Quantum Chaos—Dedicated to Professor Giulio Casati on the Occasion of His 80th Birthday

Quantum Chaos—Dedicated to Professor Giulio Casati on the Occasion of His 80th Birthday

Editor

Marko Robnik

MDPI

Basel • Beijing • Wuhan • Barcelona • Belgrade • Novi Sad • Cluj • Manchester

Editor
Marko Robnik
CAMTP - Center for Applied
Mathematics and Theoretical Physics,
University of Maribor
Maribor, Slovenia

Editorial Office
MDPI
St. Alban-Anlage 66
4052 Basel, Switzerland

This is a reprint of articles from the Special Issue published online in the open access journal *Entropy* (ISSN 1099-4300) (available at: https://www.mdpi.com/journal/entropy/special_issues/Quantum_Chaos).

For citation purposes, cite each article independently as indicated on the article page online and as indicated below:

Lastname, A.A.; Lastname, B.B. Article Title. *Journal Name* **Year**, *Volume Number*, Page Range.

ISBN 978-3-0365-8858-2 (Hbk)
ISBN 978-3-0365-8859-9 (PDF)
doi.org/10.3390/books978-3-0365-8859-9

Contents

About the Editor

Marko Robnik

Marko Robnik, the founder (1990) and director of CAMTP—Center for Applied Mathematics and Theoretical Physics, University of Maribor, Slovenia, graduated with a degree in Physics at the University of Ljubljana, Slovenia, in 1978, and a Ph.D. degree at the University of Bonn, Germany, in 1981. He was postdoc at the universities in Bonn and Bristol, a researcher at Bristol University, a researcher at the Max Planck Institute for Nuclear Physics in Heidelberg, and visiting researcher at the University of California in Santa Barbara, before returning to Slovenia in 1989. He is a Full Professor of Theoretical Physics following election at the University of Ljubljana in 1998, and he also gives lectures at the University of Maribor. He has organized and co-organized more than 55 world-class international conferences, workshops and schools since 1993, in Slovenia, Japan, Germany, Italy, Greece and Croatia. He has so far educated eight Ph.D. students from various countries (Slovenia, Greece, Brazil) and supervised a number of postdocs from Slovenia, China, Greece, Italy, Belarus, and Brazil. Robnik has published 137 original research papers in various branches of theoretical physics, predominantly in nonlinear dynamics of classical and quantum chaos, and is one of the pioneers in quantum chaos. He has (co)edited several books and proceedings of various schools, most notably for series of world-class international summer schools including Let's Face Chaos through Nonlinear Dynamics, held in Maribor, under his direction. He has been the organizer of the annual Christmas Symposia of Physicists since 2002, held in Maribor, which covers general physics. He currently has an h-index of 39 and about 6000 citations (Google Scholar). He has been a member of the European Academy of Sciences and Arts (Salzburg) since 2012. Robnik was also awarded the State Prize for Scientific Research of the Republic of Slovenia (the highest award in Slovenia for scientific work) in 1997.

Preface

This Special Issue of *Entropy* will cover quantum chaos in its broadest sense and address the general physicists as well as experts in the field. Several papers, 36 original research papers as well as 2 review papers, offer an extensive introduction to the specific topic in quantum chaos and present the most recent results. Therefore, I have written the Editorial in this spirit, in order to attract the attention of researchers from various disciplines who are interested in quantum chaos. This Special Issue is expected to be a valuable source of scientific information for newcomers in quantum chaos, including Ph.D. students, but also for experts. It encompasses a broad range of topics from various few-body and many-body dynamical systems to nano and mesoscopic systems, condensed matter physics, molecular and atomic physics, quantum field theories, high-energy physics, string theories and even cosmology. We are thus witnessing fundamental progress in understanding quantum few-body as well as many-body systems, from the dynamical aspects to statistical aspects, which also involves issues in the statistical physics of quantum systems, both in equilibrium as well as in nonequilibrium states. Similar aspects occur in many other wave systems, namely electromagnetic waves, acoustic waves, elastic waves, water surface waves, seismic waves and gravitational waves.

The motivation for this Special Issue lies in the fact that there has recently been a considerable revival in quantum chaos in the above sense. It is very much appropriate to dedicate this Special Issue to Professor Giulio Casati on the occasion of his 80th birthday. This collection of 38 papers underlines the importance of his plentiful fundamental contributions to the scientific thought and results in quantum chaos and much more.

Marko Robnik, CAMTP, Maribor, Slovenia, 1 September 2023

Marko Robnik
Editor

entropy

Editorial

Quantum Chaos—Dedicated to Professor Giulio Casati on the Occasion of His 80th Birthday

Marko Robnik

Center for Applied Mathematics and Theoretical Physics, University of Maribor, SI-2000 Maribor, Slovenia; robnik@uni-mb.si; Tel.: +38-62-235-5350 (ext. 5351)

Quantum chaos is the study of phenomena in the quantum domain which correspond to classical chaos. More generally, we speak of *wave chaos* (see below). In classically chaotic systems, we observe the sensitive dependence on initial conditions manifested in the *asymptotically (for infinite times)* exponential divergence of nearby orbits, with the rate being the largest positive Lyapunov exponent. This is a consequence of the nonlinearity of the underlying classical equations of motion. In quantum mechanics, we study the properties of the solutions of the time-dependent Schrödinger equation of motion, which is linear. Here, we observe, firstly, that the concept of orbits does not exist due to the Heisenberg uncertainty principle, and secondly, that the time-evolution of solutions (in the case of bound systems with a discrete energy spectrum) is always almost periodic due to the linearity. Consequently, an attempt to define *asymptotic* quantum Lyapunov exponents (for infinite times) results in the conclusion that they must be zero. Nevertheless, under certain conditions for *finite* times, the behavior of certain dynamical quantities, such as an out-of-time-ordered correlator (OTOC), can follow an exponential law, thereby mimicking classical behavior, but typically only up to the Heisenberg time (i.e., the ratio of the Planck constant and the mean energy spacing). In this sense, the dynamical chaos in quantum mechanics does not exist. However, there exist stationary aspects of classical chaos, namely the structure of the phase portrait which, in the general case of generic systems, consists of coexisting regular regions of ordered and stable motions and chaotic regions with positive Lyapunov exponents. It turns out that these structures are clearly revealed in precise correspondence with the classical counterpart if we study the properties of quantal eigenstates in the quantum phase space, such as those defined by the Wigner or Husimi functions. Moreover, the statistical properties of the energy spectra (and of other observables) behave in the semiclassical limit (of sufficiently short wavelengths), in correspondence with their classical counterparts. For classically integrable and regular systems, the statistics are Poissonian, while in the fully chaotic systems, the statistics of random matrix theories apply. These are universality classes, which do not involve any free parameters. In the mixed, (or generic) case, things naturally become more complicated, but we have theories to describe them. Furthermore, the statistical properties of the eigenfunctions (wave functions) are connected with the properties of the corresponding classical system.

In the strict semiclassical limit of a sufficiently small Planck constant, the Wigner or Husimi functions condense uniformly on the classical invariant regions, being either regular or chaotic, according to the Principle of Uniform Semiclassical Condensation (PUSC). For this, we have recently obtained quantitative phenomenological evidence, showing that the fraction of mixed states decays as a power law in the semiclassical limit, leaving behind only purely regular and purely chaotic eigenstates. However, before reaching the ultimate semiclassical limit, the chaotic Wigner functions may be localized in the quantum phase space, rather than occupying the entire classically available chaotic region. This phenomenon is the stationary aspect of dynamical or quantum localization, which is one of the most prominent phenomena in quantum chaos, and was discovered by Giulio Casati, Boris Chirikov, and Felix Izrailev in the late 1970s. The localization appears if the classical

Citation: Robnik, M. Quantum Chaos—Dedicated to Professor Giulio Casati on the Occasion of His 80th Birthday. *Entropy* **2023**, *25*, 1279. https://doi.org/10.3390/e25091279

Received: 28 August 2023
Accepted: 29 August 2023
Published: 31 August 2023

transport time is longer than the Heisenberg time. In the time-dependent domain, one observes the quantum diffusion typically only up to the Heisenberg time, which then stops due to the destructive interference. This phenomenon was originally studied in the quantum kicked rotator, both theoretically and numerically, which was introduced by Casati, Chirikov, and Izrailev as the main paradigm of quantum chaos, followed by analogous analyses in many other model and real physical systems, and stays an important subject of research to the present.

The Schrödinger equation for the wave function is just one of the many wave equations in the sense of mathematical physics that are linear and describe the propagation of waves. In the short wavelength approximation, the rays of the waves provide the leading order approximation of the wave dynamics. For example, Gaussian ray optics is the leading order approximation to the solutions of the Maxwell equations. The classical dynamics of orbits are the dynamics of rays of the quantum waves. It is clear that all observations of quantum chaos apply equally well to other wave equations in a variety of wave systems. These include electromagnetic, acoustic, elastic, surface water, seismic, gravitational, and other waves. Therefore, it would be more appropriate to talk about *wave chaos* than quantum chaos. However, the terminology of quantum chaos has been well established so far, but one must be aware of the broad subject of wave chaos that is encompassed by this research field.

Quantum chaos has experienced a noticeable decline in general interest and attention during a period of approximately several years around 2010, but became one of the top subjects in physics during the past few years once again, as it finds applications in solid state physics, fluid dynamics, molecular and atomic physics, quantum field theories, high energy physics, string theories, and even in cosmology. In addition to the theoretical investigations, there have been numerous experimental studies in many branches of physics, most notably the microwave experiments performed the more than three decades by Hans-Jürgen Stöckmann in Marburg, and also by Achim Richter in Darmstadt. There is hardly any important phenomenon or effect in quantum chaos which has not been experimentally studied by Stöckmann's group. Furthermore, it should be noted for the new researchers in quantum chaos that the books by Stöckmann and by Haake offer the best introduction and deeper studies of quantum chaos.

Thus, the reason and motivation for organizing and editing this Special Issue of *Entropy*, titled "Quantum Chaos", has a very solid background. As the reader can see, we have collated 36 original papers and two review papers, totaling 626 pages. Among the authors, we have many pioneers and leaders in quantum chaos, as well as some excellent researchers in the younger generation. They cover almost all important issues and aspects of quantum chaos and its developments in the last few years, covering the wide range of wave chaos in the sense described above.

Finally, I would like to stress the important reason to mark the 80th birthday of Professor Giulio Casati on this occasion, one of the most important pioneers of quantum chaos. Professor Giulio Casati has had an admirable impact on physics in a very broad sense, especially in classical and quantum chaos, where his work over more than five decades has laid down the foundations, but also in many other areas of theoretical physics and applications. Moreover, he has been, and still is, an active organizer of science, both in creating and leading elite academic and research institutions, as well as in organizing many world-top level conferences, workshops, and schools. In particular, he has been supporting young researchers all over the world in an important way. In recognition of his creative and influential life opus, and of his plentiful contributions for the good of our scientific community, this Special Issue is dedicated to him on the occasion of his 80th birthday (9 December 2022).

Conflicts of Interest: The author declares no conflict of interest.

entropy

MDPI

Article

Level Compressibility of Certain Random Unitary Matrices

Eugene Bogomolny

Université Paris-Saclay, CNRS, LPTMS, 91405 Orsay, France; eugene.bogomolny@lptms.u-psud.fr

Abstract: The value of spectral form factor at the origin, called level compressibility, is an important characteristic of random spectra. The paper is devoted to analytical calculations of this quantity for different random unitary matrices describing models with intermediate spectral statistics. The computations are based on the approach developed by G. Tanner for chaotic systems. The main ingredient of the method is the determination of eigenvalues of a transition matrix whose matrix elements equal the squared moduli of matrix elements of the initial unitary matrix. The principal result of the paper is the proof that the level compressibility of random unitary matrices derived from the exact quantisation of barrier billiards and consequently of barrier billiards themselves is equal to 1/2 irrespective of the height and the position of the barrier.

Keywords: level compressibility; barrier billiards

Citation: Bogomolny, E. Level Compressibility of Certain Random Unitary Matrices. *Entropy* **2022**, *24*, 795. https://doi.org/10.3390/e24060795

Academic Editor: Marko Robnik

Received: 26 April 2022
Accepted: 4 June 2022
Published: 7 June 2022

1. Introduction

The leading idea behind statistical descriptions of complex deterministic quantum problems is that quantum characteristics (e.g., eigenenergies) of a large variety of such problems are so erratic and irregular that their precise values are irrelevant (such as the position of a molecule in the air), and only their statistical properties are of importance. As matrices are inherent in quantum mechanics, random matrices occupy a predominant place in the application of statistics to quantum problems [1]. In a typical setting, one tries to find a random matrix ensemble whose eigenvalues have the same statistical distributions as (high-excited) eigenenergies of a given deterministic quantum problem. Until now, this query has been figured out only for two limiting classes of quantum problems: (i) models whose classical limit is integrable [2] and (ii) models whose classical limit is chaotic [3]. For generic integrable models, quantum eigenenergies are distributed as eigenvalues of diagonal matrices with independent identically distributed (i.i.d.) elements which means that their correlation functions after unfolding coincide with the ones of the Poisson distribution [2]. For generic chaotic systems, it was conjectured in [3] that their eigenenergies are distributed as eigenvalues of standard random matrix ensembles (GOE, GUE, GSE) depending only on system symmetries whose correlation functions are known explicitly [4]. The difference between these two cases is clearly seen from the limiting behaviour of their nearest-neighbour distribution $P_0(s)$, which is the probability density that two eigenvalues are separated by a distance s and there are no other eigenvalues in between. For the Poisson statistics, there is no level repulsion that means that $\lim_{s\to 0} P_0(s) \neq 0$ and for large argument $P_0(s)$ decreases exponentially with s. Standard random matrix ensembles levels repel each other, $\lim_{s\to 0} P_0(s) = 0$, and $P_0(s) \sim \exp(-as^2)$ when $s \to \infty$.

These two big conjectures form a cornerstone of quantum chaos and have been successfully applied to various problems from nuclear physics to number theory. Nevertheless, they do not cover all possible types of models. Especially intriguing is the class of pseudo-integrable billiards (see, e.g., [5]) which are two-dimensional polygonal billiards whose angles θ_j are rational multiples of π

$$\theta_j = \frac{m_j}{n_j}\pi$$

with co-prime integers m_j and n_j. A peculiarity of such billiards is seen in the fact that their classical trajectories belong to a two-dimensional surface of genus g related with angles as follows [6]

$$g = 1 + \frac{N_n}{2} \sum_j \frac{m_j - 1}{n_j}$$

where N_n is the least common multiple of all denominators n_j. Consequently, any such model with at least one numerator $m_j > 1$ is neither integrable (which would imply that trajectories belong to a two-dimensional torus with $g = 1$) nor fully chaotic (in which case trajectories should cover a three-dimensional surface of constant energy), and the aforementioned conjectures cannot be applied to such systems. Numerical calculations show that for many pseudo-integrable billiards, the spectral statistical properties of corresponding quantum problems differ from both the Poisson statistics and the random matrix statistics mentioned above (see, e.g., [7,8] and references therein). In particular, for these models, (i) $\lim_{s\to 0} P_0(s) = 0$ as for standard random matrix ensembles but (ii) $P_0(s) \sim \exp(-bs)$ for large s as for the Poisson statistics. Such hybrid statistics, labeled intermediate statistics, had been first observed in the Anderson model at the metal–insulate transition [9,10], and they constitute a special, interesting but poorly investigated class of spectral statistics.

Probably the simplest example of pseudo-integrable systems is the so-called barrier billiard, which is a rectangular billiard with a barrier inside sketched in Figure 1a. The quantum problem for this model consists of solving the Helmholtz equation

$$(\Delta + E_\alpha)\Psi_\alpha(x,y) = 0$$

imposing that eigenfunction $\Psi_\alpha(x,y)$ obeys (e.g.,) the Dirichlet conditions on the boundary of the rectangle as well as on the barrier

$$\Psi_\alpha(x,y)|_{\text{boundary}} = 0, \qquad \Psi_\alpha(x,y)|_{\text{barrier}} = 0.$$

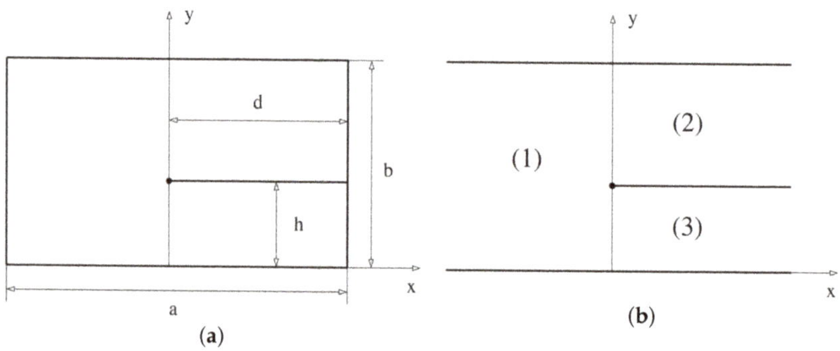

Figure 1. (**a**) Barrier billiard. (**b**) An infinite slab with a half-plane inside. Numbers indicate 3 possible channels.

Calculating the exact S-matrix for the scattering inside the infinite slab with a barrier depicted in Figure 1b, it has been demonstrated in [7,8] that spectral statistics of this model is the same as the statistics of eigenvalues of the following $N \times N$ random unitary matrix

$$B_{m,n} = e^{i\Phi_m} \frac{L_m L_n}{x_m + x_n}, \qquad m, n = 1, \ldots, N \tag{1}$$

where Φ_m are i.i.d. random variables uniformly distributed on interval $[0, 2\pi)$ and

$$L_m = \sqrt{2x_m \prod_{k\neq m} \frac{x_m + x_k}{x_m - x_k}} \tag{2}$$

where coordinates x_m depend on the position of the barrier.

Define the following quantities (momenta) of propagating modes in each of three channels indicated in Figure 1b.

$$p_m^{(1)}=\sqrt{k^2-\frac{\pi^2 m^2}{b^2}}\,,\qquad p_m^{(2)}=\sqrt{k^2-\frac{\pi^2 m^2}{(b-h)^2}}\,,\qquad p_m^{(3)}=\sqrt{k^2-\frac{\pi^2 m^2}{h^2}}\,.$$

If the ratio h/b is an irrational number, coordinates x_m have the following form

$$\vec{x}=b\Big(\underbrace{p_1^{(1)},\ldots,p_{N_1}^{(1)}}_{N_1},\underbrace{-p_1^{(2)},\ldots,-p_{N_2}^{(2)}}_{N_2},\underbrace{-p_1^{(3)},\ldots,-p_{N_3}^{(3)}}_{N_3}\Big). \tag{3}$$

Here, N_j with $j=1,2,3$ are the numbers of propagating modes in each channel

$$N_1=\left[\frac{kb}{\pi}\right],\qquad N_2=\left[\frac{k(b-h)}{\pi}\right],\qquad N_3=\left[\frac{kh}{\pi}\right] \tag{4}$$

where $[x]$ is the largest integer $\leq x$ and the total dimension of the B-matrix is $N=N_1+N_2+N_3$.

When the ratio h/b is a rational number, $h/b=p/q$ with co-prime integers p and q ($p<q$), there exist exact plane wave solutions of barrier billiard equal to zero at the whole line passing through the barrier. It is natural to disregard them and take into account only non-trivial eigenvalues. In such case, coordinates x_m have to be chosen as indicated below

$$\vec{x}=b\Big(\underbrace{p_1^{(1)},\ldots,p_k^{(1)},\ldots,p_{N_1}^{(1)}}_{k\neq 0 \mod q},\underbrace{-p_1^{(2)},\ldots,-p_{N_2}^{(2)}}_{N_2},\underbrace{-p_1^{(3)},\ldots,-p_k^{(3)},\ldots,-p_{N_3}^{(3)}}_{k\neq 0 \mod p}\Big). \tag{5}$$

The dimension of this vector is

$$N^{(\mathrm{r})}=N_1+N_2+N_3-2N_0,\qquad N_0=\left[\frac{kb}{\pi q}\right]. \tag{6}$$

The matrix B can be generalised for arbitrary vector x_m provided the following interlacing conditions are fulfilled

$$|x_1|>|x_2|>\ldots>|x_N|,\qquad x_m=(-1)^{m+1}|x_m|.$$

Exact correlation functions for the B-matrix are unknown at present. In [7,8], it was argued that an approximate Wigner-type surmise for this matrix corresponds to the so-called semi-Poisson distribution [11]. In particular, it implies that the probability density $P_n(s)$ that two levels are separated by a distance s and there are exactly n levels in between (after the standard unfolding) is given by the following expression

$$P_n(s)=\frac{2^{2n+2}}{(2n+1)!}s^{2n+1}e^{-2s},\qquad n\geq 0\,.$$

Numerical calculations presented in [7,8] agree with these simple formulas.

Despite the simplicity of the semi-Poisson distribution, it has been observed (mostly numerically) in various models ranging from certain pseudo-integrable models and quantum maps (see references in [7,8]) to the entanglement spectra of two-bits random many body quantum circuits [12].

This paper is devoted to the calculation of another important characteristic of spectral statistics, namely the level (or spectral) compressibility, which is a long-range characteristic of the spectral two-point correlation function. In particular, this quantity determines the limiting behaviour of the variance of the number of levels inside a given interval. More precisely, let $N(L)$ be the number of eigenvalues in an interval L unfolded to the unit mean density, which means that the mean number of levels in interval L equals L, $\langle N(L)\rangle = L$. By definition, the number variance is $\Sigma^{(2)}(L) \equiv \langle (N(L)-L)^2\rangle$. If for large L

$$\Sigma^{(2)}(L) \underset{L\to\infty}{\longrightarrow} \chi L \tag{7}$$

constant χ is called the level compressibility. The importance of this quantity follows from the fact that for integrable systems with the Poisson statistics, $\chi = 1$, but for standard random matrix statistics typical for chaotic models, $\chi = 0$. For all examples of intermediate statistics, it was observed that $0 < \chi < 1$.

The conventional way of determining the level compressibility for dynamical systems is the summation over all periodic orbits in the diagonal approximation initiated in [13]. For the symmetric barrier billiard with $h/b = 1/2$ and $d/a = 1/2$, it has been demonstrated in [14] that $\chi = 1/2$. The same value had been obtained in [15] for the case $h/b = 1/2$ and arbitrary barrier height d/a. Finally, for $h/b = p/q$ with co-prime integers p, q and irrational values of d/a, it has been proven in [16] that

$$\chi = \frac{1}{2} + \frac{1}{q}$$

but in the calculations, exact eigenvalues whose eigenfunctions are zero on the whole line $y = h$ (see Figure 1a) have not been excluded. When these trivial eigenvalues are removed, the answer is $\chi = 1/2$ [17]. Therefore, direct (and quite tedious) calculations suggest that for (almost) all positions and heights of the barrier, the level compressibility is the same as for the semi-Poisson statistics [11]

$$\chi = \frac{1}{2} \tag{8}$$

but the reason of this universality remains obscure.

The purpose of the paper is to find analytically the spectral compressibility for barrier billiards and for a few other models directly from the corresponding random unitary matrices. To achieve the goal, it is convenient to slightly generalise the method developed in [18] for random unitary matrices that appeared in the quantisation of quantum graphs [19]. The method is briefly explained in Section 2. In Section 3, this method is applied to random unitary matrices derived in [20] by the quantisation of a simple interval-exchange map. In this case, the transition matrix is a circulant matrix whose eigenvalues are known explicitly. The results coincide with the exact level compressibility for these models obtained in [21–24]. This example gives credit to the method and permits explaining its main features without unnecessary complications. In particular, it clarifies the situation (not covered by Refs. [21–24]) when a parameter that entered the matrix takes an irrational value. Numerically, it has been observed [22] that in such case, the spectral statistics are well described by the ones of chaotic systems (GOE or GUE), although the Lyapunov exponent of the underlying classical map is always zero.

The main part of the paper is devoted to the derivation of the level compressibility for random matrices associated with barrier billiards (1). The calculations are more complicated, as no eigenvalues (except one) are known analytically. The simplest case of the symmetric barrier billiard with $h/b = 1/2$ is investigated in Section 4. To obtain tractable expressions, a kind of paraxial approximation is developed, which permits controlling the largest terms. By using such approximation, the transition matrix is transformed into a Toeplitz matrix with a quickly decreasing symbol, which allows us to find its eigenvalues for large matrix dimension. The result of this section is that the level compressibility equals $1/2$ in accordance with periodic orbit calculations in [14,15].

In Section 5, random unitary matrices corresponding to barrier billiards with irrational ratio h/b (with coordinates given by (3)) are considered. In this case, the transition matrix in paraxial approximation contains quickly oscillating terms and, consequently, has forbidden zones in the spectrum. In spite of that, one can argue that largest moduli eigenvalues are insensitive to fast oscillations and are determined solely by a matrix averaged over such oscillations. The final matrix is also a Toeplitz type, which allows analytical calculations, proving that the level compressibility is again equal to $1/2$.

Section 6 addresses the calculation of level compressibility in the most complicated case of barrier billiards with rational ratio $h/b = p/q \neq 1/2$. The computation is cumbersome, but in the end, one comes to the conclusion that the level compressibility remains equal $1/2$. In other words, the level compressibility of barrier billiards is universal (i.e., independent of the barrier position and its height) and coincides with the semi-Poisson prediction [11].

Section 7 gives a brief summary of the results. A few technical points are discussed in Appendices A–C.

2. Generalities

It is well known that the level compressibility (7) is related with the two-point correlation form factor $K(\tau)$ as follows

$$\chi = \lim_{\tau \to 0} K(\tau). \tag{9}$$

For $N \times N$ random unitary matrices U, the form factor can, conveniently, be written in the following concise form (see, e.g., [18])

$$K(\tau) = \left\langle \frac{1}{N} |\mathrm{Tr}\, U^n|^2 \right\rangle, \qquad \tau = \frac{n}{N} \tag{10}$$

where the average is taken either on different realisations of random parameters, or over a small window of τ, or both.

Unitary matrices considered in the paper all have the product form

$$U_{j,k} = e^{i\Phi_j} w_{j,k} \tag{11}$$

where Φ_j are i.i.d. random phases uniformly distributed between 0 and 2π, and matrix $w_{j,k}$ is a fixed unitary matrix.

In [18] (see [19] for more detailed discussion), it was shown that for such unitary matrices, the averaging over random phases leads in the diagonal approximation to the following formula

$$K^{(\mathrm{diag})}\left(\frac{n}{N}\right) = g\frac{n}{N} \mathrm{Tr}\,(T^n) \tag{12}$$

where matrix elements of matrix T, called below the transition matrix, are squared moduli of matrix elements of matrix U

$$T_{j,k} = |U_{j,k}|^2 = |w_{j,k}|^2. \tag{13}$$

For systems without time-reversal invariance, $g = 1$, and for models with time-reversal invariance, $g = 2$ (here, only cases of $g = 1$ are considered). Due to the unitarily of matrix U, the T-matrix is a double stochastic matrix, $\sum_j T_{j,k} = 1$ and $\sum_k T_{j,k} = 1$, thus having the meaning of classical transition matrix.

Let Λ_β be eigenvalues of the T-matrix

$$T_{j,k} u_k(\beta) = \Lambda_\beta u_j(\beta), \qquad \beta = 0, \ldots, N-1.$$

From (12), it follows that

$$K^{(\mathrm{diag})}\left(\frac{n}{N}\right) = \frac{n}{N} \sum_{\beta=0}^{N-1} \Lambda_\beta^n. \tag{14}$$

The unitarity imposes that one eigenvalue $\Lambda_0 = 1$. Perron–Frobenius theorem states that all other eigenvalues $|\Lambda_\beta| \leq 1$.

Let the set of eigenvalues be ordered $1 = \Lambda_0 \geq \Lambda_1 \geq \ldots \geq \Lambda_{N-1}$. One has

$$K^{(\text{diag})}(\tau) = \tau(\text{Tr}\,(T^n)) = \tau + \tau \sum_{\beta=1}^{N-1} \Lambda_\beta^{N\tau}, \qquad \sum_{\beta=1}^{N-1} \Lambda_\beta^{N\tau} \leq (N-1)|\Lambda_1|^{N\tau}. \tag{15}$$

It has been noted in [18] that if

$$\lim_{N\to\infty} |\Lambda_1|^N = 0 \tag{16}$$

then the second term in (15) goes to zero for all finite τ and $K^{(\text{diag})}(\tau) = \tau$ for small τ. Consequently, it is reasonable to conjecture (as it has been done in [18]) that the whole spectral statistics of such matrices will be well described by standard random matrix formulas.

For the matrices discussed below, criterium (16) is not fulfilled. Instead, in all considered cases, the largest moduli eigenvalues of transition matrices have the form $1 - \mathcal{O}(1/N)$. To calculate the level compressibility from (14), the summation over all such eigenvalues is performed analytically, and then, the limit $\tau \to 0$ is taken.

It is well known that the form factor is not a self-average quantity. It has strong fluctuations and necessarily requires a smoothing. There exist two different sources of fluctuations. The first is related with random phases in matrices (11). Equation (12) corresponds to the averaging over these random phases in the diagonal approximation. The second has its roots in non-smoothness of $K^{(\text{diag})}(n/N)$ for different n and could be removed by a smoothing over a small interval of τ. (A trivial example is $(-1)^n$).

3. Interval-Exchange Matrices

This section is devoted to the calculation of the level compressibility for special unitary matrices derived in [20] by quantisation of a simple two-dimensional parabolic map. Slightly generalising their result [21], one can write these matrices in the following form

$$M_{n,m} = e^{i\Phi_n} \frac{1 - e^{2\pi i\alpha N}}{N\big(1 - e^{2\pi i(n-m+\alpha N)/N}\big)}, \qquad n, m = 1, \ldots, N. \tag{17}$$

Here, α is a real parameter and Φ_n are i.i.d. random variables uniformly distributed between 0 and 2π (the case with 'time-reversal symmetry' when $\Phi_{N-n+1} = \Phi_n$ requires only multiplication of the formulas below by factor $g = 2$, as indicated in (12)).

When α is a rational number $\alpha = p/q$ with co-prime integers p and q, the original classical map is an interval-exchange map and, as it was shown in detail in [21–24], spectral statistics of matrices (17) in the limit of $N \to \infty$ are unusual and peculiar. It appears that the limiting results depend on the residue of $pN \mod q$ (when $pN \equiv 0 \mod q$ matrix (17) have explicit eigenvalues not interesting for our purposes). For example, if $\alpha = 1/5$, there are two possibilities: $N \equiv \pm 1 \mod 5$ and $N \equiv \pm 2 \mod 5$. In the first case, the nearest-neighbour distribution is

$$P_0(s) = \frac{5^5}{4!} s^4 e^{-5s}$$

but for the second one, the exact result is different

$$P_0(s) = (a_2 s^2 + a_3 s^3 + a_4 s^4 + a_5 s^5 + a_6 s^6) e^{-5s}$$

where $a_2 = 625/2 - 275\sqrt{5}/2$, $a_3 = 3125/2 - 1375\sqrt{5}/2$, $a_4 = 71875/48 + 33125\sqrt{5}/48$, $a_5 = -15625/3 + 9375\sqrt{5}/4$, and $a_6 = 1015625/288 - 453125\sqrt{5}/288$.

Although the spectral correlation functions for $\alpha = p/q$ are different for different residues $pN \mod q$, the calculations show that the spectral compressibility for all residues remains the same [21,22]

$$\chi = \frac{1}{q}\,. \tag{18}$$

Matrices (17) with irrational α were investigated numerically in [22], and it was observed that their spectral statistics are well described by standard random matrix ensembles (GOE or GUE). In particular, it implies that in such case

$$\chi = 0\,. \tag{19}$$

Below, it is shown that values (18) and (19) can easily be recovered by the discussed method.

The transition matrix (13) for the discussed case has the form

$$T_{n,m} \equiv |M_{n,m}|^2 = \frac{\sin^2(\pi\alpha N)}{N^2\sin^2\left(\pi(n-m+\alpha N)/N\right)}\,. \tag{20}$$

This is a circulant matrix, and its eigenvalues are simply the Fourier transforms of its matrix elements

$$\Lambda_\beta = \frac{\sin^2(\pi\alpha N)}{N^2}\sum_{s=0}^{N-1}\frac{e^{-2\pi i\beta s/N}}{\sin^2\left(\pi(s+\alpha N)/N\right)}\,. \tag{21}$$

Differentiating both sides of the identity with integer s on z

$$\sum_{m=0}^{N-1} e^{2\pi i m(s+z)/N} = \frac{1-e^{2\pi i z}}{1-e^{2\pi i(s+z)/N}}\,,$$

after straightforward transformations, one proves that

$$\frac{\sin^2(\pi z)}{\sin^2\left(\pi(s+z)/N\right)} = -2i\sin(\pi z)e^{-\pi i z}\sum_{m=0}^{N-1}\left(m - N\frac{e^{\pi i z}}{2i\sin(\pi z)}\right)e^{2\pi i m(s+z)/N}$$

from which it follows that eigenvalues (21) are

$$\Lambda_\beta = \left(1-\frac{\beta}{N}\left(1-e^{-2\pi i\alpha N}\right)\right)e^{2\pi i\beta\alpha}\,,\qquad \beta = 0,1\,\ldots,N-1\,. \tag{22}$$

Notice that $\Lambda_0 = 1$, $\Lambda_{N-\beta} = \Lambda_\beta^*$ for $\beta = 1,\ldots,N-1$.

Consider first the case of rational $\alpha = p/q$. Assume that $N \neq 0 \mod q$ and calculate the form factor from (14) separately for $n \equiv r \mod q$ with $r = 0,1,\ldots,q-1$ (i.e., $n = qt + r$ with integer t)

$$K\left(\frac{qt+r}{N}\right) = \frac{qt+r}{N}\left[1 + 2\mathrm{Re}\sum_{\beta=1}^{N/2}\left(1-\frac{\beta}{N}\left(1-e^{-2\pi i\alpha N}\right)\right)^{qt+r}e^{2\pi i\beta pr/q}\right].$$

Here, we do not take into account that when N is even the term with $\beta = N/2$ is real and has an additional factor $1/2$.

Notice that the phase depends only on r. For large N, one can put the first factor in the exponent and sum the geometric progression from 1 to infinity. The answer is

$$K_r(\tau) = \tau + 2\tau\mathrm{Re}\left(\frac{e^{-\xi\tau+2\pi i pr/q}}{1-e^{-\xi\tau+2\pi i pr/q}}\right),\qquad \tau = \frac{qt+r}{N} \tag{23}$$

where $\xi = 1-e^{-2\pi i pN/q} = 1-e^{-2\pi i k/q}$ with $k = pN \mod q$.

When $\tau \to 0$, all terms except the one with $r = 0$ tend to zero, as they do not have a pole at $\tau = 0$. The remaining term equals

$$\lim_{\tau \to 0} K_0(\tau) = 2\,\mathrm{Re}\left(\frac{1}{\xi}\right) = \frac{2\,\mathrm{Re}\,(1 - e^{2\pi i z})}{2 - 2\cos(2\pi z)} = 1\,.$$

It means that after the averaging over random phases, the limiting value of the form factor $K_r(\tau)$ strongly depends on small changes of $\tau = (qt + r)/N$. For $r = 0$, $K_0(\tau)$ tends to 1 for small $\tau \ll 1$ but for all other terms with $r = 1, \ldots, q-1$ $K_r(\tau)$ tends to 0. The difference between these different values of τ is very small, of the order of $1/N$. Therefore, after the averaging over any small (but finite) interval of τ, one gets

$$\lim_{\tau \to 0} \langle K(\tau) \rangle = \frac{1}{q}$$

which agrees with (18) obtained in [22] by a different method.

For illustration, the results of the direct calculation of the form factor for $\alpha = 1/5$ and $N = 399$ and $N = 398$ are presented in Figure 2. First, eigenvalues of the matrix (17) were calculated numerically, and then using (10), the form factor for different n values was computed. The result is averaged over 1000 realisations of random phases. It is clear that, indeed, for different residues of n modulo 5, the results are different, and when $n \equiv 0 \bmod 5$, the form factor at small argument is close to 1, but for all other residues, it starts at 0. The average over all 5 residues begins at $1/5$, as expected.

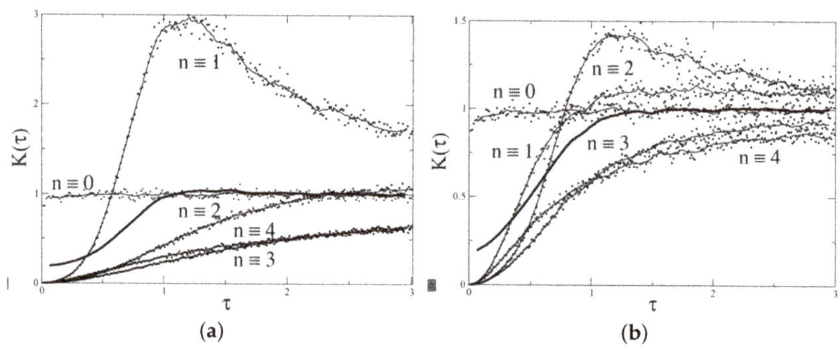

Figure 2. Form factor for the matrix (17) with $\alpha = 1/5$ and (**a**) $N = 399 \equiv -1 \bmod 5$ and (**b**) $N = 398 \equiv -2 \bmod 5$ averaged over 1000 realisations. Points are values of $K(n/N)$ for integers n with indicated residues modulo 5. Thin solid lines are a guide for the eye. Thick solid lines indicate the average over all 5 residues: $\frac{1}{5}\sum_{r=0}^{4} K((qt+r)/n)$.

Such a clear picture appears when the form factor is calculated at special values of τ, $\tau = n/N$ with integer n. Computing it at arbitrary arguments leads to an irregular plot but, of course, the average curve remains unchanged.

Exactly the same formulas can be applied for an irrational value of parameter α. In this case, one has

$$K(\tau) = \tau + 2\tau\,\mathrm{Re}\left(\frac{e^{-\zeta\tau}}{1 - e^{-\zeta\tau}}\right), \qquad \zeta = 1 - e^{-2\pi i\alpha N} - 2\pi i\alpha N\,. \tag{24}$$

The exponent $\zeta = 2\sin^2(\pi\alpha N) + i\left[\sin(2\pi\alpha N) - 2\pi\alpha N\right]$ has a large imaginary part when $N \to \infty$. It means that the above expression is a strongly oscillated function of τ. When averaged over a small interval of τ, one obtains $K(\tau) = \tau$ as it should be for the ensemble of usual random matrices (GUE). This result follows without calculations from the fact that the average of all eigenvalues $\Lambda_\beta^{\tau N}$ except $\beta = 0$ equals zero as a consequence of rapidly changing phases. (For even N, the term with $\beta = N/2$ is real, but as it tends to zero at large N, its contribution is negligible).

Notice that criterion (16) for matrix (17) with irrational α is not fulfilled. Nevertheless, the spectral statistics of such a matrix is close to GUE statistics. This example illustrates a new mechanism for the appearance of random matrix statistics. The contribution of higher eigenvalues of the transition matrix (13) decreases not because of a gap between the first and the second eigenvalues as has been proposed in (16) but due to rapid oscillations for large matrix dimensions.

4. Symmetric Barrier Billiard with *h*/*b* = 1/2

The central problem of the paper is the determination of level compressibility for the B-matrices given by (1) and (2) by employing the method proposed in [18] and used in the previous section for matrices derived from the quantisation of an interval-exchange map. The simplicity of treatment of interval-exchange matrices comes from the fact that their transition matrices are circulant matrices whose eigenvalues are known exactly. For the B-matrices, calculations are more complicated, as there are no explicit formulas for eigenvalues of the corresponding transition matrix.

$$T_{m,n} = |B_{m,n}|^2 = \frac{L_m^2 L_n^2}{(x_m + x_n)^2}, \qquad m, n = 1, \ldots, N. \tag{25}$$

This section is devoted to the investigation of the B-matrix corresponding to the symmetric barrier billiard with ratio $h/b = 1/2$. In this case, $q = 2$, $b - h = h$, and the second part of the vector $\vec{x}$ in (5) coincides with the third one. Now, trivial eigenfunctions can be removed by considering a desymmetrised rectangular billiard with height $h = b/2$ and imposing the Neumann boundary conditions for negative x and $y = h$. It is the equivalent of dropping the second part of vector (5) and taking coordinates x_m as follows [7]

$$x_m = (-1)^{m+1} b \sqrt{k^2 - \frac{\pi^2 m^2}{b^2}}, \qquad m = 1, \ldots, N, \qquad N = \left[\frac{kb}{\pi}\right]. \tag{26}$$

Odd (resp., even) indices describe the first (resp., the third) part of vector (5).

The numerically calculated spectrum of the transition matrix in this case is presented in Figure 3a.

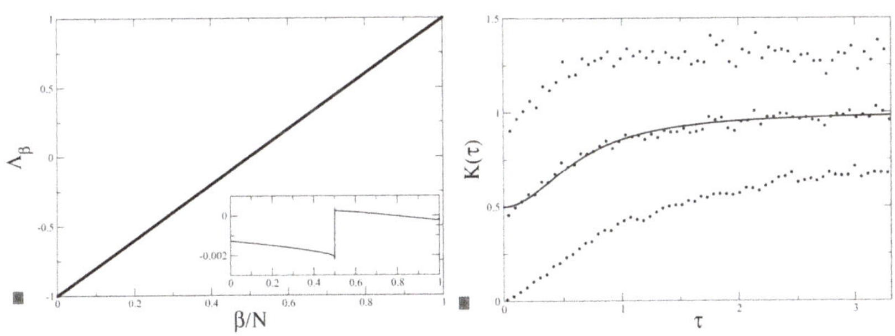

(a) (b)

Figure 3. (**a**) Transition matrix spectrum for symmetric barrier billiard with $h/b = 1/2$, $b = \pi$, $k = 1000.5$, and $N = 1000$. Insert shows the difference between true eigenvalues Λ_β and the straight line $2\beta/N - 1$. (**b**) Form factor for symmetric barrier billiard with $h/b = 1/2$, $b = \pi$, $k = 400.5$, and $N = 400$. Data are averaged over 1000 realisations of random phases. The upper dots are $K(2n/N)$, the lower dots are $K((2n-1)/N)$, and the middle dots correspond to $\Big(K(2n/N) + K((2n-1)/N)\Big)/2$. The solid line is the semi-Poisson prediction $K(\tau) = (2 + \pi^2\tau^2)/(4 + \pi^2\tau^2)$ [11].

To calculate this spectrum (or, at least, the behaviour of largest-moduli eigenvalues) analytically, a kind of paraxial approximation has been developed. It is based on the fact that the main ingredient of matrices with intermediate statistics is a linear fall-off of matrix elements from the diagonal [25,26]. In the simplest setting, it means that

$$M_{m,n} \sim \frac{R_{m,n}}{m-n+\text{const.}}\,, \qquad m,n \gg 1\,, \qquad m-n=\mathcal{O}(1)\,.$$

Therefore, it is natural to assume that the most important contributions come from the pole terms with $R_{m,n} \approx R_{m,m}$. This type of approximation can be done directly from the definition (2), as it is demonstrated in Appendix A. According to these results, the T-matrix in the paraxial approximation is a block matrix

$$T = \begin{pmatrix} 0_{\rm o,o} & t_{\rm o,e} \\ t^T_{\rm o,e} & 0_{\rm e,e} \end{pmatrix}, \qquad t_{\rm o,e} \equiv t_{2m-1,2n} = \frac{1}{\pi^2(n-m+1/2)^2}\,. \tag{27}$$

Here, subscripts 'o' and 'e' indicate odd and even indices, respectively.

It is instructive to obtain this answer without the knowledge of the exact B-matrix. One can achieve it by using the instantaneous approximation used in quantum mechanics when the interaction changes suddenly. In optics, such an approximation is analogous to the Fraunhofer diffraction. In the barrier billiard, it corresponds to the situation when a wave with large momentum quickly moving in a channel enters into another channel (cf., Figure 1b). In the instantaneous approximation, eigenfunctions in the new channel are just a re-expansion of the initial eigenfunctions into a complete set of eigenfunctions with correct boundary conditions inside the final channel.

Consider a normalised wave with the Neumann boundary conditions at $y=h=b/2$ and the Dirichlet ones at $y=0$

$$\psi^{(1)}_{2m-1}(x,y) = \frac{2}{\sqrt{b}}\sin\left(\frac{(2m-1)\pi}{b}y\right)\exp\left(ip^{(1)}_{2m-1}x\right), \qquad x<0, \qquad 0\le y\le b/2$$

propagating in the desymmetrised barrier billiard at negative x. When it penetrates into the region of positive x, it has to be expanded into correct waves propagating inside that region

$$\psi^{(1)}_{2m-1}(x,y) = \sum_{n=1}^{\infty} S_{2m-1,2n}\,\psi^{(3)}_{2n}(x,y)$$

where $\psi^{(3)}_{2n}(x,y)$ are waves obeying the Dirichlet boundary conditions at $y=0$ and $y=h=b/2$

$$\psi^{(3)}_{2n}(x,y) = \frac{2}{\sqrt{b}}\sin\left(\frac{2\pi n}{b}y\right)\exp\left(ip^{(3)}_{n}x\right), \qquad x>0, \qquad 0\le y\le b/2\,.$$

Coefficients $S_{2m-1,2n}$ are the S-matrix for this process. In the paraxial approximation, they are calculated as follows (notice that in the paraxial approximation, $p^{(1)}_{2m-1} \approx p^{(3)}_n$)

$$\begin{aligned} S^{(\text{p.a.})}_{2m-1,2n} &= \frac{4}{b}\int_0^{b/2} \sin\left(\frac{\pi(2m-1)}{b}\right)\sin\left(\frac{2\pi n}{b}y\right)dy = \\ &= \frac{(-1)^{m+n}}{\pi^2(n-m+1/2)} + \frac{(-1)^{m+n}}{\pi^2(n+m-1/2)}\,. \end{aligned}$$

Taking into account only the pole term (and symmetry of the S-matrix), one obtains for the T-matrix exactly the same expression as (A1).

Thus, the transition matrix (27) is a block Toeplitz matrix. It is plain that its eigenvalues $\Lambda = \pm\sqrt{\lambda_\beta}$ where λ_β are eigenvalues of a $N_1 \times N_1$ matrix (with $N_1 = [N/2]$).

$$P_{m,n} = \sum_{k=1}^{N_1} t_{m,k} t_{n,k}, \qquad t_{m,k} = \frac{1}{\pi^2(k-m+1/2)^2}.$$

Dominant contributions to the sum come from regions $k \sim m$ and $k \sim n$. Due to a quick decrease of the summands, the finite summation over k can safely be substituted in the limit $N_1 \to \infty$ by the sum over all integer k

$$P_{m,n} \approx \sum_{k=-\infty}^{\infty} t_{m,k} t_{n,k}.$$

Using (A6), the necessary sum is easily calculated, and the result is

$$P_{m,n} = t_{m-n}, \qquad t_0 = \frac{1}{3}, \qquad t_{r\neq 0} = \frac{2}{\pi^2 r^2}. \tag{28}$$

This formula is valid when $m, n \gg 1$ and $m - n = \mathcal{O}(1)$.

Matrix (28) is a Toeplitz matrix with quickly decreasing matrix elements. It is well known that eigenvalues of the $N \times N$ Toeplitz matrix can be asymptotically calculated as follows (see, e.g., [27–31] and references therein)

$$\lambda_\beta = f\left(\frac{\beta}{2N} + \mathcal{O}\left(\frac{1}{N}\right)\right), \qquad \beta = 1, \dots, N \tag{29}$$

where function $f(x)$ called the symbol is the Fourier series of t_r.

$$f(x) = \sum_{r=-\infty}^{\infty} t_r e^{2\pi i r x}.$$

(More precise formulas can be found in the above references).

Using (A4) and (A5) one finds that the symbol of matrix (28) is

$$f(x) = \frac{1}{3} + \frac{2}{\pi^2} \sum_{r=-\infty}^{\infty}{}' \frac{e^{2\pi i r x}}{r^2} = \frac{1}{3} + 4B_2(x) = (1-2x)^2.$$

Therefore, eigenvalues of the P-matrix for large N_1 are

$$\lambda_\beta \approx \left(1 - \frac{\beta}{N_1}\right)^2, \qquad \beta = 1, \dots, N_1.$$

Eigenvalues of block matrix (27) $\Lambda = \pm\sqrt{\lambda_\beta}$. Taking into account that the dimension of matrix (27) is $N \approx 2N_1$, one concludes that approximately its eigenvalues are

$$\Lambda = \pm\left(1 - \frac{2\beta}{N}\right), \qquad \beta = 1, \dots, \tfrac{1}{2}N. \tag{30}$$

With the corresponding redefinition of index β, these eigenvalues can be rewritten in the form

$$\Lambda_\beta \approx \frac{2\beta}{N} - 1, \qquad \beta = 1, \dots, N$$

which agrees well with numerical calculations (see Figure 3a).

The form factor in the diagonal approximation is related with transition matrix eigenvalues by (14)

$$K^{(\mathrm{diag})}\left(\frac{n}{N}\right) = \frac{n}{N}\sum_{\beta}\Lambda_{\beta}^{n}.$$

As $\Lambda_{N-\beta} = -\Lambda_{\beta}$ for $\beta = 1,\ldots,N-1$ (which is a consequence of the block structure of the transition matrix (27)), the form factor $K^{\mathrm{diag}}(n/N)$ with odd n in the diagonal approximation tends to zero when $\tau = n/N \to 0$

$$K^{(\mathrm{diag})}(\tau) \underset{\tau\to 0}{\longrightarrow} 0. \tag{31}$$

However, for even n, one obtains a different answer. Equation (30) may not be accurate for extreme eigenvalues with small β. For $\tau = 2n/N$, one can separate the contribution of small $\beta < \beta_0$ and the rest for which (30) is a good approximation

$$K^{(\mathrm{diag})}(\tau) = \tau\left(\mathrm{const} + 2\sum_{\beta=\beta_0}^{N/2}\left(1-\frac{2\beta}{N}\right)^{\tau N}\right) \underset{N\to\infty}{\longrightarrow} \tau\left(\mathrm{const} + \frac{2e^{-2j_0\tau}}{1-e^{-2\tau}}\right) \underset{\tau\to 0}{\longrightarrow} 1. \tag{32}$$

As has been discussed in the previous section, it means that the spectral compressibility of the B-matrix for symmetric barrier billiard coincides with the semi-Poisson value

$$\chi \equiv K(0) = \frac{1}{2}. \tag{33}$$

For illustration, the form factor for the symmetric billiard calculated numerically by the direct diagonalisation of 400×400 matrices (1) with coordinates given by (26) and averaged over 1000 realisations is shown in Figure 3b. Two branches corresponding to odd and even n are clearly seen. The average over odd and even values agrees well with the semi-Poisson expression for the form factor [11] and, in fortiori, the level compressibility is $1/2$, as in (33).

5. Barrier Billiard with Irrational Ratio *h*/*b*

The transition matrices for a general barrier billiard with an off-centre barrier remain the same as in (25), but coordinates x_m should have the form (3) for irrational ratio h/b and (5) for rational $h/b = m/q$. The direct calculations of eigenvalues of these matrices reveal that they are more complicated that the ones for symmetric billiard with $h/b = 1/2$ discussed in the previous section. As an example, in Figure 4, the spectra of the transition matrices with $h/b = 1/\sqrt{5}$ and $h/b = 2/5$ are presented. It is clearly seen that although eigenvalues with small moduli are quite irregular and have gaps, the largest moduli eigenvalues are well described by a straight line $\Lambda_{\beta} = 2\beta/N - 1$.

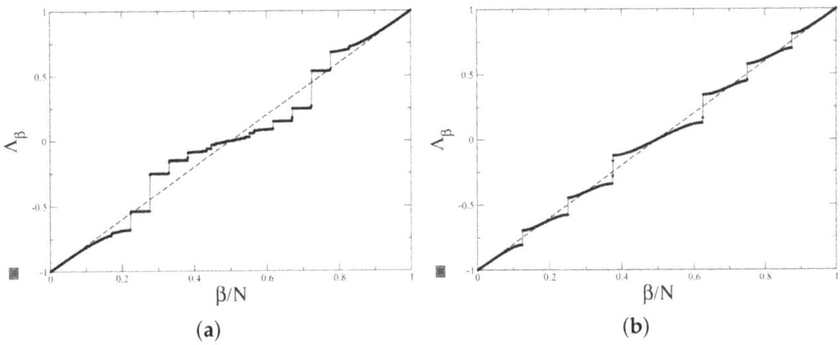

Figure 4. Spectra of the transition matrix for barrier billiard with (**a**) $h/b = 1/\sqrt{5}, b = \pi, k = 500.5$, and $N = 999$ (cf. (4)) (**b**) $h/b = 2/5, b = \pi, k = 650.5$, and $N = 1040$ (cf. (6)). The straight dashed line in the both figures is $\Lambda_\beta = 2\beta/N - 1$.

This section is concentrated on the analytical treatment of billiards with irrational ratio h/b. As in the previous section, the first step consists of the calculation of a paraxial S-matrix for the scattering inside the slab with a barrier, as shown in Figure 1b. It can easily be completed in the instantaneous approximation exactly as above. In such an approximation, only transitions from channel 1 to 2 and to 3 and their inverse are non-zero. One has

$$S_{m,n}^{1\to 3} = \frac{2}{\sqrt{bh}}\int_0^h \sin\left(\frac{\pi m}{b}y\right)\sin\left(\frac{\pi n}{h}y\right)dy = = \frac{(-1)^n \sin(\pi h m/b)}{\pi\sqrt{bh}}\left((m/b - n/h)^{-1} - (m/b + n/h)^{-1}\right).$$

Similarly,

$$S_{m,n}^{1\to 2} = \frac{2}{\sqrt{b(b-h)}}\int_h^b \sin\left(\frac{\pi m}{b}y\right)\sin\left(\frac{\pi n}{b-h}(b-y)\right)dy = = \frac{(-1)^n \sin(\pi m h/b)}{\pi\sqrt{b(b-h)}}\left((m/b - n/(b-h))^{-1} - (m/b + n/(b-h))^{-1}\right).$$

The transition matrix $T = |S|^2$ also has the same block structure. Retaining only the pole (the first) terms (and slightly changing the notations), one obtains

$$T = \begin{pmatrix} 0_{1,1} & t_{n_1,n_2}^{1\to 2} & t_{n_1,n_3}^{1\to 3} \\ t_{n_1,n_2}^{1\to 2} & 0_{2,2} & 0_{2,3} \\ t_{n_1,n_3}^{1\to 3} & 0_{3,2} & 0_{3,3} \end{pmatrix} \quad (34)$$

where $n_i = 1, \ldots, N_i$ with N_i given by (4) and

$$t_{n_1,n_2}^{1\to 2} = \frac{z_2 \sin^2(\pi n_1 h/b)}{\pi^2 (z_2 n_1 - n_2)^2}, \qquad t_{n_1,n_3}^{1\to 3} = \frac{z \sin^2(\pi n_1 h/b)}{\pi^2 (z n_1 - n_3)^2}, \qquad z_2 = 1 - z, \quad z = \frac{h}{b}. \quad (35)$$

Due to the block structure of the transition matrix (34), it follows that its eigenvalues Λ are determined by the relation $\Lambda^2 = \lambda_\beta$ where λ_β are eigenvalues of $N_1 \times N_1$ matrix

$$P_{m,n} = \sum_{k=1}^{N_2} t_{m,k}^{1\to 2} t_{n,k}^{1\to 2} + \sum_{k=1}^{N_3} t_{m,k}^{1\to 3} t_{n,k}^{1\to 3}.$$

For large matrix dimensions, the summation can be extended over all integer k, and the sums can be calculated explicitly by using (A7) from Appendix C. The results are

$$P_{m,m} = \frac{z^2 + (1-z)^2}{3}(3 - 2\sin^2(\pi m z)), \tag{36}$$

and for $m \neq n$

$$P_{m,n} = \frac{2(\sin^2(\pi z m) + \sin^2(\pi n z))}{\pi^2 (m-n)^2} - \frac{2(1-2z)\sin(\pi m z)\sin(\pi n z)\sin(\pi(m-n)z)}{\pi^3 z(1-z)(m-n)^3}. \tag{37}$$

This matrix is a combination of Toeplitz terms dependent on the difference $m-n$ and oscillating terms (which explains the existence of forbidden zones in its spectrum; see Figure 4a).

Due to the unitarity of the B-matrix, the exact transition matrix $T = |B|^2$ has the largest eigenvalue equal to 1, whose corresponding eigenvector is $(1, 1, \ldots, 1)$. It is natural (and is confirmed by calculations) that eigenvectors of the P-matrix corresponding to large moduli eigenvalues are slowly varying functions. Consequently, all oscillating terms in (36) and (37) for large m and n could be ignored. These arguments lead to the following recipe of the next step of approximation. Put $m = n + r$ and average all matrix elements of the P-matrix over quickly changing phase $\pi n z$. The calculations are straightforward and

$$\langle P_{m,n} \rangle \equiv \lim_{N \to \infty} \frac{1}{N} \sum_{n=1}^{N} P_{n+r,n} = f_{m-n} \tag{38}$$

where

$$f_0 = \frac{2(z^2 + (1-z)^2)}{3}, \qquad f_{r \neq 0} = \frac{2}{\pi^2 r^2} - \frac{(1-2z)\sin(2\pi r z)}{2\pi^3 z(1-z) r^3}.$$

Eigenvalues of such a matrix for large N are calculated by the Fourier transform of this symbol

$$\lambda_\beta = \sum_{r=-\infty}^{\infty} f_r e^{2\pi i r x}, \qquad x = \frac{\beta}{2N_1}.$$

The necessary sums are expressed through the Bernoulli polynomials (A4), (A5), and the result is

$$\lambda_\beta = \frac{2}{3}(z^2 + (1-z)^2) + 4B_2(x) - \frac{1-2z}{3z(1-z)}\Big(B_3(\{x+z\}) - B_3(\{x-z\})\Big).$$

From the beginning, one can assume that $h < h - b$, i.e., $z = h/b < 1/2$ (the case $h/b = 1/2$ was discussed in Section 4). Then

$$\lambda_\beta = \begin{cases} 1 - 4x + \dfrac{x^2}{z(1-z)}, & 0 \le x \le z \\ \dfrac{1 - 2z + 2z^2}{1-z} - \dfrac{2x(1-x)}{1-z}, & z \le x \le \frac{1}{2} \end{cases}, \qquad x = \frac{\beta}{2N_1}. \tag{39}$$

As eigenvalues of the block matrix (34) $\Lambda = \pm\sqrt{\lambda_\beta}$, it follows that close to maximum value (i.e., with small β),

$$\Lambda \approx \pm\left(1 - \frac{\beta}{N_1} + \mathcal{O}\left(\frac{1}{N_1}\right)\right) \approx \pm\left(1 - \frac{2\beta}{N}\right).$$

As in the calculation of the form factor (14) small moduli eigenvalues are irrelevant, one can ignore higher-order terms in the above expression, which gives the same expression as in (30). It means that the level compressibility of barrier billiards with irrational ratio h/b has the same value as in the preceding sections $\chi = \frac{1}{2}$.

In Figure 5a, the above formulas are compared with the results of direct calculations for the P-matrix with $h/b = 1/\sqrt{5}$. As has been demonstrated, the approximate expression (39) is tangent to the exact spectrum close to 1. The form factor computed numerically for the same ratio h/b is presented in Figure 5b. The agreement with the above result is clearly seen.

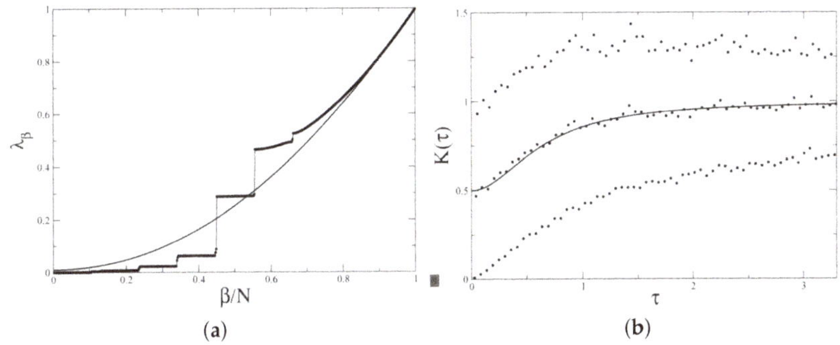

Figure 5. (**a**) Eigenvalues of the P-matrix (36), (37) for for $h/b = 1/\sqrt{5}$ for the same parameters as in Figure 3a. Solid line indicates approximate expressions (39). (**b**) Form factor for $h/b = 1/\sqrt{5}$, $b = \pi$, $k = 200.5$, $N = 399$ averaged over 1000 realisations. Other notations are as in Figure 3b.

6. Barrier Billiard with Rational Ratio *h/b* = *p/q*

The calculation of transition matrix eigenvalues when the ratio h/b is a rational number can be completed by a similar method. An additional difficulty in such case is that one has to select special combinations of states in the second and the third channels to remove trivial eigenvalues equal to zero on the whole line passing through the barrier. It has been discussed in detail in [8] and briefly reviewed in Appendix B. Combining all terms together, one concludes that the transition matrix when $h/b = p/q$ with p and q being co-prime integers has the block form similar to (34) but with one more block

$$T = \begin{pmatrix} 0_{1,1} & t^{1\to 2}_{n_1,n_2} & t^{1\to 3}_{n_1,n_3} & t^{1\to 4}_{n_1,n4} \\ t^{1\to 2}_{n_1,n_2} & 0_{2,2} & 0_{2,3} & 0_{2,4} \\ t^{1\to 3}_{n_1,n_3} & 0_{3,2} & 0_{3,3} & 0_{3,4} \\ t^{1\to 4}_{n_1,n_4} & 0_{4,2} & 0_{4,3} & 0_{4,4} \end{pmatrix}. \tag{40}$$

Here, indices $n_j = 1, \ldots N_j^{(r)}$ have the following restrictions

$$n_1 \not\equiv 0 \mod q\,, \qquad n_2 \not\equiv 0 \mod q-p\,, \qquad n_3 \not\equiv 0 \mod p$$

and

$$N_1^{(r)} = N_1 - N_0\,, \qquad N_2^{(r)} = N_2 - N_0\,, \qquad N_3^{(r)} = N_3 - N_0\,, \qquad N_4^{(r)} = N_0 \tag{41}$$

with N_1, N_2, N_3 given by (4) and N_0 is determined by (A2) or (6). The total matrix dimension is $N^{(r)} = \sum_{j=1}^4 N_j^{(r)} = N_1 + N_2 + N_3 - 2N_0$ as in (6).

Matrices $t^{1\to 2}$ and $t^{1\to 3}$ are the same as in (35) and $t^{1\to 4}$ given by (A3) from Appendix B

$$t^{1\to 4}_{n_1,n_4} = \frac{\sin^2(\pi m h/b)}{\pi^2 p(q-p)(n_1/q - n_4)^2}\,.$$

The eigenvalues of block matrix (40) Λ are $\Lambda = \lambda_\beta^2$ where λ_β with $\beta = 1, \ldots, N_1$ are eigenvalues of matrix (superscript (res), indicating that the matrix describes the resonance case $h/b = p/q$)

$$P_{m,n}^{(\mathrm{res})} = \sum_{k \neq 0 \mod (q-p)} t_{m,k}^{1\to 2} t_{n,k}^{1\to 2} + \sum_{k \neq 0 \mod p} t_{m,k}^{1\to 3} t_{n,k}^{1\to 3} + \sum_k t_{m,k}^{1\to 4} t_{n,k}^{1\to 4}.$$

Using an evident relation

$$\sum_{k \neq 0 \mod r} f(k) = \sum_k f(k) - \sum_k f(rk)$$

and (A7), the above sums can be explicitly calculated.

The results are

$$\begin{aligned} P_{m,m}^{(\mathrm{res})} &= \frac{1}{3q^2}(p^2 + (q-p)^2)(3 - 2\sin^2(\pi pm/q)) \\ &+ \frac{2\sin^4(\pi pm/q)(3 - 2\sin^2(\pi m/q))}{3p(q-p)q^2\sin^4(\pi m/q)} \end{aligned} \tag{42}$$

and when $m \neq n$

$$P_{m,n}^{(\mathrm{res})} = \frac{b_{m,n}}{(m-n)^2} - \frac{c_{m,n}\sin(\pi p(m-n)/q)}{(m-n)^3} - \frac{d_{m,n}\sin(\pi(m-n)/q)}{(m-n)^3} \tag{43}$$

where

$$\begin{aligned} b_{m,n} &= \frac{2}{\pi^2}\big(\sin^2(\pi pm/q) + \sin^2(\pi pn/q)\big) \\ &+ \frac{2\sin^2(\pi pm/q)\sin^2(\pi pn/q)}{p(q-p)\pi^2}\Big(\frac{1}{\sin^2(\pi m/q)} + \frac{1}{\sin^2(\pi n/q)}\Big), \\ c_{m,n} &= \frac{2q(q-2p)}{\pi^3 p(q-p)}\sin(\pi pm/q)\sin(\pi pn/q), \\ d_{m,n} &= \frac{4q\sin^2(\pi pm/q)\sin^2(\pi pn/q)}{\pi^3 p(q-p)\sin(\pi m/q)\sin(\pi n/q)}. \end{aligned} \tag{44}$$

Although these expressions are indexed by integers m and n, this notation is symbolic. The point is that by construction, these integers cannot be arbitrary but have to be not divisible by q. Let us order such numbers and let $\nu(k)$ with $k = 1, 2, \ldots,$ be the kth integer $\not\equiv 0 \mod q$. Then, indices of matrix $P_{m,n}^{(\mathrm{res})}$ have to be considered as follows: $m = \nu(j)$, $n = \nu(k)$ with $j, k = 1, 2, \ldots, N_1^{(\mathrm{r})}$ with $N_1^{(\mathrm{r})}$ defined in (41). In such notation, matrix $P^{(\mathrm{res})}$ is $N_1^{(\mathrm{r})} \times N_1^{(\mathrm{r})}$ matrix

$$P_{m,n}^{(\mathrm{res})} \equiv P_{\nu(j),\nu(k)}^{(\mathrm{res})}, \qquad j, k = 1, \ldots, N_1^{(\mathrm{r})}.$$

The next step, as in the previous section (cf., (38)), consists in the substitution instead of the above exact expressions, their mean values with fixed difference between the indices

$$\langle P_{m,n}^{(\mathrm{res})}\rangle(r) = \lim_{N\to\infty} \frac{1}{N}\sum_{n=1}^{N} P_{n+r,n}^{(\mathrm{res})}$$

where both integers n and $n + r$ have to be not divisible by q.

According to (42) and (43), the $P^{(\mathrm{res})}$ matrix is a mixture of functions depending explicitly on the differences of indices and certain coefficients depending on indices modulus

q. Only the latter requires the explicit averaging. Using (A8)–(A11) from Appendix C, one obtains that

$$\sum_{m=1}^{q-1} P_{m,m} = \frac{(p^2+(q-p)^2)(2q-3)}{3q^2} + \frac{2p^2(2q-3p)}{3(q-p)q^2},$$
$$\sum_{m=1}^{q-1}{}' b_{m,m+r} = \frac{2q(q-p+1-2\sin^2(\pi pr/q))}{\pi^2(q-p)},$$
$$\sum_{m=1}^{q-1}{}' c_{m,m+r} = \frac{q^2(q-2p)\cos(\pi pr/q)}{\pi^3 p(q-p)}, \qquad \sum_{m=1}^{q-1}{}' d_{m,m+r} = \frac{q^2\sin(2\pi pr/q)}{\pi^3 p(q-p)\sin(\pi r/q)}.$$

Here, it is taken into account that $p/q < 1/2$. The superscript $'$ in these sums indicates that the term with $m+r \equiv 0 \mod q$ is omitted. The latter condition implies that the number of independent terms equal $q-1$ if $r \equiv 0 \mod q$ or $q-2$ otherwise. Finally, one obtains

$$\langle P^{(\mathrm{res})}_{m,n} \rangle = \alpha_1 \delta_{m,n} + f_{m-n} \tag{45}$$

with

$$f_{m-n} = \begin{cases} \dfrac{\alpha_2}{(m-n)^2}, & (m-n) \equiv 0 \mod q \\ \dfrac{\alpha_3}{(m-n)^2} + \dfrac{\alpha_4\cos(2\pi pr/q)}{(m-n)^2} + \dfrac{\alpha_5\sin(2\pi pr/q)}{(m-n)^3}, & (m-n) \equiv r \neq 0 \mod q \end{cases}$$

where constants α_j are

$$\alpha_1 = \frac{(p^2+(q-p)^2)(2q-3)}{3q^2(q-1)} + \frac{2p^2(2q-3p)}{3(q-p)q^2(q-1)} \qquad \alpha_2 = \frac{2q(q-p+1)}{\pi^2(q-p)(q-1)},$$
$$\alpha_3 = \frac{2q}{\pi^2(q-2)}, \qquad \alpha_4 = \frac{2q}{\pi^2(q-p)(q-2)}, \qquad \alpha_5 = \frac{q^2(2p-q-2)}{2\pi^3 p(q-p)(q-2))}.$$

Although this matrix depends only on the difference of indices $m-n$, it is not a Toeplitz matrix as m and n are not arbitrary numbers but only integers not divisible by q. Nevertheless, one can argue that the largest eigenvalues for large matrix dimension can be calculated by a formula similar to Toeplitz matrices (which is a kind of variational method)

$$\lambda_\beta = \frac{1}{N_1^{(\mathrm{r})}} \sum_{j,k=1}^{N_1^{(\mathrm{r})}} \langle P^{(\mathrm{res})}_{\nu(j),\nu(k)} \rangle e^{2\pi i x(j-k)}, \qquad x = \frac{\beta}{2N_1^{(\mathrm{r})}}. \tag{46}$$

Here, as above, $\nu(k)$ is the kth integer $\not\equiv 0 \mod q$.

In Appendix C (see (A13)), it is shown that such a sum can be written as follows

$$\lambda_\beta = \alpha_1 + \alpha_2 \sum_{u=-\infty}^{\infty} \frac{e^{2\pi i x(q-1)u}}{(qu)^2} + 2\mathrm{Re}\sum_{t=1}^{q-1}\left(1 - \frac{t}{q-1}\right) \times$$
$$\times \sum_{u=-\infty}^{\infty} e^{2\pi i x((q-1)u+t)} \left(\frac{\alpha_3}{(uq+t)^2} + \frac{\alpha_4\cos(2\pi pt/q)}{(uq+t)^2} + \frac{\alpha_5\sin(2\pi pr/q)}{(uq+t)^3}\right).$$

The first sum is calculated through the Bernoulli polynomial $B_2(x)$ (see (A4)). The last sums are expressed through two functions

$$G(x,r) = \sum_{k=-\infty}^{\infty} \frac{e^{2\pi i x(kq+r)}}{(kq+r)^2}, \qquad F(x,r) = \sum_{k=-\infty}^{\infty} \frac{e^{2\pi i x(kq+r)}}{(kq+r)^3}.$$

The explicit expressions of these function can be obtained as follows.

Define one more function

$$g(x,r) = \sum_{k=-\infty}^{\infty} \frac{e^{2\pi i x(kq+r)}}{kq+r}.$$

By the differentiation over x, one has $G'(x,r) = 2\pi i g(x,r)$ and $F'(x,r) = 2\pi i G(x,r)$. As the differentiation of $g(x,r)$ over x gives the sum of δ-function, it is plain that $g(x,r)$ is the piece-wise constant function in interval $[j/q,(j+1)/q]$. Using (A6), one gets

$$g(j/q+x,r) = \frac{\pi}{q\sin(\pi r/q)} e^{i\pi r(2j+1)/q}.$$

Correspondingly, function $G(x,r)$ is a piece-wise linear function in the same intervals

$$G(j/q+x,r) = \frac{\pi^2}{q^2\sin^2(\pi r/q)} e^{2\pi i r j/q} + \frac{2\pi^2 i x}{q\sin(\pi r/q)} e^{i\pi r(2j+1)/q}. \tag{47}$$

In the same way, one proves that function $F(x,r)$ is a piece-wise quadratic function

$$\begin{aligned} F(j/q+x,r) &= \frac{\pi^3\cos(\pi r/q)}{q^3\sin^3(\pi r/q)} e^{2\pi i r j/q} + \frac{2\pi^3 i x}{q^2\sin^2(\pi r/q)} e^{2\pi i r j/q} - \\ &- \frac{2\pi^3 x^2}{q\sin(\pi r/q)} e^{i\pi r(2j+1)/q}. \end{aligned} \tag{48}$$

In all these formulas, $j = 0,\ldots,q-1$ and $0 \le x \le 1/q$.

Combining all terms together, one finds

$$\lambda_\beta = \alpha_1 + \frac{2\pi^2\alpha_2}{q^2} B_2(\{x(q-1)\}) + 2\mathrm{Re}\sum_{t=1}^{q-1}\Big(1-\frac{t}{q-1}\Big) \times$$
$$\times\; e^{ixt/q}\Big[(\alpha_3+\alpha_4\cos(2\pi pt/q))G\Big(\frac{x(q-1)}{q},t\Big) + \alpha_5\sin(2\pi pt/q)F\Big(\frac{x(q-1)}{q},t\Big)\Big].$$

The main interest for the calculation of the form factor is the behaviour of the largest eigenvalues for x close to zero. Using (47) and (48), one concludes that

$$\lambda_\beta = C_0 + C_1\frac{\beta}{N_1^{(\mathrm{r})}}, \qquad 0 \le \beta \ll N_1^{(\mathrm{r})}.$$

Here

$$\begin{aligned} C_0 &= \alpha_1 + \frac{\pi^2\alpha_2}{3q^2} + \\ &+ \frac{2\pi^2}{q^2}\sum_{t=1}^{q-1}\Big(1-\frac{t}{q-1}\Big)\Big(\frac{\alpha_3+\alpha_4\cos(2\pi pt/q)}{\sin^2(\pi t/q)} + \frac{\pi\alpha_5\sin(2\pi pr/q)\cos(\pi t/q)}{q\sin^3(\pi t/q)}\Big) \end{aligned}$$

and

$$C_1 = -\frac{2\pi^2\alpha_2(q-1)}{2q^2} - \frac{2\pi^2(q-1)}{q^2}\sum_{t=1}^{q-1}\Big(1-\frac{t}{q-1}\Big)(\alpha_3+\alpha_4\cos(2\pi pt/q)).$$

The sum over residues is of the form

$$\sum_{t=1}^{q-1}\Big(1-\frac{t}{q-1}\Big)h(t)$$

and (as it is easy to check) in the considered case $h(q-t) = h(t)$. Therefore

$$\sum_{t=1}^{q-1} t\,h(t) = \sum_{t=1}^{q-1}(q-t)\,h(q-t) = \sum_{t=1}^{q-1}(q-t)\,h(t) \longrightarrow \sum_{t=1}^{q-1} t\,h(t) = \frac{q}{2}\sum_{t=1}^{q-1} h(t)\,.$$

Consequently

$$\begin{aligned} C_0 &= \alpha_1 + \frac{\pi^2\alpha_2}{3q^2} + \\ &+ \frac{2\pi^2(q-2)}{q^2(q-1)}\sum_{t=1}^{q-1}\left(\frac{\alpha_3+\alpha_4\cos(2\pi pt/q)}{\sin^2(\pi t/q)} + \frac{\pi\alpha_5\sin(2\pi pr/q)\cos(\pi t/q)}{q\sin^3(\pi t/q)}\right) \end{aligned}$$

and

$$C_1 = -\frac{2\pi^2\alpha_2(q-1)}{2q^2} - \frac{2\pi^2(q-2)}{q^2}\sum_{t=1}^{q-1}\left(\alpha_3+\alpha_4\cos(2\pi pt/q)\right).$$

Using sums indicated in Appendix C and collecting all terms in the end, one finds that

$$C_0 = 1, \qquad C_1 = -2.$$

This result signifies that the largest moduli eigenvalues of the transition matrix for the barrier billiard with rational ratio $h/b = p/q$ are (i) independent of values of integers p and q and (ii) have the same asymptotic expression as in (30) (taking into account that $N^{(\mathrm{r})} \approx 2N_1^{(\mathrm{r})}$)

$$\Lambda = \pm\left(1 - \frac{\beta}{N_1^{(\mathrm{r})}} + \mathcal{O}\left(\frac{1}{N_1^{(\mathrm{r})}}\right)\right) \approx \pm\left(1 - \frac{2\beta}{N^{(\mathrm{r})}}\right).$$

As it has been explained above, it implies that (i) the form factor for barrier billiards is different for odd and even n and (ii) the spectral compressibility is exactly equal 1/2 for all positions of the barrier

$$\chi \equiv K(0) = \frac{1}{2}.$$

The numerical calculations exemplified in Figure 6 confirm well these results.

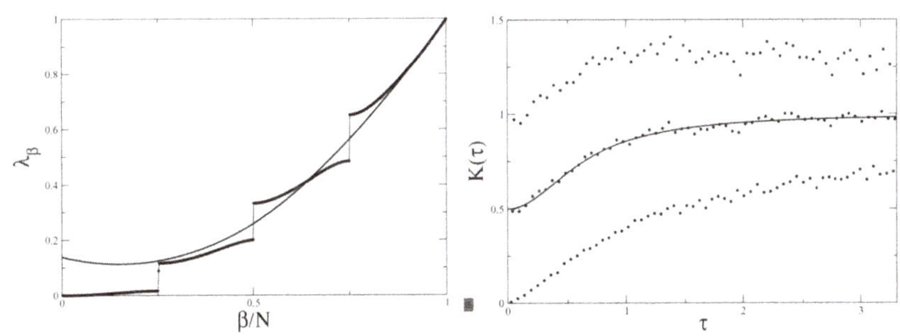

Figure 6. (**a**) Eigenvalues of the P-matrix (42), (43) for $h/b = 2/5$ with $N_1 = 520$. Solid line is the spectrum (49) of the asymptotic matrix. (**b**) Form factor for $h/b = 2/5$, $b = \pi$, $k = 250.5$, and $N = 400$ averaged over 1000 realisations. Other notations are as in Figure 3b.

7. Summary

It is demonstrated that the method of calculation of the level of compressibility proposed by G. Tanner in [18] for chaotic systems can successfully be applied for intermediate statistics models. The criterium discussed in [18] states that if the difference between the dominant eigenvalue of the transition matrix $\Lambda = 1$ and the second in magnitude eigenvalue is big enough, then only the dominant eigenvalues contributes to the form factor, and one obtains the usual value of the form factor corresponding to standard random matrix ensembles. Notably, the level compressibility is zero.

For models considered in the paper, no individual transition matrix eigenvalues dominate, and one has to sum over many of them with moduli close to 1. Two types of random unitary matrices were investigated. The first corresponding to a quantisation of an interval-exchange map [20] has been discussed in detail in [21–24]. In particular, the values of the level compressibility were derived. The application of the transition matrix approach for this case serves first of all to check the validity of Tanner's method for intermediate statistics models. It appears that interval-exchange matrices lead to circulant transition matrices whose eigenvalues are explicitly known and all necessary sums are easily estimated. In the end, one obtains the same values of the level compressibility as obtained in [21–24] but with much simpler and transparent calculations. An example is of a special interest. It corresponds to interval-exchange matrices with an irrational value of a parameter (which strictly speaking describes not an interval-exchange map but only a parabolic one). Numerically, it has observed in [22] that in such case, spectral statistics is of usual random matrix type (GOE or GUE depending on a symmetry) as for chaotic systems, which looks strange as the Lyapunov exponent of any parabolic map is zero. The transition matrix approach clearly indicates that although there is no dominant eigenvalue as has been discussed in [18], all eigenvalues except $\Lambda = 1$ for large matrix dimensions are so quickly oscillating that averaging over a small interval of the argument effectively removes their contributions, producing the standard random matrix result.

The main part of the paper contains the calculation of the level compressibility for random unitary matrices derived from the exact quantisation of barrier billiards in [7,8]. The importance of such matrices comes from the fact that they have the same spectral statistics as high-excited states of barrier billiards, which are the simplest examples of pseudo-integrable models for which very little is known analytically.

The barrier billiard transition matrices are more complicated than the ones for interval-exchange matrices. Their spectra contain forbidden zones, and their exact eigenvalues seems not to be accessible in closed form. Nevertheless, as the level compressibility requires the control only of largest moduli eigenvalues of the transition matrix, it is possible to find such eigenvalues for large matrix dimensions precisely. The main simplification comes from the fact that eigenvectors corresponding to largest moduli eigenvalues are slow oscillating functions. Therefore, quickly oscillating terms in matrix elements will give negligible contributions on these eigenvectors, and one can substitute instead of exact matrix elements their average over fast oscillations. The resulting matrices are simpler and permit finding their large moduli eigenvalues analytically. In the end, one proves that the level compressibility of barrier billiards for all positions and heights of the barrier is the same and equals $1/2$. This result strongly indicates that spectral statistics of the B-matrices associated with barrier billiards is universal (i.e., independent on the barrier position) and well described by the semi-Poisson distribution.

Funding: This research received no external funding.

Conflicts of Interest: The author declares no conflict of interest.

Appendix A. Approximate Expression for the B-Matrix for the Symmetric Billiard

The purpose of this appendix is the determination of the transition matrix for the symmetric case (i.e., $x_m = (-1)^{m+1} y_m$, $y_m = b\sqrt{k^2 - \pi^2 m^2/b^2}$) in the paraxial approximation by taking into account only the pole terms in the definition (1). From (2) with odd

$m = 2j - 1$, it follows (for simplicity, it is assumed N is even $N = 2n$ and the products is taken from $q = 1$ until $q = n$)

$$2y_{2j-1}L_{2j-1}^2 = (2y_{2j-1})^2 \prod_{q\neq j} \frac{y_{2j-1}+y_{2q-1}}{y_{2j-1}-y_{2q-1}} \prod_q \frac{y_{2j-1}-y_{2q}}{y_{2j-1}+y_{2q}}$$
$$= \prod_q \left(\frac{y_{2j-1}+y_{2q-1}}{y_{2j-1}+y_{2q}}\right)^2 \frac{\pi^2 \prod_q [q^2-(j-1/2)^2]}{\prod_{q\neq j}[(q-1/2)^2-(j-1/2)^2]} = g_j\, F^2(y_{2j-1})$$

where

$$g_j = \lim_{v\to j-1/2} \frac{\pi^2((j-1/2)^2-v^2)\prod_q[1-(j-1/2)^2/q^2]}{\prod_q[1-v^2/(q-1/2)^2]},$$
$$F(z) = \prod_q \frac{q(z+y_{2q-1})}{(q-1/2)(z+y_{2q})}.$$

Exactly in the same way, one gets

$$2y_{2j}L_{2j}^2 = -(2y_{2j})^2 \prod_{q\neq j} \frac{y_{2j}+y_{2q}}{y_{2j}-y_{2q}} \prod_q \frac{y_{2j}-y_{2q-1}}{y_{2j}+y_{2q-1}}$$
$$= -\prod_q \left(\frac{y_{2j}+y_{2q}}{y_{2j}+y_{2q-1}}\right)^2 \frac{\pi^2 \prod_q[(q-1/2)^2-j^2]}{\prod_{q\neq j}[q^2-j^2]} = \frac{f_j}{F^2(y_{2j})}$$

with

$$f_j = -\lim_{v\to j} \frac{\pi^2(j^2-v^2)\prod_q[1-j^2/(q-1/2)^2]}{\prod_q[1-v^2/q^2]}.$$

All products in the above expressions should be taken from 1 to n. If j is not too close to kb/π (i.e., the momentum is not close to the threshold of new propagating modes), the products in g_j and f_j can be extended to infinity, and these functions can easily be calculated from standard expressions

$$\sin \pi x = \pi x \prod_{q=1}^{\infty}\left(1-\frac{x^2}{q^2}\right), \qquad \cos \pi x = \prod_{q=1}^{\infty}\left(1-\frac{x^2}{(q-1/2)^2}\right).$$

In this way, one finds that $g_j = 2$ and $f_j = 2\pi^2 j^2$. Finally

$$L_{2j-1} = \frac{F(y_{2j-1})}{\sqrt{y_{2j-1}}}, \qquad L_{2j} = \frac{\pi j}{F(y_{2j})\sqrt{y_{2j}}}.$$

The T-matrix elements are

$$T_{m,n} = \frac{L_m^2 L_n^2}{(x_m+x_n)^2}.$$

In the paraxial approximation, one should take into account only the terms with x_m of different signs. For symmetric billiard, it means that $T_{2n,2m} \approx 0$, $T_{2n-1,2m-1} = 0$ and

$$T_{2m-1,2n} = \frac{L_{2m-1}^2 L_{2n}^2}{(y_{2m-1}-y_{2n})^2} \approx \frac{F^2(y_{2m-1})}{F^2(y_{2n})} \frac{\pi^2 n^2}{y_{2m-1}y_{2n}(y_{2m-1}-y_{2n})^2}.$$

As only the pole terms are important, one can put $y_{2n} \approx y_{2m-1}$ and $F^2(y_{2m-1}) \approx F^2(y_{2n})$

$$T_{2m-1,2n} \approx \frac{\pi^2 n^2}{y_{2m-1}y_{2n}(y_{2m-1}-y_{2n})^2} = \frac{n^2}{\pi^2(n^2-(m-1/2)^2)^2}\frac{(y_{2m-1}+y_{2n})^2}{y_{2m-1}y_{2n}} \approx$$
$$\approx \frac{4n^2}{\pi^2(n^2-(m-1/2)^2)^2} = \frac{1}{\pi^2(n-m+1/2)^2} + \frac{1}{\pi^2(n+m-1/2)^2}.$$

Appendix B. Instantaneous Approximation for the Resonance Case

When the ratio h/b is a rational number $h/b = p/q$ with co-prime integer p and q, it is plain that for the barrier billiard as in Figure 1a, the following three transverse momenta with integer t (and the corresponding longitudinal momenta $p_t = \sqrt{k^2 - \pi^2 w_t^2}$) are equal

$$w_t = \frac{\pi q t}{b} = \frac{\pi((q-p)t}{b-h} = \frac{\pi p t}{h}. \tag{A1}$$

Introduce the elementary solutions with these momenta in each of three regions indicated in Figure 1b

$$\begin{aligned}
\phi_{qt}^{(1)}(x,y) &= \sqrt{\frac{2}{b}}\sin\left(\frac{\pi q t}{b}y\right)\exp(ip_t x), \qquad 0 \le y \le b, \quad x<0, \\
\phi_{(q-p)t}^{(2)}(x,y) &= \sqrt{\frac{2}{b-h}}\sin\left(\frac{\pi(q-p)t}{b-h}(b-y)\right)\exp(ip_t x), \quad b-h \le y \le b, \quad x>0, \\
\phi_{pt}^{(3)}(x,y) &= \sqrt{\frac{2}{h}}\sin\left(\frac{\pi p t}{h}y\right)\exp(ip_t x), \qquad 0 \le y \le h, \quad x>0.
\end{aligned}$$

Due to the resonant conditions (A1), all these solutions represent exact solutions for the scattering inside the slab in Figure 1b. The number of such solutions is

$$N_0 = \left[\frac{kb}{\pi q}\right]. \tag{A2}$$

When spectral statistics of non-trivial eigenvalues is considered, these solutions should be removed. This has been completed in detail in [8]. Below, the derivation of the paraxial approximation for the T-matrix in such case is briefly discussed.

The paraxial approximation for the T-matrix for the scattering inside the slab in Figure 1b for non-resonant waves when in the first region $n_1 \neq 0 \mod q$, in the second region $n_2 \neq 0 \mod q-p$, and in the third one $n_3 \neq 0 \mod p$ are given by the same expression as in (35). The first step consists in removing all waves from region 1 proportional to q. However, it is not enough, as waves from the second and the third regions can diffract into waves in the first region with $n_1 \equiv 0 \mod q$. To remove them, notice that

$$\phi_{qt}^{(1)}(x,y) = \begin{cases} -(-1)^{qt}\sqrt{\dfrac{b}{b-h}}\phi_{(q-p)t}^{(2)}(x,y), & h \le y \le b \\ \sqrt{\dfrac{b}{h}}\phi_{pt}^{(3)}(x,y), & 0 \le y \le b \end{cases}.$$

Therefore, the following linear combination [8]

$$\phi_t^{(4)}(x,y) = \sqrt{\frac{h}{b}}\phi_{(q-p)t}^{(2)}(x,y) + (-1)^{qt}\sqrt{\frac{b-h}{b}}\phi_{(q-p)t}^{(3)}(x,y)$$

is orthogonal to $\phi_{qt}^{(1)}(x,y)$ and cancels undesirable waves with $n_1 \equiv 0 \mod q$.

The calculation of the scattering into such state can be performed as above

$$\begin{aligned} S^{1\to 4}_{m,t} &= 2\sqrt{\frac{h}{b^2(b-h)}}\int_h^b \sin\left(\frac{\pi m y}{b}\right)\sin\left(\frac{\pi(q-p)t(b-y)}{b-h}\right)dy + \\ &+ 2(-1)^{qt}\sqrt{\frac{b-h}{b^2 h}}\int_0^h \sin\left(\frac{\pi m y}{b}\right)\sin\left(\frac{\pi p t y}{h}\right)dy = \\ &= \frac{(-1)^{(p+q)t}\sin(\pi m h/b)b}{\sqrt{h(b-h)}}\left(\frac{1}{m-qt}-\frac{1}{m+qt}\right). \end{aligned}$$

The paraxial approximation of the corresponding transition matrix elements is given by the pole term

$$t^{1\to 4}_{m,t} = \frac{\sin^2(\pi m h/b)}{\pi^2 p(q-p)(m/q-t)^2}, \qquad m \not\equiv 0 \mod q. \tag{A3}$$

Appendix C. Divers Relations

In this appendix, a few formulas used in the text are briefly reviewed.

The known identity (see e.g., [32])

$$\sum_{r=-\infty}^{\infty}{}' \frac{1}{r^n}e^{2\pi i r x} = -\frac{(2\pi i)^n}{n!}B_n(\{x\}) \tag{A4}$$

where $\{x\}$ is the fractional part of x and $B_n(x)$ are Bernoulli polynomials. For example,

$$B_2(x) = x^2 - x + \frac{1}{6}, \qquad B_3(x) = x^3 - \frac{3}{2}x^2 + \frac{1}{2}x. \tag{A5}$$

The following identities are standard and presented for completeness.

$$\sum_{n=-\infty}^{\infty}\frac{1}{n+x} = \pi\cot(\pi x), \qquad \sum_{n=-\infty}^{\infty}\frac{1}{(n+x)(n+y)} = \frac{\pi(\cot(\pi x)-\cot(\pi y))}{y-x}. \tag{A6}$$

The first formula is simply the expansion of the right-hand side over poles. The second is a consequence of the first. Differentiating the above expressions by x and y shows that

$$\begin{aligned} &\sum_{n=-\infty}^{\infty}\frac{1}{(n+x)^2} = \frac{\pi^2}{\sin^2(\pi x)}, \qquad \sum_{n=-\infty}^{\infty}\frac{1}{(n+x)^4} = \frac{\pi^4(3-2\sin^2(\pi x))}{3\sin^4(\pi x)}, \\ &\sum_{n=-\infty}^{\infty}\frac{1}{(n+x)^2(n+y)^2} = \frac{\pi^2}{(x-y)^2}\left(\frac{1}{\sin^2(\pi x)}+\frac{1}{\sin^2(\pi y)}\right) + \\ &+ \frac{2\pi(\cot(\pi x)-\cot(\pi y))}{(x-y)^3}. \end{aligned} \tag{A7}$$

In different places of the paper, one needs to calculate finite sums $\sum_{t=1}^{q-1} f(t)$ over residues mod q. A characteristic feature of such sums is that their summands $f(t)$ can be rewritten as a ratio of two polynomials in variable $z = e^{2\pi i t/q}$. The summation over t from 1 to $q-1$ corresponds to the calculation of the integral

$$\sum_{t=1}^{q-1} f(t) = \frac{1}{2\pi i}\oint_C \frac{q z^{q-1}}{z^q - 1} f(z) dz$$

where contour C encircled all roots of $z^q - 1 = 0$ except the one with $z = 1$. By deforming the contour and calculating the necessary residues, one can obtain the necessary sums in closed form.

Below, a few formulas obtained by this manner are listed

$$\sum_{t=1}^{q-1} \frac{1}{\sin^2(\pi t/q)} = \frac{q^2-1}{3}, \qquad \sum_{t=1}^{q-1} \frac{\cos(2\pi pt/q)}{\sin^2(\pi t/q)} = \frac{q^2+6p^2-1}{3} - 2qp, \tag{A8}$$

$$\sum_{t=1}^{q-1} \frac{\sin(2\pi pt/q)\cos(\pi t/q)}{\sin^3(\pi t/q)} = \frac{2p(q^2+2p^2)}{3} - 2qp^2, \tag{A9}$$

$$\sum_{m=1}^{q-1} \frac{\sin^2(\pi pm/q)\sin^2(\pi p(m+r)/q)}{\sin^2(\pi m/q)} = \frac{pq}{2} - p^2\sin^2(\pi pr/q), \tag{A10}$$

$$\sum_{m=1}^{q-1} \frac{\sin^4(\pi pm/q)}{\sin^4(\pi m/q)} = \frac{pq(2p^2+1)}{3} - p^4. \tag{A11}$$

Here, it is assumed that $p/q < 1/2$.

In Section 6, one has to calculate the following sum where $\nu(k)$ is the kth integer not divisible by q

$$S = \frac{1}{N}\sum_{j,k=1}^{N} f(\nu(j)-\nu(k))e^{ix(j-k)}$$

in the limit $N \to \infty$ with a certain quickly decreasing function $f(x)$, $f(x) \underset{|x|\to\infty}{\longrightarrow} x^{-2}$.

To obtain an explicit expression of that sum, notice that the number of integers from 1 to n divisible by q is $[n/q]$ where $[x]$ is the largest integer less of equal n. Therefore, if $n = ql + r$ with $r = 1, \ldots, q-1$ then $k = (q-1)l + r$. It means that

$$\nu((q-1)l + r) = ql + r, \qquad r = 1, \ldots, q-1. \tag{A12}$$

As integer $\nu(k) \neq 0 \mod q$, the residue $r \neq 0$.

Writing $j = (q-1)m + s$ and $k = (q-1)n + r$ with integer m, n and $r, s = 1, \ldots, q-1$, one gets that

$$j - k = (q-1)u + t, \qquad \nu(j) - \nu(k) = qu + t, \qquad u = m - n, t = s - r.$$

The summation over integers j and k is equivalent to the summation over integers m, n and s, r. Fixing the differences $k = m - n$ and $t = s - r$, using the fact that integers with fixed residue $\mod (q-1)$ are uniformly distributed

$$\lim_{N\to\infty} \frac{1}{N}\sum_{j=1}^{N} F(j \mod (q-1)) = \frac{1}{q-1}\sum_{s=1}^{q-1} F(s)$$

and that

$$\sum_{s=1}^{q-1}\sum_{r=1}^{q-1} f(s-r) = \sum_{t=-(q-2)}^{q-2} (q-1-|t|)f(t)$$

one finds that

$$S = \sum_{u=-\infty}^{\infty}\sum_{t=-(q-2)}^{q-2}\left(1 - \frac{|t|}{q-1}\right) f(qu+t)e^{ix((q-1)u+t)}.$$

Due to a quick decrease of function $f(x)$, the summation over u can be extended to the sum over all integers. It is convenient to separate term with $t = 0$, add together terms with t and $-t$, and in the last term change $u \to -u$

$$
\begin{aligned}
S &= \sum_{u=-\infty}^{\infty} f(qu)e^{ix(q-1)u} + \\
&+ \sum_{u=-\infty}^{\infty}\sum_{t=1}^{q-1}\left(1-\frac{t}{q-1}\right)\left(f(qu+t)e^{ix((q-1)u+t)} + f(-qu-t)e^{-ix((q-1)u+t)}\right).
\end{aligned}
$$

The used function f is even $f(-x) = f(x)$, and this expression can be written as follows

$$
S = \sum_{u=-\infty}^{\infty} f(qu)e^{ix(q-1)u} + 2\mathrm{Re}\sum_{u=-\infty}^{\infty}\sum_{t=1}^{q-1}\left(1-\frac{t}{q-1}\right)f(qu+t)e^{ix((q-1)u+t)}. \qquad \text{(A13)}
$$

References

1. Wigner, E.P. Random matrices in physics. *SIAM Rev.* **1967**, *9*, 1. [CrossRef]
2. Berry, M.V.; Tabor, M. Level clustering in the regular spectrum. *Proc. R. Soc. Lond. A* **1977**, *356*, 375.
3. Bohigas, O.; Giannoni, M.J.; Schmit, C. Characterization of chaotic quantum spectra and universality of level fluctuation laws. *Phys. Rev. Lett.* **1984**, *52*, 1. [CrossRef]
4. Mehta, M.L. *Random Matrices*, 3rd ed.; Academic Press: Cambridge, MA, USA, 2014.
5. Richens, P.J.; Berry, M.V. Pseudointegrable systems in classical and quantum mechanics. *Phys. D Nonlinear Phenomena* **1981**, *2*, 495. [CrossRef]
6. Zemlyakov, A.N.; Katok, A.B. Topological transitivity in billiards in polygons. *Math. Notes* **1975**, *18*, 760. [CrossRef]
7. Bogomolny, E. Barrier billiard and random matrices. *J. Phys. A Math. Theor.* **2022**, *55*, 024001. [CrossRef]
8. Bogomolny, E. Random matrices associated with general barrier billiards. *J. Phys. A Math. Theor.* **2022**, *55*, 254002. [CrossRef]
9. Altshuler, B.L.; Zharekeshev, I.K.; Kotochigava, S.A.; Shklovskii, B.I. Repulsion between levels and the metal-insulator transition. *Sov. Phys. JETP* **1988**, *67*, 625.
10. Shklovskii, B.I.; Shapiro, B.; Sears, B.R.; Lambrianides, P.; Shore, H.B. Statistics of spectra of disordered systems near the metal-insulator transition. *Phys. Rev. B* **1993**, *47*, 11487. [CrossRef]
11. Bogomolny, E.; Gerland, U.; Schmit, C. Short-range plasma model for intermediate spectral statistics. *Eur. Phys. J.* **2001**, *19*, 121. [CrossRef]
12. Chamon, C.; Hamma, A.; Mucciolo, E.R. Emergent Irreversibility and Entanglement Spectrum Statistics. *Phys. Rev. Lett.* **2014**, *112*, 240501. [CrossRef] [PubMed]
13. Berry, M.V. Semiclassical theory of spectral rigidity. *Proc. R. Soc. Lond. A* **1985**, *400*, 229.
14. Wiersig, J. Spectral properties of quantized barrier billiards. *Phys. Rev. E* **2002**, *65*, 046217. [CrossRef]
15. Giraud, O. Spectral Statistics of Difractive Systems. Ph.D. Thesis, University Paris-Sud, Orsay, France, 2002.
16. Giraud, O. Periodic orbits and semiclassical form factor in barrier billiards. *Commun. Math. Phys.* **2005**, *260*, 183. [CrossRef]
17. Giraud, O. LPTMS, Orsay, France. Personal communication, 2021.
18. Tanner, G. Unitary-stochastic matrix ensembles and spectral statistics. *J. Phys. A Math. Gen.* **2001**, *34*, 8485. [CrossRef]
19. Gnutzmann, S.; Smilansky, U. Quantum graphs: Applications to quantum chaos and universal spectral statistics. *Adv. Phys.* **2006**, *55*, 527. [CrossRef]
20. Giraud, O.; Marklof, J.; O'Keefe, S. Intermediate statistics in quantum maps. *J. Phys. A Math. Gen.* **2004**, *37*, L303. [CrossRef]
21. Bogomolny, E.; Schmit, C. Spectral statistics of a quantum interval-exchange map. *Phys. Rev. Lett.* **2004**, *93*, 254102. [CrossRef] [PubMed]
22. Bogomolny, E.; Dubertrand, R.; Schmit, C. Spectral statistics of a pseudo-integrable map: The general case. *Nonlinearity* **2009**, *22*, 2101. [CrossRef]
23. Bogomolny, E.; Giraud, O.; Schmit, C. Random matrix ensembles associated with Lax matrices. *Phys. Rev. Lett.* **2009**, *103*, 054103. [CrossRef] [PubMed]
24. Bogomolny, E.; Giraud, O.; Schmit, C. Integrable random matrix ensembles. *Nonlinearity* **2011**, *24*, 3179. [CrossRef]
25. Levitov, L.S. Localization-delocalization transition for one-dimensional alloy potentials. *EPL* **1988**, *7*, 343. [CrossRef]
26. Altshuler, B.L.; Levitov, S. Weak chaos in a quantum Kepler problem. *Phys. Rep.* **1997**, *288*, 487. [CrossRef]
27. Grenander, U.; Szego, G. *Toeplitz Forms and Their Applications*; University of California Press: Berkeley, CA, USA; Los Angeles, CA, USA, 1958.
28. Böttcher, A.; Grudsky, S.M.; Maksimenko, E.A. Inside the eigenvalues of certain Hermitian Toeplitz band matrices. *J. Comp. Appl. Math.* **2010**, *233*, 2245. [CrossRef]
29. Deift, P.; Its, A.; Krasovsky, I. Eigenvalues of Toeplitz matrices in the bulk of the spectrum. *Bull. Inst. Math. Acad. Sinica* **2012**, *7*, 437.

30. Bogoya, J.M.; Böttcher, A.; Grudsky, S.M.; Maximenko, E.A. Eigenvectors of Hermitian Toeplitz matrices with smooth simple-loop symbols. *Linear Algebra Appl.* **2016**, *493*, 606. [CrossRef]
31. Rambour, P. Asymptotic of the eigenvalues of Toeplitz matrices with even symbol. *arXiv* **2021**, arXiv:2101.11250.
32. Bateman, H.; Erdelyi, A. *Higher Transcendental Functions*; Graw-Hill Book Company, Inc.: New York, NY, USA, 1953.

entropy

Article

Effective Field Theory of Random Quantum Circuits

Yunxiang Liao [1,2,*] and Victor Galitski [1]

1 Joint Quantum Institute, Department of Physics, University of Maryland, College Park, MD 20742, USA; galitski@umd.edu
2 Condensed Matter Theory Center, Department of Physics, University of Maryland, College Park, MD 20742, USA
* Correspondence: liao@umd.edu

Abstract: Quantum circuits have been widely used as a platform to simulate generic quantum many-body systems. In particular, random quantum circuits provide a means to probe universal features of many-body quantum chaos and ergodicity. Some such features have already been experimentally demonstrated in noisy intermediate-scale quantum (NISQ) devices. On the theory side, properties of random quantum circuits have been studied on a case-by-case basis and for certain specific systems, and a hallmark of quantum chaos—universal Wigner–Dyson level statistics—has been derived. This work develops an effective field theory for a large class of random quantum circuits. The theory has the form of a replica sigma model and is similar to the low-energy approach to diffusion in disordered systems. The method is used to explicitly derive the universal random matrix behavior of a large family of random circuits. In particular, we rederive the Wigner–Dyson spectral statistics of the brickwork circuit model by Chan, De Luca, and Chalker [Phys. Rev. X 8, 041019 (2018)] and show within the same calculation that its various permutations and higher-dimensional generalizations preserve the universal level statistics. Finally, we use the replica sigma model framework to rederive the Weingarten calculus, which is a method of evaluating integrals of polynomials of matrix elements with respect to the Haar measure over compact groups and has many applications in the study of quantum circuits. The effective field theory derived here provides both a method to quantitatively characterize the quantum dynamics of random Floquet systems (e.g., calculating operator and entanglement spreading) and a path to understanding the general fundamental mechanism behind quantum chaos and thermalization in these systems.

Keywords: quantum chaos; random quantum circuits; field theory

Citation: Liao, Y.; Galitski, V. Effective Field Theory of Random Quantum Circuits. *Entropy* **2022**, *24*, 823. https://doi.org/10.3390/e24060823

Academic Editor: Marko Robnik

Received: 3 May 2022
Accepted: 10 June 2022
Published: 13 June 2022

1. Introduction

Recent years have seen a surge of interest in random quantum circuits, which can be used to simulate various properties of interacting many-body quantum systems, including universal ergodic dynamics. Random quantum circuits [1–12] consist of qubits (or qudits) evolving under successive applications of unitary quantum gates drawn randomly from ensembles of unitaries. They capture several general properties of many-body quantum chaotic systems. In particular, the universal random matrix theory (RMT) statistics of the quasi-energy spectra has been derived for some time-periodic (Floquet) circuits [13–23]. Quantum systems with underlying classical chaotic dynamics are conjectured to exhibit level statistics identical to that of a suitably chosen random matrix ensemble [24], and the RMT statistics has been used as one of the indicators for quantum chaos. For decades, numerous efforts have been made to understand this Bohigas–Giannoni–Schmit (BGS) quantum chaos conjecture theoretically [25–36]. Random quantum circuits, due to their fine-tuned structure, allow exact analysis of the spectral statistics, which sheds light on the underlying mechanism responsible for the emergence of RMT structure. In addition to the RMT spectral statistics, random quantum circuits also exhibit other fundamental properties of many-body quantum chaotic systems, such as the decay of correlation functions of local observables [37,38], ballistic spreading of the local operators [13,39–44], ballistic growth of the entanglement [13,38,40,45–51], and Gaussian distribution of the matrix elements of observables in the energy eigenbasis (as expected from the eigenstate thermalization hypothesis [52–54]) [55,56]. Experimentally, random quantum circuits can be simulated in noisy intermediate-scale quantum (NISQ) devices [57] built with superconducting qubits [58,59], trapped ions [60,61], and Rydberg atoms [62], and some of these generic features for quantum chaotic systems have been observed [63]. Due to their high controllability, random quantum circuits provide a useful tool for exploring fundamental principles underlying chaos and thermalization.

At a high level, the approaches to studying the conventional interacting many-body systems can be broadly categorized as either (i) the microscopic approach, where the specific Hamiltonian of a particular system is studied in an ab initio or bottom-up way and where all microscopic details enter the calculation or (ii) the low-energy (usually field-theory-based) approach, where only the universal low-energy features enter the theory (e.g., gapless collective modes). In the context of the quantum circuits used to simulate many-body systems, it was mostly the former type of approach that has been used to date. In particular, theoretical studies of random quantum circuits relied heavily on fine-tuned features of specific models. Here, we construct a field-theory approach, where universal features of ergodic dynamics of a large family of random quantum circuits can be studied in a unified way.

This field-theory method generalizes a supersymmetric sigma model developed by Zirnbauer [64–66], Altland [67], and others, making use of a generalized Hubbard–Stratonovich transformation, i.e., the color–flavor transformation [64–66,68]. This model is similar to the sigma model for disordered electron systems [69,70] but is formulated for systems modeled by an evolution operator or a scattering matrix rather than a Hamiltonian. In particular, it has been used to investigate the connection between the appearance of universal RMT statistics and the underlying classical chaotic dynamics (i.e., the BGS quantum chaos conjecture [24]) in generic quantum chaotic maps [65,66], localization in quantum kicked rotors [67,71–73], eigenenergy and eigenfunction statistics of quantum graphs [74–77], and the plateau transition in quantum Hall systems [78,79]. A review of this model and a more complete list of references can be found in [80,81]. For quantum circuits, we find it convenient to reformulate this sigma model in the replica formalism (in contrast to supersymmetry), and use it to derive an ensemble-averaged effective field theory.

The spectral statistics of the Floquet operator of time-periodic systems can be extracted from this effective field theory. In contrast to the aforementioned model-specific methods, the field-theoretical approach is applicable to a wide class of Floquet systems and is particularly useful for random quantum circuits composed of local random gates. To demonstrate the usefulness of this method, we apply it to study the spectral statistics of a family of Floquet random quantum circuits, some of which are shown in Figures 1 and 2. These Floquet circuits are composed of random unitary matrices which are drawn randomly and independently from the circular unitary ensemble (CUE) and are applied to all pairs of neighboring qudits at various time substeps during one period. Among these circuits, the brickwork circuit depicted in Figure 1a was studied earlier by Chan, De Luca, and Chalker in [13] (see also the related work in [16]), and the RMT spectral statistics were derived in the limit of a large on-site Hilbert space dimension $q \to \infty$, using the generalized diagrammatic approach initially constructed in [82] for the Weingarten calculus [83–89]. Applying the field-theoretical approach, we rederive this result and also show that the universal RMT statistics is preserved under an arbitrary reordering of the two-qudit gates, for both periodic and open boundary conditions. In particular, we show that the effective field theory describing the Floquet quantum circuit of this type is identical to that of the CUE ensemble. Moreover, we consider higher-dimensional generalizations of this family of Floquet circuits and find that their quasi-energy spectra all exhibit RMT statistics (see Figure 2 for some 2D examples).

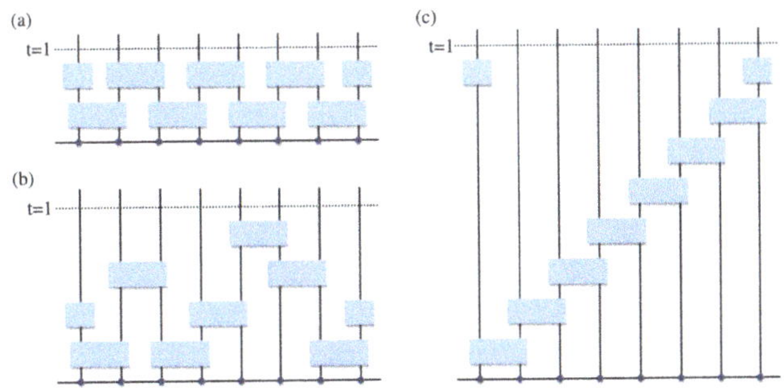

Figure 1. Floquet operators of 1D Floquet random quantum circuits with different orderings of the two-qudit unitary gates: (**a**) brickwork circuit [13], (**c**) staircase circuit, and (**b**) circuit obtained from rearranging the local gates (blue boxes) in (**a**) or (**b**). The horizontal direction represents the space coordinates, and each qudit is indicated by a black dot. The vertical direction shows the discrete time evolution within one period, and different layers represent different substeps. Each blue box represents an independent CUE matrix $w^{(n,n+1)}$ acting on the Hilbert space of two neighboring sites n and $n+1$. The two half-sized blue boxes at the boundaries constitute a random CUE matrix $w^{(L,1)}$ (identical matrix) acting on the sites L and 1, for periodic (open) boundary condition, where L is the total number of qudits. For any of these Floquet quantum circuits, which differ only by the ordering of local gates and are subject to either periodic or open boundary condition, we prove that the statistical properties of the quasi-energy spectra are identical to those of the CUE ensemble in the limit of large on-site Hilbert space dimension $q \to \infty$.

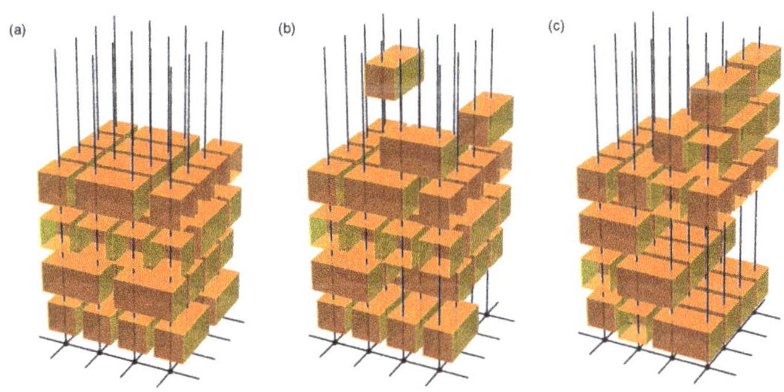

Figure 2. Floquet operators of 2D Floquet random quantum circuits with different orderings of the two-qudit gate [(**a**–**c**)]. Within one period, two-qudit unitaries (orange boxes) drawn independently from the CUE ensemble are applied to all pairs of neighboring qudits (black dots) in a 2D lattice at various time substeps (layers). Each qudit is coupled to all of its neighbors at different substeps. The half-sized box at the boundary combines with its neighbor on the opposite side to give a random unitary (identical matrix) acting on the corresponding pair of qudits, for periodic (open) boundary condition. We prove that, as in the 1D case (Figure 1), the quasi-energy spectrum for this type of Floquet random quantum circuit with arbitrary ordering of the quantum gates exhibits universal statistical properties described by the CUE ensemble in the limit of large on-site Hilbert space dimension $q \to \infty$, for both periodic and open boundary conditions. This statement extends to arbitrary dimensions (see Appendix B).

The field-theoretical approach is not only useful in the study of the spectral statistics of Floquet random quantum circuits but can also be applied to the analysis of more generic properties of a wide class of random quantum circuits (which are not necessarily time-periodic). To show this, within the same sigma model framework, we rederive the known results for Weingarten calculus, which has extensive applications in studies of quantum circuits (e.g., in [13,39,40,43,45,90,91]). When considering a quantum circuit where all or part of the quantum gates are random unitaries drawn from some ensembles with the Haar probability, one usually encounters polynomial functions of the matrix elements integrated over the relevant group with the Haar measure (or equivalently averaged over the relevant ensemble). These types of Haar integrals can be computed using the Weingarten calculus [82–88]. In the present paper, we rederive the results for the Haar integrals of products of the matrix elements of a unitary matrix over the unitary group (i.e., moments of CUE random matrices) using the field-theoretical approach. The derivation can be generalized to other compact Lie groups or compact symmetric spaces [86,92–94].

The rest of the paper is organized as follows. In Section 2, we introduce a standard diagnostic of the statistics of quasi-energies of time-periodic systems and briefly review the spectral statistical properties of the CUE ensemble. In Section 3, we present an effective field theory which is formulated as a replica sigma model and can be used to investigate the spectral statistics of a wide class of Floquet quantum systems. This effective field theory is applied in Section 4 to study a family of Floquet random quantum circuits related to the brickwork circuit in [13] by a reordering of the Haar random unitary gates (which couple the neighboring qudits) and by the higher-dimensional generalization. We show that the statistical properties of the quasi-energy spectra of the Floquet random quantum circuits of this type are universally described by the CUE ensemble, for both the periodic and open boundary conditions. Using the same sigma model method, we rederive the results for the Weingarten calculus for the unitary group in Section 5. Finally, in Section 6, we conclude with a discussion of directions for future studies. The appendices are devoted to technical details. In Appendices A and B, we provide the calculation of the second- and fourth-order moments of the Floquet operator for the family of Floquet random quantum circuits studied in the current paper, in 1D and in higher dimensions, respectively. We prove that, for arbitrary ordering of the two-qudit unitary gates, and in any dimension, these moments are identical to those of the CUE ensemble. In Appendix C, we study a non-interacting Floquet model and examine the higher-order fluctuations in the effective field theory. We show that the quartic-order fluctuations of this non-interacting model give rise to a larger contribution to the level correlation function compared with their chaotic counterparts and are therefore no longer negligible. In Appendix D, we rederive the asymptotic behavior of the Weingarten function for a unitary group of dimension q in the large $q \to \infty$ limit. Appendix E contains a derivation of the recursion relation for the Weingarten function.

2. Diagnostic of Spectral Statistics of Time-Periodic Systems

For time-periodic (Floquet) systems, the statistics of the quasi-energy spectra serve as a diagnostic of quantum chaos. The quasi-energies $\{\theta_i\}$ are the eigenphases of the time evolution operator over one period U (i.e., the Floquet operator): $U|n\rangle = e^{i\theta_n}|n\rangle$. The quasi-energy density can be expressed as

$$\rho(\phi) = \sum_{i=1}^{N} \delta_{2\pi}(\phi - \theta_i), \tag{1}$$

where N is the Hilbert space dimension and $\delta_{2\pi}(\phi)$ represents the 2π-periodic delta function defined as $\delta_{2\pi}(\phi) = \sum_{n=-\infty}^{\infty} e^{-in\phi}/2\pi$. For brevity, the subscript 2π will be omitted from now on.

For an ensemble of random Floquet systems, the two-point correlation function of the quasi-energy density is usually introduced as:

$$R_2(\phi_1, \phi_2) = \langle \rho(\phi_1)\rho(\phi_2)\rangle, \tag{2}$$

where the angular bracket denotes the ensemble averaging. The correlation function, $R_2(\phi_1, \phi_2)$ measures the probability of finding two quasi-energies at ϕ_1 and ϕ_2, and is one of the widely used probes of the spectral statistics. In the present paper, in addition to the ensemble averaging, we also perform the averaging over the entire quasi-energy spectrum:

$$\bar{R}_2(\Delta\phi) = \int_0^{2\pi} \frac{d\phi_0}{2\pi} R_2(\phi_0 + \Delta\phi/2, \phi_0 - \Delta\phi/2). \tag{3}$$

This averaging [80] is similar to the energy integration employed in [95,96] to derive a ballistic sigma model for individual Hamiltonian systems, and it allows us to extract information about the correlation function of an arbitrary pair of quasi-energy levels with the separation $\Delta\phi$, irrespective of their positions in the quasi-energy spectrum on the Floquet circle [97].

The Fourier transform of the two-level correlation function is known as the spectral form factor:

$$K(t) = \int_0^{2\pi} d\phi_1 \int_0^{2\pi} d\phi_2 R_2(\phi_1, \phi_2) e^{-i(\phi_1 - \phi_2)t} = 2\pi \int_0^{2\pi} d(\Delta\phi)\, \bar{R}_2(\Delta\phi) e^{-i\Delta\phi t}, \tag{4}$$

which is equivalent to

$$K(t) = \left\langle \left|\mathrm{Tr}(U^t)\right|^2 \right\rangle = \left\langle \sum_{i,j=1}^{N} e^{-it(\theta_i - \theta_j)} \right\rangle. \tag{5}$$

From this definition, one can immediately see that $K(t) = K(-t)$ and $K(t=0) = N^2$. Moreover, assuming no degeneracy in the quasi-energy spectrum, in the large time limit $t \to \infty$, the off-diagonal ($i \neq j$) terms in the summation in Equation (5) vanish upon ensemble averaging, due to the random phases, and the SFF acquires the value of $K(t \to \infty) = N$ (plateau).

Chaotic Floquet systems without time-reversal invariance are expected to exhibit the same spectral statistics properties as the circular unitary ensemble (CUE) [81,98,99], which is an ensemble of unitary matrices with the Haar probability measure. For the CUE ensemble, the spectrum is statistically homogeneous, and the mean level density is given by $\bar{\rho} = N/2\pi$. The two-level correlation function $R_2(\phi_1, \phi_2)$ depends only on the level separation $\Delta\phi = \phi_1 - \phi_2$ and is therefore equal to its energy-averaged value [99]:

$$R_2(\phi_1, \phi_2) = \bar{R}_2(\Delta\phi) = -\frac{1}{4\pi^2} \frac{\sin^2(N\Delta\phi/2)}{\sin^2(\Delta\phi/2)} + \frac{N^2}{4\pi^2} + \frac{N}{2\pi}\delta(\Delta\phi). \tag{6}$$

At nonzero $\Delta\phi$, the first term in the equation above corresponds to the connected part of the two-level correlation function defined as

$$R_2^{\mathrm{con}}(\phi_1, \phi_2) = \langle \rho(\phi_1)\rho(\phi_2) \rangle - \langle \rho(\phi_1) \rangle \langle \rho(\phi_2) \rangle. \tag{7}$$

In the large N limit, after rescaling the quasi-energy $\varepsilon = \bar{\rho}\Delta\phi$ and keeping it finite, the connected two-level correlation function assumes a form identical to that of the Gaussian unitary ensemble (GUE) [81,99]:

$$\frac{1}{\bar{\rho}^2} R_2^{\mathrm{con}}(\Delta\phi) = -\left(\frac{\sin(\pi\varepsilon)}{\pi\varepsilon}\right)^2. \tag{8}$$

From Equation (6), one finds that the SFF of the CUE ensemble acquires the form

$$K(t) = \min(|t|, N) + N^2 \delta_{t,0}. \tag{9}$$

It exhibits a linear ramp until a plateau sets in at $t = N$. This linear ramp reflects the repulsion between the quasi-energies and is expected to be a universal feature of quantum chaotic systems with broken time-reversal symmetry.

3. Replica Sigma Model for Generic Floquet Systems

In this section, we reformulate the sigma model initially constructed in [64–67] using the replica trick instead of the supersymmetric method, and obtain an ensemble-averaged effective field theory from which one can extract information about the statistical properties of the quasi-energy spectra for generic time-periodic quantum systems.

3.1. Generating Function for Level Correlation Function

Consider now an ensemble of Floquet systems whose Floquet operators U are from an arbitrary ensemble of unitary matrices. Note that, for many-body systems, U represents the many-body Floquet operator that acts in the many-body Hilbert space. The statistical properties of the quasi-energies of this ensemble can be obtained from the following generating function using the replica trick

$$\mathcal{Z}^{(R)}(\alpha,\beta) = \left\langle \int_0^{2\pi} \frac{d\phi}{2\pi} \left[\det\left(1-\alpha e^{i\phi}U\right)\det\left(1-\beta e^{-i\phi}U^\dagger\right)\right]^R \right\rangle. \tag{10}$$

Here, α and β are two complex numbers and R represents the replica number which will be set to zero (the replica limit) at the end of the calculation. Taking derivatives of the generating function $\mathcal{Z}^{(R)}(\alpha,\beta)$ with respect to the complex variables α and β, multiplying the result by $\alpha\beta/R^2$, and then taking the replica limit $R \to 0$, we find

$$\begin{aligned} C(\alpha,\beta) &\equiv \lim_{R\to 0}\frac{\alpha\beta}{R^2}\frac{\partial^2 \mathcal{Z}^{(R)}(\alpha,\beta)}{\partial\alpha\partial\beta} = \left\langle \int_0^{2\pi}\frac{d\phi}{2\pi}\,\mathrm{Tr}\left(\frac{\alpha e^{i\phi}U}{1-\alpha e^{i\phi}U}\right)\mathrm{Tr}\left(\frac{\beta e^{-i\phi}U^\dagger}{1-\beta e^{-i\phi}U^\dagger}\right)\right\rangle \\ &= \left\langle \int_0^{2\pi}\frac{d\phi}{2\pi}\sum_{n,m=1}^{\infty}\mathrm{Tr}\left(\alpha e^{i\phi}U\right)^n\mathrm{Tr}\left(\beta e^{-i\phi}U^\dagger\right)^m\right\rangle \\ &= \sum_{n=1}^{\infty}K(n)(\alpha\beta)^n. \end{aligned} \tag{11}$$

In this equation, the powers of U and $U^\dagger$ (i.e., n and m) need to be identical to have a non-vanishing contribution to $C(\alpha,\beta)$ after the integration over ϕ.

The two-level correlation function $\bar{R}_2(\Delta\phi)$ can be obtained from $C(\alpha,\beta)$, which is a weighted summation of the SFF $K(n)$ (Equation (5)) at discrete time n, by setting $\alpha=\beta=e^{i\Delta\phi/2}$:

$$\bar{R}_2(\Delta\phi) = \frac{1}{(2\pi)^2}\left[2\,\mathrm{Re}\,C(\alpha=\beta=e^{i\Delta\phi/2}) + N^2\right]. \tag{12}$$

Here, we have used the fact that $K(-n) = K(n)$ and $K(0) = N^2$, with N being the dimension of the Hilbert space. We note that the higher-order correlation functions of the quasi-energy density can be evaluated in an analogous manner, making use of a similar generating function [64].

3.2. Replica Sigma Model for Level Correlation Function

The generating function defined in Equation (10) can be expressed as a Grassmann path integral:

$$\begin{aligned} \mathcal{Z}^{(R)}(\alpha,\beta) = \left\langle \int_0^{2\pi}\frac{d\phi}{2\pi}\int \mathcal{D}(\bar{\psi},\psi)\exp\Big\{ -\bar{\psi}_i^{+,u}\left(\delta_{ij}-\alpha e^{i\phi}U_{ij}\right)\psi_j^{+,u} \\ -\bar{\psi}_i^{-,u}\left(\delta_{ij}-\beta e^{-i\phi}U_{ij}^\dagger\right)\psi_j^{-,u}\Big\}\right\rangle. \end{aligned} \tag{13}$$

The Grassmann field $\psi_i^{s,u}$ carries three different indices: $u = 1,2,\ldots,R$ is the replica index, $i = 1,2,\ldots,N$ labels the Hilbert space, and $s = +/-$ distinguishes the contributions from the forward (U) and backward ($U^\dagger$) evolution operators. Throughout the paper, we employ the convention that repeated indices imply the summation.

Using the color–flavor transformation [64–66,68], the integration over the center phase ϕ in Equation (13) can be converted into an integration over the $RN \times RN$ complex matrix field Z (we call matrix Z a 'field' but note that it does not have time-dependence and carries indices in the Hilbert and replica spaces only):

$$\begin{aligned}\mathcal{Z}^{(R)}(\alpha,\beta) &= c_0\Big\langle \int \mathcal{D}(\bar\psi,\psi)\exp\Big\{-\bar\psi_i^{+,u}\psi_i^{+,u} - \bar\psi_i^{-,u}\psi_i^{-,u}\Big\} \\ &\times \int \mathcal{D}(Z,Z^\dagger)\det(1+Z^\dagger Z)^{-2RN-1}\exp\Big\{\alpha\bar\psi_i^{+,u}Z_{ij}^{uv}U^\dagger_{jj'}\psi_{j'}^{-,v} - \beta\bar\psi_i^{-,u}(Z^\dagger)_{ij}^{uv}U_{jj'}\psi_{j'}^{+,v}\Big\}\Big\rangle \\ &= c_0\Big\langle \int \mathcal{D}(Z,Z^\dagger)\det(1+Z^\dagger Z)^{-2RN-1} \\ &\times \int \mathcal{D}(\bar\psi,\psi)\exp\Big\{-\begin{bmatrix}\bar\psi^+ & \bar\psi^-\end{bmatrix}\begin{bmatrix}1 & -\alpha Z U^\dagger \\ \beta Z^\dagger U & 1\end{bmatrix}\begin{bmatrix}\psi^+ \\ \psi^-\end{bmatrix}\Big\}\Big\rangle.\end{aligned} \tag{14}$$

Here, Z_{ij}^{uv} is a complex matrix which carries indices in both the replica space (labeled by u,v) and the Hilbert space (labeled by i,j), and c_0 is an indefinite unessential normalization factor, which can be determined in the $R \to 0$ limit.

Performing the Gaussian integration over ψ, we arrive at a sigma model representation of the generating function

$$\begin{aligned}\mathcal{Z}^{(R)}(\alpha,\beta) &= c_0\Big\langle \int \mathcal{D}(Z,Z^\dagger)e^{-S[Z^\dagger,Z]}\Big\rangle, \\ S[Z^\dagger,Z] &= (2RN+1)\,\mathrm{Tr}\ln\Big(1+Z^\dagger Z\Big) - \mathrm{Tr}\ln\Big(1+\alpha\beta Z U^\dagger Z^\dagger U\Big).\end{aligned} \tag{15}$$

It is sometimes convenient to make the transformation

$$Q = T\Lambda T^{-1}, \qquad T = \begin{bmatrix}1 & -Z \\ Z^\dagger & 1\end{bmatrix}, \qquad \Lambda = \begin{bmatrix}1 & 0 \\ 0 & -1\end{bmatrix}, \tag{16}$$

after which the action becomes

$$S[Q] = -2RN\,\mathrm{Tr}\ln\left(\frac{Q\Lambda+1}{2}\right) - \mathrm{Tr}\ln\left[\frac{1}{2}\left(1-\begin{bmatrix}\alpha U & 0 \\ 0 & \beta U^\dagger\end{bmatrix}\right)Q\Lambda + \frac{1}{2}\left(1+\begin{bmatrix}\alpha U & 0 \\ 0 & \beta U^\dagger\end{bmatrix}\right)\right]. \tag{17}$$

Similarly to the sigma model for a disordered system [69,70,100], the matrix field Q stays on the manifold with the constraints $\mathrm{Tr}\,Q = 0$ and $Q^2 = 1$. Equation (15) (or equivalently Equation (17)) is a replica version of the supersymmetric sigma model derived earlier to study the spectral statistics of the circular ensembles and quantum chaotic maps [64–66,80].

3.3. *Ensemble-Averaged Effective Theory*

Starting from Equation (15), we then perform the ensemble averaging and derive an effective field theory for the matrix field Z for any ensemble of Floquet systems. Note that Equation (15) can be rewritten as

$$\begin{aligned}\mathcal{Z}^{(R)}(\alpha,\beta) &= c_0\int \mathcal{D}(Z,Z^\dagger)e^{-S_{\mathrm{eff}}[Z^\dagger,Z]}, \\ S_{\mathrm{eff}}[Z^\dagger,Z] &\equiv (2RN+1)\,\mathrm{Tr}\ln\Big(1+Z^\dagger Z\Big) - \ln\Big\langle\exp\Big[\mathrm{Tr}\ln\Big(1+\alpha\beta Z U^\dagger Z^\dagger U\Big)\Big]\Big\rangle.\end{aligned} \tag{18}$$

Until now, no approximations have been made, and the expression above is exact. However, it is difficult to perform the ensemble averaging in the second term in the action S_{eff}, especially for many-body systems whose Floquet operator has a complicated structure in the many-body Hilbert space. To proceed, we expand S_{eff} around the saddle point $Z = 0$ in powers of the fluctuation Z

and carry out the ensemble averaging term by term. Up to the quartic order in Z, the action S_{eff} is given by

$$S_{\text{eff}}[Z^\dagger, Z] = S^{(2)}_{\text{eff}}[Z^\dagger, Z] + S^{(4)}_{\text{eff}}[Z^\dagger, Z] + O(Z^6), \tag{19a}$$

$$S^{(2)}_{\text{eff}}[Z^\dagger, Z] = \sum_{i_1,\ldots,i_4} \text{tr}\left(Z^\dagger_{i_1 i_2} Z_{i_3 i_4}\right)\left[(2RN+1)\delta_{i_2,i_3}\delta_{i_4,i_1} - \alpha\beta\left\langle U_{i_2 i_3} U^\dagger_{i_4 i_1}\right\rangle\right], \tag{19b}$$

$$\begin{aligned} S^{(4)}_{\text{eff}}[Z^\dagger, Z] = &-\frac{(\alpha\beta)^2}{2} \sum_{i_1,\ldots,i_8} \text{tr}\left(Z^\dagger_{i_1 i_2} Z_{i_3 i_4}\right) \text{tr}\left(Z^\dagger_{i_5 i_6} Z_{i_7 i_8}\right) \\ &\times \left[\left\langle U_{i_2 i_3} U^\dagger_{i_4 i_1} U_{i_6 i_7} U^\dagger_{i_8 i_5}\right\rangle - \left\langle U_{i_2 i_3} U^\dagger_{i_4 i_1}\right\rangle\left\langle U_{i_6 i_7} U^\dagger_{i_8 i_5}\right\rangle\right] \\ &+\frac{1}{2} \sum_{i_1,\ldots,i_8} \text{tr}\left(Z^\dagger_{i_1 i_2} Z_{i_3 i_4} Z^\dagger_{i_5 i_6} Z_{i_7 i_8}\right) \\ &\times \left[-(2RN+1)\delta_{i_2,i_3}\delta_{i_4,i_5}\delta_{i_6,i_7}\delta_{i_8,i_1} + (\alpha\beta)^2\left\langle U_{i_2 i_3} U^\dagger_{i_4 i_5} U_{i_6 i_7} U^\dagger_{i_8 i_1}\right\rangle\right]. \end{aligned} \tag{19c}$$

In the present paper, we use 'tr' to denote the trace operation that acts on the replica space only and 'Tr' to trace over both the replica and Hilbert spaces. We note that the last term in $S^{(4)}_{\text{eff}}$ (Equation (19c)) gives rise to a contribution of higher order in the replica number R compared with the remaining term in $S^{(4)}_{\text{eff}}$ and is not important in the replica limit $R \to 0$.

The specific forms of the moments of the Floquet operator in the effective theory are model dependent. Once they are known, one can insert them into the expression for the effective action in Equation (19) to obtain an effective field theory which encodes the information about the statistical properties of the quasi-energy spectrum. This field-theoretical approach is therefore applicable to a wide class of Floquet systems, including many-body and single-particle systems. One advantage of this method, compared with other model-specific methods, is that it can be used to investigate why or when universal statistical behaviors emerge. For Floquet random quantum circuits consisting of independent local unitary gates, the moments of the many-body Floquet operator are given by products of moments of the local unitaries, and therefore are usually not difficult to evaluate. This makes the current field-theoretical approach especially useful for random quantum circuits with local gates.

4. Application to Floquet Random Quantum Circuits

4.1. Floquet Random Quantum Circuits

As an example, we apply the effective field theory (Equation (19)) derived in the previous section to study the spectral statistics of a family of Floquet quantum circuits composed of random local unitary gates. Figures 1 and 2 show the Floquet operators of some of these Floquet random quantum circuits, including the brickwork circuit (Figure 1a) studied earlier in [13]. The other two circuits depicted in panels (b) and (c) of Figure 1 can be obtained from the brickwork circuit by reordering the local gates. Some examples of the 2D generalization of such Floquet circuits are depicted in Figure 2. We prove that Floquet random quantum circuits of this type, subject to either periodic or open boundary condition, are all described by the same effective field theory as the CUE ensemble in the limit of large on-site Hilbert space dimension, irrespective of the ordering of the local gates and the dimensionality of the lattice of qudits.

We first consider the Floquet quantum circuits consisting of a 1D lattice chain of L qudits, each of which contains $q \to \infty$ internal states. The dimension of the many-body Hilbert space is $N = q^L$. The time evolutions of these Floquet circuits are discrete and time-periodic, and the evolutions over one period contain $2 \leq M \leq L$ substeps (where M is model specific). For all these Floquet quantum circuits, each qudit (labeled by an integer $n = 1, 2, \ldots, L$) is coupled to its neighbors on the left hand side (at site $n-1$) and the right hand side (at site $n+1$) at two different substeps (labeled by integers $s^{(n-1,n)}$ and $s^{(n,n+1)}$, respectively). We consider both the periodic and open boundary conditions, and the qudit label n is defined modulo L for the periodic boundary condition. The local gate that couples a pair of neighboring qudits at sites n and $n+1$ is given by a $q^2 \times q^2$ random CUE matrix $w^{(n,n+1)}$ and is represented diagrammatically by a blue box in Figure 1. Unitary gates acting on different pairs of neighboring qudits are independent and uncorrelated. We consider all possible orderings of these quantum gates represented by different configurations of $\left\{1 \leq s^{(n,n+1)} \leq M | n = 1, 2, \ldots, L'\right\}$ with constraint $s^{(n,n+1)} \neq s^{(n-1,n)}$. Here, L' is defined as $L' = L$ for the periodic boundary condition and $L' = L - 1$ for the open boundary condition. The total number of substeps M is given by the total number of different integers in the set $\left\{s^{(n,n+1)}\right\}$, and we consider all possible values of $2 \le M \le L'$. For the staircase circuit in Figure 1c, $M = L'$ and $s^{(n,n+1)} = n$, while for the brickwork circuit in Figure 1a, $M = 2$ and $s^{(n,n+1)} = 2 - (n \bmod 2)$ (for the open boundary condition or the periodic boundary condition with even L).

For any of the Floquet random quantum circuits described above, the Floquet operator can be expressed as

$$U = W^{(\sigma(1),\sigma(1)+1)} W^{(\sigma(2),\sigma(2)+1)} \ldots W^{(\sigma(L'-1),\sigma(L'-1)+1)} W^{(\sigma(L'),\sigma(L')+1)}, \tag{20}$$

where $\sigma \in S_{L'}$ represents a permutation of numbers $1, 2, \ldots, L'$. W is defined as

$$W^{(n,n+1)} = w^{(n,n+1)} \otimes 1^{(n,n+1)}. \tag{21}$$

Here, $1^{(n,n+1)}$ represents an identical matrix operating in the Hilbert space of all sites except for n and $n+1$. As mentioned earlier, $w^{(n,n+1)}$ acts on the qudits at sites n and $n+1$ and is drawn randomly and independently from the CUE ensemble of dimension q^2. We consider all possible quantum circuits whose Floquet operator can be expressed in the form of Equation (20) for arbitrary permutations $\sigma \in S_{L'}$. For example, σ is the identical permutation for the staircase circuit in Figure 1c, while for the brickwork circuit in Figure 1a with the periodic boundary condition and even L, σ is given by

$$\sigma(2k-1) = k, \qquad \sigma(2k) = L/2 + k, \qquad 1 \le k \le L/2. \tag{22}$$

In Appendix A, we prove that the Floquet operator U given by Equation (20) with arbitrary permutation $\sigma \in S_{L'}$ obeys the conditions:

$$\left\langle U_{\mathbf{ij}} U^{\dagger}_{\mathbf{j'i'}} \right\rangle = \frac{1}{N} \delta_{\mathbf{ii'}} \delta_{\mathbf{jj'}}, \tag{23a}$$

$$\left\langle U_{\mathbf{i}_1\mathbf{j}_1} U_{\mathbf{i}_2\mathbf{j}_2} U^{\dagger}_{\mathbf{j}'_1\mathbf{i}'_1} U^{\dagger}_{\mathbf{j}'_2\mathbf{i}'_2} \right\rangle = \frac{1}{N^2} \left(\delta_{\mathbf{i}_1\mathbf{i}'_1} \delta_{\mathbf{j}_1\mathbf{j}'_1} \delta_{\mathbf{i}_2\mathbf{i}'_2} \delta_{\mathbf{j}_2\mathbf{j}'_2} + \delta_{\mathbf{i}_1\mathbf{i}'_2} \delta_{\mathbf{j}_1\mathbf{j}'_2} \delta_{\mathbf{i}_2\mathbf{i}'_1} \delta_{\mathbf{j}_2\mathbf{j}'_1} \right). \tag{23b}$$

Here, the L-dimensional vector $\mathbf{i} = (i^{(1)}, i^{(2)}, \ldots, i^{(L)})$ labels the many-body state of the circuits, and its n-th component $i^{(n)} = 1, 2, \ldots, q$ indexes the single-particle state of the n-th qudit. We note that Equation (23a) holds for arbitrary q, while Equation (23b) is derived in the limit of $q \to \infty$.

The above discussion can be straightforwardly generalized to higher-dimensional Floquet quantum circuits with similar configurations. In particular, let us consider now a D-dimensional cubic lattice of qudits, with L sites in each direction. The single-particle (many-body) Hilbert space dimension is $q \to \infty$ ($N = q^{L^D}$). The time evolution is again discrete and periodic, and is composed of local two-qudit gates that couple separately all pairs of neighboring qudits. During one period, each qudit is coupled to all of its neighboring qudits at different substeps by different local unitary gates drawn randomly and independently from the CUE ensemble of dimension q^2. For this type of D-dimensional Floquet circuit, with any possible ordering of these local gates and with either periodic or open boundary condition, we prove that the Floquet operator U still satisfies Equation (23). In this case, $\mathbf{i}$ can be considered as an L^D-dimensional vector whose component $i^{(\mathbf{n})} = 1, 2, \ldots, q$ labels the single-particle state of the qudit $\mathbf{n}$ in the D-dimensional lattice. The derivation is relegated to Appendix B.

Equations (23a) and (23b) are also obeyed if U is drawn randomly from a CUE ensemble of dimension $N \to \infty$. As a result, the effective field theory (Equation (19)) for the Floquet quantum circuits under consideration is equivalent to that of the CUE ensemble of dimension N, and the two-level correlation functions for the current models are given by the CUE level correlation function in Equation (6).

4.2. Quadratic Fluctuations

In the following, we present the detailed derivation of the two-level correlation function $\bar{R}_2(\Delta\phi)$ for the quasi-energies of the Floquet random quantum circuits described above (or equivalently the CUE ensemble of the same dimension N), using the effective field theory (Equation (19)) and the moments of the Floquet operator (Equation (23)).

Substituting Equation (23a) into the quadratic-order effective action $S_{\text{eff}}^{(2)}$ in Equation (19b), we obtain

$$S_{\text{eff}}^{(2)} = (2RN+1)\sum_{\mathbf{i}_1,\mathbf{i}_2} \text{tr}\left(Z^{\dagger}_{\mathbf{i}_1\mathbf{i}_2} Z_{\mathbf{i}_2\mathbf{i}_1}\right) - \frac{\alpha\beta}{N}\sum_{\mathbf{i}_1,\mathbf{i}_2} \text{tr}\left(Z^{\dagger}_{\mathbf{i}_1\mathbf{i}_1} Z_{\mathbf{i}_2\mathbf{i}_2}\right). \tag{24}$$

We now divide Z into the diagonal component $X_{\mathbf{ii}} \equiv Z_{\mathbf{ii}}$ and the off-diagonal component $Y_{\mathbf{ij}} \equiv Z_{\mathbf{ij}}$ (for $\mathbf{i} \neq \mathbf{j}$) in the Hilbert space and Fourier transform the diagonal component $X_{\mathbf{jj}}$ with respect to $\mathbf{j}$:

$$X(\mathbf{k}) = \sum_{\mathbf{j}} X_{\mathbf{jj}} e^{-i2\pi\mathbf{k}\cdot\mathbf{j}/q}. \tag{25}$$

Here, $\mathbf{k}$ is an L^D dimensional vector and the summation over each component runs over $k^{(\mathbf{n})} = 0, 1, \dots, q-1$. In terms of $X(\mathbf{k})$ and $Y_{\mathbf{ij}}$, the effective action $S_{\text{eff}}^{(2)}$ can be rewritten as

$$\begin{aligned} S_{\text{eff}}^{(2)} &= (2RN+1)\frac{1}{N}\sum_{\mathbf{k}} \text{tr}\left(X^{\dagger}(\mathbf{k})X(\mathbf{k})\right) - \frac{\alpha\beta}{N}\text{tr}\left(X^{\dagger}(0)X(0)\right) \\ &\quad + (2RN+1)\sum_{\mathbf{i}_1\neq\mathbf{i}_2} \text{tr}\left(Y^{\dagger}_{\mathbf{i}_1\mathbf{i}_2} Y_{\mathbf{i}_2\mathbf{i}_1}\right). \end{aligned} \tag{26}$$

From the equation above, one finds that the bare propagator for $X(\mathbf{k}=0)$ acquires the form

$$\left\langle X^{uv}(0)(X^{\dagger})^{v'u'}(0)\right\rangle_0 = \delta_{uu'}\delta_{vv'}\frac{N}{(2RN+1)-\alpha\beta}. \tag{27}$$

Here, the angular bracket with subscript 0 represents the averaging over the Gaussian fluctuation of matrix Z governed by the action $S_{\text{eff}}^{(2)}$ (Equation (26)). Taking the replica limit $R \to 0$ and setting $\alpha = \beta$ to $e^{i\Delta\phi/2}$, the bare propagator for $X(0)$ becomes

$$\left\langle X^{uv}(0)(X^{\dagger})^{v'u'}(0)\right\rangle_0 = \delta_{uu'}\delta_{vv'}\frac{N}{1-e^{i\Delta\phi}}, \tag{28}$$

which diverges when $\Delta\phi \to 0$. The corresponding mode is massless.

By contrast, the propagators for $X(\mathbf{k}\neq 0)$ and $Y_{\mathbf{ij}}$ are given by, respectively,

$$\begin{aligned} \left\langle X^{uv}(\mathbf{k})(X^{\dagger})^{v'u'}(\mathbf{k}')\right\rangle_0 &= \delta_{uu'}\delta_{vv'}\delta_{\mathbf{k},\mathbf{k}'}\frac{N}{2RN+1}, \\ \left\langle Y^{uv}_{\mathbf{ij}}(Y^{\dagger})^{v'u'}_{\mathbf{j}'\mathbf{i}'}\right\rangle_0 &= \delta_{uu'}\delta_{vv'}\delta_{\mathbf{ii}'}\delta_{\mathbf{jj}'}\frac{1}{2RN+1}. \end{aligned} \tag{29}$$

$X(\mathbf{k}\neq 0)$ and $Y_{\mathbf{ij}}$ are therefore massive, and give rise to an α-, β-independent contribution to the generating function $\mathcal{Z}^{(R)}$ at the quadratic order.

In summary, the Z fluctuations can be divided into two categories: the massless fluctuation $X(0)$ and the massive fluctuations $X(\mathbf{k}\neq 0)$ and $Y_{\mathbf{ij}}$. The massive modes $X(\mathbf{k}\neq 0)$ and $Y_{\mathbf{ij}}$ contribute a nonessential constant to the generating function, while the soft mode $X(0)$ governs the spectral statistics [36,101].

4.3. Quartic Fluctuations

We now investigate the contribution to the self-energy from the quartic-order fluctuations (Equation (19c)). Note that the last term in $S_{\text{eff}}^{(4)}$ (Equation (19c)) does not contribute in the replica limit $R \to 0$ due to its special structure in the replica space. We can therefore focus on the first term in $S_{\text{eff}}^{(4)}$, which can be expressed in terms of $X(\mathbf{k})$ and Y as

$$\begin{aligned} S_{\text{eff}}^{(4-1)} &= -\frac{(\alpha\beta)^2}{2N^4}\sum_{\mathbf{k}_1,\mathbf{k}_2} \text{tr}\left(X^{\dagger}(\mathbf{k}_1)X(\mathbf{k}_2)\right)\text{tr}\left(X^{\dagger}(-\mathbf{k}_1)X(-\mathbf{k}_2)\right) \\ &- \frac{(\alpha\beta)^2}{2N^2}\sum_{\mathbf{i}_1\neq\mathbf{i}_2,\mathbf{i}_3\neq\mathbf{i}_4} \text{tr}\left(Y^{\dagger}_{\mathbf{i}_1\mathbf{i}_2}Y_{\mathbf{i}_3\mathbf{i}_4}\right)\text{tr}\left(Y^{\dagger}_{\mathbf{i}_2\mathbf{i}_1}Y_{\mathbf{i}_4\mathbf{i}_3}\right) \\ &- \frac{(\alpha\beta)^2}{2N^3}\sum_{\mathbf{i}_1\neq\mathbf{i}_2}\sum_{\mathbf{k},\mathbf{k}'}\left[\text{tr}\left(Y^{\dagger}_{\mathbf{i}_1\mathbf{i}_2}X(\mathbf{k})\right)\text{tr}\left(Y^{\dagger}_{\mathbf{i}_2\mathbf{i}_1}X(-\mathbf{k})\right) + \text{tr}\left(X^{\dagger}(-\mathbf{k})Y_{\mathbf{i}_1\mathbf{i}_2}\right)\text{tr}\left(X^{\dagger}(\mathbf{k})Y_{\mathbf{i}_2\mathbf{i}_1}\right)\right]. \end{aligned} \tag{30}$$

Here, we have used Equation (23b). Comparing with the quadratic action $S_{\rm eff}^{(2)}$ in Equation (26), one can see that the quartic action $S_{\rm eff}^{(4-1)}$ is of higher order in $1/N$. Note that this counting does not hold for nonergodic circuits, and the higher-order fluctuations become important (see Appendix C).

The self-energies for the X and Y components from the quartic interactions are given by, respectively,

$$\begin{aligned}(\Sigma_X)^{ab,ba}(\mathbf{k},\mathbf{k}) &= \frac{(\alpha\beta)^2}{N^4}\left\langle X^{ab}(-\mathbf{k})(X^\dagger)^{ba}(-\mathbf{k})\right\rangle_0 = \frac{(\alpha\beta)^2}{N^3}\frac{1}{1-\alpha\beta\delta_{\mathbf{k},0}},\\ (\Sigma_Y)^{ab,ba}_{\mathbf{i}_1\mathbf{i}_2,\mathbf{i}_2\mathbf{i}_1} &= \frac{(\alpha\beta)^2}{N^2}\left\langle Y^{ab}_{\mathbf{i}_2\mathbf{i}_1}(Y^\dagger)^{ba}_{\mathbf{i}_1\mathbf{i}_2}\right\rangle_0 = \frac{(\alpha\beta)^2}{N^2}.\end{aligned} \tag{31}$$

Here, we have taken the replica limit and ignored the self-energy corrections that vanish in the limit $R \to 0$. From the equation above, we can see that the self-energy from the quartic-order fluctuations is negligible in the large N limit for the massive modes $X(\mathbf{k} \neq 0)$ and Y. For the massless mode $X(0)$, the self-energy from the quartic interactions can be ignored if we consider an energy separation $\Delta\phi \gg 1/N$ that is not too small. To recover the fine structure of the nearby quasi-energy levels, higher-order fluctuations of $X(0)$ are needed.

We note that, for integrable systems, fluctuations beyond the quadratic order are no longer negligible, even when the quasi-energy separation $\Delta\phi$ being probed is much larger than the mean level spacing. In Appendix C, we consider a non-interacting Floquet model whose single-particle dynamics within one period is generated by random CUE matrices. In particular, the single-particle Floquet operator for each particle is independently drawn from the CUE ensemble. We find that the action for the quadratic fluctuations of this non-interacting model is identical to that of the Floquet random quantum circuits considered in this section (or equivalently the CUE ensemble), and is given by Equation (26). However, the quartic fluctuations are governed by a different action which, compared with its chaotic counterpart in Equation (30), leads to a much larger contribution to the self-energy of the Z matrix field. Unlike the chaotic model, the higher-order fluctuations become important for this non-interacting model.

4.4. Two-Level Correlation Function

For the chaotic Floquet random quantum circuits under consideration here, we can focus on the quadratic fluctuations' contribution and neglect the higher-order corrections. $\mathcal{C}(\alpha,\beta)$, defined in Equation (11), is then approximately given by

$$\mathcal{C}(\alpha,\beta) = \lim_{R\to 0}\frac{\alpha\beta}{R^2}c_1\left\langle \frac{1}{N}\operatorname{tr}\left(X^\dagger(0)X(0)\right) + \frac{\alpha\beta}{N^2}\operatorname{tr}^2\left(X^\dagger(0)X(0)\right)\right\rangle_0 = \frac{\alpha\beta}{(1-\alpha\beta)^2}. \tag{32}$$

Here, the overall coefficient c_1 contains the unessential contribution from the massive modes and its value in the replica limit is determined from the fact that $\lim_{R\to 0} Z^{(R)}(\alpha,\beta) = 1$. Using Equation (12), we obtain the result for the two-level correlation function

$$\bar{R}_2(\Delta\phi) = -\frac{1}{8\pi^2}\frac{1}{\sin^2(\Delta\phi/2)} + \frac{N^2}{4\pi^2}. \tag{33}$$

It is easy to see that $\langle \operatorname{Tr} U\rangle = 0$ for the Floquet circuits under consideration. Therefore, the average quasi-energy density for any of these circuits is homogeneous and given by $\bar{\rho} = N/2\pi$. This also means that the first term in Equation (33) corresponds to the connected part of the two-level correlation function $R_2^{\rm con}(\Delta\phi \neq 0)$ defined in Equation (7). We emphasize that this equation applies to all the Floquet quantum circuits under consideration.

Comparing Equation (33) with the exact CUE result in Equation (6), one can see that the smooth part of $\bar{R}_2(\Delta\phi)$ is recovered, while an oscillatory term proportional to $\cos(N\Delta\phi)$ is missing. To recover the oscillatory term or to extract the behavior of $\bar{R}_2(\Delta\phi)$ at small energy separations $\Delta\phi \lesssim 1/N$, nonperturbative information about higher-order fluctuations is needed. This may be obtained by consideration of non-standard saddle points (see [80]), similar to the calculation of the level correlation function for the Hamiltonian systems described by the Gaussian ensembles [102–105].

5. Weingarten Calculus

To show that the field-theoretical approach described above provides access not only to the spectral statistics but also to other generic properties of quantum circuits, in this section we rederive the known results for the Weingarten calculus, which has been employed extensively in studies of quantum circuits. Let us now consider a Haar integral of a product of the matrix elements of a unitary matrix U over the unitary group in q dimensions $\mathcal{U}(q)$:

$$\begin{aligned} I &= \int_{\mathcal{U}(q)} dU U_{i_1j_1} U_{i_2j_2} ... U_{i_pj_p} U^{\dagger}_{j'_1i'_1} U^{\dagger}_{j'_2i'_2} ... U^{\dagger}_{j'_{p'}i'_{p'}} \\ &= \left\langle U_{i_1j_1} U_{i_2j_2} ... U_{i_pj_p} U^{\dagger}_{j'_1i'_1} U^{\dagger}_{j'_2i'_2} ... U^{\dagger}_{j'_{p'}i'_{p'}} \right\rangle_{\mathrm{CUE}}, \end{aligned} \tag{34}$$

where dU denotes the normalized ($\int_{\mathcal{U}(q)} dU = 1$) Haar measure. This integral can be interpreted as the product of the unitary matrix elements averaged over the CUE ensemble, denoted by the angular bracket with subscript CUE. In this section and in Appendices D and E, we use U to denote a random CUE matrix (i.e., a Haar distributed random unitary matrix).

It has been found that this integral I can be expressed in terms of the Weingarten function [82–88]:

$$I = \delta_{p,p'} \sum_{\tau,\sigma \in S_p} \mathrm{Wg}(\tau^{-1}\sigma) \prod_{k=1}^{p} \delta_{i_k, i'_{\sigma(k)}} \delta_{j_k, j'_{\tau(k)}}. \tag{35}$$

Here, the double summation runs over all permutations $\tau, \sigma \in S_p$ of the integers $1, 2, ..., p$, and the Weingarten function $\mathrm{Wg}(\tau^{-1}\sigma)$ depends only on the cycle structure of the product $\tau^{-1}\sigma$. More specifically, the permutation $\tau^{-1}\sigma$ can be divided into m disjoint cycles:

$$\left(P_1^{(1)} \to P_2^{(1)} \to ... \to P_{c_1}^{(1)}\right)\left(P_1^{(2)} \to P_2^{(2)} \to ... \to P_{c_2}^{(2)}\right) ... \left(P_1^{(m)} \to P_2^{(m)} \to ... \to P_{c_m}^{(m)}\right). \tag{36}$$

Here, $\left\{P_k^{(l)}\right\}$ are different integers from the set of integers under the permutation $\{1, 2, ..., p\}$, and they satisfy $P_k^{(l)} = (\tau^{-1}\sigma)^{k-1}(P_1^{(l)})$ (for $k \leq c_l$) and $P_1^{(l)} = (\tau^{-1}\sigma)P_{c_l}^{(l)}$. The Weingarten function $\mathrm{Wg}(\tau^{-1}\sigma)$ depends only on the lengths of these disjoint cycles $\{c_k | k = 1, 2, ..., m\}$, which obey the constraint $\sum_{k=1}^{m} c_k = p$, irrespective of their order. In the following, we will denote the Weingarten function of the permutation $\tau^{-1}\sigma$ given by Equation (36) as $\mathrm{Wg}(\tau^{-1}\sigma) = V^{(p)}_{c_1, c_2, ..., c_m}$.

The Weingarten function can be uniquely determined by the recursions relation [82,83]:

$$q V^{(p+1)}_{c_1, ..., c_m, 1} + \sum_{s=1}^{m} c_s V^{(p+1)}_{c_1, ..., c_{s-1}, c_s+1, c_{s+1}, ..., c_m} = V^{(p)}_{c_1, ..., c_m}, \tag{37a}$$

$$q V^{(p)}_{c_1, ..., c_m} + \sum_{c=1}^{c_1-1} V^{(p)}_{c_1-c, c, c_2, ..., c_m} + \sum_{s=2}^{m} c_s V^{(p)}_{c_1+c_s, c_2, ..., c_{s-1}, c_{s+1}, ..., c_m} = 0. \tag{37b}$$

Here, $\{c_i\}$ represent an arbitrary set of m positive integers that satisfy $\sum_{k=1}^{m} c_k = p$ in Equation (37a), and in Equation (37b) an additional constraint $c_1 \geq 2$ is imposed.

In the large $q \to \infty$ limit [82], the Weingarten function of the identical permutation id is given by $\mathrm{Wg}(id) = V^{(p)}_{1,1,...,1} = q^{-p} + O(q^{-p-2})$. For all the remaining permutations, the Weingarten function $V^{(p)}_{c_1, c_2, ..., c_m} = O(q^{m-2p})$ is of higher order in $1/q$ (since the number of disjoint cycles $m < p$). Using these results, one can easily see that Equation (23) is obeyed by the CUE random matrix U of dimension N in the large N limit.

These results for the Weingarten calculus have been obtained previously, using various methods [82,83,85,86,88,89]. In the following, we rederive these results using a field-theoretical method similar to the one employed earlier in the analysis of the statistics of quasi-energy spectra. We believe that this field-theoretical approach can be straightforwardly generalized and can serve as a general framework for studying quantum circuits.

5.1. Sigma Model Derivation for the Weingarten Calculus

To begin with, we rewrite the integral I in Equation (34) as a fermionic path integral

$$\begin{aligned} I &= \int_{\mathcal{U}(q)} dU \int \mathcal{D}(\bar{\psi},\psi) e^{-S_U[\bar{\psi},\psi]} F[\bar{\psi},\psi], \\ S_U[\bar{\psi},\psi] &= \sum_{u=1}^{R'} \sum_{l,k=1}^{q} \left(\bar{\psi}_k^{+,u} U_{kl} \psi_l^{+,u} + \bar{\psi}_k^{-,u} U_{kl}^{\dagger} \psi_l^{-,u} \right), \\ F[\bar{\psi},\psi] &= \prod_{k=1}^{p} \left(\psi_{i_k}^{-,k} \bar{\psi}_{j_k}^{-,k} \right) \prod_{l=1}^{p'} \left(\psi_{j_l'}^{+,l} \bar{\psi}_{i_l'}^{+,l} \right). \end{aligned} \tag{38}$$

Here, the fermionic field $\psi_k^{s,u}$ carries three indices that label, respectively, the component associated with $U/U^{\dagger}$ ($s = +/-$), the replica space ($u = 1,2,\ldots,R'$), and the Hilbert space in which the unitary matrix U acts ($k = 1,2,\ldots,q$). Note that here, the replica number R' is an integer given by $R' = \max(p,p')$ and, unlike in the earlier calculation of the level correlation function, it does not need to be set to zero at the end.

In Equation (38), we have made use of Wick's theorem and the following identities

$$\begin{aligned} &\left\langle \psi_{i_u}^{-,u} \bar{\psi}_{j_v}^{-,v} \right\rangle_{S_U} = \delta_{uv} U_{i_u j_u}, \qquad \left\langle \psi_{j_u'}^{+,u} \bar{\psi}_{i_v'}^{+,v} \right\rangle_{S_U} = \delta_{uv} U^{\dagger}_{j_u' i_u'}, \\ &\langle \psi^- \bar{\psi}^+ \rangle_{S_U} = \langle \psi^+ \bar{\psi}^- \rangle_{S_U} = 0. \end{aligned} \tag{39}$$

The angular bracket with the subscript S_U represents the functional averaging over the fermionic field ψ with the weight $e^{-S_U[\bar{\psi},\psi]}$. Note that in Equation (38), fermions with different replica indices (u) or $U/U^{\dagger}$ indices ($s = \pm$) are uncoupled. In fact, the replica space is introduced here such that, when computing the expectation value of $\langle F[\bar{\psi},\psi]\rangle_{S_U}$ using the Wick contraction, the fermionic field $\psi^{\mp}$ with index i_k (j_k') has to pair with $\bar{\psi}^{\mp}$ with index j_k (i_k'), as they share the same replica index k, leading to the factor $U_{i_k j_k}$ ($U^{\dagger}_{j_k' i_k'}$) in the integrand of I.

Applying the color–flavor transformation [64–66,68], the Haar integral over the unitary matrix U in Equation (38) can be converted into an integral over a complex $R' \times R'$ matrix field Z:

$$\begin{aligned} I =& z_1 \int \mathcal{D}(Z,Z^{\dagger}) \det(1 + Z^{\dagger}Z)^{-(2R'+q)} \int \mathcal{D}(\bar{\psi},\psi) e^{-S_Z[\bar{\psi},\psi]} F[\bar{\psi},\psi], \\ =& z_1 \int \mathcal{D}(Z,Z^{\dagger}) \det(1 + Z^{\dagger}Z)^{-(2R'+q)} \det(-ZZ^{\dagger})^q \langle F[\bar{\psi},\psi]\rangle_{S_Z}, \end{aligned} \tag{40}$$

where

$$\begin{aligned} S_Z[\bar{\psi},\psi] &= \sum_{u,v=1}^{R'} \sum_{l=1}^{q} \left(-\bar{\psi}_l^{+,u} Z_{uv} \psi_l^{-,v} + \bar{\psi}_l^{-,u} Z_{uv}^{\dagger} \psi_l^{+,v} \right), \\ z_1^{-1} &= \int \mathcal{D}(Z,Z^{\dagger}) \det(1 + Z^{\dagger}Z)^{-(2R'+q)} \det(-ZZ^{\dagger})^q. \end{aligned} \tag{41}$$

Note that here the matrix field Z acts in the replica space only. In the second equality of Equation (40), we have integrated out the fermions governed by the new action $S_Z[\bar{\psi},\psi]$, and the angular bracket with the subscript S_Z represents the functional averaging with the weight $\exp(-S_Z[\bar{\psi},\psi])$. The normalization constant z_1 is determined from the fact that $I = 1$ if we set $F = 1$.

After the color–flavor transformation, fermions with different replica indices interact through the matrix field Z, while those with different Hilbert space indices become uncoupled. The fermionic propagator now acquires the form

$$\begin{aligned} &\left\langle \psi_{i_u}^{-,u} \bar{\psi}_{i_v'}^{+,v} \right\rangle_{S_Z} = -\delta_{i_u i_v'} Z^{-1}_{uv}, \qquad \left\langle \psi_{j_u'}^{+,u} \bar{\psi}_{j_v}^{-,v} \right\rangle_{S_Z} = \delta_{j_u' j_v} (Z^{\dagger})^{-1}_{uv}, \\ &\langle \psi^- \bar{\psi}^- \rangle_{S_Z} = \langle \psi^+ \bar{\psi}^+ \rangle_{S_Z} = 0. \end{aligned} \tag{42}$$

Making use of these results, we obtain

$$\langle F[\bar{\psi},\psi]\rangle_{S_Z} = \delta_{p,p'} \sum_{\tau,\sigma \in S_p} \operatorname{sgn}(\tau^{-1}\sigma) \prod_{k=1}^{p} \delta_{i_k, i'_{\sigma(k)}} \delta_{j_k, j'_{\tau(k)}} Z^{-1}_{k\sigma(k)} (Z^{\dagger})^{-1}_{\tau(k)k}. \tag{43}$$

Inserting Equation (43) into Equation (40), and applying the transformation

$$W = Z^{-1}, \tag{44}$$

whose Jacobian leads to a contribution of $2R' \operatorname{tr}\ln(WW^\dagger)$ to the action, we find

$$I = \delta_{p,p'} \sum_{\tau,\sigma\in S_p} g(\sigma,\tau) \prod_{k=1}^{p} \delta_{i_k,i'_{\sigma(k)}} \delta_{j_k,j'_{\tau(k)}}. \tag{45}$$

Here, $g(\sigma,\tau)$ is given by

$$\begin{aligned} g(\sigma,\tau) &= \left\langle G[W,W^\dagger;\sigma,\tau]\right\rangle_W \equiv \frac{\int \mathcal{D}(W,W^\dagger) e^{-S_w[W,W^\dagger]} G[W,W^\dagger;\sigma,\tau]}{\int \mathcal{D}(W,W^\dagger) e^{-S_w[W,W^\dagger]}}, \\ S_w[W,W^\dagger] &= (2R'+q) \operatorname{tr}\ln\left(1+WW^\dagger\right), \\ G[W,W^\dagger;\sigma,\tau] &= \operatorname{sgn}(\tau^{-1}\sigma) \prod_{k=1}^{p} W_{k\sigma(k)} W^\dagger_{\tau(k)k}. \end{aligned} \tag{46}$$

The angular bracket with subscript W represents the averaging over W with the action $S_w[W,W^\dagger]$.

Using this field-theoretical approach, the Haar integral over the unitary matrix U, whose matrix elements are highly correlated due to the constraint $UU^\dagger = \mathbf{1}$, has been transformed to an integration over the complex matrix W governed by the action S_w. Note that Equation (45) is equivalent to the known result Equation (35) if $g(\sigma,\tau) = \operatorname{Wg}(\tau^{-1}\sigma)$. In the following, we will show that $g(\sigma,\tau)$, defined by Equation (46), depends only on the cycle structure of $\tau^{-1}\sigma$ and is indeed the Weingarten function $\operatorname{Wg}(\tau^{-1}\sigma)$.

5.2. General Properties of the Weingarten Function

If we apply the following transformation in Equation (46)

$$W_{ij} \to W_{i\chi(j)}, \qquad W^\dagger_{ij} \to W^\dagger_{\chi(i)j}, \tag{47}$$

where $\chi \in S_p$ represents an arbitrary permutation of p numbers, the action $S_w[W,W^\dagger]$ remains invariant, while $G[W,W^\dagger;\sigma,\tau]$ transforms to $G[W,W^\dagger;\chi\sigma,\chi\tau]$. This proves that $g(\sigma,\tau)$ obeys the condition

$$g(\sigma,\tau) = g(\chi\sigma,\chi\tau), \tag{48}$$

for an arbitrary permutation $\chi \in S_p$. Setting χ to τ^{-1}, one can immediately see that g is a function of $\tau^{-1}\sigma$ only.

In an analogous way, one can also prove that $g(\sigma,\tau)$ depends only on the cycle structure of $\tau^{-1}\sigma$:

$$g(\tau,\sigma) = V^{(p)}_{c_1,c_2,\ldots,c_m}, \tag{49}$$

where m is the total number of disjoint cycles in $\tau^{-1}\sigma$ and c_i is the length of the i-th cycle. Rearranging the W and $W^\dagger$ terms in $G[W,W^\dagger;\sigma,\tau]$, we can rewrite $g(\sigma,\tau)$ as

$$\begin{aligned} g(\sigma,\tau) &= \operatorname{sgn}(\tau^{-1}\sigma) \left\langle \prod_{k=1}^{m} \left(\prod_{l=1}^{c_k} W_{P_l^{(k)}\sigma(P_l^{(k)})} W^\dagger_{\sigma(P_l^{(k)})P_{l+1}^{(k)}} \right) \right\rangle_W \\ = g(\tau^{-1}\sigma, id) &= (-1)^{p-m} \left\langle \prod_{k=1}^{m} \left(\prod_{l=1}^{c_k} W_{P_l^{(k)}P_{l+1}^{(k)}} W^\dagger_{P_{l+1}^{(k)}P_{l+1}^{(k)}} \right) \right\rangle_W. \end{aligned} \tag{50}$$

Here, the integers $\left\{P_l^{(k)}\right\}$ denote the cycle structure of $\tau^{-1}\sigma$ (Equation (36)), and they satisfy $P_l^{(k)} = (\tau^{-1}\sigma)^{l-1}(P_1^{(k)})$. We introduced here the notation that $P_l^{(k)} = P^{(k)}_{(l \bmod c_k)}$. In the second equality, we have used $\operatorname{sgn}(\tau^{-1}\sigma) = (-1)^{p-m}$ as well as Equation (48).

Note that the action $S_w[W,W^\dagger]$ is invariant under the transformation

$$W_{ij} \to W_{\chi(i)\chi(j)}, \qquad W^\dagger_{ij} \to W^\dagger_{\chi(i)\chi(j)}, \tag{51}$$

for arbitrary $\chi \in S_p$. We now apply this transformation and choose the permutation χ, defined by

$$\chi(P_l^{(k)}) = b_k + l, \tag{52}$$

where $b_k = \sum_{h=1}^{k-1} c_h$ for $k \geq 2$ and $b_1 = 0$. After this transformation, $g(\sigma,\tau)$ becomes

$$g(\sigma,\tau) = (-1)^{p-m} \left\langle \prod_{k=1}^{m} \left(\prod_{l=1}^{c_k} W_{b_k+l, b_k+(l+1 \bmod c_k)} W^{\dagger}_{b_k+(l+1 \bmod c_k), b_k+(l+1 \bmod c_k)} \right) \right\rangle_W . \tag{53}$$

This equation shows that $g(\sigma,\tau)$ depends only on the lengths $\{c_i | i = 1,...,m\}$ of the disjoint cycles of $\tau^{-1}\sigma$.

Using these results, we prove in Appendix E that the function $g(\sigma,\tau)$ defined in Equation (46) satisfies the recursion relation (Equation (37)) and therefore is given by the Weingarten function $g(\sigma,\tau) = \mathrm{Wg}(\tau^{-1}\sigma)$ for $p \leq q$. See [83] for a discussion about extrapolating the results for the case of $p \leq q$ to $p > q$. In Appendix D, we also provide the derivation for the asymptotic behavior of the Weingarten function $g(\sigma,\tau)$ in the large $q \to \infty$ limit.

6. Conclusions

In summary, we derived an effective field theory which can be employed to investigate the spectral statistics of the Floquet operator for a large class of Floquet quantum systems. We applied it to a family of Floquet random quantum circuits whose Floquet operators are composed of two-qudit random unitaries acting on pairs of neighboring qudits in a D-dimensional lattice. Universal RMT statistics was found in the limit of infinite local Hilbert space dimensions, irrespective of the ordering of local gates, the dimensionality of the qudit lattice, and the choice of the boundary condition (open or periodic). This field-theoretical approach was also used to rederive the known results for the Weingarten calculus, which is a method of computing Haar integrals of polynomial functions of the matrix elements.

The wide applicability of the Weingarten calculus in quantum circuit studies suggests that this field-theoretical approach may be generalized to investigate other fundamental features of quantum many-body systems, which is a direction for future work. In particular, the field theory may be useful for the investigation of phase transitions in quantum circuits, such as the measurement-induced phase transition [106–109], which arises from the competition between the unitary dynamics and the projective measurements. A renormalization group analysis of the field theory may be available to carefully examine this phase transition. Moreover, a mapping has been found between the entanglement growth (or the operator spreading in the high-dimensional case) in random quantum circuits and the classical statistical mechanics problem of the surface growth in the Kardar–Parisi–Zhang universality class [39,45]. The field-theoretical approach may provide some insight into this mapping in a more generic setting.

In the present paper, we focus on the Weingarten calculus for the unitary group, having in mind random quantum circuits whose local gates are given by the CUE random matrices. The derivation can be immediately generalized to other compact groups [86,92–94], making use of the color–flavor transformation for the corresponding group [68]. This generalization may have direct applications in random quantum circuits with different symmetry classes [43]. Moreover, as a simple example, the effective field theory given by Equation (19) was employed to study the spectral statistics of Floquet quantum circuits whose local gates are drawn from the CUE ensemble. In fact, it is applicable to quantum circuits with random gates drawn from various ensembles of unitary matrices. It is a useful analytical tool to study the Thouless energy—the energy separation below which the RMT statistics appears—and also to look for the putative transitions between the ergodic phase and many-body localization phase in various Floquet random quantum circuit models [14,17,18,90,110,111].

The replica trick is used in the present paper to derive the sigma model for generic Floquet quantum systems, which was originally constructed in the supersymmetric formalism [64–67]. One advantage of using the replica method is that many terms in the action for higher-order fluctuations are irrelevant in the replica limit and therefore can be omitted. However, unlike the supersymmetric calculation, which allows for a nonperturbative analysis [64,70], the current replica calculation is perturbative and cannot recover the oscillating term in the level correlation function. Consideration of non-standard saddle points [80,102–104] may recover the missing oscillating term and is left for a future study.

Author Contributions: Conceptualization, Y.L. and V.G.; Writing—original draft, Y.L. and V.G. All authors have read and agreed to the published version of the manuscript.

Funding: This work was supported by the U.S. Department of Energy, Office of Science, Basic Energy Sciences, under Award No. DE-SC0001911. Y.L. acknowledges a postdoctoral fellowship from the Simons Foundation "Ultra-Quantum Matter" Research Collaboration.

Conflicts of Interest: The authors declare no conflict of interest.

Appendix A. Derivation of the Moments of Floquet Operator

In this appendix, we prove that the Floquet operator U given by Equation (20) with arbitrary permutation $\sigma \in S_{L'}$ follows the conditions given by Equations (23a) and (23b), in the limit $q \to \infty$. In other words, the moments of the Floquet operator U for any Floquet random quantum circuit that is related to the brickwork circuit studied in [13] by reordering the local quantum gates within one period are given by Equation (23). We make use of the fact that the moments of the Floquet operator U are given by the product of the moments of the independent random CUE matrices $w^{(n,n+1)}$, which also take the form of Equation (23) after the replacement $N \to q^2$ and $\mathbf{i} \to (i^{(n)}, i^{(n+1)})$. Here, q^2 and $(i^{(n)}, i^{(n+1)})$ are, respectively, the dimension and label of the two-qudit Hilbert space in which the CUE matrix $w^{(n,n+1)}$ operates.

For any Floquet operator U that takes the form of Equation (20), we can express its second-order moment $\left\langle U_{\mathbf{ij}} U^\dagger_{\mathbf{j'i'}} \right\rangle$ as

$$\begin{aligned}\left\langle U_{\mathbf{ij}} U^\dagger_{\mathbf{j'i'}} \right\rangle &= \sum_{\mathbf{k},\mathbf{k}'} \prod_{n=1}^{L'} \left\langle w^{(n,n+1)}_{a^{(n)},b^{(n+1)};c^{(n)},d^{(n+1)}} (w^{(n,n+1)})^\dagger_{c'^{(n)},d'^{(n+1)};a'^{(n)},b'^{(n+1)}} \right\rangle \\ &= \sum_{\mathbf{k},\mathbf{k}'} \prod_{n=1}^{L'} \frac{1}{q^2} \delta_{a^{(n)},a'^{(n)}} \delta_{b^{(n+1)},b'^{(n+1)}} \delta_{c^{(n)},c'^{(n)}} \delta_{d^{(n+1)},d'^{(n+1)}} .\end{aligned} \tag{A1}$$

As earlier, vector $\mathbf{i}$ labels the many-body state of all the qudits in the circuit and its n-th element $i^{(n)}$ indexes the single-particle state of the n-th qudit. In addition, $\mathbf{k} = (k^{(1)}, ..., k^{(L)})$ ($\mathbf{k} = (k^{(2)}, ..., k^{(L-1)})$) is an L-dimensional ($(L-2)$-dimensional) vector for the periodic (open) boundary condition. $k^{(n)}$ labels the single-particle state at site n at an intermediate substep, and one has to sum over all possible states $k^{(n)} = 1, ..., q$. Furthermore, $a^{(n)}$ ($b^{(n+1)}$) and $c^{(n)}$ ($d^{(n+1)}$) are, respectively, the row and column indices labeling the Hilbert space of the site n ($n+1$) for the CUE random matrix $w^{(n,n+1)}_{a^{(n)},b^{(n+1)};c^{(n)},d^{(n+1)}}$, and they are defined as follows. Note first that only the unitaries $w^{(n,n+1)}$ and $w^{(n-1,n)}$ act on the qudit at site n, and $\left\{a^{(n)}, b^{(n)}, c^{(n)}, d^{(n)}\right\}$ depend on the ordering of these two unitaries. In particular, if $\sigma^{-1}(n) < \sigma^{-1}(n-1)$ (i.e., unitary $w^{(n,n+1)}$ is applied before $w^{(n-1,n)}$), we have

$$a^{(n)} = i^{(n)}, \quad c^{(n)} = k^{(n)}, \quad b^{(n)} = k^{(n)}, \quad d^{(n)} = j^{(n)}, \tag{A2}$$

and otherwise,

$$a^{(n)} = k^{(n)}, \quad c^{(n)} = j^{(n)}, \quad b^{(n)} = i^{(n)}, \quad d^{(n)} = k^{(n)}. \tag{A3}$$

For the periodic boundary condition, Equations (A2) and (A3) apply to arbitrary site n, which is defined modulo L. By contrast, for the open boundary condition, these two equations hold for $1 < n < L$. At the boundaries $n = 1, L$, we instead have

$$a^{(1)} = i^{(1)}, \qquad c^{(1)} = j^{(1)}, \qquad b^{(L)} = i^{(L)}, \qquad d^{(L)} = j^{(L)}. \tag{A4}$$

The row and column indices for $(w^{(n,n+1)})^\dagger$ are defined in the same way, and $\{a'^{(n)}, b'^{(n)}, c'^{(n)}, d'^{(n)}\}$ are related to $\left\{i'^{(n)}, j'^{(n)}, k'^{(n)}\right\}$ in exactly the same way as $\left\{a^{(n)}, b^{(n)}, c^{(n)}, d^{(n)}\right\}$ are related to $\left\{i^{(n)}, j^{(n)}, k^{(n)}\right\}$.

For both the cases of $\sigma^{-1}(n) < \sigma^{-1}(n-1)$ and $\sigma^{-1}(n) > \sigma^{-1}(n-1)$, $\{a^{(n)}, b^{(n)}, c^{(n)}, d^{(n)}\}$ are related with $\left\{i^{(n)}, j^{(n)}, k^{(n)}, k^{(n)}\right\}$ by a permutation, and the identity

$$\delta_{a^{(n)},a'^{(n)}} \delta_{b^{(n)},b'^{(n)}} \delta_{c^{(n)},c'^{(n)}} \delta_{d^{(n)},d'^{(n)}} = \delta_{i^{(n)},i'^{(n)}} \delta_{j^{(n)},j'^{(n)}} (\delta_{k^{(n)},k'^{(n)}})^2, \tag{A5}$$

is always true for the periodic boundary condition (and for open boundary condition when $1 < n < L$). Making use of this property in Equation (A1), we find that the second-order moment of the Floquet operator for the periodic boundary condition is given by

$$\left\langle U_{\mathbf{ij}} U^{\dagger}_{\mathbf{j'i'}} \right\rangle = \frac{1}{q^{2L}} \sum_{\mathbf{k},\mathbf{k}'} \prod_{n=1}^{L} \delta_{i^{(n)},i'^{(n)}} \delta_{j^{(n)},j'^{(n)}} \left(\delta_{k^{(n)},k'^{(n)}}\right)^2 = \frac{1}{q^L} \prod_{n=1}^{L} \delta_{i^{(n)},i'^{(n)}} \delta_{j^{(n)},j'^{(n)}}. \tag{A6}$$

The second-order moment of U for the open boundary condition can be evaluated in a similar way:

$$\begin{aligned} \left\langle U_{\mathbf{ij}} U^{\dagger}_{\mathbf{j'i'}} \right\rangle &= \frac{1}{q^{2(L-1)}} \delta_{i^{(1)},i'^{(1)}} \delta_{j^{(1)},j'^{(1)}} \delta_{i^{(L)},i'^{(L)}} \delta_{j^{(L)},j'^{(L)}} \sum_{\mathbf{k},\mathbf{k}'} \prod_{n=2}^{L-1} \delta_{i^{(n)},i'^{(n)}} \delta_{j^{(n)},j'^{(n)}} \left(\delta_{k^{(n)},k'^{(n)}}\right)^2 \\ &= \frac{1}{q^L} \prod_{n=1}^{L} \delta_{i^{(n)},i'^{(n)}} \delta_{j^{(n)},j'^{(n)}}. \end{aligned} \tag{A7}$$

As a concrete example, consider the staircase circuit depicted in Figure 1c with the periodic boundary condition. In this case, $\left\{a^{(n)}, b^{(n)}, c^{(n)}, d^{(n)}\right\}$ are given by Equation (A3), and the second-order moment for the Floquet operator U is given by

$$\begin{aligned} \left\langle U_{\mathbf{ij}} U^{\dagger}_{\mathbf{j'i'}} \right\rangle &= \sum_{\mathbf{k},\mathbf{k}'} \left\langle w^{(1,2)}_{i^{(1)},i^{(2)};k^{(1)},k^{(2)}} \left(w^{(1,2)}\right)^{\dagger}_{k'^{(1)},k'^{(2)};i'^{(1)},i'^{(2)}} \right\rangle \\ &\times \prod_{n=2}^{L-1} \left\langle w^{(n,n+1)}_{k^{(n)},i^{(n+1)};j^{(n)},k^{(n+1)}} \left(w^{(n,n+1)}\right)^{\dagger}_{j'^{(n)},k'^{(n+1)};k'^{(n)},i'^{(n+1)}} \right\rangle \\ &\times \left\langle w^{(L,1)}_{k^{(L)},k^{(1)};j^{(L)},j^{(1)}} \left(w^{(L,1)}\right)^{\dagger}_{j'^{(L)},j'^{(1)};k'^{(L)},k'^{(1)}} \right\rangle \\ &= \frac{1}{q^L} \delta_{\mathbf{i},\mathbf{i}'} \delta_{\mathbf{j},\mathbf{j}'}. \end{aligned} \tag{A8}$$

Reordering the two-qudit quantum gates changes the row and column labels of $w^{(n,n+1)}$ and $(w^{(n,n+1)})^{\dagger}$ in the equation above but does not affect the final result. Note that this result for the second-order moment $\left\langle U_{\mathbf{ij}} U^{\dagger}_{\mathbf{j'i'}} \right\rangle$ is valid for arbitrary q.

We now evaluate the fourth-order moment of U and consider the $q \to \infty$ limit for simplicity. This can be expressed as

$$\begin{aligned} \left\langle U_{\mathbf{i}_1\mathbf{j}_1} U_{\mathbf{i}_2\mathbf{j}_2} U^{\dagger}_{\mathbf{j}'_1\mathbf{i}'_1} U^{\dagger}_{\mathbf{j}'_2\mathbf{i}'_2} \right\rangle &= \sum_{\mathbf{k}_1,\mathbf{k}'_1,\mathbf{k}_2,\mathbf{k}'_2} \prod_{n=1}^{L'} \left\langle w^{(n,n+1)}_{a_1^{(n)},b_1^{(n+1)};c_1^{(n)},d_1^{(n+1)}} w^{(n,n+1)}_{a_2^{(n)},b_2^{(n+1)};c_2^{(n)},d_2^{(n+1)}} \right. \\ &\times \left. \left(w^{(n,n+1)}\right)^{\dagger}_{c_1'^{(n)},d_1'^{(n+1)};a_1'^{(n)},b_1'^{(n+1)}} \left(w^{(n,n+1)}\right)^{\dagger}_{c_2'^{(n)},d_2'^{(n+1)};a_2'^{(n)},b_2'^{(n+1)}} \right\rangle \\ &= \sum_{\mathbf{k}_1,\mathbf{k}'_1,\mathbf{k}_2,\mathbf{k}'_2} \prod_{n=1}^{L'} \frac{1}{q^4} \sum_{s^{(n)}=1}^{2} \begin{pmatrix} \delta_{s^{(n)},1} \delta_{a_1^{(n)},a_1'^{(n)}} \delta_{b_1^{(n+1)},b_1'^{(n+1)}} \delta_{c_1^{(n)},c_1'^{(n)}} \delta_{d_1^{(n+1)},d_1'^{(n+1)}} \\ \times \delta_{a_2^{(n)},a_2'^{(n)}} \delta_{b_2^{(n+1)},b_2'^{(n+1)}} \delta_{c_2^{(n)},c_2'^{(n)}} \delta_{d_2^{(n+1)},d_2'^{(n+1)}} \\ +\delta_{s^{(n)},2} \delta_{a_1^{(n)},a_2'^{(n)}} \delta_{b_1^{(n+1)},b_2'^{(n+1)}} \delta_{c_1^{(n)},c_2'^{(n)}} \delta_{d_1^{(n+1)},d_2'^{(n+1)}} \\ \times \delta_{a_2^{(n)},a_1'^{(n)}} \delta_{b_2^{(n+1)},b_1'^{(n+1)}} \delta_{c_2^{(n)},c_1'^{(n)}} \delta_{d_2^{(n+1)},d_1'^{(n+1)}} \end{pmatrix}. \end{aligned} \tag{A9}$$

Here, $\mathbf{k}_{1,2}$ and $\mathbf{k}'_{1,2}$, like $\mathbf{k}$ and $\mathbf{k}'$, are L-dimensional ($(L-2)$-dimensional) vectors whose components label the intermediate single-particle states for all sites (sites in the bulk of the lattice $1 < n < L$) for the periodic (open) boundary condition. In addition, $\left\{a_{1/2}^{(n)}, b_{1/2}^{(n)}, c_{1/2}^{(n)}, d_{1/2}^{(n)}\right\}$ are related to $\left\{i_{1/2}^{(n)}, j_{1/2}^{(n)}, k_{1/2}^{(n)}\right\}$ in the same way as $\left\{a^{(n)}, b^{(n)}, c^{(n)}, d^{(n)}\right\}$ are related to $\left\{i^{(n)}, j^{(n)}, k^{(n)}\right\}$ (see Equations (A2)–(A4)), and similarly for $\left\{a_{1/2}'^{(n)}, b_{1/2}'^{(n)}, c_{1/2}'^{(n)}, d_{1/2}'^{(n)}\right\}$. The summation over $s^{(n)} = 1, 2$ is simply introduced here to distinguish the two terms within the bracket.

In the second equality in Equation (A9), terms involving $\delta_{k_1^{(n)},k_1'^{(n)}} \delta_{k_2^{(n)},k_2'^{(n)}} \delta_{k_1^{(n)},k_2'^{(n)}} \delta_{k_2^{(n)},k_1'^{(n)}}$, compared with those involving $\left(\delta_{k_1^{(n)},k_1'^{(n)}} \delta_{k_2^{(n)},k_2'^{(n)}}\right)^2$ or $\left(\delta_{k_1^{(n)},k_2'^{(n)}} \delta_{k_2^{(n)},k_1'^{(n)}}\right)^2$, are of higher order in $1/q$, since the number of free summations is reduced. For this reason, terms with $s^{(n)} \neq s^{(n+1)}$ for at least one n, compared with those with all $s^{(n)}$ identical, are of higher order in the large q expansion and can be ignored. As a result, we obtain

$$\begin{aligned}\left\langle U_{\mathbf{i}_1\mathbf{j}_1}U_{\mathbf{i}_2\mathbf{j}_2}U^{\dagger}_{\mathbf{j}'_1\mathbf{i}'_1}U^{\dagger}_{\mathbf{j}'_2\mathbf{i}'_2}\right\rangle &= \sum_{\mathbf{k}_1,\mathbf{k}'_1,\mathbf{k}_2,\mathbf{k}'_2}\frac{1}{q^{4L'}}\left(\begin{array}{l}\prod_{n=1}^{L'}\delta_{a_1^{(n)},a_1'^{(n)}}\delta_{b_1^{(n+1)},b_1'^{(n+1)}}\delta_{c_1^{(n)},c_1'^{(n)}}\delta_{d_1^{(n+1)},d_1'^{(n+1)}}\\ \quad\times\delta_{a_2^{(n)},a_2'^{(n)}}\delta_{b_2^{(n+1)},b_2'^{(n+1)}}\delta_{c_2^{(n)},c_2'^{(n)}}\delta_{d_2^{(n+1)},d_2'^{(n+1)}}\\ +\prod_{n=1}^{L'}\delta_{a_1^{(n)},a_2'^{(n)}}\delta_{b_1^{(n+1)},b_2'^{(n+1)}}\delta_{c_1^{(n)},c_2'^{(n)}}\delta_{d_1^{(n+1)},d_2'^{(n+1)}}\\ \quad\times\delta_{a_2^{(n)},a_1'^{(n)}}\delta_{b_2^{(n+1)},b_1'^{(n+1)}}\delta_{c_2^{(n)},c_1'^{(n)}}\delta_{d_2^{(n+1)},d_1'^{(n+1)}}\end{array}\right)\\ &= \frac{1}{q^{2L}}\left(\prod_{n=1}^{L}\delta_{i_1^{(n)},i_1'^{(n)}}\delta_{i_2^{(n)},i_2'^{(n)}}\delta_{j_1^{(n)},j_1'^{(n)}}\delta_{j_2^{(n)},j_2'^{(n)}}+\prod_{n=1}^{L}\delta_{i_1^{(n)},i_2'^{(n)}}\delta_{i_2^{(n)},i_1'^{(n)}}\delta_{j_1^{(n)},j_2'^{(n)}}\delta_{j_2^{(n)},j_1'^{(n)}}\right).\end{aligned} \tag{A10}$$

In the second equality, we used the fact that Equation (A1) reduces to Equation (A6) (Equation (A7)) upon substituting Equations (A2) and (A3) (Equations (A2)–(A4)). This proves Equation (23) for any Floquet operator described by Equation (20) with arbitrary permutation $\sigma \in S_{L'}$.

Appendix B. Generalization to Higher-Dimensional Floquet Quantum Circuits

In this appendix, we generalize the calculation in Appendix A to an arbitrary high dimension D. More specifically, we consider now a D-dimensional Floquet random quantum circuit which consists of a D-dimensional cubic lattice of qudits. During one period, each qudit is coupled to all its neighboring qudits at different time substeps by different two-qudit unitary gates. All these unitary gates are drawn randomly and independently from the CUE ensemble. We will prove that, for any Floquet quantum circuit of this type with an arbitrary ordering of the two-qudit unitary gates, and for both periodic and open boundary conditions, the second- and fourth-order moments of the Floquet operator U take the form Equation (23), same as the CUE ensemble of dimension $N = q^{L^D}$, where q is the Hilbert space dimension of each qudit and L is the number of lattice sites in each direction. For simplicity, we consider the large $q \to \infty$ limit for the calculation of the fourth-order moment.

In the following, we employ the notation that each qudit is labeled by a D-dimensional vector $\mathbf{n}$ whose α-th component n_α represents the coordinate in the lattice in dimension $\alpha = 1, ..., D$. The neighbors of qudit $\mathbf{n}$ are denoted by $\mathbf{n} \pm \mathbf{e}_\alpha$, with $\mathbf{e}_\alpha$ being the unit vector in dimension α. n_α is defined modulo L for the periodic boundary condition. The many-body state is indexed by $\mathbf{i}$ and the single-particle state at site $\mathbf{n}$ is labeled by $i^{(\mathbf{n})}$.

As in the 1D case, the second-order moment $\left\langle U_{\mathbf{ij}}U^{\dagger}_{\mathbf{j}'\mathbf{i}'}\right\rangle$ can be expressed as

$$\begin{aligned}\left\langle U_{\mathbf{ij}}U^{\dagger}_{\mathbf{j}'\mathbf{i}'}\right\rangle &= \sum_{\mathbf{k},\mathbf{k}'}\prod_{\mathbf{n}}{}'\prod_{\alpha=1}^{D}\left\langle w^{(\mathbf{n},\mathbf{n}+\mathbf{e}_\alpha)}_{a^{(\mathbf{n},\alpha)},b^{(\mathbf{n}+\mathbf{e}_\alpha,\alpha)};c^{(\mathbf{n},\alpha)},d^{(\mathbf{n}+\mathbf{e}_\alpha,\alpha)}}\left(w^{(\mathbf{n},\mathbf{n}+\mathbf{e}_\alpha)}\right)^{\dagger}_{c'^{(\mathbf{n},\alpha)},d'^{(\mathbf{n}+\mathbf{e}_\alpha,\alpha)};a'^{(\mathbf{n},\alpha)},b'^{(\mathbf{n}+\mathbf{e}_\alpha,\alpha)}}\right\rangle\\ &= \sum_{\mathbf{k},\mathbf{k}'}\prod_{\mathbf{n}}{}'\prod_{\alpha=1}^{D}\frac{1}{q^2}\delta_{a^{(\mathbf{n},\alpha)},a'^{(\mathbf{n},\alpha)}}\delta_{b^{(\mathbf{n}+\mathbf{e}_\alpha,\alpha)},b'^{(\mathbf{n}+\mathbf{e}_\alpha,\alpha)}}\delta_{c^{(\mathbf{n},\alpha)},c'^{(\mathbf{n},\alpha)}}\delta_{d^{(\mathbf{n}+\mathbf{e}_\alpha,\alpha)},d'^{(\mathbf{n}+\mathbf{e}_\alpha,\alpha)}}.\end{aligned} \tag{A11}$$

Here, $w^{(\mathbf{n},\mathbf{n}+\mathbf{e}_\alpha)}$ represents the random CUE matrix that couples the qudits at sites $\mathbf{n}$ and $\mathbf{n} + \mathbf{e}_\alpha$. $\prod'_{\mathbf{n}}$ stands for $\prod_{n_1=1}^{L'} \cdots \prod_{n_D=1}^{L'}$, where L' is defined as L $(L-1)$ for the periodic (open) boundary condition. We use $k^{(\mathbf{n},m)}$ to denote the single-particle state for qudit $\mathbf{n}$ at a intermediate substep, and it needs to be summed over $(\sum_{k^{(\mathbf{n},m)}=1}^{q})$. $\sum_{\mathbf{k}}$ denotes the product $\prod_{\mathbf{n}}\prod_{m=1}^{M_{\mathbf{n}}}\sum_{k^{(\mathbf{n},m)}=1}^{q}$, where m runs over $m = 1, 2, ..., M_{\mathbf{n}}$ and $\mathbf{n}$ runs over all sites in the lattice. $M_{\mathbf{n}} + 1$ is the number of neighbors for site $\mathbf{n}$. For the periodic boundary condition, $M_{\mathbf{n}} = 2D - 1$, while for the open boundary condition,

$$M_{\mathbf{n}} = 2D - D_{\mathbf{n}} - 1, \tag{A12}$$

where $D_{\mathbf{n}}$ denotes the number of components of vector $\mathbf{n}$ satisfying $n_\alpha = 1$ or L. $\sum_{\mathbf{k}'}$ is defined in the same manner. The row and column indices $\left\{a^{(\mathbf{n},\alpha)}, b^{(\mathbf{n},\alpha)}, c^{(\mathbf{n},\alpha)}, d^{(\mathbf{n},\alpha)}\right\}$ and $\left\{a'^{(\mathbf{n},\alpha)}, b'^{(\mathbf{n},\alpha)}, c'^{(\mathbf{n},\alpha)}, d'^{(\mathbf{n},\alpha)}\right\}$ label the single-particle state at site $\mathbf{n}$ at a certain time substep and depend on the ordering of the quantum gates that couple site $\mathbf{n}$ with all its neighbors.

Consider first the periodic boundary condition. For arbitrary ordering of the two-qudit unitary gates, the $4D$ numbers in the set $S_{\mathbf{n}}^{(1)} = \left\{a^{(\mathbf{n},\alpha)}, b^{(\mathbf{n},\alpha)}, c^{(\mathbf{n},\alpha)}, d^{(\mathbf{n},\alpha)} | \alpha = 1, 2, ..., D\right\}$ are always related to those in the set $S_{\mathbf{n}}^{(2)} = \left\{i^{(\mathbf{n})}, j^{(\mathbf{n})}, k^{(\mathbf{n},m)}, k^{(\mathbf{n},m)} | m = 1,2,...,2D-1\right\}$ by a permutation. Row and column indices for $w^\dagger$ are defined in an analogous way. In particular, the numbers in the set $S_{\mathbf{n}}^{\prime(1)} = \left\{a^{\prime(\mathbf{n},\alpha)}, b^{\prime(\mathbf{n},\alpha)}, c^{\prime(\mathbf{n},\alpha)}, d^{\prime(\mathbf{n},\alpha)} | \alpha = 1,2,...,D\right\}$ are related to those in the set $S_{\mathbf{n}}^{\prime(2)} = \left\{i^{\prime(\mathbf{n})}, j^{\prime(\mathbf{n})}, k^{\prime(\mathbf{n},m)}, k^{\prime(\mathbf{n},m)} | m = 1,2,...,2D-1\right\}$ by the same permutation that connects $S_{\mathbf{n}}^{(1)}$ to $S_{\mathbf{n}}^{(2)}$. This means that, for arbitrary ordering of the two-qudit gates, the following identity is always satisfied

$$\prod_{\alpha=1}^{D} \delta_{a^{(\mathbf{n},\alpha)},a^{\prime(\mathbf{n},\alpha)}} \delta_{b^{(\mathbf{n},\alpha)},b^{\prime(\mathbf{n},\alpha)}} \delta_{c^{(\mathbf{n},\alpha)},c^{\prime(\mathbf{n},\alpha)}} \delta_{d^{(\mathbf{n},\alpha)},d^{\prime(\mathbf{n},\alpha)}} = \delta_{i^{(\mathbf{n})},i^{\prime(\mathbf{n})}} \delta_{j^{(\mathbf{n})},j^{\prime(\mathbf{n})}} \left(\prod_{m=1}^{2D-1} \delta_{k^{(\mathbf{n},m)},k^{\prime(\mathbf{n},m)}} \right)^2. \tag{A13}$$

Inserting this equation into the second equality of Equation (A11), we obtain the second-order moment for the Floquet operator

$$\begin{aligned} \left\langle U_{\mathbf{ij}} U^\dagger_{\mathbf{j'i'}} \right\rangle &= \sum_{\mathbf{k},\mathbf{k}'} \prod_{\mathbf{n}} \frac{1}{q^{2D}} \delta_{i^{(\mathbf{n})},i^{\prime(\mathbf{n})}} \delta_{j^{(\mathbf{n})},j^{\prime(\mathbf{n})}} \left(\prod_{m=1}^{2D-1} \delta_{k^{(\mathbf{n},m)},k^{\prime(\mathbf{n},m)}} \right)^2 \\ &= \frac{1}{q^{L^D}} \prod_{\mathbf{n}} \delta_{i^{(\mathbf{n})},i^{\prime(\mathbf{n})}} \delta_{j^{(\mathbf{n})},j^{\prime(\mathbf{n})}}. \end{aligned} \tag{A14}$$

For the open boundary condition, extra care must be taken for the sites $\mathbf{n}$ at the boundaries. For any site $\mathbf{n}$ that sits in the bulk of the lattice, as in the case of the periodic boundary condition, the sets $S_{\mathbf{n}}^{(1)}$ and $S_{\mathbf{n}}^{(2)}$ ($S_{\mathbf{n}}^{\prime(1)}$ and $S_{\mathbf{n}}^{\prime(2)}$) defined above are always related by a $4D$-number permutation for any ordering of the quantum gates. Moreover, the permutation that connects $S_{\mathbf{n}}^{(1)}$ and $S_{\mathbf{n}}^{(2)}$ is identical to the one that connects $S_{\mathbf{n}}^{\prime(1)}$ and $S_{\mathbf{n}}^{\prime(2)}$. By contrast, at the boundary, for any component of the vector $\mathbf{n}$ that satisfies $n_\alpha = 1$ ($n_\alpha = L$), $b^{(\mathbf{n},\alpha)}$ and $d^{(\mathbf{n},\alpha)}$ ($a^{(\mathbf{n},\alpha)}$ and $c^{(\mathbf{n},\alpha)}$) must be removed from the set $S_{\mathbf{n}}^{(1)} = \left\{a^{(\mathbf{n},\beta)}, b^{(\mathbf{n},\beta)}, c^{(\mathbf{n},\beta)}, d^{(\mathbf{n},\beta)} | \beta = 1,2,...,D\right\}$. After applying this procedure, all the $2M_{\mathbf{n}} + 2$ numbers left in the new set $S_{\mathbf{n}}^{(1)}$ are related to those in the set $S_{\mathbf{n}}^{(2)} = \left\{i^{(\mathbf{n})}, j^{(\mathbf{n})}, k^{(\mathbf{n},m)}, k^{(\mathbf{n},m)} | m = 1,2,...,M_{\mathbf{n}}\right\}$ by a permutation. $\left\{a^{\prime(\mathbf{n},\alpha)}, b^{\prime(\mathbf{n},\alpha)}, c^{\prime(\mathbf{n},\alpha)}, d^{\prime(\mathbf{n},\alpha)}\right\}$ are defined in the same way. They are related to $\left\{i^{\prime(\mathbf{n})}, j^{\prime(\mathbf{n})}, k^{\prime(\mathbf{n},m)}\right\}$ in exactly the same way as $\left\{a^{(\mathbf{n},\alpha)}, b^{(\mathbf{n},\alpha)}, c^{(\mathbf{n},\alpha)}, d^{(\mathbf{n},\alpha)}\right\}$ are related to $\left\{i^{(\mathbf{n})}, j^{(\mathbf{n})}, k^{(\mathbf{n},m)}\right\}$. This leads to

$$\begin{aligned} &\prod_{\alpha|n_\alpha \neq L} \delta_{a^{(\mathbf{n},\alpha)},a^{\prime(\mathbf{n},\alpha)}} \delta_{c^{(\mathbf{n},\alpha)},c^{\prime(\mathbf{n},\alpha)}} \prod_{\beta|n_\beta \neq 1} \delta_{b^{(\mathbf{n},\beta)},b^{\prime(\mathbf{n},\beta)}} \delta_{d^{(\mathbf{n},\beta)},d^{\prime(\mathbf{n},\beta)}} \\ &= \delta_{i^{(\mathbf{n})},i^{\prime(\mathbf{n})}} \delta_{j^{(\mathbf{n})},j^{\prime(\mathbf{n})}} \left(\prod_{m=1}^{M_{\mathbf{n}}} \delta_{k^{(\mathbf{n},m)},k^{\prime(\mathbf{n},m)}} \right)^2. \end{aligned} \tag{A15}$$

Here, the first (second) product $\prod_{\alpha|n_\alpha \neq L}$ ($\prod_{\beta|n_\beta \neq 1}$) runs over all directions α (β) that obey $n_\alpha \neq L$ ($n_\beta \neq 1$).

Using this result, we find that Equation (A11) reduces to

$$\begin{aligned} \left\langle U_{\mathbf{ij}} U^\dagger_{\mathbf{j'i'}} \right\rangle &= \sum_{\mathbf{k},\mathbf{k}'} \prod_{\mathbf{n}} \left(\frac{1}{q^{M_{\mathbf{n}}+1}} \prod_{\alpha|n_\alpha \neq L} \delta_{a^{(\mathbf{n},\alpha)},a^{\prime(\mathbf{n},\alpha)}} \delta_{c^{(\mathbf{n},\alpha)},c^{\prime(\mathbf{n},\alpha)}} \prod_{\beta|n_\beta \neq 1} \delta_{b^{(\mathbf{n},\beta)},b^{\prime(\mathbf{n},\beta)}} \delta_{d^{(\mathbf{n},\beta)},d^{\prime(\mathbf{n},\beta)}} \right) \\ &= \prod_{\mathbf{n}} \left[\frac{1}{q^{M_{\mathbf{n}}+1}} \delta_{i^{(\mathbf{n})},i^{\prime(\mathbf{n})}} \delta_{j^{(\mathbf{n})},j^{\prime(\mathbf{n})}} \prod_{m=1}^{M_{\mathbf{n}}} \left(\sum_{k^{(\mathbf{n},m)},k^{\prime(\mathbf{n},m)}} \delta^2_{k^{(\mathbf{n},m)},k^{\prime(\mathbf{n},m)}} \right) \right] \\ &= \frac{1}{q^{L^D}} \prod_{\mathbf{n}} \delta_{i^{(\mathbf{n})},i^{\prime(\mathbf{n})}} \delta_{j^{(\mathbf{n})},j^{\prime(\mathbf{n})}}. \end{aligned} \tag{A16}$$

We have therefore proved Equation (23) for the D-dimensional Floquet random quantum circuit which has an arbitrary ordering of the two-qudit random unitary gates and is subject to either periodic or open boundary condition.

The fourth-order moment $\left\langle U_{\mathbf{i}_1\mathbf{j}_1} U_{\mathbf{i}_2\mathbf{j}_2} U^\dagger_{\mathbf{j}_1'\mathbf{i}_1'} U^\dagger_{\mathbf{j}_2'\mathbf{i}_2'} \right\rangle$, in the limit of $q \to \infty$, can be expressed in a form similar to the 1D expression in Equation (A9):

$$
\begin{aligned}
\left\langle U_{\mathbf{i}_1\mathbf{j}_1} U_{\mathbf{i}_2\mathbf{j}_2} U^{\dagger}_{\mathbf{j}_1'\mathbf{i}_1'} U^{\dagger}_{\mathbf{j}_2'\mathbf{i}_2'} \right\rangle &= \sum_{\mathbf{k}_1,\mathbf{k}_1',\mathbf{k}_2,\mathbf{k}_2'} \prod_{\mathbf{n}}' \prod_{\alpha=1}^{D} \Big\langle w^{(\mathbf{n},\mathbf{n}+\mathbf{e}_\alpha)}_{a_1^{(\mathbf{n},\alpha)},b_1^{(\mathbf{n}+\mathbf{e}_\alpha,\alpha)};c_1^{(\mathbf{n},\alpha)},d_1^{(\mathbf{n}+\mathbf{e}_\alpha,\alpha)}} w^{(\mathbf{n},\mathbf{n}+\mathbf{e}_\alpha)}_{a_2^{(\mathbf{n},\alpha)},b_2^{(\mathbf{n}+\mathbf{e}_\alpha,\alpha)};c_2^{(\mathbf{n},\alpha)},d_2^{(\mathbf{n}+\mathbf{e}_\alpha,\alpha)}} \\
&\times (w^{(\mathbf{n},\mathbf{n}+\mathbf{e}_\alpha)})^{\dagger}_{c_1'^{(\mathbf{n})},d_1'^{(\mathbf{n}+\mathbf{e}_\alpha)};a_1'^{(\mathbf{n})},b_1'^{(\mathbf{n}+\mathbf{e}_\alpha)}} (w^{(\mathbf{n},\mathbf{n}+\mathbf{e}_\alpha)})^{\dagger}_{c_2'^{(\mathbf{n})},d_2'^{(\mathbf{n}+\mathbf{e}_\alpha)};a_2'^{(\mathbf{n})},b_2'^{(\mathbf{n}+\mathbf{e}_\alpha)}} \Big\rangle \\
&= \sum_{\mathbf{k}_1,\mathbf{k}_1',\mathbf{k}_2,\mathbf{k}_2'} \prod_{\mathbf{n}}' \prod_{\alpha=1}^{D} \frac{1}{q^4} \sum_{s^{(\mathbf{n},\alpha)}=1}^{2} \begin{pmatrix} \delta_{s^{(\mathbf{n},\alpha)},1} \delta_{a_1^{(\mathbf{n},\alpha)},a_1'^{(\mathbf{n},\alpha)}} \delta_{b_1^{(\mathbf{n}+\mathbf{e}_\alpha,\alpha)},b_1'^{(\mathbf{n}+\mathbf{e}_\alpha,\alpha)}} \delta_{c_1^{(\mathbf{n},\alpha)},c_1'^{(\mathbf{n},\alpha)}} \delta_{d_1^{(\mathbf{n}+\mathbf{e}_\alpha,\alpha)},d_1'^{(\mathbf{n}+\mathbf{e}_\alpha,\alpha)}} \\ \times \delta_{a_2^{(\mathbf{n},\alpha)},a_2'^{(\mathbf{n},\alpha)}} \delta_{b_2^{(\mathbf{n}+\mathbf{e}_\alpha,\alpha)},b_2'^{(\mathbf{n}+\mathbf{e}_\alpha,\alpha)}} \delta_{c_2^{(\mathbf{n},\alpha)},c_2'^{(\mathbf{n},\alpha)}} \delta_{d_2^{(\mathbf{n}+\mathbf{e}_\alpha,\alpha)},d_2'^{(\mathbf{n}+\mathbf{e}_\alpha,\alpha)}} \\ +\delta_{s^{(\mathbf{n},\alpha)},2} \delta_{a_1^{(\mathbf{n},\alpha)},a_2'^{(\mathbf{n},\alpha)}} \delta_{b_1^{(\mathbf{n}+\mathbf{e}_\alpha,\alpha)},b_2'^{(\mathbf{n}+\mathbf{e}_\alpha,\alpha)}} \delta_{c_1^{(\mathbf{n},\alpha)},c_2'^{(\mathbf{n},\alpha)}} \delta_{d_1^{(\mathbf{n}+\mathbf{e}_\alpha,\alpha)},d_2'^{(\mathbf{n}+\mathbf{e}_\alpha,\alpha)}} \\ \times \delta_{a_2^{(\mathbf{n},\alpha)},a_1'^{(\mathbf{n},\alpha)}} \delta_{b_2^{(\mathbf{n}+\mathbf{e}_\alpha,\alpha)},b_1'^{(\mathbf{n}+\mathbf{e}_\alpha,\alpha)}} \delta_{c_2^{(\mathbf{n},\alpha)},c_1'^{(\mathbf{n},\alpha)}} \delta_{d_2^{(\mathbf{n}+\mathbf{e}_\alpha,\alpha)},d_1'^{(\mathbf{n}+\mathbf{e}_\alpha,\alpha)}} \end{pmatrix}.
\end{aligned} \tag{A17}
$$

Here, the indices $\left\{a_{1/2}^{(\mathbf{n},\alpha)}, b_{1/2}^{(\mathbf{n},\alpha)}, c_{1/2}^{(\mathbf{n},\alpha)}, d_{1/2}^{(\mathbf{n})}\right\}$ are related to $\left\{i_{1/2}^{(\mathbf{n})}, j_{1/2}^{(\mathbf{n})}, k_{1/2}^{(\mathbf{n},m)}\right\}$ in the same way as $\left\{a^{(\mathbf{n},\alpha)}, b^{(\mathbf{n},\alpha)}, c^{(\mathbf{n},\alpha)}, d^{(\mathbf{n})}\right\}$ are related to $\left\{i^{(\mathbf{n})}, j^{(\mathbf{n})}, k^{(\mathbf{n},m)}\right\}$ (and similarly for the indices of $w^{\dagger}$). As before, $s^{(\mathbf{n},\alpha)}$ is introduced to distinguish the two terms in the bracket.

Similarly to the 1D case, in the second equality in the equation above, terms that contain the factor $\delta_{k_1^{(\mathbf{n},m)},k_1'^{(\mathbf{n},m)}} \delta_{k_2^{(\mathbf{n},m)},k_2'^{(\mathbf{n},m)}} \delta_{k_1^{(\mathbf{n},m)},k_2'^{(\mathbf{n},m)}} \delta_{k_2^{(\mathbf{n},m)},k_1'^{(\mathbf{n},m)}}$ are of higher order in the large q expansion compared with those with the factor $(\delta_{k_1^{(\mathbf{n},m)},k_1'^{(\mathbf{n},m)}} \delta_{k_2^{(\mathbf{n},m)},k_2'^{(\mathbf{n},m)}})^2$ or $(\delta_{k_1^{(\mathbf{n},m)},k_2'^{(\mathbf{n},m)}} \delta_{k_2^{(\mathbf{n},m)},k_1'^{(\mathbf{n},m)}})^2$, due to the reduced number of free summations. For this reason, we only need to keep terms for which all $s^{(\mathbf{n},\alpha)}$ are identical in the leading order of the large q expansion. This leads to

$$
\begin{aligned}
&\left\langle U_{\mathbf{i}_1\mathbf{j}_1} U_{\mathbf{i}_2\mathbf{j}_2} U^{\dagger}_{\mathbf{j}_1'\mathbf{i}_1'} U^{\dagger}_{\mathbf{j}_2'\mathbf{i}_2'} \right\rangle \\
&= \sum_{\mathbf{k}_1,\mathbf{k}_1',\mathbf{k}_2,\mathbf{k}_2'} \begin{pmatrix} \prod_{\mathbf{n}}' \prod_{\alpha=1}^{D} \frac{1}{q^4} \delta_{a_1^{(\mathbf{n},\alpha)},a_1'^{(\mathbf{n},\alpha)}} \delta_{b_1^{(\mathbf{n}+\mathbf{e}_\alpha,\alpha)},b_1'^{(\mathbf{n}+\mathbf{e}_\alpha,\alpha)}} \delta_{c_1^{(\mathbf{n},\alpha)},c_1'^{(\mathbf{n},\alpha)}} \delta_{d_1^{(\mathbf{n}+\mathbf{e}_\alpha,\alpha)},d_1'^{(\mathbf{n}+\mathbf{e}_\alpha,\alpha)}} \\ \times \delta_{a_2^{(\mathbf{n},\alpha)},a_2'^{(\mathbf{n},\alpha)}} \delta_{b_2^{(\mathbf{n}+\mathbf{e}_\alpha,\alpha)},b_2'^{(\mathbf{n}+\mathbf{e}_\alpha,\alpha)}} \delta_{c_2^{(\mathbf{n},\alpha)},c_2'^{(\mathbf{n},\alpha)}} \delta_{d_2^{(\mathbf{n}+\mathbf{e}_\alpha,\alpha)},d_2'^{(\mathbf{n}+\mathbf{e}_\alpha,\alpha)}} \\ +\prod_{\mathbf{n}}' \prod_{\alpha=1}^{D} \frac{1}{q^4} \delta_{a_1^{(\mathbf{n},\alpha)},a_2'^{(\mathbf{n},\alpha)}} \delta_{b_1^{(\mathbf{n}+\mathbf{e}_\alpha,\alpha)},b_2'^{(\mathbf{n}+\mathbf{e}_\alpha,\alpha)}} \delta_{c_1^{(\mathbf{n},\alpha)},c_2'^{(\mathbf{n},\alpha)}} \delta_{d_1^{(\mathbf{n}+\mathbf{e}_\alpha,\alpha)},d_2'^{(\mathbf{n}+\mathbf{e}_\alpha,\alpha)}} \\ \times \delta_{a_2^{(\mathbf{n},\alpha)},a_1'^{(\mathbf{n},\alpha)}} \delta_{b_2^{(\mathbf{n}+\mathbf{e}_\alpha,\alpha)},b_1'^{(\mathbf{n}+\mathbf{e}_\alpha,\alpha)}} \delta_{c_2^{(\mathbf{n},\alpha)},c_1'^{(\mathbf{n},\alpha)}} \delta_{d_2^{(\mathbf{n}+\mathbf{e}_\alpha,\alpha)},d_1'^{(\mathbf{n}+\mathbf{e}_\alpha,\alpha)}} \end{pmatrix} \\
&= \frac{1}{q^{2L^D}} \left(\prod_{\mathbf{n}} \delta_{i_1^{(\mathbf{n})},i_1'^{(\mathbf{n})}} \delta_{i_2^{(\mathbf{n})},i_2'^{(\mathbf{n})}} \delta_{j_1^{(\mathbf{n})},j_1'^{(\mathbf{n})}} \delta_{j_2^{(\mathbf{n})},j_2'^{(\mathbf{n})}} + \prod_{\mathbf{n}} \delta_{i_1^{(\mathbf{n})},i_2'^{(\mathbf{n})}} \delta_{i_2^{(\mathbf{n})},i_1'^{(\mathbf{n})}} \delta_{j_1^{(\mathbf{n})},j_2'^{(\mathbf{n})}} \delta_{j_2^{(\mathbf{n})},j_1'^{(\mathbf{n})}} \right).
\end{aligned} \tag{A18}
$$

In the second equality here, we made use of the intermediate results in the derivation of the second-order moment (i.e., the fact that Equation (A11) is identical to Equations (A14) and (A16) for the periodic and open boundary conditions, respectively).

Appendix C. Effective Field Theory of a Non-interacting Floquet Model

In this appendix, we compare the effective field theory of the Floquet random quantum circuits studied earlier (or equivalently the CUE ensemble) with that of a non-interacting Floquet model and show that the quartic fluctuations become important in the non-interacting case. We consider now a time-periodic non-interacting system whose many-body Floquet operator is simply given by the tensor product of all single-particle Floquet operators $w^{(n)}$:

$$U = w^{(1)} \otimes w^{(2)} \otimes \ldots \otimes w^{(L)}. \tag{A19}$$

All $\left\{w^{(n)} | n = 1, \ldots, L\right\}$ are $q \times q$ unitary matrices drawn randomly and independently from the CUE ensemble. As before, we consider the limit of large single-particle Hilbert space dimension $q \to \infty$ for simplicity.

Since all single-particle Floquet operators $w^{(n)}$ are independent, the many-body SFF $K(t)$ is given by the product of all single-particle SFFs $k^{(n)}(t)$:

$$K(t) = \prod_{n=1}^{L} k^{(n)}(t), \qquad k^{(n)}(t) = \left\langle |\operatorname{Tr}(w^{(n)})^t|^2 \right\rangle. \tag{A20}$$

Here, $k^{(n)}(t)$ is simply the SFF for the CUE ensemble of dimension q (Equation (9) after the replacement $N \to q$). Therefore, one can easily see that the many-body SFF now exhibits a fast-growing ramp $K(t) = t^L$, which plateaus at time $t = q$. The many-body quasi-energies are uncorrelated at an energy separation of the order of the many-body level spacing $\Delta\phi \sim 1/N$, where $N = q^L$ is the dimension of the many-body Hilbert space [36].

The many-body Floquet operator U given by Equation (A19) satisfies Equation (23a) but not Equation (23b). Its fourth-order moment in the large q limit takes the form

$$\begin{aligned}\left\langle U_{\mathbf{i}_1\mathbf{j}_1} U_{\mathbf{i}_2\mathbf{j}_2} U^\dagger_{\mathbf{j}'_1\mathbf{i}'_1} U^\dagger_{\mathbf{j}'_2\mathbf{i}'_2} \right\rangle &= \prod_{n=1}^{L} \left\langle w^{(n)}_{i_1^{(n)},j_1^{(n)}} w^{(n)}_{i_2^{(n)},j_2^{(n)}} (w^{(n)})^\dagger_{j_1'^{(n)},i_1'^{(n)}} (w^{(n)})^\dagger_{j_2'^{(n)},i_2'^{(n)}} \right\rangle \\ &= \frac{1}{q^{2L}} \prod_{n=1}^{L} \left(\delta_{i_1^{(n)},i_1'^{(n)}} \delta_{j_1^{(n)},j_1'^{(n)}} \delta_{i_2^{(n)},i_2'^{(n)}} \delta_{j_2^{(n)},j_2'^{(n)}} + \delta_{i_1^{(n)},i_2'^{(n)}} \delta_{j_1^{(n)},j_2'^{(n)}} \delta_{i_2^{(n)},i_1'^{(n)}} \delta_{j_2^{(n)},j_1'^{(n)}} \right).\end{aligned} \tag{A21}$$

Note that for the chaotic Floquet circuits described by Equation (20), to have nonzero moments $\left\langle U_{\mathbf{i}_1\mathbf{j}_1} U_{\mathbf{i}_2\mathbf{j}_2} U^\dagger_{\mathbf{j}'_1\mathbf{i}'_1} U^\dagger_{\mathbf{j}'_2\mathbf{i}'_2} \right\rangle$, if $i_1^{(n)} = i_{1/2}'^{(n)}$ is obeyed by one of the sites, it must hold for all other sites as well. Otherwise, the fourth-order moment can be ignored in the leading order of the large q expansion. However, this is no longer the case for the non-interacting model. It is this difference that leads to the contrasting behavior of the many-body level statistics for the two models.

Since the second-order moment of the Floquet operator $\left\langle U_{\mathbf{ij}} U^\dagger_{\mathbf{j}'\mathbf{i}'} \right\rangle$ for the current non-interacting model is identical to that of the chaotic Floquet circuits discussed in Section 4, the quadratic fluctuations in the effective theory (Equation (19)) are therefore also governed by the action in Equation (24). On the other hand, the quartic-order effective action of the non-interacting model is different from that of the chaotic model (Equation (30)) and assumes the form

$$\begin{aligned} S^{(4-1)}_{\text{eff}} &= -\frac{(\alpha\beta)^2}{2} \sum_{\mathbf{i}_1,\ldots,\mathbf{i}_8} \operatorname{tr}\left(Z^\dagger_{\mathbf{i}_1\mathbf{i}_2} Z_{\mathbf{i}_3\mathbf{i}_4} \right) \operatorname{tr}\left(Z^\dagger_{\mathbf{i}_5\mathbf{i}_6} Z_{\mathbf{i}_7\mathbf{i}_8} \right) \\ &\times \frac{1}{q^{2L}} \left[\begin{array}{l} \prod_{n=1}^{L} \left(\delta_{i_2^{(n)},i_1^{(n)}} \delta_{i_6^{(n)},i_5^{(n)}} \delta_{i_3^{(n)},i_4^{(n)}} \delta_{i_7^{(n)},i_8^{(n)}} + \delta_{i_2^{(n)},i_5^{(n)}} \delta_{i_6^{(n)},i_1^{(n)}} \delta_{i_3^{(n)},i_8^{(n)}} \delta_{i_7^{(n)},i_4^{(n)}} \right) \\ - \prod_{n=1}^{L} \delta_{i_2^{(n)},i_1^{(n)}} \delta_{i_6^{(n)},i_5^{(n)}} \delta_{i_3^{(n)},i_4^{(n)}} \delta_{i_7^{(n)},i_8^{(n)}} \end{array} \right]. \end{aligned} \tag{A22}$$

Here, we have inserted Equation (A21) into $S^{(4)}_{\text{eff}}$ in Equation (19c) and, as before, ignored the last term in $S^{(4)}_{\text{eff}}$, which is not important in the replica limit $R \to 0$.

Rewriting Equation (A22) in terms of X and Y, which denote, respectively, the diagonal and off-diagonal components of the matrix Z in the Hilbert space, we obtain

$$\begin{aligned} S^{(4-1)}_{\text{eff}} &= -\frac{(\alpha\beta)^2}{2N^2} \sum_{\mathbf{k}_1,\ldots,\mathbf{k}_4} \operatorname{tr}\left(X^\dagger(\mathbf{k}_1) X(\mathbf{k}_2) \right) \operatorname{tr}\left(X^\dagger(\mathbf{k}_3) X(\mathbf{k}_4) \right) \\ &\times \left[\prod_{n=1}^{L} \left(\delta_{k_1^{(n)},0} \delta_{k_2^{(n)},0} \delta_{k_3^{(n)},0} \delta_{k_4^{(n)},0} + \frac{1}{q^2} \delta_{k_1^{(n)},-k_3^{(n)}} \delta_{k_2^{(n)},-k_4^{(n)}} \right) - \prod_{n=1}^{L} \left(\delta_{k_1^{(n)},0} \delta_{k_2^{(n)},0} \delta_{k_3^{(n)},0} \delta_{k_4^{(n)},0} \right) \right] \\ &- \frac{(\alpha\beta)^2}{2N^2} \sum_{\mathbf{i}_1 \neq \mathbf{i}_2, \mathbf{i}_3 \neq \mathbf{i}_4, \mathbf{i}_5 \neq \mathbf{i}_6, \mathbf{i}_7 \neq \mathbf{i}_8} \operatorname{tr}\left(Y^\dagger_{\mathbf{i}_1\mathbf{i}_2} Y_{\mathbf{i}_3\mathbf{i}_4} \right) \operatorname{tr}\left(Y^\dagger_{\mathbf{i}_5\mathbf{i}_6} Y_{\mathbf{i}_7\mathbf{i}_8} \right) \\ &\times \prod_{n=1}^{L} \left(\delta_{i_2^{(n)},i_1^{(n)}} \delta_{i_6^{(n)},i_5^{(n)}} \delta_{i_3^{(n)},i_4^{(n)}} \delta_{i_7^{(n)},i_8^{(n)}} + \delta_{i_2^{(n)},i_5^{(n)}} \delta_{i_6^{(n)},i_1^{(n)}} \delta_{i_3^{(n)},i_8^{(n)}} \delta_{i_7^{(n)},i_4^{(n)}} \right) \\ &- \frac{(\alpha\beta)^2}{2N^2} \sum_{\mathbf{i}_1 \neq \mathbf{i}_2, \mathbf{i}_5 \neq \mathbf{i}_6} \sum_{\mathbf{k}_3,\mathbf{k}_4} \left[\operatorname{tr}\left(Y^\dagger_{\mathbf{i}_1\mathbf{i}_2} X(\mathbf{k}_3) \right) \operatorname{tr}\left(Y^\dagger_{\mathbf{i}_5\mathbf{i}_6} X(\mathbf{k}_4) \right) + \operatorname{tr}\left(X^\dagger(-\mathbf{k}_3) Y_{\mathbf{i}_1\mathbf{i}_2} \right) \operatorname{tr}\left(X^\dagger(-\mathbf{k}_4) Y_{\mathbf{i}_5\mathbf{i}_6} \right) \right] \\ &\times \prod_{n=1}^{L} \left(\delta_{i_2^{(n)},i_1^{(n)}} \delta_{i_6^{(n)},i_5^{(n)}} \delta_{k_3^{(n)},0} \delta_{k_4^{(n)},0} + \frac{1}{q} \delta_{i_2^{(n)},i_5^{(n)}} \delta_{i_6^{(n)},i_1^{(n)}} \delta_{k_3^{(n)},-k_4^{(n)}} \right). \end{aligned} \tag{A23}$$

Here, $X(\mathbf{k})$ represents the Fourier transform of the diagonal component $X_{\mathbf{ii}}$ with respect to $\mathbf{i}$ (Equation (25)), and the n-th component of momentum $\mathbf{k}$ is denoted by $k^{(n)}$.

From Equation (A23), we obtain the self-energy for $X(\mathbf{k})$ and Y from the quartic interactions in the replica limit $R \to 0$,

$$
\begin{aligned}
(\Sigma_X)^{ab;ba}(\mathbf{k},\mathbf{k}) &= \frac{(\alpha\beta)^2}{N^2}\left\langle X^{ab}(-\mathbf{k})(X^{\dagger})^{ba}(-\mathbf{k})\right\rangle_0 \left[\prod_{n=1}^{L}\left(\delta_{k^{(n)},0}+\frac{1}{q^2}\right)-\delta_{\mathbf{k},0}\right] \\
&= \frac{(\alpha\beta)^2}{N}\frac{1}{1-\alpha\beta\delta_{\mathbf{k},0}}\left[\left(1+\frac{1}{q^2}\right)^{r_0}\frac{1}{q^{2(L-r_0)}}-\delta_{\mathbf{k},0}\right], \\
(\Sigma_Y)^{ab;ba}_{\mathbf{i}_1\mathbf{i}_2;\mathbf{i}'_2\mathbf{i}'_1} &= \frac{(\alpha\beta)^2}{N^2}\sum_{\mathbf{i}_3\neq\mathbf{i}_4}\left\langle Y^{ab}_{\mathbf{i}_3\mathbf{i}_4}(Y^{\dagger})^{ba}_{\mathbf{i}_4\mathbf{i}_3}\right\rangle_0 \\
&\quad\times\prod_{n=1}^{L}\left(\delta_{i_2^{(n)},i_1^{(n)}}\delta_{i_3^{(n)},i_4^{(n)}}\delta_{i_1'^{(n)},i_2'^{(n)}}+\delta_{i_2^{(n)},i_3^{(n)}}\delta_{i_1^{(n)},i_4^{(n)}}\delta_{i_2'^{(n)},i_3^{(n)}}\delta_{i_1'^{(n)},i_4^{(n)}}\right) \\
&= \frac{(\alpha\beta)^2}{N}\prod_{n=1}^{L}\left(\delta_{i_2^{(n)},i_1^{(n)}}\delta_{i_1'^{(n)},i_2'^{(n)}}+\frac{1}{q}\delta_{i_2^{(n)},i_2'^{(n)}}\delta_{i_1^{(n)},i_1'^{(n)}}\right) \\
&\quad-\frac{(\alpha\beta)^2}{N}\prod_{n=1}^{L}\left(\delta_{i_2^{(n)},i_1^{(n)}}\delta_{i_1'^{(n)},i_2'^{(n)}}+\frac{1}{q}\delta_{i_2^{(n)},i_2'^{(n)}}\delta_{i_1^{(n)},i_1'^{(n)}}\delta_{i_1^{(n)},i_2^{(n)}}\right).
\end{aligned}
\tag{A24}
$$

Here, we have used the expressions for the bare propagators (Equations (27) and (29)). r_0 represents the number of zero components in the L-dimensional vector $\mathbf{k}$. We can therefore see that the self-energy for $X(0)$ is of the order of $L/q^2N(1-\exp(i\Delta\phi))$, and the self-energy for $X(\mathbf{k})$ with $r_0 < L$ zero components in $\mathbf{k}$ is of the order of $1/q^{2(L-r_0)}N$. Furthermore, the self-energy $(\Sigma_Y)^{ab;ba}_{\mathbf{i}_1\mathbf{i}_2;\mathbf{i}'_2\mathbf{i}'_1}$ is of the order of $1/(Nq^{L-r'})$, where r' represents the number of components satisfying $i_2^{(n)} = i_1^{(n)}$ and $i_2'^{(n)} = i_1'^{(n)}$. Comparing Equation (A24) with Equation (31), we find that the self-energies for $X(\mathbf{k})$ (with at least one zero component in $\mathbf{k}$) and $Y_{\mathbf{i}_1\mathbf{i}_2}$ (when $i_1^{(n)} = i_2^{(n)}$ is satisfied for as least one site n) of the current non-interacting model are much larger than their counterparts for the chaotic Floquet circuits. Therefore, in contrast to the chaotic case, the quartic-order fluctuations for the current non-interacting model are essential.

Appendix D. Large q Expansion of the Weingarten Function

In this appendix, we study the asymptotic behavior of the Weingarten function $g(\sigma,\tau)$ (Equation (46)) for the unitary group $U(q)$ in the large $q\to\infty$ limit. Expanding the action $S_w[W,W^{\dagger}]$ in powers of W, we obtain

$$
S_w[W,W^{\dagger}] = q\sum_{n=1}^{\infty}\frac{(-1)^{n-1}}{n}\,\mathrm{tr}\left(WW^{\dagger}\right)^n, \tag{A25}
$$

where we have approximated the overall coefficient as $q+2R' \approx q$. One can see from rescaling W by $W\to W/\sqrt{q}$ that the higher-order terms in the expansion above are also of higher order in $1/q$.

At the leading order in the large q expansion, we can therefore apply the Gaussian approximation and keep only the quadratic term in $S_w[W,W^{\dagger}]$:

$$
S_w^{(2)}[W,W^{\dagger}] = q\,\mathrm{tr}\left(WW^{\dagger}\right). \tag{A26}
$$

The bare W propagator is given by

$$
\left\langle W_{uv}W^{\dagger}_{v'u'}\right\rangle_0 = \delta_{uu'}\delta_{vv'}\frac{1}{q}, \tag{A27}
$$

where the angular bracket with subscript 0 denotes the functional averaging with the Gaussian action $S_w^{(2)}$. Applying the Wick theorem to evaluate the expectation value of $\langle G[W,W^{\dagger};\sigma,\tau]\rangle_0$ (Equation (46)), we find that, in the Gaussian approximation, $g(\sigma,\tau)$ is non-vanishing only when $\sigma=\tau$:

$$
g(\sigma,\sigma) = \prod_{k=1}^{p}\left\langle W_{k\sigma(k)}W^{\dagger}_{\sigma(k)k}\right\rangle_0 = \frac{1}{q^p}, \tag{A28}
$$

which is consistent with the known result for the Gaussian approximation to the Weingarten function $\mathrm{Wg}(id) = g(\sigma,\sigma)$ [82]. The Weingarten functions for all the remaining permutations $\tau^{-1}\sigma \neq id$ are of higher order in $1/q$.

To deduce the leading-order term in the large q expansion of $g(\sigma \neq \tau)$, we should go beyond the Gaussian approximation and keep the higher-order terms in $S_w[W, W^\dagger]$. Using Equation (53), we find

$$\begin{aligned} g(\sigma,\tau) &\propto \prod_{k=1}^{m}\left\langle\left(\prod_{l=1}^{c_k} W_{b_k+l,b_k+(l+1 \bmod c_k)} W^\dagger_{b_k+(l+1 \bmod c_k),b_k+(l+1 \bmod c_k)}\right)\left[q\,\mathrm{tr}\left(WW^\dagger\right)^{c_k}\right]\right\rangle_0 \\ &\propto \prod_{k=1}^{m}\left(q\prod_{l=1}^{c_k}\left\langle W_{b_k+l,b_k+(l+1 \bmod c_k)} W^\dagger_{b_k+(l+1 \bmod c_k),b_k+l}\right\rangle_0 \right. \\ &\quad \left. \times \left\langle W^\dagger_{b_k+(l+1 \bmod c_k),b_k+(l+1 \bmod c_k)} W_{b_k+(l+1 \bmod c_k),b_k+(l+1 \bmod c_k)}\right\rangle_0\right) \\ &\propto \prod_{k=1}^{m} q^{1-2c_k}, \end{aligned} \tag{A29}$$

which leads to the known result $g(\sigma,\tau) \propto q^{m-2p}$ [82]. The extra factor of $q\,\mathrm{Tr}(WW^\dagger)^{c_k}$ in the first line of the equation above represents the interaction vertex arising from the c_k-th-order term in the expansion of $S_w[W, W^\dagger]$ (Equation (A25)).

Appendix E. Recursion Relation for the Weingarten Function

In this appendix, we will show that the function $g(\sigma,\tau)$ given by Equation (46) satisfies the recursion relations for the Weingarten function, i.e., Equation (37). To derive the first recursion relation Equation (37a), we make use of the identity $UU^\dagger = \mathbf{1}$, which leads to

$$\begin{aligned} &\int_{\mathcal{U}(q)} dU U_{i_1 j_1} U_{i_2 j_2} \ldots U_{i_p j_p} U^\dagger_{j'_1 i'_1} U^\dagger_{j'_2 i'_2} \ldots U^\dagger_{j'_p i'_p} \\ &= \sum_{i_{p+1}, i'_{p+1}, j_{p+1}=1}^{q} \delta_{i_{p+1}, i'_{p+1}} \delta_{j_{p+1}, j'_{p+1}} \\ &\quad \times \int_{\mathcal{U}(q)} dU U_{i_1 j_1} U_{i_2 j_2} \ldots U_{i_p j_p} U_{i_{p+1} j_{p+1}} U^\dagger_{j'_1 i'_1} U^\dagger_{j'_2 i'_2} \ldots U^\dagger_{j'_p i'_p} U^\dagger_{j'_{p+1} i'_{p+1}}. \end{aligned} \tag{A30}$$

Using Equation (45), the left hand side (L.H.S.) of the equation above can be expressed as

$$\text{L.H.S.} = \sum_{\tau',\sigma' \in S_p}\left(\prod_{k=1}^{p} \delta_{i_k, i'_{\sigma'(k)}} \delta_{j_k, j'_{\tau'(k)}}\right) g(\sigma',\tau'), \tag{A31}$$

while the right hand side (R.H.S.) acquires the form

$$\text{R.H.S.} = \sum_{i_{p+1}, i'_{p+1}, j_{p+1}=1}^{q} \sum_{\tau'',\sigma'' \in S_{p+1}} \delta_{i_{p+1}, i'_{p+1}} \delta_{j_{p+1}, j'_{p+1}} \left(\prod_{k=1}^{p+1} \delta_{i_k, i'_{\sigma''(k)}} \delta_{j_k, j'_{\tau''(k)}}\right) g(\sigma'',\tau''). \tag{A32}$$

We then choose an arbitrary pair of permutations $\sigma, \tau \in S_p$ and set $i_k = i'_{\sigma(k)}, j_k = j'_{\tau(k)}$, for $k = 1, 2, \ldots, p$. For simplicity, we consider the case where $p < q$ and all integers in the set $\{i'_k | k = 1, \ldots, p\}$ ($\{j'_k | k = 1, \ldots, p+1\}$) are different from each other. In this case, the L.H.S. then becomes

$$\text{L.H.S.} = g(\sigma,\tau) = V^{(p)}_{c_1, c_2, \ldots, c_m}, \tag{A33}$$

where $\{c_k | k = 1, 2, \ldots, m\}$ represent the lengths of disjoint cycles for the permutation $\tau^{-1}\sigma$ and obey the constraint $\sum_{k=1}^{m} c_k = p$. Here, we made use of the result that $g(\sigma,\tau)$ depends only on $\{c_k\}$ and expressed it as $V^{(p)}_{c_1, c_2, \ldots, c_m}$ (see Section 5.2). On the other hand, the R.H.S. becomes

$$\text{R.H.S.} = \sum_{\tau'',\sigma'' \in S_{p+1}} \sum_{i'_{p+1}=1}^{q} \left(\prod_{k=1}^{p} \delta_{i'_{\sigma(k)}, i'_{\sigma''(k)}} \delta_{j'_{\tau(k)}, j'_{\tau''(k)}}\right) \delta_{i'_{p+1}, i'_{\sigma''(p+1)}} \delta_{j'_{p+1}, j'_{\tau''(p+1)}} g(\sigma'',\tau''). \tag{A34}$$

Let us now examine all terms in the summation in Equation (A34). The contribution is nonzero for arbitrary i'_{p+1} if $\sigma'' = \sigma_p$ and $\tau'' = \tau_p$, where σ_p and τ_p are defined as

$$\sigma_p(k) = \begin{cases} \sigma(k), & k = 1, 2, \ldots, p, \\ p+1, & k = p+1, \end{cases} \qquad \tau_p(k) = \begin{cases} \tau(k), & k = 1, 2, \ldots, p, \\ p+1, & k = p+1. \end{cases} \tag{A35}$$

The permutation $\tau_p^{-1}\sigma_p$, compared with $\tau^{-1}\sigma$, contains an extra disjoint cycle of length 1. More specifically, from the earlier result (Equation (50)), we find that $g(\sigma_p, \tau_p)$ is given by

$$\begin{aligned} g(\sigma_p,\tau_p) &= (-1)^{p-m}\left\langle\left(W_{p+1,p+1} W^\dagger_{p+1,p+1}\right)\prod_{k=1}^{m}\left(\prod_{l=1}^{c_k} W_{P_l^{(k)},\sigma(P_l^{(k)})} W^\dagger_{\sigma(P_l^{(k)}),P_{l+1}^{(k)}}\right)\right\rangle_W \\ &= V^{(p+1)}_{c_1, c_2, \ldots c_m, 1}. \end{aligned} \tag{A36}$$

Compared with $g(\sigma,\tau) = V^{(p)}_{c_1,c_2,\dots,c_m}$, it contains an extra factor $W_{p+1,p+1}W^{\dagger}_{p+1,p+1}$. Here, the integers $\left\{P_l^{(k)}\right\}$ denote the cycle structure of $\tau^{-1}\sigma$ as in Equation (36). The total contribution from this pair of permutations (σ_p,τ_p) to Equation (A34) is

$$\sum_{i'_{p+1}=1}^{q} g(\sigma_p,\tau_p) = qV^{(p+1)}_{c_1,c_2,\dots,c_m,1}, \tag{A37}$$

where the factor q comes from the summation over i'_{p+1}.

Another non-vanishing contribution comes from the case when $i'_{p+1} = i'_{\sigma(l)}$ for an arbitrary positive integer $l \le p$. In this case, the contribution is non-vanishing if the permutations $\sigma'' = \sigma_l$ and $\tau'' = \tau_p$, where τ_p is defined in Equation (A35) and σ_l is given by

$$\sigma_l(k) = \begin{cases} \sigma(k), & k \neq l, p+1, \\ p+1, & k = l, \\ \sigma(l), & k = p+1. \end{cases} \tag{A38}$$

In this case, it is easy to see that $\delta_{i'_{p+1}i'_{\sigma_l(p+1)}} = \delta_{i'_{\sigma(l)},i'_{\sigma_l(l)}} = \delta_{i'_{p+1},i'_{\sigma(l)}} = 1$.

$g(\sigma_l,\tau_p)$ can be obtained from $g(\sigma,\tau)$ by replacing the factor $W_{l,\sigma(l)}W^{\dagger}_{\sigma(l),\tau^{-1}\sigma(l)}$ in Equation (50) with $W_{l,p+1}W^{\dagger}_{p+1,p+1}W_{p+1,\sigma(l)}W^{\dagger}_{\sigma(l),\tau^{-1}\sigma(l)}$. This results in adding a number to the disjoint cycle to which number l belongs (labeled by j):

$$\begin{aligned} g(\sigma,\tau) =& \left\langle W_{l,\sigma(l)}W^{\dagger}_{\sigma(l),\tau^{-1}\sigma(l)}\cdots\right\rangle_W = V^{(p)}_{c_1,\dots,c_j,\dots,c_m}, \\ g(\sigma_l,\tau_p) =& \left\langle W_{l,\sigma_l(l)}W^{\dagger}_{\sigma_l(l),\tau_p^{-1}\sigma_l(l)}W_{\tau_p^{-1}\sigma_l(l),\sigma_l\tau_p^{-1}\sigma_l(l)}W^{\dagger}_{\sigma_l\tau_p^{-1}\sigma_l(l),\tau_p^{-1}\sigma_l\tau_p^{-1}\sigma_l(l)}\cdots\right\rangle_W, \\ =& \left\langle W_{l,p+1}W^{\dagger}_{p+1,p+1}W_{p+1,\sigma(l)}W^{\dagger}_{\sigma(l),\tau^{-1}\sigma(l)}\cdots\right\rangle_W = V^{(p+1)}_{c_1,\dots,c_j+1,\dots,c_m}. \end{aligned} \tag{A39}$$

Here, "..." represents all remaining terms that stay the same in $g(\sigma_l,\tau_p)$ and $g(\sigma,\tau)$. Summing over all possibilities of l, we find the total contribution of this type is

$$\sum_l g(\sigma_l,\tau_p) = \sum_{s=1}^{m} c_s V^{(p+1)}_{c_1,\dots,c_{s-1},c_s+1,c_{s+1},\dots,c_m}. \tag{A40}$$

Combining everything (Equations (A37) and (A40)), we find the R.H.S. of Equation (A30).

$$\text{R.H.S.} = qV^{(p+1)}_{c_1,\dots,c_m,1} + \sum_{s=1}^{m} c_s V^{(p+1)}_{c_1,\dots,c_{s-1},c_s+1,c_{s+1},\dots,c_m}. \tag{A41}$$

Equating this to the L.H.S. (Equation (A33)), we arrive at the recursion relation Equation (37a), which holds for an arbitrary pair of $\sigma,\tau \in S_p$ (or equivalently, an arbitrary configuration of $\{c_j\}$ that satisfies $\sum_{j=1}^{m} c_j = p$) when $p < q$.

The recursion relation Equation (37b) can be proved in a similar way using the identity:

$$\sum_{j_p,j'_p=1}^{q} \delta_{j_p,j'_p} \int_{\mathcal{U}(q)} dU U_{i_1j_1}U_{i_2j_2}\dots U_{i_{p-1}j_{p-1}}U_{i_pj_p}U^{\dagger}_{j'_1i'_1}U^{\dagger}_{j'_2i'_2}\dots U^{\dagger}_{j'_{p-1}i'_{p-1}}U^{\dagger}_{j'_pi'_p} = 0, \tag{A42}$$

for $i_p \neq i'_p$. Using Equation (45), one can rewrite this equation as

$$\sum_{j_p,j'_p=1}^{q} \sum_{\tau',\sigma' \in S_p} \delta_{j_p,j'_p} \left(\prod_{k=1}^{p} \delta_{i_k,i'_{\sigma'(k)}} \delta_{j_k,j'_{\tau'(k)}}\right) g(\sigma',\tau') = 0. \tag{A43}$$

We now set $j_k = j'_k$ and $i_k = i'_{\chi(k)}$ for $1 \le k \le p$. $\chi \in S_p$ represents an arbitrary nonidentical permutation of numbers $1,2,\dots,p$ that satisfies $\chi(p) \neq p$. It contains m disjoint cycles with lengths $\{c_k|k = 1,2,\dots,m\}$. Since $\chi \neq id$ and $\chi(p) \neq p$, p stays in a cycle (labeled by 1) of length $c_1 \ge 2$. We consider the case $p \le q$ and choose the set $\{i'_k|k = 1,\dots,p\}$ $(\{j'_k|k = 1,\dots,p-1\})$, whose elements are all different from each other, for simplicity. Equation (A43) can be rewritten as

$$\sum_{j'_p=1}^{q} \sum_{\tau',\sigma' \in S_p} \left(\prod_{k=1}^{p} \delta_{i'_{\chi(k)},i'_{\sigma'(k)}} \delta_{j'_k,j'_{\tau'(k)}}\right) g(\sigma',\tau') = 0. \tag{A44}$$

Let us now examine all terms in the summation on the L.H.S. of Equation (A44). For arbitrary j'_p, the contribution is non-vanishing when $\sigma' = \chi$ and $\tau' = id$. The total contribution of this type is

$$\sum_{j'_p=1}^{q} g(\chi,id) = qg(\chi,id) = qV^{(p)}_{c_1,c_2,\dots,c_m}. \tag{A45}$$

Another type of non-vanishing contribution comes from the case where $j'_p = j'_l$ and l also stays within the same cycle to which p belongs (cycle 1). In this case, the contribution is non-vanishing if $\sigma' = \chi$ and $\tau = \tau_l$, where τ_l is defined as

$$\tau_l(k) = \begin{cases} k, & k \neq p, l, \\ l, & k = p, \\ p, & k = l. \end{cases} \quad \text{(A46)}$$

$g(\chi, \tau_l)$ can be obtained from $g(\chi, id)$ by replacing $W^\dagger_{ll}$ and $W^\dagger_{pp}$ in $G[W, W^\dagger; \chi, id]$ (Equation (46)) with $W^\dagger_{lp}$ and $W^\dagger_{pl}$. This breaks cycle 1 of length c_1 into two disjoint cycles of lengths $c_1 - c$ and c (c obeys $0 < c < c_1$ and depends on l):

$$\begin{aligned} g(\chi, id) &= V^{(p)}_{c_1, c_2, \ldots, c_m} \\ &= \left\langle \left(W_{p,\chi(p)} W^\dagger_{\chi(p),\chi(p)} \cdots W_{\chi^{-1}(l),l} W^\dagger_{l,l} W_{l,\chi(l)} W^\dagger_{\chi(l),\chi(l)} \cdots W_{\chi^{-1}(p),p} W^\dagger_{p,p} \right) \prod_{k \neq 1}^{m} (\ldots) \right\rangle_W, \\ g(\chi, \tau_l) &= V^{(p)}_{c_1 - c, c, c_2, \ldots, c_m} \\ &= \left\langle \left(W_{p,\chi(p)} W^\dagger_{\chi(p),\chi(p)} \cdots W_{\chi^{-1}(l),l} W^\dagger_{l,p} \right) \left(W_{l,\chi(l)} W^\dagger_{\chi(l),\chi(l)} \cdots W_{\chi^{-1}(p),p} W^\dagger_{p,l} \right) \prod_{k \neq 1}^{m} (\ldots) \right\rangle_W. \end{aligned} \quad \text{(A47)}$$

Here, $\prod_{k \neq 1}^{m} (\ldots)$ represents the contribution from all remaining cycles other than 1 and stays the same for both $g(\chi, id)$ and $g(\chi, \tau_l)$. Taking into account all possible l belonging to cycle 1, we find that the total contribution from such terms is

$$\sum_{l \in \text{cycle } 1} g(\chi, \tau_l) = \sum_{c=1}^{c_1 - 1} V^{(p)}_{c_1 - c, c, c_2, \ldots, c_m}. \quad \text{(A48)}$$

The last type of non-vanishing contribution arises from the case when $j'_p = j'_l$ and l does not belong to cycle 1 (we label the cycle which l belongs by 2). In this case, the contribution is non-vanishing if $\sigma' = \chi$ and $\tau' = \tau_l$ (Equation (A46)). As mentioned earlier, $g(\chi, \tau_l)$ can be obtained from $g(\chi, id)$ by replacing $W^\dagger_{ll}$ and $W^\dagger_{pp}$ in $G[W, W^\dagger; \chi, id]$ with $W^\dagger_{lp}$ and $W^\dagger_{pl}$, respectively. However, since now l, p belong to different cycles, this replacement combines cycle 1 and cycle 2 into one cycle of length $c_1 + c_2$:

$$\begin{aligned} g(\chi, id) &= V^{(p)}_{c_1, c_2, c_3, \ldots, c_m} \\ &= \left\langle \left(W_{p,\chi(p)} W^\dagger_{\chi(p),\chi(p)} \cdots W_{\chi^{-1}(p),p} W^\dagger_{p,p} \right) \left(W_{l,\chi(l)} W^\dagger_{\chi(l),\chi(l)} \cdots W_{\chi^{-1}(l),l} W^\dagger_{l,l} \right) \prod_{k > 2}^{m} (\ldots) \right\rangle, \\ g(\chi, \tau_0) &= V^{(p)}_{c_1 + c_2, c_3, \ldots, c_m} \\ &= \left\langle \left(W_{p,\chi(p)} W^\dagger_{\chi(p),\chi(p)} \cdots W_{\chi^{-1}(p),p} W^\dagger_{p,l} W_{l,\chi(l)} W^\dagger_{\chi(l),\chi(l)} \cdots W_{\chi^{-1}(l),l} W^\dagger_{l,p} \right) \prod_{k > 2}^{m} (\ldots) \right\rangle. \end{aligned} \quad \text{(A49)}$$

Here, $\prod_{k > 2}^{m} (\ldots)$ represents the contribution from all remaining cycles $k \neq 1, 2$, which remains the same for both cases. Summing over all possible l that do not stay in cycle 1, we find the total contribution of this type is

$$\sum_{l \notin \text{cycle } 1} g(\chi, \tau_l) = \sum_{s=2}^{m} c_s V^{(p)}_{c_1 + c_s, c_2, \ldots, c_{s-1}, c_{s+1}, \ldots, c_m}. \quad \text{(A50)}$$

From Equation (A42), one can see that, for an arbitrary choice of $\chi \in S_p$ that satisfies $\chi(p) \neq p$ (for arbitrary $\{c_k\}$ with at least one cycle length $c_1 > 1$), the sum of all three types of non-vanishing contributions (Equations (A45), (A48), and (A50)) should equal 0, which proves Equation (37b) for $p \leq q$.

References

1. Emerson, J.; Weinstein, Y.S.; Saraceno, M.; Lloyd, S.; Cory, D.G. Pseudo-random unitary operators for quantum information processing. *Science* **2003**, *302*, 2098–2100. [CrossRef] [PubMed]
2. Emerson, J.; Livine, E.; Lloyd, S. Convergence conditions for random quantum circuits. *Phys. Rev. A* **2005**, *72*, 060302. [CrossRef]
3. Oliveira, R.; Dahlsten, O.C.O.; Plenio, M.B. Generic Entanglement Can Be Generated Efficiently. *Phys. Rev. Lett.* **2007**, *98*, 130502. [CrossRef] [PubMed]
4. Harrow, A.W.; Low, R.A. Random Quantum Circuits are Approximate 2-designs. *Commun. Math. Phys.* **2009**, *291*, 257–302. [CrossRef]
5. Brandao, F.G.; Harrow, A.W.; Horodecki, M. Local Random Quantum Circuits are Approximate Polynomial-Designs. *Commun. Math. Phys.* **2016**, *346*, 397–434. [CrossRef]
6. Žnidarič, M. Optimal two-qubit gate for generation of random bipartite entanglement. *Phys. Rev. A* **2007**, *76*, 012318. [CrossRef]
7. Žnidarič, M. Exact convergence times for generation of random bipartite entanglement. *Phys. Rev. A* **2008**, *78*, 032324. [CrossRef]

8. Arnaud, L.; Braun, D. Efficiency of producing random unitary matrices with quantum circuits. *Phys. Rev. A* **2008**, *78*, 062329. [CrossRef]
9. Brown, W.G.; Viola, L. Convergence Rates for Arbitrary Statistical Moments of Random Quantum Circuits. *Phys. Rev. Lett.* **2010**, *104*, 250501. [CrossRef]
10. Nakata, Y.; Hirche, C.; Koashi, M.; Winter, A. Efficient Quantum Pseudorandomness with Nearly Time-Independent Hamiltonian Dynamics. *Phys. Rev. X* **2017**, *7*, 021006. [CrossRef]
11. Diniz, I.T.; Jonathan, D. Comment on "Random Quantum Circuits are Approximate 2-designs" by A.W. Harrow and R.A. Low (Commun. Math. Phys. 291, 257–302 (2009)). *Commun. Math. Phys.* **2011**, *304*, 281–293 [CrossRef]
12. Gross, D.; Audenaert, K.; Eisert, J. Evenly distributed unitaries: On the structure of unitary designs. *J. Math. Phys.* **2007**, *48*, 052104. [CrossRef]
13. Chan, A.; De Luca, A.; Chalker, J.T. Solution of a Minimal Model for Many-Body Quantum Chaos. *Phys. Rev. X* **2018**, *8*, 041019. [CrossRef]
14. Chan, A.; De Luca, A.; Chalker, J.T. Spectral Statistics in Spatially Extended Chaotic Quantum Many-Body Systems. *Phys. Rev. Lett.* **2018**, *121*, 060601. [CrossRef]
15. Friedman, A.J.; Chan, A.; De Luca, A.; Chalker, J.T. Spectral Statistics and Many-Body Quantum Chaos with Conserved Charge. *Phys. Rev. Lett.* **2019**, *123*, 210603. [CrossRef]
16. Garratt, S.J.; Chalker, J.T. Local Pairing of Feynman Histories in Many-Body Floquet Models. *Phys. Rev. X* **2021**, *11*, 021051. [CrossRef]
17. Garratt, S.J.; Chalker, J.T. Many-Body Delocalization as Symmetry Breaking. *Phys. Rev. Lett.* **2021**, *127*, 026802. [CrossRef]
18. Chan, A.; De Luca, A.; Chalker, J.T. Spectral Lyapunov exponents in chaotic and localized many-body quantum systems. *Phys. Rev. Res.* **2021**, *3*, 023118. [CrossRef]
19. Bertini, B.; Kos, P.; Prosen, T. Random Matrix Spectral Form Factor of Dual-Unitary Quantum Circuits. *Commun. Math. Phys* **2021**, *387*, 597–620. [CrossRef]
20. Bertini, B.; Kos, P.; Prosen, T. Exact Spectral Form Factor in a Minimal Model of Many-Body Quantum Chaos. *Phys. Rev. Lett.* **2018**, *121*, 264101. [CrossRef]
21. Kos, P.; Bertini, B.; Prosen, T. Chaos and Ergodicity in Extended Quantum Systems with Noisy Driving. *Phys. Rev. Lett.* **2021**, *126*, 190601. [CrossRef] [PubMed]
22. Moudgalya, S.; Prem, A.; Huse, D.A.; Chan, A. Spectral statistics in constrained many-body quantum chaotic systems. *Phys. Rev. Res.* **2021**, *3*, 023176. [CrossRef]
23. Chan, A.; Shivam, S.; Huse, D.A.; De Luca, A. Many-Body Quantum Chaos and Space-time Translational Invariance. *arXiv* **2021**, arXiv:2109.04475.
24. Bohigas, O.; Giannoni, M.J.; Schmit, C. Characterization of chaotic quantum spectra and universality of level fluctuation laws. *Phys. Rev. Lett.* **1984**, *52*, 1. [CrossRef]
25. Kos, P.; Ljubotina, M.; Prosen, T. Many-Body Quantum Chaos: Analytic Connection to Random Matrix Theory. *Phys. Rev. X* **2018**, *8*, 021062. [CrossRef]
26. Roy, D.; Prosen, T. Random matrix spectral form factor in kicked interacting fermionic chains. *Phys. Rev. E* **2020**, *102*, 060202. [CrossRef]
27. Roy, D.; Mishra, D.; Prosen, T. Spectral form factor in a minimal bosonic model of many-body quantum chaos. *arXiv* **2022**, arXiv:2203.05439.
28. Flack, A.; Bertini, B.; Prosen, T. Statistics of the spectral form factor in the self-dual kicked Ising model. *Phys. Rev. Res.* **2020**, *2*, 043403. [CrossRef]
29. Berry, M.V. Semiclassical theory of spectral rigidity. *Proc. R. Soc. Lond. Ser. A* **1985**, *400*, 229–251.
30. Sieber, M.; Richter, K. Correlations between Periodic Orbits and their Role in Spectral Statistics. *Phys. Scr.* **2001**, *T90*, 128. [CrossRef]
31. Müller, S.; Heusler, S.; Braun, P.; Haake, F.; Altland, A. Periodic-orbit theory of universality in quantum chaos. *Phys. Rev. E* **2005**, *72*, 046207. [CrossRef] [PubMed]
32. Altland, A.; Bagrets, D. Quantum ergodicity in the SYK model. *Nucl. Phys. B* **2018**, *930*, 45–68. [CrossRef]
33. Saad, P.; Shenker, S.H.; Stanford, D. A semiclassical ramp in SYK and in gravity. *arXiv* **2018**, arXiv:1806.06840.
34. Liao, Y.; Galitski, V. Emergence of many-body quantum chaos via spontaneous breaking of unitarity. *Phys. Rev. B* **2022**, *105*, L140202. [CrossRef]
35. Liao, Y.; Galitski, V. Universal dephasing mechanism of many-body quantum chaos. *Phys. Rev. Res.* **2022**, *4*, L012037. [CrossRef]
36. Liao, Y.; Vikram, A.; Galitski, V. Many-Body Level Statistics of Single-Particle Quantum Chaos. *Phys. Rev. Lett.* **2020**, *125*, 250601. [CrossRef]
37. Bertini, B.; Kos, P.; Prosen, T. Exact Correlation Functions for Dual-Unitary Lattice Models in $1+1$ Dimensions. *Phys. Rev. Lett.* **2019**, *123*, 210601. [CrossRef]
38. Piroli, L.; Bertini, B.; Cirac, J.I.; Prosen, T. Exact dynamics in dual-unitary quantum circuits. *Phys. Rev. B* **2020**, *101*, 094304. [CrossRef]
39. Nahum, A.; Vijay, S.; Haah, J. Operator Spreading in Random Unitary Circuits. *Phys. Rev. X* **2018**, *8*, 021014. [CrossRef]

40. Von Keyserlingk, C.W.; Rakovszky, T.; Pollmann, F.; Sondhi, S.L. Operator Hydrodynamics, OTOCs, and Entanglement Growth in Systems without Conservation Laws. *Phys. Rev. X* **2018**, *8*, 021013. [CrossRef]
41. Rakovszky, T.; Pollmann, F.; von Keyserlingk, C.W. Diffusive Hydrodynamics of Out-of-Time-Ordered Correlators with Charge Conservation. *Phys. Rev. X* **2018**, *8*, 031058. [CrossRef]
42. Khemani, V.; Vishwanath, A.; Huse, D.A. Operator Spreading and the Emergence of Dissipative Hydrodynamics under Unitary Evolution with Conservation Laws. *Phys. Rev. X* **2018**, *8*, 031057. [CrossRef]
43. Hunter-Jones, N. Operator growth in random quantum circuits with symmetry. *arXiv* **2018**, arXiv:1812.08219.
44. Bertini, B.; Piroli, L. Scrambling in random unitary circuits: Exact results. *Phys. Rev. B* **2020**, *102*, 064305. [CrossRef]
45. Nahum, A.; Ruhman, J.; Vijay, S.; Haah, J. Quantum Entanglement Growth under Random Unitary Dynamics. *Phys. Rev. X* **2017**, *7*, 031016. [CrossRef]
46. Jonay, C.; Huse, D.A.; Nahum, A. Coarse-grained dynamics of operator and state entanglement. *arXiv* **2018**, arXiv:1803.00089.
47. Bertini, B.; Kos, P.; Prosen, T. Entanglement Spreading in a Minimal Model of Maximal Many-Body Quantum Chaos. *Phys. Rev. X* **2019**, *9*, 021033. [CrossRef]
48. Bertini, B.; Kos, P.; Prosen, T. Operator Entanglement in Local Quantum Circuits I: Chaotic Dual-Unitary Circuits. *SciPost Phys.* **2020**, *8*, 67. [CrossRef]
49. Bertini, B.; Kos, P.; Prosen, T. Operator Entanglement in Local Quantum Circuits II: Solitons in Chains of Qubits. *SciPost Phys.* **2020**, *8*, 68. [CrossRef]
50. Gopalakrishnan, S.; Lamacraft, A. Unitary circuits of finite depth and infinite width from quantum channels. *Phys. Rev. B* **2019**, *100*, 064309. [CrossRef]
51. Rakovszky, T.; Pollmann, F.; von Keyserlingk, C.W. Sub-ballistic Growth of Rényi Entropies due to Diffusion. *Phys. Rev. Lett.* **2019**, *122*, 250602. [CrossRef] [PubMed]
52. Srednicki, M. Chaos and quantum thermalization. *Phys. Rev. E* **1994**, *50*, 888. [CrossRef] [PubMed]
53. Srednicki, M. The approach to thermal equilibrium in quantized chaotic systems. *J. Phys. A* **1999**, *32*, 1163. [CrossRef]
54. Deutsch, J.M. Quantum statistical mechanics in a closed system. *Phys. Rev. A* **1991**, *43*, 2046–2049. [CrossRef]
55. Chan, A.; De Luca, A.; Chalker, J.T. Eigenstate Correlations, Thermalization, and the Butterfly Effect. *Phys. Rev. Lett.* **2019**, *122*, 220601. [CrossRef]
56. Fritzsch, F.; Prosen, T. Eigenstate thermalization in dual-unitary quantum circuits: Asymptotics of spectral functions. *Phys. Rev. E* **2021**, *103*, 062133. [CrossRef]
57. Preskill, J. Quantum Computing in the NISQ era and beyond. *Quantum* **2018**, *2*, 79. [CrossRef]
58. Kjaergaard, M.; Schwartz, M.E.; Braumüller, J.; Krantz, P.; Wang, J.I.J.; Gustavsson, S.; Oliver, W.D. Superconducting Qubits: Current State of Play. *Annu. Rev. Condens. Matter Phys.* **2020**, *11*, 369–395. [CrossRef]
59. Kandala, A.; Mezzacapo, A.; Temme, K.; Takita, M.; Brink, M.; Chow, J.M.; Gambetta, J.M. Hardware-efficient variational quantum eigensolver for small molecules and quantum magnets. *Nature* **2017**, *549*, 242–246. [CrossRef]
60. Monroe, C.; Campbell, W.C.; Duan, L.M.; Gong, Z.X.; Gorshkov, A.V.; Hess, P.W.; Islam, R.; Kim, K.; Linke, N.M.; Pagano, G.; et al. Programmable quantum simulations of spin systems with trapped ions. *Rev. Mod. Phys.* **2021**, *93*, 025001. [CrossRef]
61. Blatt, R.; Roos, C.F. Quantum simulations with trapped ions. *Nat. Phys.* **2012**, *8*, 277–284. [CrossRef]
62. Browaeys, A.; Lahaye, T. Many-body physics with individually controlled Rydberg atoms. *Nat. Phys.* **2020**, *16*, 132–142. [CrossRef]
63. Mi, X.; Roushan, P.; Quintana, C.; Mandrà, S.; Marshall, J.; Neill, C.; Arute, F.; Arya, K.; Atalaya, J.; Babbush, R.; et al. Information scrambling in quantum circuits. *Science* **2021**, *374*, 1479–1483. [CrossRef]
64. Zirnbauer, M.R. Supersymmetry for systems with unitary disorder: Circular ensembles. *J. Phys. A* **1996**, *29*, 7113–7136. [CrossRef]
65. Zirnbauer, M.R. The color-flavor transformation and a new approach to quantum chaotic maps. In Proceedings of the 12th International Congress of Mathematical Physics (ICMP 97), Brisbane, Australia, 13–19 July 1997; pp. 290–297.
66. Zirnbauer, M.R. Pair correlations of quantum chaotic maps from supersymmetry. In *Supersymmetry and Trace Formulae: Chaos and Disorder*; Kluwer Academic/Plenum: New York, NY, USA, 1999; pp. 153–172.
67. Altland, A.; Zirnbauer, M.R. Field Theory of the Quantum Kicked Rotor. *Phys. Rev. Lett.* **1996**, *77*, 4536–4539. [CrossRef]
68. Zirnbauer, M.R. Color-Flavor Transformation Revisited. *arXiv* **2021**, arXiv:2109.10272.
69. Wegner, F. The mobility edge problem: Continuous symmetry and a conjecture. *Z. Phys. B* **1979**, *35*, 207–210. [CrossRef]
70. Efetov, K.B. *Supersymmetry in Disorder and Chaos*; Cambridge University Press: Cambridge, UK, 1997.
71. Tian, C.; Altland, A. Theory of localization and resonance phenomena in the quantum kicked rotor. *New J. Phys.* **2010**, *12*, 043043. [CrossRef]
72. Tian, C.; Altland, A.; Garst, M. Theory of the Anderson Transition in the Quasiperiodic Kicked Rotor. *Phys. Rev. Lett.* **2011**, *107*, 074101. [CrossRef]
73. Tian, C.; Altland, A. Field theory of Anderson transition of the kicked rotor. *Phys. Scr.* **2012**, *T151*, 014049. [CrossRef]
74. Gnutzmann, S.; Altland, A. Universal Spectral Statistics in Quantum Graphs. *Phys. Rev. Lett.* **2004**, *93*, 194101. [CrossRef] [PubMed]
75. Gnutzmann, S.; Altland, A. Spectral correlations of individual quantum graphs. *Phys. Rev. E* **2005**, *72*, 056215. [CrossRef] [PubMed]
76. Gnutzmann, S.; Keating, J.P.; Piotet, F. Quantum Ergodicity on Graphs. *Phys. Rev. Lett.* **2008**, *101*, 264102. [CrossRef] [PubMed]

77. Gnutzmann, S.; Keating, J.; Piotet, F. Eigenfunction statistics on quantum graphs. *Ann. Phys.* **2010**, *325*, 2595–2640. [CrossRef]
78. Zirnbauer, M.R. Toward a theory of the integer quantum Hall transition: Continuum limit of the Chalker–Coddington model. *J. Math. Phys.* **1997**, *38*, 2007–2036. [CrossRef]
79. Janssen, M.; Metzler, M.; Zirnbauer, M.R. Point-contact conductances at the quantum Hall transition. *Phys. Rev. B* **1999**, *59*, 15836–15853. [CrossRef]
80. Altland, A.; Gnutzmann, S.; Haake, F.; Micklitz, T. A review of sigma models for quantum chaotic dynamics. *Rep. Prog. Phys.* **2015**, *78*, 086001. [CrossRef]
81. Haake, F. *Quantum Signatures of Chaos*; Springer: Berlin/Heidelberg, Germany, 2010.
82. Brouwer, P.; Beenakker, C. Diagrammatic method of integration over the unitary group, with applications to quantum transport in mesoscopic systems. *J. Math. Phys.* **1996**, *37*, 4904–4934. [CrossRef]
83. Samuel, S. $U(N)$ Integrals, 1/N, and the De Wit-'t Hooft anomalies. *J. Math. Phys.* **1980**, *21*, 2695–2703. [CrossRef]
84. Weingarten, D. Asymptotic behavior of group integrals in the limit of infinite rank. *J. Math. Phys.* **1978**, *19*, 999–1001. [CrossRef]
85. Collins, B. Moments and cumulants of polynomial random variables on unitary groups, the itzykson-zuber integral, and free probability. *Int. Math. Res. Not.* **2003**, *2003*, 953–982. [CrossRef]
86. Collins, B.; Śniady, P. Integration with respect to the Haar measure on unitary, orthogonal and symplectic group. *Commun. Math. Phys.* **2006**, *264*, 773–795. [CrossRef]
87. Collins, B.; Matsumoto, S.; Novak, J. The weingarten calculus. *arXiv* **2021**, arXiv:2109.14890.
88. Köstenberger, G. Weingarten Calculus. *arXiv* **2021**, arXiv:2101.00921.
89. Argaman, N.; Zee, A. Diagrammatic theory of random scattering matrices for normal-metal–superconducting mesoscopic junctions. *Phys. Rev. B* **1996**, *54*, 7406–7420. [CrossRef] [PubMed]
90. Sünderhauf, C.; Pérez-García, D.; Huse, D.A.; Schuch, N.; Cirac, J.I. Localization with random time-periodic quantum circuits. *Phys. Rev. B* **2018**, *98*, 134204. [CrossRef]
91. Li, Y.; Vasseur, R.; Fisher, M.; Ludwig, A.W. Statistical Mechanics Model for Clifford Random Tensor Networks and Monitored Quantum Circuits. *arXiv* **2021**, arXiv:2110.02988.
92. Collins, B.; Matsumoto, S. On some properties of orthogonal Weingarten functions. *J. Math. Phys.* **2009**, *50*, 113516. [CrossRef]
93. Matsumoto, S. General moments of matrix elements from circular orthogonal ensembles. *Random Matrices Theory Appl.* **2012**, *1*, 1250005. [CrossRef]
94. Matsumoto, S. Weingarten calculus for matrix ensembles associated with compact symmetric spaces. *Random Matrices Theory Appl.* **2013**, *2*, 1350001. [CrossRef]
95. Andreev, A.V.; Agam, O.; Simons, B.D.; Altshuler, B.L. Quantum Chaos, Irreversible Classical Dynamics, and Random Matrix Theory. *Phys. Rev. Lett.* **1996**, *76*, 3947–3950. [CrossRef] [PubMed]
96. Andreev, A.; Simons, B.; Agam, O.; Altshuler, B. Semiclassical field theory approach to quantum chaos. *Nucl. Phys. B* **1996**, *482*, 536–566. [CrossRef]
97. Joshi, L.K.; Elben, A.; Vikram, A.; Vermersch, B.; Galitski, V.; Zoller, P. Probing Many-Body Quantum Chaos with Quantum Simulators. *Phys. Rev. X* **2022**, *12*, 011018. [CrossRef]
98. Dyson, F.J. Statistical Theory of the Energy Levels of Complex Systems. *J. Math. Phys.* **1962**, *3*, 140–156. [CrossRef]
99. Mehta, M.L. *Random Matrices*; Elsevier: Amsterdam, The Netherlands, 2004.
100. Kamenev, A. *Field Theory of Non-Equilibrium Systems*; Cambridge University Press: Cambridge, UK, 2011.
101. Winer, M.; Jian, S.K.; Swingle, B. Exponential Ramp in the Quadratic Sachdev-Ye-Kitaev Model. *Phys. Rev. Lett.* **2020**, *125*, 250602. [CrossRef]
102. Andreev, A.V.; Altshuler, B.L. Spectral Statistics beyond Random Matrix Theory. *Phys. Rev. Lett.* **1995**, *75*, 902–905. [CrossRef]
103. Kamenev, A.; Mézard, M. Wigner-Dyson statistics from the replica method. *J. Phys. A* **1999**, *32*, 4373. [CrossRef]
104. Kamenev, A.; Mézard, M. Level correlations in disordered metals: The replica σ model. *Phys. Rev. B* **1999**, *60*, 3944. [CrossRef]
105. Altland, A.; Kamenev, A. Wigner-Dyson statistics from the Keldysh σ-model. *Phys. Rev. Lett.* **2000**, *85*, 5615. [CrossRef]
106. Skinner, B.; Ruhman, J.; Nahum, A. Measurement-Induced Phase Transitions in the Dynamics of Entanglement. *Phys. Rev. X* **2019**, *9*, 031009. [CrossRef]
107. Li, Y.; Chen, X.; Fisher, M.P.A. Quantum Zeno effect and the many-body entanglement transition. *Phys. Rev. B* **2018**, *98*, 205136. [CrossRef]
108. Li, Y.; Chen, X.; Fisher, M.P.A. Measurement-driven entanglement transition in hybrid quantum circuits. *Phys. Rev. B* **2019**, *100*, 134306. [CrossRef]
109. Potter, A.C.; Vasseur, R. Entanglement dynamics in hybrid quantum circuits. *arXiv* **2021**, arXiv:2111.08018.
110. Morningstar, A.; Colmenarez, L.; Khemani, V.; Luitz, D.J.; Huse, D.A. Avalanches and many-body resonances in many-body localized systems. *arXiv* **2021**, arXiv:2107.05642.
111. Bertini, B.; Kos, P.; Prosen, T. Exact Spectral Statistics in Strongly Localising Circuits. *arXiv* **2021**, arXiv:2110.15938.

entropy

MDPI

Article

Quantum Chaos, Random Matrices, and Irreversibility in Interacting Many-Body Quantum Systems

Hans A. Weidenmüller

Max-Planck-Institut für Kernphysik, D-69029 Heidelberg, Germany; haw@mpi-hd.mpg.de

Abstract: The Pauli master equation describes the statistical equilibration of a closed quantum system. Simplifying and generalizing an approach developed in two previous papers, we present a derivation of that equation using concepts developed in quantum chaos and random-matrix theory. We assume that the system consists of subsystems with strong internal mixing. We can then model the system as an ensemble of random matrices. Equilibration results from averaging over the ensemble. The direction of the arrow of time is determined by an (ever-so-small) coupling to the outside world. The master equation holds for sufficiently large times if the average level densities in all subsystems are sufficiently smooth. These conditions are quantified in the text, and leading-order correction terms are given.

Keywords: master equation; equilibration; Markov approximation

Citation: Weidenmüller, H.A. Quantum Chaos, Random Matrices, and Irreversibility in Interacting Many-Body Quantum Systems. *Entropy* **2022**, *24*, 959. https://doi.org/10.3390/e24070959

Academic Editor: Marko Robnik

Received: 20 June 2022
Accepted: 7 July 2022
Published: 11 July 2022

1. Introduction

We investigate the connection between the irreversible statistical equilibration of closed quantum systems, quantum chaos [1], and random-matrix theory [2]. We do so by expanding and generalizing ideas formulated in Refs. [3,4].

To begin, we recall the standard theoretical description of statistical equilibration. The interaction between the constituents of an isolated (or closed) many-body quantum system drives the system toward statistical equilibrium. The process is theoretically most simply described in terms of the Pauli master equation. Given a division of the system into a finite number K of subsystems, the master equation for the time-dependent occupation probability $P_{E,\alpha}(t)$ of subsystem α at excitation energy E reads

$$\dot{P}_{E,\alpha}(t) = \sum_{\beta} \Big(R_{\beta \to \alpha} P_{E,\beta}(t) - R_{\alpha \to \beta} P_{E,\alpha}(t) \Big). \tag{1}$$

The dot indicates the time derivative, and the indices α, β range from 1 to K. Equation (1) conserves probability, $\sum_\alpha \dot{P}_{E,\alpha} = 0$. The interaction between states in subsystems α and β gives rise to the time-independent transition rates $R_{\alpha \to \beta}$ and $R_{\beta \to \alpha}$. These obey detailed balance,

$$\rho_{E,\beta} R_{\beta \to \alpha} = \rho_{E,\alpha} R_{\alpha \to \beta}. \tag{2}$$

Here, $\rho_{E,\alpha}$ is the average level density of the states in subsystem α at excitation energy E. Equation (2) implies that $P^{(0)}_{E,\alpha} = C \rho_{E,\alpha}$ with C a constant independent of α is a time-independent solution of Equation (1). That solution describes statistical equilibrium. For large times, the solutions $P_{E,\alpha}(t)$ of Equation (1) tend generically toward the equilibrium solution $P^{(0)}_{E,\alpha}$, irrespective of the initial conditions at time $t = 0$. Thus, Equation (1) describes the irreversible development in time of the system toward statistical equilibrium.

Equation (1) goes back to the early days of quantum theory. The master equation is a standard topic in textbooks on quantum statistical mechanics, and it is widely used to describe statistical equilibration in atomic, molecular, and nuclear systems. At the same time, equilibration in isolated quantum systems poses a theoretical problem that continues to cause intense discussion and research. The time-reversal-invariant Schrödinger equation furnishes a universal framework for the theoretical description of closed nonrelativistic quantum systems. How can that description be reconciled with the manifestly irreversible tendency toward statistical equilibrium encapsulated in Equation (1)? The literature on the subject is vast. For references, we confine ourselves to two review articles [5,6] summarizing a total of about 500 research papers.

The derivation of the master Equation (1) in Refs. [3,4] uses arguments that differ from the ones summarized in Refs. [5,6]. It is based upon a random-matrix approach. It is assumed that within each subsystem or class α of states, the interaction mixes the states so strongly that the resulting eigenvalues and eigenfunctions follow the statistical distribution predicted by random-matrix theory. Under that assumption, the matrix elements of the interaction connecting states in different classes $\alpha \neq \beta$ become Gaussian-distributed random variables. Averaging the differential equations in time for the occupation probabilities of states in classes $\alpha = 1, \ldots, K$ over that distribution, and using the Markov approximation, yields master Equation (1).

In the present paper, we address two issues that have remained unexplored in Refs. [3,4]. First, we address the connection between the basic statistical assumption used in Refs. [3,4] and quantum chaos (Section 2). Second, we ask for the cause of irreversibility displayed by Equation (1). Which approximation used in the derivation causes the transition from the time-reversal-invariant Schrödinger equation to the manifestly time-reversal-noninvariant master Equation (1)? Is irreversibility due to the Markov approximation, is it due to our averaging over the statistical distribution of matrix elements of the interaction, or is there another cause? To answer that question, in Section 3, we use the results of Section 2 to define the basic random-matrix ensemble. In Section 4, we use the results of Ref. [3] to work out the time evolution of average occupation probabilities in the various classes. In Section 5, we apply the Markov approximation and introduce the central element which is responsible for the violation of the time-reversal invariance. Corrections to the Markov approximation are worked out in Section 6. Time-dependent oscillatory terms are addressed in Section 7. We summarize the results and the answers to the questions raised above in Section 8.

We focus attention on the time evolution of average occupation probabilities, i.e., on the diagonal elements of the density matrix of the system as described by Equation (1). Our derivation can be extended to the equation describing the time evolution of the full density matrix, including the off-diagonal elements. That equation is often referred to as the master equation, while Equation (1) is then called a rate equation. We do not follow that custom here. It is easily seen that upon averaging and in Markov approximation, the off-diagonal elements of the density matrix decrease exponentially in time. The decrease is governed by the same rates that determine the loss term in Equation (1).

2. Quantum Chaos and Random-Matrix Theory

The connection between quantum chaos and random-matrix theory (RMT) was uncovered in the late 1970s and early 1980s. In the search for "quantum signatures of classical chaos" [1], several authors turned their attention to random-matrix theory. That theory had been introduced by Wigner [7], in 1955, in an attempt to characterize the eigenvalues of the nuclear Hamiltonian that appeared as parameters in his R-matrix theory of nuclear reactions [8]. McDonald and Kaufman [9], Casati, Valz-Gris, and Guarneri [10], and Berry [11] investigated numerically the energy spectra of two-degrees-of-freedom quantum systems (the stadium billiard and the Sinai billiard) that are chaotic in the classical limit. Their results suggested that the distribution of spacings of neighboring eigenvalues coincides with the Wigner surmise. That surmise furnishes an excellent approximation to the distribution of nearest-neighbor spacings of eigenvalues of the Gaussian orthogonal ensemble (GOE) [12] of random matrices. Following these investigations, Bohigas, Giannoni, and Schmit [13] generated numerically a considerably larger set of eigenvalues for the Sinai billiard than had been used by the earlier authors. When combined with a refined statistical analysis, the spacings of these eigenvalues showed good agreement with the GOE fluctuation measures. The agreement caused the authors to formulate the "BGS conjecture": The spectral fluctuation measures of a classically chaotic quantum system coincide with those of the canonical random-matrix ensemble in the same symmetry class (unitary, orthogonal, or symplectic).

The BGS conjecture has since been thoroughly tested numerically in several few-degrees-of-freedom systems (see the review in [2]). With the help of Gutzwiller's periodic-orbit sum for the level density, Sieber and Richter [14] and Heusler, Müller, Altland, Braun, and Haake [15] gave a formal demonstration of the validity of the conjecture for general dynamical systems, summarized in the article by Müller and Sieber [16]. The demonstration is not a mathematical proof but shows physically convincingly why the BGS conjecture holds for classically chaotic quantum systems that possess a periodic-orbit sum for the level density.

These results strongly suggest that the BGS conjecture holds quite generally for strongly interacting many-body quantum systems that are chaotic in the classical limit. Ergodic theory [17] has shown that classical deterministic chaos is ubiquitous. It is characterized by positive Lyapunov exponents. These guarantee that long periodic orbits do have the properties used in Ref. [15]. However, that connection between RMT and classical chaos is too narrow in the present context. Bosonic and Fermionic systems possess symmetries that are not part of the classical ergodic theory. Ongoing research addresses the RMT fluctuation properties in such systems (see, for example, Refs. [18,19]). Moreover, there exist strongly interacting many-body systems that do not possess a well-defined classical limit. Locally interacting spin chains are an example. And yet, the spectral form factor (the Fourier transform of the spectral pair correlation function) of these systems agrees with the RMT prediction [20]. Likewise, it is not clear that very dense quantum systems like atomic nuclei do possess a classical limit. And yet, measures of spectral fluctuations evaluated for light nuclei with a Hamiltonian containing a mean-field potential and a residual two-body interaction do show agreement with RMT predictions [21]. Thermalization and random-matrix properties in many-body systems governed by a mean field and a residual interaction are analyzed more generally in Ref. [22].

It thus seems that the BGS conjecture, while generically valid, does not cover all cases of interest. In that situation, it is useful to recall the physical reason for agreement with RMT predictions in systems such as spin chains [20] or atomic nuclei [21]. It is the strong mixing of the unperturbed states of the system by the interaction. A suitable generalization of the BGS conjecture would say: For many-body systems with a sufficiently strong generic interaction, the spectral fluctuation measures agree with RMT predictions. That formulation is unspecific as it does not say how strong the interaction has to be. It is quite specific in saying that RMT spectral fluctuation properties are attained.

In conclusion, RMT spectral fluctuation properties are generic and are expected to occur universally in strongly interacting many-body systems. Therefore, the assumption formulated in Section 1 is on firm grounds. Nevertheless, for a specific system, it is advisable to make sure that RMT spectral fluctuation properties prevail, either by demonstrating that the system is classically chaotic or by a numerical test. Attention must be paid, in particular, to one-body and many-body localization [23–25]. Either phenomenon is not compatible with RMT spectral fluctuation properties.

3. Ensemble of Random Matrices

Before we address the second issue raised in the Introduction, we give a brief summary of the assumptions and developments in Ref. [4]. We define the basic random-matrix ensemble.

It is assumed that in Hilbert space, the states of the system can be grouped into a finite number K of classes, such that within each class, the states interact strongly. The example considered in Ref. [4] is a system of Fermions with a mean-field average potential and a closed-shell ground state. Excited states are defined as particle-hole excitations out of the ground state. Classes of such states carry the same number of particle-hole states. Our theoretical development is more general, however, and is not restricted to that example. We confine ourselves, however, to the case of orthogonal symmetry.

In every class α, we use a basis of orthonormal states labeled $|\alpha m\rangle$ with $m = 1, 2, \ldots, N_\alpha$ and $N_\alpha \gg 1$. On that basis, the Hamiltonian is written as

$$H_{\alpha m, \beta n} = \delta_{\alpha\beta} H^{(\alpha)}_{mn} + \langle \alpha m | V | \beta n \rangle. \tag{3}$$

In Ref. [4], it is assumed that within each class α, the Hamiltonian $H^{(\alpha)}$ is a member of the GOE. That is in line with the arguments of Section 2. The non-statistical real matrix elements $\langle \alpha m | V | \beta n \rangle$ connect different classes and vanish for $\alpha = \beta$. They also vanish unless the states in both classes carry the same conserved quantum numbers. In Ref. [4], it was assumed that V is a two-body interaction. That interaction changes the particle-hole number by one unit. It does not connect classes of particle-hole states differing in particle-hole number by two or more units. In the present context, it is likewise admissible that V does not connect every class with every other one. For definiteness, we exclude, however, the possibility that the classes form subgroups that are not connected to each other at all by matrix elements of V. The mixing of the states in different classes due to V yields eigenfunctions that are spread over several or all classes. As we shall see, that fact ultimately causes statistical equilibration as described by Equation (1).

The diagonalization of each of the GOE Hamiltonians $H^{(\alpha)}$ in Equation (3) yields for the total Hamiltonian in Equation (3) the expression

$$H_{\alpha\mu, \beta\nu} = E_{\alpha\mu} \delta_{\alpha\beta} \delta_{\mu\nu} + \langle \alpha\mu | V | \beta\nu \rangle. \tag{4}$$

The eigenvalues $E_{\alpha\mu}$ follow Wigner–Dyson statistics. The projections of the eigenfunctions $|\alpha\mu\rangle$ onto some fixed vector $|\alpha m\rangle$ have a Gaussian distribution. It follows that the matrix elements $\langle \alpha\mu | V | \beta\nu \rangle$ have a Gaussian distribution with respect to their dependence on both μ and ν.

The central step taken in Ref. [4] consists of replacing in each class α the Hamiltonian $H^{(\alpha)}$ by an ensemble of GOE Hamiltonians. That implies that the $E_{\alpha\mu}$ in Equation (4) form an ensemble of Wigner–Dyson-distributed eigenvalues and that the matrix elements $\langle \alpha\mu | V | \beta\nu \rangle$ form an ensemble of Gaussian-distributed zero-centered random variables with second moments

$$\begin{aligned} &\Big\langle \langle \alpha\mu | V | \beta\nu \rangle \langle \alpha'\mu' | V | \beta'\nu' \rangle \Big\rangle \\ &= V^2_{\alpha\beta} \Big(\delta_{\alpha\alpha'} \delta_{\mu\mu'} \delta_{\beta\beta'} \delta_{\nu\nu'} + \delta_{\alpha\beta'} \delta_{\mu\nu'} \delta_{\beta\alpha'} \delta_{\nu\mu'} \Big). \end{aligned} \tag{5}$$

The big angular brackets denote the ensemble average. For $\alpha \neq \beta$, the states $|\alpha\nu\rangle$ and $|\beta\nu\rangle$ are statistically uncorrelated. That gives rise to the Kronecker deltas involving class labels in Equation (5). The mean square matrix elements $V^2_{\alpha\beta}$ vanish for $\alpha = \beta$. They measure the strength of the coupling of classes (α, β). Within each class α, we eventually take the GOE limit of infinite matrix dimension N_α. Then, the spectrum of $H^{(\alpha)}$ extends from -2λ to $+2\lambda$, the average level spacing $d_\alpha \propto 1/N_\alpha$ tends to zero, and the average level density $\rho_{E,\alpha}$ versus energy E is given by

$$\rho_{E,\alpha} = \frac{N}{\pi\lambda} \sqrt{1 - (E/(2\lambda))^2}. \tag{6}$$

In the following sections, we display the assumptions under which Equation (1) follows from that central step. We show that Equation (1) describes the time evolution of ensemble-averaged occupation probabilities defined for each class. Therefore, statistical equilibration does not necessarily occur for each member of the ensemble. It is an average property, characteristic of most members of the ensemble.

4. Time Evolution of Average Occupation Probabilities

To implement the approach, we use the time-dependent Schrödinger equation in the interaction representation. The wave function is expanded on the basis of states $|\alpha\mu\rangle, |\beta\nu\rangle, \ldots$ with time-dependent occupation amplitudes $c_{\alpha\mu}(t), c_{\beta\nu}(t), \ldots$. That yields

$$i\hbar\dot{c}_{\alpha\mu}(t) = \sum_{\beta\nu}\langle\alpha\mu|\tilde{V}(t)|\beta\nu\rangle c_{\beta\nu}(t). \tag{7}$$

The dot indicates the time derivative, and the time-dependent matrix element

$$\langle\alpha\mu|\tilde{V}(t)|\beta\nu\rangle = \langle\alpha\mu|V|\beta\nu\rangle \exp\{i(E_{\alpha\mu} - E_{\beta\nu})t/\hbar\} \tag{8}$$

is Hermitean, $\langle\alpha\mu|\tilde{V}(t)|\beta\nu\rangle^* = \langle\beta\nu|\tilde{V}(t)|\alpha\mu\rangle$.

Time-dependent average occupation probabilities $P_{E,\alpha}(t)$ of the states in class α are defined as

$$P_{E,\alpha}(t) = \left\langle \sum_{\mu} |c_{\alpha\mu}(t)|^2 \right\rangle. \tag{9}$$

In the GOE limit, the sum over μ encompasses states $|\alpha\mu\rangle$ with energies $E_{\alpha\mu}$ located within a very small energy interval centered on the energy E defined by the initial condition $c_{\alpha\mu}(0) = \delta_{\alpha\alpha_0}\delta_{\mu\mu_0}$ at time $t = 0$ for the coefficients $c_{\alpha\mu}(t)$. The time derivative of $P_{E,\alpha}(t)$ is

$$\dot{P}_{E,\alpha}(t) = \left\langle \sum_{\mu} (c^*_{\alpha\mu}\dot{c}_{\alpha\mu} + \dot{c}^*_{\alpha\mu}c_{\alpha\mu}) \right\rangle. \tag{10}$$

In Ref. [4], the average in Equation (10) is calculated by expanding each of the factors $c_{\alpha\mu}(t)$, $c^*_{\alpha\mu}(t)$, $\dot{c}_{\alpha\mu}(t)$, $\dot{c}^*_{\alpha\mu}(t)$ on the right-hand side perturbatively in powers of $\tilde{V}$. That results in a multiple sum over products of Gaussian-distributed matrix elements. The average is calculated for each such product separately. In each product, the average equals the sum over all ways of averaging pairs of matrix elements using Equation (5). The number of such pairwise "contractions" of matrix elements proliferates with increasing order of the perturbation expansion. Terms of leading order are obtained by suppressing all contributions that contain derivatives of the average level density $\rho_{E,\alpha}$ with respect to energy E in any class α. Such suppression allows resummation of the series, is tantamount to the Markov approximation, and yields Equation (1) for the time evolution of $P_{E,\alpha}(t)$. Leading-order contraction patterns are displayed in Section 5 of Ref. [3].

Here, we present an alternative derivation of Equation (1) which is simpler, more transparent, and more general than the one used in Ref. [4]. That allows us to answer the second question raised in Section 1.

We consider the first term on the right-hand side of Equation (10). The contraction rules given in Ref. [3] imply that the first factor V in the perturbative expansion of $\dot{c}_{\alpha\mu}$ is contracted either with the factor V immediately following it (case (i)), or with the first such factor in the perturbative expansion of $c^*_{\alpha\mu}$ (case (ii)). For case (i), we integrate Equation (7) with respect to time and reinsert the result on the right-hand side of that equation. That gives

$$\begin{aligned}\dot{c}_{\alpha\mu}(t) &= \frac{1}{i\hbar}\sum_{\beta\nu}\langle\alpha\mu|\tilde{V}(t)|\beta\nu\rangle c_{\beta\nu}(0) \\ &+ \frac{1}{(i\hbar)^2}\sum_{\beta\nu}\langle\alpha\mu|\tilde{V}(t)|\beta\nu\rangle \\ &\times \int_0^t \mathrm{d}t_1 \sum_{\gamma\rho}\langle\beta\nu|\tilde{V}(t_1)|\gamma\rho\rangle c_{\gamma\rho}(t_1).\end{aligned} \tag{11}$$

The first term on the right-hand side does not contribute because $\tilde{V}$ is not followed by a second such factor. We insert the second term into $\sum_\mu \langle c^*_{\alpha\mu}\dot{c}_{\alpha\mu}\rangle$ and contract the two factors V displayed explicitly. We obtain

$$\frac{1}{(i\hbar)^2}\sum_\beta V^2_{\alpha\beta}\sum_{\mu\nu}\int_0^t \mathrm{d}t_1 \exp\{i(E_{\alpha\mu}-E_{\beta\nu})(t-t_1)/\hbar\} \times\Big\langle c^*_{\alpha\mu}(t)c_{\alpha\mu}(t_1)\Big\rangle. \tag{12}$$

Writing $t_1 = t + (t_1 - t)$, we expand $\langle c_{\alpha\mu}(t_1)\rangle$ in a Taylor series in $(t_1 - t)$ and keep the first two terms. The leading-order term yields the Markov approximation, and the next-order term, the correction of subleading order. We define $\Delta_\nu = (E_{\alpha\mu} - E_{\beta\nu})/\hbar$, carry out the time integrations, and find

$$\frac{1}{(i\hbar)^2}\sum_\beta V^2_{\alpha\beta}\sum_{\mu\nu}\Big\{\frac{i}{\Delta_\nu}\Big(1-\exp\{i\Delta_\nu t\}\Big)\Big\langle c^*_{\alpha\mu}(t)c_{\alpha\mu}(t)\Big\rangle + \frac{1}{\Delta_\nu^2}\Big(1+(it\Delta_\nu-1)\exp\{i\Delta_\nu t\}\Big)\Big\langle c^*_{\alpha\mu}(t)\dot{c}_{\alpha\mu}(t)\Big\rangle\Big\}. \tag{13}$$

For case (ii) we proceed analogously. In the first term, on the right-hand side of Equation (10), we use Equation (7) for $\dot{c}_{\alpha\mu}$. For $c^*_{\alpha\mu}$, we integrate the complex conjugate of Equation (7) over time and obtain

$$c^*_{\alpha\mu}(t) = \frac{i}{\hbar}\sum_{\beta\nu}\int_0^t \mathrm{d}t_1\ \langle\alpha\mu|\tilde{V}(t_1)|\beta\nu\rangle^* c^*_{\beta\nu}(t_1). \tag{14}$$

The term analogous to expression (12) is

$$\frac{1}{(\hbar)^2}\sum_\beta V^2_{\alpha\beta}\sum_{\mu\nu}\int_0^t \mathrm{d}t_1 \exp\{i(E_{\alpha\mu}-E_{\beta\nu})(t-t_1)/\hbar\} \times\Big\langle c^*_{\beta\nu}(t_1)c_{\beta\nu}(t)\Big\rangle. \tag{15}$$

We expand $c^*_{\beta\nu}(t_1)$ in a Taylor series in $t_1 - t$ and keep the first two terms. We define $\Delta_\mu = (E_{\alpha\mu} - E_{\beta\nu}/\hbar$, carry out the time integrations, and obtain

$$\frac{1}{(\hbar)^2}\sum_\beta V^2_{\alpha\beta}\sum_{\mu\nu}\Big\{\frac{i}{\Delta_\mu}\Big(1-\exp\{i\Delta_\mu t\}\Big)\Big\langle c^*_{\beta\nu}(t)c_{\beta\nu}(t)\Big\rangle + \frac{1}{\Delta_\mu^2}\Big(1+(it\Delta_\mu-1)\exp\{i\Delta_\mu t\}\Big)\Big\langle \dot{c}^*_{\beta\nu}(t)c_{\beta\nu}(t)\Big\rangle\Big\}. \tag{16}$$

In evaluating expressions (13) and (16), we take the GOE limit defined on top of Equation (6). We omit the terms carrying the factor $\exp\{i\Delta_\nu t\}$ in expression (13) and the terms carrying the factor $\exp\{i\Delta_\mu t\}$ in expression (16). These terms are addressed in Section 7.

5. Markov Approximation

We focus attention on the remaining terms in the first lines of expressions (13) and (16). These are given by

$$-\frac{1}{\hbar^2}\sum_{\beta} V^2_{\alpha\beta} \sum_{\mu\nu} \left(\frac{i}{\Delta_\nu} \langle c^*_{\alpha\mu}(t) c_{\alpha\mu}(t) \rangle + c.c. \right) ,$$

$$\frac{1}{\hbar^2}\sum_{\beta} V^2_{\alpha\beta} \sum_{\mu\nu} \left(\frac{i}{\Delta_\mu} \langle c^*_{\beta\nu}(t) c_{\beta\nu}(t) \rangle + c.c. \right) . \tag{17}$$

In the summation over ν in the first line, we assume that Δ_ν carries an infinitesimal positive imaginary increment $i\varepsilon$ with $\varepsilon > 0$. In the GOE limit, that gives

$$\frac{1}{\Delta_\nu} = \mathcal{P}\frac{\hbar}{E_{\alpha\mu} - E_{\beta\nu}} - i\pi\hbar\delta(E_{\alpha\mu} - E_{\beta\nu}). \tag{18}$$

The summation over ν is replaced by an integral over $E' = E_{\beta\nu}$. The weight factor is $\rho_{E',\beta}$, the average level density in class β defined in Equation (6). The symbol $\mathcal{P}$ denotes the principal-value integral. In the first expression (17), that integral gives a purely imaginary contribution which cancels against the principal-value integral in the complex conjugate term. We are left with the contribution due to the delta function which links only states in classes α and β that pertain to the same energy. As a consequence, the process is entirely on shell. The integration over E' yields $-i\pi\hbar\rho_{E_{\alpha\mu},\beta}$. We recall that the summation over μ encompasses a set of energies $E_{\alpha\mu}$ narrowly centered at E, the energy defined by the initial value for $c_{\alpha\mu}$. Within that interval, we may replace $\rho_{E_{\alpha\mu},\beta}$ with $\rho_{E,\beta}$. We use definition (9) to perform the remaining summation over μ and obtain for the first of expressions (17)

$$-\frac{2\pi}{\hbar} P_{E,\alpha}(t) \sum_{\beta} V^2_{\alpha\beta} \rho_{E,\beta}. \tag{19}$$

That is the loss term in Equation (1), with an explicit expression for the rate $R_{\alpha\to\beta}$. Proceeding analogously for the second term in expression (17), we obtain

$$+\frac{2\pi}{\hbar} \rho_{E,\alpha} \sum_{\beta} V^2_{\alpha\beta} P_{E,\beta}(t). \tag{20}$$

That is the gain term in Equation (1), again with an explicit expression for the rate $R_{\beta\to\alpha}$. Our treatment has been very explicit because gain term plus loss term together yield Equation (1) which violates time-reversal invariance. The origin of that violation and of the resulting tendency toward equilibration is in the choice of the sign of the infinitesimal increment $i\varepsilon$ in Δ_ν and Δ_μ. That choice determines the sign of the last term in Equation (18). Choosing the opposite sign would change the signs of gain term and loss term and would result in an equation where the occupation probabilities $P_{E,\alpha}$ grow indefinitely with time. That is physically unacceptable.

Use of an imaginary increment $i\varepsilon$ in Δ_ν and in Δ_μ is a formal necessity. Without it, our approach would involve the matrix elements $\langle \alpha m | (E - H^{(\alpha)})^{-1} | \alpha m' \rangle$ of the propagators of the GOE Hamiltonians $H^{(\alpha)}$ with real energy E for $\alpha = 1, \dots, K$. Each of these quantities is a random variable that does not possess a well-defined distribution. All moments higher than the first diverge. That can easily be checked by direct calculation. Use of an imaginary increment $i\varepsilon$ in Δ_ν and in Δ_μ is also a physical necessity. It breaks time-reversal invariance. Without it, the terms in expressions (19) and (20) would vanish, and the Markov approximation would say that the time derivatives of all average occupation probabilities are equal to zero. We see that a consistent and singularity-free derivation of Equation (1) cannot be given in the present framework for $\varepsilon = 0$. These difficulties are removed by adding the increment $i\varepsilon$. Purely formally, either sign of ε is admissible. Choosing a positive value for ε reflects physical necessity and has distinct physical significance. The sign of ε determines the signs of gain and loss term and, thereby, the direction of the arrow of time. A positive value of ε causes an ever-so-small exponential decrease with time of all time-dependent exponentials in the time evolution of $c_{\alpha\mu}(t)$ and $c^*_{\alpha\mu}(t)$. The resulting small but continuous loss of occupation probability corresponds to a leakage of the system, physically caused by the evaporation of particles, by gamma emission, or by another such mechanism. That leakage determines physically the direction of time and is ultimately responsible for the loss of time-reversal invariance in Equation (1).

In summary: In the present framework, master Equation (1) cannot be derived for a strictly isolated system, i.e., for $\varepsilon = 0$. The direction of the arrow of time is determined by the sign of ε. A positive value of ε accounts for an (ever-so-small) coupling to the outside world that causes a loss of probability.

6. Corrections to the Markov Approximation

Subleading-order corrections to the Markov approximation are due to the terms in the second lines of expressions (13) and (16) and to certain contraction patterns of pairs of matrix elements of V that violate the rules laid down in Ref. [3] to leading order. We consider these in turn.

In the second lines of expressions (13) and (16), we neglect the strongly oscillating terms. The remaining expressions are proportional, respectively, to $\sum_\nu 1/\Delta^2_\nu$ and to $\sum_\mu 1/\Delta^2_\mu$. Either term can be written as a derivative with respect to energy. It follows that the said terms are proportional, respectively, to $\dot{P}_{E,\alpha} \sum_\beta V^2_{\alpha\beta} \mathrm{d}\rho_{E,\beta}/\mathrm{d}E$ and to $\mathrm{d}\rho_{E,\alpha}/\mathrm{d}E \sum_\beta V^2_{\alpha\beta} \dot{P}_{E,\beta}$. That result has two consequences. First, it contains the time derivatives of the occupation probabilities $P_{E,\alpha}$ and $P_{E,\beta}$. When statistical equilibrium is reached, all time derivatives vanish. The equilibrium solution $P^{(0)}_{E,\alpha} = C\rho_{E,\alpha}$ is determined by the interplay of gain term and loss term on the right-hand side of Equation (1). It has the same value without and with non-Markovian corrections. We conclude that the corrections to the Markov approximation in expressions (13) and (16) change the form of master Equation (1), may affect the time within which statistical equilibrium is approximately attained, but do not affect the tendency of the system to attain equilibrium or change the form of the equilibrium solution. Second, the result allows us to sharpen the criterion for the validity of the Markov approximation. The corrections to the Markov approximation are negligible if the characteristic energy interval $E_\beta = \rho_{E,\beta}/(\mathrm{d}\rho_{E,\beta}/\mathrm{d}E)$, within which the level density $\rho_{E,\beta}$ changes significantly, is large compared to the spreading width $2\pi V^2_{\alpha\beta}\rho_{E,\beta}$ for all classes α coupled to class β.

We turn to the subleading contributions due to contraction patterns that violate the rules laid down in Ref. [3]. We confine ourselves to a single example. In Equation (11), we carry the perturbation expansion of $\dot{c}_{\alpha\mu}(t)$ up to the term of fourth order in $\tilde{V}$. That term involves a time-ordered integral over times t_1, t_2, t_3. The integrand is the product of three matrix elements of $\tilde{V}$ and of $c_{\gamma\rho}(t_3)$. In the average over the four matrix elements of V displayed explicitly, the contribution of subleading order is obtained by contracting the outer pair and the inner pair. In the resulting expression, we replace $c_{\alpha\mu}(t_3)$ by $c_{\alpha\mu}(t)$ and carry out the three time integrations. As done at the end of Section 4, we suppress terms that carry rapidly oscillating exponential factors. These are addressed in the next section. That leaves us with a single term given by

$$\frac{1}{(i\hbar)^4} \sum_{\beta\gamma} V^2_{\alpha\beta} V^2_{\beta\gamma} \sum_{\mu\nu} \frac{\hbar^2}{i^2(E_{\alpha\mu} - E_{\beta\nu})^2} \times \sum_\rho \frac{\hbar}{i(E_{\gamma\rho} - E_{\alpha\mu})} \langle c^*_{\alpha\mu}(t) c_{\alpha\mu}(t) \rangle. \tag{21}$$

The sum over ν can be written as a derivative with respect to energy. We use Equation (18) and suppress the principal-value contributions. The result is proportional to

$iP_{E,\alpha}(t)\sum_\beta V^2_{\alpha\beta}\mathrm{d}\rho_{E,\beta}/\mathrm{d}E\sum_\gamma V^2_{\beta\gamma}\rho_{E,\gamma}$. It is purely imaginary and, thus, does not contribute to Equation (1). We note, however, that the form of the result shows that the criterion for the validity of the Markov approximation stated in the previous paragraph applies here as well. A term similar to expression (21) is obtained from the subleading contraction connecting a factor V in $\dot{c}_{\alpha\mu}(t)$ and three factors V in $c^*_{\alpha\mu}(t)$. Actually, neither of these two terms modifies Equation (1). The form of the equilibrium solution remains unchanged.

These results confirm that the Markov approximation is not the cause of violation of time-reversal invariance in Equation (1).

7. Oscillatory Terms

As done in Ref. [4], we have consistently omitted terms of the form

$$\sum_\nu \frac{\exp i(E - E_{\beta\nu})t/\hbar\}}{E - E_{\beta\nu} + i\varepsilon}. \tag{22}$$

Such terms are not peculiar to the Markov approximation. They arise, in general, whenever the time evolution of a quantum system is studied perturbatively. Actually, the contents of the big round brackets in Equations (13) and (16) vanish for $t = 0$. That shows that the neglect of the oscillatory terms, intuitively justified because of the rapid oscillation of the exponential, is valid only beyond some finite time t_0. At that time, the contribution of the oscillating terms becomes small compared to that of the first term in big round brackets in Equations (13) and (16). To determine t_0, we replace the infinitesimal increment ε in expression (22) by a finite width γ. That gives

$$\sum_\nu \frac{\exp i(E - E_{\beta\nu})t/\hbar\}}{(E - E_{\beta\nu})^2 + \gamma^2}\left(E - E_{\beta\nu} - i\gamma\right). \tag{23}$$

Upon summation, the real part of the big round bracket vanishes or is very small. That is true even for $t = 0$. The term proportional to $i\gamma$ involves the summation over a Lorentzian with width γ. If the remaining terms in the sum were sufficiently smooth functions of $E_{\beta\nu}$, the limit $\gamma \to 0$ may be taken. That yields the delta function on the right-hand side of Equation (18). In the present case, we cannot use that step because of the oscillation of the exponential. We assume that $\gamma \gg d_\beta$, the mean level spacing in class β. Then, the sum over the Lorentzian effectively involves a large number of terms. Because of the exponential, the sum becomes negligible for $t \gtrsim 2\pi\hbar/\gamma$. We compare that with the characteristic time scale of Equation (1) which is given by the inverse rate $\hbar/(2\pi V^2_{\alpha\beta}\rho_{E,\beta})$. The two time scales are approximately equal, and oscillating terms may be suppressed in master Equation (1) after some initial time $t_0 = 2\pi\hbar/\gamma$, if γ is of order $(2\pi)^2 V^2_{\alpha\beta}\rho_{E,\beta}$.

While an infinitesimal increment ε is sufficient to guarantee violation of time-reversal invariance in Equation (1), a finite width γ is needed to validate omission of the exponential terms for times $t \gtrsim t_0 = 2\pi\hbar/\gamma$. The width γ, caused by decay into open channels, leads to a decay in time of all exponentials in the time evolution of amplitudes, see Equation (7). To be consistent, such decay may have to be accounted for explicitly in terms of an additional loss term in Equation (1). That was done, for instance, in Ref. [26], where γ accounts for neutron evaporation in a laser-induced nuclear reaction.

8. Summary and Conclusions

We have considered a system composed of classes of states, each of which is strongly mixed internally. By recalling central results of quantum chaos and its connection to random-matrix theory, we have justified the use of a random-matrix approach to that system. Within each class, the dynamics are governed by an ensemble of GOE Hamiltonians. States in different classes are coupled by an ensemble of Gaussian-distributed matrix elements. We have simplified the derivation of Ref. [4] for the time evolution of average occupation probabilities for each class. That has enabled us to precisely define the conditions of validity of master Equation (1) and to identify the central element that causes the violation of time-reversal invariance. Irreversibility is due to an ever-so-small loss of probability (due to particle evaporation, gamma emission, or another such mechanism). The loss term defines the direction of the arrow of time. In combination with ensemble averaging and the Markov approximation, it leads to the irreversible master Equation (1).

Corrections of the subleading order to the Markov approximation may modify the speed at which the system equilibrates. They do not affect the irreversible tendency toward statistical equilibrium. The corrections are negligible, and the Markov approximation is valid, if in each class α the average level density $\rho_{E,\alpha}$ is sufficiently smooth. Quantitatively, that requires that the energy interval $E_\alpha = \rho_{E,\alpha}/(\mathrm{d}\rho_{E,\alpha}/\mathrm{d}E)$, within which $\rho_{E,\alpha}$ changes significantly, is large in comparison with every one of the spreading widths $2\pi V^2_{\beta\alpha}\rho_{E,\alpha}$ that feed states in class α from states in any other class β. For the terms that strongly oscillate in time, we have derived a lower bound t_0 on time beyond which such oscillating terms are negligible.

We have answered the questions raised in Section 1. In doing so, we have formulated precise conditions on level densities and time scales within which Equation (1) can be used with confidence.

After completion of the manuscript, I became aware of Ref. [27]. I am grateful to A. Volya and to V. Zelevinsky for drawing my attention to that paper. Investigating a number of examples, the authors demonstrate the limitations in specific cases of the universal random-matrix approach adopted in the present paper.

Funding: This research received no external funding.

Conflicts of Interest: The author declares no conflict of interest.

References

1. Haake, F. *Quantum Signatures of Chaos*; Springer: Berlin, Germany, 2000.
2. Guhr, T.; Müller-Groeling, A.; Weidenmüller, H.A. Random-Matrix Theories in Quantum Physics: Common Concepts. *Phys. Rep.* **1998**, *299*, 189. [CrossRef]
3. Agassi, D.; Weidenmüller, H.A.; Mantzouranis, G. The Statistical Theory of Nuclear Reactions for Strongly Overlapping Resonances as a Theory of Transport Phenomena. *Phys. Rep.* **1977**, *22*, 145. [CrossRef]
4. Weidenmüller, H.A. Transport Equations for Driven Many-Body Quantum Systems. *J. Phys. A Math. Theor.* **2022**, *55*, 184001. [CrossRef]
5. Eisert, J.; Friesdorf, M.; Gogolin, C. Quantum Many-Body Systems out of Equilibrium. *Nat. Phys.* **2015**, *11*, 124. [CrossRef]
6. Gogolin, C.; Eisert, J. Equilibration, Thermalization, and the Emergence of Statistical Mechanics in Closed Quantum Systems. *Rep. Prog. Phys.* **2016**, *79*, 056001. [CrossRef]
7. Wigner, E.P. Characteristic Vectors of Bordered Matrices with Infinite Dimensions. *Ann. Math.* **1955**, *62*, 548. [CrossRef]
8. Wigner, E.P.; Eisenbud, L. Higher Angular Momenta and Long Range Interaction in Resonance Reactions. *Phys. Rev.* **1947**, *72*, 29. [CrossRef]
9. McDonald, S.W.; Kaufman, A.N. Spectrum and Eigenfunctions for a Hamiltonian with Stochastic Trajectories. *Phys. Rev. Lett.* **1979**, *42*, 1189. [CrossRef]
10. Casati, G.; Valz-Gris, F.; Guarneri, I. On the Connection between Quantization of Nonintegrable Systems and Statistical Theory of Spectra. *Lett. Nuovo C. Soc. Ital. Fis.* **1980**, *28*, 279. [CrossRef]
11. Berry, M.V. Quantizing a Classicallz Ergodic System: Sinai's Billiard and the KKR Method. *Ann. Phys.* **1981**, *131*, 163. [CrossRef]
12. Dyson, F.J. Statistical Theory of the Energy Levels of Complex Systems. I. *J. Math. Phys.* **1962**, *3*, 140. [CrossRef]
13. Bohigas, O.; Giannoni, M.-J.; Schmit, C. Characterization of Chaotic Quantum Spectra and Universality of Level Fluctuation Laws. *Phys. Rev. Lett.* **1984**, *52*, 1. [CrossRef]
14. Sieber, M.; Richter, K. Correlations between Periodic Orbits and their Role in Spectral Statistics. *Phys. Scr.* **2001**, *T 90*, 128. [CrossRef]
15. Heusler, S.; Müller, S.; Altland, A.; Braun, P.; Haake, F. Periodic-Orbit Theory of Level Correlations. *Phys. Rev. Lett.* **2007**, *98*, 044103. [CrossRef] [PubMed]
16. Müller, S.; Sieber, M. Quantum Chaos and Quantum Graphs. In *The Oxford Handbook of Random Matrix Theory*; Akemann, G., Baik, J., Francesco, P.D., Eds.; Oxford University Press: Oxford, UK, 2011; pp. 683–702.
17. Eckmann, J.-P.; Ruelle, D. Ergodic Theory of Chaos and Strange Attractors. *Rev. Mod. Phys.* **1985**, *57*, 617. [CrossRef]
18. Rammensee, J.; Urbina, J.D.; Richter, K. Many-Body Quantum Interference and the Saturation of Out-of-Time-Order Correlators. *Phys. Rev. Lett.* **2018**, *121*, 124101. [CrossRef] [PubMed]
19. Geiger, B.; Urbina, J.D.; Richter, K. Emergence of a Renormalized 1/N Expansion in Quenched Critical Many-Body Systems. *Phys. Rev. Lett.* **2021**, *126*, 110602. [CrossRef]

20. Kos, P.; Ljubotina, M.; Prosen, T. Many-Body Quantum Chaos: Analytic Connection to Random-Matrix Theory. *Phys. Rev. X* **2018**, *8*, 021062. [CrossRef]
21. Zelevinsky, V.; Brown, B.A.; Frazier, N.; Horoi, M. The Nuclear Shell-Model as a Testing Ground for Many-Body Chaos. *Phys. Rep.* **1996**, *276*, 85. [CrossRef]
22. Borgonovi, F.; Izrailev, F.M.; Santos, L.F.; Zelevinsky, V.G. Quantum Chaos and Thermalization in Isolated Systems of Interacting Particles. *Phys. Rep.* **2016**, *626*, 1. [CrossRef]
23. Evers, F.; Mirlin, A.D. Anderson Transitions. *Rev. Mod. Phys.* **2008**, *80*, 1355. [CrossRef]
24. Alet, F.; Laflorencie, N. Many-Body Localization: An Introduction and Selected Topics. *C. R. Phys.* **2018**, *19*, 498. [CrossRef]
25. Abanin, D.A.; Altmann, E.; Bloch, I.; Serbyn, M. Many-Body Localization, Thermalization, and Entanglement. *Rev. Mod. Phys.* **2019**, *91*, 021001. [CrossRef]
26. Palffy, A.; Buss, O.; Hoefer, A.; Weidenmüller, H.A. Laser-Nucleus Interactions: The Quasiadiabatic Regime. *Phys. Rev. C* **2015**, *92*, 044619. [CrossRef]
27. Volya, A.; Zelevinsky, V. Time-dependent relaxation of observables in complex quantum systems. *Phys. Complex.* **2020**, *1*, 025007. [CrossRef]

entropy

MDPI

Article

Localization Detection Based on Quantum Dynamics

Kazue Kudo [1,2]

1 Department of Computer Science, Ochanomizu University, Tokyo 112-8610, Japan; kudo@is.ocha.ac.jp
2 Graduate School of Information Sciences, Tohoku University, Sendai 980-8579, Japan

Abstract: Detecting many-body localization (MBL) typically requires the calculation of high-energy eigenstates using numerical approaches. This study investigates methods that assume the use of a quantum device to detect disorder-induced localization. Numerical simulations for small systems demonstrate how the magnetization and twist overlap, which can be easily obtained from the measurement of qubits in a quantum device, changing from the thermal phase to the localized phase. The twist overlap evaluated using the wave function at the end of the time evolution behaves similarly to the one evaluated with eigenstates in the middle of the energy spectrum under a specific condition. The twist overlap evaluated using the wave function after time evolution for many disorder realizations is a promising probe for detecting MBL in quantum computing approaches.

Keywords: many-body localization; twist overlap; quantum computing

Citation: Kudo, K. Localization Detection Based on Quantum Dynamics. *Entropy* **2022**, *24*, 1085. https://doi.org/10.3390/e24081085

Academic Editor: Marko Robnik

Received: 29 June 2022
Accepted: 4 August 2022
Published: 5 August 2022
Corrected: 21 November 2022

1. Introduction

Many-body localization (MBL) has recently attracted significant interest [1–5]. MBL is a generalization of Anderson localization for disordered quantum many-body systems with interactions. When the disorder is strong enough, MBL prevents the system from thermalizing. The transition from the thermal to localized phase, i.e., MBL transition, is recognized as an eigenstate phase transition from the ergodic phase to non-ergodic phase [6–8]. The MBL transition is typically detected by quantities evaluated using eigenstates, such as entanglement, inverse participation ratio, one-particle density matrix [9–22], and spectral properties, such as level statistics [6,15–19,23] as well as local observables [6,14–16]. The random matrix theory and quantum chaos conjecture support the relationship between spectral properties and quantum chaos [24,25] as well as provide a framework for characterizing the MBL transition [26,27]. Experimental studies have also captured the signature of the MBL transition in disordered quantum systems with different architectures [28–41]. In experiments, MBL is often explored by the investigation of quantum dynamics. For example, the imbalance between the populations of even and odd sites in an atomic system is a measurable quantity for localization detection. The initial state has populations only on even sites, while odd sites are empty. Then, the imbalance is close to 1 initially and relaxes to zero as the system thermalizes. However, the imbalance maintains a finite value in the localized phase due to the initial state memory.

Recently, new approaches using quantum annealers were studied to simulate the properties of disordered quantum systems [42–49]. Probing MBL is also within the scope of quantum annealers and quantum computers. Although current quantum devices are still noisy and cannot compute exact eigenstates, some dynamical characteristics of MBL are robust against noise. In the localized phase, local quantities are conserved to some extent. For example, if the initial state is the all-spin-up state, then the magnetization maintains a large value because of the initial state memory. The magnetization, in this case, is similar to the imbalance in an atomic system in the sense that the memory effect characterizes localization. A recent experiment using a quantum annealer detected by the localization transition through magnetization measurements at the end of the time evolution [49].

This study investigated the localization detection based on quantum dynamics in a disordered quantum spin chain. Magnetization is a simple quantity that can be easily measured in quantum devices. We also employed another measurable quantity, the twist overlap. The twist overlap is a quantity proposed to detect the MBL transition and measure the extent to which an eigenstate overlaps with its twisted state [50]. A twisted state is obtained by applying a twist operator that rotates spins over the chain at gradually increasing angles. The twist overlap almost vanishes for thermal eigenstates, whereas it has a finite value for localized ones [50]. The twist overlap can also be evaluated using the state after time evolution. Using the twist overlap is convenient for localization detection in quantum devices because it can be easily obtained from the measurement of each spin.

The numerical simulations of small system sizes in this study demonstrate how the magnetization and twist overlap change from the thermal phase to the localized phase. These quantities were obtained at the end of the time evolution and averaged over different disorder realizations. They often oscillate at different frequencies for each disorder realization. Thus, we also examined the time dependence of these quantities to understand their dynamics. The time dependence, whose observation in experiments may require enormous time and effort, is also helpful in understanding the characteristics of these quantities. Although the difference between the thermal and localized phases in those quantities is evident, we cannot decide the existence of a phase transition from the limited numerical simulations. This work aims to demonstrate the effectiveness of experimentally measurable quantities, specifically twist overlap, as a localization probe.

2. Model and Methods

2.1. Model

The model used in this study is a one-dimensional transverse Ising model with local random fields, which is applied to a quantum annealer. The Hamiltonian is given by

$$H = \sum_{j=1}^{L-1} J_j \sigma_j^z \sigma_{j+1}^z + \sum_{j=1}^{L} h_j \sigma_j^z - \sum_{j=1}^{L} \Gamma_j \sigma_j^x, \quad (1)$$

where L is the system size and σ_j^x and σ_j^z are the Pauli operators of components x and z, respectively, at site j. The local field h_j consists of random numbers with a uniform distribution in the interval $[-w, w]$, wherein w denotes the disorder strength. The interactions J_j and transverse fields Γ_j are given by $1 + r_j$, where r_j is a uniform random number in the interval $[-\sigma, \sigma]$. Here, we refer to weak disorders in J_j and Γ_j as static noises. They are introduced to mimic the static noises of couplings between qubits and local fields in a quantum device.

2.2. Entanglement Entropy and Twist Overlap

The MBL transition is typically detected using quantities calculated using eigenstates. Moreover, the quantum dynamics reflects the properties of eigenstates. Before examining the quantum dynamics, we examine several quantities calculated using eigenstates. Here, we employ the half-chain entanglement entropy and twist overlap evaluated with eigenstates in the middle of the energy spectrum.

The half-chain entanglement entropy is defined by

$$S_E = -\mathrm{Tr}\rho_A \log \rho_A, \quad (2)$$

where ρ_A is the reduced density matrix for subsystem A. Subsystem A corresponds to the first half of the spin chain. Eigenstates obey the volume and area laws of entanglement in the thermal and localized phases, respectively. In other words, S_E decreases when the thermal-to-MBL transition occurs. The transition point is characterized by the variance peak of the half-chain entanglement entropy [14,18].

The twist operator is defined by

$$U_{\text{twist}} = \exp\left[\frac{i}{2}\sum_{j=1}^{L}\theta_j\sigma_j^z\right]. \tag{3}$$

It generates a spin-wave-like excitation by rotating the spins around the z axis at angles $\theta_j = 2\pi j/L$ [50,51]. The factor $1/2$ originates from spin $\frac{1}{2}$. The overlap between a state $|\psi\rangle$ and its twisted state $U_{\text{twist}}|\psi\rangle$ is the twist overlap, which is represented as

$$z = \langle\psi|U_{\text{twist}}|\psi\rangle, \tag{4}$$

where $|\psi\rangle$ denotes the eigenstate of the Hamiltonian in the original definition [50]. In the thermal phase, the twist overlap is expected to vanish because the twisted state with a spin-wave-like excitation is orthogonal to the original state. In contrast, the long-wavelength perturbation given by the twist operator has little effect on the eigenstates in the localized phase, which implies a finite twist overlap.

In quantum devices, the twist overlap is easily obtained by measuring each qubit. Writing $|\psi\rangle$ as a linear combination $|\psi\rangle = \sum_s \alpha_s|s\rangle$ of the computational basis $\{|s\rangle\}$ yields

$$z = \sum_{s,s'}\alpha_s\alpha_{s'}^*\exp\left[\frac{i}{2}\sum_{j=1}^{L}\theta_j s_j^z\right]\langle s'|s\rangle = \sum_s|\alpha_s|^2\exp\left[\frac{i}{2}\sum_{j=1}^{L}\theta_j s_j^z\right], \tag{5}$$

where $\sigma_j^z|s\rangle = s_j^z|s\rangle$. Since each measurement provides the configuration of s_j^z $(j = 1,\ldots,L)$ with probability $|\alpha_s|^2$, many measurements provide the expected value of the twist overlap.

2.3. Time Evolution

The solution of the Schrödinger equation,

$$i\frac{d}{dt}|\psi(t)\rangle = H|\psi\rangle, \tag{6}$$

is represented as

$$|\psi(t)\rangle = \exp(-iHt)|\psi_0\rangle, \tag{7}$$

where $|\psi_0\rangle$ denotes the initial state. In the numerical simulations below, the initial state is the all-spin-up state. The exact diagonalization of the Hamiltonian provides the time dependence of $|\psi(t)\rangle$. We observe the magnetization and twist overlap at the end of the time evolution, which is at the final time $t = T_{\text{fin}}$.

When the initial state is expressed as $|\psi_0\rangle = \sum_k c_k|\phi_k\rangle$ with eigenstates $|\phi_k\rangle$ of the Hamiltonian, Equation (7) becomes

$$|\psi(t)\rangle = \sum_{k=1}^{2^L} c_k\exp(-iE_k t)|\phi_k\rangle, \tag{8}$$

where E_k denotes the eigenenergy corresponding to $|\phi_k\rangle$. The z component of the magnetization defined by $M_z \equiv \langle\psi|\sum_j\sigma_j^z|\psi\rangle$ evolves as

$$M_z(t) = \sum_{k,l=1}^{2^L} c_k c_l^*\langle\phi_l|\sum_{j=1}^{L}\sigma_j^z|\phi_k\rangle e^{-i(E_k-E_l)t}. \tag{9}$$

Similarly, the twist overlap evolves as

$$z(t) = \sum_{k,l=1}^{2^L} c_k c_l^*\langle\phi_l|U_{\text{twist}}|\phi_k\rangle e^{-i(E_k-E_l)t}. \tag{10}$$

Equations (9) and (10) suggest the oscillatory behavior of $M_z(t)$ and $z(t)$. Therefore, the observed quantities depend on both the eigenstates and the final time. Since the frequency and amplitude of the oscillation differ from sample to sample, the average over disorder realizations characterizes the observed quantities if the final time is sufficiently large.

3. Results

3.1. Quantities Calculated from Eigenstates

Before investigating quantum dynamics, we confirm the localization properties characterized by eigenstates of the Hamiltonian. They help understand the quantities obtained from quantum dynamics. The exact diagonalization of the Hamiltonian without noise ($\sigma = 0$) was performed for each pair of the system size L and disorder strength w. The number of disorder realizations was 10^4 for $L = 8$ and 10^3 for $L = 10$ and 12. For each realization, the half-chain entanglement entropy and twist overlap were calculated and averaged over 20 eigenstates around the center of the energy spectrum.

Figure 1 shows the dependence on the disorder strength w of the half-chain entanglement entropy S_E and the absolute square of the twist overlap $|z|^2$ averaged over the disorder realizations. The averages of S_E and $|z|^2$ are plotted with error bars in Figure 1a and 1c, respectively, where the error bar represents the standard deviation. Figure 1b and 1d show the standard deviations of S_E and $|z|^2$, respectively.

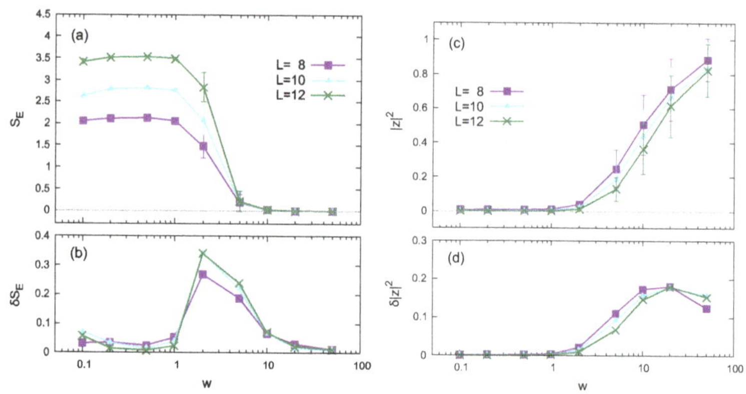

Figure 1. Disorder strength dependence of the entanglement entropy (**a**,**b**) and the twist overlap (**c**,**d**), calculated using eigenstates for different system sizes in the noiseless case ($\sigma = 0$). (**a**,**b**) plot the half-chain entanglement entropy S_E and its standard deviation δS_E as functions of disorder strength w, respectively. Similarly, (**c**,**d**) plot the absolute square of the twist overlap $|z|^2$ and its standard deviation $\delta|z|^2$, respectively. The error bars in (**a**,**c**) represent standard deviation.

The disorder strength dependence of the half-chain entanglement entropy shown in Figure 1a,b shows a behavior similar to that of Ref. [49], although the values were different because of the differences in the models. For each L, the variance (standard deviation) peaks around $w \simeq$ 1–5, where the transition or crossover between the thermal and localized phases occurs.

As expected, the twist overlap, whose absolute square is shown in Figure 1c, increases with the disorder strength. The variance is almost zero in the weak-disorder region and becomes finite in the strong-disorder region. These results indicate that both the average and variance of the twist overlap are almost zero in the thermal phase, but large in the localized phase. The disorder-strength dependence of the variance was not observed in Ref. [50], which used the random-field Heisenberg chain. The difference in the variance behavior is likely due to differences in the models.

Whereas the peaks of δS_E appear in the middle of changes in S_E, those of $\delta|z|^2$ occur as $|z|^2$ is large enough. The peaks of $\delta|z|^2$ appear probably due to the saturation of $|z|^2$.

Thus, the peak position of $\delta|z|^2$ is not related to the transition or crossover between the thermal and localized phases.

3.2. Properties Based on Quantum Dynamics

In this subsection, we investigate the magnetization and twist overlap evaluated using the wave function at the end of time evolution. The wave function $|\psi(t)\rangle$ at the final time $t = T_{\mathrm{fin}}$ was calculated from the exact diagonalization of the Hamiltonian. The final time was $T_{\mathrm{fin}} = 10$, which is sufficient to capture the difference between the thermal and localized phases, as shown in the following subsection. The initial state is taken as the all-spin-up state. $M_z = \langle\psi(T_{\mathrm{fin}})|\sum_j \sigma_j^z|\psi(T_{\mathrm{fin}})\rangle$ and $z = \langle\psi(T_{\mathrm{fin}})|U_{\mathrm{twist}}|\psi(T_{\mathrm{fin}})\rangle$ were calculated for each disorder realization. The number of disorder realizations was the same as that in the previous subsection: 10^4 for $L = 8$ and 10^3 for $L = 10$ and 12.

Figure 2 shows the dependence on the disorder strength w of the z component of the magnetization M_z and the absolute square of the twist overlap $|z|^2$ averaged over the disorder realizations. Here, the noise strength is $\sigma = 0$. The averages of M_z and $|z|^2$ are plotted with error bars in Figure 2a and 2c, respectively, where the error bar represents standard deviation. Figure 2b and 2d show the standard deviations of M_z and $|z|^2$, respectively.

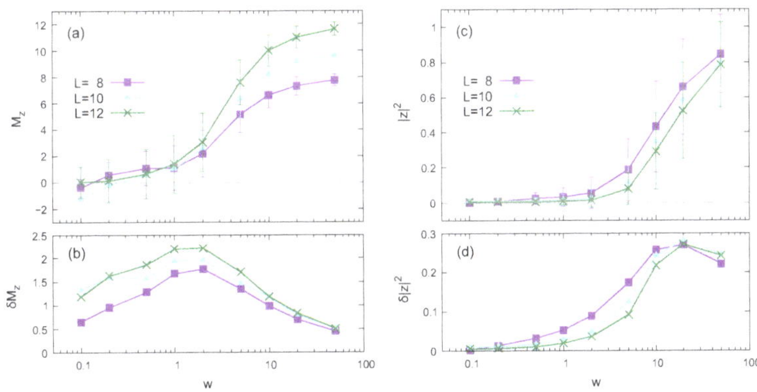

Figure 2. Disorder strength dependence based on quantum dynamics for different system sizes in the noiseless case ($\sigma = 0$). (**a**,**b**) plot the z component of the magnetization M_z and its standard deviation δM_z as functions of disorder strength w, respectively. (**c**,**d**) plot the absolute square of the twist overlap $|z|^2$ and its standard deviation $\delta|z|^2$, respectively. The error bars in (**a**,**c**) represent standard deviation.

The average magnetization is $M_z \simeq 0$ in the weak-disorder region, indicating thermalization. When the disorder is strong enough, $M_z \simeq L$, which is a signature of the memory effect because $M_z = L$ in the initial state. The memory effect is characteristic of the localized phase, which was also observed in Ref. [49]. The variance (standard deviation) of the magnetization peaks at a disorder strength slightly weaker than that of the entanglement entropy. Since magnetization fluctuates with time and can have negative values, δM_z is relatively large in the weak-disorder region. Thus, the variance peak of the magnetization in this situation cannot apply to determining the transition or overlap point.

The twist overlap also increases with the disorder strength. However, the memory effect is not the leading cause for the large value of $|z|^2$ in the strong disorder region. If the memory effect dominates the twist overlap behavior, $|z|^2$ should be close to 1, and its variance should be small. Considering the fact that the twist overlap shown in Figure 2c is similar to that in Figure 1c, we expect that the behavior of the twist overlap reflects the properties of eigenstates. The variance (standard deviation) in Figure 2d is relatively large compared with that in Figure 1d. The large variance in Figure 2d is due to the difference in the eigenstates of different disorder realizations and the oscillatory behavior of $z(t)$.

Similar results to the noiseless case ($\sigma = 0$) also appear in the presence of static noise. While several types of noises exist in quantum devices, we here consider static noises in the interaction between spins and the transverse field. As shown in Figure 3, the disorder strength of M_z and $|z|^2$ has little dependence on noise strength σ. However, the time evolution is affected by the static noises, as shown in the following subsection. The details of the time evolution are averaged out in the results in Figure 3, making the noise dependence negligible.

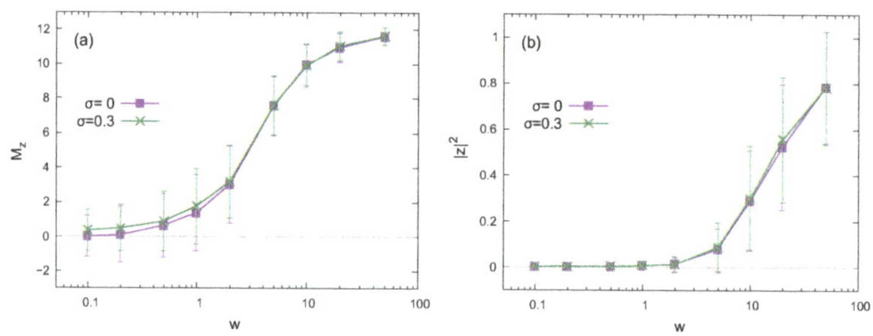

Figure 3. Disorder strength dependence of (**a**) the z component of the magnetization and (**b**) the absolute square of the twist overlap $|z|^2$ for different noise strengths. The error bars represent the standard deviation. The system size is $L = 12$.

3.3. Time Dependence

The time dependence of the magnetization and twist overlap helps us understand the characteristics of quantum dynamics in the system. Figure 4 illustrates $M_z(t)$ (in the upper row) and $|z(t)|^2$ (in the lower row) for several combinations of disorder strength w and noise strength σ. Each graph plots ten samples of the time series calculated at different disorder realizations. The graphs in the same column, for example, (a) and (e), share the same wave-function samples. That is, the curves of $M_z(t)$ and $|z(t)|^2$ with the same color in the same column are evaluated using the same wave functions.

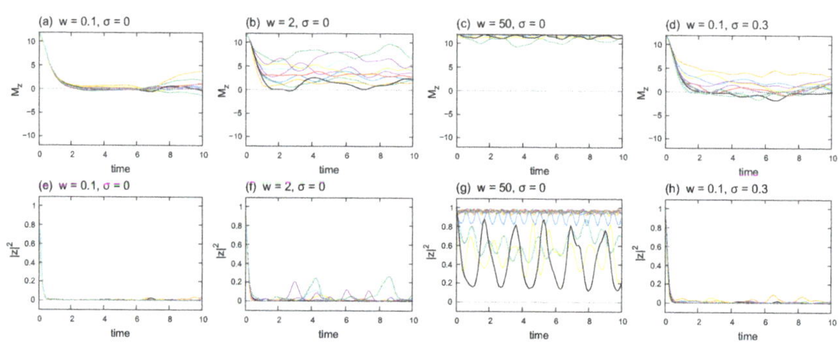

Figure 4. Time dependence of (**a–d**) the magnetization and (**e–h**) the absolute square of the twist overlap. Each graph plots ten different samples of the time series. (**a**,**e**) share the same wave-function samples, and the same applies to (**b**,**f**), (**c**,**g**), and (**d**,**h**). The system size is $L = 12$.

Figure 4a–c and 4e–g demonstrate how the behaviors of $M_z(t)$ and $|z(t)|^2$ change with disorder strength w, respectively, in the noiseless case ($\sigma = 0$). Initially, $M_z = L = 12$, which then decreases to $M_z \simeq 0$ when the disorder is weak. As the disorder strengthens, $M_z(t)$ fluctuates around positive values. Finally, $M_z(t)$ fluctuates around the initial value in the localized phase. However, $|z|^2 = 1$ at the initial time decreases rapidly to $|z|^2 \simeq 0$

when the disorder is weak. As the disorder strengthens, the fluctuations in $|z|^2$ become noticeable. In contrast to $M_z(t)$, some samples of $|z(t)|^2$ oscillate with large amplitudes around relatively low values.

Figure 4d,h show the time dependence of $M_z(t)$ and $|z(t)|^2$, respectively, for the noisy ($\sigma = 0.3$) and weak-disorder ($w = 0.1$) cases. A comparison with Figure 4a,d suggests that the time average of $M_z(t)$ is larger in the noisy case than that in the noiseless case in general. The difference in time dependence indicates that the static noise in the interaction strength and transverse field affects the quantum dynamics, which is not reflected in Figure 3.

4. Discussion

The absolute square of the twist overlap $|z|^2$ exhibits a slight variance in the weak-disorder region, as shown in Figures 2 and 3. However, the z component of the magnetization M_z has a relatively large variance in the same region. The time dependences of $M_z(t)$ and $|z(t)|^2$ also support this behavior, which seems curious. Figure 1c,d illustrates that $|z|^2 \simeq 0$ in the small-disorder region, implying that $\langle \phi_l | U_{\text{twist}} | \phi \rangle$ in Equation (10) nearly vanishes in the middle of the energy spectrum. Since the eigenstates in the high and low regions of the energy spectrum also contribute to the time-dependent $|z(t)|^2$, the variance of $|z|^2$ has a small finite value in Figures 2 and 3. However, $M_z(t)$ fluctuates around zero because of thermalization, which causes a relatively large variance.

As shown in Figure 4, $|z(t)|^2$ oscillates with a large amplitude in some strong-disorder cases, even though $M_z(t)$ remains around the initial value. Large-amplitude oscillations arise from the combination of $\langle \phi_l | U_{\text{twist}} | \phi_k \rangle$ with different eigenstates $|\phi_l\rangle$ and $|\phi_k\rangle$. As shown in Figure 1c,d, the average and variance of $|z|^2$ are significant in the strong-disorder region, which supports the variation in the combination of $\langle \phi_l | U_{\text{twist}} | \phi_k \rangle$.

5. Conclusions

We investigated a method that assumes the use of a quantum device to detect disorder-induced localization. Localization in a disordered spin chain is detected by evaluating the magnetization and twist overlap at the end of the time evolution for many disorder realizations. Numerical simulations demonstrated how the magnetization and twist overlap characteristics change between the thermal and localized phases. We found evident differences between them, although the existence of a phase transition was not decided. The disorder-strength dependence of the magnetization and twist overlap is robust against static noises in the interaction between spins and the local field.

Under the condition in this work, the twist overlap evaluated using the wave function at the end of the time evolution behaved similarly to that calculated using eigenstates in the middle of the energy spectrum. In other words, the twist overlap after time evolution can provide information on the properties of eigenstates beyond the memory effect. The twist overlap is easily obtained from the measurement of qubits in a quantum device. Although this work assumes an ideal quantum device, the results suggest that the twist overlap is a promising probe for detecting MBL in quantum computing approaches.

Funding: This research received no external funding.

Data Availability Statement: Not applicable.

Conflicts of Interest: The author declares no conflict of interest.

Abbreviation

The following abbreviation is used in this manuscript:

MBL	Many-body localization

References

1. Alet, F.; Laflorencie, N. Many-body localization: An introduction and selected topics. *Comptes Rendus Phys.* **2018**, *19*, 498–525. [CrossRef]
2. Parameswaran, S.A.; Vasseur, R. Many-body localization, symmetry and topology. *Rep. Prog. Phys.* **2018**, *81*, 082501. [CrossRef] [PubMed]
3. Abanin, D.A.; Altman, E.; Bloch, I.; Serbyn, M. Colloquium: Many-body localization, thermalization, and entanglement. *Rev. Mod. Phys.* **2019**, *91*, 021001. [CrossRef]
4. Gopalakrishnan, S.; Parameswaran, S.A. Dynamics and transport at the threshold of many-body localization. *Phys. Rep.* **2020**, *862*, 1–62. [CrossRef]
5. Tikhonov, K.S.; Mirlin, A.D. From Anderson localization on random regular graphs to many-body localization. *Ann. Phys.* **2021**, *435*, 168525. [CrossRef]
6. Pal, A.; Huse, D.A. Many-body localization phase transition. *Phys. Rev. B* **2010**, *82*, 174411. [CrossRef]
7. Huse, D.A.; Nandkishore, R.; Oganesyan, V. Phenomenology of fully many-body-localized systems. *Phys. Rev. B* **2014**, *90*, 174202. [CrossRef]
8. Pekker, D.; Clark, B.K.; Oganesyan, V.; Refael, G. Fixed Points of Wegner-Wilson Flows and Many-Body Localization. *Phys. Rev. Lett.* **2017**, *119*, 075701. [CrossRef]
9. Bardarson, J.H.; Pollmann, F.; Moore, J.E. Unbounded Growth of Entanglement in Models of Many-Body Localization. *Phys. Rev. Lett.* **2012**, *109*, 017202. [CrossRef]
10. Serbyn, M.; Papić, Z.; Abanin, D.A. Universal Slow Growth of Entanglement in Interacting Strongly Disordered Systems. *Phys. Rev. Lett.* **2013**, *110*, 260601. [CrossRef]
11. Serbyn, M.; Papić, Z.; Abanin, D.A. Local Conservation Laws and the Structure of the Many-Body Localized States. *Phys. Rev. Lett.* **2013**, *111*, 127201. [CrossRef] [PubMed]
12. Bera, S.; Schomerus, H.; Heidrich-Meisner, F.; Bardarson, J.H. Many-Body Localization Characterized from a One-Particle Perspective. *Phys. Rev. Lett.* **2015**, *115*, 046603. [CrossRef] [PubMed]
13. Bera, S.; Lakshminarayan, A. Local entanglement structure across a many-body localization transition. *Phys. Rev. B* **2016**, *93*, 134204. [CrossRef]
14. Kjäll, J.A.; Bardarson, J.H.; Pollmann, F. Many-Body Localization in a Disordered Quantum Ising Chain. *Phys. Rev. Lett.* **2014**, *113*, 107204. [CrossRef] [PubMed]
15. Enss, T.; Andraschko, F.; Sirker, J. Many-body localization in infinite chains. *Phys. Rev. B* **2017**, *95*, 045121. [CrossRef]
16. Orell, T.; Michailidis, A.A.; Serbyn, M.; Silveri, M. Probing the many-body localization phase transition with superconducting circuits. *Phys. Rev. B* **2019**, *100*, 134504. [CrossRef]
17. Luitz, D.J.; Laflorencie, N.; Alet, F. Many-body localization edge in the random-field Heisenberg chain. *Phys. Rev. B* **2015**, *91*, 081103. [CrossRef]
18. Khemani, V.; Lim, S.; Sheng, D.; Huse, D.A. Critical Properties of the Many-Body Localization Transition. *Phys. Rev. X* **2017**, *7*, 021013. [CrossRef]
19. Khemani, V.; Sheng, D.; Huse, D.A. Two Universality Classes for the Many-Body Localization Transition. *Phys. Rev. Lett.* **2017**, *119*, 075702. [CrossRef]
20. Hopjan, M.; Heidrich-Meisner, F. Many-body localization from a one-particle perspective in the disordered one-dimensional Bose-Hubbard model. *Phys. Rev. A* **2020**, *101*, 063617. [CrossRef]
21. Zhang, S.X.; Yao, H. Universal Properties of Many-Body Localization Transitions in Quasiperiodic Systems. *Phys. Rev. Lett.* **2018**, *121*, 206601. [CrossRef] [PubMed]
22. Gray, J.; Bose, S.; Bayat, A. Many-body localization transition: Schmidt gap, entanglement length, and scaling. *Phys. Rev. B* **2018**, *97*, 201105. [CrossRef]
23. Kudo, K.; Deguchi, T. Finite-size scaling with respect to interaction and disorder strength at the many-body localization transition. *Phys. Rev. B* **2018**, *97*, 220201. [CrossRef]
24. Casati, G.; Valz-Gris, F.; Guarnieri, I. On the connection between quantization of nonintegrable systems and statistical theory of spectra. *Lett. Nuovo Cimento* **1980**, *28*, 279–282. [CrossRef]
25. Bohigas, O.; Giannoni, M.J.; Schmit, C. Characterization of Chaotic Quantum Spectra and Universality of Level Fluctuation Laws. *Phys. Rev. Lett.* **1984**, *52*, 1. [CrossRef]
26. D'Alessio, L.; Kafri, Y.; Polkovnikov, A.; Rigol, M. From quantum chaos and eigenstate thermalization to statistical mechanics and thermodynamics. *Adv. Phys.* **2016**, *65*, 239–362. [CrossRef]
27. Šuntajs, J.; Bonča, J.; Prosen, T.; Vidmar, L. Quantum chaos challenges many-body localization. *Phys. Rev. E* **2020**, *102*, 062144. [CrossRef]
28. Schreiber, M.; Hodgman, S.S.; Bordia, P.; Lüschen, H.P.; Fischer, M.H.; Vosk, R.; Altman, E.; Schneider, U.; Bloch, I. Observation of many-body localization of interacting fermions in a quasirandom optical lattice. *Science* **2015**, *349*, 842–845. [CrossRef]
29. Kondov, S.; McGehee, W.; Xu, W.; DeMarco, B. Disorder-Induced Localization in a Strongly Correlated Atomic Hubbard Gas. *Phys. Rev. Lett.* **2015**, *114*, 083002. [CrossRef]
30. Smith, J.; Lee, A.; Richerme, P.; Neyenhuis, B.; Hess, P.W.; Hauke, P.; Heyl, M.; Huse, D.A.; Monroe, C. Many-body localization in a quantum simulator with programmable random disorder. *Nat. Phys.* **2016**, *12*, 907–911. [CrossRef]

31. Bordia, P.; Lüschen, H.P.; Hodgman, S.S.; Schreiber, M.; Bloch, I.; Schneider, U. Coupling Identical one-dimensional Many-Body Localized Systems. *Phys. Rev. Lett.* **2016**, *116*, 140401. [CrossRef] [PubMed]
32. Bordia, P.; Lüschen, H.; Scherg, S.; Gopalakrishnan, S.; Knap, M.; Schneider, U.; Bloch, I. Probing Slow Relaxation and Many-Body Localization in Two-Dimensional Quasiperiodic Systems. *Phys. Rev. X* **2017**, *7*, 041047. [CrossRef]
33. Lüschen, H.P.; Bordia, P.; Hodgman, S.S.; Schreiber, M.; Sarkar, S.; Daley, A.J.; Fischer, M.H.; Altman, E.; Bloch, I.; Schneider, U. Signatures of Many-Body Localization in a Controlled Open Quantum System. *Phys. Rev. X* **2017**, *7*, 011034. [CrossRef]
34. Lüschen, H.P.; Bordia, P.; Scherg, S.; Alet, F.; Altman, E.; Schneider, U.; Bloch, I. Observation of Slow Dynamics near the Many-Body Localization Transition in One-Dimensional Quasiperiodic Systems. *Phys. Rev. Lett.* **2017**, *119*, 260401. [CrossRef]
35. Wei, K.X.; Ramanathan, C.; Cappellaro, P. Exploring Localization in Nuclear Spin Chains. *Phys. Rev. Lett.* **2018**, *120*, 070501. [CrossRef]
36. Xu, K.; Chen, J.J.; Zeng, Y.; Zhang, Y.R.; Song, C.; Liu, W.; Guo, Q.; Zhang, P.; Xu, D.; Deng, H.; et al. Emulating Many-Body Localization with a Superconducting Quantum Processor. *Phys. Rev. Lett.* **2018**, *120*, 050507. [CrossRef]
37. Kohlert, T.; Scherg, S.; Li, X.; Lüschen, H.P.; Das Sarma, S.; Bloch, I.; Aidelsburger, M. Observation of Many-Body Localization in a One-Dimensional System with a Single-Particle Mobility Edge. *Phys. Rev. Lett.* **2019**, *122*, 170403. [CrossRef]
38. Rispoli, M.; Lukin, A.; Schittko, R.; Kim, S.; Tai, M.E.; Léonard, J.; Greiner, M. Quantum critical behaviour at the many-body localization transition. *Nature* **2019**, *573*, 385–389. [CrossRef]
39. Rubio-Abadal, A.; Choi, J.Y.; Zeiher, J.; Hollerith, S.; Rui, J.; Bloch, I.; Gross, C. Many-Body Delocalization in the Presence of a Quantum Bath. *Phys. Rev. X* **2019**, *9*, 041014. [CrossRef]
40. Zhu, D.; Johri, S.; Nguyen, N.H.; Alderete, C.H.; Landsman, K.A.; Linke, N.M.; Monroe, C.; Matsuura, A.Y. Probing many-body localization on a noisy quantum computer. *Phys. Rev. A* **2021**, *103*, 032606. [CrossRef]
41. Gong, M.; de Moraes Neto, G.D.; Zha, C.; Wu, Y.; Rong, H.; Ye, Y.; Li, S.; Zhu, Q.; Wang, S.; Zhao, Y.; et al. Experimental characterization of the quantum many-body localization transition. *Phys. Rev. Res.* **2021**, *3*, 033043. [CrossRef]
42. King, A.D.; Carrasquilla, J.; Raymond, J.; Ozfidan, I.; Andriyash, E.; Berkley, A.; Reis, M.; Lanting, T.; Harris, R.; Altomare, F.; et al. Observation of topological phenomena in a programmable lattice of 1800 qubits. *Nature* **2018**, *560*, 456–460. [CrossRef] [PubMed]
43. Harris, R.; Sato, Y.; Berkley, A.J.; Reis, M.; Altomare, F.; Amin, M.H.; Boothby, K.; Bunyk, P.; Deng, C.; Enderud, C.; et al. Phase transitions in a programmable quantum spin glass simulator. *Science* **2018**, *361*, 162–165. [CrossRef]
44. Kairys, P.; King, A.D.; Ozfidan, I.; Boothby, K.; Raymond, J.; Banerjee, A.; Humble, T.S. Simulating the Shastry-Sutherland Ising Model Using Quantum Annealing. *PRX Quantum* **2020**, *1*, 020320. [CrossRef]
45. Bando, Y.; Susa, Y.; Oshiyama, H.; Shibata, N.; Ohzeki, M.; Gómez-Ruiz, F.J.; Lidar, D.A.; Suzuki, S.; del Campo, A.; Nishimori, H. Probing the universality of topological defect formation in a quantum annealer: Kibble-Zurek mechanism and beyond. *Phys. Rev. Research* **2020**, *2*, 033369. [CrossRef]
46. Bando, Y.; Nishimori, H. Simulated quantum annealing as a simulator of nonequilibrium quantum dynamics. *Phys. Rev. A* **2021**, *104*, 022607. [CrossRef]
47. King, A.D.; Raymond, J.; Lanting, T.; Isakov, S.V.; Mohseni, M.; Poulin-Lamarre, G.; Ejtemaee, S.; Bernoudy, W.; Ozfidan, I.; Smirnov, A.Y.; et al. Scaling advantage over path-integral Monte Carlo in quantum simulation of geometrically frustrated magnets. *Nat. Comm.* **2021**, *12*, 1113. [CrossRef]
48. King, A.D.; Nisoli, C.; Dahl, E.D.; Poulin-Lamarre, G.; Lopez-Bezanilla, A. Qubit spin ice. *Science* **2021**, *373*, 576–580. [CrossRef]
49. Filho, J.L.C.d.C.; Izquierdo, Z.G.; Saguia, A.; Albash, T.; Hen, I.; Sarandy, M.S. Localization transition induced by programmable disorder. *Phys. Rev. B* **2022**, *105*, 134201. [CrossRef]
50. Kutsuzawa, T.; Todo, S. Nested Iterative Shift-invert Diagonalization for Many-body Localization in the Random-field Heisenberg Chain. *arXiv* **2022**, arXiv:2203.09732. [CrossRef]
51. Nakamura, M.; Todo, S. Order Parameter to Characterize Valence-Bond-Solid States in Quantum Spin Chains. *Phys. Rev. Lett.* **2002**, *89*, 077204. [CrossRef] [PubMed]

entropy

Article

Pseudoclassical Dynamics of the Kicked Top

Zhixing Zou [1] **and Jiao Wang** [1,2,*]

[1] Department of Physics and Key Laboratory of Low Dimensional Condensed Matter Physics (Department of Education of Fujian Province), Xiamen University, Xiamen 361005, China
[2] Lanzhou Center for Theoretical Physics, Lanzhou University, Lanzhou 730000, China
* Correspondence: phywangj@xmu.edu.cn

Abstract: The kicked rotor and the kicked top are two paradigms of quantum chaos. The notions of quantum resonance and the pseudoclassical limit, developed in the study of the kicked rotor, have revealed an intriguing and unconventional aspect of classical–quantum correspondence. Here, we show that, by extending these notions to the kicked top, its rich dynamical behavior can be appreciated more thoroughly; of special interest is the entanglement entropy. In particular, the periodic synchronization between systems subject to different kicking strength can be conveniently understood and elaborated from the pseudoclassical perspective. The applicability of the suggested general pseudoclassical theory to the kicked rotor is also discussed.

Keywords: kicked top model; quantum resonance; pseudoclassical limit; dynamical entanglement

Citation: Zou, Z.; Wang, J. Pseudoclassical Dynamics of the Kicked Top. *Entropy* **2022**, *24*, 1092. https://doi.org/10.3390/e24081092

Academic Editor: Marko Robnik

Received: 19 July 2022
Accepted: 5 August 2022
Published: 9 August 2022

1. Introduction

The study of quantum chaos, or quantum chaology [1], focuses on whether, how, and to what extent classical chaos may manifest itself in the quantum realm. In essence, it boils down to the general classical–quantum correspondence issue, as insightfully pointed out by Einstein at the very early development stage of quantum theory [2]. The quantum kicked rotor, presumably the best known paradigm of quantum chaos, was first introduced by Casati et al. in their seminal study that opened this field [3]. After four decades of investigation, the richness of this paradigmatic model appears to be surprising. Far beyond quantum chaos, it has also been realized that this model may play a unique role in other fundamental problems, such as Anderson localization (transition) [4–7] and the quantum Hall effect [8–10]. Centering around the kicked rotor, an expanded overlapping field encompassing all these relevant problems is emerging [11].

In contrast to its richness, another advantage of the kicked rotor lies in its simplicity, featuring only a single point particle on a circle subject to the stroboscopic external interaction, which makes the study of this model much more convenient than most others. An exception is the kicked top model [12], which has a finite Hilbert space, so that it is even more favorable for research. Interestingly and importantly, these two models usually demonstrate different aspects of quantum chaos in a complementary way. With all these advantages, they are often the first ideal candidates for probing new notions. In recent years, interesting notions having been intensively investigated range widely, from the out-of-time-order correlations [13–15], to the dynamical entanglement [15–20], the non-Hermitian properties [21,22], and so on.

The dynamical entanglement is devised to capture the decoherence process of a quantum system when being coupled to the environment. It has distinct characteristics if the system's classical counterpart is chaotic. The kicked top has the spin algebra symmetry and, as such, it can be regarded as a composite of identical qubits. An additional advantage due to such a multiqubit interpretation is that, for studying the dynamical entanglement, there is no need to introduce the environment. It has been shown both theoretically and experimentally that, indeed, the dynamical entanglement may serve as a diagnosis of quantum chaos in the kicked top model [20,23].

However, as far as we know, in most previous studies of the kicked top, only a narrower range of comparatively weak kicking strength has been investigated, leaving its properties unexplored yet in a wider range of stronger kicking strength. The reason might be that, for the classical kicked top, the transition from regular to globally chaotic motion occurs at a rather weak kicking strength. When the system is already globally chaotic, further increasing the kicking strength would not result in any qualitatively new properties. Accordingly, due to quantum–classical correspondence, it is reasonable to conjecture that this would also be the case in the quantum kicked top in the semi-classical limit. Nevertheless, as illustrated in a recent study where measures of quantum correlations were studied [18], the quantum kicked tops at certain different kicking strengths may synchronize, in clear contrast to their classical counterparts.

In fact, it is worth noting that in the kicked rotor, the similar classical–quantum non-correspondence phenomenon, termed quantum resonance [24], has been recognized and studied ever since the beginning of the quantum chaos field. Later, it has also been realized that the properties of the system when being detuned from the quantum resonance condition can even be understood in a classical way through the so-called pseudoclassical limit [25,26], rather than the conventional semi-classical limit. This reminds us to consider whether the synchronization observed in the quantum kicked top may have any underlying connections to quantum resonance and the pseudoclassical limit. In this work, we will provide a positive answer to this question. In particular, we will suggest a more general scheme of the pseudoclassical limit that involves more information of the quantum dynamics, which allows us to successfully apply it not only to the kicked top, but also to the kicked rotor. When being applied to the kicked rotor, the previously developed pseudoclassical scheme is found to be a special case of the suggested one.

In the following, we will briefly describe the kicked top model in Section 2 first. Next, in Section 3, we will discuss the quantum resonance condition for the quantum kicked top and develop the pseudoclassical theory. The properties of the system adjacent to the quantum resonance condition will be discussed in detail with two illustrating cases in Section 4. In particular, the numerical studies and the comparison with the pseudoclassical theory will be presented. In Section 5, the properties of dynamical entanglement will be investigated from the perspective of the pseudoclassical limit. Finally, we will summarize our work and discuss its extension to the kicked rotor in Section 6.

2. The Kicked Top Model

The Hamiltonian of the kicked top model is [12]

$$H = \alpha J_x + \frac{\beta}{2j} J_z^2 \sum_{n=-\infty}^{\infty} \delta(t-n),$$

where J_x, J_y, and J_z are the angular momentum operators respecting the commutations $[J_\lambda, J_\mu] = i\epsilon_{\lambda\mu\nu} J_\nu$ (the Planck constant $\hbar$ is set to be unity throughout) and j is an integer or half-integer related to the dimension of the Hilbert space N as $N = 2j+1$. The first term in H describes the procession around the x axis with angular frequency α, while the second term accounts for a periodic sequence of kicks with an overall kicking strength β (the period of kicking is set to be the unit of time). In the following, we will restrict ourselves to the case of integer j, but the discussions can be extended straightforwardly to that of half-integer j. Since the Hamiltonian is time-periodic, the evolution of the system for a unit time, or one step of evolution, can be fulfilled by applying the Floquet operator

$$U = \exp\left(-i\frac{\beta}{2j} J_z^2\right) \exp(-i\alpha J_x) \tag{1}$$

to the present state. Obviously, U does not change under the transformation $\beta \to \beta + 4j\pi$, implying that the properties of the quantum kicked top have a periodic dependence on the kicking strength β of period $4j\pi$. Thus, a better understanding of the quantum kicked top calls for investigations covering such a period.

In the semi-classical limit $j \to \infty$, following the Heisenberg equations, the one-step evolution of the system reduces to the following map [27]:

$$\begin{aligned} X' =& X\cos[\beta(Y\sin\alpha + Z\cos\alpha)] - (Y\cos\alpha \\ &- Z\sin\alpha)\cdot\sin[\beta(Y\sin\alpha + Z\cos\alpha)], \\ Y' =& X\sin[\beta(Y\sin\alpha + Z\cos\alpha)] + (Y\cos\alpha \\ &- Z\sin\alpha)\cdot\cos[\beta(Y\sin\alpha + Z\cos\alpha)], \\ Z' =& Y\sin\alpha + Z\cos\alpha, \end{aligned} \tag{2}$$

with the normalized variables $X = J_x/j$, $Y = J_y/j$, and $Z = J_z/j$. This map defines the classical kicked top. Physically, this map describes the process of rotating the top along the x axis for an angle of α first to reach the intermediate state $(\tilde{X}, \tilde{Y}, \tilde{Z})$, followed by further rotating it around $\tilde{Z}$ by $\beta\tilde{Z}$, which is the same as the quantum Floquet operator.

Note that the state (X, Y, Z) can be viewed as a point on the surface of a unit sphere. Therefore, it can be represented equivalently by two angles, denoted as Θ and Φ, via the coordinate transformation $(X, Y, Z) = (\sin\Theta\cos\Phi, \sin\Theta\sin\Phi, \cos\Theta)$. For the sake of convenience, we denote map (2) in terms of Θ and Φ as

$$(\Theta', \Phi') = \mathcal{F}(\Theta, \Phi; \alpha, \beta), \tag{3}$$

where (Θ', Φ') is the state equivalent to (X', Y', Z').

In order to make a close comparison between the quantum and the classical dynamics, we invoke the spin coherent state in the former, which has the minimum uncertainty in a spin system. A spin coherent state centered at (Θ, Φ), denoted as $|\Theta, \Phi\rangle$, can be generated from the angular momentum eigenstate $|j, j\rangle$ as

$$|\Theta, \Phi\rangle = \exp(i\Theta[J_x\sin\Phi - J_y\cos\Phi])|j, j\rangle.$$

Here, $|j, j\rangle$ satisfies that $(J_x^2 + J_y^2 + J_z^2)|j, j\rangle = j(j+1)|j, j\rangle$ and $J_z|j, j\rangle = j|j, j\rangle$. The classical counterpart of $|\Theta, \Phi\rangle$ is the point (Θ, Φ) on the unit sphere.

3. The Pseudoclassical Theory

3.1. Quantum Resonance in the Kicked Top

The concept of quantum resonance was first introduced in the kicked rotor model. The Floquet operator for the kicked rotor is

$$U_R = \exp\left(-i\frac{p^2}{2}T\right)\exp(-iK\cos\theta),$$

where T and K are two parameters, θ is the angular displacement of the rotor, and p is the corresponding conjugate angular momentum. If $T = 4\pi r/s$ with r and s as two coprime integers, except for the cases of an odd r and $s = 2$, the asymptotic growth in energy is quadratic in time, corresponding to a linear spreading of the wavepacket in the angular momentum space. This phenomenon is referred to as "quantum resonance" [24], since it is caused by the pure quantum effect, with no connections to the classical dynamics. Otherwise, the energy would undergo a linear growth stage, corresponding to the diffusive spread of the wavepacket in the angular momentum space, before it saturates due to quantum interference, which is known as the so-called dynamical localization [28]. For the special case of an odd r and $s = 2$, it follows that $U_R^2 = 1$. Namely, the quantum dynamics is periodic of period two, which is referred to as "quantum antiresonance".

There is an implicit connection between the kicked rotor and the kicked top. By assuming the scaling

$$T = \beta/j \quad \text{and} \quad K = \alpha j,$$

it has been shown that the kicked rotor emerges as a limit case of the kicked top [27]. Given this, the notion of quantum resonance can be extended to the kicked top by assigning the quantum resonance condition that

$$\beta = 4j\pi\frac{r}{s}$$

with coprime r and s. Indeed, under this condition, the kicked top has similar properties to the kicked rotor in quantum resonance. For example, for $\beta = 4j\pi$, the Floquet operator reduces to $U = \exp(-i\alpha J_x)$, implying that the top keeps rotating at a constant rate; however, when r is odd and $s = 2$, we have $U^2 = 1$, suggesting that the motion is periodic of period two as well.

3.2. The Pseudoclassical Limit of the Kicked Top

For the kicked rotor, a pseudoclassical theory has been developed to address the quantum dynamics via a classical map, the so-called pseudoclassical limit, when the system parameter is close to the resonance condition that T is an integer multiple of 2π (i.e., $s = 2$) [25,26]. In the following, we attempt to extend this theory to the kicked top and study its behavior for $\beta = 4j\pi r/s + \delta$, where δ (incommensurate to π) is a weak perturbation to the resonance condition that is unnecessarily limited to the case of $s = 2$. Suppose that the current state is $|\Theta, \Phi\rangle$ and its classical counterpart is (Θ, Φ); our task is to figure out the one-step evolution result for the latter by analogy based on the quantum evolution.

For $\beta = 4j\pi r/s + \delta$, the Floquet operator (1) can be rewritten as

$$\begin{aligned} U &= \exp\left(-i\frac{4j\pi\frac{r}{s} + \delta}{2j}J_z^2\right)\exp(-i\alpha J_x) \\ &= \exp\left(-i2\pi\frac{r}{s}J_z^2\right)\exp\left(-i\frac{\delta}{2j}J_z^2\right)\exp(-i\alpha J_x). \end{aligned} \tag{4}$$

Remarkably, the last two terms are exactly the Floquet operator of the kicked top but with the kicking strength δ instead. As shown in previous studies, when δ is small, the quantum dynamics that the last two operators represent can be well mapped to the classical kicked top (with the kicking strength δ) in the semi-classical limit. As a consequence, the classical counterpart of the last two operators is to map (Θ, Φ) into the intermediate state

$$(\tilde{\Theta}_\delta, \tilde{\Phi}_\delta) = \mathcal{F}(\Theta, \Phi; \alpha, \delta), \tag{5}$$

where the subscript δ at the l.h.s. indicates the kicking strength for the sake of clearness.

Vice versa, for the corresponding quantum evolution, due to the solid quantum–classical correspondence in the semi–classical limit, we assume that these two operators map the coherent state $|\Theta, \Phi\rangle$ into that of $|\tilde{\Theta}_\delta, \tilde{\Phi}_\delta\rangle$. Then, the remaining problem is to find out the classical counterpart of the result when the first operator at the r.h.s. of Equation (4) applies to this intermediate state $|\tilde{\Theta}_\delta, \tilde{\Phi}_\delta\rangle$. The result is (see the derivation in Appendix A)

$$\exp\left(-i2\pi\frac{r}{s}J_z^2\right)|\tilde{\Theta}_\delta, \tilde{\Phi}_\delta\rangle = \sum_{l=0}^{s-1} G_l\left|\tilde{\Theta}_\delta, \tilde{\Phi}_\delta + \frac{2\pi r}{s}l\right\rangle, \tag{6}$$

where G_l is the Gaussian sum

$$G_l = \frac{1}{s}\sum_{k=0}^{s-1}\exp\left(-i\frac{2\pi r}{s}k(k-l)\right).$$

The physical meaning of Equation (6) is clear: the intermediate coherent state $|\tilde{\Theta}_\delta, \tilde{\Phi}_\delta\rangle$ is mapped into s coherent states located along the line of $\Theta = \tilde{\Theta}_\delta$, each of which has an amplitude given by a Gaussian sum. Note that these s coherent states are not necessarily independent; some of them may correspond to the same coherent state, if their l values lead to the same angle of $\Delta = 2\pi rl/s \bmod 2\pi$. Suppose that there are $\mathcal{N}$ different such angles in total and denote them as Δ_k, $k = 1, \cdots, \mathcal{N}$; then, Equation (6) can be rewritten as

$$\exp\left(-i2\pi\frac{r}{s}J_z^2\right)|\tilde{\Theta}_\delta, \tilde{\Phi}_\delta\rangle = \sum_{k=1}^{\mathcal{N}} A_k|\tilde{\Theta}_\delta, \tilde{\Phi}_\delta + \Delta_k\rangle. \tag{7}$$

Here, for the kth component coherent state, its amplitude A_k is the sum of all G_l whose subscript l satisfying $\Delta_k = 2\pi rl/s \bmod 2\pi$.

In the semi-classical limit $j \to \infty$, a coherent state reduces to a point in the phase space. Given this, we can give Equation (7) a classical interpretation as the following: the point (Θ, Φ) is mapped into a set of $\mathcal{N}$ points and meanwhile each point is associated with a complex "amplitude". These two features make the situation here distinct from the previous pseudoclassical theory for the kicked rotor, where a point is mapped only to another point and no complex amplitude is involved. Thus, formally, the pseudoclassical map that we seek can be expressed as

$$\mathcal{M} : (\Theta, \Phi) \to \{[(\tilde{\Theta}_\delta, \tilde{\Phi}_\delta + \Delta_k); A_k], k = 1, \cdots, \mathcal{N}\}. \tag{8}$$

This is the key result of the present work. As illustrated in the next section, it does allow us to predict the quantum dynamics in such a pseudoclassical way. Here, we emphasize that the amplitudes $\{A_k\}$ are crucial to this end. Specifically, $|A_k|^2$ has to be taken as the weight of the kth point to evaluate the expected value of a given observable. Moreover, the phases encoded in these amplitudes have to be considered simultaneously to correctly trace the quantum evolution.

4. Applications of the Pseudoclassical Theory

In this section, we check the effectiveness of the pseudoclassical map by comparing its predictions with that obtained directly with the quantum Floquet operator. In general, if a point is mapped into $\mathcal{N} > 1$ points at each step, then the number of points that we have to deal with would increase exponentially. Therefore, in practice, it would be prohibitively difficult to apply it for any arbitrarily given parameters. However, fortunately, for some quantum resonance parameters, $\mathcal{N}$ could be small, and, under certain conditions, e.g., if α is an integer multiple of $\pi/2$, coherent cancellation may suppress the increase in the number of points (see the second subsection below). In such cases, the application of the pseudoclassical theory can be greatly simplified. Here, we consider two such cases as illustrating examples, i.e., $\beta = 2j\pi + \delta$ and $\beta = j\pi + \delta$, respectively.

4.1. Case I: $\beta = 2j\pi + \delta$

For this case, we can show (see Appendix B) that $\mathcal{N} = 1$, i.e., the pseudoclassical dynamics evolves the point (Θ, Φ) into another single point as

$$\mathcal{M} : (\Theta, \Phi) \to (\tilde{\Theta}_\delta, \tilde{\Phi}_\delta + \pi), \tag{9}$$

with the corresponding amplitude $A_1 = 1$.

With Equation (9) in hand, we are ready to predict the quantum properties. The most relevant quantities could be the expected values of angular momentums. In Figure 1, their dependence on time is shown for a randomly chosen initial condition. The corresponding quantum results for three different values of j are plotted together for comparison. It can be seen that the pseudoclassical results agree very well with the quantum ones, and, as expected, as j increases, the agreement improves progressively. It shows that, indeed, the pseudoclassical limit captures the quantum motion successfully.

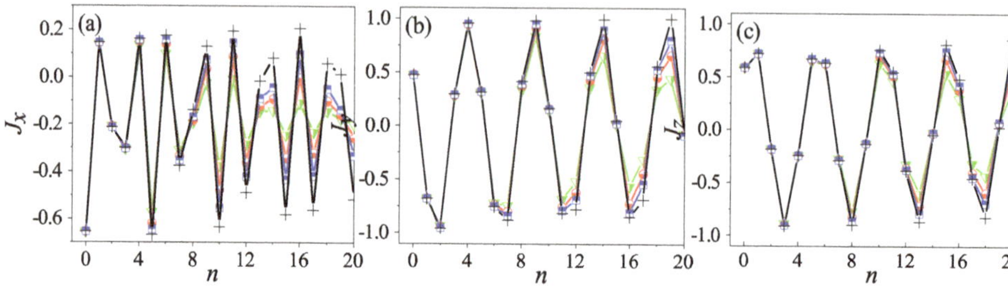

Figure 1. The time dependence of the expected value of angular momentum J_x (**a**), J_y (**b**), and J_z (**c**), respectively, for $\alpha = 1$ and $\beta = 2j\pi + 2$. The black crosses are for the results by the pseudoclassical map (Equation (9)). The green triangles, the red circles, and the blue squares are for the quantum results with $j = 100$, 200, and 400, respectively. For the initial state (Θ, Φ) and $|\Theta, \Phi\rangle$, $\Theta = 0.8\pi$ and $\Phi = 0.3\pi$.

The agreement illustrated in Figure 1 does not depend on α. However, if α is an integer multiple of $\pi/2$, the system would have an additional interesting property. Namely, its quantum entanglement entropy would remain synchronized with that of the system that has a kicking strength of $\beta = \delta$ instead [18]. Since, for such an α value, the good agreement between the pseudoclassical and the quantum evolution remains equally, we can probe this interesting phenomenon from the pseudoclassical perspective. In fact, by following Equation (9) and taking into account the extra symmetry introduced by such an α value, we can show that this synchronization in the entanglement entropy roots in the synchronization of their dynamics (see the following and the next section). The latter has a period of four (two) when α is an odd (even) multiple of $\pi/2$. In Appendix B, the pseudoclassical dynamics is detailed for the representative example where $\alpha = \pi/2$. For this case, in terms of the expected value of angular momentums, denoted as J_x, J_y, and J_z as well, without confusion, the connection between these two systems at a given time n can be made explicitly as follows:

$$
\begin{aligned}
J_x(n;\beta) &= \begin{cases} J_x(n;\delta), & \mathrm{mod}\ (n,4) = 0 \text{ or } 2; \\ -J_x(n;\delta), & \mathrm{mod}\ (n,4) = 1 \text{ or } 3, \end{cases} \\
J_y(n;\beta) &= \begin{cases} J_y(n;\delta), & \mathrm{mod}\ (n,4) = 0 \text{ or } 3; \\ -J_y(n;\delta), & \mathrm{mod}\ (n,4) = 1 \text{ or } 2, \end{cases} \\
J_z(n;\beta) &= \begin{cases} J_z(n;\delta), & \mathrm{mod}\ (n,4) = 0 \text{ or } 1; \\ -J_z(n;\delta), & \mathrm{mod}\ (n,4) = 2 \text{ or } 3. \end{cases}
\end{aligned}
\tag{10}
$$

Note that the angular momentum values at the l.h.s. and the r.h.s. are for the system with kicking strength $\beta = 2j\pi + \delta$ and δ, respectively.

To check this prediction, we compare the numerical results of the quantum evolution of the two systems. The results are presented in Figure 2, where not only the four-step synchronization but also the details of the intermediate states can be recognized immediately. We can also make a close comparison of these two systems by visualizing their quantum evolution in the phase space with the Husimi distribution [29]. At a given point (Θ, Φ) in the phase space, the Husimi distribution $P(\Theta, \Phi)$ is defined as the expectation value of the density matrix ρ with respect to the corresponding spin coherent state, i.e.,

$$P(\Theta, \Phi) = \frac{2j+1}{4\pi} \langle \Theta, \Phi | \rho | \Theta, \Phi \rangle.$$

The results for $\beta = 2$ and $\beta = 2j\pi + 2$ at four different times are shown in Figures 3 and 4, respectively. It can be seen that, when $n = 1$, the centers of the two wavepackets only differ by an angle of π in Φ, while, when n = 2, they become symmetric with respect to $(\Theta, \Phi) = (\pi/2, \pi)$. When $n = 4$ and 8, the two wavepackets are indistinguishable, which is a sign that the two systems are synchronized.

Obviously, all these numerical checks have well corroborated the effectiveness of our pseudoclassical analysis.

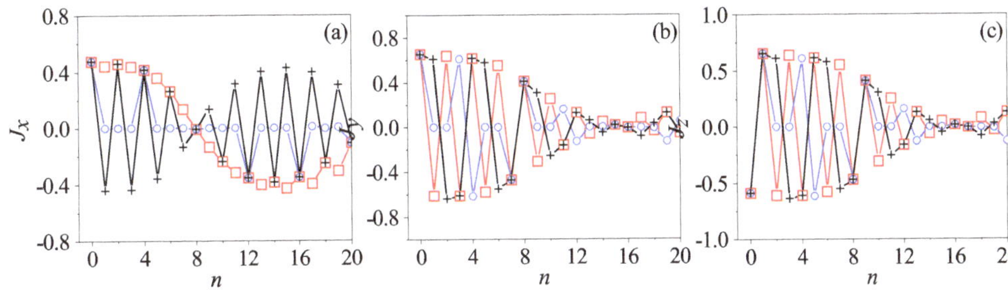

Figure 2. The time dependence of the expected value of angular momentum J_x (**a**), J_y (**b**), and J_z (**c**), respectively. The same as Figure 1 but for the quantum kicked top with kicking strength $\beta = 2j\pi + 2$ (red squares), $\beta = j\pi + 2$ (blue circles), and $\beta = 2$ (black crosses), respectively. Here, $\alpha = \pi/2$, $j = 400$, and the initial state is $|\Theta, \Phi\rangle = |0.7\pi, 0.3\pi\rangle$.

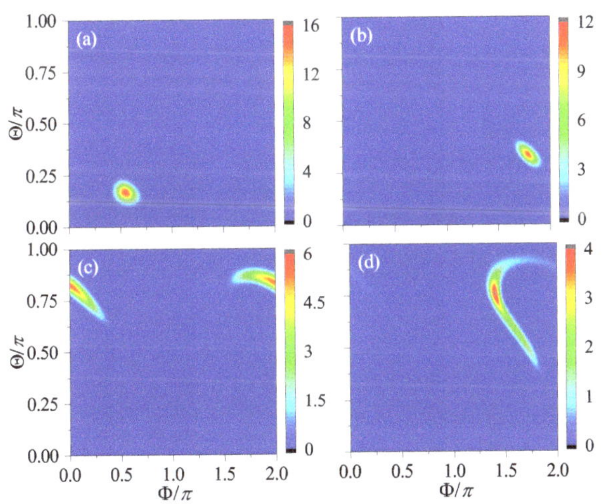

Figure 3. The Husimi distribution for $\beta = 2$ at the time $n = 1$ (**a**), $n = 2$ (**b**), $n = 4$ (**c**), and $n = 8$ (**d**), respectively. Here, $\alpha = \pi/2$, $j = 100$, and the initial state is $|\Theta, \Phi\rangle = |\pi/2, \pi/3\rangle$.

4.2. Case II: $\beta = j\pi + \delta$

Now, let us consider a more complex case, i.e., $\beta = j\pi + \delta$. For this case, $\mathcal{N} = 2$ and the pseudoclassical map is (see Appendix C)

$$\mathcal{M} : (\Theta, \Phi) \to \begin{cases} (\tilde{\Theta}_\delta, \tilde{\Phi}_\delta + \pi); \;\; A_1, \\ (\tilde{\Theta}_\delta, \tilde{\Phi}_\delta); \;\; A_2, \end{cases} \tag{11}$$

with $A_1 = (1+i)/2$ and $A_2 = (1-i)/2$. It implies that, after each step, a point will be mapped into two points at the same probability but with different phases. This map looks simple, but as $\mathcal{N} = 2$, if we use it to predict the quantum evolution, the points will proliferate in time so that, in practice, we can trace the quantum motion for a few steps only. Interestingly, this fact might explain why the quantum motion would be complicated from a new perspective.

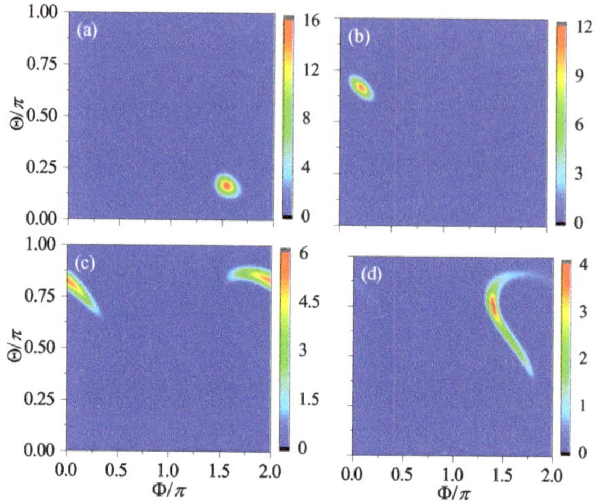

Figure 4. The Husimi distribution for $\beta = 2$ at the time $n = 1$ (**a**), $n = 2$ (**b**), $n = 4$ (**c**), and $n = 8$ (**d**), respectively. The same as Figure 3 but for $\beta = 2j\pi + 2$ instead (other parameters remain unchanged).

Nevertheless, for some special values of α, due to the coherent effect, the newly generated points after a step of iteration may overlap and cancel each other out, making the number of points increase more slowly. An intriguing example is that discussed in the previous subsection, i.e., where α is an integer multiple of $\pi/2$. Again, for such an α value, the system is brought to synchronization with the system of kicking strength $\beta = \delta$ as well, but with instead a period of eight (four) if α is an odd (even) multiple of $\pi/2$. To be explicit, for $\alpha = \pi/2$, the connections between the two systems are presented in Appendix C. In terms of the expected value of angular momentums, we have

$$
\begin{aligned}
J_x(n;\beta) &= \begin{cases} J_x(n;\delta), & \text{mod } (n,8) = 0 \text{ or } 4; \\ 0, & \text{else,} \end{cases} \\
J_y(n;\beta) &= \begin{cases} J_y(n;\delta), & \text{mod } (n,8) = 0 \text{ or } 7; \\ -J_y(n;\delta), & \text{mod } (n,8) = 3 \text{ or } 4; \\ 0, & \text{else,} \end{cases} \\
J_z(n;\beta) &= \begin{cases} J_z(n;\delta), & \text{mod } (n,8) = 0 \text{ or } 1; \\ -J_z(n;\delta), & \text{mod } (n,8) = 4 \text{ or } 5; \\ 0, & \text{else.} \end{cases}
\end{aligned}
\tag{12}
$$

The simulation results of the quantum angular momentums for $\beta = j\pi + \delta$ are shown in Figure 2 as well; they support this derivation convincingly.

Based on the pseudoclassical dynamics, we find that the initial point will be mapped into two and then four points after the first and the second iteration, respectively. However, after the third iteration, the points do not become eight as expected; rather, these eight points can be divided into four pairs and the two points in each pair overlap with each other. Moreover, two of these four points disappear in effect as the resultant total amplitude for each of them turns out to be zero (see Appendix C). Thus, only two points remain, and after the fourth iteration, these two points further merge into one. As a consequence, the number of points varies in time with a period of four. The results for the Husimi distribution given in Figure 5 are in good agreement with this analysis. Comparing this with the results for $\beta = \delta$ in Figure 3, we can see that after the fourth iteration, there is only one wavepacket of the same shape in both cases, but their positions are different. Only after the eighth iteration, the two wavepackets are identical, which explains why the synchronization period should be eight.

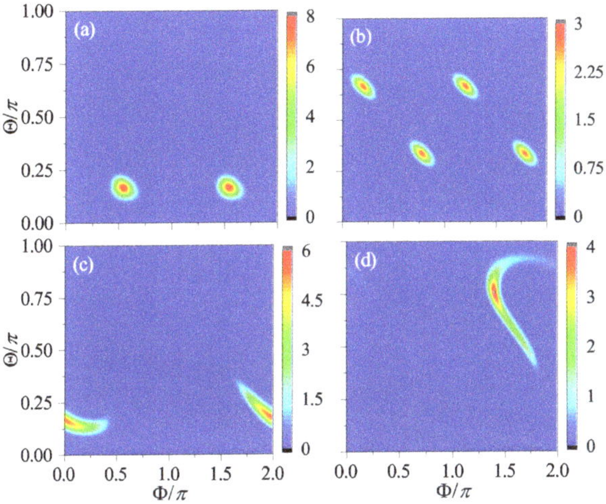

Figure 5. The Husimi distribution for $\beta = 2$ at the time $n = 1$ (**a**), $n = 2$ (**b**), $n = 4$ (**c**), and $n = 8$ (**d**), respectively. The same as Figures 3 and 4 but for $\beta = j\pi + 2$. All other parameters are the same as in the former two figures.

5. The Dynamical Entanglement

The dynamical entanglement of the quantum kicked top has been studied carefully in recent years. We discuss this issue in this section by taking advantage of the pseudoclassical results obtained in the previous section.

For the quantum kicked top, its momentum can be represented by $2j$ qubits, or a collection of $2j$ spin-1/2 identical particles. If the initial state of the system is symmetric under permutations for identical qubits, this permutation symmetry will be preserved, as it is respected by the action of the Floquet operator of the kicked top. As a consequence, the expected spin value for any single qubit of these $2j$ identical qubits is $s_\gamma = J_\gamma/(2j)$, where J_γ is the expected momentum value of the top and $\gamma = x, y$, and z [30].

On the other hand, though various bipartite entanglement measures have been suggested, it has been shown that they are qualitatively equivalent. Thus, we adopt the measure considered the most frequently, i.e., the bipartite entanglement between any qubit and the subsystem made up of the remaining $2j - 1$ qubits. This entanglement is usually quantified by computing the linear entropy $S = 1 - \mathrm{Tr}(\rho_1^2)$, where ρ_1 denotes the reduced density operator for a single qubit. As ρ_1 is a 2×2 operator, it can be expressed as $\rho_1 = 1/2 + \sum_\gamma s_\gamma \sigma_\gamma$, given the expected spin value s_γ. Here, σ_γ is the Pauli operator. By substituting $s_\gamma = J_\gamma/(2j)$, we have $\rho_1 = 1/2 + \sum_\gamma J_\gamma \sigma_\gamma/(2j)$ in terms of J_γ instead [30]. It follows that

$$S = \frac{1}{2}\left(1 - \frac{[J_x]^2 + [J_y]^2 + [J_z]^2}{j^2}\right), \tag{13}$$

which has a well-defined classical counterpart and is easy to compute numerically. Here, $[J_\gamma]^2$ represents the square of the expected value of angular momentum J_γ.

For the two cases close to the quantum resonance condition discussed in the previous section, we can see immediately how their linear entropy is related to that of the case $\beta = \delta$ based on the pseudoclassical analysis. First, for $\beta = 2j\pi + \delta$, from Equation (10), we have that $[J_\gamma(n;\beta)]^2 = [J_\gamma(n;\delta)]^2$ at any time n; hence, $S(n)$ must coincide with that for the system of $\beta = \delta$ throughout. However, for $\beta = j\pi + \delta$, from Equation (12), we know that $\sum_\gamma [J_\gamma(n;\beta)]^2 = \sum_\gamma [J_\gamma(n;\delta)]^2$ only when $\mod(n, 8) = 0$ or 4. Therefore, we may expect a synchronization of period four in the linear entropy. In addition, from Equation (12), we also know that $\sum_\gamma [J_\gamma(n;\beta)]^2 = 0$ when $\mod(n, 8) = 2$ or 6, suggesting that the linear entropy should reach its maximal value repeatedly in a period of four as well. As for the case $\beta = \delta$ itself, because the quantum motion can be approximated by the semi-classical limit if δ is small, the time dependence of the linear entropy can be predicted based on the classical dynamics [15]. In a chaotic region of the phase space, $S(n)$ should increase linearly before saturation; otherwise, it proceeds in a logarithmic law that features the regular motion. The numerical results of the linear entropy for these three cases and two representative initial states are presented in Figure 6, which meet all these expectations.

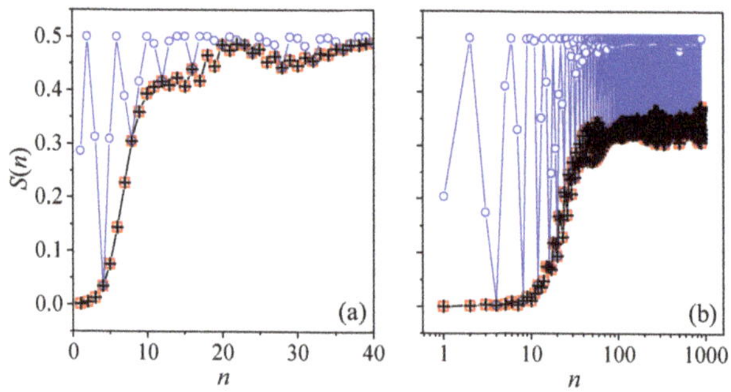

Figure 6. The linear entropy as a function of time for the initial condition $|\Theta, \Phi\rangle = |0.7\pi, 0.3\pi\rangle$ (**a**) and $|0.7\pi, 0.6\pi\rangle$ (**b**), respectively. The classical counterparts of these two states are in the chaotic and regular region of the phase space, respectively, for the classical kicked top of $\beta = 2$. In both panels, the red squares, the blue circles, and the black crosses are for, respectively, $\beta = 2j\pi + 2$, $\beta = j\pi + 2$, and $\beta = 2$. For all the cases, $\alpha = \pi/2$ and $j = 400$.

Another interesting and related quantity that has been intensively investigated is the time-averaged entanglement entropy. If the semi-classical limit exists, it is used to estimate the equilibrium value that the corresponding classical system tends to. For the linear entropy, it is defined as

$$S_\tau = \frac{1}{\tau} \sum_{n=1}^{\tau} S(n), \tag{14}$$

where τ is a sufficiently long time. Interestingly, it was found that the contour plot of S_τ can well capture the characteristics of the phase space portrait of the corresponding classical system [19,20]. The reason is that, given the semi-classical limit, for all initial coherent states centered on the same classical trajectory, by definition, S_τ should be the same in the long average time limit. As such, S_τ can be used to distinguish the trajectories. As an illustration, in Figure 7a, the contour plot of S_τ for the case $\beta = 2$, where the semi-classical limit holds well, is shown. The corresponding phase space portrait of its classical counterpart is shown in Figure 8. The similarity between them is easy to recognize.

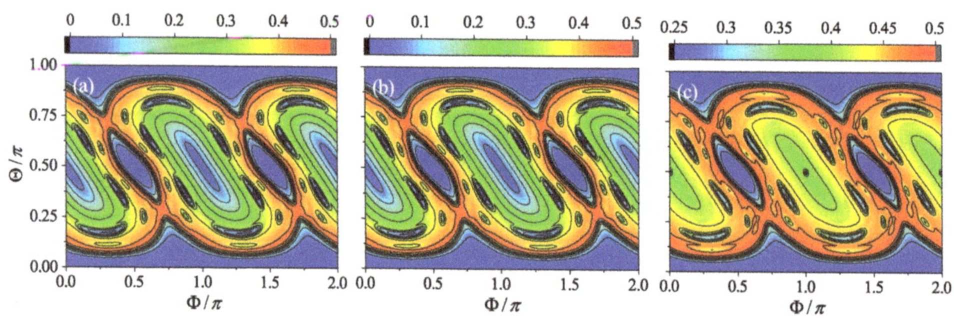

Figure 7. The contour plot of the time-averaged entanglement entropy, S_τ for kicking strength $\beta = 2$ (**a**), $\beta = 2j\pi + 2$ (**b**), and $\beta = j\pi + 2$ (**c**), respectively. Here, $\alpha = \pi/2$, $j = 400$, $\tau = 300$, and a grid of 201×201 initial coherent states is simulated for each case.

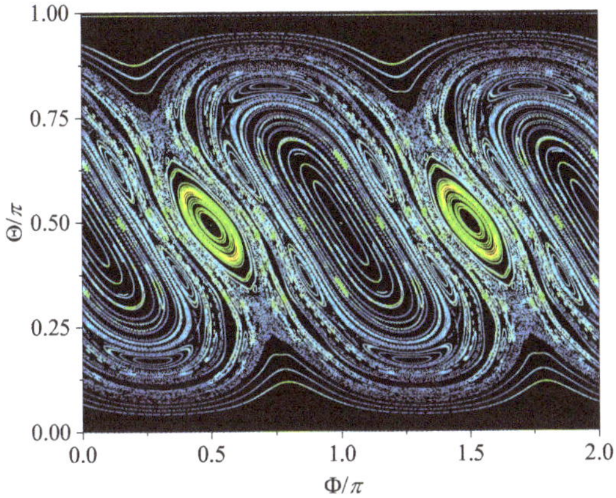

Figure 8. The phase space portrait of the classical kicked top with $\alpha = \pi/2$ and $\beta = 2$.

For the two cases close to the quantum resonance condition, the semi-classical limit breaks. However, as the pseudoclassical limit exists, it allows us to use S_T to probe the corresponding pseudoclassical systems. In Figure 7b, the contour plot for $\beta = 2j\pi + 2$ is shown. It is identical to that for $\beta = 2$, because the two cases share the same linear entropy at every step. However, it is worth noting that, although, from Equation (9), we know that any trajectory of the pseudoclassical system for $\beta = 2j\pi + \delta$ is related to one of the semi-classical system for $\beta = \delta$, they are not located at the same positions in their respective phase spaces (see Figures 3 and 4, for example). This suggests that S_T is still equally helpful for obtaining an overall sense of the pseudoclassical dynamics, but some details could be missed inevitably by definition.

More interesting is the case for $\beta = j\pi + \delta$. For the pseudoclassical dynamics, the concept of the conventional trajectory does not apply any longer, because, at a given time, the number of points in the phase space can be multiple, and they have their respective complex amplitudes. Regardless of this fact, as Figure 7c shows, S_T works well again for schematizing the pseudoclassical dynamics. For example, due to the periodic synchronization with the case of $\beta = \delta$, we may expect that the chaotic and regular regions of the former are exactly those of the latter, respectively. This is indeed the case, which can be seen by comparing Figure 7a,c. Alternatively, if we compute S_T by taking the average of $S(n)$ once every four steps (the period of synchronization), the results should be the same exactly for both cases.

6. Summary and Discussion

In summary, we have introduced the quantum resonance condition into the kicked top model. In order to study the behavior of the quantum kicked top detuned from the quantum resonance condition, we have established the corresponding pseudoclassical theory. By analytical and numerical studies, we have shown that this theory is effective. In particular, when being applied to discuss the dynamical entanglement, the properties of the quantum kicked top are successfully predicted based on the pseudoclassical dynamics. Our results also suggested that the time-averaged entanglement entropy is still a powerful tool for grasping the pseudoclassical dynamics.

The suggested pseudoclassical scheme is distinct from the one originally introduced in the kicked rotor that works only near the special quantum resonance condition that T is an integer multiple of 2π. To make this explicit, let us extend our scheme to the kicked rotor and compare it with the original pseudoclassical limit. The result is similar to Equation (8):

$$\mathcal{M} : (p, \theta) \to \{[(\tilde{p}_\delta, \tilde{\theta}_\delta + \Delta_k); A_k], k = 1, \cdots, \mathcal{N}\}, \tag{15}$$

but, here, (p,θ) is the classical counterpart and the center of the coherent state $|p,\theta\rangle$ for the kicked rotor, and $(\tilde{p}_\delta,\tilde{\theta}_\delta)$ is that (p,θ) is mapped to by the classical kicked rotor dynamics with the kicking strength δK. For $T=2k\pi+\delta$ (k is an integer), it gives exactly the original pseudoclassical result. However, this scheme also applies when the system is close to other quantum resonances, making it more general than the original one.

In practice, the main challenge for applying this scheme is the same as that encountered in the kicked top, i.e., the rapid proliferation of phase space points. However, if we introduce an additional symmetry into the system, i.e., the translation invariance for $\theta \to \theta + 2\pi/w$, where $w=s/2$ for an even s and $w=s$ for an odd s, respectively, it is found that the coherent cancellation mechanism works efficiently so that the proliferation can be greatly suppressed. The translation invariance can be fulfilled by replacing the potential $\cos\theta$ in the Floquet operator U_R with $\cos(w\theta)$. Such a favorable property makes the kicked rotor even more advantageous than the kicked top for demonstrating the pseudoclassical dynamics. A detailed discussion will be published elsewhere [31].

With the pseudoclassical theory in hand, some interesting problems could be investigated further. For example, it may be applied to study the entanglement in the kicked top with weak measurements of one or several qubits [32] to identify the measurement effect from a different perspective. Moreover, the proliferation of the phase space points is in clear contrast with the conventional classical dynamics. As it is inherited from the quantum evolution, $\log\mathcal{N}$ might be taken as a complexity measure of the quantum dynamics. For some previously studied problems in the double kicked rotor and top, such as Hofstadter's butterfly spectrum [33–35] and exponential and superballistic wavepacket spreading [36,37], it may help us to gain a deeper understanding. For some frontier topics mentioned in the Introduction, this theory may find applications as well.

Author Contributions: Conceptualization, Z.Z. and J.W.; writing—original draft preparation, Z.Z. and J.W.; funding acquisition, J.W. All authors have read and agreed to the published version of the manuscript.

Funding: This work is supported by the National Natural Science Foundation of China (Grants No. 12075198 and No. 12047501).

Institutional Review Board Statement: Not applicable.

Informed Consent Statement: Not applicable.

Data Availability Statement: Not applicable.

Conflicts of Interest: The authors declare no conflict of interest.

Appendix A. Derivation of Equation (6)

First, we expand the coherent state $|\tilde{\Theta}_\delta,\tilde{\Phi}_\delta\rangle$ over the eigenstates $\{|m\rangle\}$ of J_z, i.e., $|\tilde{\Theta}_\delta,\tilde{\Phi}_\delta\rangle = \sum_{m=-j}^{j} c_m|m\rangle$. Applying the operator $\exp(-i\frac{2\pi r}{s}J_z^2)$ to both sides, it results in

$$\exp\left(-i\frac{2\pi r}{s}J_z^2\right)|\tilde{\Theta}_\delta,\tilde{\Phi}_\delta\rangle = \sum_{k=0}^{s-1}\exp\left(-i\frac{2\pi r}{s}k^2\right)\sum_{\mathrm{mod}\,(m,s)=k} c_m|m\rangle. \tag{A1}$$

Note that by the translation operator $\exp(-iJ_z\phi)$, coherent state $|\tilde{\Theta}_\delta,\tilde{\Phi}_\delta\rangle$ is shifted into $|\tilde{\Theta}_\delta,\tilde{\Phi}_\delta+\phi\rangle$, and we thus have

$$\begin{aligned}\left|\tilde{\Theta}_\delta,\tilde{\Phi}_\delta+\frac{2\pi r}{s}l\right\rangle &= \exp\left(-iJ_z\frac{2\pi r}{s}l\right)|\tilde{\Theta}_\delta,\tilde{\Phi}_\delta\rangle \\ &= \sum_{k=0}^{s-1}\exp\left(-i\frac{2\pi kr}{s}l\right)\sum_{\mathrm{mod}\,(m,s)=k} c_m|m\rangle\end{aligned} \tag{A2}$$

by setting $\phi = 2\pi lr/s$, where l is an integer. Next, multiplying both sides with $\exp(i2\pi\lambda lr/s)$, where λ is an integer, $0 \leq \lambda \leq s-1$, and taking summation over l from $l = 0$ to $s-1$, we can obtain that

$$\sum_{l=0}^{s-1} \frac{1}{s} \exp\left(i\frac{2\pi r}{s}\lambda l\right) \left|\tilde{\Theta}_\delta, \tilde{\Phi}_\delta + \frac{2\pi r}{s} l\right\rangle = \sum_{\text{mod }(m,s)=\lambda} c_m |m\rangle. \tag{A3}$$

Finally, by replacing λ in Equation (A3) with k and substituting this equation into Equation (A1), we have the result of Equation (6).

Appendix B. Pseudoclassical Dynamics for $\beta = 2j\pi + \delta$

For this case, $r = 1$ and $s = 2$; we have $G_0 = 0$, $G_1 = 1$, $A_1 = 1$, and $\mathcal{N} = 1$, so that the pseudoclassical map Equation (8) reduces to

$$\mathcal{M} : (\Theta, \Phi) \to (\tilde{\Theta}_\delta, \tilde{\Phi}_\delta + \pi). \tag{A4}$$

Namely, a point is mapped into another with unity amplitude.

On the other hand, note that for the particular case that $\alpha = \frac{\pi}{2}$, the classical map given by Equation (3) has the following properties:

$$\begin{aligned}
(\pi - \Theta', \pi - \Phi') &= \mathcal{F}(\Theta, \Phi + \pi, \frac{\pi}{2}, \beta), \\
(\Theta', \Phi' + \pi) &= \mathcal{F}(\pi - \Theta, \pi - \Phi, \frac{\pi}{2}, \beta), \\
(\pi - \Theta', 2\pi - \Phi') &= \mathcal{F}(\pi - \Theta, 2\pi - \Phi, \frac{\pi}{2}, \beta).
\end{aligned} \tag{A5}$$

If we use $(\Theta(0), \Phi(0))$ to denote the initial state and $(\tilde{\Theta}_\delta(n), \tilde{\Phi}_\delta(n))$ the state after n kicks following the classical map with kicking strength δ, i.e.,

$$(\tilde{\Theta}_\delta(n), \tilde{\Phi}_\delta(n)) = \mathcal{F}^n(\Theta_\delta(0), \Phi_\delta(0), \frac{\pi}{2}, \delta), \tag{A6}$$

we can write the results of the pseudoclassical map step by step as follows:

$n = 0 \to n = 1$:

$$(\Theta(0), \Phi(0)) \to (\tilde{\Theta}_\delta(1), \tilde{\Phi}_\delta(1) + \pi);$$

$n = 1 \to n = 2$:

$$(\tilde{\Theta}_\delta(1), \tilde{\Phi}_\delta(1) + \pi) \to (\pi - \tilde{\Theta}_\delta(2), 2\pi - \tilde{\Phi}_\delta(2));$$

$n = 2 \to n = 3$:

$$(\pi - \tilde{\Theta}_\delta(2), 2\pi - \tilde{\Phi}_\delta(2)) \to (\pi - \tilde{\Theta}_\delta(3), \pi - \tilde{\Phi}_\delta(3));$$

$n = 3 \to n = 4$:

$$(\pi - \tilde{\Theta}_\delta(3), \pi - \tilde{\Phi}_\delta(3)) \to (\tilde{\Theta}_\delta(4), \tilde{\Phi}_\delta(4)).$$

At each step, the amplitude is unity. It shows that, after every four steps, the pseudoclassical dynamics coincides with the classical dynamics of kicking strength δ. Based on these results, we can write the results given in Equation (10) straightforwardly.

Appendix C. Pseudoclassical Dynamics for $\beta = j\pi + \delta$

For this case, $r = 1$ and $s = 4$; we have $G_1 = 0$, $G_2 = \frac{\sqrt{2}}{2}\exp(i\frac{\pi}{4})$, $G_3 = 0$, and $G_4 = \frac{\sqrt{2}}{2}\exp(-i\frac{\pi}{4})$. As a result, $\mathcal{N} = 2$; $A_1 = G_2 = \frac{\sqrt{2}}{2}\exp(i\frac{\pi}{4})$ and $A_2 = G_4 = \frac{\sqrt{2}}{2}\exp(-i\frac{\pi}{4})$, respectively.

The pseudoclassical map Equation (8) can be written as

$$\mathcal{M}:(\Theta,\Phi)\rightarrow\begin{cases}(\tilde{\Theta}_\delta,\tilde{\Phi}_\delta+\pi);\ \frac{\sqrt{2}}{2}\exp(i\frac{\pi}{4}),\\(\tilde{\Theta}_\delta,\tilde{\Phi}_\delta);\ \frac{\sqrt{2}}{2}\exp(-i\frac{\pi}{4}).\end{cases} \tag{A7}$$

For the particular case that $\alpha=\frac{\pi}{2}$, by taking into account the properties of (A5), we can write the pseudoclassical map step by step as follows. Note that the notation is the same as in Appendix B and the constant factor $\frac{\sqrt{2}}{2}$ of A_1 and A_2 is dropped for the sake of clearness and convenience.
$n=0\rightarrow n=1$:

$$(\Theta(0),\Phi(0))\rightarrow\begin{cases}(\tilde{\Theta}_\delta(1),\tilde{\Phi}_\delta(1));\ \exp(-i\frac{\pi}{4}),\\(\tilde{\Theta}_\delta(1),\tilde{\Phi}_\delta(1)+\pi);\ \exp(i\frac{\pi}{4});\end{cases}$$

$n=1\rightarrow n=2$:

$$(\tilde{\Theta}_\delta(1),\tilde{\Phi}_\delta(1));\ \exp(-i\frac{\pi}{4})\rightarrow\begin{cases}(\tilde{\Theta}_\delta(2),\tilde{\Phi}_\delta(2));\ -i,\\(\tilde{\Theta}_\delta(2),\tilde{\Phi}_\delta(2)+\pi);\ 1;\end{cases}$$

$$(\tilde{\Theta}_\delta(1),\tilde{\Phi}_\delta(1)+\pi);\ \exp(i\frac{\pi}{4})\rightarrow\begin{cases}(\pi-\tilde{\Theta}_\delta(2),\pi-\tilde{\Phi}_\delta(2));\ 1,\\(\pi-\tilde{\Theta}_\delta(2),2\pi-\tilde{\Phi}_\delta(2));\ i;\end{cases}$$

$n=2\rightarrow n=3$:

$$(\tilde{\Theta}_\delta(2),\tilde{\Phi}_\delta(2));\ -i\rightarrow\begin{cases}(\tilde{\Theta}_\delta(3),\tilde{\Phi}_\delta(3));\ \exp(-i\frac{3\pi}{4}),\\(\tilde{\Theta}_\delta(3),\tilde{\Phi}_\delta(3)+\pi);\ \exp(-i\frac{\pi}{4});\end{cases}$$

$$(\tilde{\Theta}_\delta(2),\tilde{\Phi}_\delta(2)+\pi);\ 1\rightarrow\begin{cases}(\pi-\tilde{\Theta}_\delta(3),\pi-\tilde{\Phi}_\delta(3));\ \exp(-i\frac{\pi}{4}),\\(\pi-\tilde{\Theta}_\delta(3),2\pi-\tilde{\Phi}_\delta(3));\ \exp(i\frac{\pi}{4});\end{cases}$$

$$(\pi-\tilde{\Theta}_\delta(2),\pi-\tilde{\Phi}_\delta(2));\ 1\rightarrow\begin{cases}(\tilde{\Theta}_\delta(3),\tilde{\Phi}_\delta(3));\ \exp(i\frac{\pi}{4}),\\(\tilde{\Theta}_\delta(3),\tilde{\Phi}_\delta(3)+\pi);\ \exp(-i\frac{\pi}{4});\end{cases}$$

$$(\pi-\tilde{\Theta}_\delta(2),2\pi-\tilde{\Phi}_\delta(2));\ i\rightarrow\begin{cases}(\pi-\tilde{\Theta}_\delta(3),\pi-\tilde{\Phi}_\delta(3));\ \exp(i3\frac{\pi}{4}),\\(\pi-\tilde{\Theta}_\delta(3),2\pi-\tilde{\Phi}_\delta(3));\ \exp(i\frac{\pi}{4});\end{cases}$$

Note that at $n=3$, the total amplitude of the point $(\tilde{\Theta}_\delta(3),\tilde{\Phi}_\delta(3))$ vanishes and so does that of $(\pi-\tilde{\Theta}_\delta(3),\pi-\tilde{\Phi}_\delta(3))$ as a consequence of coherence cancellation, while the other two points $(\tilde{\Theta}_\delta(3),\tilde{\Phi}_\delta(3)+\pi)$ and $(\pi-\tilde{\Theta}_\delta(3),2\pi-\tilde{\Phi}_\delta(3))$ remain.
$n=3\rightarrow n=4$:

$$(\Theta(3),\Phi(3)+\pi);\exp(-i\frac{\pi}{4})\rightarrow\begin{cases}(\pi-\tilde{\Theta}_\delta(4),2\pi-\tilde{\Phi}_\delta(4));\ 1,\\(\pi-\tilde{\Theta}_\delta(4),\pi-\tilde{\Phi}_\delta(4));\ -i;\end{cases}$$

$$(\pi-\tilde{\Theta}(3),2\pi-\tilde{\Phi}(3));\exp(i\frac{\pi}{4})\rightarrow\begin{cases}(\pi-\tilde{\Theta}_\delta(4),\pi-\tilde{\Phi}_\delta(4));\ i,\\(\pi-\tilde{\Theta}_\delta(4),2\pi-\tilde{\Phi}_\delta(4));\ 1;\end{cases}$$

Note that at $n = 4$, the total amplitude of $(\pi - \tilde{\Theta}_\delta(4), \pi - \tilde{\Phi}_\delta(4))$ turns out to be zero, but the other point $(\pi - \tilde{\Theta}_\delta(4), 2\pi - \tilde{\Phi}_\delta(4))$ survives.
$\boldsymbol{n = 4 \to n = 5}$:

$$(\pi - \tilde{\Theta}_\delta(4), 2\pi - \tilde{\Phi}_\delta(4)); 1 \to \begin{cases} (\pi - \tilde{\Theta}_\delta(5), \pi - \tilde{\Phi}_\delta(5)); & \exp(i\frac{\pi}{4}), \\ (\pi - \tilde{\Theta}_\delta(5), 2\pi - \tilde{\Phi}_\delta(5)); & \exp(-i\frac{\pi}{4}); \end{cases}$$

$\boldsymbol{n = 5 \to n = 6}$:

$$(\pi - \tilde{\Theta}_\delta(5), \pi - \tilde{\Phi}_\delta(5)); \quad \exp(i\frac{\pi}{4}) \to \begin{cases} (\tilde{\Theta}_\delta(6), \tilde{\Phi}_\delta(6)); \ i, \\ (\tilde{\Theta}_\delta(6), \tilde{\Phi}_\delta(6) + \pi); \ 1; \end{cases}$$

$$(\tilde{\pi} - \Theta_\delta(5), 2\pi - \tilde{\Phi}_\delta(5)); \quad \exp(-i\frac{\pi}{4}) \to \begin{cases} (\pi - \tilde{\Theta}_\delta(6), \pi - \tilde{\Phi}_\delta(6)); \ 1 \\ (\pi - \tilde{\Theta}_\delta(6), 2\pi - \tilde{\Phi}_\delta(6)); \ -i; \end{cases}$$

$\boldsymbol{n = 6 \to n = 7}$:

$$(\tilde{\Theta}_\delta(6), \tilde{\Phi}_\delta(6)); \ i \to \begin{cases} (\tilde{\Theta}_\delta(7), \tilde{\Phi}_\delta(7)); & \exp(i\frac{\pi}{4}), \\ (\tilde{\Theta}_\delta(7), \tilde{\Phi}_\delta(7) + \pi); & \exp(i3\frac{\pi}{4}); \end{cases}$$

$$(\tilde{\Theta}_\delta(6), \tilde{\Phi}_\delta(6) + \pi); \ 1 \to \begin{cases} (\pi - \tilde{\Theta}_\delta(7), \pi - \tilde{\Phi}_\delta(7)); & \exp(-i\frac{\pi}{4}), \\ (\pi - \tilde{\Theta}_\delta(7), 2\pi - \tilde{\Phi}_\delta(7)); & \exp(i\frac{\pi}{4}); \end{cases}$$

$$(\pi - \tilde{\Theta}_\delta(6), \pi - \tilde{\Phi}_\delta(6)); \ 1 \to \begin{cases} (\tilde{\Theta}_\delta(7), \tilde{\Phi}_\delta(7)); & \exp(i\frac{\pi}{4}), \\ (\tilde{\Theta}_\delta(7), \tilde{\Phi}_\delta(7) + \pi); & \exp(-i\frac{\pi}{4}); \end{cases}$$

$$(\pi - \tilde{\Theta}_\delta(6), 2\pi - \tilde{\Phi}_\delta(6)); \ -i \to \begin{cases} (\pi - \tilde{\Theta}_\delta(7), \pi - \tilde{\Phi}_\delta(7)); & \exp(-i\frac{\pi}{4}), \\ (\pi - \tilde{\Theta}_\delta(7), 2\pi - \tilde{\Phi}_\delta(7)); & \exp(-i\frac{3\pi}{4}); \end{cases}$$

Similar to that at $n = 3$, at $n = 7$, the points eventually emerge are $(\tilde{\Theta}_\delta(7), \tilde{\Phi}_\delta(7))$ and $(\pi - \tilde{\Theta}_\delta(7), \pi - \tilde{\Phi}_\delta(7))$. The other two vanish.
$\boldsymbol{n = 7 \to n = 8}$:

$$(\Theta(7), \Phi(7)); \exp(i\frac{\pi}{4}) \to \begin{cases} (\tilde{\Theta}_\delta(8), \tilde{\Phi}_\delta(8)); \ 1, \\ (\tilde{\Theta}_\delta(8), \tilde{\Phi}_\delta(8) + \pi); \ i; \end{cases}$$

$$(\pi - \tilde{\Theta}(7), \pi - \tilde{\Phi}(7)); \exp(-i\frac{\pi}{4}) \to \begin{cases} (\tilde{\Theta}_\delta(8), \tilde{\Phi}_\delta(8) + \pi); \ 1, \\ (\tilde{\Theta}_\delta(8), \tilde{\Phi}_\delta(8) + \pi); \ -i. \end{cases}$$

Finally, after eight kicks, the pseudoclassical dynamics brings the initial condition to $(\tilde{\Theta}_\delta(8), \tilde{\Phi}_\delta(8))$, the same as the classical map does. We thus have Equation (12) immediately.

References

1. Berry, M. Quantum chaology, not quantum chaos. *Phys. Scr.* **1989**, *40*, 335. [CrossRef]
2. Einstein, A. Zum Quantensatz von Sommerfeld und Epstein. *Verh. Dtsch. Phys. Ges.* **1917**, *19*, 82.
3. Casati, G.; Chirikov, B.V.; Izraelev, F.M.; Ford, J. Stochastic behavior of a quantum pendulum under a periodic perburbation. *Lect. Notes Phys.* **1979**, *93*, 334. [CrossRef]
4. Fishman, S.; Grempel, D.R.; Prange, R.E. Chaos, Quantum Recurrences, and Anderson Localization. *Phys. Rev. Lett.* **1982**, *49*, 509. [CrossRef]

5. Casati, G.; Guarneri, I.; Shepelyansky, D.L. Anderson transition in a one-dimensional system with three incommensurate frequencies. *Phys. Rev. Lett.* **1989**, *62*, 345. [CrossRef] [PubMed]
6. Chabé, J.; Lemarié, G.; Grémaud, B.; Delande, D.; Szriftgiser, P.; Garreau, J.-C. Experimental observation of the Anderson metal-insulator transition with atomic matter waves. *Phys. Rev. Lett.* **2008**, *101*, 255702. [CrossRef]
7. Cherroret, N.; Vermersch, B.; Garreau, J.-C.; Delande, D. How nonlinear interactions challenge the three-dimensional Anderson transition. *Phys. Rev. Lett.* **2014**, *112*, 170603. [CrossRef]
8. Chen, Y.; Tian, C. Planck's quantum-driven integer quantum Hall effect in chaos. *Phys. Rev. Lett.* **2014**, *113*, 216802. [CrossRef]
9. Tian, C.; Chen, Y.; Wang, J. Emergence of integer quantum Hall effect from chaos. *Phys. Rev. B* **2016**, *93*, 75403. [CrossRef]
10. Guarneri, I.; Tian, C.; Wang, J. Self-duality triggered dynamical transition. *Phys. Rev. B* **2020**, *102*, 45433. [CrossRef]
11. Garreau, J.-C. Quantum simulation of disordered systems with cold atoms. *C. R. Phys.* **2017**, *18*, 31. [CrossRef]
12. Haake, F.; Kuś, M.; Scharf, R. Classical and quantum chaos for a kicked top. *Z. Phys. B Condensed Matter* **1987**, *65*, 381. [CrossRef]
13. Maldacena, J.; Shenker, S.H.; Stanford, D. A bound on chaos. *J. High Energy Phys.* **2016**, *2016*, 106. [CrossRef]
14. Rozenbaum, E.B.; Ganeshan, S.; Galitski, V. Lyapunov exponent and out-of-time-ordered correlator's growth rate in a chaotic system. *Phys. Rev. Lett.* **2017**, *118*, 86801. [CrossRef]
15. Lerose, A.; Pappalardi, S. Bridging entanglement dynamics and chaos in semiclassical systems. *Phys. Rev. A* **2020**, *102*, 32404. [CrossRef]
16. Zurek, W.H.; Paz, J.P. Decoherence, chaos, and the second law. *Phys. Rev. Lett.* **1994**, *72*, 2508. [CrossRef]
17. Ruebeck, J.B.; Lin, J.; Pattanayak, A.K. Entanglement and its relationship to classical dynamics. *Phys. Rev. E* **2017**, *95*, 62222. [CrossRef]
18. Bhosale, U.T.; Santhanam, M.S. Periodicity of quantum correlations in the quantum kicked top. *Phys. Rev. E* **2018**, *98*, 52228. [CrossRef]
19. Zarum, R.; Sarkar, S. Quantum–classical correspondence of entropy contours in the transition to chaos. *Phys. Rev. E* **1998**, *57*, 5467. [CrossRef]
20. Wang, X.; Ghose, S.; Sanders, B.C.; Hu, B. Entanglement as a signature of quantum chaos. *Phys. Rev. E* **2004**, *70*, 16217. [CrossRef]
21. Longhi, S. Localization, quantum resonances, and ratchet acceleration in a periodically kicked PT-symmetric quantum rotator. *Phys. Rev. A* **2017**, *95*, 12125. [CrossRef]
22. Mudute-Ndumbe, S.; Graefe, E. A non-Hermitian PT-symmetric kicked top. *New J. Phys.* **2020**, *22*, 103011. [CrossRef]
23. Chaudhury, S.; Smith, A.; Anderson, B.E.; Ghose, S.; Jessen, P.S. Quantum signatures of chaos in a kicked top. *Nature* **2009**, *461*, 768. [CrossRef] [PubMed]
24. Izrailev, F.M.; Shepelyanskii, D.L. Quantum resonance for a rotator in a nonlinear periodic field. *Theor. Math. Phys.* **1980**, *43*, 553. [CrossRef]
25. Fishman, S.; Guarneri, I.; Rebuzzini, L. Stable quantum resonances in atom optics. *Phys. Rev. Lett.* **2002**, *89*, 84101.
26. Fishman, S.; Guarneri, I.; Rebuzzini, L. A theory for quantum accelerator modes in atom optics. *J. Stat. Phys.* **2003**, *110*, 911. [CrossRef]
27. Haake, F.; Shepelyansky, D.L. The kicked rotator as a limit of the kicked top. *Europhys. Lett.* **1988**, *5*, 671. [CrossRef]
28. Casati, G.; Chirikov, B.V. *Quantum Chaos: Between Order and Disorder*; Cambridge University Press, Cambridge, UK, 1995.
29. Agarwal, G.S. Relation between atomic coherent-state representation, state multipoles, and generalized phase-space distributions. *Phys. Rev. A* **1981**, *24*, 2889. [CrossRef]
30. Ghose, S.; Stock, R.; Jessen, P.; Lal, R.; Silberfarb, A. Chaos, entanglement, and decoherence in the quantum kicked top. *Phys. Rev. A* **2008**, *78*, 42318. [CrossRef]
31. Zou, Z.; Wang, J. The Pseudoclassical Theory for a Class of Kicked Rotors. (To be published).
32. Schomerus, H. Noisy monitored quantum dynamics of ergodic multi-qubit systems. *J. Phys. A* **2022**, *55*, 214001. [CrossRef]
33. Wang, J.; Gong, J.B. Proposal of a cold-atom realization of quantum maps with Hofstadter's butterfly spectrum. *Phys. Rev. A* **2008**, *77*, 31405(R); Erratum in *Phys. Rev. A* **2011**, *84*, 39904(E). [CrossRef]
34. Wang, J.; Gong, J.B. Butterfly Floquet spectrum in driven SU(2) systems. *Phys. Rev. Lett.* **2009**, *102*, 244102. [CrossRef] [PubMed]
35. Wang, J.; Gong, J.B. Generating a fractal butterfly Floquet spectrum in a class of driven SU(2) systems. *Phys. Rev. E* **2010**, *81*, 26204.
36. Wang, J.; Guarneri, I.; Casati, G.; Gong, J.B. Long-lasting exponential spreading in periodically driven quantum systems. *Phys. Rev. Lett.* **2011**, *107*, 234104. [CrossRef] [PubMed]
37. Fang, P.; Wang, J. Superballistic wavepacket spreading in double kicked rotors. *Sci. China Phys. Mech. Astron.* **2016**, *59*, 680011. [CrossRef]

entropy

Article

Elliptic Flowers: New Types of Dynamics to Study Classical and Quantum Chaos

Hassan Attarchi [1,*] and Leonid A. Bunimovich [2,†]

1 Department of Mathematics, University of California Riverside, Riverside, CA 92521, USA
2 School of Mathematics, Georgia Institute of Technology, Atlanta, GA 30332, USA
* Correspondence: hassan.attarchi@ucr.edu; Tel.: +1-470-553-5653
† To Giulio—A pioneer of Milano-Siberia track.

Abstract: We construct examples of billiards where two chaotic flows are moving in opposite directions around a non-chaotic core or vice versa; the dynamics in the core are chaotic but flows that are moving in opposite directions around it are non-chaotic. These examples belong to a new class of dynamical systems called elliptic flowers billiards. Such systems demonstrate a variety of new behaviors which have never been observed or predicted to exist. Elliptic flowers billiards, where a chaotic (non-chaotic) core coexists with the same (chaotic/non-chaotic) type of dynamics in flows were recently constructed. Therefore, all four possible types of coexisting dynamics in the core and tracks are detected. However, it is just the beginning of studies of elliptic flowers billiards, which have already extended the imagination of what may happen in phase spaces of nonlinear systems. We outline some further directions of investigation of elliptic flowers billiards, which may bring new insights into our understanding of classical and quantum dynamics in nonlinear systems.

Keywords: elliptic flowers billiards; chaotic and non-chaotic core; chaotic and non-chaotic tracks

Citation: Attarchi, H.; Bunimovich, L.A. Elliptic Flowers: New Types of Dynamics to Study Classical and Quantum Chaos. *Entropy* **2022**, *24*, 1223. https://doi.org/10.3390/e24091223

Academic Editor: Marko Robnik

Received: 1 August 2022
Accepted: 30 August 2022
Published: 1 September 2022

1. Introduction

The chaos theory, as well as nonlinear dynamics as a whole, was essentially built from some "simple" basic examples. The simplicity of these examples is often questionable, especially from the point of mathematicians, who still failed to investigate a fundamental Chirikov standard map, as well as some other basic examples of chaotic systems. That is why it is important, especially for physicists, to have a decent collection of completely (rigorously) studied examples that demonstrate the enormous richness of nonlinear dynamics.

Recently found elliptic flowers billiards [1] provide yet another type of behavior that not only had not been observed but even never imagined before. Indeed, there are essentially very (extremely) few nontrivial well-understood, and rigorously investigated examples with the coexistence of several ergodic components in the phase space of a system under study. In elliptic flowers (EF) billiards, there are three ergodic components. One of them is called a core. The two other ergodic components consist of orbits that move around the core clockwise and counter-clockwise, respectively. Therefore, EF billiards have much richer and more interesting dynamics than track billiards [2], where the entire phase space consists of two tracks, and orbits move in opposite directions to each other.

A large class of EF billiards was constructed in [1], and it was shown that in some simple sub-classes of EF, the dynamics in the core and tracks are chaotic, or the dynamics in all these three parts of the phase space can also be non-chaotic. By chaotic, we mean dynamics with an exponential divergence of nearby orbits and mixing (i.e., decay of correlations).

In the present paper, we give examples of EF billiards with opposite types of dynamics in tracks and the core. In other words, in these examples, a chaotic core coexists with non-chaotic tracks and vice versa. Therefore, all four possible types of coexistence of dynamics in the core and tracks are possible.

It is worth mentioning that EF billiards can have even more rich behavior. Particularly, EF billiards demonstrate a complicated evolution of the wavefronts (for example, see the videos by Nils Berglund on YouTube (https://www.youtube.com/c/nilsberglund) 7 June, 9 June, 11 June and 25 June 2021) and certainly could be used for studies of various types of the evolution of such waves (and wavefronts) with time in a lesson of dynamics. Another natural direction of studies is the analysis of global bifurcations that occur under a change of parameters as well as of local bifurcations of wavefronts. In addition, it turned out that track billiards are useful for some problems in quantum chaos, for example, in the studies of the Schnirelman's peak [3]. Further, a class of billiards that is more general and abstract than tracks was considered in [4,5] to study quantum ergodicity. Thus, it seems reasonable to assume that EF billiards will help to understand something new about quantum chaos, especially in relation to the Berry–Robnik theory.

2. Construction/Definition of Elliptic Flowers Billiards

The general construction of a (multilayered) elliptic flower billiard (EF-billiard) starts with a choice of any convex (base) polygon A on the Euclidean plane. Let $A_1, A_2, \ldots, A_n$ be the vertices of A. The first elliptic layer over A is formed by all arcs of ellipses, which do not intersect A and have focuses at the points A_i and A_{i+1}, where i varies between 1 and n, and $i+1$ is taken as the modulus n. In other words, the first layer of ellipses over polygon A consists of all ellipses with focuses at the ends of one and the same side of A. The second layer of ellipses over A is a union of all ellipses that have focuses at the points A_i and A_{i+2}, where again $i+2$ is taken $\mod n$. Thus the second layer of ellipses over polygon A consists of all ellipses with focuses at the ends of the small diagonals of A, which connect the ends of two neighboring sides of A.

Analogously, the mth layer of ellipses, where $m < \frac{n+1}{2}$, over a given polygon is defined in the same way, where instead of $(i+2) \mod n$, one should take $(i+m) \mod n$. Therefore, all ellipses in all layers have a focus at some vertex of A. It is easy to see that a convex polygon with n vertices has $\frac{n}{2}$ layers if n is an even number, and $\frac{n-1}{2}$ layers if n is odd. If a base polygon A is a triangle, then the sides also play the role of small diagonals. Therefore, in this case, there is only one layer.

Definition 1 ([1]). *A simply connected billiard table $Q(A)$ is called an elliptic flower over a convex base polygon A if the following conditions are satisfied:*

1. *A is a subset of the billiard table, which does not intersect the boundary $\partial Q(A)$.*
2. *the boundary $\partial Q(A)$ consists of pieces of ellipses that belong to the layers of ellipses over A.*

We say that an invariant subset of an EF billiard is a track if it consists of all orbits that move clockwise or counter-clockwise in a billiard table of the corresponding EF billiard. Clearly, if there is a clockwise track, then also exists a counter-clockwise track. The orbits in tracks move around some subset Co of the EF billiard table, and they never intersect it. A core consists of all orbits that intersect Co between any two consecutive reflections off the boundary of an EF billiard table. Clearly, if A is a triangle or a rectangle, then the lines passing through the sides of A never intersect outside this polygon. It is easy to see [1] that interesting dynamics of elliptic flowers billiards appear when they are built over convex polygons with at least five sides. The following definition singles out a special subclass of EF billiards.

Definition 2 ([1]). *Consider a base polygon A with the vertices $A_1, A_2, \ldots, A_n$. A simply connected billiard table Q is called a structural elliptic flower (SEF) over A if all regular components of the boundary ∂Q are the arcs of the ellipses from the layers over A, which are not allowed to cross the lines passing through the sides of A, but they may have their endpoints on these lines.*

Clearly, not all boundary components of general EF billiards may have their endpoints on the straight lines passing through the sides of A. However, here for the sake of simplicity, we are considering much more narrow classes of EF, for which, particularly, the construction of elliptic flowers over a base polygon A becomes trivial. A natural difficulty in the construction of general-type elliptic flowers is how to finish (close) the boundary by the last elliptic arc when all other regular components of the boundary were already built. Such an ellipse, which passes through two fixed points of the boundary, and has focuses at some vertices of A, may just not exist.

Although we expect that general elliptic flowers billiards may well demonstrate a large variety of dynamics, our concern here is to present the (hopefully) simplest examples of EF billiards with interesting nontrivial dynamics, which, to the best of our knowledge, have never been seen before.

We will start with considering an even simpler class of EF billiards, which are called the special one-layer (SOL) elliptic flowers [1]. It was proven in [1] that in some classes of SOL EF billiards a chaotic core coexists with chaotic tracks, and in some other classes of such billiards, a non-chaotic core coexists with non-chaotic tracks.

To construct EF billiards, where the coexisting core and tracks demonstrate not similar but different types of dynamics, which is a goal of the present paper, we need to move beyond SOL elliptic flowers billiards. However, our examples will be obtained as some kinds of perturbations of the SOL EF billiards.

Construction of special one-layer elliptic flowers: Let A be a regular convex polygon with n vertices. Consider the straight semi-lines, which have the ends at the centers of the sides of A, are orthogonal to the corresponding sides and do not intersect A. Then a special one-layer elliptic flower over A is a billiard table with n identical boundary components, which are the arcs of ellipses with the focuses located at the ends of the small diagonals of A. Moreover, the endpoints of each elliptic arc lie on two semi-lines orthogonal to a pair of neighboring sides of A, i.e., the sides of A that have a common vertex (See Figure 1).

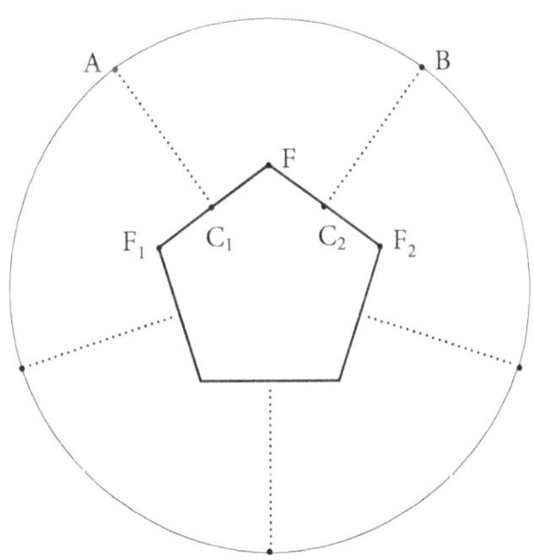

Figure 1. Demonstration of an SOL EF on a pentagon. The boundary component AB is a piece of an ellipse with focuses at F_1 and F_2, where they are endpoints of a small diagonal of the pentagon. Note that, in the SOL EF structure, line segments C_1A and C_2B are perpendicular bisectors of the sides F_1F and FF_2, respectively.

3. Non-Chaotic Core and Chaotic Tracks

In this section, we construct examples of elliptic flowers billiards, such that the dynamics in the core are non-chaotic while the dynamics in the tracks are chaotic.

Consider a special one-layer (SOL) elliptic flower built over a regular hexagon. Let a, b, and c denote the lengths of the semi-major axis, semi-minor axis, and linear eccentricity of the ellipses, containing the boundary components, respectively. In this EF structure, if we assume that the lengths of the hexagon's sides equal 2, then $c = \sqrt{3}$. If $b = 3$, then the centers of all maximal osculating circles of the boundary components coincide with the center of the hexagon. Recall that an osculating circle to a curve γ at a point x is the circle tangent to γ at x and, the radius of this circle is $1/|\kappa(x)|$, where $\kappa(x)$ is the curvature of γ at point x. Clearly, the maximal osculating circles of an ellipse are tangent to the boundary at the endpoints of the minor axis. Let b be (slightly) greater than 3. This assumption guarantees that the core and tracks are chaotic (see Theorem 3.9 in [1]).

To get a non-chaotic core, we change one of the boundary components of the EF billiard by moving its major axis toward the center O of the hexagon. In Figure 2, component AB has focal points F_1 and F_2 instead of endpoints of its corresponding short diagonal in the usual SOL EF structure.

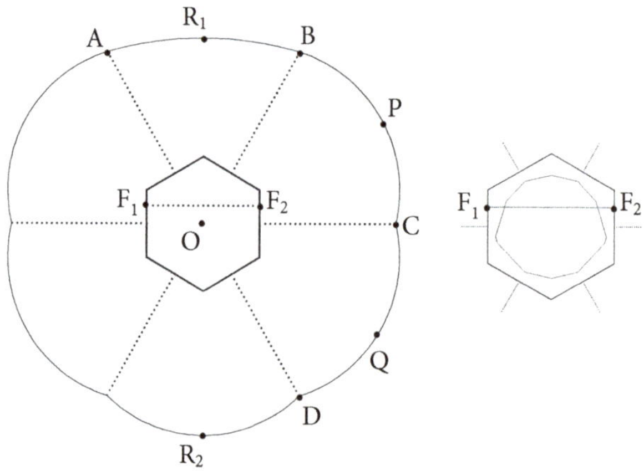

Figure 2. Deformed SOL EF on a hexagon (**left**), and its core (**right**).

Let a_1 and b_1 be the lengths of a semi-major and a semi-minor axis of AB, respectively. Note that the linear eccentricity of AB remains equal to $c = \sqrt{3}$. Now, we fix parameter b, and check for which range of parameter b_1 the 2-periodic orbit R_1R_2 is stable (Figure 2), while all periodic orbits in the tracks are unstable.

Consider a Cartesian coordinate with the center at O. Then

$$F_1 = (-\sqrt{3}, y_0), \text{ and } F_2 = (\sqrt{3}, y_0),$$

where $0 \leq y_0 < 1$. By fixing $b = 4$, we also fix the points A and B in our coordinate system. Therefore, we can express a_1 (also, b_1) as a function of y_0, where

$$2a_1 = |AF_1| + |AF_2|. \tag{1}$$

Under these assumptions, the distance between R_1 and R_2 is given by

$$|R_1R_2| = (b+1) + (b_1 + y_0) = 5 + b_1 + y_0. \tag{2}$$

Let r_1 and r_2 be the radii of osculating circles at R_1 and R_2, respectively. Then,

$$r_1 = \frac{a_1^2}{b_1} = b_1 + \frac{c^2}{b_1} = b_1 + \frac{3}{b_1}, \qquad r_2 = \frac{a^2}{b} = b + \frac{c^2}{b} = 4.75\,. \tag{3}$$

Let $y_0 = 1$ in (1). Then we have $a_1 = a$. Further, it is a direct consequence of the geometry of the deformed SOL EF that $a_1 > a$, when $0 \leq y_0 < 1$. Therefore, $b_1 > b = 4$ if $0 \leq y_0 < 1$. Hence, (2) and (3) imply that

$$|R_1R_2| > r_1, \qquad |R_1R_2| > r_2. \tag{4}$$

The Jacobians of the billiard maps $T_{R_1R_2}$ and $T_{R_2R_1}$, corresponding to the trajectories from R_1 to R_2 and vice versa, are equal to [6]:

$$dT_{R_1R_2} = \begin{bmatrix} |R_1R_2|/r_1 - 1 & -|R_1R_2| \\ -|R_1R_2|/(r_1r_2) + 1/r_1 + 1/r_2 & |R_1R_2|/r_2 - 1 \end{bmatrix},$$

and

$$dT_{R_2R_1} = \begin{bmatrix} |R_1R_2|/r_2 - 1 & -|R_1R_2| \\ -|R_1R_2|/(r_1r_2) + 1/r_1 + 1/r_2 & |R_1R_2|/r_1 - 1 \end{bmatrix}.$$

Therefore,

$$trace(dT_{R_1R_2}dT_{R_2R_1}) = 2 - 4\left(\frac{|R_1R_2|}{r_1} + \frac{|R_1R_2|}{r_2} - \frac{|R_1R_2|^2}{r_1r_2}\right).$$

For all values of $y_0 \in [0,\ 0.414569)$, we obtain

$$|trace(dT_{R_1R_2}dT_{R_2R_1})| < 2, \tag{5}$$

and

$$r_1 + r_2 > |R_1R_2|. \tag{6}$$

Inequality (5) shows that the 2-periodic orbit between R_1 and R_2 is linearly stable. In order to demonstrate that this 2-periodic orbit is, in fact, elliptic (nonlinearly stable), we must check that the first Birkhoff coefficient is nonzero.

In the ellipse $E = \{(a\cos\theta,\ b\sin\theta) \mid 0 \le \theta < 2\pi\}$, the radius of curvature $r(\theta)$ can be expressed as

$$r(\theta) = \frac{(\sqrt{a^2\sin^2\theta + b^2\cos^2\theta})^3}{ab}.$$

Let s be the arc length parameter of the ellipse E. Then

$$\frac{dr}{ds}(\theta) = \frac{3(a^2-b^2)\sin\theta\cos\theta}{ab}, \tag{7}$$

and

$$\frac{d^2r}{ds^2}(\theta) = \frac{3(a^2-b^2)(\cos^2\theta - \sin^2\theta)}{ab\sqrt{a^2\sin^2\theta + b^2\cos^2\theta}}. \tag{8}$$

Let r_1' and r_1'' (also, r_2' and r_2'') be the first and second derivatives of the radius of curvature at the point R_1 (also, R_2), respectively. From Equations (7) and (8), we have

$$r_1' = r_2' = 0, \qquad r_1'' = \frac{-3(a_1^2 - b_1^2)}{a_1^2 b_1}, \qquad r_2'' = \frac{-3(a^2-b^2)}{a^2 b}. \tag{9}$$

Denote by τ_1 the first Birkhoff coefficient (see [7] for the explicit formula for τ_1 corresponding to an elliptic 2-periodic orbit) of the 2-periodic orbit between R_1 and R_2. Then, by making use of (3) and (9), we obtain that

$$\tau_1 = -\frac{1}{8}\frac{r_1+r_2}{r_1r_2} - \frac{1}{8}\frac{|R_1R_2|}{|R_1R_2| - r_1 - r_2}\left(\frac{|R_1R_2| - r_1}{|R_1R_2| - r_2}r_2'' + \frac{|R_1R_2| - r_2}{|R_1R_2| - r_1}r_1''\right). \tag{10}$$

It follows from (3), (4), (6), and (9) that τ_1 is nonzero because both terms on the right-hand side of (10) are strictly negative. Therefore, the 2-periodic orbit R_1R_2 is elliptic (nonlinearly stable). Hence, the global dynamics in the core are non-chaotic if $y_0 \in [0, 0.414569)$ because an elliptic island with a positive measure exists. One should expect, though, that there are some subsets of the core where dynamics are chaotic.

We will check for what subrange of parameter y_0, with the same choice of other parameters, the dynamics in tracks is chaotic. First, we check that all boundary components are absolutely focusing curves. The boundary consists of six pieces of ellipses. Five of them have the parameters $b = 4$, $c = \sqrt{3}$, and $a = \sqrt{19}$. Let the projections of these five pieces to the corresponding major axis of the ellipses have the length L. Then

$$L < 2(b+1)\tan\frac{\pi}{6} = \frac{10}{\sqrt{3}}.$$

It implies that

$$L < \frac{10}{\sqrt{3}} < \sqrt{38} = a\sqrt{2}. \quad (11)$$

It was proven in [8] that a piece of an ellipse (symmetric with respect to its minor axis) is absolutely focusing if its projection to the major axis has a length less than $a\sqrt{2}$. Thus, (11) shows that the similar five pieces of ellipses are absolutely focusing. The projection of the boundary component AB to the major axis of the corresponding ellipse also has the length L. Indeed, it immediately follows from our construction, since A and B are endpoints of the boundary components in a SOL EF. Because $a_1 > \sqrt{19} = a$, the boundary component AB is also absolutely focusing (i.e., $L < a_1\sqrt{2}$). Thus, all six boundary components are absolutely focusing in this deformed SOL EF. Moreover, the angles between adjacent (focusing) boundary components are greater than π. Because of these facts, we only need to show that the boundary components are sufficiently far apart from each other, which will imply that the dynamics in the tracks are chaotic.

On the other hand, if $y_0 \in [0.40357, 0.414569)$, then

$$r_1\cos\phi_1 + r_2\cos\phi_2 < |R_1P|, \quad (12)$$

and

$$r_1\cos\phi_3 + r_2\cos\phi_4 < |R_1Q|, \quad (13)$$

where P and Q are points at the end of minor axes of components BC and CD, respectively, and

- ϕ_1: the angle between the normal line at R_1 and the line segment R_1P,
- ϕ_2: the angle between the normal line at P and the line segment R_1P,
- ϕ_3: the angle between the normal line at R_1 and the line segment R_1Q,
- ϕ_4: the angle between the normal line at Q and the line segment R_1Q.

Inequalities (12) and (13) imply that the tangent circles of radius $\frac{r_1}{2}$ and $\frac{r_2}{2}$ at R_1 and P (or Q), respectively, do not cover the entire segment R_1P (or R_1Q). Therefore, the boundary components are sufficiently far apart from each other to have hyperbolicity (generated by the defocusing mechanism) of trajectories traveling between them [8–10]. We also showed before that boundary components are absolutely focusing. Therefore, it follows from the general theory of billiards [6,9,10] that the dynamics in tracks are hyperbolic and chaotic thanks to the mechanism of defocusing.

4. Chaotic Core and Non-Chaotic Tracks

In this section, we construct examples of elliptical flowers with a chaotic core and non-chaotic tracks.

Consider a SOL EF billiard table built over a regular pentagon. Let a, b, and c denote the semi-major axis, semi-minor axis, and linear eccentricity of the elliptical components of the boundary, respectively. Let the lengths of the sides of the pentagon equal 2. Then $c = \cos\frac{\pi}{5}$. Moreover, if we choose

$$b = c\tan\frac{2\pi}{5} = \cos\frac{\pi}{5}\tan\frac{2\pi}{5} \simeq 2.489898,$$

then the centers of the osculating circles with the maximum radius of all regular components of the boundary coincide with the center of the pentagon. Choose now b to be slightly smaller than $c \tan \frac{2\pi}{5} \simeq 2.489898$. More precisely, we assume that

$$2.484 < b < 2.489898. \tag{14}$$

Under this assumption, we will show that the (pentagon shape) 5-periodic orbit $ABCDE$ (Figure 3) is a linearly stable orbit in a track. Clearly, then the 5-periodic orbit, which traces the same points in the opposite direction, is also linearly stable. We have

$$\tau := |AB| = |BC| = |CD| = |DE| = |EA| = 2\left(b + \cos\frac{\pi}{5}\tan\frac{\pi}{10}\right)\cos\frac{3\pi}{10}.$$

Moreover, the angles of reflection φ and the curvatures κ of the boundary at points A, B, C, D, and E are equal to $\frac{3\pi}{10}$ and $\frac{b}{a^2}$, respectively.

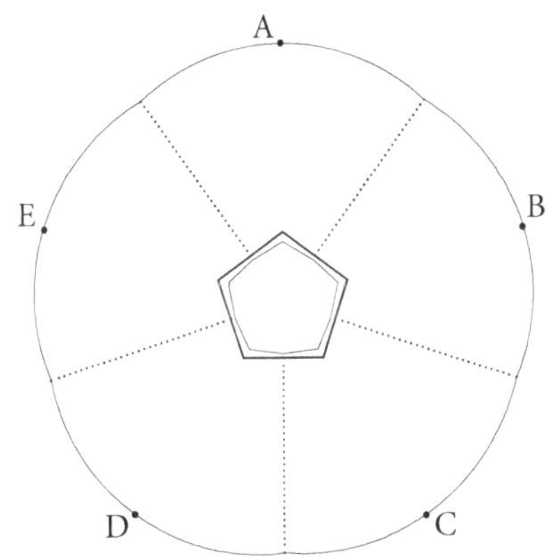

Figure 3. A SOL EF on a pentagon, and its actual core is a decagon.

The jacobians of the billiard map corresponding to consecutive reflections along this periodic orbit are equal to

$$dT = \begin{bmatrix} X & Y \\ Z & X \end{bmatrix} = \frac{-1}{\cos\varphi}\begin{bmatrix} -\tau\kappa + \cos\varphi & \tau \\ \tau\kappa^2 - 2\kappa\cos\varphi & -\tau\kappa + \cos\varphi \end{bmatrix},$$

and

$$trace((dT)^5) = 2X^5 + 10XY^2Z^2 + 20X^3YZ.$$

Under the assumption (14), we obtain that

$$|trace((dT)^5)| < 2.$$

Hence, if b is slightly smaller than $c \tan \frac{2\pi}{5}$, then there are linearly stable periodic orbits in the tracks. Therefore, the dynamics in the tracks are not globally chaotic, although generically, some subset(s) with chaotic dynamics (chaotic seas) should be present there. However, at this stage of analysis of EF billiards, we are only concerned with global chaos (or with the absence of global chaos) in the tracks and in the core.

We will show now that the dynamics in the core (which is a decagon in this EF billiard) are chaotic. The way how a decagon core appears from a EF billiard table, with a pentagon as a base polygon, is demonstrated in Figure 4 (this process is described in detail in [1]). The reason is that the lines connecting the ends of the boundary components to the corner points of the pentagon (as point K is connected to point L in Figure 4) cut out pieces of the base pentagon. The remaining subset of the pentagon, which forms a core of the EF billiard, is a decagon.

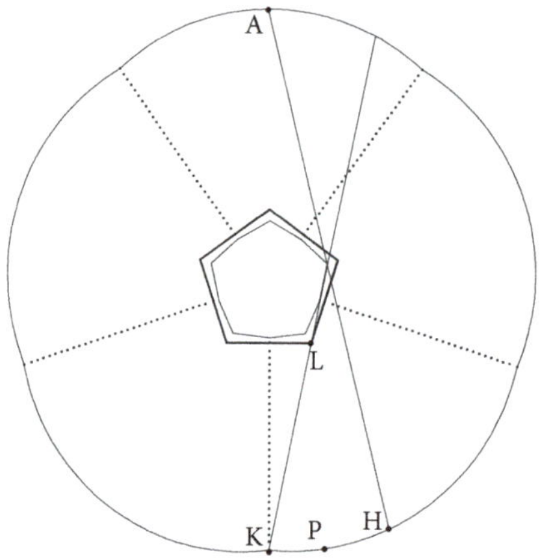

Figure 4. Continuation of all segments similar to KL cuts out from the pentagon a smaller polygon, which becomes a core.

Let point A, see Figure 4, be the endpoint of the minor axis of a boundary component of this billiard table. All straight lines, which start at point A and pass through the core (decagon) form a cone with vertex A, where AH belongs to the boundary of this cone (Figure 4). A simple computation for any point such as P on the boundary between H and K shows that

$$r_A \cos \phi_A + r_P \cos \phi_P < |AP|, \tag{15}$$

where r_A and r_P are radii of curvature at the points A and P, respectively, and

- ϕ_A: the angle between the normal line at A and the line segment AP,
- ϕ_P: the angle between the normal line at P and the line segment AP.

Inequality (15) implies that the (osculating) tangent circles with the radii $\frac{r_A}{2}$ and $\frac{r_P}{2}$ at A and P, respectively, do not cover segment AP. This observation, together with the fact that trajectories passing through the core cannot have two consecutive reflections off one and the same boundary component, prove that all orbits in the core are hyperbolic (due to the defocusing mechanism). Therefore, the dynamics in the core are hyperbolic and chaotic.

5. Concluding Remarks

This paper, together with [1], shows that all four possible types of coexistence of dynamics in the core and in tracks are realized in the elliptic flowers billiards. Therefore, dynamics could be chaotic in the core and tracks, it could be chaotic in the core and non-chaotic in tracks, it could be non-chaotic in the core and chaotic in tracks, and, finally, the dynamics could be non-chaotic both in the core and in tracks.

However, these results just scratch the surface of this new area. General multilayered elliptic flowers billiards, especially the ones built over non-regular polygons, would likely demonstrate a more rich variety of behaviors. The analysis of standard (without a core) track billiards allowed some results in classical and quantum chaos to be obtained [3,11]. There is little doubt that studies of general EF billiards will bring something new, interesting, and surprising. A great challenge is to construct such billiards (or other dynamical systems similar to EF billiards dynamics) in dimensions greater than two. It has actually been performed for track billiards [2]. However, the construction of high-dimensional elliptic flowers seems to be a real challenge.

Author Contributions: Formal analysis, H.A.; Supervision, L.A.B. All authors have read and agreed to the published version of the manuscript.

Funding: This research received no external funding.

Institutional Review Board Statement: Not applicable.

Informed Consent Statement: Not applicable.

Data Availability Statement: Not applicable.

Conflicts of Interest: The authors declare no conflict of interest.

References

1. Bunimovich, L.A. Elliptic flowers: Simply connected billiard tables where chaotic (non-chaotic) flows move around chaotic (non-chaotic) cores. *Nonlinearity* **2022**, *35*, 3245–3260. [CrossRef]
2. Bunimovich, L.A.; Del Magno, G. Track Billiards. *Commun. Math. Phys.* **2009**, *288*, 699–713. [CrossRef]
3. Veble, G.; Prosen, T.; Robnik, M. Expanded boundary integral method and chaotic time-reversal doublets in quantum billiards. *New J. Phys.* **2007**, *9*, 15–27. [CrossRef]
4. Gutkin, B. Dynamical 'breaking' of time reversal symmetry. *J. Phys. A Math. Theor.* **2007**, *40*, F761. [CrossRef]
5. Gutkin, B. Note on Converse Quantum Ergodicity. *Proc. Am. Math. Soc.* **2009**, *137*, 2795–2800. [CrossRef]
6. Chernov, N.; Markarian, R. Chaotic Billiards. In *Mathematical Surveys and Ponographs*; AMS: Providence, RI, USA, 2006; Volume 127.
7. Kamphorst, S.O.; Pinto-de-Carvalho, S. The First Birkhoff coefficient and the stability of 2-periodic orbits on billiards. *Exp. Math.* **2005**, *14*, 299–306. [CrossRef]
8. Wojtkowski, M. Principles for the Design of Billiards with Nonvanishing Lyapunov Exponents. *Commun. Math. Phys.* **1986**, *105*, 391–414. [CrossRef]
9. Bunimovich, L.A. On absolutely focusing mirrors. In *Ergodic Theory and Related Topics III*; Springer: Berlin/Heidelberg, Germany, 1992; pp. 62–82.
10. Bunimovich, L.A. Mechanisms of chaos in billiards: dispersing, defocusing and nothing else. *Nonlinearity* **2018**, *31*, R78–R92. [CrossRef]
11. Horvat, M.; Prosen, T. Uni-directional transport properties of a serpent billiard. *J. Phys. A Math. Gen.* **2004**, *37*, 3133–3145. [CrossRef]

entropy

MDPI

Article

Quantum Chaos in the Extended Dicke Model

Qian Wang [1,2]

[1] CAMTP-Center for Applied Mathematics and Theoretical Physics, University of Maribor, SI-2000 Maribor, Slovenia; qwang@zjnu.edu.cn
[2] Department of Physics, Zhejiang Normal University, Jinhua 321004, China

Abstract: We systematically study the chaotic signatures in a quantum many-body system consisting of an ensemble of interacting two-level atoms coupled to a single-mode bosonic field, the so-called extended Dicke model. The presence of the atom–atom interaction also leads us to explore how the atomic interaction affects the chaotic characters of the model. By analyzing the energy spectral statistics and the structure of eigenstates, we reveal the quantum signatures of chaos in the model and discuss the effect of the atomic interaction. We also investigate the dependence of the boundary of chaos extracted from both eigenvalue-based and eigenstate-based indicators on the atomic interaction. We show that the impact of the atomic interaction on the spectral statistics is stronger than on the structure of eigenstates. Qualitatively, the integrablity-to-chaos transition found in the Dicke model is amplified when the interatomic interaction in the extended Dicke model is switched on.

Keywords: quantum chaos; extended Dicke model; spectral statistics; eigenstate structure

Citation: Wang, Q. Quantum Chaos in the Extended Dicke Model. *Entropy* **2022**, *24*, 1415. https://doi.org/10.3390/e24101415

Academic Editor: Lawrence Horwitz

Received: 15 September 2022
Accepted: 30 September 2022
Published: 4 October 2022

1. Introduction

In recent years, the study of quantum chaos in many-body systems has attracted much attention, both theoretically and experimentally in different fields of physics, such as statistical physics [1–5], condensed matter physics [6–13], and high-energy physics [14–19], as well as quantum information science [20–26]. To some extent, this great interest in quantum many-body chaos is due to the close connections of chaos to several fundamental questions that arise in current theoretical and experimental studies. Although a full understanding of quantum many-body chaos is still lacking, much progress has been achieved. It is known that chaos in interacting quantum many-body systems can lead to thermalization [1–3], the fast scrambling of quantum information [14,27–29], and an exponential growth of quantum complexities [18,30–35], and diffusive transport [36].

The notion of chaos in the classical regime is usually defined by the so-called butterfly effect, namely the exponential separation of infinitesimally nearby trajectories for initial perturbations [37,38]. However, as the concept of trajectory is ill-defined in quantum mechanics, the definition of quantum chaos remains an open question. Therefore, probing the signatures of chaos in quantum many-body systems becomes a central task in the studies of quantum many-body chaos. To date, many complementary detectors of quantum chaos and the limits of their usefulness have been widely investigated in the literature [28,31–33,39–54]. Important model systems in this context are billiards [40,55]. Another task, which has recently also drawn great interest, is to unveil different factors that affect the chaotic properties of quantum many-body systems. While the impacts of the strength of disorder and the choice of initial states on the development of quantum chaos in various many-body systems have been extensively explored [4,56–61], more works are required in order to obtain deeper insights into the universal aspects of quantum many-body chaos.

In the present work, we analyze the emergence of chaos in the extended Dicke model. There are several different versions of the extended Dicke model [62–65]. Here, we focus on the one that has been discussed in ref. [63]. Different from the original Dicke model [66], which consists of an ensemble of noninteracting two-level atoms interacting with a single bosonic mode, the atoms in our considered extended Dicke model are permitted to interact. This allows us to analyze the effects of the atomic interaction on the degree of chaos of the model. Previously, the role of the atom–field coupling in the Dicke model for the emergence of quantum chaos has been investigated [67–70], while here we explore how this transition is affected by additional atomic interaction. By performing a detailed analysis of the energy spectral statistics and the structure of eigenstates, we systematically study both the chaotic signatures of the extended Dicke model and examine the effect of the atomic interaction on the chaotic features in the model. We show how the atomic interaction affects the spectral statistics and the structure of eigenstates, respectively.

The article is structured as follows. The model is introduced in Section 2. The influences of the atomic interaction on the energy spectral statistics is discussed in Section 3. A detailed investigation of the consequence of the atomic interaction on the structure of eigenstates is presented in Section 4. Finally, we conclude in Section 5 with a brief summary of our results and outlook.

2. Extended Dicke Model

As an extension of the original Dicke model [66–68], the extended Dicke model studied here consists of N mutual interacting two-level atoms with energy gap ω_0 coupled to a single cavity mode with frequency ω. By employing the collective spin operators $J_{x,y,z} = \sum_{i=1}^{N} \hat{\sigma}_{x,y,z}^{(i)}$ ($\hat{\sigma}_{x,y,z}$ are the Pauli matrices), the Hamiltonian of the extended Dicke model can be written as (hereafter, we set $\hbar = 1$) [63,71]

$$H = \omega a^{\dagger} a + \omega_0 J_z + \frac{2\lambda}{\sqrt{N}} J_x (a + a^{\dagger}) + \frac{\kappa}{N} J_z^2, \tag{1}$$

where $a(a^{\dagger})$ denotes the bosonic annihilation (creation) operator, λ is the coupling strength between atom and field, and κ represents the strength of the atomic interaction.

The conservation of total spin operator $\mathbf{J}^2 = J_x^2 + J_y^2 + J_z^2$ for the Hamiltonian (1) leads to the Hamiltonian matrix being block diagonal in $\mathbf{J}^2$ representation. In this work, we focus on the maximum spin sector $j = N/2$, which involves the experimental realizations and includes the ground state. Moreover, the commutation between Hamiltonian (1) and the parity operator $\Pi = e^{i\pi(j + J_z + a^{\dagger} a)}$ enables us to further separate the Hamiltonian matrix into even- and odd-parity subspaces. Here, we restrict our study to the even-parity subspace.

To numerically diagonalize the Hamiltonian (1), we work in the usual Fock–Dicke basis $\{|n, m\rangle\} = \{|n\rangle \otimes |j, m\rangle\}$. Here, $|n\rangle$ are the Fock states of bosonic mode with $n = 0, 1, 2, \ldots, \infty$, and $|j, m\rangle$ represent the so-called Dicke states with $m = -j, -j+1, \ldots, j$. Then, the elements of the Hamiltonian matrix are given by

$$\begin{aligned} \langle n', m' | H | n, m \rangle = & (n\omega + m\omega_0)\delta_{n',n}\delta_{m',m} + \frac{\kappa}{N} m^2 \delta_{m',m} + \frac{\lambda}{\sqrt{N}} \left[\sqrt{n}\delta_{n',n-1} + \sqrt{n+1}\delta_{n',n+1} \right] \\ & \times \left[\sqrt{j(j+1) - m(m+1)}\delta_{m',m+1} + \sqrt{j(j+1) - m(m-1)}\delta_{m',m-1} \right]. \end{aligned} \tag{2}$$

We remark that the value of n is unbounded from above, the actual dimension of the Hilbert space is infinite, regardless of the value of j. In practice, we need to cut off the bosonic number states at a larger but finite value $\mathcal{N}_c$. Moreover, the dependence of the chaoticity in the Dicke model on the energy [63,72,73] further implies that it is also necessary to cut off the energy in order to obtain the finite number of considered states. In our numerical simulations, we set $\mathcal{N}_c = 320$ and restrict our analysis to the eigenstates with energies $E/N \in [0.4, 4]$; the convergence of our results is carefully examined. For our selected energy interval, we checked that our main results still hold for other choices of $\mathcal{N}_c$, as long as $N \geq 16$ and the convergence of the Fock–Dicke basis is fulfilled.

The extended Dicke model exhibits both ground-state and excited-state quantum phase transitions and displays a transition from integrable to chaotic behavior with increasing system energy, such as in the original Dicke model. The features of these transitions have been extensively investigated in the semiclassical regime [63]. It is worth mentioning that several possible experimental realizations of the extended Dicke model have been pointed out in [63,71,74].

3. Energy Spectrum Statistics

In this section, we explore the transition from integrability to chaos in the extended Dicke model by analyzing its energy-level spacing distribution. In our study, we focus on the energy levels with energies changing from $E/N = 0.4$ to $E/N = 4$. We compare our results to the level distributions of fully integrable and chaotic cases, respectively. We are aiming to characterize the quantum signatures of chaos in the model and unveil the impact of the atomic interaction on its chaotic properties.

3.1. Level Spacing Statistics

As the most frequently used probe of quantum chaos, the distribution $P(s)$ of the spacings s of the consecutive unfolded energy levels quantifies the degree of correlations between levels. For integrable systems, where the energy levels are allowed to cross, the distribution $P(s)$ is given by the Poissonian distribution [75],

$$P_P(s) = \exp(-s). \tag{3}$$

On the other hand, the energy levels in chaotic systems exhibit level repulsion, and the distribution $P(s)$ follows the Wigner–Dyson distribution [39–41]. For the systems with symmetric and real Hamiltonian matrices, as in the extended Dicke model, the Wigner–Dyson distribution has the following expression:

$$P_{WD}(s) = \frac{\pi s}{2} \exp\left(-\frac{\pi}{4}s^2\right). \tag{4}$$

In Figure 1, we show $P(s)$ of the extended Dicke model with $\omega = \omega_0 = 1$ and $j = N/2 = 16$ for different values of λ and κ. Here, the level spacings s are obtained from the unfolded eigenlevels $\tilde{E}_\mu = E_\mu/\Delta E$, with $1/\Delta E$ being the local density of states. One can clearly see that $P(s)$ undergoes a transition to Wigner–Dyson distribution as λ increases, regardless of the value of κ. The case $\kappa = 0$ is the original Dicke model as in our previous work [68]. However, we find that, with increasing κ, the Poissonian distribution at $\lambda = 0.1$, and $\kappa = 0$ in Figure 1a turns into an intermediate case, as observed in Figure 1g. This suggests that the degree of chaos in the extended Dicke model can be tuned by the atomic interaction.

To quantitatively characterize the effect of the atomic interaction on the degree of chaos, we consider the proximity of $P(s)$ to Wigner–Dyson or to Poissonian distributions. There are several ways to measure the distance between two distributions, here, the difference between $P(s)$ and $P_{WD}(s)[P_P(s)]$ is quantified by the chaos indicators η and β. The indicator η is defined as [7,67,76]

$$\eta = \left| \frac{\int_0^{s_0}[P(s) - P_{WD}(s)]ds}{\int_0^{s_0}[P_P(s) - P_{WD}(s)]ds} \right|, \tag{5}$$

where $s_0 = 0.4729\ldots$ is the first intersection point of $P_{WD}(s)$ and $P_P(s)$. For $P(s) = P_{WD}$, we have $\eta = 0$, while $P(s) = P_P(s)$ leads to $\eta = 1$. The indicator β is the level repulsion exponent and can be obtained by fitting $P(s)$ to the Brody distribution [77]

$$P_B(s) = b_\beta(\beta+1)s^\beta \exp[-b_\beta s^{\beta+1}], \tag{6}$$

where the factor b_β is given by

$$b_\beta = \left[\Gamma\left(\frac{\beta+2}{\beta+1}\right)\right]^{\beta+1}, \tag{7}$$

with $\Gamma(x)$ being the gamma function. The value of β varies in the interval $\beta \in [0,1]$. When $\beta = 0$, it means the level spacing distribution $P(s)$ is Poisson. On the other hand, for chaotic systems, we would expect $P(s) = P_{WD}(s)$, and therefore $\beta = 1$.

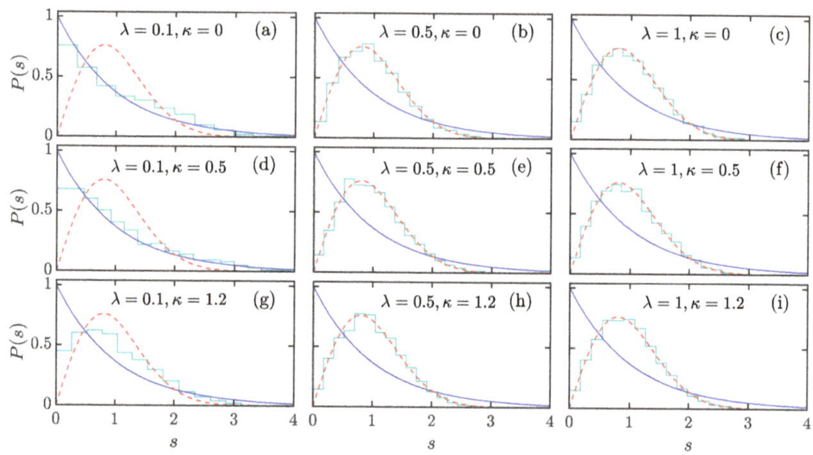

Figure 1. Level spacing distribution of the extended Dicke model for several combinations of λ and κ, see the legends in panels (**a**–**i**). The considered energy levels are the ones that have energies $E/N \in [0.4, 4]$. The total atom number is $N = 2j = 32$, and the cut off in bosinic Hilbert space is $\mathcal{N}_c = 320$. The Poissonian (blue solid lines) and Wigner–Dyson distributions (red dashed lines) are, respectively, plotted in each panel for comparison. Other parameters are: $\omega = \omega_0 = 1$. All quantities are dimensionless.

In Figure 2a, we plot η as a function of λ for various values κ. It can be seen that irrespective of the strength of the atomic interaction, the extended Dicke model undergoes a transition from integrability to chaos as the coupling strength λ increases. However, as the strength of the atomic interaction increases, the onset of chaos happens for smaller values of coupling strength λ. This is more evident from Figure 2b, where we show how η evolves as a function κ and λ. Clearly, the width of the region with larger values of η decreases with increasing κ, suggesting that the location of the crossover to quantum chaos moves towards the lower values of λ with increasing κ. The statement above is further confirmed by the boundary of chaotic region, which is plotted as the black curve in Figure 2b. Here, we determine the boundary of chaos by the condition $\eta \leq \eta_d = 0.3$ [76]. We set the threshold $\eta_d = 0.3$ as it implies that the model has already departed from the integrablity and is tending to the chaotic regime.

Figure 2c shows β as a function of λ with increasing κ. As observed in Figure 2a, while the chaotic behavior at higher values λ is independent of κ, the coupling strength λ that needs to be for the transition to chaos decreases with increasing κ. Figure 2d illustrates β as a function of κ and λ. One can see that with increasing κ, the region with $\beta \approx 1$ extends to smaller values of λ. By identifying the chaotic region as $\beta \geq 0.7$, we show that the boundary of chaos strongly depends on the strength of the atomic interaction, see the black curve in Figure 2d. Notice that the boundary extracted from η behaves in a similar way to the one extracted from β.

3.2. Level Spacing Ratio

The study of level spacing distribution requires the so-called unfolding procedure [78]. It proceeds by rescaling the original eigenlevels to ensure that the local density of states of the resulting spectrum is 1. It is usually a nontrivial task, in particular for quantum many-body systems. To circumvent this disadvantage, one can resort to another chaotic probe based on the ratio of adjacent level spacings [79], which is free from the unfolding procedure.

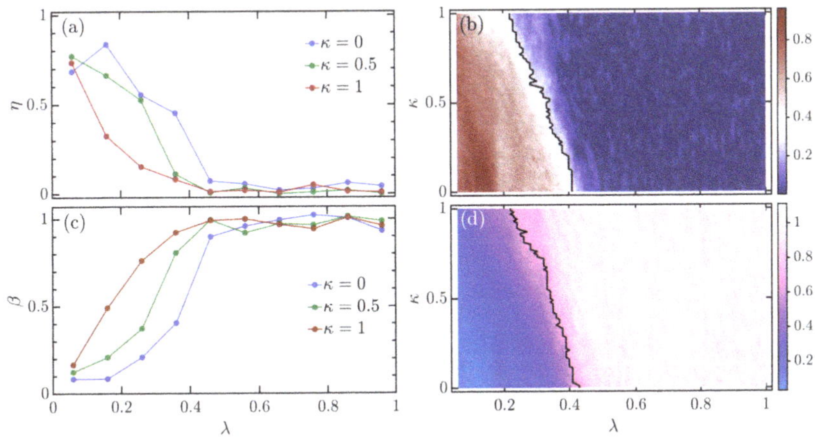

Figure 2. (**a**) Chaos indicator η as a function of λ for several values of κ. (**b**) η as a function of κ and λ. (**c**) Chaos indicator β as a function of λ for different values of κ. (**d**) β as a function of κ and λ. These results are obtained from the energy levels with energies $E/N \in [0.4, 4]$. In panels (**b**,**d**), the boundaries of the chaotic region are marked by the black curves, which are defined as $\eta \leq 0.3$ and $\beta \geq 0.7$, respectively. Other parameters are: $\omega = \omega_0 = 1$ and $N = 2j = 32$. All quantities are dimensionless.

For a given set of level spacings $\{s_\mu = E_{\mu+1} - E_\mu\}$, the ratio of adjacent level spacings is defined as [79,80]

$$r_\mu = \min\left(\delta_\mu, \frac{1}{\delta_\mu}\right), \tag{8}$$

where $\delta_\mu = s_{\mu+1}/s_\mu$ is the ratio between two adjacent level spacings. Obviously, r_μ is defined in the interval $r_\mu \in [0,1]$. The distribution of r_μ for both integrable and chaotic systems has been analytically investigated [80–82]. It is known that for the chaotic systems with Hamitonian from the Gaussian orthogonal ensemble (GOE), the level spacing ratio distribution is given by

$$P_{GOE}(r) = \frac{1}{Z_1}\frac{2(r+r^2)}{(1+r+r^2)^{5/2}}, \tag{9}$$

where $Z_1 = 8/27$ is the normalization constant. On the other hand, as the eigenlevels in the integrable systems are uncorrelated (independent Poisson levels), one can simply find that the ratio distribution is

$$P_{Poi}(r) = \frac{2}{(r+1)^2}. \tag{10}$$

Due to $r_\mu \in [0,1]$, the ratio distribution $P_{GOE/Poi}(r)$ vanishes outside the range $[0,1]$.

Figure 3a–f show how the level spacing ratio distribution $P(r)$ evolves for different combinations of atomic interaction strength κ and coupling strength λ. Similarly to what we observe for the level spacing distribution $P(s)$ in Figure 1, the spacing ratio distribution $P(r)$ tends to P_{GOE} with increasing coupling strength λ, independent of κ value. However, as evident from Figure 3a,d, increasing κ leads to the enhancement in degree of chaos of the model at smaller values of λ. Therefore, as mentioned above, the atomic interaction can be used to tune the level of chaoticity in the model. By switching on the interatomic interaction $\kappa > 0$, the regularity-to-chaos transition of the original Dicke model [68,70] is amplified.

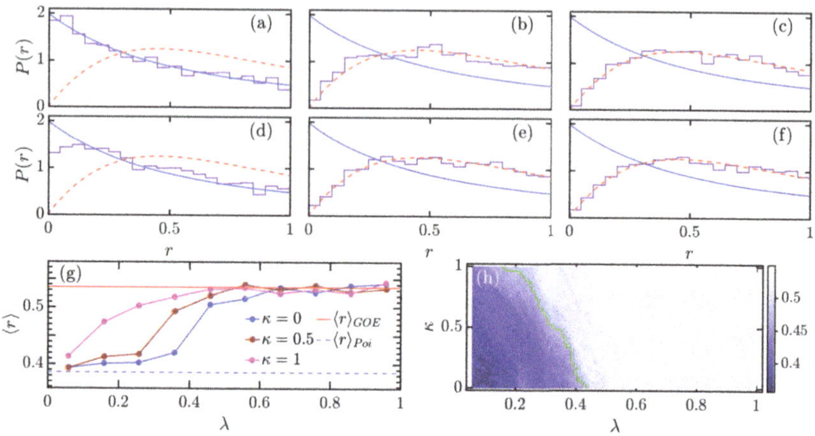

Figure 3. Level spacing ratio distributions $P(r)$ for various combinations of κ and λ: (**a**) $\kappa = 0, \lambda = 0$, (**b**) $\kappa = 0, \lambda = 0.5$, (**c**) $\kappa = 0, \lambda = 1$, (**d**) $\kappa = 0.7, \lambda = 0$, (**e**) $\kappa = 0.7, \lambda = 0.5$, and (**f**) $\kappa = 0.7, \lambda = 1$. In each panel, the Poisson distribution $P_{Poi}(r)$ is plotted as a blue solid curve, while the red dashed line denotes $P_{GOE}(r)$. (**g**) Averaged level spacing ratio $\langle r \rangle$ as a function of λ for several values of κ. (**h**) $\langle r \rangle$ as a function of κ and λ. The green line indicates the chaotic boundary, which is determined by $\langle r \rangle \geq 0.48$. The energy levels used in our numerical calculation have energies $E/N \in [0.4, 4]$. Other parameters: $\omega = \omega_0 = 1$ and $N = 2j = 40$. All quantities are dimensionless.

A more stringent analysis of the effect of the atomic interaction is made with the average level spacing ratio, defined as

$$\langle r \rangle = \int_0^1 r P(r) dr. \tag{11}$$

It takes the value $\langle r \rangle_{Poi} = 2 \ln 2 - 1 \approx 0.386$ for integrable systems with $P(r) = P_{Poi}(r)$, while for chaotic systems with $P(r) = P_{GOE}(r)$, one has $\langle r \rangle_{GOE} = 4 - 2\sqrt{3} \approx 0.536$. Hence, $\langle r \rangle$ acts as a detector to diagnose whether the studied system is in the integrable or chaotic regime and has been widely used to track the crossover from integrability to chaos.

Figure 3g demonstrates $\langle r \rangle$ as a function of λ for three different values κ. We see that, regardless of the value of κ, the transition from integrability to chaos is well captured by the behavior of $\langle r \rangle$, which varies from $\langle r \rangle_{Poi}$ to $\langle r \rangle_{GOE}$ with increasing λ. We further observe that the chaotic phase is robust with respect to the variation of κ, but the integrable phase exhibits a strong dependence on κ. For the integrable phase with smaller values of λ, we find that increasing κ gives rise to an increase in $\langle r \rangle$. As a consequence, the location of transition to chaos can be varied by the atomic intraction. This effect is more clearly observed in Figure 3h, where the evolution of $\langle r \rangle$ as a function of κ and λ is illustrated. Again, we define the boundary of chaotic region by the condition $\langle r \rangle \geq \langle r \rangle_c = 0.48$, meaning that the region with value $\langle r \rangle \geq 0.48$ is considered as chaos. We checked that our main result still holds for other choices of $\langle r \rangle_c$, as long as $\langle r \rangle_c \in (0.45, 0.5)$. The green line in Figure 3h denotes the obtained boundary of chaos. As expected, the behavior of the boundary line confirms the extension of the chaotic region to lower values of λ as κ increases.

4. Structure of Eigenstates

The onset of chaos also bears a remarkable change in the structure of eigenstates. In this section, we explore the impact of atomic interaction on the transition to chaos by investigating the variation in the structure of eigenstates.

It is known that the eigenstates of chaotic systems are uncorrelated and are well described by random matrix theory (RMT) [3,83–85]. For the model studied in this work, one can expect that in the chaotic phase the eigenstates of the model will have the same structure as those of random GOE matrices. The GOE eigenstates are fully delocalized random vectors with real components consisting of independent Gaussian random numbers. Hence, the deviation of the eigenvector structure from Gaussian behavior is an alternative benchmark to certify quantum chaos [42,86–89].

The analysis of the structure of eigenstates requires expansion of the eigenstates in a chosen basis. The choice of basis is usually decided by the physical problem and the system under consideration. Here, we use the Fock–Dicke bases, $\{|n,m\rangle\}$, which are the eigenstates of $a^\dagger a$ and J_z, as mentioned in Section 2. The decomposition of the kth eigenstate, $|k\rangle$, of the Hamiltonian (1) in the selected basis is given by

$$|k\rangle = \sum_{\nu=1}^{\mathcal{D}} c_k^\nu |\nu\rangle, \tag{12}$$

where $|\nu\rangle = |n,m\rangle$, $\mathcal{D}$ is the dimension of the Hilbert space, and $c_k^\nu = \langle\nu|k\rangle$ are the kth eigenstate components in the basis $\{|\nu\rangle\}$, satisfying the normalization condition $\sum_\nu |c_k^\nu|^2 = 1$. The characterizations of the eigenstate structure are provided by the statistical properties of eigenstate coefficients $\{c_k^\nu\}$.

To analyze the fingerprint of chaos in the properties of the eigenstates of H in Equation (1), as well as to examine whether the atomic interaction has an effect on the eigenstate structure, we explore the coefficients' distribution in comparison with the corresponding GOE result. As mentioned above, the eigenstates for the chaotic systems with real and symmetric $\mathcal{D}$-dimensional Hamiltonian matrices are consistent with the GOE eigenstates. The components of a GOE eigenstate, $\{c_\nu\}$, are random numbers that are uniformly distributed on a unit sphere with dimension $\mathcal{D}-1$. In the limit $\mathcal{D} \gg 1$, the dependence between components vanishes, and the distribution of components can be well described by a Gaussian distribution with zero mean and variance $1/\mathcal{D}$ [77,90–92],

$$P_{GOE}(c) = \sqrt{\frac{\mathcal{D}}{2\pi}} e^{-\mathcal{D}c^2/2}. \tag{13}$$

In the chaotic systems, it has been known that the coefficients of the mid-spectrum eigenstates are distributed as a near-Gaussian distribution [77,78,93–95], while the coefficients' distribution for the eigenstates of nonchaotic systems and the edge eigenstates of chaotic systems is significantly different from Gaussian distribution [95–97]. As increasing λ leads to the onset of chaos in the model; one would expect that the distribution of mid-spectrum eigenstates coefficients should be turned from non-Gaussian into near-Gaussian.

Figure 4a–f show the evolution of the eigenstate coefficient distributions, denoted by $P(c)$, for several (κ, λ) combinations in comparsion with the Gaussian distribution provided by Equation (13). We see that, regardless of the value of κ, the eigenstate coefficients' distribution tends to Gaussian as λ increases. The larger the value of λ is, the higher the degree of chaos in the model, and, therefore, the closer the coefficients' distribution to Gaussian, as expected. Another prominent feature, that is also independent of κ, observed in the behaviors of $P(c)$ is the larger peak around $c_k^\nu \sim 0$. Even in the most chaotic regime, $P(c)$ still exhibits a high peak near zero, as shown in Figure 4c,f. This excessive number of zero coefficients is mainly due to the mixed feature of the model. This means the regular and chaotic behavior coexist in the model for the considered parameters. Detailed analysis of the mixture of regular and chaotic behaviors in the extended Dicke model is beyond the scope of the present work; we leave this investigation for a future work.

Let us now turn to discuss how the atomic interaction κ affects the eigenstate coefficients' distribution. As evident from Figure 4a–f, the presence of atomic interaction has almost no effect on the behavior of the eigenstate coefficients' distribution, but just a reduction in the height of peak in the regular phase [compare Figure 4a to Figure 4d]. This implies that the impact of the atomic interaction on the structure of eigenstates is not as strong as on the eigenlevels.

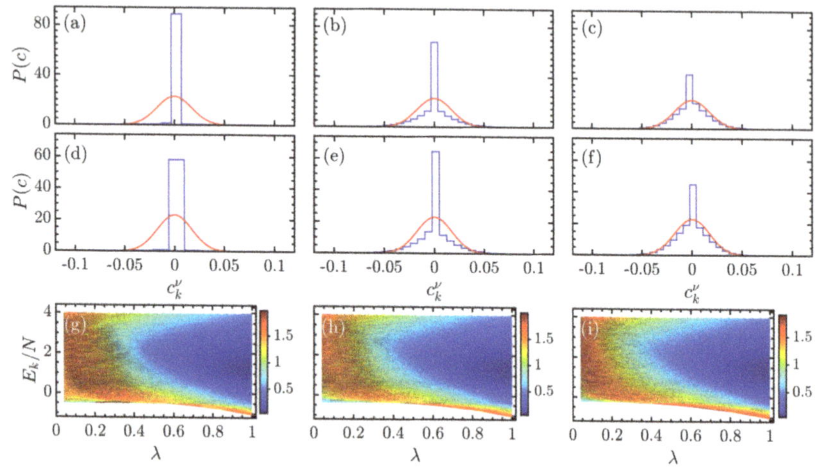

Figure 4. Panels (**a–f**): Histogramas of the coefficients c_k^ν of eigenstates with energy $E_k/N \in [1.75, 2.25]$, in the Fock–Dicke basis. Each panel corresponds to different (κ, λ) combinations: (**a**) $\kappa = 0, \lambda = 0$, (**b**) $\kappa = 0, \lambda = 0.5$, (**c**) $\kappa = 0, \lambda = 1$, (**d**) $\kappa = 0.7, \lambda = 0$, (**e**) $\kappa = 0.7, \lambda = 0.5$, and (**f**) $\kappa = 0.7, \lambda = 1$. The red solid line in each panel denotes the Gaussian distribution in Equation (13). Panels (**g–i**): Kullback–Leibler divergence D_{KL} [cf. Equation (14)] as a function of rescaled energy E_k/N and λ for $\kappa = 0$ (**g**), $\kappa = 0.5$ (**h**), and $\kappa = 1$ (**i**). Other parameters: $\omega = \omega_0 = 1$ and $N = 2j = 40$. All quantities are dimensionless.

To measure the difference between $P(c)$ and Gaussian distribution (13), we consider Kullback–Leibler (KL) divergence [98], which is commonly used to measure how close an observed distribution is to a predicted distribution. In our study, $P(c)$ and $P_{GOE}(c)$ are, respectively, identified as the observed and predicted distributions. Then, the KL divergence between them is given by

$$D_{KL} = \int_{c_{min}}^{c_{max}} P(c) \ln\left[\frac{P(c)}{P_{GOE}(c)}\right] dc, \tag{14}$$

where c_{min} and c_{max} denote, respectively, the minimum and maximum values in $\{c_k^\nu\}$. The KL divergence is non-negative, so that $D_{KL} \geq 0$, and it vanishes only when $P(c) = P_{GOE}(c)$. Qualitatively, a larger D_{KL} value indicates a larger difference between $P(c)$ and $P_{GOE}(c)$.

In Figure 4g–i, we plot the KL divergence for the extended Dicke model as a function of E_k/N and λ with $\kappa = \{0, 0.5, 1\}$. We see that the behaviors of D_{KL} versus E_k/N and λ for different κ values are very similar. As expected, the eigenstates in the non-chaotic phase ($\lambda \lesssim 0.45$) and the eigenstates at the spectrum edge in the chaotic phase ($\lambda \gtrsim 0.5$) have larger values of D_{KL}, suggesting that the corresponding coefficients' distributions strongly deviate from Gaussian, as illustrated in Figure 4a,d for the integrable case. Here, it is worth pointing out that the infinite dimension of the Hilbert space for the extended Dicke model leads to its spectrum has only one edge, namely the ground state. On the other hand, the lower values of D_{KL} for mid-spectrum eigenstates in the chaotic phase imply that their coefficients' distributions are closer to Gaussian, as shown in Figure 4c,f. The similarity between behaviors of D_{KL} observed in the bottom row of Figure 4 prompts a more detailed investigation of the impact of the atomic interaction on the eigenstate coefficient distribution.

To determine whether the eigenstate coefficients' distribution is robust with respect to the variation of the atomic interaction κ, we calculate D_{KL} for the mid-spectrum eigenstates with energies $E_k/N \in [1.75, 2.25]$. The final result for various cases is shown in Figure 5. The evolution of D_{KL} as a function of λ for different values of κ is plotted in Figure 5a.

One can see that the behavior of D_{KL} for different κ is very similar. The KL divergence varies slowly for smaller λ until $\lambda \approx 0.25$, after which it rapidly decreases to small values as the coupling strength is increased beyond $\lambda_c \approx 0.5$. The change in D_{KL} behavior is a manifestation of the transition to chaos resulting from the reduction in eigenstates correlation with increasing λ. We also observe that the dependence of D_{KL} on κ in the integrable phase is very different from that of the chaotic phase. The explicit dependence of D_{KL} on κ for several values of λ is shown in Figure 5b. As expected from Figure 5a, in the chaotic regime with $\lambda \gtrsim 0.5$, D_{KL} is independent of κ, whereas the KL divergence strongly depends on the atomic interaction for the cases of smaller λ. Overall, the KL divergence decreases with increasing κ in the integrable regime. This indicates that the proximity of the eigenstate coefficients' distribution to the Gaussian distribution can be improved by the atomic interaction. Therefore, the atomic interaction can vary the degree of chaos in the extended Dicke model, in agreement with the results obtained from eigenvalue statistics. An overall evolution of D_{KL} as a function of κ and λ is depicted in Figure 5c. We see that the chaotic region remains almost unchanged as κ increases, in contrast to the extension behavior revealed by the eigenvalue-based detectors of quantum chaos [see Figure 2b,d,h]. This suggests that the level repulsion is more sensitive to the effect of atomic interaction than the structure of eigenstates.

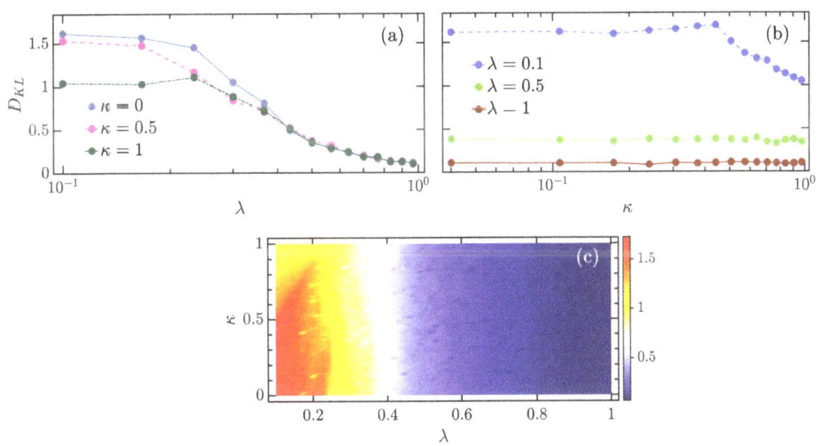

Figure 5. (**a**) KL divergence D_{KL} (14) as a function of λ for several κ values. (**b**) D_{KL} as a function of κ for several coupling strengths λ. (**c**) Color-scaled plot of D_{KL} in the $\kappa - \lambda$ plane. In all panels, the coefficients distribution $P(c)$ is obtained from the eigenstates with energies $E_k/N \in [1.75, 2.25]$. Other parameters: $\omega = \omega_0 = 1$ and $N = 2j = 40$. All quantities are dimensionless.

5. Conclusions

In this article, we performed a detailed analysis of quantum chaotic characters of the extended Dicke model through the statistical properties of eigenvalues and eigenstates. The presence of interaction between atoms in the model further allows us to explore the dependence of chaotic properties of the model on the atomic interaction. It was shown that the integrability-to-chaos transition of the original Dicke model as a function of the atom–field coupling is amplified by the interatomic interaction.

We demonstrated that as the model moves from regular phase to chaotic phase, both the level spacing and level spacing ratio distributions undergo a crossover from Poisson to Wigner distribution, regardless of the strength of atomic interaction. However, the presence of atomic interaction can lead to a notable deviation of level spacing and spacing ratio distributions from Poisson distribution. To quantify this deviation and to measure the degree of chaos in the model, we consider three different chaos indictors to probe the transition from integrablity to chaos. All of these indicators are complementary to each other and are able to capture the crossover from Poisson to Wigner for both level spacing and level spacing ratio distributions. We have also shown that the behaviors of these indicators as a function of control parameters are very similar. In particular, we found that the degree of chaos of the model can be controlled by tuning the strength of atomic interaction. This result highlights the role of interaction in the development of chaos in quantum many-body systems and opens up the possibility to tune the degree of chaos in the extended Dicke model.

Additional quantum chaotic signatures showing how the atomic interaction affects the degree of chaos in the extended Dicke model are unveiled in the structure of the eigenstates. To analyze the eigenstate structure, we expand each eigenstate in the Fock–Dicke basis and focus on the expansion coefficients' distribution for mid-spectrum eigenstates. For fully chaotic systems with Hamitonian from GOE, such distribution is well described by Gaussian distribution. We showed that the transition to chaos can be detected by the deviation of coefficients' distribution from Gaussian distribution. However, we note that even within the chaotic phase, the coefficients' distribution is still different from Gaussian, indicating the existence of correlations between them. By using the KL divergence to measure the distance between the coefficients' distribution and Gaussian distribution, we illustrated that the onset of chaos corresponds to the rapid decrease in the behavior of KL divergence as a function of coupling strength. Although the atomic interaction leads to the decrease in KL divergence in the regular phase, the transition to chaos revealed by KL divergence is almost independent of atomic interaction. This is different from the results obtained by the eigenvalue-based chaos indicators and implies that unlike the eigenlevels, the eigenstate structure is robust with respect to the change in atomic interaction.

A natural extension of the present work is to investigate the dynamical role played by the atomic interaction in the development of chaos. It would also be interesting to analyze the effect of atomic interaction on the level of chaoticity through long-range spectral correlations, which can be detected by the spectral form factor [41]. In addition, understanding the emergence of chaos and the impact of atomic interaction from the dynamics of the classical counterpart of the model would be another interesting topic. Very recently, the critical phenomena in the extended Dicke model have been thoroughly analyzed in this direction [65]. Finally, we would like to mention that a direct demonstration of level spacing distribution in an ultracold-atom system has been realized in a recent experiment [99]. Hence, we expect that the spectral statistics of our studied extended Dicke model can be verified by state-of-the-art experimental platforms.

Funding: This research was funded by the Slovenian Research Agency (ARRS) under the grant number J1-9112. Q.W. acknowledges support from the National Science Foundation of China under grant number 11805165, Zhejiang Provincial Nature Science Foundation under grant number LY20A050001.

Institutional Review Board Statement: Not applicable.

Informed Consent Statement: Not applicable.

Data Availability Statement: Not applicable.

Conflicts of Interest: The author declares no conflict of interest.

References

1. Altland, A.; Haake, F. Quantum Chaos and Effective Thermalization. *Phys. Rev. Lett.* **2012**, *108*, 073601. [CrossRef] [PubMed]
2. D'Alessio, L.; Kafri, Y.; Polkovnikov, A.; Rigol, M. From quantum chaos and eigenstate thermalization to statistical mechanics and thermodynamics. *Adv. Phys.* **2016**, *65*, 239–362. [CrossRef]
3. Borgonovi, F.; Izrailev, F.; Santos, L.; Zelevinsky, V. Quantum chaos and thermalization in isolated systems of interacting particles. *Phys. Rep.* **2016**, *626*, 1–58. [CrossRef]
4. Nandkishore, R.; Huse, D.A. Many-Body Localization and Thermalization in Quantum Statistical Mechanics. *Ann. Rev. Condens. Matter Phys.* **2015**, *6*, 15–38. [CrossRef]
5. Deutsch, J.M. Eigenstate thermalization hypothesis. *Rep. Prog. Phys.* **2018**, *81*, 082001. [CrossRef]
6. Chan, A.; De Luca, A.; Chalker, J.T. Solution of a Minimal Model for Many-Body Quantum Chaos. *Phys. Rev. X* **2018**, *8*, 041019. [CrossRef]

7. Garcia-March, M.A.; van Frank, S.; Bonneau, M.; Schmiedmayer, J.; Lewenstein, M.; Santos, L.F. Relaxation, chaos, and thermalization in a three-mode model of a Bose–Einstein condensate. *New J. Phys.* **2018**, *20*, 113039. [CrossRef]
8. Friedman, A.J.; Chan, A.; De Luca, A.; Chalker, J.T. Spectral Statistics and Many-Body Quantum Chaos with Conserved Charge. *Phys. Rev. Lett.* **2019**, *123*, 210603. [CrossRef]
9. Ray, S.; Cohen, D.; Vardi, A. Chaos-induced breakdown of Bose-Hubbard modeling. *Phys. Rev. A* **2020**, *101*, 013624. [CrossRef]
10. Rautenberg, M.; Gärttner, M. Classical and quantum chaos in a three-mode bosonic system. *Phys. Rev. A* **2020**, *101*, 053604. [CrossRef]
11. Kobrin, B.; Yang, Z.; Kahanamoku-Meyer, G.D.; Olund, C.T.; Moore, J.E.; Stanford, D.; Yao, N.Y. Many-Body Chaos in the Sachdev-Ye-Kitaev Model. *Phys. Rev. Lett.* **2021**, *126*, 030602. [CrossRef] [PubMed]
12. Fogarty, T.; García-March, M.Á.; Santos, L.F.; Harshman, N.L. Probing the edge between integrability and quantum chaos in interacting few-atom systems. *Quantum* **2021**, *5*, 486. [CrossRef]
13. Wittmann W., K.; Castro, E.R.; Foerster, A.; Santos, L.F. Interacting bosons in a triple well: Preface of many-body quantum chaos. *Phys. Rev. E* **2022**, *105*, 034204. [CrossRef] [PubMed]
14. Maldacena, J.; Shenker, S.H.; Stanford, D. A bound on chaos. *J. High Energy Phys.* **2016**, *2016*, 106. [CrossRef]
15. Stanford, D. Many-body chaos at weak coupling. *J. High Energy Phys.* **2016**, *2016*, 9. [CrossRef]
16. Magán, J.M. Black holes, complexity and quantum chaos. *J. High Energy Phys.* **2018**, *2018*, 43. [CrossRef]
17. Jahnke, V. Recent Developments in the Holographic Description of Quantum Chaos. *Adv. High Energy Phys.* **2019**, *2019*, 9632708. [CrossRef]
18. Ali, T.; Bhattacharyya, A.; Haque, S.S.; Kim, E.H.; Moynihan, N.; Murugan, J. Chaos and complexity in quantum mechanics. *Phys. Rev. D* **2020**, *101*, 026021. [CrossRef]
19. Rabinovici, E.; Sánchez-Garrido, A.; Shir, R.; Sonner, J. Operator complexity: A journey to the edge of Krylov space. *J. High Energy Phys.* **2021**, *2021*, 62. [CrossRef]
20. Schack, R.; Caves, C.M. Information-theoretic characterization of quantum chaos. *Phys. Rev. E* **1996**, *53*, 3257–3270. [CrossRef] [PubMed]
21. Vidmar, L.; Rigol, M. Entanglement Entropy of Eigenstates of Quantum Chaotic Hamiltonians. *Phys. Rev. Lett.* **2017**, *119*, 220603. [CrossRef] [PubMed]
22. Piga, A.; Lewenstein, M.; Quach, J.Q. Quantum chaos and entanglement in ergodic and nonergodic systems. *Phys. Rev. E* **2019**, *99*, 032213. [CrossRef] [PubMed]
23. Bertini, B.; Kos, P.; Prosen, T. Entanglement Spreading in a Minimal Model of Maximal Many-Body Quantum Chaos. *Phys. Rev. X* **2019**, *9*, 021033. [CrossRef]
24. Lerose, A.; Pappalardi, S. Bridging entanglement dynamics and chaos in semiclassical systems. *Phys. Rev. A* **2020**, *102*, 032404. [CrossRef]
25. Lantagne-Hurtubise, E.; Plugge, S.; Can, O.; Franz, M. Diagnosing quantum chaos in many-body systems using entanglement as a resource. *Phys. Rev. Res.* **2020**, *2*, 013254. [CrossRef]
26. Anand, N.; Styliaris, G.; Kumari, M.; Zanardi, P. Quantum coherence as a signature of chaos. *Phys. Rev. Res.* **2021**, *3*, 023214. [CrossRef]
27. Hosur, P.; Qi, X.L.; Roberts, D.A.; Yoshida, B. Chaos in quantum channels. *J. High Energy Phys.* **2016**, *2016*, 4. [CrossRef]
28. Chenu, A.; Molina-Vilaplana, J.; del Campo, A. Work Statistics, Loschmidt Echo and Information Scrambling in Chaotic Quantum Systems. *Quantum* **2019**, *3*, 127. [CrossRef]
29. Prakash, R.; Lakshminarayan, A. Scrambling in strongly chaotic weakly coupled bipartite systems: Universality beyond the Ehrenfest timescale. *Phys. Rev. B* **2020**, *101*, 121108. [CrossRef]
30. Balasubramanian, V.; DeCross, M.; Kar, A.; Parrikar, O. Quantum complexity of time evolution with chaotic Hamiltonians. *J. High Energy Phys.* **2020**, *2020*, 134. [CrossRef]
31. Bhattacharyya, A.; Haque, S.S.; Kim, E.H. Complexity from the reduced density matrix: A new diagnostic for chaos. *J. High Energy Phys.* **2021**, *2021*, 28. [CrossRef]
32. Bhattacharyya, A.; Chemissany, W.; Haque, S.S.; Murugan, J.; Yan, B. The Multi-faceted Inverted Harmonic Oscillator: Chaos and Complexity. *SciPost Phys. Core* **2021**, *4*, 2. [CrossRef]
33. Parker, D.E.; Cao, X.; Avdoshkin, A.; Scaffidi, T.; Altman, E. A Universal Operator Growth Hypothesis. *Phys. Rev. X* **2019**, *9*, 041017. [CrossRef]
34. Dymarsky, A.; Gorsky, A. Quantum chaos as delocalization in Krylov space. *Phys. Rev. B* **2020**, *102*, 085137. [CrossRef]
35. Caputa, P.; Magan, J.M.; Patramanis, D. Geometry of Krylov complexity. *Phys. Rev. Res.* **2022**, *4*, 013041. [CrossRef]
36. Bertini, B.; Heidrich-Meisner, F.; Karrasch, C.; Prosen, T.; Steinigeweg, R.; Žnidarič, M. Finite-temperature transport in one-dimensional quantum lattice models. *Rev. Mod. Phys.* **2021**, *93*, 025003. [CrossRef]
37. Cvitanovic, P.; Artuso, R.; Mainieri, R.; Tanner, G.; Vattay, G.; Whelan, N.; Wirzba, A. Chaos: Classical and quantum. In *ChaosBook. Org*; Niels Bohr Institute: Copenhagen, Denmark, 2005; Volume 69, p. 25.
38. Schuster, H.G.; Just, W. *Deterministic Chaos: An Introduction*; John Wiley & Sons: Hoboken, NJ, USA, 2006.
39. Bohigas, O.; Giannoni, M.J.; Schmit, C. Characterization of Chaotic Quantum Spectra and Universality of Level Fluctuation Laws. *Phys. Rev. Lett.* **1984**, *52*, 1–4. [CrossRef]
40. Stöckmann, H.J. *Quantum Chaos: An Introduction*; Cambridge University Press: Cambridge, UK, 1999. [CrossRef]

41. Haake, F. *Quantum Signatures of Chaos*; Springer: Berlin/Heidelberg, Germany, 2010.
42. Zyczkowski, K. Indicators of quantum chaos based on eigenvector statistics. *J. Phys. A* **1990**, *23*, 4427. [CrossRef]
43. Emerson, J.; Weinstein, Y.S.; Lloyd, S.; Cory, D.G. Fidelity Decay as an Efficient Indicator of Quantum Chaos. *Phys. Rev. Lett.* **2002**, *89*, 284102. [CrossRef] [PubMed]
44. Rozenbaum, E.B.; Ganeshan, S.; Galitski, V. Lyapunov Exponent and Out-of-Time-Ordered Correlator's Growth Rate in a Chaotic System. *Phys. Rev. Lett.* **2017**, *118*, 086801. [CrossRef] [PubMed]
45. García-Mata, I.; Saraceno, M.; Jalabert, R.A.; Roncaglia, A.J.; Wisniacki, D.A. Chaos Signatures in the Short and Long Time Behavior of the Out-of-Time Ordered Correlator. *Phys. Rev. Lett.* **2018**, *121*, 210601. [CrossRef]
46. Chen, X.; Ludwig, A.W.W. Universal spectral correlations in the chaotic wave function and the development of quantum chaos. *Phys. Rev. B* **2018**, *98*, 064309. [CrossRef]
47. Kos, P.; Ljubotina, M.; Prosen, T. Many-Body Quantum Chaos: Analytic Connection to Random Matrix Theory. *Phys. Rev. X* **2018**, *8*, 021062. [CrossRef]
48. Bertini, B.; Kos, P.; Prosen, T. Exact Spectral Form Factor in a Minimal Model of Many-Body Quantum Chaos. *Phys. Rev. Lett.* **2018**, *121*, 264101. [CrossRef]
49. Gietka, K.; Chwedeńczuk, J.; Wasak, T.; Piazza, F. Multipartite entanglement dynamics in a regular-to-ergodic transition: Quantum Fisher information approach. *Phys. Rev. B* **2019**, *99*, 064303. [CrossRef]
50. Xu, T.; Scaffidi, T.; Cao, X. Does Scrambling Equal Chaos? *Phys. Rev. Lett.* **2020**, *124*, 140602. [CrossRef] [PubMed]
51. Cao, Z.; Xu, Z.; del Campo, A. Probing quantum chaos in multipartite systems. *Phys. Rev. Res.* **2022**, *4*, 033093. [CrossRef]
52. Zonnios, M.; Levinsen, J.; Parish, M.M.; Pollock, F.A.; Modi, K. Signatures of Quantum Chaos in an Out-of-Time-Order Tensor. *Phys. Rev. Lett.* **2022**, *128*, 150601. [CrossRef]
53. Raúl González Alonso, J.; Shammah, N.; Ahmed, S.; Nori, F.; Dressel, J. Diagnosing quantum chaos with out-of-time-ordered-correlator quasiprobability in the kicked-top model. *arXiv* **2022**, arXiv:2201.08175.
54. Joshi, L.K.; Elben, A.; Vikram, A.; Vermersch, B.; Galitski, V.; Zoller, P. Probing Many-Body Quantum Chaos with Quantum Simulators. *Phys. Rev. X* **2022**, *12*, 011018. [CrossRef]
55. Lozej, Č.; Lukman, D.; Robnik, M. Phenomenology of quantum eigenstates in mixed-type systems: Lemon billiards with complex phase space structure. *arXiv* **2022**, arXiv:2207.07197.
56. Abanin, D.A.; Altman, E.; Bloch, I.; Serbyn, M. Colloquium: Many-body localization, thermalization, and entanglement. *Rev. Mod. Phys.* **2019**, *91*, 021001. [CrossRef]
57. Turner, C.J.; Michailidis, A.A.; Abanin, D.A.; Serbyn, M.; Papić, Z. Weak ergodicity breaking from quantum many-body scars. *Nat. Phys.* **2018**, *14*, 745–749. [CrossRef]
58. Sinha, S.; Sinha, S. Chaos and Quantum Scars in Bose-Josephson Junction Coupled to a Bosonic Mode. *Phys. Rev. Lett.* **2020**, *125*, 134101. [CrossRef]
59. Turner, C.J.; Desaules, J.Y.; Bull, K.; Papić, Z. Correspondence Principle for Many-Body Scars in Ultracold Rydberg Atoms. *Phys. Rev. X* **2021**, *11*, 021021. [CrossRef]
60. Mondragon-Shem, I.; Vavilov, M.G.; Martin, I. Fate of Quantum Many-Body Scars in the Presence of Disorder. *PRX Quantum* **2021**, *2*, 030349. [CrossRef]
61. Serbyn, M.; Abanin, D.A.; Papić, Z. Quantum many-body scars and weak breaking of ergodicity. *Nat. Phys.* **2021**, *17*, 675–685. [CrossRef]
62. Kloc, M.; Stránský, P.; Cejnar, P. Quantum phases and entanglement properties of an extended Dicke model. *Ann. Phys.* **2017**, *382*, 85–111. [CrossRef]
63. Rodriguez, J.P.J.; Chilingaryan, S.A.; Rodríguez-Lara, B.M. Critical phenomena in an extended Dicke model. *Phys. Rev. A* **2018**, *98*, 043805. [CrossRef]
64. Guerra, C.A.E.; Mahecha-Gómez, J.; Hirsch, J.G. Quantum phase transition and Berry phase in an extended Dicke model. *Eur. Phys. J. D* **2020**, *74*, 200. [CrossRef]
65. Herrera Romero, R.; Bastarrachea-Magnani, M.A.; Linares, R. Critical Phenomena in Light-Matter Systems with Collective Matter Interactions. *Entropy* **2022**, *24*, 1198. [CrossRef] [PubMed]
66. Dicke, R.H. Coherence in Spontaneous Radiation Processes. *Phys. Rev.* **1954**, *93*, 99–110. [CrossRef]
67. Emary, C.; Brandes, T. Chaos and the quantum phase transition in the Dicke model. *Phys. Rev. E* **2003**, *67*, 066203. [CrossRef] [PubMed]
68. Wang, Q.; Robnik, M. Statistical properties of the localization measure of chaotic eigenstates in the Dicke model. *Phys. Rev. E* **2020**, *102*, 032212. [CrossRef]
69. Bastarrachea-Magnani, M.A.; del Carpio, B.L.; Lerma-Hernández, S.; Hirsch, J.G. Chaos in the Dicke model: Quantum and semiclassical analysis. *Phys. Scr.* **2015**, *90*, 068015. [CrossRef]
70. Bastarrachea-Magnani, M.A.; López-del Carpio, B.; Chávez-Carlos, J.; Lerma-Hernández, S.; Hirsch, J.G. Delocalization and quantum chaos in atom-field systems. *Phys. Rev. E* **2016**, *93*, 022215. [CrossRef]
71. Robles Robles, R.A.; Chilingaryan, S.A.; Rodríguez-Lara, B.M.; Lee, R.K. Ground state in the finite Dicke model for interacting qubits. *Phys. Rev. A* **2015**, *91*, 033819. [CrossRef]
72. Cejnar, P.; Stránský, P.; Macek, M. Regular and Chaotic Collective Modes in Nuclei. *Nucl. Phys. News* **2011**, *21*, 22–27. [CrossRef]

73. Chávez-Carlos, J.; López-del Carpio, B.; Bastarrachea-Magnani, M.A.; Stránský, P.; Lerma-Hernández, S.; Santos, L.F.; Hirsch, J.G. Quantum and Classical Lyapunov Exponents in Atom-Field Interaction Systems. *Phys. Rev. Lett.* **2019**, *122*, 024101. [CrossRef]
74. Rodríguez-Lara, B.M.; Lee, R.K. Classical dynamics of a two-species condensate driven by a quantum field. *Phys. Rev. E* **2011**, *84*, 016225. [CrossRef]
75. Berry, M.V.; Tabor, M. Level clustering in the regular spectrum. *Proc. R. Soc. A* **1977**, *356*, 375–394. [CrossRef]
76. Jacquod, P.; Shepelyansky, D.L. Emergence of Quantum Chaos in Finite Interacting Fermi Systems. *Phys. Rev. Lett.* **1997**, *79*, 1837–1840. [CrossRef]
77. Brody, T.A.; Flores, J.; French, J.B.; Mello, P.A.; Pandey, A.; Wong, S.S.M. Random-matrix physics: Spectrum and strength fluctuations. *Rev. Mod. Phys.* **1981**, *53*, 385–479. [CrossRef]
78. Guhr, T.; Müller-Groeling, A.; Weidenmüller, H.A. Random-matrix theories in quantum physics: Common concepts. *Phys. Rep.* **1998**, *299*, 189–425. [CrossRef]
79. Oganesyan, V.; Huse, D.A. Localization of interacting fermions at high temperature. *Phys. Rev. B* **2007**, *75*, 155111. [CrossRef]
80. Atas, Y.Y.; Bogomolny, E.; Giraud, O.; Roux, G. Distribution of the Ratio of Consecutive Level Spacings in Random Matrix Ensembles. *Phys. Rev. Lett.* **2013**, *110*, 084101. [CrossRef] [PubMed]
81. Atas, Y.Y.; Bogomolny, E.; Giraud, O.; Vivo, P.; Vivo, E. Joint probability densities of level spacing ratios in random matrices. *J. Phys. A Math. Theor.* **2013**, *46*, 355204. [CrossRef]
82. Giraud, O.; Macé, N.; Vernier, E.; Alet, F. Probing Symmetries of Quantum Many-Body Systems through Gap Ratio Statistics. *Phys. Rev. X* **2022**, *12*, 011006. [CrossRef]
83. Berry, M.V. Regular and irregular semiclassical wavefunctions. *J. Phys. A* **1977**, *10*, 2083–2091. [CrossRef]
84. Porter, C.E.; Thomas, R.G. Fluctuations of Nuclear Reaction Widths. *Phys. Rev.* **1956**, *104*, 483–491. [CrossRef]
85. Mehta, M.L. *Random Matrices*; Elsevier: Amsterdam, The Netherlands, 2004.
86. Izrailev, F.M. Simple models of quantum chaos: Spectrum and eigenfunctions. *Phys. Rep.* **1990**, *196*, 299–392. [CrossRef]
87. Haake, F.; Życzkowski, K. Random-matrix theory and eigenmodes of dynamical systems. *Phys. Rev. A* **1990**, *42*, 1013–1016. [CrossRef]
88. Leboeuf, P.; Voros, A. Chaos-revealing multiplicative representation of quantum eigenstates. *J. Phys. A Math. Gen.* **1990**, *23*, 1765–1774. [CrossRef]
89. Wang, Q.; Robnik, M. Multifractality in Quasienergy Space of Coherent States as a Signature of Quantum Chaos. *Entropy* **2021**, *23*, 1347. [CrossRef]
90. Torres-Herrera, E.J.; Karp, J.; Távora, M.; Santos, L.F. Realistic Many-Body Quantum Systems vs. Full Random Matrices: Static and Dynamical Properties. *Entropy* **2016**, *18*, 359. [CrossRef]
91. Bäcker, A.; Haque, M.; Khaymovich, I.M. Multifractal dimensions for random matrices, chaotic quantum maps, and many-body systems. *Phys. Rev. E* **2019**, *100*, 032117. [CrossRef] [PubMed]
92. Nakerst, G.; Haque, M. Chaos in the three-site Bose-Hubbard model—Classical vs. quantum. *arXiv* **2022**, arXiv:2203.09953.
93. Khaymovich, I.M.; Haque, M.; McClarty, P.A. Eigenstate Thermalization, Random Matrix Theory, and Behemoths. *Phys. Rev. Lett.* **2019**, *122*, 070601. [CrossRef] [PubMed]
94. Pausch, L.; Carnio, E.G.; Rodríguez, A.; Buchleitner, A. Chaos and Ergodicity across the Energy Spectrum of Interacting Bosons. *Phys. Rev. Lett.* **2021**, *126*, 150601. [CrossRef]
95. Beugeling, W.; Bäcker, A.; Moessner, R.; Haque, M. Statistical properties of eigenstate amplitudes in complex quantum systems. *Phys. Rev. E* **2018**, *98*, 022204. [CrossRef] [PubMed]
96. Luitz, D.J.; Bar Lev, Y. Anomalous Thermalization in Ergodic Systems. *Phys. Rev. Lett.* **2016**, *117*, 170404. [CrossRef] [PubMed]
97. Luitz, D.J.; Khaymovich, I.M.; Lev, Y.B. Multifractality and its role in anomalous transport in the disordered XXZ spin-chain. *SciPost Phys. Core* **2020**, *2*, 6. [CrossRef]
98. Kullback, S.; Leibler, R.A. On Information and Sufficiency. *Ann. Math. Stat.* **1951**, *22*, 79–86. [CrossRef]
99. Frisch, A.; Mark, M.; Aikawa, K.; Ferlaino, F.; Bohn, J.L.; Makrides, C.; Petrov, A.; Kotochigova, S. Quantum chaos in ultracold collisions of gas-phase erbium atoms. *Nature* **2014**, *507*, 475–479. [CrossRef] [PubMed]

entropy MDPI

Article

Mather β-Function for Ellipses and Rigidity

Michael Bialy

School of Mathematical Sciences, Raymond and Beverly Sackler Faculty of Exact Sciences, Tel-Aviv University, Tel Aviv 6997801, Israel; bialy@tauex.tau.ac.il

Abstract: The goal of the first part of this note is to get an explicit formula for rotation number and Mather β-function for ellipse. This is done here with the help of non-standard generating function of billiard problem. In this way the derivation is especially simple. In the second part we discuss application of Mather β-function to rigidity problem.

Keywords: Birkhoff billiard; invariant curve; action minimizers: rigidity, integrable billiards

1. Introduction

Consider the confocal family of ellipses

$$E_\lambda = \left\{ \frac{x^2}{a^2-\lambda} + \frac{y^2}{b^2-\lambda} = 1 \right\}, \quad 0 < \lambda < b^2 < a^2.$$

The initial ellipse is $E = E_0$. Polygonal lines with the vertices on E circumscribed about confocal caustic E_λ correspond to billiard trajectories of the billiard in E. A caustic E_λ is called rational [1], of rotation number $\rho = m/n$, if a billiard trajectory circumscribing E_λ closes after n reflections making m rotations. These closed billiard trajectories are called Poncelet polygons. By famous Poncelet theorem if one billiard trajectory tangent to E_λ is closed with $\rho = m/n$, then all of them are closed with the same ρ. Given a caustic E_λ, all Poncelet polygons have the same perimeter. Mather β-function assigns to the rotation number ρ of E_λ the value of this perimeter divided by the number of vertices. Let me remark that traditionally Mather β-function is negative of ours. However we prefer, for convenience, sign + for generating function and hence for Mather β-function as well.

Citation: Bialy, M. Mather β-Function for Ellipses and Rigidity. *Entropy* **2022**, *24*, 1600. https://doi.org/10.3390/e24111600

Academic Editor: Marko Robnik

Received: 25 September 2022
Accepted: 1 November 2022
Published: 3 November 2022

Example 1. *It is not difficult to compute the perimeter and the corresponding λ for 4-gons. Namely the perimeter equals $4\sqrt{a^2+b^2}$ and hence $\beta(1/4) = \sqrt{a^2+b^2}$, $\sqrt{\lambda} = ab/\sqrt{a^2+b^2}$. The perimeter of Poncelet triangles and the corresponding λ can be geometrically found, but this requires solution of cubic equation. We leave this as an exercise.*

In this note we show how to compute the perimeter of the Poncelet polygons for a given caustic E_λ. By a different method a similar formula was discovered in a recent paper [2]. Notice that the straightforward computation of the lengths of the edges seems to be difficult. The main idea of this paper is to use *non-standard generating function* of the billiard. This function was found in [3] for ellipsoids, and in [4,5] for general convex billiard tables. Using this idea we replace the straightforward computation of the action functional by expressing this functional via the non-standard generating function. This approach leads immediately to a formula containing pseudo-elliptic integral, which can be further reduced to elliptic integrals, using [6]. The question of existence of such a formula was explicitly addressed by S. Zelditch in [7,8]. We also get by our method the known formulas for rotation number and the invariant measure [1,9] in a very direct way. In addition, we derive in Section 6 a simple formula for the so-called Lazutkin parameter of the caustic E_λ.

There is an extensive literature on Poncelet porism, formulas for invariant measure and the rotation number. I refer to the incomplete list of papers on the subject [1,9–13].

The non-standard generating function for convex billiards has been already used in our paper [14], explaining conservation laws for elliptical billiards discovered recently by Dan Reznik [15,16] et al., see also [11,12,17]. Additionally, the non-standard generating function is a key ingredient in the recent proof of a part of Birkhoff conjecture for centrally symmetric billiard tables [18].

Mather β-function is very important function related both to classical dynamics inside the domain as well as to the spectral problems. In this paper, we shall discuss in Section 8 the relation of Mather β-function to the rigidity questions. The idea to use Mather β-function for rigidity in billiards belongs to K.F. Siburg [19]. We refer [1,8,20–22] for further developments and other approaches.

2. Results

In this section, we formulate our main contributions. Other results are placed in the corresponding sections.

Theorem 1. *Consider the invariant curve of rotation number ρ corresponding to the caustic E_λ. Mather β-function corresponding to the caustic E_λ is given by the following formula:*

$$\beta(\rho) = \frac{2ce\sqrt{e^2 - f^2}}{e^2 - 1} - \frac{2cf}{K(k)}[K(k)E(\phi, k) - E(k)F(\phi, k)], \quad k = 1/f,\ \phi = \arcsin\frac{\sqrt{\lambda}}{b},$$

where $E(\phi, k)$ is elliptic integral of the second kind, $K(k), E(k)$ are complete elliptic integrals of first and second kind, and e, f are eccentricities of the ellipses E, E_λ.

Corollary 1. *The following formula holds*

$$\beta(\rho) = \frac{2a\sqrt{\lambda}}{b} - 2\sqrt{a^2 - \lambda}E(\phi, k) + \rho|E_\lambda|,$$

$$\phi = \arcsin\frac{\sqrt{\lambda}}{b},\ k = 1/f,$$

where $|E_\lambda|$ denotes the circumference of the ellipse E_λ.

We give a proof of these formulas Section 1.

Example 2. *(1) One can see from this formula that for $\rho = 0$, that is when $f \to e$ (confocal ellipse coincides with the boundary, i.e., $\lambda = 0$), it follows that $\phi \to 0$ and hence $\beta \to 0$.*

(2) When $f \to 1$ (corresponding to the confocal ellipse shrinking to the segment between the focii), $\beta \to 2a$—the diameter (only the first summand of the formula remains, the second one tends to zero).

We shall discuss now the relation of Mather β-function to the rigidity questions. The following problem is important. Let Ω_1, Ω_2 be two strictly convex domains having the same Mather β-functions $\beta_1 = \beta_2$, can one state that the domains are isometric. It is especially important in view of its applications to spectral rigidity.

Remarkably, if Ω_1 is an ellipse then there are many approaches leading to the affirmative answer. In this paper, we do not consider infinitesimal behavior of Mather β-function at 0 (cf. [21,22]), but rather study this function on a finite neighborhood of 0. Our contribution is based on the recent paper with a partial resolution of Birkhoff conjecture for centrally symmetric convex billiards [18]. The result of [18] can be formulated in terms of Mather β-function as follows:

Theorem 2. *Let Ω_1, Ω_2 be two strictly convex C^2-smooth centrally symmetric planar domains such that Ω_1 is an ellipse. Suppose that Mather β-functions β_1, β_2 of the domains satisfy*

$$\beta_1(\rho) = \beta_2(\rho),\ \forall \rho \in (0, \frac{1}{4}].$$

Then Ω_2 is an ellipse isometric to Ω_1.

In Section 8, we shall give the proof of this result and discuss further application of Mather β-function to rigidity problems.

3. Preliminaries and Methods

3.1. Non-Standard Generating Function

Consider the space of oriented lines in the plane $\mathbb{R}^2(x, y)$. A line can be written as

$$\cos\varphi \cdot x + \sin\varphi \cdot y = p,$$

where φ is the direction of the right normal to the oriented line, and p is the signed distance to the line from the origin. Thus, (p, φ) are coordinates in the space of oriented lines, see Figure 1. The 2-form $\rho = dp \wedge d\varphi$ is the area (symplectic) form on the space of oriented lines used in geometrical optics and integral geometry.

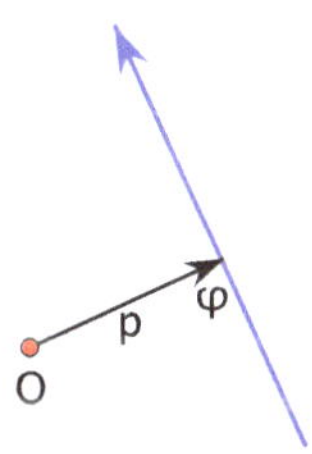

Figure 1. Coordinates in the space of oriented lines.

Consider a smooth strictly convex billiard curve γ, and let $h(\varphi)$ be its support function, that is, the distance from the origin (supposed to be inside γ) to the tangent line to γ at the point where the outer normal has direction φ. The sub-space **A** of the oriented lines intersecting the curve γ is the phase space cylinder of the billiard map. The billiard transformation acts on **A** as an exact symplectic map.

$$T : (p_1, \varphi_1) \mapsto (p_2, \varphi_2)$$

sends the incoming trajectory to the outgoing one. Let

$$\psi = \frac{\varphi_1 + \varphi_2}{2},\ \delta = \frac{\varphi_2 - \varphi_1}{2},$$

where ψ is the direction of the outer normal at the reflection point and δ is the reflection angle.

Proposition 1. *The function*

$$S(\varphi_1, \varphi_2) = 2h\left(\frac{\varphi_1 + \varphi_2}{2}\right)\sin\left(\frac{\varphi_2 - \varphi_1}{2}\right) = 2h(\psi)\sin\delta$$

is a generating function of the billiard transformation, that is, $T(p_1, \varphi_1) = (p_2, \varphi_2)$ if and only if

$$-\frac{\partial S_1(\varphi_1, \varphi_2)}{\partial \varphi_1} = p_1,\quad \frac{\partial S_2(\varphi_1, \varphi_2)}{\partial \varphi_2} = p_2.$$

Proof. We refer to Figure 2.

One has

$$-\frac{\partial S_1(\varphi_1,\varphi_2)}{\partial \varphi_1} = -h'(\psi)\sin\delta + h(\psi)\cos\delta.$$

The position vector of the point of the curve γ with the outer normal having direction ψ is

$$\gamma(\psi) = h(\psi)(\cos\psi, \sin\psi) + h'(\psi)(-\sin\psi, \cos\psi)$$

(this formula is well known in convex geometry). Then, using some trigonometry,

$$p_1 = \gamma(\psi)\cdot(\cos\varphi_1, \sin\varphi_1) = h(\psi)\cos\delta - h'(\psi)\sin\delta,$$

as needed. The argument for p_2 is similar. □

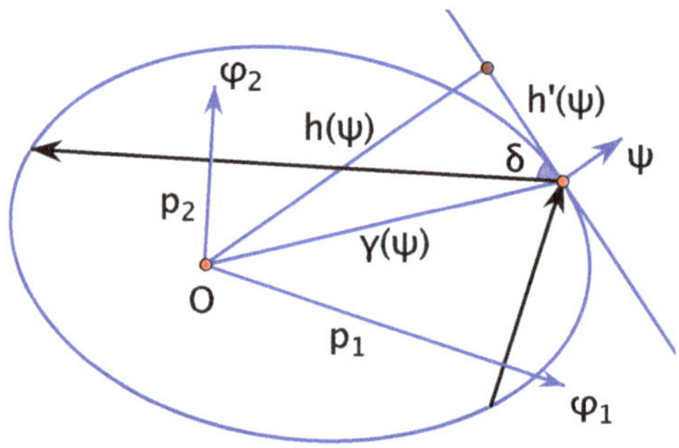

Figure 2. To Proposition 1.

In order to use the function S for ellipse let me remind the computation of the support function, with respect to the center of the ellipse, as a function of ψ which is the angle made by the outer normal with the positive x-axes.

Lemma 1. *Let E be the ellipse* $\{\frac{x^2}{a^2} + \frac{y^2}{b^2} = 1\}$. *One has:*

$$h(\psi) = \sqrt{a^2\cos^2\psi + b^2\sin^2\psi}.$$

Proof. Consider a point (ξ, η) of the ellipse. A normal vector is given by

$$N = \left(\frac{\xi}{a^2}, \frac{\eta}{b^2}\right) = \ell(\cos\psi, \sin\psi), \quad \ell = |N|.$$

and the tangent line at this point has the equation

$$\frac{\xi x}{a^2} + \frac{\eta y}{b^2} = 1.$$

The distance from the origin to this line is

$$\frac{1}{\sqrt{\frac{\xi^2}{a^4} + \frac{\eta^2}{b^4}}} = \frac{1}{\ell}.$$

On the other hand,

$$\xi = a^2\ell\cos\psi,\ \eta = b^2\ell\sin\psi,$$

and the equation of the ellipse implies that

$$\ell^2 = \frac{1}{a^2\cos^2\psi + b^2\sin^2\psi}.$$

Therefore $h(\psi) = 1/\ell = \sqrt{a^2\cos^2\psi + b^2\sin^2\psi}$, as claimed. □

3.2. *Integral for Elliptic Billiard in Various Forms*

Billiard in ellipse is integrable. The integral can be understood at least in three ways.

1. Jacobi-Chasles integral λ.

Given an oriented line not intersecting the segment between the focii. Consider the confocal ellipse

$$E_\lambda = \left\{ \frac{x^2}{a^2-\lambda} + \frac{y^2}{b^2-\lambda} = 1 \right\}$$

tangent to this line then λ is an integral of the billiard, i.e., it remains constant under the reflections.

2. Joachimsthal integral $J := \frac{\sin\delta}{h}$.

This corresponds to the conservation of $< v, \nabla q >$, where v the unit vector of the line, and q is the quadratic form $q(x) = \frac{<Qx,x>}{2}$, with the diagonal matrix $Q = diag(\frac{1}{a^2}, \frac{1}{b^2})$, see Figure 3.

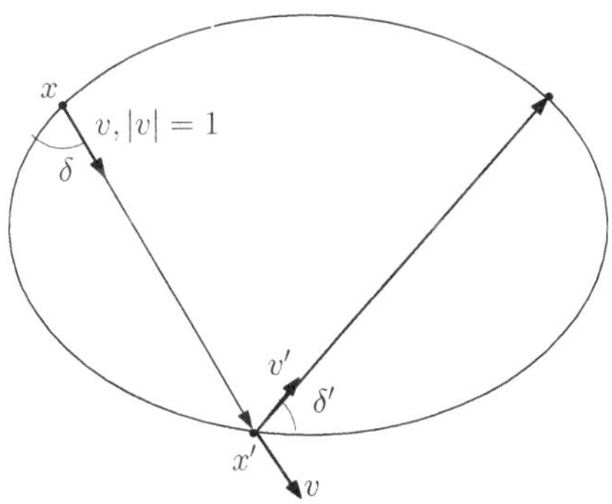

Figure 3. Joachimsthal integral.

Indeed, the conservation follows from the following implications:

$$< Q(x+x'), x'-x >= 0 \quad \Rightarrow \quad < Qx, v >= - < Qx', v > .$$

Additionally,

$$< Qx', v+v' >= 0 \Rightarrow < Qx', v' >= - < Qx', v > \Rightarrow < Qx, v >=< Qx', v' >.$$

One can express this conservation law in terms of h, δ as follows:

$$- < v, \nabla f >= -|\nabla f| \sin\delta = |Qx| \sin\delta = \frac{\sin\delta}{h} = J.$$

Here we used that $|Qx| = \frac{1}{h}$ as we explained in the proof of Lemma 1.

3. Product of two momenta F.

Let us consider a segment of the billiard trajectory tangent to a confocal ellipse E_λ with the semi-axes $\sqrt{b^2-\lambda} < \sqrt{a^2-\lambda}$. Let d_1, d_2 be the distances from the foci to the line and α be the direction of its normal. Then we define $F := d_1 d_2$ (This definition of the integral we learned from Michael Berry). It then follows from the next theorem that F is indeed an integral.

3.3. The Relations between Conserved Quantities

Theorem 3. *The following relations hold true:*

1. $J = \sqrt{\lambda}/ab$
2. $F = \sqrt{b^2 - \lambda}$
3. *In terms of the eccentricities* $e = \frac{a}{c}, f = \frac{\sqrt{a^2-\lambda}}{c}$ *of* E, E_λ *we have the formulas:*

$$\lambda = c^2(e^2 - f^2),\ J = \frac{\sqrt{e^2 - f^2}}{c\,e\sqrt{e^2 - 1}},\ \textit{where}\ c = \sqrt{a^2 - b^2}.$$

Proof. (1) Consider an oriented line passing through the point $(a, 0)$ with right normal having angle δ (see Figure 4). Then for p of this line we have

$$p = \sqrt{(a^2 - \lambda)\cos^2\delta + (b^2 - \lambda\sin^2\delta)},$$

and hence

$$p = a\ \cos\delta,$$

see Figure 4. Therefore these two give

$$b\ \sin\delta = \sqrt{\lambda}.$$

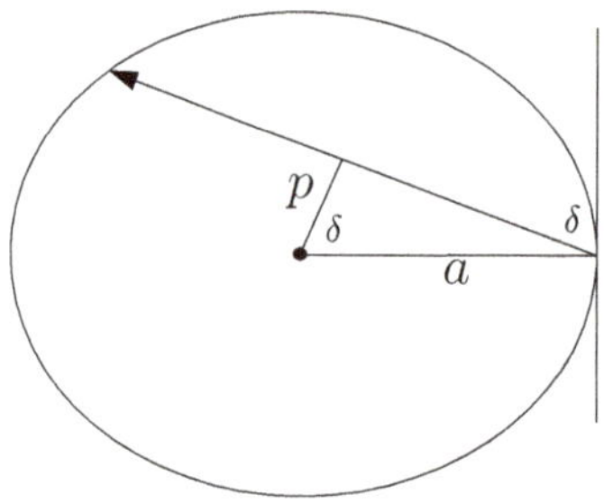

Figure 4. Relation of J and λ.

On the other hand from the definition of J we have:

$$J = \frac{\sin\delta}{a}.$$

Thus,

$$J = \frac{\sqrt{\lambda}}{ab}.$$

(2) Given a line with coordinates (p, φ), we have

$$d_1 = p - c\cos\phi,\ d_2 = p + c\cos\varphi,$$

where $c^2 = a^2 - b^2$ (see Figure 5).

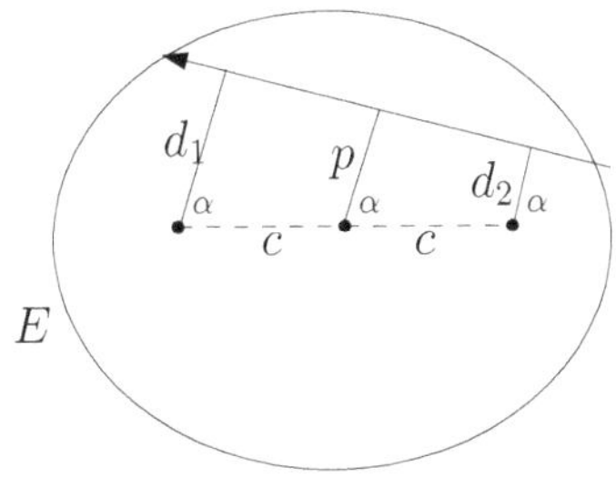

Figure 5. Integral $F = d_1 d_2$.

$$F := d_1 d_2 = p^2 - c^2 \cos^2 \varphi. \tag{1}$$

If the line is tangent to E_λ, then $p = h_\lambda(\varphi)$, where h_λ is the support function of E_λ. Hence, using Lemma 1 we rewrite (1)

$$F = (a^2 - \lambda)\cos^2 \varphi + (b^2 - \lambda)\sin^2 \varphi - c^2 \cos^2 \varphi = b^2 - \lambda.$$

(3) Follows from item (1) and the definition of eccentricities. □

4. Invariant Measure on an Invariant Curve

Suppose we have a curve on the phase cylinder **A** which is invariant under the billiard map T. Suppose this curve is a graph and lies in the level set of the integral $F(p, \varphi) = const$. Then there is a natural measure $d\mu$ on the curve which is invariant under T. According to V.I. Arnold this is called Gelfand-Leray form, which by another Arnold' principle was probably discovered earlier. Next we compute this measure explicitly.

Theorem 4. *The invariant measure on the invariant curve corresponding to the value J of Joachimsthal integral and other related quantities given by Theorem 3 is given by the formula:*

$$d\mu = \frac{d\psi}{\sqrt{a^2 - c^2 \sin^2 \psi}\sqrt{(1 - J^2 a^2) + J^2 c^2 \sin^2 \psi}}.$$

Therefore the measure of the arc $[0, \psi]$ equals

$$\mu([0, \psi]) = \frac{1}{cf} F\left(\varphi, \frac{1}{f}\right), \quad \varphi = \arcsin \sqrt{\frac{(d+1)\tan^2 \psi}{(d+1)\tan^2 \psi + d}},$$

$$d = \frac{1 - J^2 a^2}{J^2 c^2} = \frac{(b^2 - \lambda)c^2}{\lambda} = \frac{f^2 - 1}{e^2 - f^2} e^2 > 0$$

The measure of the whole invariant curve equals

$$U = \frac{4}{cf} F\left(\frac{\pi}{2}, \frac{1}{f}\right).$$

Here and below e, f are the eccentricities of E, E_λ and $F(\varphi, k) = \int_0^\varphi \frac{dt}{\sqrt{1 - k^2 \sin^2 t}}$ is the elliptic integral of the first kind.

Proof. The invariant measure on the curve $\{F = const\}$ can be written as:

$$d\mu = \frac{1}{F_p} d\varphi.$$

Due to explicit form of F in (p, φ) coordinates (Theorem 3) we have:

$$d\mu = \frac{1}{p} d\varphi$$

We compute using the change of variable on the invariant curve $\varphi \to \psi$ (see Figure 6):

$$\varphi = \psi + \delta(\psi).$$

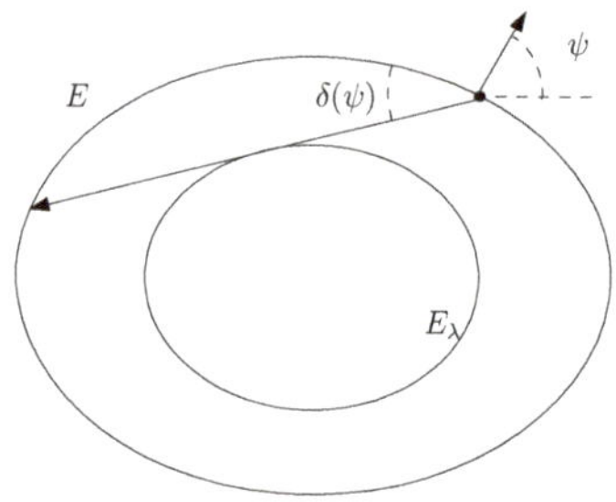

Figure 6. Change of variable on the invariant curve $\varphi \to \psi$ $\varphi = \psi + \delta(\psi)$.

Using the formula

$$p = h(\psi)\cos\delta(\psi) + h'(\psi)\sin\delta(\psi)$$

We have

$$d\mu = \frac{1}{p} d\varphi = \frac{(1 + \delta'(\psi))d\psi}{h(\psi)\cos\delta(\psi) + h'(\psi)\sin\delta(\psi)} \tag{2}$$

Next we use the explicit form of Joachimsthal integral:

$$Jh(\psi) = \sin\delta(\psi)$$

and hence also

$$Jh'(\psi) = \cos\delta(\psi)\delta'(\psi).$$

Substituting into (2) we get

$$d\mu = \frac{1}{p} d\varphi = \frac{J(1 + \delta'(\psi))d\psi}{\sin\delta(\psi)\cos\delta(\psi)(1 + \delta'(\psi))} = \frac{Jd\psi}{\sin\delta(\psi)\cos\delta(\psi)} = \frac{d\psi}{h\sqrt{1 - h^2}}. \tag{3}$$

thus we compute using Lemma 1:

$$d\mu = \frac{d\psi}{\sqrt{a^2\cos^2\psi + b^2\sin^2\psi}\sqrt{1 - J^2(a^2\cos^2\psi + b^2\sin^2\psi)}} =$$

$$\frac{d\psi}{\sqrt{a^2 - (a^2 - b^2)\sin^2\psi}\sqrt{(1 - J^2a^2) + J^2(a^2 - b^2)\sin^2\psi}} =$$

$$\frac{d\psi}{\sqrt{a^2 - c^2\sin^2\psi}\sqrt{(1 - J^2a^2) + J^2c^2\sin^2\psi}}.$$

Therefore the measure of the segment $[0, \psi]$

$$\mu([0, \psi]) = \int_0^\psi \frac{d\psi}{\sqrt{a^2 - c^2\sin^2\psi}\sqrt{(1 - J^2a^2) + J^2c^2\sin^2\psi}}.$$

Changing to $x = \sin^2 \psi$ we get

$$\mu([0,\psi]) = \frac{1}{2}\int_0^{\sin^2\psi} \frac{dx}{\sqrt{x(1-x)}\sqrt{a^2-c^2x}\sqrt{(1-J^2a^2)+J^2c^2x}} =$$

$$\frac{1}{2Jc^2}\int_0^{\sin^2\psi} \frac{dx}{\sqrt{x(1-x)}\sqrt{e^2-x}\sqrt{d+x}} = \frac{g}{2Jc^2}F(\varphi,k),$$

where in the last step we used the reduction of the pseudo-elliptic integral to the elliptic integral of the first kind [6] (p. 112; integral 254.00). In the last formula $e = a/c$ is the eccentricity of the ellipse, $c = \sqrt{a^2-b^2}, d = \frac{1-J^2a^2}{J^2c^2}$. Now we need to compute parameters of the elliptic integral given in [6] (p. 112). In order to compute d we use Theorem 3

$$d = \frac{1-J^2a^2}{J^2c^2} = \frac{(b^2-\lambda)e^2}{\lambda} = \frac{f^2-1}{e^2-f^2}e^2 > 0, \tag{4}$$

$$J = \frac{\sqrt{e^2-f^2}}{ce\sqrt{e^2-1}}, \ \lambda = c^2(e^2-f^2),$$

where f is the eccentricity of E_λ. Moreover, we compute the parameters k, φ, g for the elliptic integral.

$$k = \sqrt{\frac{e^2+d}{e^2(1+d)}} = 1/f, \quad g = \frac{2}{\sqrt{e^2(1+d)}} = \frac{2\sqrt{e^2-f^2}}{fe\sqrt{e^2-1}}.$$

The angle φ is computed by the formula:

$$\sin^2\varphi = \frac{(d+1)(\sin^2\psi)}{\sin^2\psi+d} = \frac{(d+1)\tan^2\psi}{(d+1)\tan^2\psi+d}, \tag{5}$$

where d is given in (4). Next we see that the coefficient

$$\frac{g}{2Jc^2} = \frac{1}{cf}.$$

Thus, finally we have

$$\mu([0,\psi]) = \frac{1}{cf}F\left(\varphi,\frac{1}{f}\right), \quad U = \frac{4}{cf}F\left(\frac{\pi}{2},\frac{1}{f}\right),$$

where U is the measure of the whole curve. □

5. Mather β-Function

Now we are in position to find Mather β-function for ellipse stated in Theorem 1 and Corollary 1. We shall use the invariant measure and non-standard generating function S. Consider the invariant curve of the rotation number $\rho = \frac{m}{n}$ corresponding to the rational caustic E_λ and to the value J of Joachimsthal integral. We shall give a proof of the formula for rational rotation number ρ, but it is easy to see that it remains valid for irrational ρ.

Proof of Theorem 1. Let $\rho = m/n$ and $(p_i, \varphi_i), i = 1, \ldots, n$ denote the coordinates of the edges l_i of a Poncelet polygon. Set

$$\psi_i = \frac{\varphi_{i-1}+\varphi_i}{2}, \ \delta_i = \frac{\varphi_i-\varphi_{i-1}}{2}.$$

The perimeter of the Poncelet polygon can be computed by means of the generating function S given in Proposition 1 as follows (see [14]):

$$\beta\left(\frac{m}{n}\right) = \frac{1}{n}\sum_{i=1}^{n} S(\varphi_{i-1}, \varphi_i) = \frac{2}{n}\sum_{i=1}^{n} h(\psi_i)\sin\delta_i, \tag{6}$$

Next we integrate both sides of (6) with respect to the measure $d\mu$ and using the invariance of the measure we get:

$$\beta\left(\frac{m}{n}\right)U = 2\int h(\psi)\sin\delta(\psi)d\mu,$$

where U is the measure of the whole curve. Thus, we have using the explicit expression of the measure (3):

$$\beta(\rho) = \frac{2}{U}\int_0^{2\pi}\frac{Jh\sin\delta}{\sin\delta\cos\delta}d\psi = \frac{2J}{U}\int_0^{2\pi}\frac{h}{\sqrt{1-J^2h^2}}d\psi.$$

Substitute the explicit formula for h we obtain:

$$\beta(\rho) = \frac{8J}{U}\int_0^{\pi/2}\frac{\sqrt{a^2\cos^2\psi + b^2\sin^2\psi}}{\sqrt{1-J^2(a^2\cos^2\psi + b^2\sin^2\psi)}}d\psi =$$

$$\frac{8J}{U}\int_0^{\pi/2}\frac{\sqrt{a^2 - c^2\sin^2\psi}}{\sqrt{1-J^2a^2+J^2c^2\sin^2\psi}}d\psi =$$

$$\frac{8Jc}{UJc}\int_0^{\pi/2}\frac{\sqrt{e^2-\sin^2\psi}}{\sqrt{d+\sin^2\psi}}d\psi = \frac{8}{U}\int_0^{\pi/2}\frac{\sqrt{e^2-\sin^2\psi}}{\sqrt{d+\sin^2\psi}}\frac{d\sin^2\psi}{2\sin\psi\cos\psi} =$$

$$\frac{4}{U}\int_0^1\frac{\sqrt{e^2-x}}{\sqrt{d+x}\sqrt{x(1-x)}}dx = \frac{4e^2g}{U}\left(\frac{k^2}{\alpha^2}F(\pi/2,k) - \left(\frac{k^2}{\alpha^2}-1\right)\Pi(\pi/2,\alpha^2,k)\right).$$

where we used the values $\alpha^2 = \frac{1}{1+d}$, $k = \frac{1}{f}$ and g, φ as above. This reduction to the complete elliptic integral of the third kind is given in [6] (p.112 integral 254.13 then 339.01). Next we use [6] (integral 414.01) for the complete integral $\Pi(\pi/2,\alpha^2,k) =: \Pi(\alpha^2,k)$ and finally obtain:

$$\beta(\rho) = \frac{4e^2g}{U}\left(\frac{k^2}{\alpha^2}K(k) - \left(\frac{k^2-\alpha^2}{\alpha^2}\right)\left(K(k) + \frac{\alpha[K(k)E(\phi,k) - E(k)F(\phi,k)]}{\sqrt{(1-\alpha^2)(k^2-\alpha^2)}}\right)\right),$$

where $\phi = \arcsin(\alpha/k)$. Simplifying we get:

$$\beta(\rho) = \frac{4e^2g}{U}\left(K(k) - \left(\frac{\sqrt{k^2-\alpha^2}}{\alpha\sqrt{1-\alpha^2}}\right)[K(k)E(\phi,k) - E(k)F(\phi,k)]\right).$$

Substituting the values of parameters

$$g = \frac{2\sqrt{e^2-f^2}}{fe\sqrt{e^2-1}},\ U = \frac{4}{cf}K(k),\ k = 1/f,\ \alpha^2 = \frac{e^2-f^2}{f^2(e^2-1)},\ k^2-\alpha^2 = \frac{f^2-1}{f^2(e^2-1)}$$

we get:

$$\beta(\rho) = \frac{2ce\sqrt{e^2-f^2}}{e^2-1} - \frac{2cf}{K(k)}[K(k)E(\phi,k) - E(k)F(\phi,k)],$$

$$\phi = \arcsin\sqrt{\frac{e^2 - f^2}{e^2 - 1}} = \arcsin\frac{\sqrt{\lambda}}{b}.$$

□

Proof of the Corollary 1. This follows immediately from Theorem 1 using the following relations. The first is on the perimeter of the ellipse $|E_\lambda|$, and the second for the rotation number ρ which we shall prove in Theorem 2 in Section 7.

$$|E_\lambda| = 4\int_0^{\pi/2}\sqrt{(a^2-\lambda) - c^2\sin^2 t}\,dt = 4\sqrt{a^2-\lambda}\,E(k),\ k = 1/f,\ cf = \sqrt{a^2-\lambda},$$

where f is the eccentricity of E_λ as above.

$$\rho = \frac{F(\phi, k)}{2K(k)}.$$

□

6. Mather β-Function and the Lazutkin Parameter

Let me remind the notion of the Lazutkin parameter. Given a convex caustic $\mathcal{C}$ of convex billiard curve γ (not necessarily ellipse), one has a conservation law stating that for any point $P \in \gamma$ the Lazutkin parameter

$$L := |PX| + |PY| - |\widehat{XY}|$$

does not depend on the point P (see [23]). Here $X, Y \in \mathcal{C}$ are the tangency points of tangent lines to $\mathcal{C}$ from P and overarc denotes the arc between the indicated points.

Suppose $P_i, i = 1, \ldots, n$ are the vertices of billiard n-periodic trajectory $\mathcal{P}$ making m turns. For any vertex P_i we write the Lazutkin parameter:

$$L = |P_iX_i| + |P_iY_i| - |\widehat{X_iY_i}|, \quad i = 1, \ldots, n.$$

Summing these identities we get

$$nL = |\mathcal{P}| - m|\mathcal{C}|.$$

Dividing by n, we obtain the general formula (see [19]), valid for any billiard with convex caustic $\mathcal{C}$:

$$\beta(\rho) = L + \rho|\mathcal{C}|,$$

where L is Lazutkin parameter, $\rho = \frac{m}{n}$ is the rotation number and $|\cdot|$ is the perimeter. Comparing the last formula with one of Corollary 1 we get the following:

Corollary 2. *For the Lazutkin parameter L of the caustic E_λ of the elliptic billiard E we have the following formula:*

$$L(E_\lambda) = \frac{2a\sqrt{\lambda}}{b} - 2\sqrt{a^2-\lambda}\,E(\phi, k).$$

7. Rotation Number ρ

In this section, we give another derivation of the formula for the rotation number ρ corresponding to caustic E_λ [1,9].

Theorem 5. *For the invariant curve corresponding to caustic E_λ having eccentricity f the rotation number is:*

$$\rho = \frac{F(\phi, k)}{2K(k)},\ k = \frac{1}{f},\ \phi = \arcsin\sqrt{\frac{e^2 - f^2}{e^2 - 1}} = \arcsin\frac{\sqrt{\lambda}}{b},$$

where $K(k) = F(\pi/2, k)$ is the complete elliptic integral.

Example 3. *We see from this Theorem that for $\lambda \to 0$ the $\phi \to 0$ and hence $\rho \to 0$. On the other hand if $\lambda \to b$ that is $f \to 1$ we have $\phi \to \pi/2$ and hence $\rho \to 1/2$.*

Remark 1. *This formula is given in [1,9]. A beautiful method to get formula for rotation number is given in [24]. Unfortunately there is a computational mistake for the integrals at the end of page 298. Another formula for the rotation number is given without proof in [13]. However, in that formula $f \to 1$ does not imply to $\rho \to 1/2$.*

Proof of Theorem 5. We shall use the formula for rotation number:

$$\rho = \mu[\psi, T(\psi)]/U,$$

where ψ is a point on the curve and $T(\psi)$ its image. This is independent on the choice of ψ since measure μ is invariant. Here and below we use ψ as a coordinate on the invariant curve related to the angle φ by the formula $\varphi = \psi + \delta(\psi)$ as before. Now we shall choose ψ in this formula in such a way that the segment $[\psi, T(\psi)]$ is vertical and tangent to E_λ (see Figure 7):

$$\psi = -\theta \quad and \quad T(\psi) = \theta.$$

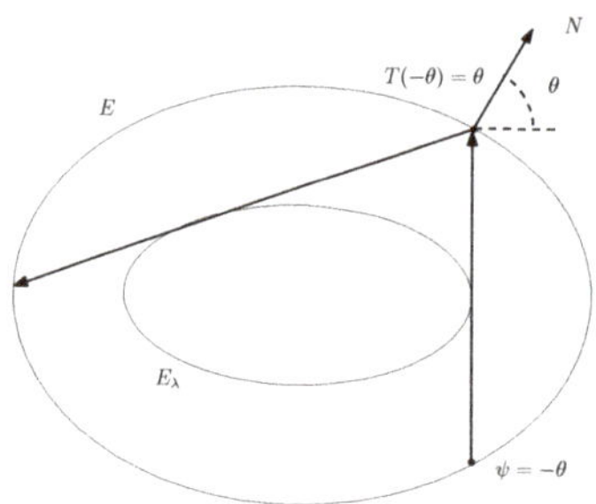

Figure 7. For computation of ρ.

We can easily compute θ using the normal vector $N = \left(\frac{x}{a^2}, \frac{y}{b^2}\right)$, where by the definition of θ, we have

$$x = \sqrt{a^2 - \lambda}, \quad y = \frac{\sqrt{\lambda} b}{a}.$$

Hence we get:

$$\tan\theta = \frac{a^2}{b^2}\frac{\sqrt{\lambda}b}{a\sqrt{a^2-\lambda}} = \frac{a}{b}\frac{\sqrt{\lambda}}{\sqrt{a^2-\lambda}} = \frac{e\sqrt{e^2-f^2}}{f\sqrt{e^2-1}}. \tag{7}$$

It then follows from Theorem 4 that

$$\mu[-\theta, \theta] = 2\mu[0, \theta] = \frac{2}{cf}F\left(\phi, \frac{1}{f}\right),$$

where

$$\sin^2\phi = \frac{(d+1)\tan^2\theta}{(d+1)\tan^2\theta + d}.$$

Substituting $d, d+1$ from (1) and $\tan\theta$ from (7) we obtain:

$$\sin\phi = \sqrt{\frac{e^2 - f^2}{e^2 - 1}} = \frac{\sqrt{\lambda}}{b}.$$

Thus, we have for the rotation number:

$$\rho = \frac{2}{U}\mu[0,\theta] = \frac{2}{cfU}F\left(\phi, \frac{1}{f}\right) = \frac{1}{2F\left(\frac{\pi}{2}, \frac{1}{f}\right)}F\left(\phi, \frac{1}{f}\right),$$

$$\phi = \arcsin\sqrt{\frac{e^2 - f^2}{e^2 - 1}} = \arcsin\frac{\sqrt{\lambda}}{b}.$$

□

Remark 2. *Analogously to the proof of Theorem 1 the following relation can be derived:*

$$\rho = \frac{4}{\pi U}\int_0^{\pi/2} \frac{\arcsin\left(J\sqrt{a^2\cos^2\psi + b^2\sin^2\psi}\right)\, d\psi}{\sqrt{a^2\cos^2\psi + b^2\sin^2\psi}\sqrt{1 - J^2(a^2\cos^2\psi + b^2\sin^2\psi)}}. \quad (8)$$

Indeed, by the the following formula holds for (m, n)*-periodic:*

$$2\pi m = \sum_{i=1}^{n} 2\delta_i,$$

because $2\delta_i$ *is the angle between the edges* l_{i-1} *and the* l_i*. Integrating this with respect to the invariant measure* $d\mu$ *we get:*

$$2\pi m\, U = 2n \int \delta d\mu.$$

Thus, we have

$$\rho = \frac{1}{\pi U}\int \delta d\mu = \frac{1}{\pi U}\int_0^{2\pi} \frac{J\delta}{\sin\delta\cos\delta} d\psi.$$

The last integral gives formula (8). Notice, that unlike Theorem (5), integral (8), cannot be reduced to elliptic integrals.

8. Mather β-Function and Rigidity

We start this section with the proof of Theorem 2.

Proof. The first step is based on a combination of several powerful results. By a Theorem of John Mather [25] the function β is differentiable at a rational point ρ if and only if there is an invariant curve consisting of periodic orbits with rotation number ρ. Moreover, all the orbits lying on these invariant curves are action minimizing. It then follows from Aubry-Mather theory and theorem of Mather on differentiability of β-function that there exist invariant curves of all rotation numbers $\rho \in (0, \frac{1}{4}]$, and these curves foliate the domain between the curve for $\rho = 1/4$ and the boundary of the phase cylinder **A** (see [19] for the argument in the case of circular billiard).

Therefore, billiard in Ω_2 meets the assumptions of [18] and hence must be an ellipse.

The last step is to show that this ellipse is an isometric copy of Ω_1. Indeed let $a_i > b_i$, $i = 1, 2$ are semi-axes of the two ellipses. First, take the value of the rotation number $\frac{1}{4}$ and use the equality of the β-functions at the value $1/4$. This yields

$$a_1^2 + b_1^2 = a_2^2 + b_2^2. \quad (9)$$

Second, mention that by the definition $\beta(0) = 0$ holds true for any domain. However, the derivative $\beta'(0)$ gives the circumference of the domain. Therefore, by the assumption of Theorem 2, we have $\beta_1'(0) = \beta_2'(0)$ and hence $|\Omega_1| = |\Omega_2|$, where $|\Omega|$ is the circumference of Ω. Next we use classical formula for $|\Omega|$ of arbitrary convex domain via the support function:

$$|\Omega| = \int_0^{2\pi} h(\psi) d\psi.$$

Therefore for the ellipses $\Omega_{1,2}$ we write

$$|\Omega_i| = 4\int_0^{\pi/2}\sqrt{\frac{a_i^2+b_i^2}{2}+\frac{a_i^2-b_i^2}{2}\cos 2\psi}\quad d\psi =$$

$$2\sqrt{2}\int_0^{\pi}\sqrt{(a_i^2+b_i^2)+(a_i^2-b_i^2)\cos t}\ dt = 2\sqrt{2}\int_0^{\pi}\sqrt{A+c_i^2\cos t}\ dt,$$

where we introduced $A := a_1^2+b_1^2 = a_2^2+b_2^2$. Consider now the last integral as a function of the parameter $C := c^2 = a^2-b^2$, while A is fixed.

$$f(C) := 2\sqrt{2}\int_0^{\pi}\sqrt{A+C\cos t}\,dt$$

Differentiating f with respect to C we obtain:

$$f' = \sqrt{2}\int_0^{\pi}\frac{\cos t}{\sqrt{A+C\cos t}}\,dt = \sqrt{2}\int_0^{\pi/2}\left[\frac{\cos t}{\sqrt{A+C\cos t}} - \frac{\cos t}{\sqrt{A-C\cos t}}\right]dt.$$

It is easy to see that the for $t \in (0,\pi/2)$ the integrand is negative, hence f is strictly monotone decreasing in C. Therefore, the equality $|\Omega_1| = |\Omega_2|$ is possible only when $C_1 = C_2$. This together with (9) implies that the ellipses are isometric. □

The second part of the given proof leads naturally to the following.

Question. How many values of β-function determine the ellipse in the class of ellipses. More precisely we ask if ellipse is determined by any two values of β-function $\beta(\rho_1), \beta(\rho_2)$ for the rotation numbers $\rho_{1,2} \in (0, \frac{1}{2}]$.

In order to prove this one needs more analysis of the formula of minimal action of Theorem 1. Notice that in [22] the reconstruction of ellipse is given by means of infinitesimal data of the β-function near 0.

A partial result in the direction of this question is the following

Theorem 6. *Ellipse can be determined by two values of $\beta(\rho_1), \beta(\rho_2)$ where $\rho_1 = \frac{1}{2}$ and $\rho_2 = \frac{m}{n}$ is any rational in $(0, \frac{1}{2})$.*

Proof. Notice first that $\beta(\frac{1}{2}) = 2a$ is the diameter of ellipse. We argue by contradiction. Suppose Ω_1, Ω_2 are two ellipses with the same diameter $2a$, satisfying $\beta_1(\frac{m}{n}) = \beta_2(\frac{m}{n})$, but $b_1 < b_2$, see Figure 8. In this case we can introduce a linear map A which is the expansion map along the y-axes transforming Ω_1 to Ω_2. Notice that A increases perimeter of any polygon.

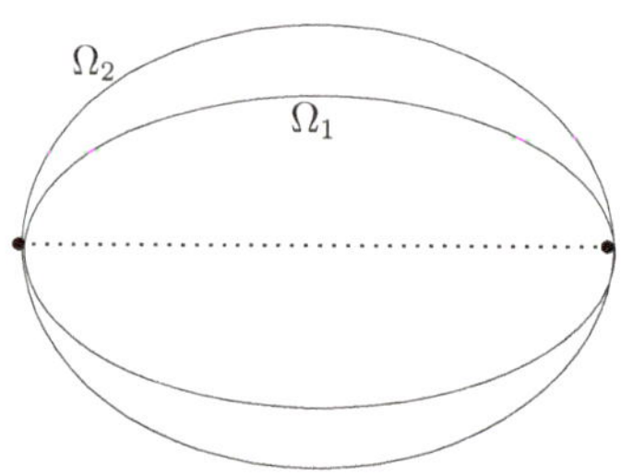

Figure 8. Ellipses with the same diameter.

Denote by P_1, P_2 two Poncelet polygons of the rotation number $\frac{m}{n}$ for Ω_1 and Ω_2, respectively. Obviously, the polygons $A(P_1)$ and P_2 have the same rotation number. The condition $\beta_1(\frac{m}{n}) = \beta_2(\frac{m}{n})$ implies that the perimeters of $P_{1,2}$ are equal:

$$|P_1| = |P_2|.$$

Hence we have the inequality

$$|A(P_1)| > |P_2|,$$

since A is expanding. However, this contradicts the fact P_2 is a Poncelet polygon is a length maximizer in its homotopy class. □

Remark. It is plausible that the result of Theorem 6 remains valid when the rotation number ρ_2 is irrational.

9. Discussion

Let me pose here most natural problems related the results of this paper:

1. Is it possible to relax symmetry assumption in the main Theorem 1? Our method of proof of Theorem 1 relies on the approach related to the so-called E.Hopf type rigidity phenomenon from [18]. This method is very robust and it is not clear at the moment how it can be generalized.
2. Another problem is to adopt our approach to a smaller neighborhood of the boundary of the phase cylinder.
3. All known approaches to rigidity in billiards, are based on the properties of orbits near the boundary. We believe there are rigidity results based on the behavior far from the boundary.
4. It would be interesting to prove that ellipse is determined by any two values of Mather β function $\beta(\rho_1), \beta(\rho_2)$ where ρ_1, ρ_2 are any two rotation numbers in $(0, \frac{1}{2})$.

Funding: M.B. was partially supported by ISF grant 580/20 and DFG grant MA-2565/7-1 within the Middle East Collaboration Program.

Institutional Review Board Statement: Not applicable.

Informed Consent Statement: Not applicable.

Data Availability Statement: Not applicable.

Acknowledgments: This paper is a continuation of our previous paper with Sergei Tabachnikov [14]. I am grateful to them for useful discussions and providing references.

Conflicts of Interest: The author declares no conflict of interest. The funders had no role in the design of the study; in the collection, analyses, or interpretation of data; in the writing of the manuscript; or in the decision to publish the results.

References

1. Kaloshin, V.; Sorrentino, A. On the local Birkhoff conjecture for convex billiards. *Ann. Math.* **2018**, *188*, 315–380. [CrossRef]
2. Garcia, R.; Koiller, J.; Reznik, D. Estimating Elliptic Billiard Invariants with Spatial Integrals. *J. Dyn. Control Syst.* **2022**, 1–11. [CrossRef]
3. Suris, Y.B. Billiards in confocal quadrics as a pluri-Lagrangian system. *Theor. Appl. Mech.* **2016**, *43*, 221–228. [CrossRef]
4. Bialy, M.; Mironov, A.E. Angular billiard and algebraic Birkhoff conjecture. *Adv. Math.* **2017**, *313*, 102–126. [CrossRef]
5. Bialy, M. Gutkin billiard tables in higher dimensions and rigidity. *Nonlinearity* **2018**, *31*, 2281–2293. [CrossRef]
6. Byrd, P.F.; Friedman, M.D. *Handbook of Elliptic Integrals for Engineers and Scientists*, 2nd ed.; Die Grundlehren der Mathematischen Wissenschaften; Springer: Berlin/Heidelberg, Germany, 1971; pp. xvi+358.
7. Zelditch, S. (Northwestern University, Evanston, IL 60208, USA). Personal communication, August 2022.
8. Zelditch, S. Survey on the inverse spectral problem. *Not. Int. Congr. Chin. Math.* **2014**, *2*, 1–20. [CrossRef]
9. Chang, S.-J.; Friedberg, R. Elliptical billiards and Poncelet us theorem. *J. Math. Phys.* **1988**, *29*, 1537. [CrossRef]
10. Dragović, V.; Radnović, M. *Poncelet Porisms and Beyond. Integrable Billiards, Hyperelliptic Jacobians and Pencils of Quadrics*; Frontiers in Mathematics; Birkhäuser/Springer Basel AG: Basel, Switzerland, 2011; pp. viii+293.
11. Stachel, H. On the motion of billiards in ellipses. *Eur. J. Math.* **2022**, 1–21. [CrossRef]
12. Stachel, H. The geometry of billiards in ellipses and their Poncelet grids. *J. Geom.* **2021**, *112*, 1–29. [CrossRef]
13. Tabanov, M.B. New ellipsoidal confocal coordinates and geodesics on an ellipsoid. *Math. Sci.* **1996**, *82*, 3851–3858. [CrossRef]
14. Bialy, M.; Tabachnikov, S. Dan Reznik's identities and more. *Eur. J. Math.* **2020**, 1–14. [CrossRef]
15. Reznik, D.; Garcia, R.; Koiller, J. Can the elliptic billiard still surprise us? *Math. Intell.* **2020**, *42*, 6–17. [CrossRef]
16. Reznik, D.; Garcia, R.; Koiller, J. Fifty new invariants of N-periodics in the elliptic billiard. *Arnold Math. J.* **2021**, *7*, 341–355. [CrossRef]

17. Akopyan, A.; Schwartz, R.; Tabachnikov, S. Billiards in ellipses revisited. *Eur. J. Math.* **2020**, 1–15. [CrossRef]
18. Bialy, M.; Mironov, A.E. The Birkhoff-Poritsky conjecture for centrally-symmetric billiard tables. *Ann. Math.* **2022**, *196*, 389–413. [CrossRef]
19. Siburg, K.F. *The Principle of Least Action in Geometry and Dynamics*; Lecture Notes in Mathematics, 1844; Springer: Berlin/Heidelberg, Germany, 2004.
20. Hezari, H.; Zelditch, S. One can hear the shape of ellipses of small eccentricity. *arXiv* **2019**, arXiv:1907.03882.
21. Huang, G.; Kaloshin, V.; Sorrentino, A. On the marked length spectrum of generic strictly convex billiard tables. *Duke Math. J.* **2018**, *167*, 175–209. [CrossRef]
22. Sorrentino, A. Computing Mather's β-function for Birkhoff billiards. *Discrete Contin. Dyn. Syst.* **2015**, *35*, 5055–5082. [CrossRef]
23. Tabachnikov, S. *Geometry and Billiards*; American Mathematical Society: Providence, RI, USA, 2005.
24. Kołodziej, R. The rotation number of some transformation related to billiards in an ellipse. *Studia Math.* **1985**, *81*, 293–302. [CrossRef]
25. Mather, J.N. Differentiability of the minimal average action as a function of the rotation number. *Bol. Soc. Brasil. Mat.* **1990**, *21*, 59–70. [CrossRef]

entropy

MDPI

Article

Ray–Wave Correspondence in Microstar Cavities

Julius Kullig * and Jan Wiersig

Institut für Physik, Otto-von-Guericke-Universität Magdeburg, Postfach 4120, D-39016 Magdeburg, Germany
* Correspondence: julius.kullig@ovgu.de

Abstract: In a previous work published by the authors in 2020, a novel concept of light confinement in a microcavity was introduced which is based on successive perfect transmissions at Brewster's angle. Hence, a new class of open billiards was designed with star-shaped microcavities where rays propagate on orbits that leave and re-enter the cavity. In this article, we investigate the ray–wave correspondence in microstar cavities. An unintuitive difference between clockwise and counterclockwise propagation is revealed which is traced back to nonlinear resonance chains in phase space.

Keywords: microcavities; quantum chaos; ray–wave correspondence

Citation: Kullig, J.; Wiersig, J. Ray–Wave Correspondence in Microstar Cavities. *Entropy* **2022**, *24*, 1614. https://doi.org/10.3390/e24111614

Academic Editor: Marko Robnik

Received: 14 October 2022
Accepted: 3 November 2022
Published: 5 November 2022

1. Introduction

Optical microcavities are fascinating systems to confine light on a very small scale [1,2]. Traditional whispering-gallery cavities rely on successive total internal reflections at the cavity's dielectric interface which prevent a ray from leaving the cavity. Correspondingly, long-lived whispering-gallery modes exist as solutions of Maxwell's equations. Such whispering-gallery cavities appear, e.g., as microdisks [3–6], microtoroid [7–9], microspheres [10], or microjet cavities [11]. They have attracted an immense amount of research interest, e.g., for quantum chaos [12–15] or exceptional points in non-Hermitian systems [16–20] and, additionally, offer a lot of applications, e.g., the generation of optical frequency combs [21], as microlasers with directional emission [22–25], optical sensors for nanoparticles [26–28] or rotating motion [29], or orbital angular momentum lasers [30].

In a former article [31], an alternative concept for light confinement was introduced which is completely different from the traditional whispering-gallery design. The perfect transmissions at Brewster's angle are utilized such that a ray can leave (and reenter) the cavity without partial back reflection. Thus, via successive transmissions through the dielectric interface of the cavity, rays are guided along a periodic orbit without loss of intensity. This idea can be implemented in a star-shaped cavity as shown in Figure 1a or in Brewster-notched cavities [32].

In this article, we further investigate the ray–wave correspondence in the microstar cavity. In particular we look at (asymmetric) deformed microstar cavities where the formation of the modes is governed by the regular dynamics around an elliptical fixed point in phase space. In the semiclassical regime, we observe modes localizing along the regular structures in phase space, while this localization might be expected from quantum-chaos theory ,we can also identify an unexpected difference between clockwise (CW) and counterclockwise (CCW) propagation in the star cavity. In the previous literature, a general imbalance of CW and CCW propagation due to asymmetric backscattering was discussed for microcavities [19,33–35]. However, in this article, we study the difference pattern between CW and CCW propagation that is due to a nonuniform loss across a resonance chain. Therefore, the revealed mechanism is a general feature of open systems and not restricted to asymmetric boundary deformations. In addition, a Frobenius–Perron operator (FPO) formalism is used as a connection between ray dynamics and optical modes.

The paper is organized as follows. In Section 2, the ray dynamics of the microstar cavity are issued with a special focus on the dynamics of CW and CCW propagation in phase space. The corresponding impacts on the wave dynamics are discussed in Section 3. A conclusion is given in Section 4.

2. Ray Dynamics and Phase-Space Description for Microstar Cavities

A microstar is a quasi-two-dimensional cavity with homogeneous refractive index n [31]. It is characterized by the number of spikes ν. In order to guide a ray via successive transmissions at Brewster's angle $\chi_B = \arctan(1/n)$ along a closed orbit, both the refractive index and the opening angle α of the spikes need to be adjusted as

$$n = \frac{1+\sin(\pi/\nu)}{\cos(\pi/\nu)} \tag{1}$$

$$\alpha = \frac{\nu-2}{2\nu}\pi. \tag{2}$$

In the following, a microstar with $\nu = 9$ spikes is considered with a refractive index of $n \approx 1.43$ and spike opening angle $\alpha = 70°$. The microstar with such a polygonal shape supports a family of periodic orbits propagating without loss of intensity. However, these periodic orbits are only marginally stable. A slight boundary deformation of the microstar can be used to stabilize the dynamics around one central orbit as shown in Figure 1a. In polar coordinates (r, ϕ), the boundary of the most right deformed spike can be expressed as

$$r(\phi) = \frac{1}{n\sin(\phi)+\cos(\phi)} - \varepsilon_1\phi(\phi - \tfrac{\pi}{\nu}) \quad \text{for } \phi \in [0, \pi/\nu] \tag{3}$$

$$r(\phi) = \frac{1}{n\sin(\phi)+\cos(\phi)} + \varepsilon_2\phi(\phi + \tfrac{\pi}{\nu}) \quad \text{for } \phi \in [-\pi/\nu, 0] \tag{4}$$

with ϵ_1 (ϵ_2) being the deformation parameter of the upper (lower) segment of the spike. Thus, the boundary of the deformed microstar is given by periodic continuation. Choosing $\varepsilon_1 \neq \varepsilon_2$ allows for an asymmetric boundary deformation without a mirror reflection symmetry of the cavity.

The ray dynamics in the microstar are conveniently expressed in a Poincaré phase-space section, in the following phase space for short, which is introduced in Figure 1. Whenever a ray propagating along the spikes is passing the abscissa, its position $x > 0$ and the transversal momentum $p = \sin\theta$ are recorded. For a microstar cavity without deformed boundary, i.e., $\epsilon_1 = \epsilon_2 = 0$, a family of marginally stable orbits exists at $p = 0$, whereas in the deformed case, an elliptical fixed point with a region of regular dynamics can be achieved. For a boundary deformation with a mirror-reflection symmetry ($\epsilon_1 = \epsilon_2$), the fixed point is either at $p = 0$ or occurs pairwise at $\pm|p|$. Contrary, for an asymmetric deformation ($\epsilon_1 \neq \epsilon_2$), a single elliptical fixed point can occur at a finite nonzero momentum, as in Figure 1c with $(\epsilon_1, \epsilon_2) = (0.76, 1.2)$. For this deformation, a 5:1 resonance chain exists in the regular region, which is of further interest below. In comparison, the deformation parameters $(\epsilon_1, \epsilon_2) = (0.78, 1.212)$ lead to an island of regular dynamics with a similar shape but without a (macroscopic) resonance chain, see Figure 1d.

Note that the phase space represents CCW propagating rays. The corresponding rays propagating in CW direction are given by time reversal of the (geometrical) dynamics. For the phase-space variable p describing the momentum along the x-axis, this relates to a change in the sign, cf. Figure 1b. A mirror reflection, however, would not change the sign of p. Therefore, even in a cavity with a mirror-reflection symmetry, CW and CCW dynamics can have differences, namely in the direction along the symmetry line.

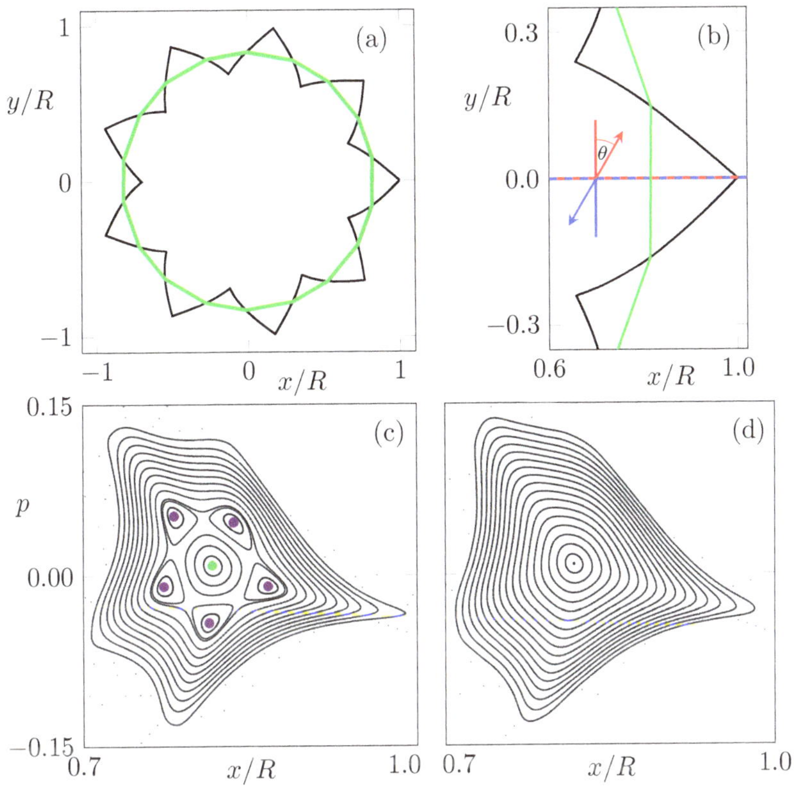

Figure 1. In (**a**), a microstar cavity with $\nu = 9$ asymmetric spikes is shown as a black curve. The simple periodic orbit is shown by the green curve. In (**b**), a magnification of one spike is shown. Propagation in CCW (CW) direction is illustrated by red (blue) arrows. Figure (**c**) shows a phase-space portrait of the microstar with deformation parameters $(\epsilon_1, \epsilon_2) = (0.76, 1.2)$, as in (**a**,**b**). The simple periodic orbit is shown as a green dot. A period-5 orbit is shown as magenta dots. (**d**) Phase-space portrait for deformation parameters $(\epsilon_1, \epsilon_2) = (0.78, 1.212)$.

In the following, we analyze the difference between CW and CCW propagation. Therefore, initial conditions (x_0, p_0) on a uniform grid in phase space with unit intensity $I_0 = 1$ are iterated N_{turns} turns in CW and in CCW direction. The remaining fractions of the intensity I_{CCW} in CCW direction are shown in Figure 2a–d for $N_{\text{turns}} = 1, 15, 50, 100$, respectively. As can be seen, the loss introduced by partial back-reflections from Fresnel formulas is not uniform in the regular region, while the central fixed point suffers almost no loss due to Brewster-angle transmissions, the border of the regular region has more loss, which results in a loss gradient over the resonance chain. A consequence is a distortion of the intensity pattern, which can be seen in Figure 2b,c. This distortion is different for both propagation directions, resulting in a remarkable difference $I_{\text{CCW}} - I_{\text{CW}}$ around the resonance chain; see Figure 2e–h. Note that the effect becomes considerably weaker for long times where the only remaining intensity is centered around the central fixed point.

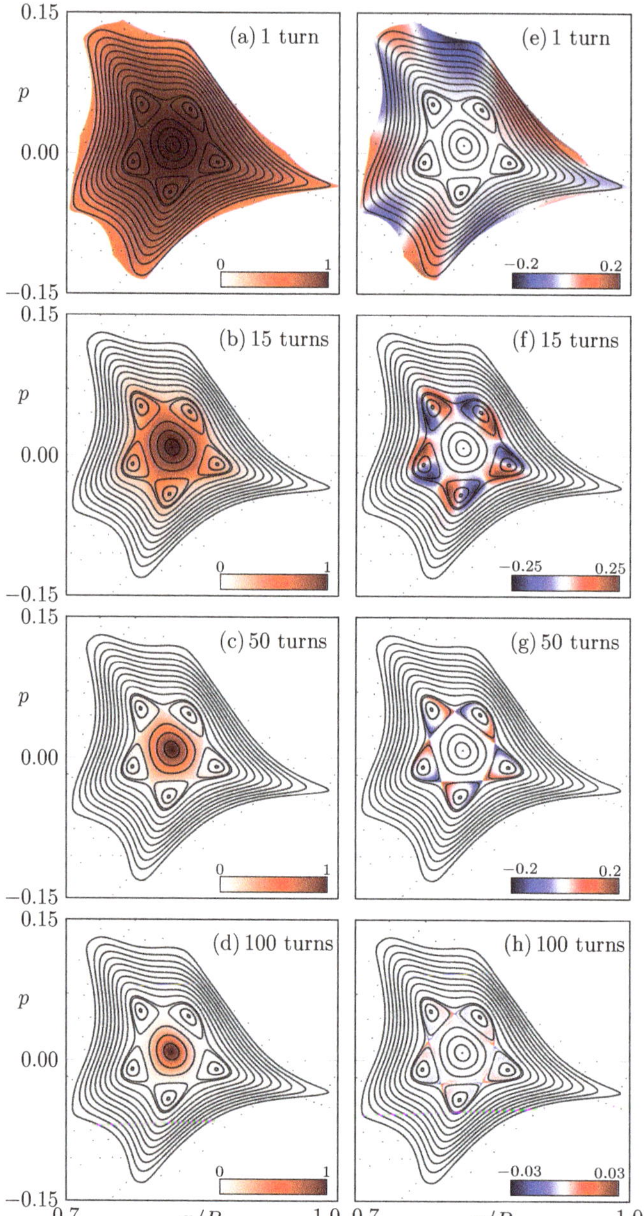

Figure 2. The intensity distribution (x_0, p_0, I_{CCW}) of rays with initial conditions $(x_0, p_0, I_0 = 1)$ after $1, 15, 50, 100$ turns in CCW direction is shown in (**a–d**). Correspondingly, (**e–h**) show the distribution $(x_0, p_0, I_{CCW} - I_{CW})$ of the difference between propagation in CCW and CW direction. Note the adjusted scale in the colormap for (**e–h**).

Intuitively, the difference between both propagation directions can be understood by analyzing the dynamics around the resonance chain, as shown in Figure 3. Focusing on one unstable orbit of the resonance chain (see red dot in Figure 3), the dynamics on the left side push the orbits closer to the center of the regular region, whereas on the right side, the orbits are pushed away from the regular region's center. Hence, orbits on the left side of the unstable orbit will suffer less loss then orbits on the right side in the next iterations. This behavior is inverted when the dynamics are reversed from CCW to CW, which leads to the characteristic pattern of the intensity difference in Figure 2f–h.

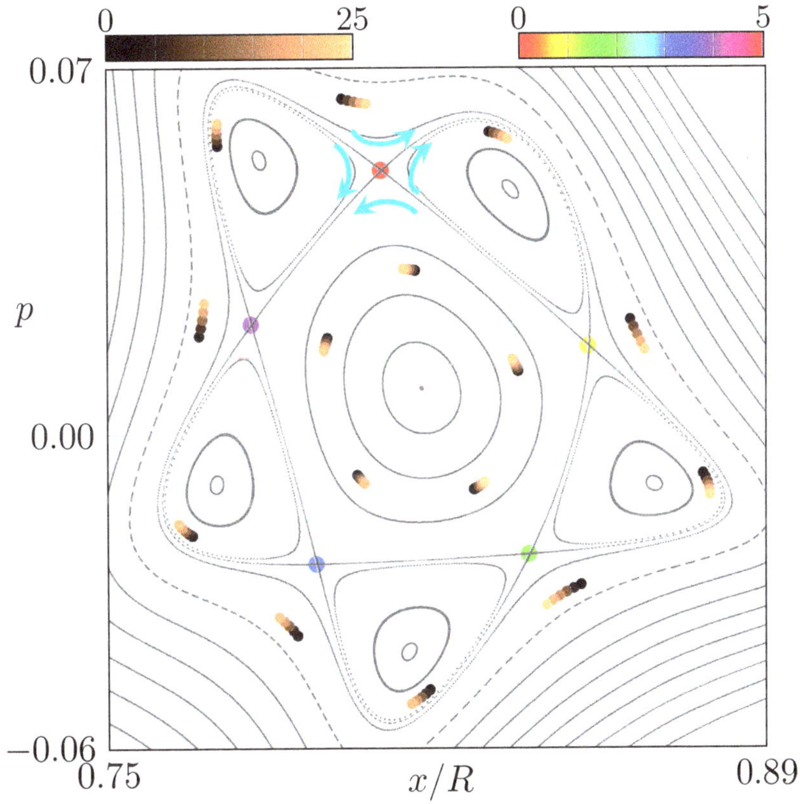

Figure 3. Magnification of the phase-space portrait from Figure 1c around the 5:1 resonance chain. Iterations for selected (nonperiodic) orbits are shown as dots colored from black to orange. The unstable period-5 orbit is represented by dots with a cyclic colormap. The arrows illustrate the dynamics around one of the unstable fixed points. The deformation parameters of the microstar are $(\epsilon_1, \epsilon_2) = (0.76, 1.2)$.

For comparison, in Figure 4, the system without (macroscopic) resonance chain is investigated. Note that here the difference between intensities I_{CCW} and I_{CW} is much weaker and not structured in phase space.

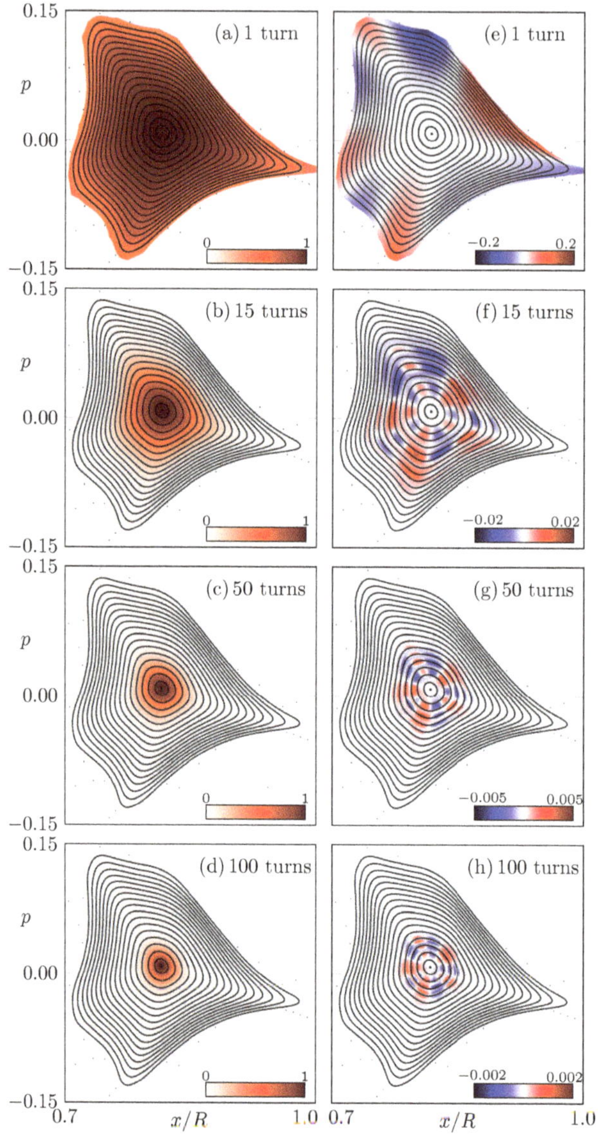

Figure 4. Similar as in Figure 2, **(a–d)** show the intensity distributions $(x_0, p_0, I_{\text{CCW}})$ and **(e–h)** show the distributions $(x_0, p_0, I_{\text{CCW}} - I_{\text{CW}})$, but for deformation parameters $(\epsilon_1, \epsilon_2) = (0.78, 1.212)$, where no macroscopic resonance chain occurs in the phase space. Note the adjusted scale in the colormap for **(e–h)**.

Frobenius–Perron Operator

The Frobenius–Perron operator (FPO) describes the evolution of density (intensity) distributions in phase space [36]. It has been used for, e.g., maps [37–41], billiard systems [37], and the asymmetric backscattering in deformed microcavities [42]. Here, the aim is to construct classical intensity eigenstates for CCW and CW propagation in the deformed microstar via the FPO. For this purpose, Ulam's method [43] is used to construct a matrix approximation $\mathcal{F}$ of the FPO as follows. The part of the phase space between $x \in [r(\phi/\nu), r(0)]$ and $p \in [-0.2, 0.2]$ covering the regular island is divided into a lattice of

$N \times N$ uniform rectangular cells. In the cell i, random initial conditions with total intensity I_i are iterated once in (C)CW direction. The transferred intensity I_f to the phase-space cell f is then recorded. Thus, the two $N^2 \times N^2$ matrices with the elements $\mathcal{F}_{fi}^{(C)CW} = I_f / I_i$ are an approximation for the intensity transport in CW and CCW direction, respectively. As such, the eigenvalues are inside the unit circle, as shown in Figure 5a,b. Of special interest are the eigenstates of the FPO with the eigenvalue of largest magnitude, as they represent states with a long lifetime. For the system with a resonance chain, these long-lived eigenstates for (C)CW propagation are shown in Figure 5c–k, in addition to their differences. As it can be seen, the localization of the eigenstates around the central fixed point depends on the fineness of the discretization given by N. This is consistent with the previous literature on FPO eigenstates of maps with a mixed phase space [39,44]. For an infinite discretization, the intensity is localized on the stable periodic orbit, which is a single point in phase space. However, for a more rough discretization, i.e., $N \lesssim 200$, the eigenstates extend over the resonance chain leading to a local intensity distortion, which is different for CW and CCW eigenstates. Consequently, the difference between CW and CCW eigenstates is significant around the resonance chain and consistent with the ray dynamics shown in Figure 2f,g. In particular, the pattern with alternating preference of CW or CCW propagation along the stable and unstable points of the resonance chain is visible.

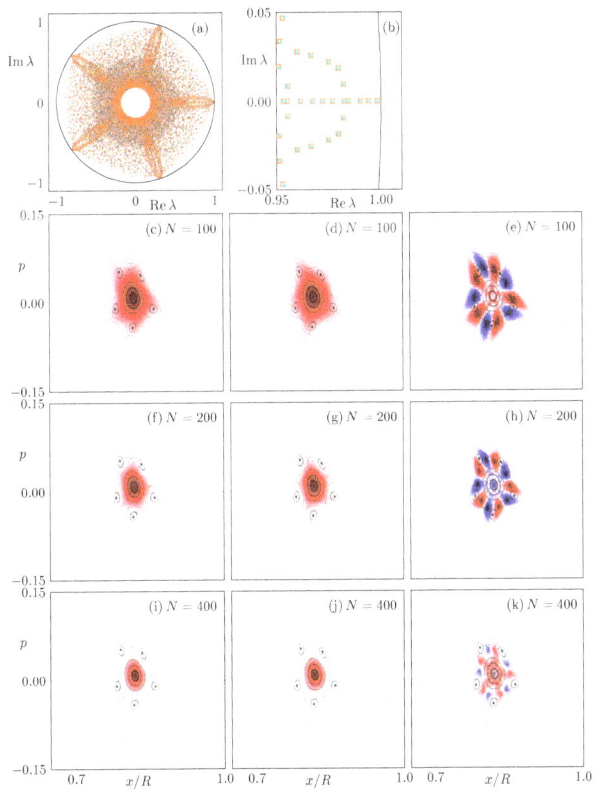

Figure 5. (**a**) The 10,000 largest-magnitude eigenvalues λ of the FPO with $N = 200$ are shown in the complex plane. Orange squares (blue crosses) represent the eigenvalues of the CCW (CW) FPO. Note that they are often on top of each other. (**b**) Shows a magnification of (**a**) around unity. (**c**–**e**) show the eigenstates corresponding to the eigenvalue of the largest modulus for CCW direction, CW direction, and the difference between both eigenstates for $N = 100$. The same sequence is shown in (**f**–**h**) for $N = 200$ and in (**i**–**k**) for $N = 400$.

Note that the shown difference pattern between CW and CCW propagation is not due to an asymmetry of the cavity's boundary. In fact, for deformation parameters $\epsilon_1 = \epsilon_2 = 0.98$, the microstar has a mirror-reflection symmetry, a similar 5:1 resonance chain in the phase space, and a similar alternating pattern of the eigenstate difference (not shown).

In Figure 6, the eigenstates corresponding to the second, third, and fourth eigenvalues next to unity are shown. As such an eigenstate has positive and negative parts, it does not represent a proper intensity distribution by itself. However, for the dynamics of an arbitrary (positive) phase-space intensity on intermediate timescales, these eigenstates still provide useful information. Especially, for the second (Figure 6a,b) and fourth (Figure 6e,f) eigenstates, the chirality around the resonance chain as well as the difference between CW and CCW propagation can be seen clearly.

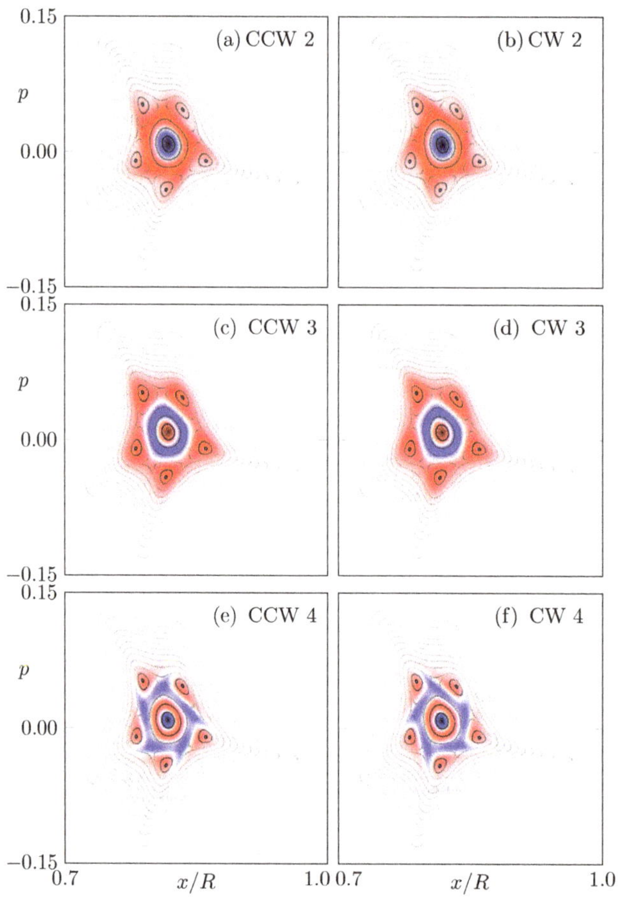

Figure 6. Eigenstates of the FPO ($N = 200$) with (**a**,**b**) second, (**c**,**d**) third, and (**e**,**f**) fourth eigenvalues next to unity. The left (right) panel shows the eigenstates for CCW (CW) propagation.

For a comparison, the eigenvalues and eigenstates of the FPO for the system without a resonance chain are shown in Figure 7. The eigenstates also localize at the center of the regular region, but they do not show a significant chiral behavior, i.e., the intensity distribution along the regular orbits is uniform.

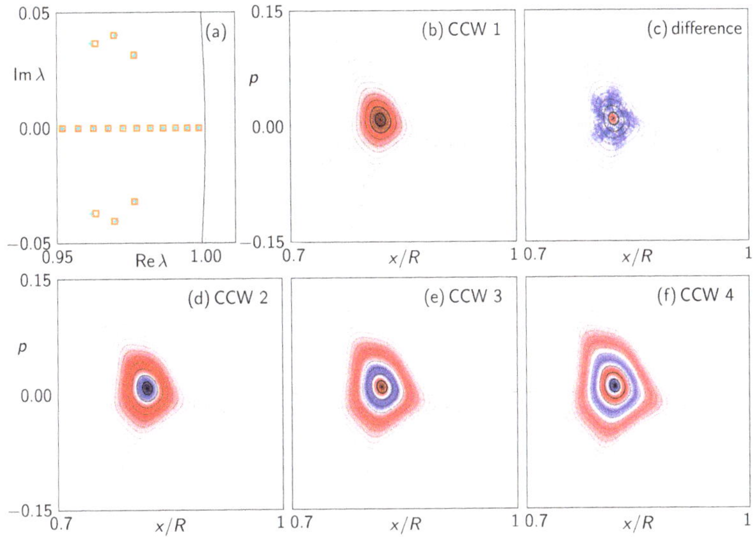

Figure 7. (**a**) Eigenvalues close to unity of the FPO ($N = 200$) for the system without a resonance chain characterized by $(\epsilon_1, \epsilon_2) = (0.78, 1.212)$ (cf. Figure 4). (**b**) Eigenstate of the FPO for the CCW dynamics with largest-modulus eigenvalue. (**c**) Difference between CCW and CW eigenstate. (**d**–**f**) Next eigenstates of the FPO in CCW direction with eigenvalues close to unity.

Note that in contrast to ref. [42], here, the true-time mapping is not included in the FPO. However, this incorporation of different time scales for the iteration is not relevant here, as the rays propagating through the spikes of the microstar have almost the same optical path length regardless of whether they propagate with a large or small radius.

3. Wave Dynamics

In this section, the wave dynamics of the microstar are analyzed with a focus on the semiclassical regime and the difference between CW and CCW propagation. For quasi-two-dimensional microcavities, Maxwell's equations reduce to the scalar mode Equation [45]

$$\left[\Delta + n^2k^2\right]\psi = 0 \tag{5}$$

where $k = \omega/c$ is the complex frequency which determines the Q-factor (lifetime) of a mode via $Q = -\mathrm{Re}\,k/(2\mathrm{Im}\,k)$. To utilize Brewster's angle, here, transverse electrical polarization is chosen such that the wave function ψ represents the magnetic field as $\vec{H} = (0, 0, \mathrm{Re}\,[\psi e^{-i\omega t}])$. Hence, along the cavities interface ψ, and its scaled normal derivative $n^{-2}\partial_{\vec{v}}\psi$ are continuous. In addition, ψ is required to fulfill Sommerfeld's outgoing wave condition. For the numerical calculation of the long-lived modes in the microstar, the boundary element method [46] and FEM software COMSOL [47] are used. In Figure 8, the long-lived modes in the complex frequency plane in the semiclassical regime with $350 \leq \mathrm{Re}\,kR \leq 600$ are shown. As in the wave regime discussed in ref. [31], the most long-lived mode localizes along the stable periodic orbit passing through the spikes of the microstar; see Figure 9d. However, additional long-lived modes exist in the semiclassical regime that also pass through the spikes but have more intensity maxima in radial direction; see Figure 9a–c. Thus, a classification of these modes with a radial mode number l is suitable.

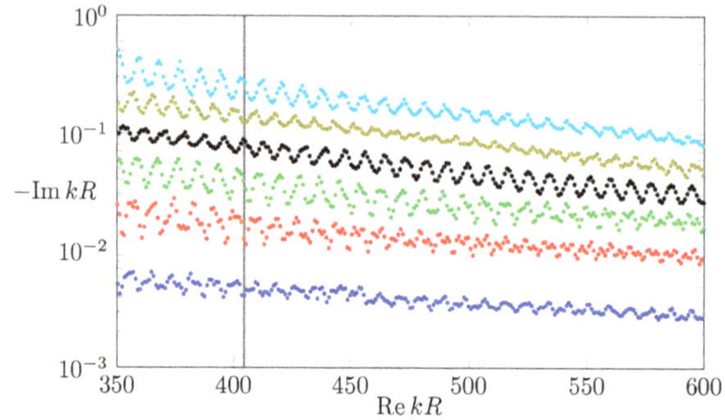

Figure 8. Dimensionless frequencies kR of the long-lived modes in the microstar cavity that can be classified by their mode number in radial direction by (blue) $l = 1$, (red) $l = 2$, (green) $l = 3$, (black) $l = 4$, (yellow) $l = 5$, and (cyan) $l = 6$. The gray line indicates the frequency of the modes shown in Figure 9.

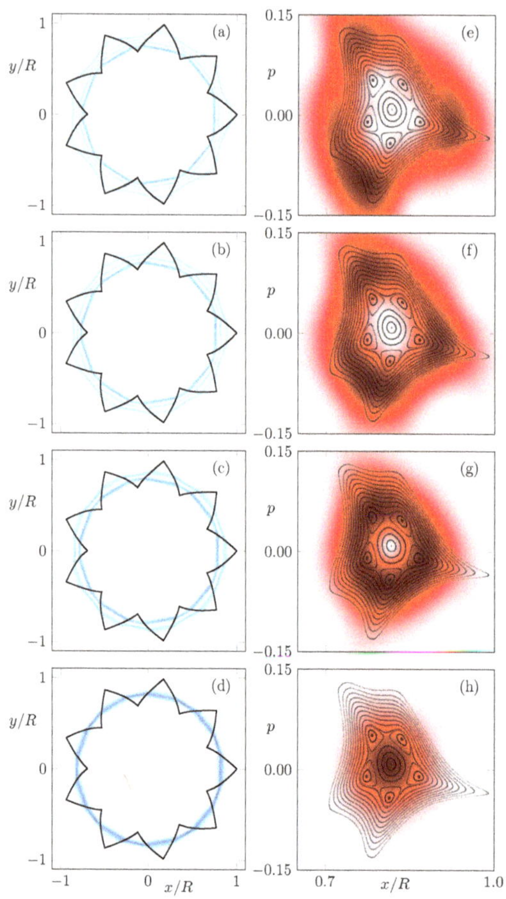

Figure 9. (**a–d**) shows the intensity pattern of the modes with radial mode number $l = 4$ to $l = 1$ and $\mathrm{Re}\,kR \approx 404$ (see gray line in Figure 8). (**e–h**) represents the Husimi projection of the corresponding mode.

To compare the wave and ray dynamics in phase space, the Husimi function $\mathcal{H}(x,p;\psi,\partial_\nu\psi)$ of a mode ψ is used. Here, we adapt the boundary Husimi function [48] to the phase space taken along the x-axis (see Figure 1). Thus, for $x \in [0.5R, 1.1R]$, the Husimi function (for CCW propagation) is constructed via

$$\mathcal{H}(x,p;\psi,\partial_\nu\psi) = \frac{1}{\mathcal{N}}\left|kFh_\psi(x,p) + \frac{i}{F}h_{\partial_\nu\psi}(x,p)\right|^2 \quad (6)$$

where $F = \sqrt{n\sqrt{1-p^2}}$ and $\sigma = \sqrt{R/(nk)}$. The functions h_ψ and $h_{\partial_\nu\psi}$ are defined via an overlap with a Gaussian wave packet with

$$h_f(x,p) = \int_{x_{min}}^{x_{max}} \zeta(t;x,p) f(t)\,dt \quad (7)$$

$$\zeta(t;x,p) = \exp\left(\frac{(x-t)^2}{2\sigma^2} - ipt\right). \quad (8)$$

The normalization constant $\mathcal{N}$ in Equation (6) is chosen such that $\max_{x,p}\mathcal{H} = 1$. Note that for the magnetic field, the normal derivative along the x-axis can be computed via $\partial_\nu\psi = \partial_y H_z = -i\omega\varepsilon_0\varepsilon_r E_x$.

In Figure 9e–h, the Husimi functions of the corresponding modes are shown in the right panel. As an expected result, the Husimi function of the most long-lived mode localizes in the center of the regular island, see Figure 9h, with a maximum at the finite momentum $p \approx 0.01$, which nicely reflects the stable periodic orbit. Furthermore, as an expected result, the modes with higher radial mode numbers l have their Husimi functions localized on outer tori of the regular island. Hence, these states can be seen as higher-order quantized states on the regular region. However, in this regime of $\mathrm{Re}\,kR \approx 404$, the modes cannot localize on the smaller sub-islands of the resonance chain itself. Consequently, the Husimi functions of the related modes interpolate through the resonance chain.

To verify the different intensity distortion for CCW and CW propagation, a few individual modes with a larger dimensionless frequency $\mathrm{Re}\,kR > 1200$ are calculated. Then, for a mode, the Husimi function for CCW propagation $\mathcal{H}(x,p;\psi,\partial_\nu\psi)$ and the corresponding Husimi function for CW propagation $\mathcal{H}(x,-p;\psi,-\partial_\nu\psi)$ are calculated to obtain the difference.

$$\mathcal{I}(x,p;\psi) = \mathcal{H}(x,p;\psi,\partial_\nu\psi) - \mathcal{H}(x,-p;\psi,-\partial_\nu\psi). \quad (9)$$

Note that in Equation (9), each Husimi function is normalized as discussed above. Hence, a preferred sense of rotation in the mode ψ does not impact the difference $\mathcal{I}(x,p;\psi)$. In Figure 10a–i, the Husimi function and the difference $\mathcal{I}(x,p;\psi)$ are shown for the long-lived modes in the microstar with a resonance chain. It is remarkable that over the range of $kR \approx 1200$, $kR \approx 1280$ and $kR \approx 1360$, each mode and its partner mode with a quasi-degenerate frequency show almost the same pattern in $\mathcal{I}(x,p;\psi)$. This pattern is organized around the stable and unstable fixed point of the resonance chain and in a very good agreement with the pattern of the FPO eigenstates, e.g., in Figure 5e.

To highlight the importance of the resonance chain for this phenomenon, a mode pair in the microstar without macroscopic resonance chain is shown in Figure 10j–l. Already within the mode pair, the pattern $\mathcal{I}(x,p;\psi)$ is different and not organized around an obvious phase-space structure.

A simplified model system that explains the effect in an open wave (quantum) system is discussed in the Appendix A.

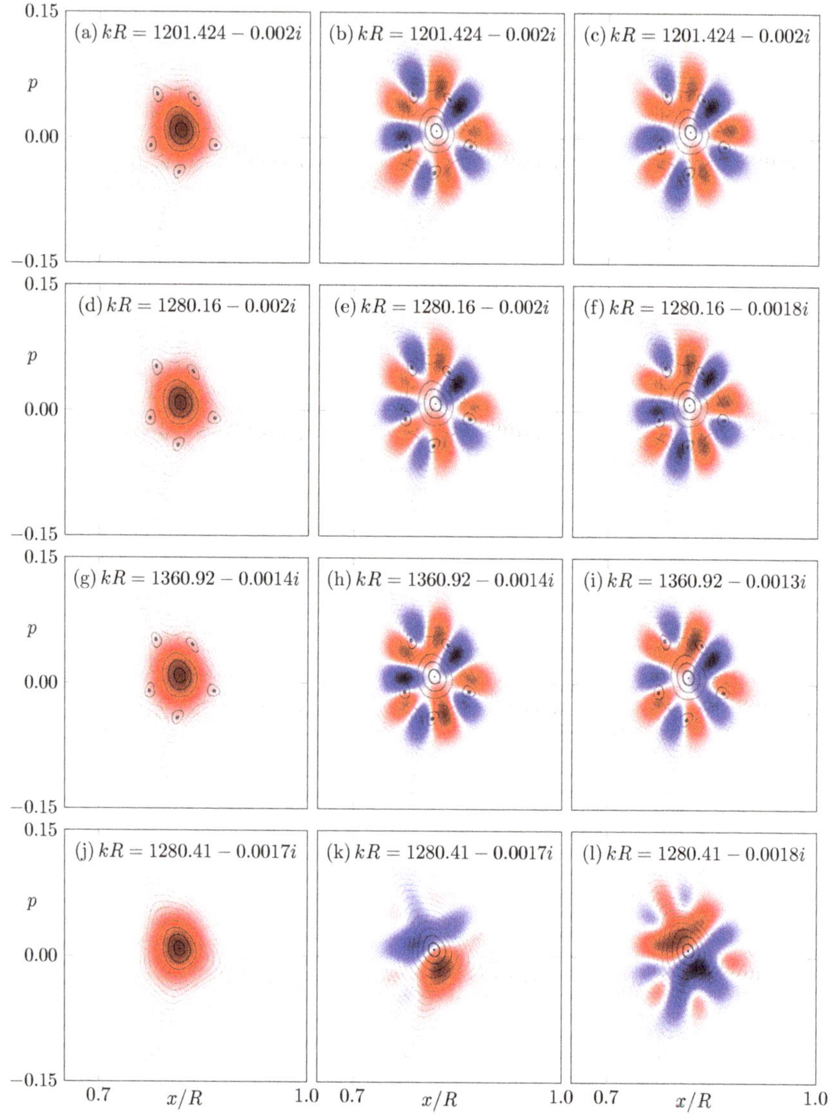

Figure 10. (**a**) Husimi function for CCW propagation, (**b**) difference $\mathcal{I}(x, p; \psi)$ between Husimi functions for CCW and CW propagation, see Equation (9). (**c**) Same as (**b**) but for a partner mode with almost the same frequency kR as in (**b**). The panels (**d–i**) follow the same sequence as (**a**,**b**) but for modes with different frequency kR in the same system given by deformation parameters $(\epsilon_1, \epsilon_2) = (0.76, 1.2)$. Modes for the microstar given by $(\epsilon_1, \epsilon_2) = (0.78, 1.212)$ are shown in (**j–l**).

4. Conclusions

In this article, we discussed the ray–wave correspondence in microstar cavities. From the ray dynamics, a (asymmetric) boundary deformation is used to manipulate the light around a periodic orbit with a regular island in phase space. Correspondingly, in the semiclassical regime, optical modes localize hierarchically on the regular island. Depending on the deformation of the microstar resonance, chains might appear in the regular island. Since the intensity loss is not uniform in the regular region, those resonance chains have an unusual effect on the dynamics. The propagation in CW and CCW direction shows an intensity distortion with a pattern that organizes along the resonance chain. This distortion is explained by the ray dynamics, and therefore manifests in the eigenstates of the FPO. Furthermore, the difference between CW and CCW propagation is revealed in the Husimi functions of the optical modes in the semiclassical regime.

Such an interplay between nonuniform loss and nonlinear dynamics is a rather general effect, as it can also be observed in a simple quantum model system. Hence, the effect is not restricted to microstar cavities. However, microstar cavities are ideal systems to study such dynamics, as a nontrivial connection between forward and backward dynamics in time manifests naturally in terms of CW and CCW propagation. Therefore, we believe our studies highlight microstar cavities as an interesting class of systems to unveil novel phenomena.

Author Contributions: Investigation, J.K.; Supervision, J.W.; Writing—original draft, J.K.; Writing—review and editing, J.W. All authors have read and agreed to the published version of the manuscript.

Funding: This research received no external funding.

Institutional Review Board Statement: Not applicable.

Data Availability Statement: Not applicable.

Acknowledgments: We acknowledge support for the Book Processing Charge by the Open Access Publication Fund of Magdeburg University.

Conflicts of Interest: The authors declare no conflict of interest.

Appendix A. Difference in Forward and Backward Dynamics in a Generalized Pendulum Model

In this appendix, a simple model system is constructed to illustrate the interplay between a resonance chain and a loss gradient to cause a difference between two directions of dynamical propagation. In the microstar cavity, such two directions are given naturally by CW and CCW propagation. In the simplified one-dimensional model, the two directions are given by forward and backward propagation in time under Hamiltonian dynamics. The Hamiltonian is given by a generalized pendulum or normal form ansatz for a $r{:}s$ resonance chain as [49–51]

$$H_{r:s}(\theta, I) = H_0(I) + 2V_{r:s}\left(\frac{I}{I_{r:s}}\right)^{\frac{r}{2}} \cos(r\theta). \tag{A1}$$

Here, $I \geq 0$ and $\theta \in [0, 2\pi]$ are action-angle variables of the unperturbed part $H_0(I) = (I - I_{r:s})^2/(2M_{r:s})$. The parameters are fixed to $(I_{r:s}, M_{r:s}, V_{r:s}) = (0.8, 1, 0.005)$ and $r = 3$. Using a canonical transformation

$$\mathcal{M}: \ (\theta, I) \mapsto (Q, P) = (\sqrt{2I}\cos\theta, -\sqrt{2I}\sin\theta), \tag{A2}$$

the Hamiltonian given in the coordinates (θ, I) can be related to a regular island in the phase space (Q, P) with an elliptical fixed point at the origin and a resonance chain around it; see Figure A1a,b.

Next, the Hamiltonian (A1) is quantized. Since the angle coordinate θ is 2π-periodic, the action eigenstates $|I_n\rangle$ are a discrete basis of the Hilbert space with $I_n = \hbar(n + 1/2)$. Thus, the Hamiltonian (A1) has the matrix elements [52,53]

$$\langle I_m|H_{r:s}|I_n\rangle = \frac{(I_n - I_{r:s})^2}{2M_{r:s}}\delta_{nm} + V_{r:s}\left(\frac{\hbar}{I_{r:s}}\right)^{\frac{r}{2}}\left(\sqrt{\frac{n!}{(n-r)!}}\delta_{m,n-r} + \sqrt{\frac{(n+r)!}{n!}}\delta_{m,n+r}\right) \tag{A3}$$

and is therefore represented as an infinite matrix. In the numerical simulations, the reduced Planck's constant is set to $\hbar = 0.3$, and a cutoff for the action indices at $N_{\max} = 90$ is used. Hence, the (forward) time evolution operator $\mathcal{U}_{\text{for}}$ over one unit time step is given as a finite matrix

$$\mathcal{U}_{\text{for}} = \exp\left(-\frac{i}{\hbar}H_{r:s}\right). \tag{A4}$$

Correspondingly, the operator for the dynamic one unit time step backward in time yields

$$\mathcal{U}_{\text{back}} = \exp\left(\frac{i}{\hbar}H_{r:s}\right). \tag{A5}$$

Next, a nonuniform loss is introduced which has a gradient across the resonance chain and therefore mimics the situation in the microstar cavity. In the basis of action eigenstates, the loss is described by a reflection operator $\mathcal{R}$ with matrix elements

$$R(I_n) = \exp(-I_n/I_{r:s}) \tag{A6}$$

in the diagonal, cf. Figure A1c. Hence, states with small I_n suffer less loss than the states with large I_n. The dynamics of the open system is therefore described by the sub-unitary time-evolution operator

$$\mathcal{U}^{(\text{open})}_{\text{back, for}} = \mathcal{R}\mathcal{U}_{\text{back, for}}. \tag{A7}$$

Note that $\mathcal{U}^{(\text{open})}_{\text{back}}$ is not the time-reversed operator of $\mathcal{U}^{(\text{open})}_{\text{for}}$ as loss is not turned into gain. This is consistent with the microstar cavity, where CCW is the time-reversed CW propagation, but both directions of the dynamics suffer loss from radiation and partial reflections. The eigenstates $\Psi^{(\text{open})}_{\text{back, for}}$ of $\mathcal{U}^{(\text{open})}_{\text{back, for}}$ are calculated numerically. In particular, we are interested in the eigenstates with eigenvalues of the largest magnitude. Due to the loss profile, these eigenstates localize near $I = 0$. Their Husimi function can be calculated as

$$H_{\text{back, for}}(\theta, I) = \left|\sum_{n=0}^{N_{\max}} \langle \alpha(\theta, I)|I_n\rangle\langle I_n|\Psi^{(\text{open})}_{\text{back, for}}\rangle\right|^2 \tag{A8}$$

with $\alpha(\theta, I)$ being a minimal uncertainty wave packet at the phase-space point (θ, I) given by

$$\langle \alpha(\theta, I)|I_n\rangle = \mathcal{N}(I)\exp\left(-\frac{(I_n - I)^2}{4\sigma^2} + \frac{i}{\hbar}(I_n - I)\theta\right). \tag{A9}$$

Here, $\sigma = \sqrt{\hbar/2}$ is a parameter and $\mathcal{N}(I)$ is a normalization factor ensuring $\|\alpha(\theta, I)\|^2 = \sum_{n=0}^{N_{\max}} |\langle \alpha(\theta, I)|I_n\rangle|^2 = 1$. The Husimi function for the long-lived eigenstates corresponding to the forward/backward dynamics are shown in Figure A1d,e. As it can be seen in the magnifications, the Husimi function's contour lines are curved sinusoidally with a phase shift between forward and backward dynamics. Exactly this phase shift leads to the oscillatory pattern for the difference between both Husimi functions, which is organized along the stable and unstable fixed point of the resonance chain; see Figure A1f.

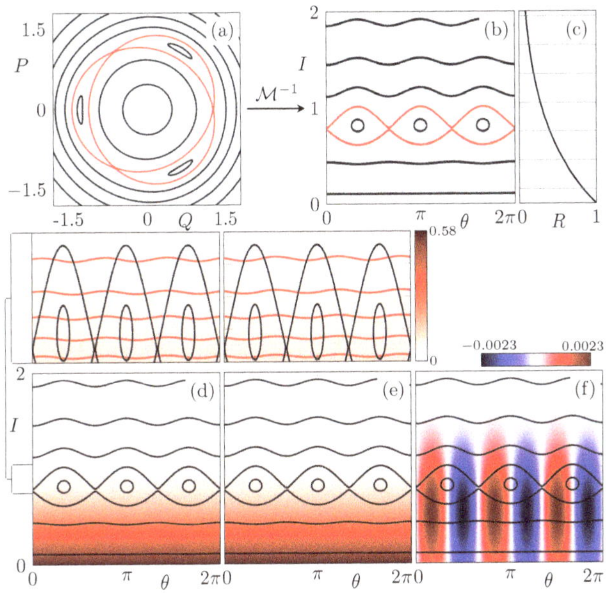

Figure A1. (**a**,**b**) Contour lines of the generalized pendulum Hamiltonian in the phase-space (Q, P) and (θ, I) are shown as black curves. The separatrix of the resonance chain is highlighted via a red curve. (**c**) The reflectivity function which defines the loss profile is shown via a thick black curve. Thin horizontal lines indicate the position of action eigenstates at I_n. (**d**,**e**) The Husimi function for eigenstates of $U_{\text{for}}^{(\text{open})}$ ($U_{\text{back}}^{(\text{open})}$) localizing predominantly at I_0 are shown via the red color gradient. Contour lines of the Husimi function are shown as red curves in the magnification plots. (**f**) Difference between the Husimi function for forward and backward propagation.

References

1. Vahala, K.J. *Optical Microcavities*; Advanced Series in Applied Physics; World Scientific: Singapore, 2004; Volume 5.
2. Chang, R.K.; Campillo, A.J. *Optical Processes in Microcavities*; *Advanced Series in Applied Physics*; World Scientific: Singapore, 1996; Volume 3.
3. Kippenberg, T.J.; Kalkman, J.; Polman, A.; Vahala, K.J. Demonstration of an erbium-doped microdisk laser on a silicon chip. *Phys. Rev. A* **2006**, *74*, 051802(R). [CrossRef]
4. McCall, S.L.; Levi, A.F.J.; Slusher, R.E.; Pearton, S.J.; Logan, R.A. Whispering-gallery mode microdisk lasers. *Appl. Phys. Lett.* **1992**, *60*, 289–291. [CrossRef]
5. Levi, A.F.J.; Slusher, R.E.; McCall, S.L.; Glass, J.L.; Pearton, S.J.; Logan, R.A. Directional light coupling from microdisk lasers. *App. Phys. Lett.* **1993**, *62*, 561–563. [CrossRef]
6. Tamboli, A.C.; Haberer, E.D.; Sharma, R.; Lee, K.W.; Nakamura, S.; Hu, E.L. Room-temperature continuous-wave lasing in GaN/InGaN microdisks. *Nat. Photonics* **2007**, *1*, 61–64. [CrossRef]
7. Ilchenko, V.S.; Gorodetsky, M.L.; Yao, X.S.; Maleki, L. Microtorus: A high-finesse microcavity with whispering-gallery modes. *Opt. Lett.* **2001**, *26*, 256–258. [CrossRef] [PubMed]
8. Armani, D.K.; Kippenberg, T.J.; Spillane, S.M.; Vahala, K.J. Ultra-high-*Q* toroid microcavity on a chip. *Nature* **2003**, *421*, 925–928. [CrossRef]
9. Zhang, X.; Choi, H.S.; Armani, A.M. Ultimate quality factor of silica microtoroid resonant cavities. *Appl. Phys. Lett.* **2010**, *96*, 153304. [CrossRef]
10. Collot, L.; Lefevre-Seguin, V.; Brune, M.; Raimond, J.; Haroche, S. Very high-*Q* whispering-gallery modes observed on fused silica microspheres. *Europhys. Lett.* **1993**, *23*, 327–334. [CrossRef]
11. Yang, J.; Moon, S.; Lee, S.B.; Lee, J.H.; An, K.; Shim, J.B.; Lee, H.W.; Kim, S.W. Development of a deformation-tunable quadrupolar microcavity. *Rev. Sci. Instrum.* **2006**, *77*, 083103. [CrossRef]
12. Stöckmann, H.J. *Quantum Chaos*; Cambridge University Press: Cambridge, UK, 2000
13. Cao, H.; Wiersig, J. Dielectric microcavities: Model systems for wave chaos and non-Hermitian physics. *Rev. Mod. Phys.* **2015**, *87*, 61–111. [CrossRef]
14. Nöckel, J.U.; Stone, A.D. Ray and wave chaos in asymmetric resonant optical cavities. *Nature* **1997**, *385*, 45–47. [CrossRef]

15. Sunada, S.; Shinohara, S.; Fukushima, T.; Harayama, T. Signature of Wave Chaos in Spectral Characteristics of Microcavity Lasers. *Phys. Rev. Lett.* **2016**, *116*, 203903. [CrossRef] [PubMed]
16. Peng, B.; Özdemir, S.K.; Lei, F.; Monifi, F.; Gianfreda, M.; Long, G.L.; Fan, S.; Nori, F.; Bender, C.M.; Yang, L. Parity–time-symmetric whispering-gallery microcavities. *Nat. Phys.* **2014**, *10*, 394–398. [CrossRef]
17. Lee, S.B.; Yang, J.; Moon, S.; Lee, S.Y.; Shim, J.B.; Kim, S.W.; Lee, J.H.; An, K. Observation of an Exceptional Point in a Chaotic Optical Microcavity. *Phys. Rev. Lett.* **2009**, *103*, 134101. [CrossRef]
18. Zhu, J.; Özdemir, Ş.K.; He, L.; Yang, L. Controlled manipulation of mode splitting in an optical microcavity by two Rayleigh scatterers. *Opt. Express* **2010**, *18*, 23535–23543. [CrossRef]
19. Peng, B.; Özdemir, Ş.K.; Liertzer, M.; Chen, W.; Kramer, J.; Yilmaz, H.; Wiersig, J.; Rotter, S.; Yang, L. Chiral modes and directional lasing at exceptional points. *Proc. Natl. Acad. Sci. USA* **2016**, *113*, 6845. [CrossRef]
20. Richter, S.; Zirnstein, H.G.; Zúñiga Pérez, J.; Krüger, E.; Deparis, C.; Trefflich, L.; Sturm, C.; Rosenow, B.; Grundmann, M.; Schmidt-Grund, R. Voigt Exceptional Points in an Anisotropic ZnO-Based Planar Microcavity: Square-Root Topology, Polarization Vortices, and Circularity. *Phys. Rev. Lett.* **2019**, *123*, 227401. [CrossRef]
21. Kippenberg, T.J.; Holzwarth, R.; Diddams, S.A. Microresonator-Based Optical Frequency Combs. *Science* **2011**, *332*, 555–559. [CrossRef]
22. Nöckel, J.U.; Stone, A.D.; Chen, G.; Grossman, H.L.; Chang, R.K. Directional emission from asymmetric resonant cavities. *Opt. Lett.* **1996**, *21*, 1609–1611. [CrossRef]
23. Wiersig, J.; Hentschel, M. Combining directional light output and ultralow loss in deformed microdisks. *Phys. Rev. Lett.* **2008**, *100*, 033901. [CrossRef]
24. Wang, Q.J.; Yan, C.; Yu, N.; Unterhinninghofen, J.; Wiersig, J.; Pflügl, C.; Diehl, L.; Edamura, T.; Yamanishi, M.; Kan, H.; et al. Whispering-gallery mode resonators for highly unidirectional laser action. *Proc. Natl. Acad. Sci. USA* **2010**, *107*, 22407. [CrossRef] [PubMed]
25. Albert, F.; Hopfmann, C.; Eberspächer, A.; Arnold, F.; Emmerling, M.; Schneider, C.; Höfling, S.; Forchel, A.; Kamp, M.; Wiersig, J.; et al. Directional whispering gallery mode emission from Limaçon-shaped electrically pumped quantum dot micropillar lasers. *Appl. Phys. Lett* **2012**, *101*, 021116. [CrossRef]
26. Armani, A.M.; Kulkarni, R.P.; Fraser, S.E.; Flagan, R.C.; Vahala, K.J. Label-Free, Single-Molecule Detection with Optical Microcavities. *Science* **2007**, *317*, 783–787. [CrossRef] [PubMed]
27. Vollmer, F.; Arnold, S. Whispering-gallery-mode biosensing: Label-free detection down to single molecules. *Nat. Methods* **2008**, *5*, 591. [CrossRef]
28. Lu, T.; Lee, H.; Chen, T.; Herchak, S.; Kim, J.H.; Fraser, S.E.; Flagan, R.C.; Vahala, K. High sensitivity nanoparticle detection using optical microcavities. *Proc. Nat. Acad. Sci. USA* **2011**, *108*, 5976–5979. [CrossRef] [PubMed]
29. Sunada, S.; Harayama, T. Design of resonant microcavities: Application to optical gyroscopes. *Opt. Express* **2007**, *15*, 16245–16254. [CrossRef]
30. Miao, P.; Zhang, Z.; Sun, J.; Walasik, W.; Longhi, S.; Litchinitser, N.M.; Feng, L. Orbital angular momentum microlaser. *Science* **2016**, *353*, 464–467. [CrossRef]
31. Kullig, J.; Jiang, X.; Yang, L.; Wiersig, J. Microstar cavities: An alternative concept for the confinement of light. *Phys. Rev. Res.* **2020**, *2*, 012072. [CrossRef]
32. Kullig, J.; Wiersig, J. Microdisk cavities with a Brewster notch. *Phys. Rev. Res.* **2021**, *3*, 023202. [CrossRef]
33. Wiersig, J.; Kim, S.W.; Hentschel, M. Asymmetric scattering and nonorthogonal mode patterns in optical microspirals. *Phys. Rev. A* **2008**, *78*, 053809. [CrossRef]
34. Wiersig, J.; Eberspächer, A.; Shim, J.B.; Ryu, J.W.; Shinohara, S.; Hentschel, M.; Schomerus, H. Nonorthogonal pairs of copropagating optical modes in deformed microdisk cavities. *Phys. Rev. A* **2011**, *84*, 023845. [CrossRef]
35. Wiersig, J. Non-Hermitian Effects Due to Asymmetric Backscattering of Light in Whispering-Gallery Microcavities. In *Parity-Time Symmetry and Its Applications*; Christodoulides, D., Yang, J., Eds.; Springer: Singapore, 2018; pp. 155–184.
36. Beck, C.; Schögl, F. *Thermodynamics of Chaotic Systems: An Introduction*; Cambridge Nonlinear Science Series; Cambridge University Press: Cambridge, UK, 1993.
37. Altmann, E.G.; Portela, J.S.E.; Tél, T. Leaking chaotic systems. *Rev. Mod. Phys.* **2013**, *85*, 869–918. [CrossRef]
38. Frahm, K.M.; Shepelyansky, D.L. Ulam method for the Chirikov standard map. *Eur. Phys. J. B* **2010**, *76*, 57–68. [CrossRef]
39. Weber, J.; Haake, F.; Šeba, P. Frobenius-Perron Resonances for Maps with a Mixed Phase Space. *Phys. Rev. Lett.* **2000**, *85*, 3620–3623. [CrossRef]
40. Ermann, L.; Shepelyansky, D.L. Ulam method and fractal Weyl law for Perron-Frobenius operators. *Eur. Phys. J. B* **2010**, *75*, 299–304. [CrossRef]
41. Carlo, G.G.; Rivas, A.M.F.; Spina, M.E. Classical to quantum correspondence in dissipative directed transport. *Phys. Rev. E* **2015**, *92*, 052907. [CrossRef]
42. Kullig, J.; Wiersig, J. Frobenius–Perron eigenstates in deformed microdisk cavities: Non-Hermitian physics and asymmetric backscattering in ray dynamics. *New J. Phys.* **2016**, *18*, 015005. [CrossRef]
43. Ulam, S.M. *A Collection of Mathematical Problems*; Interscience Publishers: New York, NY, USA, 1960.
44. Weber, J.; Haake, F.; Braun, P.A.; Manderfeld, C.; Seba, P. Resonances of the Frobenius-Perron operator for a Hamiltonian map with a mixed phase space. *J. Phys. A Math. Gen.* **2001**, *34*, 7195–7211. [CrossRef]

45. Jackson, J.D. *Classical Electrodynamics*; John Wiley and Sons: New York, NY, USA, 1962.
46. Wiersig, J. Boundary element method for resonances in dielectric microcavities. *J. Opt. A Pure Appl. Opt.* **2003**, *5*, 53. [CrossRef]
47. *COMSOL Multiphysics*® *v. 6.0*; COMSOL AB: Stockholm, Sweden, 2022. Available online: www.comsol.com (accessed on 4 November 2022).
48. Hentschel, M.; Schomerus, H.; Schubert, R. Husimi functions at dielectric interfaces: Inside-outside duality for optical systems and beyond. *Europhys. Lett.* **2003**, *62*, 636–642. [CrossRef]
49. Lebœuf, P.; Mouchet, A. Normal Forms and Complex Periodic Orbits in Semiclassical Expansions of Hamiltonian Systems. *Ann. Phys.* **1999**, *275*, 54–112. [CrossRef]
50. Lichtenberg, A.J.; Liebermann, M.A. *Regular and Stochastic Motion*; Springer: New York, NY, USA, 1983.
51. Keshavamurthy, S.; Schlagheck, P.E. *Dynamical Tunneling: Theory and Experiment*, 1st ed.; CRC Press: Boca Raton, FL, USA, 2011.
52. Löck, S.; Bäcker, A.; Ketzmerick, R.; Schlagheck, P. Regular-to-Chaotic Tunneling Rates: From the Quantum to the Semiclassical Regime. *Phys. Rev. Lett.* **2010**, *104*, 114101. [CrossRef] [PubMed]
53. Mertig, N.; Kullig, J.; Löbner, C.; Bäcker, A.; Ketzmerick, R. Perturbation-free prediction of resonance-assisted tunneling in mixed regular-chaotic systems. *Phys. Rev. E* **2016**, *94*, 062220. [CrossRef] [PubMed]

entropy

MDPI

Article

Universal Single-Mode Lasing in Fully Chaotic Billiard Lasers

Mengyu You [1], Daisuke Sakakibara [1], Kota Makino [1], Yonosuke Morishita [1], Kazutoshi Matsumura [1], Yuta Kawashima [1], Manao Yoshikawa [1], Mahiro Tonosaki [1], Kazutaka Kanno [2], Atsushi Uchida [2], Satoshi Sunada [3], Susumu Shinohara [4] and Takahisa Harayama [1,*]

[1] Department of Applied Physics, School of Advanced Science and Engineering, Waseda University, 3-4-1 Okubo, Shinjuku-ku, Tokyo 169-8555, Japan
[2] Department of Information and Computer Sciences, Saitama University, 255 Shimo-okubo, Sakura-ku, Saitama City 338-8570, Saitama, Japan
[3] Faculty of Mechanical Engineering, Institute of Science and Engineering, Kanazawa University, Kanazawa 920-1192, Ishikawa, Japan
[4] Department of Production Systems Engineering and Sciences, Komatstu University, Nu 1-3 Shicho-machi, Komatsu 923-8511, Ishikawa, Japan
* Correspondence: harayama@waseda.jp

Abstract: By numerical simulations and experiments of fully chaotic billiard lasers, we show that single-mode lasing states are stable, whereas multi-mode lasing states are unstable when the size of the billiard is much larger than the wavelength and the external pumping power is sufficiently large. On the other hand, for integrable billiard lasers, it is shown that multi-mode lasing states are stable, whereas single-mode lasing states are unstable. These phenomena arise from the combination of two different nonlinear effects of mode-interaction due to the active lasing medium and deformation of the billiard shape. Investigations of billiard lasers with various shapes revealed that single-mode lasing is a universal phenomenon for fully chaotic billiard lasers.

Keywords: quantum chaos; billiard lasers; microcavity lasers

Citation: You, M.; Sakakibara, D.; Makino, K.; Morishita, Y.; Matsumura, K.; Kawashima, Y.; Yoshikawa, M.; Tonosaki, M.; Kanno, K.; Uchida, A.; et al. Universal Single-Mode Lasing in Fully Chaotic Billiard Lasers. *Entropy* **2022**, *24*, 1648. https://doi.org/10.3390/e24111648

Academic Editor: Marko Robnik

Received: 13 October 2022
Accepted: 11 November 2022
Published: 14 November 2022

1. Introduction

Two-dimensional (2D) billiard lasers have been widely investigated over the past decades with various billiard shapes [1–39]. From the viewpoint of the property of ray dynamics which is equivalent to the motion of a point particle on a billiard table, 2D billiard lasers can be classified into three categories: integrable, partially chaotic, and fully chaotic billiard lasers [40]. On the basis of this classification of dynamical systems, the universality of energy level statistics was discovered by Casati et al. for the first time for a quantum billiard, which is a quantum particle on a billiard table where the motion of the corresponding classical point particle is fully chaotic [41]. This discovery led to the Bohigas–Gianonni–Schmidt (BGS) conjecture [42], and made significant contributions to the establishment of the research field of quantum chaos [43,44].

Although optical billiards can be viewed as a realization of quantum billiards, there is a noticeable difference between them. The light field inside an optical billiard is confined by the refractive index difference between inside and outside the optical billiard, and the corresponding rays obey Fresnel's law at the billiard boundary. Meanwhile, quantum particles are completely confined inside the billiard table by the 2D rigid wall potential, and the corresponding classical point particles obey full reflection at the billiard boundary. As a result, eigenstates of a quantum billiard are bound states with real eigenenergies, whereas those of an optical billiard are quasibound states, or resonances, with complex eigenfrequencies because of non-Hermitian boundary conditions [1,2], where their real parts represent oscillation frequencies and their imaginary parts decay or loss rates of modes. In spite of the difference mentioned above, low-loss (i.e., well-confined) resonance wave functions in an optical billiard have properties similar to those of the eigenfunctions of the corresponding quantum billiard. Hence, these phenomena which were found for chaotic quantum billiards have been also observed in chaotic billiard lasers [3,4,7,10–12,17,20,21,26]. In this sense, chaotic billiard lasers have provided a valuable platform where the phenomena theoretically predicted in the research field of quantum chaos can be experimentally investigated.

When optical billiards are used as laser cavities (billiard lasers, in short), they exhibit essentially different properties from those of quantum billiards. That is, the resonance modes interact with each other through the active lasing medium in billiard lasers, which causes nonlinear dynamics of the light field [1,6,8,9,14,16]. In the spirit of the quantum chaos theory pioneered by Casati, it is important to turn our attention to universal properties of chaotic billiard lasers that resulted from the synergistic effect of the deformation of the billiard shape and nonlinear dynamics of resonance modes such as mode competition.

As an example of such a universal phenomenon for fully chaotic billiard lasers, it has been reported for the experiments of semiconductor billiard lasers with high pumping conditions that only single-mode lasing was observed for fully chaotic billiard lasers with stadium shapes with various aspect ratios, whereas only multi-mode lasing for integrable billiard lasers with elliptic shapes of different aspect ratios [28,32]. In addition, by assuming the similarity of wave functions for fully chaotic billiard lasers, it has been theoretically shown that single-mode lasing states are stable, whereas multi-mode lasing states are unstable when the billiard size is much larger than the wavelength and the external pumping power is sufficiently large [33]. Accordingly, in analogy with the universality that Casati found for quantum billiards [41], it can be expected that single-mode lasing is a universal phenomenon for fully chaotic billiard lasers which does not depend on the details of the billiard lasers. Although lasing characteristics of 2D billiard lasers with various 2D shapes have been intensively studied, universality has not been much pursued so far. Therefore, it is important to verify this universality by numerically studying fully chaotic billiard lasers with different shapes in the same way as the BGS conjecture was numerically examined for various systems [42].

In this paper, we numerically investigate the laser dynamics of three different shapes of fully chaotic billiard lasers; cardioid, D-shaped, and stadium billiard lasers shown in Figure 1, whose ray-dynamical trajectories have been exactly proven to be fully chaotic [45,46]. These billiards have already been studied in the research fields of quantum chaos and chaotic billiard lasers [8,9,13,15,17,19,28,30,32,37,41,42]. It is numerically shown that the single-mode lasing states are stable in all these three billiards in high pumping regimes, which verifies the universal single-mode lasing in fully chaotic billiard lasers [28,33]. We explain the detailed transition mechanism from multi to single-mode lasing, revealing the mode-pulling and the mode-pushing interaction among lasing modes. We also experimentally show that the numbers of lasing modes of the semiconductor cardioid, D-shaped, and stadium lasers are clearly different from those of semiconductor microdisk lasers, which also experimentally verifies the universality of single-mode lasing in fully chaotic billiard lasers.

The subsequent sections of this paper are organized as follows. In Section 2, we review the theoretical model to describe the dynamics of the light field and the lasing medium in billiard lasers and show numerical results of the resonances and wave functions for the cardioid billiard. In Section 3, the single-mode lasing is achieved in the course of time evolution of a numerically calculated light field in the cardioid billiard laser, whereas multi-mode lasing in the elliptic billiard laser when the pumping power is much larger than the lasing threshold. In Section 4, it is shown that the properties of the resonance wave functions of the cardioid and elliptic billiards are different from each other. We numerically obtain phase diagrams for the cardioid, D-shaped, and stadium billiard lasers, which verify the universal single-mode lasing. In Section 5, we provide experimental data for semiconductor billiard lasers with the cardioid, D-shaped, stadium, and circular shapes, which also validate the universal single mode lasing for fully chaotic billiard lasers. In Section 6, we numerically demonstrate that the occasional appearance of a couple of peaks experimentally observed in the spectra of the semiconductor fully-chaotic billiard lasers is attributed to the fluctuation of the temperature. Summary and discussion are given in Section 7.

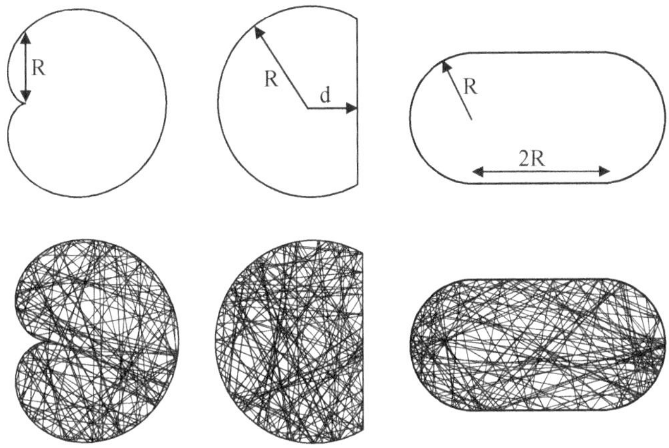

Figure 1. Fully chaotic billiards. From left to right: cardioid, D-shaped, and stadium.

2. The Schrödinger–Bloch Model and the Resonances of the Cardioid Billiard Laser

We assume that the billiard laser is wide in the xy-directions and thin in the z-direction. As a result, the electromagnetic fields are separated into transverse magnetic (TM) and transverse electric (TE) modes. Here, we focus only on TM modes, whose electric field vector is expressed as $\boldsymbol{E} = (0, 0, E_z)$. The atoms in the lasing medium are assumed to have spherical symmetry and two energy levels. The relaxation due to the interaction with the reservoir can be described phenomenologically with decay constants $\gamma_\perp$ for the microscopic polarization ρ and $\gamma_\parallel$ for the population inversion W. The effect of the external energy injected into the lasing medium is introduced phenomenologically and represented by the pumping power W_∞.

The slowly varying envelope approximation reduces the Maxwell equations as follows:

$$\frac{\partial \tilde{E}}{\partial t} = \frac{i}{2}\left(\nabla_{xy}^2 + \frac{n^2}{n_{\text{in}}^2}\right)\tilde{E} - \alpha_L(x,y)\tilde{E} + \frac{2\pi N\kappa\hbar}{\varepsilon}\tilde{\rho}, \quad (1)$$

where $\tilde{E}$ and $\tilde{\rho}$ are, respectively, the slowly varying envelopes of the z-component of the electric field and the microscopic polarization, and N is the number density of the atoms, κ is the coupling strength, ε is the permittivity, and α_L represents the losses describing absorption inside the billiard and it equals to zero outside the billiard. The refractive index $n(x,y)$ equals to n_{in} when the position (x,y) is inside the billiard and n_{out} outside the billiard. In the above, space and time are made dimensionless by the scale transformation $((n_{in}\omega_s/c)x, (n_{in}\omega_s/c)y) \rightarrow (x,y)$, $t\omega_s \rightarrow t$, respectively, where ω_s is the oscillation frequency of the fast oscillation part of the electric field.

The optical Bloch equations describing the active gain medium in interaction with the light field are written as

$$\frac{\partial}{\partial t}\tilde{\rho} = -\tilde{\gamma}_\perp\tilde{\rho} - i\Delta_0\tilde{\rho} + \tilde{\kappa}W\tilde{E}, \quad (2)$$

and

$$\frac{\partial}{\partial t}W = -\tilde{\gamma}_\parallel(W - W_\infty) - 2\tilde{\kappa}\left(\tilde{E}\tilde{\rho}^* + \tilde{E}^*\tilde{\rho}\right), \quad (3)$$

where $\tilde{\gamma}_\perp \equiv \gamma_\perp/\omega_s$ and $\tilde{\gamma}_\parallel \equiv \gamma_\parallel/\omega_s$ respectively represent the dimensionless transversal and longitudinal relaxation rates, and $\Delta_0 \equiv [\omega_0 - \omega_s]/\omega_s$ represents the gain center, and ω_0 is the transition frequency of the two-level atoms, and $\tilde{\kappa} \equiv \kappa/\omega_s$ represents the dimensionless couple strength. W_∞ represents the external pumping power. The set of Equations (1)–(3) is called the Schrödinger–Bloch (SB) model [1,6,8,9,14]. The SB model can be numerically solved by the split-operator method (or so-called symplectic-integrator method). Therefore, it is more convenient for numerical simulations than the Maxwell–Bloch model.

By assuming the stationary oscillation of the electric and polarization field, respectively, $\tilde{E} = \tilde{E}_s e^{-i\Delta_s t}$, $\tilde{\rho} = \tilde{\rho}_s e^{-i\Delta_s t}$, and the constant population inversion, we obtain the time-independent SB equation,

$$\left[\nabla_{xy}^2 + \left(2\Delta_s + \frac{n^2}{n_{in}^2}\right)\right]\tilde{E}_s = 2i\alpha(\Delta_s)\frac{W_\infty}{1+\eta(\Delta_s)|\tilde{E}_s|^2}\tilde{E}_s - 2i\alpha_L\tilde{E}_s, \tag{4}$$

where the complex Lorentzian gain $\alpha(\Delta_s)$ and the strength of the nonlinearity $\eta(\Delta_s)$ are defined as follows:

$$\alpha(\Delta_s) \equiv \alpha_0 \frac{1}{\tilde{\gamma}_\perp^2 + (\Delta_s - \Delta_0)^2}(\tilde{\gamma}_\perp + i(\Delta_s - \Delta_0)), \tag{5}$$

and

$$\eta(\Delta_s) \equiv \frac{4\tilde{\kappa}^2}{\tilde{\gamma}_{//}\tilde{\gamma}_\perp}\frac{\tilde{\gamma}_\perp^2}{\tilde{\gamma}_\perp^2 + (\Delta_0 - \Delta_s)^2}, \tag{6}$$

where $\alpha_0 \equiv 2\pi N\kappa\tilde{\kappa}\hbar/\varepsilon$. The right-hand side of Equation (4) represents the effect of the lasing medium and vanishes outside the billiard.

The resonance wave function corresponding to the resonance (i.e., complex eigenfrequency) Δ_j is the solution of Equation (4) with $W_\infty = 0$ and $\alpha_L = 0$. When the pumping power W_∞ is sufficiently increased from zero, the low-loss resonance closest to the gain center Δ_0 changes into a stationary lasing state of the SB model. By linearizing the SB model equations, the lasing condition for the resonance Δ_j is obtained from the energy balance condition that the gain due to the pumping exceeds the losses [14]. Its mathematical expression is given by

$$\alpha_0 \frac{\tilde{\gamma}_\perp W_\infty}{\tilde{\gamma}_\perp^2 + (\Delta_0 - \mathrm{Re}\,\Delta_j)^2} > -\mathrm{Im}\,\Delta_j + \alpha_L, \tag{7}$$

where the left-hand side represents the gain, and the right-hand side the losses. The frequency of the lasing mode is given by

$$\Delta_j' = \frac{\tilde{\gamma}_\perp \mathrm{Re}\,\Delta_j + (-\mathrm{Im}\,\Delta_j + \alpha_L)\Delta_0}{-\mathrm{Im}\,\Delta_j + \alpha_L + \tilde{\gamma}_\perp}, \tag{8}$$

which is generally shifted from the frequency of the resonance $\mathrm{Re}\,\Delta_j$.

Figure 2 shows the distribution of the resonances of the cardioid optical billiard numerically obtained by the boundary element method [47] and the gain function for the SB model defined by $g(\Delta) = \alpha_0\tilde{\gamma}_\perp/\{\tilde{\gamma}_\perp^2 + (\Delta_0 - \mathrm{Re}\,\Delta)^2\}$. The shape of a cardioid optical billiard is defined as $r = R(1+\cos\theta)$ in the 2D polar coordinates with $R = 30$ in our numerical simulations as shown in Figure 1. Due to the symmetry of the cardioid billiard, the resonances are classified into two different classes of even and odd parities. The refractive index n_{in} is 3.3 inside the billiard and $n_{\mathrm{out}} = 1$ outside the billiard. The other parameters are set as follows: $\Delta_0 = 0.04$, $\tilde{\gamma}_\perp = 0.02$, $\alpha_L = 0.004$, $\tilde{\kappa} = 0.5$, $\epsilon = 3.3$, and $N\kappa\hbar = 0.5$. From Figure 2, one can see that there exist a number of resonances with low decay rates (i.e., $|\mathrm{Im}\Delta| < 0.01$).

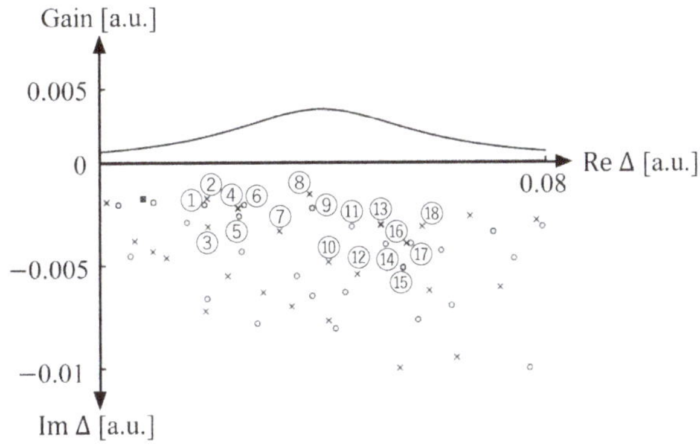

Figure 2. Gain and resonances in a cardioid billiard. Circles (○) denote the resonances with odd parity while crosses (×) those with even parity.

It has been theoretically shown that in a fully chaotic billiard laser, at least one single-mode lasing state is stable, whereas all multi-mode lasing states are unstable, when the external pumping power is sufficiently large and the billiard size is much larger than the wavelength so that $\tilde{\gamma}_{\parallel} \gg |\Delta_{ij}|$ holds, where Δ_{ij} is the difference between adjacent lasing frequencies [33]. This result provides a theoretical ground for recent experimental observations of universal single-mode lasing in fully chaotic billiard lasers [28,32]. We emphasize that the cardioid billiard laser numerically studied in this paper is much larger than those billiard lasers which have been studied so far. In addition, the gain band is much wider than those of previous studies. Therefore, the simulation condition shown in Figure 2 is more appropriate to numerically verify the universal single-mode lasing of fully chaotic billiard lasers. Figure 3 shows the spatial patterns of the resonance wave functions corresponding to the resonances labeled in Figure 2. All of these low-loss wave functions are distributed over the billiard domain and significantly overlap with one another reflecting fully chaotic ray-dynamical trajectories in the cardioid billiard [31–33,48].

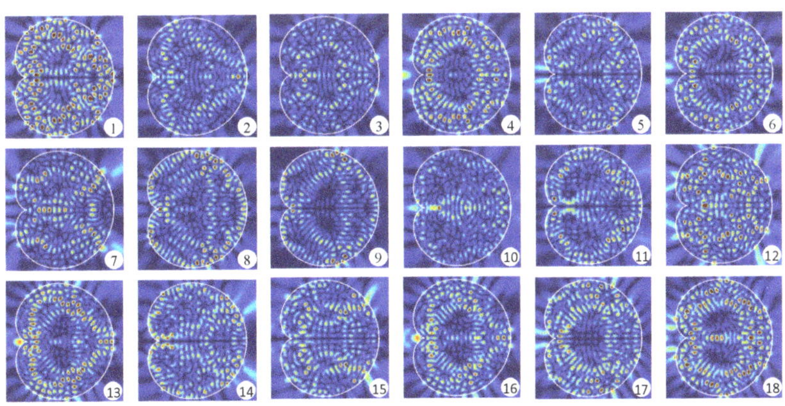

Figure 3. Spatial intensity patterns of eighteen low-loss resonances labeled in Figure 2.

3. Dynamical Simulation of Large Fully Chaotic and Integrable Billiard Lasers

Typical examples of the time evolution of the total light intensity inside the cardioid billiard laser are shown in Figure 4a,d,g. The longitudinal relaxation rate $\tilde{\gamma}_{\parallel}$ is set to 0.01, which is much larger than the frequency differences of adjacent resonances, i.e., $\tilde{\gamma}_{\parallel} \gg |\Delta_{ij}|$ (see also Figure 2). The lasing threshold W_{th} is found to be 0.000866. When the pumping power W_∞ is 0.001 just above the threshold, the intensity converges to a certain constant after initial transient behavior as shown in Figure 4a. The power spectrum is obtained by the Fourier transformation of the time series in the stationary lasing regime in Figure 4a and has a single peak as shown in Figure 4b, which means that the stationary lasing state includes only one lasing mode. The spatial intensity pattern of this stationary single-mode lasing state is shown in Figure 4c. One can see that this intensity pattern is almost the same as that of the resonance wave function labeled by ⑧ in Figure 3.

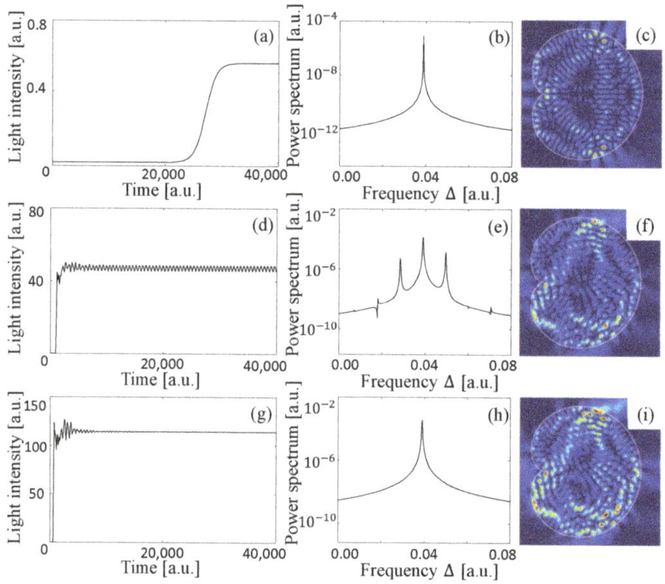

Figure 4. Dynamics of a cardioid billiard laser numerically simulated by the SB model. Time evolution of the total intensity inside the billiard with the pumping power $W_\infty =$ (**a**) 0.001, (**d**) 0.005, and (**g**) 0.010. (**b**,**e**,**h**) The spectrum obtained from the stationary oscillation regime with the corresponding pumping powers. (**c**,**f**,**i**) The intensity patterns of the final lasing states with the corresponding pumping powers.

As the pumping power W_∞ is increased, the other resonances become the stationary lasing modes which interact with one another due to the nonlinear effect of the lasing medium and the time evolution of the light intensity becomes oscillatory around a certain constant value as shown in Figure 4d where $W_\infty = 0.005$. In this multi-mode lasing regime, the spectrum has several peaks as shown in Figure 4e and the final lasing state is not a stationary state but a limit cycle or chaotic oscillation. Figure 4f shows a snapshot of the intensity pattern of the multi-mode lasing state.

When the pumping power W_∞ is increased further, the intensities of the lasing modes increase much more, and hence, they strongly interact with one another, and finally, only one lasing mode wins the mode competition. The light intensity takes a constant value and the spectrum has a single peak as shown in Figure 4g,h as in the case of just above the threshold. However, typically, the stationary lasing state in this high pumping power regime is not a simple lasing mode corresponding to one resonance as in the case of just above the threshold but a locked state of a couple of resonances as shown in Figure 4i. One can see that the spatial pattern of this final stationary lasing state is asymmetric, violating the symmetry of the cardioid shape due to the spontaneous symmetry breaking caused by locking of resonance wave functions with different parities [9]. To summarize, the spectral peak in Figure 4b is attributed to a single resonance, but that in Figure 4h corresponds to a locked state of resonances.

The intensity pattern in Figure 4i is almost the same as that in Figure 5 which is obtained by superposing the resonance wave functions corresponding to the two resonances denoted by ⑧ and ⑨ in Figure 3. Accordingly, one can see that the final lasing state in the highly pumping regime is composed of two different resonance wave functions whose frequencies are originally different from each other. The frequency difference of these two modes vanishes because they are locked due to the mode pulling effect of the lasing medium. Consequently, the spectrum has a single peak and the light intensity takes a constant value instead of beating oscillatory behavior with two peaks in the spectrum. This is a typical transition process from multi-mode to single-mode lasing as the pumping power is increased in fully chaotic billiard lasers.

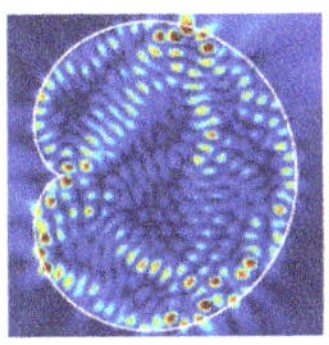

Figure 5. The spatial intensity pattern obtained by the superposition of the two resonances denoted by ⑧ and ⑨ in Figure 3, which have even and odd parity, respectively. The superposition of different-parity wave functions yields an asymmetric intensity pattern (see Ref. [9] for a detailed explanation).

Next, we show laser dynamics of an elliptic billiard laser. The elliptic billiard shape for our numerical simulation is defined by $x^2 + y^2/0.7^2 = R_0^2$, where $R_0 = 43.9155$, so that this elliptic billiard has the same area as the cardioid billiard. Hence, the numbers of resonances are almost the same because of the Weyl formula. The longitudinal relaxation rate $\tilde{\gamma}_{\parallel}$ is set to 0.01, which is much larger than the frequency differences of adjacent resonances, i.e., $\tilde{\gamma}_{\parallel} \gg |\Delta_{ij}|$, that is, the same condition as that of the SB model simulation for the cardioid billiard lasers.

The wave functions corresponding to the bouncing ball trajectories have much larger decay rates than the whispering gallery modes which become stationary lasing states. Therefore, the resonance wave functions of an elliptic billiard which comprise the stationary lasing states are regularly localized inside the billiard according to the quantum numbers in the radial and rotational directions, which systematically determine the high-intensity positions in the spatial patterns of the resonance wave functions. Consequently, many lasing modes coexist without interacting with one another by avoiding the overlap of high-intensity regions of the wave functions, and hence, many peaks are observed in the spectrum as shown in Figure 6e. For the elliptic billiard laser, single-mode lasing is only observed just above the threshold $W_{\text{th}} = 0.000666$. Figures 6a–c show data for such single-mode lasing at $W_{\infty} = 0.001$.

When the pumping power is increased further, those resonance wave functions which do not overlap with the dominant lasing modes become new stationary lasing modes and the number of the stationary lasing modes increases as shown in Figure 6g–i. Consequently, multi-mode lasing is always observed in the high pumping regime in integrable billiard lasers.

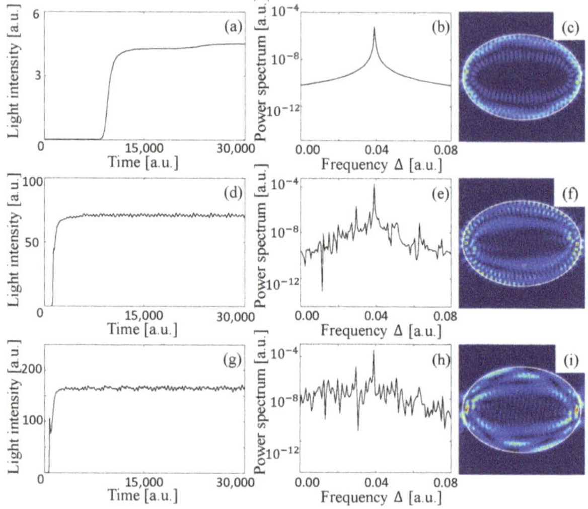

Figure 6. Dynamics of an elliptic billiard laser numerically simulated by the SB model. Time evolution of the total intensity inside the billiard with the pumping power $W_\infty =$ (**a**) 0.001, (**d**) 0.005, and (**g**) 0.010. (**b**,**e**,**h**) The spectrum obtained from the stationary oscillation regime with the corresponding pumping powers. (**c**,**f**,**i**) The intensity patterns of the final lasing states with the corresponding pumping powers.

4. Spatial Overlap between Two Modes and Phase Diagram

Long-lived eigenstates in open chaotic mapping systems have been shown to be localized on the forward trapped set of the corresponding classical dynamics in the short-wavelength limit [49–53]. This localization is a manifestation of quantum ergodicity in open quantized classically fully-chaotic systems. Similar phenomena have also been observed in fully chaotic 2D optical billiards numerically [32], where resonance wave functions of low-loss modes are localized around the forward trapped set of the corresponding ray-dynamics with Fresnel's law in the short wavelength limit as shown in Figure 7 [31,48,54]. In addition, assuming that all of the spatial patterns of the wave functions for low-loss modes in a fully chaotic billiard lasers are similar to one another, one can analytically prove the stability of a single-mode lasing state and instability of a multi-mode lasing state [33]. The overlap is an important quantity to measure the wave function similarity. It is defined by the following cross-correlation of the amplitude distributions between an arbitrary pair of resonance wave functions,

$$C = \frac{\int |\phi(\boldsymbol{r})||\psi(\boldsymbol{r})|)d\boldsymbol{r}}{\sqrt{(\int |\phi(\boldsymbol{r})|^2 d\boldsymbol{r})(\int |\psi(\boldsymbol{r})|^2 d\boldsymbol{r})}}, \tag{9}$$

where ϕ and ψ are the resonance wave functions and the integrals are taken over the 2D billiard [32].

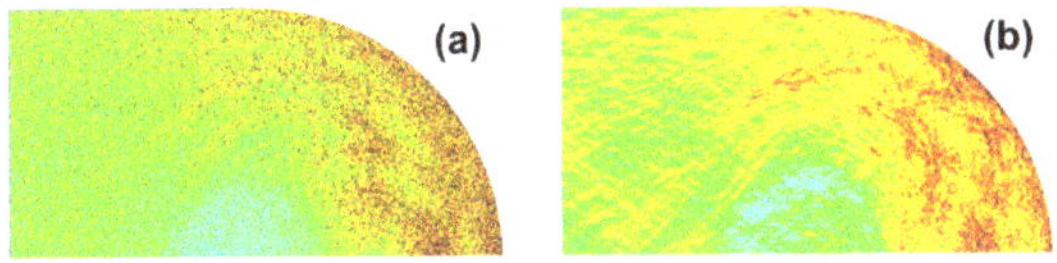

Figure 7. Numerically computed light intensity distributions inside the stadium optical billiard with refractive index 3.3 (because of the C_{2v} symmetry of the stadium shape, only a quarter domain is shown). (**a**) Conditionally invariant measure obtained by ray simulation incorporating Fresnel's law with 25 million initial conditions (See Ref. [48] for a detailed explanation). (**b**) Average of 30 low-loss resonance wave functions.

Figure 8 shows the histogram of the overlap C for the cardioid billiard, where most of the overlap values are found between 0.6 and 0.8. The functional form of the histogram is found to be universal for fully chaotic billiards [32,48]. For the elliptic billiard, the overlap C can take a smaller value and is distributed over a wider range. These results indicate that almost all resonance wave functions of the cardioid billiard lasers are similarly distributed inside the billiard, whereas the resonance wave functions of the elliptic billiard lasers exhibit various spatial patterns inside the billiard [31,33,48]. On the basis of quantum-classical correspondence in the semiclassical regimes, it is expected that the shorter the wavelength, the more distinct the histograms between fully chaotic and integrable billiards. This means that the effects caused by the spatial distributional difference of the resonances between fully chaotic and integrable billiards are expected to be more prominent for larger billiard lasers.

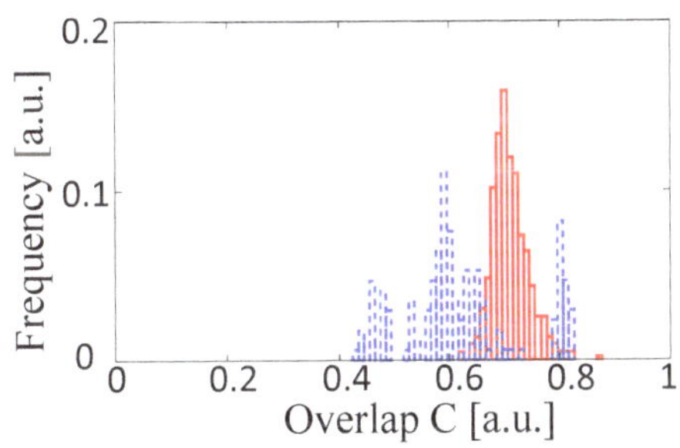

Figure 8. The histogram of the overlap C for the cardioid billiard (red solid line) and the elliptic billiard (blue dotted line) calculated by using 30 and 19 low-loss resonance wave functions, respectively.

In addition, as the size of the billiard laser becomes larger, the frequency difference Δ_{ij} between adjacent resonances is decreased and becomes much smaller than the longitudinal relaxation rate $\tilde{\gamma}_{\parallel}$. Therefore, large billiards satisfy the condition $\tilde{\gamma}_{\parallel} \gg |\Delta ij|$. In order to verify the importance of this condition, we change the value of the longitudinal relaxation rate $\tilde{\gamma}_{\parallel}$ with fixing the size and pumping power of the cardioid billiard laser. As we have already shown in Figure 4g–i, single-mode lasing is observed with $\tilde{\gamma}_{\parallel} = 0.01$ and $W_{\infty} = 0.01$. Here, we decrease the value of $\tilde{\gamma}_{\parallel}$ without changing the remaining parameter values. When its value is decreased to 0.009, the other two peaks appear in the spectrum as shown in Figure 9a, where the left and right peaks respectively correspond to the locked state of the resonances ④ and ⑤ and the locked state of the resonances ⑬ and ⑭ (resonances are labeled in Figure 3). As the value of $\tilde{\gamma}_{\parallel}$ is decreased gradually to 0.007, more and more peaks appear as shown in Figure 9b,c. All of these peaks can also be assigned to the resonances in Figure 3 and some of them are the locked states of adjacent different parity resonance pairs. We confirmed that even with these smaller values of $\tilde{\gamma}_{\parallel}$, those peaks in the spectrum disappear and change into a single peak when the pumping power W_{∞} is increased. This way, we obtain the phase diagram of the lasing state in the $\tilde{\gamma}_{\parallel}$-$W_{\infty}$ plane shown in Figure 10a for the cardioid billiard laser. One can see that only single-mode lasing is always observed as far as $\tilde{\gamma}_{\parallel}$ and W_{∞} are sufficiently large.

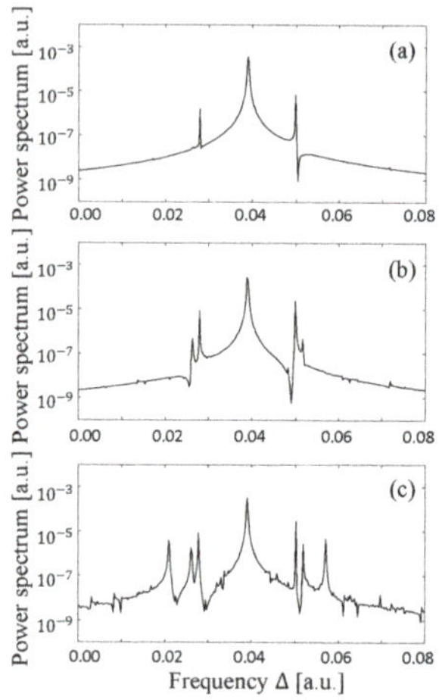

Figure 9. Power spectra obtained from the lasing states for the cardioid billiard laser with the fixed pumping power $W_{\infty} = 9.1W_{th}$ and the values of the longitudinal relaxation rate $\tilde{\gamma} =$ (**a**) 0.009 (**b**) 0.008 (**c**) 0.007.

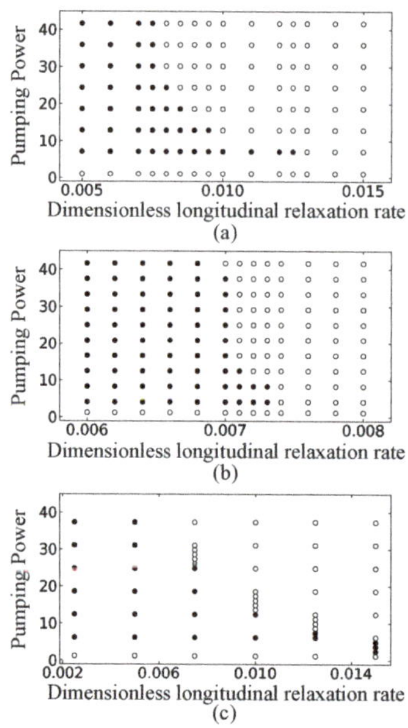

Figure 10. Phase diagrams of lasing states with various longitudinal relaxation rates $\tilde{\gamma}_{\parallel}$ and pumping powers$W_{\infty}/W_{\mathrm{th}}$ for (**a**) the cardioid billiard laser, (**b**) the D-shaped billiard laser, and (**c**) the stadium billiard laser. The white and black circles correspond to single-mode and multi-mode lasing states, respectively.

Our simulations of the cardioid billiard laser revealed the existence of two types of single-mode lasing regimes in the $\tilde{\gamma}_{\parallel}$-$W_{\infty}$ plane. One regime appears just above the lasing threshold, where the lasing mode corresponds to a single resonance with the largest net gain. Another regime appears for large W_∞ values, where the lasing mode corresponds to a locked mode of different symmetry-class resonances. We confirmed that these two types of single-mode lasing regimes were observed in the SB model simulations of the D-shaped billiard ($R = 30$ and $d = 15$) and the stadium billiard ($R = 30$), as shown in Figure 10b,c. Figure 10 is our main result of this paper, showing that single-mode lasing of a locked mode is commonly observed for the three different fully chaotic billiard lasers when the pumping power is sufficiently large.

We note that although the SB model simulations reported in this paper are performed for larger sizes of billiards than those previously numerically studied, they are still not so large as the typical semiconductor lasers used in experiments. The sizes of billiards for the SB model simulation in this paper are estimated less than 10 μm if the wavelength is assumed 1 μm, which is typical for conventional semiconductor lasers while their typical sizes are much larger than 10 μm.

Therefore, in relatively large billiard lasers with linear dimensions more than 100 μm, the frequency differences Δ_{ij} between adjacent resonances are significantly smaller than those in the simulation of this paper, which implies that the transition threshold from multi-mode to single-mode lasing is much smaller than those in Figure 10. Consequently, it is expected that the universal single-mode lasing can be observed in the experiments of semiconductor fully-chaotic billiard lasers with typical sizes.

5. Experiments of Semiconductor Billiard Lasers

By applying a reactive-ion-etching technique to a graded index separate-confinement-heterostructure single quantum well GaAs/$Al_xGa_{1-x}As$ structure, we fabricated semiconductor cardioid, D-shaped, and stadium billiard lasers of which microscope images are shown in Figure 11a–c. We also fabricated the circular billiard laser, that is, the microdisk laser as the representative example of integrable billiard lasers shown in Figure 11d.

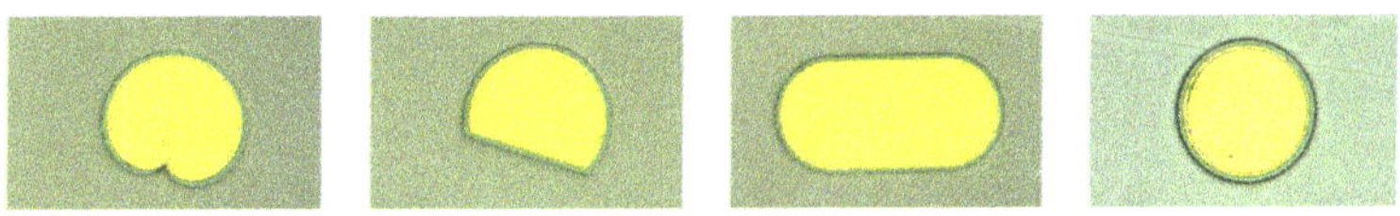

(a) (b) (c) (d)

Figure 11. Optical microscope images of the fabricated lasers of (**a**) the cardioid billiard (R = 35 μm), (**b**) the D-shaped billiard (R = 40 μm, d = 20 μm), (**c**) the stadium billiard (R = 50 μm), and (**d**) the circular billiard, i.e., microdisk (R = 40 μm).

The lasers were soldered on aluminum nitride submounts, and electrically driven with current injection under continuous-wave (cw) operation. The temperature was kept at room temperature. The difference between cw and pulse operations is crucial for observing the universal single-mode lasing, because relaxation from initial transient multi-mode lasing to single-mode lasing needs a constant pumping with a certain duration. Thus, for pulse operations with pulse widths smaller than the above duration, one cannot observe the universal single-mode lasing [13,28,55].

To measure the optical spectra of the semiconductor billiard lasers, the emitted light was collected with anti-reflection-coated lenses and coupled to a multimode optical fiber via a 30 dB optical isolator. The multimode optical fiber was coupled to a spectrum analyzer (Advantest Q8347). Figure 12 shows the measured spectra of the semiconductor billiard lasers. Every spectrum of three different fully chaotic billiard lasers has a single peak with the injection currents much larger than their thresholds as shown in Figure 12a–c, whereas the spectrum of the integrable billiard (microdisk) laser has more than 10 peaks as shown in Figure 12d. The appearance of a single lasing peak in the spectra of the semiconductor fully-chaotic billiard lasers can be more convincingly confirmed with the high-resolution mode of the spectrum analyzer with wavelength interval of 0.002 nm. As shown in Figure 13, a narrow linewidth peak was measured in the semiconductor D-shaped billiard lasers.

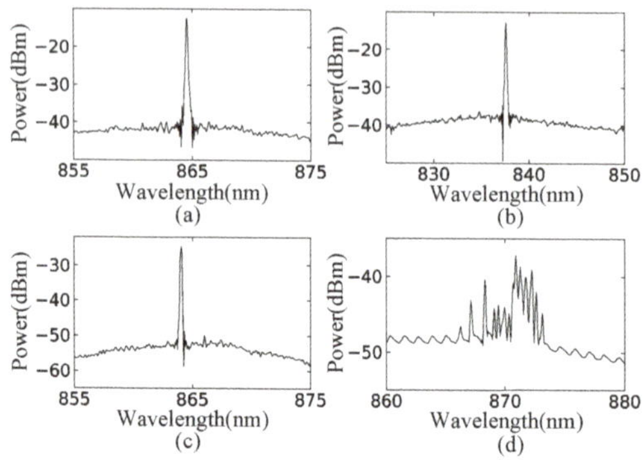

Figure 12. Spectra observed in the experiments of (**a**) the cardioid laser with the injection current 160 mA, (**b**) the D-shaped laser with the injection current 200 mA, (**c**) the stadium laser with the injection current 200 mA, and (**d**) the microdisk laser with the injection current 120 mA. The thresholds of these lasers are around 30 mA, 45 mA, 40 mA, and 20 mA, respectively.

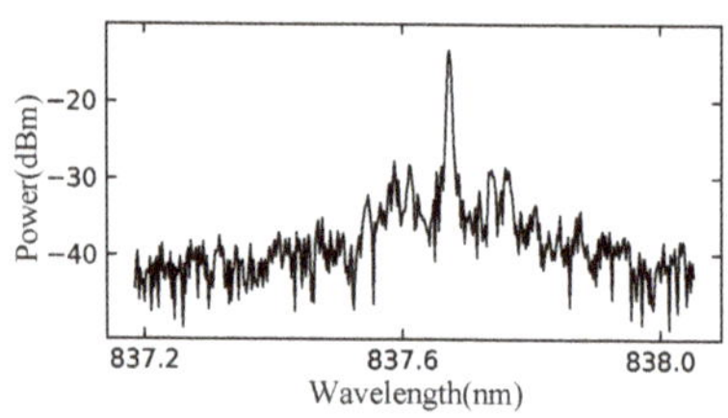

Figure 13. A spectrum obtained by using the spectrum analyzer in high-resolution mode with wavelength interval of 0.002 nm in the experiment of the semiconductor D-shaped billiard laser with the injection current 200 mA.

To gain further insight in the spectral characteristics, we counted the number of the peaks whose intensities were larger than −35 dB of the maximum peak intensity in each spectrum. Figure 14 shows the injection–current dependence of the numbers of peaks in the spectra. From Figure 14, one can see that the number of the lasing modes in a fully chaotic billiard laser is distinctly different from that of the integrable one when the injection currents are much larger than their thresholds. Similar experimental results were reported for the semiconductor stadium and elliptic billiard lasers with various aspect ratios [32]. Accordingly, we conclude that this phenomenon is universal and independent of the details of billiard lasers. Thus, the universal single-mode lasing in fully chaotic billiard lasers is verified experimentally.

The spectra observed in the experiments of fully chaotic billiard lasers occasionally contain a couple of peaks instead of one single peak even when the injection current is much larger than the threshold as one can see in Figure 14. In the next section, we show by the SB model simulation that this phenomenon can be explained by the temperature fluctuation of the billiard lasers.

In the experiments, we always try to keep the temperature of the billiard lasers to a fixed value by the Peltier cooler with the thermostat. However, even if we set the temperature of the temperature control device to a fixed value, it is impossible to make the actual temperature of the laser strictly kept to be this value, and the fluctuation of temperature is practically unavoidable during the experiments. Temperature fluctuations affect the states of the lasing medium and change the gain profile. Since the sizes of the semiconductor fully-chaotic billiard lasers are large and the gain band contains a number of resonances, a slight change of the gain profile can change the stationary lasing states significantly.

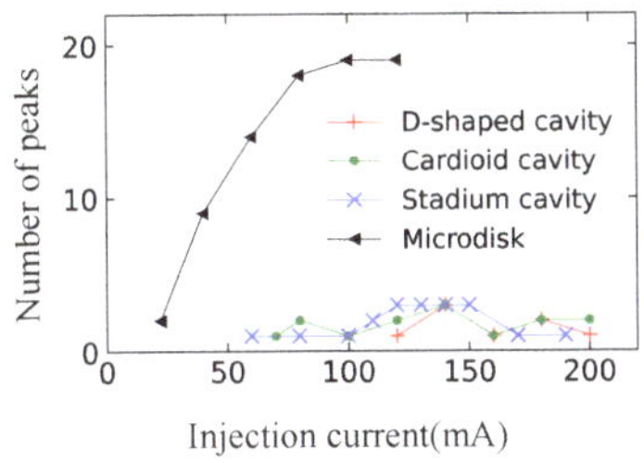

Figure 14. Injection current dependence of the number of the spectral peaks of fully chaotic and integrable billiard lasers.

In order to elucidate the temperature dependence of the stationary lasing states of the semiconductor stadium billiard laser, we investigate the change of spectral peaks by increasing the temperature from 20.1 °C to 21.0 °C by 0.1 °C. The result is shown in Figure 15, where we find that the gradual temperature increase from 20.4 °C to 20.8 °C results in a drastic change of the lasing state (i.e., the appearance of a peak at 862 nm in addition to a peak at around 857 nm), whereas the other temperature changes do not cause significant spectral changes. The temperature increase shifts the gain center to the right, i.e., longer wavelengths, which explains the appearance of the peak at around 862 nm. What is interesting here is that only this peak is observed, despite there are a number of resonances between the two peaks, and the gain center is gradually moved from left to right. In the following section, we show that this experimental result can be reproduced by the SB model simulations. Namely, we numerically demonstrate that low-loss resonances very close to the gain center do not necessarily contribute to the formation of a final lasing state obtained for a high pumping power, and that there is a selected set of low-loss resonances that can contribute to the final lasing state.

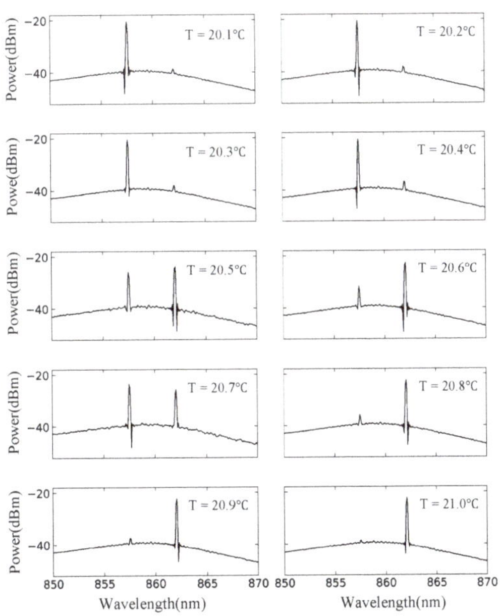

Figure 15. Spectra of the semiconductor stadium billiard laser with the injection current 80 mA at various temperatures between 20.1 °C and 21.0 °C.

6. Effect of Temperature Fluctuation

We investigate the dynamics of the D-shaped billiard laser by changing the gain center Δ_0 in the SB model simulation with the fixed pumping power W_∞. The size and deformation parameters of this billiard are $R = 30$ and $d = 15$, respectively, and the distribution of the resonances in the complex frequency plane is shown in Figure 16. When the gain centers are 0.181 and 0.22, the power spectra have single peaks, respectively corresponding to the locked state of the resonances A and B and the locked state of the resonances C and D as shown in Figure 17a,c. When the gain center is 0.206, the two locked states simultaneously appear in the power spectrum as shown in Figure 17b. This is explained by the fact that the net gains of two locked states are larger than those of the resonances around 0.206. Consequently, two distinct peaks are simultaneously observed in the spectra, which correspond to the experimental results shown in Figure 15.

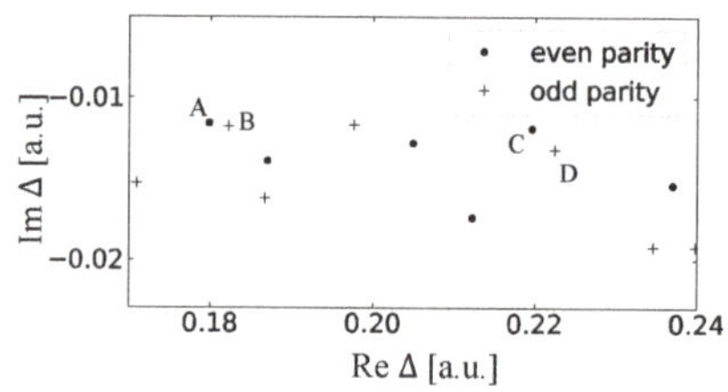

Figure 16. The resonance distribution for the D-shaped billiard.

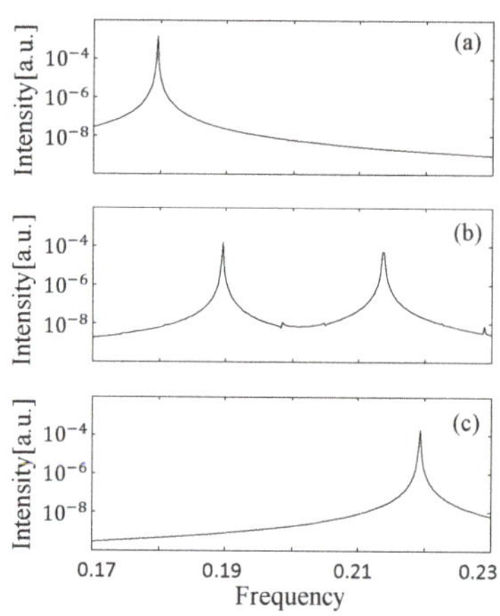

Figure 17. Power spectra of the lasing states with different gain centers Δ_0 ($\tilde{\gamma}_\parallel = 0.008$). (**a**) $\Delta_0 = 0.181$. (**b**) $\Delta_0 = 0.206$. (**c**) $\Delta_0 = 0.220$.

Finally, the pumping power W_∞ as well as the gain center Δ_0 are increased. Even if the position of the gain center is located between the real parts of the resonances B and C, one of the two locked states corresponding to the two peaks in Figure 17b wins the mode competition and becomes a single-mode lasing state with large pumping power. We define this transition threshold W_t as the value of the pumping power where the lasing state changes from the multi-mode to the single-mode lasing state. The gain center dependence of W_t is shown in Figure 18. W_t is small when the gain center is near the low-loss resonances such as 0.181 and 0.22. However, it becomes extremely large when the gain center is the midpoint between these two values, and we could not obtain the single-mode lasing state by the SB model simulation because the net gains of the locked state of A and B and the locked state of C and D are equally large. If the size of the billiard is much larger, the resonance distribution is much more dense. Therefore, the two-mode lasing regime in the pumping power is expected to be small in large fully-chaotic billiard lasers. However, when the injection current is large, the temperature of the upper contact layer might be very high due to the resistance there, and hence, the lower Peltier device might operate to cool the laser from the bottom and the whole situation might become a highly non-equilibrium state, which could result in the fluctuation of the gain center. Therefore, it might be inevitable to observe a couple of peaks occasionally in the spectra of the semiconductor fully-chaotic billiard lasers, which is the effect of the combination of the universal single-mode lasing and the fluctuation of the gain center.

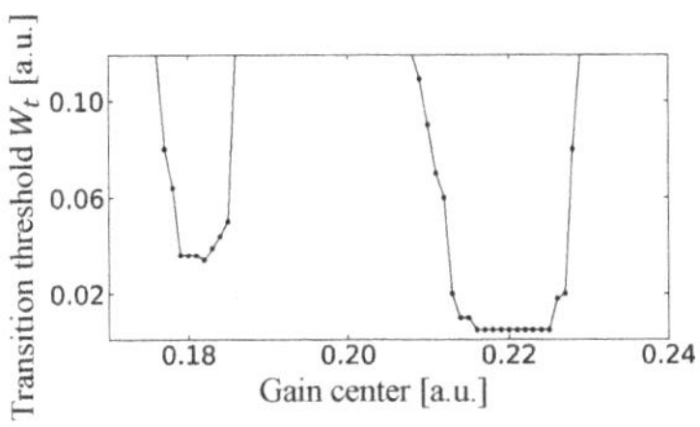

Figure 18. The gain center dependence of the transition threshold W_t from multi-mode to single-mode lasing.

7. Summary and Discussion

By the SB model simulation of the cardioid, D-shaped, and stadium billiard lasers, we verified the universal single mode lasing in fully chaotic billiard lasers: at least one single-mode lasing state is stable, whereas all multi-mode lasing states are unstable, when the external pumping power is sufficiently large and the billiard size is much larger than the wavelength so that $\tilde{\gamma}_{\parallel} \gg |\Delta_{ij}|$ is satisfied, where Δ_{ij} is the difference between adjacent lasing frequencies. The stability of multi-mode lasing states in integrable billiard lasers is also demonstrated by the SB model simulation of the elliptic billiard laser.

We also verified the universal single-mode lasing by the experiments of the semiconductor cardioid, D-shaped, stadium, and circular billiard lasers. The occasional observation of a couple of peaks instead of a single peak in the spectra of the semiconductor fully-chaotic billiard lasers was attributed to the fluctuation of the temperature on the basis of the SB model simulation. We expect that the occasional appearance of a couple of peaks in the spectra would vanish and only a single peak would be observed if the semiconductor fully-chaotic billiard lasers have much lower resistances and the control of the temperature is more accurate.

Author Contributions: Conceptualization, M.Y. (Mengyu You) and T.H.; methodology, M.Y. (Mengyu You), T.H., S.S. (Satoshi Sunada) and S.S. (Susumu Shinohara); investigation, D.S., K.M. (Kota Makino), Y.M., K.M. (Kazutoshi Matsumura), Y.K., M.Y. (Manao Yoshikawa) and M.T.; device fabrication, S.S.(Satoshi Sunada), A.U. and K.K.; original draft preparation, M.Y. (Mengyu You) and T.H.; review and editing. S.S. (Satoshi Sunada) and S.S. (Susumu Shinohara); funding acquisition, T.H., S.S. (Satoshi Sunada) and A.U. All authors have read and agreed to the published version of the manuscript.

Funding: JSPS KAKENHI Grant Numbers JP21K03410, JP20H04255, JP22H05198, Waseda University (2022C-136), JST CREST Grant Number JPMJCR17N2.

Data Availability Statement: Not applicable.

Acknowledgments: T.H. is deeply grateful to Giulio Casati for having inspired him to study quantum chaos. He also thanks Roland Ketzmerick, A. Douglas Stone, Hui Cao, and Stefan Bittner for valuable discussions.

Conflicts of Interest: The authors declare no conflict of interest.

References

1. Harayama, T.; Shinohara, S. Two-dimensional microcavity lasers. *Laser Photonics Rev.* **2011**, *5*, 247–271. [CrossRef]
2. Cao, H.; Wiersig, J. Dielectric microcavities: Model systems for wave chaos and non-Hermitian physics. *Rev. Mod. Phys.* **2015**, *87*, 61–111. [CrossRef]
3. Nöckel, J.U.; Stone, A.D. Ray and wave chaos in asymmetric resonant optical cavities. *Nature* **1997**, *385*, 45–47. [CrossRef]
4. Gmachl, C.; Capasso, F.; Narimanov, E.; Nöckel, J.U.; Stone, A.D.; Faist, J.; Sivco, D.L.; Cho, A.Y. High-power directional emission from microlasers with chaotic resonators. *Science* **1998**, *280*, 1556–1564. [CrossRef]
5. Fukushima, T.; Biellak, S.A.; Sun, Y.; Siegman, A.E. Beam propagation behavior in a quasi-stadium laser diode. *Opt. Express* **1998**, *2*, 21–28. [CrossRef]
6. Harayama, T.; Davis, P.; Ikeda, K.S. Nonlinear Whispering Gallery Modes. *Phys. Rev. Lett.* **1999**, *82*, 3803–3806. [CrossRef]
7. Lee, S.B.; Lee, J.H.; Chang, J.S.; Moon, H.J.; Kim, S.W.; An, K. Observation of Scarred Modes in Asymmetrically Deformed Microcylinder Lasers. *Phys. Rev. Lett.* **2002**, *88*, 033903. [CrossRef]
8. Harayama, T.; Davis, P.; Ikeda, K.S. Stable Oscillations of a Spatially Chaotic Wave Function in a Microstadium Laser. *Phys. Rev. Lett.* **2003**, *90*, 063901. [CrossRef]
9. Harayama, T.; Fukushima, T.; Sunada, S.; Ikeda, K.S. Asymmetric Stationary Lasing Patterns in 2D Symmetric Microcavities. *Phys. Rev. Lett.* **2003**, *91*, 073903. [CrossRef]
10. Lee, S.Y.; Rim, S.; Ryu, J.W.; Kwon, T.Y.; Choi, M.; Kim, C.M. Quasiscarred Resonances in a Spiral-Shaped Microcavity. *Phys. Rev. Lett.* **2004**, *93*, 164102. [CrossRef]
11. Schwefel, H.G.L.; Rex, N.B.; Tureci, H.E.; Chang, R.K.; Stone, A.D.; Ben-Messaoud, T.; Zyss, J. Dramatic shape sensitivity of directional emission patterns from similarly deformed cylindrical polymer lasers. *J. Opt. Soc. Am. B* **2004**, *21*, 923–934. [CrossRef]
12. Podolskiy, V.A.; Narimanov, E.; Fang, W.; Cao, H. Chaotic microlasers based on dynamical localization. *Proc. Natl. Acad. Sci. USA* **2004**, *101*, 10498–10500. [CrossRef] [PubMed]
13. Fukushima, T.; Harayama, T. Stadium and quasi-stadium laser diodes. *IEEE J. Sel. Top. Quantum Electron.* **2004**, *10*, 1039–1051. [CrossRef]
14. Harayama, T.; Sunada, S.; Ikeda, K.S. Theory of two-dimensional microcavity lasers. *Phys. Rev. A* **2005**, *72*, 013803. [CrossRef]
15. Lebental, M.; Lauret, J.S.; Hierle, R.; Zyss, J. Highly directional stadium-shaped polymer microlasers. *Appl. Phys. Lett.* **2006**, *88*, 031108. [CrossRef]
16. Türeci, H.E.; Stone, A.D.; Collier, B. Self-consistent multimode lasing theory for complex or random lasing media. *Phys. Rev. A* **2006**, *74*, 043822. [CrossRef]
17. Wiersig, J. Formation of Long-Lived, Scarlike Modes near Avoided Resonance Crossings in Optical Microcavities. *Phys. Rev. Lett.* **2006**, *97*, 253901. [CrossRef]
18. Tanaka, T.; Hentschel, M.; Fukushima, T.; Harayama, T. Classical Phase Space Revealed by Coherent Light. *Phys. Rev. Lett.* **2007**, *98*, 033902. [CrossRef]
19. Dubertrand, R.; Bogomolny, E.; Djellali, N.; Lebental, M.; Schmit, C. Circular dielectric cavity and its deformations. *Phys. Rev. A* **2008**, *77*, 013804. [CrossRef]
20. Bogomolny, E.; Dubertrand, R.; Schmit, C. Trace formula for dielectric cavities: General properties. *Phys. Rev. E* **2008**, *78*, 056202. [CrossRef] [PubMed]
21. Wiersig, J.; Hentschel, M. Combining Directional Light Output and Ultralow Loss in Deformed Microdisks. *Phys. Rev. Lett.* **2008**, *100*, 033901. [CrossRef] [PubMed]
22. Shinohara, S.; Hentschel, M.; Wiersig, J.; Sasaki, T.; Harayama, T. Ray-wave correspondence in limaçon-shaped semiconductor microcavities. *Phys. Rev. A* **2009**, *80*, 031801. [CrossRef]
23. Yan, C.; Wang, Q.J.; Diehl, L.; Hentschel, M.; Wiersig, J.; Yu, N.; Pflügl, C.; Capasso, F.; Belkin, M.A.; Edamura, T.; et al. Directional emission and universal far-field behavior from semiconductor lasers with limaçon-shaped microcavity. *Appl. Phys. Lett.* **2009**, *94*, 251101. [CrossRef]
24. Song, Q.; Fang, W.; Liu, B.; Ho, S.T.; Solomon, G.S.; Cao, H. Chaotic microcavity laser with high quality factor and unidirectional output. *Phys. Rev. A* **2009**, *80*, 041807. [CrossRef]
25. Yi, C.H.; Kim, M.W.; Kim, C.M. Lasing characteristics of a Limaçon-shaped microcavity laser. *Appl. Phys. Lett.* **2009**, *95*, 141107. [CrossRef]
26. Shinohara, S.; Harayama, T.; Fukushima, T.; Hentschel, M.; Sasaki, T.; Narimanov, E.E. Chaos-Assisted Directional Light Emission from Microcavity Lasers. *Phys. Rev. Lett.* **2010**, *104*, 163902. [CrossRef]
27. Song, Q.; Ge, L.; Redding, B.; Cao, H. Channeling Chaotic Rays into Waveguides for Efficient Collection of Microcavity Emission. *Phys. Rev. Lett.* **2012**, *108*, 243902. [CrossRef]
28. Sunada, S.; Fukushima, T.; Shinohara, S.; Harayama, T.; Adachi, M. Stable single-wavelength emission from fully chaotic microcavity lasers. *Phys. Rev. A* **2013**, *88*, 013802. [CrossRef]
29. Sarma, R.; Ge, L.; Wiersig, J.; Cao, H. Rotating Optical Microcavities with Broken Chiral Symmetry. *Phys. Rev. Lett.* **2015**, *114*, 053903. [CrossRef]
30. Redding, B.; Cerjan, A.; Huang, X.; Lee, M.L.; Stone, A.D.; Choma, M.A.; Cao, H. Low spatial coherence electrically pumped semiconductor laser for speckle-free full-field imaging. *Proc. Natl. Acad. Sci. USA* **2015**, *112*, 1304–1309. [CrossRef]

31. Harayama, T.; Shinohara, S. Ray-wave correspondence in chaotic dielectric billiards. *Phys. Rev. E* **2015**, *92*, 042916. [CrossRef] [PubMed]
32. Sunada, S.; Shinohara, S.; Fukushima, T.; Harayama, T. Signature of Wave Chaos in Spectral Characteristics of Microcavity Lasers. *Phys. Rev. Lett.* **2016**, *116*, 203903. [CrossRef] [PubMed]
33. Harayama, T.; Sunada, S.; Shinohara, S. Universal single-mode lasing in fully chaotic two-dimensional microcavity lasers under continuous-wave operation with large pumping power. *Photon. Res.* **2017**, *5*, B39–B46. [CrossRef]
34. Kawashima, Y.; Shinohara, S.; Sunada, S.; Harayama, T. Self-adjustment of a nonlinear lasing mode to a pumped area in a two-dimensional microcavity. *Photon. Res.* **2017**, *5*, B47–B53. [CrossRef]
35. Cao, Q.T.; Wang, H.; Dong, C.H.; Jing, H.; Liu, R.S.; Chen, X.; Ge, L.; Gong, Q.; Xiao, Y.F. Experimental Demonstration of Spontaneous Chirality in a Nonlinear Microresonator. *Phys. Rev. Lett.* **2017**, *118*, 033901. [CrossRef]
36. Lee, J.W.; Yi, C.H.; Kim, M.W.; Ryu, J.; Oh, K.R.; Kim, C.M. Unidirectional emission of high-Q scarred modes in a rounded D-shape microcavity. *Opt. Express* **2018**, *26*, 34864–34871. [CrossRef]
37. Bittner, S.; Guazzotti, S.; Zeng, Y.; Hu, X.; Yılmaz, H.; Kim, K.; Oh, S.S.; Wang, Q.J.; Hess, O.; Cao, H. Suppressing spatiotemporal lasing instabilities with wave-chaotic microcavities. *Science* **2018**, *361*, 1225–1231. [CrossRef]
38. Kim, Y.; Lee, S.Y.; Ryu, J.W.; Kim, I.; Han, J.H.; Tae, H.S.; Choi, M.; Min, B. Designing whispering gallery modes via transformation optics. *Nat. Photonics* **2016**, *10*, 647–652. [CrossRef]
39. Kim, K.; Bittner, S.; Zeng, Y.; Guazzotti, S.; Hess, O.; Wang, Q.J.; Cao, H. Massively parallel ultrafast random bit generation with a chip-scale laser. *Science* **2021**, *371*, 948–952. [CrossRef]
40. Stone, A.D. Chaotic billiard lasers. *Nature* **2010**, *465*, 696–697. [CrossRef]
41. Casati, G.; Valz-Gris, F.; Guarnieri, I. On the connection between quantization of nonintegrable systems and statistical theory of spectra. *Lett. Al Nuovo Cimento* **1980**, *28*, 279–282. [CrossRef]
42. Bohigas, O.; Giannoni, M.J.; Schmit, C. Characterization of Chaotic Quantum Spectra and Universality of Level Fluctuation Laws. *Phys. Rev. Lett.* **1984**, *52*, 1–4. [CrossRef]
43. Haake, F. *Quantum Signatures of Chaos*; Springer: Berlin/Heidelberg, Germany, 2001.
44. Nakamura, K.; Harayama, T. *Quantum Chaos and Quantum Dots*; Oxford University Press on Demand: Oxford, UK, 2004; Volume 3.
45. Bunimovich, L.A. On the ergodic properties of nowhere dispersing billiards. *Commun. Math. Phys.* **1979**, *65*, 295–312. [CrossRef]
46. Wojtkowski, M. Principles for the design of billiards with nonvanishing Lyapunov exponents. *Commun. Math. Phys.* **1986**, *105*, 391 – 414. [CrossRef]
47. Wiersig, J. Boundary element method for resonances in dielectric microcavities. *J. Opt. A Pure Appl. Opt.* **2002**, *5*, 53–60. [CrossRef]
48. Ketzmerick, R.; Clauß, K.; Fritzsch, F.; Bäcker, A. Chaotic Resonance Modes in Dielectric Cavities: Product of Conditionally Invariant Measure and Universal Fluctuations. *Phys. Rev. Lett.* **2022**, *129*, 193901. [CrossRef]
49. Schomerus, H.; Tworzydło, J. Quantum-to-Classical Crossover of Quasibound States in Open Quantum Systems. *Phys. Rev. Lett.* **2004**, *93*, 154102. [CrossRef] [PubMed]
50. Nonnenmacher, S.; Zworski, M. Fractal Weyl laws in discrete models of chaotic scattering. *J. Phys. A Math. Gen.* **2005**, *38*, 10683–10702. [CrossRef]
51. Keating, J.P.; Novaes, M.; Prado, S.D.; Sieber, M. Semiclassical Structure of Chaotic Resonance Eigenfunctions. *Phys. Rev. Lett.* **2006**, *97*, 150406. [CrossRef]
52. Shepelyansky, D.L. Fractal Weyl law for quantum fractal eigenstates. *Phys. Rev. E* **2008**, *77*, 015202. [CrossRef]
53. Novaes, M. Resonances in open quantum maps. *J. Phys. A Math. Theor.* **2013**, *46*, 143001. [CrossRef]
54. Altmann, E.G. Emission from dielectric cavities in terms of invariant sets of the chaotic ray dynamics. *Phys. Rev. A* **2009**, *79*, 013830. [CrossRef]
55. Cerjan, A.; Bittner, S.; Constantin, M.; Guy, M.; Zeng, Y.; Wang, Q.J.; Cao, H.; Stone, A.D. Multimode lasing in wave-chaotic semiconductor microlasers. *Phys. Rev. A* **2019**, *100*, 063814. [CrossRef]

entropy

Article

Canonical Density Matrices from Eigenstates of Mixed Systems §

Mahdi Kourehpaz, Stefan Donsa, Fabian Lackner, Joachim Burgdörfer and Iva Březinová *

Institute for Theoretical Physics, Vienna University of Technology, Wiedner Hauptstraße 8-10/136, 1040 Vienna, Austria
* Correspondence: iva.brezinova@tuwien.ac.at
§ Dedicated to Professor Giulio Casati on the occasion of his 80th birthday.

Abstract: One key issue of the foundation of statistical mechanics is the emergence of equilibrium ensembles in isolated and closed quantum systems. Recently, it was predicted that in the thermodynamic ($N \to \infty$) limit of large quantum many-body systems, canonical density matrices emerge for small subsystems from almost all pure states. This notion of canonical typicality is assumed to originate from the entanglement between subsystem and environment and the resulting intrinsic quantum complexity of the many-body state. For individual eigenstates, it has been shown that local observables show thermal properties provided the eigenstate thermalization hypothesis holds, which requires the system to be quantum-chaotic. In the present paper, we study the emergence of thermal states in the regime of a quantum analog of a mixed phase space. Specifically, we study the emergence of the canonical density matrix of an impurity upon reduction from isolated energy eigenstates of a large but finite quantum system the impurity is embedded in. Our system can be tuned by means of a single parameter from quantum integrability to quantum chaos and corresponds in between to a system with mixed quantum phase space. We show that the probability for finding a canonical density matrix when reducing the ensemble of energy eigenstates of the finite many-body system can be quantitatively controlled and tuned by the degree of quantum chaos present. For the transition from quantum integrability to quantum chaos, we find a continuous and universal (i.e., size-independent) relation between the fraction of canonical eigenstates and the degree of chaoticity as measured by the Brody parameter or the Shannon entropy.

Keywords: thermal state; isolated many-body system; quantum chaos; quantum integrability; canonical density matrix

Citation: Kourehpaz, M.; Donsa, S.; Lackner, F.; Burgdörfer, J.; Březinová, I. Canonical Density Matrices from Eigenstates of Mixed Systems. *Entropy* **2022**, *24*, 1740. https://doi.org/10.3390/e24121740

Academic Editor: Marko Robnik

Received: 3 October 2022
Accepted: 4 November 2022
Published: 29 November 2022

1. Introduction

As first recognized by Ludwig Boltzmann [1,2] "molecular" chaos lies at the core of the foundation of classical statistical mechanics. Only when the phase space of an isolated mechanical system is structureless can the motion be safely assumed to be ergodic and the equal a priori probability for phase space points on the energy hypersurface, the basic tenet of the microcanonical ensemble, is realized. Moreover, chaotic dynamics is "mixing", thereby enforcing the approach to the thermal equilibrium state from "almost all" out-of-equilibrium initial conditions. While any large isolated system is expected to be described by a microcanonical ensemble, any well-defined small subsystem thereof that is only allowed to exchange energy with the remainder of the large system (referred to as bath or environment in the following) is described by the canonical ensemble. The phase-space density of the subsystem is weighted by the Boltzmann factor $e^{-\beta H_s}$, where H_s is the Hamilton function of the subsystem, $\beta = 1/k_B T$ with T the temperature imprinted by the environment and k_B the Boltzmann constant. However, when the phase space of the system is not chaotic but rather dominated by regular motion on KAM tori [3,4], neither ergodicity nor mixing is a priori assured, and thermalization of an initial non-equilibrium state may be elusive. The implicit assumption of classical equilibrium statistical mechanics is that in the limit of a large number of degrees of freedom, chaos is generic for any interacting many-particle system.

How those concepts translate into quantum physics has remained a topic of great conceptual interest and lively debate [5–21]. Renewed interest is stimulated by the experimental accessibility of ultracold quantum gases [22–28], trapped ions [29], and nanosystems [30,31], where many of the underlying concepts became quantitatively accessible in large but finite quantum systems in unprecedented detail. The foundation of thermalization of quantum systems has been pioneered by von Neumann in terms of the quantum ergodic theorem [32–37]. Accordingly, the entropy is an increasing function of time and expectation values of generic macroscopic observables for pure states formed by coherent superposition of states within microscopic energy shells converge to that of the microcanonical ensemble provided that the energy spectrum of the system is strictly non-degenerate. Recently, this description of thermal equilibrium states was extended to the notion of canonical typicality [37–40]. Accordingly, starting from almost any pure state formed by a coherent superposition of energy eigenstates of a large isolated many-body system with eigenenergies within a given energy shell $[E, E + \Delta E]$ of macroscopically small thickness ΔE, the reduction to a small subsystem by tracing out the degrees of freedom of the bath will yield the same reduced density matrix one would obtain from the reduction of the microcanonical density matrix for the entire system. If the bath is sufficiently large and the interactions between the bath and the subsystem sufficiently weak, the reduced density matrix corresponds to the standard canonical density matrix $\hat{\rho}_s = e^{-\beta \hat{H}_s}/\mathrm{Tr}[e^{-\beta \hat{H}_s}]$ with $\hat{H}_s$ the Hamilton operator of the subsystem. The proof of this canonical typicality invokes the intrinsic randomness of the expansion coefficients of the pure state in terms of entangled subsystem–bath states. The latter assumption goes back to the notion of intrinsic quantum complexity of entangled states in large systems put forward already by Schrödinger [41].

An alternative approach to thermalization is tied to the eigenstate thermalization hypothesis (ETH) [5–8] first put forward by Landau and Lifshitz [42], stating that basic properties of statistical mechanics can emerge not only from ensemble averages but from typical single wavefunctions. However, the condition under which such an equivalence may emerge has remained open. The more recent formulation of the ETH [5–8] invokes the notion of quantum chaos and Berry's conjecture. Characteristics of quantum chaos were originally identified in few-degrees of freedom systems whose classical limit exhibits chaos [43–49]. Nowadays, the notion of quantum chaos is invoked more generally for systems that display the same signatures such as energy level distributions predicted by random matrix theory (RMT) [43,48,50,51] or randomness of wavefunction amplitudes [5,52] even when a well-defined classically chaotic counterpart is not known. The ETH conjectures that for chaotic systems, the diagonal matrix elements of any generic local observable taken in the energy eigenstate basis are smooth functions of the total energy while the off-diagonal elements are exponentially decreasing randomly fluctuating variables with zero mean [6–8]. If the ETH is valid for a specific system, individual eigenstates show thermal properties upon reduction to a small subsystem. The ETH has been shown to hold for a large variety of systems without a classical analogue [24,53–64]. Deviations from the ETH have been observed for local observables in finite systems of hard-core bosons and spin-less fermions [57,58,65] when the energy level distribution deviates from the Wigner–Dyson level statistics of RMT characteristic for chaotic systems.

In the present paper, we explore the quantitative relationship between thermal properties of reduced density matrices (RDMs) emerging from single isolated eigenstates of the entire system and quantum chaos. More specifically, we want to address the question: Is for large but finite systems quantum chaos a conditio sine qua non for the emergence of the Gibbs ensemble, i.e., the canonical ensemble of the subsystem, from eigenstates of the entire system? Or is quantum entanglement and complexity in these systems itself sufficient to render the reduced density matrix of a small subsystem canonical? To this end, we determine the fraction of canonical density matrices emerging upon reduction from the entire set of eigenstates. We explore the existence of a quantitative relationship between the fraction of eigenstates that upon reduction lead to canonical eigenstates, also termed fraction of canonical eigenstates, and the degree of quantum chaos of the entire system. We unravel the connection between this eigenstate canonicity and quantum chaos by exact diagonalization of a large yet finite mesoscopic quantum system. We emphasize that this measure addresses isolated energy eigenstates of the many-body system, in contrast to coherent superpositions of energy eigenstates from a given energy shell of finite width with random expansion coefficients as invoked in the well-established notion of canonical typicality [37–40].

As a prototypical case in point, we consider an itinerant impurity embedded in a spin-polarized Fermi–Hubbard system. Unlike impurity models for disordered systems [66], our model is fully deterministic. All key ingredients for the realization of the present system, i.e., discrete lattice, tunable interactions, and impurity can be experimentally realized with ultracold fermionic atoms (see, e.g., [67–72]). In the present scenario, the impurity serves as a probe or "thermometer" in the isolated many-body quantum system providing an unambiguous subsystem–bath decomposition with tunable coupling strength between subsystem and bath. Moreover, our system features a tunable transition from quantum chaos to quantum integrability without invoking any extrinsic stochasticity or disorder [66]. The fact that the subsystem consists of a distinguishable particle has a number of distinct advantages: The reduced density matrix of the probe is uniquely defined and its properties are basis-independent. No choice of a specific basis for the probe such as the independent-particle basis is involved. Moreover, its canonical RDM approaches a Maxwell–Boltzmann rather than a Fermi–Dirac distribution for indistinguishable fermions. Its thermal state is thus characterized by a single equilibrium parameter, the temperature T, without the need for introducing a chemical potential μ, thereby improving the numerical accuracy of the test of canonicity. We measure the proximity of the reduced density matrix of the impurity to the canonical density matrix and identify a direct and size-independent correlation between the fraction of canonical eigenstates and quantum chaos.

The paper is structured as follows. In Section 2, we introduce our impurity-Fermi–Hubbard model which serves as a prototypical (sub)system–environment model system. Quantitative measures for quantum chaos are introduced in Section 3. The mapping of spectral properties of this isolated many-body system onto thermal states of the impurity within the framework of the microcanonical and canonical ensembles are discussed in Section 4. The distance in Liouville space between the reduced density matrix of the impurity and a generic canonical density matrix is analyzed and the relation between the fraction of canonical eigenstates and quantum chaos is established in Section 5. Concluding remarks are given in Section 6.

2. The Fermi–Hubbard Model with Impurity

We investigate a variant of the single-band one-dimensional Fermi–Hubbard model which is particularly well suited to study entanglement and quantum correlations between subsystem and its environment or bath. The bath is represented by spin-polarized fermions enforcing single occupancy of sites by bath particles while the distinguishable impurity can occupy any site. Accordingly, the Hamiltonian of the total system is given by

$$\hat{H} = \hat{H}_{\mathrm{I}} + \hat{H}_{\mathrm{B}} + \hat{H}_{\mathrm{IB}}, \tag{1}$$

where the Hamiltonian of the subsystem, i.e., the impurity (I), is

$$\hat{H}_{\mathrm{I}} \;\; = -J_{\mathrm{I}} \textstyle\sum_{j=1}^{M_s-1} \left[\hat{a}_{j+1}^{\dagger}\hat{a}_j + c.c.\right] + \sum_{j=1}^{M_s} V(j)\hat{n}_j, \tag{2}$$

while the Hamiltonian of the bath is

$$\hat{H}_{\mathrm{B}} = -J_{\mathrm{B}} \sum_{j=1}^{M_s-1} \left[\hat{b}_{j+1}^{\dagger}\hat{b}_j + c.c.\right] + W_{\mathrm{BB}} \sum_{j=1}^{M_s-1} \hat{N}_{j+1}\hat{N}_j + \sum_{j=1}^{M_s} V(j)\hat{N}_j. \tag{3}$$

The interaction between the subsystem and the bath is given by

$$\hat{H}_{\text{IB}} = W_{\text{IB}} \sum_{j=1}^{M_s} \hat{n}_j \hat{N}_j. \tag{4}$$

The operators $\hat{a}_j$ and $\hat{a}_j^\dagger$ ($\hat{b}_j$ and $\hat{b}_j^\dagger$) are the creation and annihilation operators of the impurity (bath particles) on site j with the anticommutation relations $\{a_i, a_j\} = 0$, $\{a_i^\dagger, a_j^\dagger\} = 0$, $\{a_i, a_j^\dagger\} = \delta_{ij}$, and $\{a_i^{(\dagger)}, b^{(\dagger)}\} = 0$. The operators $\hat{n}_j = \hat{a}_j^\dagger \hat{a}_j$ and $\hat{N}_j = \hat{b}_j^\dagger \hat{b}_j$ correspond to the number operators of impurity and bath $\hat{n}_j|j\rangle = n_j|j\rangle$ and $\hat{N}_j|j\rangle = N_j|j\rangle$ with occupation numbers n_j and N_j of site j, respectively. J_{I} (J_{B}) describes the hopping matrix elements of the impurity (bath particles). The bath particles interact with each other by a nearest-neighbor interaction with strength W_{BB} while the impurity interacts with the bath particles via an on-site interaction with strength W_{IB}. The Hubbard chain has M_s sites with Dirichlet boundary conditions imposed at the edges. An additional very weak external background potential ($V \ll J_{\text{I}}, J_{\text{B}}$) with on-site matrix element $V(j)$ ($j = 1, \dots, M_s$) is applied,

$$V(j) = 0.01\left[-0.5 + \frac{(j-1)^n}{(M_s-1)^n}\right], \tag{5}$$

for which we use a linear ($n = 1$) or quadratic ($n = 2$) function in order to remove residual geometric symmetries such that the irreducible state space coincides with the entire state space and symmetry related degeneracies are lifted. Alternative impurity models were recently suggested for the investigation of the ETH [64].

We solve the system via exact diagonalization to determine all eigenstates and eigenenergies of the entire system. The dimension of the Hilbert space of the system is $d_{\text{H}} = M_s\binom{M_s}{N_{\text{B}}}$, where N_{B} is the number of bath particles. We consider typical half-filling configurations with $N_{\text{B}} \approx M_s/2$. The largest M_s considered is $M_s = 15$ resulting in a Hilbert space dimension of $d_{\text{H}} = 96{,}525$ for $N_{\text{B}} = 7$. We set $J_{\text{I}} = J_{\text{B}} = J$ which also defines the unit of energy ($J = 1$) in the following. The key advantage of the present model is that it allows to control and tune the properties of the bath separately by varying W_{BB} while keeping fixed the properties of the subsystem whose reduced density matrix we probe. This clear-cut subsystem–bath decomposition allows for the unambiguous probing of the emergence of canonical density matrices, thereby avoiding any ad hoc separation by "cutting out" of the subsystem which then requires the grand canonical density matrix for an open quantum system since both energy and particles can be exchanged [65]. Moreover, its thermal state is unambiguously characterized by T rather than by T and μ as for indistinguishable fermions, thereby improving the numerical reliability of the performed tests.

The present system should be realizable for ultracold fermionic atoms trapped in optical lattices [28,69–74]. All key ingredients required for its realization including tunable interactions and impurity–bath mixtures are available in the toolbox of ultracold atomic physics. We note that tuning the nearest-neighbor interaction W_{BB} between the atoms in optical lattices to large values in the regime of strong correlations, $W_{\text{BB}}/J_{\text{B}} \gtrsim 1$, still poses an experimental challenge which might be overcome in the near future.

3. Measures of Quantum Chaos

The present single-band Fermi–Hubbard model does not possess an obvious classical counterpart whose phase space consists of regions of regular and/or chaotic motion. Lacking such direct quantum–classical correspondence, quantum integrability and quantum chaos in the present system is identified by signatures of the quantum system that have been shown to probe chaotic and regular motion in systems where quantum–classical correspondence does prevail. Several measures of quantum chaos have been proposed that are based on either properties of eigenstates or of the spectrum [14,49,58,75–79]. As will be shown below, by tuning W_{BB}, we can continuously tune the entire system from the limit of quantum integrability to the limit of quantum chaos across the transition region of a mixed quantum system in which integrable and chaotic motion coexist and explore its impact on the fraction of eigenstates which upon reduction lead to canonical density matrices. The influence of the continuous transition from quantum integrability to quantum chaos on the thermal state of the subsystem will be explored with the help of the present prototypical system.

3.1. Spectral Measures

Starting point for analyzing and quantifying quantum chaos by means of spectral statistics is the cumulative spectral function also called the staircase function

$$N(E) = \sum_{\alpha} \Theta(E - E_\alpha), \tag{6}$$

where E_α are the energy eigenvalues of the entire system, and Θ is the Heaviside step function. Its spectral derivative is the density of states (DOS)

$$\Omega(E) = \frac{d}{dE} N(E). \tag{7}$$

Examples for $N(E)$ and $\Omega(E)$ of the present system are shown in Figure 1.

The smoothed "average" spectral staircase function $\bar{N}(E)$ fitted to a polynomial of order 10, also shown in Figure 1, provides the reference for spectral unfolding required for certain measures of quantum fluctuations about the (classical) mean. Accordingly, the unfolded energy spectrum is given by $e_\alpha = \bar{N}(E_\alpha)$. For systems for which quantum–classical correspondence holds, $\bar{N}(E)$ corresponds to the classical phase space volume in units of Planck's constant h and $\Omega(E)$ to the microcanonical energy shell. We note that the saturation of $N(E)$ observed with increasing E (Figure 1a) or, likewise, the bell-shaped curve for the DOS (Figure 1b) decreasing at large E is in the present case a consequence of the single-band approximation of the Fermi–Hubbard model (Equation (1)) and, more generally, appears for systems with a spectrum bounded from above. For realistic macroscopic systems, $N(E)$ and $\Omega(E)$ should generically increase monotonically with E. As discussed in more detail below, this non-generic decrease of the density of states observed for the present as well as for other finite and mesoscopic systems has implications for the ensuing thermal properties.

The probability density $P(s)$ of the nearest-neighbor level spacings (NNLS), $s = e_{\alpha+1} - e_\alpha$, features distinctively different shapes for quantum integrable and quantum chaotic systems. While for integrable systems, the NNLS have been predicted by Berry and Tabor [50] to feature an exponential (or Poissonian) distribution $P_\mathrm{P}(s) = \exp(-s)$, for chaotic systems it closely follows random matrix theory [43]. In our case of a time-reversal symmetric system, the corresponding random-matrix ensemble is the Gaussian orthogonal ensemble (GOE) which has been shown (see e.g., [46]) to closely follow the Wigner–Dyson distribution (or Wigner surmise) given by

$$P_\mathrm{WD}(s) = \frac{\pi s}{2} e^{-\pi s^2/4}. \tag{8}$$

A complementary spectral measure first proposed by Gurevich and Pevzner [80] and applied to quantum chaos [81,82] has the advantage that it does not require spectral unfolding but can be applied to the spectral raw data, i.e., the restricted gap ratios r_α

$$r_\alpha = \min\left(r_\alpha, \frac{1}{r_\alpha}\right), \tag{9}$$

where $r_\alpha = (E_{\alpha+1} - E_\alpha)/(E_\alpha - E_{\alpha-1})$. The distribution of restricted gap ratios has been shown to obey for 3×3 GOE matrices the analytical prediction

$$W_{\mathrm{GOE}}(r) = \frac{27}{4} \frac{r + r^2}{(1 + r + r^2)^{5/2}}. \tag{10}$$

For chaotic systems, this prediction remains very accurate even for large systems [82]. In the limit of quantum integrable systems, the distribution of restricted gap ratios is given by (see [81])

$$W_{\mathrm{P}}(r) = \frac{2}{(1 + r)^2}. \tag{11}$$

The search for generic spectral measures for the transition regime between the quantum integrable and quantum chaotic limit has remained an open problem. For systems possessing a classical counterpart with a mixed phase space in which integrable and chaotic motion coexist, several models for the NNLS have been proposed [83–87]. Empirically, one of the best fits to spectral data for mixed systems has been provided by a heuristic ansatz suggested by Brody [88] which allows for a one-parameter smooth interpolation of the NNLS distribution in the transition region between the quantum integrable and quantum chaotic limit,

$$P_{\mathrm{B}}(s) = (\gamma + 1) b s^\gamma e^{-b s^{\gamma+1}}, \tag{12}$$

where the Brody parameter γ characterizes the transition from the integrable ($\gamma = 0$) to the chaotic limit ($\gamma = 1$) and b follows from the normalization as

$$b = \left[\Gamma\left(\frac{\gamma + 2}{\gamma + 1}\right)\right]^{\gamma+1}. \tag{13}$$

The Brody parameter can be viewed as a measure of the strength of level repulsion between neighboring levels of the quantum system. For mixed few-degrees of freedom systems with a classical analogue, γ could be identified as a measure for the chaotic fraction of classical phase space [86,89]. Moreover, γ has also been found to be directly proportional to the degree of phase-space (de)localization of eigenstates as measured by their Husimi distribution [77]. The parameterization of the transition from quantum integrability to quantum chaos in terms of a variable exponent γ has the salient feature that even for very small but finite γ, $0 < \gamma \ll 1$, $P_{\mathrm{B}}(0) = 0$, reflecting the fact that any perturbation of quantum integrability immediately causes level repulsion and suppresses the probability density for any exact degeneracy. We recall that non-degeneracy is one of the key prerequisites of von Neumann's quantum ergodic theorem [32]. We further note that the Hasegawa distribution [84] sometimes provides an even more accurate fit to the NNLS distribution (see, e.g., [90]), however, at the price of a second adjustable parameter.

To determine γ, we fit Equation (12) to the data for $P(s)$ (Figure 2). The quality of the fit is evaluated through the χ^2-function

$$\chi^2 = \sum_i [P(s_i) - P_{\mathrm{B}}(s_i)]^2 \Delta s, \tag{14}$$

which measures the deviation of the distribution of nearest-neighbor spacings $P(s)$ from the Brody distribution $P_{\mathrm{B}}(s)$ using a bin size of Δs. As an additional measure for the uncertainty of γ, we use the fact that the Brody parameter can be alternatively determined from a fit to the integral $\int ds' P(s')$ rather than to $P(s)$ itself. The small differences found between the two fits can be used as a measure for the numerical error.

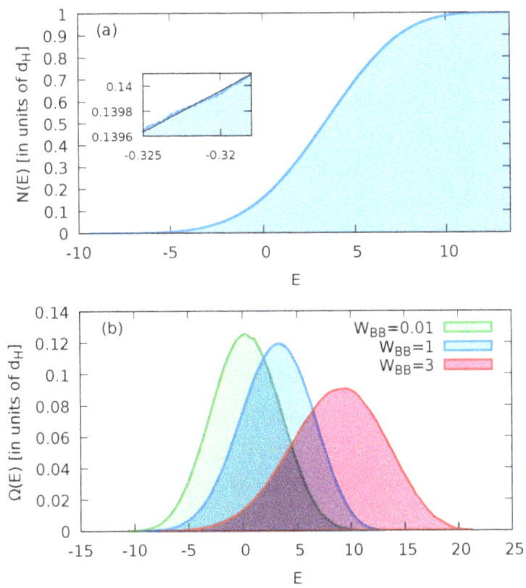

Figure 1. (**a**) The spectral staircase function $N(E)$ for the Fermi–Hubbard model with impurity with $W_{BB} = 1$ and a total number of states $d_H = 96{,}525$ ($M_s = 15$, $N_B = 7$). The inset shows a magnification of $N(E)$ with a fit for the smoothed "average" staircase function $\bar{N}(E)$ entering the spectral unfolding. (**b**) The normalized density of states (DOS), $\Omega(E)$, using a bin size of $\Delta E = 0.4$ for different interaction strengths of the bath particles. The impurity–bath interaction in (**a**,**b**) is $W_{IB} = 1$.

For $W_{BB} = 0$ and for the linear tilt of the external potential $V(j)$ (Equation (5)), we observe an excess of (near-) degenerate states as compared to the prediction of the exponential (Poisson) distribution in the first bin at $s = 0$ with $\Delta s = 0.01$. This hints at the presence of an only weakly broken symmetry which disappears when using a quadratic tilt. For reasons of consistency, we employ for all W_{BB} a linear tilt in the following. Neglecting the first bin in the fitting procedure for $W_{BB} = 0$, we obtain $\gamma = 0.005$ and, overall, a very good agreement with the Poisson distribution (Figure 2a). As the intra-bath interaction is varied from $W_{BB} = 0$ to $W_{BB} = 1$, we observe a continuous transition from a near Poissonian to an approximate Wigner–Dyson NNLS distribution (Figure 2a–c). The Brody parameter monotonically increases from $\gamma \simeq 0.005$ at $W_{BB} = 0$ to $\gamma \simeq 0.9$ at $W_{BB} = 1$. We note that after reaching a plateau at $\gamma \simeq 0.93$ near $W_{BB} = 3$, the Brody parameter decreases again for $W_{BB} > 5$ and vanishes in the strongly correlated limit of $W_{BB} \gg 1$. The decrease of the Brody parameter for large W_{BB} results from clustering of the energy spectrum in the strongly interacting regime. The bath fragments into clusters of particles with the interactions between separate clusters suppressed. Thus, a partially ordered system emerges reducing the degree of quantum chaoticity. We will focus in the following on the parameter range $W_{BB} \leq 1$ within which the transition from a nearly quantum integrable to a nearly fully quantum chaotic system occurs.

For the two limiting cases of quantum integrability ($W_{BB} \to 0$) and quantum chaos ($W_{BB} \to 1$) of the present Fermi–Hubbard system, we can also apply the predictions for the restricted gap ratio distribution (Equations (10) and (11)) . We find for these two limiting cases very good agreement between the prediction and the data (Figure 3), confirming that the identification of quantum integrability and quantum chaos is independent of the particular choice of the spectral measure.

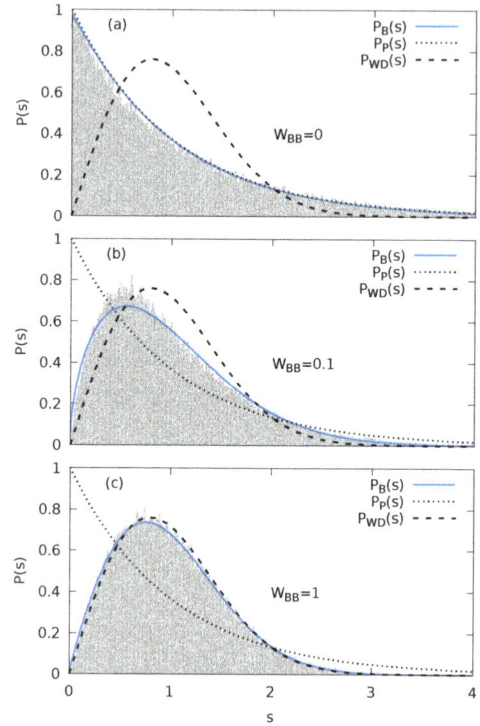

Figure 2. The numerically determined nearest-neighbor level statistics $P(s)$ for the Fermi–Hubbard model with impurity (Equations (1)–(4)) **(a)** $W_{\rm BB} = 0$, **(b)** $W_{\rm BB} = 0.1$ and **(c)** $W_{\rm BB} = 1$ compared to the Poisson (exponential) distribution $P_{\rm P}(s)$, the Wigner–Dyson distribution $P_{\rm WD}(s)$, as well as the fit to the Brody distribution $P_{\rm B}(s)$. The bin size used is $\Delta s = 0.01$. Other parameters are $M_s = 15$, $N_{\rm B} = 7$, and $W_{\rm IB} = 1$.

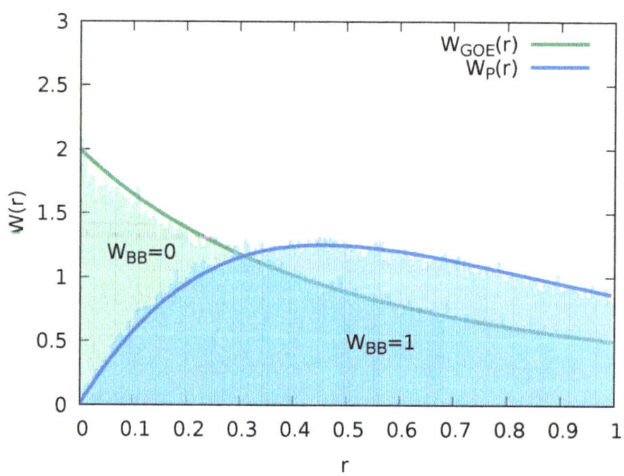

Figure 3. Statistical distribution function of restricted gap ratios for the Fermi–Hubbard model with impurity ($M_s = 15$, $N_{\rm B} = 7$) for different $W_{\rm BB}$ and $W_{\rm IB} = 1$ compared to the analytical predictions for random matrices within the GOE ensemble (Equation (10)) and for integrable spectra (Equation (11)).

For the first moment of the restricted gap ratio distribution, we find $\langle r \rangle = 0.5284$ for $W_{\mathrm{BB}} = 1$ agreeing to within 0.5% with the GOE expectation value for asymptotically large matrices $\langle r \rangle_{\mathrm{GOE}} = 0.5307$ [82]. Conversely, for $W_{\mathrm{BB}} = 0$, we find $\langle r \rangle = 0.3811$ in very good agreement with the prediction for a Poisson distribution $\langle r \rangle_{\mathrm{P}} = 0.3863$. As there is presently no interpolation function $W(r)$ available for the transition between the quantum integrable limit (Equation (11)) and the quantum chaotic limit (Equation (10)), we will focus in the following on the Brody distribution for the NNLS as spectral measure for the transition regime.

3.2. Measures for Wavefunctions

As an alternative to spectral measures, on can also explore and quantify chaos through the complexity of the eigenstates. According to Berry's conjecture, the eigenstates of a chaotic system feature randomly distributed amplitudes over an appropriate basis, e.g., in quantum billiards, they correspond to randomly distributed plane waves [75]. Following this conjecture, a large number of such measures have been proposed. They include the statistical distribution of eigenvectors [9,12,46,91], the configuration-space probability distribution [92], the configuration-space self-avoiding path correlation function [52], the Wigner function-based wavefunction autocorrelation function [75], the inverse participation ratio [93], the Shannon entropy [94], and the phase space localization measured in terms of the information entropy encoded in the Husimi distribution [77]. One limitation for the quantitative significance of most of these measures (with the possible exception of [77]) is their dependence on the chosen basis of representation. For systems that can be continuously tuned from integrable to chaotic, the eigenstates of the integrable limit suggest themselves as a convenient basis to monitor the transition to chaos [9,12,58]. For many-body systems, the eigenstates of the mean-field Hamiltonian often provide the reference basis for measuring quantum chaoticity [14]. In the following, we use the eigenstates $|\psi_\alpha^0\rangle$ of the integrable system with $W_{\mathrm{BB}} = 0$ as a basis for determining the statistical distribution of eigenvectors. From the amplitudes $c_\alpha^{\alpha'} = \langle \psi_{\alpha'}^0 | \psi_\alpha \rangle$ and probabilities $|c_\alpha^{\alpha'}|^2$, we calculate the Shannon entropy [94] for each eigenstate $|\psi_\alpha\rangle$

$$S_\alpha = -\sum_{\alpha'=1}^{d_{\mathrm{H}}} |c_\alpha^{\alpha'}|^2 \ln |c_\alpha^{\alpha'}|^2. \tag{15}$$

We observe that for $W_{\mathrm{BB}} = 1$, the Shannon entropy as a function of E_α forms an inverted parabola-like function with remarkably small eigenstate-to-eigenstate fluctuations (Figure 4). At the apex near the center of the spectrum, S_α reaches a maximum S_{max} close to the GOE limit $S_{\mathrm{GOE}} \approx \ln 0.48 d_{\mathrm{H}}$ [58] with d_{H} the dimension of the Hilbert space (Figure 4a). States in the tails of the spectrum show strong deviations from this limit as the eigenstates in this region are less complex and do not fulfill the ETH [58]. Best agreement with GOE predictions can therefore be expected near the center of the spectrum at $\alpha \approx d_{\mathrm{H}}/2$ with the highest density of states.

For smaller W_{BB} (Figure 4b–d), the Shanon entropy reveals a significantly diminished complexity of the eigenstates indicated by a reduced S_{max} and, at the same time, drastically increased state-to-state fluctuations. Probing the generic features of the wavefunctions, we will use the dependence of the scaled Shannon entropy

$$\bar{S} = S_{\mathrm{max}}(W_{\mathrm{BB}}) / S_{\mathrm{GOE}} \tag{16}$$

as an alternative wavefunction-based measure of quantum chaoticity complementing the Brody parameter γ as spectral measure. Numerically, we determine S_{max} by averaging over small intervals of energy and calculating the maximum of the resulting smooth curve.

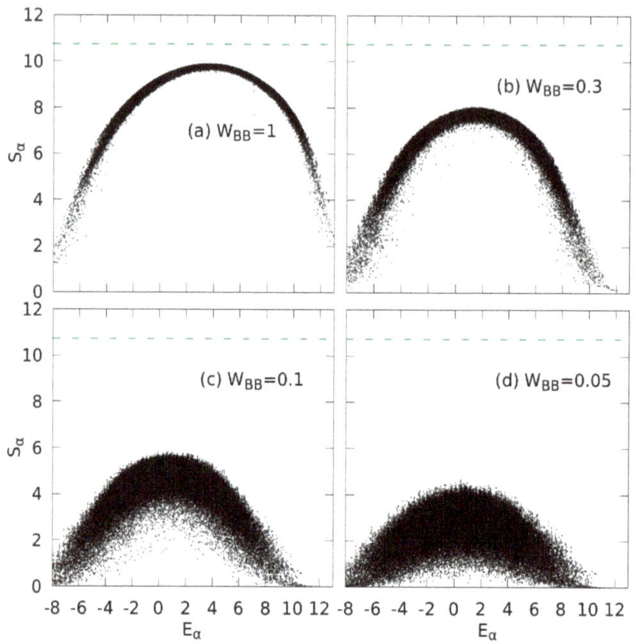

Figure 4. Distribution of Shannon entropies (Equation (15)) as a measure of the complexity of eigenstates of the system for different W_{BB}, (**a**) $W_{BB} = 1$, (**b**) $W_{BB} = 0.3$, (**c**) $W_{BB} = 0.1$, and (**d**) $W_{BB} = 0.05$. The horizontal lines mark the value $S_{GOE} \approx \ln 0.48 d_H$ expected for the GOE ensemble. All other parameters as in Figure 2.

Empirically, we find that the dependence of the Brody parameter γ, i.e., the degree of quantum chaoticity on the interaction parameter of the bath particles, $\gamma(W_{BB})$ (Figure 5) can be accurately approximated by

$$\gamma(W_{BB}) \approx \gamma_0 \tanh(W_{BB}/W_{BB}^0) \tag{17}$$

with $\gamma_0 = 0.88$ and $W_{BB}^0 = 0.15$. While a monotonic increase is intuitively expected, the origin of this particularly simple functional form remains to be understood. Remarkably, the evolutions of γ and $\bar{S}$ as a function of W_{BB} closely mirror each other, thereby representing two independent measures of the degree of quantum chaoticity during the transition from integrability to chaos. Overall, the agreement between γ and $\bar{S}$ is very good. Residual differences can be viewed as a measure for the residual uncertainty in the quantitative determination of the degree of the eigenstate quantum chaoticity.

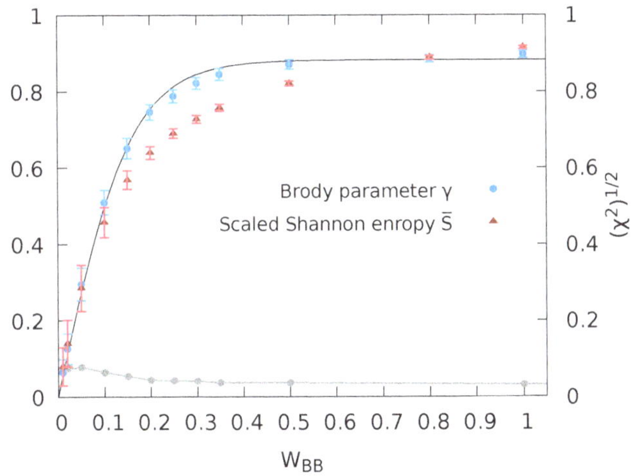

Figure 5. The Brody parameter γ (Equation (12)) or scaled Shannon entropy $\bar{S}$ (Equation (16), left y-axis) as a function of $W_{\rm BB}$. The error bars for γ correspond to the standard deviation by comparison between the fits to $P(s)$ with fits to $\int_0^s ds' P(s')$, and the black line corresponds to a fit to a tanh function $\gamma(W_{\rm BB}) \approx \gamma_0 \tanh(W_{\rm BB}/W^0_{\rm BB})$ with the parameters $\gamma_0 = 0.88$ and $W^0_{\rm BB} = 0.15$. The error bars in $\bar{S}$ reflect the width of S in Figure 4 and correspond to the scaled standard deviation around $S_{\max}$. Error of the fit to the Brody distribution as measured by the square root of the χ^2 function (Equation (14)) (gray line and right y-axis).

4. The Reduced Density Matrix of the Impurity

The impurity embedded in the Fermi–Hubbard system serves as a "thermometer", i.e., as a sensitive probe of the thermal state of the interacting many-body system. We aim at exploring the emergence of thermal properties of the impurity when the entire (subsystem and bath) system is in a given pure and stationary eigenstate of $\hat{H}$ with energy E_α and vanishing state entropy (or von Neumann entropy $S_{\rm vN} = 0$). Such an isolated large quantum system can be viewed as the limiting case of the quantum microcanonical ensemble where the width of the energy shell ΔE vanishes, i.e., $\Delta E \to 0$. Unlike other approaches, it does not invoke any coarse-graining over a macroscopically small but finite width of the energy shell nor any random interactions. For such a quantum system without any a priori built-in statistical randomness, we pose the following question: Starting from a given isolated eigenstate of the entire system, under which conditions will the reduced density matrix of the impurity correspond to a canonical density matrix, i.e., the thermometer will be accurately represented by a Gibbs ensemble or, for short, be in a Gibbs state? If such a thermal state emerges, what will be its temperature T, or its inverse temperature $\beta = 1/k_B T$? We refer to this process as emergence of a thermal equilibrium state rather than the frequently used term "thermalization" as the latter (implicitly) implies a time-dependent approach to an equilibrium state starting from an out-of-equilibrium (statistical or pure) initial state that represents a coherent superposition of different energy eigenstates. We neither invoke any ensemble average over states from the microcanonical energy shell of finite thickness ΔE nor do we invoke wave packet dynamics of a non-stationary state of the entire system.

For finite isolated systems, in particular, systems with a bounded spectrum such as the present Fermi–Hubbard model, the extraction of proper thermodynamic (or thermostatic) variables from the microcanonical ensemble requires special care. As has been recently demonstrated [95,96], the alternative definitions of the entropy used as the fundamental thermodynamic potential for the microcanonical ensemble yield, in general, inequivalent results. The standard definition [97] attributed to Boltzmann

$$S_{\mathrm{Boltzmann}} = k_B \ln \Omega(E) = k_{\mathrm{B}} \ln N'(E), \tag{18}$$

with $\Omega(E)$ the DOS of the entire closed system, implies an inverse temperature

$$\beta_{\mathrm{Boltzmann}}(E) = \frac{1}{k_{\mathrm{B}}}\frac{\partial S_{\mathrm{Boltzmann}}(E)}{\partial E} = \frac{\partial \ln \Omega(E)}{\partial E} = \frac{\Omega'(E)}{\Omega(E)} = \frac{N''(E)}{N'(E)}, \tag{19}$$

that may violate certain thermodynamic relations for mesoscopic systems with a bounded spectrum [95,96]. As shown more than 100 years ago [97,98], the Gibbs entropy defined by

$$S_{\mathrm{Gibbs}} = k_B \ln N(E) \tag{20}$$

results in an inverse temperature

$$\beta_{\mathrm{Gibbs}}(E) = \frac{\partial \ln N(E)}{\partial E} = \frac{N'(E)}{N(E)} = \frac{\Omega(E)}{N(E)} \tag{21}$$

that is free of such inconsistencies. From Equations (19) and (21), it follows that the two inverse temperature definitions are interrelated through the specific heat C [95]

$$\beta_{\mathrm{Boltzmann}} = (1 - k_B/C)\beta_{\mathrm{Gibbs}} \tag{22}$$

with $C = (\partial T_{\mathrm{Gibbs}}/\partial E)^{-1}$ and $T_{\mathrm{Gibbs}} = \beta_{\mathrm{Gibbs}}^{-1}/k_B$. Only for systems with a small specific heat of the order of k_{B} or smaller, differences between $\beta_{\mathrm{Boltzmann}}$ and β_{Gibbs} become noticeable. This is in particular the case for systems with a bounded spectrum. While $\beta_{\mathrm{Boltzmann}}(E)$ features negative values as soon as the density of states $\Omega(E) = N'(E)$ decreases (Equation (19)), $\beta_{\mathrm{Gibbs}}(E)$ remains always positive semi-definite (Equation (21)). Figure 6 presents a comparison between β_{Gibbs} and $\beta_{\mathrm{Boltzmann}}$ for the present Fermi–Hubbard model with an impurity where we have applied the microcanonical thermodynamic relations for $\beta_{\mathrm{Boltzmann}}$ and β_{Gibbs} (Equations (19) and (21)) to the numerically determined spectral data (Figure 1) of the entire system over a wide range of energies E. The two inverse temperatures closely follow each other in parallel with β_{Gibbs} shifted upwards relative to $\beta_{\mathrm{Boltzmann}}$ as long as $\Omega'(E) > 0$. For larger E when $\beta_{\mathrm{Boltzmann}}$ turns negative, the discrepancies increase as β_{Gibbs} remains positive for all E.

Alternatively, the entire system can be assigned an inverse temperature β_c by treating the system as a canonical ensemble. Accordingly, the energy E can be expressed in terms of the canonical expectation value

$$E = \frac{\mathrm{Tr}\left[\hat{H}e^{-\beta_c \hat{H}}\right]}{\mathrm{Tr}\left[e^{-\beta_c \hat{H}}\right]} = \frac{\partial \ln Z_c}{\partial \beta_c}, \tag{23}$$

where $Z_c = \mathrm{Tr}\left[\exp\left(-\beta_c \hat{H}\right)\right]$ is the canonical partition function and $\hat{H}$ is the Hamiltonian of the entire system (see Equation (1)). For a given E, Equation (23) yields an implicit relation for β_c also shown in Figure 6. Obviously, for this finite system, β_c is close to $\beta_{\mathrm{Boltzmann}}$. In the thermodynamic limit, we would expect $\beta_c = \beta_{\mathrm{Boltzmann}}$. In spite of the fact that the size of our system is still far from the thermodynamic limit ($N \to \infty$), the agreement between different thermodynamic ensembles is already remarkably close. Deviations appear primarily near the tails of the density of states and are larger in the region of negative $\beta_{\mathrm{Boltzmann}}$ where the DOS decreases rather than increases with E.

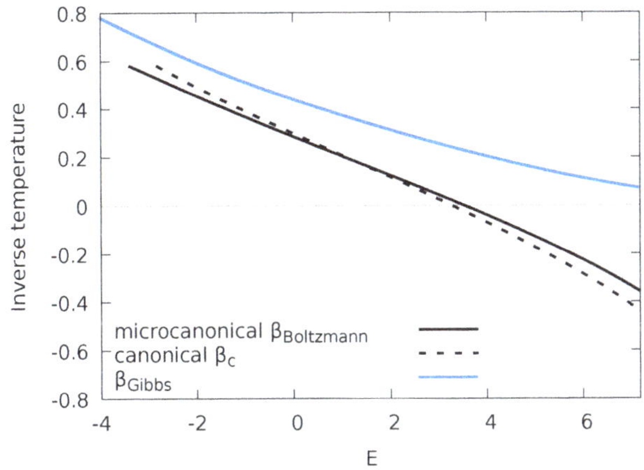

Figure 6. The inverse temperature as a function of the energy of the entire system predicted by the microcanonical Boltzmann entropy (Equation (18), solid black) and the Gibbs entropy (Equation (20), blue) as well as the canonical expectation value (Equation (23), dashed black). The energy is restricted to the interval $[E_{\rm min}, E_{\rm peak} + E_{\rm FWHM}/2,]$ with $E_{\rm min}$ the lower bound where the DOS of the entire system is $\geq 15\%$ of its peak value at $E_{\rm peak}$, and $E_{\rm FWHM}$ the full-width-at-half-maximum of the DOS. Bath–bath interaction strength $W_{\rm BB} = 1$ and impurity–bath interaction $W_{\rm IB} = 1$ (see Figure 1b).

The conceptually interesting question now arises which of these temperatures, if any, will be imprinted on the impurity upon an exact calculation of its reduced density matrix by tracing out all bath degrees of freedom from a given single exact eigenstate of a the isolated many-body system, and without invoking any a priori assumption of the microcanonical ensemble.

To address this question, we start from the density operator for any pure energy eigenstate $|\psi_\alpha\rangle$ of the entire system given by the projector $|\psi_\alpha\rangle\langle\psi_\alpha|$. Consequently, the reduced density matrix (RDM) of the impurity follows from tracing out all bath degrees of freedom,

$$D_\alpha^{(\rm I)} = {\rm Tr}_{N_{\rm B}}[|\psi_\alpha\rangle\langle\psi_\alpha|], \tag{24}$$

which will, in general, depend on the parent state $|\psi_\alpha\rangle$ it is derived from. We explore now the generic properties of $D_\alpha^{(\rm I)}$ independent of the particular parent state. Specifically, we investigate whether a given $D_\alpha^{(\rm I)}$ emerging from an individual eigenstate $|\psi_\alpha\rangle$ approaches a canonical density matrix. To this end, we diagonalize the RDM

$$D_\alpha^{(\rm I)} = \sum_{m=1}^{M_s} n_{m,\alpha}|\eta_{m,\alpha}\rangle\langle\eta_{m,\alpha}|, \tag{25}$$

yielding natural orbitals $|\eta_{m,\alpha}\rangle$ with natural occupation numbers $n_{m,\alpha}$ [99]. We emphasize that within the present approach, the RDMs $D_\alpha^{(\rm I)}$ and their eigenvalues, the occupation numbers $n_{m,\alpha}$, which characterize the thermal state, are a priori uniquely determined and not influenced by the choice of any (approximate) basis. Compared to previous investigations, this is one distinguishing feature of the present study of the thermal state emerging from an isolated deterministic many-body system. RDMs have been previously employed in studies of disordered fermionic systems [100–102].

Canonicity is reached when $n_{m,\alpha}$ is given by the Boltzmann factor $e^{-\beta\epsilon_{m,\alpha}^{(\rm I)}}$ with $\epsilon_{m,\alpha}^{(\rm I)}$ the expectation value of the Hamilton operator $H_{\rm I}$ of the impurity alone evaluated in the basis of natural orbitals, $\epsilon_{m,\alpha}^{(\rm I)} = \langle\eta_{m,\alpha}|\hat{H}_{\rm I}|\eta_{m,\alpha}\rangle$, which, in turn, should be close to the eigenstates of $\hat{H}_{\mathrm{I}}$. Moreover, the resulting value for β extracted from the fit to the exponential distribution allows the identification of the inverse temperature uniquely characterizing the thermal distribution.

For a finite-size system with an impurity and a bath with an order of magnitude of 10 particles and finite impurity–bath coupling, the residual interaction of the impurity with the bath is not negligible and should therefore be included to improve the numerical accuracy. We account for the residual impurity–bath interaction on the level of the mean-field (MF) or Hartree approximation [14]. Accordingly, the energies $\epsilon^{(\mathrm{I})}$ of the impurity appearing in the Boltzmann factor include a correction term

$$\bar{\epsilon}^{(\mathrm{I})}_{m,\alpha} = \langle \eta_{m,\alpha} | \hat{H}_{\mathrm{I}} + \hat{W}^{(\mathrm{IB})}_{\mathrm{MF},\alpha} | \eta_{m,\alpha} \rangle, \tag{26}$$

where the MF interaction operator in site-representation reads

$$W^{(\mathrm{IB})}_{\mathrm{MF},\alpha}(j) = W_{\mathrm{IB}}\, \rho_{\mathrm{B},\alpha}(j) \tag{27}$$

with

$$\rho_{\mathrm{B},\alpha}(j) = \langle j | \mathrm{Tr}_{N_{\mathrm{B}}-1,\mathrm{I}}[|\Psi_\alpha\rangle\langle\Psi_\alpha|] | j \rangle \tag{28}$$

the reduced one-body density of residual bath particles at the site j when the entire system is in state $|\psi_\alpha\rangle$. In Equation (28), the partial trace over all but one ($N_{\mathrm{B}} - 1$) bath particles and the impurity (I) is denoted by $\mathrm{Tr}_{N_{\mathrm{B}}-1,\mathrm{I}}$. The energy fluctuations

$$\Delta\bar{\epsilon}^{(\mathrm{I})}_{m,\alpha} = \sqrt{\langle \eta_{m,\alpha} | (\hat{H}_{\mathrm{I}} + \hat{W}^{(\mathrm{IB})}_{\mathrm{MF},\alpha})^2 | \eta_{m,\alpha} \rangle - \bar{\epsilon}^{(\mathrm{I})2}_{m,\alpha}} \tag{29}$$

provide a measure for the proximity of the natural orbitals of the RDM to the eigenstates of the (perturbed) single-particle Hamilton operator of the subsystem, $\hat{H}_{\mathrm{I,eff}} = \hat{H}_{\mathrm{I}} + \hat{W}^{(\mathrm{IB})}_{\mathrm{MF},\alpha}$. The energy fluctuations (Equation (29)) vanish only when the natural orbitals $|\eta_{m,\alpha}\rangle$ with which the matrix elements in Equation (29) are evaluated do coincide with the eigenstates of $\hat{H}_{\mathrm{I,eff}}$. Therefore, the variance $\Delta\bar{\epsilon}^{(\mathrm{I})}_{m,\alpha}$ can serve as a distance measure of the natural orbitals from eigenstates of the impurity Hamiltonian operator. The MF correction in Equation (27) follows from the Liouville–von Neumann equation for the reduced system where the interaction with the bath consists of the MF term and a collision operator. The collision operator describes the correlations between the impurity and the bath particles and contains the so-called two-particle (subsystem–bath) cumulant Δ_{12}. We numerically monitor the validity of the MF approximation through the magnitude of the two-particle correlation energy determined by Δ_{12}. Consistently, we find that for all many-particle states $|\psi_\alpha\rangle$ which reduce to a near-canonical RDM for the impurity, the correlation energy is negligible compared to the MF energy thereby justifying Equation (26). Of course, in the limit of weak impurity–bath coupling, the MF correction (Equation (27)) becomes negligible as well.

A representative example for the spectrum of the impurity RDM, i.e., the occupation number distribution of natural orbitals of the impurity RDM emerging from a single energy eigenstate of the entire system with state index $\alpha = 4364$ (with α sorted by energy) and energy eigenvalue $E_\alpha = -2.396$ lying on the tail of the DOS with positive β for $W_{\mathrm{BB}} = 1$, is shown in Figure 7.

Indeed, a Boltzmann distribution $\propto e^{-\beta\bar{\epsilon}^{(\mathrm{I})}_{m,\alpha}}$ characterizing the canonical density matrix is observed. Moreover, the fit to an exponential yields $\beta \approx 0.58$ in close agreement with $\beta_{\mathrm{Boltzmann}} = 0.58$ predicted by Equation (19) for the inverse temperature within the microcanonical ensemble (see also Figure 6) and reproduces the distribution of occupation numbers very well. It also agrees with β_c predicted by Equation (23) where the entire system is treated as a canonical ensemble. We note that the Boltzmann-like decay of the diagonal elements would remain qualitatively unchanged when neglecting the MF correction in Equation (26) but the fit to β would deteriorate. Thus, from the reduction of state $\alpha = 4364$, we have verified that a canonical density matrix emerges.

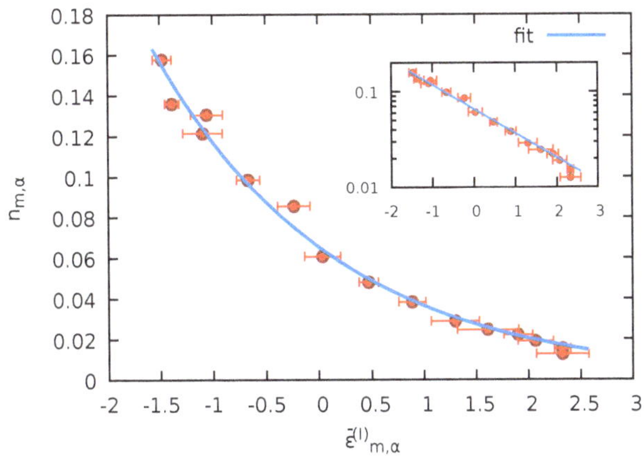

Figure 7. The occupation numbers $n_{m,\alpha}$ of the natural orbitals as a function of their energies $\bar{\epsilon}_{m,\alpha}$ (Equation (26)) for the eigenstate number $\alpha = 4364$ of the total system with energy $E_\alpha \approx -2.396$ and $W_{\rm BB} = 1$ ($M_s = 15$, $N_{\rm B} = 7$). The impurity–bath coupling strength is $W_{\rm IB} = 1$. The horizontal error bars indicate the fluctuations $\Delta\bar{\epsilon}_{m,\alpha}$ (Equation (29)). The blue solid line corresponds to the best exponential fit yielding the exponent $\beta \approx 0.58$ in agreement with $\beta_{\rm Boltzmann}$ deduced for this state from Equation (19). The inset shows the same plot on a logarithmic scale.

On a conceptual level, the present results confirm the analysis by Dunkel and Hilbert [95] who showed that the recently observed experimental single-particle population distribution in an isolated finite cold-atom system [22] is governed by $\beta_{\rm Boltzmann}$. Thus, the canonical density matrix of a small system emerging from tracing out bath variables is characterized by the inverse Boltzmann temperature $\beta_{\rm Boltzmann}$ rather than by $\beta_{\rm Gibbs}$. Consequently, level inversion in a small system in thermal contact with a bath, in particular, spin systems [103,104], can be properly characterized by negative $\beta_{\rm Boltzmann}$. The point to be noted is that while $\beta_{\rm Boltzmann}$ describes the canonical density matrix, the use of $\beta_{\rm Gibbs}$ is required for consistency in thermodynamic relations such as the Carnot efficiency [95,96]. In the following, we present the numerical results for the canonical density matrix of the impurity in terms of $\beta_{\rm Boltzmann}$ which we denote, from now on, for notational simplicity by β. We point out that β can be straightforwardly transformed into $\beta_{\rm Gibbs}$ using Equation (22) and that none of the conclusions to be drawn in the following are altered by this transformation.

5. Eigenstate Canonicity and Degree of Quantum Chaoticity

The demonstration of the emergence of a canonical density matrix from a particular eigenstate $|\psi_\alpha\rangle$ ($\alpha = 4364$) of the entire system invites now the following questions: Is the reduction to a canonical density matrix generic, i.e., will it emerge for almost all $|\psi_\alpha\rangle$? Is this appearance related to the quantum chaos present in the underlying many-body system? On a more quantitative footing: For how many of the eigenstates will a canonical density matrix emerge and does this number depend on the degree of quantum chaos of the system?

We explore these questions by determining the fraction of many-body eigenstates reducing to a canonical density matrix of the impurity, referred to in the following as eigenstate canonicity, as a function of the exact total energy E_α for the complete set of eigenstates α of the entire system and for varying bath–bath interaction $W_{\rm BB}$. The corresponding degree of quantum chaoticity of the entire system is measured by either the Brody parameter (Equation (12)) or the Shannon entropy (Equation (16)). Striking differences in the approach to the thermal state with inverse temperature β appear which are controlled by the Brody parameter γ (or Shannon entropy $\bar{S}$): At $W_{\rm BB} = 1$, when the system is chaotic as indicated by a Brody parameter $\gamma \approx 0.9$ (or scaled Shannon entropy $\bar{S} = 0.9$), a thermal distribution with a well-defined inverse temperature β, consistent with the (micro)canonical ensemble prediction (Equations (19) and (23)), emerges for an overwhelming fraction of states with the exception of states in the tails of the spectrum where the DOS is strongly suppressed (Figure 8a). The large deviations in the tails are consistent with the corresponding deviations of $\bar{S}$ in the same spectral region (Figure 4a). With decreasing $W_{\rm BB}$ and, correspondingly, decreasing γ or $\bar{S}$, an increasing fraction of states yields values of β that are far from the thermal ensemble prediction. Moreover, the quality of the fit to a canonical density matrix measured by the variance of $\Delta\beta$ and indicated by the color coding of Figure 8 drastically deteriorates. In other words, for a significant fraction of states, the emerging RDMs do not conform with the constraints of a canonical density matrix.

In order to quantify the decomposition of the Hilbert space into the subspace of states $|\psi_\alpha\rangle$ whose reduction to the subsystem yields a canonical density matrix and into the complement whose reduction fails to yield such a thermal state, we introduce a threshold for the variance of the inverse temperature $\Delta\beta_{\rm th}$ above which we consider the eigenstate canonicity to be failing. We then calculate for all states $|\psi_\alpha\rangle$ the fraction of emerging canonical density matrices satisfying $\Delta\beta \leq \Delta\beta_{\rm th}$. Of course, the resulting fraction of states will depend on the precise value of $\Delta\beta_{\rm th}$ chosen. We have determined these fractions for thresholds ranging from $\Delta\beta_{\rm th} = 5 \times 10^{-3}$ to 1.5×10^{-2}. Changes of the fractions due to variation of $\Delta\beta_{\rm th}$ are indicated by the vertical error bars in Figure 9. An unambiguous trend of a monotonic increase of the fraction of canonical density matrices with chaoticity is emerging, obviously unaffected by the choice of $\Delta\beta_{\rm th}$. This fraction representing Gibbs states, denoted in the following by G, monotonically increases with quantum chaoticity as parameterized by either the Brody parameter γ, $G(\gamma)$, or alternatively by the scaled Shannon entropy, $G(\bar{S})$ (Figure 9). Since γ and $\bar{S}$ both increase monotonically with the bath interaction $W_{\rm BB}$ (see Figure 5), this implies also a monotonic relationship $G(W_{\rm BB})$. The conceptually important observation emerging from Figure 9 is that the degree of canonicity of the RDM, $G(\gamma)$, undergoes a continuous transition from the quantum-integrable ($\gamma \to 0$) to the quantum-chaotic limit ($\gamma \to 1$). The strength of level repulsion in the NNLS parameterized by γ directly determines the probability of finding the RDM of the impurity represented by a Gibbs ensemble.

The approach of the RDM of the impurity to the Gibbs ensemble

$$D_\alpha^{\rm Gibbs} = \frac{1}{Z_{c,\alpha}} e^{-\beta_\alpha \left(\hat{H}_I + \hat{W}_{{\rm MF},\alpha}^{\rm (IB)}\right)}, \tag{30}$$

with $Z_{c,\alpha} = {\rm Tr}[e^{-\beta_\alpha \left(\hat{H}_I + W_{{\rm MF},\alpha}^{\rm (IB)}\right)}]$ can be also directly observed in the spatial site representation (j_1, j_2) of the RDM of the impurity (Figure 10b).

We illustrate the RDM in the site representation for two energetically nearest-neighbor states ($\alpha = 13{,}637$ and $\alpha = 13{,}638$) when the system is in the transition regime between integrable and non-integrable (in the present case, $W_{\rm BB} = 0.1$). We quantify the approach to $D_\alpha^{\rm Gibbs}$ through the density matrix site correlation function

$$C_\alpha(\Delta j) = \sum_{j=1}^{M_s - \Delta j} \langle j | D_\alpha^{\rm (I)} | j + \Delta j \rangle, \tag{31}$$

where $\langle j | D_\alpha^{\rm (I)} | j' \rangle$ is the RDM of the impurity (Equation (24)) in the site basis. While the state $\alpha = 13{,}638$ results in a nearly diagonal RDM in the site basis (Figure 10a) with rapidly decaying site correlations closely following the prediction for a Gibbs ensemble (Equation (30)), the adjacent state $\alpha = 13{,}637$ yields a RDM with significant off-diagonal entries, extended site correlations, and strong deviations from Equation (30). Thus, the emergence of a thermal density matrix in the transition regime between quantum integrability and quantum chaos displays strong state-to-state fluctuations and is not a smooth function of the energy E_α.

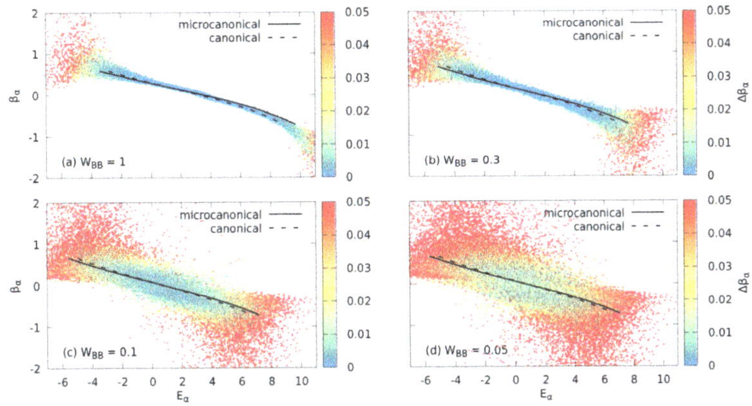

Figure 8. The inverse Boltzmann temperature β_α of the impurity as a function of energy E_α for the eigenstates of the entire system as obtained from fits to the RDM of the impurity for varying interaction strengths W_{BB}, (**a**) $W_{\mathrm{BB}} = 1$ with $\gamma \approx 0.9$, (**b**) $W_{\mathrm{BB}} = 0.3$ with $\gamma \approx 0.8$, (**c**) $W_{\mathrm{BB}} = 0.1$ with $\gamma \approx 0.5$ and (**d**) $W_{\mathrm{BB}} = 0.05$ with $\gamma \approx 0.3$. The color bar on the right-hand side represents the variance of β_α, $\Delta\beta_\alpha$, obtained from the fit. Variances above $\Delta\beta > 0.05$ are shown in red. The lines correspond to $\beta(E)$ obtained from the microcanonical ensemble Equation (19) (solid) and the canonical ensemble Equation (23) (dashed), respectively. Other parameters are $M_s = 15$, $N_{\mathrm{B}} = 7$, $W_{\mathrm{IB}} = 1$.

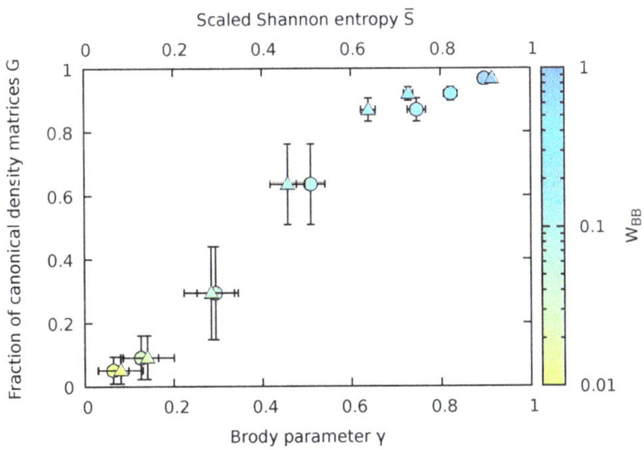

Figure 9. The fraction of canonical density matrices G obtained for the RDM of the impurity as a function of the Brody parameter γ (lower horizontal axis, dots) or as a function of the scaled Shannon entropy $\bar{S}$ (upper horizontal axis, triangles). The dots are color-coded by the interaction strength W_{BB} between the bath particles. Horizontal error bars for γ indicate the uncertainty in the extraction of the Brody parameter, horizontal error bars in $\bar{S}$ indicate the standard deviation of the Shannon entropy (see Figure 5). The vertical error bars give the variation of G under variation of the threshold $\Delta\beta$. Other parameters are $M_s = 15$, $N_{\mathrm{B}} = 7$, and $W_{\mathrm{IB}} = 1$.

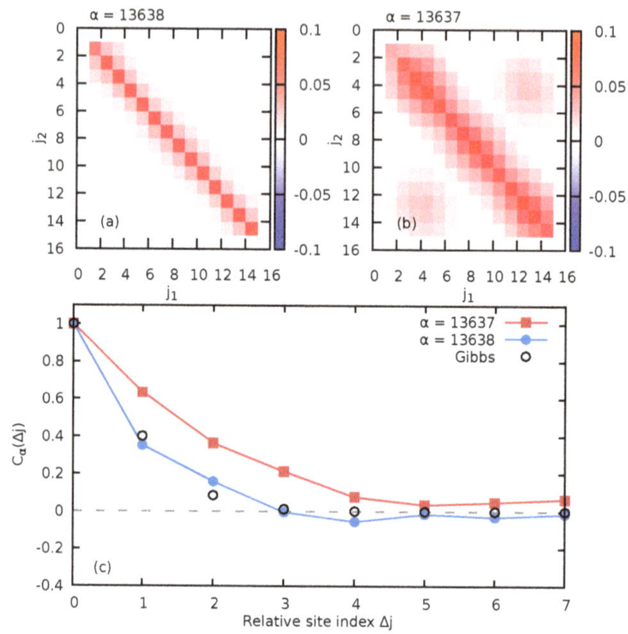

Figure 10. Site representation of the impurity RDM, $\langle j_2|D_\alpha^{(\mathrm{I})}|j_1\rangle$, resulting from the reduction of two adjacent states (**a**) $\alpha = 13{,}638$ and (**b**) $\alpha = 13{,}637$ of the entire system. In (**c**), we compare the site correlation function $C_\alpha(\Delta j)$ (Equation (31)) for these two states with the prediction for the ideal Gibbs state (Equation (30)) (black open circles). The system is in the transition regime between quantum integrable and quantum chaotic ($W_{\mathrm{BB}} = 0.1$). Other parameters are $M_s = 15$, $N_{\mathrm{B}} = 7$, $W_{\mathrm{IB}} = 1$.

As quantitative measure for the distance of a given RDM from the Gibbs ensemble, we use the trace-class norm

$$\Delta D_\alpha = ||D_\alpha^{(\mathrm{I})} - D_\alpha^{\mathrm{Gibbs}}||_1 \tag{32}$$

with $||M||_1 = \mathrm{Tr}\sqrt{M^\dagger M}$ the largest of the Schatten p-norms ($p = 1$). For hermitian positive-semidefinite matrices of unit trace, the Schatten 1-norm is bounded by $0 \leq ||M_1 - M_2||_1 \leq 2$. We observe for RDMs derived from all eigenstates of the entire system (Figure 11) an overall reduction of distances ΔD_α from a canonical density matrix with increasing W_{BB}. For $W_{\mathrm{BB}} = 1$ (Figure 11a) in the (near) quantum chaotic limit, the vast majority of impurity RDMs have a distance of $\lesssim 0.15$ from an ideal Gibbs ensemble (apart from those reduced from many-body states in the tail regions of the spectrum with low DOS). The distribution of ΔD_α mirrors the distribution of Shannon entropies (Figure 4). We note that for the present finite quantum system, we find that the distance measured by the Schatten 1-norm has a lower bound of $\Delta D_\alpha \gtrsim 0.05$. As the Schatten 1-norm is sensitive to small deviations in both diagonal and off-diagonal elements, these deviations are due to residual fluctuations (Equation (29), Figure 7) of the natural orbitals of the impurity which are expected to vanish in the thermodynamic limit $N \to \infty$. Indeed, plotting the value of the smallest distance $(\Delta D_\alpha)_{\min}$ as a function of the dimension of the Hilbert space of the system d_{H} for three numerically feasible system sizes indicates that the minimal distance vanishes in the thermodynamic limit as $d_{\mathrm{H}}^{-1/3}$ (inset Figure 11a). With decreasing W_{BB}, e.g., $W_{\mathrm{BB}} = 0.1$ in Figure 11b, the mean distance of RDMs from a Gibbs state significantly increases and, moreover, the spread becomes much larger reflecting, again, the behavior of the Shannon entropy (Figure 4c).

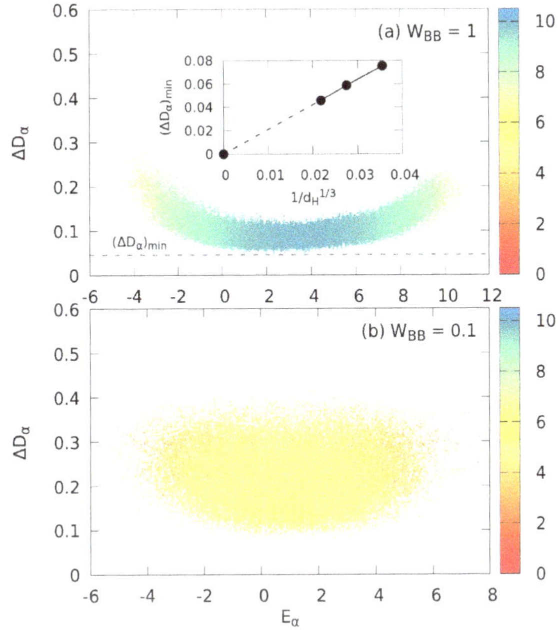

Figure 11. Distribution of distances ΔD_α (Equation (32)) from the (ideal) Gibbs state of the impurity density matrices reduced from the eigenstates $|\psi_\alpha\rangle$ of the large system with energy E_α. Shown are only those states with variance $\Delta\beta_\alpha < 0.01$. The points are color-coded by the Shannon entropy of their parent state $|\psi_\alpha\rangle$. (**a**) Near the quantum chaotic limit ($W_{\rm BB} = 1$); (**b**) in the transition regime between quantum integrability and quantum chaos ($W_{\rm BB} = 0.1$). The dashed horizontal line in (**a**) marks the minimal distance $(\Delta D_\alpha)_{\rm min}$ plotted in the inset of (**a**) as a function of the dimension of the Hilbert space $d_{\rm H}$ for $W_{\rm BB} = 1$. Other parameters are $M_s = 15$, $N_{\rm B} = 7$, $W_{\rm IB} = 1$.

The emergence of canonical density matrices, i.e., of Gibbs states for almost all $|\psi_\alpha\rangle$ in the quantum chaotic limit ($\gamma \simeq 1$ or $\bar{S} \simeq 1$) can be viewed as a rather specific manifestation and extension of the ETH [5–8]. The local observable in this case is the RDM of the impurity, $D_\alpha^{(\rm I)}$, itself. Its diagonal elements are, indeed, a smooth function of the total energy E_α as predicted by ETH but now, more specifically, Boltzmann-distributed $\propto e^{-\beta_\alpha \bar{\epsilon}_{m,\alpha}^{(\rm I)}}$ over impurity states with the inverse temperature imprinted by E_α. The present analysis covers, in addition, also the transition regime between quantum integrability and quantum chaos ($0 < \gamma < 1$) where, in general, the ETH does not apply. A canonical density matrix may still emerge but now only for a decreasing fraction of eigenstates of the finite large system. The size of this fraction G is predicted by the degree of quantum chaoticity as measured by the Brody parameter γ or Shannon entropy $\bar{S}$ (Figure 9).

The direct relation between the emergence of the canonical density matrix for a small subsystem from eigenstate reduction and the quantum chaoticity of the large system it is embedded in, established here for a finite quantum system, raises the conceptual question as to the extension of this connection to the thermodynamic ($N \to \infty$) limit. Clearly, this question cannot be conclusively addressed by the present method of exact diagonalization. Nevertheless, we can provide evidence to this effect by exploring the scaling with system size still within computational reach. We first establish that the degree of quantum chaoticity as measured by the Brody parameter γ (or the Shannon entropy $\bar{S}$) indeed increases with system size at fixed strength of the interaction $W_{\rm BB}$ that breaks quantum integrability (Figure 12).

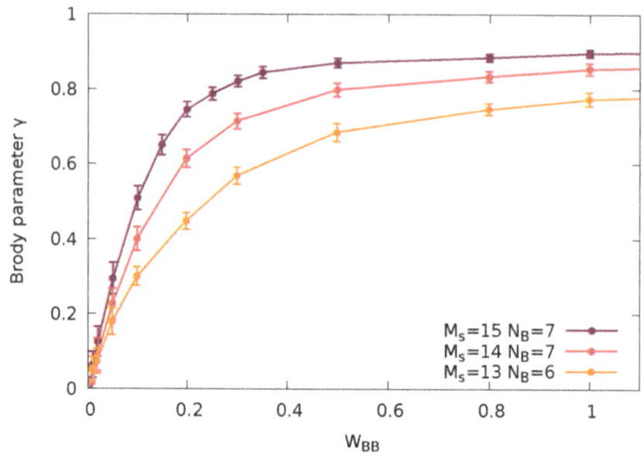

Figure 12. Variation of the Brody parameter γ characterizing the transition from quantum integrability to quantum chaos as a function of the bath–bath interaction W_{BB} breaking quantum integrability shown for different system sizes. The impurity–bath interaction is $W_{\mathrm{IB}} = 1$.

We vary the system size by increasing the total number of sites while keeping the system at (approximate) half-filling of bath particles. The corresponding Hilbert space increases from $d_{\mathrm{H}} = 22{,}308$ ($M_s = 13$, $N_{\mathrm{B}} = 6$) to $d_{\mathrm{H}} = 96{,}525$ ($M_s = 15$, $N_{\mathrm{B}} = 7$). The observed increase of quantum chaoticity with system size is qualitatively in line with properties of classical chaos: In a mixed phase space with surviving local regular structures such as tori, their influence on phase space dynamics is rapidly diminishing with increasing phase space dimension a prominent example of which is Arnold diffusion [3,4]. This increase of quantum chaoticity with system size at fixed interaction strength turns out to be key for the emergence of a universal, i.e., (nearly) system-size-independent, interrelation between the fraction of canonical eigenstates and the degree of quantum chaoticity. Both the Brody parameter γ as well as the fraction of density matrices complying with the Gibbs ensemble increase with system size at fixed bath interaction strength. As a consequence, a near universal, i.e., size-independent, relation $G(\gamma)$ between the fraction of (approximate) Gibbs states and the degree of quantum chaos as measured by γ emerges (Figure 13).

The data for different combinations of values of W_{BB} and M_s fall on the same curve. A very similar relation would emerge for $G(\bar{S})$ as a function of the scaled Shannon entropy. We have thus established the remarkable feature that the fraction of canonical eigenstates, i.e., the likelihood that a subsystem is in a Gibbs state when the large but finite system is in a pure energy eigenstate with zero von Neumann entropy is controlled and can be tuned by γ (or $\bar{S}$) and, in turn, by the degree of level repulsion in the quantum many-body system which is controlled by γ.

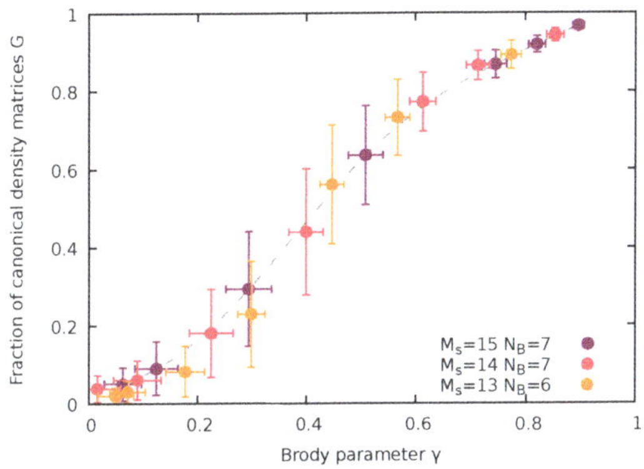

Figure 13. Universal relation between the fraction G of RDMs of the impurity converging to a Gibbs state and the Brody parameter γ for different combinations of system sizes (M_s and N_B) and bath interaction strengths W_{BB}. Impurity–bath interaction in all systems considered is $W_{IB} = 1$. Dashed line to guide the eye.

6. Conclusions and Outlook

In this work, we have explored the emergence of a thermal state (or Gibbs ensemble) of a small (sub)system in contact with a bath when the combined large but finite deterministic quantum system is isolated and in a well-defined energy eigenstate. As prototypical case, we have considered an impurity embedded in an interacting spin-polarized Fermi–Hubbard many-body bath which facilitates a clear-cut subsystem–bath decomposition and a tunable transition of the entire system from quantum integrability to quantum chaos. By tracing out the bath degrees of freedom, we have investigated how many of the resulting reduced density matrices of the subsystem represent a canonical density matrix. We have shown that the probability for finding a canonical density matrix monotonically increases with the degree of quantum chaos. The degree of quantum chaos is identified here by both the energy-level statistics as well as by the randomness of the eigenstates as measured by the Shannon entropy. The likelihood for the emergence of thermal states is thus found to be controlled by the degree of quantum chaoticity as parameterized by the Brody parameter or the Shannon entropy. Even though our simulations are limited to finite-size systems, the present results for varying system sizes suggest that the relation between the fraction of eigenstates of the isolated many-body system whose reduction to a small subsystem yields a reduced canonical density matrix and the degree of quantum chaoticity is universal, i.e., size-independent.

Each many-body eigenstate represents the fine-grained version of the energy shell of the microcanonical ensemble of the entire impurity–bath system. This connection between the fraction of canonical eigenstates and quantum chaoticity thus offers a direct quantum analogue to the role of classical chaos which Boltzmann invoked in deducing the classical (micro-)canonical ensemble. One can view this as an example of classical–quantum correspondence to this cornerstone of the foundation of statistical mechanics. The statistical ensemble properties can already emerge for isolated energy eigenstates without invoking any randomness, e.g., coarse-graining over a macroscopically thin energy shell or superposition of many eigenstates of the isolated large system as frequently employed. The emergence of statistical ensemble properties from the reduction of pure states was already early anticipated by Landau and Lifshitz [42] and later related to quantum chaos [14]. The present study establishes a direct quantitative relationship between the degree of canonicity and the degree of quantum chaos, in particular, also covering the transition regime from quantum integrability to quantum chaos.

The present results are also expected to have implications for the topical issue of thermalization in finite quantum systems [24,26,27,30]. In this paper, we intentionally avoided this notion and, instead, focused on thermal equilibrium states as we deduce the canonical density matrix from stationary energy eigenstates bypassing any explicit time dependence of the dynamics. Thermalization of an initial non-equilibrium state is, by contrast, a fundamental probe of the time evolution of quantum many-body systems. Up to now, one primary focus has been on quantum quenches, the relaxation of out-off equilibrium initial states. Their time evolution has typically shown a transition from an exponential decay for weakly perturbed many-body systems to a Gaussian decay in the strongly coupled limit, however, without an unambiguous correlation to quantum chaos [105,106]. The Shannon entropy was found to linearly increase with time before reaching saturation [107]. For disordered systems, an initial rapid decay followed by a slow power-law relaxation of occupation numbers has been observed [66,102]. The extension of the present study to a non-equilibrium initial state of a deterministic many-body system would yield the time evolution of the entire one-body RDM, and its eigenvalues and eigenvectors, the time dependence of which remains to be explored. Moreover, the dependence of the relaxation dynamics of the RDM on the choice of the initial state for systems in the transition regime between quantum integrability and quantum chaos (i.e., for intermediate values of the Brody parameter γ) is of particular interest. Most importantly, will quantum chaos play an analogous role for the process of mixing as classical chaos does for classical non-equilibrium dynamics and the relaxation to equilibrium? The origin and properties of such "quantum mixing" remain a widely open question.

Author Contributions: Data curation, M.K. and I.B.; Formal analysis, J.B. and I.B.; Funding acquisition, J.B. and I.B.; Investigation, M.K., J.B. and I.B.; Methodology, I.B.; Software, M.K., S.D., F.L. and I.B.; Supervision, I.B.; Validation, I.B.; Visualization, I.B.; Writing—original draft, I.B.; Writing—review and editing, S.D., F.L., J.B. and I.B. All authors have read and agreed to the published version of the manuscript.

Funding: This research was funded by the WWTF grant MA-14002, the Austrian Science Fund (FWF) grant P 35539-N, the FWF doctoral college Project No. FWF- W1243 (Solids4Fun), as well as the International Max Planck Research School of Advanced Photon Science (IMPRS-APS). Calculations were performed on the Vienna Scientific Cluster (VSC3 and VSC4).

Institutional Review Board Statement: Not applicable.

Informed Consent Statement: Not applicable.

Data Availability Statement: The presented data can be obtained from the corresponding author upon a reasonable request.

Acknowledgments: We thank Sebastian Bichelmaier for helpful discussions.

Conflicts of Interest: The authors declare no conflict of interest.

Abbreviations

The following abbreviations are used in this manuscript:

DOS	Density of states
ETH	Eigenstate thermalization hypothesis
GOE	Gaussian orthogonal ensemble
MF	Mean field
NNLS	Nearest-neighbor level spacing
RDM	Reduced density matrix

References

1. Boltzmann, L. *Vorlesungen über Gastheorie*; Verlag Johann Ambrosius Barth: Leipzig, Germany, 1896.
2. Huang, K. *Statistical Mechanics*; Wiley: New York, NY, USA, 1988.
3. Arnold, V.I. *Mathematical Methods of Classical Mechanics*; Springer: New York, NY, USA, 1979.
4. Lichtenberg, A.J.; Lieberman, M.A. *Regular and Chaotic Dynamics*; Springer: New York, NY, USA, 1991.
5. Deutsch, J.M. Quantum statistical mechanics in a closed system. *Phys. Rev. A* **1991**, *43*, 2046–2049. [CrossRef]
6. Srednicki, M. Chaos and quantum thermalization. *Phys. Rev. E* **1994**, *50*, 888–901. [CrossRef] [PubMed]

7. Srednicki, M. Thermal fluctuations in quantized chaotic systems. *J. Phys. A Math. Gen.* **1996**, *29*, L75–L79. [CrossRef]
8. Srednicki, M. The approach to thermal equilibrium in quantized chaotic systems. *J. Phys. A Math. Gen.* **1999**, *32*, 1163–1175. [CrossRef]
9. Zelevinsky, V.; Brown, B.A.; Frazier, N.; Horoi, M. The nuclear shell model as a testing ground for many-body quantum chaos. *Phys. Rep.* **1996**, *276*, 85–176. [CrossRef]
10. Flambaum, V.V.; Izrailev, F.M.; Casati, G. Towards a statistical theory of finite Fermi systems and compound states: Random two-body interaction approach. *Phys. Rev. E* **1996**, *54*, 2136–2139. [CrossRef]
11. Flambaum, V.V.; Izrailev, F.M. Statistical theory of finite Fermi systems based on the structure of chaotic eigenstates. *Phys. Rev. E* **1997**, *56*, 5144–5159. [CrossRef]
12. Borgonovi, F.; Guarneri, I.; Izrailev, F.; Casati, G. Chaos and thermalization in a dynamical model of two interacting particles. *Phys. Lett. A* **1998**, *247*, 140–144. [CrossRef]
13. Borgonovi, F.; Celardo, G.; Izrailev, F.M.; Casati, G. Semiquantal Approach to Finite Systems of Interacting Particles. *Phys. Rev. Lett.* **2002**, *88*, 054101. [CrossRef]
14. Borgonovi, F.; Izrailev, F.; Santos, L.; Zelevinsky, V. Quantum chaos and thermalization in isolated systems of interacting particles. *Phys. Rep.* **2016**, *626*, 1–58. [CrossRef]
15. Müller, M.P.; Adlam, E.; Masanes, L.; Wiebe, N. Thermalization and Canonical Typicality in Translation-Invariant Quantum Lattice Systems. *Commun. Math. Phys.* **2015**, *340*, 499–561. [CrossRef]
16. Reimann, P.; Gemmer, J. Why are macroscopic experiments reproducible? Imitating the behavior of an ensemble by single pure states. *Phys. A Stat. Mech. Its Appl.* **2020**, *552*, 121840. [CrossRef]
17. Brenes, M.; Pappalardi, S.; Goold, J.; Silva, A. Multipartite Entanglement Structure in the Eigenstate Thermalization Hypothesis. *Phys. Rev. Lett.* **2020**, *124*, 040605. [CrossRef] [PubMed]
18. Brenes, M.; Pappalardi, S.; Mitchison, M.T.; Goold, J.; Silva, A. Out-of-time-order correlations and the fine structure of eigenstate thermalization. *Phys. Rev. E* **2021**, *104*, 034120. [CrossRef] [PubMed]
19. Sugimoto, S.; Hamazaki, R.; Ueda, M. Eigenstate Thermalization in Long-Range Interacting Systems. *Phys. Rev. Lett.* **2022**, *129*, 030602. [CrossRef]
20. Wang, J.; Benenti, G.; Casati, G.; Wang, W.G. Complexity of quantum motion and quantum-classical correspondence: A phase-space approach. *Phys. Rev. Res.* **2020**, *2*, 043178. [CrossRef]
21. Wang, J.; Benenti, G.; Casati, G.; Wang, W.g. Quantum chaos and the correspondence principle. *Phys. Rev. E* **2021**, *103*, L030201. [CrossRef]
22. Braun, S.; Ronzheimer, J.P.; Schreiber, M.; Hodgman, S.S.; Rom, T.; Bloch, I.; Schneider, U. Negative Absolute Temperature for Motional Degrees of Freedom. *Science* **2013**, *339*, 52–55. [CrossRef]
23. Kaufman, A.M.; Tai, M.E.; Lukin, A.; Rispoli, M.; Schittko, R.; Preiss, P.M.; Greiner, M. Quantum thermalization through entanglement in an isolated many-body system. *Science* **2016**, *353*, 794–800. [CrossRef]
24. Jansen, D.; Stolpp, J.; Vidmar, L.; Heidrich-Meisner, F. Eigenstate thermalization and quantum chaos in the Holstein polaron model. *Phys. Rev. B* **2019**, *99*, 155130. [CrossRef]
25. Abanin, D.A.; Altman, E.; Bloch, I.; Serbyn, M. *Colloquium*: Many-body localization, thermalization, and entanglement. *Rev. Mod. Phys.* **2019**, *91*, 021001. [CrossRef]
26. Prüfer, M.; Kunkel, P.; Strobel, H.; Lannig, S.; Linnemann, D.; Schmied, C.M.; Berges, J.; Gasenzer, T.; Oberthaler, M.K. Observation of universal dynamics in a spinor Bose gas far from equilibrium. *Nature* **2018**, *563*, 217–220. [CrossRef] [PubMed]
27. Erne, S.; Bücker, R.; Gasenzer, T.; Berges, J.; Schmiedmayer, J. Universal dynamics in an isolated one-dimensional Bose gas far from equilibrium. *Nature* **2018**, *563*, 225–229. [CrossRef] [PubMed]
28. Scherg, S.; Kohlert, T.; Sala, P.; Pollmann, F.; Hebbe Madhusudhana, B.; Bloch, I.; Aidelsburger, M. Observing non-ergodicity due to kinetic constraints in tilted Fermi-Hubbard chains. *Nat. Commun.* **2021**, *12*, 4490. [CrossRef]
29. Neyenhuis, B.; Zhang, J.; Hess, P.W.; Smith, J.; Lee, A.C.; Richerme, P.; Gong, Z.X.; Gorshkov, A.V.; Monroe, C. Observation of prethermalization in long-range interacting spin chains. *Sci. Adv.* **2017**, *3*, e1700672. [CrossRef]
30. Neill, C.; Roushan, P.; Fang, M.; Chen, Y.; Kolodrubetz, M.; Chen, Z.; Megrant, A.; Barends, R.; Campbell, B.; Chiaro, B.; et al. Ergodic dynamics and thermalization in an isolated quantum system. *Nat. Phys.* **2016**, *12*, 1037–1041. [CrossRef]
31. Trotzky, S.; Chen, Y.A.; Flesch, A.; McCulloch, I.P.; Schollwöck, U.; Eisert, J.; Bloch, I. Probing the relaxation towards equilibrium in an isolated strongly correlated one-dimensional Bose gas. *Nat. Phys.* **2012**, *8*, 325–330. [CrossRef]
32. Neumann, J.V. Beweis des Ergodensatzes und desH-Theorems in der neuen Mechanik. *Z. Phys.* **1929**, *57*, 30–70. [CrossRef]
33. Goldstein, S.; Lebowitz, J.L.; Tumulka, R.; Zanghì, N. Long-time behavior of macroscopic quantum systems: Commentary accompanying the English translation of John von Neumann's 1929 article on the quantum ergodic theorem. *Eur. Phys. J. H* **2010**, *35*, 173–200. [CrossRef]
34. D'Alessio, L.; Kafri, Y.; Polkovnikov, A.; Rigol, M. From quantum chaos and eigenstate thermalization to statistical mechanics and thermodynamics. *Adv. Phys.* **2016**, *65*, 239–362. [CrossRef]
35. Siskens, T.J.; Bongaarts, P.J.M. Ergodicity of operators in finite quantum systems. *Physica* **1973**, *68*, 315–341. [CrossRef]
36. Bongaarts, P.J.M.; Siskens, T.J. Observables, constants of the motion and ergodicity in quantum-statistical mechanics of finite systems. *Physica* **1974**, *71*, 529–559. [CrossRef]

37. Mori, T.; Ikeda, T.N.; Kaminishi, E.; Ueda, M. Thermalization and prethermalization in isolated quantum systems: A theoretical overview. *J. Phys. At. Mol. Opt. Phys.* **2018**, *51*, 112001. [CrossRef]
38. Popescu, S.; Short, A.J.; Winter, A. Entanglement and the foundations of statistical mechanics. *Nat. Phys.* **2006**, *2*, 754–758. [CrossRef]
39. Goldstein, S.; Lebowitz, J.L.; Tumulka, R.; Zanghì, N. Canonical Typicality. *Phys. Rev. Lett.* **2006**, *96*, 050403. [CrossRef] [PubMed]
40. Reimann, P. Typicality for Generalized Microcanonical Ensembles. *Phys. Rev. Lett.* **2007**, *99*, 160404. [CrossRef]
41. Schrödinger, E. *Statistical Thermodynamics*; Cambridge University Press: Cambridge, UK, 1952.
42. Landau, L.D.; Lifshitz, E.M. *Statistical Physics*; Pergamon Press: Oxford, UK, 1958.
43. Bohigas, O.; Giannoni, M.J.; Schmit, C. Characterization of Chaotic Quantum Spectra and Universality of Level Fluctuation Laws. *Phys. Rev. Lett.* **1984**, *52*, 1–4. [CrossRef]
44. Casati, G.; Chirikov, B.V.; Izraelev, F.M.; Ford, J. Stochastic behavior of a quantum pendulum under a periodic perturbation. In *Stochastic Behavior in Classical and Quantum Hamiltonian Systems*; Series Title: Lecture Notes in Physics; Casati, G., Ford, J., Eds.; Springer: Berlin/Heidelberg, Germany, 1979; Volume 93, pp. 334–352. [CrossRef]
45. Casati, G. Relevance of classical chaos in quantum mechanics: The hydrogen atom in a monochromatic field. *Phys. Rep.* **1987**, *154*, 77–123. [CrossRef]
46. Haake, F. *Quantum Signatures of Chaos*; Springer: Berlin/Heidelberg, Germany, 2001.
47. Gutzwiller, M. *Chaos in Classical and Quantum Mechanics*; Springer: New York, NY, USA, 1990.
48. Berry, M. Quantum chaology, not quantum chaos. *Phys. Scr.* **1989**, *40*, 335–336. [CrossRef]
49. Prosen, T.; Robnik, M. Energy level statistics in the transition region between integrability and chaos. *J. Phys. A Math. Gen.* **1993**, *26*, 2371–2387. [CrossRef]
50. Berry, M.V.; Tabor, M. Level Clustering in the Regular Spectrum. *Proc. R. Soc. A Math. Phys. Eng. Sci.* **1977**, *356*, 375–394. [CrossRef]
51. Mehta, M.L. *Random Matrices*; Elsevier Academic Press: Amsterdam, The Netherlands, 2004.
52. Shapiro, M.; Goelman, G. Onset of Chaos in an Isolated Energy Eigenstate. *Phys. Rev. Lett.* **1984**, *53*, 1714–1717. [CrossRef]
53. Rigol, M.; Dunjko, V.; Olshanii, M. Thermalization and its mechanism for generic isolated quantum systems. *Nature* **2008**, *452*, 854–858. [CrossRef] [PubMed]
54. Rigol, M. Breakdown of Thermalization in Finite One-Dimensional Systems. *Phys. Rev. Lett.* **2009**, *103*, 100403. [CrossRef]
55. Rigol, M. Quantum quenches and thermalization in one-dimensional fermionic systems. *Phys. Rev. A* **2009**, *80*, 053607. [CrossRef]
56. Rigol, M.; Srednicki, M. Alternatives to Eigenstate Thermalization. *Phys. Rev. Lett.* **2012**, *108*, 110601. [CrossRef]
57. Santos, L.F.; Rigol, M. Localization and the effects of symmetries in the thermalization properties of one-dimensional quantum systems. *Phys. Rev. E* **2010**, *82*, 031130. [CrossRef]
58. Santos, L.F.; Rigol, M. Onset of quantum chaos in one-dimensional bosonic and fermionic systems and its relation to thermalization. *Phys. Rev. E* **2010**, *81*, 036206. [CrossRef]
59. Genway, S.; Ho, A.F.; Lee, D.K.K. Dynamics of Thermalization in Small Hubbard-Model Systems. *Phys. Rev. Lett.* **2010**, *105*, 260402. [CrossRef]
60. Genway, S.; Ho, A.F.; Lee, D.K.K. Thermalization of local observables in small Hubbard lattices. *Phys. Rev. A* **2012**, *86*, 023609. [CrossRef]
61. Kim, H.; Ikeda, T.N.; Huse, D.A. Testing whether all eigenstates obey the eigenstate thermalization hypothesis. *Phys. Rev. E* **2014**, *90*, 052105. [CrossRef]
62. Garrison, J.R.; Grover, T. Does a Single Eigenstate Encode the Full Hamiltonian? *Phys. Rev. X* **2018**, *8*, 021026. [CrossRef]
63. Schlagheck, P.; Shepelyansky, D.L. Dynamical thermalization in Bose-Hubbard systems. *Phys. Rev. E* **2016**, *93*, 012126. [CrossRef]
64. Brenes, M.; LeBlond, T.; Goold, J.; Rigol, M. Eigenstate Thermalization in a Locally Perturbed Integrable System. *Phys. Rev. Lett.* **2020**, *125*, 070605. [CrossRef]
65. Santos, L.F.; Polkovnikov, A.; Rigol, M. Weak and strong typicality in quantum systems. *Phys. Rev. E* **2012**, *86*, 010102. [CrossRef] [PubMed]
66. Krause, U.; Pellegrin, T.; Brouwer, P.W.; Abanin, D.A.; Filippone, M. Nucleation of Ergodicity by a Single Mobile Impurity in Supercooled Insulators. *Phys. Rev. Lett.* **2021**, *126*, 030603. [CrossRef]
67. Gaunt, A.L.; Schmidutz, T.F.; Gotlibovych, I.; Smith, R.P.; Hadzibabic, Z. Bose-Einstein Condensation of Atoms in a Uniform Potential. *Phys. Rev. Lett.* **2013**, *110*, 200406. [CrossRef] [PubMed]
68. Fukuhara, T.; Kantian, A.; Endres, M.; Cheneau, M.; Schauß, P.; Hild, S.; Bellem, D.; Schollwöck, U.; Giamarchi, T.; Gross, C.; et al. Quantum dynamics of a mobile spin impurity. *Nat. Phys.* **2013**, *9*, 235–241. [CrossRef]
69. Haller, E.; Hudson, J.; Kelly, A.; Cotta, D.A.; Peaudecerf, B.; Bruce, G.D.; Kuhr, S. Single-atom imaging of fermions in a quantum-gas microscope. *Nat. Phys.* **2015**, *11*, 738–742. [CrossRef]
70. Parsons, M.F.; Huber, F.; Mazurenko, A.; Chiu, C.S.; Setiawan, W.; Wooley-Brown, K.; Blatt, S.; Greiner, M. Site-Resolved Imaging of Fermionic Li 6 in an Optical Lattice. *Phys. Rev. Lett.* **2015**, *114*, 213002. [CrossRef]
71. Cheuk, L.W.; Nichols, M.A.; Okan, M.; Gersdorf, T.; Ramasesh, V.V.; Bakr, W.S.; Lompe, T.; Zwierlein, M.W. Quantum-Gas Microscope for Fermionic Atoms. *Phys. Rev. Lett.* **2015**, *114*, 193001. [CrossRef] [PubMed]
72. Greif, D.; Parsons, M.F.; Mazurenko, A.; Chiu, C.S.; Blatt, S.; Huber, F.; Ji, G.; Greiner, M. Site-resolved imaging of a fermionic Mott insulator. *Science* **2016**, *351*, 953–957. [CrossRef] [PubMed]

73. Schreiber, M.; Hodgman, S.S.; Bordia, P.; Lüschen, H.P.; Fischer, M.H.; Vosk, R.; Altman, E.; Schneider, U.; Bloch, I. Observation of many-body localization of interacting fermions in a quasirandom optical lattice. *Science* **2015**, *349*, 842–845. [CrossRef] [PubMed]
74. Boll, M.; Hilker, T.A.; Salomon, G.; Omran, A.; Nespolo, J.; Pollet, L.; Bloch, I.; Gross, C. Spin- and density-resolved microscopy of antiferromagnetic correlations in Fermi-Hubbard chains. *Science* **2016**, *353*, 1257–1260. [CrossRef]
75. Berry, M.V. Regular and irregular semiclassical wavefunctions. *J. Phys. A Math. Gen.* **1977**, *10*, 2083–2091. [CrossRef]
76. Pandey, M.; Claeys, P.W.; Campbell, D.K.; Polkovnikov, A.; Sels, D. Adiabatic Eigenstate Deformations as a Sensitive Probe for Quantum Chaos. *Phys. Rev. X* **2020**, *10*, 041017. [CrossRef]
77. Wang, Q.; Robnik, M. Statistical properties of the localization measure of chaotic eigenstates in the Dicke model. *Phys. Rev. E* **2020**, *102*, 032212. [CrossRef]
78. Lozej, C.; Casati, G.; Prosen, T. Quantum chaos in triangular billiards. *Phys. Rev. Res.* **2022**, *4*, 013138. [CrossRef]
79. Li, B.; Robnik, M. Statistical properties of high-lying chaotic eigenstates. *J. Phys. A Math. Gen.* **1994**, *27*, 5509–5523. [CrossRef]
80. Gurevich, I.I.; Pevzner, M.I. "Repulsion" of Nuclear Levels. *J. Exp. Theor. Phys.* **1957**, *4*, 278.
81. Oganesyan, V.; Huse, D.A. Localization of interacting fermions at high temperature. *Phys. Rev. B* **2007**, *75*, 155111. [CrossRef]
82. Atas, Y.Y.; Bogomolny, E.; Giraud, O.; Roux, G. Distribution of the Ratio of Consecutive Level Spacings in Random Matrix Ensembles. *Phys. Rev. Lett.* **2013**, *110*, 084101. [CrossRef] [PubMed]
83. Berry, M.V.; Robnik, M. Semiclassical level spacings when regular and chaotic orbits coexist. *J. Phys. A Math. Gen.* **1984**, *17*, 2413–2421. [CrossRef]
84. Hasegawa, H.; Mikeska, H.J.; Frahm, H. Stochastic formulation of energy-level statistics. *Phys. Rev. A* **1988**, *38*, 395–399. [CrossRef]
85. Caurier, E.; Grammaticos, B.; Ramani, A. Level repulsion near integrability: A random matrix analogy. *J. Phys. A Math. Gen.* **1990**, *23*, 4903–4909. [CrossRef]
86. Lenz, G.; Haake, F. Reliability of small matrices for large spectra with nonuniversal fluctuations. *Phys. Rev. Lett.* **1991**, *67*, 1–4. [CrossRef]
87. Izrailev, F.M. Quantum localization and statistics of quasienergy spectrum in a classically chaotic system. *Phys. Lett. A* **1988**, *134*, 13–18. [CrossRef]
88. Brody, T.A.; Flores, J.; French, J.B.; Mello, P.A.; Pandey, A.; Wong, S.S.M. Random-matrix physics: Spectrum and strength fluctuations. *Rev. Mod. Phys.* **1981**, *53*, 385–479. [CrossRef]
89. Yang, X.; Burgdörfer, J. Molecular-dynamics approach to the statistical properties of energy levels. *Phys. Rev. Lett.* **1991**, *66*, 982–985. [CrossRef]
90. Libisch, F.; Stampfer, C.; Burgdörfer, J. Graphene quantum dots: Beyond a Dirac billiard. *Phys. Rev. B* **2009**, *79*, 115423. [CrossRef]
91. Izrailev, F.M. Simple models of quantum chaos: Spectrum and eigenfunctions. *Phys. Rep.* **1990**, *196*, 299–392. [CrossRef]
92. McDonald, S.W.; Kaufman, A.N. Wave chaos in the stadium: Statistical properties of short-wave solutions of the Helmholtz equation. *Phys. Rev. A* **1988**, *37*, 3067–3086. [CrossRef] [PubMed]
93. Bell, R.J.; Dean, P. Atomic vibrations in vitreous silica. *Discuss. Faraday Soc.* **1970**, *50*, 55. [CrossRef]
94. Blümel, R.; Smilansky, U. Suppression of classical stochasticity by quantum-mechanical effects in the dynamics of periodically perturbed surface-state electrons. *Phys. Rev. A* **1984**, *30*, 1040–1051. [CrossRef]
95. Dunkel, J.; Hilbert, S. Consistent thermostatistics forbids negative absolute temperatures. *Nat. Phys.* **2014**, *10*, 67–72. [CrossRef]
96. Hilbert, S.; Hänggi, P.; Dunkel, J. Thermodynamic laws in isolated systems. *Phys. Rev. E* **2014**, *90*, 062116. [CrossRef]
97. Hertz, P. Über die mechanischen Grundlange der Thermodynamik. *Ann. Phys. (Leipz.)* **1910**, *338*, 225–274. [CrossRef]
98. Gibbs, J.W. *Elementary Principles in Statistical Mechanics*; Reprint of the 1902 edition; Dover: Mineola, NY, USA, 2014.
99. Löwdin, P.O.; Shull, H. Natural Orbitals in the Quantum Theory of Two-Electron Systems. *Phys. Rev.* **1956**, *101*, 1730–1739. [CrossRef]
100. Bera, S.; Schomerus, H.; Heidrich-Meisner, F.; Bardarson, J.H. Many-Body Localization Characterized from a One-Particle Perspective. *Phys. Rev. Lett.* **2015**, *115*, 046603. [CrossRef]
101. Bera, S.; Martynec, T.; Schomerus, H.; Heidrich-Meisner, F.; Bardarson, J.H. One-particle density matrix characterization of many-body localization. *Ann. Phys.* **2017**, *529*, 1600356. [CrossRef]
102. Lezama, T.L.M.; Bera, S.; Schomerus, H.; Heidrich-Meisner, F.; Bardarson, J.H. One-particle density matrix occupation spectrum of many-body localized states after a global quench. *Phys. Rev. B* **2017**, *96*, 060202. [CrossRef]
103. Purcell, E.M.; Pound, R.V. A Nuclear Spin System at Negative Temperature. *Phys. Rev.* **1951**, *81*, 279–280. [CrossRef]
104. Ramsey, N.F. Thermodynamics and Statistical Mechanics at Negative Absolute Temperatures. *Phys. Rev.* **1956**, *103*, 20–28. [CrossRef]
105. Torres-Herrera, E.J.; Santos, L.F. Local quenches with global effects in interacting quantum systems. *Phys. Rev. E* **2014**, *89*, 062110. [CrossRef] [PubMed]
106. Torres-Herrera, E.J.; Kollmar, D.; Santos, L.F. Relaxation and thermalization of isolated many-body quantum systems. *Phys. Scr.* **2015**, *T165*, 014018. [CrossRef]
107. Santos, L.F.; Borgonovi, F.; Izrailev, F.M. Onset of chaos and relaxation in isolated systems of interacting spins: Energy shell approach. *Phys. Rev. E* **2012**, *85*, 036209. [CrossRef]

entropy

MDPI

Article

On the Quantization of AB Phase in Nonlinear Systems

Xi Liu [1], Qing-Hai Wang [2] and Jiangbin Gong [2,3,*]

[1] NUS Graduate School—Integrative Sciences and Engineering Programme (ISEP), National University of Singapore, Singapore 119077, Singapore
[2] Department of Physics, National University of Singapore, Singapore 117551, Singapore
[3] Center for Quantum Technologies, National University of Singapore, Singapore 117543, Singapore
* Correspondence: phygj@nus.edu.sg

Abstract: Self-intersecting energy band structures in momentum space can be induced by nonlinearity at the mean-field level, with the so-called nonlinear Dirac cones as one intriguing consequence. Using the Qi-Wu-Zhang model plus power law nonlinearity, we systematically study in this paper the Aharonov–Bohm (AB) phase associated with an adiabatic process in the momentum space, with two adiabatic paths circling around one nonlinear Dirac cone. Interestingly, for and only for Kerr nonlinearity, the AB phase experiences a jump of π at the critical nonlinearity at which the Dirac cone appears and disappears (thus yielding π-quantization of the AB phase so long as the nonlinear Dirac cone exists), whereas for all other powers of nonlinearity, the AB phase always changes continuously with the nonlinear strength. Our results may be useful for experimental measurement of power-law nonlinearity and shall motivate further fundamental interest in aspects of geometric phase and adiabatic following in nonlinear systems.

Keywords: AB phase; Berry phase; power-law nonlinearity; Dirac cone; adiabatic dynamics; quantization

Citation: Liu, X.; Wang, Q.-H.; Gong, J. On the Quantization of AB Phase in Nonlinear Systems. *Entropy* **2022**, *24*, 1835. https://doi.org/10.3390/e24121835

Academic Editor: Marko Robnik

Received: 30 November 2022
Accepted: 12 December 2022
Published: 16 December 2022

1. Introduction

The dynamics depicted by a nonlinear discretized Schödinger equation (NDSE) can be extremely rich, including the emergence of many-dimensional chaos, solitons, and breathers, etc. The problem can be much reduced by assuming the translational invariance of a wave under consideration. With this assumption, the main physics is about the features of Bloch waves, the associated energy bands, and how they respond to changes in the parameters of a nonlinear system. Interestingly, the nonlinear Bloch bands of NDSE can induce gapless band structures absent in linear systems, such as 2-dimensional (2D) nonlinear Dirac cones [1] induced by Kerr nonlinearity [2]. Even more peculiarly, such nonlinear Dirac cones are formed by exotic nonlinear energy bands in a subregime of the Brillouin zone [1,3–8].

As a close analog to a setting in real space to measure the Aharonov–Bohm (AB) phase around a singularity point with magnetic flux, let us now imagine two adiabatic paths, in the momentum space, circling around a band-crossing point. If we adiabatically change the Bloch momentum, so as to guide the Bloch wave to evolve along the two adiabatic paths, the final phase difference thus generated between the two adiabatic paths is termed the nonlinear AB phase [1]. One may naïvely think of the following: provided that the dynamical phases between the two adiabatic paths are identical and hence have zero contribution to the phase difference of interest, the obtained AB phase would be just the Berry phase associated with the band degeneracy point. The actual physics turns out to be more interesting than just a Berry phase. As a result of nonlinearity, any small deviation of the adiabatically following state from the instantaneous Bloch wave causes a tiny correction to the dynamical phase, and accumulation of such tiny corrections over the entire adiabatic protocol yields an unfamiliar geometrical phase on top of the expected Berry phase. Remarkably, as a possible means of topological characterization of nonlinear

Dirac cones, it is found in Ref. [1] that the nonlinear AB phase around nonlinear Dirac cones induced by Kerr nonlinearity added to the so-called Qi-Wu-Zhang (QWZ) model [9] is quantized in π, whereas the Berry phase is not quantized (thus in sharp contrast to a variety of linear systems, where the Berry phase around a Dirac cone is quantized in π [10–13]). Echoing with the finding in [1], Ref. [3] found π-quantization of a nonlinear Zak phase and Ref. [14] further confirmed the π-quantization of the nonlinear AB phase around a nodal line induced by Kerr nonlinearity.

The goal of this work is entirely focused on aspects of the nonlinear AB phase around Dirac cones induced by general power law nonlinearity [15–23]. In this way, it becomes possible to answer whether the previously obtained AB phase quantization is unique to Kerr nonlinearity and if so, why there is such uniqueness. Using the QWZ model [9] as the linear limit, we are able to analytically show that Kerr nonlinearity happens to be a critical case among all kinds of power law nonlinearity. Specifically, for any nonlinearity other than the cubic order, the π-quantization of nonlinear AB phase does not exist. Our analytical results are further confirmed by direct numerical simulations.

2. Hamiltonian and Energy Spectrum

The momentum-space Hamiltonian is composed of a QWZ model with power law nonlinearity characterized by a parameter p:

$$\hat{H}(\psi) = J_1 \sin k_1 \sigma_1 + J_2 \sin k_2 \sigma_2 + \beta(k_1, k_2)\sigma_3 + g \begin{bmatrix} |\psi_1|^{2p} & 0 \\ 0 & |\psi_2|^{2p} \end{bmatrix}, \tag{1}$$

where σ_i are Pauli matrices and ψ_a are two components of the wavefunction, $\psi = \begin{bmatrix} \psi_1 \\ \psi_2 \end{bmatrix}$. The normalization of the wavefunction means that $|\psi_1|^2 + |\psi_2|^2 = 1$. The nonlinearity parameter p is a non-negative real number. The Kerr nonlinearity corresponds to $p = 1$. The parameters k_1 and k_2 are two quasimomenta, whose values will be adiabatically tuned in order to implement an actual adiabatic protocol to generate the nonlinear AB phase.

To solve the nonlinear eigenvalue problem,

$$\hat{H}(\psi) |\psi\rangle = E |\psi\rangle, \tag{2}$$

we introduce a real parameter x as

$$\psi_1 = \sqrt{\frac{1+x}{2}}, \qquad \psi_2 = \sqrt{\frac{1-x}{2}} e^{i\varphi}. \tag{3}$$

We will see later that the angular variable φ is the same as in the Figure 1. It turns out that x is the central quantity for expressing energy, dynamical phase, Berry phase, and nonlinear AB phase. It can be shown that the instantaneous eigenenergy is

$$E = \frac{\beta}{x} + \frac{g}{x}\left[\left(\frac{1+x}{2}\right)^{p+1} - \left(\frac{1-x}{2}\right)^{p+1}\right], \tag{4}$$

where x satisfies the following algebraic equation,

$$\frac{1-x^2}{x^2}\left\{\beta + \frac{g}{2}\left[\left(\frac{1+x}{2}\right)^{p} - \left(\frac{1-x}{2}\right)^{p}\right]\right\}^2 = |\gamma|^2, \tag{5}$$

with $\gamma := J_1 \sin k_1 - iJ_2 \sin k_2$.

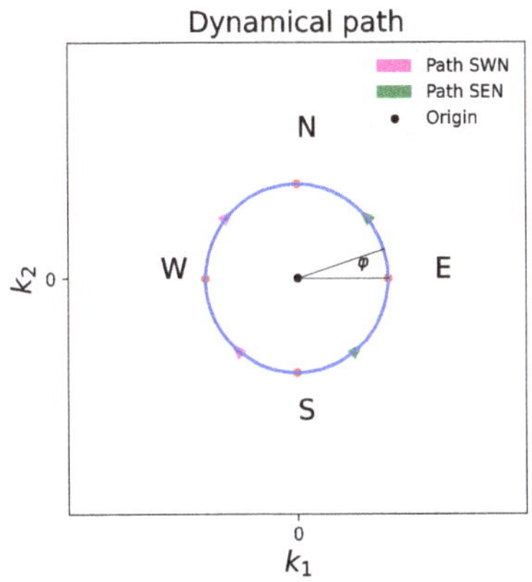

Figure 1. Dynamical paths in the momentum space. Path SWN and path SEN are symmetric halves of the perimeter of the circle. The system starts its adiabatic following at point S, and ends at point N. The two paths are parameterized by φ in the main text.

In order to have a Dirac point in the energy spectrum, the energy must be doubly degenerate at $k_1 = k_2 = 0$. Since $\gamma = 0$ at this point, x must satisfy

$$\beta(0,0) + \frac{g}{2}\left[\left(\frac{1+x}{2}\right)^p - \left(\frac{1-x}{2}\right)^p\right] = 0. \tag{6}$$

For simplicity, we choose

$$J_1 = J_2 := B, \tag{7}$$

$$\beta(k_1, k_2) = B(-1 + \cos k_1 + \cos k_2). \tag{8}$$

Hence, $\beta(0,0) = B$. It is clear that the nonlinearity strength g and energy E can be scaled in terms of B. Energy spectra with $p = 1, 1.5, 2$ and $g = 2.5B$ are shown in Figure 2, where the Dirac cone is clearly visible around the origin. A perturbative analysis of energy spectrum near the Dirac cone can be found in Appendix A.1.

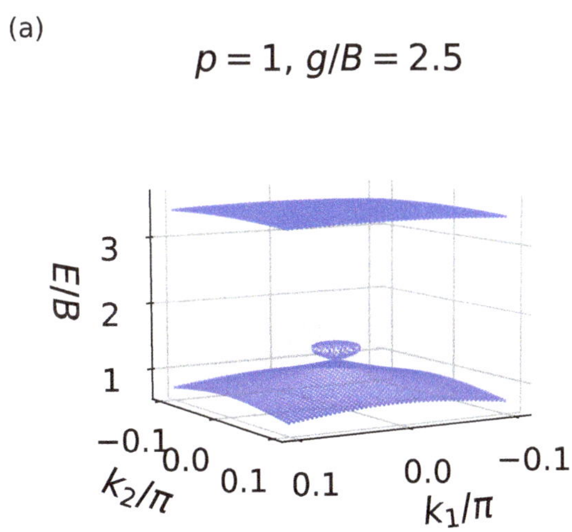

Figure 2. *Cont.*

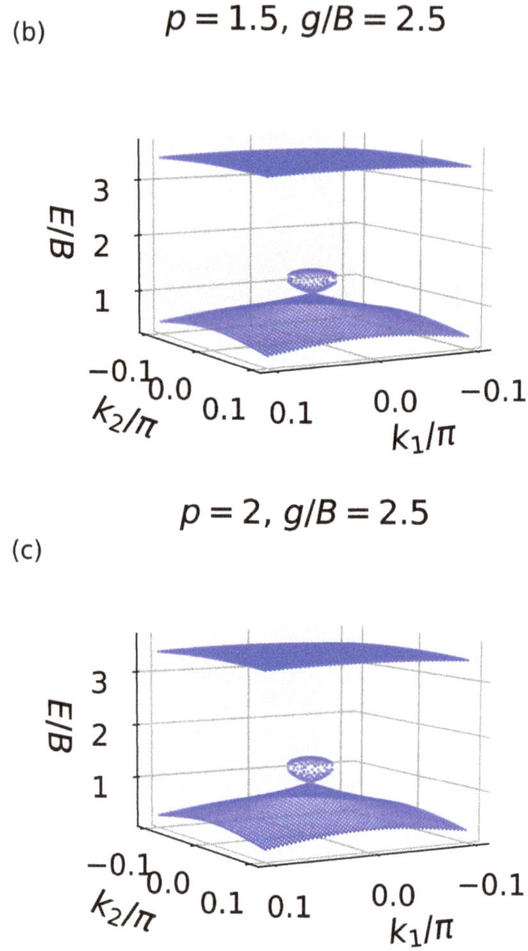

Figure 2. Nonlinear band structure for small momenta in the vicinity of the origin, i.e., for small values of $|k_1|$ and $|k_2|$, with the nonlinear strength parameter $g = 2.5B$, and with the power-law nonlinearity parameter (**a**) $p = 1$, (**b**) $p = 1.5$, (**c**) $p = 2$. See the main text for details of the system parameters. The Dirac cone emerges from the lower energy band.

3. Dynamics of Adiabatic Following

To obtain the nonlinear AB phase, let us consider two adiabatic paths along a small circle around the origin $k_1 = k_2 = 0$. As shown in Figure 1, starting at the same point S, along each path, the system is guided to move along one half of the perimeter of the circle using the same amount of time. The two adiabatic paths are "recombined" at the end of the evolution at point N. As introduced in Section 1, the phase difference acquired by the system between two adiabatic paths is called the nonlinear AB phase. Clearly, the nonlinear AB phase here is the sum of the dynamical phase difference and the Berry phase associated with the closed loop around the band-degeneracy point. We shall study below the possible AB phase quantization for a varying nonlinearity strength g and for different nonlinear parameters p. The quasimomenta k_1 and k_2 associated with two spatial dimensions are parameterized by φ and will be made to adiabatically change.

At the starting point S, the system is assumed to be prepared in the Bloch eigenstate at momentum space location S. As the system adiabatically evolves along the path SEN or SWN, the time-evolving state deviates from the instantaneous eigenstate along the path, with the tiny deviation at the order of the adiabatic parameter ε. The slower the rate of adiabatic change is, the smaller ε is, and the smaller the deviation. Here, nonlinearity plays a key role. That is, the dynamical phase also obtains a correction at the order of ε. Since the total evolution time is of order $O(\varepsilon^{-1})$, the $O(\varepsilon)$ term in this phase correction will contribute an ε-independent term through accumulation, yielding a geometric phase term out of the dynamical phase. This will not occur in linear terms because such correction accumulated over the entire adiabatic process is at most of the order of ε, which vanishes for sufficiently slow adiabatic protocols.

The dynamics of the states is governed by the time-dependent Schrödinger equation,

$$i\,|\dot{\Psi}\rangle = \widehat{H}(\Psi)\,|\Psi\rangle\,, \tag{9}$$

where the Hamiltonian is given by Equation (1) with ψ being replaced by Ψ. Here, the overhead dot denotes the time derivative. We will solve this equation up to the order of ε as described above. Through the lengthy computation, as illustrated in Appendix A.2, we obtain the instantaneous change rate of the overall phase of a time-evolving state as

$$\dot{\theta} \sim -E - \frac{1-x}{2}\dot{\varphi} + gp\frac{x(1-x^2)}{4\Delta}\left[\left(\frac{1+x}{2}\right)^p - \left(\frac{1-x}{2}\right)^p\right]\dot{\varphi}, \tag{10}$$

with

$$\Delta := \beta + \frac{g}{2}\left[(1-px+px^2)\left(\frac{1+x}{2}\right)^p - (1+px+px^2)\left(\frac{1-x}{2}\right)^p\right]. \tag{11}$$

We recognize that the circular integration of the second term in Equation (10) is nothing but the Berry phase θ_B, because it assumes the same form as in the linear limit. The rest of the phase is from the dynamical phase θ_D, which contains two parts: the first part comes from the instantaneous eigenenergy E and the second part from the third term in Equation (10) as a new contribution from the nonlinearity. Specifically,

$$\theta_B := -\oint \frac{1-x}{2}d\varphi, \tag{12}$$

$$\theta_D := -\int Edt + gp\int \frac{x(1-x^2)}{4\Delta}\left[\left(\frac{1+x}{2}\right)^p - \left(\frac{1-x}{2}\right)^p\right]d\varphi. \tag{13}$$

In the event that the Dirac cone does exist at the point $k_1 = k_2 = 0$, the obtained phase difference between the two adiabatic paths described in Figure 1 then becomes the nonlinear AB phase θ_{AB}. Since the two adiabatic paths are symmetric by construction and they take the same amount of time, the leading term in Equation (13) contributes the same in each of the two paths. Thus, the difference of the dynamical phases between two paths comes from the second term of Equation (13) only. Thus, the total nonlinear AB phase is

$$\begin{aligned}\theta_{AB} &:= \theta_B + \delta\theta_D \\ &\sim -\pi(1-x) + \pi gp\frac{x(1-x^2)}{2\Delta}\left[\left(\frac{1+x}{2}\right)^p - \left(\frac{1-x}{2}\right)^p\right].\end{aligned} \tag{14}$$

Note that we take into account that the paths are chosen to be close to the Dirac cone (so that the cones indeed have linear dispersion relations), namely, $|k_1|$ and $|k_2|$ are small at all times. The leading behavior of the dynamical phase difference term is then found to be

$$\delta\theta_D \sim \pi gp\frac{x_0(1-x_0^2)}{2\Delta_0}\left[\left(\frac{1+x_0}{2}\right)^p - \left(\frac{1-x_0}{2}\right)^p\right], \tag{15}$$

where Δ_0 is Δ evaluated at $x = x_0$ and $k_1 = k_2 = 0$, x_0 is the solution of Equation (6), and

$$\Delta_0 = -\frac{gpx_0}{2}\left[(1-x_0)\left(\frac{1+x_0}{2}\right)^p + (1+x_0)\left(\frac{1-x_0}{2}\right)^p\right]. \tag{16}$$

For the Berry phase, the leading behavior is

$$\theta_B \sim -\pi(1-x) \sim -\pi(1-x_0). \tag{17}$$

As detailed in Appendix A.1, For $|g| > 2B$, a nonlinear Dirac cone is located at the origin. For $|g| < 2B$, the only possible solutions to Equation (5) are $x = \pm 1$ and there is no Dirac cone. For $g \in (0, 2B)$, we can hence assign $x_0 = -1$, and for $g \in (-2B, 0)$, we may assign $x_0 = 1$. With this convention, it is clear to see that θ_B is constantly 0 (mod 2π) for $g \in (-2B, 2B)$. The Berry phase θ_B becomes nonzero and changes continuously for $|g| > 2B$. For each p, as we continuously tune g, x_0 can be easily solved numerically using Equation (6), thus obtaining the theoretical values of the leading terms of the dynamical phase, Berry phase and AB phase around the origin. We also numerically solve the evolution using the Schrödinger equation Equation (9) along the two paths, and compute the dynamical phase, AB phase and Berry phase using numerical solutions of the evolution. The evolution is computed using an operator-splitting algorithm. The results are presented in Figure 3.

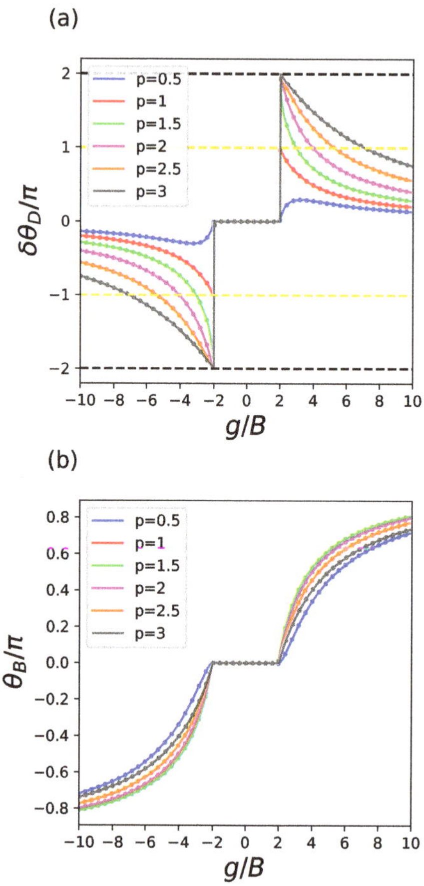

Figure 3. *Cont.*

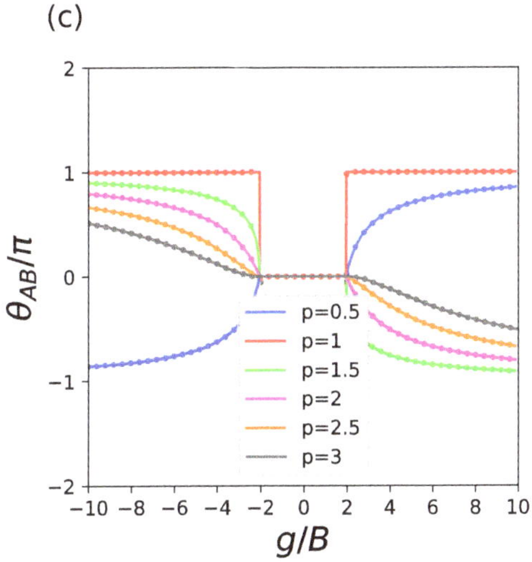

Figure 3. Dynamical, Berry and the nonlinear AB phase plotted against nonlinearity strength g for different values of power-law nonlinearity parameter p. The solid lines are theoretical values, and the dots are numerical verifications. (**a**) For $p > 1$, the jump of dynamical phase at $g = \pm 2B$ is $\pm 2\pi$; for $p = 1$, the jump of dynamical phase at $g = \pm 2B$ is $\pm\pi$; and for $p < 1$, the dynamical phase changes continuously. (**b**) The Berry phases change continuously for any p. (**c**) Only for Kerr nonlinearity $p = 1$, the AB phase has a quantized jump of π at the critical value $g = \pm 2B$ and stays at π for $|g| > 2B$.

In each plot, solid lines are theoretical values, while dots on the solid lines are computed from numerical evolutions. In Figure 3a, for any $p = 0.5, 1, 1.5, 2, 2.5, 3$, the dynamical phase around the origin is 0 for $g \in (-2B, 2B)$. At the critical value $g = \pm 2B$ where the Dirac cone appears, for $p = 0.5$, the Dirac cone changes continuously with respect to g. For $p = 1$, there is a quantized jump of $\pm\pi$ at $g = \pm 2B$. For $p = 1.5, 2, 2.5, 3$, there is a quantized jump of $\pm 2\pi$ at the critical value $g = \pm 2B$ (so this is equivalent to no change). In Figure 3b, the Berry phase (modulo 2π) is identically 0 for $g \in (-2B, 2B)$, and changes continuously with respect to g. In Figure 3c, the AB phase (modulo 2π) is the sum of the dynamical phase in Figure 3a and the Berry phase in Figure 3b. Only for $p = 1$, the AB phase has a quantized jump of π at the critical value $g = \pm 2B$ and stays at π for $|g| > 2B$, as discovered by Ref. [1]. For all other values of p, the AB phase changes continuously with respect to g. The special behavior of $p = 1$ is a result of the fact that $p = 1$ is a critical value for the limit $\lim_{g \to \pm 2B^{\pm}} \delta\theta_D$, as will be explained in the next section.

4. Mechanism of the Jump of AB Phase at $g = \pm 2B$ for Kerr Nonlinearity

For $p > 1$, we can factor out a factor $(1 - x_0^2)$ from Δ_0 which cancels the same factor in the numerator of $\delta\theta_D$,

$$\delta\theta_D(p > 1) \sim -\pi \frac{(1 + x_0)^p - (1 - x_0)^p}{(1 + x_0)^{p-1} + (1 - x_0)^{p-1}}, \tag{18}$$

which equals $\mp 2\pi$ or equivalently zero since $x_0 = \pm 1$, for $|g| = 2B$ or when the Dirac cone starts to appear.

Likewise, for $p = 1$, we have

$$\delta\theta_D(p = 1) \sim -x_0\pi, \tag{19}$$

which equals $\mp\pi$ since $x_0 = \pm 1$, for $|g| = 2B$.

Finally, for $0 < p < 1$,

$$\delta\theta_D(0 < p < 1) \sim -\pi(1 - x_0^2)^{1-p} \frac{(1 + x_0)^p - (1 - x_0)^p}{(1 + x_0)^{1-p} + (1 - x_0)^{1-p}}, \tag{20}$$

which vanishes for $x_0 = \pm 1$, for $|g| = 2B$.

The calculations above make it clear that the nonlinear AB phase associated with Kerr nonlinearity ($p = 1$) is most special as the extra nonlinearity-induced correction to dynamical phase experiences a π jump when the Dirac cone appears. What is intriguing for Kerr nonlinearity is that the nonlinear AB phase stays quantized at π for $|g| > 2B$, as θ_B and $\delta\theta_D$ happen to be complementary to each other, as shown in Equations (17) and (19). For all other forms of power-law nonlinearity, there is no such jump, π-quantization is thus absent, and consequently, the nonlinear AB phase only changes continuously with respect to g. This finally explains why in Figure 3 only the nonlinear AB phase for Kerr nonlinearity ($p = 1$) displays a quantization plateau for $|g| > 2B$.

5. Conclusions

In this paper, we analytically and computationally examined the so-called nonlinear AB phase around Dirac cones induced by power-law nonlinearity added to the QWZ model often used for studies of topological band structures. With our analytical results, we are able to explain why the nonlinear AB phase has a quantized jump of π when Dirac cone starts to appear or disappear, for and only for Kerr nonlinearity. In the context of nonlinear AB phase that can be in principle measured in experiments, Kerr nonlinearity is thus identified as a critical form of nonlinearity. As seen from our theoretical considerations above, our result will not be restricted to the QWZ model alone since it is based only on the asymptotic dispersion relation in the vicinity of the nonlinear Dirac cone. It is thus of considerable interest to investigate the generality of our results in other models with nonlinearity.

Author Contributions: J.G. designed this project. Guided by Q.-H.W. and J.G., X.L. carried out analytical and numerical calculations. Q.-H.W. simplified some of the derivations. All participated in the writing of the manuscript. J.G. finalized this draft. All authors have read and agreed to the published version of the manuscript.

Funding: This research received no external funding.

Institutional Review Board Statement: This research did not require ethical approval.

Data Availability Statement: The codes for reproducing the results of this work can be found at https://github.com/saschapojot/diracCone (accessed on 16 November 2022).

Acknowledgments: J.G. is grateful to Giulio Casati for his many years of guidance, interaction and collaboration. J.G. would also like to thank Chushun Tian for very useful discussions. The computational work for this article was fully performed on resources of the National Supercomputing Centre, Singapore (https://www.nscc.sg, accessed on 16 November 2022).

Conflicts of Interest: The authors declare no conflict of interest.

Appendix A

Appendix A.1. Eigenvalue Problem

The instantaneous eigenenergy and eigenstate satisfy the Schrödinger equation,

$$\begin{bmatrix} \beta(k_1,k_2) + g|\psi_1|^{2p} & \gamma(k_1,k_2) \\ \gamma^*(k_1,k_2) & -\beta(k_1,k_2) + g|\psi_2|^{2p} \end{bmatrix} \begin{bmatrix} \psi_1 \\ \psi_2 \end{bmatrix} = E \begin{bmatrix} \psi_1 \\ \psi_2 \end{bmatrix}. \tag{A1}$$

In terms of the two components of the state, we have

$$\gamma\psi_2 = (E - \beta - g|\psi_1|^{2p})\psi_1, \tag{A2}$$

$$\gamma^*\psi_1 = (E + \beta - g|\psi_2|^{2p})\psi_2. \tag{A3}$$

Plugging the expressions in Equation (3), we see that

$$\gamma e^{i\varphi} = \left[E - \beta - g\left(\frac{1+x}{2}\right)^p\right]\sqrt{\frac{1+x}{1-x}}. \tag{A4}$$

Since the right-hand side of the above equation is real, we recognize that the phase variable φ is simply the opposite of the phase of γ,

$$\varphi = -\arg(\gamma). \tag{A5}$$

Recall that $\gamma = B(\sin k_1 - i \sin k_2)$ in our choice, this means that φ is the same angle illustrated in Figure 1 for sufficiently small $|k_1|$ and $|k_2|$.

Multiplying ψ_1^* on both sides of Equation (A2), multiplying ψ_2^* on Equation (A3) and taking complex conjugate, then subtracting the two equations, one obtains

$$\beta + g(|\psi_1|^{2p+2} - |\psi_2|^{2p+2}) = E(|\psi_1|^2 - |\psi_2|^2). \tag{A6}$$

In terms of the parameter x defined in Equation (3), we obtain the instantaneous eigenenergy as in Equation (4).

One can then multiply the two equations in (A2) and (A3) together. Eliminating the common factor $\psi_1\psi_2$, we arrive at

$$|\gamma|^2 = E^2 - gE(|\psi_1|^{2p} + |\psi_2|^{2p}) - \beta^2 - \beta g(|\psi_1|^{2p} - |\psi_2|^{2p}) + g^2|\psi_1|^{2p}|\psi_2|^{2p}. \tag{A7}$$

Further using Equation (3), we obtain the equation satisfied by the variable x in Equation (5).

Apparently, if the Dirac cone exists, at the Dirac point $k_1 = k_2 = 0$, the energy is doubly degenerate. As a result, x is also doubly degenerate. Namely, it must satisfy Equation (6) with $\beta(0,0) = B$. That is,

$$2^{p+1}B + g[(1+x)^p - (1-x)^p] = 0. \tag{A8}$$

We denote the solution of the above equation as x_0, i.e.,

$$\left(\frac{1+x_0}{2}\right)^p - \left(\frac{1-x_0}{2}\right)^p = -\frac{2B}{g}. \tag{A9}$$

Note that the left hand side of the above equation is a monotonically increasing function of x_0 as $x_0 \in [-1,1]$, with a minimum of -1 and a maximum of $+1$. Therefore,

$$-1 \leq -\frac{2B}{g} \leq 1. \tag{A10}$$

This means that

$$|g| \geq 2B. \tag{A11}$$

This is the necessary condition for a Dirac cone to exist.

It is also of interest to use the perturbation theory to solve the eigenenergies near the Dirac cone. For sufficiently small $|k_1|$ and $|k_2|$, we let

$$x \sim x_0 + \chi, \tag{A12}$$

$$\beta \sim B + \rho, \tag{A13}$$

$$|\gamma|^2 \sim 0 + \eta, \tag{A14}$$

where χ is at least in the first order in k_1 and k_2, and ρ and η are at least in the second order in k_1 and k_2. Plugging Equations (A12)–(A14) into Equation (5), we have

$$\frac{g^2p^2}{2^{2p+2}}\frac{1-x_0^2}{x_0^2}\left[(1+x_0)^{p-1}+(1-x_0)^{p-1}\right]^2\chi^2\sim\eta. \tag{A15}$$

To this order, we obtain the correction to the parameter x,

$$\chi\sim\pm\frac{2^{p+1}}{gp}\frac{x_0}{(1+x_0)^{p-1}+(1-x_0)^{p-1}}\frac{\sqrt{J_1^2k_1^2+J_2^2k_2^2}}{\sqrt{1-x_0^2}}. \tag{A16}$$

Plugging this into Equation (4), we find the expression for the eigenenergy,

$$E\sim E_0\left(1+p\frac{\chi}{x_0}\right), \tag{A17}$$

where the nonperturbed eigenenergy is

$$E_0=\frac{g}{2}\left[\left(\frac{1+x_0}{2}\right)^p+\left(\frac{1-x_0}{2}\right)^p\right]. \tag{A18}$$

We can see clearly from the expansion of E that there is a Dirac cone structure at the origin, provided $|x_0|<1$, which corresponds to $|g|>2B$. For $|g|<2B$, the system contains two smooth energy bands. At the critical value $g=2B$ ($g=-2B$), a kink will develop on the lower (upper) band at $k_1=k_2=0$. Once $g>2B$ ($g<-2B$), a 2D self-intersection structure, i.e., a nonlinear Dirac cone, will appear from the lower (upper) band, whose vertex is at $k_1=k_2=0$. This is true for any $p>0$. We show five plots with different values of nonlinearity in Figure A1, along section $k_1=0$ and with $-0.1\pi\le k_2\le 0.1\pi$. In each plot, the red dots are perturbative eigenenergies around the Dirac point (or at the origin for $|g|\le 2B$), while the blue lines are numerical solutions by solving Equation (5) exactly. We can see that the perturbative solutions perfectly match the numerical solutions for sufficiently small $|k_2|$.

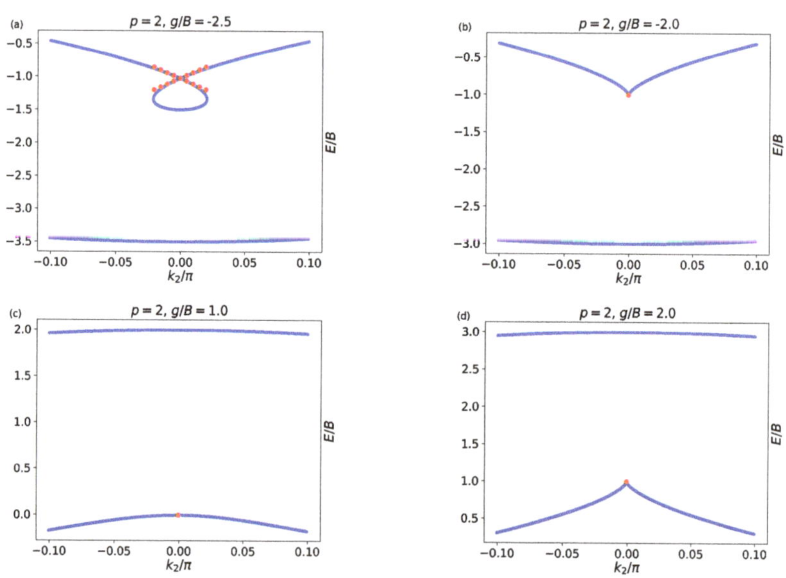

Figure A1. *Cont.*

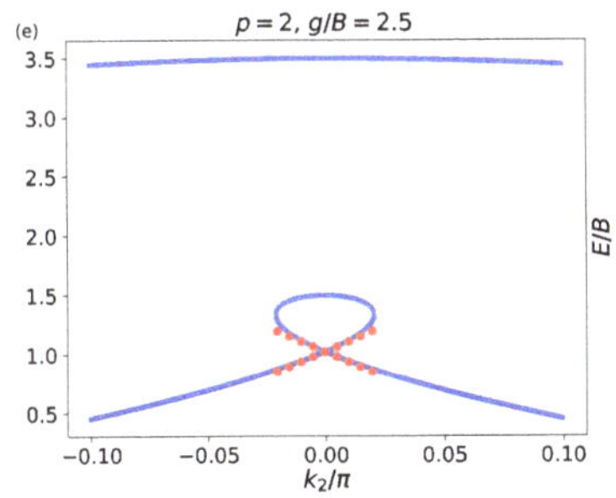

Figure A1. Numerical and perturbative solutions of E for $p = 2$ with $g/B = -2.5, -2, 1, 2, 2.5$, $B = 2$ along section $k_1 = 0$. In these plots, the red dots are perturbative eigenenergies near the Dirac cone and the blue lines are numerical solutions.

Appendix A.2. Dynamics around the Dirac Cone

We solve the following Schrödinger equation perturbatively in terms of adiabatic parameter ε [1],

$$i\partial_t \begin{bmatrix} \Psi_1 \\ \Psi_2 \end{bmatrix} = \begin{bmatrix} \beta & \gamma \\ \gamma^* & -\beta \end{bmatrix} \begin{bmatrix} \Psi_1 \\ \Psi_2 \end{bmatrix} + g \begin{bmatrix} |\Psi_1|^{2p}\Psi_1 \\ |\Psi_2|^{2p}\Psi_2 \end{bmatrix}. \tag{A19}$$

During the adiabatic following process, the quasimomenta are tuned adiabatically,

$$k_1 = k_1(\varepsilon t), \qquad k_2 = k_2(\varepsilon t), \tag{A20}$$

with $0 < \varepsilon \ll 1$. Let

$$|\Psi\rangle = \begin{bmatrix} \Psi_1 \\ \Psi_2 \end{bmatrix} = e^{i\theta}(|\psi\rangle + \varepsilon\,|\phi\rangle) = e^{i\theta} \begin{bmatrix} \psi_1 \\ \psi_2 \end{bmatrix} + \varepsilon e^{i\theta} \begin{bmatrix} \phi_1 \\ \phi_2 \end{bmatrix}, \tag{A21}$$

with

$$\dot{\theta} \sim -E + \varepsilon\alpha, \tag{A22}$$

where ψ_a are the solutions to the eigenvalue problem in Equation (2), and $\varepsilon\phi_a$ are the first-order corrections. As we shall see, the dynamical phase comes from both E and $\varepsilon\alpha$, and geometric phase comes from only $\varepsilon\alpha$.

The solution to the adiabatic process deviates from the instantaneous eigensolution by the order of ε, multiplied by a phase term. Note that the increment of phase θ over a small duration of time dt may deviate from the contribution of dynamical phase by $O(\varepsilon)$, but the accumulation of the $O(\varepsilon)$ term over the total time $O(\varepsilon^{-1})$ has a contribution of $O(1)$.

The Hamiltonian can be expand in the power of ε accordingly,

$$\widehat{H}(\Psi) \sim \widehat{H}(\psi) + \varepsilon\hat{h}(\psi,\phi), \tag{A23}$$

where $\widehat{H}(\psi)$ is given in Equation (1) and $\hat{h}$ depends on both ψ and ϕ with a diagonal form,

$$\hat{h}(\psi,\phi) = gp \begin{bmatrix} |\psi_1|^{2p-2}(\psi_1^*\phi_1 + \psi_1\phi_1^*) & 0 \\ 0 & |\psi_2|^{2p-2}(\psi_2^*\phi_2 + \psi_2\phi_2^*) \end{bmatrix}. \tag{A24}$$

To compute $\hat{h}$, we need to expand $|\Psi|^{2p}$. To the first order in ε, we have

$$\begin{aligned} |\psi_a + \varepsilon\phi_a|^{2p} &\sim |\psi_a|^{2p}\left[1 + 2\varepsilon p\,\mathrm{Re}\left(\frac{\phi_a}{\psi_a}\right)\right] \\ &= |\psi_a|^{2p} + \varepsilon p|\psi_a|^{2p-2}(\psi_a^*\phi_a + \psi_a\phi_a^*). \end{aligned} \tag{A25}$$

Plugging Equations (A21), (A22) and (A25) into the time-dependent Schrödinger equation in Equation (A19), up to the first order in ε, we obtain

$$E|\psi\rangle - \varepsilon\alpha|\psi\rangle + \varepsilon E|\phi\rangle + i|\dot{\psi}\rangle + i\varepsilon|\dot{\phi}\rangle \sim \widehat{H}(\psi)|\psi\rangle + \varepsilon\hat{h}|\psi\rangle + \varepsilon\widehat{H}(\psi)|\phi\rangle. \tag{A26}$$

Note that the time derivative brings a factor of ε because we are in the adiabatic regime, thus, the term $\varepsilon|\dot{\phi}\rangle$ is actually in the order of ε^2 and it can be discarded. Apply the instantaneous eigenvalue equation in Equation (2), we obtain the equation for $|\phi\rangle$,

$$\left[E - \widehat{H}(\psi)\right]|\phi\rangle \sim \alpha|\psi\rangle - \frac{i}{\varepsilon}|\dot{\psi}\rangle + \hat{h}|\psi\rangle. \tag{A27}$$

Multiplying $\langle\psi|$ from the left to Equation (A27), we obtain

$$\varepsilon\alpha = i\langle\psi|\dot{\psi}\rangle - \varepsilon\langle\psi|\hat{h}|\psi\rangle. \tag{A28}$$

After a lengthy calculation, we find the solution to the correction of wavefunction as

$$\varepsilon\phi_1 = -\frac{x(1-x)\sqrt{1+x}}{4\sqrt{2}}\frac{\dot{\varphi}}{\Delta} - i\frac{x}{4\sqrt{2(1+x)}}\frac{\dot{x}}{\Delta'}, \tag{A29}$$

$$\varepsilon\phi_2 = \frac{x(1+x)\sqrt{1-x}}{4\sqrt{2}}\frac{\dot{\varphi}}{\Delta}e^{i\varphi} + i\frac{x}{4\sqrt{2(1-x)}}\frac{\dot{x}}{\Delta'}e^{i\varphi}, \tag{A30}$$

where we introduce two quantities, Δ in Equation (11) and Δ' is defined as

$$\Delta' := \beta + \frac{g}{2}\left[\left(\frac{1+x}{2}\right)^p - \left(\frac{1-x}{2}\right)^p\right]. \tag{A31}$$

It turns out that the $\dot{x}/\Delta'$ terms do not contribute to $\hat{h}$,

$$\hat{h} = gp\frac{x(1-x^2)}{4\Delta}\frac{\dot{\varphi}}{\varepsilon}\begin{bmatrix} -\left(\frac{1+x}{2}\right)^{p-1} & 0 \\ 0 & \left(\frac{1-x}{2}\right)^{p-1} \end{bmatrix}. \tag{A32}$$

Putting all together, the change rate of the overall phase is given in Equation (10).

References

1. Bomantara, R.; Zhao, W.; Zhou, L.; Gong, J. Nonlinear Dirac cones. *Phys. Rev. B* **2017**, *96*, 121406. [CrossRef]
2. New, G. *Introduction to Nonlinear Optics*; Cambridge University Press: Cambridge, UK, 2011.
3. Tuloup, T.; Bomantara, R.; Lee, C.; Gong, J. Nonlinearity induced topological physics in momentum space and real space. *Phys. Rev. B* **2020**, *102*, 115411. [CrossRef]
4. Wu, B.; Niu, Q. Nonlinear Landau-Zener tunneling. *Phys. Rev. A* **2000**, *61*, 023402. [CrossRef]
5. Liu, J.; Fu, L.; Ou, B.; Chen, S.; Choi, D.; Wu, B.; Niu, Q. Theory of nonlinear Landau-Zener tunneling. *Phys. Rev. A* **2002**, *66*, 023404. [CrossRef]
6. Witthaut, D.; Graefe, E.; Korsch, H. Towards a generalized Landau-Zener formula for an interacting Bose-Einstein condensate in a two-level system. *Phys. Rev. A* **2006**, *73*, 063609. [CrossRef]
7. Zhang, Q.; Hänggi, P.; Gong, J. Two-mode Bose-Einstein condensate in a high-frequency driving field that directly couples the two modes. *Phys. Rev. A* **2008**, *77*, 053607. [CrossRef]
8. Zhang, Q.; Hänggi, P.; Gong, J. Nonlinear Landau–Zener processes in a periodic driving field. *New J. Phys.* **2008**, *10*, 073008. [CrossRef]
9. Qi, X.; Wu, Y.; Zhang, S. Topological quantization of the spin Hall effect in two-dimensional paramagnetic semiconductors. *Phys. Rev. B* **2006**, *74*, 085308. [CrossRef]
10. Novoselov, K.; Geim, A.; Morozov, S.; Jiang, D.; Katsnelson, M.; Grigorieva, I.; Dubonos, S.; Firsov, A. Two-dimensional gas of massless Dirac fermions in graphene. *Nature* **2005**, *438*, 197–200. [CrossRef]
11. Zhang, Y.; Tan, Y.; Stormer, H.; Kim, P. Experimental observation of the quantum Hall effect and Berry's phase in graphene. *Nature* **2005**, *438*, 201–204. [CrossRef]

12. Ando, T.; Nakanishi, T.; Saito, R. Berry's Phase and Absence of Back Scattering in Carbon Nanotubes. *J. Phys. Soc. Jpn.* **1998**, *67*, 2857–2862. [CrossRef]
13. Mikitik, G.; Sharlai, Y. Manifestation of Berry's Phase in Metal Physics. *Phys. Rev. Lett.* **1999**, *82*, 2147–2150. [CrossRef]
14. Tuloup, T.; Bomantara, R.; Gong, J. Topological characteristics of gap closing points in nonlinear Weyl semimetals. *Phys. Rev. B* **2022**, *106*, 195411. [CrossRef]
15. Milovanov, A.; Rasmussen, J.; Dif-Pradalier, G. Self-consistent model of the plasma staircase and nonlinear Schrödinger equation with subquadratic power nonlinearity. *Phys. Rev. E* **2021**, *103*, 052218. [CrossRef] [PubMed]
16. Wazwaz, A. Exact solutions for the fourth order nonlinear Schrodinger equations with cubic and power law nonlinearities. *Math. Comput. Model.* **2006**, *43*, 802–808. [CrossRef]
17. Kilic, B.; Inc, M. Optical solitons for the Schrödinger–Hirota equation with power law nonlinearity by the Bäcklund transformation. *Optik* **2017**, *138*, 64–67. [CrossRef]
18. Sulem, C.; Sulem, P. *The Nonlinear Schrödinger Equation: Self-Focusing and Wave Collapse*; Springer Science & Business Media: Berlin/Heidelberg, Germany, 2004.
19. Osman, M.; Lu, D.; Khater, M. A study of optical wave propagation in the nonautonomous Schrödinger-Hirota equation with power-law nonlinearity. *Results Phys.* **2019**, *13*, 102157. [CrossRef]
20. Mirzazadeh, M.; Ekici, M.; Zhou, Q.; Biswas, A. Exact solitons to generalized resonant dispersive nonlinear Schrödinger's equation with power law nonlinearity. *Optik* **2017**, *130*, 178–183. [CrossRef]
21. Dai, C.; Zhang, X.; Fan, Y.; Chen, L. Localized modes of the (n + 1)-dimensional Schrödinger equation with power-law nonlinearities in PT-symmetric potentials. *Commun. Nonlinear Sci. Numer. Simul.* **2017**, *43*, 239–250. [CrossRef]
22. Mirzazadeh, M.; Eslami, M.; Vajargah, B.; Biswas, A. Optical solitons and optical rogons of generalized resonant dispersive nonlinear Schrödinger's equation with power law nonlinearity. *Optik* **2014**, *125*, 4246–4256. [CrossRef]
23. Biswas, A.; Konar, S. *Introduction to Non-Kerr Law Optical Solitons*; Chapman and Hall/CRC: New York, NY, USA, 2006.

entropy

Article

Chaos and Thermalization in the Spin-Boson Dicke Model

David Villaseñor [1,2], Saúl Pilatowsky-Cameo [3], Miguel A. Bastarrachea-Magnani [4], Sergio Lerma-Hernández [5,*], Lea F. Santos [6,*] and Jorge G. Hirsch [1,*]

1 Instituto de Ciencias Nucleares, Universidad Nacional Autónoma de México, Apdo. Postal 70-543, Mexico City 04510, Mexico
2 Instituto de Investigaciones en Matemáticas Aplicadas y en Sistemas, Universidad Nacional Autónoma de México, Mexico City 04510, Mexico
3 Center for Theoretical Physics, Massachusetts Institute of Technology, Cambridge, MA 02139, USA
4 Departamento de Física, Universidad Autónoma Metropolitana-Iztapalapa, Av. Ferrocarril San Rafael Atlixco 186, Mexico City 09340, Mexico
5 Facultad de Física, Universidad Veracruzana, Circuito Aguirre Beltrán s/n, Xalapa 91000, Mexico
6 Department of Physics, University of Connecticut, Storrs, CT 06269, USA
* Correspondence: slerma@uv.mx (S.L.-H.); lea.santos@uconn.edu (L.F.S.); hirsch@nucleares.unam.mx (J.G.H.)

Abstract: We present a detailed analysis of the connection between chaos and the onset of thermalization in the spin-boson Dicke model. This system has a well-defined classical limit with two degrees of freedom, and it presents both regular and chaotic regions. Our studies of the eigenstate expectation values and the distributions of the off-diagonal elements of the number of photons and the number of excited atoms validate the diagonal and off-diagonal eigenstate thermalization hypothesis (ETH) in the chaotic region, thus ensuring thermalization. The validity of the ETH reflects the chaotic structure of the eigenstates, which we corroborate using the von Neumann entanglement entropy and the Shannon entropy. Our results for the Shannon entropy also make evident the advantages of the so-called "efficient basis" over the widespread employed Fock basis when investigating the unbounded spectrum of the Dicke model. The efficient basis gives us access to a larger number of converged states than what can be reached with the Fock basis.

Keywords: quantum chaos; eigenstate thermalization hypothesis; quantum entanglement

Citation: Villaseñor, D.; Pilatowsky-Cameo, S.; Bastarrachea-Magnani, M.A.; Lerma-Hernández, S.; Santos, L.F.; Hirsch, J.G. Chaos and Thermalization in the Spin-Boson Dicke Model. *Entropy* **2023**, *25*, 8. https://doi.org/10.3390/e25010008

Academic Editor: Marko Robnik

Received: 14 November 2022
Revised: 13 December 2022
Accepted: 14 December 2022
Published: 21 December 2022

1. Introduction

The onset of thermalization in isolated quantum systems, which evolve unitarily and are described by pure states, was discussed in von Neumann's 1929 paper on the quantum ergodic theorem [1–4]. In Pechukas's 1984 paper "Remarks on Quantum Chaos" [5], one finds a brief summary of von Neumann's work and the subsequent developments against it, particularly by Bocchieri and Loinger [6], which, Pechukas claims, actually make von Neumann's results sharper [7]. In Pechukas's words,

> if one selects a state "at random" from an energy shell and determines, as a function of time, the probability that it lies in a "typical" subspace of the shell, the time average of this probability is liable to be much closer to the statistical expectation than is its instantaneous value, the more so the less degenerate the spectrum of the Hamiltonian.

The ideas in this quotation are connected with the notion of "typicality" [3,4], which has been one of the directions in studies of thermalization.

In retrospect, one can identify in the 1985 work by Jensen and Shakar [8] the seeds for what later became known as the eigenstate thermalization hypothesis (ETH) [9,10]. There, the authors study the infinite-time average and the eigenstate expectation value of the magnetization of a finite spin-1/2 chain for different initial states and different energy eigenstates, and show that both exhibit very good agreement with the microcanonical average when the system is chaotic. They state that in the chaotic system, "the magnetization is a fairly smooth and monotonic function of energy", and in this case,

> even an initial energy eigenstate will exhibit a constant value for the observable which is very close to that predicted by the statistical theory,

while for the integrable Hamiltonian, the "deviations from equilibrium tend to be large". This means that an observable O should thermalize when its expectation value $\langle E_k|\hat{O}|E_k\rangle$ is a smooth function of energy, since in this situation, its infinite-time average, $\overline{O}$, will be very close to the microcanonical average, $O_{\rm mic}$; that is [11]

$$\overline{O} = \sum_k |c_k|^2 O_{k,k} \simeq O_{\rm mic} = \frac{1}{W_{E,\Delta E}} \sum_k O_{k,k} \ , \tag{1}$$

where $|E_k\rangle$ are the eigenstates of the system's Hamiltonian, $O_{k,k} = \langle E_k|\hat{O}|E_k\rangle$ are the diagonal elements of the operator $\hat{O}$, $c_k = \langle E_k|\Psi(0)\rangle$ are the coefficients of the initial state $|\Psi(0)\rangle = \sum_k c_k|E_k\rangle$, and $W_{E,\Delta E}$ is the number of eigenstates $|E_k\rangle$ contained in the energy window $E_k \in [E - \Delta E, E + \Delta E]$ with $|E - E_k| < \Delta E$. Under the assumption of the smooth behavior in energy, even a single eigenstate inside the microcanonical window should give $O_{k,k}$ very close to the microcanonical average; therefore, the term "eigenstate thermalization hypothesis" (ETH) was coined by Srednicki in his 1994 paper [12]. Notions of quantum chaos [8], random matrices [13], and Berry's conjecture [12] have been invoked to justify the validity of Equation (1). Starting with Rigol et al's 2008 paper [11], several numerical studies have confirmed Equation (1) for chaotic systems. Since then, various studies further elaborated the framework of the ETH to take into account the convergence of $\overline{O}$ and $O_{\rm mic}$ as the system size increases [9,14], the analysis of the off-diagonal elements $O_{k,k'} = \langle E_k|\hat{O}|E_{k'}\rangle$, known as off-diagonal ETH [9,15], the dependence on the energy of the initial state [16,17], and the structure of the eigenstates in realistic many-body quantum systems [18–20].

The analysis in Ref. [18] was inspired by the 1990s papers from previous members of the Novosibirsk school, including Casati [21], Flambaum [22,23], Izrailev [24,25], Shepelyansky [26], and Zelevinsky [27,28], who focused on the structure of the eigenstates to define quantum chaos and explain the onset of thermalization [29]. In realistic interacting many-body quantum systems, the energy eigenstates can be rather complicated, but they are not random vectors, as the eigenstates of Gaussian random matrices or the random superpositions of plane waves with random phases and Gaussian random amplitude stated by the Berry's conjecture [30]. For such random vectors, Equation (1) is trivially satisfied, but they are associated with unphysical models. In the above mentioned 1990s papers, chaotic eigenstates are those that, when written in the mean field basis, exhibit coefficients that are random variables following a Gaussian distribution around the envelope defined by the energy shell. They emerge away from the edges of the spectrum of quantum systems with many strongly interacting particles and ensure the validity of the Fermi–Dirac [24] and Bose–Einstein [31] distributions. These ideas were employed in Ref. [18] to corroborate the validity of the ETH for few-body observables when the eigenstates are chaotic.

The present article is dedicated to Professor Giulio Casati on the occasion of his 80th birthday in 2022 and to Professor Felix Izrailev's 80th birthday in 2021. Not only their past, but also their recent works [31–38] continue to have an enormous impact in the developments of quantum chaos and its connections with thermalization, quantum statistical mechanics, and the quantum-classical correspondence.

Here, we investigate the validity of the diagonal and off-diagonal ETH, in connection with the structure of the eigenstates, for the spin-boson Dicke model [39–41]. This system has a well-defined classical limit and two-degrees of freedom, so it does not quite fall within the requirement for thermalization of a large number of coupled degrees of freedom [13].

Depending on the parameters and energy region, the Dicke model may be regular or chaotic. Since its introduction in the 1950s to explain superradiance in spontaneous radiation processes [39], it has been employed in a variety of theoretical studies that include quantum phase transitions [42–44], classical and quantum chaos [45–49], non-equilibrium quantum dynamics [41,50–54], the evolution of out-of-time-ordered correlators (OTOCs) [55–57], quantum scarring [58–62], and quantum localization measures in phase space [61,63]. Experimentally, the model can be realized with superconducting circuits [64], cavity assisted Raman transitions [65,66], trapped ions [67,68], and other systems [69]. In the particular context of the relationship between chaos and thermalization, the Dicke model was studied in Ref. [70] with emphasis on the behavior of the fidelity OTOC and the agreement between the long-time average of the collective spin observable and the microcanonical ensemble. The chaos–thermalization connection was also explored for the kicked Dicke model in Ref. [71].

We analyze the diagonal and off-diagonal ETH for the number of photons and the number of excited atoms in the regular and chaotic regions of the Dicke model, and compare the results with the structure of the eigenstates analyzed with the entanglement entropy and the Shannon entropy. The latter is a delocalization measure that depends on the basis in which the eigenstates are written. The Hilbert space of the Dicke model is infinite, because its number of bosons is unbounded. Our results show that the values of the Shannon entropy grows rapidly with the eigenvalues when the eigenstates are written in the Fock basis, but are more restricted when the "efficient basis" is used, which makes evident the advantages of the latter when one wants to study large systems and high energies.

The paper is organized as follows. We introduce the Dicke model in Section 2 and briefly review the onset of classical and quantum chaos in Section 3. In Section 4, we study the diagonal and off-diagonal ETH and analyze the eigenstates in Section 5. Our conclusions are summarized in Section 6.

2. Dicke Model

The Dicke model [39] describes a system of $\mathcal{N}$ two-level atoms coupled with a single mode of a quantized radiation field. Setting $\hbar = 1$, the Hamiltonian is given by

$$\hat{H}_{\mathrm{D}} = \hat{H}_{\mathrm{F}} + \hat{H}_{\mathrm{A}} + \hat{H}_{\mathrm{I}}, \tag{2}$$

$$\hat{H}_{\mathrm{F}} = \omega \hat{a}^{\dagger}\hat{a}, \tag{3}$$

$$\hat{H}_{\mathrm{A}} = \omega_0 \hat{J}_z, \tag{4}$$

$$\hat{H}_{\mathrm{I}} = \frac{\gamma}{\sqrt{\mathcal{N}}}(\hat{a}^{\dagger} + \hat{a})(\hat{J}_+ + \hat{J}_-), \tag{5}$$

where $\hat{H}_{\mathrm{F}}$ defines the field's energy, $\hat{H}_{\mathrm{A}}$ the energy of the two-level atoms, and $\hat{H}_{\mathrm{I}}$ the atom–field interaction energy. In the equations above, $\hat{a}^{\dagger}$ ($\hat{a}$) is the bosonic creation (annihilation) operator of the single field mode and $\hat{J}_+$ ($\hat{J}_-$) is the raising (lowering) collective pseudo-spin operator, where $\hat{J}_{\pm} = \hat{J}_x \pm i\hat{J}_y$ and $\hat{J}_{x,y,z} = (1/2)\sum_{k=1}^{\mathcal{N}} \hat{\sigma}^k_{x,y,z}$ for the Pauli matrices $\hat{\sigma}^k_{x,y,z}$ acting on the kth two-level atom. Since the squared total pseudo-spin operator $\hat{\mathbf{J}}^2 = \hat{J}_x^2 + \hat{J}_y^2 + \hat{J}_z^2$ commutes with the Hamiltonian, $[\hat{H}_D, \hat{\mathbf{J}}^2] = 0$, the eigenvalues of $\hat{\mathbf{J}}^2$, given by $j(j+1)$, label the invariant subspaces in the Hilbert space. We work within the totally symmetric subspace, which includes the collective ground state and is defined by the maximum pseudo-spin value $j = \mathcal{N}/2$.

The Dicke Hamiltonian $\hat{H}_{\mathrm{D}}$ commutes with the parity operator

$$\hat{\Pi} = \exp(i\pi\hat{\Lambda}), \tag{6}$$

where

$$\hat{\Lambda} = \hat{a}^{\dagger}\hat{a} + \hat{J}_z + j\hat{1} = \hat{n} + \hat{n}_{\mathrm{ex}}, \tag{7}$$

the number of photons is $\hat{n} = \hat{a}^{\dagger}\hat{a}$, and the number of excited atoms is $\hat{n}_{\mathrm{ex}} = \hat{J}_z + j\hat{1}$. Because of this symmetry, the eigenstates have one of two different parities, $\hat{\Pi}|E_k\rangle = \pm|E_k\rangle$.

The three parameters of the Dicke Hamiltonian, ω, ω_0, and γ, determine the energy scales of the system and the onset of chaos. The radiation frequency of the single-mode electromagnetic field (the boson) is given by ω, the energy splitting of each two-level atom is ω_0, and γ is the coupling strength modulating the atom–field interaction. In the thermodynamic limit, $\mathcal{N} \to \infty$, the light-matter coupling divides the parameter space in the normal phase, when the strength γ in smaller than the critical value $\gamma_c = \sqrt{\omega\omega_0}/2$, and the superradiant phase, when $\gamma > \gamma_c$ [42,72–74]. The Dicke model also presents regular and chaotic regions depending on the parameters and excitation energies [47]. We fix the coupling strength in the superradiant phase, $\gamma = 2\gamma_c = 1$, and the resonant frequencies at $\omega = \omega_0 = 1$, so that we ensure that both the classical dynamics and the quantum eigenspectrum are chaotic at excitation energies $\epsilon \geq -0.8$ [47], where $\epsilon = E/j$ denotes the energy scaled to the system size j.

The Hilbert space of the Dicke model is infinite. In Appendix A, we provide details on how to diagonalize the system's Hamiltonian on an efficient basis. We consider three system sizes, $j = 30, 60, 100$, whose associated Hilbert space dimensions in the efficient basis are given by $d_D^{EB} = 8\,601, 30\,371, 80\,601$, ensuring the convergence of the set of eigenstates from the ground-state energy $\epsilon_{GS} = -2.125$ until a truncated value ϵ_T.

3. Chaos in the Dicke Model

The classical Hamiltonian of the Dicke model is presented in Appendix B. The generation of Poincaré sections and the computation of Lyapunov exponents for trajectories in phase space [47,75] reveal the onset of chaos for high energies and strong couplings. In Figure 1a, we show a classical map of the percentage of chaos as a function of excitation energies and coupling strengths. The percentage of chaos is defined as the ratio of the number of chaotic initial conditions over the total number of initial conditions used in the sample and is illustrated with a color gradient, where dark indicates that the majority of the initial conditions are regular and light indicates that most initial conditions are chaotic.

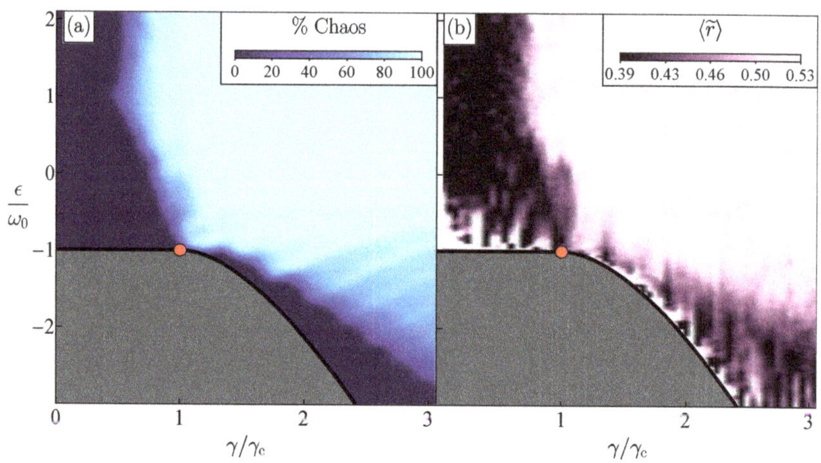

Figure 1. Panel (**a**): Map of the percentage of chaos for the classical trajectories as a function of the excitation energy ϵ and the coupling strength γ. Panel (**b**): Map of the average values of the ratio of consecutive energy levels $\langle\tilde{r}\rangle$ (see Equation (8)) obtained for moving windows of eigenenergies, which contain approximately 1000 energy levels. The low energy regions contain few energy levels and were averaged using windows of eigenenergies with the available energy levels. Because of this, the fluctuations of $\langle\tilde{r}\rangle$ are more pronounced in these regions. In both panels (**a**) and (**b**), the red dot represents the critical point for the transition from the normal to the superradiant phase for the ground state. The system size in panel (**b**) is $j = 100$.

Signatures of the onset of chaos in classical systems are found in the quantum domain, as suggested by Casati et al's 1980 paper [76] and Bohigas et al's 1984 work [77], and became known as "quantum chaos". The eigenvalues of a quantum system become correlated when its classical counterpart is chaotic. The degree of correlations between neighboring levels can be detected with the level spacing distribution [78] or the ratio of consecutive energy levels [79,80],

$$\widetilde{r}_k = \min(r_k, r_k^{-1}) = \frac{\min(s_k, s_{k-1})}{\max(s_k, s_{k-1})}, \tag{8}$$

where $s_k = E_{k+1} - E_k$ is the nearest-neighbor spacing between energy levels and $r_k = s_k/s_{k-1}$; while both short- and long-range correlations are measured with quantities such as the level number variance [81] or the spectral form factor [78]. In the case of the Dicke model, level spacing distributions [75,82,83] and spectral form factors [53,54] in agreement with random matrix theory (RMT) were verified for the energies and parameters associated with the onset of chaos in the classical limit.

Other tests of quantum chaos include Peres lattices [84] and the evolution of OTOCs [85–87]. The Peres lattice is a plot of the eigenstate expectation values of an observable as a function of the eigenvalues. It has been widely employed (not always under this name) in studies of ETH and provides visual evidence of the loss of integrability [75,88]. As for the OTOCs, their exponential growth rates are seen as quantum analogs of the classical Lyapunov exponents. This was indeed confirmed for the Dicke model in the chaotic region [55], although its exponential growth happens also in regular regions due to instability [57].

In Figure 1, we compare the onset of classical chaos, investigated in Figure 1a, with the degree of correlations between neighboring energy levels, as captured by the ratio of consecutive energies in Figure 1b. To smooth spectral fluctuations, an average is performed and denoted by $\langle \widetilde{r} \rangle$. In the regular regime, where the eigenvalues are uncorrelated and the level spacing distribution is Poissonian, $\langle \widetilde{r} \rangle_{\mathrm{P}} \approx 0.39$, while in the chaotic region, where the eigenvalues are correlated as in RMT and the level spacing distribution follows the Wigner–Dyson distribution, $\langle \widetilde{r} \rangle_{\mathrm{WD}} \approx 0.53$. Our results in Figure 1 exhibit an evident classical-quantum correspondence. There is a visible relationship between the classical map of Figure 1a and the quantum map of Figure 1b, when the percentage of chaos is large (small) in Figure 1a; $\langle \widetilde{r} \rangle$ approaches 0.53 (0.39) in Figure 1b.

4. Thermalization in the Dicke Model

For an isolated quantum system initially in a pure state, the evolution is governed by the unitary time operator as

$$|\Psi(t)\rangle = e^{-i\hat{H}t}|\Psi(0)\rangle = \sum_k c_k e^{-iE_k t}|E_k\rangle, \tag{9}$$

where $\hat{H}|E_k\rangle = E_k|E_k\rangle$ and $c_k = \langle E_k|\Psi(0)\rangle$. Thus, the expectation value of a given operator $\hat{O}$ under states evolved in time can be calculated as follows

$$O(t) = \langle \Psi(t)|\hat{O}|\Psi(t)\rangle = \sum_{k\neq k'} c_k^* c_{k'} e^{i(E_k - E_{k'})t} O_{k,k'} + \sum_k |c_k|^2 O_{k,k} \quad , \tag{10}$$

where $O_{k,k'} = \langle E_k|\hat{O}|E_{k'}\rangle$ are the matrix elements of the operator expressed in the energy eigenbasis.

The onset of thermalization according to the ETH happens for local observables if two assumptions are satisfied:

(i) The infinite-time average,

$$\overline{O} = \lim_{t\to+\infty} \frac{1}{t}\int_0^t dt' O(t') = \sum_k |c_k|^2 O_{k,k}, \tag{11}$$

coincides with the microcanonical average,

$$O_{\text{mic}} = \frac{1}{W_{E,\Delta E}} \sum_k O_{k,k} \approx \overline{O}, \tag{12}$$

or more precisely, $\overline{O}$ approaches the thermodynamic average as the system size grows. This is sometimes referred to as "diagonal ETH". In the equation above, $W_{E,\Delta E}$ is the number of eigenstates $|E_k\rangle$ contained in the energy window $E_k \in [E - \Delta E, E + \Delta E]$ with $|E - E_k| < \Delta E$.

As explained above, the ETH should be valid when the eigenstates are chaotic (ergodic), filling the energy shell. In this case, one obtains Gaussian distributions for few-body observables in many-body quantum systems, which can be understood as follows. Assume that the few-body observable $\hat{O}$ has N nonzero eigenvalues O_n in $\hat{O}|O_n\rangle = O_n|O_n\rangle$, where N is large. We can project the energy eigenstates in the basis $|O_n\rangle$ as $|E_k\rangle = \sum_n C^k_{O_n}|O_n\rangle$, and write

$$O_{kk} = \langle E_k|\hat{O}|E_k\rangle = \sum_{n=1}^{N} |C^k_{O_n}|^2 \langle O_n|\hat{O}|O_n\rangle = O_{1,1}|C^k_{O_1}|^2 + \ldots + O_{N,N}|C^k_{O_N}|^2. \tag{13}$$

If the eigenstates are fully chaotic (ergodic), then $C^k_{O_n}$'s are independent Gaussian random numbers. According to the central limit theorem, a large sum of independent random numbers $O_{n,n}|C^k_{O_n}|^2$ will also follow a Gaussian distribution. Therefore, in the region where the eigenstates are chaotic, the distribution of $O_{k,k}$ should be Gaussian.

(ii) The fluctuations around $\overline{O}$, which are determined by the phases $e^{i(E_k - E_{k'})t}$, the coefficients c_k, and the off-diagonal elements $O_{k,k'}$ in Equation (10), decrease with system size and cancel out on average. This is sometimes referred to as "off-diagonal ETH". A Gaussian distribution of $O_{k,k'}$ indicates that the eigenstates are strongly chaotic (ergodic), so condition (ii) should be satisfied if the energy of the initial state (of a system perturbed far from equilibrium) is in the chaotic region.

The relationship between the Gaussian distribution of the off-diagonal elements of a few-body observable and ergodicity can be understood as above, using $\hat{O}|O_n\rangle = O_n|O_n\rangle$ and $|E_k\rangle = \sum_n C^k_{O_n}|O_n\rangle$ in

$$O_{kk'} = \langle E_k|\hat{O}|E_{k'}\rangle = \sum_{n=1}^{N} (C^k_{O_n})^* C^{k'}_{O_n} \langle O_n|\hat{O}|O_n\rangle = O_{1,1}(C^k_{O_1})^* C^{k'}_{O_1} + \ldots + O_{N,N}(C^k_{O_N})^* C^{k'}_{O_N}. \tag{14}$$

For chaotic eigenstates, $C^k_{O_n}$'s are independent Gaussian random numbers, and the product of two independent Gaussian random numbers, $(C^k_{O_n})^* C^{k'}_{O_n}$, which appears in each term of Equation (14), is also an independent random number. According to the central limit theorem, a large sum of independent random numbers follows a Gaussian distribution, so in the region where the eigenstates are chaotic, the distribution of $O_{k,k'}$ should be Gaussian. This has been confirmed for different chaotic systems with many degrees of freedom [15,89,90], but not in chaotic systems with one [91] or few particles [92], few degrees of freedom [93] or in many-body systems when $\hat{O}$ is not few-body [94].

4.1. Diagonal ETH

To test the validity of the diagonal ETH, we compute the deviation of the eigenstate expectation values with respect to the microcanonical value [19,20]

$$\Delta^{\text{mic}}[O] = \frac{\sum_k |O_{k,k} - O_{\text{mic}}|}{\sum_k O_{k,k}}, \tag{15}$$

and also use a stronger test that takes into account the normalized extremal fluctuations and is given by [19]

$$\Delta_e^{\mathrm{mic}}[O] = \left| \frac{\max(O) - \min(O)}{O_{\mathrm{mic}}} \right|, \tag{16}$$

where $\max(O)$ and $\min(O)$ are taken from the same energy window $E_k \in [E - \Delta E, E + \Delta E]$ used in Equation (12). We consider as observables the number of excited atoms, $\hat{n}_{\mathrm{ex}} = \hat{J}_z + j\hat{1}$, and the number of photons, $\hat{n} = \hat{a}^\dagger \hat{a}$, of the Dicke model.

In Figure 2, we show the Peres lattices for the expectation values of the number of photons, $n_{k,k} = \langle E_k | \hat{n} | E_k \rangle$ (Figure 2a), and of the number of excited atoms, $(n_{\mathrm{ex}})_{k,k} = \langle E_k | \hat{n}_{\mathrm{ex}} | E_k \rangle$ (Figure 2b), for all eigenstates of the Dicke model with positive parity that range from the ground state energy, $\epsilon_{\mathrm{GS}} = -2.125$, up to a maximal converged eigenstate with eigenenergy $\epsilon_{\mathrm{T}} = 1.755$. In the regular region of low energies, the Peres lattices present a clear pattern related with the quasi-conserved quantities. Above $\epsilon \approx -0.8$, where the system becomes chaotic, the lattices become smoother in energy. It is visible that the spread of the expectation values in the chaotic region decreases as the system size increases; that is, for high excitation energies, the fluctuations are larger for $j = 30$ than for $j = 100$. It is also noticeable that the number of excited atoms fluctuate less than the number of photons.

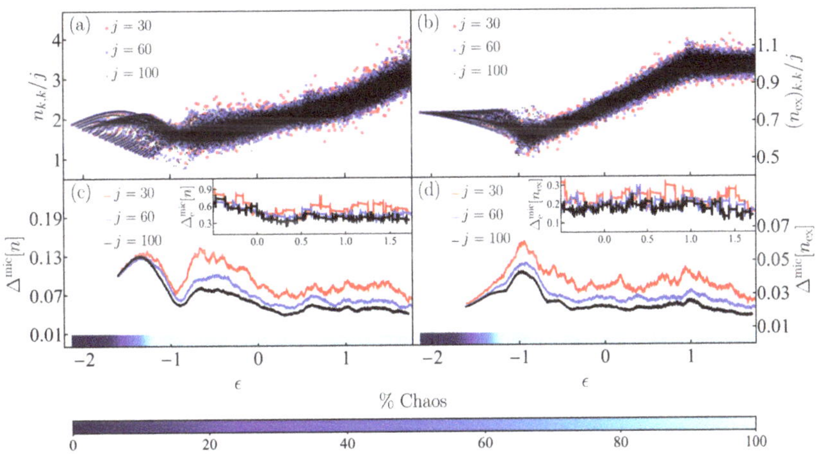

Figure 2. Panels (**a**) and (**b**): Peres lattice of the eigenstate expectation values of the number of photons $n_{k,k}$ (**a**) and the number of excited atoms $(n_{\mathrm{ex}})_{k,k}$ (**b**) scaled to the system size j for eigenstates $|E_k\rangle$ with positive parity of the Dicke model. Panels (**c**) and (**d**): Deviations of the expectation values of the same observables $n_{k,k}$ (**c**) and $(n_{\mathrm{ex}})_{k,k}$ (**d**) with respect to their microcanonical value (see Equation (15)). The insets in (**c**) and (**d**) show the extremal deviations of the respective quantities in the chaotic energy regime (see Equation (16)). The color scales contained within the panels (**c**) and (**d**) correspond to the values of the percentage of classical chaos shown at the bottom in the numbered color scale. The system size in all panels (**a**–**d**) is indicated with a given color for three values $j = 30, 60, 100$.

The reduction of the fluctuations with the increase of system size, which is a necessary condition for the validity of the ETH, is better quantified in Figure 2c,d, where we show $\Delta^{\mathrm{mic}}[n]$ and $\Delta^{\mathrm{mic}}[n_{\mathrm{ex}}]$, respectively. To compute these quantities, we used moving windows of eigenenergies for each system size $j = 30, 60, 100$, which contain 100, 350, 900 energy levels. In the chaotic region, both deviations clearly decrease as j increases, while at low energies ($\epsilon < -0.8$), where chaos and regularity coexist, the fluctuations decrease very slowly or do not decrease at all.

The insets of Figure 2c,d contain the results for the normalized extremal fluctuation in Equation (16). Only the chaotic energy interval defined by $\epsilon \in [-0.5, 1.755]$ is shown. The reduction of the extremal fluctuations as j increases is also perceptible in these insets, thus confirming the validity of the diagonal ETH in the chaotic region of the Dicke model.

In Figure 3, we consider the energy interval $\epsilon \in [0.5, 1]$, where hard chaos manifests itself classically, and use only the positive parity sector of eigenstates. In Figure 3a,b, we show, respectively, the distribution of the diagonal elements of the number of photons and the number of excited atom, which are presented in a semi-logarithmic scale. The figures confirm that the distribution shape is Gaussian, as explained in Equation (13).

4.2. Off-Diagonal ETH

We now analyze the off-diagonal elements of the number of photons in Figure 3c and the number of excited atoms in Figure 3d. As discussed in Equation (14), a Gaussian distribution validates the off-diagonal ETH. This is indeed the case of Figure 3c,d.

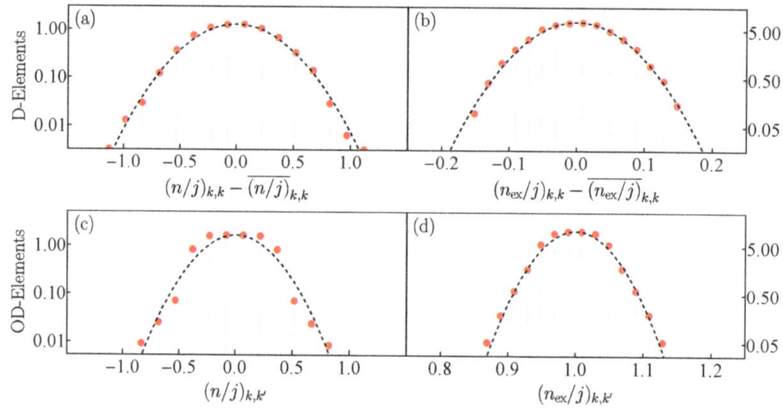

Figure 3. Panels (**a**) and (**b**): Statistical distribution (red solid dots) in semi-logarithmic scale of the diagonal matrix elements of number of photons $n_{k,k}$ (**a**) and excited atoms $(n_{ex})_{k,k}$ (**b**) scaled to the system size j for eigenstates $|E_k\rangle$ with positive parity of the Dicke model contained in the chaotic energy interval $\epsilon_k \in (0.5, 1.0)$. The term $\overline{O}_{k,k}$ represents the average of the expectation value $O_{k,k}$ in the same energy region for each operator $\hat{n}$ and $\hat{n}_{ex}$. The black dashed line depicts a Gaussian fit. Panels (**c**) and (**d**): The same as panels (**a**) and (**b**) for the off-diagonal matrix elements of the same observables $n_{k,k'}$ (**c**) and $(n_{ex})_{k,k'}$ (**d**). The system size in all panels (**a**–**d**) is $j = 30$.

5. Entropies of the Eigenstates of the Dicke Model

In the last section, we verified numerically the validity of the ETH in the chaotic region of the Dicke model using the matrix elements of the number of photons and the number of excited atoms. In this section, we use the von Neumann entanglement entropy and the Shannon entropy to analyze the structure of the eigenstates of the model. The von Neumann entanglement entropy can be regarded as the limit of higher-order observables in a replicated Hilbert space, and one may generalize the ETH to include higher-order statistical moments [95] arising from these replicated spaces, which allows one to explain the so-called Page correction [96] to the volume law. The von Neumann entanglement entropy has been linked to the onset of chaos in quantum systems [97–101]. The Shannon entropy is a basis-dependent quantity. We study this entropy with respect to both the Fock and an efficient basis.

The Dicke model is a bipartite system, whose Hilbert space is a tensor product of the atomic $\mathcal{H}_A$ and bosonic $\mathcal{H}_B$ sectors, $\mathcal{H}_D = \mathcal{H}_A \otimes \mathcal{H}_B$. The bosonic sector is infinite-dimensional, while the atomic one has dimension $d_A = 2j + 1$. For a pure state expanded in the Fock basis $|n; j, m_z\rangle$,

$$|\Psi\rangle = \sum_{n=0}^{\infty} \sum_{m_z=-j}^{j} c_{n,m_z} |n; j, m_z\rangle, \tag{17}$$

the density matrix is the projector operator $\hat{\rho} = |\Psi\rangle\langle\Psi|$, and the reduced density matrix in the atomic sector is calculated as,

$$\hat{\rho}_{\mathrm{A}} = \mathrm{Tr}_{\mathrm{B}}[\hat{\rho}] = \sum_{m_z=-j}^{j} \sum_{m_z'=-j}^{j} \left(\sum_{n=0}^{\infty} c_{n,m_z} c^*_{n,m_z'} \right) |j, m_z\rangle\langle j, m_z'|. \tag{18}$$

The von Neumann entanglement entropy is given by

$$S_{\mathrm{En}} = -\,\mathrm{Tr}[\hat{\rho}_{\mathrm{A}} \ln(\hat{\rho}_{\mathrm{A}})]. \tag{19}$$

For numerical convenience, we trace out the infinite bosonic sector first, but from the Schmidt decomposition [102], the result for S_{En} is the same if we would instead trace out the atomic sector first. (See Appendix C for a generalization of quantum entanglement to multipartite systems.)

The standard basis to compute the entanglement entropy is the Fock basis, which is defined as a decoupled basis between photons and atoms $|n; j, m_z\rangle = |n\rangle \otimes |j, m_z\rangle$ (see Appendix A.1). Nevertheless, alternative bases can be used to calculate this entropy by selecting the correct partition of the system. In this work, in addition to the Fock basis, we use an efficient basis (see Appendix A.2 for a full description) that allows us to reach larger system sizes ($j > 30$) than we can reach with the Fock basis [103,104]. The efficient basis is defined as the tensor product $|N; j, m_x\rangle = |N\rangle_{m_x} \otimes |j, m_x\rangle$, where $|j, m_x\rangle$ are the atomic pseudo-spin states rotated by an angle $-\pi/2$ around the y axis and $|N\rangle_{m_x}$ are displaced Fock states, whose displacing depends explicitly on the atomic pseudo-spin eigenvalue m_x. In the following general notation for a pure state,

$$|\Psi\rangle = \sum_{x=0}^{\infty} \sum_{y=-j}^{j} c_{x,y} |x; j, y\rangle, \tag{20}$$

$(x, y) = (n, m_z)$ stands for the Fock basis and $(x, y) = (N, m_x)$ for the efficient basis.

To select the correct atomic sector starting with the efficient basis, we need to perform an adequate trace over the modified bosonic sector related with this basis, such that it is equivalent to the trace over the bosonic sector of the Fock basis. This can be accomplished by mapping one basis into the other (see Appendix A.3). As the Hilbert space associated with the atomic pseudo-spin states $|j, m_z\rangle$ is the same as the Hilbert space of the rotated ones $|j, m_x\rangle$, the atomic sector for both subspaces is the same and the entanglement entropy can be computed properly. The Shannon entropy of the pure state in Equation (20) is given by

$$S_{\mathrm{Sh}} = -\sum_{x=0}^{\infty} \sum_{y=-j}^{j} |c_{x,y}|^2 \ln(|c_{x,y}|^2). \tag{21}$$

5.1. Results for the Entanglement Entropy

We start by comparing the von Neumann entanglement entropy for the Dicke model, which is nonintegrable, with the results for one of the integrable limits of the model, namely the Tavis–Cummings model (see Appendix D for the derivation of this model). This comparison was also made in Ref. [105].

In Figure 4, we plot the Peres lattice of the exponential of the von Neumann entanglement entropy in Equation (19) for the integrable Tavis–Cummings model (Figure 4a) and the nonintegrable Dicke model (Figure 4b). Results for all states from both parity sectors, from the ground-state of the Tavis–Cummings model up to a maximal converged eigenstate with eigenenergies $\epsilon_k \in [\epsilon_{\mathrm{GS}} = -2.136, \epsilon_{\mathrm{T}} = 2.5]$, are shown. As evident in Figure 4a, the values of the entropy for the integrable case show large fluctuations and patterns associated with regularity are visible. The fluctuations indicate that even states very close in energy may have very different structures, so ETH should not be satisfied. In contrast, S_{En} becomes a smoother function of energy in the chaotic region of the Dicke model ($\epsilon > -0.8$), as seen in Figure 4b, which indicates the states close in energy are very similar. In the case of the Dicke model, regular patterns are restricted to the low energies, where the model is not chaotic.

We stress, however, that some isolated eigenstates located in the chaotic regime of the Dicke model present low entanglement. They should be related with strongly scarred states, which are known to exist in this model [61,62,106].

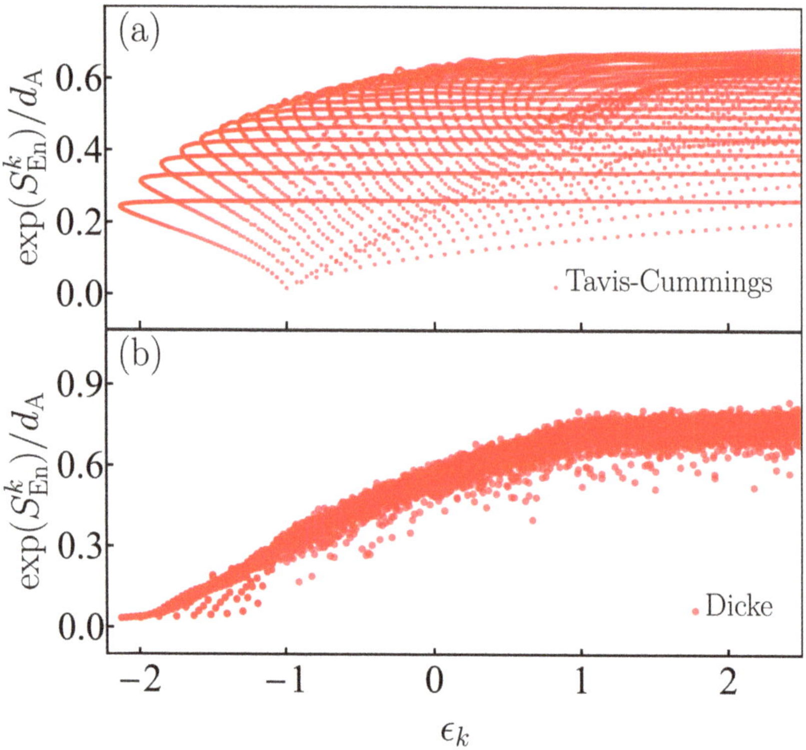

Figure 4. Panels (**a**) and (**b**): Peres lattice of the exponential of the von Neumann entanglement entropy S^k_{En} (see Equation (19)) scaled to the atomic Hilbert-space dimension $d_A = 2j + 1$ for eigenstates $|E_k\rangle$ from both parity sectors of the integrable Tavis–Cummings model (**a**) (see Equation (A23)) and the nonintegrable Dicke model (**b**) (see Equation (2)). The system size in both panels (**a**) and (**b**) is $j = 30$.

In Figure 5a, we analyze the von Neumann entanglement entropy for the eigenstates of the Dicke model, ranging from the ground state until a maximal converged eigenstate with eigenenergies $\epsilon_k \in [\epsilon_{\text{GS}} = -2.125, \epsilon_{\text{T}} = 1.841]$, for three values of the system size $j = 30$, 60, 100. As discussed in Section 4 and in Figure 2, thermalization requires the convergence of the infinite-time average of a few-body observable towards the microcanonical average in the thermodynamic limit, so scaling analysis needs to be performed. As seen in Figure 5a, the patterns at the low energies of the regular region of the Dicke model do not disappear as the system size increases, but in the chaotic region, the fluctuations clearly shrink as the system size increases. This behavior is quantified in Figure 5b, where we present the deviation of the entanglement entropy from the microcanonical average, as computed in Equation (15) for the three system sizes $j = 30, 60, 100$. The fluctuations decay to values close to zero for energies above $\epsilon \approx -1.2$, indicating the transition to a region where the majority of the eigenstates are chaotic. In the inset of Figure 5b, we present the extremal fluctuations calculated with Equation (16) for the chaotic energy interval only, $\epsilon \in [-0.5, 1.841]$. We avoid the regular region, where the changes in the extremal values are abrupt. The inset confirms the decrease in fluctuations with the increase in system size.

By comparing Figure 5b with Figure 3c,d, we observe that the deviation from the microcanonical average of the von Neumann entanglement entropy decreases to zero more abruptly as compared to the the number of photons and excited atoms. This different behavior between entropies and expectation values of observables is not well understood yet and motivates future work.

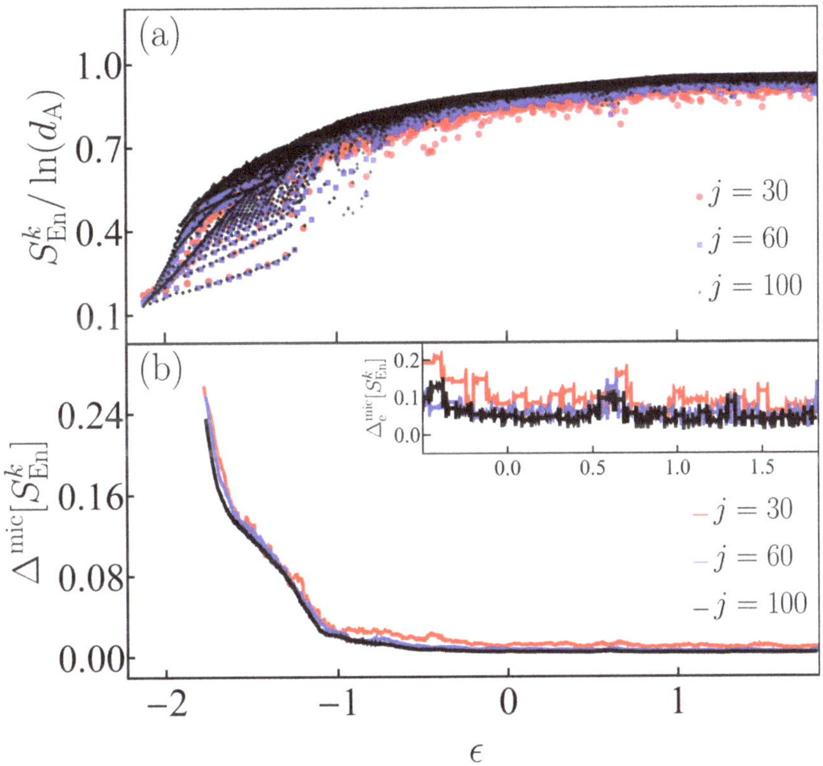

Figure 5. Panel (**a**): Peres lattice of the von Neumann entanglement entropy $S^k_{\rm En}$ scaled to the atomic Hilbert-space dimension $d_{\rm A} = 2j + 1$ for eigenstates $|E_k\rangle$ from both parity sectors of the Dicke model. Panel (**b**): Deviations of the von Neumann entanglement entropy with respect to its microcanonical value, computed with Equation (15). The inset shows the extremal deviations in the chaotic energy regime computed with Equation (16). The system size in both panels (**a**) and (**b**) is indicated with a given color for three values $j = 30, 60, 100$.

5.2. Results for the Shannon Entropy

We proceed with the analysis of the structure of the eigenstates of the Dicke model making use now of the basis-dependent Shannon entropy. In Figure 6a, we plot the Peres lattice of the Shannon entropy for the eigenstates of the Dicke model in the Fock and efficient bases. The eigenstates range from the ground state until a maximal converged eigenstate with eigenenergies $\epsilon_k \in [\epsilon_{\rm GS} = -2.125, \epsilon_{\rm T} = 2.356]$. The inset in Figure 6a contains the same data, but shows the exponential of the entropy. The values of the Shannon entropy computed in the Fock basis are larger than in the efficient basis. For the Fock basis, the entropy grows unboundedly with energy, while for the efficient basis, $S_{\rm Sh}$ saturates at high energies. These features make evident the advantages of using the efficient basis, since it requires fewer basis states to build a given eigenstate than what is needed by the Fock basis. While in Figure 6, our system size was restricted to $j = 30$, when we employ the efficient basis, the same computational resources allow us to go up to $j = 100$, as used in Figures 2, 5 and 7 below.

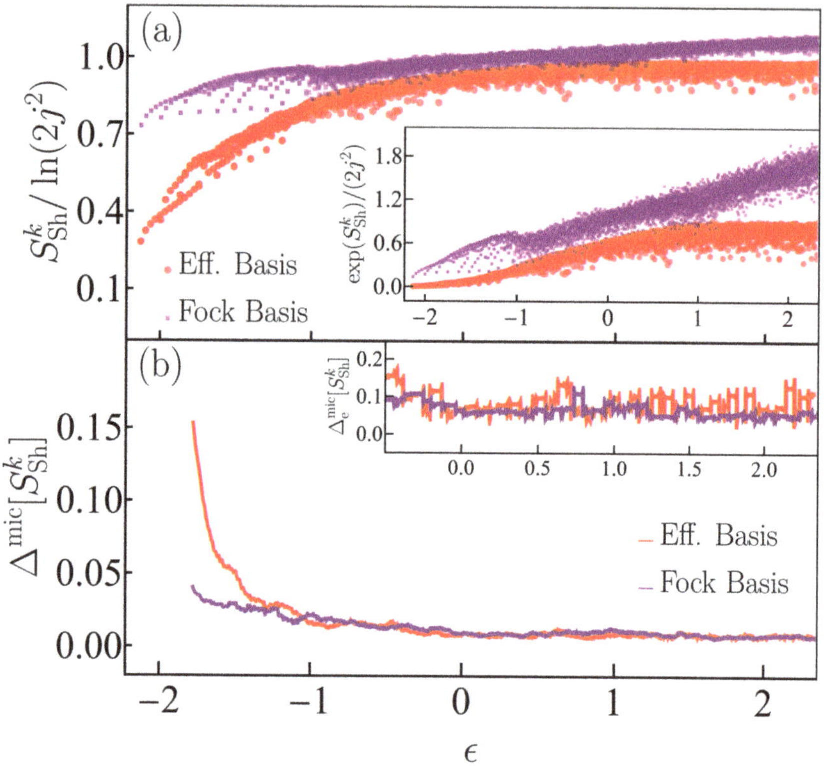

Figure 6. Panel (**a**): Peres lattices of the Shannon entropy $S^k_{\rm Sn}$ (see Equation (21)) scaled to the system-size dependence of the density of states, $\nu(\epsilon) \propto 2j^2$ (see Appendix B), for eigenstates $|E_k\rangle$ from both parity sectors of the Dicke model written in the efficient basis (red dots) and the Fock basis (purple dots). The inset shows the exponential values of the Shannon entropy for both bases. Panel (**b**): Deviation of Shannon entropy with respect to its microcanonical value (see Equation (15)). The inset shows the extremal deviation of the same quantity in the chaotic energy regime (see Equation (16)). The system size in both panels (**a**) and (**b**) is $j = 30$.

The analysis of the fluctuations of the Shannon entropy in Figure 6b and its inset shows that, similarly to what we observed for the entanglement entropy in Figure 5, they approach zero in the chaotic region. The results are comparable for both the Fock and efficient bases. In what follows, we examine the fluctuations of the Shannon entropy for different system sizes obtained for eigenstates written in the efficient basis only.

In Figure 7a, we plot the Peres lattice of the Shannon entropy for the eigenstates of the Dicke model written in the efficient basis. These eigenstates are contained in the energy interval $\epsilon_k \in [\epsilon_{\rm GS} = -2.125, \epsilon_{\rm T} = 1.841]$ for the same values of the system size which were considered previously, $j = 30, 60, 100$. Similarly to what we observe for the entanglement entropy in Figure 5a, the regular patterns and large fluctuations are visible at low energies, where we do not expect the validity of ETH. The fluctuations decrease as we move to high energies and as the system size is increased. This behavior is quantified with the deviations from the microcanonical average shown in Figure 7b and its inset. The decay of $\Delta^{\text{mic}}[S_{\text{En}}]$ to values close to zero for the entanglement in Figure 5b is more abrupt than what we see for $\Delta^{\text{mic}}[S_{\text{Sh}}]$ in Figure 7b, but both cases signal the transition to chaos.

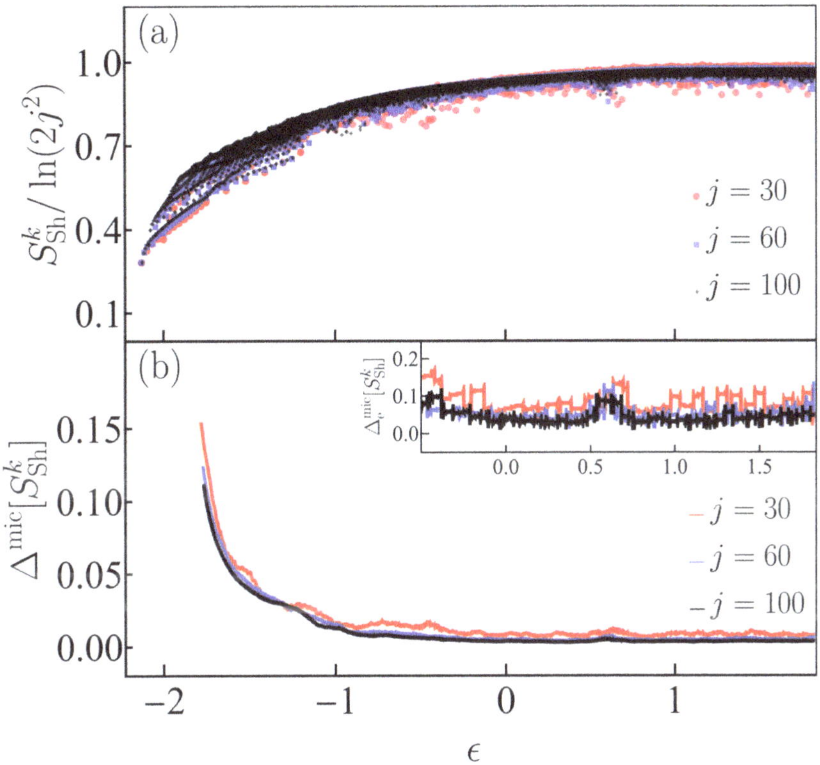

Figure 7. Panel (**a**): Peres lattice of Shannon entropy S^k_{Sh} scaled to the system-size dependence of the density of states $\nu(\epsilon) \propto 2j^2$ for eigenstates $|E_k\rangle$ from both parity sectors of the Dicke model. The Shannon entropy for these eigenstates was built in the efficient basis (see Appendix A.2). Panel (**b**): Deviations of the Shannon entropy with respect to its microcanonical value, computed with Equation (15). The inset shows the extremal deviations in the chaotic energy regime computed with Equation (16). The system size in both panels (**a**) and (**b**) is indicated with a given color for three values $j = 30, 60, 100$.

We close this section mentioning that the regular patterns seen at low energies in Figures 2 and 5–7 reflect the existence of families of periodic orbits, which were studied in previous works [61,62,106]. We also note that for high energies, the behavior of the Shannon entropy computed in the efficient basis and of the von Neumann entanglement entropy are similar in the sense that both measures nearly saturate. In contrast, the Shannon entropy in the Fock basis grows rapidly with energy, because of the infinite bosonic Hilbert subspace of the Dicke model. The modified bosonic sector of the efficient basis is responsible for the smaller values of the Shannon entropy when compared to the values for the Fock basis.

6. Conclusions

In this work, we confirmed the validity of the ETH in the chaotic region of the Dicke model. This was done by analyzing the diagonal and off-diagonal elements of the number of photons and the number of excited atoms for different system sizes. We corroborated that the validity of the ETH stems from the presence of chaotic eigenstates, which we showed by analyzing their components and measures of entanglement and delocalization.

The Shannon entropy, used to quantify the level of delocalization of the eigenstates in a given basis, made evident the advantages of using the efficient basis over the Fock basis. For high energies, the first leads to a slower growth of the entropy than the Fock basis, allowing us to reach converged states for larger system sizes than accessible with the Fock basis.

Author Contributions: All authors have contributed equally to the conceptualization, development, and writing of the work. All authors have read and agreed to the published version of the manuscript.

Funding: This research was funded by the DGAPA- UNAM project number IN104020, the Mexican CONACyT project number CB2015-01/255702, and the United States NSF, Grant No. DMR-1936006. L.F.S. had support from the MPS Simons Foundation Award ID: 678586.

Institutional Review Board Statement: Not applicable.

Data Availability Statement: All the data that support the results and plots showed within this work are available from the corresponding authors upon request.

Acknowledgments: We acknowledge the support of the Computation Center—ICN, in particular to Enrique Palacios, Luciano Díaz, and Eduardo Murrieta.

Conflicts of Interest: The authors declare no conflict of interest.

Abbreviations

The following abbreviations are used in this manuscript:

ETH	eigenstate thermalization hypothesis
RMT	random matrix theory
OTOC	out-of-time-ordered correlator

Appendix A. Diagonalization Bases

Appendix A.1. Fock Basis

The Fock basis is the natural choice to diagonalize the Dicke Hamiltonian, since it is composed by the tensor product between bosonic Fock states $|n\rangle$ and atomic pseudo-spin states $|j, m_z\rangle$, $|n; j, m_z\rangle = |n\rangle \otimes |j, m_z\rangle$. On the one hand, the eigenvalues $n = 0, 1, 2, \ldots$ of the number operator $\hat{n} = \hat{a}^\dagger \hat{a}$ (with eigenvalue equation $\hat{n}|n\rangle = n|n\rangle$) provide an infinite bosonic Hilbert subspace. On the other hand, the eigenvalues $m_z = -j, -j+1, \ldots, j-1, j$ of the pseudo-spin operator $\hat{J}_z$ (with eigenvalue equation $\hat{J}_z|j, m_z\rangle = m_z|j, m_z\rangle$) provide a finite atomic Hilbert subspace with dimension $d_{\mathrm{A}}^{\mathrm{FB}} = 2j+1$. In order to diagonalize the Dicke Hamiltonian a truncation value n_{max} of the bosonic Hilbert subspace has to be chosen, which allows to define a finite bosonic dimension $d_{\mathrm{B}}^{\mathrm{FB}} = n_{\mathrm{max}} + 1$. In this way, the global Hilbert-space dimension of the Dicke model is given by the product $d_{\mathrm{D}}^{\mathrm{FB}} = d_{\mathrm{A}}^{\mathrm{FB}} \times d_{\mathrm{B}}^{\mathrm{FB}}$.

Appendix A.2. Efficient Basis

The truncation required by the Fock basis for the convergence of the high-energy eigenstates rapidly increases with the system size j. This can be circumvented by using the so-called efficient basis, which was originally obtained by studying the integrable limit $\omega_0 \to 0$ [107,108]. This basis can be written in terms of a displaced bosonic annihilation operator $\hat{A} = \hat{a} + G\hat{J}_x$ with $G = 2\gamma/(\omega\sqrt{\mathcal{N}})$ and the eigenstates $|j, m_x\rangle$ of $\hat{J}_x$ ($\hat{J}_x|j, m_x\rangle = m_x|j, m_x\rangle$). The efficient basis states are defined by

$$|N; j, m_x\rangle = \frac{(\hat{A}^\dagger)^N}{\sqrt{N!}}|\alpha_{m_x}\rangle \otimes |j, m_x\rangle, \tag{A1}$$

where $|\alpha_{m_x}\rangle = D(\alpha_{m_x})|0\rangle$ is a coherent state centered at $\alpha_{m_x} = \alpha^*_{m_x} = -Gm_x$, and $\hat{D}(\alpha_{m_x}) = \exp(\alpha_{m_x}\hat{a}^\dagger - \alpha^*_{m_x}\hat{a})$ is the displacement operator. By commuting $\hat{A}^\dagger$ with the displacement operator and using that $G\hat{J}_x|j,m_x\rangle = -\alpha_{m_x}|j,m_x\rangle$, one may further write

$$\begin{aligned}|N;j,m_x\rangle &= \frac{(\hat{a}^\dagger - \alpha_{m_x})^N}{\sqrt{N!}}|\alpha_{m_x}\rangle \otimes |j,m_x\rangle \\ &= \hat{D}(\alpha_{m_x})|N\rangle \otimes |j,m_x\rangle \\ &= |N\rangle_{m_x} \otimes |j,m_x\rangle\end{aligned} \tag{A2}$$

where $|N\rangle_{m_x} = \hat{D}(\alpha_{m_x})|N\rangle$ are displaced Fock states, also called generalized coherent states.

The eigenvalues $N = 0,1,2,\dots$ of the displaced number operator $\hat{N} = \hat{A}^\dagger\hat{A}$ (with eigenvalue equation $\hat{N}|N\rangle_{m_x} = N|N\rangle_{m_x}$) label a modified bosonic Hilbert subspace. The eigenvalues $m_x = -j,-j+1,\dots,j-1,j$ of the pseudo-spin operator $\hat{J}_x$ label the same finite atomic Hilbert sector with dimension $d_{\mathrm{A}}^{\mathrm{EB}} = 2j+1$. As for the Fock basis, the modified bosonic sector must be truncated to some $N_{\max}$ for diagonalization. This yields a finite modified bosonic dimension $d_{\mathrm{B}}^{\mathrm{EB}} = N_{\max}+1$ and a global Hilbert-space dimension $d_{\mathrm{D}}^{\mathrm{EB}} = d_{\mathrm{A}}^{\mathrm{EB}} \times d_{\mathrm{B}}^{\mathrm{EB}}$.

In contrast to the Fock basis, where convergence of high-energy eigenstates is infeasible for large system sizes (usually $j > 30$), the efficient basis allows to get thousands of converged eigenstates in high-energy regimes, even beyond $j = 100$ [103,104].

Appendix A.3. Mapping from Efficient Basis to Fock Basis

To compute the entanglement entropies shown in Figures 6 and 7, we perform a partial trace over the usual atomic sector of the Fock basis, but we diagonalize in the efficient basis. Thus, we need to map the wave function of the eigenstates from the efficient basis to the Fock basis. This is done as follows.

A general pure quantum state $|\Psi\rangle$ can be expanded in the efficient basis and the rotated Fock basis ($|n;j,m_x\rangle$), respectively

$$|\Psi\rangle = \sum_{x=0}^{x_{\max}} \sum_{m_x=-j}^{j} C_{x,m_x}|x;j,m_x\rangle, \tag{A3}$$

where $x = N$ defines the efficient basis and $x = n$ the Fock basis. Moreover, $C_{x,m_x} = \langle x;j,m_x|\Psi\rangle$ are the coefficients of the arbitrary state in each basis, which must satisfy the normalization condition $\sum_{x,m_x}|C_{x,m_x}|^2 = \hat{1}$.

Note that

$$\begin{aligned}C_{n,m_x} &= \langle n;j,m_x|\Psi\rangle \\ &= \sum_{N=0}^{N_{\max}} \sum_{m'_x=-j}^{j} C_{N,m'_x}\langle n;j,m_x|N;j,m'_x\rangle \\ &= \sum_{N=0}^{N_{\max}} \sum_{m'_x=-j}^{j} C_{N,m'_x}\langle n|N\rangle_{m'_x} \otimes \langle j,m_x|j,m'_x\rangle r \\ &= \sum_{N=0}^{N_{\max}} C_{N,m_x}\langle n|\hat{D}(\alpha_{m_x})|N\rangle,\end{aligned} \tag{A4}$$

using that $\langle j,m_x|j,m'_x\rangle = \delta_{m_x,m'_x}$. The term $\langle n|\hat{D}(\alpha_{m_x})|N\rangle$ for $n > N$ is given by [109–111]

$$\langle n|\hat{D}(\alpha_{m_x})|N\rangle = \sqrt{\frac{N!}{n!}}\alpha_{m_x}^{n-N}e^{-|\alpha_{m_x}|^2/2}\mathcal{L}_N^{n-N}(|\alpha_{m_x}|^2), \tag{A5}$$

where $\mathcal{L}_{n_0}^{n_1}(x)$ is an associated Laguerre polynomial given by the Rodrigues formula

$$\mathcal{L}_{n_0}^{n_1}(x) = \frac{x^{-n_1}e^x}{n_0!}\frac{d^{n_0}}{dx^{n_0}}(e^{-x}x^{n_0+n_1}). \tag{A6}$$

We numerically found that, in order to ensure a correct convergence of the coefficients C_{n,m_x} given by Equation (A4), the truncation value must be chosen as $n_{\max} \approx 3N_{\max}$. The associated Laguerre polynomials can be efficiently calculated to arbitrary precision with a package included in the Wolfram Mathematica software [112].

Appendix B. Classical Limit of the Dicke Model

The classical limit of the Dicke model can be obtained taking the expectation value of the quantum Hamiltonian $\hat{H}_\text{D}$ under the tensor product of Glauber and Bloch coherent states $|\mathbf{x}\rangle = |q,p\rangle \otimes |Q,P\rangle$, and dividing it by the system size j [47,54,58,75,82,113,114]

$$h_\text{D}(\mathbf{x}) = \frac{\langle\mathbf{x}|\hat{H}_\text{D}|\mathbf{x}\rangle}{j} = h_\text{F}(\mathbf{x}) + h_\text{A}(\mathbf{x}) + h_\text{I}(\mathbf{x}), \tag{A7}$$

$$h_\text{F}(\mathbf{x}) = \frac{\omega}{2}(q^2+p^2), \tag{A8}$$

$$h_\text{A}(\mathbf{x}) = \frac{\omega_0}{2}(Q^2+P^2) - \omega_0, \tag{A9}$$

$$h_\text{I}(\mathbf{x}) = 2\gamma q Q\sqrt{1-\frac{Q^2+P^2}{4}}, \tag{A10}$$

where $h_\text{F}(\mathbf{x})$ and $h_\text{A}(\mathbf{x})$ represent the Hamiltonians of two classical harmonic oscillators, and $h_\text{I}(\mathbf{x})$ the coupling between them. The bosonic Glauber and the atomic Bloch coherent states, represented by the canonical variables (q,p) and (Q,P) respectively, are given explicitly by

$$|q,p\rangle = e^{-(j/4)(q^2+p^2)}e^{\left[\sqrt{j/2}(q+ip)\right]\hat{a}^\dagger}|0\rangle, \tag{A11}$$

$$|Q,P\rangle = \left(1-\frac{Q^2+P^2}{4}\right)^j e^{\left[(Q+iP)/\sqrt{4-Q^2-P^2}\right]\hat{J}_+}|j,-j\rangle, \tag{A12}$$

where $|0\rangle$ is the photon vacuum and $|j,-j\rangle$ is the state with all the atoms in the ground state.

The classical Dicke Hamiltonian $h_\text{D}(\mathbf{x})$, obtained with the latter method, has an infinite four-dimensional phase space $\mathcal{M}$ in the canonical variables $\mathbf{x} = (q,p;Q,P)$, where the atomic variables are bounded ($Q^2+P^2 \leq 4$). A useful property of this phase space is that it can be partitioned into a family of classical energy shells with finite volume $\mathcal{V}(\epsilon) < \infty$, given by

$$\mathcal{M}(\epsilon) = \{\mathbf{x} \in \mathcal{M} \mid h_\text{D}(\mathbf{x}) = \epsilon\}, \tag{A13}$$

where $\epsilon = E/j$ is the classical energy of the shell scaled to the system size j, which defines an effective Planck constant $\hbar_\text{eff} = 1/j$ [115]. The finite volume $\mathcal{V}(\epsilon)$ of the classical energy shells $\mathcal{M}(\epsilon)$ is obtained with a semiclassical approximation to the quantum density of states $\nu(\epsilon)$, using the Gutzwiller trace formula [114,116,117]. The explicit expression is given by

$$\mathcal{V}(\epsilon) = \int_\mathcal{M} d\mathbf{x}\,\delta(h_\text{D}(\mathbf{x})-\epsilon) = (2\pi\hbar_\text{eff})^2\nu(\epsilon), \tag{A14}$$

where the density of states is proportional to the system size $\nu(\epsilon) \propto 2j^2$, and can be derived explicitly following Ref. [114].

Appendix C. Quantum Entanglement

An arbitrary pure state $|\Psi\rangle$ of a multipartite system composed of $\mathcal{S}$ subsystems, whose Hilbert space is given by the tensor product $\mathcal{H} = \mathcal{H}_1 \otimes \mathcal{H}_2 \otimes \ldots \otimes \mathcal{H}_\mathcal{S}$, can be expanded due to the superposition principle in a tensor-product basis $\{|\psi^1_{k_1}\rangle \otimes |\psi^2_{k_2}\rangle \otimes \ldots \otimes |\psi^\mathcal{S}_{k_\mathcal{S}}\rangle\}$ as

$$|\Psi\rangle = \sum_{k_1,k_2,\ldots,k_\mathcal{S}}^{d_1,d_2,\ldots,d_\mathcal{S}} c_{k_1,k_2,\ldots,k_\mathcal{S}} |\psi^1_{k_1}\rangle \otimes |\psi^2_{k_2}\rangle \otimes \ldots \otimes |\psi^\mathcal{S}_{k_\mathcal{S}}\rangle, \tag{A15}$$

where $d_1, d_2, \ldots, d_\mathcal{S}$ are the dimensions of each subspace and the coefficients $c_{k_1,k_2,\ldots,k_\mathcal{S}}$ satisfy the normalization condition

$$\sum_{k_1,k_2,\ldots,k_\mathcal{S}}^{d_1,d_2,\ldots,d_\mathcal{S}} |c_{k_1,k_2,\ldots,k_\mathcal{S}}|^2 = \hat{1}. \tag{A16}$$

Thus, the state $|\Psi\rangle$ is called separable (entangled) if it can (cannot) be written as a tensor product of states corresponding to each subspace

$$|\Psi\rangle = |\Psi_1\rangle \otimes |\Psi_2\rangle \otimes \ldots \otimes |\Psi_\mathcal{S}\rangle. \tag{A17}$$

For a mixed state $\hat{\rho}$ the definition of entanglement is no longer equivalent to that of a pure state. In this way, the mixed state $\hat{\rho}$ is called separable (entangled) if it can (cannot) be written as a convex combination of tensor-product states [102,118], corresponding to each subspace

$$\hat{\rho} = \sum_{k_1,k_2,\ldots,k_\mathcal{S}}^{d_1,d_2,\ldots,d_\mathcal{S}} p_{k_1,k_2,\ldots,k_\mathcal{S}} \hat{\rho}^1_{k_1} \otimes \hat{\rho}^2_{k_2} \otimes \ldots \otimes \hat{\rho}^\mathcal{S}_{k_\mathcal{S}}, \tag{A18}$$

where $d_1, d_2, \ldots, d_\mathcal{S}$ are the dimensions of each subspace and the probabilities $p_{k_1,k_2,\ldots,k_\mathcal{S}}$ satisfy the normalization condition

$$\sum_{k_1,k_2,\ldots,k_\mathcal{S}}^{d_1,d_2,\ldots,d_\mathcal{S}} p_{k_1,k_2,\ldots,k_\mathcal{S}} = \hat{1}. \tag{A19}$$

When multipartite systems are studied, a useful tool to work with is the reduced density matrix corresponding to a given subspace, which is obtained by taking the partial trace of the whole density matrix $\hat{\rho}$. For example, the reduced density matrix of the i-th subspace is given by

$$\hat{\rho}_i = \mathrm{Tr}_{(1,2,\ldots,\mathcal{S})\neq i}[\hat{\rho}] = \sum_{k_i=1}^{d_i} P_{k_i}\hat{\rho}^i_{k_i}, \tag{A20}$$

where $\sum_{k_i=1}^{d_i} P_{k_i} = \hat{1}$ and

$$P_{k_i} = \sum_{(k_1,k_2,\ldots,k_\mathcal{S})\neq k_i}^{(d_1,d_2,\ldots,d_\mathcal{S})\neq d_i} p_{k_1,k_2,\ldots,k_\mathcal{S}}. \tag{A21}$$

Appendix D. Integrable Dicke Model: Tavis–Cummings Model

The integrable limit of the Dicke model is known as the Tavis–Cummings model [119], and can be obtained by applying the rotating wave approximation (RWA) to the Dicke Hamiltonian (2). The RWA consists of ignoring the interacting terms in $\hat{H}_\mathrm{I}$ (see Equation (5)) that oscillates very fast, that is, $\hat{a}^\dagger \hat{J}_+$ and $\hat{a}\hat{J}_-$. The last results in a modified interacting Hamiltonian given by

$$\hat{H}_\mathrm{I}^\mathrm{RWA} = \frac{\gamma}{\sqrt{\mathcal{N}}}(\hat{a}^\dagger \hat{J}_- + \hat{a}\hat{J}_+), \tag{A22}$$

such that, the complete Tavis–Cummings Hamiltonian is given by

$$\hat{H}_{\mathrm{TC}} = \hat{H}_{\mathrm{F}} + \hat{H}_{\mathrm{A}} + \hat{H}_{\mathrm{I}}^{\mathrm{RWA}}, \tag{A23}$$

where the Hamiltonians $\hat{H}_{\mathrm{F}}$ and $\hat{H}_{\mathrm{A}}$ are the same terms given in Equations (3) and (4). The particularity of the Tavis–Cummings Hamiltonian is that it commutes with the operator $\hat{\Lambda}$ (see Equation (7)), which is a conserved quantity that defines the number of excitations. The last feature allows to diagonalize the Tavis–Cummings Hamiltonian in finite subspaces of such operator [119].

References

1. von Neumann, J. Beweis des Ergodensatzes und des H-Theorems in der neuen Mechanik. *Zeitschrift für Physik* **1929**, *57*, 30–70. [CrossRef]
2. von Neumann, J. Proof of the ergodic theorem and the H-theorem in quantum mechanics. *Eur. Phys. J. H* **2010**, *35*, 201–237. [CrossRef]
3. Goldstein, S.; Lebowitz, J.L.; Tumulka, R.; Zanghì, N. Long-time behavior of macroscopic quantum systems. *Eur. Phys. J. H* **2010**, *35*, 173–200. [CrossRef]
4. Goldstein, S.; Lebowitz, J.L.; Mastrodonato, C.; Tumulka, R.; Zanghì, N. Normal typicality and von Neumann's quantum ergodic theorem. *Proc. R. Soc. A Math. Phys. Eng. Sci.* **2010**, *466*, 3203–3224. [CrossRef]
5. Pechukas, P. Remarks on "quantum chaos". *J. Phys. Chem.* **1984**, *88*, 4823. [CrossRef]
6. Bocchieri, P.; Loinger, A. Ergodic Theorem in Quantum Mechanics. *Phys. Rev.* **1958**, *111*, 668–670. [CrossRef]
7. Pechukas, P. Sharpening an inequality in quantum ergodic theory. *J. Math. Phys.* **1984**, *25*, 532–534. [CrossRef]
8. Jensen, R.V.; Shankar, R. Statistical Behavior in Deterministic Quantum Systems with Few Degrees of Freedom. *Phys. Rev. Lett.* **1985**, *54*, 1879–1882. [CrossRef]
9. D'Alessio, L.; Kafri, Y.; Polkovnikov, A.; Rigol, M. From quantum chaos and eigenstate thermalization to statistical mechanics and thermodynamics. *Adv. Phys.* **2016**, *65*, 239–362. [CrossRef]
10. Deutsch, J.M. Eigenstate thermalization hypothesis. *Rep. Prog. Phys.* **2018**, *81*, 082001. [CrossRef]
11. Rigol, M.; Dunjko, V.; Olshanii, M. Thermalization and its mechanism for generic isolated quantum systems. *Nature* **2008**, *452*, 854. [CrossRef] [PubMed]
12. Srednicki, M. Chaos and quantum thermalization. *Phys. Rev. E* **1994**, *50*, 888–901. [CrossRef] [PubMed]
13. Deutsch, J.M. Quantum statistical mechanics in a closed system. *Phys. Rev. A* **1991**, *43*, 2046–2049. [CrossRef] [PubMed]
14. Beugeling, W.; Moessner, R.; Haque, M. Finite-size scaling of eigenstate thermalization. *Phys. Rev. E* **2014**, *89*, 042112. [CrossRef]
15. LeBlond, T.; Mallayya, K.; Vidmar, L.; Rigol, M. Entanglement and matrix elements of observables in interacting integrable systems. *Phys. Rev. E* **2019**, *100*, 062134. [CrossRef]
16. Torres-Herrera, E.J.; Santos, L.F. Effects of the interplay between initial state and Hamiltonian on the thermalization of isolated quantum many-body systems. *Phys. Rev. E* **2013**, *88*, 042121. [CrossRef]
17. He, K.; Rigol, M. Initial-state dependence of the quench dynamics in integrable quantum systems. III. Chaotic states. *Phys. Rev. A* **2013**, *87*, 043615. [CrossRef]
18. Santos, L.F.; Rigol, M. Onset of quantum chaos in one-dimensional bosonic and fermionic systems and its relation to thermalization. *Phys. Rev. E* **2010**, *81*, 036206. [CrossRef]
19. Santos, L.F.; Rigol, M. Localization and the effects of symmetries in the thermalization properties of one-dimensional quantum systems. *Phys. Rev. E* **2010**, *82*, 031130. [CrossRef]
20. Rigol, M.; Santos, L.F. Quantum chaos and thermalization in gapped systems. *Phys. Rev. A* **2010**, *82*, 011604(R). [CrossRef]
21. Benenti, G.; Casati, G.; Shepelyansky, D. Emergence of Fermi–Dirac thermalization in the quantum computer core. *Eur. Phys. J. D* **2001**, *17*, 265. [CrossRef]
22. Flambaum, V.V.; Gribakina, A.A.; Gribakin, G.F.; Kozlov, M.G. Structure of compound states in the chaotic spectrum of the Ce atom: Localization properties, matrix elements, and enhancement of weak perturbations. *Phys. Rev. A* **1994**, *50*, 267–296. [CrossRef] [PubMed]
23. Flambaum, V.V.; Izrailev, F.M.; Casati, G. Towards a statistical theory of finite Fermi systems and compound states: Random two-body interaction approach. *Phys. Rev. E* **1996**, *54*, 2136–2139. [CrossRef] [PubMed]
24. Flambaum, V.V.; Izrailev, F.M. Statistical theory of finite Fermi systems based on the structure of chaotic eigenstates. *Phys. Rev. E* **1997**, *56*, 5144–5159. [CrossRef]
25. Borgonovi, F.; Guarneri, I.; Izrailev, F.; Casati, G. Chaos and thermalization in a dynamical model of two interacting particles. *Phys. Lett. A* **1998**, *247*, 140–144. [CrossRef]
26. Jacquod, P.; Shepelyansky, D.L. Emergence of Quantum Chaos in Finite Interacting Fermi Systems. *Phys. Rev. Lett.* **1997**, *79*, 1837. [CrossRef]
27. Horoi, M.; Zelevinsky, V.; Brown, B.A. Chaos vs Thermalization in the Nuclear Shell Model. *Phys. Rev. Lett.* **1995**, *74*, 5194–5197. [CrossRef]

28. Zelevinsky, V.; Brown, B.A.; Frazier, N.; Horoi, M. The nuclear shell model as a testing ground for many-body quantum chaos. *Phys. Rep.* **1996**, *276*, 85–176. [CrossRef]
29. Borgonovi, F.; Izrailev, F.M.; Santos, L.F.; Zelevinsky, V.G. Quantum chaos and thermalization in isolated systems of interacting particles. *Phys. Rep.* **2016**, *626*, 1. [CrossRef]
30. Berry, M.V. Regular and irregular semiclassical wavefunctions. *J. Phys. A Math. Gen.* **1977**, *10*, 2083–2091. [CrossRef]
31. Borgonovi, F.; Mattiotti, F.; Izrailev, F.M. Temperature of a single chaotic eigenstate. *Phys. Rev. E* **2017**, *95*, 042135. [CrossRef] [PubMed]
32. Wang, J.; Benenti, G.; Casati, G.; Wang, W.G. Complexity of quantum motion and quantum-classical correspondence: A phase-space approach. *Phys. Rev. Res.* **2020**, *2*, 043178. [CrossRef]
33. Balachandran, V.; Benenti, G.; Casati, G.; Poletti, D. From the eigenstate thermalization hypothesis to algebraic relaxation of OTOCs in systems with conserved quantities. *Phys. Rev. B* **2021**, *104*, 104306. [CrossRef]
34. Wang, J.; Casati, G.; Benenti, G. Classical Physics and Blackbody Radiation. *Phys. Rev. Lett.* **2022**, *128*, 134101. [CrossRef]
35. Wang, J.; Benenti, G.; Casati, G.; ge Wang, W. Statistical and dynamical properties of the quantum triangle map. *J. Phys. A* **2022**, *55*, 234002. [CrossRef]
36. Santos, L.F.; Borgonovi, F.; Izrailev, F.M. Chaos and Statistical Relaxation in Quantum Systems of Interacting Particles. *Phys. Rev. Lett.* **2012**, *108*, 094102. [CrossRef]
37. Borgonovi, F.; Izrailev, F.M.; Santos, L.F. Exponentially fast dynamics of chaotic many-body systems. *Phys. Rev. E* **2019**, *99*, 010101. [CrossRef]
38. Borgonovi, F.; Izrailev, F.M.; Santos, L.F. Timescales in the quench dynamics of many-body quantum systems: Participation ratio versus out-of-time ordered correlator. *Phys. Rev. E* **2019**, *99*, 052143. [CrossRef]
39. Dicke, R.H. Coherence in Spontaneous Radiation Processes. *Phys. Rev.* **1954**, *93*, 99. [CrossRef]
40. Garraway, B.M. The Dicke model in quantum optics: Dicke model revisited. *Philos. Trans. Royal Soc. A* **2011**, *369*, 1137. [CrossRef]
41. Kirton, P.; Roses, M.M.; Keeling, J.; Dalla Torre, E.G. Introduction to the Dicke Model: From Equilibrium to Nonequilibrium, and Vice Versa. *Adv. Quantum Technol.* **2019**, *2*, 1800043. [CrossRef]
42. Emary, C.; Brandes, T. Chaos and the quantum phase transition in the Dicke model. *Phys. Rev. E* **2003**, *67*, 066203. [CrossRef] [PubMed]
43. Emary, C.; Brandes, T. Quantum Chaos Triggered by Precursors of a Quantum Phase Transition: The Dicke Model. *Phys. Rev. Lett.* **2003**, *90*, 044101. [CrossRef] [PubMed]
44. Brandes, T. Excited-state quantum phase transitions in Dicke superradiance models. *Phys. Rev. E* **2013**, *88*, 032133. [CrossRef] [PubMed]
45. Furuya, K.; Nemes, M.C.; Pellegrino, G.Q. Quantum Dynamical Manifestation of Chaotic Behavior in the Process of Entanglement. *Phys. Rev. Lett.* **1998**, *80*, 5524–5527. [CrossRef]
46. Lóbez, C.M.; Relaño, A. Entropy, chaos, and excited-state quantum phase transitions in the Dicke model. *Phys. Rev. E* **2016**, *94*, 012140. [CrossRef]
47. Chávez-Carlos, J.; Bastarrachea-Magnani, M.A.; Lerma-Hernández, S.; Hirsch, J.G. Classical chaos in atom–field systems. *Phys. Rev. E* **2016**, *94*, 022209. [CrossRef]
48. Sinha, S.; Sinha, S. Chaos and Quantum Scars in Bose-Josephson Junction Coupled to a Bosonic Mode. *Phys. Rev. Lett.* **2020**, *125*, 134101. [CrossRef]
49. Valencia-Tortora, R.J.; Kelly, S.P.; Donner, T.; Morigi, G.; Fazio, R.; Marino, J. Crafting the dynamical structure of synchronization by harnessing bosonic multi-level cavity QED. *arXiv* **2022**, arXiv:2210.14224.
50. Altland, A.; Haake, F. Equilibration and macroscopic quantum fluctuations in the Dicke model. *New J. Phys.* **2012**, *14*, 073011. [CrossRef]
51. Kloc, M.; Stránský, P.; Cejnar, P. Quantum quench dynamics in Dicke superradiance models. *Phys. Rev. A* **2018**, *98*, 013836. [CrossRef]
52. Lerma-Hernández, S.; Chávez-Carlos, J.; Bastarrachea-Magnani, M.A.; Santos, L.F.; Hirsch, J.G. Analytical description of the survival probability of coherent states in regular regimes. *J. Phys. A Math. Theor.* **2018**, *51*, 475302. [CrossRef]
53. Lerma-Hernández, S.; Villaseñor, D.; Bastarrachea-Magnani, M.A.; Torres-Herrera, E.J.; Santos, L.F.; Hirsch, J.G. Dynamical signatures of quantum chaos and relaxation time scales in a spin-boson system. *Phys. Rev. E* **2019**, *100*, 012218. [CrossRef] [PubMed]
54. Villaseñor, D.; Pilatowsky-Cameo, S.; Bastarrachea-Magnani, M.A.; Lerma-Hernández, S.; Santos, L.F.; Hirsch, J.G. Quantum vs classical dynamics in a spin-boson system: Manifestations of spectral correlations and scarring. *New J. Phys.* **2020**, *22*, 063036. [CrossRef]
55. Chávez-Carlos, J.; López-del Carpio, B.; Bastarrachea-Magnani, M.A.; Stránský, P.; Lerma-Hernández, S.; Santos, L.F.; Hirsch, J.G. Quantum and Classical Lyapunov Exponents in Atom-Field Interaction Systems. *Phys. Rev. Lett.* **2019**, *122*, 024101. [CrossRef] [PubMed]
56. Lewis-Swan, R.J.; Safavi-Naini, A.; Bollinger, J.J.; Rey, A.M. Unifying thermalization and entanglement through measurement of fidelity out-of-time-order correlators in the Dicke model. *Nat. Commun.* **2019**, *10*, 1581. [CrossRef]

57. Pilatowsky-Cameo, S.; Chávez-Carlos, J.; Bastarrachea-Magnani, M.A.; Stránský, P.; Lerma-Hernández, S.; Santos, L.F.; Hirsch, J.G. Positive quantum Lyapunov exponents in experimental systems with a regular classical limit. *Phys. Rev. E* **2020**, *101*, 010202(R). [CrossRef]
58. de Aguiar, M.; Furuya, K.; Lewenkopf, C.; Nemes, M. Chaos in a spin-boson system: Classical analysis. *Ann. Phys.* **1992**, *216*, 291–312. [CrossRef]
59. Furuya, K.; de Aguiar, M.; Lewenkopf, C.; Nemes, M. Husimi distributions of a spin-boson system and the signatures of its classical dynamics. *Ann. Phys.* **1992**, *216*, 313–322. [CrossRef]
60. Bakemeier, L.; Alvermann, A.; Fehske, H. Dynamics of the Dicke model close to the classical limit. *Phys. Rev. A* **2013**, *88*, 043835. [CrossRef]
61. Pilatowsky-Cameo, S.; Villaseñor, D.; Bastarrachea-Magnani, M.A.; Lerma-Hernández, S.; Santos, L.F.; Hirsch, J.G. Ubiquitous quantum scarring does not prevent ergodicity. *Nat. Commun.* **2021**, *12*, 852. [CrossRef] [PubMed]
62. Pilatowsky-Cameo, S.; Villaseñor, D.; Bastarrachea-Magnani, M.A.; Lerma-Hernández, S.; Santos, L.F.; Hirsch, J.G. Quantum scarring in a spin-boson system: Fundamental families of periodic orbits. *New J. Phys.* **2021**, *23*, 033045. [CrossRef]
63. Wang, Q.; Robnik, M. Statistical properties of the localization measure of chaotic eigenstates in the Dicke model. *Phys. Rev. E* **2020**, *102*, 032212. [CrossRef]
64. Jaako, T.; Xiang, Z.L.; Garcia-Ripoll, J.J.; Rabl, P. Ultrastrong-coupling phenomena beyond the Dicke model. *Phys. Rev. A* **2016**, *94*, 033850. [CrossRef]
65. Baden, M.P.; Arnold, K.J.; Grimsmo, A.L.; Parkins, S.; Barrett, M.D. Realization of the Dicke Model Using Cavity-Assisted Raman Transitions. *Phys. Rev. Lett.* **2014**, *113*, 020408. [CrossRef]
66. Zhang, Z.; Lee, C.H.; Kumar, R.; Arnold, K.J.; Masson, S.J.; Grimsmo, A.L.; Parkins, A.S.; Barrett, M.D. Dicke-model simulation via cavity-assisted Raman transitions. *Phys. Rev. A* **2018**, *97*, 043858. [CrossRef]
67. Cohn, J.; Safavi-Naini, A.; Lewis-Swan, R.J.; Bohnet, J.G.; Gärttner, M.; Gilmore, K.A.; Jordan, J.E.; Rey, A.M.; Bollinger, J.J.; Freericks, J.K. Bang-bang shortcut to adiabaticity in the Dicke model as realized in a Penning trap experiment. *New J. Phys.* **2018**, *20*, 055013. [CrossRef]
68. Safavi-Naini, A.; Lewis-Swan, R.J.; Bohnet, J.G.; Gärttner, M.; Gilmore, K.A.; Jordan, J.E.; Cohn, J.; Freericks, J.K.; Rey, A.M.; Bollinger, J.J. Verification of a Many-Ion Simulator of the Dicke Model Through Slow Quenches across a Phase Transition. *Phys. Rev. Lett.* **2018**, *121*, 040503. [CrossRef]
69. Chelpanova, O.; Lerose, A.; Zhang, S.; Carusotto, I.; Tserkovnyak, Y.; Marino, J. Intertwining of lasing and superradiance under spintronic pumping. arXiv **2021**, arXiv:2112.04509.
70. Kirkova, A.V.; Ivanov, P.A. Quantum chaos and thermalization in the two-mode Dicke model. arXiv **2022**, arXiv:2207.03825.
71. Ray, S.; Ghosh, A.; Sinha, S. Quantum signature of chaos and thermalization in the kicked Dicke model. *Phys. Rev. E* **2016**, *94*, 032103. [CrossRef] [PubMed]
72. Hepp, K.; Lieb, E.H. On the superradiant phase transition for molecules in a quantized radiation field: The Dicke maser model. *Ann. Phys. (N.Y.)* **1973**, *76*, 360–404. [CrossRef]
73. Hepp, K.; Lieb, E.H. Equilibrium Statistical Mechanics of Matter Interacting with the Quantized Radiation Field. *Phys. Rev. A* **1973**, *8*, 2517–2525. [CrossRef]
74. Wang, Y.K.; Hioe, F.T. Phase Transition in the Dicke Model of Superradiance. *Phys. Rev. A* **1973**, *7*, 831–836. [CrossRef]
75. Bastarrachea-Magnani, M.A.; Lerma-Hernández, S.; Hirsch, J.G. Comparative quantum and semiclassical analysis of atom–field systems. II. Chaos and regularity. *Phys. Rev. A* **2014**, *89*, 032102. [CrossRef]
76. Casati, G.; Valz-Gris, F.; Guarnieri, I. On the connection between quantization of nonintegrable systems and statistical theory of spectra. *Lett. Nuov. Cim.* **1980**, *28*, 279–282. [CrossRef]
77. Bohigas, O.; Giannoni, M.J.; Schmit, C. Characterization of Chaotic Quantum Spectra and Universality of Level Fluctuation Laws. *Phys. Rev. Lett.* **1984**, *52*, 1–4. [CrossRef]
78. Mehta, M.L. *Random Matrices*; Academic Press: Boston, MA, USA, 1991.
79. Oganesyan, V.; Huse, D.A. Localization of interacting fermions at high temperature. *Phys. Rev. B* **2007**, *75*, 155111. [CrossRef]
80. Atas, Y.Y.; Bogomolny, E.; Giraud, O.; Roux, G. Distribution of the Ratio of Consecutive Level Spacings in Random Matrix Ensembles. *Phys. Rev. Lett.* **2013**, *110*, 084101. [CrossRef]
81. Guhr, T.; Müller-Groeling, A.; Weidenmüller, H.A. Random Matrix Theories in Quantum Physics: Common Concepts. *Phys. Rep.* **1998**, *299*, 189. [CrossRef]
82. Bastarrachea-Magnani, M.A.; del Carpio, B.L.; Lerma-Hernández, S.; Hirsch, J.G. Chaos in the Dicke model: Quantum and semiclassical analysis. *Phys. Scr.* **2015**, *90*, 068015. [CrossRef]
83. Wang, J.; Wang, W.-G. Characterization of random features of chaotic eigenfunctions in unperturbed basis. *Phys. Rev. E* **2018**, *97*, 062219. [CrossRef] [PubMed]
84. Peres, A. New Conserved Quantities and Test for Regular Spectra. *Phys. Rev. Lett.* **1984**, *53*, 1711–1713. [CrossRef]
85. Larkin, A.I.; Ovchinnikov, Y.N. Quasiclassical Method in the Theory of Superconductivity. *Sov. Phys. JETP* **1969**, *28*, 1200.
86. Maldacena, J.; Stanford, D. Remarks on the Sachdev-Ye-Kitaev model. *Phys. Rev. D* **2016**, *94*, 106002. [CrossRef]
87. Maldacena, J.; Shenker, S.H.; Stanford, D. A bound on chaos. *J. High Energy Phys.* **2016**, *2016*, 106. [CrossRef]
88. Bastarrachea-Magnani, M.A.; Hirsch, J.G. Peres lattices and chaos in the Dicke model. *J. Phys. Conf. Ser.* **2014**, *512*, 012004. [CrossRef]

89. Beugeling, W.; Moessner, R.; Haque, M. Off-diagonal matrix elements of local operators in many-body quantum systems. *Phys. Rev. E* **2015**, *91*, 012144. [CrossRef]
90. Santos, L.F.; Pérez-Bernal, F.; Torres-Herrera, E.J. Speck of chaos. *Phys. Rev. Res.* **2020**, *2*, 043034. [CrossRef]
91. Łydżba, P.; Zhang, Y.; Rigol, M.; Vidmar, L. Single-particle eigenstate thermalization in quantum-chaotic quadratic Hamiltonians. *Phys. Rev. B* **2021**, *104*, 214203. [CrossRef]
92. Zisling, G.; Santos, L.F.; Lev, Y.B. How many particles make up a chaotic many-body quantum system? *SciPost Phys.* **2021**, *10*, 88. [CrossRef]
93. Wittmann W., K.; Castro, E.R.; Foerster, A.; Santos, L.F. Interacting bosons in a triple well: Preface of many-body quantum chaos. *Phys. Rev. E* **2022**, *105*, 034204. [CrossRef] [PubMed]
94. Khaymovich, I.M.; Haque, M.; McClarty, P.A. Eigenstate Thermalization, Random Matrix Theory, and Behemoths. *Phys. Rev. Lett.* **2019**, *122*, 070601. [CrossRef] [PubMed]
95. Kaneko, K.; Iyoda, E.; Sagawa, T. Characterizing complexity of many-body quantum dynamics by higher-order eigenstate thermalization. *Phys. Rev. A* **2020**, *101*, 042126. [CrossRef]
96. Page, D.N. Average entropy of a subsystem. *Phys. Rev. Lett.* **1993**, *71*, 1291–1294. [CrossRef] [PubMed]
97. Miller, P.A.; Sarkar, S. Signatures of chaos in the entanglement of two coupled quantum kicked tops. *Phys. Rev. E* **1999**, *60*, 1542–1550. [CrossRef]
98. Lakshminarayan, A. Entangling power of quantized chaotic systems. *Phys. Rev. E* **2001**, *64*, 036207. [CrossRef]
99. Bandyopadhyay, J.N.; Lakshminarayan, A. Testing Statistical Bounds on Entanglement Using Quantum Chaos. *Phys. Rev. Lett.* **2002**, *89*, 060402. [CrossRef]
100. Bandyopadhyay, J.N.; Lakshminarayan, A. Entanglement production in coupled chaotic systems: Case of the kicked tops. *Phys. Rev. E* **2004**, *69*, 016201. [CrossRef]
101. Wang, X.; Ghose, S.; Sanders, B.C.; Hu, B. Entanglement as a signature of quantum chaos. *Phys. Rev. E* **2004**, *70*, 016217. [CrossRef]
102. Horodecki, R.; Horodecki, P.; Horodecki, M.; Horodecki, K. Quantum entanglement. *Rev. Mod. Phys.* **2009**, *81*, 865–942. [CrossRef]
103. Bastarrachea-Magnani, M.A.; Hirsch, J.G. Efficient basis for the Dicke model: I. Theory and convergence in energy. *Phys. Scr.* **2014**, *T160*, 014005. [CrossRef]
104. Hirsch, J.G.; Bastarrachea-Magnani, M.A. Efficient basis for the Dicke model: II. Wave function convergence and excited states. *Phys. Scr.* **2014**, *T160*, 014018. [CrossRef]
105. Kloc, M.; Stránský, P.; Cejnar, P. Quantum phases and entanglement properties of an extended Dicke model. *Ann. Phys.* **2017**, *382*, 85–111. [CrossRef]
106. Pilatowsky-Cameo, S.; Villaseñor, D.; Bastarrachea-Magnani, M.A.; Lerma-Hernández, S.; Santos, L.F.; Hirsch, J.G. Identification of quantum scars via phase-space localization measures. *Quantum* **2022**, *6*, 644. [CrossRef]
107. Chen, Q.H.; Zhang, Y.Y.; Liu, T.; Wang, K.L. Numerically exact solution to the finite-size Dicke model. *Phys. Rev. A* **2008**, *78*, 051801. [CrossRef]
108. Bastarrachea-Magnani, M.A.; Hirsch, J.G. Numerical solutions of the Dicke Hamiltonian. *Rev. Mex. Fis. S* **2011**, *57*, 69.
109. Cahill, K.E.; Glauber, R.J. Ordered Expansions in Boson Amplitude Operators. *Phys. Rev.* **1969**, *177*, 1857–1881. [CrossRef]
110. Cahill, K.E.; Glauber, R.J. Density Operators and Quasiprobability Distributions. *Phys. Rev.* **1969**, *177*, 1882–1902. [CrossRef]
111. de Oliveira, F.A.M.; Kim, M.S.; Knight, P.L.; Buek, V. Properties of displaced number states. *Phys. Rev. A* **1990**, *41*, 2645–2652. [CrossRef]
112. Wolfram Research, Inc. *Mathematica, Version 13.1*; Wolfram Research, Inc.: Champaign, IL, USA, 2022.
113. de Aguiar, M.A.M.; Furuya, K.; Lewenkopf, C.H.; Nemes, M.C. Particle-Spin Coupling in a Chaotic System: Localization-Delocalization in the Husimi Distributions. *EPL* **1991**, *15*, 125–131. [CrossRef]
114. Bastarrachea-Magnani, M.A.; Lerma-Hernández, S.; Hirsch, J.G. Comparative quantum and semiclassical analysis of atom–field systems. I. Density of states and excited-state quantum phase transitions. *Phys. Rev. A* **2014**, *89*, 032101. [CrossRef]
115. Ribeiro, A.D.; de Aguiar, M.A.M.; de Toledo Piza, A.F.R. The semiclassical coherent state propagator for systems with spin. *J. Phys. A Math. Gen.* **2006**, *39*, 3085–3097. [CrossRef]
116. Gutzwiller, M.C. Periodic Orbits and Classical Quantization Conditions. *J. Math. Phys.* **1971**, *12*, 343–358. [CrossRef]
117. Gutzwiller, M.C. *Chaos in Classical and Quantum Mechanics*; Springer: New York, NY, USA, 1990.
118. Werner, R.F. Quantum states with Einstein-Podolsky-Rosen correlations admitting a hidden-variable model. *Phys. Rev. A* **1989**, *40*, 4277–4281. [CrossRef]
119. Tavis, M.; Cummings, F.W. Exact Solution for an *N*-Molecule—Radiation-Field Hamiltonian. *Phys. Rev.* **1968**, *170*, 379–384. [CrossRef]

entropy

Article

Relaxation Exponents of OTOCs and Overlap with Local Hamiltonians

Vinitha Balachandran [1,*] and Dario Poletti [1,2,3,4,5,*]

1 Science, Mathematics and Technology Cluster, Singapore University of Technology and Design, 8 Somapah Road, Singapore 487372, Singapore
2 EPD Pillar, Singapore University of Technology and Design, 8 Somapah Road, Singapore 487372, Singapore
3 MajuLab, CNRS-UCA-SU-NUS-NTU International Joint Research Unit, Singapore 117543, Singapore
4 Centre for Quantum Technologies, National University of Singapore, Singapore 117543, Singapore
5 The Abdus Salam International Centre for Theoretical Physics, Strada Costiera 11, 34151 Trieste, Italy
* Correspondence: vinitha_balachandaran@ihpc.a-star.edu.sg (V.B.); dario_poletti@sutd.edu.sg (D.P.)

Abstract: OTOC has been used to characterize the information scrambling in quantum systems. Recent studies have shown that local conserved quantities play a crucial role in governing the relaxation dynamics of OTOC in non-integrable systems. In particular, the slow scrambling of OTOC is seen for observables that have an overlap with local conserved quantities. However, an observable may not overlap with the Hamiltonian but instead with the Hamiltonian elevated to an exponent larger than one. Here, we show that higher exponents correspond to faster relaxation, although still algebraic, and such exponents can increase indefinitely. Our analytical results are supported by numerical experiments.

Keywords: OTOC; information scrambling; relaxation dynamics

Citation: Balachandran, V.; Poletti, D. Relaxation Exponents of OTOCs and Overlap with Local Hamiltonians. *Entropy* **2023**, *25*, 59. https://doi.org/10.3390/e25010059

Academic Editor: Marko Robnik

Received: 25 November 2022
Accepted: 20 December 2022
Published: 28 December 2022

1. Introduction

For generic many-body quantum systems, information initially encoded in a few local degrees of freedom can spread in time over the entire accessible space. This process is called information scrambling and can be characterized by out-of-time ordered correlators (OTOCs) [1–21]. For quantum systems with a classical limit, OTOCs can be mapped to Lyapunov exponents [22–33]. Because of this, OTOCs have been applied to understand the thermalization in many-body quantum systems [3–13].

Recent studies have pointed out the relevance of local conserved quantities in the relaxation dynamics of OTOCs [34,34–40]. In particular, in [39] it was shown that the emergence of algebraic relaxation can stem from the locality of the Hamiltonian, i.e., the ensuing presence of a Lieb–Robinson bound [41], and the eigestate thermalization hypothesis (ETH) [42,43]. Importantly, with the approach developed in [39] it was also possible to show that the algebraic relaxation of the OTOC is typical.

In the scenarios considered until now, the operators in the OTOC had non-zero overlap with the Hamiltonian or a local conserved quantity (i.e., total magnetization). Here, we investigate how the relaxation dynamics would be affected if the operators in the OTOC, e.g., A, do not overlap with the Hamiltonian H but only with one of its powers, i.e., $\mathrm{tr}(AH) = 0$, but $\mathrm{tr}(AH^m) \neq 0$ for m, being an integer larger than one. We show that depending on the exponent m at which the overlap becomes non-zero, we expect an algebraic relaxation of the OTOC in time with an exponent proportional to m. To obtain this result, we also show the relation between the first non-zero derivative of the diagonals of an operator in the energy basis, with the exponent m at which $\mathrm{tr}(AH^m) \neq 0$.

The paper is organized as follows. In Section 2, we introduce the definition of OTOCs and explain the relaxation dynamics of OTOCs from the knowledge of the matrix elements of the observables in the eigenenergy basis. In Section 3, we show analytically our main result, i.e., that any different exponents can emerge in the relaxation of the OTOC, depending on the order at which the operators in the OTOC overlap with the Hamiltonian. Our numerical results are presented in Section 4. We draw our conclusions in Section 5.

2. Emergence of Slow Scrambling

2.1. Definition

Consider the infinite-temperature out-of-time-ordered correlator (OTOC) between two local observables A and B defined as

$$O^{AB}(t) = \frac{1}{2}\langle [A(t), B][A(t), B]^{\dagger} \rangle \tag{1}$$

where $A(t) = U^{\dagger} A U$ is the time evolved operator A due to the unitary evolution $U = \mathcal{T} e^{-\mathrm{i} \int_0^t H(\tau) d\tau}$ from the time-ordered integration of the (generically) time-dependent Hamiltonian $H(t)$. Expanding the commutators, we can rewrite Equation (1) as

$$\begin{aligned} \frac{1}{2}\langle [A(t), B][A(t), B]^{\dagger} \rangle &= \langle B^2 A(t)^2 \rangle - \langle A(t) B A(t) B \rangle \\ &= G^{AB}(t) - F^{AB}(t), \end{aligned} \tag{2}$$

where $G^{AB}(t) = \langle B^2 A(t)^2 \rangle$ is the time-ordered part of OTOC and $F^{AB}(t) = \langle A(t) B A(t) B \rangle$ is the not-time-ordered part. We consider only unitary and Hermitian observables for which $G(t) = 1$, and hence we restrict ourselves to $F(t)$ in the remaining part. Taking energy eigenstates as the basis of the Hilbert space, the time evolution of OTOC can be written in the eigenenergy basis $|p\rangle$ as

$$F^{AB}(t) = \frac{1}{\mathcal{V}} \sum_{p,q,k,l} e^{\mathrm{i}(E_p - E_q + E_k - E_l)t} A_{pq} B_{qk} A_{kl} B_{lp} \tag{3}$$

where E_p is the eigenenergy, $A_{pq} = \langle p|A|q \rangle$, and $B_{qk} = \langle q|B|k \rangle$. We work in units for which $\hbar = 1$.

As $t \to \infty$, dominant terms in the above expression are those for which $E_p - E_q + E_k - E_l = 0$. Hence, for generic systems [44,45], the infinite-time value of $F^{AB}(t)$ is given by

$$\begin{aligned} F^{AB}(\infty) = \frac{1}{\mathcal{V}} \Bigg(\sum_p A_{pp}^2 B_{pp}^2 + \sum_{p,q \neq p} (A_{pp} B_{pq} A_{qq} B_{qp} \\ + A_{pq} B_{qq} A_{qp} B_{pp}) \Bigg). \end{aligned} \tag{4}$$

Equation (4) highlights the importance of diagonal elements of A and B in the eigenenergy basis in the infinite-time value of OTOC. Indeed, a non-zero diagonal element in A or B is necessary to guarantee a non-zero value of $F^{AB}(\infty)$.

2.2. Conditions for Algebraic Relaxation of OTOC

Two sufficient conditions for the emergence of algebraic relaxation of OTOC [39,40] are

- A Lieb–Robinson bound (or even an algebraic spreading of correlation that occurs in systems with power-law interactions),
- The algebraic scaling of the infinite-time value of the OTOC with the system size.

In local and bounded Hamiltonians, the speed of propagation of the correlations is limited by Lieb–Robinson bound [41,46]. Hence, an accurate description of the evolution of OTOC of a thermodynamically large system can be obtained simply considering a finite portion of it. Assuming that the system is maximally scrambled within the region of size L, the decay of $F^{AB}_{L=\infty}(t)$ is bounded by the Lieb–Robinson velocity $v_{\rm LR}$ as

$$F^{AB}_{L=\infty}(t) \approx F^{AB}_{L=s\, v_{\rm LR}\, t}(\infty), \tag{5}$$

where s is a real number larger than 1. Hence, L increases with time and is a time-dependent quantity. Therefore, the scaling of $F^{AB}_L(\infty)$ is crucial to predict the bound for the relaxation of OTOC. In particular, when $F^{AB}_L(\infty)$ decays algebraically with the system size, e.g., $F^{AB}_L(\infty) \propto L^{-\alpha}$, then the OTOC of the thermodynamic size system cannot decay faster than algebraically in time, or more precisely from Equation (5) one can write that it cannot be faster than

$$F^{AB}_{L=\infty}(t) \propto \frac{1}{t^{\alpha}} \tag{6}$$

because $L = s\, v_{\rm LR}\, t$.

The actual decay of the OTOC may even be slower, for example, considering cases in which the system goes through prethermalization [47] or in which the system is many-body localized [48]. However, the relaxation cannot be faster; hence, the OTOC will have a slow, non-exponential relaxation. A comprehensive analysis of this is presented in [39].

3. Generic Algebraic Relaxation in Short-Ranged Systems

3.1. Estimate of the Infinite Time Value of OTOC

In this section, we show how to obtain the approximate value of the infinite-time, finite-size, OTOC $F^{AB}_L(\infty)$

$$\begin{aligned}
F^{AB}_L(\infty) =& \frac{1}{\mathcal{V}}\sum_p A^2_{pp}B^2_{pp} + \frac{1}{\mathcal{V}}\sum_{p,q\neq p} A_{pp}A_{qq}|B_{pq}|^2 \\
&+ \frac{1}{\mathcal{V}}\sum_{p,q\neq p} B_{pp}B_{qq}|A_{pq}|^2 \\
\approx& \frac{1}{\mathcal{V}}\sum_p A^2_{pp}B^2_{pp} + \frac{1}{\mathcal{V}}\sum_p A_{pp}A_{pp}\left[(BB^\dagger)_{pp} - B^2_{pp}\right] \\
&+ \frac{1}{\mathcal{V}}\sum_p B_{pp}B_{pp}\left[(AA^\dagger)_{pp} - A^2_{pp}\right] \\
\approx& \frac{1}{\mathcal{V}}\sum_p A^2_{pp}B^2_{pp} + \frac{1}{\mathcal{V}}\sum_p \left[\mathrm{tr}(BB^\dagger) - B^2_{pp}\right]A^2_{pp} \\
&+ \frac{1}{\mathcal{V}}\sum_p \left[\mathrm{tr}(AA^\dagger) - A^2_{pp}\right]B^2_{pp} \\
\approx& \frac{1}{\mathcal{V}}\sum_p A^2_{pp}B^2_{pp} + \frac{1}{\mathcal{V}}\sum_p \left[1 - B^2_{pp}\right]A^2_{pp} \\
&+ \frac{1}{\mathcal{V}}\sum_p \left[1 - A^2_{pp}\right]B^2_{pp} \\
\approx& \frac{1}{\mathcal{V}}\sum_p \left[A^2_{pp} + B^2_{pp} - A^2_{pp}B^2_{pp}\right] \\
\approx& \frac{1}{\mathcal{V}}\sum_p \left[A^2_{pp} + B^2_{pp}\right],
\end{aligned} \tag{7}$$

where we have used steps similar to [39,45], and a similar discussion can be found in [40]. Thus, the main contribution of the infinite-time finite-size OTOC comes from the A^2_{pp} and B^2_{pp} terms, which we will be discussing in the following.

3.2. Structure of the Diagonal Elements

In short, the diagonal element A_{pp} can be approximated by a function of eigenenergy E_p

$$|A_{pp} - f_A(E_p/L)| \leq e^{-(\Omega(L))}, \tag{8}$$

where $f_A(E_p/L)$ can be expanded as

$$\begin{aligned} f_A(E_p/L) &= f_A(0) + f_A^{(1)}(0)E_p/L + \frac{1}{2}f_A^{(2)}(0)E_p^2/L^2 + ... \\ &= \sum_q \frac{f_A^{(q)}}{q!}\left(\frac{E_p}{L}\right)^q \end{aligned} \tag{9}$$

with $f_A^{(q)}$ being the q-th derivative of f_A. We also note that, using Lemma 1 in [45], one can write

$$\frac{1}{\mathcal{V}}\sum_p E_p^q = \langle H^q \rangle = O(L^{q/2}). \tag{10}$$

In [45] it was shown that, for traceless operators $f_A(0) = 0$, and if $f_A^{(1)}(0) \neq 0$, then we can write

$$\begin{aligned} \mathrm{tr}(AH) &= \frac{1}{\mathcal{V}}\sum_p A_{pp}E_p \\ &\approx \frac{1}{\mathcal{V}L}\sum_p E_p^2 f_A^{(1)}(0) \\ &\approx \frac{\langle H^2 \rangle}{L} f_A^{(1)}(0) \end{aligned} \tag{11}$$

and thus

$$f_A^{(1)}(0) \approx \frac{\mathrm{tr}(AH)L}{\langle H^2 \rangle}. \tag{12}$$

Hence, the first derivative of a local observable A is independent of the system size. From Equation (12), we obtain

$$\begin{aligned} F_L^{AB}(\infty) &\approx \frac{1}{\mathcal{V}}\sum_p (A_{pp}^2 + B_{pp}^2) \\ &\approx \frac{1}{\mathcal{V}}\sum_p \frac{E_p^2}{L^2}\left[\left(f_A^{(1)}(0)\right)^2 + \left(f_B^{(1)}(0)\right)^2\right] \\ &\approx \frac{1}{\mathcal{V}}\sum_p \frac{E_p^2}{L^2} \frac{\left[\mathrm{tr}(AH)^2 + \mathrm{tr}(BH)^2\right]L^2}{\langle H^2 \rangle^2} \\ &\approx \frac{\mathrm{tr}(AH)^2 + \mathrm{tr}(BH)^2}{\langle H^2 \rangle} \\ &\propto \frac{1}{L}. \end{aligned} \tag{13}$$

The last step stems from the fact that $\mathrm{tr}(AH)$ and $\mathrm{tr}(BH)$ are independent of the system size, while $\langle H^2 \rangle \propto L$ from Equation (10).

If $\mathrm{tr}(AH) = 0$ but, for instance, $\mathrm{tr}(AH^p) \neq 0$ only for $p \geq p_c$ then one can generalize the previous result. Considering $f_A^{(q)}(0)$ as the smalles non-zero derivative of f_A at zero energy (with the same parity as p_c), then we can write

$$\mathrm{tr}(AH^{p_c}) = \sum_n \frac{f_A^{(q)}}{q!} \frac{E_n^{q+p_c}}{L^q} \tag{14}$$

which implies that

$$f_A^{(q)} = \frac{q!\mathrm{tr}(AH^{p_c})L^q}{\langle H^{p_c+q} \rangle}. \tag{15}$$

Now, if $q < p_c$ then $f_A^{(q)}$ would decay as $L^{-(p_c-q)/2}$, which implies that they are 0, and the non-size dependent $f_A^{(q)}(0)$ would occur exactly at $q = p_c$. This implies that the first non-zero derivative of $f_A(0)$ is the p_c-th one. Thus, when $\mathrm{tr}(AH^p) \neq 0$ only for $p \geq p_c$ we can write

$$\begin{aligned} F_L^{AB}(\infty) &\approx \frac{1}{\mathcal{V}} \sum_n \left(\frac{E_n}{L}\right)^{2p_c} \left[\left(f_A^{(p_c)}(0)\right)^2 + \left(f_B^{(p_c)}(0)\right)^2 \right] \\ &\approx \frac{1}{\mathcal{V}} \sum_n \left(\frac{E_n}{L}\right)^{2p_c} \frac{\left[\mathrm{tr}(AH^{p_c})^2 + \mathrm{tr}(BH^{p_c})^2\right] L^{2p_c}}{\langle H^{2p_c}\rangle^2} \\ &\approx \frac{\mathrm{tr}(AH^{p_c})^2 + \mathrm{tr}(BH^{p_c})^2}{\langle H^{2p_c}\rangle} \\ &\propto \frac{1}{L^{p_c}}. \end{aligned} \tag{16}$$

Building on Equation (16), and combining it with the Lieb–Robinson bound $L = s\, v_{\mathrm{LR}}\, t$, we can thus guarantee that F^{AB} cannot relax faster than t^{-p_c}. Furthermore, for systems in which correlations mostly spread diffusively, i.e., proportional to $t^{1/2}$, we can can expect F^{AB} to relax as $t^{-p_c/2}$. Hence, the structure of the diagonal elements of the observables, which is the first non-zero derivative at 0 energy, i.e., which is the first exponent of the Hamiltonian that has non-zero overlap with the operators A and B considered, plays an important role in the relaxation dynamics of the OTOC in the system. This is numerically verified in detail in the following section.

4. Results

4.1. Model

We consider a prototypical non-integrable model, the tilted Ising chain with Hamiltonian

$$H = \sum_{l=1}^{L-1} J_z \sigma_l^z \sigma_{l+1}^z + \sum_{l=1}^{L} (h_x \sigma_l^x + h_z \sigma_l^z), \tag{17}$$

where J_z is the coupling constant in the z direction, while h_x and h_z are the transverse and the longitudinal field strengths. The model is integrable when either $h_x = 0$ or $h_z = 0$. This can be verified by studying the level spacing statistics, which typically follows a Poisson distribution for integrable systems and a Wigner–Dyson distribution for non-integrable ones [49,50]. In particular, $\delta_n = E_{n+1} - E_n$, the level spacing between two consecutive energy levels E_n and E_{n+1} within a single symmetry sector, define the ratio $r_n = \max(\delta_n, \delta_{n+1})/\min(\delta_n, \delta_{n+1})$ and take an average $r = \sum_n r_n/N$, where N is the number of energy level differences considered. For a Poisson distribution, r can be computed analytically, and it gives $r = 2\ln 2 - 1 \approx 0.386$, while for a Wigner–Dyson distribution r can be evaluated numerically to be $r \approx 0.529$ [51]. In the current work, we use parameters $J_z = 1$, $h_z = 0.809$, and $h_x = 0.9$, which result in $r \approx 0.53$ already for a system size of $L = 12$ spins.

4.2. Observables and Structure of Their Diagonal Elements

To span over a variety of different structures, and to have operators A, which have $\mathrm{tr}(AH^p) \neq 0$ only for $p \leq p_c$ with p_c, which can be different from 1, we analyze both single-site and multi-site observables in our study. In particular, we consider the following four types of observables:

$$\text{single}-\text{site} \rightarrow \sigma_l^\alpha \tag{18}$$

$$\text{double}-\text{site} \rightarrow \sigma_l^\alpha \sigma_{l+1}^\alpha \tag{19}$$

$$\text{triple}-\text{site} \rightarrow \sigma_{l-1}^\alpha \sigma_l^\alpha \sigma_{l+1}^\alpha \tag{20}$$

$$\text{quadruple}-\text{site} \rightarrow \sigma_{l-2}^\alpha \sigma_{l-1}^\alpha \sigma_l^\alpha \sigma_{l+1}^\alpha \tag{21}$$

where $\alpha = x, y$ or z. The diagonal elements of these operators in the eigenbasis of Hamiltonian Equation (17) are shown in Figure 1. The left column is for $\alpha = x$, the center column is for $\alpha = y$, and the right column is for $\alpha = z$. The rows are for increasing the range of operators from top to bottom, with the top row for single-site operators and the fourth row for four-site operators. In all of the panels, the dashed lines represent the expected algebraic energy dependence of f_A near energy zero from Section 3.2. We note that these fits are evaluated directly from calculating $f_A^{(n)}(0)$ with Equation (15) along with eigenenergies E_n for the system Hamiltonian in Equation (17) with no fitting parameters.

For single-site observables $A = \sigma_l^x$ and σ_l^z, and for the non-integrable Ising chain $\mathrm{tr}(AH) \neq 0$ and so $f_A^{(1)}(0) \neq 0$. However, with $A = \sigma_l^y$, $\mathrm{tr}(AH^n) = 0$ for any n. Hence, we expect a linear variation of the diagonal elements of σ_l^x and σ_l^z with energy density E_n/L and a flat profile for σ_l^y. This can be seen in Figure 1a–c. To conform our analytical predictions, we plot $f_A^{(1)}(0)E_n/L$ where $f_A^{(1)}(0)$ is calculated explicity form Equation (12).

The two-site observables $A = \sigma_{L/2}^\alpha \sigma_{L/2+1}^\alpha$, are shown in the panels (d–f). For $A = \sigma_{L/2}^x \sigma_{L/2+1}^x$, $\mathrm{tr}(AH^2) \neq 0$, whereas $\mathrm{tr}(AH) = 0$ and, as predicted in Section 3.2, we thus observe that f_A can be fitted by a parabola $f_A^{(2)}(0)E_n^2/(2!L^2)$ indicated by the dashed black lines. Since $\mathrm{tr}(AH) \neq 0$ for $A = \sigma_{L/2}^z \sigma_{L/2+1}^z$, we see a linear scaling of A_{nn} with E_n/L. For the $A = \sigma_{L/2}^y \sigma_{L/2+1}^y$ observable, $\mathrm{tr}(AH^m) \neq 0$ for $m \geq 3$. Hence, we see a cubic structure of the diagonal elements with a fitting of the form $f_A^{(3)}(0)E_n^3/(3!L^3)$.

We also consider triple-site observables $A = \sigma_{L/2-1}^\alpha \sigma_{L/2}^\alpha \sigma_{L/2+1}^\alpha$. These are depicted in the panels (g–i). Here, $\mathrm{tr}(AH^m) \neq 0$ for $m \geq 3$ for σ_l^x observables, and we clearly see a cubic structure for the diagonal elements that can be fitted with lines of the form $f_A^{(3)}(0)E_n^3/(3!L^3)$. Since there are no diagonal elements for any power of H for the σ^y observable, a flat profile is seen. With the σ^z observable, a parabolic structure is seen since $\mathrm{tr}(AH^2) \neq 0$, whereas $\mathrm{tr}(AH) = 0$. This is also nicely fitted by $f_A^{(?)}(0)E_n^2/(2!L^2)$ in panel (i).

For the four site observable, we study $A = \sigma_{L/2-2}^\alpha \sigma_{L/2-1}^\alpha \sigma_{L/2}^\alpha \sigma_{L/2+1}^\alpha$. For $A = \sigma_{L/2-2}^x \sigma_{L/2-1}^x \sigma_{L/2}^x \sigma_{L/2+1}^x$, a quartic structure can be seen as $\mathrm{tr}(AH^m) \neq 0$ only for $m \geq 4$. This is fitted by $f_A^{(4)}(0)E_n^4/(4!L^4)$ (black dashed lines). With σ_l^y observables, the expected structure is hexic (polynomial of sixth degree) because $\mathrm{tr}(AH^m) \neq 0$ only for $m \geq 6$. Though it is less clear, we fit it with the expected scaling $f_A^{(6)}(0)E_n^6/(6!L^6)$ using the black dashed lines. For the σ^z observable, we find a parabolic structure in accordance with our prediction as $\mathrm{tr}(AH^2) \neq 0$, whereas $\mathrm{tr}(AH) = 0$. Since we consider systems of size $L = 14$, the results in Figure 1j–l are partially affected by finite-size effects. Despite this, the numerics are aligned with our theoretical predictions.

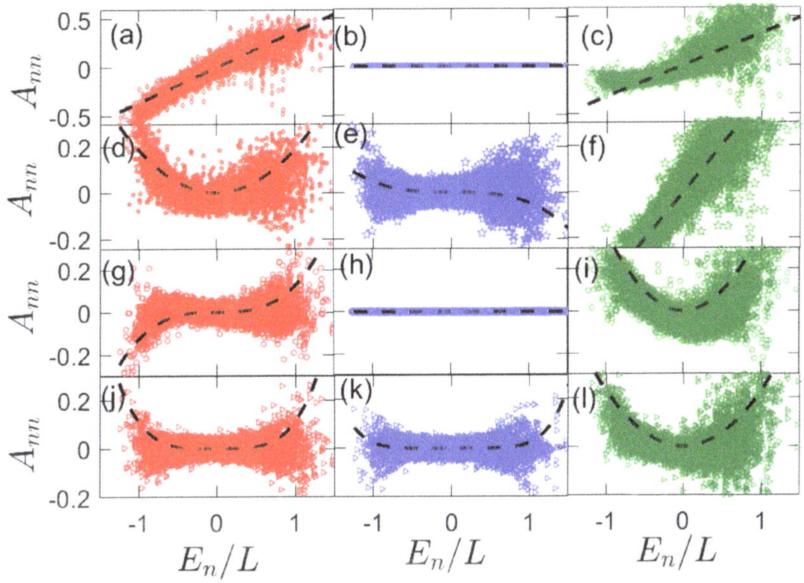

Figure 1. Diagonal elements of the observable in the energyeigen basis for single-site observables $A = \sigma^{\alpha}_{L/2}$ panel (**a–c**), double-site observables $A = \sigma^{\alpha}_{L/2}\sigma^{\alpha}_{L/2+1}$ (**d–f**), triple-site observables $A = \sigma^{\alpha}_{L/2-1}\sigma^{\alpha}_{L/2}\sigma^{\alpha}_{L/2+1}$ (**g–i**), and quadruple-site observables $A = \sigma^{\alpha}_{L/2-2}\sigma^{\alpha}_{L/2-1}\sigma^{\alpha}_{L/2}\sigma^{\alpha}_{L/2+1}$ (**j–l**). Left panels are for σ^x_l ($\alpha = x$) observables, middle panels are for σ^y_l ($\alpha = y$) observables, and right panels are for σ^z_l ($\alpha = z$) observables. Dashed lines are the lowest order fits in the Taylor expansion of the observable in Equation (9). Here, $L = 14$, $J_z = 1$, $h_x = 0.9$, and $h_z = 0.809$.

To summarize this section, we observe clearly that the diagonal elements of operators can have a very different dependence as a function of energy near zero. In particular, we have numerically verified the prediction that $A_{nn} \sim 1/L^p$, where p is the lowest positive integer such that $\mathrm{tr}(AH^p) \neq 0$.

4.3. Scaling of the Infinite Time Value of OTOC

In Figure 2, we show numerical confirmation that, given the minimum positive integer p_c such that $\mathrm{tr}(AH^{p_c}) \neq 0$ or $\mathrm{tr}(BH^{p_c}) \neq 0$, then $F^{AB}_{L=\infty}(t) \propto 1/L^{p_c}$. In each of the panels, we show how the infinite time value of the OTOC $F^{AB}_L(t = \infty)$ varies as a function of the system size L. In the different panels, we will focus on single-site, panel (a); two-site, panel (b), three-site, panel (c); and four-site, panel (d), observables. In each panel, the red line with circles corresponds to $\alpha = x$, blue with stars to $\alpha = y$, and green with diamonds to $\alpha = z$. In panel Figure 2a, we plot the infinite time values of OTOC with single-site observables $A, B = \sigma^{\alpha}_l$, where $l = L/2$ for observable B and $l = L/2 - 1$ for observable A. We see that these observables have $p_c = 1$, and hence they follow $1/L$ scaling, as shown by dashed line. σ^y_l has no overlap with any local conserved quantities, and hence the diagonal elements as well as the infinite time values of OTOC are zero. Figure 2b is for double-site observables Equation (19), where $l = L/2$ for observable B and $l = L/2 - 2$ for observable A. We compare the numerical results with fitted lines, in particular with $1/L^2$ (dotted), $1/L^3$ (dashed dotted lines), and $1/L$ (dashed), respectively, corresponding to operators with $p_c = 2, 3$ and 1. We note that due to the small value of the overlap of $\sigma^y_j\sigma^y_{j+1}$ with the Hamiltonian, the expected scaling is followed only at larger system sizes. In panel (c), we plot the triple-site observables Equation (20), where $j = L/2 - 3$ for A and $j = L/2$ for B. Fitted lines are for $1/L^2$ and $1/L^3$ scalings, as expected, since $\alpha = 2$ and 3, respectively. Since the diagonal elements of $\sigma^y_j\sigma^y_{j+1}\sigma^y_{j+2}$ are zero, the infinite time value of the OTOC $F_L^{AB}(t=\infty)$ is zero. Panel (d) is for quadruple-site observables Equation (21), where $j = L/2-4$ for A and $j = L/2$ for B. The expected scalings are $p_c = 4, 6, 2$. However, due to the fact that the observables have a large support at initial time, we see that the correct scaling of $1/L^4, 1/L^6, 1/L^2$ is followed only at large system sizes.

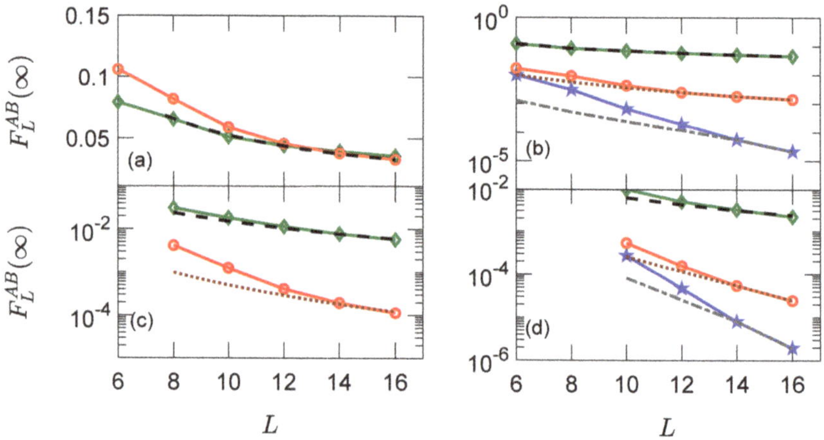

Figure 2. Infinite time values of OTOC corresponding to the single-site observables with $A = \sigma^\alpha_{L/2-1}$, $B = \sigma^\alpha_{L/2}$ panel (**a**), double-site observables with $A = \sigma^\alpha_{L/2-2}\sigma^\alpha_{L/2-1}$, $B = \sigma^\alpha_{L/2}\sigma^\alpha_{L/2+1}$ (**b**), triple-site observables with $A = \sigma^\alpha_{L/2-3}\sigma^\alpha_{L/2-2}\sigma^\alpha_{L/2-1}$, $B = \sigma^\alpha_{L/2}\sigma^\alpha_{L/2+1}\sigma^\alpha_{L/2+2}$ (**c**), and quadruple-site observables with $A = \sigma^\alpha_{L/2-4}\sigma^\alpha_{L/2-3}\sigma^\alpha_{L/2-2}\sigma^\alpha_{L/2-1}$, $B = \sigma^\alpha_{L/2}\sigma^\alpha_{L/2+1}\sigma^\alpha_{L/2+2}\sigma^\alpha_{L/2+3}$ (**d**). Green lines with diamonds are for observables involving only σ^z_l ($\alpha = z$) operators, red lines with circles are for σ^x_l ($\alpha = x$), and blue lines with stars are for σ^y_l ($\alpha = y$) operators, respectively. Black-dashed, brown-dotted, and grey-dashed dotted lines are the fits for σ^z_l, σ^x_l, and σ^y_l observables. Here, $J_z = 1$, $h_x = 0.9$, and $h_z = 0.809$.

4.4. Dynamics of OTOCs

We study the dynamics of OTOC in Figure 3, where each panel reflects the same case analyzed in the corresponding panel of Figure 2. Green lines are for observables involving only σ^z_l operators, red lines are for σ^x_l, and blue lines are for σ^y_l operators, respectively. In these plots, we need to study the long-time evolution. We thus need to disregard initial transients. At the same time, though, our results are affected by finite size, so we would need to concentrate on long yet intermediate times to evaluate the relaxation of the OTOC over time. Light shades are for $L = 14$, and dark shades for $L = 12$. Black-dashed, brown-dotted, and grey-dashed dotted lines are the fits for σ^z_l, σ^x_l, and σ^y_l observables. Figure 3a is for single-site observables, as in Equation (18). We have already seen that since $\mathrm{tr}(OH) \neq 0$, for $(O = A, B)$, then the infinite time value of OTOC $F_L^{AB}(t=\infty)$ scales as $1/L$. From our discussion at the end of Section 3.2, we thus expect that $F^{AB}(t) \propto 1/t^{1/2}$, and the numerical result of the dynamics, is well fitted by the black dashed line proportional to $t^{1/2}$.

In Figure 3b, we study the two-site observables of Equation (19) with $l = L/2$ for observable B and $l = L/2-2$ for observable A. As already discussed, the lowest order terms that have non zero values in the Taylor expansion for these observables are 1, 2 and 3, respectively, for the $\sigma^z_l, \sigma^x_l, \sigma^y_l$ observables. In Figure 2b, we showed the scaling of the infinite-time OTOC for these observables as $1/L$, $1/L^2$, $1/L^3$. Here, we would thus expect a scaling with times of $1/t^{1/2}$, $1/t$, and $1/t^{3/2}$, as shown in the plots by dotted, dashed dotted, and dashed lines, respectively. We study the evolution of three-site observables of Equation (20) in Figure 3c. Here, $l = L/2-3$ for A and $l = L/2$ for B. Fitted lines are for $1/t$ and $1/t^{3/2}$ scaling as expected since $\mathrm{tr}(OH) \neq 0$ for σ^z_l and $\mathrm{tr}(OH^2) \neq 0$ for the σ^x_l observable. Panel (d) shows the dynamics for four site observables with $l = L/2-4$ for A and $l = L/2$ for B. The expected scaling is $1/t$, $1/t^2$, and $1/t^3$ for $\alpha = z, x$ and y, respectively, whose operators for the corresponding critical exponent p_c that gives non-zero overlap are 2, 4, and 6 .

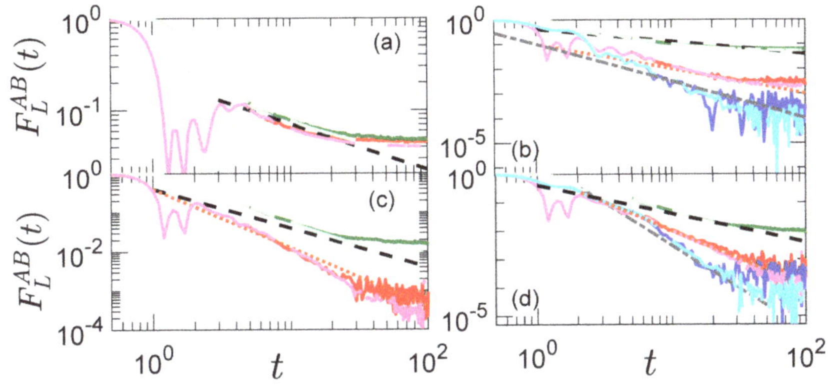

Figure 3. Time evolution of OTOC corresponding to the single-site observables with $A = \sigma^{\alpha}_{L/2-1}$, $B = \sigma^{\alpha}_{L/2}$ panel (**a**), double-site observables with $A = \sigma^{\alpha}_{L/2-2}\sigma^{\alpha}_{L/2-1}$, $B = \sigma^{\alpha}_{L/2}\sigma^{\alpha}_{L/2+1}$ (**b**), triple-site observables $A = \sigma^{\alpha}_{L/2-3}\sigma^{\alpha}_{L/2-2}\sigma^{\alpha}_{L/2-1}$, $B = \sigma^{\alpha}_{L/2}\sigma^{\alpha}_{L/2+1}\sigma^{\alpha}_{L/2+2}$ (**c**), and quadruple-site observables $A = \sigma^{\alpha}_{L/2-4}\sigma^{\alpha}_{L/2-3}\sigma^{\alpha}_{L/2-2}\sigma^{\alpha}_{L/2-1}$, $B = \sigma^{\alpha}_{L/2}\sigma^{\alpha}_{L/2+1}\sigma^{\alpha}_{L/2+2}\sigma^{\alpha}_{L/2+3}$ (**d**). Green lines are for observables involving only σ^z_l $(\alpha = z)$ operators, red lines are for σ^x_l $(\alpha = x)$, and blue lines are for σ^y_l $(\alpha = y)$ operators respectively. Black-dashed, brown-dotted, and grey-dashed dotted lines are the fits for σ^z_l, σ^x_l, and σ^y_l observables discussed in the text. Here, $L = 14$ for lighter shades and $L = 12$ for darker shades and $J_z = 1$, $h_x = 0.9$, and $h_z = 0.809$.

5. Conclusions

OTOCs have been studied as a probe for quantum information scrambling. Slow, algebraic scrambling has been reported in systems with local conserved quantities [34,38–40].

In this paper, we showed that the higher the exponent at which ones elevates the Hamiltonian in order to have a non-zero overlap with the operators in the OTOC, the faster is the relaxation of the OTOC over time. Furthermore, if there is an exponent such that the overlap is non-zero, then the relaxation, even if it appears to be fast, is bounded to be, at the fastest, algebraic, and only if there is no overlap with any power of the Hamiltonian (or other conserved quantities), then the relaxation can be exponential.

From our results, it follows that considering single-site operators in the OTOC, and a local Hamiltonian with only a single site and nearest neighbours term, relaxation can only take a limited set of exponents. It is thus necessary to consider operators with larger support, such as two-site, three-site, and four-site operators, to observe a larger variety and magnitude of relaxation exponents. This, however, leads to the difficulty of studying the relaxation numerically due to more pronounced finite-size effects when studying operators with larger support. Future developments in numerical methods could help to test our results for larger systems.

In order to derive these results, we also found a relation between the first non-zero derivative of the function representing the diagonals of an operator in the energy basis and the first non-zero exponent of the Hamiltonian (which has non-zero overlap with the operators of the OTOC). Future works could extend these results to time-dependent systems with other types of conserved quantities.

Author Contributions: V.B. and D.P. have contributed equally to Conceptualization, Methodology, Formal Analysis, and Writing (original draft and review/editing). All authors have read and agreed to the published version of the manuscript.

Funding: D.P. acknowledges support from Ministry of Education Singapore, grant T2EP50120-0041.

Institutional Review Board Statement: Not applicable.

Informed Consent Statement: Not applicable.

Data Availability Statement: The data that support the findings of this study are available from the corresponding author upon reasonable request.

Acknowledgments: We acknowledge C. von Keyserlingk who asked a critical and insightful question which started this project. The computational work for this article were partially performed on the National Supercomputing Centre, Singapore [52]. D.P. and V.B. are extremely grateful to G. Casati, to whom this volume is dedicated, for his guidance and example throughout the many years we had the fortune to learn from, interact and collaborate with.

Conflicts of Interest: The authors declare no conflict of interest.

References

1. Witten, E. Anti-de Sitter space and holography. *Adv. Theor. Math. Phys.* **1998**, *2*, 253–291. [CrossRef]
2. Maldacena, J. The Large-N Limit of Superconformal Field Theories and Supergravity. *Int. J. Theor. Phys.* **1999**, *38*, 1113–1133. [CrossRef]
3. Hayden, P.; Preskill, J. Black holes as mirrors: Quantum information in random subsystems. *J. High Energy Phys.* **2007**, *2007*, 120. [CrossRef]
4. Sekino, Y.; Susskind, L. Fast scramblers. *J. High Energy Phys.* **2008**, *2008*, 65. [CrossRef]
5. Shenker, S.H.; Stanford, D. Black holes and the butterfly effect. *J. High Energy Phys.* **2014**, *2014*, 67. [CrossRef]
6. Sachdev, S.; Ye, J. Gapless spin-fluid ground state in a random quantum Heisenberg magnet. *Phys. Rev. Lett.* **1993**, *70*, 3339. [CrossRef]
7. Kitaev, A. A Simple Model of Quantum Holography. *Talks at KITP*. 2015. Available online: https://online.kitp.ucsb.edu/online/entangled15/kitaev/ (accessed on 23 November 2022).
8. Lashkari, N.; Stanford, D.; Hastings, M.; Osborne, T.; Hayden, P. Towards the fast scrambling conjecture. *J. High Energy Phys.* **2013**, *2013*, 22. [CrossRef]
9. Roberts, D.A.; Stanford, D. Diagnosing Chaos Using Four-Point Functions in Two-Dimensional Conformal Field Theory. *Phys. Rev. Lett.* **2015**, *115*, 131603. [CrossRef]
10. Cotler, J.S.; Gur-Ari, G.; Hanada, M.; Polchinski, J.; Saad, P.; Shenker, S.H.; Stanford, D.; Streicher, A.; Tezuka, M. Black holes and random matrices. *J. High Energy Phys.* **2017**, *2017*, 118. [CrossRef]
11. Roberts, D.A.; Stanford, D.; Susskind, L. Localized shocks. *J. High Energy Phys.* **2015**, *2015*, 51. [CrossRef]
12. Hosur, P.; Qi, X.L.; Roberts, D.A.; Yoshida, B. Chaos in quantum channels. *J. High Energy Phys.* **2016**, *2016*, 4. [CrossRef]
13. Borgonovi, F.; Izrailev, F.M.; Santos, L.F. Timescales in the quench dynamics of many-body quantum systems: Participation ratio versus out-of-time ordered correlator. *Phys. Rev. E* **2019**, *99*, 052143. [CrossRef] [PubMed]
14. Li, J.; Fan, R.; Wang, H.; Ye, B.; Zeng, B.; Zhai, H.; Peng, X.; Du, J. Measuring Out-of-Time-Order Correlators on a Nuclear Magnetic Resonance Quantum Simulator. *Phys. Rev. X* **2017**, *7*, 031011. [CrossRef]
15. Gärttner, M.; Bohnet, J.G.; Safavi-Naini, A.; Wall, M.L.; Bollinger, J.J.; Rey, A.M. Measuring out-of-time-order correlations and multiple quantum spectra in a trapped-ion quantum magnet. *Nat. Phys.* **2017**, *13*, 781–786. [CrossRef]
16. Landsman, K.A.; Figgatt, C.; Schuster, T.; Linke, N.M.; Yoshida, B.; Yao, N.Y.; Monroe, C. Verified quantum information scrambling. *Nature* **2019**, *567*, 61–65. [CrossRef]
17. Niknam, M.; Santos, L.F.; Cory, D.G. Sensitivity of quantum information to environment perturbations measured with a nonlocal out-of-time-order correlation function. *Phys. Rev. Res.* **2020**, *2*, 013200. [CrossRef]
18. Joshi, M.K.; Elben, A.; Vermersch, B.; Brydges, T.; Maier, C.; Zoller, P.; Blatt, R.; Roos, C.F. Quantum Information Scrambling in a Trapped-Ion Quantum Simulator with Tunable Range Interactions. *Phys. Rev. Lett.* **2020**, *124*, 240505. [CrossRef]
19. Blok, M.S.; Ramasesh, V.V.; Schuster, T.; O'Brien, K.; Kreikebaum, J.M.; Dahlen, D.; Morvan, A.; Yoshida, B.; Yao, N.Y.; Siddiqi, I. Quantum Information Scrambling on a Superconducting Qutrit Processor. *Phys. Rev. X* **2021**, *11*, 021010. [CrossRef]
20. Mi, X.; Roushan, P.; Quintana, C.; Mandra, S.; Marshall, J.; Neill, C.; Arute, F.; Arya, K.; Atalaya, J.; Babbush, R.; et al. Information Scrambling in Computationally Complex Quantum Circuits. *Science* **2021**, *374*, 1479–1483. [CrossRef]
21. Braumüller, J.; Karamlou, A.H.; Yanay, Y.; Kannan, B.; Kim, D.; Kjaergaard, M.; Melville, A.; Niedzielski, B.M.; Sung, Y.; Vepsäläinen, A.; et al. Probing quantum information propagation with out-of-time-ordered correlators. *Nat. Phys.* **2021**, *18*, 172–178. [CrossRef]
22. Rozenbaum, E.B.; Ganeshan, S.; Galitski, V. Lyapunov Exponent and Out-of-Time-Ordered Correlator's Growth Rate in a Chaotic System. *Phys. Rev. Lett.* **2017**, *118*, 086801. [CrossRef] [PubMed]
23. Hashimoto, K.; Murata, K.; Yoshii, R. Out-of-time-order correlators in quantum mechanics. *J. High Energy Phys.* **2017**, *2017*, 138. [CrossRef]
24. Cotler, J.S.; Ding, D.; Penington, G.R. Out-of-time-order operators and the butterfly effect. *Ann. Phys.* **2018**, *396*, 318–333. [CrossRef]

25. García-Mata, I.; Saraceno, M.; Jalabert, R.A.; Roncaglia, A.J.; Wisniacki, D.A. Chaos Signatures in the Short and Long Time Behavior of the Out-of-Time Ordered Correlator. *Phys. Rev. Lett.* **2018**, *121*, 210601. [CrossRef] [PubMed]
26. Chávez-Carlos, J.; López-del Carpio, B.; Bastarrachea-Magnani, M.A.; Stránský, P.; Lerma-Hernández, S.; Santos, L.F.; Hirsch, J.G. Quantum and Classical Lyapunov Exponents in Atom-Field Interaction Systems. *Phys. Rev. Lett.* **2019**, *122*, 024101. [CrossRef] [PubMed]
27. Fortes, E.M.; García-Mata, I.; Jalabert, R.A.; Wisniacki, D.A. Gauging classical and quantum integrability through out-of-time-ordered correlators. *Phys. Rev. E* **2019**, *100*, 042201. [CrossRef]
28. Rammensee, J.; Urbina, J.D.; Richter, K. Many-Body Quantum Interference and the Saturation of Out-of-Time-Order Correlators. *Phys. Rev. Lett.* **2018**, *121*, 124101. [CrossRef]
29. Prakash, R.; Lakshminarayan, A. Scrambling in strongly chaotic weakly coupled bipartite systems: Universality beyond the Ehrenfest timescale. *Phys. Rev. B* **2020**, *101*, 121108. [CrossRef]
30. Bergamasco, P.D.; Carlo, G.G.; Rivas, A.M.F. Out-of-time ordered correlators, complexity, and entropy in bipartite systems. *Phys. Rev. Res.* **2019**, *1*, 033044. [CrossRef]
31. Rozenbaum, E.B.; Bunimovich, L.A.; Galitski, V. Early-Time Exponential Instabilities in Nonchaotic Quantum Systems. *Phys. Rev. Lett.* **2020**, *125*, 014101. [CrossRef]
32. Wang, J.; Benenti, G.; Casati, G.; Wang, W.G. Complexity of quantum motion and quantum-classical correspondence: A phase-space approach. *Phys. Rev. Res.* **2020**, *2*, 043178. [CrossRef]
33. Wang, J.; Benenti, G.; Casati, G.; Wang, W.G. Quantum chaos and the correspondence principle. *Phys. Rev. E* **2021**, *103*, L030201. [CrossRef] [PubMed]
34. Rakovszky, T.; Pollmann, F.; von Keyserlingk, C.W. Diffusive Hydrodynamics of Out-of-Time-Ordered Correlators with Charge Conservation. *Phys. Rev. X* **2018**, *8*, 031058. [CrossRef]
35. Nahum, A.; Ruhman, J.; Vijay, S.; Haah, J. Quantum Entanglement Growth under Random Unitary Dynamics. *Phys. Rev. X* **2017**, *7*, 031016. [CrossRef]
36. Nahum, A.; Vijay, S.; Haah, J. Operator Spreading in Random Unitary Circuits. *Phys. Rev. X* **2018**, *8*, 021014. [CrossRef]
37. von Keyserlingk, C.W.; Rakovszky, T.; Pollmann, F.; Sondhi, S.L. Operator Hydrodynamics, OTOCs, and Entanglement Growth in Systems without Conservation Laws. *Phys. Rev. X* **2018**, *8*, 021013. [CrossRef]
38. Khemani, V.; Vishwanath, A.; Huse, D.A. Operator Spreading and the Emergence of Dissipative Hydrodynamics under Unitary Evolution with Conservation Laws. *Phys. Rev. X* **2018**, *8*, 031057. [CrossRef]
39. Balachandran, V.; Benenti, G.; Casati, G.; Poletti, D. From the eigenstate thermalization hypothesis to algebraic relaxation of OTOCs in systems with conserved quantities. *Phys. Rev. B* **2021**, *104*, 104306. [CrossRef]
40. Balachandran, V.; Santos, L.F.; Rigol, M.; Poletti, D. Effect of symmetries in out-of-time ordered correlators in interacting integrable and nonintegrable many-body quantum systems. *arXiv* **2022**, arXiv:2211.07073.
41. Lieb, E.H.; Robinson, D.W. The finite group velocity of quantum spin systems. *Commun. Math. Phys.* **1972**, *28*, 251–257. [CrossRef]
42. Srednicki, M. Thermal fluctuations in quantized chaotic systems. *J. Phys. Math. Gen.* **1996**, *29*, L75–L79. [CrossRef]
43. Deutsch, J.M. Quantum statistical mechanics in a closed system. *Phys. Rev. A* **1991**, *43*, 2046–2049. [CrossRef] [PubMed]
44. Srednicki, M. The approach to thermal equilibrium in quantized chaotic systems. *J. Phys. A Math. Gen.* **1999**, *32*, 1163–1175. [CrossRef]
45. Huang, Y.; Brandão, F.G.S.L.; Zhang, Y.L. Finite-Size Scaling of Out-of-Time-Ordered Correlators at Late Times. *Phys. Rev. Lett.* **2019**, *123*, 010601. [CrossRef] [PubMed]
46. Cheneau, M.; Barmettler, P.; Poletti, D.; Endres, M.; Schauß, P.; Fukuhara, T.; Gross, C.; Bloch, I.; Kollath, C.; Kuhr, S. Light-cone-like spreading of correlations in a quantum many-body system. *Nature* **2012**, *481*, 484 – 487. [CrossRef] [PubMed]
47. Luitz, D.J.; Moessner, R.; Sondhi, S.L.; Khemani, V. Prethermalization without Temperature. *Phys. Rev. X* **2020**, *10*, 021046. [CrossRef]
48. Lee, J.; Kim, D.; Kim, D.H. Typical growth behavior of the out-of-time-ordered commutator in many-body localized systems. *Phys. Rev. B* **2019**, *99*, 184202. [CrossRef]
49. Bohigas, O.; Giannoni, M.J.; Schmit, C. Characterization of Chaotic Quantum Spectra and Universality of Level Fluctuation Laws. *Phys. Rev. Lett.* **1984**, *52*, 1–4. [CrossRef]
50. Casati, G.; Valz-Gris, F.; Guarnieri, I. Connection between quantization of nonintegrable systems and statistical theory of spectra. [CrossRef]
51. Oganesyan, V.; Huse, D.A. Localization of interacting fermions at high temperature. *Phys. Rev. B* **2007**, *75*, 155111. [CrossRef]
52. Available online: https://www.nscc.sg/ (accessed on 30 September 2022).

entropy

MDPI

Review

Quantum Chaos in the Dynamics of Molecules

Kazuo Takatsuka

Fukui Institute for Fundamental Chemistry, Kyoto University, Kyoto 606-8103, Japan; kaztak@fukui.kyoto-u.ac.jp

Abstract: Quantum chaos is reviewed from the viewpoint of "what is molecule?", particularly placing emphasis on their dynamics. Molecules are composed of heavy nuclei and light electrons, and thereby the very basic molecular theory due to Born and Oppenheimer gives a view that quantum electronic states provide potential functions working on nuclei, which in turn are often treated classically or semiclassically. Therefore, the classic study of chaos in molecular science began with those nuclear dynamics particularly about the vibrational energy randomization within a molecule. Statistical laws in probabilities and rates of chemical reactions even for small molecules of several atoms are among the chemical phenomena requiring the notion of chaos. Particularly the dynamics behind unimolecular decomposition are referred to as Intra-molecular Vibrational energy Redistribution (IVR). Semiclassical mechanics is also one of the main research fields of quantum chaos. We herein demonstrate chaos that appears only in semiclassical and full quantum dynamics. A fundamental phenomenon possibly giving birth to quantum chaos is "bifurcation and merging" of quantum wavepackets, rather than "stretching and folding" of the baker's transformation and the horseshoe map as a geometrical foundation of classical chaos. Such wavepacket bifurcation and merging are indeed experimentally measurable as we showed before in the series of studies on real-time probing of nonadiabatic chemical reactions. After tracking these aspects of molecular chaos, we will explore quantum chaos found in nonadiabatic electron wavepacket dynamics, which emerges in the realm far beyond the Born-Oppenheimer paradigm. In this class of chaos, we propose a notion of Intra-molecular Nonadiabatic Electronic Energy Redistribution (INEER), which is a consequence of the chaotic fluxes of electrons and energy within a molecule.

Keywords: quantum chaos; electronic-state chaos; electron dynamics; nonadiabatic dynamics; chemical dynamics; semiclassical mechanics; quantum tunneling; stochasticity and determinicity

Citation: Takatsuka, K. Quantum Chaos in the Dynamics of Molecules. *Entropy* **2023**, *25*, 63. https://doi.org/10.3390/e25010063

Academic Editor: Marko Robnik

Received: 7 November 2022
Revised: 22 December 2022
Accepted: 22 December 2022
Published: 29 December 2022

1. Introductory Remarks: What Makes Molecules Special in Quantum Science and Chaos

Molecules consist of heavy nuclei and light electrons, which give rise to quite interesting and characteristic properties. All of the living bodies and biological systems on the globe are made up with molecules and are subject to the beautiful chemical laws. In the biological activities most of those molecular dynamics and properties should appear to be *deterministic* in the molecular levels, whereas the general dynamics of molecules often work in *stochastic* and *statistical* manners. The origin of such statistical properties arising even from small molecular systems are widely ascribed to classical [1] and quantum [2–12] chaos. This in turn suggests that one may find novel chaos in the dynamics of molecules.

Light electrons in a molecule move very fast and yet are accompanied with long wavelength in their matter waves, while the nuclear motion is much slower, but the relevant wavelengths are relatively far shorter due to the heavy masses. Thus, the two intrinsic hierarchical structures emerge in time-scale and length-scale within the dynamics of a molecule. Consequently the general picture of a molecule are: (i) Nuclei move on mean-field potential functions that are generate by the electrons. (ii) The nuclear motion can bear classical and semiclassical nature due to the short wavelength, while electrons are utterly quantum. Therefore, in contrast to classical chaos of the nuclear motion, it is often claimed that there is no chaos in electronic states because of discreteness in the energy levels with large gaps among them. Physics of the mapping of classical chaos onto quantum systems is often referred to as quantum chaology [13]. As a matter of fact most of the chaos studies in molecular science have been focused on the vibrational spectroscopy and nonlinear mixing of the vibrational modes and the theoretical foundation of statistical laws in chemical reaction dynamics. All these are about the chaoticity arising from the nuclear motion alone.

However, the theoretical separation of the electronic and nuclear motions due to the fast and slow modes, which is known as the Born-Oppenheimer separation, can sometimes break down through large kinematic interaction between them, which cause a mixing of the relevant electronic states. This interaction is generally and vaguely referred to as nonadiabatic interaction [14–20]. This interaction is extremely important in molecular science, since the electronic states and the relevant molecular properties can change instantaneously and drastically when a molecule passes through those nonadiabatic interaction regions. Moreover, we will show that nonadiabatic mixing among many electronic states involved in the manifold of densely quasi-degenerate electronic states can result in strong chaos, which can manifest as a diffusion motion of electronic states in the Hilbert space. This article will hence discuss chaos in what we call nonadiabatic electron dynamics [21–27].

Chaos theory for Hamilton mechanics was initiated (resumed from the view point of Poincaré) around 1950s [28]. Many beautiful theoretical models, elegant mathematics of geometry along with sophisticated numerical studies on ordinary differential equations have been studied since [1]. As for molecular chaos, there are very many pieces of theoretical studies in the literature, yet, due to the complexity of molecular nature most of the theoretical developments and applications are limited to small dimensional systems, typically two-dimensional in the early days. On the other side, the present author and his group have long been studying the chemical dynamics ranging from nuclear classical and semiclassical dynamics (see ref. [29,30] and more cited there) to full quantum electron dynamics [25–27]. Such molecular studies on chemical dynamics often guide us to chaotic phenomena that appear only in molecules. Conversely, tracking the path of our chaos studies, along with some data added herewith, would give partly an idea about what the dynamics of molecule are. We therefore apologize in advance that this review article covers only a very biased aspect of chaos studies. We hence do not cover the topics of the main-stream theories and well-established materials for quantum chaos such as level statistics, the random matrix theory, chaos in external fields and dissipative systems, and so on. Extensive reviews about general quantum chaos theory [11,12] are now widely available in numerous monographs [2–10] and review articles published elsewhere. The special issue of this journal Entropy, "Quantum Chaos—Dedicated to Professor Giulio Casati on the Occasion of His 80th Birthday", is certainly among the newest ones.

We begin this review with the very basic structure of molecular science in Section 2, along with the so-called Born-Oppenheimer approximation, which separates nuclear and electronic dynamics in molecules [31].

In Section 3, we briefly touch upon the aims and implications of chaos studies in molecular science, particularly for chemical dynamics. For this purpose only, we take a couple of examples from previous studies on classical chaos.

Section 4 resumes with the full quantum chaos using the modified Hénon-Heiles potential as a model system of two-dimensional vibrational motions. We track the quantum wavepackets running within the quasi-separatrix and construct some of the associated eigenfunctions. A class of dynamical tunneling that exists only in the quasi-separatrix will be presented. The long-time energy spectra and the dependence of the chaotic spectra on the magnitude of the Planck constant are briefly examined.

Multidimensional semiclassical mechanics is indispensable in the treatment of the dynamics of heavy nuclei in molecules. We review our developed semiclassics [29,30] in Section 5, which we call the method of Action Decomposed Function (ADF) [32,33]. With ADF we analyze the mechanism of quantization not only of forming the energy peaks but of nullification of the off-resonance components. As a consequence of our theoretical and numerical studies, we proceed to the idea of amplitude free quantization without use of the annoying amplitude factors such as the van Vleck determinant [34] and those inherent to the semiclassical kernels (Greens functions) [35–37]. Semiclassical tunneling is also touched upon.

Then we study the characteristic molecular science, in which the nuclear dynamics kinematically couples with the electronic states, where the electronic-nuclear separation is broken down. In Section 6, we first concentrate on the nonadiabatic nuclear dynamics, which are subject to coupled equations of motion on multiple potential energy surfaces. The electronic states are still represented in time-independent wavefunctions. We show that the phenomenon of "bifurcation and merging (remixing) of quantum wavepackets", a fundamental mechanism of quantum chaos, is indeed observable experimentally.

The final part of this review is devoted to the nonadiabatic electron dynamics in Section 7. Electronic wavepacket description is particularly useful in the stage where the adiabatic electronic states behind are embedded in a quasi-degenerate manifold and thereby nonadiabatic mixing among them is very intense and continual. Those strong and enduring electronic state mixing result in quantum chaos for the molecular electronic states.

This review concludes with some remarks in Section 8, stressing that the theory of quantum chaos and philosophy behind is indeed fundamental and useful also in practical science like chemistry.

2. Brief Introduction to the Theoretical Framework of Molecules: Born-Oppenheimer Approximation to Separate Electronic and Nuclear Motions

We first outline briefly the very basic frame of quantum description of molecules suggesting how chaos studies of molecules proceed.

2.1. The Born-Oppenheimer (BO) Approximation

The nonrelativistic molecular Hamiltonian is given as

$$\begin{aligned} H(\mathbf{r},\mathbf{R}) &= T_N + H^{el}(\mathbf{r};\mathbf{R}) \\ &= \frac{1}{2}\sum_k \frac{\hat{P}_k^2}{M_k} + H^{(el)}(\mathbf{r};\mathbf{R}), \end{aligned} \tag{1}$$

and

$$i\hbar\frac{\partial}{\partial t}\Psi(\mathbf{r},\mathbf{R},t) = H(\mathbf{r},\mathbf{R}), \tag{2}$$

where $H^{(el)}(\mathbf{r};\mathbf{R})$ is the electronic Hamiltonian defined as

$$\begin{aligned} H^{(el)}(\mathbf{r};\mathbf{R}) &= T_e + V_c(\mathbf{r};\mathbf{R}) \\ &= \frac{1}{2m}\sum_j \hat{p}_j^2 + V_c(\mathbf{r};\mathbf{R}) \end{aligned} \tag{3}$$

with $\mathbf{r}$ and $\mathbf{R}$ representing the electronic and nuclear coordinates, respectively, while $\hat{p}_j$ and $\hat{P}_k$ denote the operators for their conjugate momenta of the jth and kth component of $\mathbf{r}$ (specifically written as r_j) and $\mathbf{R}$ (R_k), respectively. We here consider only the Coulombic interactions $V_c(\mathbf{r};\mathbf{R})$ among electrons and nuclei, that is

$$V_c(\mathbf{r};\mathbf{R}) = \sum_{a<b}\frac{e^2}{|\mathbf{r}_a - \mathbf{r}_b|} - \sum_a\sum_A \frac{Z_A e^2}{|\mathbf{r}_a - \mathbf{R}_A|} + \sum_{A<B}\frac{Z_A Z_B e^2}{|\mathbf{R}_A - \mathbf{R}_B|}, \tag{4}$$

where $\mathbf{r}_a$ and $\mathbf{R}_A$ are the ath electron and Ath nuclei, respectively, and Z_A indicate the nuclear charge on the Ath nuclei.

Due to the great difference in the time scales between electrons and nuclei, their motion is assumed to be separated at each time, as the so-called the Born-Oppenheimer (BO) approximation represents the total wavefunction $\Psi(\mathbf{r}, \mathbf{R}, t)$ to be

$$\Psi(\mathbf{r}, \mathbf{R}, t) \simeq \chi_I(\mathbf{R}, t)\Phi_I(\mathbf{r}; \mathbf{R}), \tag{5}$$

where $\Phi_I(\mathbf{r}; \mathbf{R})$ and $\chi_I(\mathbf{R}, t)$ are the electronic and nuclear wavefunctions, respectively. Note that electrons are assumed to move so fast that it can adjust themselves as a stationary wave at any nuclear position $\mathbf{R}$, and thereby the electronic wavefunction $\Phi_I(\mathbf{r}; \mathbf{R})$ is assumed to be free of time coordinate t. This in turn requires that $\Phi_I(\mathbf{r}; \mathbf{R})$ is to be attained at each $\mathbf{R}$ such that

$$H^{el}(\mathbf{r}; \mathbf{R})\Phi_I(\mathbf{r}; \mathbf{R}) = V_I(\mathbf{R})\Phi_I(\mathbf{r}; \mathbf{R}). \tag{6}$$

Notice that $\mathbf{R}$ in $\Phi_I(\mathbf{r}; \mathbf{R})$ and in Equation (6) is thereby treated as a parameter. The electronic energy $V_I(\mathbf{R})$ as a function of nuclear coordinates serves as a potential energy function $(I = 1, 2, \ldots)$ for the nuclear wavefunction in such a manner that

$$i\hbar\frac{\partial}{\partial t}\chi_I(\mathbf{R}, t) = \left[\sum_k \frac{\hat{P}_k^2}{2M_k}\chi + V_I(\mathbf{R})\right]\chi_I(\mathbf{R}, t). \tag{7}$$

The time independent figure $V_I(\mathbf{R})$ is often called the potential energy hyper-surface (PES) in the chemical literature [31].

The lowest-order correction to the Born–Oppenheimer energy is found to be in the order of

$$(m/M)^{6/4}. \tag{8}$$

This analysis [38] clearly demonstrates that the Born–Oppenheimer approximation is far better than would be expected from a simple mass ratio m/M. For example, if $m/M \sim 10^{-4}$, which is typical in molecular systems, the error term is $(m/M)^{1.5} \sim 10^{-6}$.

2.2. The Born-Huang Expansion

To approach the exact solution of Equation (2) from the BO approximation, we need to expand $\Psi(\mathbf{r}, \mathbf{R}, t)$ in the stationary electronic basis functions $\{\Phi_I(\mathbf{r}; \mathbf{R})\}$ at each nuclear position $\mathbf{R}$ as

$$\Psi(\mathbf{r}, \mathbf{R}, t) = \sum_I \chi_I(\mathbf{R}, t)\Phi_I(\mathbf{r}; \mathbf{R}), \tag{9}$$

where unknown time-dependent coefficients $\chi_I(\mathbf{R}, t)$ are supposed to describe the *nuclear wavepackets*. This is referred to as the Born-Huang expansion [39]. The electronic basis functions $\{|\Phi_I\rangle\}$ are assumed to be orthonormal at each nuclear configuration as

$$\langle\Phi_I(\mathbf{R})|\Phi_J(\mathbf{R})\rangle = \delta_{IJ}. \tag{10}$$

The nuclear wavefunctions are subject to the following coupled equations

$$i\hbar\frac{\partial}{\partial t}\chi_I = \frac{1}{2}\sum_k \frac{\hat{P}_k^2}{M_k}\chi_I + \sum_J H_{IJ}^{el}\chi_J - i\hbar\sum_k\sum_J \frac{X_{IJ}^k}{M_k}\hat{P}_k\chi_J - \frac{\hbar^2}{2}\sum_k\sum_J \frac{Y_{IJ}^k}{M_k}\chi_J, \tag{11}$$

where

$$X_{IJ}^k = \left\langle \Phi_I \middle| \frac{\partial\Phi_J}{\partial R_k} \right\rangle, \quad Y_{IJ}^k = \left\langle \Phi_I \middle| \frac{\partial^2\Phi_J}{\partial {R_k}^2} \right\rangle, \tag{12}$$

and

$$H_{IJ}^{el}(\mathbf{R}) = \left\langle \Phi_I \middle| H^{el} \middle| \Phi_J \right\rangle. \tag{13}$$

Since X_{IJ}^k and Y_{IJ}^k serve as the mixing elements among the adiabatic electronic states, they are collectively called nonadiabatic interactions. Tough difficulties in an attempt to solve the coupled equations of motion (11) are due to the many dimensionality of $\mathbf{R}$, the oscillatory nature (due to the short wavelength) of $\{\chi_I\}$, and many channels to be taken into account. Besides, some of $\{\chi_I\}$ can extend to asymptotic regions in reaction (scattering) dynamics.

Note that the above Born-Huang expansion can be performed with use of other basis functions that do not satisfy Equation (6). In such a case $H_{IJ}^{el}(\mathbf{R})$ in Equation (13), have significant amount of nonzero off-diagonal elements. $\{\Phi_I(\mathbf{r};\mathbf{R})\}$ satisfying Equation (6) are referred to as the *adiabatic* wavefunctions, while others are to *diabatic* wavefunctions collectively. There are many ways to determine the diabatic functions [16–18,20], a typical set being those that minimize $\left|X_{IJ}^k\right|$. We will use the adiabatic representation in what follow unless otherwise noted.

3. Implication of Chaos in Chemical Dynamics

3.1. Determinicity versus Stochasticity in Molecules

It was Arrhenius who established the well-known thermodynamical reaction rate k from initial molecules to products in the form

$$k \propto \exp(-E_a/RT). \tag{14}$$

This formula is indeed revolutionary in that the *rate or speed* has been derived based on the Maxwell-Boltzmann distribution for *equilibrium states*. The physical insight of this rate expression was uncovered by Eyring [40–42]. He studied a potential energy surface for simple atomic exchange reaction

$$\mathrm{A} + \mathrm{BC} \rightarrow [\mathrm{ABC}]^{\ddagger} \rightarrow \mathrm{AB} + \mathrm{C} \tag{15}$$

where $[\mathrm{ABC}]^{\ddagger}$ is a transient state, called the transition state, at the bottleneck area on the way of reaction. He assumed a statistical distribution for the vibrational modes transversal to the translational one and associated the partition functions with them. He thus clarified the quantum mechanical meaning of the barrier height E_a along with the preexponential factors to k (Polanyi and Wigner also stressed the geometrical roles of the potential energy surface $V_I(\mathbf{R})$ of Equation (6)) [40–42]. It is now known that the transition state theory is approximately valid even in binary collisions in vacuum. Even a quantum mechanical extension is one of the important subjects [43,44]. On the other hand, rather simple reactions for small systems can now well be treated in full quantum mechanics in a completely deterministic manner. (I do never mean that the full quantum scattering calculation for chemical reactions is an easy task. See refs. [45,46] for examples.) Yet the fundamental question is aimed at the theoretical ground of the statistical assumption made in the transition state.

While being subject to statistical laws on one hand, the dynamics of molecular motions are generally deterministic on the other hand. Let us see such a rather intricate example. See Figure 1, which is an ab initio (first principle) dynamics of 5-methyltropolone (5MTR). Ushiyama found that the proton shift from on O_2 in the panel (a) to the position next to O_1 (panel (c)) induces the rotation of the methyl group -$C_8H_aH_bH_c$ at the bottom. The reciprocal motion of proton turns to rotational motion of the methyl group through such a long distance. Careful inspection at the panel (a) and (c) shows that this determinicity is supported by the shift of double bonds (see ref. [47] for the electronic state mechanism). This example denies the simple minded view that a molecule can be always regarded as a set of points of nuclear point masses connected by "springs" among them. We therefore need a careful thought when we study the chaos of molecular vibrational motions.

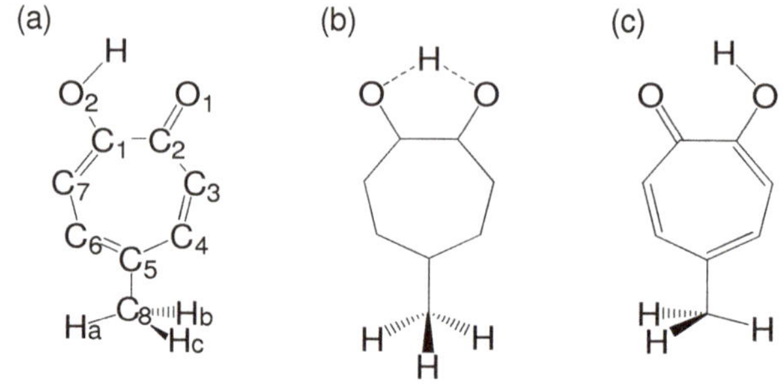

Figure 1. (**a**) Stable structure of 5-methyltropolone (5MTR). Hydrogen atoms attached on C_3, C_4, C_6, and C_7 are not shown. (**b**) The transition state on the way. (**c**) Proton has ben shifted from O_2 to O_1 inducing the internal rotation of the methyl group at the bottom. [Taken from Ushiyama, H.; Takatsuka, K. *Angew. Chem. Int. Ed.* **2005,** *44*, 2 with permission].

Another example of statistical reaction rate theory was given by Marcus [48], which augmented the successors' formalism due to Rice, Ramsperger, and Kassel, thereby now called the RRKM theory for unimolecular dissociation, which predicts the rate of slow unimolecular dissociation of a molecule energized high enough to surmount the potential barrier. This process is schematically represented as

$$[MA]^* \rightarrow M + A, \tag{16}$$

where an energized molecule $[MA]^*$ may have been produced by collision and/or photo-excitation, and so on. From the view point of quantum scattering theory, $[MA]^*$ can be regarded as a core (temporally) excited resonance, and the unimolecular dissociation may be viewed as a half-collision process of the Feshbach resonance [49,50]. In case where those decaying vibrational energy levels are densely packed having high density of states, a state in the vibrational-state manifold wanders among many nonlinearly coupled vibrational states with a long life-time. It is intuitively clear that statistical prediction should work well in such a situation, since the corresponding microcanonical ensemble is expected to be highly mixing. The years of 1970s have been devoted to prove that the ergodicity and mixing in the relevant classical phase space of $[MA]^*$ is indeed realized by classical chaos. Such stochastic energy redistribution within a molecule is now generally termed Intramolecular Vibrational energy Redistribution (IVR), which represents one of the most important concepts in chemical dynamics and will be briefly touched upon below. In short, statistical mechanics works surprisingly well even in such small molecules composed only of several atoms, not of the Avogadro number. Those dynamics deviating from the RRKM theory are now placed under special focus for their possible peculiar phase-space structures, which are often termed as non-RRKM dynamics.

The key quantity that brings about stochastic and statistical nature into small molecular dynamical systems is the density of state (DOS), not the number of particles (atoms and molecules). In case where the DOS is low, the quantum dynamics should be subject to the Schrödinger equation and naturally deterministic. The higher is DOS, the harder become the relevant coupled equations to solve, and statistical treatments turn out to be more feasible. Due to the heavy masses of nuclei, the vibrational and rotational degrees of freedom of a molecule tend to have very high DOS, which can readily amount to the order 10^{10} eV^{-1} even for several atomic systems. This is not as large as the Avogadro number but sufficiently large to allow for statistical treatment. On the other hand, the DOS for the typical low-lying *electronic states* is less than 1.0 eV^{-1}, and quantum mechanics is almost only one faithful approach. For this reason, the theory of stationary molecular electronic states is widely called quantum chemistry. For these low DOS and discrete states, it is a natural assertion that there is no chaos. Yet, in the end of this review, we will penetrate the *electronic state manifold* of densely quasi-degenerate states, ending up with finding chaos.

3.2. *Intramolecular Vibrational Energy Redistribution (IVR)*

Given the statistical assumption of molecular vibrational energy distribution, the RRKM theory actually estimates the rate of unimolecular dissociation based on statistical inference at the bottle neck position on the way to products in phase space [48], using the transition state [40–42]. Yet, chemical dynamics has made a further step to the physical processes of intramolecular vibrational (along with rotational) energy flow (redistribution) within a molecule.

Such experimental studies were made possible by the advent of short pulse laser, which has unified two then separated subfields in molecular science, namely, spectroscopy of molecular structures and chemical reaction dynamics. As such the so-called stimulated emission pumping (SEP) spectroscopy applied to acetylene molecule H-C≡C-H (linear in shape having a triple bond) has given a huge impact [51–54]. Just briefly is outlined the essence of the experiment below. Preparing beforehand an electronically excited state of H-Ċ=Ċ-H (bent structure and longer C-C distances, double bond), they de-excited it to the high vibrational excited state on the electronic ground state by induced emission. Varying the time-width of the pumping laser, they observed very interesting vibrational emission spectra in the various hierarchical levels of resolution. Reflecting the highly vibrational excited states of polyatomic molecule, the attained spectra were extremely complicated, manifesting the chaotic feature behind. Surprisingly nevertheless, Yamanouchi succeeded to assign the spectral origins of most of the relevant peaks [52]. Among others, the most striking spectral feature is the so-called nested clump structure: Each of clump-like spectral peaks of a rather broad width is found to be composed of finer peaks attained when a longer width pulse is applied. Each of these fine peaks in turn shows another clump structure and is observed to consist of the even finer peaks with the pulse of further long width. They ascribed the nested structures to the stepwise intramolecular energy redistribution triggered by the mode change from the vibrational structure of H-Ċ=Ċ-H to that H-C≡C-H. For example, the elongated Ċ=Ċ bond releases much vibrational energy to the shorter C≡C in the first stage, which are transferred to the C-H stretching modes, which in turn coupled with the bending motion of ∠HCC in due time ordering. More complicated energy redistribution can follow such as a very large amplitude motion of the hydrogen atoms, but we do not pursue them here. Since the spectra of the longer time-width can track the long physical process of energy redistribution. The related experiments and precise analyses have vividly shown the real-time and actual energy flow within a molecule. The notion of chaos has thus physical reality in chemical dynamics.

The energy transfer among the different vibrational modes are induced by high anharmonicity of the relevant potential functions, leading to nonlinear canonical equations, and are also due to the overlap among the Fermi resonances [55] or the Feshbach resonance [50]. (see ref. [5] for the overlap criterion of chaos.) Therefore, the relevant dynamics can readily result in chaos leading to ergodicity [56]. Many classical mechanical studies have been devoted to IVR. Yet, only recently become available extensive reviews on quantum ergodicity and intramolecular energy flow due to Leitner [57] and quantum ergodicity transition in relation to IVR in a phase space perspective by Karmakar and Keshavemurthy [58].

Incidentally, energy transfer, relaxation, and redistribution of intra- and inter-molecular systems in condensed molecular systems [55,59] and biological molecules [60,61] are also among the basic processes of many of chemical phenomena, irrespective of the extent of possible contribution by quantum chaos. Currently, molecular dynamics studies are among the main streams in these fields, but those are beyond the scope of this review.

3.3. *Onset of Statistical Properties: Liquid-Like Clusters*

Beck and Steve Berry are the first, to the best of out knowledge, who found the critical roles of classical chaos in the statistical behavior of "melting and phase transition" appearing in structural isomerization dynamics of small atomic clusters [62]. Berry and their coworkers have achieved many fundamental works about cluster dynamics. See ref. [63] as a representative paper. We also refer to the work by Chandler et al. [64] as a pioneering study on stochastic dynamics of molecular isomerization.

Chaos not only makes the relevant phase space ergodic and mixing but also gives interesting qualitative features to the dynamics of molecules. Such properties can be typically seen in the structural isomerization dynamics of atomic clusters. Take an example from Ar_7 cluster on the Lennard-Jones potential, which has 4 isomers as depicted in Figure 2. (Permutation symmetry counts more than 5000 isomers.) Each isomer is supported by its own potential energy basin. In a low energy, each isomer can stay within its potential basin keeping its structure as a *solid*, while at some high energy the isomerization begins to take place by surmounting the potential barriers, and further in considerably high energy frequent isomerization keeps taking place, which can be viewed as extensive large scale vibrational motions over the entire molecule or as a *liquid*-phase analog [63]. Even higher energy induces evaporation of one atom and two atomic molecule [65,66]. This entire process is hence regarded as a molecular level prototype of the first-order phase transition [62,63,67].

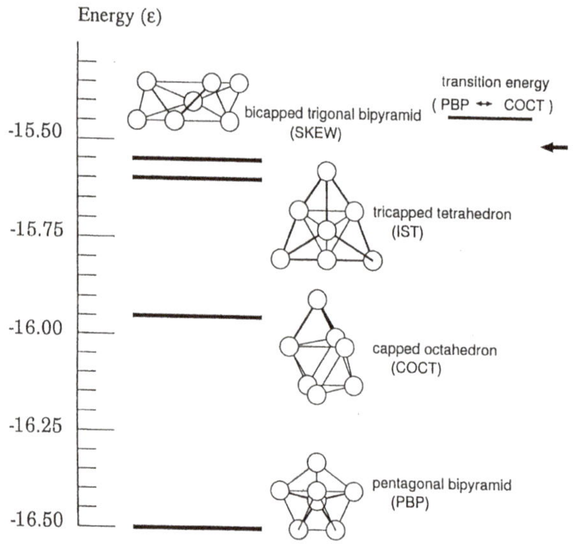

Figure 2. Four isomers on the single Lennard-Jones potential energy surface. The names and the bottom energy of each potential basins are presented. [Drawn by Dr. C. Seko].

In the liquid-like phase, chaotic dynamics manifest themselves in the following phenomena. (1) Each classical path wanders among the potential basins for individual isomers in a random fashion (see Figure 2). The time-series with respect to the names of the isomers such as {PBP→COCT→IST→COCT→PBOP→SKEW→ ...} is stochastic in the Newtonian dynamics, yet a path stays for somewhat long time in each basin [68,69]. (2) Memory loss of the classical paths occurs. Despite mutually different potential barriers among the individual potential basins, they behave as though they forget which basin they came from and to which basin they are going to visit next. They seem not to care about the presence of the so-called transition state, which is the lowest energy saddle between two basins [68,69]. (3) The time scale for a trajectory to penetrate into a mixing area after getting into a potential basin is so short that trajectories readily lose the memory in the mixing area in phase space. We call this situation inter-basin mixing [70]. (4) Microcanonical temperature can be defined with which the lifetimes of the isomers are subject to an Arrhenius relation. This in turn suggests the existence of the free energy despite the nonequilibrium dynamics [71].

Behind the above chaotic dynamics we can imagine geometrical structures in phase space, referred to as the reaction tubes. Suppose that a trajectory enters a PBP basin from the neighboring COCT by passing through a divided surface (many dimensional). Then there should exist a large bundle of similar trajectories flowing from the COCT to this PBP that includes this particular trajectory. See Figure 3, which illustrates a part of the unit piece of reaction tube. Into this PBP basin schematically represented as a circle, for instance, many other reaction tubes come in from other neighboring potential basins such as those of SKEW and IST. Soon after the tubes come in, each branches frequently as schematically shown in the figure. The successive branching actually reduces the diameter exponentially as time. Indeed one can measure the exponent of the decay in the form $\exp(-\beta t)$ [70,72]. This is a geometrical background for classical paths to lose their memory from which basin they came in. Those extremely thin tubes entangle with other tubes of different histories, and they repeatedly merge with one another to thicker tubes and each eventually gets out of this basin to the neighboring ones (take the time reversal dynamics of the branching). This is how they lose the future "memory" of which basin they are going to visit next [70,72]. By construction it is obvious that the initial fat trunk of the reaction tube depicted in Figure 3 is already a mixture of finer tubes of different histories, each of those finer tubes being composed of even finer ones. Thus, the cross-section of the tube should exhibit very complicated fractal structures, along with globally periodic orbits of many different symbolic dynamics and different periods. In summary, repeated *branching and remerging* of the reaction tubes exist behind the memory loss.

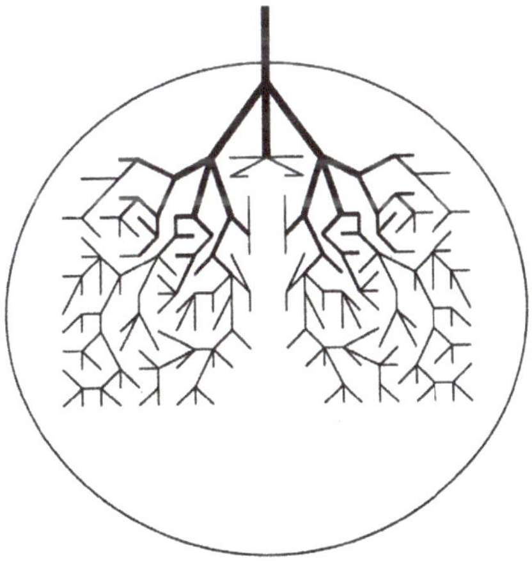

Figure 3. Schematic presentation of a unit of reaction tube in a potential basin and its bifuraction structure. The circle schematically models a potential basin, each of which supports one isomer. Only one of tubes is depicted and truncated at some points. The reaction tubes lie behind classical chaos in structural isomerization dynamics, which is one of the so-called many-valey problems. The similar tube units are extended from every neighboring potential basins, and one basin is filled with many tubes. Time-revesal symmetry gives an idea of how the reaction tubes are reconnected with one another to form thicker tubes and get out of the basin to next ones. [Drawn by Dr. C. Seko].

The original idea of reaction tube is due to De Leon and their colleague, which they called the cylindrical manifold in the study of two dimensional two potential basin dynamics [73,74]. "Bifurcation and remerging (reconnection)" of the reaction tubes can be regarded as a basic mechanism of complex isomerization dynamics besides the baker's transformation. There are interesting studies about the fine structure of the reaction tubes relevant to the stochastic dynamics. However we skip this aspect to quantum chaos.

4. Quantum Dynamics in the Quasi-Separatrix of Two Dimensional Molecular Vibration

In the early 1980s the studies of quantum chaos in terms of quantum wavepacket propagation were made rather extensively in the community of chemical physics [75–78]. In particular, Feit and Fleck [78] are among the first, to the best of our knowledge, who numerically integrated the Schrödinger equation for wavepackets on the Hénon-Heiles Hamiltonian [28]. The scientific insights attained thus far with the accurate and the efficient methodologies developed for those studies are indispensable to date.

We herein study pure quantum chaos in terms of wavepackets and eigenfunctions confined in the separatrix on the modified Hénon-Heiles potential. This potential function is regarded as a two dimensional potential energy surface (PES) on which for molecular vibrational motion to develop.

4.1. *Quantum Wavepackets Embedded in the Quasi-Separatrix*

The Hamiltonian we adopt is

$$H = \frac{1}{2m_x} p_x^2 + \frac{1}{2m_y} p_y^2 + \frac{1}{2}(x^2 + y^2) + x^2\left(ay^2 + y\right) + \frac{1}{3}y^3(by - 1), \tag{17}$$

where the masses are chosen as $m_x = 1.0087$ and $m_y = 1.0$ and $a = 0.6$, $b = 0.2$ with the Planck constant $\hbar = 0.005$. The parameters have been set so as to close the dissociation channel inherent to the original Hénon-Heiles potential. The potential functions are drawn in the upper panels of Figure 4, while the Poincaré surface of section at $x = 0$ with $p_x > 0$ for a selected energy is drawn at the left of the lower panel. One of the regular vibrational modes lying in a torus, marked with "lib", is drawn in the (x, y)-space right to the Poincaré surface of section. An appropriate energy is chosen so that the quasi-separatrix has a small to medium size so as to coexist with the set of tori of significant scales. These altogether represent *weak chaos*.

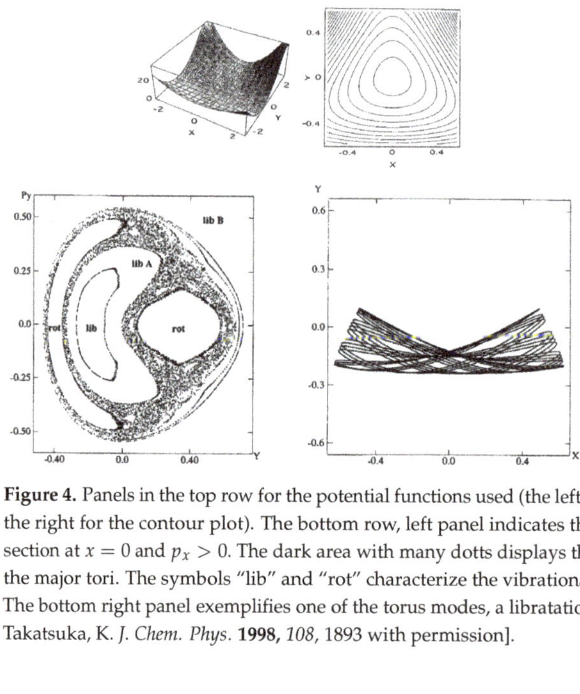

Figure 4. Panels in the top row for the potential functions used (the left one for a perspective view and the right for the contour plot). The bottom row, left panel indicates the classical Poincaré surface of section at $x = 0$ and $p_x > 0$. The dark area with many dotts displays the quasi separatorix separating the major tori. The symbols "lib" and "rot" characterize the vibrational modes of the individual tori. The bottom right panel exemplifies one of the torus modes, a libratation. [Taken from Hashimoto, N.; Takatsuka, K. *J. Chem. Phys.* **1998,** *108*, 1893 with permission].

We first track a wavepacket propagating in time within the major quasi-separatrix. To do so, we extended the Poincaré surface of section to a quantum version as follows [79,80]. Given a two-dimensional wavefunction $\psi(x,y,t)$, we first take a momentum representation only in the x-coordinate such that

$$\tilde{\psi}(p_x,y,t) = \int_{-\infty}^{\infty} dx \exp\left(-\frac{i}{\hbar}p_x x\right)\psi(x,y,t). \tag{18}$$

Then Fourier back it using only the positive p_x such that

$$\psi^+(x,y,t) = \int_0^{\infty} dp_x \exp\left(\frac{i}{\hbar}p_x x\right)\tilde{\psi}(p_x,y,t) \tag{19}$$

and put $x = 0$ there for making a cross section at $x = 0$. Bring it then into the Husimi representation in such a way

$$\begin{aligned}\rho_H(q_y,p_y,t) &= (2\pi\hbar)^{-1}(\pi\hbar)^{-1/4} \\ &\times \left|\int_{-\infty}^{\infty} dy \exp\left(-\frac{1}{2\hbar}(y-p_y)^2 + \frac{i}{\hbar}p_y y\right)\psi^+(0,y,t)\right|^2.\end{aligned} \tag{20}$$

We then plot $\rho_H(q_y,p_y,t)$ in the (q_y,p_y)-plane corresponding to the classical counterpart. This is rather a faithful extension of the classical Poincaré section.

An example of the quantum Poincaré section is exhibited in Figure 5, the upper row indicates $|\psi(x,y,t)|^2$ at selected times, while the lower row shows the corresponding quantum Poincaré section. It is clear in the quantum Poincaré section that the present packet is running on the quasi-separatrix. It is noteworthy that in upper panels of Figure 5a,c, the wavepacket stays around the unstable fixed points for a longer time, thereby leaving two significant strips in the wavefunction in the corresponding (x,y) space. These dense strips are an example of the so-called quantum scar first found by Heller [81] (see also E. J. Heller, page 547 in ref. [4]).

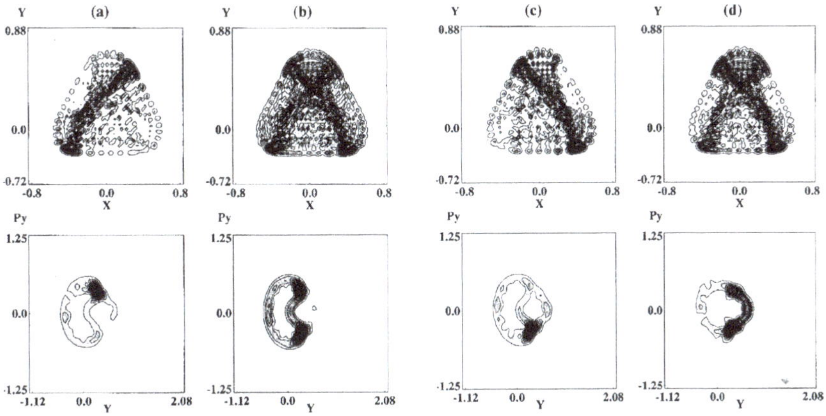

Figure 5. Time propagation of a quantum wavepacket in the quasi-separatrix. Time passes from panel (**a–d**). The upperrow indicates $|\psi(x,y,t)|^2$ at selected times, while the lower row shows the corresponding quantum Poincaré section, in which the wavepacket is seen to be running on the classical quasi-separatrix. [Taken from Hashimoto, N.; Takatsuka, K. *J. Chem. Phys.* **1998,** *108*, 1893 with permission].

4.2. Energy Spectra and Eigenfunctions in the Quasi-Sparatrix

The next question is whether the eigenfunctions can be built in the quasi-separatrix without decaying to the stable tori. We next explore this aspect.

4.2.1. Energy Screening of Quantum Wavepackets

Prior to the numerical presentation we need to show how only (or at least mostly) the separatrix eigenfunctions are extracted from the very many eigenfunctions. This is a difficult task if we resort to the straightforward method of diagonalization of the Hamiltonian matrix, which may be expanded in arbitrary basis functions. Fortunately, those wavepackets we have shown above in Figure 5 are certainly running on the quasi-separatrix, and it would be an appropriate idea to quantize those packets. Our way to do so was as follows [82].

As a standard way to obtain eigenfunctions at any chosen energy E from a wave packet, one may want to calculate the time-energy Fourier transform [83,84]

$$\psi(x,E) = (2\pi\hbar)^{-1} \int_{-\infty}^{+\infty} e^{+\frac{i}{\hbar}Et}\psi(x,t)\,dt\,. \tag{21}$$

However, this approach needs to pick extremely accurate $E's$ beforehand, usually by means of the Fast Fourier Transform of a correlation functions. (Incidentally see ref. [85] to extract very accurate spectrum from the FFT beyond a given grid size.) This is because the integral in Equation (21) simply vanishes in use of an inaccurate E. Moreover one may need to know in advance whether E should indeed belong to one of the quasi-separatrix eigenfunctions.

An efficient alternative to calculate eigenfunctions directly without the prior knowledge on the energy spectrum is the screening method [82]. Suppose we seek for an eigenfunction in an arbitrary narrow energy range $[\epsilon_1, \epsilon_2]$. A wavepacket residing in the interval can be generated as

$$\begin{aligned} |\Phi[\epsilon_1,\epsilon_2]\rangle &= \int_{\epsilon_1}^{\epsilon_2} dE\,\delta(E-\mathcal{H})|\phi(0)\rangle \\ &= \frac{1}{2\pi}\int_{-\infty}^{\infty} dt\,\frac{e^{i\epsilon_2 t}-e^{i\epsilon_1 t}}{it}|\phi(t)\rangle, \end{aligned} \tag{22}$$

with the help of

$$\delta(E-\hat{H}) = (2\pi\hbar)^{-1}\int_{-\infty}^{+\infty} dt\, e^{\frac{i}{\hbar}(E-\hat{H})t}\,, \tag{23}$$

where $|\phi(t)\rangle = e^{-i\mathcal{H}t}|\phi(0)\rangle$, and $|\phi(0)\rangle$ is an initial wavepacket supposed to run on the quasi-separatrix in the present purpose. If there is only one eigenstate in $[\epsilon_1, \epsilon_2]$, this wavefunction $|\Phi[\epsilon_1,\epsilon_2]\rangle$ should converge to to the exact one, aside from the norm. Its energy, if needed, is to be readily estimated as an expectation value. If the target energy range does not contain any eigenfunction, $|\Phi[\epsilon_1,\epsilon_2]\rangle$ converges to the null. When several eigenstates are included in the range, one can step to a further elaboration, which is not difficult at all [82,86]. Incidentally, notice the faster convergence in Equation (22) than in Equation (21) due to the presence of t in the denominator.

4.2.2. Chaotic Eigenfunctions and Spectra

Figure 6 represents an almost degenerate pair of eigenfunctions belonging to the quasi-separatrix wavefunctions. The left (right) panel represents an odd (even) function with respect to the mirror line at the $x = 0$, with energy $E = 0.152094$ ($E = 0.152500$). The two energies are close to the one for the classical Poincaré section of Figure 4. As seen in the nodal structures, these are actually highly excited states. The classical motions at the two major unstable fixed pints are separated by symmetry as a left-handed (right-handed) librating motions (suggested by panel (a) and (c) in the upper row of Figure 5). However, the quantum wavepackets mix together by floating from and into those fixed point areas.

Looking back at the structure of the quasi-separatrix in Figure 4, it is noticed that two wide areas around the fixed points (chaotic seas) are connected through thin canals, thus forming constrictions (see at $p_y = 0$ in the section for instance). Therefore the wavepackets must pass through those thin channels to move from one wide chaotic sea to the other to become a global eigenfunction. In other words, those canals serve as "dynamical barriers", which suppress the mobility of the wavepackets. The abovc pair of eigenfunctions in Figure 6 are generated through such dynamical barriers, and the slight difference in their energies and the odd-even symmetry separation are all the consequences of that kind of blockade or *quasi-tunneling*.

Davis and Heller found a quantum mechanical tunneling between two (or more) tori, which are separated by separatrix(es). This phenomenon is referred to as the dynamical tunneling through dynamical barrier [87–89]. In contrast, the pair of eigenstates observed in Figure 6 are not literally dynamical tunneling, since the canal-passing is allowed both quantum- and classical-mechanically. However, the view of the Poincaré section in Figure 5 suggests that the chaotic seas around the major fixed points reserve two major wavepackets to stay long, which are weakly separated by the thin canals thereby serving as dynamical barriers effectively. We therefore referred to this tunneling-like motion as dynamical tunneling of the second kind [80]. This phenomenon seems to be peculiar to quantum chaos, and yet we are not aware whether it has been found in actual molecular spectra.

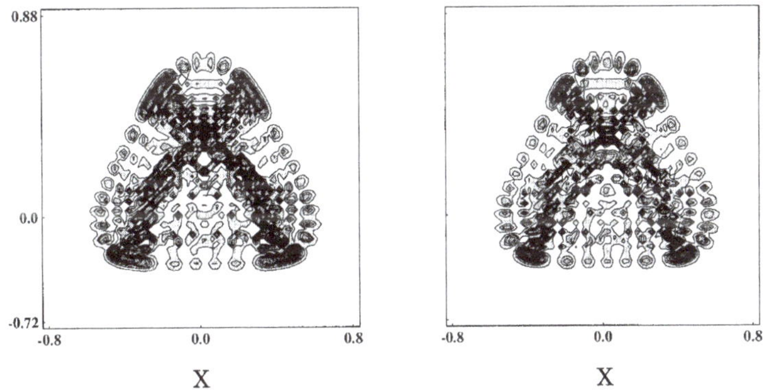

Figure 6. A pair of almost degenerate eigenfunctions lying in the quasi-separatrix. In the left panel is an odd function with respect to the symmetry axis $x = 0$ [energy $E = 0.152094$], while the right being even [$E = 0.152500$]. The slight energy difference and the separation of the symmetry are due to what we call dynamical tunneling of the second kind. See the text. [Taken from Hashimoto, N.; Takatsuka, K. *J. Chem. Phys.* **1998,** *108*, 1893 with permission].

4.2.3. Dependence on the Magnitude of the Planck Constant

We are curious about the possible effects of the magnitude of the Planck constant on quantum chaos. To save the space, we refer this aspect to our original paper [79]. Yet we hereby briefly touch upon the smoothing of chaos in the larger value of $\hbar$ as demonstrated in terms of the Poincaré section. The above two panels in Figure 7 come from a separatrix eigenfunction with $\hbar = 0.005$ ($[E_1, E_2] = [0.150.0.153]$), while the lower panel displays the section $\hbar = 0.010$ ($[E_1, E_2] = [0.150.0.156]$). Such quantum smoothing is not at all surprising, but the corresponding energy spectra are found not as simple (not shown here) chematic [79]. As a chemist, I suspect that the biological worlds on the globe may not be robust against a small change of the Planck constant.

4.3. Time-Dependent Spectrum of Very Long-Time Dynamics

The classical trajectories within the thin quasi-separatrix are generally not periodic nor quasi-periodic by definition (meaning a zero measure for the exception), and moreover they wander around circulating near one tori to another one from time to time, each torus of which represents different mode of vibration (see Figure 4). Therefore those trajectories seem to continue to undergo different vibrational motions mostly in an intermittent way [90]. Therefore it can take extremely long time to extract the associated vibrational quantum spectra from the correlation functions based on those trajectories. The resultant spectrum can be time dependent, meaning that it is dependent on which timing we begin to calculate the correlation function and on when we finish. Such an example of time-dependent spectra are seen in [79,80] This is a serious issue in vibrational spectroscopy for complicated molecules of chaotic behavior. (Recall the nested clump structure in the SEP spectrum, mentioned in Section 3.2) The structural isomerization dynamics in the liquid-like phase discussed in the previous section (Section 3.3) is actually a large amplitude vibrational motion of the extremely long time-scale associated with chaos. It is still left unresolved how to take vibrational spectra for such wandering motions from one potential basin to another, spending some long time at each basin. Proteins in molten globule state must have the similar nature to the liquid-like clusters.

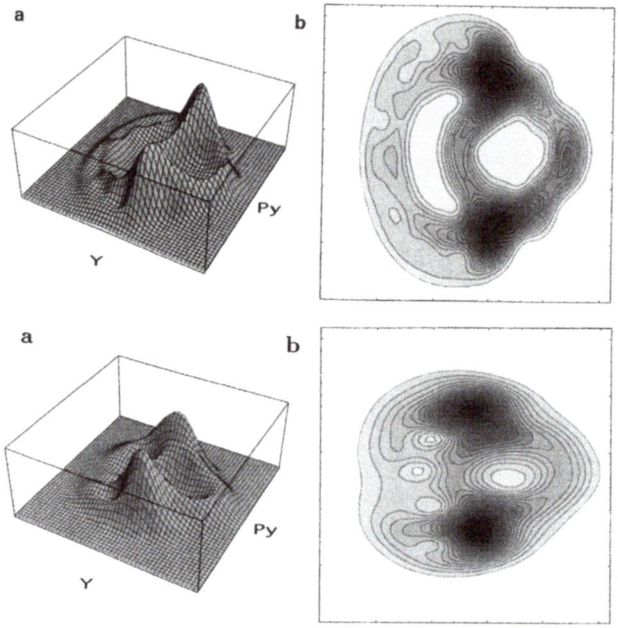

Figure 7. Quantum Poincaré surface of section. Panels at the top row reprsent an quasi-separatorix eigenfunction given by $\hbar = 0.005$ ($[E_1, E_2] = [0.150.0.153]$). The bottom panels are the eigenfunction at $\hbar = 0.010$ ($[E_1, E_2] = [0.150.0.156]$). Left column (**a**) for the perspective views and the right (**b**) for contour plots. [Taken from Hashimoto, N.; Takatsuka, K. *J. Chem. Phys.* **1995**, *103*, 6914 with permission].

5. Quantization of Chaos with Semiclassical Wavepackets

The mechanism of energy quantization is a landmark lying in between classical and quantum mechanics. It is therefore quite natural for semiclassical mechanics to play a major role there. Besides, the chemical dynamics quite often demands semiclassical treatment for the nuclear motion apart from chaos. There are many beautiful studies in the literature on the EBK semiclassical quantization for regular systems (see ref. [91,92] for representative examples). As for quantization of chaotic dynamics the periodic orbit theory, or the trace formula, due to Gutzwiller [35–37] is the central work [2,4–10,12]. One of the most intriguing aspects of the trace formula is its relation to the Riemann and Selberg zeta functions. However, reviewing it is by far beyond the scope of this article [93].

5.1. The Gutzwiller Periodic Orbit Theory

The periodic orbit theory [35–37] treats the semiclassical estimate of the density of states with the Feynman kernel K as

$$D(E) = \int dq \int dt K(q,q,t) \exp\left(\frac{i}{\hbar}Et\right). \tag{24}$$

The theory applies the (first-order) semiclassical approximation to the Feynman kernel or its energy representation (the energy Green function) to the density of states, thereby extracting the essential roles of phase-space periodic orbits [94]. The successive applications of the stationary phase approximations give rise to the famous expression

$$D(E) \simeq \frac{1}{(2\pi\hbar)^N} \int\int dq dp \delta(E - H_{cl}(q,p)) + \frac{1}{\pi\hbar}\sum_{\gamma}\sum_{k=1}^{\infty} \frac{T_\gamma \cos\left(k\left(\frac{W_\gamma}{\hbar} - \frac{\pi}{2}\lambda_\gamma\right)\right)}{\sqrt{\left|Det\left(M_\gamma^k - I\right)\right|}}, \tag{25}$$

where the first term of the right hand side is the Thomas–Fermi (classical) density of states with the classical Hamiltonian $H_{cl}(q,p)$ arising from the periodic orbits of zero length and the second oscillatory terms represent the quantum interference among the finite periodic orbits (specified by γ) in terms of the period T_γ, the action integral W_γ, and the Monodromy matrix M_γ^k in the transversal direction of phase-space periodic motion. These terms should be summed up with respect to the number of circulations k over all the possible periodic orbits. However, it is very difficult particularly in many-dimensional systems to find the necessary periodic orbits (see ref. [95] for instance) and count up their contributions accurately. Moreover, it is well known that the periodic orbits proliferate exponentially as the length of the period T_γ. Another intrinsic difficulty in the path summation appears in the denominator in the second term of $D(E)$. This is a direct reflection that most of the semiclassical expressions bear the serious theoretical difficulty related to divergence of the monodromy matrix, which explodes at the caustics or the turning points. The divergence is particularly prominent in chaotic systems. It is therefore often claimed that the series of periodic orbit sum in Equation (25) may not be absolutely convergent, which sometimes leaves the trace formula to be obscure in practice. Furthermore, a naive question may be cast at whether the oscillatory terms in the trace formula can indeed cancel the Thomas–Fermi density at any energy of not-to-be quantized (off-resonant energy).

It seems that there are two major opposite directions to cope with the fundamental difficulty of the trace formula. One is an attempt to recount the periodic orbit contributions in Equation (25) in such a way to attain a conditional convergence [96]. Furthermore, the convergence of the trace formula can be better considered in the complex energy plane [97]. Semiclassical mechanics of higher order $\hbar$ is obviously a canonical candidate to improve the trace formula [98]. Deep mathematics are involved in these approaches.

The other one is to take an easier, simpler and more tractable approach, mainly based on the semiclassical propagation of wavepackets. Our approach below is along this line. Besides, semiclassical wavepacket theory is one of the main methodologies in chemical reaction dynamics. As such, we here present our own approach, which is intended to step further beyond semiclassical approximation to the exact one. We then attain some insights about the mechanism of energy quantization, that is, the explicit roles of not only constructive but destructive interferences in forming discrete energy levels. We also see that the energy levels in chaos can be determined without the tedious pre-exponential factor (or the Monodromy matrix) of the semiclassical kernel.

5.2. Wavepacket Semiclassics with Action Decomposed Function

Many semiclassical theories have been proposed in the literature [7,14,15,34,94,99–104], either starting from the Feynman kernel, WKB theory or others. Yet, we consider the following theory to clarify the dynamical hierarchical structure in the range extending from classical and semiclassical to full-quantum dynamics.

5.2.1. Short Time Dynamics of the Maslov Type Wavefunction

We represent a wavefunction $\Psi(q,t)$ in the so-called Maslov type [102], such that

$$\Psi(q,t) = F(q,t)\exp\left(\frac{i}{\hbar}S_{cl}(q,t)\right), \tag{26}$$

for a short time interval $[t, t+\Delta t]$ on a multidimensional coordinate q in configuration space, where S_{cl} is assumed to satisfy the classical Hamilton–Jacobi (HJ) equation

$$\frac{\partial S_{cl}}{\partial t} + \frac{1}{2m}(\nabla S_{cl})^2 + V = 0, \tag{27}$$

which defines relevant classical paths during the interval, with m being the mass. The classical factors are thus factored out for this short interval and thereby the purely quantum factors are thus supposed to be included only in the function $F(q,t)$ [32,33]. $F(q,t)$ is thus referred to as Action Decomposed Function (ADF) [29,30]. Then in this interval, the Schrödinger equation for $\Psi(q,t)$ is transformed to a linear equation of motion for the complex valued amplitude function $F(q,t)$ as

$$\frac{\partial F(q,t)}{\partial t} = \left(-p\cdot\nabla - \frac{1}{2}(\nabla\cdot p)\right)F(q,t) + \frac{i\hbar}{2}\nabla^2 F(q,t), \tag{28}$$

where p is a momentum at (q,t), which is $q(t)$, as

$$p = \nabla S_{cl}(q,t). \tag{29}$$

We use the mass weighted coordinates so that all the masses (m) are scaled to unity $m = 1$, and thereby p is set to be numerically equivalent to the corresponding velocity v. In transforming the Euler picture to the Lagrange by defining

$$\frac{D}{Dt} = \frac{\partial}{\partial t} + v\cdot\nabla \tag{30}$$

Equation (28) reads

$$\frac{D}{Dt}F(q,t) = \left[-\frac{1}{2}(\nabla\cdot p) + \frac{i\hbar}{2}\nabla^2\right]F(q,t). \tag{31}$$

$F(q - q_{cl}(t), t)$ is hence carried along a classical path $q_{cl}(t)$ in this interval. Then the Trotter decomposition [94] valid for a very short time–interval Δt gives

$$\begin{aligned} &F(q - q_{cl}(t+\Delta t), t+\Delta t) \\ &\simeq \exp\left[\frac{i\hbar}{2}\Delta t\nabla^2\right]\exp\left[-\frac{1}{2}(\nabla\cdot p)\Delta t\right]F(q - q_{cl}(t), t) \\ &\simeq \exp\left[-\frac{1}{2}(\nabla\cdot p)\Delta t\right]\exp\left[\frac{i\hbar}{2}\Delta t\nabla^2\right]F(q - q_{cl}(t), t) \end{aligned} \tag{32}$$

and represents the quantum effects in a product between the momentum gradient ($\exp\left[-\frac{1}{2}(\nabla\cdot p)\Delta t\right]$) and the pure quantum diffusion ($\exp\left[\frac{i\hbar}{2}\Delta t\nabla^2\right]$). The theory is exact except for the Trotter decomposition, provided that the classical action $S_{cl}(q,t)$ exists.

5.2.2. Semiclassical Approximation

Equation (32) is first reduced to a semiclassical level of approximation by considering the momentum gradient part only, which results in

$$
\begin{aligned}
&F(q - q_{cl}(t+\Delta t), t+\Delta t) \\
&\simeq \exp\left[-\frac{1}{2}(\nabla \cdot p)\Delta t\right] F(q - q_{cl}(t), t).
\end{aligned}
\tag{33}
$$

Notice that the Planck constant does not yet appear in this level. The time evolution of it is approximately given by estimating the momentum gradient term as follows [29]

$$
\begin{aligned}
&F(q - q(t+\Delta t), t+\Delta t)|_{q=q_{cl}(t+\Delta t)} \\
&= \exp\left[-\frac{1}{2}\int_t^{t+\Delta t} dt \sum_k \frac{\partial p_k}{\partial q_k}\right] F(q - q_{cl}(t), t)\Bigg|_{q=q_{cl}(t)} \\
&= \left(\frac{\sigma(t)}{\sigma(t+\Delta t)}\right)^{1/2} F(q - q_{cl}(t), t)\Bigg|_{q=q_{cl}(t)},
\end{aligned}
\tag{34}
$$

or

$$
\sigma(t+\Delta t)^{1/2} F(q - q_{cl}(t+\Delta t), t+\Delta t)\Big|_{q=q_{cl}(t+\Delta t)} = \sigma(t)^{1/2} F(q - q_{cl}(t), t)\Big|_{q=q_{cl}(t)}, \tag{35}
$$

where the deviation determinant $\sigma(t)$ is expressed as

$$
\sigma(t) = \prod_{i=1}^{N} \wedge \left(q^i_{cl}(t) - q_{cl}(t)\right), \tag{36}
$$

which is an N-dimensional orientable tiny volume surrounding the point $q_{cl}(t)$ in configuration space, with $q^i_{cl}(t)$ $(i = 1, \ldots, N)$ being the configuration space point of an ith classical trajectories nearby the reference path $q_{cl}(t)$. See Figure 8 for geometrical definition of $\sigma(t)$. Since $\sigma(t)$ arises from the orientable volume, the square root of it gives rise to the part of the so-called Maslov phase [105]. (see ref. [106] for another geometrical meaning of the Maslov phase.)

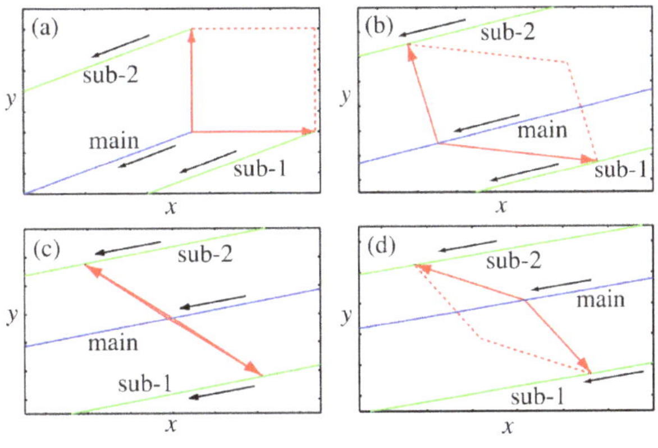

Figure 8. An illustration for geometrical implication of $\sigma(t)$ defined in Equation (36). A deviation determinant $\sigma(t)$, which is an orientable volume, withrespect to the main path is formed with the help of nearby classical paths. Time passes from panel (**a–d**). [Taken from Takahashi, S.; Takatsuka, K. *Phys. Rev. A* **2004**, *69*, 022110 with permission].

The above results can be represented in a symmetric manner by defining "square root" of the volume element [107]. Suppose that an arbitrary function $f(q_t)$, where q_t is the coordinate system in configuration space, is mapped from the original point q_0 at time 0 by the classical flow $q_0 \rightarrow q_t$. An integral also mapped is represented as

$$I = \int f(q_t, t) dq_t = \int f(q_t, t) dq_t^{1/2} dq_t^{1/2*} \tag{37}$$

with

$$dq_t^{1/2} \equiv \exp\left[\frac{i\pi M(q_0, q_t)}{2}\right] |dq_t|^{1/2}, \tag{38}$$

where $dq_t^{1/2*}$ is intended to be the complex conjugate to $dq_t^{1/2}$, and $M(q_0, q_t)$ is the Maslov index counting the number of the change of sign of $\sigma(t)$ of Equation (36). The semiclassical conservation rule is then written up to the phase as [107]

$$F(q_t, t) dq_t^{1/2} = F(q_0, 0) dq_0^{1/2}. \tag{39}$$

This is rather similar to the law of the Poincaré-Cartan absolute invariance in phase space [108].

The momentum gradient $\sum_k \partial p_k / \partial q_k$ can diverge, when it happens to hold $\sigma(t) = 0$ in Equation (35), where a classical path come to a focal point either in configuration or momentum space. This is the origin of the difficulty of the theories of semiclassical level. Yet, such a singularity can be removed by the quantum diffusion below [30].

5.2.3. Gaussian Wavepacket Approximation

We next proceed to take account of the quantum diffusion term. In doing so, we here survey the dynamics of a Gaussian wavepacket among other possible choices of $F(q, t)$. Heller introduced the dynamics of a Gaussian function as early as in 1975 by assuming that the potential function at the Gaussian center is truncated up to the quadratic terms [109].

We insert a Gaussian of the following form into the general scheme of ADF Equation [30],

$$G(q - q_{cl}(t), t) = g(t) \exp\left[-\frac{1}{c(t) + id(t)} (q - q_{cl}(t))^2\right], \tag{40}$$

with both $c(t)$ and $d(t)$ being real-valued. Then some (rigorous) manipulations give

$$c(t + \Delta t) = c(t) \left(\frac{\sigma(t + \Delta t)}{\sigma(t)}\right)^2 \tag{41}$$

and

$$d(t + \Delta t) = d(t) \left(\frac{\sigma(t + \Delta t)}{\sigma(t)}\right)^2 + \frac{2\hbar \Delta t}{m}. \tag{42}$$

or shows they satisfy the following differential equation, respectively,

$$\dot{c}(t) = 2 \frac{\dot{\sigma}(t)}{\sigma(t)} c(t). \tag{43}$$

and

$$\dot{d}(t) = 2 \frac{\dot{\sigma}(t)}{\sigma(t)} d(t) + \frac{2\hbar}{m}. \tag{44}$$

Note that the Planck constant appears only in this level of dynamical hierarchy. This normalized Gaussian form are propagated in time quite accurately, under the dynamical circumstances where the Gaussian may remain Gaussian. Yet, the theory needs additional procedures when the dynamics under study faces difficult situations like wavepacket bifurcation [110].

Further, one can find an interesting internal structure in the Gaussian packet. Define $\zeta(t)$ then such that

$$d(t) = \zeta(t)\sigma(t), \tag{45}$$

and insert this back into Equation (44) to obtain

$$\sigma(t)\dot{\zeta}(t) - \dot{\sigma}(t)\zeta(t) = \frac{2\hbar}{m}, \tag{46}$$

which is a Wronskian relation between $\sigma(t)$ and $\zeta(t)$. This suggests that *something* is rotating to maintain the packet in the quantum realm [30]. However, we step aside from this subject to be back to the energy quantization problem [110].

5.3. *On the Quantization Condition: Creation and Elimination of the Spectral Peaks by the Phases*

It is quite often that only the to-be-quantized (resonant) energy peaks are under focus without paying much attention to to-be-nullified part of $D(E)$ at off-resonant energy areas. We therefore reconsider here the basic mechanisms of evolution of the resonant peaks and those of suppression of the off-resonant counterparts.

5.3.1. Prior Quantization Conditions Based on the Periodic Orbits

To reconsider the mechanism of quantization from the view point of wavepacket dynamics, we get back to a little more elementary stage of the theory. Consider a spectral density function

$$P(E) = \int_{PO} \frac{dq_0 dp_0}{\hbar^N} P_{tr}(E_{cl}(q_0, p_0); E), \tag{47}$$

for an N-dimensional system, where the pre-spectral density is defined as

$$\begin{aligned} &P_{tr}(E_{cl}(q_0, p_0); E) \\ &= \lim_{L \to \infty} \frac{1}{L} \int_0^L dt R(q_t, q_0, t) \exp\left[\frac{i}{\hbar} S(q_t, q_0, t) - i\frac{\pi}{2} M(q_0, q_t)\right] \exp\left(\frac{i}{\hbar} Et\right), \end{aligned} \tag{48}$$

which is a correlation function starting from a classical point (q_0, p_0) on a manifold of energy E_{cl}. $M(q_0, q_t)$ is the Maslov index. $R(q_t, q_0, t)$ represents an overlap integral between a wavepacket at $t = 0$ and that of later time t. We implicitly assume that the wavepacket adopted here has been taken from the above ADF framework, Equation (26), yet we do not specify the functional form of it in this stage and it does not have to be a Gaussian. The Fourier transform in P_{tr} eventually gives rise to the energy spectrum in Equation (47). In the following analysis, we assume that (1) only the periodic orbits are to be considered in $P_{tr}(E_{cl}(q_0, p_0); E)$ as the symbol $\int_{PO}$ indicates, which is a natural consequence of the stationary-phase condition, (2) the phases in Equation (47) are just made up with the standard action integral and the associated Maslov phases alone, and (3) the amplitude factor $R(q_t, q_0, t)$ is always finite and smooth along the trajectories [111,112].

5.3.2. Peaks Arising from Individual Orbits and an Extended Quantization Condition

In the time-integration of $P(E)$ in Equation (48), one can apply the stationary phase approximation (SPA) to obtain

$$\frac{\partial}{\partial t}[S(q, q_0; t) + Et] = -E_{cl} + E = 0, \tag{49}$$

which indicates that classical trajectories with E_{cl} should make a major contribution to $P_{tr}(E_{cl}(q_0, p_0); E)$, which is already well-known [35–37,113,114]. And $P_{tr}(E_{cl}(q_0, p_0); E)$ can be viewed as a function of both E and $E_{cl}(q_0, p_0)$.

We first survey the oscillatory property of the integrand of $P_{tr}(E_{cl}(q_0,p_0);E)$. With the classical action integral

$$S(q_t,q_0,t) = \int_{q_0}^{q_t} p(q;E_{cl}(q_0,p_0))dq - E_{cl}(q_0,p_0)t, \tag{50}$$

we define

$$\Lambda(t;q_0,p_0) = R(q_t,q_0,t)\exp\left[\frac{i}{\hbar}\int_{q_0}^{q_t} p(q;E_{cl}(q_0,p_0))dq\bigg|_{\mathrm{PO}} - i\frac{\pi}{2}M(q_0,q_t)\right], \tag{51}$$

to rewrite $P_{tr}(E_{cl}(q_0,p_0);E)$ in a simplified fashion as

$$P_{tr}(E_{cl}(q_0,p_0);E) = \int dt\Lambda(t;q_0,p_0)\exp\left[\frac{i}{\hbar}(E-E_{cl}(q_0,p_0))t\right]. \tag{52}$$

Note that both $\int_{q_0}^{q_t} p(q;E_{cl}(q_0,p_0))dq$ and $\frac{\pi}{2}M(q_0,q_t)$ have the following time dependence on the periodic orbits

$$(\text{constant}) \times t + (\text{oscillatory part}). \tag{53}$$

Here the constant part is calculated out as an average of $\int_{q_0}^{q_t} p(q;E_{cl}(q_0,p_0))dq$ and $\frac{\pi}{2}M(q_0,q_t)$ in $\Lambda(t;q_0,p_0)$ of Equation (51) for the interval of, say, $[q_0,q_T]$. Let the period of the periodic orbit under consideration be $T(q_0,p_0)$. Then we define a quantity $X(q_0,p_0)$ such that

$$\frac{1}{T(q_0,p_0)}\left[\frac{1}{\hbar}\oint_{q_0}^{q_T} p(q;E_{cl}(q_0,p_0))dq - \frac{\pi}{2}M(q_0,q_T)\right] = X(q_0,p_0), \tag{54}$$

and extract $X(q_0,p_0)\times t$ from the phase part of (51) (that is Equation (53)) in such a way that

$$\frac{1}{\hbar}\int_{q_0}^{q_t} p(q;E_{cl}(q_0,p_0))dq - \frac{\pi}{2}M(q_0,q_t) - X(q_0,p_0)t. \tag{55}$$

Then this function should appears to be an exactly periodic function. Furthermore, since the preexponential factor $R(q_t,q_0,t)$ is also assumed to be periodic of $T(q_0,p_0)$,

$$\Lambda(t;q_0,p_0)\exp[-iX(q_0,p_0)t] \tag{56}$$

is periodic with the period $T(q_0,p_0)$ as a whole. Then one can expand the function $\Lambda(t;q_0,p_0)$ in a Fourier series such that

$$\Lambda(t;q_0,p_0) = \sum_{n=-\infty}^{\infty} c_n\exp\left(-i\frac{2n\pi}{T(q_0,p_0)}t + iX(q_0,p_0)t\right), \tag{57}$$

where

$$c_n = \frac{1}{T(q_0,p_0)}\int_0^T dt\Lambda(t;q_0,p_0)\exp\left[i\frac{2n\pi}{T(q_0,p_0)}t - iX(q_0,p_0)t\right]. \tag{58}$$

Here we take a limit in the time integral for $P_{tr}(E_{cl}(q_0,p_0);E)$ in Equation (48) up to sufficiently long time L and rewrite it with the help of Equation (57) as

$$\begin{aligned}
&P_{tr}(E_{cl}(q_0,p_0);E)\\
&\equiv \frac{1}{L}\int_0^L dt\Lambda(t;q_0,p_0)\exp\left[\frac{i}{\hbar}(E-E_{cl}(q_0,p_0))t\right]\\
&= \frac{1}{L}\sum_{-\infty}^{\infty} c_n\frac{-i\hbar T(q_0,p_0)}{-2n\pi\hbar + (E-E_{cl}(q_0,p_0)+X\hbar)T(q_0,p_0)}\\
&\times\left(\exp\left[\frac{iL}{\hbar T(q_0,p_0)}(-2n\pi\hbar + (E-E_{cl}(q_0,p_0)+X\hbar)T(q_0,p_0))\right] - 1\right).
\end{aligned} \tag{59}$$

The correlation function $R(q_t, q_0, t)$ is assumed to be so smoothly varying (or much less oscillatory than the phase factor) that it can be practically factored out from the Fourier expansion with respect to the phases.

It is now clear from Equation (59) that only the denominator in this expression satisfying the condition

$$(E - E_{cl}(q_0, p_0) + X\hbar)T(q_0, p_0) = 2n\pi\hbar \quad (60)$$

will lead to a peak in $P_{tr}(E_{cl}(q_0, p_0); E)$ of the height of c_n. The substitution of $X(q_0, p_0)$ with that of Equation (54) gives rise to a condition

$$(E - E_{cl}(q_0, p_0))T(q_0, p_0) + \left[\int_{q_0}^{q_t} p(q; E_{cl}(q_0, p_0))dq \Big|_{PO} - \frac{\pi}{2} M(q_0, q_T)\hbar \right] = 2n\pi\hbar. \quad (61)$$

This is a quantization (resonance) condition for $P_{tr}(E_{cl}(q_0, p_0); E)$ (but not for $P(E)$), which we call the *prior* quantization condition [111,115]. Not only E but (q_0, p_0) has to fulfill this condition to make a peak in $P_{tr}(E_{cl}(q_0, p_0); E)$. Although $P_{tr}(E_{cl}(q_0, p_0); E)$ has far more peaks than $P(E)$ itself, many of these peaks are to be eliminated by *destructive interference* upon integration over the trajectory space, which will be discussed below.

Yet, the true peaks should be contained in the set of peaks fulfilling Equation (61). Note also that the additional condition of Equation (49) nullifies

$$(E - E_{cl}(q_0, p_0))T(q_0, p_0) \quad (62)$$

in Equation (61), thereby reducing it directly to the EBK-like condition.

5.3.3. Destructive Interference Suppressing the Spurious Energy Peaks: Another Essential Role of the Phases for Quantization

The off-resonant trajectories should suppress the values of $P(E)$ to the order of $O(\hbar)$ due to the Riemann-Lebesgue lemma. However, there can exist many quasi-resonant trajectories, which would form spurious peaks in $P_{tr}(E_{cl}(q_0, p_0); E)$. These peaks are anticipated to be cancelled away by destructive interference with many other anonymous trajectories. As such there are two basic case in which to suppress virtual peaks as follows. (We refer to the original papers, ref. [111,112], for more details along with numerical presentations.)

Case in which E is Slightly Shifted from a True Eigenvalue (Near-Resonance Peaks)

Suppose there is a peak satisfying

$$\left[\int_{q_0}^{q_t} p(q; E_{cl}(q_0, p_0))dq - \frac{\pi}{2} M(q_0, q_T)\hbar \right] = 2n\pi\hbar \quad (63)$$

at an energy $E_{cl}(q_0, p_0) = E_x$. It is expected theoretically that the contribution from nearby orbits having $E'_{cl}(q'_0, p'_0)$ should make $P_{tr}(E'_{cl}(q'_0, p'_0); E_x)$ very small to the order of $O(\hbar)$. However, if $E'_{cl}(q'_0, p'_0)$ happens to be sufficiently close to the $E_x = E_{cl}(q_0, p_0)$, the peak is split into "*a pair of* very *high spurious* peaks" that contain the peak of $P_{tr}(E'_{cl}(q'_0, p'_0); E_x)$ in between. These peaks constitute a continuous width near the true peak. (For numerical and graphical examples, see [111].)

Case of Generation of Harmonics in E

There is yet another mechanism to generate spurious peaks. A trajectory with $E_{cl}(q_0, p_0)$ happens to satisfy the resonance condition of Equation (63) with some "quantum number" n and the energy E_x. Then, it is possible that the prior quantization condition of Equation (61) may be satisfied by another set of (n', E'_x). They must be simply the harmonics of the correct peak of (n, E_x) [111]. These harmonics also should be eliminated by off-resonant trajectories in terms of the appropriate destructive interference.

Having presented the specific roles of off-resonant trajectories to suppress the spurious peaks, we are concerned whether the trace formula implements such a function in it.

5.4. Amplitude-Free Energy Quantization of Classical Chaos

One of the theoretical and numerical difficulties of the trace formula manifests itself explicitly in the denominator of Equation (25) of the second term, which comes from the amplitude factor of the energy Greens function. In the above analysis of the energy spectrum, from Equations (48)–(61), no significant role was found for the amplitude factor $R(q_t, q_0, t)$ to play. We hence try to pursue a possible energy quantization disregarding the absolute value of the amplitude factor, but taking the Maslov phase into account.

To explicitly monitor the sensitivity of the amplitude factor to the variation of the preexponential factor, we first make the following preliminary calculations [116], in which the amplitude factor is rescaled intentionally as follows.

$$C_s^{ADF}(t) = \int dq_0 F^*(q_t, 0) F(q_0, 0) \left| \frac{\partial q_t}{\partial q_0} \right|^{1-s} \times \exp\left[\frac{i}{\hbar} S_1(q_t, q_0, t) + \frac{i}{\hbar} p_0 \cdot (q_0 - q_t) - \frac{i\pi M}{2} \right], \tag{64}$$

where $C_s^{ADF}(t)$ is a correlation function using a primitive Gaussian function, and an artificial scaling parameter s has been introduced. The value $s = 0.5$ reduces C_s^{ADF} to the original semiclassical correlation function, while there is no physical justification to choose other values. Yet, it numerically turns out that values up to at least $s = 1.5$ give correct energy positions in the resultant spectra [116]. In particular the choice of $s = 1.0$ implies that one can abandon the semiclassical amplitude factor as $\left| \frac{\partial q_t}{\partial q_0} \right|^0$, aside from the Maslov phase, to obtain the spectral positions.

After somewhat extensive analysis and numerical studies, we have confirmed that the spectrum positions are accurately reproduced through an amplitude correlation function [117,118], of the form

$$C_{p_0}^{III} \equiv \int dq_0 F^*(q_t) F(q_0) \exp\left(-\frac{i}{\hbar} p_0 q_t \right) \exp\left[\frac{i}{\hbar} S_2(q_t, p_0, t) - \frac{i\pi M}{2} \right], \tag{65}$$

where p_0 is an initial momentum to be scanned to cover the entire spectrum, and $F(q_t)$ is a (frozen) Gaussian function to be carried from the initial position q_0 [115,116]. We only need the way of calculation of the Maslov index as exemplified in Figure 8. It turns out that the amplitude-free correlation function $C_{p_0}^{III}$ works quite well even in chaotic regime, provided that the semiclassical dynamics under study is essentially dominated by the phases. An example of chaotic spectrum in chaotic region for the Hamiltonian

$$H = \frac{1}{2m_x} p_x^2 + \frac{1}{2m_y} p_y^2 + \frac{1}{2}(x^2 + y^2) + x^2\left(0.6y^2 + y\right) + \frac{1}{3} y^3 (0.2y - 1) + 0.1x \tag{66}$$

with $m_x = 1.0087$ and $m_y = 1.0$ is shown in Figure 9, where we have adopted $\hbar = 0.005$ and $E_{cl} \simeq 0.157$. (Note that for a large $\hbar$, the quality of semiclassical quantization is accordingly deteriorated and should not be used.)

The partial success of the amplitude-free quantization without paying full attention to the semiclassical amplitude factors, except for the Maslov phase, suggests that one does not necessarily have to faithfully track the every fine complexity of the Lagrange manifold behind classical chaos in order to perform semiclassical mechanics, partly due to the quantum smoothing.

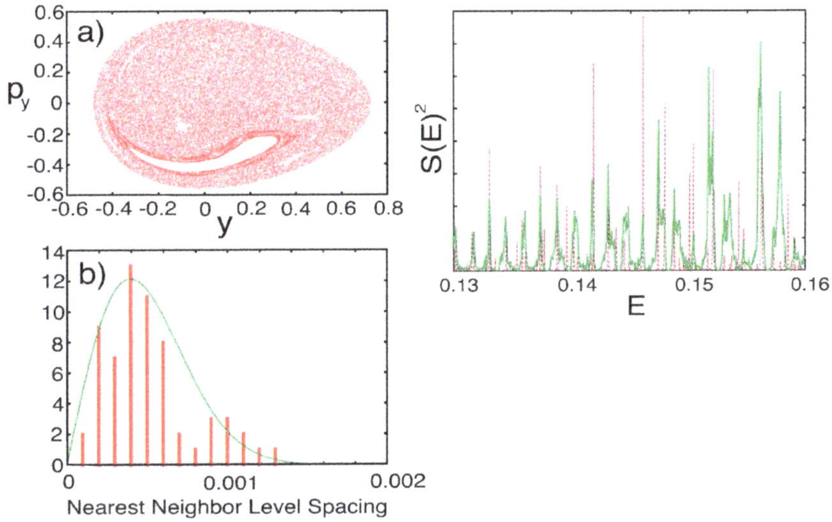

Figure 9. (**a**) Classical Poincaré surface of section at $x = 0$, $p_x = 0$ for an ensemble of trajectories with the total energy $E_{cl} = 0.157$. $\hbar = 0.005$. (**b**) The nearest neighbor level spacing of the full quantum spectrum in this energy region, which is subject to the Wigner distribution, an indication of quantum chaos. The right panel shows the spectrum taken from the amplitude free correlation function (green solid) and full quantum counterpart (red). Only the spectral positions should be compared. [Taken from Takahashi, S.; Takatsuka, K. *J. Chem. Phys.* **2007**, *127*, 084112 with permission].

5.5. Chaos Mediated by Dynamical Tunneling

Tunneling effect is one of the key phenomena to characterize chemical reactions not only in very low temperature circumstances but even in excited states. It is now widely accepted that complex semiclassical theory is very effective in the study of quantum tunneling [119]. However, classical chaos generally makes the structures of the Lagrange manifold and associated caustics very complicated in multidimensional complex space. Besides, the choice of the stationary paths there is not necessarily achieved as in the Feynman kernel over the real paths. Shudo and Ikeda have developed a pioneering work to lay out and cope with the intricate yet beautiful tunneling paths in the complex domain [120,121]. We here present a simple-minded idea on semiclassical tunneling by introducing the parity of motion, which splits phase space into many mutually exclusive nonclassical sub-spaces [122,123].

5.5.1. Semiclassical Tunneling Theory

We first need to formulate a tunneling theory in a semiclassical context [122,123]. Let us begin with Feynman's path integral representation of the kernel

$$K(x, \Delta t; y) = \left(\frac{m}{2\pi i\hbar\Delta t}\right)^{N/2} \exp\left(\frac{i\Delta t}{\hbar}\left[\frac{m}{2}(\frac{x-y}{\Delta t})^2 - V(\frac{x+y}{2})\right]\right) \tag{67}$$

or in a more limited alternative

$$K(q + \Delta q, \Delta t; q) = \left(\frac{m}{2\pi i\hbar\Delta t}\right)^{N/2} \exp\left(\frac{i\Delta t}{\hbar}\left[\sum_k \frac{m_k}{2}(\frac{\Delta q_k}{\Delta t})^2 - V(q)\right]\right), \tag{68}$$

where Δq is not necessarily as short as Δt, and $V(q)$ is an arbitrary potential function. In the path integration, the Euler variational principle

$$\frac{d}{dt}\left(\frac{\partial L}{\partial \dot{q}_k}\right) - \frac{\partial L}{\partial q_k} = 0, \tag{69}$$

with the Lagrangian

$$\begin{aligned} L &= \sum_k \frac{m_k}{2}(\frac{\Delta q_k}{\Delta t})^2 - V(q) \\ &= \sum_k \frac{m_k}{2}(\dot{q}_k)^2 - V(q) \quad \text{as } \Delta q \to 0 \text{ with } \Delta t \to 0 \end{aligned} \tag{70}$$

brings about the stationarity in the integral [94]. It is seen though that the kernel of Equation (68) is invariant to a set of parameters (constant) σ_k that are introduced in a manner that

$$K(q+\Delta q, \Delta t; q) = \left(\frac{m}{2\pi i\hbar\Delta t}\right)^{N/2} \exp\left(\frac{i\Delta t}{\hbar}\left[\sum_k \frac{m_k}{2}(\sigma_k \frac{\Delta q_k}{\Delta t})^2 - V(q)\right]\right) \tag{71}$$

as long as

$$\sigma_k^2 = 1. \tag{72}$$

Define accordingly the corresponding Lagrangian $L(\sigma)$ as

$$L(\sigma) = \sum_k \frac{m_k}{2}(\sigma_k \dot{q}_k)^2 - V(q) \tag{73}$$

which should satisfy the stationary condition

$$\frac{d}{dt}\left(\frac{\partial L(\sigma)}{\partial \dot{q}_k}\right) - \frac{\partial L(\sigma)}{\partial q_k} = 0. \tag{74}$$

An additional set of the Newtonian equations then results

$$m_k \ddot{q}_k = -\sigma_k \frac{\partial V(q)}{\partial q_k}. \tag{75}$$

Nothing happens when all $\sigma_k = 1$. If all the σ_k is -1, the paths emerged from Equation (75) are equivalent to the so-called instanton path [94]. We refer to σ_k as the parity of motion. A variety of the set $\{\sigma_k\}$ gives the various combinations of non-tunneling and tunneling paths in the real time. Thus, the sets of $\{\sigma_k\}$ defines the individual phase spaces, which we call sheets. No communication exists among them as long as the parity sets are frozen.

We can move to the Hamiltonian framework by defining

$$p_k = \sqrt{\sigma_k}\dot{q}_k, \tag{76}$$

where p_k is pure imaginary for the negative parity. The corresponding canonical equations of motion

$$\sigma_k \dot{\bar{p}}_k = -\frac{\partial H(\sigma)}{\partial q_k} \quad \text{and} \quad \sigma_k \dot{q}_k = \frac{\partial H(\sigma)}{\partial \bar{p}_k}, \tag{77}$$

is defined with the modified Hamiltonian

$$H(\sigma) = \sum_k \frac{\sigma_k}{2m_k}\bar{p}_k{}^2 + V(q). \tag{78}$$

Setting $\sigma_k = -1$ is effectively equivalent to flipping the mass m_k from positive to negative value.

To see the invariance property inherent to the Hamilton system, the following coordinates are convenient, that is

$$Q_k = \sqrt{\sigma_k} q_k \qquad \text{and} \qquad P_k = \sqrt{\sigma_k} \bar{p}_k, \tag{79}$$

in which the new coordinates Q_k's in configuration space are also pure imaginary when their parities are negative. Then Equation (77) are transformed to

$$\dot{P}_k = -\frac{\partial H(\sigma)}{\partial Q_k} \qquad \text{and} \qquad \dot{Q}_k = \frac{\partial H(\sigma)}{\partial P_k}. \tag{80}$$

Due to the Poincaré-Cartan theorem of absolute invariance [108], the symplectic structure the 2-form

$$\omega^2 = \sum_k dP_k \wedge dQ_k \tag{81}$$

is conserved and the Stokes theorem allows to define the path-action S_{cl}, that is,

$$S_{cl} = \int \sum_k P_k dQ_k - H(Q, P)dt. \tag{82}$$

In order to adopt these paths in semiclassical mechanics, the Q-coordinates should be rotated back to q-coordinates keeping P in such a way that

$$(Q, P) \to (q, P). \tag{83}$$

Then we have an action integral

$$\begin{aligned} S_{cl}(Q, P) \to S_{\text{tnl}}(q, P) &= \int \sum_k P dq_k - H(\sigma)dt \\ &= \sum_k \int \sqrt{\sigma_k} \bar{p}_k dq_k - H(\sigma)dt, \end{aligned} \tag{84}$$

and a semiclassical wavefunction for tunneling is formally written as

$$\psi = A \exp\left(\frac{i}{\hbar} S_{\text{tnl}}\right), \tag{85}$$

where the pre-exponential factor A represents an amplitude. Since S_{tnl} in Equation (84), is generally complex-valued, we rewrite S_{tnl} in the standard form as

$$S_{\text{tnl}} = S_R + iS_I. \tag{86}$$

Insertion of Equation (86) into Equation (85) gives

$$\psi = A \exp\left(-\frac{1}{\hbar}|S_I|\right) \exp\left(\frac{i}{\hbar} S_R\right), \tag{87}$$

where the phase convention to S_{tnl} is taken so that S_I becomes positive. Thus, the damping of the norm due to tunneling is estimated as

$$\exp\left(-2\frac{|\text{Im } S_{\text{tnl}}|}{\hbar}\right), \tag{88}$$

which gives a natural definition of the relevant tunneling probability.

The tunneling paths induce additional phases when they are incorporated into the semiclassical Feynman kernel such that [124]

$$K(q_t, t; q_s, s : \sigma) = \sum_{\alpha} \left(\frac{1}{-2\pi i\hbar} \right)^{\frac{N}{2}} \frac{1}{\sqrt{\left| \frac{\partial q_t}{\partial p_s} \right|}} \exp\left(\frac{i}{\hbar} S_{\text{tnl}} + i\phi \right), \tag{89}$$

where N is the number of dimensions and the additional phase ϕ is given by

$$\phi = -\frac{\pi}{2} N_f + \frac{\pi}{4} N_p \tag{90}$$

at an entrance of a tunneling path and

$$\phi = -\frac{\pi}{2} N_f - \frac{\pi}{4} N_p \tag{91}$$

at an exit of a tunneling path. Here N_p is the number of the negative parities while N_f is the sum of the multiplicities of the zeros of $\partial q_f / \partial \bar{p}_i = 0$ along the trajectory, which gives the Maslov index.

Connection Problem

Tunneling paths can arise only upon transferring from $\{\sigma_k = 1\}$ to another, typically only one of σ_k is negative. This is because as the number of negative parities increases the damping probability in Equation (88) also become large. Those paths on the different sheets should be connected smoothly. To ensure smoothness, we change the sign of one of the parities at a caustic in the direction normal to the caustic line. The caustics, at which the density of the paths becomes very high (actually infinite in the primitive semiclassical approximation), are defined with the so-called Jacobi field [94] in such a way that

$$det|\frac{\partial \vec{q}_f}{\partial \vec{p}_i}| = 0, \tag{92}$$

where $\vec{p}_i$ is a real-valued (quasi-)momentum vector for a given parity set $\{\sigma_k = \pm 1\}$ at the initial point on each sheet and $\vec{q}_f$ is the final points of a trajectory on the same sheet. The tunneling path can get back to the Newtonian sheet of $\{\sigma_k = 1\}$ in a similar manner [125].

The stationary phase condition in the semiclassical approximation to the Feynman kernel requires that the quantum phase S_{HJ} in Equation (89) should be connected smoothly at the connection point. This requirement is expressed as

$$\frac{\partial S_{HJ}}{\partial q} \text{ (in classical region)} = \frac{\partial S_{\text{tnl}}}{\partial q} \text{ (in tunnel region)}, \tag{93}$$

which simply implies

$$\bar{p}_k = -\bar{p}_k \tag{94}$$

at the connection point. Obviously,

$$\bar{p}_k = 0 \tag{95}$$

is the unique solution [126]. Therefore, to connect the wavefunctions smoothly at the turning points in the relevant direction k, we may change the parity there (or parities depending on the degeneracy of turning points) to the direction of zero momentum.

Tunneling phenomena are quite important in many chemical reaction dynamics, and some applications of the present treatment are found in Refs. [124,126–128].

5.5.2. Statistical Redistribution of Tunneling Paths

In view of the nature of this review article, we briefly show an example of chaos that is induced by tunneling dynamics, just for curiosity. The Hamiltonian is

$$H = \frac{1}{2m_x}p_x^2 + \frac{1}{2m_y}p_y^2 + \frac{1}{2}(x^2 + y^2) + x^2y - \frac{1}{3}y^3, \tag{96}$$

where the masses are chosen as $m_x = 1.0087$ and $m_y = 1.0$ so as to break the symmetry of C_3. The vibrational modes and the corresponding tori in the Poincaré surface of section are depicted in Figure 10. The bottom panel displays a long time Poincaré section arising from a single trajectory, which repeats dynamical tunneling from time to time among tori, thus giving rise to chaos effectively. Here, however, we have intensionally disregarded the damping of the amplitude induced by tunneling. Careful inspection will find the faint trace of a couple of tori [125]. See ref. [125] for details about the present tunneling dynamics.

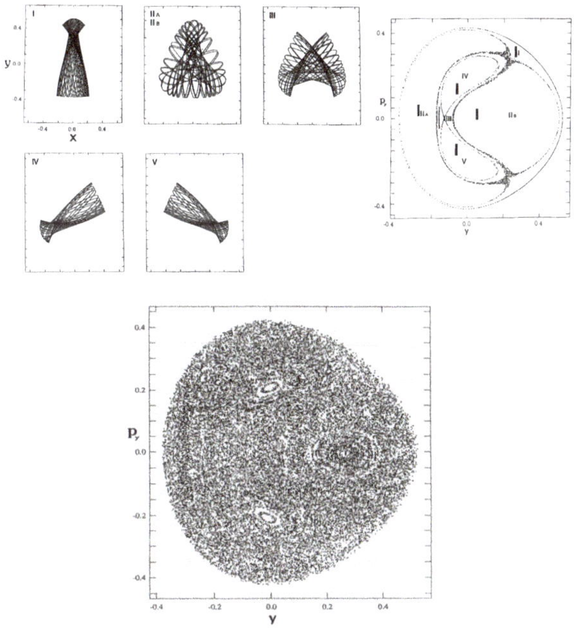

Figure 10. The right panel of the upper row shows the Poincaré section ($x = 0$ and $p_x > 0$) along with the vibrational modes (left) belonging to the tori. $E = 0.09$. The botton row shows a long time plot at the Poincaré section, which wanders from one torus to another through tunneling. Statistical distribution through tunneling is recognized. [Taken from Ushiyama, H.; Takatsuka, K. *Phys. Rev. E* **1996**, *53*, 115 with permission].

6. Chaos Arising from Repeated Bifurcation and Merge of the Quantum Wavepackets on Nonadiabatically Coupled Potential Basins

One of the characteristic features of quantum chaos is reflected on the energy level statistics such as the patterns in the nearest neighbor level spacing distribution (NNLSD) [114,129]. The extensive level repulsions among the states are widely recognized to be the origin of the Wigner distribution in NNLSD. It is therefore natural to pay attention to the *nonadiabatic interactions* as one of the key factors to modulate the structure of the energy levels. As such, the theory of level dynamics [9,130–132] sets a focus on nonadiabatic interactions as a factor that can modulate the energy levels in an anomalous fashion. Shudo [133,134] explicitly clarified the effects of an avoided crossing on the level distributions, the avoided crossing of which is induced by a variation of a part (s) of interaction potential but not by those interactions presented in Equation (12). The nonadiabatic interactions of Equation (12) are basically about electronic-state mixing due to the *nuclear kinematic coupling* on the occasion of the variation of nuclear configuration (molecular geometry) and are one of the most critical interactions in chemical dynamics [14,16–20]. Therefore, the beautiful methods in the level dynamics does not seem to apply to the dynamics of molecules.

Besides, we are quite often concerned with the wavepacket dynamics of electrons rather than the electronic energy levels at a given nuclear configuration. In other words, the essential characteristics of molecular nonadiabatic dynamics is seen when the relevant quantum wavepackets *bifurcate* and coherently *merge* (remix) many times at the molecular geometry of significant nonadiabatic interactions. It is quite natural therefore to suspect that the wavepacket bifurcation and merging should be the dynamical origin of purely quantum chaos in place of "stretching and folding" in the baker's transformation for classical chaos. We hence survey the nuclear wavepacket dynamics of bifurcation and remixing in this section first and then proceed to the nonadiabatic electron wavepacket dynamics in the next section.

6.1. Experimental Observation of Bifurcation and Merge of the Quantum Wavepackets

We here show that "bifurcation and merging" of quantum wavepackets can be observed with the modern real-time experiments. Among the cutting-age experimental methodologies to track the femtosecond-scale nuclear motions in chemical dynamics, the time- and angle-resolved photoelectron spectroscopy (TRPES) has been very well established since 1999 [135–137], when the group of this author happened to start the first complete ab initio full quantum mechanical simulation to track nuclear wavepackets in TRPES [138–140]. Among others, we showed for the first time that the instance of bifurcation and merging of nuclear wavepackets in passing across the nonadiabatic regions can be directly observed [141,142]. That is, such bifurcation and merging are of physical reality. Now other novel experimental methodologies are available.

For readers who might be interested in the recent experimental progress since then, which are related to observation nonadiabatic dynamics, we refer to the following works (we apologize for not a complete list). T. Suzuki is the first who used photoelectron imaging technique to observe the dynamics passing through an intersystem crossing of pyradine [143,144]. Wörner and his group observed the passage of NO_2 molecule across the conical intersection by means of the high harmonic spectroscopy [145] and time-resolve photoelectron spectroscopy [146]. Further novel methods have been introduced, such as attosecond stimulated X-ray Raman spectroscopy [147], time-resolved fluorescent spectroscopy [148], ultrafast electron diffraction technique [149], and so on. We also found that the wavepacket bifurcation can be directly observed not only by pump-probe technique by means of pulse lasers but with induced photoemission spectroscopy from molecules placed under a CW laser [150].

The basic experimental setting for the time-resolved photoelectron spectroscopy is as follows: (1) First excite a ground state wavepacket to an electronically excited state by a pump pulse-laser. This laser is of a sufficiently short width so that the nuclear wavepacket thus excited begins to run spontaneously, which eventually passes across the nonadiabatic region(s). Then, (2) shine another laser (probe laser) to ionize the excited state. (3) Measure the energy and angular distributions of photoelectrons as a function of the time delay (the time lag between the probe and pump), and thereby retrieve the information of the dynamics of the nuclear (chemical) dynamics. This TRPES technique is quite successful in that there is no closed channel to be probed.

The theoretical framework of TRPES is outlined below [138–140]. Briefly, the total wavefunction is expanded in terms of the three relevant electronic states: Φ_1 (with a potential function V_1), Φ_2 (with V_2), and $\Phi_{\mathbf{k}}^{(-)}$ (ion state), as

$$\Psi(\mathbf{r}, R, t) = \chi_1(R,t)\Phi_1(\mathbf{r};R) + \chi_2(R,t)\Phi_2(\mathbf{r};R) + \int d\mathbf{k}\chi_{\mathbf{k}}(R,t)\Phi_{\mathbf{k}}^{(-)}(\mathbf{r};R), \quad (97)$$

where $\Phi_{\mathbf{k}}^{(-)}(\mathbf{r};R)$ is a scattering state having the in-going boundary condition [50] with $\mathbf{k}$ labeling the wave vector of the photoelectron. The time-dependent coupled Schrödinger equations for the vibrational wavepackets χ_1, χ_2, and $\chi_{\mathbf{k}}$ are to be solved in a coupled Schrödinger equations (note that $\chi_{\mathbf{k}}$ is continuous with respect to $\mathbf{k}$) under all the matrix elements being evaluated in full quantum mechanics [138–140]. For instance, we need the matrix elements with respect to the electronic continuum representing the electronic ionization process. This is not an easy task. Once they are all solved through, the photoelectron spectra are calculated from the vibrational wavepackets after the probe interaction as

$$P(\varepsilon_k) = k \int dR d\hat{k} |\chi_{\mathbf{k}}(R)|^2. \tag{98}$$

Figure 11 shows the photoelectron spectra of NaI molecule exhibiting the wavepacket bifurcation and merging, taken for the pump-probe delays of 675, 700, 725, and 750 fs. Snapshots of the absolute square of the wavepackets on the diabatic potentials are shown in the upper column. The red and blue ones represent $\left|\chi_1^{(dia)}(R,t)\right|^2$ on $V_1^{(dia)}(R)$ and $\left|\chi_2^{(dia)}(R,t)\right|^2$ on $V_2^{(dia)}(R)$, respectively. These wavepackets are passing across the crossing point (nonadiabatic region) from the right to the left. The photoelectron signals, the kinetic energy distribution of photoelectrons, at each time are correspondingly draw in the lower column. It is clearly seen that the peaks of photoelectron signals are modulated in real time from one to two and three peaks, highlighting the wavepacket bifurcation to those components. These dynamics can be comprehended primarily with the Condon principle: The wavepackets running on the lower potential energy make the higher photoelectron kinetic energy signals. The wavepacket bifurcation and merging and can thus be physically observed in real time.

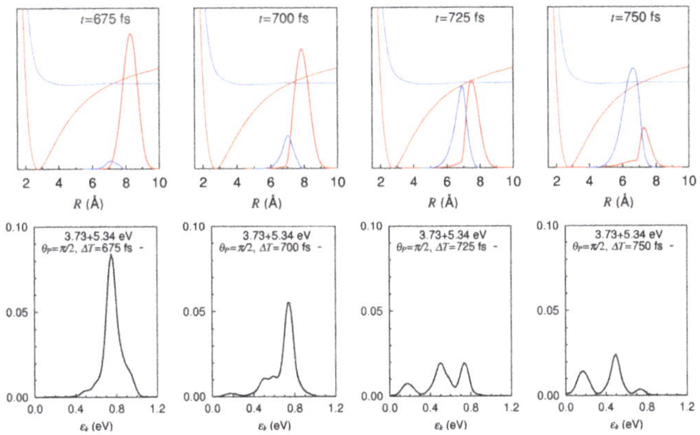

Figure 11. Photoelectron kinetic energy spectrum as the wavepackets bifurcate and merge at the avoided crossing. (The potential curves in this figure are graphcally represented in the *diabatic* representation.) The polarization vector of the probe laser light is perpendicular to that of the pump, while the latter is set parallel to the line between Na and I. In the upper row are shown the absolute square of the wavepackets at selected times (red (blue) one on the red (blue) potential curve). The bottom row represents the corresponding energy-resolved photelectron signals. Figure drawn by Dr. Y. Arasaki. [Taken from Takatsuka, K. *Bull. Chem. Soc. Jpn.* **2021**, *94*, 1421 with permission].

6.2. Chaotic Eigenfunctions in Nonadiabatically Coupled Potential Functions

We next survey how the infinite repetition of bifurcation and merge of the quantum wavepackets results in the specific signature in eigenfunctions. Fujisaki systematically investigated this aspect by placing a focus on the transition from regular to chaotic states of the wavefunctions as functions of energy and the magnitude of nonadiabatic coupling elements [151,152]. The system used was the so-called Heller model, which was defined in their pioneering work [153]. It is defined as (see also Figure 12)

$$\mathcal{H} = \begin{pmatrix} T + V_A & J \\ J & T + V_B \end{pmatrix},$$
$$T = \frac{1}{2}(p_x^2 + p_y^2), \quad V_A = \frac{1}{2}(\omega_x^2 x^2 + \omega_y^2 y^2) + \epsilon_A, \quad V_B = \frac{1}{2}(\omega_x^2 \xi^2 + \omega_y^2 \eta^2) + \epsilon_B \tag{99}$$

with $\xi = (x + 2a\sin\theta)\cos 2\theta + (y - 2a\cos\theta)\sin 2\theta$ and $\eta = -(x + 2a\sin\theta)\sin 2\theta + (y - 2a\cos\theta)\cos 2\theta$, where T is a nuclear kinetic energy (nuclear masses are assumed to be unity), V_i $(i = A, B)$ two-dimensional diabatic harmonic potentials of each electronic state i, J the nonadiabatic coupling constant between the electronic states induces an interaction between the two otherwise uncoupled harmonic potentials. ϵ_i $(i = A, B)$ indicates the bottoms of each harmonic potential. The geometrical meanings of a and θ are shown in Figure 12. The angle 2θ referred to as the Duschinsky angle cause a coupling between x and y. Unlike the original Heller model, we chose the frame for V_A to be 'diagonalized' (no crossing terms like xy in V_A). The location of the crossing manifold (usually called crossing seam) should be presented in the original paper [152].

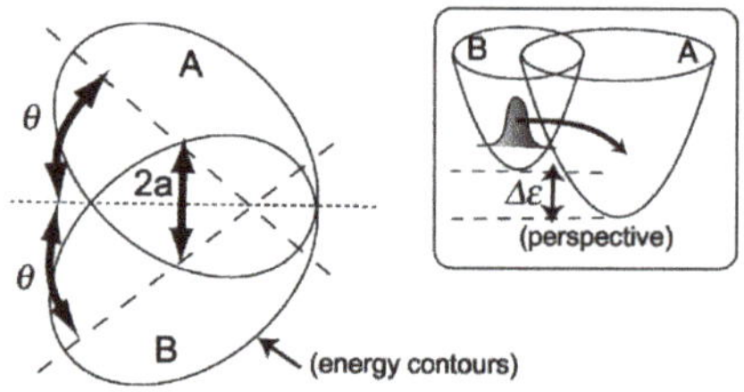

Figure 12. A schematic representation of Heller's model system. The distance between the centers of the potentials is 2a, and the angle between the relevant crossing seam (dotted line) and the primary axis of each potential (dashed line) is θ. Inset: The perspective view of the present system. The energy difference of the potential bottoms is $\Delta\epsilon = \epsilon_B - \epsilon_A$. [Taken from Fujisaki, H.; Takatsuka, K. *J. Chem. Phys.* **2001**, *114*, 3497 with permission].

Since the Heller model is simply an uncoupled pair of harmonic potentials at $J = 0$, it can have merely regular states depending on the values on J and θ. Yet, the intense mixing induced by these coupling elements lead the wavefunctions and energy spectra to exhibit chaotic patterns. As an example, Figure 13 demonstrates the spatial distribution of some (selected) very chaotic wavefunctions. The chaoticity was quantified in terms of a transition from the Poisson to Wigner distributions in the nearest neighbor level spacing distribution (NNLSD), Δ_3 statistics, and so on [151]. Here in the graph of Figure 13, only the so-called amplitude distributions of the eigenfunctions, proposed by Berry [154], on one of the surfaces (A) are shown. The definition of the amplitude distributions is

$$P(\phi) = \frac{1}{\sqrt{2\pi\sigma^2}} \exp\left(-\frac{\phi^2}{2\sigma^2}\right), \tag{100}$$

where $P(\phi)$ is the frequency distribution of the amplitude of the eigenfunction that is randomly sampled at points that are energetically accessible in configuration space on each potential. About 6000 points are randomly sampled in Figure 13, and it turns out that $P(\phi)$ is found to be fitted with a large σ, reflecting the spatially wide distribution due to chaos. Although not shown here, we found that the regular wavefunctions give much smaller σ (far narrower distributions) [151,152]. Thus, we have seen a regular to chaotic transition of eigenfunctions (and eigenvalues as well, not shown here) in nonadiabatically coupled quantum system.

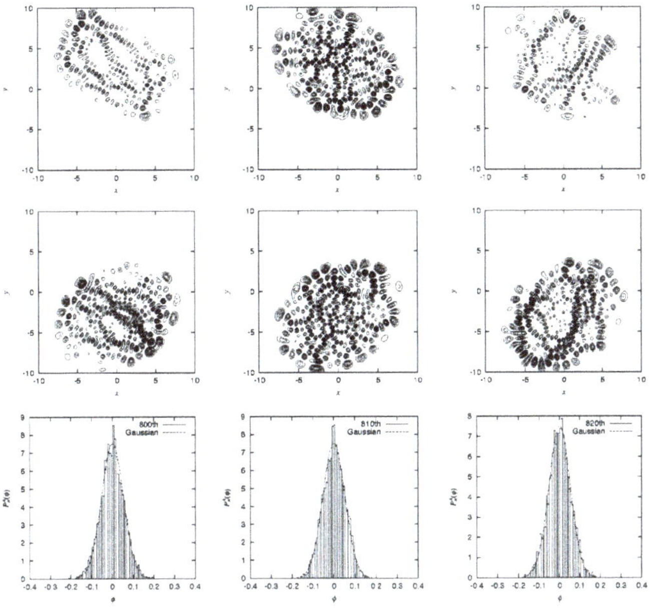

Figure 13. First row from the top: Eigenfunctions squared (from left to right, 800th, 810th, and 820th) of Heller's model system on surface A (see Figure 12) with $J = 1.5$ (intermediate coupling) and $\theta = \pi/6$. Second row: The corresponding eigenfunctions (squared) on surface B. Third row: The corresponding amplitude distributions of the eigenfunctions onsurface A. The dashed envelope curves represent Gaussian functions. [Taken from Fujisaki, H.; Takatsuka, K. *Phys. Rev. E* **2001**, *63*, 066221 with permission].

6.3. Need for Measures of Chaoticity in Quantum Wavepackets

One of the most interesting issues in quantum chaos is how to measure the extent of chaoticity. This would depend on the definition of quantum chaos [2–12]. Characterization of chaoticity embedded in discrete quantum *energy levels* has been extensively studied (see Refs. [5,9,11] for instance). For example, the measures such as NNLSD, Δ_3 statistics, and so on, work well to capture chaos. On the other hand, we need to characterize chaoticity in the *dynamics* of wavefunctions of molecules. Berry's amplitude distributions of the eigenfunctions $P(\phi)$ in Equation (100) can be used as one of such measures by applying at each time of the time-propagation of a packet. However, this measure seems to be based on a rather intuitive expectation that the chaotic wavefunction should be always widely distributed in space. Another concern is whether it can distinguish well between regular systems and those of quantum localization and the quantum scar.

The studies on classical chaos have already established such measures like the Lyapunov exponent [155], the real-complex valuedness of the eigenvalues of the stability matrix, KS entropy, and so on [1]. In practical studies on chemical dynamics we often need a crisp measure to detect and/or characterize chaoticity in the propagation of wavefunctions. However, it is known that quantum coherence working in quantum wavepackets particularly make a significant difference from the classical one. Suppose we have two wavepackets $\phi_\mu(t)$ and $\phi_{\mu'}(t)$, which are slightly different from each other in their initial conditions μ and μ', the initial condition of two wavepackets for instance. In marked contrast to the exponential deviation of two classical trajectories of slightly different initial conditions, the following overlap between the two packet

$$O_{\mu\mu'}(t) = \left| \left\langle \phi_\mu(t) | \phi_{\mu'}(t) \right\rangle \right|^2 \tag{101}$$

is kept constant as long as they are subject to the same Hamiltonian. For the same reason, quantities like the Shannon entropy

$$S(t) = -\int dq|\phi_\mu(q,t)|^2 \log|\phi_\mu(q,t)|^2 \tag{102}$$

is conserved as well. Then, a primitive idea to break the coherence is as follows [156]. Prepare a projection operator

$$P_i = \int_{A_i} dr|r\rangle\langle r| \tag{103}$$

where A_i is a subset (local area) of the entire space and satisfies $A_i \cap A_j = \emptyset$ for $i \neq j$. In place of the overlap integral $O_{\mu\mu'}(t)$

$$O_{\mu\mu'}(t) = \left|\left\langle \phi_\mu(t) \middle| \sum_i P_i \middle| \phi_\mu(t) \right\rangle\right|^2 \tag{104}$$

we consider the following modulation

$$D_{\mu\mu'}(t) = \sum_i \left|\left\langle \phi_\mu(t)|P_i|\phi_{\mu'}(t) \right\rangle\right|^2, \tag{105}$$

which obviously destroys the coherence among the different regions. In chaos $\phi_\mu(t)$ and $\phi_{\mu'}(t)$ deviate from each other in configuration space, and $D_{\mu\mu'}(t)$ is anticipated to exhibit a quick damp from a slight difference in the initial condition [156]. A good compromise to balance the number of A_i and size is necessary to sensitive yet economical detection of chaos. It has been shown numerically that the method works very well even for the nonadiabatic dynamics as in the above Heller model system [156]. However, this decoherence method costs much labor in many dimensional systems. We hence need general and tractable alternatives. We will be back to this aspect in the next section.

7. Chaos in Nonadiabatic Electron Dynamics of Molecules

In this section, we review chaotic dynamics in electronic state manifolds, those states involved in which are mixed frequently and extensively by the nuclear kinematic couplings, or as vaguely termed, nonadiabatic interactions [21,23–27]. (We do not review the traditional and canonical theories of nonadiabatic transitions for nuclear wavefunctions, essentially originating from the spirit of the Landau-Zener-Stuekelberg theory [14,16,17,19,20]).

7.1. *The Hamiltonian Studied for Electron Dynamics*

7.1.1. Electron Dynamics

The coupled Equations (11) based on the Born-Huang expansion, Equation (9), set the theoretical foundation of molecular science. Yet, various drawbacks arise coming from the time-independent representation of the electronic wavefunctions $\{\Phi_I(\mathbf{r};\mathbf{R})\}$. First of all, it is technically hard to prepare a set of potential functions $\{V_I(\mathbf{R})\}$, on which the nuclear wavepackets $\{\chi_I(\mathbf{R},t)\}$ propagate in time. In particular, the most serious situation appears when the electronic states are densely quasi-degenerate and are strongly coupled to one another through the nonadiabatic interactions. In such a case a huge time-dependent fluctuation over the electronic states can appear. To cope with those difficulties, we resume the dynamics with time-dependent electronic wavefunctions from the beginning.

We hereby restart our basic framework from the following representation of the total Hamiltonian operator [21,157]

$$\begin{aligned} &H(\mathbf{R}, elec) \\ &= \frac{1}{2}\sum_k \left(\hat{P}_k - i\hbar \sum_{IJ} |\Phi_I\rangle X^k_{IJ} \langle \Phi_J|\right)^2 + \sum_{IJ} |\Phi_I\rangle H^{(\mathrm{el})}_{IJ} \langle \Phi_J|, \end{aligned} \tag{106}$$

where $\hat{P}_k$ is the nuclear momentum operator. $\mathbf{R}$ indicates the nuclear coordinates, and we represent the electronic states in the Hilbert space, not in the configuration space $\mathbf{r}$. Note that $\hat{P}_k$ are to be operated only on the nuclear wavefunctions, not on the electronic wavefunctions $\Phi_I(\mathbf{r};\mathbf{R})$, since such kinematic interactions have already been taken into account in X^k_{IJ} of Equation (106). The sum over the electronic states suffixed with I and J may include continuum states.

No approximation has been made so far, and, we here introduce an approximation that the nuclear momentum operators $\hat{P}_k$ are replaced with their classical counterparts P_k as

$$\tilde{H}(\mathbf{R},\mathbf{P},elec) \equiv \frac{1}{2}\sum_k\left(P_k - i\hbar\sum_{IJ}|\Phi_I\rangle X^k_{IJ}\langle\Phi_J|\right)^2 + \sum_{IJ}|\Phi_I\rangle H^{(\mathrm{el})}_{IJ}\langle\Phi_J|. \tag{107}$$

Then the total wavefunction subject to the Hamiltonian of Equation (106) is to be reduced to an electron wavepacket state running on paths $\mathbf{R}(t)$ (but not necessarily Newtonian trajectories). More explicitly, we expand it as

$$|\Phi_{\mathbf{elec}}(t;\mathbf{R}(t))\rangle = \sum_I C_I(t)|\Phi_I(\mathbf{r};\mathbf{R})\rangle\Big|_{\mathbf{R}=\mathbf{R}(t)}, \tag{108}$$

along each $\mathbf{R}(t)$, where $\{\Phi_I(\mathbf{r};\mathbf{R})\}$ is a basis functions taken at each nuclear coordinate $\mathbf{R}$, and $\Phi_I(\mathbf{r};\mathbf{R})$ do not have to be the eigenfunctions of the electronic Hamiltonian. Equations (107) and (108) are widely referred to as mixed quantum-classical representation. The time-dependent coefficients $C_I(t)$ are to be evaluated with the coupled equations of motion for electron wavepackets in such a way that

$$i\hbar\frac{dC_I}{dt} = \sum_J\left[H^{(\mathrm{el})}_{IJ} - i\hbar\sum_k \dot{R}_k X^k_{IJ} - \frac{\hbar^2}{4}\sum_k(Y^k_{IJ} + Y^{k*}_{JI})\right]C_J, \tag{109}$$

where $H^{(\mathrm{el})}_{IJ}$, X^k_{IJ}, and Y^k_{IJ} have been defined in Equations (12) and (13), respectively. Again, the electronic basis $\{\Phi_I(\mathbf{r};\mathbf{R})\}$ are not demanded to be adiabatic ones. The bra-ket notation used here indicates integration over the electronic coordinates $\mathbf{r}$ only. Furthermore, the second order derivative terms Y^k_{IJ} in Equation (109) are quite often neglected in practice because they are always accompanied by the small quantity $\hbar^2$ [25,26], although it is not always negligible in general.

7.1.2. Nuclear Dynamics: Path Branching

The nuclear paths $\mathbf{R}(t)$ are driven by the force field provided by the electronic states, more explicitly in terms of the force matrix F^k_{IJ} (see ref. [21]) defined as

$$F^k_{IJ} = -\frac{\partial H^{(\mathrm{el})}_{IJ}}{\partial R_k} - \sum_K\left(X^k_{IK}H^{(\mathrm{el})}_{KJ} - H^{(\mathrm{el})}_{IK}X^k_{KJ}\right) + i\hbar\sum_l \dot{R}_l\left[\frac{\partial X^l_{IJ}}{\partial R_k} - \frac{\partial X^k_{IJ}}{\partial R_l}\right], \tag{110}$$

which arises from the variational principle or the Hamilton canonical equations of motion for $(\mathbf{R},\mathbf{P})$ [21].

The non-zero off-diagonal elements of F^k_{IJ} can induce branching of the nuclear paths $\mathbf{R}(t)$ [22]. If, on the other hand, the force matrix is diagonal, that is, $F^k_{IJ} = 0$ for $I \neq J$, each of F^k_{II} gives a Newtonian force on each potential function I, which gives the theoretical foundation for the so-called ab initio dynamics, molecular dynamics, the first principle dynamics, and so on.

7.1.3. Nuclear Dynamics: Mean-Field Approximation (Semiclassical Ehrenfest Theory)

To save the work needed for the path branching calculations [22], we often adopt a mean-field approximation, the so-called semiclassical Ehrenfest theory (SET), in which the force matrix of Equation (110) is to be averaged over the electron wavepacket as

$$\begin{aligned}\langle F_k \rangle &= \sum_{IJ} C_I^* F_{IJ}^k C_J \\ &= -\sum_{IJ} C_I^* \frac{\partial H_{IJ}^{(\mathrm{el})}}{\partial R_k} C_J - \sum_{IJ}\sum_{K} C_I^* \left(X_{IK}^k H_{KJ}^{(\mathrm{el})} - H_{IK}^{(\mathrm{el})} X_{KJ}^k \right) C_J, \end{aligned} \tag{111}$$

which gives rise to a single force for a single path. If the basis set $\{\Phi_I(\mathbf{r};\mathbf{R})\}$ happens to be complete, Equation (111) is reduced to the form of Hellmann–Feynman force

$$\langle F_k \rangle = -\left\langle \Psi_{\mathrm{elec}} \left| \frac{\partial \hat{H}^{(\mathrm{el})}}{\partial R_k} \right| \Psi_{\mathrm{elec}} \right\rangle. \tag{112}$$

The use of the mean-field approximation is justified for short-time propagation, typically shorter than 100 fs. Indeed, it has been numerically shown [22] that the electronic-state mixing is very well (accurately) reproduced by SET for a short time, provided that the nonadiabatic coupling elements X_{IK}^k and electronic matrix elements $H_{IK}^{(\mathrm{el})}$ are both correctly evaluated. As the nuclear positions proceed to some extent, the Ehrenfest paths should be modified accordingly [22].

7.2. Longuet–Higgins (Berry) Phase and Lorentz-like Force

Before proceeding to the electronic-state chaos, we make a detour to a specific force working on the nuclei, which can cause breaking the mirror symmetry of molecular shape, namely molecular chirality.

Let us review the molecular Hamiltonian of Equation (107) now in classical electromagnetic field

$$\begin{aligned}&\tilde{\mathcal{H}}(\mathbf{R},\mathbf{P},elec,\mathbf{A}) \equiv \\ &\frac{1}{2}\sum_k \left(P_k - \frac{Z_k e}{c} A_k - i\hbar \sum_{IJ} |\Phi_I\rangle X_{IJ}^k \langle \Phi_J| \right)^2 + \sum_{IJ} |\Phi_I\rangle \mathcal{H}_{IJ}^{(el)}(\mathbf{A}^{ele}) \langle \Phi_J|, \end{aligned} \tag{113}$$

where $A_k(\mathbf{R})$ are the (classical) electromagnetic vector potentials on nuclei. The electronic Hamiltonian also includes the vector potential $\mathbf{A}^{ele}$. In this expression we notice that the nonadiabatic couplings mechanically stand parallel with the vector field $\mathbf{A}$, which suggests the qualitative roles of the nonadiabatic couplings working on the nuclear momentum may be akin to those of the vector potentials.

Indeed it is well-known that there is a mathematical similarity between the Aharonov–Bohm phase [158] and the Longuet–Higgins phase [159], the latter being well known in chemical dynamics relevant to the so-called conical intersection, which is a special case of nonadiabatic interaction. (see ref. [18] for the interesting properties of conical intersection.) These two phases have been unified as the so-called geometrical phase, often called Berry's phase [160,161]. In light of the above parallelism in Equation (113), we suspect another similarity to exist between them: As the electromagnetic Lorentz force works on a charged particle running in a magnetic field, a novel force, which we call the Lorentz-like nonadiabatic force, appears to work on nuclei running in a field created by the nonadiabatic interaction field [21,162].

To be more specific, let us imagine a molecular system as illustrated in Figure 14.

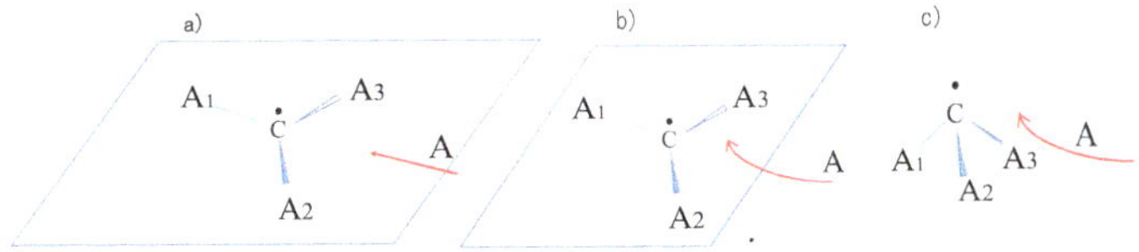

Figure 14. A symmetry breaking by pyramidalization of a coplanar molecular system, leading to molecular chirality. (**a**) Atom A is approaching the planar molecule on the same plane. (**b**) The Lorentz-like nonadiabatic force triggers to guide the incident atom to the direction out of the plane. (**c**) The Hellmann-Feynman force follows to make the pyramidalization complete. [Taken from K. Takatsuka, J. Chem. Phys. **146**, 084312 (2017) with permission].

At the center we have an atom C, say carbon atom, and around it lie three atomic or molecular elements A_1, A_2, A_3. All the elements A_1, A_2, A_3 in the molecule $CA_1A_2A_3$ are mutually different and sit stably on a common plane. Suppose an external atom A is coming in on the same plane of an $x-y$ coordinates. (A_1, A_2, A_3, and A should be all different species from each other.) The z-direction is set to be perpendicular to this plane and the plus or minus of this direction determines the right or left with respect to the plane. Let (x_a, y_a, z_a) be a point of the incident atom A. Then the force matrix in z_a-direction (perpendicular to the plane) [162] is given such that

$$F_{IJ}^{z_a} = -\left(\frac{\partial H^{(el)}}{\partial R_{z_a}}\right)_{IJ} - i\hbar\left(\dot{R}_{x_a}\left[\frac{\partial X_{IJ}^{x_a}}{\partial R_{z_a}} - \frac{\partial X_{IJ}^{z_a}}{\partial R_{x_a}}\right] + \dot{R}_{y_a}\left[\frac{\partial X_{IJ}^{y_a}}{\partial R_{z_a}} - \frac{\partial X_{IJ}^{z_a}}{\partial R_{y_a}}\right]\right)$$
$$- i\hbar \sum_{b\neq a}^{A_1,A_2,A_3} \sum_{j_b}^{x,y,z} \dot{R}_{j_b}\left[\frac{\partial X_{IJ}^{j_b}}{\partial R_{z_a}} - \frac{\partial X_{IJ}^{z_a}}{\partial R_{j_b}}\right], \tag{114}$$

where the suffix b specifies A_1, A_2, and A_3, and j_b indicates the (x,y,z)-coordinate of each element. This is a force while the state undergoes nonadiabatic transition from an Ith to the Jth one. Just for simplicity, we assume that all the atoms other than the incident atom A stay still, thus neglecting the possible vibrational motion. For the sake of clear presentation, we simply set to

$$\dot{R}_{j_b} = 0 \qquad (j = x, y, z \text{ and } b \neq a). \tag{115}$$

Hence only the atom A is supposed to be moving on the plane with a nonzero value of $(\dot{R}_{x_a}, \dot{R}_{y_a})$ and $\dot{R}_{z_a} = 0$. $F_{IJ}^{z_a}$ is then reduced to

$$F_{IJ}^{z_a} = -\left(\frac{\partial H^{(el)}}{\partial R_{z_a}}\right)_{IJ} - i\hbar\left(\dot{R}_{x_a}\left[\frac{\partial X_{IJ}^{x_a}}{\partial R_{z_a}} - \frac{\partial X_{IJ}^{z_a}}{\partial R_{x_a}}\right] + \dot{R}_{y_a}\left[\frac{\partial X_{IJ}^{y_a}}{\partial R_{z_a}} - \frac{\partial X_{IJ}^{z_a}}{\partial R_{y_a}}\right]\right). \tag{116}$$

We here define the Lorentz-like nonadiabatic force with the second term of the right hand side of this equation as [162]

$$f_{IJ}^{z_a} \equiv -i\hbar\left(\dot{R}_{x_a}\left[\frac{\partial X_{IJ}^{x_a}}{\partial R_{z_a}} - \frac{\partial X_{IJ}^{z_a}}{\partial R_{x_a}}\right] + \dot{R}_{y_a}\left[\frac{\partial X_{IJ}^{y_a}}{\partial R_{z_a}} - \frac{\partial X_{IJ}^{z_a}}{\partial R_{y_a}}\right]\right) \tag{117}$$

and simply write as

$$F_{IJ}^{z_a} = -\left(\partial_{z_a} H^{(el)}\right)_{IJ} + f_{IJ}^{z_a}, \tag{118}$$

where $-\left(\partial_{z_a} H^{(el)}\right)_{IJ}$ is the short-hand notation of the off-diagonal Hellmann-Feynman term. (Note that the diagonal elements of the nonadiabatic interaction X_{II} are zero if the function used as the basis functions are real. However, this is not the case for complex-valued wavefunctions.)

The expression in Equation (117) can be reexpressed in a more familiar manner: Let us consider a vector of the force matrix elements as

$$\vec{f}^a_{IJ} = \left(f^{x_a}_{IJ}, f^{y_a}_{IJ}, f^{z_a}_{IJ} \right). \quad (119)$$

In terms of the vector $\vec{X}^a_{IJ}(\mathbf{R})$ of the nonadiabatic interaction fields

$$\vec{X}^a_{IJ}(\mathbf{R}) = \left(X^{x_a}_{IJ}(\mathbf{R}), X^{y_a}_{IJ}(\mathbf{R}), X^{z_a}_{IJ}(\mathbf{R}) \right) \quad (120)$$

we define a vector field

$$\vec{Z}^a_{IJ}(\mathbf{R}) = -i\hbar \vec{\nabla}^a \times \vec{X}^a_{IJ}(\mathbf{R}), \quad (121)$$

in which the dependence on the nuclear coordinates $\mathbf{R}$ is explicitly denoted. Then $\vec{f}^a_{IJ}$ is expressed as

$$\vec{f}^a_{IJ}(\mathbf{R}) = \vec{v}^a \times \vec{Z}^a_{IJ}(\mathbf{R}), \quad (122)$$

where $\vec{v}^a$ is the velocity vector of the atom A, and $\vec{\nabla}^a \times$ indicates the curl on the coordinates for the atom A only [162].

It is immediately noticed that this expression is essentially of the same mathematical form as that behind the Fleming *left-handed* rule in classical electromagnetic theory. The Lorentz force distinguishes the front and back sides of a plane made up by an electron current and a magnetic field. See Figure 15. Thus, a simple relationship is found between the electromagnetic Lorentz force and the Lorentz-like nonadiabatic force in such a way that

$$\vec{A}^a \leftrightarrow \vec{X}^a_{IJ}, \quad \vec{B}^a \leftrightarrow \vec{Z}^a_{IJ}, \quad \dot{\mathbf{R}}_a \times q_a e \mathbf{B}_a \leftrightarrow \vec{v}^a \times \vec{Z}^a_{IJ}, \quad (123)$$

where $\mathbf{B}$ is a magnetic field.

We thus recognize that the nonadiabatic force can distinguish right and left with respect to the plane of the vector field of nonadiabatic interactions, which can serve as one of the fundamental mechanisms of the origin of molecular chirality. Yet, further study is needed to quantify the actual magnitude of anisotropic force.

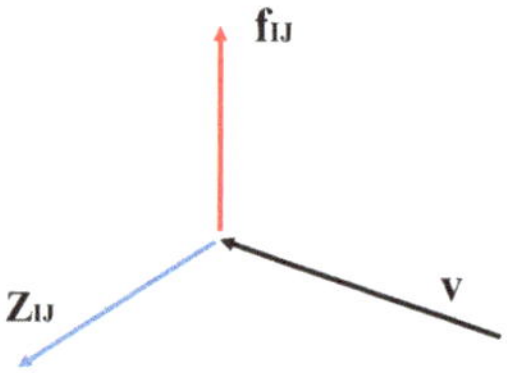

Figure 15. A particle coming in with a velocity vector $\mathbf{v}$ into the $\mathbf{Z}_{IJ}$ field feels a force $\mathbf{f}_{IJ}$, which is orthogonal to both $\mathbf{v}$ and $\mathbf{Z}_{IJ}$, according to $\vec{f}^a_{IJ}(\mathbf{R}) = \vec{v}^a \times \vec{Z}^a_{IJ}(\mathbf{R})$. See the text for the definition of $\mathbf{Z}_{IJ}$.

7.3. *Chaotic Electron Dynamics in Densely Quasi-Degenerate Electronic-State Manifold; B_{12} Cluster as an Example*

As a candidate among molecules possibly having electronic-state chaos, we study some of the highly excited states of a cluster consisting of 12 boron atoms, B_{12}. The nonadiabatic interactions working among the excited states of this cluster cause very frequent transitions among adiabatic states [163–165], resulting in frequent and continual state-mixing as represented as (recall Equation (108))

$$\sum_I^{\text{adiabatic}} C_I(t) \Phi^{\text{ad}}_I(\boldsymbol{r}; \boldsymbol{R}(t)). \quad (124)$$

In contrast to classical chaos, there is no crisp criterion with which to define or identify chaos in quantum dynamics, to the best of our knowledge. Below we only quantify the extent of chaoticity of the relevant electron dynamics represented as in Equation (124). A special attention is paid to the random mixing among the deterministic electronic states in the state space. We then discuss the possible manifestation of those strong chaoticity in chemical and physical properties of the clusters.

7.3.1. Nearest Neighbor Level Spacing Distribution (NNLSD)

At any selected time t, or at any molecular shape $\boldsymbol{R}$, one can diagonalize the electronic Hamiltonian matrix $H_{IJ}^{(\mathrm{el})}$ in Equation (109). The energy levels are therefore geometry dependent. We have surveyed the NNLSD for the B_{12} cluster (see ref. [164] and its supplemental material stored in J. Chem. Phys.). It turns out that whereas the NNLSD shows a clear Wigner distribution in case of low symmetry of molecular shape, level clustering occurs to some extent when the molecule is of higher symmetry, actually C_{3v} symmetry. This is natural because the systems of higher symmetry tend to bear accidental degeneracy more often. On the other hand, the possibility (measure) for this system to take C_{3v} symmetry is actually zero in the dynamical studies of variable $\boldsymbol{R}(t)$. Thus, we can judge that the present system has a chaotic structure in its energy level. However, our goal is to clarify and characterize the chaotic feature possibly hidden in the individual electronic wavepackets.

7.3.2. Diffusive Dynamics in the State Space: Fractional Brown Motion

It is experimentally hard to pick and prepare a pure adiabatic excited state out of the densely quasi-degenerate electronic state. Even if one could successfully pump the ground state to one of the highly adiabatic states, the resultant state would quickly begin to undergo continual nonadiabatic mixing with other states, each of which in turn further mixes with other states. These enduring state-mixings look like a diffusion as a whole in the electronic Hilbert space. Figure 16 shows the magnitude of $|C_I(t)|^2$ of Equation (124) in a color code in the (I,t) space. The brighter color indicates the larger $|C_I(t)|^2$. It visibly demonstrates that diffusive dynamics has been certainly realized in the electronic-state space.

To quantify the extent of the diffusive motion of the wavepackets in the electronic state-space explicitly, we survey the scaling property of the dynamics in terms of the fractional Brownian motion (fBm) of Mandelbrot [166]. Consider

$$\sum_{J=1}(J-I(0))^2|C_J(t)|^2 = D_{I(0)}t^{\alpha}, \tag{125}$$

where $I(0)$ indicates the state number of the initial state (that is $C_I(0)$), and $C_J(t)$ is the mixing coefficient at time t starting from this initial $I(0)$ state. Again $C_J(t)$ have been evaluated with the SET electron wavepackets. Recall that the exponent $\alpha = 1.0$ implies the ordinary diffusion and $D_{I(0)}$ is just the ordinary diffusion constant. States of the initial adiabatic state $I(0) = 3$, 40, 100, 300, and 500 have been numerically examined [164]. Furthermore, it turns out that the nonadiabatic electron dynamics of $I(0) \geq 40$ are indeed the fractional Brownian motion (fBm) in the sense of Equation (125). The exponent of the present fBm is found to be about $\alpha = 0.15$, which suggests a milder stochastic motion than ordinary diffusion. Interestingly, the values of the exponent is mostly independent of the initial states chosen, whereas the diffusion constants $D_{I(0)}$ significantly depend on the initial states, with the higher states having the larger diffusion constant.

Thus, Figure 16 convinces enough that the concept of adiabatic state in the Born-Oppenheimer approximation and the associated the energy level (potential energy hypersurface) all lose sense in the present situation. Besides, the dynamical aspect of chaos is far more directly relevant to the present huge electronic-state fluctuation than the static signatures of chaos like the level statistics.

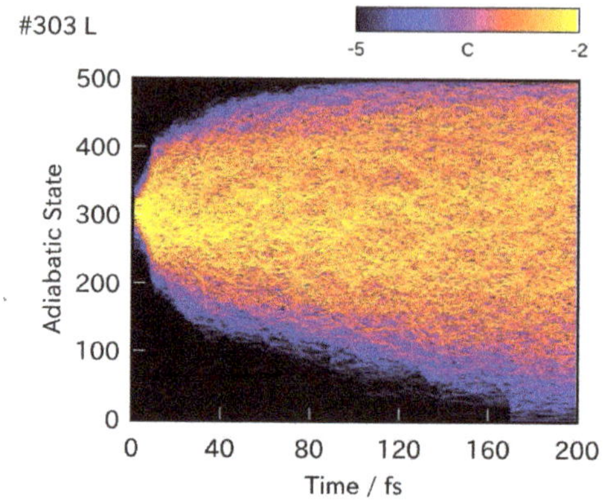

Figure 16. Diffusion–like redistribution of the electronic states in the excited electronic state manifold of B_{12} cluster. It is actually a fractional Brownian motion [164] in the state space. The dynamics of the $\log|C_I(t)|^2$ of Equation (108) is tracked with the SET and is charted in color code. An initial state has been prepared at the 300th adiabatic state. The extensive diffusive state is realized well before 10 fs. [Figure drawn by Dr. Y. Arasaki. Taken from Takatsuka, K. *Bull. Chem. Soc. Jpn.* **2021**, *94*, 1421 with permission].

7.3.3. Shannon Entropy

We next calculate the Shannon entropy in terms of $|C_I(t)|^2$ taken from Equation (124), which is the probability for the electronic wavepacket to occupy the Ith adiabatic state at time t. Hence,

$$S = -\sum_I |C_I(t)|^2 \log|C_I(t)|^2 \tag{126}$$

quantifies the information about randomness in the electronic Hilbert space. Note this entropy is different from Equation (102). Incidentally, the present system is too large to apply the spirit of decoherence as in Equation (105). Figure 17 shows S for the three wavepackets, those starting from the adiabatic state #100, #200, and #300. A very quick rise from $S = 0$ is observed for all the wavepackets up to about 5 fs, and much slower uprises follow since then. The absolute magnitude of the entropy is found to be larger for the higher state, which seems natural. In case of the adiabatic dynamics, where no electronic-state mixing is induced, the entropy is kept zero since $|C_I(t)|^2 = 1.0$. We thus see that the present electron dynamics keep distributing a large amount of the population over the extensive range in the Hilbert space.

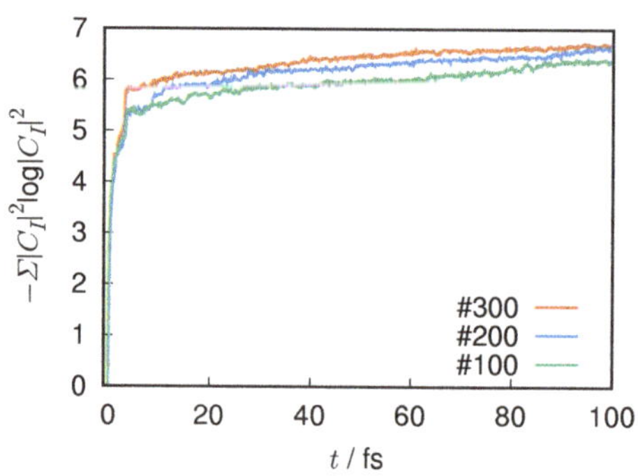

Figure 17. Time evolution of the Shannon entropy with respect to the adiabatic state population $|C_I(t)|^2$ involved in the wavepackets, each starting from the adiabatic state #100, #200, and #300, respectively. [Provided by Dr. Yasuki Arasaki].

7.3.4. Lyapunov Exponents for the Loss of Electronic State Memory

In the present nonadiabatic electron wavepacket dynamics, the main origin of chaoticity is not nonlinearity in the nuclear path dynamics but the nonadiabatic electronic state-mixing. To quantify the chaoticity due to the nonadiabatic interactions directly, we compare two dynamics that start from the same initial condition and one running on an *adiabatic potential* and the other to run on the *nonadiabatic averaged potential*. Suppose we have the Ith pure adiabatic state $\Phi_I^{ad}(\boldsymbol{r};\boldsymbol{R}(t))$ at the nuclear position $\boldsymbol{R}(t)$ at time t. Let a classical path run on this potential surface $V_I(\boldsymbol{R})$ for a short time Δt with an arbitrary momentum $\mathbf{P}(t)$, reaching a point at $\boldsymbol{R}^{ad}(t+\Delta t)$ with the electronic state $\Phi_I^{ad}(\boldsymbol{r};\boldsymbol{R}^{ad}(t+\Delta t))$. Likewise we propagate a nonadiabatic electronic state starting from the same set of $\left(\Phi_I^{ad}(\boldsymbol{r};\boldsymbol{R}(t)),\boldsymbol{R}(t),\mathbf{P}(t)\right)$ but the nonadiabatic interactions are taken account this time. The nuclear path then arrives at $\boldsymbol{R}^{SET}(t+\Delta t)$, which is slightly different from $\boldsymbol{R}^{ad}(t+\Delta t)$, but a significant different electronic wavepacket $\Phi_{(I)}^{SET}(\boldsymbol{r};\boldsymbol{R}^{SET}(t+\Delta t))$ is to be attained. The process is summarized as

$$\begin{cases} \Phi_I^{ad}(\boldsymbol{r};\boldsymbol{R}(t)) \rightarrow \Phi_I^{ad}(\boldsymbol{r};\boldsymbol{R}^{ad}(t+\Delta t)) & \text{by adiabatic path} \\ \Phi_I^{ad}(\boldsymbol{r};\boldsymbol{R}(t)) \rightarrow \Phi_{(I)}^{SET}(\boldsymbol{r};\boldsymbol{R}^{SET}(t+\Delta t)) & \text{by nonadiabatic path.} \end{cases} \tag{127}$$

The deviation of $\Phi_{(I)}^{SET}(\boldsymbol{r};\boldsymbol{R}^{SET}(t+\Delta t))$ from $\Phi_I^{ad}(\boldsymbol{r};\boldsymbol{R}^{ad}(t+\Delta t))$ should reflect a memory loss of the initial state, and we quantify this the process by regarding it as an exponential decay of

$$\begin{aligned} &\left|\left\langle \Phi_I^{ad}(\boldsymbol{r};\boldsymbol{R}^{ad}(t+\Delta t)) \middle| \Phi_{(I)}^{SET}(\boldsymbol{r};\boldsymbol{R}^{SET}(t+\Delta t))\right\rangle\right|^2 \\ &= \left|\left\langle \Phi_I^{ad}(\boldsymbol{r};\boldsymbol{R}(t)) \middle| \Phi_I^{ad}(\boldsymbol{r};\boldsymbol{R}(t))\right\rangle\right|^2 \exp(-\xi_I \Delta t), \end{aligned} \tag{128}$$

where t is chosen arbitrary. Unfortunately the integral in the left hand side of Equation (128) is not easy to estimate. However, for a sufficiently short Δt, the following approximation holds very accurately

$$\boldsymbol{R}^{SET}(t+\Delta t) \simeq \boldsymbol{R}^{ad}(t+\Delta t), \tag{129}$$

since their deviation is far slower and smaller. In fact, the deviation is numerically found to be about 0.005 Å in 1.0 fs (not shown graphically). Thus, we may define

$$\xi_I(t;\Delta t) = -\frac{1}{\Delta t}\log\left|\left\langle \Phi_I^{ad}(\boldsymbol{r};\boldsymbol{R}^{SET}(t+\Delta t)) \middle| \Phi_{(I)}^{SET}(\boldsymbol{r};\boldsymbol{R}^{SET}(t+\Delta t))\right\rangle\right|^2, \tag{130}$$

and with use of Equation (124) it is estimated by

$$\xi_I = -\frac{1}{\Delta t}\log|C_I(t+\Delta t)|^2. \tag{131}$$

Figure 18 represents the process of the above memory-loss process as observed in the selected states. Panel (a) shows that the nonadiabatic electronic state starting from the adiabatic #300 one quickly loses its memory about which adiabatic state it emerged from [167]. Actually it takes only 0.5 fs for the nonadiabatic state to lose most of the memory in $|C_I(t)|^2$. The relevant exponents defined in Equation (128) is shown in panel (b). In this calculation, Δt was set to 0.5 fs and the time t was scanned. Panel (b) displays two exponents ξ_I of Equation (131) for the nonadiabatic states starting from the #100 and #300 adiabatic states, respectively. They fluctuate from time to time with the average value around 5.0 fs^{-1}. This implies that the adiabatic path can immediately lose its memory as soon as the nonadiabatic interactions are switched on, no matter where the paths run. In other words, the adiabatic states are entirely embedded in a wide chaotic sea.

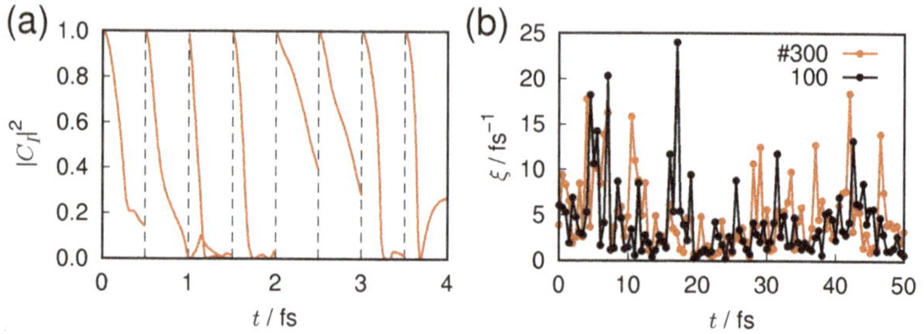

Figure 18. Loss of memory of nonadiabatic states in the Hilbert space. (**a**) Quick decay of $|C_I(t)|^2$ from 1.0 to $C_I(t + \Delta t)$ with $\Delta t = 0.5$ fs at selcted points of t for B_{12} cluster SET dynamics starting from the 300th adiabatic state. Dynamics starting at $t = 0.0$, 0.5, 1.0, 1.5, 2.0, 2.5, 3.0, and 3.5 fs. (**b**) Exponent of memory loss, defined in Equation (128), for the initial adiabatic component weight, $\xi(t, \Delta t)$, starting from either the 300th (red) or the 100th (black) adiabatic state. $\Delta t = 0.5$ fs. Lines connect calculated points only for visualization. [Provided by Dr. Yasuki Arasaki].

7.3.5. Turbulent Electron Flow in the Cluster

To observe one of the physical consequences from the present stochastic electron dynamics, we next survey the internal electron current driven by the nonadiabatic interactions. The probability current density of electrons $\vec{j}(\mathbf{r}, t)$, which is also termed electron flux, quantifies and visualizes the flow pattern. $\vec{j}(\mathbf{r}, t)$ is supposed to satisfy the continuity equation

$$\frac{\partial \rho(\mathbf{r}, t)}{\partial t} + \nabla \vec{j}(\mathbf{r}, t) = 0, \tag{132}$$

and in quantum mechanics $\vec{j}(\mathbf{r}, t)$ is defined as

$$\vec{j}(\mathbf{r}, t) = \frac{\hbar}{2im_e}[\psi^* \nabla \psi - \psi \nabla \psi^*], \tag{133}$$

where m_e and ψ are the mass and wave function of the involved particles [168]. For many-electron systems it is reduced to

$$\begin{aligned} \vec{j}(\mathbf{r}, t) &= \frac{\hbar}{2im_e}[\nabla_{\mathbf{r}} \rho(\mathbf{r}', \mathbf{r}) - \nabla_{\mathbf{r}'} \rho(\mathbf{r}', \mathbf{r})]\Big|_{\mathbf{r}' \to \mathbf{r}} \\ &= \frac{\hbar}{2im_e} \sum_i^{\mathrm{NO}} n_i \left[\lambda_i^*(\mathbf{r}') \nabla_{\mathbf{r}} \lambda_i(\mathbf{r}) - \nabla_{\mathbf{r}'} \lambda_i^*(\mathbf{r}') \lambda_i(\mathbf{r})\right]\Big|_{\mathbf{r}' \to \mathbf{r}}, \end{aligned} \tag{134}$$

where $\nabla_{\mathbf{r}}$ and $\nabla_{\mathbf{r}'}$ are the nabla with respect to $\mathbf{r}$ and $\mathbf{r}'$, respectively, and $\{\lambda_i\}$ are the natural orbitals that are eigenfunction of the density operator $\hat{\rho}(t)$ ($\rho(\mathbf{r}', \mathbf{r}) = \langle \mathbf{r}' | \hat{\rho} | \mathbf{r} \rangle$) with occupation numbers n_i. Note that only the complex-valued natural orbitals can make contributions to Equation (134). Stationary-state electronic wavefunctions like most of the eigenfunctions of the electronic Hamiltonian $\hat{H}^{(el)}$ are real-valued (with exceptions with respect to angular momentum eigenfunctions) and thereby give zero flux only (see [169,170] for the general electronic and nuclear flux in general molecules, refs. [171–174] for the early applications of electron flux in chemical dynamics, ref. [175] for an extensive review, and ref. [176] for one of the latest).

Figure 19 shows the electron flux induced by the electron wavepacket dynamics starting from the 74th and 75th states, individually, with no nuclear kinetic energy but released spontaneously from the initial geometry. The snapshots have been taken at two very short times at 0.6 fs and 1.0 fs. These flux vectors are viewed from the top. The packet starting from the 74th state happens to exhibit the electron current in the counterclockwise direction at 0.6 fs, while the 75th counterpart shows the current mostly in the direction clockwise. Although the cluster seems to be in C_{3v} symmetry, it is relaxed to the lower one, and consequently, these circular patters are not conserved and shortly transformed to a more complicated flow. Indeed, only shortly after 1.0 fs both fluxes begin to fall into very complicated in spatial patterns. After all, those electron currents become turbulent or random flow (not shown graphically). The most exciting chemical study is then to identify the consequences of electronic turbulence in chemical reactivity and molecular properties.

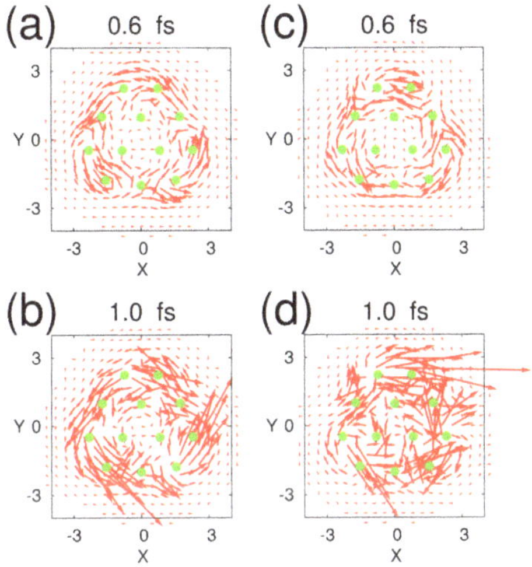

Figure 19. Spontaneous electron flux arising from the wavepackets in the SET dynamics starting from the 74th adiabatic state at (**a**) 0.6 fs and (**b**) 1.0 fs, and from the 75th adiabatic states at (**c**) 0.6 fs and (**d**) 1.0 fs. dynamics. The flux vectors are systematically multiplied by 2.0×10^4 for visualization. These fluxes are driven by nonadiabatic interactions. Soon after 1.0 fs, the regular-looking fluxes become turbulent. [Taken from Yonehara,T; Takatsuka, K. *J. Chem. Phys.* **2016**, *144*, 164304 with permission].

Incidentally, one can define the flux of electronic energy [177], which is of course different from the above electron flux. It is quite interesting to track real-time energy flow in molecules. We have observed the turbulent flux of the electronic energy too in the above excited-state cluster dynamics [177].

We have thus observed the clear signatures of quantum chaos in molecular nonadiabatic electron dynamics.

7.4. The Long Life-Time of Dynamical Chemical Bonding: Hyper Resonance

The above diffusive and chaotic phenomena due to the very frequent nonadiabatic interactions prevent molecular dissociation and bring about a notion of quantum resonance we have not seen before. As resonance phenomena the Pauling resonance in the theory of chemical electronic structure [178] and the Feshbach resonance [49] in quantum scattering theory [50] are most relevant to the present study. The Pauling resonance in the so-called valence bond theory is regarded as a configuration mixing (interaction) among the so-called resonance structures, each of which corresponds intuitively to a valence bonding. The Feshbach resonance, or core-excited resonance, is schematically represented as

$$\begin{aligned} &M + A \to (MA)^* \\ &\to \text{dissociations eventually with a long lifetime,} \end{aligned} \tag{135}$$

where a state of a target nuclei, atom, or molecule (denoted as M) is pumped to the higher levels by borrowing the kinetic energy of an incident particle A. The Feshbach resonances and their overlapping set the foundation of Intramolecular Vibrational energy Redistribution (IVR) in the unimolecular dissociation dynamics, as discussed in Section 3.2. Thus, the dissociation channels are temporarily closed by a temporal compound state $(MA)^*$ surviving a long life-time. Both the Pauling and Feshbach resonances can be regarded basically as events on a single adiabatic potential surface.

During the continual state-mixing in the diffusive motion in B_{12}, the electronic and nuclear energies are mutually exchanged through the nonadiabatic interactions. This phenomenon can be regarded as another kind of quantum mechanical resonance. The present dynamics found in B_{12} may be termed as *hyper-resonance* (resonance among the resonant states), since each adiabatic state is already in a Feshbach and/or Pauling resonance state. We refer to such a very long life-time tentative chemical bonding as *dynamical bonding* [165].

7.5. *Intra-Molecular Nonadiabatic Electronic Energy Redistribution*

7.5.1. Huge Inflation of Phase-Space Volume

Despite the very high energy of those excited-state boron clusters, they do not readily break apart, since the dissociation channels of them tend to be closed by very frequent nonadiabatic transitions. This is a natural consequence of hyper-resonance. We then survey a possible mathematical ground of the long life-time in terms of a drastically simplified statistical theory based on the phase space theory, originally proposed by J. C. Light [179]. This theory has been extensively applied to and modified before in the context of cluster dynamics [180–182], Calvo [183–187], and Fujii [65,66].

We consider only the simplest dissociation of B_{12} cluster in which only one atom is evaporated from N-body cluster ($N = 12$). The phase space theory under no external field predicts the dissociation rate k as

$$k = \frac{W_{N-1}(E)}{\Omega_N(E)}, \tag{136}$$

where $\Omega_N(E)$ is the density of the states of the N-particle cluster at the total energy E, and $W_{N-1}(E)$ denotes the total flux induced by the dissociation of a single atom [179]. As for a single adiabatic potential energy surface, say $V_I(\boldsymbol{R})$ at a nuclear configuration $\boldsymbol{R}$, $\Omega_N(E)$ is approximately estimated with the expression

$$\Omega_N^{(I)}(E) = \hbar^{-3N} \int d\boldsymbol{R}d\boldsymbol{P}\, \delta(E - H_I(\boldsymbol{R},\boldsymbol{P})), \tag{137}$$

where

$$H_I(\boldsymbol{R},\boldsymbol{P}) = \frac{\boldsymbol{P}\cdot\boldsymbol{P}}{2} + V_I(\boldsymbol{R}), \tag{138}$$

with the obvious notation of the total nuclear momentum $\boldsymbol{P}$. We assume that the total energy is high above the potential barrier of the dissociation channels and that there are not significant centrifugal barriers for the present non-rotating clusters. In such systems, the so-called bottle-neck volume, say $\partial\Omega_N^{(I)}$, of the phase-space volume $\Omega_N^{(I)}$ to the dissociation channel cannot be appropriately defined. In place of resorting to $\partial\Omega_N^{(I)}$ therefore, we use the phase space volume of the daughter cluster of $N-1$ particles, denoted as $\Omega_{N-1}^{(I)}(E - \Delta E)$, where ΔE is the energy for the dissociating particle to take away with the relative velocity $\vec{v}$. The flux is approximately expressed as

$$W_{N-1}^{(I)}(E) = \int \Omega_{N-1}^{(I)}(E - \Delta E)\vec{v}(\Delta E) \cdot d\vec{v}. \tag{139}$$

(The reduced mass is set to be unity for simplicity.) Thus, the critical quantity for the statistical rate of the dissociation probability turns out to be the ratio

$$\frac{\Omega_{N-1}^{(I)}(E - \Delta E)}{\Omega_N^{(I)}(E)} \tag{140}$$

for a given ΔE.

We further suppose that the phase space of a system under study has the mixing property and is completely statistical due to chaos. Then we may regard the quantity

$$\Omega_N^{(I)}(E) - \Omega_{N-1}^{(I)}(E - \Delta E) \quad (141)$$

as a phase space volume for trajectories to "tentatively" remain bound as N-body cluster. Then the related entropy $s^{(I)}(E, \Delta E)$ can then be defined as

$$s^{(I)}(E, \Delta E) = -\log\left(\frac{\Omega_N^{(I)}(E) - \Omega_{N-1}^{(I)}(E - \Delta E)}{\Delta\Omega}\right), \quad (142)$$

where $\Delta\Omega$ represents the size of a unit cell of the phase space volume, such as $\hbar^{3N}$, which is used to count the number of cells in the volume $\left[\Omega_N^{(I)}(E) - \Omega_{N-1}^{(I)}(E - \Delta E)\right]$.

Although the state mixing is caused by the interaction

$$-i\hbar \sum_{k=1}^{3N} \dot{R}_k(t) X_{IJ}^k(\boldsymbol{R}(t)) \quad (143)$$

and/or $H_{IJ}^{(\mathrm{el})}(\boldsymbol{R}(t))$ as in Equation (109), we herein assume *for simplicity* that the transition among the electronic states occur freely. Then the entropy confined in the individual state $s^{(I)}(E, \Delta E)$ should be represented by the sum of them as

$$\begin{aligned} s^{(\max)}(E, \Delta E) &= \sum_I s^{(I)}(E, \Delta E) \\ &= -\sum_I \log\left(\frac{\Omega_N^{(I)}(E) - \Omega_{N-1}^{(I)}(E - \Delta E)}{\Delta\Omega}\right). \end{aligned} \quad (144)$$

In such an extreme case the connected phase-space volume within ΔE is effectively extended and can be approximated as

$$\prod_I \left[\Omega_N^{(I)}(E) - \Omega_{N-1}^{(I)}(E - \Delta E)\right]. \quad (145)$$

On the other hand, the phase space volume directly connected to the dissociation channel should be represented as the simple sum of each as

$$\sum_I \Omega_{N-1}^{(I)}(E - \Delta E). \quad (146)$$

Hence the the dissociation probability can be dramatically lowered from the order of Equation (140) to

$$\frac{\sum_I \Omega_{N-1}^{(I)}(E - \Delta E)}{\prod_J \left[\Omega_N^{(J)}(E) - \Omega_{N-1}^{(J)}(E - \Delta E)\right]}, \quad (147)$$

indicating that the relative phase-space volume towards the dissociation is by far smaller than the phase space volume of the original N-body cluster, and the probability for the cluster to come across the dissociation channel is extremely small. Consequently, the lifetime of the clusters is expected to be far longer than in case of no nonadiabatic transitions.

7.5.2. Intra-Molecular Nonadiabatic Electronic Energy Redistribution (INEER): Nonadiabatic Interaction to Close Dissociation Channels

Nonadiabatic electronic-state chaos thus inflates the volume of phase-space of nuclear motion of clusters to a very large extent. Due to this inflation, the molecular states lose chances to reach one of their dissociation channels. That is, a dissociating molecule can be brought back to the bound state by nonadiabatic interactions, which would dissociate otherwise. Here is an example of such an event [165]. In Figure 20, a B_{12} of initial hindered geometry is prepared, which is close to a dissociation channel. (Such dissociation channels can be readily found by colliding a B_{11} cluster and a B atom) In panel (a) is shown a trajectory for the 47th adiabatic state at this geometry to run on its adiabatic potential surface (no nonadiabatic interaction). After 110 fs of running, one of the atoms leaves the cluster leaving B_{11} cluster behind. Likewise, the same 47th adiabatic state in panel (b) begins to run from the same geometry with the same momentum but with the nonadiabatic interactions explicitly included in the electron dynamics. The trajectory clearly demonstrates that an atom being about to leave the cluster is brought back again to the B_{12}, thus evidencing that the dissociation channel is closed. This phenomenon is what we call intra-molecular nonadiabatic electronic energy redistribution (INEER) due to hyper-resonance [188]. This makes a contrast to intramolecular vibrational energy redistribution (IVR) caused by classical chaos on a single potential basin. INEER is a crucial consequence of the intense chaos due to the continual nonadiabatic electronic-state mixing.

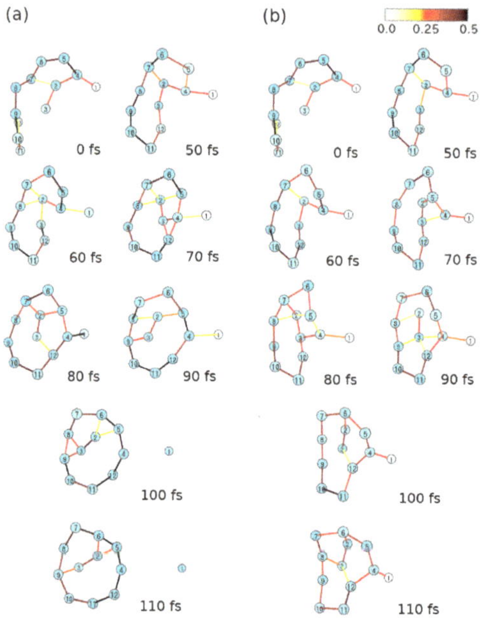

Figure 20. B_{12} bonding structures along a trajectory under (**a**) adiabatic dynamics, and (**b**) nonadiabatic dynamics, starting from the same initial conditions (initial adiabatic state 47). Time t in fs is indicated beside each structure. At $t = 110$ fs, atom 1 is in the dissociation channel in the adiabatic dynamics, while this atom is pulled back to the cluster site in the nonadiabatic dynamics. [Taken from Arasaki, Y.; Takatsuka, K. *J. Chem. Phys.* **2019**, *150*, 114101 with permission].

8. Concluding Remarks

To review "quantum chaos" in the scope of chemical dynamics, we herein have tracked the following subjects. We began with full quantum dynamics, wavepackets and eigenfunctions localized in the quasi-separatrix in two-dimensional potential function.

This study was made in search for a novel molecular vibrational mode. Indeed, in this small dimensional system lie several vibrational modes, which give birth to very long-time wandering motions in the quasi-separatrix. A pair of the eigenstates of the dynamical tunneling of the second kind have been identified.

Semiclassical theory and energy quantization thereof, which is indispensable in the study of molecular vibration were next reviewed. We took a wavepacket approach in studying semiclassical dynamics rather than the direct semiclassical reduction of the Feynman kernel. With this approach, we discussed the mechanism of energy quantization for not only making the peaks but suppressing the off-resonant components. One of the most annoying parts in semiclassical kernel and related quantities is the divergence at caustics and/or turning points, which casts an inherent difficulty to the Gutzwiller trace formula. To avoid the difficulty, we have shown that the amplitude-free energy quantization actually works quite well. Chaotic paths that appear only through quantum tunneling have been shown along with a semiclassical tunneling formalism based on the notion of the parity of motion.

One of the main aims of the present review has been to guide the readers to chaos in molecular nonadiabatic dynamics, which is induced by the kinematic couplings between nuclei and electrons. The entanglement between electrons and nuclei is one of the most important subjects in general chemical science. We have emphasized the critical roles of "bifurcation and merging" of quantum wavefunctions. The most characteristic molecular chaos is found in nonadiabatic electron dynamics in the densely quasi-degenerate electronic-state manifold, every state of which mixes together with other members of the manifold. We have shown that the dynamical properties of electronic-state chaos can be measured in terms of the quantities related to diffusion and memory loss in the Hilbert space, turbulent flow of electron current, and so on. The chaos of the electronic excited state should become more and more vital in future molecular science, which will proceed far beyond the Born-Oppenheimer paradigm of the current theoretical framework of chemistry.

Some of the characteristics found for molecular quantum chaos are summarized below:

The basic mechanism of chaos in nonadiabatic wavepacket dynamics is "bifurcation and merging", which is a quantum mechanical synonym of "stretching and folding" in the classical baker's (or the Smale horse-shoe) transformation. Indeed repeated occurrences of "bifurcation and merging" of quantum wavepackets result in genuine chaos in quantum mechanics not only in energy spectra but more importantly in the dynamics of wavepackets. We would like to particularly emphasize again that the bifurcation and merging of wavepackets are indeed physically observable. Interestingly, "bifurcation (branching) and merging of the reaction tubes" sets also a geometrical foundation of classical chaos in structural isomerization dynamics. Extension of the tube geometry to quantum mechanics has not yet been explored.

"Loss of memory" is one of the central concepts to characterize chaos in general. The positive Lyapunov exponent in classical systems indicates an exponential separation of nearby orbits, leading to the loss of the memory of their initial stage. In our quantum chaos, the exponential decay of the memory of an electronic state in the time-propagation against the nonadiabatic (kinematic) interactions can take place in the Hilbert space. Again, we have observed that classical chaos in the structural isomerization dynamics of liquid-like clusters is associated with a characteristic feature of memory loss, which leads to a random yet statistical appearance of the isomers in the time series.

The advent of the ultrafast pulse laser has revolutionized molecular science since the 1970s. The femtosecond laser chemistry [189] to date has made it possible to real-time track the molecular deformation and reactions, an example of which can be seen in Figure 11 of this review. We are now in the age of technical development and applications of attosecond pulse lasers, the width of which is compatible with the time scales of electronic motions in molecules. This situation has been driving us in nonadiabatic electron wavepacket dynamics in Section 7. Nevertheless, we also call for attention for the study of extremely long-time spectroscopy. As we noted in this article, the structural isomerization of not only clusters but also large molecules like proteins can wander from one shape (potential valley, potential basin) to another taking a very long time at each local structure. These many-dimensional dynamics are often accompanied by large dimensional chaotic spaces, which makes the entire dynamics even longer and more complicated. It is extremely challenging therefore to figure out spectroscopic means to capture the essential feature of the long-time dynamics. The time-dependent spectroscopy, which was proposed in Section 4 and discussed briefly in the section of semiclassical mechanics Section 5, is just one primitive idea in an attempt to cope with this matter. A parallel computation scheme with many different initial conditions should be a good candidate to overcome long-time quantum dynamics in ergodic phase space, if a method to appropriately connect the pieces of spectra taken individually is devised.

A final remark before closing. The essential nature of classical and quantum chaos may lie in mathematical beauty and intricate geometry behind the complicated-looking phenomena. However, the complexity of molecules often makes such elegant mathematical structures obscure and even less mighty. Moreover, We are afraid that sticking to the too rigorous definitions of quantum chaos would isolate itself from other natural science. Therefore, I would like to conclude this review by stressing that more studies on quantum chaos made mathematically easy and conceptually useful are demanded than ever, since there are so many unknowns left (not only) in the molecular phenomena, notions, and laws that need an understanding of quantum chaos.

Funding: JSPS KAKENHI, grant number JP20H00373.

Institutional Review Board Statement: Not applicable.

Informed Consent Statement: Not applicable.

Data Availability Statement: All data included in this manuscript are available upon reasonable request by contacting the corresponding author.

Acknowledgments: The author thanks their former students and postdoctral rsearch fellows, who are coauthors of the papers with the present author. In particular, he is grateful for long standing collaboration with Hiroshi Ushiyama, Satoshi Takahashi, and Yasuki Arasaki. Special thanks go to Yasuki Arasaki, who kindly provided Figures 17 and 18 prior to publication. He is also grateful to Marko Robnik for their invitation to make a contribution to the special issue of quantum chaos celebrating Professor Giulio Casati. This work has been supported by JSPS KAKENHI (Japan), the grant number JP20H00373.

Conflicts of Interest: The author declares no conflict of interest.

References

1. Lichtenberg, A.J.; Lieberman M.A. *Regular and Chaotic Dynamics*, 2nd ed.; Springer: Berlin, Germany, 1992.
2. Ozorio de Almeida, A.M. *Hamiltonian Systems, Chaos and Quantization;* Cambridge University Press: Cambridge, UK, 1988.
3. Tabor, M. *Chaos and Integrability in Nonlinear Dynamics*; John Wiley: New York, NY, USA, 1989.
4. Giannoni, M.-J.; Voros, A.; Zinn-Justin, J. *Chaos and Quantum Physics*; North Holland Publisher: Amsterdam, The Netherlands, 1991.
5. Reichl, L.E. *The Transition to Chaos;* Springer: New York, NY, USA, 1992.
6. Nakamura K. *Quantum Chaos*; Cambridge University Press: Cambridge, UK, 1993.
7. Gaspard, P.; Alonso, D.; Burghardt, I. New Way of Understanding Semiclassical Quantization. *Adv. Chem. Phys.* **1995**, *XC*, 105.
8. Casati, G.; Chirikov, B. (Eds.) *Quantum Chaos: Between Order and Disorder;* Cambridge University Press: Cambrdige, UK, 1995.
9. Haake, F. *Quantum Signatures of Chaos*; Springer: Berlin/Heidelberg, Germany, 2010.
10. Berggren K.-F.; Åberg, S. (Eds.) *Quantum Chaos Y2K*; Physica Scripta; The Royal Swedish Academy of Sciences: Stockholm, Sweden, 2001.
11. Robnik, M. Fundamental concepts of quantum chaos. *Eur. Phys. J. Spec. Top.* **2016**, *225*, 959–976. [CrossRef]
12. Wang, J.; Benenti, G.; Casati, G.; Wang, W. Quantum Chaos and the Correspondence Principle. *Phys. Rev. E* **2021**, *103*, L030201. [CrossRef] [PubMed]
13. Berry, M. Quantum chaology, not quantum chaos. *Phys. Scr.* **1989**, *40*, 335. [CrossRef]
14. Child, M.S. *Molecular Collision Theory*; Academic Press: London, UK, 1974.
15. Child, M.S. *Semiclassical Mechanics with Molecular Applications*; Clarendon Press: Oxford, UK, 1991.

16. Baer, M. Introduction to the theory of electronic non-adiabatic coupling terms in molecular systems. *Phys. Rep.* **2002**, *358*, 75–142. [CrossRef]
17. Jasper, A.W.; Kendrick, B.K.; Mead, C.A.; Truhlar, D.G. *Modern Trends in Chemical Reaction Dynamics Part I, Chapter 8*; Yang, X., Liu, K. Eds.; World Scientific: Singapore, 2004.
18. Yarkony, W.D.D.R.; Köppel, H. (Eds.), *Conical Intersections: Electronic Structure, Dynamics and Spectroscopy*; Advanced Series in Physical Chemistry; World Scientific: Hackensack, NJ, USA, 2004; Volume 15.
19. Nakamura, H. *Nonadiabatic Transition: Concepts, Basic Theories and Applications*, 2nd ed.; World Scientific: Singapore, 2012.
20. Baer, M. *Beyond Born–Oppenheimer: Electronic Nonadiabatic Coupling Terms and Conical Intersections;* Wiley: Hoboken, NJ, USA, 2006.
21. Takatsuka, K. Generalization of classical mechanics for nuclear motions nonadiabatically coupled with electron wavepacket dynamics and in quantum-classical mixed representation. *J. Phys. Chem. A* **2007**, *111*, 10196. [CrossRef]
22. Yonehara, T.; Takatsuka, K. Phase-space averaging and natural branching of nuclear paths for nonadiabatic electron wavepacket dynamics. *J. Chem. Phys.* **2008**, *129*, 134109. [CrossRef]
23. Takatsuka, K.; Yonehara, T. Nonadiabatic chemical dynamics in intermediate and intense laser fields. *Adv. Chem. Phys.* **2010**, *144*, 93.
24. Takatsuka, K.; Yonehara, T. Exploring dynamical electron theory beyond the Born-Oppenheimer framework: From chemical reactivity to non-adiabatically coupled electronic and nuclear wavepackets on-the-fly under laser field. *Phys. Chem. Chem. Phys.* **2011**, *13*, 4987. [CrossRef]
25. Yonehara, T.; Hanasaki, K.; Takatsuka, K. Fundamental approaches to nonadiabaticity: Towards a chemical theory beyond the Born-Oppenheimer paradigm. *Chem. Rev.* **2012**, *112*, 499–542. [CrossRef]
26. Takatsuka, K.; Yonehara, T.; Hanasaki, K.; Arasaki, Y. *Chemical Theory beyond the Born—Oppenheimer Paradigm*; World Scientific: Singapore, 2015.
27. Takatsuka, K. Electron dynamics in molecular elementary processes and chemical reactions. *Bull. Chem. Soc. Jpn.* **2021**, *94*, 1421. [CrossRef]
28. Henon, M.; Heiles, C. The applicability of the third integral of motion. *Astron. J.* **1964**, *69*, 73. [CrossRef]
29. Takahashi, S.; Takatsuka, K. Towards many-dimensional real-time quantum theory for heavy-particle dynamics. I. Semiclassics in the Lagrange picture of classical phase flow. *Phys. Rev. A* **2014**, *89*, 012108. [CrossRef]
30. Takatsuka, K.; Takahashi, S. Towards many-dimensional real-time quantum theory for heavy particle dynamics. II. Beyond semiclassics by quantum smoothing of singularity in quantum-classical correspondence. *Phys. Rev. A* **2014**, *89*, 012109. [CrossRef]
31. Helgaker, T.; Jørgensen, P.; Olsen, J. *Molecular Electronic-Structure Theory;* John Wiley: New York, NY, USA, 2013.
32. Takatsuka, K.; Inoue, A. Global Representation of Maslov-type Semiclassical Wavefunction and Its Spectrum in a Small Number of Classical Trajectories. *Phys. Rev. Lett.* **1997**, *78*, 1404–1407. [CrossRef]
33. Inoue-Ushiyama, A.; Takatsuka, K. Semiclassical Theory for Maslov-type Wave Packet. Hierarchy below the Semiclassical Feynman Kernel. *Phys. Rev. A* **1999**, *59*, 3256–3269. [CrossRef]
34. Miller, W.H. Classical-Limit Quantum Mechanics and the Theory of Molecular Collisions. *Adv. Chem. Phys.* **1974**, *25*, 69–177.
35. Gutzwiller, M.C. Energy Spectrum According to Classical Mechanics. *J. Math. Phys.* **1970**, *11*, 1791–1806. [CrossRef]
36. Gutzwiller, M.C. Periodic Orbits and Classical Quantization Conditions. *J. Math. Phys.* **1971**, *12*, 343–358. [CrossRef]
37. Gutzwiller, M.C. *Chaos in Classical and Quantum Mechanics;* Springer: Berlin/Heidelberg, Germany, 1990.
38. Takahashi, S.; Takatsuka, K. On the validity range of the Born-Oppenheimer approximation: A semiclassical study for all-particle quantization of three-body Coulomb systems. *J. Chem. Phys.* **2006**, *124*, 144101. [CrossRef]
39. Born, M.; Huang, K. *Dynamical Theory of Crystal Lattices*; Oxford University Press: Oxford, UK, 1954.
40. Laidler, K.J. *Chemical Kinetics*, 3rd ed; Harpeer & ROW: New York, NY, USA, 1987.
41. Steinfeld, J.I.; Francisco, J.S.; Hase, W.L. *Chemical Kinetics and Dynamics;* Prentice Hall: Englewood Cliffs, NJ, USA, 1989.
42. Fueno, T. (Ed.) *The Transition State: A Theoretical Approach;* Kodansha: Tokyo, Japan, 1999.
43. Miller, W.H. Quantum mechanical transition state theory and a new semiclassical model for reaction rate constants. *J. Chem. Phys.* **1974**, *61*, 1823–1834. [CrossRef]
44. Miller, W.H. Beyond Transition-State Theory: A Rigorous Quantum Theory of Chemical Reaction Rates. *Acc. Chem. Res.* **1993**, *26*, 174–181. [CrossRef]
45. Zhang, P.; Han, K. Adiabatic/Nonadiabatic State-to-State Reactive Scattering Dynamics Implemented on Graphics Processing Units. *J. Phys. Chem. A* **2013**, *117*, 8512–8518. [CrossRef] [PubMed]
46. Naskar, K.; Ghosh, S.; Adhikari, S. Accurate Calculation of Rate Constant and Isotope Effect for the F + H_2 Reaction by the Coupled 3D Time-Dependent Wave Packet Method on the Newly Constructed Ab Initio Ground Potential Energy Surface. *J. Phys. Chem. A* **2022**, *126*, 3311–3328. [CrossRef] [PubMed]
47. Ushiyama, H.; Takatsuka, K. Methyl group rotation driven by proton transfer through a long-range chemical interaction. *Angew. Chem. Int. Ed.* **2005**, *44*, 2–5. [CrossRef]
48. Marcus, R.A. Unimolecular Dissociations and Free Radical Recombination Reactions. *J. Chem. Phys.* **1952**, *20*, 359–364. [CrossRef]
49. Feshbach, H. A unified theory of nuclear reactions. II. *Ann. Phys.* **1962**, *19*, 287–313. [CrossRef]
50. Newton, R.G. *Scattering Theory of Waves and Particles*; Springer: Berlin/Heidelberg, Germany, 1982.
51. Abramson, E.; Field, R.W.; Imre, D.; Innes, K.K.; Kinsey, J.L. Fluorescence and stimulated emission $S_1 \rightarrow S_0$ spectra of acetylene: Regular and ergodic regions. *J. Chem. Phys.* **1985**, *83*, 453–465. [CrossRef]

52. Yamanouchi, K.; Ikeda, N.; Tsuchiya, S. Vibrationally highly excited acetylene as studied by dispersed fluorescence and stimulated emission pumping spectroscopy: Vibrational assignment of the feature states. *J. Chem. Phys.* **1991**, *95*, 6330–6342. [CrossRef]
53. Jonas, D.M.; Solina, S.A.B.; Rajaram, B.; Silbey, R.J.; Field, R.W.; Yamanouchi, K.; Tsuchiya, S. Intramolecular vibrational relaxation and forbidden transitions in the SEP spectrum of acetylene. *J. Chem. Phys.* **1992**, *97*, 2813–2816. [CrossRef]
54. Jonas, D.M.; Solina, S.A.B.; Rajaram, B.; Silbey, R.J.; Field, R.W. Intramolecular vibrational redistribution of energy in the stimulated emission pumping spectrum of acetylene. *J. Chem. Phys.* **1993**, *99*, 7350–7370. [CrossRef]
55. Leitner, D.M.; Wolynes, P.G. Vibrational relaxation and energy localization in polyatomics: Effects of high order resonances on flow rates and the quantum ergodicity transition. *J. Chem. Phys.* **1996**, *105*, 11226. [CrossRef]
56. Karmakar, S.; Yadav, P.K.; Keshavamurthy, S. Stable chaos and delayed onset of statisticality in unimolecular dissociation reactions. *Comm. Chem.* **2020**, *3*, 1–11. [CrossRef]
57. Leitner, D.M. Quantum ergodicity and energy flow in molecules. *Adv. Phys.* **2015**, *64*, 445–517. [CrossRef]
58. Karmakar, S.; Keshavamurthy, S. Intramolecular vibrational energy redistribution and the quantum ergodicity transition: A phase space perspective. *Phys. Chem. Chem. Phys.* **2020**, *22*, 11139–11173. [CrossRef]
59. Leitner, D.M. Influence of Quantum Energy Flow and Localization on Molecular Isomerization in Gas and Condensed Phases. *Intern. J. Quant. Chem.* **1999**, *75*, 523–531. [CrossRef]
60. Fujisaki, H.; Straub, J.E. Vibrational energy relaxation in proteins. *Proc. Natl. Acad. Sci. USA* **2005**, *102*, 6726–6731. [CrossRef]
61. Fujisaki, H.; Zhang, Y.; Straub, J.E. Time-dependent perturbation theory for vibrational energy relaxation and dephasing in peptides and proteins. *J. Chem. Phys.* **2006**, *124*, 144910. [CrossRef]
62. Beck, T.L.; David M. Leitner, D.M.; Berry, R.S. Melting and phase space transitions in small clusters: Spectral characteristics, dimensions, and K entropy. *J. Chem. Phys.* **1988**, *89*, 1681–1694. [CrossRef]
63. Berry, R.S.; Jellinek, J.; Natanson, G. Melting of clusters and melting. *Phys. Rev. A.* **1984**, *30*, 911–931. [CrossRef]
64. Kuharski, R.A.; Chandler, D.; Montgomery, J.A., Jr.; Rabii, F.; Singer, S.J. Stochastic Molecular Dynamics Study of Cyclohexane Isomerization. *J. Phys. Chem.* **1988**, *92*, 3261–3267. [CrossRef]
65. Fujii, M.; Takatsuka, K. Nonempirical statistical theory for atomic evaporation from nonrigid clusters: Applications to the absolute rate constant and kinetic energy release. *J. Phys. Chem. A* **2007**, *111*, 1389–1402. [CrossRef] [PubMed]
66. Fujii, M.; Takatsuka, K. Nonempirical statistical theory for molecular evaporation from nonrigid clusters. *J. Chem. Phys.* **2008**, *128*, 114318. [CrossRef] [PubMed]
67. Takatsuka, K. Temperature, geometry, and variational statistical structure in microcanonical ensemble for structural isomerization dynamics of clusters: A multichannel chemical reaction beyond the transition-state concept. *Adv. Chem. Phys.* **2005**, *130*, 25–85.
68. Seko, C.; Takatsuka, K. Non-Ergodicity and Two Sub-Phases in the Coexistence Region in Isomerization Dynamics of Ar7-like Molecules. *J. Chem. Phys.* **1996**, *104*, 8613–8626. [CrossRef]
69. Seko, C.; Takatsuka, K. A Scrutiny of the Premise of the Rice-Ramsberger-Kassel-Marcus Theory in Isomerization Reaction of an Ar7-type Molecule. *J. Chem. Phys.* **1996**, *105*, 10356–10366.
70. Takatsuka, K.; Seko, C. Concept of Inter-basin Mixing and Extension of the Liapunov Exponen in Multiple Potential-basin Dynamics as Structural Isomerization of Clusters. *J. Chem. Phys.* **1999**, *110*, 3263–3266. [CrossRef]
71. Takatsuka, K.; Yanao, T. Microcanonical Temperature and Its Arrhenius Relation to Lifetimes in Isomerization Dynamics of Clusters. *J. Chem. Phys.* **2000**, *113*, 2552–2562. [CrossRef]
72. Seko, C.; Takatsuka, K. Origin of the complex dynamics in structural isomerization of small clusters: The effects of potential topography. *J. Chem. Phys.* **1998**, *109*, 4768–4781. [CrossRef]
73. De Leon, N.; Mehta, M.A.; Robert, Q. Topper, R.Q. Cylindrical manifolds in phase space as mediators of chemical reaction dynamics and kinetics. I. Theory. *J. Chem. Phys.* **1991**, *94*, 8310–8328. [CrossRef]
74. De Leon, N.; Mehta, M.A.; Robert Q. Topper, R.Q. Cylindrical manifolds in phase space as mediators of chemical reaction dynamics and kinetics. II. Numerical considerations and applications to models with two degrees of freedom. *J. Chem. Phys.* **1991**, *94*, 8329–8341. [CrossRef]
75. Davis, J.; Stechel, E.B.; Heller, E.J. Quantum dynamics in classically integrable and non-integrable regions. *Chem. Phys. Lett.* **1980**, *76*, 21–26. [CrossRef]
76. Hutchinson, J.S.; Wyatt, R.E. Quantum ergodicity for time-dependent wave-packet dynamics. *Phys. Rev. A* **1980**, *23*, 1567–1584. [CrossRef]
77. Bixon, M.; Jortner, J. Quantum dynamics of the Hénon–Heiles system. *J. Chem. Phys.* **1982**, *77*, 4175–4187. [CrossRef]
78. Feit, M.D.; Fleck, J.A., Jr. Wave packet dynamics and chaos in the Héon–Heiles system. *J. Chem. Phys.* **1984**, *80*, 2578–2584. [CrossRef]
79. Hashimoto, N.; Takatsuka, K. Expected Significance of Weakly Chaotic Vibrational Motions in Single Molecule Spectroscopy. *J. Chem. Phys.* **1995**, *103*, 6914–6929. [CrossRef]
80. Hashimoto, N.; Takatsuka, K. Quantum Localization and Dynamic Tunneling ofQuasi-separatrix Wavefunctions for Molecular Vibration. *J. Chem. Phys.* **1998**, *108*, 1893–1903. [CrossRef]
81. Heller, E.J. Bound-State Eigenfunctions of Classically Chaotic Hamiltonian Systems: Scars of Periodic Orbits. *Phys. Rev. Lett.* **1984**, *53*, 1515–1518. [CrossRef]
82. Takatsuka, K.; Hashimoto, N. A novel method to calculate eigenfunctions and eigenvalues in a given energy range. *J. Chem. Phys.* **1995**, *103*, 6057. [CrossRef]

83. Feit, M.D.; Fleck, J.A.; Steiger, A. Solution of the Schröinger equation by a spectral method. *J. Comput. Phys.* **1982**, *47*, 412–433. [CrossRef]
84. Viswanathan, R.; Shi, S.;Villalonga, E.; Rabitz, H. Calculation of scattering wave functions by a numerical procedure based on the MØller wave operator. *J. Chem. Phys.* **1989**, *91*, 2333. [CrossRef]
85. Takatsuka, K. Extraction of Accurate Frequencies from the Fast-Fourier-Transform Spectra. *J. Compt. Phys.* **1992**, *102*, 374–380. [CrossRef]
86. Wall, M.R.; Neuhauser, D. Extraction, through filter-diagonalization, of general quantum eigenvalues or classical normal mode frequencies from a small number of residues or a short-time segment of a signal. I. Theory and application to a quantum-dynamics model. *J. Chem. Phys.* **1995**, *102*, 8011. [CrossRef]
87. Davis, M.J.; Heller, E.J. Quantum dynamical tunneling in bound states. *J. Chem. Phys.* **1981**, *75*, 246. [CrossRef]
88. Keshavamurthy, S. Dynamical tunneling in molecules: Role of the classical resonances and chaos. *J. Chem. Phys.* **2003**, *119*, 161. [CrossRef]
89. Karmakar, S.; Keshavamurthy, S. Relevance of the Resonance Junctions on the Arnold Web to Dynamical Tunneling and Eigenstate Delocalization. *J. Phys. Chem. A* **2018**, *122*, 8636–8649. [CrossRef]
90. Takatsuka, K. Concept of Phase-Space Large Amplitude Motion. A Classical Study. *Chem. Phys. Lett.* **1993**, *204*, 491–495. [CrossRef]
91. Noid, D.W.; Koszykowski, M.L.; Marcus, R.A. Quasiperiodic and Stochastic Behavior in Molecules. *Ann. Rev. Phys. Chem.* **1981**, *32*, 267–309. [CrossRef]
92. Heller, E.J. *The Semiclassical Way to Dynamics and Spectroscopy: Structure, Dynamics and Quantum Chaos in Atoms and Molecules under Strong Magnetic Fields*; Princeton University Press: Princeton, NJ, USA, 2018.
93. Bogomolony, E. Riemann Zeta function and quantum chaos. *Prog. Theoret. Phys. Suppl.* **2007**, *166*, 19–36. [CrossRef]
94. Schulman, L.S. *Techniques and Applications of Path Integration*; Wiley: New York, NY, USA, 1981.
95. Koh, Y.-W.; Takatsuka, K. Finding periodic orbits of higher-dimensional flows by including tangential components of trajectory motion. *Phys. Rev. E* **2007**, *76*, 066205. [CrossRef]
96. Berry, M.V.; Howls, C.J. High orders of the Weyl expansion for quantum billiards: Resurgence of periodic orbits, and the Stokes phenomenon. *Proc. R. Soc. Lond. A* **1994**, *447*, 527–555.
97. Eckhardt, B.; Aurell, E. Convergence of the Semi-Classical Periodic Orbit Expansion. *Europhys. Lett.* **1989**, *9*, 509–512. [CrossRef]
98. Alonso, D.; Gaspard, P. $\hbar$ expansion for the periodic orbit quantization of chaotic systems. *CHAOS* **1993**, *3*, 601–612. [CrossRef]
99. Bohm, D. *Quantum Theory*; Dover: New York, NY, USA, 1989.
100. Messiah, A. *Quantum Mechanics*; Dover: New York, NY, USA, 2017.
101. Berry, M.V.; Mount, K.E. *Rep. Prog. Phys.* **1972**, *35*, 315. [CrossRef]
102. Maslov, V.P.; Fedoriuk, M.V. *Semi-Classical Approximation in Quantum Mechanics*; Reidel: Dordrecht, The Netherlands, 1981.
103. Blrack, M.; Bhaduri, R. *Semicalssical Physics*; Taylor & Francis: Boca Raton, FL, USA, 2003.
104. Tannor, D.J. *Introduction to Quantum Mechanics*; University Science Books: Melville, NY, USA, 2007.
105. Takahashi, S.; Takatsuka, K. Geometrical evaluation of the Maslov index. *Phys. Rev. A* **2004**, *69*, 022110. [CrossRef]
106. Takatsuka, K. Dynamics and Quantization of Hamiltonian Chaos. Density of States in Phase-Space Semiclassical Mechanics. *Phys. Rev. A* **1992**, *45*, 4326–4339. [CrossRef]
107. Takatsuka, K. An Amplitude-free Correlation Function Based on an Algebra for Coordinate Transformation in Semiclassical Integrals. *Phys. Rev. E* **2001**, *64*, 016224. [CrossRef]
108. Arnold, V.I. *Mathematical Methods of Classical Mechanics*, 2nd ed.; Springer: New York, NY, USA, 1989.
109. Heller, E.G. Time-dependent approach to semiclassical dynamics. *J. Chem. Phys.* **1975**, *62*, 1544–1555. [CrossRef]
110. Takatsuka, K. Huygens-like proliferation of the secondary coherent-state wavelets from a single Gaussian wavepacekt to fit in the length-scale hierarchical Schrödinger dynamics. 2023. *to be published.*
111. Ushiyama, H.; Takatsuka, K. Extended quantization condition for constructive and destructive interferences and trajectories dominating molecular vibrational eigenstates. *J. Chem. Phys.* **2005**, *122*, 224112. [CrossRef]
112. Inoue-Ushiyama, A.; Takatsuka, K. Roles of Phase Cancellation Removing Spuriously Quantized States in Semiclassical Mechanics. *Phys. Rev. A* **1999**, *60*, 112–120. [CrossRef]
113. Berry, M.V.; Tabor, M. Closed orbits and the regular bound spectrum. *Proc. R. Soc. Lond. Ser. A* **1976**, *349*, 101.
114. Berry, M.V.; Tabor, M. Calculating the bound spectrum by path summation in action-angle variables. *J. Phys. A* **1977**, *10*, 371. [CrossRef]
115. Takahashi, S.; Takatsuka, K. Phase quantization of chaos in semiclassical regime. *J. Chem. Phys.* **2007**, *127*, 084112. [CrossRef]
116. Yamashita, T.; Takatsuka, K. Phase quantization of chaos and the role of the semiclassical amplitude factor. *Prog. Theor. Phys. Suppl.* **2007**, *166*, 56. [CrossRef]
117. Hotta, K.; Takatsuka, K. Semiclassical quantization of chaos in terms of an amplitude-free quasi-correlation function. *J. Phys. A Gen. Math.* **2003**, *36*, 4785–4803. [CrossRef]
118. Takatsuka, K.; Takahashi, S.; Koh, Y.W.; Yamashita, T. Energy quantization of chaos with the semiclassical phases alone. *J. Chem. Phys.* **2007**, *126*, 021104. [CrossRef]
119. Adachi, S. A numerical evaluation of the semiclassical coherent state path integral. *Ann. Phys.* **1989**, *195*, 45–93. [CrossRef]
120. Shudo, A.; Ikeda, K.S. Complex classical trajectories and chaotic tunneling. *Phys. Rev. Lett.* **1995**, *74*, 682–685. [CrossRef]

121. Shudo, A.; Ikeda, K.S. Tunneling effect and the natural boundary of invariant tori. *Phys. Rev. Lett.* **2012**, *109*, 154102. [CrossRef]
122. Takatsuka, K.; Ushiyama, H. Tunneling Solutions of Hamilton–Jacobi Equation for Multi-dimensional Semiclassical Theory. *Phys. Rev. A* **1995**, *51*, 4353–4364. [CrossRef]
123. Takatsuka, K.; Ushiyama, H.; Inoue-Ushiyama, A. Tunneling Paths in Multi-dimensional Semiclassical Dynamics. *Phys. Rep.* **1999**, *322*, 347–417. [CrossRef]
124. Ushiyama, H.; Takatsuka, K. Semiclassical Study on Multidimensional Effects in Tunneling Chemical Reactions: Tunneling Paths and Tunneling Tubes. *J. Chem. Phys.* **1997**, *106*, 7023–7035. [CrossRef]
125. Ushiyama, H.; Takatsuka, K. Statistical Redistribution of Trajectories from a Torus to Tori by Chaotic Dynamical-Tunneling. *Phys. Rev. E* **1996**, *53*, 115–123. [CrossRef]
126. Ushiyama, H.; Takatsuka, K. Quasi-semiclassical Approach to Tunneling Chemical Reactions. *J. Chem. Phys.* **1998**, *109*, 9664–9673. [CrossRef]
127. Ushiyama, H.; Takatsuka, K. Time-dependent probability of quantum tunneling in terms of quasi-semiclassical method. *J. Chem. Phys.* **2004**, *120*, 4561–4572. [CrossRef]
128. Ushiyama, H.; Takatsuka, K. Very fast tunneling in the early stage of reaction dynamics. *J. Phys. Chem. A* **2005**, *51*, 11807–11814. [CrossRef]
129. Berry, M.V.; Robnik, M. Semiclassical level spacings when regular and chaotic orbits coexist. *J. Phys. A Math. Gen.* **1984**, *17*, 2413–2421. [CrossRef]
130. Pechukas, P. Distribution of Energy Eigenvalues in the Irregular Spectrum. *Phys. Rev. Lett.* **1983**, *51*, 943. [CrossRef]
131. Yukawa, T. New Approach to the Statistical Properties of Energy Levels. *Phys. Rev. Lett.* **1985**, *54*, 1883–1886. [CrossRef]
132. Nakamura, K.; Lakshmanan, M. Complete Integrability in a Quantum Description of Chaotic Systems. *Phys. Rev. Lett.* **1986**, *56*, 1661. [CrossRef] [PubMed]
133. Shudo, A.; Saito, N. Level Spacing Distribution and Avoided Crossing in Quantum Chaos. *J. Phys. Soc. Jpn.* **1987**, *56*, 2641–2652. [CrossRef]
134. Wang, S.-J.; Chu, S.Y. Level dynamics: An approach to the study of avoided level crossings and transition to chaos. *Phys. Rev. A* **1993**, *47*, 3546–3553. [CrossRef] [PubMed]
135. Suzuki, T.; Wang, L.; Kohguchi, K. Femtosecond time-resolved photoelectron imaging on ultrafast electronic dephasing in an isolated molecule. *J. Chem. Phys.* **1999**, *111*, 4859. [CrossRef]
136. Blanchet, V.; Zgierski, M.Z.; Seideman, T.; Stolow, A. Discerning vibronic molecular dynamics using time-resolved photoelectron spectroscopy. *Nature* **1999**, *401*, 52. [CrossRef]
137. Davies, J.A.; LeClaire, J.E.; Continetti, R.E.; Hayden, C.C. Femtosecond time-resolved photoelectron—Photoion coincidence imaging studies of dissociation dynamics. *J. Chem. Phys.* **1999**, *111*, 1–4. [CrossRef]
138. Arasaki, Y.; Takatsuka, K.; Wang, K.; McKoy, V. Femtosecond Energy- and Angle-Resolved Photoelectron Spectra. *Chem. Phys. Lett.* **1999**, *302*, 363–374. [CrossRef]
139. Arasaki, Y.; Takatsuka, K.; Wang, K.; McKoy, V. Femtosecond Energy- and Angle-resolved Photoelectron Spectroscopy. *J. Chem. Phys.* **2000**, *112*, 8871. [CrossRef]
140. Takatsuka, K.; Arasaki, Y.; Wang, K.; McKoy, V. Probing Wavepacket Dynamics with Femtosecond Energy- and Angl-Resolved Photoelectron Spectroscopy. *Faraday Discuss.* **2000**, *115*, 1–15. [CrossRef]
141. Arasaki, Y.; Takatsuka, K.; Wang, K.; McKoy, V. Pump-probe photoionization study of the passage and bifurcation of a quantum wave packet across an avoided crossing. *Phys. Rev. Lett.* **2003**, *90*, 248303. [CrossRef]
142. Arasaki, Y.; Takatsuka, K.; Wang, K.; McKoy, V. Studies of electron transfer in NaI with pump-probe femtosecond photoelectron spectroscopy. *J. Chem. Phys.* **2003**, *119*, 7913–7923. [CrossRef]
143. Horio, T.; Fuji, T.; Suzuki, Y.; Suzuki, T. Probing ultrafast internal conversion through conical intersection via time-energy Map of photoelectron angular anisotropy. *J. Am. Chem. Soc.* **2009**, *131*, 10392–10393. [CrossRef]
144. Suzuki, Y.; Fuji, T.; Horio, T.; Suzuki, T. Time-resolved photoelectron imaging of ultrafast $S_2 \rightarrow S_1$ internal conversion through conical intersection in pyrazine. *J. Chem. Phys.* **2010**, *132*, 174302. [CrossRef]
145. Wörner, H.J.; Bertrand, J.B.; Fabre, B.; Higuet, J.; Ruf, H.; Dubrouil, A.; Patchkovskii, S.; Spanner, M.; Mairesse, Y.; Blanchet, V.; et al. Conical Intersection Dynamics in NO2 Probed by Homodyne High-Harmonic Spectroscopy. *Science* **2011**, *334*, 208–212. [CrossRef]
146. von Contal, A.; Tehlar, A.; Schletter, A.; Arasaki, Y.; Takatsuka, K.; Wörner, H.J. Conical-intersection dynamics and ground-state chemistry probed by extreme-ultraviolet time-resolved photoelectron spectroscopy. *Nat. Comm.* **2018**, *9*, 3162. [CrossRef]
147. Kowalewski, M.; Bennett, K.; Dorfman, K.E.; Mukamel, S. Catching Conical Intersections in the Act: Monitoring Transient Electronic Coherences by Attosecond Stimulated X-Ray Raman Signals. *Phys. Rev. Lett.* **2015**, *115*, 193003. [CrossRef]
148. Brazard, J.; Bizimana, L.A.; Gellen, T.; Carbery, W.P.; Turner, D.B. Experimental Detection of Branching at a Conical Intersection in a Highly Fluorescent Molecule. *J. Phys. Chem. Lett.* **2016**, *7*, 14–19 [CrossRef]
149. Yang, J.; Zhu, X.; Wolf, T.J.A.; Li, Z.; Nunes, J.P.F.; Coffee, R.; Cryan, J.P.; Gühr, M.; Hegazy, K.; Heinz, T.F.; et al. CF_3I conical intersection and photodissociation dynamics with ultrafast electron diffraction. *Science* **2018**, *361*, 64–68. [CrossRef]
150. Mizuno, Y.; Arasaki, Y.; Takatsuka, K. Real-time observation of wavepacket bifurcation on nonadiabatically coupled field-dressed potential energy curves by means of spectrogram of induced photon-emission from molecules driven by CW laser. *J. Chem. Phys.* **2016**, *145*, 184305. [CrossRef]

151. Fujisaki, H.; Takatsuka, K. Chaos Induced by Quantum Effect Due to Breakdown of the Born-Oppenheimer Nonadiabaticity. *Phys. Rev. E* **2001**, *63*, 066221. [CrossRef]
152. Fujisaki, H.; Takatsuka, K. Highly Excited Vibrational Eigenfunctions in a Multimode Nonadiabatic System with Duschinsky Rotation. *J. Chem. Phys.* **2001**, *114*, 3497. [CrossRef]
153. Heller, E.J. Mode mixing and chaos induced by potential surface crossings. *J. Chem. Phys.* **1990**, *92*, 1718–1727. [CrossRef]
154. Berry, M.V. Regular and irregular semiclassical wavefunctions. *J. Phys. A* **1977**, *10*, 2083. [CrossRef]
155. Benettin, G.; Galgani, L.; Strelcyn, J.M. Kolmogorov entroyy anti numerical exyeriments. *Phy. Rev. A* **1976**, *14*, 2338–2345. [CrossRef]
156. Higuchi, H.; Takatsuka, K. Quantum chaos induced by nonadiabatic coupling in wavepacket dynamics. *Phys. Rev. E* **2002**, *66*, 035203. [CrossRef]
157. Takatsuka, K. Non-Born-Oppenheimer paths in anti-Hermitian dynamics for nonadiabatic transition. *J. Chem. Phys.* **2006**, *124*, 064111. [CrossRef]
158. Aharonov, Y.; Bohm, D. Significance of electromagnetic potentials in the quantum theory. *Phys. Rev.* **1959**, *115*, 485–491. [CrossRef]
159. Herzberg, G.; Longuet-Higgins, H.C. Intersection of potential energy surfaces in polyatomic molecules. *Discuss. Faraday Soc.* **1963**, *35*, 77. [CrossRef]
160. Berry, M. Quantal phase factors accompanying adiabatic changes. *Proc. R. Soc. Lond. A* **1984**, *392*, 45–57.
161. Simon, B. Holonomy, the Quantum Adiabatic Theorem, and Berry's. *Phase. Phys. Rev. Lett.* **1984**, *51*, 2167–2170. [CrossRef]
162. Takatsuka, K. Lorentz-like force emerging from kinematic interactions between electrons and nuclei in molecules: A quantum mechanical origin of symmetry breaking that can trigger molecular chirality. *J. Chem. Phys.* **2017**, *146*, 084312. [CrossRef]
163. Yonehara, T.; Takatsuka, K. Electron wavepacket dynamics in highly quasi-degenerate coupled electronic states: A theory for chemistry where the notion of adiabatic potential energy surface loses the sense. *J. Chem. Phys.* **2012**, *137*, 22A520. [CrossRef]
164. Yonehara, T.; Takatsuka, K. Nonadiabatic electron dynamics in densely quasidegenerate states in highly excited boron cluster. *J. Chem. Phys.* **2016**, *144*, 164304. [CrossRef]
165. Arasaki, Y.; Takatsuka, K. Chemical bonding and nonadiabatic electron wavepacket dynamics in densely quasi-degenerate excited state manifold of boron clusters. *J. Chem. Phys.* **2019**, *150*, 114101. [CrossRef]
166. Mandelbrot, B. *The Fractal Geomery of Nature*; Freeman: San Francisco, CA, USA, 1982.
167. Arasaki, Y.; Takatsuka, K. Quantum chaos in nonediabatic electron dynamics. 2023. *to be published*.
168. Schiff, L.I. *Quantum Mechanics*; McGraw-Hill: New York, NY, USA, 1968.
169. Matsuzaki, R.; Takatsuka, K. Electronic and nuclear flux analysis on nonadiabatic electron transfer reaction: A view from single-configuration adiabatic Born-Oppenheimer representation. *J. Comput. Chem.* **2019**, *40*, 148–163. [CrossRef]
170. Matsuzaki, R.; Takatsuka, K. Electronic and nuclear fluxes induced by quantum interference in the adiabatic and nonadiabatic dynamics in the Born-Huang representation. *J. Chem. Phys.* **2019**, *150*, 014103. [CrossRef]
171. Okuyama, M.; Takatsuka, K. Electron flux in molecules induced by nuclear motions. *Chem. Phys. Lett.* **2009**, *476*, 109. [CrossRef]
172. Nagashima, K.; Takatsuka, K. Electron-wavepacket reaction dynamics in proton transfer of formamide. *J. Phys. Chem. A* **2009**, *113*, 15240. [CrossRef]
173. Nagashima, K.; Takatsuka, K. Early-stage dynamics in coupled proton-electron transfer from $\pi - \pi^*$ state of phenol to solvent ammonia clusters: An electron dynamics study. *J. Phys. Chem. A* **2012**, *116*, 11167. [CrossRef]
174. Okuyama, M.; Takatsuka, K. Dynamical electron mechanism of double proton transfer in formic acid dimer. *Bull. Chem. Soc. Jpn.* **2012**, *85*, 217. [CrossRef]
175. Bredtmann, T.; Diestler, D.J.; Li, S.-D.; Manz, J.; Pérez-Torrés, J.F.; Tian, W.-J.; Wu, Y.-B.; Yang, Y.; Zhai, H.-J. Quantum theory of concerted electronic and nuclear fluxes associated with adiabatic intramolecular processes. *Phys. Chem. Chem. Phys.* **2015**, *17*, 29421. [CrossRef]
176. Hanasaki, K.; Takatsuka, K. Spin current in chemical reactions. *Chem. Phys. Lett.* **2022**, *793*, 139462. [CrossRef]
177. Takatsuka, K.; Arasaki, Y. Real-time electronic energy current and quantum energy flux in molecules. *J. Chem. Phys.* **2022**, *in press*. [CrossRef]
178. Pauling, L. *The Nature of the Chemical Bond and the Structure of Molecules and Crystals*, 3rd ed.; Cornell University Press: Ithaca, NY, USA, 1960.
179. Light, J.C. Phase-Space Theory of Chemical Kinetics. *J. Chem. Phys.* **1964**, *40*, 3221–3229. [CrossRef]
180. Weerasinghe, S.; Amar, F.G. Absolute classical densities of states for very anharmonic systems and applications to the evaporation of rare gas clusters. *J. Chem. Phys.* **1993**, *98*, 4967–4983. [CrossRef]
181. Parneix, P.; Bréchignac, P.; Amar, F.G. Isomer specific evaporation rates: The case of aniline–Ar_2. *J. Chem. Phys.* **1996**, *104*, 983–991. [CrossRef]
182. Parneix, P.; Amar, F.G.; Bréchignac, P. Structure, On the use of evaporation dynamics to characterize phase transitions in van der Waals clusters: Investigations in aniline–$(argon)_n$ up to $n = 15$. *Chem. Phys.* **1998**, *239*, 121–138. [CrossRef]
183. Calvo, F.; Labastie, P. Monte-Carlo simulations of rotating clusters.*Eur. Phys. J. D* **1998**, *3*, 229–236. [CrossRef]
184. Calvo, F. Thermal Stability of the Solidlike and Liquidlike Phases of $(C_{60})_n$ Clusters. *J. Phys. Chem. B* **2001**, *105*, 2183–2190. [CrossRef]
185. Calvo, F.; Parneix, P. Statistical evaporation of rotating clusters. I. Kinetic energy released. *J. Chem. Phys.* **2003**, *119*, 256–264. [CrossRef]

186. Parneix, P.; Calvo, F. Statistical evaporation of rotating clusters. II. Angular momentum distribution. *J. Chem. Phys.* **2003**, *119*, 9469–9475. [CrossRef]
187. Calvo, F.; Parneix, P. Statistical evaporation of rotating clusters. III. Molecular clusters. *J. Chem. Phys.* **2004**, *120*, 2780–2787. [CrossRef]
188. Takatsuka, K.; Arasaki, Y. Intra-molecular nonadiabatic electronic energy redistribution. 2023. *to be published*.
189. Zewail, A.H. *Femtochemistry: Ultrafast Dynamics of the Chemical Bond*; World Scientific: Singapore, 1994.

entropy

MDPI

Article

Extreme Eigenvalues and the Emerging Outlier in Rank-One Non-Hermitian Deformations of the Gaussian Unitary Ensemble

Yan V. Fyodorov [1,*], Boris A. Khoruzhenko [2,*] and Mihail Poplavskyi [2]

[1] Department of Mathematics, King's College London, London WC2R 2LS, UK
[2] School of Mathematical Sciences, Queen Mary University of London, London E1 4NS, UK
* Correspondence: yan.fyodorov@kcl.ac.uk (Y.V.F.); b.khoruzhenko@qmul.ac.uk (B.A.K.)

Abstract: Complex eigenvalues of random matrices $J = \text{GUE} + i\gamma\,\text{diag}(1, 0, \ldots, 0)$ provide the simplest model for studying resonances in wave scattering from a quantum chaotic system via a single open channel. It is known that in the limit of large matrix dimensions $N \gg 1$ the eigenvalue density of J undergoes an abrupt restructuring at $\gamma = 1$, the critical threshold beyond which a single eigenvalue outlier ("broad resonance") appears. We provide a detailed description of this restructuring transition, including the scaling with N of the width of the critical region about the outlier threshold $\gamma = 1$ and the associated scaling for the real parts ("resonance positions") and imaginary parts ("resonance widths") of the eigenvalues which are farthest away from the real axis. In the critical regime we determine the density of such extreme eigenvalues, and show how the outlier gradually separates itself from the rest of the extreme eigenvalues. Finally, we describe the fluctuations in the height of the eigenvalue outlier for large but finite N in terms of the associated large deviation function.

Keywords: non-Hermitian random matrices; complex eigenvalues; extreme eigenvalues; eigenvalue outlier; resonances; resonance trapping

Citation: Fyodorov, Y.V.; Khoruzhenko, B.A.; Poplavskyi, M. Extreme Eigenvalues and the Emerging Outlier in Rank-One Non-Hermitian Deformations of the Gaussian Unitary Ensemble. *Entropy* **2023**, *25*, 74. https://doi.org/10.3390/e25010074

Academic Editor: Marko Robnik

Received: 13 November 2022
Revised: 28 December 2022
Accepted: 28 December 2022
Published: 30 December 2022

1. Introduction

Rank-one non-normal deformations of the Gaussian and Circular Unitary Ensembles are a useful analytic tool for studying statistics of resonances in quantum scattering from a chaotic domain via a single channel [1,2]. As surveyed in [2,3], these random matrix ensembles are integrable in the sense that the joint probability density of their complex eigenvalues and, in some spectral scaling limits of interest, the eigenvalue correlation functions can be determined in a closed form. Such integrability, which also proves to be useful in other physics contexts, see, e.g., [4], extends to a certain degree to the deformed $\beta-$Gaussian and $\beta-$circular ensembles [5,6], especially to the classical values $\beta = 1, 4$ [7,8], but is lost if the underlying normal random matrix ensemble (Hermitian or unitary) is not integrable, as is the case with, e.g., finite rank non-Hermitian deformations of Wigner matrices [9–11] or band matrices [12]. Still, the latter matrices are found to share, in appropriate parameter ranges, some statistical characteristics of their complex eigenvalues and eigenvectors with their integrable counterparts.

In this paper, we aim to investigate complex eigenvalues with extreme imaginary parts for the rank-one non-Hermitian deformations of the Gaussian Unitary Ensemble (GUE) by exploiting the above-mentioned integrability. The latter feature gives access to the asymptotics of the eigenvalue density in the complex plane on mesoscopic scales and allows us to carry out a quantitative analysis of the separation of the eigenvalue outlier (which is known to exist in this model [9,10]) from the rest of the eigenvalues. Eigenvalue outliers in the complex plane have recently attracted renewed interest [11,13–15]. Our analysis refines and complements the existing knowledge about the outliers of nearly Hermitian matrices [9–11] albeit for arguably the simplest model of its type. As we will demonstrate, despite the simplicity of the model, its extreme eigenvalues exhibit an interesting transition at a certain value of the deformation parameter, with rich critical behaviour which deserves to be studied in more detail.

The non-Hermitian matrices that we consider are of the form

$$J = H + i\Gamma, \tag{1}$$

where H is a GUE matrix and Γ is a diagonal matrix with all diagonal entries being zero except the first one,

$$\Gamma = \gamma \,\mathrm{diag}(1, 0, \ldots, 0). \tag{2}$$

Denoting the matrix dimension by N, we fix the global spectral scale by the condition that the expected value of $\mathrm{Tr}\, H^2$ is N. Then the joint probability density function (JDPF) of matrix elements of the GUE matrix H is

$$f_N(H) = const \times \exp\left\{-\frac{N}{2}\,\mathrm{Tr}\, H^2\right\}. \tag{3}$$

With this normalisation, the limiting eigenvalue distribution of H, as the matrix dimension is approaching infinity, is supported on the interval $[-2, 2]$, and, inside this interval, the eigenvalue density is $\nu(X) = \frac{1}{2\pi}\sqrt{4 - X^2}$.

Note that due to the invariance of the JPDF (3) with respect to unitary rotations $H \to UHU^{-1}$ one may equivalently replace Γ in (2) with any other rank-one Hermitian matrix. Without loss of generality we may also assume γ to be positive. Then the eigenvalues $X_j + iY_j$ of matrices J (1)–(3) are all in the upper half of the complex plane and for N large they all, except possibly one outlier, lie just above the interval $[-2, 2]$ of the real line. Whether such an outlier is present or not is determined by the value of γ. For fixed values of $\gamma < 1$, almost surely, for N sufficiently large, *all* N eigenvalues lie within distance $c_N N^{-1}$ from the real line, with $c_N = o(N^\epsilon)$ for every $\epsilon > 0$ [9]. Furthermore, if $\gamma > 1$ then the same is true of all but one eigenvalue. This outlier lies much higher in the complex plane: to leading order in N, its imaginary part (the "height") is $\gamma - \gamma^{-1}$ [9,10,14]. For precise statements and proofs we refer the reader to [9,10] where these and similar facts were established for finite rank non-Hermitian deformations of real symmetric matrices with independent matrix entries.

For finite but large matrix dimensions, one would expect to find a transition region of infinitesimal width Ω about the outlier threshold value $\gamma = 1$ which captures the emergence of the outlier from the sea of low lying eigenvalues. Questions about the scaling of Ω with N and the corresponding characteristic height and distribution of the eigenvalues that lie farthest away from the real line are natural and interesting in this context. These are open questions in the mathematics and mathematical physics literature on the subject.

Apart from the mathematical curiosity, there is also motivation coming from physics. In the physics literature, the eigenvalues of J are associated with the *zeroes* of a scattering matrix in the complex energy plane, and their complex conjugates with the *poles* of the same scattering matrix, known as "resonances". The latter are obviously the eigenvalues of matrices (1)–(2) with γ replaced by $-\gamma$. In that context the absolute value of the eigenvalue's imaginary part is associated with the "resonance width". The eigenvalues close to real axis are called "narrow resonances" and the outlier is called the "broad resonance". The use of the Gaussian Unitary Ensemble for H is justified by invoking the so-called Bohigas-Giannoni-Schmidt conjecture [16] describing spectral statistics of highly excited energy levels of some classes of systems whose classical counterparts are chaotic. The resulting ensemble J is then an important ingredient in characterising statistical properties of scattering matrices in systems with quantum chaos and no time-reversal invariance, see [1] for description of the associated framework going back to the pioneering paper [17]. In that framework, the phenomenon of the outlier separation and the simultaneous movement of the rest of the eigenvalues towards the real axis was first discussed, albeit at a heuristic level, already in early theoretical works [18,19], the latter work even establishing the correct asymptotic position of the outlier. Later on, this phenomenon got considerable attention under the name "resonance trapping" and eventually was observed in experiments [20].

Very recently, Dubach and Erdős [11] performed a detailed analysis of the eigenvalue trajectories, with respect to changing the parameter γ, in the random matrix ensemble $H + i\gamma vs.v^*$ in the settings when H is assumed to be a Wigner matrix and v a column vector of unit length. It turned out that the evolution of the eigenvalues is governed by a system of *deterministic* first-order differential equations subject to random initial conditions, with the initial positions and velocities expressed in terms of the eigenvalues and eigenvectors of H. In addition, under suitable conditions on the distribution of matrix entries of H ensuring the validity of the uniform isotropic local law (Theorem 5 in [11]), Dubach and Erdős proved that with high probability the eigenvalue outlier is distinctly separated from the rest of the eigenvalues for all

$$\gamma > 1 + \frac{N^{\varepsilon}}{\sqrt[3]{N}}, \quad \varepsilon > 0. \tag{4}$$

Moreover, if $\varepsilon < 1/3$, i.e., if $N^{-1/3+\varepsilon}$ is asymptotically small, the outlier's height is $2N^{-1/3+\varepsilon}$ and its real part is in the window of width $N^{-1/3-\varepsilon/4}$ around the origin, whereas all other eigenvalues are no higher than $N^{-1/3-\varepsilon}$. In addition, with high probability, for all

$$\gamma < 1 - \frac{N^{\varepsilon}}{\sqrt[3]{N}}, \quad \varepsilon > 0, \tag{5}$$

no eigenvalue reaches the heights

$$Y = \frac{m}{\sqrt[3]{N}}, \quad m > 0. \tag{6}$$

These findings suggest that the width Ω of the transition region around $\gamma = 1$ scales with as $N^{-1/3}$ for N large. Naturally, for γ inside this region one would expect to find several eigenvalues, including the emerging "atypical" outlier, with imaginary parts on the critical scale (6) much exceeding the height $O(N^{-1})$ of low lying eigenvalues, as illustrated in Figure 1. One might call such eigenvalues "typical extremes" to emphasise atypicality of the emerging outlier.

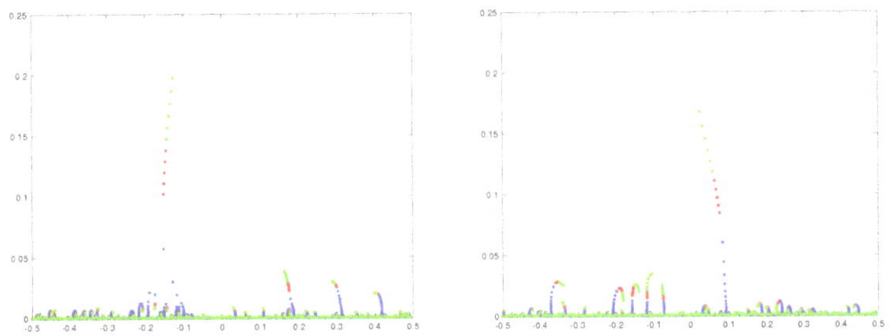

Figure 1. γ-trajectories of eigenvalues of matrices (1)–(3) of dimension $N = 1000$ near the origin. Each plot represents a different sample of H from the GUE (3). The parameter γ is varying in the interval $[0, 0.5]$ in the increments of 0.05 (blue dots), in the interval $[0.5, 1]$ in the increments of 0.1 (red dots), and in the interval $[1, 1.5]$ in the increments of 0.1 (green dots).

To a large extent our paper is motivated by [11] and aims to provide quantitative insights into this picture of the outlier emerging from the cloud of extreme eigenvalues. Whilst the approach of Dubach and Erdős is dynamical (fix matrix H and study eigenvalue trajectories as the magnitude γ of the deformation increases), our approach is statistical (fix a scale for γ and count the number of eigenvalues on characterisitc spectral scales in the complex plane averaged over the distribution of H which, for technical reasons, we assume to be GUE). Our present approach is limited to the expected values; analysing higher order moments is left as an interesting problem for future investigations. However, even with such a basic tool we are able to develop rather detailed quantitative understanding of the outlier separation and the associated restructuring transition in the spectra of matrices J.

As such, the two approaches complement each other very well. For example, we prove that for

$$\gamma = 1 + \frac{\alpha}{\sqrt[3]{N}}, \quad \alpha \in \mathbb{R}, \tag{7}$$

the expected number of eigenvalues whose height exceeds the level (6) is asymptotically given by the integral $\int_m^\infty p_\alpha^{(\mathrm{Im})}(m')dm'$ with density

$$\widetilde{p}_\alpha^{(\mathrm{Im})}(m) = \frac{1}{2\sqrt{\pi}} \frac{\frac{3}{2m} + \left(\frac{3m}{2} - \alpha\right)^2}{m^{3/2}} e^{-m\left(\alpha - \frac{m}{2}\right)^2}, \quad m > 0.$$

This density is the average density of the extreme eigenvalues at height (6). Together with findings in [11] this result establishes that the width Ω of the transition region around $\gamma = 1$ indeed scales with $N^{-1/3}$. Similarly, we are able to determine the average density of extreme eigenvalues $Z_j = X_j + iY_j$ of J near the origin in the complex plane in the critical scaling regime when when $q + im = \sqrt[3]{N}Z = O(1)$. As a function of coordinates q and m, this density, when appropriately rescaled, is given by

$$\widetilde{p}_\alpha(q,m) = \frac{1}{4\pi m}\left[\frac{1}{m} + \frac{q^2}{4} + \left(\frac{3m}{2} - \alpha\right)^2\right] e^{-m\left[\frac{q^2}{4} + \left(\alpha - \frac{m}{2}\right)^2\right]}, \quad q \in \mathbb{R}, m > 0.$$

It can be verified that $\int_{-\infty}^{+\infty} \widetilde{p}_\alpha(q,m)dq = \widetilde{p}_\alpha^{(\mathrm{Im})}(m)$, implying that the population of extreme eigenvalues at the critical height (6) which generates the eventual outlier (as α is approaching infinity) is constrained to a narrow vertical strip of width $O(N^{-1/3})$ about the origin (the centre of the eigenvalue band of H). Thus, our results both confirm and complement the analysis in [11], and show that it indeed touched the optimal scales in γ (7), both along the real and imaginary axes.

We would like to conclude this section with a short description of the structure of our paper. In Section 2 we develop quantitative heuristic analysis of the outlier separation. This analysis elucidates the emerging critical scaling in γ and the critical spectral scalings in the complex plane and provides a useful background for rigorous calculations later on. This section also offers our outlook on the universality of the exponent $-1/3$ in (7). Section 3 contains the statement of our main results and discussion. In Section 4 we express the expected density of eigenvalues of J and the density of their imaginary parts at finite matrix dimensions in terms of, respectively, Hermite and Laguerre polynomials. These expressions are then used in Sections 5 and 6 for asymptotic analysis of eigenvalue densities in various scaling limits. The two appendices contain derivations of technical auxiliary results.

2. Low Lying Eigenvalues and Their Extremes: A Heuristic Outlook

Before presenting our main results in the next Section, we would like to offer our quantitative heuristic insights into the outlier separation elucidating the emerging scalings and mechanisms behind them and providing a useful background for rigorous calculations later on.

With $z_j = X_j + iY_j$ standing for the eigenvalues of matrices $J = H + i\Gamma$, the angular brackets $\langle \ldots \rangle$ standing for averaging over the GUE matrix H (3), and $\delta(X)$ for the Dirac delta-function, the expected number of eigenvalues of J in domain D can be computed by integrating the mean eigenvalue density

$$\rho_N(X,Y) = \left\langle \frac{1}{N} \sum_{j=1}^{N} \delta(X - X_j)\delta(Y - Y_j) \right\rangle \tag{8}$$

over D and multiplying the result by N. For example, the expected number $\mathcal{N}_\gamma(Y)$ of the eigenvalues of J which lie above the line $\operatorname{Im} z = Y$ in the complex plane is given by the integral

$$\mathcal{N}_\gamma(Y) = N \int_{-\infty}^{\infty} \int_{Y}^{\infty} \rho_N(X,Y')\, dX dY' = N \int_{Y}^{\infty} \rho_N^{(\mathrm{Im})}(Y')\, dY', \tag{9}$$

where $\rho_N^{(\mathrm{Im})}(Y)$ is the mean density of the imaginary parts *irrespective of the value of the real part*,

$$\rho_N^{(\mathrm{Im})}(Y) = \left\langle \frac{1}{N} \sum_{j=1}^{N} \delta(Y - Y_j) \right\rangle. \tag{10}$$

Guided by the eigenvalue perturbation theory one can expect that the typical height Y of the eigenvalues whose real part is close to a point $X \in (-2,2)$ in the spectral bulk scales with the mean separation $\Delta = (N\nu(X))^{-1}$ between neighbouring real eigenvalues of the GUE matrix H in the limit $N \to \infty$. On a more formal level, introducing the scaled version of $\rho_N(X,Y)$ [21]

$$\widetilde{\rho}_N(X,y) := \frac{1}{\nu(X)} \left\langle \frac{1}{N} \sum_{j=1}^{N} \delta(X - X_j)\delta\big(y - 2\pi\nu(X)NY_j\big) \right\rangle, \quad -2 < X < 2, \tag{11}$$

one finds that such scaled density is well-defined in the limit of large matrix dimensions [1,2,21,22]: for every $y > 0$

$$\widetilde{\rho}(X,y) := \lim_{N\to\infty} \widetilde{\rho}_N(X,y) = -\frac{d}{dy}\left[e^{-yg(X)} \frac{\sinh y}{y} \right], \quad g(X) = \frac{\gamma + \frac{1}{\gamma}}{2\pi\nu(X)}, \tag{12}$$

confirming that locally the typical height of low lying eigenvalues scales with $\Delta = (N\nu(X))^{-1}$.

Globally, the typical height of low lying eigenvalues scales with N^{-1}. Intuitively, this can be understood from the exact sum rule

$$\sum_{j=1}^{N} Y_j = \gamma \tag{13}$$

which follows from the obvious relation $\operatorname{Tr} J = i\gamma + \operatorname{Tr} H$. On a more formal level, consider the expected fraction of the eigenvalues of J which lie above the level $\operatorname{Im} z = Y$, and set $y = NY$. In the limit $N \to \infty$,

$$\frac{1}{N}\mathcal{N}_\gamma\left(\frac{y}{N}\right) \sim \int_{-2}^{2} dX\, \nu(X) \int_{2\pi\nu(X)y}^{\infty} dy'\, \widetilde{\rho}(X,y') \tag{14}$$

$$= \frac{e^{-y\left(\gamma+\frac{1}{\gamma}\right)}}{y} \int_{-2}^{2} \frac{dX}{4\pi} \left(e^{y\sqrt{4-X^2}} - e^{-y\sqrt{4-X^2}} \right). \tag{15}$$

The integral in (15) is the modified Bessel function $I_1(2y)$. Therefore,

$$\lim_{N\to\infty}\frac{1}{N}\mathcal{N}_\gamma\left(\frac{y}{N}\right)=\frac{e^{-y\left(\gamma+\frac{1}{\gamma}\right)}}{y}I_1(2y)\,. \tag{16}$$

From this,

$$\widetilde{\rho}^{(\mathrm{Im})}(y):=\lim_{N\to\infty}\frac{1}{N}\rho_N^{(\mathrm{Im}\,z)}\left(\frac{y}{N}\right)=-\frac{d}{dy}\left[\frac{e^{-y\left(\gamma+\frac{1}{\gamma}\right)}}{y}I_1(2y)\right] \tag{17}$$

$$=\frac{e^{-y\left(\gamma+\frac{1}{\gamma}\right)}}{y}\left[\left(\gamma+\frac{1}{\gamma}-2\right)I_1(2y)-I_0(2y)-I_2(2y)\right]. \tag{18}$$

The density $\widetilde{\rho}^{(\mathrm{Im})}(y)$ is the mean density of the scaled imaginary parts $y_j=NY_j$ in the limit of large matrix dimensions. Even though it describes low lying eigenvalues it contains some useful information about eigenvalues higher up in the complex plane.

As an example, consider the expected value of the sum of the imaginary parts of low lying eigenvalues. Using definition (10), the sum rule (13) implies that

$$N\int_0^\gamma Y\rho_N^{(\mathrm{Im})}(Y)dY=\gamma\,. \tag{19}$$

Upon rescaling $y=NY$, one could naively jump to the conclusion that $\int_0^\infty y\,\widetilde{\rho}^{(\mathrm{Im})}(y)\,dy\ =\gamma$. However, by making use of (17) and integral 6.623(3) in [23], one actually finds that

$$\int_0^\infty y\,\widetilde{\rho}^{(\mathrm{Im})}(y)\,dy=\int_0^\infty\frac{e^{-y\left(\gamma+\frac{1}{\gamma}\right)}}{y}I_1(2y)\,dy=\frac{\gamma+\frac{1}{\gamma}-\sqrt{\left(\gamma+\frac{1}{\gamma}\right)^2-4}}{2}=\begin{cases}\gamma, & \text{if } \gamma<1,\\ \frac{1}{\gamma}, & \text{if } \gamma>1.\end{cases}$$

Thus, if $\gamma<1$ then the imaginary parts of low lying eigenvalues indeed add up to γ, in full agreement with the sum rule (19), whereas if $\gamma>1$ they add up only to $\frac{1}{\gamma}<\gamma$. The sum rule deficit $\gamma-\frac{1}{\gamma}$ is exactly the imaginary part of the outlier, and suggests that the rescaled limiting density of low lying eigenvalues, $\widetilde{\rho}^{(\mathrm{Im})}(y)$, precisely misses the delta-functional mass $\frac{1}{N}\delta\left(y-\left(\gamma-\frac{1}{\gamma}\right)\right)$.

As another example, consider the asymptotic form of $\widetilde{\rho}^{(\mathrm{Im})}(y)$ when $y\gg1$. It is markedly different depending on whether $\gamma=1$ or not. In the later case, using in (17) the asymptotic expansion for the modified Bessel function of large argument, $I_p(x)\sim\frac{e^x}{\sqrt{2\pi x}}(1-\frac{4p^2-1}{8x}+\ldots)$ one finds an exponential decay, whilst in the former case the decay is algebraic:

$$\widetilde{\rho}^{(\mathrm{Im})}(y)=\begin{cases}\dfrac{e^{-y\frac{(1-\gamma)^2}{\gamma}}}{2\sqrt{\pi}\,y^{3/2}}\left[\dfrac{(1-\gamma)^2}{\gamma}+\dfrac{30-3(\gamma+\gamma^{-1})}{16y}+O\left(\dfrac{1}{y^2}\right)\right] & \text{if } \gamma\neq1,\\[2ex] \dfrac{3}{4\sqrt{\pi}}\dfrac{1}{y^{5/2}}+O\left(\dfrac{1}{y^{7/2}}\right) & \text{if } \gamma=1\,.\end{cases} \tag{20}$$

It is instructive to return to the unscaled imaginary part Y and take a closer look at the expected number of the eigenvalues of J exceeding the level $\mathrm{Im}\,z=Y$ in the limit $N\to\infty$. It is evident from (16) that

$$\mathcal{N}_\gamma(Y)\sim\frac{e^{-NY\left(\gamma+\frac{1}{\gamma}\right)}}{Y}I_1(2NY)\,, \tag{21}$$

provided $NY = O(1)$. Extending this asymptotic relation to large values of NY allows one to get insights, even if only heuristically, about the characteristic scale of the highest placed among the low lying eigenvalues. Along these lines, we define the *characteristic scale of the height of typical extreme eigenvalues* as such level Y_e that the expected number of eigenvalues with imaginary part exceeding Y_e is of order of unity:

$$\mathcal{N}_\gamma(Y_e) = O(1). \tag{22}$$

We add the word typical to exclude the atypical eigenvalue (the outlier) which is known to exist when $\gamma > 1$. Now, assuming NY_e to be large (but still anticipating $Y_e \ll 1$) one can replace the Bessel function in (21) by its corresponding asymptotic expression and approximate:

$$\mathcal{N}_\gamma(Y_e) \approx \frac{e^{-NY_e\frac{(1-\gamma)^2}{\gamma}}}{2\sqrt{\pi N}\,Y_e^{3/2}}, \qquad 1 \ll NY_e \ll N. \tag{23}$$

The condition in (22) then leads to two essentially different scenarios depending on the value of γ. Namely, for every fixed positive $\gamma \neq 1$ the characteristic scale of the typical extreme values is, to leading order in N, $O(N^{-1}\ln N)$. On the other hand, if $\gamma = 1$ then the typical extreme values raise from the sea of low lying eigenvalues to a much higher height of $O(N^{-1/3})$. This change of scale for extreme values is easy to trace back to the emerging power-law decay in the vicinity of $\gamma = 1$ which is evident in (20).

In fact, as evident from (23), the typical extreme values scale as $Y_e = O(N^{-1/3})$ not only at $\gamma = 1$, but also as long as $|1-\gamma| \propto N^{-1/3}$. Actually, by setting simultaneously $\gamma = 1 + \alpha N^{-1/3}$ and $Y_e = mN^{-1/3}$ the asymptotic relation (23) is converted into

$$\mathcal{N}_{1+\frac{\alpha}{\sqrt[3]{N}}}\left(\frac{m}{\sqrt[3]{N}}\right) \approx \frac{e^{-m\alpha^2}}{2\sqrt{\pi}\,m^{3/2}}, \tag{24}$$

an expression that is indeed of order of unity for all fixed values of α and $m > 0$. Thus, the width of the transition region about $\gamma = 1$ must scale with $N^{-1/3}$. Combined with the existence of a distinct outlier at height $\gamma - \gamma^{-1} \gg Y_e$ one may indeed see that our heuristic argument perfectly agrees with the conjecture of Dubach and Erdős about the critical scaling $\gamma = 1 + O(N^{-1/3})$ where the separation of typical and atypical extreme values happens.

Before continuing our exposition of the heuristics behind the restructuring of the density of complex eigenvalues we would like to make two remarks.

Remark 1. *To make further contact with the standard subject of extreme value statistics, it is useful to recourse to the classical theory of extreme values for i.i.d. sequences of random variables* $y_1, \ldots, y_N$*, a succinct albeit informal summary of which can be found in, e.g., [24]. In that case the probability law of extreme values is characterised by the tail behaviour of the "parent" probability density function (pdf)* $p(y)$ *of* y_j *and is essentially universal in the limit* $N \to \infty$*. In our context, the pertinent case for comparison is that of non-negative continuous i.i.d. random variables with the parent distribution supported on the entire semi-axis* $[0, \infty)$*. Then only two possibilities may arise. Those sequences which are characterised by the power-law decaying pdf* $p(y) \sim Ay^{-(1+\alpha)}$*,* $\alpha > 0$*, as* $y \to \infty$ *have their extreme values scaling with* $(AN/\alpha)^{1/\alpha}$ *and the distribution of their maximum,* $y_{max} = \max(y_1, \ldots, y_N)$*, after rescaling converges to the so-called Fréchet law in the limit* $N \to \infty$*. In contrast, if the parent pdf decays faster than any power, e.g., if* $\ln p(y) \sim -y^\delta$*,* $\delta > 0$*, then, to leading order, extreme values scale with* $(\ln N)^{1/\delta}$*, and the distribution of the largest value* y_{max}*, converges, after a shift and further rescaling, to the so-called Gumbel law. Although, the imaginary parts of complex eigenvalues in the random matrix ensemble (1)–(3) are not at all independent (as is evident from their JPDF (51) resulting in a non-trivial determinantal two-point and higher order correlation functions at the scale* N^{-1}*, see [22]), our scaling predictions for the typical extreme eigenvalues are in formal correspondence with the i.i.d. picture: a Gumbel-like* *scaling (with $\delta = 1$) if $\gamma \neq 1$ and a Fréchet-like scaling (with $\alpha = 3/2$) if $\gamma = 1$. This is exactly as would have been implied in the i.i.d. picture by the tail behaviour of the mean eigenvalue densities in the two cases in (20). This fact naturally suggests to conjecture Gumbel statistics for the typical largest imaginary part (excluding possible outlier) for any $\gamma \neq 1$, changing to a Fréchet-like law for $\gamma = 1$, with a possible family of $\alpha-$ dependent nontrivial extreme value statistics in the crossover critical regime $\gamma = 1 + \alpha N^{-1/3}$. Although we are not able to shed light on the distribution of typical extreme eigenvalues in the random matrix ensemble (1)–(3), we will discuss some results in that direction for a somewhat related model at the end of the next section.*

Remark 2. *The phenomenon of resonance width restructuring with increasing the coupling to continuum (controlled in the present model by the parameter γ) and the emergence of the broad resonance has many features in common with the so-called super-radiant phenomena in optics. This is well known in the physics literature, see [25] and references therein. Here, we would like to point to a similarity of the spectral restructure in the random matrix ensemble (1)–(3) to a process in a different physics context, the so-called "condensation transition" which occurs in models of mass transport when the globally conserved mass M exceeds a critical value, see, e.g., [26] for a review. In such a regime, the excess mass forms a localised in space condensate coexisting with a background fluid in which the remaining mass is evenly distributed over the rest of the system. A particularly simple case for analysing the condensation phenomenon is when the system has a stationary state such that probability of observing a configuration of masses m_i factorises into the form $\prod_i f(m_i)\delta(\sum_i m_i - M)$. In that context again the tail behaviour of the "parent" mass density $f(m)$ plays important role. Although we would like to stress again that in our model the imaginary parts of the complex eigenvalues are not independent, the analogy with the condensation phenomenon is quite evident.*

Essentially the same heuristic analysis as in the above helps to clarify the numerically observed fact of the outlier emerging mostly close to the origin of the spectrum $\mathrm{Re}\, z = 0$. From this angle it is instructive to ask what should be the scale of extreme values for eigenvalues satisfying $|\mathrm{Re}\, z| < W$, that are sampled in a window of a small widths $W \ll 1$ around the origin (still assuming typically many eigenvalues in the window, so that $W \gg \Delta \sim 1/N$). The total mean number of eigenvalues in the window W whose imaginary parts exceed the level Y (but still formally remain of the order of $1/N$) is now given by

$$\mathcal{N}_{\gamma,W}(Y) = \frac{e^{-NY\left(\gamma+\frac{1}{\gamma}\right)}}{4\pi Y}[T_W(NY) - T_W(-NY)], \quad T_W(NY) = 2\int_0^W e^{NY\sqrt{4-X^2}}dX. \quad (25)$$

For $NY \gg 1$ the term $T_W(-NY)$ is exponentially suppressed, while the integral in $T_W(NY)$ is dominated by $X \ll 1$ and with required accuracy yields the leading-order expression in the form:

$$\mathcal{N}_{\gamma,W}(Y \gg 1/N) \approx \frac{e^{-NY\left(\frac{(1-\gamma)^2}{\gamma}\right)}}{2\pi Y^{3/2}}\sqrt{\frac{2}{N}}\int_0^{W\sqrt{NY/2}} e^{-\frac{t^2}{2}}dt. \quad (26)$$

Now, let us assume that both the width W of the window and the parameter γ scale with N in this non-trivial way as

$$W \sim N^{-1+\kappa}, \quad 0 < \kappa \leq 1, \qquad \text{and} \qquad \gamma = 1 - \alpha N^{-\delta}, \quad 0 < \delta \leq \infty, \ \alpha \in \mathbb{R}, \quad (27)$$

and again apply the same heuristic procedure to determine the scale of extreme values $Y_e(\kappa, \delta)$ in the window as $N \to \infty$ for given values of exponents κ and δ. A straightforward computation shows that the arising scale of extreme values very essentially depends on whether the parameter δ satisfies $0 < \delta < 1/3$ or $1/3 \leq \delta < 1$. In the former case we find

$$Y_e(\kappa, 0 < \delta < 1/3) \approx \begin{cases} N^{-1+\kappa}, & \text{if } 0 < \kappa < 2\delta, \\ \frac{\kappa - 2\delta}{\alpha} N^{-1+2\delta} \ln N, & \text{if } 2\delta < \kappa < 1 - \delta, \\ \frac{1 - 3\delta}{\alpha} N^{-1+2\delta} \ln N, & \text{if } 1 - \delta < \kappa < 1. \end{cases} \tag{28}$$

whereas in the latter case

$$Y_e(\kappa, 1/3 \leq \delta < 1) \approx \begin{cases} N^{-1+\kappa}, & \text{if } 0 < \kappa < 2/3, \\ N^{-1/3}, & \text{if } 2/3 < \kappa < 1. \end{cases} \tag{29}$$

One may say that as long as $\delta < 1/3$ the system is not fully in the well-developed "critical regime", and the extreme value scale is growing with the window width, saturating at the Gumbel-like scale $N^{-1+2\delta} \ln N$. At the same time, as long as δ exceeds the threshold value $\delta = 1/3$, the typical extreme values reach the scale $Y_e = O\left(N^{-1/3}\right)$ as long as they are sampled in a window of width exceeding the scale $W_c = O\left(N^{-1/3}\right)$, thus containing $O(N^{2/3})$ eigenvalues. This heuristics suggests that only eigenvalues satisfying $|X| < W_c$ typically have a nonvanishing probability to reach to the maximum height in the complex plane, and eventually to generate an outlier as α increases. It would be also natural to expect the corresponding extreme eigenvalues to follow the Fréchet-type statistics for their imaginary parts, as opposed to the Gumbel statistics in the former case.

We would like to end our heuristic considerations with a brief heuristic outlook on the universality of the scaling factor $N^{-1/3}$ which is key to the correct description of the transition in question. As is evident from (23) the exponent $-1/3$ is implied by the scaling law

$$\mathcal{N}_{\gamma=1}(Y) \propto \frac{1}{N^{1/2} Y^{3/2}} \tag{30}$$

in the limit $NY \gg 1$ for the expected number of eigenvalues exceeding the level line $\mathrm{Im}\, z = Y$. Thus, to investigate the extent of universality of this exponent one needs to trace the origin of the scaling law (30). This can be readily done by returning to the asymptotic relation (14) and (15) which was used to obtain (30). On evaluating the integral in (15) for large values of $y = Y/N$ by the Laplace method it becomes immediately apparent that the power $Y^{-3/2}$ on the right-hand side in (30) and, hence, the exponent in question stems from the quadratic shape of the limiting GUE eigenvalue density function $\nu(X) = (2\pi)^{-1}\sqrt{4 - X^2}$ in the vicinity of its maximum. It is natural to conjecture that had one started from a random Hermitian matrix H taken from the broad class of invariant ensembles characterised by joint probability density function $\propto \exp\{-N \,\mathrm{Tr}\, V(H)\}$ with a suitable potential $V(H)$ (or from the class of Wigner matrices with suitable conditions on the iid entries), the asymptotic expression (12) for the scaled eigenvalue density $\widetilde{\rho}_N(X, y)$ would retain its validity after replacing $\nu(X)$ in (11) and (12) by the corresponding limiting eigenvalue density of H. For example, as was shown albeit not fully rigorously in [27], such universality of the scaled eigenvalue density near the real line is exhibited by almost Hermitian random matrices which are morally similar to finite rank non-Hermitian deviations as in (1) and (2). Since asymptotic relation (14) and (15) is the immediate corollary of (12), one then concludes that as long as the limiting eigenvalue density of H has a single global parabolic-shaped maximum, an additive rank-one non-Hermitian deformation will demonstrate the same type of critical scaling for its extreme complex eigenvalues, and, most probably, after appropriate rescaling, the same type of critical behaviour of the density of imaginary parts as described in the next section. One can however imagine invariant ensembles where the mean eigenvalue density would have a non-parabolic behaviour close to the maximum point.

From this point of view, the noticed in [11] resemblance of the $N^{-1/3}$ critical scaling in the present model and the edge scaling of extreme real eigenvalues of GUE, which, e.g., manifests itself in the so-called BBP [28] transition under additive rank-one Hermitian perturbation of the GUE, looks to us purely coincidental. Indeed, the latter is known to have its origin in the square root behaviour of the mean density $\nu(X)$ at the spectral edges where $\nu(X)$ vanishes, and as such seems to have nothing to do with the behaviour of the same density close to its maximal point.

3. Main Results and Discussion

Our first result concerns the mean density of imaginary parts $\rho_N^{(\mathrm{Im})}(Y)$ (10) in the large deviation regime $Y \gg N^{-1}$. We note that no eigenvalue of J has imaginary part equal or greater than γ. This is a consequence of the sum rule (13). Therefore we only consider the range of values $Y \in [0, \gamma)$.

Theorem 1. *Consider the random matrix ensemble (1)–(3) in the scaling regime*

$$N^{1-\epsilon} Y = y > 0, \quad 0 < \epsilon \leq 1, \quad N \to \infty. \tag{31}$$

Then for every fixed $\gamma > 0$ and $\epsilon \in (0,1]$

$$\rho_N^{(\mathrm{Im})}(Y) \sim \frac{1}{\sqrt{N}} \Psi_\gamma(Y) \exp\{-N\Phi_\gamma(Y)\}, \tag{32}$$

with

$$\Phi_\gamma(Y) = Y(\gamma - Y) - \ln \frac{\gamma - Y}{\gamma} - Y r_*(Y) + 2 \ln r_*(Y), \tag{33}$$

$$\Psi_\gamma(Y) = \frac{1}{\sqrt{2\pi}} \frac{\gamma}{(\gamma - Y)^2} \frac{[1 - r_*(Y)(\gamma - Y)]^2}{Y^{3/2}(Y^2 + 4)^{1/4}}, \tag{34}$$

and

$$r_*(Y) = \frac{\sqrt{Y^2 + 4} - Y}{2}. \tag{35}$$

The rate function $\Phi_\gamma(Y)$ is a smooth non-negative function of Y on the interval $[0, \gamma)$ vanishing at $Y = 0$. The rate function is monotone increasing on this interval if $\gamma \leq 1$, whereas if $\gamma > 1$ then it has two local extrema: a local minimum at $Y_* = \gamma - \gamma^{-1}$ where it vanishes, and a local maximum at $Y_{**} = \frac{2(\gamma - \gamma^{-1})}{3 + \sqrt{1 + 8\gamma^{-2}}} < Y_*$.

By the way of discussion of the above Theorem a few remarks are in order.

Remark 3. *The two distinct profiles of the rate function are illustrated in Figure 2. If $\gamma > 1$, the point $Y_* = \gamma - \gamma^{-1}$ where the Large Deviation Rate function $\Phi_\gamma(Y)$ vanishes can be identified as the most probable value of the imaginary part in the region $Y \gg N^{-1}$, converging in the limit $N \to \infty$ to (the height of) the outlier, see next comment. At the same time, the other extremal point, Y_{**}, can be interpreted as the* true boundary, *along the imaginary axis in the complex plane, between the bulk of eigenvalues and the spectral outlier. This is because the pre-exponential factor $\Psi_\gamma(Y)$ in (32) vanishes at $Y = Y_{**}$ too. Hence, $\rho_N^{(\mathrm{Im})}(Y_{**}) \to 0$ in the scaling limit (31).*

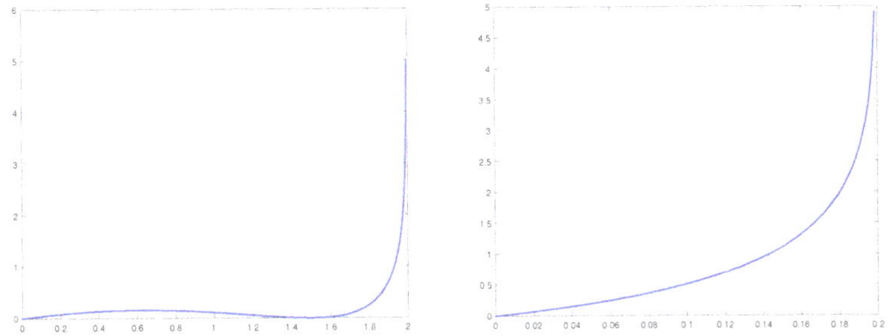

Figure 2. Plots of the rate function $\Phi_\gamma(Y)$ for $\gamma = 2$ (plot on the **left**) and $\gamma = 0.2$ (plot on the **right**).

Remark 4. *The Large Deviation approximation (32) for $\gamma > 1$ describes fluctuations of the imaginary part of the outlier around its most probable value $Y_* = \gamma - \gamma^{-1}$. The law of these fluctuations in the limit $N \to \infty$ can be easily determined from (32). To this end, we first note that for N large the magnitude of fluctuations about Y_* scales with $1/(\sqrt{N}|\Phi''_\gamma(Y_*)|)$. Calculating the second derivative and rescaling the density $\rho_N^{(\mathrm{Im})}(Y)$ correspondingly, one finds (in the limit $N \to \infty$) that*

$$N\,\rho_N^{(\mathrm{Im})}\Big(Y_* + \frac{\sigma u}{\sqrt{N}}\Big) \sim \frac{1}{\sqrt{2\pi}} e^{-\frac{u^2}{2}}, \quad \sigma^2 = \frac{1}{\gamma^2}\frac{\gamma^2+1}{\gamma^2-1}. \tag{36}$$

The integral of the rescaled density on the left-hand side over the entire range of values of u counts the expected number of eigenvalues in the $\frac{\sigma}{\sqrt{N}}$-neighbourhood of Y_*. Evidently, this integral is approaching unity as $N \to \infty$, confirming that the rescaled density on the left-hand side in (36) describes the law of fluctuations of a single eigenvalue - the outlier. Thus, we recover one of the results of [10] where laws of outlier fluctuations were established in greater generality than our assumptions (2) and (3). We note that for finite but large values of N the function

$$p_N(Y) := \sqrt{N}\,\Psi_\gamma(Y)\exp\{-N\Phi_\gamma(Y)\} \tag{37}$$

provides an approximation of the probability density function of the outlier $Y_{max} = \max Y_j$ in the interval $0 < \varepsilon < Y < \gamma, \gamma > 1$.

In Figure 3, we plot histograms of the imaginary parts Y_j of the eigenvalues and of their maximal value $Y_{max} = \max Y_j$ in the random matrix ensemble (1)–(3) and make comparison with the corresponding Large Deviation approximations. Although the value of $N = 50$ is only moderately large, one can observe a good agreement. Furthermore, one can observe that the large-N approximation (37) of the probability density of Y_{max} captures well the skewness of the distribution of Y_{max} for finite matrix dimensions. This skewness disappears in the limit $N \to \infty$, see Equation (36).

Remark 5. *Consider now the scales $Y = O(N^{-1+\varepsilon})$ with $\varepsilon \in (0,1)$. The expected number of eigenvalues with $N^{1-\varepsilon}Y \in [y_1, y_2]$ is given by the integral*

$$N\int_{y_1}^{y_2} \frac{1}{N^{1-\varepsilon}}\,\rho_N^{(\mathrm{Im})}\Big(\frac{y}{N^{1-\varepsilon}}\Big)dy. \tag{38}$$

The rescaled density in this integral can be found from (32)–(34):

$$\frac{1}{N^{1-\varepsilon}}\,\rho_N^{(\mathrm{Im})}\Big(\frac{y}{N^{1-\varepsilon}}\Big) \sim \frac{1}{N^{\varepsilon/2}}\frac{1}{2\sqrt{\pi}}\frac{(1-\gamma)^2}{\gamma}\frac{1}{y^{3/2}}e^{-N^\varepsilon y\frac{(1-\gamma)^2}{\gamma}}, \quad \varepsilon \in (0,1). \tag{39}$$

Evidently, if $\gamma \neq 1$ then, away from the boundary point $y = 0$, the integral in (38) vanishes in the limit $N \to \infty$. Therefore for every fixed $\gamma \neq 1$ and $0 < \varepsilon < 1$ there are no eigenvalues of J whose imaginary part is scaling with $N^{-1+\varepsilon}$. On the other hand, according to the heuristics of Section 2, *one should expect finite numbers of eigenvalues whose imaginary part is scaling with $N^{-1}\ln N$. These would be the extremes of the eigenvalues with the typical imaginary part $Y = O(N^{-1})$.*

By formally letting $\varepsilon \to 0$ in (39) one obtains

$$\frac{1}{N}\rho_N^{(\mathrm{Im})}\Big(\frac{y}{N}\Big) \sim \frac{1}{2\sqrt{\pi}}\frac{(1-\gamma)^2}{\gamma}\frac{1}{y^{3/2}}e^{-y\frac{(1-\gamma)^2}{\gamma}}.$$

This relation reproduces the leading order of the asymptotic form of the density of the rescaled imaginary parts $y = NY$ in the region $y \gg 1$, see the top line in (20). Thus, for a fixed value of $\gamma \neq 1$ *Theorem 1 describes a crossover of the density of imaginary parts from the characteristic scale of low lying eigenvalues to larger scales, including $Y = O(1)$ which is the scale of the outlier.*

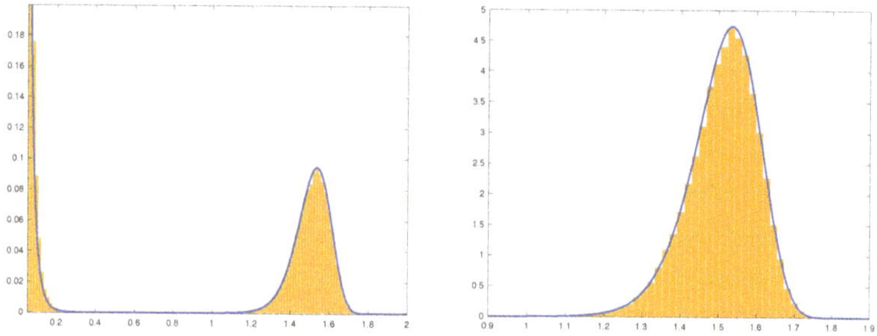

Figure 3. Histograms of the imaginary parts Y_j of the eigenvalues in the random matrix ensemble (1)–(3) of dimension $N = 50$ with $\gamma = 2$. Plot on the **left**: Histogram of Y_j's versus the large deviation approximation of density of the imaginary parts given by (32) (solid line). Plot on the **right**: Histogram of the largest imaginary part $Y_{max} = \max Y_j$ versus the large deviation approximation $p_N(Y)$ (37) of the p.d.f. of Y_{max} (solid line). Each plot was produced using 100,000 samples from the GUE distribution (3).

Whereas the picture described by Theorem 1 is quite complete for a fixed γ, it is not detailed enough to accurately describe the typical extreme eigenvalues in the situation when the parameter γ approaches its critical value $\gamma = 1$ as N is approaching infinity. For example, from the heuristics of Section 2 we know that both the width of the transition region about $\gamma = 1$ and the height of the typical extreme eigenvalues scale with $N^{-1/3}$. The Large Deviation approximation (32), if applied formally in the transition region parametrised by $\gamma = 1 + \alpha N^{-1/3}$, yields the following approximate expression for the rescaled density of imaginary parts:

$$\frac{1}{N^{1/3}}\rho_N^{(\mathrm{Im})}\Big(\frac{m}{N^{1/3}}\Big) \approx \frac{1}{N}\frac{1}{2\sqrt{\pi}}\frac{\left(\frac{3m}{2}-\alpha\right)^2}{m^{3/2}}e^{-m(\alpha-\frac{m}{2})^2}. \tag{40}$$

Evidently, in the limit of small values of m which corresponds to approaching the scale $Y = O(N^{-1})$ from above, this expression does not reproduce the correct power $5/2$ of algebraic decay (20) characteristic of this scale when $\gamma = 1$. In contrast, the heuristics based on (21), see the approximations in (23) and (24), do reproduce the correct power. Indeed, by taking the derivative in m of the expression on the right-hand side in (24), one gets

$$\frac{1}{N^{1/3}}\rho_N^{(\mathrm{Im})}\Big(\frac{m}{N^{1/3}}\Big) \approx \frac{1}{N}\frac{1}{2\sqrt{\pi}}\frac{\frac{3}{2m}+\alpha^2}{m^{3/2}}e^{-m\alpha^2}. \tag{41}$$

In the limit of small values of m the expression on the right-hand side agrees with the bottom line in (20). One can also arrive at (41) by making the formal substitution $\gamma = 1 + \frac{\alpha}{N^{1/3}}$ and $y = NY = mN^{2/3}$ in (20).

Our next Theorem is a refinement of Theorem 1 in that it provides an accurate description of the density of the typical extreme eigenvalues in the transition region between the sea of low lying eigenvalues and the eigenvalue outlier.

Theorem 2. *Consider the random matrix ensemble (1)–(3) in the scaling regime*

$$\gamma = 1 + \frac{\alpha}{N^{1/3}}, \; Y = \frac{m}{N^{1/3}}, \quad N \to \infty . \tag{42}$$

Then, for every fixed $\alpha \in \mathbb{R}$ and $m > 0$,

$$\frac{1}{N^{1/3}} \rho_N^{(\mathrm{Im})}\Big(\frac{m}{N^{1/3}}\Big) \sim \frac{1}{N} \frac{1}{2\sqrt{\pi}} \frac{\frac{3}{2m} + \left(\frac{3m}{2} - \alpha\right)^2}{m^{3/2}} \, e^{-m\left(\alpha - \frac{m}{2}\right)^2} . \tag{43}$$

This theorem confirms that the characteristic scale of the height of the typical extreme eigenvalues of matrix J is $O\Big(N^{-1/3}\Big)$. Indeed, the expected number of eigenvalues with imaginary part exceeding the level $Y = \frac{m}{N^{1/3}}$ is given by

$$N \int_m^\infty \frac{1}{N^{1/3}} \rho_N^{(\mathrm{Im})}\Big(\frac{m}{N^{1/3}}\Big) \, dm ,$$

which is a finite number in the limit $N \to \infty$.

Theorem 2 also describes the density $\rho_N^{(\mathrm{Im})}(Y)$ in the cross-over from the characteristic scale of low lying eigenvalues to the Large Deviation regime of Theorem 1. Indeed, for small values of m the asymptotic expression (43) matches the one in (41), whilst in the limit of large values of m it matches (40).

The emerging outlier is captured by (43) when *both* m and $\alpha > 0$ are large. Intuitively this is clear from the comparison of (43) and (40). On a more formal level, one can come to the same conclusion by analysing the limiting density of extreme values

$$\widetilde{p}_\alpha^{(\mathrm{Im})}(m) = \frac{1}{2\sqrt{\pi}} \frac{\frac{3}{2m} + \left(\frac{3m}{2} - \alpha\right)^2}{m^{3/2}} \, e^{-m\left(\alpha - \frac{m}{2}\right)^2}, \quad m > 0 . \tag{44}$$

Using Wolfram Mathematica one finds

$$\frac{d}{dm} \widetilde{p}_\alpha^{(\mathrm{Im})}(m) = \frac{e^{-m\left(\alpha - \frac{m}{2}\right)^2}}{32\sqrt{\pi}\, m^{7/2}} Q_6(\alpha, m) ,$$

where

$$Q_6(\alpha, m) = -60 - 48\alpha^2 m + 72\alpha m^2 - 16\alpha^4 m^2 + 80\alpha^3 m^3 - 144\alpha^2 m^4 + 108\alpha m^5 - 27m^6 .$$

Evidently, $Q_6(\alpha, m) < 0$ for all $m > 0$ if α is negative. Therefore, if $\alpha < 0$ (subcritical values of γ) then the limiting density $\widetilde{p}_\alpha^{(\mathrm{Im})}(m)$ is a monotonically decreasing function of m on the entire interval $m > 0$. One can interpret this profile as a population of extreme eigenvalues without an obvious "leader". By continuity, this profile persevere for small positive α. Indeed, at $\alpha = 0$ the polynomial $Q_6(0, m)$ has three pairs of complex conjugated roots, none are real. Since the roots of polynomials depend continuously on its coefficients, there exists an $\alpha_0 > 0$ such that for all $\alpha \in [0, \alpha_0]$ the polynomial $Q_6(\alpha, m)$ in m will still have no real roots and, hence, will take only negative values, implying that $\widetilde{p}_\alpha^{(\mathrm{Im})}(m)$ is a monotonically decreasing function of m. By computing the roots of $Q_6(\alpha, m)$ in variable m, we can show that $0.6485 < \alpha_0 < 0.649$.

Once $\alpha > \alpha_0$, the polynomial $Q_6(\alpha, m)$ in m acquires real roots. In the limit of large positive α there are two real roots: to leading order these are

$$m_1 = 2\alpha\left(1 + \frac{3}{8\alpha^3} + o\left(\frac{1}{\alpha^3}\right)\right) \quad \text{and} \quad m_2 = \frac{2}{3}\alpha\left(1 + \frac{15}{8\alpha^3} + o\left(\frac{1}{\alpha^3}\right)\right).$$

The larger root, m_1, is the point of local maximum of $\widetilde{p}_\alpha^{(\mathrm{Im})}(m)$, where $\widetilde{p}_\alpha^{(\mathrm{Im})}(m_1) \propto \alpha^{1/2} \gg 1$, and the smaller root, m_2, is the point of local minimum $\widetilde{p}_\alpha^{(\mathrm{Im})}(m)$, where $\widetilde{p}_\alpha^{(\mathrm{Im})}(m_2) \propto \alpha^{-5/2} \ll 1$. In fact, in the limit $\alpha \to \infty$ the larger root is transitioning into Y_*, the most probable value of imaginary parts, and, hence, it can be interpreted as the emerging spectral outlier. At the same time, the smaller root is transitioning into the true boundary Y_{**} between the sea of low lying eigenvalues and the outlier. This cross-over can be validated by noticing that in the scaling limit (42) $Y_* = \gamma - \gamma^{-1} \sim 2\alpha$ and $Y_{**} = \frac{2(\gamma-\gamma^{-1})}{3+\sqrt{1+8\gamma^{-2}}} \sim \frac{2}{3}\alpha$.

Further insights into the restructuring of the spectrum of J can be obtained by looking at the γ-dependence of the expected number of the eigenvalues of J with imaginary parts exceeding the level $Y = mN^{-1/3}$. In the scaling limit (42) this number converges to

$$\widetilde{\mathcal{N}}_\alpha(m) = \int_m^\infty \widetilde{p}_\alpha^{(\mathrm{Im})}(m')dm'\,.$$

In Figure 4, we plot $\widetilde{\mathcal{N}}_\alpha(m)$ as function of α for several values of m. One can observe that for any fixed $m > 0$ the population of the extreme eigenvalues of J that exceed the level $Y = mN^{-1/3}$ is, on average, growing as γ is approaching the critical value $\gamma = 1$ from below. For γ on the other side of $\gamma = 1$, this population peaks a some point and then it starts to decline as γ increases further, to a single eigenvalue which is the outlier. All the other extreme eigenvalues are getting closer and closer to the real line with the increase of γ. One can think of them as being trapped in the sea of low lying eigenvalues. This picture is consistent with the eigenvalue trajectories of Figure 1 and provides a more quantitative description of the "resonance trapping" phenomenon [20] in the framework of random matrix theory.

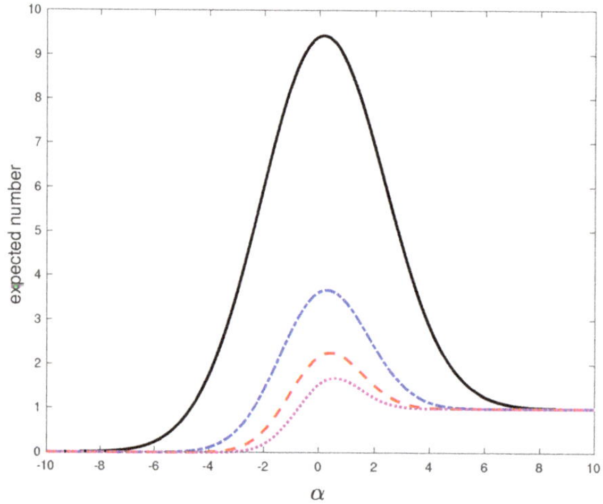

Figure 4. Plot of the expected number $\widetilde{\mathcal{N}}_\alpha(m)$ of the eigenvalues of J with imaginary parts exceeding the level $Y = mN^{-1/3}$ as function of α when $m = 0.1$ (black solid line), $m = 0.2$ (blue dashdotted line), $m = 0.3$ (red dashed line), and $m = 0.4$ (magenta dotted line).

Our final result aims to clarify the length of the central part of the spectrum of J supporting nontrivial scaling behaviour of the extreme eigenvalues in the vicinity of the separation transition. To this end, let us consider eigenvalues $z_j = X_j + iY_j$ of J in the scaling regime when

$$\gamma = 1 + \frac{\alpha}{N^{1/3}}, \; X = \frac{q}{N^{1/3}}, Y = \frac{m}{N^{1/3}}, \quad N \to \infty. \tag{45}$$

On average, eigenvalue numbers in this regime can be counted using the rescaled density

$$\widetilde{p}_N(q,m) := \left\langle \sum_{i=1}^{N} \delta\left(q - N^{1/3}X_j\right) \delta\left(m - N^{1/3}Y_j\right) \right\rangle = \frac{N}{N^{2/3}} \rho_N\left(\frac{q}{N^{1/3}}, \frac{m}{N^{1/3}}\right),$$

where, as before, the angle brackets stand for the averaging over the GUE matrix H in (1) and $\rho_N(X,Y)$ is the mean eigenvalue density (8).

Theorem 3. *Consider the random matrix ensemble (1)–(3) in the scaling regime (45). Then, for every fixed $\alpha \in \mathbb{R}$, $q \in \mathbb{R}$ and $m > 0$,*

$$\widetilde{p}_\alpha(q,m) := \lim_{N\to\infty} \widetilde{p}_N(q,m) = \frac{1}{4\pi m}\left[\frac{1}{m} + \frac{q^2}{4} + \left(\frac{3m}{2} - \alpha\right)^2\right] e^{-m\left[\frac{q^2}{4} + \left(\alpha - \frac{m}{2}\right)^2\right]}. \tag{46}$$

It is easy to see from (46) that $\int_{-\infty}^{\infty} \widetilde{p}(q,m)dq = \widetilde{p}_\alpha^{(\mathrm{Im})}(m)$. Thus, Theorem 3 confirms the heuristics of Section 2 in that the population of extreme eigenvalues which generates the eventual outlier (as α is approaching infinity) is constrained to a narrow vertical strip of width $O(N^{-1/3})$ about the origin.

Our results demonstrate that despite being one of the simplest tools available, the mean eigenvalue density captures the eigenvalue and parameter scales associated with the spectral restructuring in the random matrix ensemble (1)–(3). However, it gives no information about finer details, such as the probability distribution of the extreme eigenvalues during the restructure. Calculating all the higher order eigenvalue correlation functions in the scaling regime (45) would be a significant step towards describing such finer details. Unfortunately, the eigenvalue point process in the random matrix ensemble (1)–(3) is not determinantal at finite matrix dimensions and such a calculation is a considerably more difficult analytic task compared to the mean eigenvalue density.

At this point we want to mention that the probability distribution of extreme eigenvalues can be determined in a related but different random matrix ensemble exhibiting a spectral restructuring not unlike one in (1)–(3). This ensemble consists of subunitary matrices of the form

$$J_{CUE} = U\,\mathrm{diag}(\sqrt{1-T}, 1, \ldots, 1), \tag{47}$$

where the matrix U is taken from the Circular Unitary Ensemble (CUE) of complex unitary matrices uniformly distributed over $U(N)$ with the Haar's measure and $T \in [0,1]$ is a parameter. The ensemble was originally introduced in [29] and various statistical aspects of their spectra and eigenvectors were addressed in [2,6,30,31] and most recently in [15].

Obviously, if $T = 0$ then the matrix J_{CUE} is unitary and all of its eigenvalues lie on the unit circle $|z| = 1$. If $T > 0$ and is fixed in the limit $N \gg 1$ then, typically, the eigenvalues of J_{CUE} lie at a distance $O(N^{-1})$ from the unit circle with the farthest away being at a distance $O(\frac{\log N}{(1-T)N})$ with probability close to one. On the other hand, for $T = 1$ one of the eigenvalues becomes identically zero, and the rest are distributed inside the unit circle in the same way as eigenvalues of the so-called "truncated" CUE [32].

The similarity between the random matrix ensembles (47) and (1)–(3) can be exemplified by analysing the mean density of the eigenvalue moduli $r_j = |z_j|$

$$\rho_N(r) = \left\langle \frac{1}{N} \sum_{j=1}^{N} \delta(r - r_j) \right\rangle_{CUE}$$

in the limit of large matrix dimensions $N \to \infty$. One finds [29] that for every fixed $T \in [0,1]$

$$\lim_{N\to\infty} N\rho_N(r) = \begin{cases} \delta(r) + \frac{2r}{(1-r^2)^2}, & \text{if } T = 1, \\ 0, & \text{if } 0 < T < 1, \end{cases}$$

whereas, on rescaling the radial density near the unit circle [2,32],

$$\widetilde{\rho}_{CUE}(y) := \lim_{N\to\infty} \frac{1}{N}\rho_N\left(1 - \frac{y}{N}\right) = -\frac{d}{dy}\left[e^{-gy}\frac{\sinh y}{y}\right], \quad \text{with } g = \frac{2}{T} - 1. \tag{48}$$

Equation (48) is identical, with the obvious correspondence

$$\frac{1}{2}\left(\gamma + \frac{1}{\gamma}\right) = \frac{2}{T} - 1, \tag{49}$$

to Equation (12) considered at the centre of the GUE spectrum. In the limit of large values of y,

$$\widetilde{\rho}_{CUE}(y) \sim \begin{cases} \dfrac{1-T}{T}\dfrac{1}{y}e^{-2y\frac{1-T}{T}}, & \text{if } 0 < T < 1, \\ \dfrac{1}{y^2} & \text{if } T = 1. \end{cases} \tag{50}$$

The rescaled radial density has an exponentially light tail if $0 < T < 1$, and it is heavy-tailed if $T = 1$ which hints at markedly different behaviour of the extreme eigenvalues in the two cases. Reflecting on (50), one can convince themselves that this change occurs in an infinitesimal region near $T = 1$ of width N^{-1}. Such a scaling regime was earlier identified and analysed from a somewhat different angle in [15]. The precise relation of our analysis to one in [15] will be given in a separate paper [33].

On setting $T = 1 - \frac{t}{N}$, $t > 0$, one can investigate this transition region in much detail [33]. For example, the smallest eigenvalue modulus of the subunitary matrices J_{CUE},

$$x_{min} = \min_{j=1,\ldots,N} |z_j|,$$

converges in the limit $N \to \infty$ to a random variable X whose cumulative probability distribution function is given by the series

$$\Pr\{X \le x\} = \sum_{n=1}^{\infty} (-1)^{n+1} \frac{x^{n(n-1)}}{\prod_{k=1}^{n}(1 - x^{2k})} e^{t\left(1 - \frac{1}{x^{2n}}\right)}, \quad 0 < x < 1.$$

This family of probability distributions interpolates between the Fréchet and Gumbel distributions and is different from the standard family of probability distributions that characterise the extreme values in long sequence of i.i.d. random variables. In the limit of small values of t

$$\lim_{t\to 0+} \Pr\left\{X < y\sqrt{t}\right\} = e^{-y^{-2}}, \quad y > 0,$$

whereas

$$\lim_{t\to+\infty} \Pr\{2t(1-X) - \log t + \log(\log t) < y\} = e^{-e^{-y}}, \quad y > 0.$$

4. Mean Density of Eigenvalues at Finite Matrix Dimensions

Our analysis of various scaling regimes of the random matrix ensemble (1)–(3) is based on finite-N expressions for the mean eigenvalue density and the mean density of imaginary parts in terms of orthogonal polynomials, see Equations (56)–(64). These representations are new and the current Section contains their derivations.

4.1. Joint Eigenvalue Density and Correlation Functions

Our starting point is a closed form expression for the joint density $P_N(z_1,\ldots,z_n)$ of the eigenvalues $z_k = X_k + iY_k$ of J (1)–(3):

$$P_N(z_1,\ldots,z_N) = \tag{51}$$
$$\frac{N^{N^2/2}}{(2\pi)^{N/2} N! G(N) \gamma^{N-1}} \exp\left\{-\frac{N}{2}\left(\gamma^2 + \sum_{k=1}^{N} \mathrm{Re}\left(z_k^2\right)\right)\right\} \delta\left(\gamma - \sum_{k=1}^{N} \mathrm{Im}\, z_k\right) \prod_{j<k}^{N} |z_j - z_k|^2,$$

where $G(N)$ is the Barnes G-function. This expression was derived in [22] (see also [5]) and, for the obvious reason, it holds for $(z_1,\ldots,z_N) \in \mathbb{C}_+^N$, where $\mathbb{C}_+$ is the upper half of the complex plane $\mathbb{C}_+ = \{z = X + iY : Y \geq 0\}$

The first key fact that makes our analysis possible is that the eigenvalue correlation functions

$$R_{N,n}(z_1,\ldots,z_n) = \frac{N!}{(N-n)!} \int_{\mathbb{C}_+^{N-n}} P_N(z_1,\ldots,z_n,z_{n+1},\ldots,z_N) \prod_{k=n+1}^{N} dX_k dY_k,$$

can be expressed in terms of averages of products of characteristic polynomials of random matrices $J(\widehat{\gamma})$ having the same structure as (1)–(3) but of smaller dimension and with a different parameter γ. The relevance of this to our investigation is in that the mean eigenvalue density $\rho_N(X,Y)$ (8) which is the main object of our interest is

$$\rho_N(X,Y) = \frac{1}{N} R_{N,1}(X+iY). \tag{52}$$

It has been shown in [22] that

$$R_{N,n}(z_1,\ldots,z_n) = \frac{1}{(2\pi)^{n/2}\gamma^n}\left(1 - \frac{\sum_{k=1}^{n} Y_k}{\gamma}\right)^{N-n-1} \frac{N^{\frac{n^2}{2}} (N-n)^{Nn-n^2}}{\prod_{j=1}^{n}(N-j-1)!} \prod_{1\leq j<k\leq n} |z_j - z_k|^2 \times$$
$$\exp\left\{-\frac{N}{2}\sum_{k=1}^{n} X_k^2 - N\sum_{k=1}^{n} Y_k(\gamma - Y_k)\right\} \left\langle \prod_{k=1}^{n} \left|\det\left[\widehat{z}_k \mathbf{1}_{N-n} - J_{\widehat{\gamma} - \sum_{k=1}^{n} \widehat{Y}_k}\right]\right|^2 \right\rangle_{H_{N-n}},$$

where

$$\widehat{\gamma} = \left(\frac{N}{N-n}\right)^{1/2}\gamma, \quad \widehat{z}_k = \left(\frac{N}{N-n}\right)^{1/2}(X_k + iY_k), \quad \widehat{Y}_k = \left(\frac{N}{N-n}\right)^{1/2} Y_k,$$

and $J_{\widehat{\gamma}-\sum_{k=1}^{n}\widehat{Y}_k}$ are the random matrices (1)–(3) of dimension $N-n$ with N in (3) replaced by $N-n$ and γ in (2) replaced by $\widehat{\gamma}-\sum_{k=1}^{n}\widehat{Y}_k$,

$$J_{\widehat{\gamma}-\sum_{k=1}^{n}\widehat{Y}_k} = H_{N-n} + i\left(\widehat{\gamma} - \sum_{k=1}^{n}\widehat{Y}_k\right)\mathrm{diag}(1,0,\ldots,0).$$

The GUE average $\langle\ldots\rangle_{H_{N-n}}$ of the product of the characteristic polynomials of $J_{\widehat{\gamma}-\sum_{k=1}^{n}\widehat{Y}_k}$ can be performed with the help of the following proposition which we prove in Appendix A.

Proposition 1. *Let*

$$F_{\gamma}(z_1,z_2,\ldots,z_n) = \left\langle \prod_{j=1}^{n}|\det(z_j\mathbf{1}_N - J_{\gamma})|^2 \right\rangle,$$

where J_γ are the rank-one deviations from the GUE of dimension N defined by (1)–(3) and the average is taken over the GUE distribution (3). Then

$$F_{\gamma}(z_1,z_2,\ldots,z_n) =$$
$$\frac{1}{2^n}\left(\frac{N}{\pi}\right)^{2n^2}\int D[S_{2n}]\exp\left\{-\frac{N}{2}\mathrm{Tr}\,S_{2n}^2\right\}\det{}^{N-1}(Z_{2n}+iS_{2n})\,\det(Z_{2n}+iS_{2n}-i\gamma L_{2n}),$$

where the integration is over the space of $2n\times 2n$ Hermitian matrices S_{2n}, $D[S_{2n}]$ is the standard volume element in this space and

$$Z_{2n} = \mathrm{diag}(z_1,z_2,\ldots,z_n,\bar{z}_1,\bar{z}_2,\ldots,\bar{z}_n), \qquad L_{2n} = \mathrm{diag}(1,-1)\otimes\mathbf{1}_n.$$

Using this Proposition one arrives, after rescaling $S_{2n} = \left(\frac{N}{N-n}\right)^{1/2}\widehat{S}_{2n}$ in the resulting matrix integral, at a useful integral representation for the eigenvalue correlation functions in the random matrix ensemble (1)–(3):

$$R_{N,n}(z_1,\ldots,z_n) = \tag{53}$$
$$\frac{c_N}{\gamma^n}\left(1-\frac{\sum_{k=1}^{n}Y_k}{\gamma}\right)^{N-n-1}\exp\left\{-\frac{N}{2}\sum_{k=1}^{n}X_k^2 - N\sum_{k=1}^{n}Y_k(\gamma-Y_k)\right\}\prod_{1\le j<k\le n}|z_j-z_k|^2\times$$
$$\int d[\widehat{S}_{2n}]\exp\left\{-\frac{N}{2}\mathrm{Tr}\,\widehat{S}_{2n}^2\right\}\det{}^{N-n-1}\left[Z_{2n}+i\widehat{S}_{2n}\right]\det\left[Z_{2n}+i\widehat{S}_{2n}-i\left(\gamma-\sum_{k=1}^{n}Y_k\right)L_{2n}\right]$$

with

$$c_{N,n}(\gamma) = \frac{N^{3n^2/2+Nn}}{(2\gamma)^n\,(2\pi)^{n/2}\,\pi^{2n^2}\prod_{j=1}^{n}(N-j-1)!}.$$

4.2. Mean Density of Complex Eigenvalues

Setting $n=1$ and $z_1 = X+iY$ in (53) and then shifting the variable of integration by making the substitution $\widehat{S}_2 = S_2 - YL_2$ in the matrix integral, one obtains the following integral representation for the mean density of eigenvalues (52) in the random matrix ensemble (1)–(3):

$$\rho_N(X,Y) = \frac{1}{2\gamma\pi^2\sqrt{2\pi}}\frac{N^{N+1/2}}{(N-2)!}\left(1-\frac{Y}{\gamma}\right)^{N-2}\exp\left\{-\frac{N}{2}X^2 - NY\gamma\right\}\times \tag{54}$$
$$\int D[S_2]\exp\left\{-\frac{N}{2}\operatorname{Tr} S_2^2 + NY\operatorname{Tr} S_2L_2\right\}\det{}^{N-2}(X\mathbf{1}_2 + iS_2)\det\left(X\mathbf{1}_2 + iS_2 - i(\gamma-Y)L_2\right).$$

It is convenient to parametrise the hermitian matrix S_2 by diagonalising it:

$$S_2 = U_2\Sigma_2U_2^*, \quad \Sigma_2 = \operatorname{diag}(\sigma_1,\sigma_2), \quad \sigma_1 \geq \sigma_2 \in \mathbb{R},$$

where U_2 is a 2×2 unitary matrix, which can be parametrised as

$$U_2 = \begin{pmatrix} \cos\theta & \sin\theta e^{i\phi} \\ -\sin\theta e^{-i\phi} & \cos\theta \end{pmatrix}, \quad \theta \in \left[0,\frac{\pi}{2}\right], \phi \in [0,2\pi].$$

Noting that

$$D[S_2] = (\sigma_1-\sigma_2)^2\,\frac{\sin(2\theta)}{2}\,d\sigma_1\,d\sigma_2\,d\theta\,d\phi.$$

one arrives, on making the substitution $S_2 = U_2\Sigma_2U_2^*$ in (54), at

$$\rho_N(X,Y) = \frac{c_N}{\gamma}\left(1-\frac{Y}{\gamma}\right)^{N-2}\exp\left\{-\frac{N}{2}\left(X^2+2Y\gamma\right)\right\}\times$$
$$\int_0^{\frac{\pi}{2}}d\theta\int_{-\infty}^{+\infty}d\sigma_1\int_{-\infty}^{\sigma_1}d\sigma_2\,(\sigma_1-\sigma_2)^2\sin(2\theta)\exp\left\{-\frac{N}{2}\left(\sigma_1^2+\sigma_2^2\right)+NY(\sigma_1-\sigma_2)\cos(2\theta)\right\}\times$$
$$(X+i\sigma_1)^{N-2}(X+i\sigma_2)^{N-2}\Big[(X+i\sigma_1)(X+i\sigma_2)+(\gamma-Y)^2-(\gamma-Y)(\sigma_1-\sigma_2)\cos(2\theta)\Big],$$

where we have introduced

$$c_N = \frac{1}{(2\pi)^{3/2}}\frac{N^{N+1/2}}{(N-2)!} \sim \frac{N^2e^N}{2\pi^2} \qquad (N\to\infty). \tag{55}$$

The integral over θ can be performed by the substitution $t = (\sigma_1-\sigma_2)\cos(2\theta)$. This yields

$$\rho_N(X,Y) = \frac{c_N}{2NY\gamma}\left(1-\frac{Y}{\gamma}\right)^{N-2}\exp\left\{-\frac{N}{2}\left(X^2+2\gamma Y-2Y^2\right)\right\}J_N(X,Y), \tag{56}$$

where

$$J_N(X,Y) = \int_{-\infty}^{+\infty}d\sigma_1\int_{-\infty}^{+\infty}d\sigma_2\,e^{-\frac{N}{2}(\sigma_1^2+\sigma_2^2)}(z+i\sigma_1)^{N-2}(\overline{z}+i\sigma_2)^{N-2}\frac{(z+i\sigma_1)-(\overline{z}+i\sigma_2)}{i}\times$$
$$\left[(z+i\sigma_1)(\overline{z}+i\sigma_2)+(\gamma-Y)^2+\frac{\gamma-Y}{Ny}-(\gamma-Y)\frac{(z+i\sigma_1)-(\overline{z}+i\sigma_2)}{i}\right],$$

with $z = X+iY$.

Further, introducing functions

$$\pi_m(z) = \int_{-\infty}^{+\infty}d\sigma\,e^{-\frac{N}{2}\sigma^2}\,(z+i\sigma)^{N-m}, \quad m = 0,1,\ldots,N, \tag{57}$$

one can rewrite the integral $J_N(X,Y)$ in the following form

$$\begin{aligned} J_N(X,Y) = & -i[\pi_0(Z)\pi_1(\overline{z}) - \pi_0(\overline{z})\pi_1(z)] \\ & -i\left((\gamma-Y)^2 + \frac{\gamma-Y}{Ny}\right)[\pi_1(z)\pi_2(\overline{z}) - \pi_1(\overline{z})\pi_2(z)] \\ & +(\gamma-Y)[\pi_0(z)\pi_2(\overline{z}) + \pi_0(\overline{z})\pi_2(z) - 2\pi_1(z)\pi_1(\overline{z})], \end{aligned} \tag{58}$$

Now one observes that $\pi_m(z)$ are actually a rescaled version of Hermite polynomials. We have that

$$\pi_m(z) = \sqrt{\pi}\left(\frac{2}{N}\right)^{\frac{N-m+1}{2}} \widetilde{H}_{N-m}\left(z\sqrt{\frac{N}{2}}\right) = \sqrt{2\pi}\pi^{1/4}\sqrt{\frac{(N-m)!}{N^{N-m+1}}}\, p_{N-m}\left(z\sqrt{\frac{N}{2}}\right), \tag{59}$$

where $\widetilde{H}_k(z)$ are the monic Hermite polynomials

$$\widetilde{H}_k(z) = \left(-\frac{1}{2}\right)^k e^{z^2}\frac{d}{dz}e^{-z^2}$$

and $p_k(z)$ are the orthonormal Hermite polynomials

$$p_k(z) = \sqrt{\frac{2^k}{k!\sqrt{\pi}}}\,\widetilde{H}_k(z)$$

satisfying the orthogonality relations

$$\int_{-\infty}^{+\infty} dz\, p_k(z)p_m(z)\, e^{-z^2}dz = \delta_{k,m}.$$

The polynomials $p_k(z)$ also satisfy the recurrence relation

$$p_{k+1}(z) = z\sqrt{\frac{2}{k+1}}p_k(z) - \sqrt{\frac{k}{k+1}}p_{k-1}(z).$$

Using the above definitions and the expression for the eigenvalue density $\rho_N(X,Y)$ in (56) and with the notation $z = X + iY$ we obtain

$$\begin{aligned} \rho_N(X,Y) = & \frac{N-1}{\sqrt{2N}Y\gamma}\left(1-\frac{Y}{\gamma}\right)^{N-2}\exp\left\{-\frac{N}{2}X^2 - NY(\gamma-Y)\right\}\times \\ & \left\{\operatorname{Im} p_N\left(z\sqrt{\frac{N}{2}}\right)p_{N-1}\left(\overline{z}\sqrt{\frac{N}{2}}\right) - (\gamma-Y)\left|p_{N-1}\left(z\sqrt{\frac{N}{2}}\right)\right|^2\right. \\ & +\sqrt{\frac{N}{N-1}}\left[\left((\gamma-Y)^2 + \frac{\gamma-Y}{NY}\right)\operatorname{Im} p_{N-1}\left(z\sqrt{\frac{N}{2}}\right)p_{N-2}\left(\overline{z}\sqrt{\frac{N}{2}}\right)\right. \\ & \left.\left.+(\gamma-Y)\operatorname{Re} p_N\left(z\sqrt{\frac{N}{2}}\right)p_{N-2}\left(\overline{z}\sqrt{\frac{N}{2}}\right)\right]\right\}. \end{aligned}$$

which, by using the recurrence relation, can be further rewritten as

$$\rho_N(X,Y) = \frac{1}{Y\gamma}\sqrt{\frac{N}{2}}\left(1-\frac{Y}{\gamma}\right)^{N-2}\exp\left\{-\frac{N}{2}X^2 - NY(\gamma-Y)\right\}\times$$
$$\left\{\mathrm{Im}\, p_N\left(z\sqrt{\frac{N}{2}}\right)p_{N-1}\left(\bar{z}\sqrt{\frac{N}{2}}\right)\left[1-\frac{1}{N}+(\gamma-Y)\left(\gamma+\frac{1}{NY}\right)\right]\right.$$
$$-\left|p_{N-1}\left(z\sqrt{\frac{N}{2}}\right)\right|^2\left[Y(\gamma-Y)^2+(\gamma-Y)-Y(\gamma-Y)\right]$$
$$\left.-\left|p_N\left(z\sqrt{\frac{N}{2}}\right)\right|^2(\gamma-Y)+\mathrm{Re}\, p_N\left(z\sqrt{\frac{N}{2}}\right)p_{N-1}\left(\bar{z}\sqrt{\frac{N}{2}}\right)X(\gamma-Y)\right\}. \tag{60}$$

4.3. Density of the Imaginary Parts

In this section, we present the derivation of the density for the imaginary parts of the eigenvalues, irrespective of their real parts, as defined in (10). We start with an observation, see integral 7.377 in [23]:

Lemma 1. *Let $\beta \geq \alpha$ be two non-negative integers and $z = X + iY$. Then*

$$\int_{-\infty}^{\infty} e^{-\frac{N}{2}X^2} p_{N-\alpha}\left(z\sqrt{\frac{N}{2}}\right)p_{N-\beta}\left(\bar{z}\sqrt{\frac{N}{2}}\right)dX =$$
$$i^{\beta-\alpha}\sqrt{\frac{2}{N}\frac{N^{\beta-\alpha}}{(N-\alpha)\dots(N-\beta+1)}}Y^{\beta-\alpha}L_{N-\beta}^{(\beta-\alpha)}\left(-NY^2\right),$$

where $L_M^{(\alpha)}$ is a standard Laguerre polynomial.

Integrating with respect to X expression for the density $\rho_N(X,Y)$ in (60) one gets the probability density of imaginary parts in the form

$$\rho_N^{(\mathrm{Im})}(Y) = \frac{1}{Y\gamma}\left(1-\frac{Y}{\gamma}\right)^{N-2}e^{-NY(\gamma-Y)}F_N(Y) \tag{61}$$

with

$$F_N(Y) = \frac{N-1}{N}YL_{N-1}^{(1)}\left(-NY^2\right) - \frac{N-1}{N}(\gamma-Y)L_{N-1}^{(0)}\left(-NY^2\right) \tag{62}$$
$$+Y\left[(\gamma-Y)^2+\frac{\gamma-Y}{NY}\right]L_{N-2}^{(1)}\left(-NY^2\right) - (\gamma-Y)Y^2L_{N-2}^{(2)}\left(-NY^2\right)$$
$$= -2\gamma L_{N-1}^{(0)}\left(-NY^2\right) + \left(\frac{N-1}{N}3Y+\frac{2\gamma}{N}\right)L_{N-1}^{(1)}\left(-NY^2\right) \tag{63}$$
$$+\left[-2Y+Y(\gamma-Y)^2\right]L_{N-2}^{(1)}\left(-NY^2\right)$$
$$= \frac{N-1}{N}(3Y-2\gamma)L_{N-1}^{(1)}\left(-NY^2\right) + \left[2\gamma-2Y+Y(\gamma-Y)^2\right]L_{N-2}^{(1)}\left(-NY^2\right) \tag{64}$$

where we systematically used the recursion relations:

$$L_{N-1}^{(0)}\left(-NY^2\right) = L_{N-1}^{(1)}\left(-NY^2\right) - L_{N-2}^{(1)}\left(-NY^2\right)$$

and

$$-Y^2 L_{N-2}^{(2)}\left(-NY^2\right) = L_{N-2}^{(1)}\left(-NY^2\right) - \frac{N-1}{N} L_{N-1}^{(1)}\left(-NY^2\right).$$

5. Proof of Theorems 1 and 2

In both proofs we use the following integral representation for the Laguerre polynomials in terms of the modified Bessel functions $I_\alpha(x)$ (see, e.g., Equation 4.19.13 in [34]):

$$L_{N-k}^{(\alpha)}\left(-NY^2\right) = \frac{2N^{N-k+1}}{(N-k)!} \frac{e^{-NY^2}}{|Y|^\alpha} \int_0^\infty \tau^{2N-2k+\alpha+1} e^{-N\tau^2} I_\alpha(2\tau|Y|N)\, d\tau \quad (\alpha > -1). \tag{65}$$

The integral in (65) can be evaluated in the limit $N \to \infty$ in various scaling regimes for Y using the Laplace method, see Appendix B. The resulting asymptotic expression depends on the scaling of the variable $Y > 0$ with N.

Proof of Theorem 1. Consider the scaling regime (31) with $\gamma > 0$ being fixed. In this regime the asymptotic form of the mean density of the imaginary parts can be found using the leading order form of $L_{N-k}^{(1)}(-NY^2)$ which can be read from (A8) as

$$L_{N-k}^{(1)}\left(-NY^2\right) \sim \frac{e^{NYr_*}}{\sqrt{2\pi N}} \frac{r_*(Y)^{-2(N-k+1)}}{Y^{3/2}(Y^2+4)^{1/4}}, \quad r_*(Y) = \frac{\sqrt{Y^2+4}-Y}{2}. \tag{66}$$

On substituting (66) into (64) one gets an asymptotic expression for the density (61) precisely in the Large Deviation form (32) with the rate function (33) and the pre-exponential factor in the form

$$\frac{1}{\sqrt{N}} \Psi_\gamma(Y) = \frac{1}{\sqrt{2\pi N}} \frac{\gamma}{(\gamma - Y)^2} \frac{3Y - 2\gamma + r_*(Y)^2(\gamma - Y)(2 + Y(\gamma - Y))}{Y^{5/2}(Y^2+4)^{1/4}}. \tag{67}$$

Finally, by exploiting the relation $1 - r_*(Y)^2 = Yr_*(Y)$,

$$\begin{aligned} 3Y - 2\gamma + r_*(Y)^2(\gamma - Y)(2 + Y(\gamma - Y)) =& Y - (\gamma - Y)\left[2(1 - r_*(Y)^2) - r_*(Y)^2 Y(\gamma - Y)\right] \\ =& Y - (\gamma - Y)\left[2r_*(Y)Y - r_*(Y)^2 Y(\gamma - Y)\right] \\ =& Y[1 - r_*(Y)(\gamma - Y)(2 - r_*(Y)(\gamma - Y))]. \end{aligned}$$

This brings the function $\Psi_\gamma(Y)$ in (67) to the form as given in (34).

To analyse the shape of the rate function $\Phi_\gamma(Y)$ in (33) it is convenient to parametrise

$$Y = e^\theta - e^{-\theta}, \quad \theta > 0. \tag{68}$$

In this parametrisation, the rate function transforms to

$$\widetilde{\Phi}_\gamma(\theta) := \Phi_\gamma\left(e^\theta - e^{-\theta}\right) = \gamma\left(e^\theta - e^{-\theta}\right) + 1 - e^{2\theta} - 2\theta - \ln\left(1 - \frac{e^\theta - e^{-\theta}}{\gamma}\right),$$

and its derivative in θ factorises as follows:

$$\begin{aligned} \widetilde{\Phi}_\gamma'(\theta) =& \gamma\left(e^\theta + e^{-\theta}\right) - 2\left(e^{2\theta} + 1\right) + \frac{e^\theta + e^{-\theta}}{\gamma - \left(e^\theta - e^{-\theta}\right)} \\ =& \left(e^\theta + e^{-\theta}\right)\left(\gamma - e^\theta\right)\left[1 - \frac{e^\theta}{\gamma - \left(e^\theta - e^{-\theta}\right)}\right]. \end{aligned}$$

Therefore, the stationary points of $\widetilde{\Phi}_\gamma(\theta)$ solve the equations

$$e^{\theta} = \gamma \tag{69}$$

and

$$e^{\theta} = \gamma - \left(e^{\theta} - e^{-\theta}\right). \tag{70}$$

These equations yields two stationary points $e^{\theta_*} = \gamma$ and $e^{\theta_{**}} = \frac{\gamma+\sqrt{8+\gamma^2}}{4}$. Correspondingly, the rate function $\Phi_\gamma(Y)$ has two stationary points

$$Y_* = \gamma - \gamma^{-1} \quad \text{and} \quad Y_{**} = \frac{3\gamma - \sqrt{8+\gamma^2}}{4} = \frac{2\left(\gamma - \frac{1}{\gamma}\right)}{3 + \sqrt{1 + 8/\gamma^2}}.$$

It is evident that if $0 < \gamma < 1$ both stationary points Y_* and Y_{**} are negative. One can easily check that in this case $\Phi_\gamma(Y)$ is monotonically increasing on the interval $Y > 0$ and is positive on this interval.

If $\gamma > 1$ then taking the second derivative in θ one can easily show that

$$\widetilde{\Phi}''_\gamma(\theta_*) = (\gamma^2 - 1)(\gamma^2 + 1) > 0, \quad \widetilde{\Phi}''_\gamma(\theta_{**}) = -\frac{\left(e^{2\theta_{**}} - e^{-2\theta_{**}}\right)\left(1 + \gamma e^{\theta_{**}}\right)}{\left[\gamma - \left(e^{\theta_{**}} - e^{-\theta_{**}}\right)\right]^2} < 0,$$

so that Y_* is the point of local minimum of the rate function $\Phi_\gamma(Y)$, and Y_{**} is the point of local maximum. It is also easy to verify that the rate function $\Phi_\gamma(Y)$ vanishes in the limit $Y \to 0$ and also at $Y = Y_*$, staying positive at all other $Y > 0$, so that that the point $Y = Y_*$ is the point of absolute minimum. Finally, to verify that the pre-exponential factor (34) vanishes at $Y = Y_{**}$ it suffices to show that $r_*(Y_{**})(\gamma - Y_{**}) = 1$. On noticing that

$$r_*(Y) = \frac{\sqrt{Y^2+4} - Y}{2} = e^{-\theta}.$$

this relation evidently follows from (68) and (70). □

Proof of Theorem 2. In the scaling regime (42) the variable Y scales with $N^{-1/3}$. As $NY \gg 1$ in this case, the required asymptotic expressions for Laguerre polynomials can be read from (A8). It turns out that in order to calculate the density of imaginary parts to leading order in this regime, one has to retain the subleading term in the pre-exponential factor as specified in (A8). On substituting $Y = mN^{-1/3}$ in (A8) we obtain that with the required precision

$$L^{(1)}_{N-1}\left(-N^{1/3}m^2\right) = \frac{e^{N\mathcal{L}_0(Y)}}{\sqrt{2\pi m^3}\,(Y^2+4)^{1/4}}\left(1 - \frac{3}{16}\frac{1}{mN^{2/3}}\right) \tag{71}$$

$$L^{(1)}_{N-2}\left(-N^{1/3}m^2\right) = \frac{e^{N\mathcal{L}_0(Y)}}{\sqrt{2\pi m^3}\,(Y^2+4)^{1/4}}\,r_*(Y)^2\left(1 - \frac{3}{16}\frac{1}{mN^{2/3}}\right) \tag{72}$$

$$L^{(0)}_{N-1}\left(-N^{1/3}m^2\right) = \frac{e^{N\mathcal{L}_0(Y)}}{N^{1/3}\sqrt{2\pi m^3}\,(Y^2+4)^{1/4}}\left(1 + \frac{1}{16}\frac{1}{mN^{2/3}}\right), \tag{73}$$

where $\mathcal{L}_0(Y) = Yr_*(Y) - 2\ln r_*(Y)$ with r_* (35) and $\ln r_*$ expanded in powers of $Y \ll 1$:

$$r_*(Y) = 1 - \frac{Y}{2} + \frac{Y^2}{8} + O(Y^4), \quad \ln r_*(Y) = -\frac{Y}{2} + \frac{Y^3}{48} + O(Y^4). \tag{74}$$

It is easy to see that the overall exponential behaviour of the mean density (61) will still be given by (33) duly expanded:

$$\Phi_\gamma(Y) = Y\left(\gamma + \frac{1}{\gamma} - 2\right) - \frac{Y^2}{2}\left(1 - \frac{1}{\gamma^2}\right) + \frac{Y^3}{3}\left(\frac{1}{\gamma^3} - \frac{1}{4}\right) + O(Y^4). \tag{75}$$

Putting in here the scaling form $\gamma = 1 + \frac{\alpha}{N^{1/3}}$ and recalling $Y = \frac{m}{N^{1/3}}$ we find from (75), assuming that the parameters $\alpha \in \mathbb{R}$ and $m > 0$ are fixed, that

$$N\Phi_{1+\frac{\alpha}{N^{1/3}}}\left(\frac{m}{N^{1/3}}\right) = m\alpha^2 - m^2\alpha + \frac{m^3}{4} = m\left(\alpha - \frac{m}{2}\right)^2 := \Phi_\alpha(m).$$

This verifies the exponent in (43). To find the pre-exponential terms we find it most convenient to use Equation (63). Substituting there (71)–(73) we first get

$$\begin{aligned} F_N(m) = &\frac{e^{N\mathcal{L}_0(Y)}}{\sqrt{2\pi m}\,(Y^2+4)^{1/4}}\left\{-\frac{2\gamma}{N^{1/3}} r_*(Y)\left(1 + \frac{1}{16}\frac{1}{mN^{2/3}}\right)\right. \\ &+\frac{1}{m}\left(1 - \frac{3}{16}\frac{1}{mN^{2/3}}\right)\left(\frac{3m}{N^{1/3}} + \frac{2\gamma}{N} - \frac{3m}{N^{4/3}}\right) \\ &\left.+\left[-\frac{2}{N^{1/3}} + \frac{1}{N^{1/3}}\left(\gamma^2 - 2\gamma\frac{m}{N^{1/3}} + \frac{m^2}{N^{2/3}}\right)\right] r_*(Y)^2\left(1 - \frac{3}{16}\frac{1}{mN^{2/3}}\right)\right\}. \end{aligned} \tag{76}$$

After rearranging and collecting the relevant terms in the above expression we arrive at

$$\begin{aligned} F_N(m) = &\frac{e^{N\mathcal{L}_0(Y)}}{\sqrt{2\pi m}\,(Y^2+4)^{1/4}}\left\{\frac{(r_*(Y)\gamma - 1)^2 + 2(1 - r_*(Y)^2)}{N^{1/3}} - \frac{2\gamma m r_*(Y)^2}{N^{2/3}}\right. \\ &\left.+\frac{1}{N}\left[-\frac{9}{16m} + m^2 r_*(Y)^2 - \frac{3}{16}\frac{(\gamma^2-2)r_*(Y)^2}{m} + \frac{2\gamma}{m} - \frac{\gamma r_*(Y)}{8m}\right]\right\}. \end{aligned} \tag{77}$$

The expansion (74) together with $\gamma = 1 + \frac{\alpha}{N^{1/3}}$ give the relations

$$\frac{(r_*(Y)\gamma - 1)^2 + 2(1 - r_*(Y)^2)}{N^{1/3}} = \frac{2m}{N^{2/3}} + \frac{1}{N}\left[\left(\alpha - \frac{m}{2}\right)^2 - m^2\right] \tag{78}$$

and

$$-\frac{2\gamma m r_*(Y)^2}{N^{2/3}} = -\frac{2m}{N^{2/3}} - \frac{2m(\alpha - m)}{N}, \tag{79}$$

which are exact to the subleading order. We can now see that the leading order terms inside the curly brackets in (77) cancel. This also implies that at the leading order it is enough to replace the factor $(Y^2+4)^{1/4}$ in (77) with $\sqrt{2}$. Finally, adding the leading order contribution from

$$\frac{1}{N}\left[-\frac{9}{16m} + m^2 r_*(Y)^2 - \frac{3}{16}\frac{(\gamma^2-2)r_*(Y)^2}{m} + \frac{2\gamma}{m} - \frac{\gamma r_*(Y)}{8m}\right] = \frac{1}{N}\left(m^2 + \frac{3}{2m}\right)$$

to the $1/N$ terms in (78) and (79) results in

$$F_N(m) = \frac{e^{N\mathcal{L}_0(Y)}}{2\sqrt{\pi m}}\frac{1}{N}\left[\frac{3}{2m} + \left(\alpha - 3\frac{m}{2}\right)^2\right], \tag{80}$$

thus verifying the pre-exponential factors in (43). □

Let us finally present the derivation of the marginal density of imaginary parts (18) pertinent to keeping the product $y = YN$ fixed as $N \to \infty$. This task is straightforwardly achieved by performing the limit $N \to \infty$ in (61) via substituting the corresponding asymptotics of Laguerre polynomials (A3) into the Formula (63) and using the identity $\frac{d}{dy}I_1(2y) = I_0(2y) - I_2(2y)$.

6. Proof of Theorem 3

Proof. We will use Equations (56)–(58) which express the mean density of eigenvalues $\rho_N(X,Y)$ in terms of the rescaled Hermite polynomials $\pi_k(X+iY)$ (59).

Using the integral representation in (57) it can be shown that in the scaling limit

$$z = X + iY, \quad X = \frac{q}{N^{1/3}}, \quad Y = \frac{m}{N^{1/3}} > 0 \tag{81}$$

the rescaled Hermite polynomials $\pi_k(z)$ are given by the asymptotic equations

$$\pi_k(z) \sim \sqrt{\frac{2\pi}{N(1+\sigma_+^2)}}(-i\sigma_+)^k e^{-\frac{N}{2}(1+iz\sigma_+ + 2\ln(-i\sigma_+))}, \tag{82}$$

$$\pi_k(\bar{z}) \sim \sqrt{\frac{2\pi}{N(1+\sigma_-^2)}}(-i\sigma_-)^k e^{-\frac{N}{2}(1+i\bar{z}\sigma_- + 2\ln(-i\sigma_-))}, \tag{83}$$

where we have introduced the notations

$$\sigma_+ = \frac{iz + \sqrt{4-z^2}}{2}, \quad \sigma_- = \frac{i\bar{z} - \sqrt{4-\bar{z}^2}}{2}. \tag{84}$$

This implies for $J_N(X,Y)$ (58) that

$$J_N(X,Y) \sim \frac{2\pi}{N}e^{-N}\frac{(\sigma_+ - \sigma_-)}{\sqrt{(1+\sigma_+^2)(1+\sigma_-^2)}}e^{-N\left(\frac{i}{2}(z\sigma_+ + \bar{z}\sigma_-) + \ln(-\sigma_+\sigma_-)\right)} \tag{85}$$

$$\times\left[(1-\sigma_+(\gamma-y))(1+\sigma_-(\gamma-Y)) - \sigma_+\sigma_-\frac{\gamma-Y}{NY}\right]. \tag{86}$$

We are here interested in the limit of small $|z|$ (81), and, hence, can use the expansions

$$\sigma_+ = 1 + \frac{iz}{2} - \frac{z^2}{8} + \ldots, \quad \sigma_- = -1 + \frac{i\bar{z}}{2} + \frac{\bar{z}^2}{8} + \ldots$$

and, consequently,

$$\begin{aligned} 1 - \sigma_+(\gamma - Y) &= 1 - \gamma + Y + \frac{Y}{2}\gamma - \frac{iX}{2}\gamma + O(|z|^2), \\ 1 - \sigma_-(\gamma - Y) &= 1 - \gamma + Y + \frac{Y}{2}\gamma + \frac{iX}{2}\gamma + O(|z|^2). \end{aligned}$$

Hence,

$$(1-\sigma_+(\gamma-Y))(1+\sigma_-(\gamma-Y)) = \left[1-\gamma+Y\left(1+\frac{\gamma}{2}\right)\right]^2 + \frac{X^2}{4}\gamma^2 + O\left(X^2+Y^2\right).$$

Setting here $X = \frac{q}{N^{1/3}}$, $Y = \frac{m}{N^{1/3}}$ and $\gamma = 1 + \frac{\alpha}{N^{1/3}}$ one obtains that to leading order in N

$$(1-\sigma_+(\gamma-Y))(1+\sigma_-(\gamma-Y)) = \frac{1}{N^{2/3}}\left[\left(\frac{3}{2}m - \alpha\right)^2 + \frac{q^2}{4}\right].$$

With the same precision we have

$$-\sigma_+\sigma_-\frac{\gamma - Y}{NY} = \frac{1}{N^{2/3}}\frac{1}{m},$$

and, consequently,

$$\left[(1-\sigma_+(\gamma-Y))(1+\sigma_-(\gamma-Y))-\sigma_+\sigma_-\frac{\gamma-Y}{NY}\right]=\frac{1}{N^{2/3}}\left[\frac{1}{m}+\left(\frac{3m}{2}-\alpha\right)^2+\frac{q^2}{4}\right]. \quad (87)$$

On inspecting (56) and (85), one concludes that the overall exponential factor in (56) is given by $e^{-N\widetilde{\Phi}_\gamma}$, where

$$\widetilde{\Phi}_\gamma=\frac{i}{2}(z\sigma_+ + \bar{z}\sigma_-)+\ln(-\sigma_+\sigma_-)+\frac{X^2}{2}+\gamma Y-Y^2-\ln\left(1-\frac{Y}{\gamma}\right).$$

The leading order form of $\widetilde{\Phi}_\gamma$ can be found by expanding in powers of X and Y, in a similar way as before:

$$\begin{aligned}
\frac{i}{2}(z\sigma_+ + \bar{z}\sigma_-) &= -Y-\frac{X^2-Y^2}{2}+\frac{Y}{8}(3X^2-Y^2)+O(|z|^4),\\
\ln(-\sigma_+\sigma_-) &= -Y+\frac{Y^3}{24}-\frac{YX^2}{8}+O(|z|^4),\\
-\ln\left(1-\frac{Y}{\gamma}\right) &= \frac{Y}{\gamma}+\frac{Y^2}{2\gamma^2}+\frac{Y^3}{3\gamma^3}+O(Y^4).
\end{aligned}$$

Adding all contributions,

$$\widetilde{\Phi}_\gamma = Y\frac{(\gamma-1)^2}{\gamma}+\frac{Y^2}{2\gamma^2}(1-\gamma^2)+\frac{Y^3}{3}\left(\frac{1}{\gamma^3}-\frac{1}{4}\right)+\frac{YX^2}{4}+O(|z|^4).$$

Setting here $X=\frac{q}{N^{1/3}}$, $Y=\frac{m}{N^{1/3}}$ and $\gamma=1+\frac{\alpha}{N^{1/3}}$, one obtains that to leading order

$$\widetilde{\Phi}_\gamma=\frac{m}{N}\left(\left(\frac{m}{2}-\alpha\right)^2+\frac{q^2}{4}\right). \quad (88)$$

Combining (88) with (87), and trivially taking into account asymptotic expressions for the remaining multiplicative factors in (56) and (86), one arrives at (46). □

Author Contributions: Conceptualization, Y.V.F. and B.A.K.; Methodology, Y.V.F., B.A.K. and M.P.; Formal analysis, Y.V.F., B.A.K. and M.P.; Investigation, Y.V.F. and B.A.K.; Writing—original draft, Y.V.F. and B.A.K.; Writing—review & editing,Y.V.F. and B.A.K. All authors have read and agreed to the published version of the manuscript.

Funding: Y.V.F. acknowledges financial support from EPSRC Grant EP/V002473/1 "Random Hessians and Jacobians: theory and applications".

Acknowledgments: The authors are grateful to Guillaume Dubach for highly useful comments on the revised version of the manuscript. The authors are also grateful to Bertrand Lacroix A Chez Toine for bringing their attention to paper [26] and the similarity between the spectral restructure in the random matrix ensemble (1)–(3) and the condensation transition in models of mass transport.

Conflicts of Interest: The authors declare no conflict of interest. The funders had no role in the design of the study; in the collection, analyses, or interpretation of data; in the writing of the manuscript; or in the decision to publish the results.

Abbreviations

The following abbreviations are used in this manuscript:

GUE	Gaussian Unitary Ensemble
JPDF	Joint Probability Density Function

Appendix A. Proof of Proposition 1

We prove here a more general version of Proposition 1 which holds for rank-M deviations $J = H + i\Gamma$, $\Gamma = \mathrm{diag}\{\gamma_1, \ldots, \gamma_M, 0, \ldots, 0\}$ from the GUE (3) with arbitrary real parameters γ_j, $j = 1, \ldots, M < N$. The Proposition 1 follows as the special case $M = 1$.

Proposition A1. *Let $\Gamma = \mathrm{diag}(\gamma_1, \ldots, \gamma_M, 0, \ldots, 0)$ be a diagonal matrix of dimension N with $M < N$ non-zero real entries γ_j and*

$$\mathcal{F}_\Gamma(z_1, z_2, \ldots, z_n) = \left\langle \prod_{j=1}^{n} |\det(z_j \mathbf{1}_N - H - i\Gamma)|^2 \right\rangle_H ,$$

where the average is taken over the GUE distribution (3). Then

$$\mathcal{F}_\Gamma(z_1, z_2, \ldots, z_n) = \frac{1}{2^n}\left(\frac{N}{\pi}\right)^{2n^2} \int D[S_{2n}]\, e^{-\frac{N}{2}\mathrm{Tr}\, S_{2n}^2} \det{}^{N-M}(Z_{2n} + iS_{2n}) \prod_{j=1}^{M} \det(Z_{2n} + iS_{2n} - i\gamma_j L_{2n}), \tag{A1}$$

where the integration is over the space of $2n \times 2n$ Hermitian matrices S_{2n}, $D[S_{2n}]$ is the standard volume element in this space and

$$Z_{2n} = \mathrm{diag}(z_1, z_2, \ldots, z_n, \bar{z}_1, \bar{z}_2, \ldots, \bar{z}_n), \qquad L_{2n} = \mathrm{diag}(1, -1) \otimes \mathbf{1}_n.$$

Proof of Proposition A1. The average of the product of the characteristic polynomials over the GUE in in (A1) can be calculated using Grassmann integration. First we use the well-known identity

$$\int (\mathrm{d}\,\overline{\Psi} \mathrm{d}\,\Psi)_{\mathrm{ent}} \exp\{-\langle \overline{\Psi}, M\Psi \rangle\} = \det M,$$

where M is $N \times N$ matrix, and $\overline{\Psi}, \Psi$ are Grassmann variables vectors of length N and $(\mathrm{d}\,\overline{\Psi} \mathrm{d}\,\Psi)_{\mathrm{ent}} = \prod_{j=1}^{N} \mathrm{d}\,\overline{\psi}_j \mathrm{d}\,\psi_j$. We also write each square of determinant in the form

$$|\det(z_k - H - i\Gamma)|^2 = \det\begin{pmatrix} z_k - H - i\Gamma & 0 \\ 0 & \bar{z}_k - H + i\Gamma \end{pmatrix}.$$

Combining the above relations,

$$\mathcal{F}_\Gamma(z_1, z_2, \ldots, z_n) = \left\langle \int \prod_{k=1}^{n} \left(\mathrm{d}\,\overline{\Psi}^{(k)} \mathrm{d}\,\Psi^{(k)}\right)_{\mathrm{ent}} \exp\left\{ -\left\langle \overline{\Psi}^{(k)}, \begin{pmatrix} z_k - H - i\Gamma & 0 \\ 0 & \bar{z}_k - H + i\Gamma \end{pmatrix} \Psi^{(k)} \right\rangle \right\} \right\rangle_H .$$

Now we interchange the order of integrations and perform the GUE average first:

$$
\begin{aligned}
F_\Gamma =& \left\langle \prod_{k=1}^{n} \exp\left\{ \left\langle \overline{\Psi}^{(k)}, \begin{pmatrix} H & 0 \\ 0 & H \end{pmatrix} \Psi^{(k)} \right\rangle \right\} \right\rangle_H \\
=& 2^{-N/2} \left(\frac{N}{\pi}\right)^{N^2/2} \int \exp\left\{ -N \sum_{p<q}^{N} (\operatorname{Re} h_{p,q})^2 + (\operatorname{Im} h_{p,q})^2 - \frac{N}{2} \sum_{p=1}^{N} h_{p,p}^2 \right\} \\
&\times \exp\left\{ \sum_{p<q}^{N} \operatorname{Re} h_{p,q} \sum_{k=1}^{n} \left(\overline{\psi}_p^{(k)} \psi_q^{(k)} + \overline{\psi}_{N+p}^{(k)} \psi_{N+q}^{(k)} + \overline{\psi}_q^{(k)} \psi_p^{(k)} + \overline{\psi}_{N+q}^{(k)} \psi_{N+p}^{(k)} \right) \right\} \\
&\times \exp\left\{ i \sum_{p<q}^{N} \operatorname{Im} h_{p,q} \sum_{k=1}^{n} \left(\overline{\psi}_p^{(k)} \psi_q^{(k)} + \overline{\psi}_{N+p}^{(k)} \psi_{N+q}^{(k)} - \overline{\psi}_q^{(k)} \psi_p^{(k)} - \overline{\psi}_{N+q}^{(k)} \psi_{N+p}^{(k)} \right) \right\} \\
&\times \exp\left\{ \sum_{p=1}^{N} h_{p,p} \sum_{k=1}^{n} \left(\overline{\psi}_p^{(k)} \psi_p^{(k)} + \overline{\psi}_{N+p}^{(k)} \psi_{N+q}^{(k)} \right) \right\}, \\
=& \exp\left\{ \frac{1}{4N} \sum_{p<q}^{N} \left(\sum_{k=1}^{n} \left(\overline{\psi}_p^{(k)} \psi_q^{(k)} + \overline{\psi}_{N+p}^{(k)} \psi_{N+q}^{(k)} + \overline{\psi}_q^{(k)} \psi_p^{(k)} + \overline{\psi}_{N+q}^{(k)} \psi_{N+p}^{(k)} \right) \right)^2 \right\} \\
&\times \exp\left\{ -\frac{1}{4N} \sum_{p<q}^{N} \left(\sum_{k=1}^{n} \left(\overline{\psi}_p^{(k)} \psi_q^{(k)} + \overline{\psi}_{N+p}^{(k)} \psi_{N+q}^{(k)} - \overline{\psi}_q^{(k)} \psi_p^{(k)} - \overline{\psi}_{N+q}^{(k)} \psi_{N+p}^{(k)} \right) \right)^2 \right\} \\
&\times \exp\left\{ \frac{1}{2N} \sum_{p=1}^{N} \left(\sum_{k=1}^{n} \left(\overline{\psi}_p^{(k)} \psi_p^{(k)} + \overline{\psi}_{N+p}^{(k)} \psi_{N+p}^{(k)} \right) \right)^2 \right\} \\
=& \exp\left\{ \frac{1}{2N} \sum_{p,q}^{N} \left(\sum_{k=1}^{n} \left(\overline{\psi}_p^{(k)} \psi_q^{(k)} + \overline{\psi}_{N+p}^{(k)} \psi_{N+q}^{(k)} \right) \right) \left(\sum_{k=1}^{n} \left(\overline{\psi}_q^{(k)} \psi_p^{(k)} + \overline{\psi}_{N+q}^{(k)} \psi_{N+p}^{(k)} \right) \right) \right\}.
\end{aligned}
$$

In the last expression one can see quartic terms in Grassmann variables. To deal with these terms, we use the so-called Hubbard-Stratonovich transformation. Let

$$
a_{k,k'} = \sum_{j=1}^{N} \overline{\psi}_j^{(k)} \psi_j^{(k')}, \quad b_{k,k'} = \sum_{j=1}^{N} \overline{\psi}_{N+j}^{(k)} \psi_{N+j}^{(k')}, \quad c_{k,k'} = \sum_{j=1}^{N} \overline{\psi}_j^{(k)} \psi_{N+j}^{(k')}, \quad d_{k,k'} = \sum_{j=1}^{N} \overline{\psi}_{N+j}^{(k)} \psi_j^{(k')},
$$

and

$$
A = \left(\begin{array}{c|c} \{a_{k,k'}\}_{k,k'=1}^{n} & \{c_{k,k'}\}_{k,k'=1}^{n} \\ \hline \{d_{k,k'}\}_{k,k'=1}^{n} & \{b_{k,k'}\}_{k,k'=1}^{n} \end{array} \right).
$$

Then

$$
\widehat{F}_\Gamma = \exp\left\{ -\frac{1}{2N} \operatorname{Tr} A^2 \right\}.
$$

The quadratic term in the $2n \times 2n$ matrix A can be linearised at the expense of the additional integration over $2n \times 2n$ hermitian matrices S_{2n} (the Hubbard-Stratonovich transformation):

$$
\widehat{F}_\Gamma = 2^{-n} \left(\frac{N}{\pi}\right)^{2n^2} \int D[S_{2n}] \exp\left\{ -\frac{N}{2} \operatorname{Tr} S_{2n}^2 - i \operatorname{Tr} S_{2n} A \right\}.
$$

Now, we can integration over the Grassmann variables. We have

$$F_\Gamma(z_1,z_2,\ldots,z_n) = \frac{1}{2^n}\left(\frac{N}{\pi}\right)^{2n^2}\int D[S_{2n}]\exp\left\{-\frac{N}{2}\operatorname{Tr}S_{2n}^2\right\}\int\prod_{k=1}^{n}\left(d\,\overline{\Psi}^{(k)}d\,\Psi^{(k)}\right)_{\text{ent}}$$
$$\times\exp\left\{-\sum_{j=1}^{N}\left(\sum_{k=1}^{n} z_k\overline{\psi}_j^{(k)}\psi_j^{(k)}+\sum_{k=1}^{n}\overline{z_k}\overline{\psi}_{N+j}^{(k)}\psi_{N+j}^{(k)}\right)\right\}$$
$$\times\exp\left\{i\sum_{j=1}^{N}\gamma_j\left(\sum_{k=1}^{n}\overline{\psi}_j^{(k)}\psi_j^{(k)}-\overline{\psi}_{N+j}^{(k)}\psi_{N+j}^{(k)}\right)\right\}$$
$$\times\exp\left\{-i\sum_{j=1}^{N}\left(\sum_{k,k'=1}^{n} s_{k',k}\overline{\psi}_j^{(k)}\psi_j^{(k')}+s_{n+k',k}\overline{\psi}_j^{(k)}\psi_{N+j}^{(k')}+s_{k',n+k}\overline{\psi}_{N+j}^{(k)}\psi_j^{(k')}+s_{n+k',n+k}\overline{\psi}_{N+j}^{(k')}\psi_{N+j}^{(k)}\right)\right\}.$$

By manipulating terms in the exponentials,

$$F_\Gamma(z_1,z_2,\ldots,z_n) = \frac{1}{2^n}\left(\frac{N}{\pi}\right)^{2n^2}\int D[S_{2n}]\exp\left\{-\frac{N}{2}\operatorname{Tr}S_{2n}^2\right\}\int\prod_{j=1}^{N}\left(d\,\overline{\psi}_j^{(\cdot)}d\,\psi_j^{(\cdot)}\right)_{\text{ent}}\left(d\,\overline{\psi}_{N+j}^{(\cdot)}d\,\psi_{N+j}^{(\cdot)}\right)_{\text{ent}}$$
$$\times\exp\left\{-\left\langle\begin{pmatrix}\overline{\psi}_j^{(\cdot)}\\ \overline{\psi}_{N+j}^{(\cdot)}\end{pmatrix},\left[\begin{pmatrix}Z-i\gamma_j\mathbf{1}_n & 0\\ 0 & \overline{Z}-i\gamma_j\mathbf{1}_n\end{pmatrix}+iS_{2n}\right]\begin{pmatrix}\psi_j^{(\cdot)}\\ \psi_{N+j}^{(\cdot)}\end{pmatrix}\right\rangle\right\}$$
$$=\frac{1}{2^n}\left(\frac{N}{\pi}\right)^{2n^2}\int D[S_{2n}]\exp\left\{-\frac{N}{2}\operatorname{Tr}S_{2n}^2\right\}\prod_{j=1}^{N}\det(Z_{2n}+iS-i\gamma_jL_{2n}).$$

Now, recalling that $\gamma_j = 0$ for $j = M+1,\ldots,N$ we obtain the statement of the Proposition. □

Appendix B. Various Asymptotic Regimes for Laguerre Polynomials

Asymptotic behaviour of the Laguerre polynomials $L_{N-k}^{(\alpha)}(-NY^2)$ in the limit when $N\to\infty$ and k and α are fixed depends on the scale of the variable $Y>0$ compared to N. For our investigation we need two scales: (i) $YN = y > 0$ is fixed and (ii) $YN \gg 1$. In both cases the desired approximations can be obtained from the integral representation (65) which we rewrite as

$$L_{N-k}^{(\alpha)}\left(-NY^2\right) = \frac{2N^{N-k+1}}{(N-k)!}\frac{e^{-NY^2}}{Y^\alpha}\int_0^\infty \tau^{-2k+\alpha+1}e^{-N(\tau^2-2\ln\tau)}I_\alpha(2\tau YN)\,d\tau, \qquad Y>0. \tag{A2}$$

We start with simpler case of $YN = y > 0$ being fixed in the limit $N\to\infty$. In this case significant contributions to the integral in (A2) are coming from a neighbourhood of the point $\tau = 1$ which is the point of minimum the function $\tau^2 - 2\ln\tau$ inside the interval of integration. Straightforward evaluation of the integral by the Laplace method together with the Stirling approximation $(N-k)! \sim \sqrt{2\pi}e^{-N}N^{N-k+1/2}$ yields that

$$L_{N-k}^{(\alpha)}\left(-\frac{y^2}{N}\right)\sim\frac{N^\alpha}{y^\alpha}I_\alpha(2y), \qquad y>0. \tag{A3}$$

In the other regime of interest for us, $YN \gg 1$, one can use the following asymptotic expansion for the modified Bessel function $I_\alpha(z)$ (see, e.g., Formula 5.11.10 in [34]):

$$I_\alpha(z) = \frac{e^z}{\sqrt{2\pi z}}\sum_{p=0}^{n}\frac{(-1)^p}{(2z)^p}A_p^{(\alpha)} + O\left(|z|^{-n-1}\right), \quad A_p^{(\alpha)} = \frac{\Gamma(\alpha+p+1/2)}{\Gamma(\alpha-p+1/2)}. \tag{A4}$$

It reduces the asymptotic analysis of $L^{(\alpha)}_{N-k}(-NY^2)$ to analysis of the following expression:

$$\frac{e^{-NY^2}}{2Y^{\alpha}\sqrt{\pi YN}}\int_0^{\infty}\tau^{-2k+\alpha+1/2}e^{-N\mathcal{L}(\tau)}\sum_p\frac{(-1)^p}{(4YN)^p}A_p^{(\alpha)}\,d\tau,\quad \mathcal{L}(\tau)=\tau^2-2\ln\tau-2\tau Y. \quad \text{(A5)}$$

In this case significant contributions to the integral in (A5) are coming from a neighbourhood of the point $\tau=\tau_*(Y)$ which is the point of minimum the function $\mathcal{L}(\tau)$ inside the interval of integration.

$$\tau_*(Y)=\frac{Y+\sqrt{Y^2+4}}{2}=\frac{1}{r_*(Y)},\quad r_*(Y)=\frac{\sqrt{Y^2+4}-Y}{2}>0, \quad \text{(A6)}$$

Using the relations $\tau_*(Y)=r_*(Y)+Y$ and $1+r_*(Y)^2=r_*(Y)(Y^2+4)^{1/2}$ we find that

$$\mathcal{L}(\tau_*(Y))=1+2\ln r_*(Y)-Y(r_*(Y)+Y),\quad \mathcal{L}''(\tau_*)=2r_*(Y)\left(Y^2+4\right)^{1/2}. \quad \text{(A7)}$$

Expanding the integrand in the standard way around $\tau=\tau_*(Y)$ and collecting the leading and subleading order terms we get asymptotic expressions for Laguerre polynomials with the precision sufficient for our purposes:

$$L^{(\alpha)}_{N-k}(-NY^2)=\begin{cases}\dfrac{e^{NYr_*(Y)}}{\sqrt{2\pi N}}\dfrac{r_*(Y)^{-2(N-k)-\alpha-1}}{Y^{\alpha+\frac{1}{2}}(Y^2+4)^{1/4}}\left[1+O\left(\dfrac{1}{N}\right)\right], & Y=O(1),\\[2ex] \dfrac{e^{NYr_*(Y)}}{\sqrt{2\pi N}}\dfrac{r_*(Y)^{-2(N-k)-\alpha-1}}{Y^{\alpha+\frac{1}{2}}(Y^2+4)^{1/4}}\left[1-\dfrac{(4\alpha^2-1)r_*(Y)}{16YN}+O\left(\dfrac{1}{N}\right)\right], & Y\ll 1\ll NY.\end{cases} \quad \text{(A8)}$$

References

1. Fyodorov, Y.V.; Sommers, H.-J. Statistics of resonance poles, phase shifts and time delays in quantum chaotic scattering: Random matrix approach for systems with broken time-reversal invariance. *J. Math. Phys.* **1997**, *38*, 1918–1981 [CrossRef]
2. Fyodorov, Y.V.; Sommers, H.-J. Random Matrices Close to Hermitian or Unitary: Overview of Methods and Results. *J. Phys. A Math. Gen.* **2003**, *36*, 3303–3347. [CrossRef]
3. Forrester, P.J. Rank-1 perturbations in random matrix theory—A review of exact results. *arXiv* **2022**, arXiv:2201.00324.
4. Poplavskyi, M.; Schehr, G. Exact Persistence Exponent for the 2*D*-Diffusion Equation and Related Kac Polynomials. *Phys. Rev. Lett.* **2018**, *121*, 150601. [CrossRef] [PubMed]
5. Kozhan, R. Rank One Non-Hermitian Perturbations of Hermitian β-Ensembles of Random Matrices. *J. Stat. Phys.* **2017**, *168*, 92–108. [CrossRef]
6. Killip, R.; Kozhan, R. Matrix Models and Eigenvalue Statistics for Truncations of Classical Ensembles of Random Unitary Matrices. *Commun. Math. Phys.* **2017**, *349*, 99–1027. [CrossRef]
7. Fyodorov, Y.V.; Osman, M. Eigenfunction non-orthogonality factors and the shape of CPA-like dips in a single-channel reflection from lossy chaotic cavities. *J. Phys. A Math. Theor.* **2022**, *55*, 224013. [CrossRef]
8. Fyodorov, Y.V.; Osman, M.; Tublin, R. A Few Results and Conjectures about Rank-One Non-Hermitian Deformations of β-Hermite Ensembles. Manuscript in preparation.
9. O'Rourke, S.; Wood, P. Spectra of nearly Hermitian random matrices. *Ann. l'Institut Henri Poincare* **2017**, *53*, 1241–1279. [CrossRef]
10. Rochet, J. Complex Outliers of Hermitian Random Matrices. *J. Theor. Probab.* **2017**, *30*, 1624–1654. [CrossRef]
11. Dubach, G.; Erdős, L. Dynamics of a rank-one perturbation of a Hermitian matrix. *arXiv* **2022**, arXiv:2108.13694
12. Shcherbina, M.; Shcherbina, T. Finite-rank complex deformations of random band matrices: Sigma-model approximation. *arXiv* **2022**, arXiv:2112.04455
13. Tao, T. Outliers in the spectrum of iid matrices with bounded rank perturbations. *Probab. Theory Relat. Fields* **2013**, *155*, 231–263. [CrossRef]
14. O'Rourke, S.; Renfrew, D. Low rank perturbations of large elliptic random matrices. *Electron. J. Probab.* **2014**, *19*, 1–65. [CrossRef]
15. Forrester, P.J.; Ipsen, J.R. A generalisation of the relation between zeros of the complex Kac polynomial and eigenvalues of truncated unitary matrices. *Probab. Theory Relat. Fields* **2019**, *175*, 833–847. [CrossRef]
16. Bohigas, O.; Giannoni, M.J.; Schmit, C. Characterization of chaotic quantum spectra and universality of level fluctuation laws. *Phys. Rev. Lett.* **1984**, *52*, 1–4. [CrossRef]
17. Verbaarschot, J.J.M.; Weidenmüller, H.A.; Zirnbauer, M.R. Grassmann integration in stochastic quantum physics: The case of compound-nucleus scattering. *Phys. Rep.* **1985**, *129*, 367–438. [CrossRef]

18. Sokolov, V.V.; Zelevinsky, V.G. Dynamics and statistics of unstable quantum states. *Nucl. Phys. A* **1989**, *504*, 562–588. [CrossRef]
19. Dittes, F.-M.; Harney, H.L.; Rotter, I. Formation of fast and slow decay modes in N-level systems coupled to one open channel. *Phys. Lett. A* **1991**, *153*, 451–455. [CrossRef]
20. Persson, E.; Rotter, I.; Stoeckmann, H.-J.; Barth, M. Observation of Resonance Trapping in an Open Microwave Cavity. *Phys. Rev. Lett.* **2000**, *85*, 2478–2481. [CrossRef]
21. Fyodorov, Y.V.; Sommers, H.-J. Statistics of S-matrix poles in few-channel chaotic scattering: Crossover from isolated to overlapping resonances. *JETP Lett.* **1996**, *63*, 1026–1030. [CrossRef]
22. Fyodorov, Y.V.; Khoruzhenko, B.A. Systematic Analytical Approach to Correlation Functions of Resonances in Quantum Chaotic Scattering. *Phys. Rev. Lett.* **1999**, *83*, 65–68. [CrossRef]
23. Gradshteyn, I.S.; Ryzhik, I.M. *Table of Integrals, Series, and Products*, 7th ed.; Elsevier/Academic Press: Amsterdam, The Netherlands, 2007.
24. Majumdar, S.N.; Pal, A.; Schehr, G. Extreme value statistics of correlated random variables: A pedagogical review. *Phys. Rep.* **2020**, *840*, 1–32. [CrossRef]
25. Auerbach, N.; Zelevinsky, V. Super-radient dynamics, doorways, and resonances in nuclei and other open mesoscopic systems. *Rep. Progr. Phys.* **2011**, *74*, 106301. [CrossRef]
26. Majumdar, S.N. Real-space Condensation in Stochastic Mass Transport Models. In *Exact Methods in Low-dimentional Statistical Physics and Quantum Computing*; (Lecture Notes of the Les Houches Summer School); Jacobsen, J., Ed.; Oxford University Press: New York, NY, USA, 2009; Volume 89.
27. Fyodorov, Y.V.; Khoruzhenko, B.A.; Sommers, H.-J. Universality in the random matrix spectra in the regime of weak non-Hermiticity. *Ann. Inst. H Poincaré Phys. Théor.* **1998**, *68*, 449–489.
28. Baik, J.; Ben Arous, G.; Peche, S. Phase transition of the largest eigenvalue for nonnull complex sample covariance matrices. *Ann. Probab.* **2005**, *33*, 1643–1697. [CrossRef]
29. Fyodorov, Y.V. Spectra of random matrices close to unitary and scattering theory for discrete-time systems. *arXiv* **2001**, arXiv:nlin/0002034.
30. Fyodorov, Y.V.; Mehlig, B. Statistics of resonances and nonorthogonal eigenfunctions in a model for single-channel chaotic scattering. *Phys. Rev. E* **2002**, *66*, 045202(R). [CrossRef]
31. Fyodorov, Y.V.; Khoruzhenko, B.A. On absolute moments of characteristic polynomials of a certain class of complex random matrices. *Commun. Math. Phys.* **2007**, *273*, 561–599. [CrossRef]
32. Życzkowski, K.; Sommers, H.-J. Truncations of random unitary matrices. *J. Phys. A Math. Gen.* **2000**, *33*, 2045–2057. [CrossRef]
33. Fyodorov, Y.V.; Khoruzhenko, B.A. Extreme eigenvalues of random sub-unitary matrices: From Fréchet to Gumbel. Manuscript in preparation.
34. Lebedev, N.N. *Special Functions and Their Applications*; Dover Publications: New York, NY, USA, 1972.

entropy

MDPI

Article

Universality and beyond in Optical Microcavity Billiards with Source-Induced Dynamics

Lukas Seemann and Martina Hentschel *

Institute of Physics, Technische Universität Chemnitz, D-09107 Chemnitz, Germany
* Correspondence: martina.hentschel@physik.tu-chemnitz.de

Abstract: Optical microcavity billiards are a paradigm of a mesoscopic model system for quantum chaos. We demonstrate the action and origin of ray-wave correspondence in real and phase space using far-field emission characteristics and Husimi functions. Whereas universality induced by the invariant-measure dominated far-field emission is known to be a feature shaping the properties of many lasing optical microcavities, the situation changes in the presence of sources that we discuss here. We investigate the source-induced dynamics and the resulting limits of universality while we find ray-picture results to remain a useful tool in order to understand the wave behaviour of optical microcavities with sources. We demonstrate the source-induced dynamics in phase space from the source ignition until a stationary regime is reached comparing results from ray, ray-with-phase, and wave simulations and explore ray–wave correspondence.

Keywords: microcavity billiards; sources; ray–wave correspondence; phase-space dynamics; Husimi function; quantum chaos

1. Introduction

Two-dimensional (2D) systems have inspired the field of quantum chaos for many years [1–3]. The origins of studying the quantum mechanical pendants of classically non-integrable systems trace back to the 1980s when universality was established as a common property of very different chaotic systems [4,5], in particular in the energy-level statistics in the 1980 paper of Casati et al. [4], and a variety of studies have been initiated focusing on the statistical properties based, e.g., on Random Matrix Theory [6].

The impressive properties of this new class of mesoscopic model systems [7] for both electrons and photons soon initiated interest in possible applications. Besides the ballistic quantum dots [3], the optical microcavities [8–14] have received a lot of interest, and lately also in non-euclidean [15] and Dirac Fermion optics [16]. One practical motivation was certainly the realization of microcavity lasers with directional emission, and plenty of solutions were found and investigated [17]. One realization involves deformed microdisk cavities of various shapes [18], including the Limaçon cavity [19–25]. Besides the experimental verification of the predicted [19] directional and universal, resonance-independent far-field emission originating from the cavity's invariant manifold, a remarkable ray–wave correspondence was seen. While all results were obtained for very different wavelengths λ—the ray modeling in the $\lambda \to 0$-limit, the wave simulation for λ larger than the experimentally relevant values—the agreement between all three approaches was convincing with slight, interference-inspired deviations between the three curves. Shinohara et al. [22] complemented this interpretation nicely by showing that averaging over a large number of resonances (42 in Ref. [22]) improves ray–wave correspondence by averaging out the resonance-specific features.

The reason for the universality of the observed far-field emission properties of whispering-gallery (WG)-type modes with high Q-factors is that the so-called natural measure (or Fresnel-weighted unstable manifold or steady probability distribution) [26]

Citation: Seemann, L.; Hentschel, M. Universality and beyond in Optical Microcavity Billiards with Source-Induced Dynamics. *Entropy* **2023**, *25*, 95. https://doi.org/10.3390/e25010095

Academic Editor: Marko Robnik

Received: 10 December 2022
Revised: 22 December 2022
Accepted: 29 December 2022
Published: 3 January 2023

determines the emission characteristics. Assuming a light ray to be initially captured by total internal reflection, it will, in a chaotic cavity, undergo a number of reflections where this condition remains fulfilled. However, at one point it will be violated and the angle of incidence χ will cross the critical lines $\sin \chi_c = \pm 1/n$ in phase space (n is refractive index of the cavity and we assume $n_0 = 1$ outside), and ray splitting occurs with the amount of light remaining in the cavity determined by Fresnel's reflection coefficient. The crossing of the critical line will be ruled by the unstable manifold of the system, weighted by the Fresnel reflection coefficient for our open optical system. Notice that this quantity describes the expanding directions along which the light will escape the cavity (actually, in the wave picture, by evanescent or tunneling escape from high-Q modes). This implies that the unstable manifold, as an important and central, yet abstract quantity of nonlinear dynamics is directly accessible and visible in experiments and the corresponding simulations. We point out that, therefore, simulations of the passive, non-lasing cavity can successfully describe even lasing cavities as long as mode interactions [27] do not play a role. In terms of ray picture modeling, the initial conditions are homogeneously distributed in phase space and the far-field characteristics are recorded when an initial transient regime is lapsed.

In this paper, we will explore another situation in optical microcavities that is induced by the presence of sources (or, similarly, relevant in microlasers with non-uniform pumping conditions [28,29]). This setting can capture situations where not all initial conditions are homogeneously populated, in contrast to the microlaser case discussed above. Source-induced phenomena can be relevant, for example, due to a specific distribution of fluorescent particles, a local pumping scheme, or even due to the coupling of two or more systems that effectively change the initial conditions to be non-uniform in phase space. The base element of any source can be described as a point-like emitter.

The paper is organized as follows. We will consider point-like sources and study their impact on the far-field emission in Section 2. We then investigate the source-initiated dynamics in phase space and introduce a ray picture extended by the phase information in Section 3 before we end with a conclusion and summary in Section 4.

2. Optical Microcavity Billiards with Sources

We start our investigation for a Limaçon cavity [19] with the shape given in polar coordinates (r, ϕ) as $r(\phi) = R_0(1 + \epsilon \cos \phi)$ where we set $R_0 = 1$ and choose the deformation parameter $\epsilon = 0.43$, such that the phase space of the cavity is known to be almost fully chaotic. We use Birkhoff coordinates, i.e., the arclength s along the boundary starting at its intersection with the positive x axis, and the sine of the angle of incidence χ of light traveling inside the cavity to specify the position in phase space. The far-field angle φ_{ff} is measured to be mathematically positive with respect to the positive x axis. We will consider TM polarized light (electric field transverse to the resonator plane, i.e. along the z axis, $\vec{E} = E_z \vec{e_z}$), a refractive index $n = 3.3$, and vary the position x_s of the source along the x axis. We use simulations with the open-source software package meep [30] and so-called meep units with the velocity of light set to 1, such that frequency f and (vacuum) wavelength λ are reciprocal to each other, as is the period $T = 1/f$. We use the built-in continuous source function of meep to describe a point-like coherent source that oscillates as $\exp(-2\pi i f t)$. Such a point-like source can be considered the basic module of any other source distribution.

Variation of the source position. For a mode at resonance frequency $f = 1.2$ (corresponding to $nkR \approx 25$), the far-field emission depends critically on the source position x_s as is visible in Figure 1. In general, more central source positions relate to more isotropic emissions, and the far-field emission characteristics of the uniformly pumped cavity can be completely lost. Similar results were found in a study of graphene and optical billiards in Ref. [16], where the importance of lensing effects in particular for single-layer graphene cavities was discussed.

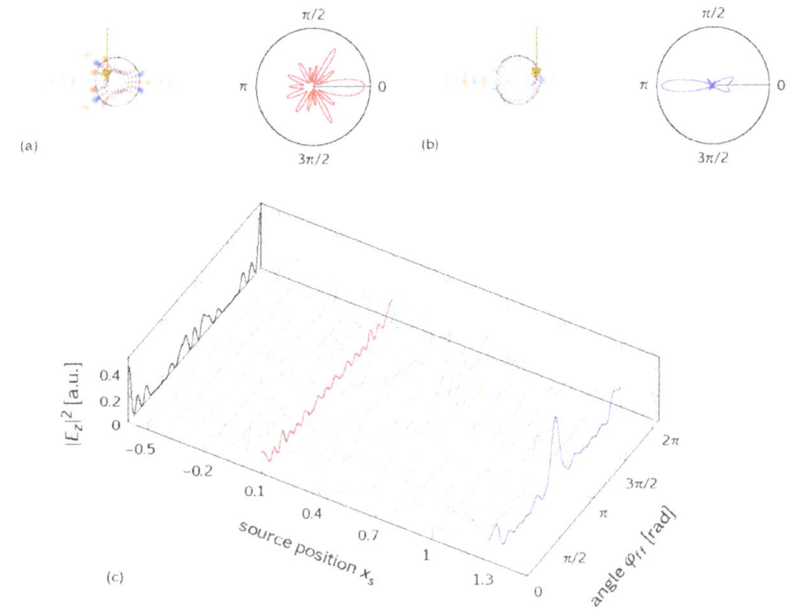

Figure 1. (**a**) Amplitude distribution E_z in real space for a mode at resonant frequency $f = 1.2$ and source position $x_s = 0.1$ (marked by arrow), and corresponding far-field emission as polar plot φ_{ff}. (**b**) Same for $x_s = 1.3$. (**c**) Far-field emission $|E_z|^2$ for source positions x_s varied along the x-axis, revealing a high sensitivity on x_s. The black line on the left marks the far-field emission "without source", i.e., the initial conditions uniformly distributed in phase space, and far-field data are taken when a stationary regime with exponential total internal intensity decay was reached, i.e., omitting the contribution of short rays.

Despite this deviation from the universality seen in the uniform pumping case, ray–wave correspondence still holds as illustrated in Figure 2 for $x_s = 1.3$. The far-field wave intensity $I = |E_z|^2$ and the ray-simulated intensities agree reasonably well. In particular, we find the wave intensity I to be reproduced by the short rays with trajectory lengths $l < 3$ corresponding to typically very few reflections at the system boundary. In other words, in the presence of sources, the far field is mainly determined by refractive escape of rays leaving the source and dwelling of very few reflections in the cavity. This indicates the relevance of lensing effects when the cavity acts similar to a thick lens. However, longer rays are needed to establish a semiquantitative agreement with the wave result for all far-field angles φ_{ff}. These long trajectories carry the information of the cavity geometry as a whole, namely in terms of the unstable manifold.

Variation of the source frequency. It is worthwhile to characterize the far-field sensitivity against variations in the source frequency, cf. Figure 3. In Figure 3a,b, mode patterns (amplitude E_z) are shown for two different frequencies f: resonant ($f = 1.2$) in Figure 3a and off-resonant ($f = 1.6$) in Figure 3b. The source position is fixed again on the x axis, here at $x_s = -0.42$ (marked by arrows). While the intra-cavity patterns deviate a lot—as expected upon a change in wavelength—the emission characteristics are less affected, cf. Figure 3c.

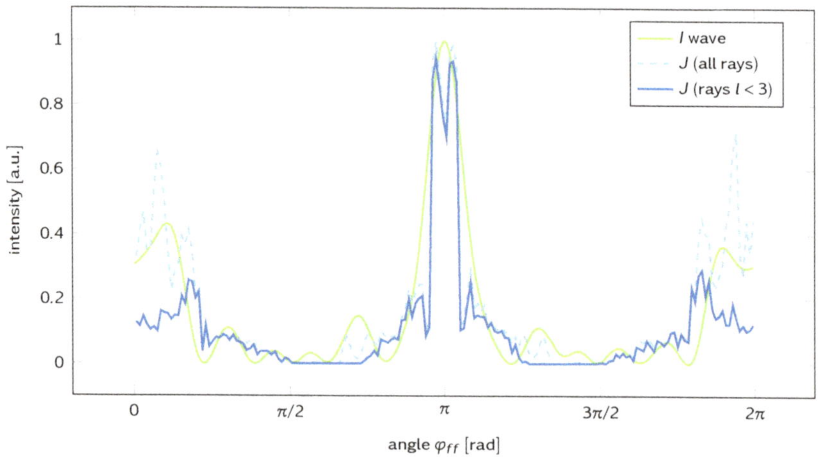

Figure 2. Far-field emission for source position $x_s = 1.3$. Wave intensity $I = |E_z|^2$ for the resonance at frequency $f = 1.2$ (full green line) and for ray-simulated intensities J including the far-field contribution of short rays only (full blue line) and of all rays (dashed line). Evidently, short rays with a trajectory length $l < 3$ contribute significantly to the far-field emission.

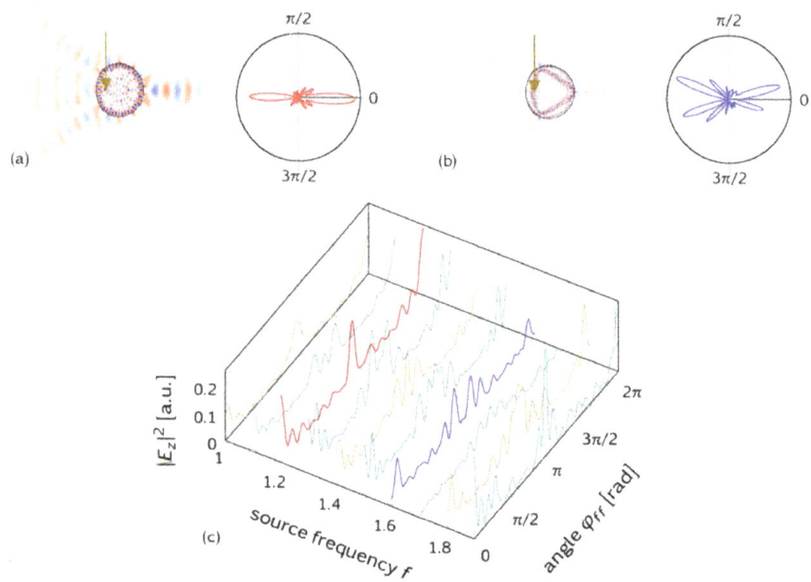

Figure 3. Far field emission depending on the source frequency f, while the source position is fixed at $x_s = -0.42$ (marked by arrow). (**a**) At the resonance frequency $f = 1.2$, the real space amplitude E_z and the far-field emission (polar plot) are shown. (**b**) An off-resonant frequency $f = 1.6$ yields a different mode pattern and a far field that differs in the details, but preserves the generic emission characteristics towards $\varphi_{ff} = 0$ and π. (**c**) Comparison of different source frequencies f confirms that the details are frequency-dependent, while the overall far-field emission is rather robust and dominated by the position of the source.

This result may seem surprising at first glance. However, it can be straightforwardly interpreted on the basis of ray-wave correspondence. Our starting point is the ray–wave correspondence for resonant frequencies, as illustrated in Figure 2 and confirmed in numerous other situations. As the naive ray picture does not know about frequencies or wavelengths, this would suggest no frequency dependence of the far-field emission at all. Of course, this cannot be correct as we know that the details of the far-field emission will be resonance-, or more generally, frequency-dependent [22]. This is precisely what can be seen in Figure 3c. This interpretation is also consistent with the fact that the far-field patterns are determined by short ray trajectories, i.e., before the formation of resonances can be expected.

3. Source-Induced Dynamics: Ray–Wave Correspondence in Phase Space

So far, we have discussed the implications of the presence of sources inside billiards for light, mainly in real space and in terms of far fields, and for the stationary situation. We will now complement the discussion in phase space, discussing both the Husimi function [31] and the ray signature of the source-induced dynamics. To this end, we will discuss the dynamics initiated when a source is turned on and follow it until a stationary state (with the source constantly emitting) is reached.

We will consider a Limaçon-shaped cavity with an intermediate deformation parameter $\epsilon = 0.25$ where a rich, mixed-phase space is present [32,33] (see the gray structure in Figure 4a). We will compare two source positions on the x axis, namely a rather central position $x_s = 0.6$ and an outer position $x_s = 1.0$, thereby manipulating the excitability of WG-type modes and trajectories. We choose a source frequency of $f = 0.64$ and use meep units as before, such that the period of the oscillation $T = 1/f \approx 1.56$. We will consider 80 time steps per period T and use the time frame number t as our variable of time. The time to travel across the cavity, i.e., to travel the optical distance $2nR_0$, is found to be 6.6 in meep units, so it will take about 4.23 T or approximately $t = 338$ frames (time steps) to travel the cavity's diameter.

Initial dynamics. What is the signature of the light emitted from the source? We start our study in the wave picture and excite a source placed at $x_s = 0.6$ with a (resonant) frequency of $f = 0.64$, cf. Figure 5. The real space evolution of the electromagnetic field amplitude E_z is shown in Figure 5a for $t = 186$, just before emitted light from the source reaches the far cavity interface. The corresponding phase–space representation is shown in Figure 5b, and we use the incoming Husimi function H^1_{in} inside the cavity [31] to characterize its signature at the interface boundary where we will also take the Poincaré surface of the section. We see that H^1_{in} contains the signature of light that has reached the cavity boundary at and around $s \approx 0$.

The snapshots in Figure 5c,d are taken about one period T later when all light emitted from the cavity at $t = 0$ has reached the boundary. There is a distinct extra signature that must characterize light emitted from the source at its first reflection at the cavity interface where the Husimi function H^1_{in} is recorded (see also the yellow line in Figure 4a and the discussion there).

This signature is discussed in more detail in Figure 6 for two different source positions and in the wave- and phase-information extended ray picture, respectively. In addition to the naive ray model considered before, we now include the ray's phase ϕ_p. This phase changes along the trajectory path according to $\phi_p = 2\pi l/\lambda$, with l denoting the optical trajectory length traveled. Upon the reflection at the boundary, an additional phase shift would have to be taken into account; however, we will limit our study to just the time before the first reflection. Note that the wavelength enters the ray picture via the phase, and thus resonance-specific properties can, in principle, become accessible within the ray model.

The Husimi functions shown in Figure 6a,b show comparable signatures and reach higher $|\sin\chi|$ for the outer source position $x_s = 1.0$ in (c) as a direct geometrical consequence of x_s being placed closer to the boundary. Notice that from frame to frame t, the location of the intensity maxima varies somewhat (not shown), indicating the importance of interference effects. This is straightforwardly confirmed qualitatively in Figure 6c,d where the phase-modulated ray intensity is shown at the first reflection point (in agreement with the time frames chosen for the Husimi plots) and clearly seen to possess a rather similar structure.

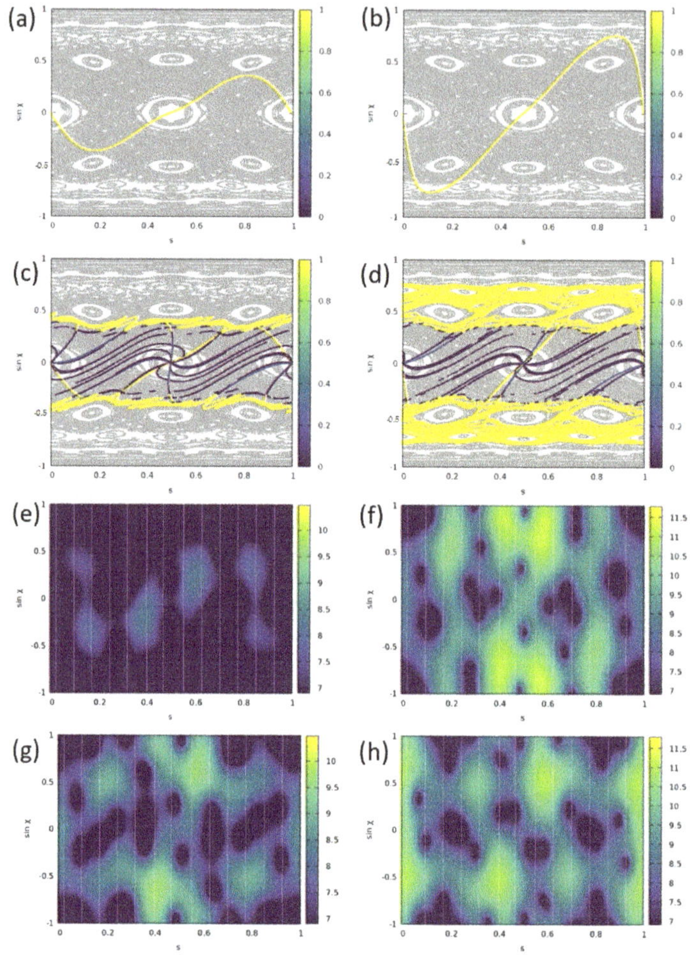

Figure 4. Dynamics induced by the source represented in phase space for two different source positions $x_s = 0.6$ in (**a**,**c**,**e**,**g**) and $x_s = 1$ in (**b**,**d**,**f**,**h**). In the ray simulation results (**a**–**d**), the Poincaré surface of section for a Limaçon cavity with $\epsilon = 0.25$ is indicated as gray background. (**a**,**b**) Characteristiics of a homogeneously emitting source at the first boundary reflection. Shown is the Fresnel-weighted intensity inside the cavity. (**c**,**d**) Same as (**a**,**b**) but after 30 reflections when stationarity is reached. In (**e**–**h**), the wave-simulation results are visualized in phase space in terms of the Husimi function H^1_{in} when a stationary regime was reached. The Husimi patterns are evolving periodically with period approximately $T/2$, and typical patterns are shown at time frames (**e**) $t = 16$, (**f**) $t = 9$, (**g**) $t = 37$, and (**h**) $t = 30$.

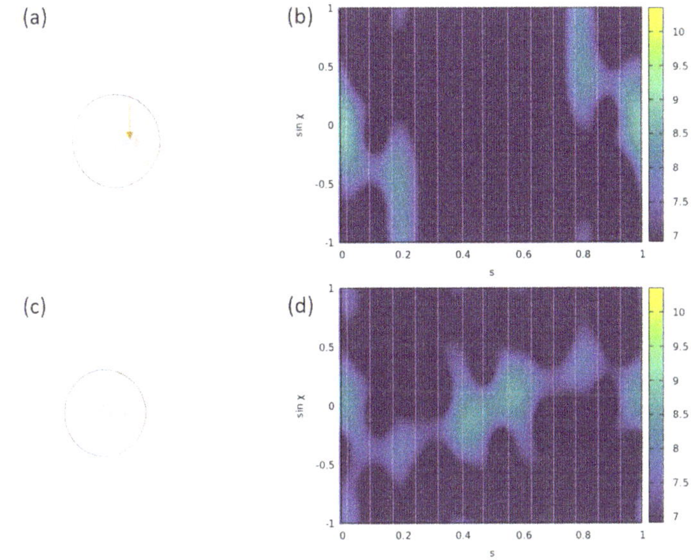

Figure 5. Initial temporal evolution of an electromagnetic wave emitted from a source with $f = 0.64$ at $x_s = 0.6$ (marked by arrow in (**a**)), $\epsilon = 0.25$. (**a**) Real-space E_z and (**b**) phase–space portrait in terms of the Husimi function H^1_{in} at time frame $t = 186$, and similarly one period later at $t = 272$ in (**c**,**d**). Note the extra signature in the center of (**d**) that can be attributed to the source signature after the first reflection at the boundary.

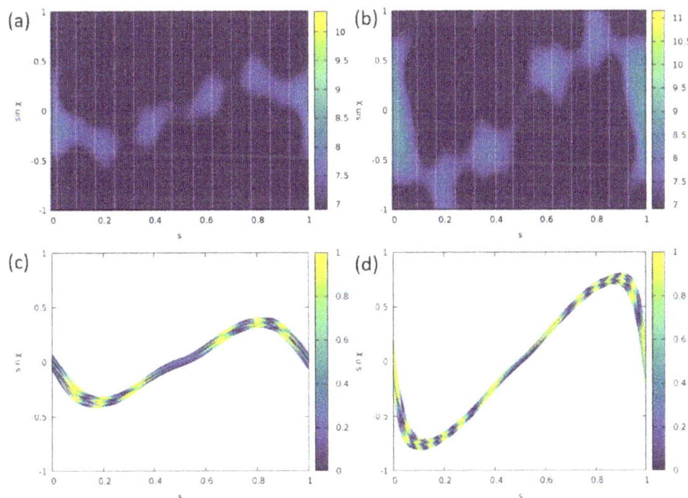

Figure 6. Phase–space representation of light emitted from a source in (**a**,**b**) wave and (**c**,**d**) ray simulations where the ray picture is extended by the phase information. (**a**) Husimi function H^1_{in} for $x_s = 0.6$ and $t = 253$. (**b**) H^1_{in} for $x_s = 1.0$ and $t = 314$. (**c**) Ray-with-phase simulation of the intensity at the first reflection point, 20 sources were randomly placed in a square with side length 0.05 around $x_s = 0.6$ and 2500 rays were started isotropically from each source. (**d**) Same as (**c**) for $x_s = 1.0$.

Stationary dynamics. After having discussed the initial source dynamics, we will now consider the stationary state reached after the transient regime. The results are presented in Figure 4, again for the source positions $x_s = 0.6$ (left column) and for $x_s = 1.0$ (right column). We start our considerations in the naive ray picture (without phase information) for which we revisit the initial source dynamics in Figure 4a,b. To this end, 10,000 rays are started uniformly from the point-like source and traced to the first reflection with the boundary, yielding the yellow curves. We point out that the time needed for the first reflection point will depend on the starting direction of the ray, especially for non-central x_s. Although the reflection-number based Poincaré map representation will thus differ from the time frame t-based study used in the wave simulations, it still provides a useful tool that can be directly superimposed on the Poincaré surface of section (PSOS). The mixed structure of the PSOS is indicated as the gray background in Figure 4a–d.

In Figure 4c,d, the source has been followed over several reflections at the boundary until stationarity (i.e., intensity saturation) was reached. The phase–space distribution of the Fresnel-weighted intensity emitted by the source is indicated in color scale. The area between the critical lines $\sin\chi = \pm 1/n$ carries a lower intensity due to refractive escape. It is evident that the spread in the phase space depends on the source position—the closer to the boundary x_s is, the larger $|\sin\chi|$ can be reached. In addition, we see that with the source positions chosen here, the three-island orbits cannot be excited within a ray simulation.

The results of wave simulations in the stationary regime (after 50 T) are displayed in Figure 4e–h. For both x_s, we find the pattern of the Husimi function H^1_{in} to periodically (with about $T/2$) vary. For each of the evolutions, we pick two characteristic patterns. For $x_s = 0.6$, we find a typical pattern that represents the source characteristics, cf. Figure 4e. Another one, cf. Figure 4g, displays intensity structured by the three-island chains. Although these islands cannot be populated in the ray-based counterpart model, it may well be possible within wave simulations due to a finite wavelength and when taking semiclassical corrections to the ray picture into account [34–41].

In particular, semiclassical arguments can explain why the intensity maxima in the Husimi function H^1_{in} seem to be placed at somewhat larger $|\sin\chi|$ in comparison to the ray model expectation. In the wave description, the evanescent wave associated with a WG-type mode will penetrate a distance of the order λ into the outer space. The corrected ray picture analogue deploys the Goos–Hänchen shift that causes the reflection to take place at an effective interface [42,43] such that the cavity appears effectively larger. However, then, the Husimi function is determined at a radius R_0 that is too small, thereby making the associated angle of incidence (somewhat) too large, as illustrated in Ref. [36], and thus explaining the deviation.

Eventually, for the other source position $x_s = 1$, we illustrate two typical Husimi patterns in Figure 4f,h. It is evident that now the H^1_{in} reaches larger values $|\sin\chi|$, in agreement with the ray-model expectation. One may speculate about the population of the four-island orbit in Figure 4f or a mode beating interaction induced by the source. We will investigate this in further studies.

4. Conclusions

We have investigated optical microcavities in the presence of sources and discussed and explained source-related (non-)universalities in the ray and wave pictures, and on the grounds of ray–wave correspondence. We showed that the position of a localized source crucially influences the properties of the optical microcavity in comparison to the source-free or uniform situation. These findings can be useful in the context of non-uniformly pumped microlasers or when light is coupled into an optical microcavity at certain positions, for example along the boundary.

We found a good agreement between the ray and wave picture results in terms of far fields and phase–space representations, and identified the signatures of the source in the ray and wave dynamics. We illustrated that a ray picture extended by the phase information can improve the agreement by capturing interference effects and introducing a wavelength into the ray picture. The phase information can also be used to distinguish regular and chaotic orbits as presented in Figure 7: a chaotic trajectory can be associated with an unpredictable phase value at the reflection points, whereas the phase evolves regularly for a periodic orbit. However, possible source-induced mode (interaction) dynamics [27] that might be suggested by Figure 4 in the stationary regime is beyond the scope of ray–wave correspondence.

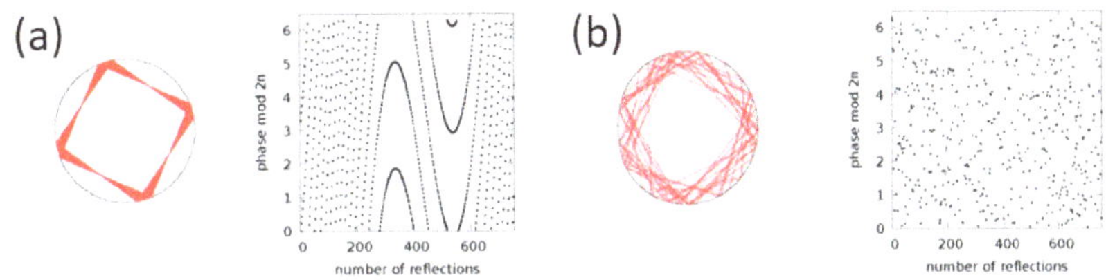

Figure 7. Phase evolution in a ray picture with phase information taken into account for (**a**) a regular orbit ($\epsilon = 0.25$) and (**b**) a chaotic trajectory ($\epsilon = 0.43$).

Author Contributions: Conceptualization, M.H.; Software, L.S.; Investigation, L.S.; Data curation, M.H.; Writing—original draft, M.H.; Visualization, L.S.; Supervision, M.H. All authors have read and agreed to the published version of the manuscript.

Funding: The publication of this article was funded by Chemnitz University of Technology and by the Deutsche Forschungsgemeinschaft (DFG, German Research Foundation)—491193532.

Institutional Review Board Statement: Not applicable.

Data Availability Statement: The data that support the findings of this study are available upon reasonable request from the authors.

Acknowledgments: We thank Tom Rodemund and Marika Carmen Federer for discussions. The special thanks of M.H. goes to Giulio Casati, whom she had the chance to meet several times as a young PhD student when the very basics of this work were laid.

Conflicts of Interest: There is no conflict of interest.

References

1. Stöckmann, H.J. *Quantum Dots: An Introduction*; Cambridge University Press: Cambridge, MA, USA, 1999.
2. Haake, F. *Quantum Signatures of Chaos*; Springer Series in Synergetics; Springer: Berlin, Germany, 2001.
3. Nakamura, K.; Harayama, T. *Quantum Chaos and Quantum Dots*; Oxford University Press: Oxford, UK, 2004.
4. Casati, G.; Val-Gris, F.; Guarnieri, I. On the connection between quantization of nonintegrable systems and statistical theory of spectra. *Lett. Nuovo C. (1971–1985)* **1980**, *28*, 279–282. [CrossRef]
5. Bohigas, O.; Giannoni, M.J.; Schmit, C. Characterization of Chaotic Quantum Spectra and Universality of Level Fluctuation Laws. *Phys. Rev. Lett.* **1984**, *52*, 1–4. [CrossRef]
6. Mehta, M.L. *Random Matrix Theory*; Elsevier: Amsterdam, The Netherlands, 2004.
7. Imry, Y. *Introduction to Mesoscopic Physics*; Oxford University Press: Oxford, UK, 1997.
8. Vahala, K. *Optical Microcavities*; World Scientific: Singapore, 2004.
9. Nöckel, J.U.; Stone, A.D. Ray and wave chaos in asymmetric resonant optical cavities. *Nature* **1997**, *385*, 45–47. [CrossRef]
10. Gmachl, C.; Capasso, F.; Narimanov, E.E.; Nöckel, J.U.; Stone, A.D.; Faist, J.; Sivco, D.L.; Cho, A.Y. High-Power Directional Emission from Microlasers with Chaotic Resonators. *Science* **1998**, *280*, 1556–1564. [CrossRef] [PubMed]
11. Doya, V.; Legrand, O.; Mortessagne, F.; Miniatura, C. Light scarring in an optical fiber. *Phys. Rev. Lett.* **2001**, *88*, 014102. [CrossRef]
12. Schäfer, R.; Kuhl, U.; Stöckmann, H.J. Directed emission from a dielectric microwave billiard with quadrupolar shape. *New J. Phys.* **2006**, *8*, 46. [CrossRef]
13. Bäcker, A.; Ketzmerick, R.; Löck, S.; Robnik, M.; Vidmar, G.; Höhmann, R.; Kuhl, U.; Stöckmann, H.J. Dynamical Tunneling in Mushroom Billiards. *Phys. Rev. Lett.* **2008**, *100*, 174103. [CrossRef]
14. Bittner, S.; Dietz, B.; Günther, U.; Harney, H.L.; Miski-Oglu, M.; Richter, A.; Schäfer, F. PT Symmetry and Spontaneous Symmetry Breaking in a Microwave Billiard. *Phys. Rev. Lett.* **2012**, *108*, 024101. [CrossRef] [PubMed]
15. Song, Y.; Monceaux, Y.; Bittner, S.; Chao, K.; Reynoso de la Cruz, H.M.; Lafargue, C.; Decanini, D.; Dietz, B.; Zyss, J.; Grigis, A.; et al. Möbius Strip Microlasers: A Testbed for Non-Euclidean Photonics. *Phys. Rev. Lett.* **2021**, *127*, 203901. [CrossRef]
16. Schrepfer, J.K.; Chen, S.C.; Liu, M.H.; Richter, K.; Hentschel, M. Dirac fermion optics and directed emission from single- and bilayer graphene cavities. *Phys. Rev. B* **2021**, *104*, 155436. [CrossRef]

17. Cao, H.; Wiersig, J. Dielectric microcavities: Model systems for wave chaos and non-Hermitian physics. *Rev. Mod. Phys.* **2015**, *87*, 61–111. [CrossRef]
18. Schermer, M.; Bittner, S.; Singh, G.; Ulysse, C.; Lebental, M.; Wiersig, J. Unidirectional light emission from low-index polymer microlasers. *Appl. Phys. Lett.* **2015**, *106*, 101107. [CrossRef]
19. Wiersig, J.; Hentschel, M. Combining Directional Light Output and Ultralow Loss in Deformed Microdisks. *Phys. Rev. Lett.* **2008**, *100*, 033901. [CrossRef] [PubMed]
20. Song, Q.; Fang, W.; Liu, B.; Ho, S.T.; Solomon, G.S.; Cao, H. Chaotic microcavity laser with high quality factor and unidirectional output. *Phys. Rev. A* **2009**, *80*, 041807. [CrossRef]
21. Yi, C.H.; Kim, M.W.; Kim, C.M. Lasing characteristics of a Limaçon-shaped microcavity laser. *Appl. Phys. Lett.* **2009**, *95*, 141107. [CrossRef]
22. Shinohara, S.; Hentschel, M.; Wiersig, J.; Sasaki, T.; Harayama, T. Ray-wave correspondence in limaçon-shaped semiconductor microcavities. *Phys. Rev. A* **2009**, *80*, 031801. [CrossRef]
23. Yan, C.; Wang, Q.J.; Diehl, L.; Hentschel, M.; Wiersig, J.; Yu, N.; Pflügl, C.; Capasso, F.; Belkin, M.A.; Edamura, T.; et al. Directional emission and universal far-field behavior from semiconductor lasers with Limaçon-shaped microcavity. *Appl. Phys. Lett.* **2009**, *94*, 251101. [CrossRef]
24. Wang, Q.J.; Yan, C.; Diehl, L.; Hentschel, M.; Wiersig, J.; Yu, N.; Pflügl, C.; Belkin, M.A.; Edamura, T.; Yamanishi, M.; et al. Deformed microcavity quantum cascade lasers with directional emission. *New J. Phys.* **2009**, *11*, 125018. [CrossRef]
25. Albert, F.; Hopfmann, C.; Eberspächer, A.; Arnold, F.; Emmerling, M.; Schneider, C.; Höfling, S.; Forchel, A.; Kamp, M.; Wiersig, J.; et al. Directional whispering gallery mode emission from Limaçon-shaped electrically pumped quantum dot micropillar lasers. *Appl. Phys. Lett.* **2012**, *101*, 021116. [CrossRef]
26. Lee, S.Y.; Ryu, J.W.; Kwon, T.Y.; Rim, S.; Kim, C.M. Scarred resonances and steady probability distribution in a chaotic microcavity. *Phys. Rev. A* **2005**, *72*, 061801. [CrossRef]
27. You, M.; Sakakibara, D.; Makino, K.; Morishita, Y.; Matsumura, K.; Kawashima, Y.; Yoshikawa, M.; Tonosaki, M.; Kanno, K.; Uchida, A.; et al. Universal Single-Mode Lasing in Fully Chaotic Billiard Lasers. *Entropy* **2022**, *24*, 1648. [CrossRef] [PubMed]
28. Michel, C.; Doya, V.; Legrand, O.; Mortessagne, F. Selective amplification of scars in a chaotic optical fiber. *Phys. Rev. Lett.* **2007**, *99*, 224101. [CrossRef]
29. Hentschel, M.; Kwon, T.Y. Designing and understanding directional emission from spiral microlasers. *Opt. Lett.* **2009**, *34*, 163–165. [CrossRef]
30. Oskooi, A.F.; Roundy, D.; Ibanescu, M.; Bermel, P.; Joannopoulos, J.D.; Johnson, S.G. MEEP: A flexible free-software package for electromagnetic simulations by the FDTD method. *Comput. Phys. Commun.* **2010**, *181*, 687–702. [CrossRef]
31. Hentschel, M.; Schomerus, H.; Schubert, R. Husimi functions at dielectric interfaces: Inside-outside duality for optical systems and beyond. *Europhys. Lett.* **2003**, *62*, 636. [CrossRef]
32. Berry, M.V.; Robnik, M. Semiclassical level spacings when regular and chaotic orbits coexist. *J. Phys. A Math. Gen.* **1984**, *17*, 2413. [CrossRef]
33. Prosen, T.; Robnik, M. Semiclassical energy level statistics in the transition region between integrability and chaos: Transition from Brody-like to Berry-Robnik behaviour. *J. Phys. A Math. Gen.* **1994**, *27*, 8059. [CrossRef]
34. Tureci, H.E.; Stone, A.D. Deviation from Snell's law for beams transmitted near the critical angle: Application to microcavity lasers. *Opt. Lett.* **2002**, *27*, 7–9. [CrossRef]
35. Rex, N.B.; Tureci, H.E.; Schwefel, H.G.L.; Chang, R.K.; Stone, A.D. Fresnel filtering in lasing emission from scarred modes of wave-chaotic optical resonators. *Phys. Rev. Lett.* **2002**, *88*, 094102. [CrossRef]
36. Hentschel, M.; Schomerus, H. Fresnel laws at curved dielectric interfaces of microresonators. *Phys. Rev. E* **2002**, *65*, 045603. [CrossRef]
37. Schomerus, H.; Hentschel, M. Correcting Ray Optics at Curved Dielectric Microresonator Interfaces: Phase-Space Unification of Fresnel Filtering and the Goos-Hänchen Shift. *Phys. Rev. Lett.* **2006**, *96*, 243903. [CrossRef] [PubMed]
38. Unterhinninghofen, J.; Wiersig, J.; Hentschel, M. Goos-Hänchen shift and localization of optical modes in deformed microcavities. *Phys. Rev. E* **2008**, *78*, 016201. [CrossRef] [PubMed]
39. Harayama, T.; Shinohara, S. Ray-wave correspondence in chaotic dielectric billiards. *Phys. Rev. E Stat. Nonlinear Soft Matter Phys.* **2015**, *92*, 042916. [CrossRef] [PubMed]
40. Stockschläder, P.; Kreismann, J.; Hentschel, M. Curvature dependence of semiclassical corrections to ray optics: How Goos-Hänchen shift and Fresnel filtering deviate from the planar case result. *EPL* **2014**, *107*, 64001. [CrossRef]
41. Stockschläder, P.; Hentschel, M. Consequences of a wave-correction extended ray dynamics for integrable and chaotic optical microcavities. *J. Opt.* **2017**, *19*, 125603. [CrossRef]
42. Goos, F.; Hänchen, H. Ein neuer und fundamentaler Versuch zur Totalreflexion. *Ann. Phys.* **1947**, *436*, 333–346. [CrossRef]
43. Goos, F.; Lindberg-Hänchen, H. Neumessung des Strahlversetzungseffektes bei Totalreflexion. *Ann. Phys.* **1949**, *440*, 251–252. [CrossRef]

entropy

MDPI

Article

Entanglement Dynamics and Classical Complexity

Jiaozi Wang [1,*], Barbara Dietz [2], Dario Rosa [2,3] and Giuliano Benenti [4,5,6]

1 Department of Physics, University of Osnabrück, D-49069 Osnabrück, Germany
2 Center for Theoretical Physics of Complex Systems, Institute for Basic Science (IBS), Daejeon 34126, Republic of Korea
3 Basic Science Program, Korea University of Science and Technology (UST), Daejeon 34113, Republic of Korea
4 Center for Nonlinear and Complex Systems, Dipartimento di Scienza e Alta Tecnologia, Università degli Studi dell'Insubria, Via Valleggio 11, 22100 Como, Italy
5 Istituto Nazionale di Fisica Nucleare, Sezione di Milano, Via Celoria 16, 20133 Milano, Italy
6 NEST, Istituto Nanoscienze-CNR, 56126 Pisa, Italy
* Correspondence: jiaozi.wang@uos.de

Abstract: We study the dynamical generation of entanglement for a two-body interacting system, starting from a separable coherent state. We show analytically that in the quasiclassical regime the entanglement growth rate can be simply computed by means of the underlying classical dynamics. Furthermore, this rate is given by the Kolmogorov–Sinai entropy, which characterizes the dynamical complexity of classical motion. Our results, illustrated by numerical simulations on a model of coupled rotators, establish in the quasiclassical regime a link between the generation of entanglement, a purely quantum phenomenon, and classical complexity.

Keywords: quantum complexity; quantum to classical transition

Citation: Wang, J.; Dietz, B.; Rosa, D.; Benenti, G. Entanglement Dynamics and Classical Complexity. *Entropy* **2023**, 25, 97. https://doi.org/10.3390/e25010097

Academic Editor: Marko Robnik

Received: 25 November 2022
Revised: 30 December 2022
Accepted: 30 December 2022
Published: 3 January 2023

1. Introduction

The characterization of complexity in quantum systems is a key problem, not only for fundamental reasons but also for the development of quantum technologies [1–3]. While for classical dynamical systems a well-established notion of complexity exists, based on Kolmogorov–Sinai (KS) entropy [4], which in turn is related to the exponential instability of orbits, in the quantum realm the measure of complexity has proven to be an elusive problem.

First, we cannot *sic et simpliciter* use trajectories, due the Heisenberg uncertainty principle. To circumvent such problem, phase-space approaches have been proposed [5–20], based on the evolution of phase space distributions. Second, entanglement, the key resource in the quest for quantum advantage, is peculiar to quantum composite systems and therefore is a source of quantum complexity without a classical analogue. Since for pure bipartite systems the reduced von Neumann entropy, known as entanglement entropy, is the well-established measure of entanglement [21], it is interesting to investigate whether its growth in a dynamical system is related to the KS entropy of the underlying classical dynamics [22].

For bosonic systems with an unstable quadratic Hamiltonian, entanglement entropy grows linearly in time, with a rate upper bounded by the KS entropy, the bound being saturated under suitable conditions on the size of the bipartitions [23]. The question then arises, whether the entanglement growth of chaotic quantum systems in the quasiclassical regime is also determined by the KS entropy. This issue was investigated more than two decades ago, with numerical results suggesting that the entanglement generation rate is given by the KS entropy [24]. On the other hand, such results were obtained in the weakly chaotic regime, with coexistence of chaotic seas and tori, while another study in the strongly chaotic region, where the effect of tori is negligible, showed instead no increase of the entanglement production rate upon an increase of the maximum Lyapunov exponents [25]. This apparent contradiction was explained by a quasiclassical calculation for the linear entropy, approximating the entanglement entropy, under the condition of weak coupling between the subsystems in the underlying classical dynamics [26,27]. This work showed that the entanglement growth rate is determined by the minimal value of the three rates given by the standard one deduced from the interaction term and the largest Lyapunov exponents of the two subsystems, respectively.

In this paper, we remove the above restriction on the coupling strength and compare the quantum evolution starting from separable coherent states with the classical evolution of initially Gaussian distributions, of a size determined by the effective Planck constant of the corresponding quantum dynamics. We show that in the quasiclassical regime quantum and classical linear entropy are in agreement and grow with the rate given by the KS entropy of classical dynamics. Our analytical results are illustrated by numerical simulations for a model of kicked-coupled rotators.

This work is dedicated to our friend and colleague Giulio Casati, who has always had a deep interest in understanding the complexity of quantum motion.

2. Analytical Results

In this section, we connect, for an overall pure bipartite system, the growth rate of linear entropy to the KS entropy of the classical underlying dynamics. We consider a two body system, whose Hamiltonian reads

$$\hat{H} = \hat{H}_1(\hat{q}_1, \hat{p}_1) + \hat{H}_2(\hat{q}_2, \hat{p}_2) + \hat{H}_{12}(\hat{q}_1, \hat{p}_1, \hat{q}_2, \hat{p}_2). \tag{1}$$

The corresponding classical Hamiltonian is written as

$$H(q_1, p_1, q_2, p_2) = H_1(q_1, p_1) + H_2(q_2, p_2) + H_{12}(q_1, p_1, q_2, p_2). \tag{2}$$

We compute as entanglement measure the linear entropy (also known as *second Rényi entropy*) of a subsystem (for example, system 1), which is defined as

$$S(\hat{\rho}_1) = -\ln(\mathrm{Tr}(\hat{\rho}_1^2)). \tag{3}$$

Here $\hat{\rho}_1$ is the reduced density matrix of the system 1, $\hat{\rho}_1 = \mathrm{Tr}_2(\hat{\rho})$, where the partial trace is taken over system 2 and $\hat{\rho}$ is the density matrix of the composite system. Note that equivalently we could have considered system 2, since $S(\hat{\rho}_2) = S(\hat{\rho}_1)$, with $\hat{\rho}_2 = \mathrm{Tr}_1(\hat{\rho})$. The linear entropy is much more convenient for numerical and analytical investigations than the reduced von Neumann entropy $S_{\mathrm{vN}}(\hat{\rho}_1) = -\mathrm{Tr}(\hat{\rho}_1 \ln \hat{\rho}_1)$. At the same time, it is a very useful entanglement probe. Indeed, if the linear entropy of a part is larger than the linear entropy of the overall system, bipartite entanglement exists between that part and the rest of the system. Moreover, for maximally mixed states of a subsystem of dimension N, the linear entropy and the reduced von Neumann entropy are both maximized and equal to $\ln N$.

In order to obtain the classical analog of the linear entropy, we make use of the Husimi function [28] of the density matrix $\hat{\rho}$, given by

$$W_H(\gamma) = \frac{1}{(2\pi\hbar)^2}\langle\gamma|\hat{\rho}|\gamma\rangle, \tag{4}$$

where $\gamma = (q_1, p_1, q_2, p_2)$, $|\gamma\rangle$ denotes the coherent state of the composite system centered at γ, and $\hbar$ is the effective Planck constant. In the quasiclassical limit $\hbar \to 0$, the trace of $\hat{\rho}_1^2$ can be carried out by making use of the Husimi function W_H^1 of $\hat{\rho}_1$ as

$$\mathrm{Tr}(\hat{\rho}_1^2) = \int d\boldsymbol{\gamma_1}[W_H^1(\boldsymbol{\gamma_1})]^2, \tag{5}$$

where $\gamma_1 = (q_1, p_1)$, $|\gamma_1\rangle$ denotes the coherent state of system 1 centered at γ_1, and

$$W_H^1(\gamma_1) = \frac{1}{2\pi\hbar}\langle\gamma_1|\hat{\rho}_1|\gamma_1\rangle. \tag{6}$$

Furthermore, the reduced density matrix $\hat{\rho}_1$ can also be obtained in terms of the coherent states of the system 2, denoted by $|\gamma_2\rangle$, as

$$\hat{\rho}_1 = \mathrm{Tr}_2(\hat{\rho}) = \frac{1}{2\pi\hbar}\int d\gamma_2\langle\gamma_2|\hat{\rho}|\gamma_2\rangle. \tag{7}$$

Substituting Equation (7) into Equation (6), we have

$$W_H^1(\gamma_1) = \int d\gamma_2 W_H(\gamma), \tag{8}$$

yielding with Equation (5)

$$\mathrm{Tr}(\hat{\rho}_1^2) = \int d\gamma_1\left|\int W_H(\gamma)d\gamma_2\right|^2. \tag{9}$$

Hence, we obtain

$$S(\hat{\rho}_1) = -\ln\left[\int d\gamma_1\left(\int d\gamma_2 W_H(\gamma)\right)^2\right]. \tag{10}$$

After replacing the Husimi function $W_H(\gamma)$ with the classical distribution function $\rho(\gamma)$, the classical analog of linear entropy can be written as

$$S_{cl}(\rho_1) = -\ln\left[\int d\gamma_1(\rho_{re}^1(\gamma_1))^2\right], \tag{11}$$

where $\rho_{re}^1(\gamma_1)$ indicates the marginal distribution function of γ_1,

$$\rho_{re}^1(\gamma_1) = \int d\gamma_2\rho(\gamma). \tag{12}$$

It is expected that

$$S(\hat{\rho}_1) \approx S_{cl}(\rho_1) \tag{13}$$

holds in the quasiclassical limit in which the effective Planck constant $\hbar \to 0$.

An explicit expression can be derived for the classical entropy $S_{cl}(\rho_1)$ as follows. We consider the initial state as the "most classical" state, that is, a coherent state $|\gamma\rangle$, whose corresponding classical distribution function can be written as

$$\rho_0(\gamma) = \frac{1}{(\pi\hbar_c)^2}\exp\left(-\frac{1}{\hbar_c}|\gamma - \gamma^0|^2\right), \tag{14}$$

which has a Gaussian form whose center is denoted by γ^0, $\hbar_c = \hbar$ is chosen to be the same as the effective Planck constant in the quantum case, and $|\gamma - \gamma^0|$ indicates the norm of the vector $\delta\gamma = \gamma - \gamma^0$. In the quasiclassical limit, one has $\hbar_c \to 0$, which means that, for times smaller than the Ehrenfest time scale t_E (with $t_E \to \infty$ as $\hbar_c \to 0$), almost all the states in the ensemble remain close to the center $\gamma^0(t)$. This implies that the distribution of states at time t, $\rho_t(\gamma)$, is significantly different from zero only for small $|\delta\gamma|$. In this case, the time evolution of $\delta\gamma$ is determined by the so-called stability matrix

$$\boldsymbol{M}_t^{ij} = \left.\frac{\partial(\delta\gamma_i(t))}{\partial(\delta\gamma_j(0))}\right|_{\delta\gamma(0)=0}, \tag{15}$$

with

$$\delta\gamma(t) = \boldsymbol{M}_t \delta\gamma(0). \tag{16}$$

As the classical linear entropy is independent of the coordinate's origin, for the convenience of the following discussion, we choose the position of the center $\gamma^0(t)$ as the origin of coordinates. In this local coordinate system along $\gamma^0(t)$, we can replace $\delta\gamma(t) = \gamma(t) - \gamma^0(t)$ by $\gamma(t)$.

Then making use of Liouville's theorem, the distribution at time t can be written as

$$\rho_t(\gamma) = \rho_0(\boldsymbol{M}_t^{-1}\gamma), \tag{17}$$

and therefore

$$\rho_t(\gamma) = \frac{1}{(\pi\hbar_c)^2} \exp\left(-\frac{1}{\hbar_c}|\boldsymbol{M}_t^{-1}\gamma|^2\right). \tag{18}$$

Using the positive definite symmetric matrix

$$\boldsymbol{A}_t \equiv (\boldsymbol{M}_t^{-1})^T \boldsymbol{M}_t^{-1}, \tag{19}$$

the density distribution at time t can be written as

$$\rho_t(\boldsymbol{x}) = \frac{1}{(\pi\hbar_c)^2} \exp\left(-\frac{1}{\hbar_c}\sum_{i,j=1}^{4} x_i A_t^{ij}(t) x_j\right), \tag{20}$$

which is a Gaussian distribution, with $\boldsymbol{x}$ corresponding to γ, that is, $(x_1, x_2, x_3, x_4) = (q_1, p_1, q_2, p_2)$. In order to calculate the classical linear entropy, we first calculate the marginal distribution function of $\rho_t(\boldsymbol{x})$ for system 1:

$$\rho_t^1(x_1, x_2) = \int \rho_t(x_1, x_2, x_3, x_4) dx_3 dx_4. \tag{21}$$

Then the classical linear entropy at time t can be written as

$$S_{cl}(\rho_t) = -\ln\left[\int dx_1 dx_2 \left(\rho_t^1(x_1, x_2)\right)^2\right]. \tag{22}$$

As outlined in the Appendix A, by writing $\boldsymbol{A}_t$ in block form,

$$\boldsymbol{A}_t = \begin{pmatrix} \hat{\boldsymbol{a}} & \hat{\boldsymbol{b}} \\ \hat{\boldsymbol{b}}^T & \hat{\boldsymbol{d}} \end{pmatrix}, \tag{23}$$

we obtain our first main result

$$S_{cl}(\rho_t) = \ln(2\pi\hbar) + \frac{1}{2}\ln[\det(\hat{\boldsymbol{d}})]. \tag{24}$$

In the derivation we restrict to the case of a two-particle system, however, the generalization to N particles is straightforward.

In order to compute $\det \hat{\boldsymbol{d}}$, we sum the eigenvalues of the operator $\hat{\boldsymbol{d}}$ (denoted by d_k, in order of descending energy), which are in close relation to the eigenvalues of $\boldsymbol{A}_t$. We diagonalize the symmetric matrix $\boldsymbol{A}_t$ as

$$\boldsymbol{A}_t = \boldsymbol{V}\text{diag}\{A_1, A_2, A_3, A_4\}\boldsymbol{V}^T, \tag{25}$$

where diag indicates a diagonal matrix, A_k is the k-th eigenvalue of $\boldsymbol{A}_t$, and $\boldsymbol{V}$ is an orthogonal matrix. If the system is chaotic, $A_k \propto e^{2\lambda_k t}$, where λ_k is the k-th Lyapunov exponent, with $\lambda_1 > \lambda_2 > 0 > \lambda_3 > \lambda_4$, and $\lambda_3 = -\lambda_2$, $\lambda_4 = -\lambda_1$.

Hence, in the typical case in which the eigenvectors (denoted by $|A_k\rangle, k = 1, 2$) of $\boldsymbol{A}_t$ corresponding to the eigenvalues A_1 and A_2 have non-zero components within the Hilbert space of system 2, we have

$$d_1 \propto e^{2\lambda_1 t}, \quad d_2 \propto e^{2\lambda_2 t}. \tag{26}$$

As a result,

$$\det \hat{\boldsymbol{d}} \propto \exp 2(\lambda_1 + \lambda_2)t, \tag{27}$$

which directly leads to

$$S_{cl}(\rho_t) - S_{cl}(\rho_0) = (\lambda_1 + \lambda_2)t, \tag{28}$$

indicating that the growth rate of the linear entropy is given by the Kolmogorov–Sinai entropy of the overall system, which is the second main result of our work.

3. Numerical Results

In this section, we numerically illustrate the prediction of equivalence between the classical and quantum growth of linear entropies, Equation (13), as well as the growth as predicted in Equation (28), by means of a two-body system which has a clearly defined classical counterpart. More specifically, we consider two coupled rotators (or coupled tops) [29,30], with respective angular momentum operators $\hat{\boldsymbol{S}} = (\hat{S}_x, \hat{S}_y, \hat{S}_z)^T$ and $\hat{\boldsymbol{L}} = (\hat{L}_x, \hat{L}_y, \hat{L}_z)^T$, and a time-dependent Hamiltonian with kicked interaction:

$$\hat{\mathcal{H}} = \frac{a}{j}(\hat{S}_z + \hat{L}_z) + \frac{c}{j^2}\hat{S}_x\hat{L}_x \sum_{n=-\infty}^{\infty} \delta(t - n), \tag{29}$$

where j is the (half-integer or integer) total angular momentum quantum number of both tops. The Hamiltonian possesses constants of motion, $\hat{\boldsymbol{S}}^2$ and $\hat{\boldsymbol{L}}^2$. The Hilbert space is expanded by making use of $|s, m_s, l, m_l\rangle \equiv |s, m_s\rangle \otimes |l, m_l\rangle$, which are the joint eigenvectors of $\hat{\boldsymbol{S}}^2, \hat{S}_z, \hat{\boldsymbol{L}}^2, \hat{L}_z$,

$$\begin{aligned} \hat{\boldsymbol{S}}^2|s, m_s, l, m_l\rangle &= s(s+1)|s, m_s, l, m_l\rangle, \\ \hat{S}_z|s, m_s, l, m_l\rangle &= m_s|s, m_s, l, m_l\rangle, \\ \hat{\boldsymbol{L}}^2|s, m_s, l, m_l\rangle &= l(l+1)|s, m_s, l, m_l\rangle, \\ \hat{L}_z|s, m_s, l, m_l\rangle &= l_s|s, m_s, l, m_l\rangle. \end{aligned} \tag{30}$$

where $m_s \in \{-s, -s+1, \cdots, s-1, s\}$, and $l_s \in \{-l, -l+1, \cdots, l-1, l\}$. Here we choose $s = l = j$.

The Floquet operator, that is the unitary evolution operator between consecutive kicks, can be written as

$$\hat{F} = \exp[-ia(\hat{S}_z + \hat{L}_z)] \exp[-i\frac{c}{j}\hat{S}_x\hat{L}_x]. \tag{31}$$

The classical counterpart can be obtained by taking the quasiclassical limit $\hbar = \frac{1}{j} \to 0$. Introducing the rescaled angular momenta $\hat{\mathcal{S}}_k = \frac{\hat{S}_k}{j}$ and $\hat{\mathcal{L}}_k = \frac{\hat{L}_k}{j}$, and considering the quasiclassical limit $j \to \infty$, yields the classical analog of the model,

$$\mathcal{H}_c = a(\mathcal{S}_z + \mathcal{L}_z) + c\mathcal{S}_x\mathcal{L}_x \sum_{n=-\infty}^{\infty} \delta(t - n), \tag{32}$$

where $\mathcal{S}_x^2 + \mathcal{S}_y^2 + \mathcal{S}_z^2 = \mathcal{L}_x^2 + \mathcal{L}_y^2 + \mathcal{L}_z^2 = 1$. Depending on the coupling strength the classical motion can be either chaotic or nearly-integrable, as shown by the three-dimensional Poincaré surfaces of sections of Figure 1.

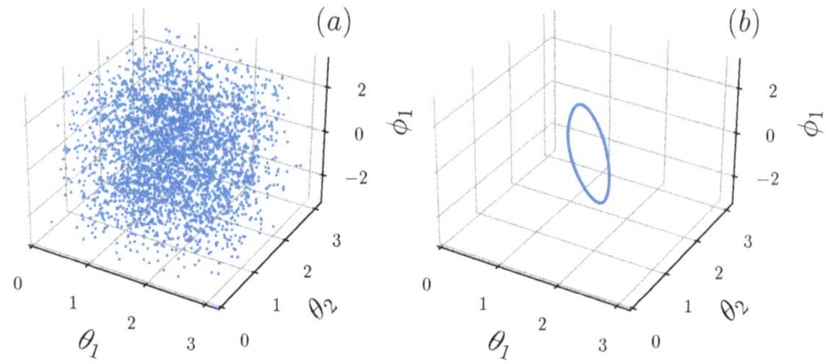

Figure 1. Three-dimensional Poincaré surface of section for the chaotic case (**a**): $a = 5, c = 3$ and the near-integrable case (**b**): $a = 5, c = 0.5$, where we fix $\phi_2 = 0$. Here we only consider a single trajectory starting from $(\theta_1, \phi_1, \theta_2, \phi_2) = (\frac{\pi}{4}, 0, \frac{3\pi}{4}, 0)$ (see text for the definition of the angles θ_k and $\phi_K, k = 1, 2$).

In the numerical simulations of both the quantum and classical cases, the linear entropy is averaged over N_p different initial states. In the quantum case, we consider the initial states $|\theta_1, \phi_1, \theta_2, \phi_2\rangle \equiv |\theta_1, \phi_1\rangle \otimes |\theta_2, \phi_2\rangle$, where $|\theta_1, \phi_1\rangle$ and $|\theta_2, \phi_2\rangle$ indicate the spin coherent state of the first rotator,

$$|\theta_1, \phi_1\rangle = e^{i\theta_1 \hat{S}_z} e^{i\phi_1 \hat{S}_y} |j, j\rangle, \tag{33}$$

and an analogous expression holds for the second rotor. Then, the quantum averaged linear entropy is calculated as follows,

$$\overline{S}_q(t) = \frac{1}{N_p} \sum_p \mathrm{Tr}((\hat{\rho}_1^p(t))^2), \tag{34}$$

where

$$\hat{\rho}_1^p(t) = \mathrm{Tr}_2(\hat{F}^t |\theta_1^p, \phi_1^p, \theta_2^p, \phi_2^p\rangle\langle\theta_1^p, \phi_1^p, \theta_2^p, \phi_2^p| (\hat{F}^\dagger)^t), \tag{35}$$

and $(\theta_1^p, \phi_1^p, \theta_2^p, \phi_2^p)$ are chosen randomly. In the classical case, we consider an initial ensemble of Gaussian states,

$$\rho_0(\theta_1', \phi_1', \theta_2', \phi_2') = A \exp\left(-\frac{(\theta_1' - \theta_1)^2}{\hbar_c} - \frac{\sin^2(\theta_1)(\phi_1' - \phi_1)^2}{\hbar_c}\right) \times \exp\left(-\frac{(\theta_2' - \theta_2)^2}{\hbar_c} - \frac{\sin^2(\theta_2)(\phi_2' - \phi_2)^2}{\hbar_c}\right), \tag{36}$$

which in case of $\hbar_c \to 0$ can be written in terms of canonical variables $(q_1, p_1, q_2, p_2) = (\phi_1, \cos\theta_1, \phi_2, \cos\theta_2)$ as

$$\rho_0(q_1', p_1', q_2', p_2') = A' \exp\left(-\frac{(q_1' - q_1)^2}{\hbar_c(1 - p_1^2)} - \frac{(1 - p_1^2)(p_1' - p_1)^2}{\hbar_c}\right) \times \exp\left(-\frac{(q_2' - q_2)^2}{\hbar_c(1 - p_2^2)} - \frac{(1 - p_2^2)(p_2' - p_2)^2}{\hbar_c}\right). \tag{37}$$

Here A and A' are normalization constants. Then the classical averaged linear entropy is calculated as

$$\overline{S}_{cl}(t) = \frac{1}{N_p} \sum_p S_{cl}(\rho_t^p), \tag{38}$$

where $S_{cl}(\rho_t^p)$ indicates the classical linear entropy (defined in Equation (11)), starting from the initial ensemble, centered at $(q_1^k, p_1^k, q_2^k, p_2^k)$. In our numerical simulations, we considered 10^7 trajectories for each initial ensemble, and the integral in Equation (11) is calculated by summing over the whole phase space with respect to system 1, which is divided into 4×10^6 phase cells.

Results for the chaotic regime are shown in Figure 2. Note that λ_2 is comparable to λ_1, and the behavior $S(t) - S(0) = (\lambda_1 + \lambda_2)t$ predicted in Equation (28) on the basis of a purely classical calculation, can be clearly seen both for quantum and classical simulations. The growth rate, in very good agreement with the KS entropy $\lambda_1 + \lambda_2$, is clearly distinguished from the growth rate given by the largest Lyapunov exponent λ_1 alone. Note that by increasing the coupling strength c the entanglement growth rate increases, in accordance with the increase of the classical KS entropy. Moreover, it can be clearly seen that the agreement between the classical and quantum linear entropy extends to longer times as $\hbar = \hbar_c$ is reduced.

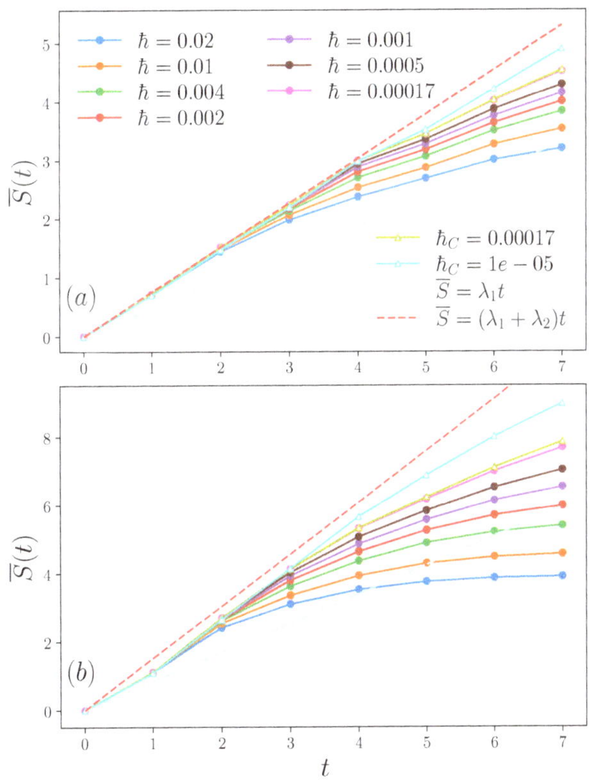

Figure 2. Quantum (circles with solid line) and classical (triangles with solid line) averaged linear entropy for different $\hbar$ and $\hbar_c$ in the kicked coupled tops model defined in Equation (29), for (**a**): $a = 5, c = 3$ and (**b**): $a = 5, c = 5$. The dashed lines indicate the functions $\overline{S} = (\lambda_1 + \lambda_2)t$ (red) and $\overline{S} = \lambda_1 t$ (light blue). The initial values of $S(t = 0)$ are subtracted.

In Figure 3, we show data in the regular regime with weaker coupling strength, for which invariant tori of the integrable model at $c = 0$ are deformed but survive. The volume occupied by the tori is the largest portion of the phase space and this affects the growth of the linear entropy, which is logarithmic rather than linear. Our numerical results show that, for large enough $\hbar$, the entropy $\overline{S}(t) \propto \log t^{\alpha}$, with $\alpha \approx 1$, while α slowly increases with reducing $\hbar$. Note that the separation between nearby trajectories increases linearly in time for integrable dynamics. Therefore, the number of cells of area $\hbar$ occupied in the two-dimensional phase-space for system 1 is proportional to t^2, leading to the expected growth $\overline{S}(t) \propto \log t^2$. We therefore conjecture that such growth would be achieved in the limit $\hbar \to 0$.

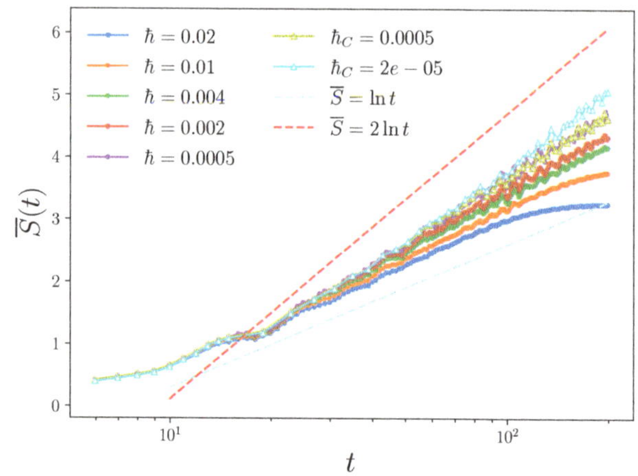

Figure 3. Same as in Figure 2, but for weaker coupling strength $c = 0.5$, for which motion is quasi-integrable. The lines $\overline{S}(t) \propto \log t$ and $\overline{S}(t) \propto \log t^2$ are also drawn.

4. Conclusions

We have shown that in the quasiclassical regime the entanglement growth rate is given by the Kolmogorov–Sinai entropy of the underlying classical dynamics. Note that we are considering initial separable coherent states, so that the quantum wave packet closely follows the underlying classical phase space distribution up to the Ehrenfest time, which diverges as the effective Planck constant $\hbar \to 0$. In spite of the lack of entanglement in classical mechanics, our results prove, in the quasiclassical regime, the close connection between entanglement generation and complexity of classical motion. Moreover, our derivation based on purely classical grounds provides an intuitive picture that could hardly be obtained on the basis of purely quantum calculations. Finally, the entanglement growth is linear in the classically chaotic and logarithmic in the regular regime, thus showing the entangling power of chaos.

Author Contributions: J.W. developed analytical calculations and performed numerical simulations. The work was supervised by G.B., with inputs from B.D. and D.R. All authors discussed the results and contributed to writing and revising the manuscript. All authors have read and agreed to the published version of the manuscript

Funding: J.W. is supported by the Deutsche Forschungsgemeinschaft (DFG) within the Research Unit FOR 2692 under Grant No. 397107022 (GE 1657/3-2)and No. 397067869 (STE 2243/3-2), B.D. and D.R. acknowledge support from the Institute for Basic Science in Korea (IBSR024-D1). G.B. acknowledges the financial support of the INFN through the project QUANTUM.

Institutional Review Board Statement: Not applicable.

Data Availability Statement: The data are contained within the article.

Conflicts of Interest: The authors declare no conflict of interest. The funders had no role in the design of the study; in the collection, analyses, or interpretation of data; in the writing of the manuscript, or in the decision to publish the results.

Appendix A. Derivation of Equation (24)

We write A_t in block form,

$$A_t = \begin{pmatrix} \hat{a} & \hat{b} \\ \hat{b}^T & \hat{d} \end{pmatrix}, \tag{A1}$$

where $\hat{a} = \hat{a}^T$, $\hat{d} = \hat{d}^T$, and $\hat{b}$ are 2×2 matrices. Furthermore, we introduce two-dimensional vectors $q_1^T = (x_1, x_2)$ and $q_2^T = (x_3, x_4)$. The matrix Equation (A1) can be brought to the form

$$\left(q_1^T, q_2^T\right)\begin{pmatrix} \hat{a} & \hat{b} \\ \hat{b}^T & \hat{d} \end{pmatrix}\begin{pmatrix} q_1 \\ \tilde{q}_2 \end{pmatrix} = \left(q_1{}^T, \tilde{q}_2^T\right)\begin{pmatrix} \hat{\hat{a}} & \hat{0} \\ \hat{0} & \hat{d} \end{pmatrix}\begin{pmatrix} q_1 \\ \tilde{q}_2 \end{pmatrix}, \tag{A2}$$

with

$$\tilde{q}_2 = \hat{Y} q_1 + q_2, \hat{Y} = \hat{d}^{-1}\hat{b}^T, \tag{A3}$$

and

$$\hat{\hat{a}} = \hat{a} - \hat{b}\hat{d}^{-1}\hat{b}^T. \tag{A4}$$

Then, Equation (20) becomes

$$\rho_t^1(q_1) = \frac{1}{(\pi\hbar_c)^2}\exp\left(-\frac{1}{\hbar_c} q_1^T \hat{\hat{a}} q_1\right)\int d\tilde{q}_2 \exp\left(-\frac{1}{\hbar_c}\tilde{q}_2^T \hat{d}\tilde{q}_2\right). \tag{A5}$$

The integrals can be performed after introducing the integration variable transformation $\tilde{q}_2 \to \hat{R}_2\tilde{q}_2$, with

$$\hat{d} = \hat{R}_2^T \Lambda_2 \hat{R}_2, \quad \Lambda_2 = \begin{pmatrix} d_1 & 0 \\ 0 & d_2 \end{pmatrix}, \tag{A6}$$

yielding

$$\rho_t^1(q_1) = \frac{1}{\pi\hbar_c}\frac{1}{\sqrt{\det\left(\hat{d}\right)}}\exp\left(-\frac{1}{\hbar_c} q_1^T \hat{\hat{a}} q_1\right). \tag{A7}$$

Using that $\det(A_t) = 1 = \det\left(\hat{d}\right)\det\left(\hat{\hat{a}}\right)$, one has

$$\int dq_1 \left(\rho_t^1(q_1)\right)^2 = \frac{\det(\hat{\hat{a}})}{(\pi\hbar_c)^2}\int dq_1 \exp\left(-\frac{2}{\hbar_c} q_1^T \hat{\hat{a}} q_1\right) = \frac{1}{2\pi\hbar_c}\sqrt{\det(\hat{\hat{a}})} = \frac{1}{2\pi\hbar_c}\frac{1}{\sqrt{\det(\hat{d})}}, \tag{A8}$$

which leads to Equation (24).

References

1. Dowling, J.P.; Milburn, G.J. Quantum technology: The second quantum revolution. *Phil. Trans. R. Soc. A* **2003**, *361*, 1655–1674. [CrossRef] [PubMed]
2. Wang, J.; Sciarrino, F.; Laing, A.; Thompson, M.G. Integrated photonic quantum technologies. *Nat. Photonics* **2020**, *14*, 273–284. [CrossRef]
3. Benenti, G.; Casati, G.; Rossini, D.; Strini, G. *Principles of Quantum Computation and Information (A Comprehensive Textbook)*; World Scientific Singapore: Singapore, 2019.
4. Cornfeld, I.P.; Fomin, S.V.; Sinai, Y.G. *Ergodic Theory*; Springer: New York, NY, USA, 1982. [CrossRef]
5. Gu, Y. Evidences of classical and quantum chaos in the time evolution of nonequilibrium ensembles. *Phys. Lett.* **1990**, *149*, 95–100. [CrossRef]
6. Ford, J.; Mantica, G.; Ristow, G.H. The Arnol'd cat: Failure of the correspondence principle. *Phys. Nonlinear Phenom.* **1991**, *50*, 493–520. [CrossRef]
7. Gu, Y.; Wang, J. Time evolution of coarse-grained entropy in classical and quantum motions of strongly chaotic systems. *Phys. Lett.* **1997**, *229*, 208–216. [CrossRef]
8. Pattanayak, A.K.; Brumer, P. Chaos and Lyapunov exponents in classical and quantal distribution dynamics. *Phys. Rev. E* **1997**, *56*, 5174–5177. [CrossRef]

9. Sokolov, V.V.; Zhirov, O.V.; Benenti, G.; Casati, G. Complexity of quantum states and reversibility of quantum motion. *Phys. Rev. E* **2008**, *78*, 046212. [CrossRef]
10. Benenti, G.; Casati, G. How complex is quantum motion? *Phys. Rev. E* **2009**, *79*, 025201. [CrossRef]
11. Balachandran, V.; Benenti, G.; Casati, G.; Gong, J. Phase-space characterization of complexity in quantum many-body dynamics. *Phys. Rev. E* **2010**, *82*, 046216. [CrossRef]
12. Prosen, T. Complexity and nonseparability of classical Liouvillian dynamics. *Phys. Rev. E* **2011**, *83*, 031124. [CrossRef]
13. Benenti, G.; Carlo, G.G.; Prosen, T. Wigner separability entropy and complexity of quantum dynamics. *Phys. Rev. E* **2012**, *85*, 051129. [CrossRef] [PubMed]
14. Qin, P.; Wang, W.G.; Benenti, G.; Casati, G. Complexity and instability of quantum motion near a quantum phase transition. *Phys. Rev. E* **2014**, *89*, 032120. [CrossRef] [PubMed]
15. Rozenbaum, E.B.; Ganeshan, S.; Galitski, V. Lyapunov Exponent and Out-of-Time-Ordered Correlator's Growth Rate in a Chaotic System. *Phys. Rev. Lett.* **2017**, *118*, 086801. [CrossRef]
16. Rammensee, J.; Urbina, J.D.; Richter, K. Many-Body Quantum Interference and the Saturation of Out-of-Time-Order Correlators. *Phys. Rev. Lett.* **2018**, *121*, 124101. [CrossRef] [PubMed]
17. García-Mata, I.; Saraceno, M.; Jalabert, R.A.; Roncaglia, A.J.; Wisniacki, D.A. Chaos Signatures in the Short and Long Time Behavior of the Out-of-Time Ordered Correlator. *Phys. Rev. Lett.* **2018**, *121*, 210601. [CrossRef]
18. Bergamasco, P.D.; Carlo, G.G.; Rivas, A.M.F. Out-of-time ordered correlators, complexity, and entropy in bipartite systems. *Phys. Rev. Res.* **2019**, *1*, 033044. [CrossRef]
19. Prakash, R.; Lakshminarayan, A. Scrambling in strongly chaotic weakly coupled bipartite systems: Universality beyond the Ehrenfest timescale. *Phys. Rev. B* **2020**, *101*, 121108. [CrossRef]
20. Wang, J.; Benenti, G.; Casati, G.; Wang, W.G. Complexity of quantum motion and quantum-classical correspondence: A phase-space approach. *Phys. Rev. Res.* **2020**, *2*, 043178. [CrossRef]
21. Bennett, C.H.; Bernstein, H.J.; Popescu, S.; Schumacher, B. Concentrating partial entanglement by local operations. *Phys. Rev. A* **1996**, *53*, 2046–2052. [CrossRef]
22. Lerose, A.; Pappalardi, S. Bridging entanglement dynamics and chaos in semiclassical systems. *Phys. Rev. A* **2020**, *102*, 032404. [CrossRef]
23. Bianchi, E.; Hackl, L.; Yokomizo, N. Linear growth of the entanglement entropy and the Kolmogorov-Sinai rate. *J. High Energy Phys.* **2018**, *2018*, 25. [CrossRef]
24. Miller, P.A.; Sarkar, S. Signatures of chaos in the entanglement of two coupled quantum kicked tops. *Phys. Rev. E* **1999**, *60*, 1542–1550. [CrossRef] [PubMed]
25. Fujisaki, H.; Miyadera, T.; Tanaka, A. Dynamical aspects of quantum entanglement for weakly coupled kicked tops. *Phys. Rev. E* **2003**, *67*, 066201. [CrossRef] [PubMed]
26. Jacquod, P. Semiclassical Time Evolution of the Reduced Density Matrix and Dynamically Assisted Generation of Entanglement for Bipartite Quantum Systems. *Phys. Rev. Lett.* **2004**, *92*, 150403. [CrossRef]
27. Petitjean, C.; Jacquod, P. Lyapunov Generation of Entanglement and the Correspondence Principle. *Phys. Rev. Lett.* **2006**, *97*, 194103. [CrossRef] [PubMed]
28. Husimi, K. Some formal properties of the density matrix. *Proc. Phys. Math. Soc. Jpn.* **1940**, *22*, 264.
29. Emerson, J.; Ballentine, L. Characteristics of quantum-classical correspondence for two interacting spins. *Phys. Rev. A* **2001**, *63*, 052103. [CrossRef]
30. Haake, F.; Gnutzmann, S.; Kuś, M. *Quantum Signatures of Chaos*; Springer: Heidelberg, Germany, 2018.

entropy

MDPI

Article

Implications of Spectral Interlacing for Quantum Graphs

Junjie Lu [1], Tobias Hofmann [2], Ulrich Kuhl [1,2,*] and Hans-Jürgen Stöckmann [2,*]

[1] Institut de Physique de Nice, CNRS, Université Côte d'Azur, 06108 Nice, France
[2] Fachbereich Physik, Philipps-Universität Marburg, 35032 Marburg, Germany
* Correspondence: ulrich.kuhl@univ-cotedazur.fr (U.K.); stoeckma@physik.uni-marburg.de (H.-J.S.)

Abstract: Quantum graphs are ideally suited to studying the spectral statistics of chaotic systems. Depending on the boundary conditions at the vertices, there are Neumann and Dirichlet graphs. The latter ones correspond to totally disassembled graphs with a spectrum being the superposition of the spectra of the individual bonds. According to the interlacing theorem, Neumann and Dirichlet eigenvalues on average alternate as a function of the wave number, with the consequence that the Neumann spectral statistics deviate from random matrix predictions. There is, e.g., a strict upper bound for the spacing of neighboring Neumann eigenvalues given by the number of bonds (in units of the mean level spacing). Here, we present analytic expressions for level spacing distribution and number variance for ensemble averaged spectra of Dirichlet graphs in dependence of the bond number, and compare them with numerical results. For a number of small Neumann graphs, numerical results for the same quantities are shown, and their deviations from random matrix predictions are discussed.

Keywords: quantum graphs; interlacing theorem; random matrix theory

1. Motivation

Quantum graphs are composed of bonds which are connected with each other at vertices. Along the bonds wave propagation is governed by the Schrödinger equation without potential and boundary conditions depending on the details of the coupling at the vertices. Quantum graphs were first introduced by Pauling [1] in the context of free electron models of organic molecules. Later, they were studied intensely in physics [2] and mathematics [3], and experimentally implemented in correspondingly-shaped microwave networks [4]. They are conceptually simple, but still complex, and there is a straightforward symbolic alphabet to classify the periodic orbits. Casati and coworkers [5] suggested that the universal features of the spectra of chaotic systems might be described by random matrix theory (RMT), which later was expressed by Bohigas, Giannoni, and Schmit [6] in the form of a conjecture. Using supersymmetry techniques, Gnutzmann and Altland [7] proved the conjecture for the two-point correlation function for fully connected graphs with incommensurate bond lengths. Their result was generalized to all correlation functions by Pluhař and Weidenmüller [8]. Just as for billiard systems [9], there is a one-to-one correspondence between a quantum graph and the corresponding microwave networks, which has been used, in particular, by Sirko and coworkers in numerous experiments to study spectral and scattering properties of microwave graphs (see e.g., Ref. [4]).

In a recent microwave experiment in tetrahedral graphs [10], however, we noticed that one important aspect is missing in the above scenario. It is hidden in the structure of the equation system determining the graph spectrum. Using energy and current conservation (the Kirchhoff rules in experimental networks), one arrives at a secular equation [2]

$$\sum_{m=1}^{V} h_{nm}\varphi_m = 0\,, \tag{1}$$

Citation: Lu, J.; Hofmann, T.; Kuhl, U.; Stöckmann, H.-J. Implications of Spectral Interlacing for Quantum Graphs. *Entropy* **2023**, *25*, 109. https://doi.org/10.3390/e25010109

Academic Editor: Marko Robnik

Received: 21 November 2022
Revised: 20 December 2022
Accepted: 27 December 2022
Published: 4 January 2023

where the sum runs over all vertices V, and φ_m is the potential at vertex m. In the experiment, the bonds are connected by ordinary T junctions corresponding to Neumann boundary conditions. For this situation, the elements of the secular matrix h are given by

$$h_{nm} = -\delta_{nm} \sum_{m'} \cot kl_{nm'} + \frac{1}{\sin kl_{nm}}, \tag{2}$$

where the l_{nm} are the lengths of the bonds connecting vertices n and m, and k is the wave number. For the homogeneous equation system (1) to have non-trivial solutions, the determinant of $h(k)$ has to vanish,

$$|h(k)| = 0. \tag{3}$$

The roots k_n of the equation generate the spectrum of the graph. It will be called "Neumann" spectrum in the following since the T junctions at all vertices obey Neumann boundary conditions. On the other hand, h_{nm} becomes singular, whenever kl_{nm} is an integer multiple of π. This situation corresponds to a totally disassembled graph with a spectrum being the sum of the spectra of all individual bonds with Dirichlet boundary conditions at both ends, thus the vertices have no influence any longer. This "Dirichlet" spectrum hence appears via the *poles* of $|h(k)|$, whereas the Neumann spectrum is given by the *zeros* of $|h(k)|$. In the following, all lengths will be assumed to be incommensurable to avoid degeneracies of the Dirichlet spectrum.

This "spectral duality", as we termed it in our previous publication [10], has important consequences for the spectral statistics. The cause is the interlacing theorem (see, e.g., Chapter 3.11 of Ref. [3]): *If the boundary conditions at one vertex of a graph are changed from Neumann to Dirichlet, or somewhere in between, the eigenvalues of the original and the new graph appear strictly alternating.*

To move from the Neumann to the Dirichlet spectrum for a complete graph, the boundary conditions have to be changed one after the other at *all* vertices, not just at one of them. Now, there is no longer a strict alternation in the sequence of the respective eigenvalues, but still a strong correlation remains—the maximum number of Neumann eigenvalues confined between two successive Dirichlet ones is given by the number of vertices V, and vice versa.

The mean density of states for a graph of a total length of l_{tot} is given by

$$\bar{\rho}(k) = \frac{l_{\text{tot}}}{\pi}. \tag{4}$$

For a graph with B bonds and a given total length, the maximum level spacing is found for the limiting case where all bonds are equal. For this case, the Dirichlet spectrum is B-fold degenerate, and the maximum distance between neighboring eigenvalues, in units of the mean level spacing, is just $s_{\text{max}} = B$. Due to the interlacing theorem, the same must be true for the Neumann resonances. There is hence a cut-off in the level spacing distribution $p(s)$ at $s_{\text{max}} = B$, at the latest, both for the Dirichlet and the Neumann spectrum. Consequences of spectral interlacing for the number variance have been discussed already in our previous paper [10].

Thus, there are clear deviations from the RMT expectation for small graphs. This is not in contradiction with the proofs mentioned in the beginning that the spectra of graphs with irrational length ratio *do* exhibit RMT behavior, since these proofs work in the limit of infinitely large graphs only. From the practical point of view this is of little help since numerical, as well as experimental, studies are necessarily restricted to comparatively small graphs.

Therefore, an understanding of the impact of Dirichlet–Neumann interlacing is mandatory for the correct interpretation of the spectral statistics in small graphs. Since, in the moment, a good idea to approach Neumann spectral statistics is still missing, we start with a more modest task—the interpretation of Dirichlet spectral statistics. Analytic results are given for level spacing distribution and number variance for a random superposition of lattice fence spectra and compared with numerical results. For the Neumann spectra, we restrict ourselves to an illustration of the fingerprints of spectral interlacing in level spacing distribution and number variance, but have to leave the theoretical interpretation to future papers. We do not discuss experimental results from microwave graphs in the present paper. This remark may be necessary since probably this is exactly what readers do expect from our group.

2. Dirichlet Graphs

For Dirichlet graphs, there are Dirichlet boundary conditions at each end of all bonds, thus the bonds are not coupled at the vertices and the spectrum corresponds to a superposition of B separated bonds. Here, we present analytical and numerical results of the spectral statistics for ensemble-averaged Dirichlet graphs. Following the usual practice, the mean density of states $\bar{\rho}(k) = l_{\text{tot}}/\pi$ was kept constant and normalized to one, meaning a total length of $l_{\text{tot}} = \pi$ for all graphs entering the average. For the numerics, the lengths had been created by generating $B-1$ random numbers r_n between 0 and π, and by taking the appearing B segments as lengths l_n. The procedure yields $p_B(l_1,\ldots,l_B) = \frac{1}{\pi^{B-1}}\delta(\sum l_n - \pi)$ for the joint length probability. The l_n are hence uniformly distributed on the interval 0 to π with the constraint $\sum l_n = \pi$. Integrating out all l_n but one obtains the distribution

$$p_B(l) = \frac{B-1}{\pi^{B-1}}(\pi - l)^{B-2} \tag{5}$$

for the remaining l, being constant only for $B = 2$. The derivation and a plot of the length distributions can be found in Appendix A.

Alternatively, one could think of taking B lengths l'_n from an interval between 0 and 1, and afterward, normalizing each length via $l_n = \pi l'_n / \sum_{n=1}^{B} l'_n$ to a mean density of one, i.e., $\sum l_n = \pi$. The resulting joint length probability is non-uniform, in contrast to the one above. For the sake of conciseness, we shall refer in the following to the two respective ensembles as the uniform and the non-uniform one. The non-uniform approach would be more in the spirit of the usual unfolding technique used in quantum chaos to make spectra taken from different systems comparable. For the non-uniform ensemble again, numerical length distributions are presented in Appendix A. Since it would be hard to obtain analytical results for the non-uniform ensemble, all analytics and numerics, if not explicitly stated differently, are for the uniform one.

In the next two subsections, theoretical expressions for nearest neighbor spacing distribution $p(s)$ and number variance Σ^2 are given and compared with numerical data.

2.1. Nearest Neighbor Spacing Distribution for Dirichlet Graphs

To calculate the distribution of nearest neighbors spacings $p(s)$ for the Dirichlet spectrum of a graph, we apply a strategy that had been used already by Berry and Robnik [11] to calculate $p(s)$ for an uncorrelated superposition of two spectra, one associated with the chaotic part, the other with the regular part of a mixed phase-space system. A key element in the calculation is the gap probability $e(s)$ describing the probability for a spectral range of length s to be empty of eigenvalues. The gap probability is related to the level spacing distribution via

$$p(s) = e''(s), \tag{6}$$

where a mean level spacing of one has been assumed. Expression (6) is well-known to those working in the field, but for readers not familiar with the subject, a didactic derivation is given in Appendix B. For a picket fence spectrum with a mean level spacing of $\Delta s = 1$, the gap probability is given by

$$e(s) = \begin{cases} 1-s, & \text{if } 0 \le s \le 1, \\ 0, & s > 1. \end{cases} \tag{7}$$

$e(s)$ is, in contrast to $p(s)$, multiplicative for superimposed uncorrelated spectra,

$$e(s) = \prod_n e_n(s)\,, \tag{8}$$

whence follows for the Dirichlet spectrum of a graph with B bonds of lengths l_n, $n = 1, \ldots, B$

$$e_B(k) = \prod_{n=1}^{B} e\left(\frac{l_n}{\pi}k\right). \tag{9}$$

From Equation (9) for $e_B(k)$, the Dirichlet level spacing distribution can now be obtained by taking the second derivative, see Equation (6). The Dirichlet level spacing distribution has already been calculated by Barra and Gaspard [12], who did not follow, however, the approach of Berry and Robnik [11]. Their derivation therefore is much less concise and considerably longer then the present one, see the appendix of [12].

Expression (9) has to be averaged over all different realizations of l_n with the constraint that the total length $l_{\rm tot}$ is constant

$$\sum_{n=1}^{B} l_n = l_{\rm tot}\,. \tag{10}$$

with the substitution $s_n = \frac{l_n}{\pi}k$, the constraint becomes

$$s = \sum_{n=1}^{B} s_n = \frac{l_{\rm tot}}{\pi}k\,. \tag{11}$$

Thus, s is the wave number in units of the mean level spacing $\pi/l_{\rm tot}$. In the following we shall use the letter s exclusively for spectra with a mean level spacing of $\Delta s = 1$. Now the average can written as

$$\begin{aligned}
\langle e_B(s)\rangle &= \frac{(B-1)!}{s^{B-1}}\int_0^s ds_1 e(s_1)\int_0^{s-s_1} ds_2 e(s_2)\cdots\int_0^{s-s_1-\cdots-s_{B-2}} ds_{B-1} e(s_{B-1}) \\
&\quad \cdot e(s - s_{B-1} - \cdots - s_1) \\
&= \frac{(B-1)!}{s^{B-1}} w_B(s)\,,
\end{aligned} \tag{12}$$

where $w_B(s)$ is given by

$$\begin{aligned}
w_B(s) &= \int_0^s ds_1 e(s_1)\int_0^{s-s_1} ds_2 e(s_2)\cdots \\
&= \int_0^s ds_1 e(s_1) w_{B-1}(s - s_1)\,, \text{ with } w_1(s) = e(s)\,.
\end{aligned} \tag{13}$$

The factorial in Equation (12) reflects the number of possible l sequences to do the average. Equation (13) can be used to calculate $w_B(s)$ iteratively. For $B = 2$, e.g., one obtains

$$w_2(s) = \begin{cases}
\int\limits_0^s ds_1 e(s_1) e(s - s_1) = s - s^2 + \frac{1}{6}s^3\,, & s < 1\,, \\
\int\limits_{s-1}^1 ds_1 e(s_1) e(s - s_1) = \frac{4}{3} - 2s + s^2 - \frac{1}{6}s^3\,, & 1 < s < 2\,, \\
0\,, & s > 2\,,
\end{cases} \tag{14}$$

where the limits of integration in the different s windows take care of the cut-off of $e(s)$. With help of Equations (12) and (6), we now obtain for the level spacing distribution

$$p_2(s) = \begin{cases} \frac{1}{3}, & 0 < s < 1, \\ \frac{8 - x^3}{3x^3}, & 1 < s < 2, \\ 0, & s > 2. \end{cases} \quad (15)$$

In this way, the $p_B(s)$ may be obtained iteratively, resulting in formulas with a complexity increasing step by step.

A more direct approach takes advantage of the fact that the integral in Equation (13) is nothing but a convolution. In such a situation, Laplace transform techniques are the method of choice. Applying a Laplace transform to Equation (13), the convolution theorem yields

$$\hat{w}_B(\lambda) = \hat{e}(\lambda)\hat{w}_{B-1}(\lambda), \quad (16)$$

where

$$\hat{w}_B(\lambda) = \mathcal{L}[w_B(s)] = \int_0^\infty w_B(s)e^{-\lambda s}\mathrm{d}s \quad (17)$$

and

$$\hat{e}(\lambda) = \mathcal{L}[e(s)] = \int_0^\infty e(x)e^{-\lambda x}\mathrm{d}x = \frac{1}{\lambda^2}\left[e^{-\lambda} - 1 + \lambda\right] \quad (18)$$

are the Laplace transforms of $w(s)$, and $e(s)$, respectively. Iterating Equation (16), one gets

$$\hat{w}_B(\lambda) = [\hat{e}(\lambda)]^B, \quad (19)$$

whence $w_B(s)$ is obtained via an inverse Laplace transform

$$w_B(x) = \mathcal{L}^{-1}(\hat{w}_B(\lambda)). \quad (20)$$

The inverse Laplace transform can be done with the result

$$w_B(s) = \begin{cases} 0, & s > B, \\ \sum_{m=0}^{\lfloor s \rfloor} \sum_{l=m}^{B} c_{ml}^B (-1)^{l-m}(s-m)^{l+B-1}, & s < B, \end{cases} \quad (21)$$

where $\lfloor s \rfloor$ denotes the largest integer $\leq s$ and

$$c_{ml}^B = \frac{B!}{m!(l-m)!(B-l)!(B+l-1)!}. \quad (22)$$

Further details can be found in Appendix C.

To verify our results, we compare the analytical results with numerical simulations. In Figure 1, the histograms for $B = 2$ to 6 and 100 are shown together with the corresponding theoretical predictions in a linear, and in Figure 2 in a logarithmic scale. All distributions show the expected cut-off at $s_{\max} = B$. There is a perfect agreement between numerics and theory. Note the discontinuity for $B = 2$ at $s = 1$, which for larger B is smoothed out and vanishes for $B \to \infty$, where an an exponential decay is expected, corresponding to a Poisson distribution. This can be seen in Figures 1f and 2f, showing the results for $B = 100$. There are still deviations from the exponential behavior as can be seen in the inset of Figure 2f, showing the same results over a larger s range. Still the analytic solution matches better. In Appendix D, numerical findings are presented for the non-uniform ensemble.

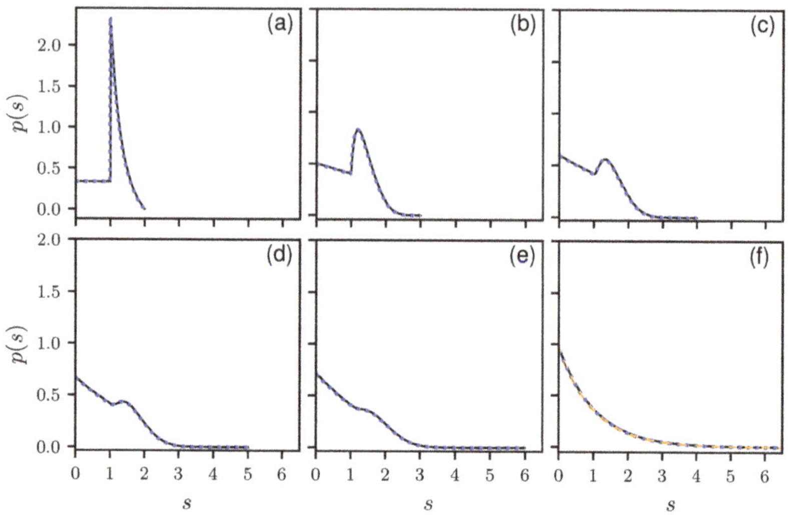

Figure 1. The distribution of nearest neighbor spacings of the Dirichlet graphs for different numbers of bonds $B = 2$ (**a**), 3 (**b**), 4 (**c**) , 5 (**d**), 6 (**e**) , 100 (**f**) in linear scale. The solid lines correspond to numerical simulations taking into account 10^9 realization, each of them containing about 900 spacings. The blue dotted lines corresponds to the theoretical prediction, Equations (6) and (12). In (**f**), the orange dashed line corresponds to a Poisson distribution, i.e., an exponential.

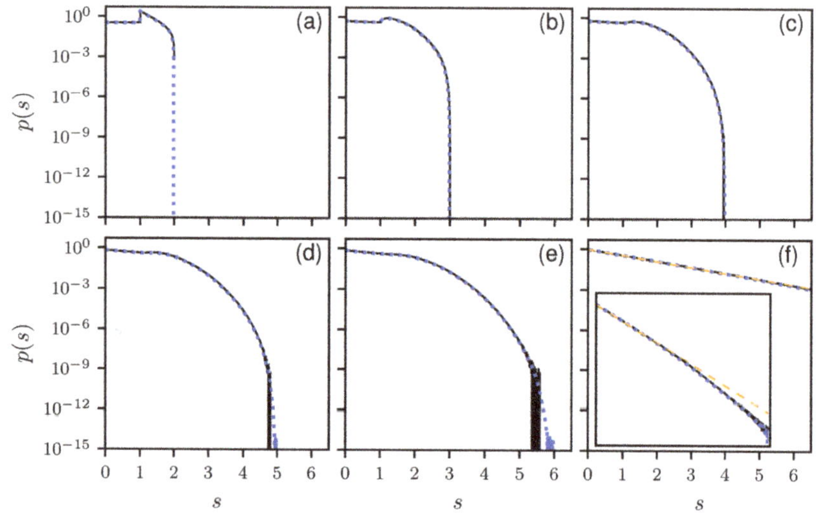

Figure 2. The same as Figure 1 but in logarithm scale. In the inset in (**f**), the abscissa ranges from 0 to 16, and the ordinate from 10^{-9} to 5.

2.2. Number Variance for Dirichlet Graphs

The number variance, defined as

$$\Sigma^2(s) = \langle n^2 \rangle - (\langle n \rangle)^2, \tag{23}$$

where n is the number of eigenvalues in an interval of length s, yields for the lattice fence spectrum of a single bond of length l

$$\Sigma^2(s) = \{s\}[1 - \{s\}] \,, \text{with } s = \frac{kl}{\pi} \text{ and } \{s\} = s - \lfloor s \rfloor \,. \tag{24}$$

It is convenient to express $\Sigma^2(s)$ in terms of its Fourier transform,

$$\Sigma^2(s) = \frac{1}{6} - \frac{1}{\pi^2} \sum_{m=1}^{\infty} \frac{\cos(2\pi m s)}{m^2} \,. \tag{25}$$

For B bonds with independent bond lengths l_n, the spectrum is just the superposition of the B spectra with bond lengths $l_1, l_2, \ldots, l_B$. The number variation $\Sigma^2(s)$ is additive for uncorrelated spectra leading to

$$\langle \Sigma^2(s) \rangle_l = B \left\{ \frac{1}{6} - \frac{1}{\pi^2} \sum_{m=1}^{\infty} \frac{\langle \cos(2\pi m s_k) \rangle_l}{m^2} \right\} \tag{26}$$

for the ensemble averaged number variance, where $s_k = \frac{k l_k}{\pi}$, and $\langle \ldots \rangle_l$ means the average over all l_n with the constraint $\sum l_k = l_{\text{tot}}$, i.e.,

$$\langle \cos(\alpha l_k) \rangle_l = \int_0^{\pi} dl \; p_B(l) \cos(\alpha l) \,, \tag{27}$$

with $\alpha = 2mk$ and $p_B(l)$ given by Equation (5).

In Figure 3, the ensemble averaged number variance for Dirichlet graphs are shown for a number of different bonds. For a single bond ($B = 1$), there is just one lattice fence spectrum with a spacing of one. Hence, one observes a periodic modulation with an average value of 1/6, as described by Equation (25). With increasing B, these oscillations are damped out more and more, until $\Sigma^2(s)$ eventually approaches the linear increase expected for a Poissonian ensemble. A good agreement between the simulations and the analytical predictions given by Equation (27) is found.

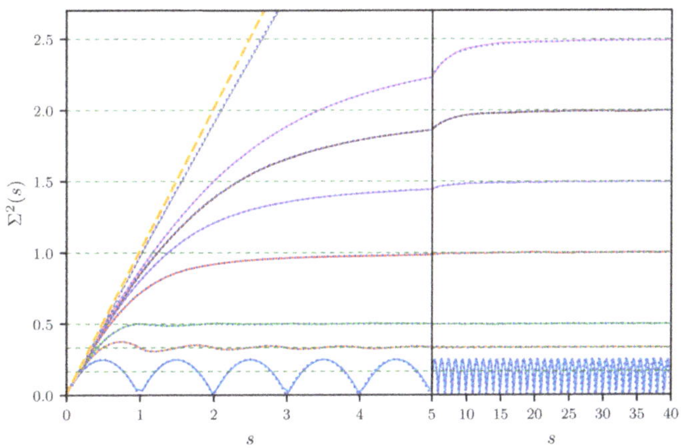

Figure 3. Ensemble averaged number variance $\Sigma^2(s)$ for the Dirichlet graphs with $B = 1, 2, 3, 6, 9, 12, 15, 100$ bonds. The solid lines correspond to the numerical simulations and the blue dotted lines to the analytical result given by Equation (26). The horizontal green dashed lines mark the limit $\Sigma^2(s) \to B/6$ for $s \to \infty$. The straight orange dashed line represents the number variance for integrable systems given by $\Sigma^2(s) = s$. Note the change of the abscissa scale at $s = 5$.

3. Neumann Graphs

Here, we present the results of the Neumann graphs shown in Table 1. We restricted ourselves to graphs where bonds are connected at least to two other bonds and where there are no disconnected parts. In addition the verticity V_B, the number of bonds that connect at a vertex, has been assumed to be the same for all vertices. The smallest graph of interest is the tetrahedron which has been used repeatedly for RMT studies in numerics [13], as well as in experiments [4,10].

Table 1. The investigated graphs for Neumann boundary conditions at the vertices. The lower part of the table shows the results of the numerics: (i) s_c is the s value, where $p(s)$ drops below 10^{-5}, i. e., $p(s_c) = 10^{-5}$, (ii) s_m is the s value, where $\Sigma^2(s)$ takes its maximal value, and (iii) Σ^2_{sat} is the limit of $\Sigma^2(s)$ for $s \to \infty$, obtained by taking the average of $\Sigma^2(s)$ in the range of s between 10 and 20.

Name	Tetrahedron	f. c. with $V = 5$	Octahedron	Hexahedron
	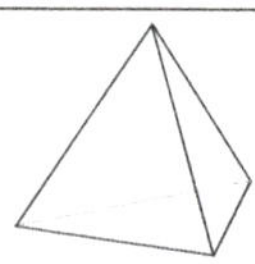	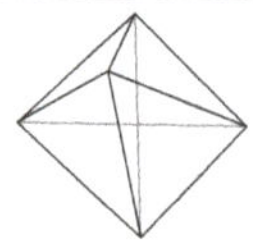	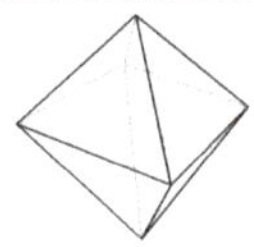	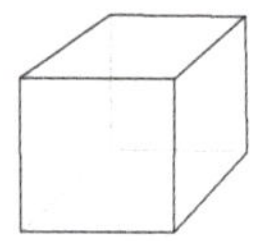
#Bonds B	6	10	12	12
#Vertices V	4	5	6	8
Valency of vertices	3	4	4	3
Fully connected (f. c.)	Yes	Yes	No	No
s_c	3.19	3.75	3.96	3.56
s_m	1.64	2.56	2.88	2.08
Σ^2_{sat}	0.40	0.48	0.51	0.44

3.1. Nearest Neighbor Spacing Distribution for Neumann Graphs

Figure 4 shows level spacing distributions for the graphs presented in Table 1. In addition, the exact RMT nearest neighbor distribution is shown [14,15]. Whereas the linear plot suggests a reasonable agreement with the RMT prediction, in the logarithmic plot for all four graphs, a suppression for large values becomes obvious, with the strongest suppression for the tetrahedron having the smallest number of bonds. This is in accordance with the expectation; due to the interlacing theorem, the largest possible distance is given by the number of bonds, $B = 6$ for the tetrahedron, and 10 or 12 for the other graphs. The decay, however, does not increase monotonously with B—it is faster for the hexahedron than for the octahedron, though the number of bonds is the same, and the decay for the fully connected five-vertices graph is as fast as for the octahedron, though the number of bonds is not the same.

Similar deviations of $p(s)$ from RMT have also been observed by Barra and Gaspard [12] but not discussed in detail.

To quantify these findings, we determined s_c, the value where $p(s)$ drops below 10^{-5}, i.e., $p(s_c) = 10^{-5}$, close to the limit of our statistical precision. In Figure 4, this s value reflects the point where $p(s)$ crosses the abscissa. The extracted values are presented in Table 1. Regrettably, it is impossible to fix the cut-off point by the numerics, which should be, e.g., $s_{\max} = 6$ for the tetrahedron. This would need more than 10^{11} spacings, by far beyond our computer resources.

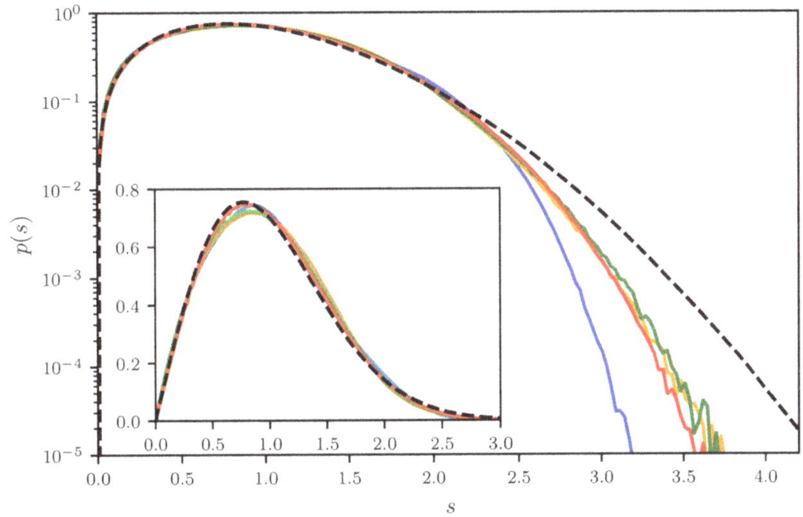

Figure 4. Level spacing distributions for the graphs shown in Table 1 in a logarithmic scale, with the tetrahedron (blue), the fully connected graph with 5 vertices (orange), the octahedron (green), and the hexahedron (red). The plots were generated by superimposing the results from $52.2 \cdot 10^6$, $14.6 \cdot 10^6$, $6.7 \cdot 10^6$, $35.0 \cdot 10^6$ spacings. The inset shows the same data in a linear scale. In addition, the exact RMT distribution is plotted (dashed black).

3.2. Number Variance for Neumann Graphs

In Figure 5, the ensemble-averaged number variances for the graphs shown in Table 1 are plotted, exhibiting a saturation at about $s = 2$ in contrast to the behavior predicted by RMT. Similar deviations from RMT predictions for the number variance in graphs have been reported in Refs. [16,17], and for the spectral rigidity in Ref. [4]. For non-experts, we mention that the spectral rigidity may be looked upon as a smoothed version of the number variance (the exact definition is technical and not of relevance here [14]). Already in 1985, Casati and coworkers [18] discovered a saturation of the spectral rigidity in the spectra of rectangular billiards, which could be traced back by Berry to the influence of the shortest periodic orbit [19]. In the present case this can not be the explanation. From the shortest periodic orbit, there should be a saturation of $\Sigma^2(s)$ at about $s_{\text{sat}} = \pi/l_{\text{min}} = l_{\text{tot}}/l_{\text{min}}$. Since necessarily $l_{\text{min}} \leq l_{\text{tot}}/B$, periodic orbit theory predicts a saturation of the number variance not until $s_{\text{sat}} \geq B$, whereas actually for all graphs the saturation is observed much earlier at about $s = 2$. In the lower part of Table 1, the saturation values are given, as well as s_m corresponding to the s value where $\Sigma^2(s)$ is maximal. s_m is a convenient tool to quantify the point of cross-over from a linear increase of $\Sigma^2(s)$ for small s values to a saturation for $s \to \infty$. There is a clear correlation between s_c, s_m, and Σ^2_{sat}, collected in the lower part of Table 1. From the interlacing theorem one would expect a correlation of these quantities with the bond number, which is observed for the first three graphs presented in the table, the tetrahedron, the totally connected graphs with five vertices, and the octahedron, but the hexahedron does not fit into the sequence. In fact, there is a stronger correlation with the vertex valency, the number of bonds meeting at a vertex. Obviously, the interlacing theorem alone is not sufficient to describe all these features.

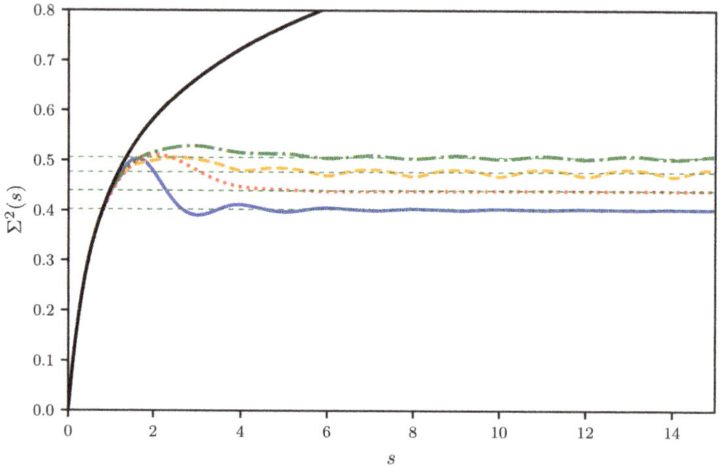

Figure 5. Number variance $\Sigma^2(s)$ for the tetrahedron (solid blue), the fully connected graph with 5 vertices (dashed orange), the octahedron (dashed dotted green), and the hexahedron (dotted red). The solid black line correspond to the RMT prediction, the horizontal thin dashed green lines mark the limiting values obtained from an average of $\Sigma^2(s)$ over the range $s = 10$ to 20.

A quantitative explanation, in particular, of the saturation values, has to be postponed to further studies, but there is a qualitative explanation. Semiclassical theory relates RMT to periodic orbits [20]. Essential ingredients are correlations between orbits and their time-reversed partners, in case there is time-reversal symmetry [19], and between various types of self-intersecting orbits with their non-intersecting partners [21,22]. Apart from this, all orbit lengths are assumed as uncorrelated. This assumption is severely violated in graphs, where all orbits are composed from a finite number of elements.

4. Conclusions

The implications of spectral interlacing in quantum graphs have been discussed. For Dirichlet graphs, explicit analytic expressions have been obtained for level spacing distribution and number variance, and compared with numerical results. For Neumann graphs, numerical results for the same quantities have been presented, showing clear deviations from RMT predictions due to spectral interlacing. For Neumann graphs, an analytic description of these features is still missing and has to be left to future work.

Author Contributions: U.K. and H.-J.S. developed concept and theory, J.L. and T.H. performed the numerics and analyzed the data. All authors participated in the discussions and in writing the paper. All authors have read and agreed to the published version of the manuscript.

Funding: This research received no external funding.

Data Availability Statement: The data supporting the findings of this study are available from the corresponding authors upon request.

Acknowledgments: J.L. acknowledges financial support from the China Scholarship Council via Grant No. 202006180008.

Conflicts of Interest: The authors declare no conflict of interest.

Abbreviations

The following abbreviation is used in this manuscript:

RMT	Random Matrix Theory

Appendix A. Length Distributions

In most of our numerical studies, the lengths were created by generating $B-1$ random numbers r_n between 0 and π, and taking the appearing B segments as lengths l_i, resulting in the uniform ensemble introduced in Section 2. This results in a joint probability

$$p(l_1,\ldots,l_B) = \langle \delta(l_1 - r_1)\delta(l_2 - r_2 + r_1)\cdots\delta(l_B - \pi + r_{B-1})\rangle_r \quad \text{(A1)}$$

for the lengths $l_1,\ldots,l_B$. The brackets denote an average over the r_i,

$$\langle \ldots \rangle_r = \prod_{i=0}^{B-1}\left(\frac{1}{\pi}\int_0^{\pi} dr_i\right)\ldots \quad \text{(A2)}$$

with the constraint $0 \le r_1 \le \cdots \le r_{B-1} \le \pi$. Due to the delta functions, the integrations are done easily step-by-step with the result

$$p(l_1,\ldots,l_B) = \frac{1}{\pi^{B-1}}\delta\left(\pi - \sum_{i=0}^{B} l_i\right). \quad \text{(A3)}$$

The l_n are thus equally distributed over the interval 0 to π with the constraint $\sum l_n = \pi$. Integrating over all l_n but one, the distribution for a single length l is obtained,

$$p_B(l) = \frac{(B-1)!}{\pi^{B-1}}\int_0^{\pi-l} dl_1 \int_0^{\pi-l-l_1} dl_2 \cdots \int_0^{\pi-l-l_1\cdots-l_{B-2}} dl_{B-1}\,\delta\left(\pi - l - \sum_{i=0}^{B-1} l_i\right), \quad \text{(A4)}$$

where the factorial takes account of the number of possible sequences to do the integrations. The integrations can be performed iteratively, the result is Equation (5). Figure A1 shows numerically obtained length distributions together with the theoretical curves from Equation (5). A perfect agreement is found.

Figure A2 shows the same for the non-uniform ensemble but without the analytic. There are clear differences to the uniform ensemble. In particular, for $B = 2$, there is a uniform length distribution for the first ensemble, whereas for the second one, there is a cusp in the distribution for $l = \pi/2$. We did not try to calculate analytic expressions for the length distribution for the latter case, they are not needed in the present context.

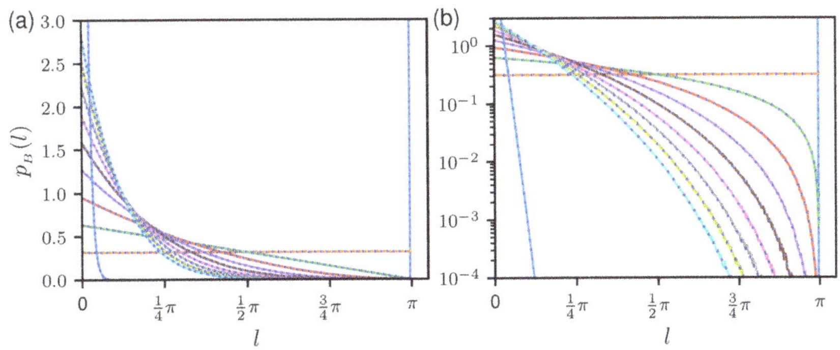

Figure A1. Distribution of lengths for the uniform ensemble in a linear (**a**) and a logarithmic (**b**) scale, for $B = 1,\ldots 10$, and 100 numbers of bonds, using 10^7 realizations, shown as colored solid lines. The blue dotted lines correspond to the analytic expression (5). With increasing B the distribution concentrates more and more at l values close to 0.

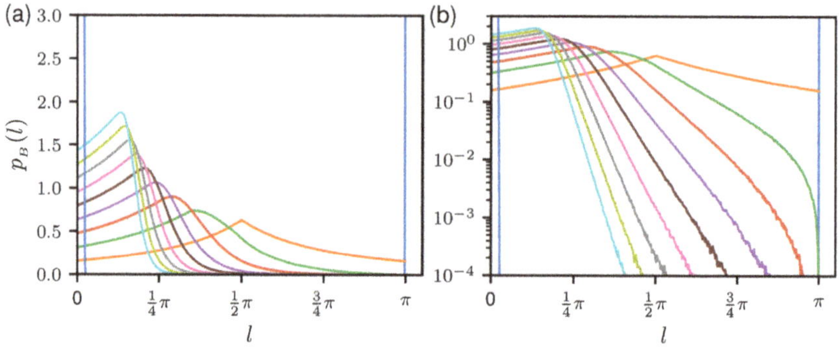

Figure A2. The same as in Figure A1, but for the non-uniform ensemble, and without analytical expression.

Appendix B. Derivation of Equation (6)

The relation between gap probability $e(k)$ and level spacing distribution $p(k)$ is obtained as follows (see, e.g., Section 3.2.2 of Ref. [23]):

The difference $(e(k) - e(k + \Delta k))$ may be interpreted as

$$e(k) - e(k + \Delta k) = h(k)\rho\Delta k\,. \tag{A5}$$

Here, $h(k)$ is the one-sided gap probability, namely the probability for an interval of length s to be empty of eigenvalue, and an eigenvalue at one end of the interval, let us assume the lower one. $\rho\Delta k$ is the probability to have an eigenvalue in the interval Δs, where ρ is the mean density of states. It follows

$$h(k) = -\frac{1}{\rho}e'(k)\,. \tag{A6}$$

Furthermore, $[h(k) - h(k + \Delta k)]$ may be interpreted as the probability to have one eigenvalue at the lower end and the next nearest one between k and $k + \Delta k$, i.e.,

$$h(k) - h(k + \Delta k) = p(k)\Delta k\,, \tag{A7}$$

whence follows

$$p(k) = -h'(k)\,, \tag{A8}$$

where $p(k)$ is the probability density for an interval to be empty of eigenvalues, and an eigenvalue on each side, i.e., the nearest neighbor spacing distribution.

Combining Equations (A6) and (A8), one obtains the wanted relation between $e(k)$ and $p(k)$,

$$p(k) = \frac{1}{\rho}e''(k)\,, \tag{A9}$$

reducing to Equation (6) for a spectrum with a mean density $\rho = 1$. For a graphical illustration see Figure A3.

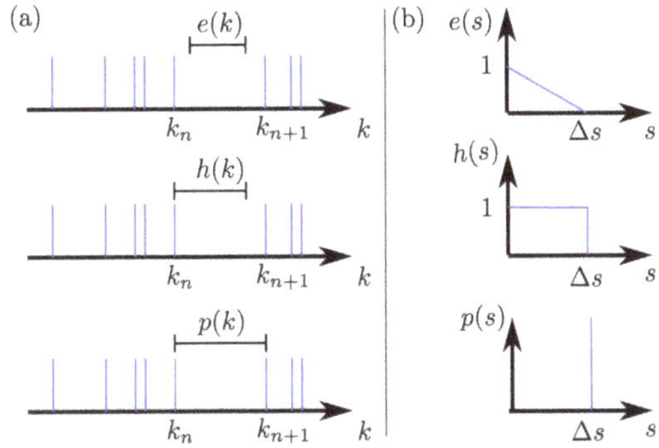

Figure A3. (**a**) Graphical illustration of $e(k)$, $h(k)$, $p(k)$ (from top to bottom). (**b**) $e(s)$, $h(s)$, $p(s)$ for a lattice fence spectrum with spacing $\Delta s = 1$.

Appendix C. Calculation of the Inverse Laplace Transform (20)

Using the standard formula for the inverse Laplace transform Equation (20) yields

$$w_B(x) = \frac{1}{2\pi i}\int_{-i\infty+a}^{i\infty+a} d\lambda \frac{1}{\lambda^{2B}}\left(e^{-\lambda} - 1 + \lambda\right)^B e^{\lambda x} \quad \text{(A10)}$$
$$= \frac{1}{2\pi i}\sum_{k=0}^{B}\binom{B}{k}\int_{-i\infty+a}^{i\infty+a} \frac{d\lambda}{\lambda^{2B}} e^{\lambda(x-k)}(\lambda - 1)^{B-k}.$$

The path of integration has to be performed to the right of all singularities of the integrand. The only possible singularity is at $\lambda = 0$, hence a has to be positive real. The path of integration can be closed at infinity by a large semi-circle either to the left, for $x > k$, or to the right, for $x < k$.

For $x < k$, the integration loop encloses no singularities, i.e., $\oint d\lambda \cdots = 0$, for $x > k$ the integration can be performed by means of the residuum method resulting in

$$w_B(x) = \sum_{k=0}^{k_{\max}}\binom{B}{k}\frac{1}{(2B-1)!}\frac{\partial^{2B-1}}{\partial\lambda^{2B-1}}\left[e^{\lambda(x-k)}(\lambda - 1)^{B-k}\right]\bigg|_{\lambda=0}, \quad \text{(A11)}$$

where $k_{\max} = \mathrm{Min}(B, \lfloor x \rfloor)$. For $x > B$, Equation (A11) yields

$$w_B(x) = \frac{1}{(2B-1)!}\frac{\partial^{2B-1}}{\partial\lambda^{2B-1}}\sum_{k=0}^{B}\binom{B}{k}\left[e^{\lambda(x-k)}(\lambda - 1)^{B-k}\right]\bigg|_{\lambda=0} \quad \text{(A12)}$$
$$= \frac{1}{(2B-1)!}\frac{\partial^{2B-1}}{\partial\lambda^{2B-1}}\left[\left(e^{-\lambda} - 1 + \lambda\right)^B e^{\lambda x}\right]\bigg|_{\lambda=0} = 0,$$

since $(e^{-\lambda} - 1 + \lambda) = \mathcal{O}(\lambda^2)$. For $x < B$, the sum in Equation (A11) runs from 0 to $\lfloor x \rfloor$, and after a number of elementary steps one arrives at Equation (21).

Appendix D. Dirichlet Level Spacing Distributions for the Non-Uniform Ensemble

To illustrate the difference between the two ensembles introduced in Section 2, we present here numerical results for the level spacing distribution for the non-uniform ensembles. Figures A4 and A5 show the results in linear and logarithmic scale. In addition, the analytical curves for the uniform ensemble are plotted. Not surprisingly, there are deviations in detail, but the qualitative behavior, in particular, the discontinuities for integer values of s, and the cut-off at $s = B$ is the same in both cases. Note that the non-uniform ensemble approaches the Poisson limit faster than the uniform one, see inset of Figure A5f.

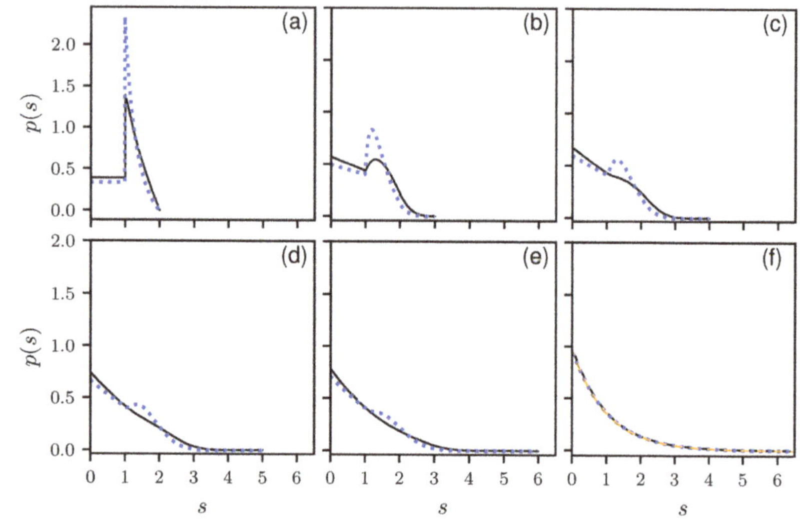

Figure A4. Level spacing distribution for the non-uniform ensemble in linear scale for different number of bonds $B = 2$ (**a**), 3 (**b**), 4 (**c**) , 5 (**d**), 6 (**e**) , 100 (**f**). The blue dotted lines correspond to the analytic results for the uniform ensemble.

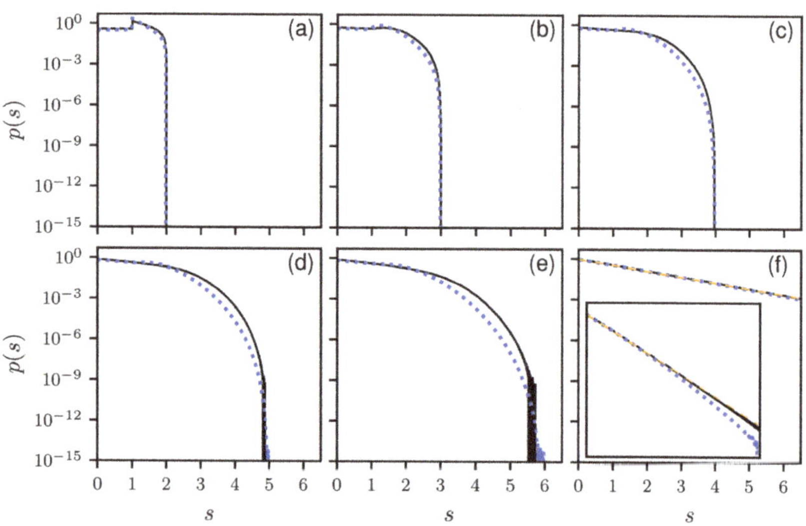

Figure A5. The same as Figure A4, but in a logarithmic scale. In the inset in (**f**), the abscissa ranges from 0 to 16, and the ordinate from 10^{-9} to 5.

References

1. Pauling, L. The Diamagnetic Anisotropy of Aromatic Molecules. *J. Chem. Phys.* **1936**, *4*, 673. [CrossRef]
2. Kottos, T.; Smilansky, U. Periodic orbit theory and spectral statistics for quantum graphs. *Ann. Phys.* **1999**, *274*, 76. [CrossRef]
3. Berkolaiko, G.; Kuchment, P. Introduction to Quantum Graphs. In *Mathematical Surveys and Monographs*; American Mathematical Society: Providence, RI, USA, 2013; Volume 186. [CrossRef]
4. Hul, O.; Bauch, S.; Pakoński, P.; Savytskyy, N.; Życzkowski, K.; Sirko, L. Experimental simulation of quantum graphs by microwave networks. *Phys. Rev. E* **2004**, *69*, 056205. [CrossRef] [PubMed]
5. Casati, G.; Valz-Gris, F.; Guarnieri, I. On the connection between quantization of nonintegrable systems and statistical theory of spectra. *Lett. Nuov. Cim.* **1980**, *28*, 279. [CrossRef]

6. Bohigas, O.; Giannoni, M.J.; Schmit, C. Characterization of chaotic spectra and universality of level fluctuation laws. *Phys. Rev. Lett.* **1984**, *52*, 1. [CrossRef]
7. Gnutzmann, S.; Altland, A. Universal Spectral Statistics in Quantum Graphs. *Phys. Rev. Lett.* **2004**, *93*, 194101. [CrossRef] [PubMed]
8. Pluhař, Z.; Weidenmüller, H.A. Universal quantum graphs. *Phys. Rev. Lett.* **2014**, *112*, 144102.[CrossRef] [PubMed]
9. Stöckmann, H.J.; Stein, J. "Quantum" chaos in billiards studied by microwave absorption. *Phys. Rev. Lett.* **1990**, *64*, 2215. [CrossRef] [PubMed]
10. Hofmann, T.; Lu, J.; Kuhl, U.; Stöckmann, H.J. Spectral duality in graphs and microwave networks. *Phys. Rev. E* **2021**, *104*, 045211. [CrossRef] [PubMed]
11. Berry, M.V.; Robnik, M. Semiclassical level spacings when regular and chaotic orbits coexist. *J. Phys. A* **1984**, *17*, 2413. [CrossRef]
12. Barra, F.; Gaspard, P. On the Level Spacing Distribution in Quantum Graphs. *J. Stat. Phys.* **2000**, *101*, 283. [CrossRef]
13. Kottos, T.; Smilansky, U. Quantum chaos on graphs. *Phys. Rev. Lett.* **1997**, *79*, 4794. [CrossRef]
14. Mehta, M.L. *Random Matrices (Revised and Enlarged Second Edition)*; Academic Press: San Diego, CA, USA, 1991. [CrossRef]
15. Dietz, B.; Haake, F. Taylor and Padé analysis of the level spacing distributions of random-matrix ensembles. *Z. Phys. B* **1990**, *80*, 153. [CrossRef]
16. Dietz, B.; Yunko, V.; Białous, M.; Bauch, S.; Ławniczak, M.; Sirko, L. Nonuniversality in the spectral properties of time-reversal-invariant microwave networks and quantum graphs. *Phys. Rev. E* **2017**, *95*, 052202. [CrossRef] [PubMed]
17. Lu, J.; Che, J.; Zhang, X.; Dietz, B. Experimental and numerical investigation of parametric spectral properties of quantum graphs with unitary or symplectic symmetry. *Phys. Rev. E* **2020**, *102*, 022309. [CrossRef] [PubMed]
18. Casati, G.; Chirikov, B.V.; Guarneri, I. Energy-level statistics of integrable quantum systems. *Phys. Rev. Lett.* **1985**, *54*, 1350. [CrossRef] [PubMed]
19. Berry, M.V. Semiclassical theory of spectral rigidity. *Proc. R. Soc. Lond. A* **1985**, *400*, 229. [CrossRef]
20. Gutzwiller, M.C. Periodic orbits and classical quantization conditions. *J. Math. Phys.* **1971**, *12*, 343. [CrossRef]
21. Sieber, M.; Richter, K. Correlations between Periodic Orbits and their Rôle in Spectral Statistics. *Phys. Scr.* **2001**, *T90*, 128. [CrossRef]
22. Müller, S.; Heusler, S.; Altland, A.; Braun, P.; Haake, F. Periodic-orbit theory of universal level correlations in quantum chaos. *New J. Phys.* **2009**, *11*, 103025. [CrossRef]
23. Stöckmann, H.J. *Quantum Chaos—An Introduction*; University Press: Cambridge, UK, 1999. [CrossRef]

entropy

Article

Quantum Manifestation of the Classical Bifurcation in the Driven Dissipative Bose–Hubbard Dimer

Pavel Muraev [1,2,3,*], Dmitrii Maksimov [1,3] and Andrey Kolovsky [1,2]

1 Kirensky Institute of Physics, Federal Research Center KSC SB RAS, 660036 Krasnoyarsk, Russia
2 School of Engineering Physics and Radio Electronics, Siberian Federal University, 660041 Krasnoyarsk, Russia
3 IRC SQC, Siberian Federal University, 660041 Krasnoyarsk, Russia
* Correspondence: muraev.pavel@mail.ru

Abstract: We analyze the classical and quantum dynamics of the driven dissipative Bose–Hubbard dimer. Under variation of the driving frequency, the classical system is shown to exhibit a bifurcation to the limit cycle, where its steady-state solution corresponds to periodic oscillation with the frequency unrelated to the driving frequency. This bifurcation is shown to lead to a peculiarity in the *stationary* single-particle density matrix of the quantum system. The case of the Bose–Hubbard trimer, where the discussed limit cycle bifurcates into a chaotic attractor, is briefly discussed.

Keywords: open quantum system; non-linear dynamics; chaotic attractors

Citation: Muraev, P.; Maksimov, D.; Kolovsky, A. Quantum Manifestation of the Classical Bifurcation in the Driven Dissipative Bose–Hubbard Dimer. *Entropy* **2023**, *25*, 117. https://doi.org/10.3390/e25010117

Academic Editor: Marko Robnik

Received: 28 September 2022
Revised: 28 December 2022
Accepted: 1 January 2023
Published: 5 January 2023

1. Introduction

In the present work, we analyze the dynamics of the two-site driven dissipative Bose–Hubbard (BH) model. Similar to the conservative two-site BH model, which provides a model for the Josephson oscillations and the phenomenon of self-trapping [1–4], the driven dissipative BH systems model a number of phenomena in open quantum systems. For example, the one-site system, which is nothing else than the driven dissipative non-linear oscillators, is the paradigm system for quantum bistability (see Ref. [5] and references therein). Extending the system to two sites enriches its dynamics and drastically complicates the classical bifurcation diagram [6,7], which poses the problem of a quantum signature of these bifurcations [8,9]. Finally, the classical three-site BH system can show the chaotic attractor that brings us to the problem of the dissipative Quantum Chaos [10,11]. We mention that nowadays the few-site open BH model can be and has been realized experimentally by using different physical platforms, among which the most successful are exciton–polariton semi-conductor systems [4,12,13] and super-conducting circuits [14–17]. In what follows, we theoretically analyze the two- and three-site driven dissipative BH model by keeping in mind laboratory experiments on photon transport in the chain of transmons, which are micro-cavities coupled to Josephson's junctions. The presence of Josephson's junction introduces an effective inter-particle interaction for photons in the cavity and, thus, each transmon can be viewed as a quantum non-linear oscillator, see Figure 1. We mention that in the present work we do not try to relate the model parameters to the system parameters used in one or the other laboratory experiment. In this sense, the model depicted in Figure 1 captures only the general scheme of these experiments, where quantum non-linear oscillators are arranged in the 'transmission line' and one measures the amplitude of the transmitted signal, i.e., the current of the microwave photons.

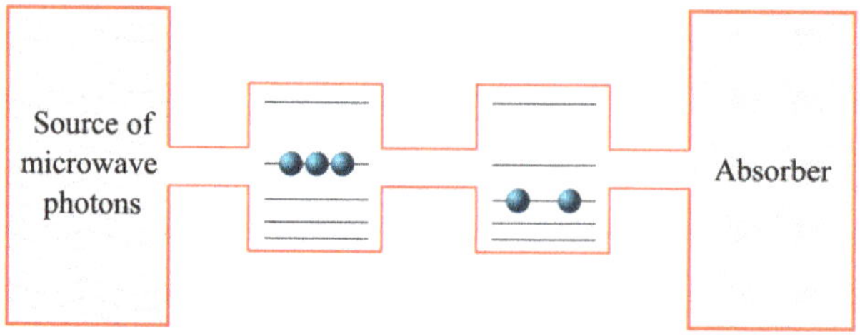

Figure 1. Pictorial presentation of the considered model. The figure shows two coupled non-linear micro-cavities with three photons in the first cavity and two photons in the second cavity. It is assumed that the microwave photons are injected into the first cavity by using a generator of the microwave field, and the transmitted photons are absorbed with some probability by a measurement device.

2. Quantum Dynamics

We consider two coupled transmons where the first transmon is excited by a microwave generator and the transmitted signal is read from the second transmon. The governing equation for the system density matrix $\widehat{\mathcal{R}}$ reads [13,16]

$$\frac{\partial \widehat{\mathcal{R}}}{\partial t} = -\frac{i}{\hbar}[\widehat{\mathcal{H}}, \widehat{\mathcal{R}}] - \frac{\gamma}{2}\left(\hat{a}_2^\dagger \hat{a}_2 \widehat{\mathcal{R}} - 2\hat{a}_2 \widehat{\mathcal{R}} \hat{a}_2^\dagger + \widehat{\mathcal{R}} \hat{a}_2^\dagger \hat{a}_2\right), \tag{1}$$

where γ is the decay constant proportional to the absorption rate. Using the rotating wave approximation, the Hamiltonian $\widehat{\mathcal{H}}$ in Equation (1) has the form

$$\widehat{\mathcal{H}} = -\hbar\Delta \sum_{\ell=1}^{2} \hat{n}_\ell - \frac{\hbar J}{2}\left(\hat{a}_2^\dagger \hat{a}_1 + \text{h.c.}\right) + \frac{\hbar^2 U}{2} \sum_{\ell=1}^{2} \hat{n}_\ell(\hat{n}_\ell - 1) + \frac{\sqrt{\hbar}\Omega}{2}(\hat{a}_1^\dagger + \hat{a}_1), \tag{2}$$

where $\hat{a}^\dagger$ and $\hat{a}$ are the creation and annihilation operators with the commutation relation $[\hat{a}, \hat{a}^\dagger] = 1$, $\hat{n}_\ell$ is the number operator, Ω is the Rabi frequency, Δ the detuning defined as the difference between the driving frequency and the cavity eigen-frequency, J the coupling constant, U the microscopic interaction constant, and $\hbar$ is the dimensionless Planck constant which determines how close is the system to its classical counterpart. For quantum systems with the conserved number of particles N, one can define the dimensionless Planck constant as $\hbar = 1/N$. In our case, where the number of particles is not conserved, $\hbar$ is just the scaling parameter that leaves invariant the classical dynamics. We focus on the case $\hbar \ll 1$ where the quantum dynamics shows similarities with the classical dynamics. In the opposite limit $\hbar > 1$ the system dynamics is dominated by the multi-photon resonances which have no classical analog.

Our main object of interest is the single-particle density matrix (SPDM) $\hat{\rho}$ which is defined as follows,

$$\rho_{\ell,m}(t) = \text{Tr}[\hat{a}_\ell^\dagger \hat{a}_m \widehat{\mathcal{R}}(t)]. \tag{3}$$

The diagonal elements of the SPDM obviously determine the populations of the sites, while off-diagonal elements determine the current of Bose particles (photons) between the sites. We find the stationary SPDM for different Δ, which will be our control parameter, by using two methods: (i) by evolving the system for fixed Δ for a long time sufficient to reach the steady-state regime, and (ii) by sweeping Δ in the *negative* direction starting from a large positive Δ. In both cases, the initial condition corresponds to the empty system, i.e., $\hat{\rho}(t=0) = 0$. The obtained results are depicted in Figure 2. It is seen in Figure 2 that stationary occupations of the chain sites (i.e., the mean number of photons in transmons) show a kind of plateau in the certain interval of Δ. Notice that this plateau is absent for the one-site BH model. It is argued in the next section that this peculiarity is a signature of the attractor bifurcation which one finds in the classical counterpart of the system (2).

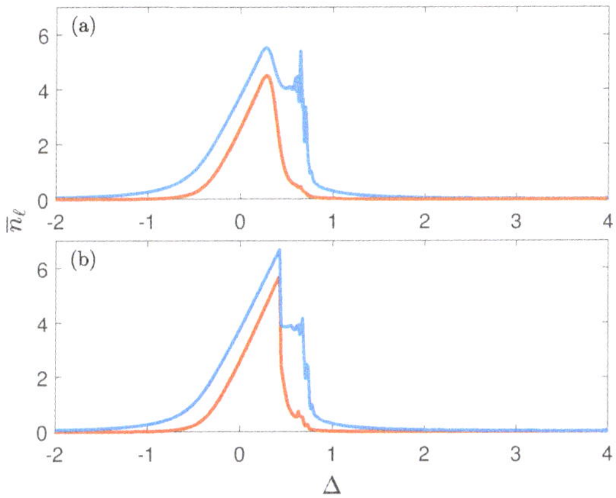

Figure 2. The mean number of bosons in the BH dimer as the function of the detuning Δ. The blue and red lines correspond to the first and second sites, respectively. (**a**) Quasi-adiabatic passage with the sweeping rate $d\Delta/dt = 0.0012$. (**b**) Steady-state solution for different Δ. The system parameters are $J = 0.5$, $U = 0.5$, $\Omega = 0.5$, $\hbar = 1/4$, $\gamma = 0.2$.

3. Classical Dynamics

The classical (mean-field) dynamics of the system is governed by the equations

$$\begin{aligned} i\dot{a}_1 &= (-\Delta + U|a_1|^2)a_1 - \frac{J}{2}a_2 + \frac{\Omega}{2} \\ i\dot{a}_2 &= (-\Delta + U|a_2|^2)a_2 - \frac{J}{2}a_1 - i\frac{\gamma}{2}a_2 \end{aligned} \tag{4}$$

where a_ℓ are complex amplitudes of the local oscillators. The numerical simulations were performed by using the fourth-order Runge–Kutta method. It is found that for non-zero γ the system (4) relaxes in course of time to some attractor which determines the system's stationary response to the external driving. In what follows, we shall be interested only in attractors whose basin contains the point $\mathbf{a} = 0$. For $\Delta < 0.48$ and $\Delta > 0.77$, we found these attractors to be simple focuses (in the rotating frame), where the populations $|a_\ell(t)|^2$ approach their stationary values depicted in Figure 3a by the blue and red solid lines. However, in the interval $0.48 < \Delta < 0.77$ these simple attractors bifurcate into the limit cycle, where the steady state solution of Equation (4) corresponds to periodic oscillations of the oscillator amplitudes with the frequency $\nu = \nu(\Delta)$ not related to the driving frequency, see Figure 3b. (Bifurcation of a simple attractor into a limit cycle in the driven dissipative two-site BH system was discussed earlier in Refs. [8,9] where the authors considered a specific model with *non-local* driving and dissipation. Additionally, when addressing the quantum system, the authors focussed on the transient dynamics for particular initial conditions but not on the stationary regime). In addition to this characteristic frequency, we also introduce the mean squared amplitudes $\overline{|a_\ell|^2}$ where the bar denotes the time average. We depict these quantities in Figure 3a by using the same line styles. Comparing now Figures 2b and 3a we conclude that bifurcation of the classical attractor is well reflected in the quantum dynamics of the BH dimer already for $\hbar = 1/4$.

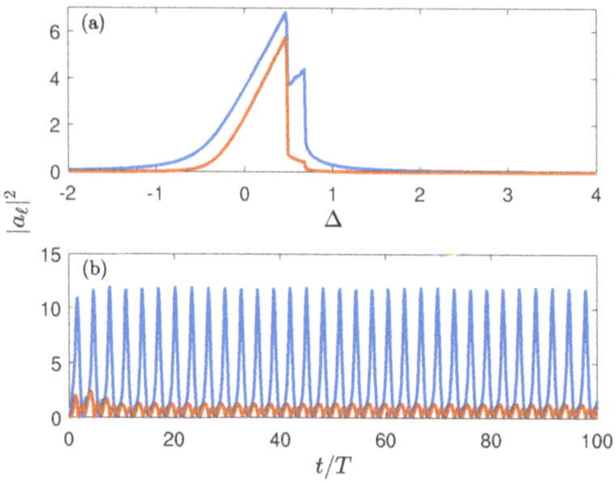

Figure 3. (**a**) The mean number of bosons in the BH dimer averaged over time as a function of detuning Δ. (**b**) System dynamics for $\Delta = 0.5$. Shown are the squared amplitudes of the local oscillators as the function of time.

4. Dissipative BH Trimer

We repeated calculations for the BH trimer. The dynamics of the trimer is controlled by the following set of equations of motion

$$\begin{aligned} i\dot{a}_1 &= (-\Delta + U|a_1|^2)a_1 - \frac{J}{2}a_2 + \frac{\Omega}{2} \\ i\dot{a}_2 &= (-\Delta + U|a_2|^2)a_2 - \frac{J}{2}(a_1 + a_3) \\ i\dot{a}_3 &= (-\Delta + U|a_3|^2)a_2 - \frac{J}{2}a_2 - i\frac{\gamma}{2}a_3 \end{aligned} \tag{5}$$

In the trimer, the new feature is that the above-discussed limit cycle bifurcates into a chaotic attractor, see the upper panel in Figure 4, which shows the Lyapunov exponent of the steady-state solution as the function of the control parameter Δ. An example of this 'stationary' solution is given in the lower panel in Figure 4 for $\Delta = 0.406$ where the Lyapunov exponent is maximal.

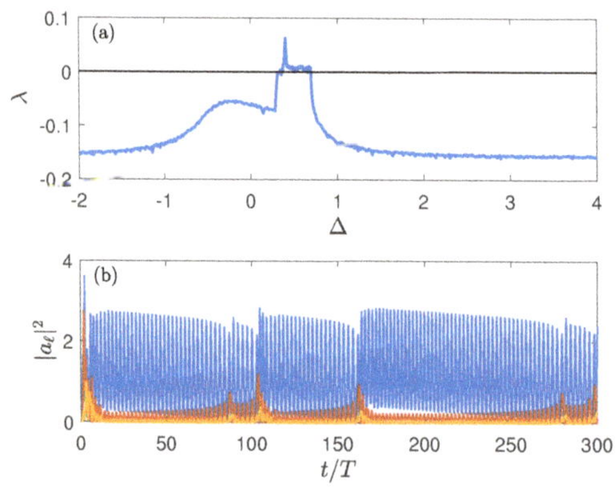

Figure 4. (**a**) Lyapunov exponent of the 'stationary' solution of the driven dissipative BH trimer as a function of the detuning Δ. (**b**) System dynamics for $\Delta = 0.406$. Shown are the squared amplitudes of the local oscillators as the function of time.

5. Summary

We showed that attractor bifurcations in the driven dissipative BH system can be well observed already for the value of the effective Planck constant $\hbar = 1/4$. In the laboratory experiment, the value of this effective constant is determined by the ratio of the interaction constant U (non-linearity of the transmon spectrum) to the Rabi frequency Ω, which is proportional to the amplitude of the microwave field. Clearly, the larger the Rabi frequency is, the more photons are simultaneously present in the system. The presented in this work results indicate that the quantum BH dimer reproduces the dynamics of the classical BH dimer when the mean number of photons is of the order of 10.

To conclude, we would like to briefly comment on the experiments where the detuning Δ is monotonically swept in time. In the present work, we restricted ourselves by the case where Δ is swept in the negative direction. If the sweeping direction is inverted, the result may strongly deviate from that shown in Figure 2a due to quantum hysteresis. Within the classical approach, the positive sweeping populates the other attractor whose basin *excludes* the point $\mathbf{a} = 0$ [18]. In the quantum approach, however, this attractor is a metastable state with a finite lifetime. Thus, unlike the case of negative sweeping, the result of a quasi-adiabatic passage in the positive direction strongly depends on the sweeping rate $d\Delta/dt$.

Author Contributions: Methodology, A.K.; Software, P.M.; Formal analysis, A.K.; Data curation, P.M., D.M. and A.K.; Writing—original draft, P.M. and A.K.; Writing—review & editing, D.M.; Visualization, P.M.; Supervision, A.K. All authors have read and agreed to the published version of the manuscript.

Funding: This work has been supported by Russian Science Foundation through grant N19-12-00167.

Conflicts of Interest: The authors declare no conflict of interest. The funders had no role in the design of the study; in the collection, analyses, or interpretation of data; in the writing of the manuscript; or in the decision to publish the results.

Abbreviations

The following abbreviations are used in this manuscript:

BH	Bose–Hubbard
SPDM	Single-particle density matrix

References

1. Smerzi, A.; Fantoni, S.; Giovanazzi, S.; Shenoy, S.R. Quantum Coherent Atomic Tunneling between Two Trapped Bose-Einstein Condensates. *Phys. Rev. Lett.* **1997**, *79*, 4950–4953. [CrossRef]
2. Albiez, M.; Gati, R.; Fölling, J.; Hunsmann, S.; Cristiani, M.; Oberthaler, M.K. Direct Observation of Tunneling and Nonlinear Self-Trapping in a Single Bosonic Josephson Junction. *Phys. Rev. Lett.* **2005**, *95*, 010402. [CrossRef] [PubMed]
3. Levy, S.; Lahoud, E.; Shomroni, I.; Steinhauer, J. The a.c. and d.c. Josephson effects in a Bose–Einstein condensate. *Nature* **2007**, *449*, 579–583. [CrossRef] [PubMed]
4. Abbarchi, M.; Amo, A.; Sala, V.G.; Solnyshkov, D.D.; Flayac, H.; Ferrier, L.; Sagnes, I.; Galopin, E.; Lemaître, A.; Malpuech, G.; et al. Macroscopic quantum self-trapping and Josephson oscillations of exciton polaritons. *Nat. Phys.* **2013**, *9*, 275–279. [CrossRef]
5. Kolovsky, A. Bistability in the dissipative quantum systems I: Damped and driven nonlinear oscillator. *arXiv* **2020**, arXiv:2002.11373. https://doi.org/10.48550/arXiv.2002.11373.
6. Giraldo, A.; Krauskopf, B.; Broderick, N.G.R.; Levenson, J.A.; Yacomotti, A.M. The driven-dissipative Bose–Hubbard dimer: Phase diagram and chaos. *New J. Phys.* **2020**, *22*, 043009. [CrossRef]
7. Giraldo, A.; Masson, S.J.; Broderick, N.G.R.; Krauskopf, B. Semiclassical bifurcations and quantum trajectories: A case study of the open Bose–Hubbard dimer. *Eur. Phys. J. Spec. Top.* **2022**, *231*, 385–401. [CrossRef]
8. Ledó, C.; Mavrogordatos, T.K.; Szymańska, M.H. Driven Bose-Hubbard dimer under nonlocal dissipation: A bistable time crystal. *Phys. Rev. B* **2019**, *100*, 054303. [CrossRef]
9. Lledó, C.; Szymańska, M.H. A dissipative time crystal with or without $\mathbb{Z}_2$ symmetry breaking. *New J. Phys.* **2020**, *22*, 075002. [CrossRef]
10. Braun, D. *Dissipative Quantum Chaos and Decoherence*; Springer: Berlin/Heidelberg, Germany, 2001; Volume 172. [CrossRef]
11. Sá, L.; Ribeiro, P.; Prosen, T. Complex Spacing Ratios: A Signature of Dissipative Quantum Chaos. *Phys. Rev. X* **2020**, *10*, 021019. [CrossRef]

12. Lagoudakis, K.G.; Pietka, B.; Wouters, M.; André, R.; Deveaud-Plédran, B. Coherent Oscillations in an Exciton-Polariton Josephson Junction. *Phys. Rev. Lett.* **2010**, *105*, 120403. [CrossRef] [PubMed]
13. Rodriguez, S.R.K.; Amo, A.; Sagnes, I.; Gratiet, L.L.; Galopin, E.; Lemaître, A.; Bloch, J. Interaction-induced hopping phase in driven-dissipative coupled photonic microcavities. *Nat. Commun.* **2016**, *7*, 1–6. [CrossRef] [PubMed]
14. Eichler, C.; Salathe, Y.; Mlynek, J.; Schmidt, S.; Wallraff, A. Quantum-Limited Amplification and Entanglement in Coupled Nonlinear Resonators. *Phys. Rev. Lett.* **2014**, *113*, 110502. [CrossRef] [PubMed]
15. Raftery, J.; Sadri, D.; Schmidt, S.; Türeci, H.; Houck, A. Observation of a Dissipation-Induced Classical to Quantum Transition. *Phys. Rev. X* **2014**, *4*, 031043. [CrossRef]
16. Fitzpatrick, M.; Sundaresan, N.M.; Li, A.C.Y.; Koch, J.; Houck, A. Observation of a Dissipative Phase Transition in a One-Dimensional Circuit QED Lattice. *Phys. Rev. X* **2017**, *7*, 011016. [CrossRef]
17. Fedorov, G.; Remizov, S.; Shapiro, D.; Pogosov, W.; Egorova, E.; Tsitsilin, I.; Andronik, M.; Dobronosova, A.; Rodionov, I.; Astafiev, O.; et al. Photon Transport in a Bose-Hubbard Chain of Superconducting Artificial Atoms. *Phys. Rev. Lett.* **2021**, *126*, 180503. [CrossRef] [PubMed]
18. Muraev, P.S.; Maksimov, D.N.; Kolovsky, A.R. Ballistic transport of interacting Bose particles in a tight-binding chain. *Phys. Rev. E* **2022**, *106*, 064203. [CrossRef]

entropy

Article

Ray-Stretching Statistics and Hot-Spot Formation in Weak Random Disorder

Sicong Chen † and Lev Kaplan *,†

Department of Physics and Engineering Physics, Tulane University, New Orleans, LA 70118, USA
* Correspondence: lkaplan@tulane.edu
† These authors contributed equally to this work.

Abstract: Weak scattering in a random disordered medium and the associated extreme-event statistics are of great interest in various physical contexts. Here, in the context of non-relativistic particle motion through a weakly correlated random potential, we show how extreme events in particle densities are strongly related to the stretching exponents, where the 'hot spots' in the intensity profile correspond to minima in the stretching exponents. This strong connection is expected to be valid for different random potential distributions, as long as the disorder is correlated and weak, and is also expected to apply to other physical contexts, such as deep ocean waves.

Keywords: random scattering; random potential; extreme event statistics

1. Introduction

1.1. Rogue Waves

In the past few decades, many encounters with extreme oceanic waves have been publicized and documented. These historical records include a 25.6 m wave that hit the Draupner oil platform in the North Sea in 1995, two ships that suffered damage at 30 m above sea level from a single wave in the South Atlantic in 2001, and the cruise liner Norwegian Dawn that met a series of three 21-m waves off the coast of Georgia in 2005. Such waves, known as freak ocean waves, or rogue waves, are of extreme height relative to the typical wave in a given sea state. Satellite images taken in 2001 and analyzed as part of the European MaxWave project [1] detected ten waves above 25 m in height, suggesting that such waves commonly occur in the world's oceans.

Freak waves have attracted much interest over the years, especially for quantitative predictions for freak wave probability distributions. Commonly known approaches to model freak waves include the Longuet–Higgins random sea model [2], the Nonlinear Schrödinger Equation (NLS) and its extension in the Dys equation [3,4], and ray dynamics in the work of White and Fornberg [5]. The Longuet–Higgins random sea model is based on random linear superposition of many plane waves with different directions and wavelengths, where (unlucky) constructive interference leads to an extreme event. By the central limit theorem, in this model the sea surface height at any spatial location behaves as a Gaussian random variable with some standard deviation σ. In the limit of a narrow frequency spectrum, the crest height then follows a Rayleigh distribution: the probability of crest height exceeding H is given by

$$P_{\text{Rayleigh}}(H) = e^{-H^2/2\sigma^2}. \tag{1}$$

According to observational data [1], this purely stochastic Rayleigh model significantly underestimates the actual probabilities of freak waves. There are also several alternative theories of the freak wave phenomenon [6].

The NLS or the Dysthe equation works well to incorporate nonlinear effects in the regime of small or moderate values of wave steepness kH, where k is the wave number.

Citation: Chen, S.; Kaplan, L. Ray-Stretching Statistics and Hot-Spot Formation in Weak Random Disorder. *Entropy* **2023**, *25*, 161. https://doi.org/10.3390/e25010161

Academic Editor: Marko Robnik

Received: 4 December 2022
Revised: 11 January 2023
Accepted: 11 January 2023
Published: 13 January 2023

Since nonlinear effects scale as a power of the wave steepness, strong nonlinear evolution is more likely to come from initial conditions with already unusually high waves. In other words, the tail of the crest height distribution is likely to be influenced by linear triggering mechanisms, even if subsequent nonlinear development is also significant.

One model that combines such a linear triggering mechanism with nonlinear evolution is the focusing or refraction of incoming plane waves by random current eddies [5], the study of which was motivated by the fact that many freak waves have been observed in regions of strong currents. Whenever a focusing current is present, an incoming plane wave evolves into caustics or singularities with infinite ray density, which—when smoothed due to nonzero wavelength and/or nonzero spread in the initial angle or frequency—forms a repeated and reproducible branched flow pattern. Consequently, freak waves appear within the 'branches' of enhanced intensity. Statistics regarding the distribution of crest heights can be obtained by combining the stochastic random seas picture (given by a local Rayleigh distribution) and the statistics of ray focusing in the presence of random currents.

Besides extreme waves in the ocean, extreme waves are known to occur in many other physical systems, governed by different equations of motion, where the waves or rays are scattered by a weak random potential. Examples of extreme waves and branched flow have been reported on a wide range of length scales, including the branching of electron flow [7–11]; amplification of tsunami waves [12–14]; branching of light traveling through a soap film [15]; and freak waves in optical [16,17], acoustic [18,19], and microwave propagation [20,21]. These systems share similarities in statistics and scaling relations, suggesting that a universal theory of scattering in weak random potentials may describe these different phenomena. Indeed, a search for universality in branched flows through potentials with differing correlation structures has obtained success [22,23]. The theory extends naturally to the case of an anisotropic scattering potential [24]. Very recently, a one-parameter model was shown to describe the behavior of classical branched flow in a time-dependent, one-dimensional random potential (equivalent to a time-independent two-dimensional potential for weak scattering), and a two-parameter phase diagram was shown to describe the corresponding quantum branched flow [25]. A recent overview of the theory and applications of branched flow appeared in reference [26].

The starting point of our work is similar to the model employed by White and Fornberg [5], as developed further in reference [27], so here we begin with a brief review of that model. The ray dynamics of deep-water surface gravity waves are governed by the dispersion relation:

$$\omega(\vec{k},\vec{r}) = \sqrt{g|\vec{k}|} + \vec{k}\cdot\vec{U}(\vec{r}), \quad (2)$$

where $\vec{U}(\vec{r})$ is the time-independent current velocity of root mean square strength u_0 and spatial correlation length ξ. An incoming wave field undergoes weak scattering when $u_0 \ll v$ (wave speed $\vec{v} = \partial\omega/\partial\vec{k}$) and the scattering angle scales as u_0/v after one correlation length ξ in the forward direction. Eventually, singularities appear as local focusing of the manifold of initial conditions. The first cusp singularities or 'hot spots' will appear after a median travel distance:

$$y = L \sim \xi(u_0/v)^{-2/3} \gg \xi, \quad (3)$$

and at that distance scale, the angular spread of trajectories due to scattering is characterized by $\delta\theta \sim (u_0/v)^{2/3}$. The dimensionless 'freak index' γ that describes the strength of scattering is defined as $\gamma = \delta\theta/\Delta\theta \sim (u_0/v)^{2/3}/\Delta\theta$, where $\Delta\theta$ is the initial angular spread of trajectories [28].

Ray dynamics is applicable for understanding wave behavior in the semiclassical regime, $k\xi \ll 1$, where the potential slowly varies on the scale of a wavelength. In that regime, the wave-height distribution may be obtained as a convolution of the Rayleigh distribution, Equation (1), with the distribution of classical ray densities, depending only on the freak index γ [27].

In the following, we apply analogous methods to the study of Schrödinger evolution in a two-dimensional random potential, where the ray limit is given by classical Newtonian mechanics. The choice of dispersion relation for weak scattering is not important, as the locations of the first generation of hot spots follow a universal scaling law across various dispersion relations. This will be further discussed in Ref. [29].

1.2. Density and Stretching Exponents

For ray dynamics in weak random disorder, the relationships between density distributions and the properties of the disorder have attracted much research interest, and the location of extreme densities has been the major focus. However, while a random superposition of waves as in the Longuet–Higgins model [2] corresponds to a swarm of trajectories, properties of the ray density have not been directly connected to statistics of the individual propagating ray trajectories. Therefore, we introduce the 'stretching exponent'—which in the case of parallel ray bundles in one dimension coincides with the rarefaction exponent introduced by Shaw and Heller [30]—as a way to quantify the degree of exponential convergence or divergence among nearby ray trajectories, and demonstrate a strong correlation between stretching exponents and the density. The region with the highest density is shown to correspond to the strongest focusing as measured by the stretching exponent distribution, and the strong correlation that first appears in this region is shown to persist everywhere further along the forward direction.

The strong negative correlation allows the density distribution to be studied from the perspective of the distribution of stretching exponents. Furthermore, the stretching exponents are directly connected with individual ray trajectories via the evolving monodromy matrices. The evolving monodromy matrix of an individual ray trajectory can be obtained from the second derivatives of the random disorder along the trajectory. The detailed mathematics are discussed in Section 2.

A relationship between extreme ray densities and minimal stretching exponents is consistent with intuitive expectations—indeed, if one considers the paraxial approximation and the case where the density at a given point is generated solely by one neighborhood of rays, then that density will simply be inversely proportional to the stretching factor. However, in Section 3, we also show that in the region of initial hot spot formation, the average stretching exponent becomes *negative*, and this is accompanied by a peak in the stretching exponent variance, or scintillation index. For a chaotic system with positive maximal Lyapunov exponent, a negative dip in the average stretching exponents is quite surprising and demands further investigation. Possible explanations involve the relationship between the statistics of the stretching exponent and the distribution of second derivatives of the disorder, which is further discussed in Ref. [31]. In any case, the non-monotonic behavior of the stretching-exponent statistics with distance greatly enhances the appearance of hot spots and rogue waves on distance scales given by Equation (3), beyond what would be expected based on the assumption of a simple exponential stretching with time.

2. Materials and Methods

In the following study, we focus on a simple example of a Hamiltonian system and establish a general connection between the ray-stretching exponents and intensity. Instead of ocean waves refracted by random currents, our model is based on a single-particle Schrödinger wave function (or, in the ray limit, a non-relativistic particle) scattered by a random potential field. In the region of weak potential or weak currents, despite the different dispersion relations, the statistics of the two models should resemble each other closely [27,29].

A non-relativistic particle moving in a 2D potential follows classical Newtonian mechanics:

$$\frac{d\vec{r}}{dt} = \vec{k}, \quad \frac{d\vec{k}}{dt} = -\frac{\partial \vec{V}(\vec{r})}{\partial \vec{r}}, \tag{4}$$

where energy $E = \frac{1}{2}|\vec{k}|^2 + V(\vec{r})$ (without loss of generality, taking $\hbar = m = 1$) is conserved. In the following analysis, $V(\vec{r})$ is considered to be a random time-independent potential with zero average value, variance $\sigma^2 = \overline{V(\vec{r})^2}$, and some two-point spatial correlation function $C(\vec{r} - \vec{r}')$ characterized by correlation length ξ. In practice, the potential $V(\vec{r})$ may be constructed by a convolution between the shape of one potential bump $U(\vec{r})$ and a random noise $g(\vec{r})$:

$$V \sim \mathcal{F}^{-1}[\mathcal{F}[g] \cdot \mathcal{F}[U]], \quad (5)$$

and then normalized to have variance σ^2. Here $g(\vec{r}) = \sum_i h_i \delta(\vec{r}_i)$ (with h_i random, independent, and having zero mean), $\mathcal{F}$ represents a Fourier transform, and the bump shape is related to the correlation function C as $|\mathcal{F}[U]|^2 \sim \mathcal{F}[C]$ [29]. A sample random potential constructed using $(256)^2$ Gaussian bumps is shown in Figure 1a.

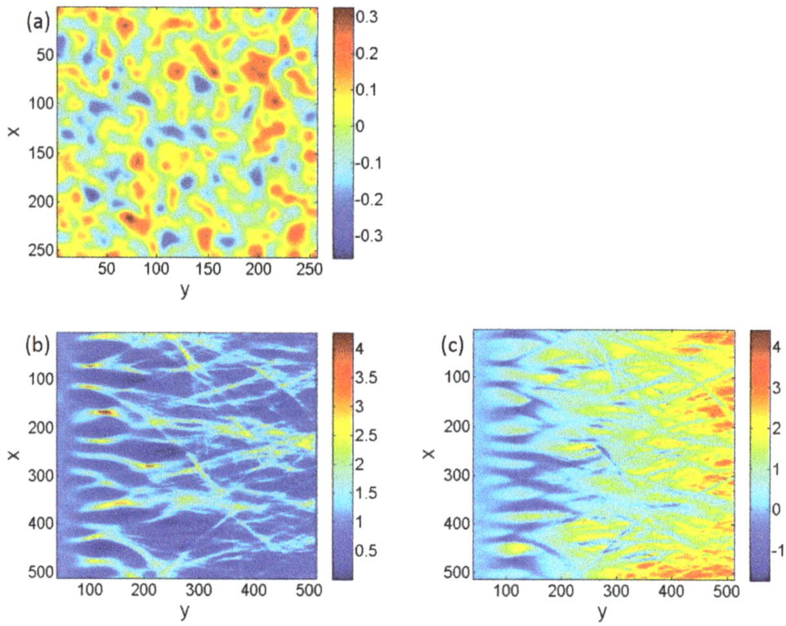

Figure 1. (**a**) A Gaussian random potential generated on a 256 by 256 grid with zero mean, standard deviation $\sigma = 0.1$, and a Gaussian two-point correlation function with correlation length $\xi = 5$. The potential is constructed by starting with $(256)^2$ delta-shaped peaks on a square grid with spacing unity, and then convolving with a Gaussian, as described in Equation (5). (**b**,**c**) The ray density and stretching exponent of particles scattered in the random potential field shown in (**a**), but with x and y rescaled to a 512 by 512 grid. Here (**b**) shows the particle density $I(x, y)$, with intensity normalized to unity before scattering, and (**c**) shows the normalized average stretching exponent $\alpha(x, y)$. The correlation is clearly negative; i.e., higher densities in panel (**b**) are associated with a lower stretching exponent in panel (**c**).

The initial particles of total energy E are distributed uniformly in the x direction at $y = y_0$, with initial angles θ_0 relative to the y direction drawn from a Gaussian distribution $P(\theta_0) \sim e^{-\theta_0{}^2/2(\Delta\theta)^2}$, with a small but finite angular spread $\Delta\theta$. (The choice of y as the propagation coordinate is consistent with the convention adopted, for example, in Ref. [27]). For the equations of motion (4), the scattering strength is determined by the dimensionless quantity σ/E in place of u_0/v. The particle trajectories will form a definite pattern of particle density for a given realization of the random potential field. Some regions ('hot spots') will have above-average particle probability density, and others ('cold spots') will have below-average density. Intuitively, the hot spots are more likely to be associated with local focusing of nearby particle trajectories, and the defocusing of trajectories will tend to produce cold spots.

To quantitatively monitor the degree of stretching or focusing, we define the stretching exponent α as the logarithm of the stretching ratio between nearby trajectories in the x direction, i.e., in the direction transverse to the main flow direction y:

$$\alpha(t) = \log\left(\left|\frac{x_1(t) - x_2(t)}{x_1(0) - x_2(0)}\right|\right) = \log\left(\left|\frac{x_1(t) - x_2(t)}{\delta x(0)}\right|\right), \tag{6}$$

where trajectories 1 and 2 are initially parallel but separated by infinitesimal $\delta x(0)$ in the x direction. The value of $\alpha(t)$ describes the cumulative divergence or convergence up to time t, and its time derivative $d\alpha(t)/dt$ gives the rate of exponential divergence or convergence at time t. The large-time limit of the stretching exponent in this 2D system is also the maximal Lyapunov exponent. While the concept of Lyapunov exponents is well established and studied in dynamical systems, the stretching exponent is employed here to study the behavior of chaotic systems for short or intermediate time scales (on scales around one Lyapunov time) and in a particular direction in phase space. For parallel ray bundles in one dimension, α coincides with the rarefaction exponent introduced by Shaw and Heller [30]. We are interested specifically in focusing or defocusing in the transverse position coordinate, which is most relevant for fluctuations in the position space density. Quantities of this type are widely applicable, for example, in semiclassical approximations, e.g., in the Van Vleck or Gutzwiller semiclassical propagator [32].

Very generally, dynamics in an N-dimensional space are described by the evolution of the phase-space vector: $\phi = (\phi^{[1]}, \phi^{[2]}, \ldots, \phi^{[2N]}) = (r^{[1]}, r^{[2]}, \ldots, r^{[N]}, k^{[1]}, k^{[2]}, \ldots, k^{[N]})$; and the continuous flow may be considered as the limit of a discrete-time map where the time step approaches zero. The following analysis makes use of the discrete-time picture. Consider a map F from time step n to the next step $n+1$,

$$\phi_{n+1} = F(\phi_n). \tag{7}$$

The iteration of the tangent space is given by the Jacobian matrix:

$$K_{ij}(\phi_n) = \frac{\partial F_i}{\partial \phi^{(j)}}|_{\phi=\phi_n}, \tag{8}$$

so that the shift δ_n in the phase vector ϕ_n is mapped to the next time step $K(\phi_n)$ as

$$\delta_{n+1} = K(\phi_n)\delta_n. \tag{9}$$

Therefore, the initial perturbation δ_0 evolves by a product of Jacobian matrices to δ_n at time step n:

$$\delta_n = M_n\delta_0 = [K(\phi_{n-1})K(\phi_{n-2})\ldots K(\phi_0)]\delta_0, \tag{10}$$

where M_n is the monodromy or stability matrix. Similarly, in the continuous-time limit, $d\phi/dt = f(\phi)$, the stability matrix $M(t)$ is given by $dM(t)/dt = J(\phi(t))M(t)$ where $J_{ij} = df_i/d\phi_j|_{\phi=\phi(t)}$ or $d\delta(t)/dt = J(\phi(t))\delta(t)$.

In the large-time limit, the $2N$ eigenvalues of M can be written in an exponential form: $\{e^{\lambda_1 t}, e^{\lambda_2 t}, \ldots, e^{\lambda_{2N} t}\}$, where the spectrum of Lyapunov exponents $\{\lambda_1, \lambda_2, \ldots, \lambda_{2N}\}$ is independent of time. The matrix M has several important properties. It generates a linear, canonical transformation, and the effective dimension of the spectrum is reduced from $2N$ to N, with a determinant equal to unity. Moreover, every independent constant of motion causes one pair of eigenvalues to become unity or one pair of exponents to vanish. Thus, in our 2D Hamiltonian model, with energy conserved, the eigenvalues of M are $\{e^{\lambda t}, e^{-\lambda t}, 1, 1\}$, where a single Lyapunov exponent λ completely captures the large-time behavior of the system.

On the other hand, the stretching exponent α defined by Equation (6) can also be viewed as

$$\alpha(t) = \log(|M_{11}(t)|), \tag{11}$$

where $\phi = \begin{pmatrix} x \\ y \\ k_x \\ k_y \end{pmatrix}$, and for large time periods, the exponent $\alpha(t)$ is expected to grow linearly with time:

$$\alpha(t) = \lambda t \quad (t \to \infty). \tag{12}$$

However, the first generation of hot spots happens at intermediate time periods, $t \sim \lambda^{-1}$, well before the stretching exponent $\alpha(t)$ begins to behave linearly. It is this intermediate behavior of $\alpha(t)$ that is of greatest interest for explaining the formation mechanism of the most extreme events.

The evolution of the displacement from time t to $t + \delta t$ can be written out explicitly as

$$\begin{aligned} \delta \begin{pmatrix} x \\ y \\ k_x \\ k_y \end{pmatrix}_{t+\delta t} &= K(t) \cdot \delta \begin{pmatrix} x \\ y \\ k_x \\ k_y \end{pmatrix}_t \\ &= \begin{bmatrix} 1 - V_{xx}\delta t^2 & -V_{xy}\delta t^2 & \delta t & 0 \\ -V_{xy}\delta t^2 & 1 - V_{yy}\delta t^2 & 0 & \delta t \\ -V_{xx}\delta t & -V_{xy}\delta t & 1 & 0 \\ -V_{xy}\delta t & -V_{yy}\delta t & 0 & 1 \end{bmatrix} \cdot \delta \begin{pmatrix} x \\ y \\ k_x \\ k_y \end{pmatrix}, \end{aligned} \tag{13}$$

where $V_{xx} = \frac{\partial^2 V}{\partial x^2}$, $V_{yy} = \frac{\partial^2 V}{\partial y^2}$, and $V_{xy} = \frac{\partial^2 V}{\partial x \partial y}$ are the second derivatives of the potential field $V(x, y)$. Then, Equation (10) gives the monodromy matrix M at time t as $M(t) = K(t - \delta t)K(t - 2\delta t) \ldots K(\delta t)K(0)$. For a specific trajectory, computing $M(t)$ requires the positions of the particle at all the time points, which can only be obtained by integrating the equations of motion. However, upon ensemble averaging over homogeneous random potentials for a given potential variance and given a two-point potential correlation function, we can consider V_{xx}, V_{yy}, and V_{xy} to be (correlated) random numbers drawn from appropriate distributions.

Moreover, when the initial angular spread $\Delta\theta$ in the forward y direction is small, $\Delta\theta \ll 1$ rad, the 2D model is analogous to a 1D model where the particle evolves in phase space (x, k) via a time-dependent random potential line $V(x, t)$ with correlation time $\xi/\sqrt{2E}$. In the 1D model, the stretching exponent α is likewise defined via the exponential divergence of two trajectories with neighboring initial positions $x(0)$ and $x(0) + \delta x(0)$.

3. Results and Discussion

For numerically computing particle trajectories and stretching exponents in the 2D Hamiltonian model, we generated a Gaussian random potential $V(\vec{r})$ with Gaussian spatial correlations:

$$\overline{V(\vec{r})V(\vec{r}')} \sim C(\vec{r} - \vec{r}') = e^{-(\vec{r}-\vec{r}')^2/2\xi^2}, \tag{14}$$

by Fourier convolution, as in Equation (5), and normalized $V(\vec{r})$ to satisfy $\overline{V(\vec{r})} = 0$ and $\overline{V^2(\vec{r})} = \sigma^2$. The choice of a Gaussian-correlated random potential was made for convenience. The above theoretical discussion does not depend on any specific choice of a random ensemble, but only on the correlation length scale ξ and strength σ. The effect of varying the correlation function C is addressed in detail in Ref. [29].

The evolution was performed on a potential field of size 512 by 512 (in arbitrary units), with correlation length $\xi = 10$ and a periodic boundary condition in the transverse (x) direction. A sample potential on a 256 by 256 grid, which was subsequently rescaled to a 512 by 512 grid for performing trajectory evolution, is shown in Figure 1a. The specific value of ξ was arbitrary and served merely to set the scale for the simulation. Without loss of generality, we set $E = 1$ (energy). The strength of scattering γ can be controlled either by varying the potential strength, σ, or by controlling the angular spread, $\Delta\theta$, in the initial conditions. To avoid boundary effects, particles were launched from $y_0 = 40$ inside the potential field and uniformly distributed in the transverse x direction, with initial angles θ_0. The initial phase-space vector for each trajectory was $(x, y_0, v_0 \sin\theta_0, v_0 \cos\theta_0)$, for which the initial velocity v_0 was calculated based on its starting position $v_0 = \sqrt{\frac{2(E-V(x,y_0))}{m}} = \sqrt{2(1-V(x,y_0))}$ so that energy was fixed ($E = 1$) for all trajectories.

Each trajectory was evolved by integrating the equations of motion, Equation (4), using a fourth order Runge–Kutta integration method. Cubic interpolation was used for the potential $V(\vec{r})$ when running trajectories. The trajectories were weighted by the angular spread $P(\theta_0) \sim e^{-\theta_0{}^2/2(\Delta\theta)^2}$, and then points along each trajectory were binned using Gaussian-shaped windows of size $\tilde{\xi}$ to generate a ray density map $I(x,y)$. Gaussian-shaped binning eliminates any artificial discontinuity in the binned density and effectively smooths the density data $I(x,y)$ on the scale $\tilde{\xi}$, which must be chosen to be small compared to the physical correlation scale ξ. A spacing of $\Delta x = 2$ in the initial trajectory positions x, an increment of 1° in the initial angle θ_0, and Gaussian intensity bins of width $\tilde{\xi} = 1$ with spacing 2 on the 512 by 512 grid were seen to be sufficient to achieve convergence in all the density data. Using initial positions $0 \le x_0 < 512$ with spacing $\Delta x = 2$ and initial angles $-15^\circ \le \theta_0 \le 15^\circ$ with spacing $\Delta\theta = 1^\circ$ requires 7936 trajectories; see Figure 1b. For stronger scattering (larger γ), structures appear at a smaller scale so that the convergence of the density data requires a greater number of trajectories.

A typical density map for initial angular spread $\Delta\theta = 5^\circ$ with a potential of strength $\sigma = 0.1$ is shown in Figure 1b, where the first generation of caustics forms around $80 < y < 160$, and the corresponding freak index is $\gamma = 2.5$. Note that the particle density $I(x,y)$ is normalized to unity, $I(x,y) = 1$, before scattering.

Next, we demonstrate the connection between the density and the stretching exponent $\alpha(t)$. For every original trajectory launched at (x, y_0), we launched a 'twin' trajectory in the same potential at $(x + \delta x(0), y_0)$. Then, $\alpha(t)$ was computed for each trajectory according to Equation (6); the trajectories were weighted by the initial angular spread $P(\theta_0) \sim e^{-\theta_0{}^2/2(\Delta\theta)^2}$; and the stretching exponents were eventually mapped into the same grid that was used for density data $I(x,y)$ to produce an average position-dependent stretching rate $\alpha(x,y)$. Again, the binning was performed using a Gaussian window function, with a width chosen appropriately for data smoothing. Due to the long-term exponential stretching trend (Equation (12)), the initial separation $\delta x(0)$ must be chosen sufficiently small so that the separation between twin trajectories remains small during the whole time evolution. In the following, we used $\delta x(0) = 10^{-5}$. We have confirmed that our results are independent of $\delta x(0)$ as long as $\epsilon_m \ll \delta x(0) \ll \xi$, where $\xi = 10$ is the correlation scale of the potential and ϵ_m represents machine precision.

First, we notice the obvious connection between the density map Figure 1b and the stretching exponent map Figure 1c. The correlation is clearly negative; i.e., higher densities in Figure 1b are associated with a lower stretching exponent in Figure 1c. Indeed, every major hot spot (maximum) of the density map corresponds visually to a minimum at the same location in the stretching exponent map. This is consistent with our prediction that extreme density events would occur where the trajectories focus most significantly. Indeed, as noted above in Section 1.2, if we assume the paraxial approximation and further assume that the density at any given point (x,y) is all coming from parallel rays originating in the neighborhood of one initial point (x_0, y_0), then the proportionality $I(x,y) \sim 1/M = e^{-\alpha}$ will hold exactly. Of course, in reality, chaotic dynamics leads at sufficient time scales to caustics and folds in the time evolution, so that the density $I(x,y)$ is given by a sum of contributions originating at different initial points with different exponents α.

In Figure 2, we show scatter plots of the relationship between $I(x,y)$ and $\alpha(x,y)$ before ($y = 50$), during ($y = 100$ and $y = 125$), and after ($y = 400$) the region with the strongest density fluctuations. Clearly, the stretching exponent is negatively correlated with density: The strong correlation grows as the rays encounter the first generation of caustics and eventually dies off after a few Lyapunov lengths. In the region of the first caustics, extremely high densities are always associated with most negative stretching exponents, and the largest stretching exponents lead to the lowest density levels. More specifically, the particle density scales as

$$I(x,y) \sim e^{-b\alpha(x,y)}, \tag{15}$$

with constant coefficient b around the first caustics, as seen in Figure 2b,c. This relationship relies only on ray dynamics, and apart from the coefficient b, it does not depend on any specific dispersion relation. At larger time scales ($t \gg \lambda^{-1}$), the stretching exponents α grow linearly with time, as described by Equation (12), and the density probability distribution gradually collapses to a Gaussian one. In this regime, the correlation between the intensity and the stretching exponent declines and eventually disappears. The crossover to the large-time regime is illustrated in Figure 2d. Nevertheless, in the regime of greatest interest, i.e., in the first caustic region $t \sim \lambda^{-1}$ where the strongest hot spots are present, the relationship is very robust.

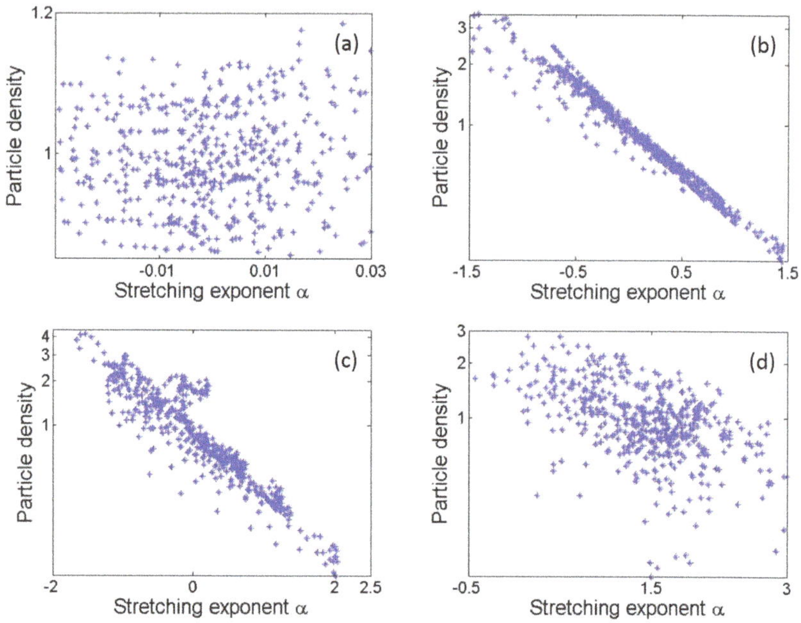

Figure 2. Scatter plots of density vs. stretching exponent before, during, and after the first generation of hot spots (around y = 80–130). (**a–d**) represent data for $y = 50$ (before), $y = 100$ (during), $y = 125$ (during), and $y = 400$ (after), respectively.

The distribution of the ray density is shown in Figure 3a, where all the density data points in the region $60 < y < 500$ from five realizations of the ensemble are included (while the trajectories are launched at $y = 40$ and the computational area extends through $y = 512$, a slightly smaller region was used for collecting statistics to avoid possible edge effects). As mentioned earlier, the bulk of the distribution is close to a Gaussian distribution, whereas the fatter tail represents events of extreme densities. Then, for Figure 3b, we averaged the stretching exponents for all the spatial cells whose density values fall within each bin in Figure 3a. Not only in the area of the first caustics but the density distribution $I(x,y)$ and stretching exponent distribution $\alpha(x,y)$ in general have a strong negative correlation; the locations with smaller or more negative stretching exponents are very likely to exhibit higher ray densities. Note that the fluctuations after very modest ensemble averaging in Figure 3b are remarkably small compared to fluctuations for one realization in Figure 2.

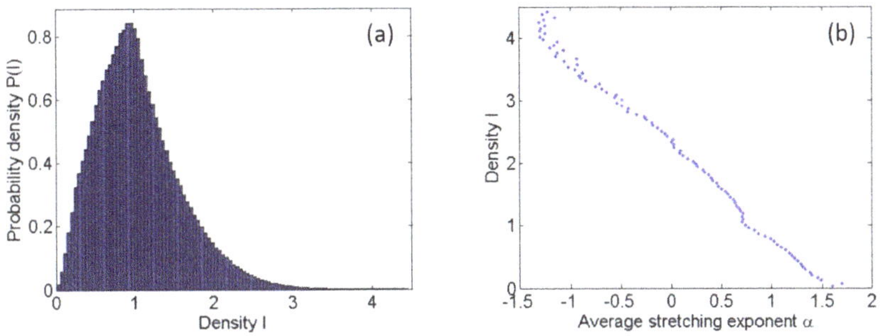

Figure 3. (**a**) Probability distribution of particle density I in the region $60 < y < 500$. (**b**) Average stretching exponent α in the same region, with each data point corresponding to one bin in panel (**a**).

To further investigate the relationship between stretching exponent and ray density, we show in Figure 4 the average and variance of these two quantities as functions of the forward distance y. Here, for each value of y, we collected data over all transverse positions x and again used five different realizations of the random potential to reduce statistical noise.

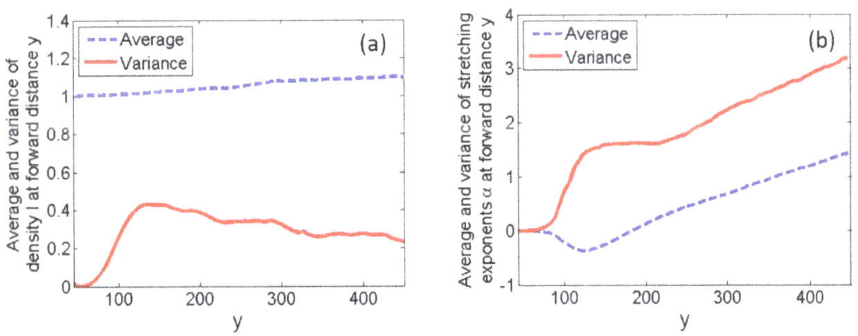

Figure 4. Evolution of intensity and stretching-exponent statistics as a function of forward travel distance. (**a**) Average and variance of the density $I(x,y)$ at forward travel distance y. (**b**) Average and variance of the stretching exponent $\alpha(x,y)$ at forward travel distance y.

The average density $\bar{I}(y)$ increases with the factor $\overline{(\cos\theta)^{-1}}$ as required by probability conservation of the ray dynamics, where θ is the angle measured from the forward y direction. In the region of small angular spread, $\bar{I}(y) \sim \overline{(\cos\theta)^{-1}} \approx 1 + \overline{\theta^2}/2 = 1 + (\Delta\theta)^2/2 + (\delta\theta)^2/2$, where $(\Delta\theta)^2$ and $(\delta\theta)^2$ are the variances associated with the initial angular spread and the scattering, respectively. As $(\delta\theta)^2$ increases linearly with forward distance y, the average intensity grows linearly with the forward distance, as observed in Figure 4a.

We now turn to the stretching exponent. In the regime of small-angle scattering in the forward y direction, we have $y \approx v_0 t$, and the stretching exponent $\bar{\alpha}(y)$ averaged over the transverse direction must grow linearly with the forward distance y at large distances, in accordance with Equation (12), with slope λ / v_0. Furthermore, at large time scales, we can consider the process of focusing or defocusing in the transverse direction as a random walk or diffusive process, which gives rise to fluctuations in the stretching exponent α around its average value. Thus, the variance in the stretching exponents is also expected to grow linearly at large distances. The linear growth in both the average stretching exponent and its variance is observed in Figure 4b.

Of greater interest, however, is the behavior on the scale of a Lyapunov length v_0/λ, corresponding to $100 < y < 150$ in Figure 4b. The marked dip in the average exponent $\overline{\alpha}(y)$ at short time scales indicates substantial local focusing, which is consistent with the visual evidence in Figure 1b. In this same region, we see in Figure 4b strong fluctuations in the stretching exponent (as measured by the variance), and correspondingly, in Figure 4a, large fluctuations in the ray density associated with the formation of the first and strongest hot spots. Here, we note that the density variance is closely related to the scintillation index, defined as $Var(I(y))/(\overline{I}(y))^2$, and in our case, $\overline{I}(y) \approx 1$ throughout. Subsequently, the variance in the density and the scintillation index decline as the number of independent trajectories contributing to the intensity at a given point grows exponentially when y is larger than a Lyapunov length, gradually washing out the pattern of hot spots and cold spots associated with extreme events.

The negativity of the average exponent $\overline{\alpha}$ at short time scales is rather surprising. This unexpected behavior can be confirmed analytically using perturbation theory over a small t, where it turns out that $\overline{\alpha}$ scales as t^3, with a prefactor depending on the correlation function of the random potential [29].

As the most significant fluctuations in both density and the stretching exponent were detected in the same spatial region, we confirm that local focusing of trajectories is directly correlated with the extreme high densities. Future work [29] extends the scaling relationship of the form (3) based on the scaling of the stretching-exponent statistics.

4. Conclusions and Outlook

We have seen that scattering of non-relativistic particles in a random weak potential field generates density patterns very similar to those observed for deep water ocean waves in the work of White and Fornberg [5]. The similarity in the intensity distributions in different physical contexts suggests the value of a universal perspective on the topic of scattering or refraction in a weak random disorder. Consequently, we explored the general connection between the density and stretching exponent in the context of non-relativistic particle motion. We conclude that the stretching exponent directly correlates with the density, and the negative correlation arises everywhere, including in the region of the strongest density fluctuations, and also further along the forward direction. By explicitly connecting these two quantities, we can treat the stretching exponent as a quantitative mirror for the intensity.

We also note that particle dynamics in a 2D ray model is analogous to a 1D model with a time-dependent potential in the regime of small angular spread. Whereas the full 4 by 4 monodromy matrix in the 2D model may be challenging for numerical evolution and even more so when it comes to an analytical treatment, the 1D model may be simple enough to obtain quantitative predictions for the distribution of the stretching exponent. Therefore, we are able to further explore the mechanism and statistics of freak wave events by studying monodromy matrix statistics in the 1D model [29].

Our calculations here are based on a model where nonlinear effects are absent. The linear model may be regarded a starting point for a more sophisticated nonlinear analysis. Nevertheless, nonlinear wave effects become significant only where the intensity produced by linear mechanisms is already large. It is therefore unlikely that nonlinear effects would substantially change the strong negative correlation between the density and stretching exponent.

While the long time behavior of the stretching exponent displays simple linear growth, the behavior on the scale of the Lyapunov time is of greater interest due to the appearance of the first generation of hot spots. A surprising minimum in the average stretching exponent is seen to correspond to a peak in the intensity variance in this regime. Future work [29] will investigate in much greater detail the statistics of the stretching exponent as a function of system parameters and distance from the origin, and explore the robustness of the stretching-exponent statistics with respect to changes in the equations of motion and the correlation properties of the random scattering potential.

Author Contributions: Conceptualization, S.C. and L.K.; methodology, S.C. and L.K.; investigation, S.C. and L.K.; writing—original draft preparation, S.C., writing—review and editing: L.K.; supervision: L.K. All authors have read and agreed to the published version of the manuscript.

Funding: This research was supported in part using high performance computing (HPC) resources and services provided by Technology Services at Tulane University, New Orleans, LA. This work was supported in part by the NSF under Grant No. PHY-1205788.

Institutional Review Board Statement: Not applicable.

Informed Consent Statement: Not applicable.

Data Availability Statement: Data available from the authors on request.

Conflicts of Interest: The authors declare no conflict of interest.

References

1. Dankert, H.; Hortsmann, J.; Lehner, S.; Rosenthal, W. Detection of wave groups in SAR images and radar image sequences. *IEEE Trans. Geosci. Remote Sens.* **2003**, *41*, 1437–1446. [CrossRef]
2. Longuet-Higgins, M.S. Detection of wave groups in SAR images and radar image sequences. *Philos. Trans. R. Soc. Ser. A* **1957**, *249*, 321–387.
3. Trulsen, K.; Dysthe, K. B. A modified nonlinear Schrödinger equation for broader bandwidth gravity waves on deep water. *Wave Motion* **1996**, *24*, 281–289. [CrossRef]
4. Dysthe, K.B. Note on a modification to the nonlinear Schrödinger equation for application to deep water waves. *Proc. R. Soc. Ser. A* **1979**, *369*, 105–114.
5. White, B.S.; Fornberg, B. On the chance of freak waves at sea. *J. Fluid Mech.* **1998**, *355*, 113–138. [CrossRef]
6. Kharif, C.; Pelinovsky, E. Physical mechanisms of the rogue wave phenomenon. *Eur. J. Mech. B* **2003**, *22*, 603–634. [CrossRef]
7. Topinka, M.A.; LeRoy, B.J.; Westervelt, R.M.; Shaw, S.E.J.; Fleischmann, R.; Heller, E.J.; Maranowski, K.D.; Gossard, A.C. Coherent branched flow in a two-dimensional electron gas. *Nature* **2001**, *410*, 183–186. [CrossRef] [PubMed]
8. Jura, M.P.; Topinka, M.A.; Urban, L.; Yazdani, A.; Shtrikman, H.; Pfeiffer, L.N.; West, K.W.; Goldhaber-Gordon, D. Unexpected features of branched flow through high-mobility two-dimensional electron gases. *Nat. Phys.* **2007**, *3*, 841–845. [CrossRef]
9. Aidala, K.E.; Parrott, R.E.; Kramer, T.; Heller, E.J.; Westervelt, R.M.; Hanson, M.P.; Gossard, A.C. Imaging magnetic focusing of coherent electron waves. *Nat. Phys.* **2007**, *3*, 464–468. [CrossRef]
10. Maryenko, D.; Ospald, F.; von Klitzing, K.; Smet, J.; Metzger, J.J.; Fleischmann, R.; Geisel, T.; Umansky, V. How branching can change the conductance of ballistic semiconductor devices. *Phys. Rev. B* **2012**, *85*, 195329. [CrossRef]
11. Kaplan, L. Statistics of branched flow in a weak correlated random potential. *Phys. Rev. Lett.* **2002**, *89*, 184103. [CrossRef]
12. Berry, M.V. Tsunami asymptotics. *New J. Phys.* **2005**, *7*, 129. [CrossRef]
13. Berry, M.V. Focused tsunami waves. *Proc. R. Soc. A* **2007**, *463*, 3055-3071. [CrossRef]
14. Dobrokhotov, S.Y.; Sekerzh-Zenkovich, S.Y.; Tirozzi, B.; Tudorovskii, T.Y. Description of tsunami propagation based on the Maslov canonical operator. *Doklady Math.* **2006**, *74*, 592–596. [CrossRef]
15. Patsyk, A.; Sivan, U.; Segev, M; Bandres, M.A. Observation of branched flow of light. *Nature* **2020**, *583*, 60–65. [CrossRef]
16. Arecchi, F.T.; Bortolozzo, U.; Montina, A.; Residori, S. Granularity and inhomogeneity are the joint generators of optical rogue waves. *Phys. Rev. Lett.* **2011**, *106*, 153901. [CrossRef]
17. Bonatto, C.; Feyereisen, M.; Barland, S.; Giudici, M.; Masoller, C.; Rios Leite, J.R.; Tredicce, J.R. Deterministic optical rogue waves. *Phys. Rev. Lett.* **2011**, *107*, 053901. [CrossRef]
18. Wolfson, M.A.; Tomsovic, S. On the stability of long-range sound propagation through a structured ocean. *J. Acoust. Soc. Am.* **2001**, *109*, 2693–2703. [CrossRef]
19. Colosi, J.A.; Baggeroer, A.B.; Birdsall, T.G.; Clark, C.; Cornuelle, B.D.; Costa, D.; Dushaw, B.D.; Dzieciuch, M.A.; Forbes, A.M.G.; Howe, B.M.; et al. A review of recent results on ocean acoustic wave propagation in random media: Basin scales. *IEEE J. Ocean. Eng.* **1999**, *24*, 138. [CrossRef]
20. Höhmann, R.; Kuhl, U.; Stöckmann, H.J.; Kaplan, L.; Heller, E.J. Freak waves in the linear regime: A microwave study. *Phys. Rev. Lett.* **2010**, *104*, 093901. [CrossRef]
21. Barkhofen, S.; Metzger, J.J.; Fleischmann, R.; Kuhl, U.; Stöckmann, H.-J. Experimental observation of a fundamental length scale of waves in random media. *Phys. Rev. Lett.* **2013**, *111*, 183902. [CrossRef] [PubMed]
22. Metzger, J.J.; Fleischmann, R.; Geisel, T. Universal statistics of branched flows. *Phys. Rev. Lett.* **2010**, *105*, 020601. [CrossRef]
23. Metzger, J.J.; Fleischmann, R.; Geisel, T. Statistics of extreme waves in random media. *Phys. Rev. Lett.* **2014**, *112*, 203903. [CrossRef]
24. Degueldre, H.; Metzger, J.J.; Schultheis, E.; Fleischmann, R. Channeling of branched flow in weakly scattering anisotropic media. *Phys. Rev. Lett.* **2017**, *118*, 024301. [CrossRef] [PubMed]
25. Šťavina, J.; Bokes, P. Quantum and classical branching flow in space and time. *Phys. Rev. A* **2022**, *106*, 052215. [CrossRef]
26. Heller, E.J.; Fleischmann, R; Kramer T. Branched flow. *Phys. Today* **2021**, *74*, 44–51. [CrossRef]
27. Heller, E.J.; Kaplan, L.; Dahlen, A. Refraction of a Gaussian seaway. *J. Geophys. Res.* **2008**, *113*, C09023. [CrossRef]

28. Heller, E.J. Freak waves: Just bad luck, or avoidable? *Europhys. News* **2005**, *35*, 159–162. [CrossRef]
29. Chen, S.; Kaplan, L. Universality of ray focusing statistics in two-dimensional and one-dimensional potential flows. *in preparation*.
30. Heller E. J.; Shaw S. Branching and fringing in microstructure electron flow. *Int. J. Mod. Phys. B* **2003**, *17*, 3977–3987. [CrossRef]
31. Chen, S.; Kaplan, L. Ray divergence statistics as a function of disorder in weak semi-classical scattering. *in preparation*.
32. Gutzwiller, M.C. *Chaos in Classical and Quantum Mechanics*; Springer: Berlin, Germany, 1990.

entropy

MDPI

Article

Scarring in Rough Rectangular Billiards

Felix M. Izrailev [1,2,†,*], German A. Luna-Acosta [1,†] and J. A. Mendez-Bermudez [1,†]

1 Instituto de Física, Benemérita Universidad Autónoma de Puebla, Apartado Postal J-48, Puebla 72570, Mexico
2 Department of Physics and Astronomy, Michigan State University, East Lansing, MI 48824-1321, USA
* Correspondence: felix.izrailev@gmail.com
† These authors contributed equally to this work.

Abstract: We study the mechanism of scarring of eigenstates in rectangular billiards with slightly corrugated surfaces and show that it is very different from that known in Sinai and Bunimovich billiards. We demonstrate that there are two sets of scar states. One set is related to the bouncing ball trajectories in the configuration space of the corresponding classical billiard. A second set of scar-like states emerges in the momentum space, which originated from the plane-wave states of the unperturbed flat billiard. In the case of billiards with one rough surface, the numerical data demonstrate the repulsion of eigenstates from this surface. When two horizontal rough surfaces are considered, the repulsion effect is either enhanced or canceled depending on whether the rough profiles are symmetric or antisymmetric. The effect of repulsion is quite strong and influences the structure of all eigenstates, indicating that the symmetric properties of the rough profiles are important for the problem of scattering of electromagnetic (or electron) waves through quasi-one-dimensional waveguides. Our approach is based on the reduction of the model of one particle in the billiard with corrugated surfaces to a model of two artificial particles in the billiard with flat surfaces, however, with an effective interaction between these particles. As a result, the analysis is conducted in terms of a two-particle basis, and the roughness of the billiard boundaries is absorbed by a quite complicated potential.

Keywords: scars; localization effects; quantum billiards; rough billiards

Citation: Izrailev, F.M.; Luna-Acosta, G.A.; Mendez-Bermudez, J.A. Scarring in Rough Rectangular Billiards. *Entropy* **2023**, *25*, 189. https://doi.org/10.3390/e25020189

Academic Editor: Marko Robnik

Received: 22 December 2022
Revised: 11 January 2023
Accepted: 11 January 2023
Published: 18 January 2023

1. Introduction

Recently, there has been a sharp increase in articles discussing the occurrence of ergodicity in physical systems, both classical and quantum [1–8]. This interest is due to the practical problem of statistical description of the behavior of a closed many-particle system whose equations of motion are strictly deterministic, that is, do not contain any random parameters. For a long time, this problem was considered fundamental for the justification of statistical mechanics, and it was assumed that its solution is associated with the so-called ergodicity of the motion of the system under consideration. The classical trajectory of ergodic systems densely and uniformly covers the entire phase space of the system, as a result of which the time average value of any observable is equal to the phase space average as time tends to infinity.

As a result of numerous attempts to prove this property in general terms for real physical systems, it was realized that the proof is possible only for special systems. The most well-known of such systems are the so-called scattering billiards, in particular, Sinai and Bunimovich billiards, in which the trajectory of a particle due to multiple elastic reflections from the walls is unstable for any initial condition except for special values whose measure is equal to zero. Thus, the behavior of the system is determined by a single trajectory, which makes it possible to speak of an ergodic type of motion.

It should be noted that the instability of motion in such billiards is exponentially strong, at which the distance between two close trajectories grows, on average, exponentially fast with time. When trajectories are reflected from the walls of the billiard, the so-called mixing occurs, as a result of which the motion becomes chaotic and indistinguishable from random. Indeed, as already discussed by the Soviet physicist N. S. Krylov in his book [9], the mixing is the main mechanism for the emergence of the most characteristic property of the statistical behavior, namely, the relaxation of a system to statistical equilibrium. Conversely, examples can be given when a given system is ergodic but its movement is regular, meaning that the relaxation of the system occurs at infinite times, which does not correspond to physical reality.

For quantum systems, the situation is much more complicated, since there is no rigorous and generally accepted definition of quantum ergodicity. The only reliable situation when one can speak of quantum ergodicity is for the systems that are fully ergodic in the classical limit. Needless to say, there are just a few such systems. As an example of the "true" quantum ergodicity, one can consider completely random matrices of large size for which the theory is well developed. For such matrices, the components of the eigenfunctions are Gaussian distributed, which is a property that is a consequence of the uniform distribution of the eigenfunction vector over the surface of the N-dimensional sphere with $N \gg 1$. It can be seen that in this case, the eigenfunctions are not only ergodic but also maximally chaotic (random), see e.g., [10].

However, for realistic physical systems, many-body matrix elements in a physically chosen basis, as a rule, are not completely random and fill only a band of finite width (determined by a finite radius of interaction in the energy space). Therefore, one can speak of ergodic eigenfunctions occupying only some energy shell of finite size. Clearly, determining the shape and size of this shell is extremely difficult [11]. In any case, at the moment, there is no unified theory of quantum ergodicity of isolated systems which takes into account the influence of the finiteness of the energy shell.

Returning to systems with a small number of particles (small number of degrees of freedom), for example chaotic billiards for which classical ergodicity is rigorously proven, after several years of their study, it was found that many of the eigenfunctions are non-ergodic even for very high energies [12,13]. In principle, such non-ergodic states do not contradict Shnirel'man's theorem [14], according to which their number must decrease with increasing energy. Numerical experiments with ergodic billiards have shown that in many cases, these "scarring" states in the configuration space can be associated with unstable periodic orbits, the number of which increases with energy. However, since these trajectories are isolated, when averaged over infinite time, they do not contribute to the mean values of the observables.

To date, there have been many papers that investigate the properties of many-particle scarring states in various physical systems (see, for example, [15–21] and references therein). The main interest is related to the mechanism for the emergence of such states immersed in a set of ergodic states. In the case of quantum systems that do not have a corresponding classical analogue, this issue becomes extremely complicated. It was found that the main characteristic of such scars in the quantum description is a small number of components of the corresponding eigenstates treated on some physically justified basis in comparison with the large number of states that can be considered as ergodic. Thus, scars can be defined as the localized states, associated with unstable periodic orbits when applicable, and embedded into the set of the ergodic states. As a result of numerous studies of quantum many-body systems, it became clear that the typical mechanism for scars is the presence of local symmetries of the system, which can correspond to the presence of local integrals of motion [20,21].

Our interest in this article is to study the mechanism for the emergence of scar states in rectangular billiards with rough horizontal surfaces. In these billiards, it is relatively easy to trace their emergence since such systems can be considered close to integrable, despite the fact that they are seem to be rigorously ergodic even for any weak boundary roughness. Such billiards of finite size were under close investigation in view of surface scattering in quasi-one-dimensional waveguides. The conventional theory of scattering in the waveguides with a weak roughness was developed long ago; however, recently, researchers found a new mechanism of surface scattering which is due to the so-called square-gradient roughness [22–25]. This mechanism was neglected in previous studies; however, it should be taken into account when the correlation length computed along the scattering profiles is very small. In particular, it was analytically shown that the scattering properties strongly depend on the correlations between upper and lower horizontal profiles. Specifically, the scattering is very different whether the two random profiles are symmetric or antisymmetric. Our study also shows a quite strong dependence of the degree of localization of the scar states on the type of correlations between the scattering profiles.

In Section 2, we describe the model under study, giving exact expressions for the matrix elements of an effective Hamiltonian in close correspondence with the model. In Section 3, we explain how we measure the degree of localization of the eigenstates in the energy basis corresponding to non-interacting artificial particles and therefore to the billiard with flat horizontal surfaces. In this representation, one easily identifies one set of scar states related to classical trajectories which are parallel to the rough boundaries of the billiard. We also show what these eigenstates look like in the configuration representation for one rough surface. In addition, we demonstrate that there is another set of scar states related to bouncing ball classical trajectories which are perpendicular to the rough boundaries of the billiard. These states consist of many "unperturbed eigenstates", which is in contrast with those from the first set. In Section 5, we analytically show how the number of strongly localized states from the first set, which corresponds to classical plane waves with the longest perpendicular wavelength, decreases with energy. In Section 6, we study billiards with two horizontal rough profiles and show how the structure of localized eigenstates depends on whether the profiles are symmetric or antisymmetric. The main attention here is paid to the effect of the repulsion of eigenstates from the corrugated surfaces. In Section 7, we draw our conclusions.

2. Model of Rough Billiards

We consider billiards that are periodic in the longitudinal coordinate x, with Dirichlet boundary conditions on the upper f_1 and lower f_2 surfaces,

$$\begin{aligned} f_1 &= L_y + W_1\xi_1(x)\,, \\ f_2 &= W_2\xi_2(x)\,. \end{aligned} \tag{1}$$

Here, L_y is the average width of the billiard and $\xi_{1,2}(x + L_x) = \xi_{1,2}(x)$ with $\langle \xi_{1,2}(x) \rangle = 0$. The angular brackets stand for the average over one period L_x or, in the case of a random profile, over different realizations of $\xi_{1,2}(x)$.

This model has been thoroughly studied in refs. [26–35] for the specific case $\xi_1(x) = \cos(2\pi x/L_x)$ and $\xi_2(x) = 0$, the so-called "cosine" or "ripple" billiard [36]. The main interest was in the properties of the energy spectrum [26,28] and in the quantum–classical correspondence for the shape of eigenstates (SE) and local density of states (LDOS) [29–32]. It was shown that for highly excited states, the global properties of the SES and LDOS in the quantum regime are similar to those described by the equations of motion for a classical particle moving inside the billiard. On the other hand, quite strong quantum effects have been revealed for individual eigenstates in a deep semiclassical region [31,32].

In this paper, we address the case of billiards with rough surfaces,

$$\xi_{1,2}(x) = \sum_{k=1}^{N_T} A_k \cos\left(\frac{2\pi k x}{L_x}\right), \tag{2}$$

focusing on the structure of eigenstates, see also [22–25,37–39]. The surfaces are modeled by a large sum of harmonics A_k drawn from a flat random distribution defined in the interval $[-\mathcal{A}, \mathcal{A}]$ (with $\mathcal{A}$ such that $|\xi_{1,2}(x)| \leq 1$). With the increase of N_T, the degree of complexity of $\xi_{1,2}(x)$ also increases, and for $N_T \gg 1$, the surfaces can be treated as random: see an example in Figure 1. In what follows, we focus on rough billiards with $N_T = 100$. For simplicity, we choose $\xi_1(x) = \xi_2(x)$. In the first part of our paper, we review the case of billiards with one flat boundary; that is $W_2 = 0$, see, e.g., Figure 1. Below, in Section 6, we consider the billiards with two antisymmetric, $W_1 = W_2$, and symmetric surfaces, $W_1 = -W_2$.

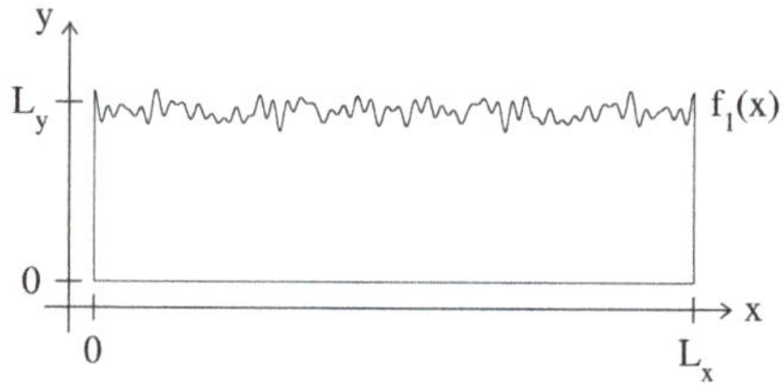

Figure 1. Example geometry for the rough billiard. $f_1(x) = L_y + W_1\xi_1(x)$ and $f_2(x) = 0$; see Equation (1). Here, $\xi_1(x)$ is given by Equation (2) with $N_T = 30$ and $W_1 = L_y/10$. The harmonics A_k used for $\xi_1(x)$ are drawn from a flat random distribution defined in the interval $[-\mathcal{A}, \mathcal{A}]$ (with $\mathcal{A}$ such that $|\xi_1(x)| \leq 1$).

Originally, the model is described by the Hamiltonian

$$\hat{H} = \frac{1}{2m}(\hat{P}_x^2 + \hat{P}_y^2) = -\frac{\hbar^2}{2m}(\partial_x^2 + \partial_y^2), \quad (3)$$

for a free particle inside the billiard with rough boundaries (such that the corresponding wave function $\Psi(x,y)$ obeys Dirichlet boundary conditions, $\Psi(x,y) = 0$ at $y = f_1(x)$ and $y = f_2(x)$). However, in order to solve it numerically, it is useful to make a canonical transformation to new variables in which the new Hamiltonian incorporates surface-scattering effects into an effective interaction potential between artificial particles identified with the two new degrees of freedom. This can be achieved by the transformation to new canonical coordinates,

$$\begin{aligned} u &= x\,, \\ v &= \frac{f_2(x) - y}{f_2(x) - f_1(x)}\,. \end{aligned} \quad (4)$$

As a result, the boundary conditions for the new wave function are trivial: $\Psi(u,v) = 0$ at $v = 0$ and $v = 1$.

The Schrödinger equation in new coordinates can be obtained from the covariant expression for a particle moving (in the absence of potentials) in a Riemannian curved space [40],

$$-\frac{\hbar^2}{2m}\Delta_{cov}\Psi(u,v) = \frac{\hbar^2}{2m}g^{-1/2}\partial_\alpha g^{\alpha\beta}g^{1/2}\partial_\beta\Psi(u,v)\,, \quad (5)$$

where the quantum Hamiltonian in covariant form [41] is

$$\hat{H} = \frac{1}{2m}g^{-1/4}\hat{P}_\alpha g^{\alpha\beta}g^{1/2}\hat{P}_\beta g^{-1/4}\,, \quad (6)$$

and the covariant momenta are

$$\hat{P}_\alpha = -i\hbar\left[\partial_\alpha + \frac{1}{4}\partial_\alpha ln(g)\right] = -i\hbar g^{-1/4}\partial_\alpha g^{1/4}\,. \quad (7)$$

Here, $\alpha, \beta = u, v$, the operator Δ_{cov} is the Laplace–Beltrami operator, g is the metric, and $g^{\alpha\beta}$ is the metric tensor. The wave functions $\Psi(u,v)$ are normalized as

$$\int_0^{L_x} du \int_0^{L_y} dv\sqrt{g}\Psi^\dagger\Psi = 1\,. \quad (8)$$

Even though (5) is still the kinetic energy, the resulting differential equation takes a much more complicated form than the ordinal Laplacian. This is the price to pay when transferring the effect of the boundaries onto the operator. Then, by substituting in (6) the explicit expressions for the metric tensor

$$g^{\alpha\beta} = \begin{pmatrix} 1 & -\dfrac{f_1'v + f_2'(1-v)}{f_1 - f_2} \\ -\dfrac{f_1'v + f_2'(1-v)}{f_1 - f_2} & \dfrac{1 + [f_1'v + f_2'(1-v)]^2}{(f_1 - f_2)^2} \end{pmatrix},$$

where $f_{1,2}' \equiv \partial f_{1,2}(u)/\partial u$, and the metric

$$g = Det(g_{\alpha\beta}) = (f_1 - f_2)^2 , \tag{9}$$

we obtain

$$\begin{aligned} \hat{H} &= \frac{1}{2\mathrm{m}} g^{-1/4} \Big\{ \hat{P}_u (f_1 - f_2) \hat{P}_u + \hat{P}_v \frac{1 + [f_1'v + f_2'(1-v)]^2}{(f_1 - f_2)^2} \hat{P}_v \\ &\quad - \big[\hat{P}_u (f_1'v + f_2'(1-v)) \hat{P}_v + \hat{P}_v (f_1'v + f_2'(1-v)) \hat{P}_u \big] \Big\} g^{-1/4} . \end{aligned} \tag{10}$$

Note that the rough boundaries $f_{1,2}$ and their derivatives $f_{1,2}'$ are fully incorporated into the Hamiltonian operator.

We remark that this representation allows us to treat the original model of one free particle in the rough billiard as a model of two interacting "particles" identified with the two degrees of freedom u and v, where the Hamiltonian (10) is separated as

$$\hat{H} = \hat{H}^0 + \hat{V}(u, v, \hat{P}_u, \hat{P}_v) , \tag{11}$$

with

$$\hat{H}^0 = \frac{1}{2\mathrm{m}} (\hat{P}_u^2 + \hat{P}_v^2). \tag{12}$$

Here, $\hat{P}_u = -i\hbar[\partial_u + (1/4)\partial_u \ln(g)]$, $\hat{P}_v = -i\hbar\partial_v$, and $\hat{V}$ stands for an effective interaction potential between the "particles" u and v. In the following, we treat $\hat{H}^0$ as the Hamiltonian of two *non-interacting particles*. The eigenstates of $\hat{H}^0$ define the unperturbed basis in which the eigenstates of the total Hamiltonian $\hat{H}(u, v, \hat{P}_u, \hat{P}_v)$ are expanded. Such a representation turns out to be convenient for the study of the chaotic properties of the model, since one can use the tools and concepts developed in the theory of interacting particles (see, for example, [42,43]).

Now, we expand the αth eigenstate of energy E^α as

$$\Psi^\alpha(u, v) = \sum_m \sum_n C_{mn}^\alpha \phi_{mn}(u, v) , \tag{13}$$

where $\phi_{mn}(u, v) = \langle u, v \mid m, n \rangle$ are the eigenstates of the unperturbed Hamiltonian $\hat{H}^0(u, v)$, and the indexes n and m are the quantum numbers corresponding to the u and v coordinates, respectively. The function $\phi_{mn}(u, v)$ has to satisfy the boundary conditions of the problem in order to form a Galerkin basis [44]. The eigenstates of $\hat{H}^0$ are

$$\phi_{mn}(u, v) = \frac{1}{\pi^{1/2} g^{1/4}} \sin\left(\frac{m\pi v}{L_y}\right) \exp\left[i\left(k + \frac{2\pi n}{L_x}\right) u\right], \tag{14}$$

with eigenvalues

$$E_{mn}^0(k) = \frac{\hbar^2}{2\mathrm{m}} \left[\left(k + \frac{\pi n}{L_x}\right)^2 + \left(\frac{\pi m}{L_y}\right)^2 \right] . \tag{15}$$

Here, formally, $-\infty < n < \infty$ and $1 \leq m < \infty$, but in practice, for the numerical computations, the unperturbed basis is truncated such that $-N_{max} < n < N_{max}$ and $1 \leq m < M_{max}$. The factor $\pi^{-1/2}g^{-1/4}$ in (14) arises from the orthonormality condition (8).

Then, we can solve the eigenvalue problem by diagonalizing the Hamiltonian (10) on the basis $\phi_{mn}(u,v)$:

$$\sum_{m'}\sum_{n'} H_{mn,m'n'}\phi_{m'n'}(u,v) = E\phi_{mn}(u,v)\,, \tag{16}$$

where the matrix elements are

$$H_{mn,m'n'} \equiv \langle mn|\hat{H}|m'n'\rangle$$
$$= -\frac{\hbar^2}{2\mathrm{m}}\int_0^{L_x} du \int_0^{L_y} dv\, g^{1/2}\left[\phi^{\dagger}_{mn}\right]\Delta_{cov}[\phi_{m'n'}] = -\frac{\hbar^2}{2\mathrm{m}}\int_0^{L_x} du \int_0^{L_y} dv\, \partial_\alpha\left[\phi^{\dagger}_{mn}\right] g^{1/2} g^{\alpha\beta}\partial_\beta[\phi_{m'n'}]\,.$$

Since the Hamiltonian (10) is periodic in u, its eigenstates are Bloch states. This allows us to write the solution of the Schrödinger equation in the form $\Psi(u,v) = \exp(iku)\,\Psi_k(u,v)$, with $\Psi_k(u+L_x,v) = \Psi_k(u,v)$. Here, the Bloch wave vector $k(E)$ is in the first Brillouin zone, $(-\pi/L_x \leq k \leq \pi/L_x)$. The statistical properties of eigenstates do not depend on a specific value of the Bloch index k inside the band, except at $k=0$ and $k = \pm\pi/L_x$, so we avoid these values of k in our numerical calculations.

Finally after some algebra, we obtain

$$H_{mn,m'n'} = \frac{\hbar^2}{2\mathrm{m}}\left\{\left(k+\frac{\pi n}{L_x}\right)^2\delta_{nn'}\delta_{mm'} + \frac{m^2\pi^2}{L_x}\left[J_1 + J_2 + \left(\frac{1}{3}+\frac{1}{4m^2\pi^2}\right)J_3\right]\delta_{mm'}\right.$$
$$\left. +i\,\frac{4mm'}{(m^2-m'^2)L_x}\left(k+\frac{\pi(n+n')}{L_x}\right)\left[J_4 - (-1)^{m+m'}J_5\right] + \frac{4mm'(m^2+m'^2)}{(m^2-m'^2)^2L_x}\left[(-1)^{m+m'}J_6 - J_7\right]\right\} \tag{17}$$

where

$$J_1 = \int_0^{L_x} dx\,\exp\left[-i\frac{2\pi}{L_x}(n-n')x\right]\frac{f_1'f_2'}{(f_1-f_2)^2}\,,$$
$$J_2 = \int_0^{L_x} dx\,\exp\left[-i\frac{2\pi}{L_x}(n-n')x\right]\frac{1}{(f_1-f_2)^2}\,,$$
$$J_3 = \int_0^{L_x} dx\,\exp\left[-i\frac{2\pi}{L_x}(n-n')x\right]\frac{(f_1'-f_2')^2}{(f_1-f_2)^2}\,,$$
$$J_4 = \int_0^{L_x} dx\,\exp\left[-i\frac{2\pi}{L_x}(n-n')x\right]\frac{f_2'}{f_1-f_2}\,,$$
$$J_5 = \int_0^{L_x} dx\,\exp\left[-i\frac{2\pi}{L_x}(n-n')x\right]\frac{f_1'}{f_1-f_2}\,,$$
$$J_6 = \int_0^{L_x} dx\,\exp\left[-i\frac{2\pi}{L_x}(n-n')x\right]\frac{(f_1'-f_2')f_1'}{(f_1-f_2)^2}\,,$$
$$J_7 = \int_0^{L_x} dx\,\exp\left[-i\frac{2\pi}{L_x}(n-n')x\right]\frac{(f_1'-f_2')f_2'}{(f_1-f_2)^2}\,.$$

Notice that depending on the billiard geometry, some of the integrals above may vanish. For example, in the case $W_2 = 0$ (i.e., for the billiard of Figure 1), we have $J_1 = J_4 = J_7 = 0$.

3. Structure of the Hamiltonian Matrix

In order to study the structure of eigenstates of the total Hamiltonian $\hat{H}(u,v)$ one needs, first, to choose a way of ordering the unperturbed basis in which we represent the Hamiltonian matrix $H_{l,l'}(k) = \langle l \mid \hat{H} \mid l'\rangle_k$. Specifically, we have to assign an index l, labeling the basis state $\mid l\rangle_k \equiv \mid m,n\rangle_k$ to each pair of indecies (m,n) (note that although the energy spectra are independent of the assignment $(m,n)\to l$, the structure of the eigenstates is not). In addition, note that the size of the Hamiltonian matrix is determined by the maximum values of n and m: $-N_{max} \leq n \leq N_{max}$ and $1 \leq m \leq M_{max}$.

A natural assignment is the following one [30]. Let us fix the lowest value of n $(-N_{max})$ and sweep all values of m $(1 \leq m \leq M_{max})$. This gives $l = 1, 2, \ldots, M_{max}$. Then, do the same for $n = -N_{max} + 1$, which gives $M_{max} + 1 \leq l \leq 2M_{max}$, and so on, until finally, we have $1 \leq l \leq L_{max}$, where $L_{max} = (2N_{max} + 1)M_{max}$ defines the Hamiltonian matrix size, $L_{max} \times L_{max}$. This rule results in a block structure of the Hamiltonian matrix with a block size equal to M_{max}. Figure 2 (left) shows the lower part of a 4030×4030 matrix with $(N_{max}, M_{max}) = (32, 62)$. Here, we can see a number of blocks of size 62×62 corresponding to $n, n' = -32, \ldots, -17$. In this representation, the matrix is clearly a block matrix.

The above way of ordering the unperturbed basis corresponds to the "channel representation" (or momentum representation), since the index m labels a specific transverse channel for the propagation of the wave through the billiard; see Equation (14). However, for our purposes, it is essential to use the "energy representation", according to which the unperturbed basis is ordered in increasing energy, $E^0_{l_{new}+1}(k) \geq E^0_{l_{new}}(k)$. This defines a new rule $l \to l_{new} = l_{new}(n, m)$. In Figure 2 (right), we show the lower part of the Hamiltonian matrix of Figure 2 (left) but now in the "energy representation". Notice that in contrast to the block structure of the Hamiltonian matrix in the "channel representation", in the "energy representation", the Hamiltonian matrix shows a band-like structure.

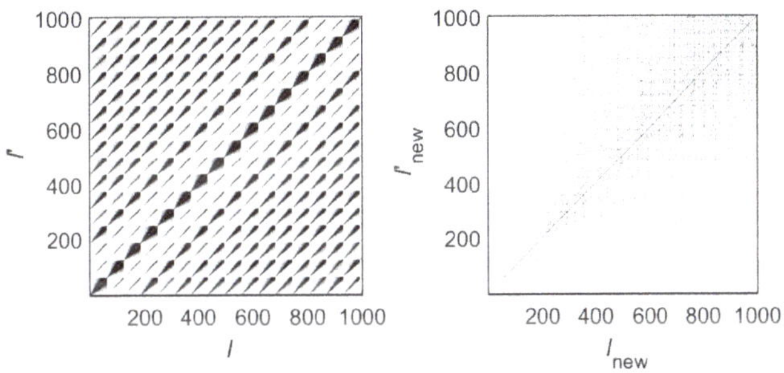

Figure 2. (**left**) Lower part of the Hamiltonian matrix in the "channel representation" $| H_{l,l'} |$. (**right**) Lower part of the Hamiltonian matrix in the "energy representation" $| H_{l_{new},l'_{new}} |$. $L_x = 2\pi$, $L_y = 2\pi$, $W_1 = 0.06L_y$, $k = 0.1$, $N_T = 100$, and $(N_{max}, M_{max}) = (32, 62)$ were used. With this choice of N_{max} and M_{max}, we obtain $L_{max} = 4030$. Both matrices are shown in grayscale where darker means higher amplitude of the matrix element.

In order to analyze the structure of the eigenstates of $\hat{H}(u, v)$ in detail, we diagonalize Hamiltonian matrices in the "energy representation" and construct the "state matrices" $|C^{\alpha}_{l_{new}}|^2$. In Figure 3, we show the lower part of the state matrix corresponding to the Hamiltonian matrix of Figure 2 (right). Here, $C^{\alpha}_{l_{new}}$ are the amplitudes of the eigenstates in the (energy-ordered) basis representation given by the index l_{new}. Namely, the index l_{new} refers to unperturbed basis states that correspond to the unperturbed Hamiltonian $\hat{H}^0$. The index α refers to a specific exact eigenstate. All eigenstates are ordered in increasing energy, with $\alpha = 1$ the ground state. Therefore, to understand how strongly localized/extended the exact eigenstates in the unperturbed basis are, one should fix the value of α and explore the dependence of $|C^{\alpha}_{l_{new}}|^2$ on l_{new}.

A crucial point in our study is that the eigenstates of the Hamiltonian in the "energy representation" have a very convenient form for the analysis. The advantage of the "energy representation" over the "channel representation" (i.e., when the Hamiltonian matrix has a block structure) is clearly seen in Figure 4 where an arbitrarily chosen eigenstate is given in the two representations. One can see that in the "channel representation", the eigenstate has a kind of regular and extended structure, while in the "energy representation", the eigenstate is compressed. In the latter case, one may use a statistical approach to describe the global properties of such eigenstates; see refs. [43,45]. Specifically, these eigenstates can be characterized by introducing an envelope around which the components are expected to fluctuate in a pseudo-random way. We stress that by using this energy ordering, it is possible to relate the global form of eigenstates in the energy representation with its classical counterpart; see, e.g., refs. [29–32].

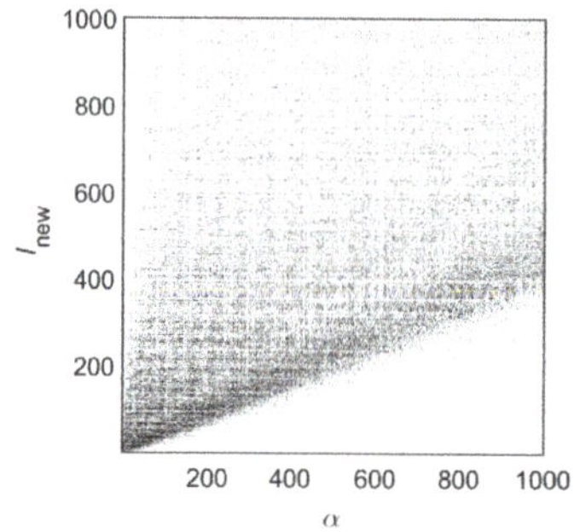

Figure 3. Lower part of the state matrix $|C^{\alpha}_{l_{new}}|^2$ from the Hamiltonian matrix of Figure 2 (right); that is, $L_x = 2\pi$, $L_y = 2\pi$, $W_1 = 0.06L_y$, $k = 0.1$, $N_T = 100$, and $(N_{max}, M_{max}) = (32, 62)$ were used. The matrix is shown in grayscale where darker means higher amplitude of the matrix element.

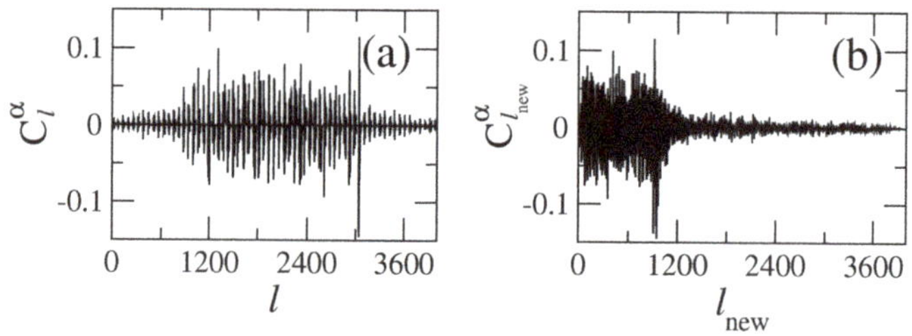

Figure 4. Example of an eigenstate obtained from the Hamiltonian in (**a**) the "channel representation" and (**b**) the "energy representation". $L_x = 2\pi$, $L_y = 2\pi$, $W_1 = 0.06L_y$, $k = 0.1$, and $N_T = 100$ were used. (**a**) The eigenstate $\alpha = 392$ as a function of l and (**b**) the same eigenstate as a function of l_{new}.

4. Eigenstates in Energy Representation

In Figure 5, we present two typical pairs of consecutive eigenstates ($\alpha = 390$ and 391 and $\alpha = 407$ and 408). The difference between the eigenstates on the left panels ($\alpha = 390$ and 407) and the other two is clearly qualitative. More specifically, while the states $\alpha = 390$ and 407 are *extended* (in energy) eigenstates, constituted by practically all basis states within the shown energy range, the eigenstates $\alpha = 391$ and 408 are mostly unperturbed: they are extremely *localized* in energy. Indeed, by neglecting all small amplitude components surrounding the main component (see left panels of Figure 5), we can determine the basis state l_{new}, defined by the pair (m, n), that most closely resembles the exact eigenstates. We find that this always corresponds to the lowest values of the transversal mode m. This fact can be understood by the following physical argument. Consider an eigenstate of the flat billiard $\phi_{m,n}(X, Y, k = 0) \propto \sin(m\pi Y/L_y)\exp(iK_x X)$ with energy $E^0 = (\hbar^2/2m_e)(K_x^2 + K_y^2)$, where $K_y = m\pi/L_y = 2\pi/\Lambda_y$. Turning on the perturbation (flat to rough billiard) will affect the high energy unperturbed states differently depending mainly on the value of Λ_y. For example, for $m = 1$, the ratio Λ_y/W_1 is $2L_y/W_1 \approx 33$ (with $W_1 = 0.06L_y$, the value of W_1 we use throughout this work), which is so large that the state cannot "see" the roughness and thus will remain essentially unperturbed. In contrast, for unperturbed states with the same (or about the same) energy but with large values of m (say, $m = 62 = M_{max}$ and correspondingly small K_x), their Λ_y is sufficiently small compared to the amplitude of the roughness ($\Lambda_y/W_1 \approx 0.5$), so that the rough boundary produces a strong mixing of unperturbed levels. The resulting exact eigenstate will consist of many components extended over the energy. Note that we may treat strongly localized states in the channel representation as a kind of scar state.

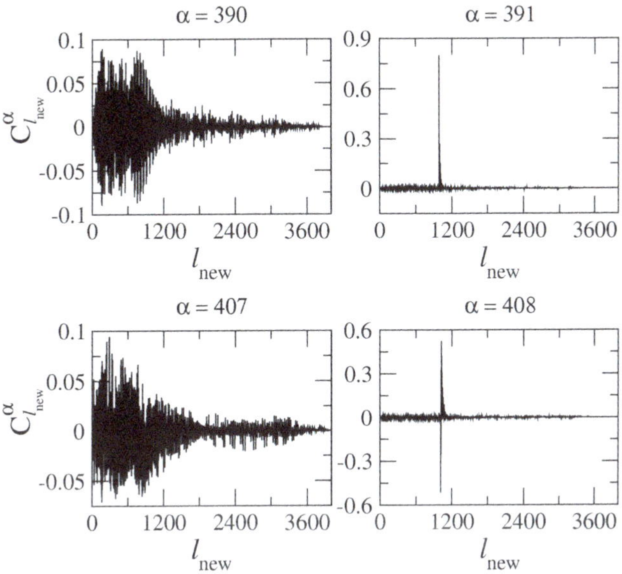

Figure 5. Typical pairs of consecutive eigenstates ($\alpha = 390, 391, 407$ and 408) in the energy representation for $L_x = 2\pi$, $L_y = 2\pi$, $W_1 = 0.06L_y$, $k = 0.1$, and $N_T = 100$.

The fact that the extremely localized (in energy) eigenstates can be identified with the plane waves $\phi_{m,n}(x,y)$ with small m, proper of the flat billiard, makes us expect that they will also be similar to plane waves when presented in configuration representation $|\Psi^{\alpha}(x,y)|^2$. Then, in Figure 6 (right), the two localized (in energy) eigenstates of Figure 5 (right) are shown in configuration representation. It is quite unexpected that these eigenstates are very different from the unperturbed ones, even though they are similar in energy representation. Figure 6 (right) shows that the rough boundary "pushes" the probability $|\Psi^{\alpha}(x,y)|^2$ away from it. Thus, the eigenstate $\alpha = 391$, whose main component in energy representation is identified with $m = 1$, differs importantly from the unperturbed mode with $m = 1$ whose maximum is at the billiard center $y = L_y/2$ (as in the case of the lowest eigenmode of a box of width L_y with hard walls). Similar repulsion occurs for the eigenstate $\alpha = 408$ identified with $m = 2$. We stress that this repulsion effect occurs only for rough billiards and is stronger the more rough the billiard boundary is, see, e.g., ref. [39]. A detailed analysis of this repulsion effect will be performed in Section 6. Finally, for comparison purposes, in Figure 6 (left), we also present the two extended (in energy) eigenstates of Figure 5 (left).

In order to characterize quantitatively the eigenstates, we compute various localization measures. The first one is the so-called *entropy localization length* l_H,

$$l_H = \exp\left[-(\mathcal{H} - \mathcal{H}_{\mathcal{GOE}})\right] \approx 2.08 \exp(-\mathcal{H}). \tag{18}$$

Here, $\mathcal{H}$ stands for the Shannon entropy of an eigenstate in a given basis,

$$\mathcal{H} = \sum_{l_{new}=1}^{N} w^{\alpha}_{l_{new}} \ln w^{\alpha}_{l_{new}}, \tag{19}$$

and $\mathcal{H}_{\mathcal{GOE}}$ is the entropy of a completely chaotic state which is characterized by Gaussian fluctuations (for $N \to \infty$) of all components $C^{\alpha}_{l_{new}}$ with the same variance $\langle w^{\alpha}_{l_{new}} \rangle = 1/N$, where $w^{\alpha}_{l_{new}} = |C^{\alpha}_{l_{new}}|^2$. The latter property occurs for completely random matrices belonging to a Gaussian Orthogonal Ensemble (GOE). Defined in this way, the quantity l_H gives the measure of the effective number of components in an eigenstate. For example, the eigenstates of Figure 5 have $l_H(\alpha = 390) = 1189.7$, $l_H(\alpha = 391) = 5.75$, $l_H(\alpha = 407) = 1517.5$, and $l_H(\alpha = 408) = 13.16$; that is, extended eigenstates have large values of l_H, while localized eigenstates are characterized by small values of l_H.

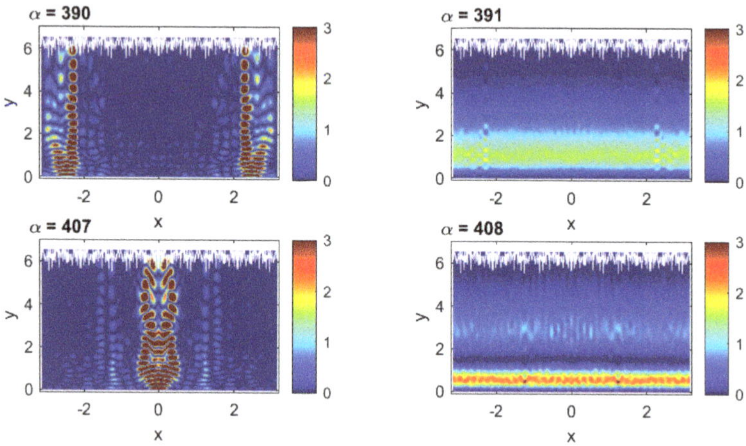

Figure 6. The eigenstates of Figure 5 in the configuration representation $|\Psi^{\alpha}(x,y)|^2$. The scale of the color code on the right of the panels should be multiplied by 10^{-3}.

The second quantity, l_{ipr}, which gives another measure of the effective number of components in an eigenstate, is expressed via the *inverse participation ratio* $\mathcal{P}$,

$$l_{ipr} = \left[\frac{\mathcal{P}_{\mathcal{GOE}}}{\mathcal{P}}\right] \approx \frac{3}{\mathcal{P}} \tag{20}$$

with

$$\mathcal{P} = \sum_{l=1}^{N} (w^{\alpha}_{l_{new}})^2. \tag{21}$$

where $\mathcal{P}_{\mathcal{GOE}} \approx 3$ is chosen in order to obtain $l_{ipr} = N$ in the GOE limit case. Correspondingly, the eigenstates of Figure 5 have $l_{ipr}(\alpha = 390) = 1061.2$, $l_{ipr}(\alpha = 391) = 6.32$, $l_{ipr}(\alpha = 407) = 1464.4$, and $l_{ipr}(\alpha = 408) = 14.99$, where a high correlation with l_H can be seen. The above two definitions of localization lengths are the most frequently used when describing the global structure of eigenstates. One should note that these quantities provide an estimate of the effective number of large components, independently on the location of these components, in the unperturbed basis.

To obtain a complete panorama, in Figure 7a,b we plot these two measures, l_H and l_{ipr}, for the eigenstates $|\alpha\rangle$ of the rough billiard. The strong fluctuations of the localization measures are evident in these figures. We can see that neighboring high-energy eigenstates may have drastically different localization measures, which is in agreement with the discussion above about the existence of localized and extended eigenstates. Moreover, these figures give us information about the relative number of each type (localized and extended eigenstates) to be found in a given energy range.

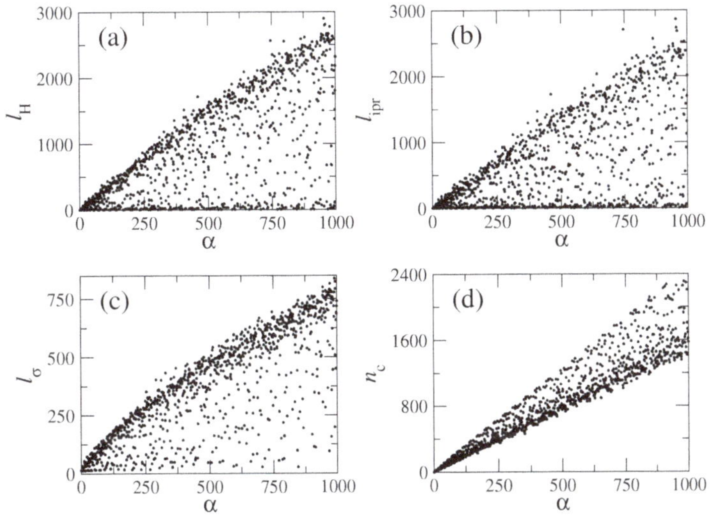

Figure 7. Localization measures for the eigenstates α of the rough billiard. (**a**) Entropy localization length $l_{\mathcal{H}}$, (**b**) inverse participation ratio l_{ipr}, (**c**) mean square root l_σ, and (**d**) centroid n_c. Here, $L_x = 2\pi$, $L_y = 2\pi$, $W_1 = 0.06L_y$, $k = 0.1$, and $N_T = 100$ were used.

Additional information about the structure of eigenstates can be obtained from the "width" or mean square root l_σ of an eigenstate, which is computed as

$$l_\sigma = \left[\sum_{l_{new}=1}^{N} w^{\alpha}_{l_{new}} \left[l_{new} - n_c(\alpha) \right]^2 \right]^{1/2}, \tag{22}$$

where $n_c = \sum_{l_{new}} l_{new}\, w^{\alpha}_{l_{new}}$ determines the centroid of an eigenstate in the unperturbed basis.

Comparison of the width l_σ with l_H and l_{ipr} gives the possibility of detecting the so-called *sparsity* of eigenstates. Indeed, small values of the ratio l_H/l_σ (or l_{ipr}/l_σ) indicate that there are "holes" in the structure of the eigenstates; therefore, such eigenstates are *sparse*; see Figure 7c. A detailed analysis shows the existence of sparsed eigenstates. As for the centroid, $n_c > \alpha$ observed in Figure 7d indicates that the interaction strength $\hat{V}$ is relatively strong compared to the unperturbed part $\hat{H}_0$.

The data of Figure 7a,b show the existence of a wide range of values of both localization measures even at high energies; moreover, there are eigenstates with small localization lengths (visible as clusters of points in the lower part of the plots) along the entire energy range. This situation is also present for the same quantities computed for the local density of states (LDOS): that is, for the basis eigenstates $\mid l_{new}\rangle$ expanded in the exact basis $\mid \alpha\rangle$; see Figure 8. Note that to compute the localizations lengths, as well as the centroid, for the LDOS, one must interchange $\alpha \leftrightarrow l_{new}$ in expressions (19), (21) and (22). It is relevant to stress that the localization lengths for the LDOS show well-defined patterns, see Figure 8, which are not present for the exact eigenstates, see Figure 7. Inspection of Figure 8a–d clearly demonstrates that there is a kind of regularity in the structure of the LDOS: the same type of states appear repeatedly, almost periodically as a function of the basis number l_{new}. These figures show the repetition of extremely localized states and of states with different values of the localization measures. The physical origin of all these types of states (localized, intermediate and extended) was explained above, and their appearance can be decoded by examining the structure of the "channel representation" of the Hamiltonian matrix; see Figure 2 (left). A detailed inspection of that matrix shows that the coupling between unperturbed states depends strongly on the values of the index m, labeling the transversal modes of the flat billiard. An unperturbed state specified by a large value of m (an m close to $M_{max} = 62$) couples strongly to several other unperturbed states. In contrast, the state with $m = 1$ has practically no coupling to other states. In particular, the extremely localized states, corresponding to the first position on the left line of each brach of Figure 8c occur because of the negligible coupling of the diagonal elements of the $H_{l,l'}$ matrix with $m = 1$, the states on the second position of each branch occur for $m = 3$, and so on with m odd. Similarly, the states on the right side of the branches result from elements of the $H_{l,l'}$ matrix with even values of m.

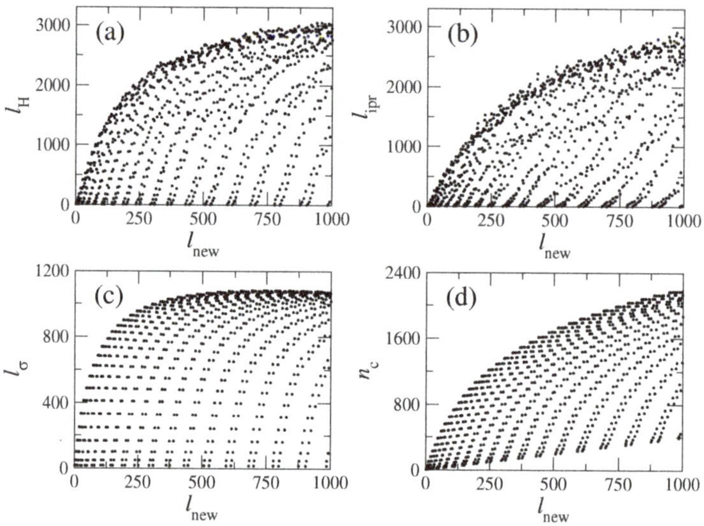

Figure 8. Localization measures for the LDOS, l_{new}, of the rough billiard. (**a**) $l_{\mathcal{H}}$, (**b**) l_{ipr}, (**c**) l_σ, and (**d**) n_c. $L_x = 2\pi$, $L_y = 2\pi$, $W_1 = 0.06L_y$, $k = 0.1$, and $N_T = 100$ were used.

This structure is expected to prevail at all energy ranges, since in any sufficiently large range of energies, there are unperturbed states with all values of $m \in [1, M_{max}]$. Even deep in the semiclassical regime, extremely localized and sparse states will appear but less and less frequently since the energy differences between states of the same type increases with energy.

5. Scar-Like States in Energy Space

In this section, we discuss the origin of the scar-like states emerging in the energy representation; they seem to be generic for plane billiards with rough surfaces. The starting point is the observation that for a weak roughness, these states can be associated with those unperturbed states that have a small value of the index m, especially for $m = 1$. Since the localized eigenstates identified above with $m = 1$ exist at all energies, the important question is about the fraction of these states relative to the total number of states $N(E) \equiv \sum_i \{i|E_i \leq E\}$ as a function of energy. This question is important in view of Shnirel'man's theorem [14] stating that the eigenfunctions of a classically ergodic system are equidistributed over the energy shell in the classical limit. In other words, we expect that

$$\lim_{E\to\infty} \frac{N_{m=1}(E)}{N(E)} = 0, \tag{23}$$

where $N_{m=1}(E)$ is the number of $m = 1$ eigenstates up to some energy E:

$$N_{m=1}(E) \equiv \sum_i \{i|E_i \leq E; \Psi_i \cong \phi_{1n}(u,v)\}. \tag{24}$$

Neither Shnirel'man's theorem nor Equation (23) says anything about how the limit is approached. Moreover, Shnirel'man's theorem does not consider the possible existence of parabolic fixed points. These occur in our billiard; they are the bouncing ball orbits or the continuous set of all horizontal trajectories ($P_y = 0$) which do not hit the two boundaries. Quantum mechanically, there are no $P_y = 0$ states, but the eigenstates of the flat channel with a minimum value of P_y are the $m = 1$ states. Thus, we have referred to the $m = 1$ eigenstates as "bouncing ball states" in [38].

For $W_1/L_y \ll 1$, one can use the unperturbed spectra given by Equation (15) to obtain a good estimate for $N_{m=1}(E)$ and $N(E)$. For large N_{max} and M_{max}, Equation (15) represents, ignoring the k^2 term, half of the ellipse

$$\frac{n^2}{\frac{2m_e}{\hbar^2}\left(\frac{L_x}{2\pi}\right)^2 E} + \frac{m^2}{\frac{2m_e}{\hbar^2}\left(\frac{L_y}{\pi}\right)^2 E} = 1, \tag{25}$$

in the $n - m$ plane (recall that $m \in [1, M_{max}]$ and $n \in [-N_{max}, N_{max}]$); above $E \equiv E^{(0)}_{n,m}(k) \cong E^{(0)}_{n,m}(0)$. In Figure 9, we show the $n - m$ plane for the case of $N_{max} = 32$ and $M_{max} = 62$; there, the dots represent basis states and the red ellipse is Equation (25) with $E = 400(\hbar^2/2m_e)$ and $L_x = Ly = 2\pi$. Then, for large E, the number $N(E)$ equals the area of the half ellipse, and the number $N_{m=1}(E)$ of localized eigenstates equals twice the size or its minor axis:

$$N(E) = \left(\frac{2m_e}{\hbar^2}\right)\frac{\sigma}{4\pi}E, \tag{26}$$

$$N_{m=1} = \left(\frac{2m_e}{\hbar^2}\right)^{1/2}\frac{\sigma}{\pi}\frac{\sqrt{E}}{L_y}; \tag{27}$$

therefore,

$$\frac{N_{m=1}(E)}{N(E)} = \left(\frac{\hbar^2}{2m_e}\right)^{1/2}\frac{4L_x}{\sigma\sqrt{E}}, \tag{28}$$

where $\sigma = L_x L_y$ is the area of one period of the billiard.

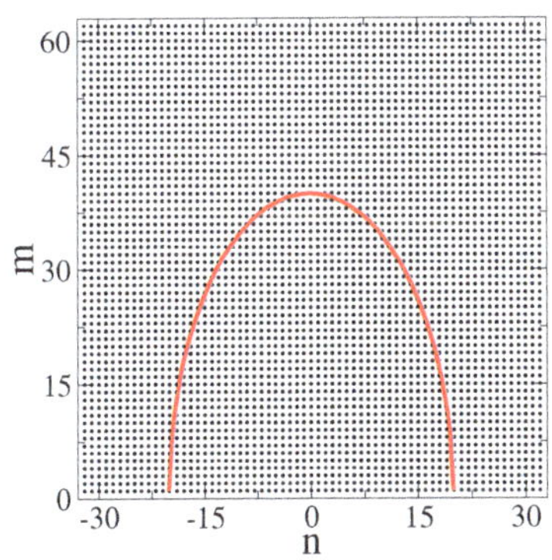

Figure 9. The $n - m$ plane for the case of $N_{max} = 32$ and $M_{max} = 62$. Dots represent basis states $| \, l\rangle \equiv | \, m, n\rangle$ or $| \, l_{new}\rangle \equiv | \, m, n\rangle$. The red ellipse is Equation (25) with $E = 400(\hbar^2/2m_e)$ and $L_x = Ly = 2\pi$.

We see that Equation (26) is precisely the first term in the Weyl series for the integrated density of states (see for example [46], Section 7). Equation (28) indicates that (i) for fixed E and σ, the portion of $m = 1$ states, relative to the total number of states, is larger the narrower the billiard is and (ii) the convergence of $N_{m=1}(E)/N(E)$ to zero is rather slow; it decreases as a power-law of E.

It is important to stress that in the case of a rough billiard having a highly modulated boundary composed of many harmonics, the expression for $N(E)$ is not simply given by (26), since now, the perimeter γ of the boundary, as well as its curvature, contribute significantly to $N(E)$ as [47]

$$N(E) = \left(\frac{2m_e}{\hbar^2}\right)\frac{\sigma}{4\pi}E - \frac{\gamma}{4\pi}\left(\frac{2m_e}{\hbar^2}\right)^{1/2}\sqrt{E} - \left(\frac{2m_e}{\hbar^2}\right)^{1/2}\frac{\sqrt{E}}{\pi}\sum_{r=1}^{\infty}\frac{(-1)^r}{(r-1/2)E^r}C_{2r+1}. \quad (29)$$

The coefficients C_{2r+1}, $r \geq 1$, in the sum of the Weyl series (29), depend on the curvature (and its derivative) of the modulated boundary (see Table 3 of [47]). As is shown in [47], the higher the index r of the Weyl coefficient C_{2r+1}, the more complex the expansions for C_r are, involving higher and higher powers of the curvature and its derivatives. Clearly, the larger the number of harmonics N_T, the more important the perimeter and curvature terms become.

In addition, it is also instructive to look at the location of the eigenstates on the $n-m$ plane. Thus, in Figure 10, we present the eigenstates of Figure 5 also in energy representation but now on the $n-m$ plane. From this figure, we can observe that: (i) the components of the energy-extended eigenstates of Figure 5 (left) are not equidistributed on the $n-m$ plane; instead, their main components are concentrated around the region with $n \sim 0$ and large m. That the components are concentrated around $n \sim 0$ supports the fact that these eigenstates are not extended over the x-axis when plotted in configuration representation; see Figure 6 (left). (ii) The components of the energy-localized eigenstates of Figure 5 (right) are also localized on the $n-m$ plane; they are indeed characterized by a single value of n and several but small values of m (a detailed decoding of this m dependence will be performed in the next section).

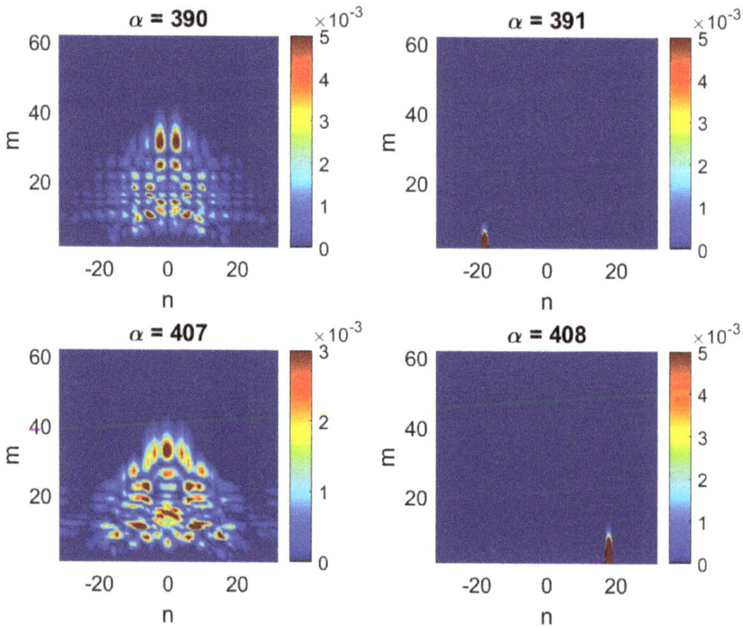

Figure 10. The eigenstates $\alpha = 390, 391, 407$ and 408 in energy representation (see Figure 5) on the $n-m$ plane.

6. Scars and Billiard Symmetries; Repulsion Effect

In this section, we study in detail the repulsion, in configuration representation, that suffer the energy-localized eigenstates of rough billiards from the rough boundaries. This effect, already shown in Figure 6 right panels, was first reported in [38,39] but also observed in [24,48,49]. Moreover, in addition to the one-flat-boundary billiard of Figure 11a, we now consider two additional billiard geometries: the antisymmetric billiard, Figure 11b, and the symmetric billiard, Figure 11c; see also [24,25].

We start by presenting in Figure 12 few typical extremely energy-localized eigenstates for the billiards of Figure 11 in three different energy regions. The main component of all those eigenstates can be identified with an unperturbed state with $m = 1$ (see also Figure 13) where we present again the eigenstates of Figure 12 but now in configuration representation). Close inspection of Figure 12 shows that for the one-flat-boundary billiard and the symmetric billiard, panels (a) and (c) of Figure 12, respectively, the localized eigenstates have several components of appreciable magnitude. This is in contrast with the eigenstates of the antisymmetric billiard, see panel (b) of Figure 12, which has only one effective component with magnitude very close to unity.

The decoding of the main components of the energy-localized eigenstates of the one–flat–boundary billiard, corresponding to panels (a) of Figure 12, shows that the second most important component of each of the localized states corresponds to $m = 2$, the next one to $m = 3$ and so on, all with the same value of n. That is, there is no mixing between different values of n. Moreover, the amplitudes of these eigenstate components decay exponentially as a function of m; see Figure 14 (left).

On the other hand, by decoding the main components of the energy-localized eigenstates of the symmetric billiard, corresponding to panels (c) of Figure 12, we see that the second most important component of each of the localized states corresponds to $m = 3$, the next one to $m = 5$ and so on, all with the same value of n but with alternating signs. Again, there is no mixing between different values of n. In addition, the amplitudes of these eigenstate components show an exponential decay as a function of m, which is in this case modulated by a sinus function; see Figure 14 (right).

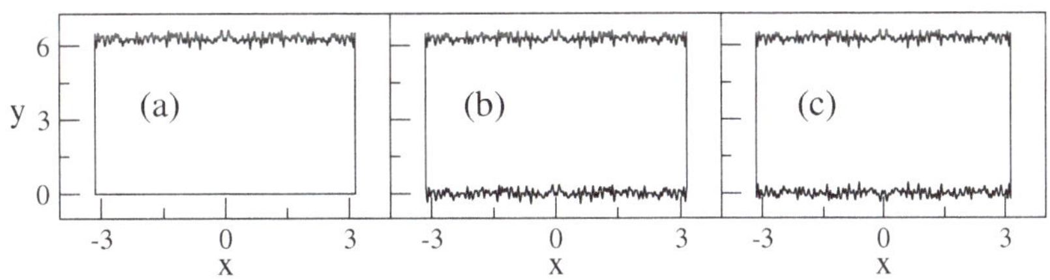

Figure 11. The rough billiards analyzed below: (**a**) a billiard with one flat boundary, $W_2 = 0$; (**b**) an antisymmetric billiard, $W_1 = W_2$; and (**c**) a symmetric billiard $W_1 = -W_2$. $N_T = 100$, $W_1/L_y = 0.06$, $L_x/L_y = 1$, $L_x = 2\pi$, and $\xi_1(x) = \xi_2(x)$.

Following the above observations, one can substitute in Equation (13) the following expressions:

$$C^{\alpha}_{mn}[a] \cong S_a \exp[-\beta_a(m-1)]\delta_{nn_\alpha}\,, \tag{30}$$

$$C^{\alpha}_{mn}[b] \cong \delta_{m,1}\delta_{nn_\alpha}\,, \tag{31}$$

$$C^{\alpha}_{mn}[c] \cong S_c \sin(m\pi/2)\exp[-\beta_c(m-1)]\delta_{nn_\alpha}\,, \tag{32}$$

for the energy-localized eigenstates of [a] the one–flat–boundary billiard, [b] the antisymmetric billiard, and [c] the symmetric billiard, respectively. Here, n_α corresponds to the value of n characterizing all the main components of a given energy-localized eigenstate.

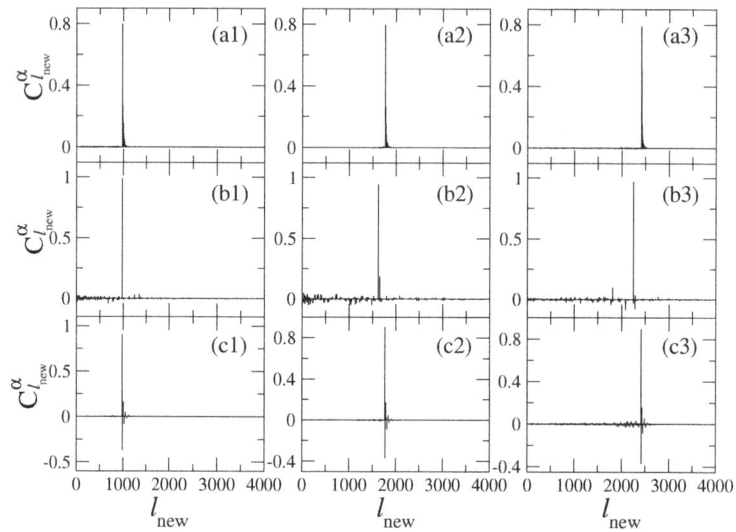

Figure 12. Localized eigenstates in the energy representation $C^{\alpha}_{l_{new}}$ for (**a**) a one-flat-boundary billiard, (**b**) an antisymmetric billiard, and (**c**) a symmetric billiard; see Figure 11. Specifically, (**a1**) $\alpha = 391$, (**a2**) $\alpha = 705$, (**a3**) $\alpha = 968$, (**b1**) $\alpha = 170$, (**b2**) $\alpha = 283$, (**b3**) $\alpha = 401$, (**c1**) $\alpha = 389$, (**c2**) $\alpha = 704$, and (**c3**) $\alpha = 962$. The main components of these localized eigenstates in energy representation correspond to unperturbed states characterized by $m = 1$ and (**a1**) $n = -18$, (**a2**) $n = -24$, (**a3**) $n = -28$, (**b1**) $n = -18$, (**b2**) $n = -23$, (**b3**) $n = -27$, (**c1**) $n = -18$, (**c2**) $n = -24$, and (**c3**) $n = -28$. $N_T = 100$, $W_1/L_y = 0.06$, $L_x/L_y = 1$, $L_x = 2\pi$, $k = 0.1$, and $\xi_1(x) = \xi_2(x)$ were used.

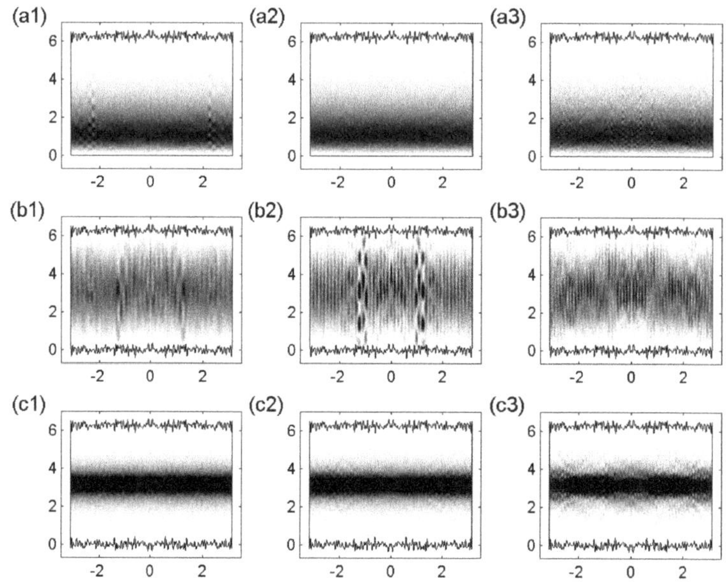

Figure 13. Localized eigenstates of Figure 12 in the configuration representation $|\Psi^{\alpha}(x,y)|^2$. Eigenstates for (**a1–a3**) a one-flat-boundary billiard, (**b1–b3**) an antisymmetric billiard, and (**c1–c3**) a symmetric billiard; same labeling as in Figure 12.

The parameters $S_a = 0.8$, $\beta_a = 0.53$, $S_c = 0.9$, and $\beta_c = 0.426$ are obtained by fitting the data to the dependence given by the expressions for C^{α}_{mn} above; see the fittings in Figure 14. It is important to remark that for a fixed N_T, all energy-localized states are characterized by the same values of $S_{a,b}$ and $\beta_{a,b}$. Substituting the dependence of C^{α}_{mn} into Equation (13) gives

$$\begin{aligned}
\Psi^{\alpha}_{loc}(u,v)[a] &\cong C_a S_a \frac{\exp[i(k+n_\alpha)u]}{\pi^{1/2}g^{1/4}} \sum_{m=1}^{\infty} \sin\left(\frac{m\pi v}{L_y}\right) \exp[-\beta(m-1)]\,, \\
\Psi^{\alpha}_{loc}(u,v)[b] &\cong C_c \frac{\exp[i(k+n_\alpha)u]}{\pi^{1/2}g^{1/4}} \sin\left(\frac{\pi v}{L_y}\right), \\
\Psi^{\alpha}_{loc}(u,v)[c] &\cong C_c S_c \frac{\exp[i(k+n_\alpha)u]}{\pi^{1/2}g^{1/4}} \sum_{m=1}^{\infty} \sin\left(\frac{m\pi}{2}\right) \sin\left(\frac{m\pi v}{L_y}\right) \exp[-\beta(m-1)]\,,
\end{aligned}$$

where $C_{a,b,c}$ arise to satisfy the orthonormality condition in curvilinear coordinates,

$$\int_0^{L_y}\int_0^{L_x} dudv\sqrt{g}\Psi^{\alpha\,*}_{loc}\Psi^{\alpha}_{loc} = 1\,. \tag{33}$$

Then, it follows that

$$\begin{aligned}
&\mid \Psi^{\alpha}_{loc}(y) \mid^2 [a] \cong \frac{1}{\pi}\left(\sum_{m=1}^{\infty} \exp[-2\beta_a(m-1)]\right)^{-1} \left[\sum_{m=1}^{\infty} \sin\left(\frac{m\pi y}{L_y}\right) \exp[-\beta_a(m-1)]\right]^2, \\
&\mid \Psi^{\alpha}_{loc}(y) \mid^2 [b] \cong \frac{1}{\pi} \sin\left(\frac{\pi y}{L_y}\right), \\
&\mid \Psi^{\alpha}_{loc}(y) \mid^2 [c] \cong \\
&\quad \frac{1}{\pi}\left(\sum_{m=1}^{\infty} \sin^2\left(\frac{m\pi}{2}\right) \exp[-2\beta_c(m-1)]\right)^{-1} \left[\sum_{m=1}^{\infty} \sin\left(\frac{m\pi}{2}\right) \sin\left(\frac{m\pi y}{L_y}\right) \exp[-\beta_c(m-1)]\right]^2.
\end{aligned}$$

These latter expressions give the average shape of the localized eigenstates in the configuration representation, projected onto the $y- \mid \Psi^{\alpha} \mid^2$ plane. In addition, note that $\Psi^{\alpha}_{loc}(y)$ does not depend on the parameters $S_{a,c}$, so the relevant parameters are $\beta_{a,b}$. To obtain $\mid \Psi^{\alpha}_{loc}(y) \mid^2$ above, we have made the approximation $v \approx y$ and $g \approx 1$ since $W_1/L_y = 0.06 \ll 1$.

Finally, after some algebra, we obtain

$$\mid \Psi^{\alpha}_{loc}(y) \mid^2 [a] \cong \frac{[\exp(2\beta_a)-1]\sin^2(\pi y/L_y)}{4\pi\big[\cos(\pi y/L_y)-\cosh(\beta_a)\big]^2}, \tag{34}$$

$$\mid \Psi^{\alpha}_{loc}(y) \mid^2 [b] \cong \frac{1}{\pi}\sin\left(\frac{\pi y}{L_y}\right), \tag{35}$$

$$\mid \Psi^{\alpha}_{loc}(y) \mid^2 [c] \cong \frac{\sinh(2\beta_c)\sinh^2(\beta_c)\sin^2(\pi y/L_y)}{2\pi\Big[\cosh^2(\beta_c)-\sin^2(\pi y/L_y)\Big]^2}. \tag{36}$$

Note that when $\beta_{a,b} \to \infty$, i.e., when the energy-localized eigenstates have only one single component, $\mid \Psi^{\alpha}_{loc}(y) \mid^2 [a] = \mid \Psi^{\alpha}_{loc}(y) \mid^2 [c] = (1/\pi)\sin(\pi y/L_y)$, as required.

In Figure 15, we plot Equations (34)–(36) together with the numerical data from the eigenstates of Figure 13. From this figure, it is clear that (i) in the one-flat-boundary billiard, the energy-localized eigenstates are repelled from the rough boundary toward the flat boundary; see the curves labeled with (a), as already observed in Figure 6; (ii) in the symmetric billiard, the energy-localized eigenstates are repelled from both rough boundaries toward the billiard center; see the curves labeled with (c); while (iii) in the antisymmetric billiard, the repulsion effect is completely absent, so the eigenstates correspond to those of the flat billiard; see the curves labeled with (b). A detailed study [24] revealed that the absence of repulsion in the antisymmetric billiard is due to the fact that in this billiard, there is no square-gradient scattering between different modes or channels, in contrast to the one-flat-boundary billiard and the symmetric billiard. Thus, one can conclude that the main contribution to the repulsion is due to the intermode square-gradient scattering terms of the Hamiltonian matrix (i.e., the integrals J_3, J_6, and J_7 in Equation (17), which vanish for the antisymmetric billiard). This conclusion supports the observation made above according to which the effect of repulsion can be explained as due to a strong localization in the channel space. This localization occurs due to a relatively strong interaction between different conducting channels (or billiard modes).

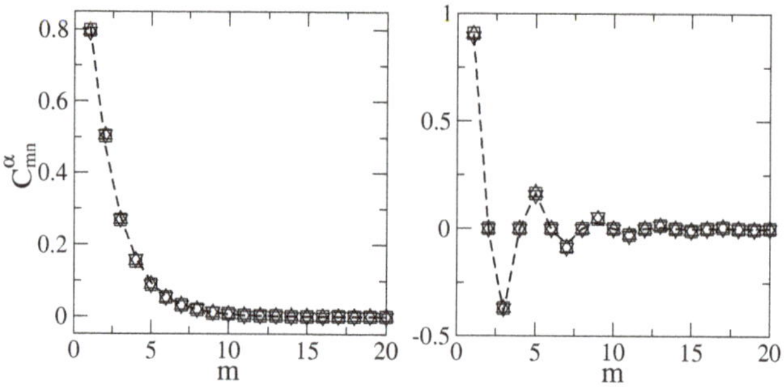

Figure 14. Main components C^{α}_{mn} as the functions of m from the localized eigenstates in the energy representation of Figure 13a (right) and Figure 13c (left). Different symbols correspond to different eigenstates. The thick dashed curves are the best fits to Equation (30) (right) and Equation (32) (left) with $S_a = 0.8$ and $\beta_a = 0.53$ and $S_c = 0.9$ and $\beta_c = 0.426$, respectively.

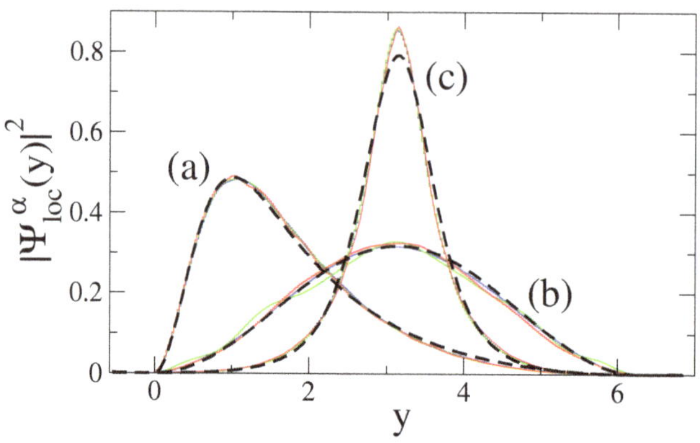

Figure 15. Projection of the eigenstate profiles of Figure 13 (color curves) onto the y coordinate together with the analytical expressions for $|\ \Psi^{\alpha}_{loc}(y)\ |^2$ (thick dashed lines) given by Equations (34)–(36) with $\beta_a = 0.53$ and $\beta_c = 0.426$. Profiles of eigenstates for (**a**) a one-flat-boundary billiard, (**b**) an antisymmetric billiard, and (**c**) a symmetric billiard.

Note that since the parameters $S_{a,b}$ and $\beta_{a,b}$ are the same for all $m = 1$ eigenstates for a fixed N_T, one can find $S_{a,b}$ and $\beta_{a,b}$ for some low-energy eigenstate (so that the matrix to be diagonalized is small) and with this one can infer the global shape of all $m = 1$ eigenstates. Conversely, knowing the shape of an $m = 1$ eigenstate within any energy range, one may assess by looking at the amount of repulsion the degree of complexity, presumably unknown, of the boundaries.

The excellent agreement between Equations (34)–(36) and the numerical data from the eigenstates of Figure 13 indicates that for extremely localized eigenstates with $m = 1$, there is no need to diagonalize the whole matrix since, there is no mixing between different values of n. In order to obtain a good approximation of these eigenstates, one only needs to diagonalize the block of size $M_{max} \times M_{max}$ corresponding to the quantum number n_α, which is a good quantum number in the presence of perturbation.

Within this approach, it is also possible to study energy-localized eigenstates with $m > 1$ using blocks of the Hamiltonian (17) of size $M_{max} \times M_{max}$. So, in Figure 16, we plot the 2nd, 3rd, and 4th lowest eigenstates in configuration representation of a block of the Hamiltonian (17) of size 100×100 for the one-flat-boundary billiard (left panels), an antisymmetric billiard (central panels), and the symmetric billiard (right panels). They correspond to eigenstates characterized by $m = 2$, 3, and 4, respectively. As expected, they have 2, 3, and 4 maxima in the transverse direction, respectively. As well as for the eigenstates characterized by $m = 1$, see Figures 13 and 15, the eigenstates of Figure 16 suffer repulsion from the rough boundaries in the one-flat-boundary billiard and the symmetric billiard, while in the antisymmetric billiard, the repulsion is absent.

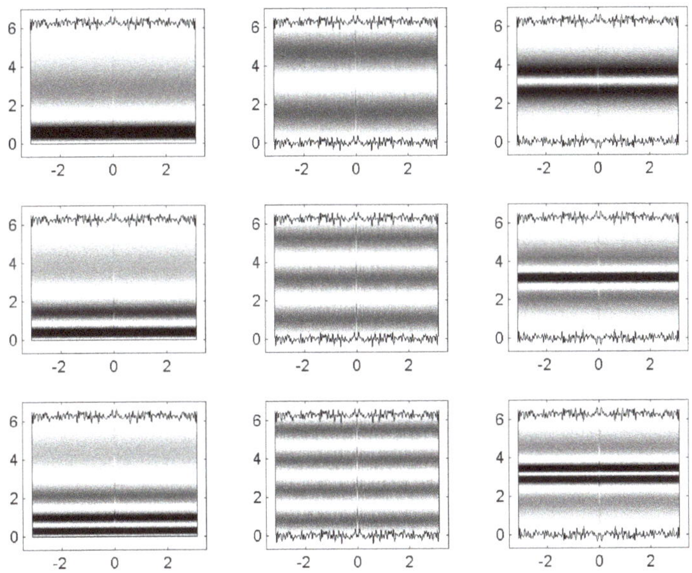

Figure 16. Left column: 2nd, 3rd, and 4rd lowest eigenstates in the configuration representation $|\Psi^\alpha(x,y)|^2$ for a block of the Hamiltonian (17) of size $M_{max} \times M_{max}$ for the billiards of Figure 11a. Middle and right column: the same but for the billiards of Figure 11b,c, respectively. $N_T = 100$, $W_1/L_y = 0.06$, $L_x/L_y = 1$, $L_x = 2\pi$, $\xi_1(x) = \xi_2(x)$, $M_{max} = 100$, $n_\alpha = 0$, and $k = 0$ were used.

7. Conclusions

In this paper, we have studied the properties of rectangular billiards with one and two rough boundaries. In the classical description, such billiards can be considered as completely chaotic, although this has not been rigorously proven. Our interest was to understand the typical structure of the eigenfunctions in the quantum description and to discover the mechanism of occurrence of scar states. For numerical simulation, we have chosen the method of reducing the original model with one particle moving inside the billiard to a model of two artificial particles in a billiard with flat boundaries, however with an effective interaction between the particles. Thus, the complexity of the rough boundary is embedded into a quite complicated interaction potential. In this representation, the Hamiltonian of the system can naturally be represented as the sum of two terms, one of which corresponds to the motion of two particles in a billiard with flat walls, and the second term describes the interaction between particles. As can be seen, such a scheme is similar to that which is often used in physics when describing isolated systems with many interacting particles. Thus, the question arises about the correspondence between the properties of eigenfunctions in a many-particle basis (in our case, a two-particle basis) and the properties of eigenfunctions in the configuration space. As was found, this correspondence is not simple, especially if we are interested in the degree of localization of the eigenfunctions and, in particular, in the occurrence of scar states.

As a result of our study, we came to the conclusion that one can roughly speak of three different types of eigenstates. The first type includes states whose density in the configuration basis is concentrated in the vicinity of bouncing ball trajectories that are perpendicular to two horizontal surfaces. Such highly localized functions (with some degree of chaos) are analogous to well-studied scar states in stadium billiards. The second type of eigenfunctions, which are strongly localized in the momentum space, can be associated with the plane waves in the billiard with flat boundaries. To some extent, they can also be termed as scar states because of their strong correspondence to unperturbed eigenstates. There is a third type of eigenfunctions that consists of many components in both the energy space of the Hamiltonian and in the configuration space. These eigenstates are quite complicated and can be treated as partially chaotic. Thus, it can be assumed that these eigenstates quickly become completely chaotic with increasing energy in comparison with those functions that are strongly localized either in energy or in configuration space.

We have considered in detail how the number of eigenstates strongly localized in the energy space decreases with increasing energy. This question is related to Shnirel'man's theorem, according to which, in the classical limit, the measure (number) of scar functions decreases to zero. It is worth noting that this theorem does not predict at what rate (in energy) all eigenfunctions become ergodic. In the general case of ergodic billiards, this question remains open, but in our case, the estimate indicates that the number of scar states (in the energy space) decreases rather slowly, namely, inversely proportional to the square of the energy.

We were also interested in the "repulsion" effect, discovered in [22], according to which for a billiard with one corrugated surface, some of the eigenstates suffer a shift of the maximal density toward the surface, which is flat. Our numerical data showed that this effect is highly pronounced for those eigenfunctions that are close to plane waves traveling in the horizontal direction. For a subset of such eigenstates, we have derived an approximate expression that clearly indicates an exponential localization in the momentum space. Even more interesting was to compare the structure of this type of eigenstate for the billiard with two corrugated surfaces. We have found that for symmetric rough surfaces, the repulsion effect is strongly enhanced. On the other hand, for antisymmetric profiles, the repulsion is absent. The origin of this phenomenon can be explained by analyzing the structure of the Hamiltonian matrix, which is strongly influenced by the form of the off-diagonal matrix elements that depend on the amplitude, first derivative, and the second derivative of the billiard profiles. Apart from that, there are terms that describe the inter-correlations between profiles. Our analysis explains both the repulsion enhancement and the disappearance of repulsion. Due to this analysis, one can predict the impact of the type of symmetry between the profiles on the scattering properties of quasi-one-dimensional waveguides with corrugated surfaces.

This article is dedicated to Professor Giulio Casati on the occasion of his 80th birthday. We wish him to be healthy and scientifically productive for the next 20 years.

Author Contributions: The authors contributed equally to this work. F.M.I., G.A.L.-A. and J.A.M.-B. conceived, designed and performed the numerical experiments; F.M.I., G.A.L.-A. and J.A.M.-B. analyzed the data and wrote the paper. All authors have read and agreed to the published version of the manuscript.

Funding: This research was funded by CONACyT grant number 286633.

Institutional Review Board Statement: Not applicable.

Data Availability Statement: The data presented in this study are available on request from the corresponding author.

Acknowledgments: J.A.M.-B. thanks support from CONACyT-Fronteras (Grant No. 425854), VIEP-BUAP (Grant No. 100405811-VIEP2022), and Laboratorio Nacional de Supercómputo del Sureste de México (Grant No. 202201007C), Mexico.

Conflicts of Interest: The authors declare no conflict of interest. The founding sponsors had no role in the design of the study; in the collection, analyses, or interpretation of data; in the writing of the manuscript, or in the decision to publish the results.

References

1. Zelditch, S.; Zworski, M. Ergodicity of eigenfunctions for ergodic billiards. *Commun. Math. Phys.* **1996**, *175*, 673–682. [CrossRef]
2. Vikram, A.; Galitski, V. Dynamical quantum ergodicity from energy level statistics. *arXiv* **2022**, arXiv:2205.05704.
3. Zhang, D.; Quan, H.T.; Wu, B. Ergodicity and mixing in quantum dynamics. *Phys. Rev. E* **2016**, *94*, 022150. [CrossRef] [PubMed]
4. Barnett, A. Asymptotic rate of quantum ergodicity in chaotic Euclidean billiards. *Comm. Pure Appl. Math.* **2006**, *59*, 1379–1521. [CrossRef]
5. Luitz, D.J.; Lev, Y.B. The ergodic side of the many-body localization transition. *Ann. Phys.* **2017**, *529*, 1600350. [CrossRef]
6. Kravtsov, V.E.; Altshuler, B.L.; Ioffe, L.B. Non-ergodic delocalized phase in Anderson model on Bethe lattice and regular graph. *Ann. Phys.* **2018**, *389*, 148–191. [CrossRef]
7. Avetisov, V.; Gorsky, A.; Nechaev, S.; Valba, O. Localization and non-ergodicity in clustered random networks. *J. Complex Netw.* **2020**, *8*, cnz026. [CrossRef]
8. Huang, W.J.; Wu, Y.B.; Guo, G.C.; Zou, X.B. Ergodic-nonergodic transition with cold spinless fermions in a cavity. *Phys. Rev. A* **2022**, *105*, 033315. [CrossRef]
9. Krylov, N.S. *Works on the Foundations of Statistical Physics by Nikolai Sergeevich Krylov*; Migdal, A.B.; Sinai, Y.G.; Zeeman, Y.L., Translators; [Translated from: Raboty po Obosnovaniiu Statisticheskoj Fiziki, Moscow, 1950]; Princeton University Press: Princeton, NJ, USA, 1979.
10. O'Rourke, S.; Vu, V.; Wang, K. Eigenvectors of random matrices: A survey. *J. Comb. Theo. Series A* **2016**, *144*, 361–442. [CrossRef]
11. Borgonovi, F.; Izrailev, F.M.; Santos, L.F.; Zelevinsky, V.G. Quantum chaos and thermalization in isolated systems of interacting particles. *Phys. Rep.* **2016**, *626*, 1–58. [CrossRef]
12. Heller, E.J. Bound-state eigenfunctions of classically chaotic hamiltonian systems: Scars of periodic orbits. *Phys. Rev. Lett.* **1984**, *53*, 1515. [CrossRef]
13. Berry, M.V. Quantum scars of classical closed orbits in phase space. *Proc. R. Soc. A* **1989**, *423*, 219.
14. Shnirel'man, A.I. Ergodic properties of eigenfunctions. *Usp. Math. Nauk.* **1974**, *29*, 181–182.
15. Serbyn, M.; Abanin, D.A.; Papic, Z. Quantum many-body scars and weak breaking of ergodicity. *Nat. Phys.* **2021**, *17*, 675–685. [CrossRef]
16. Mondal, D.; Sinha, S.; Ray, S.; Kroha, J.; Sinha, S. Classical route to ergodicity and scarring phenomena in a two-component Bose-Josephson junction. *Phys. Rev. A* **2022**, *106*, 043321. [CrossRef]
17. Turner, C.J.; Desaules, J.-Y.; Bull, K.; Papic, Z. Correspondence principle for many-body scars in ultracold rydberg atoms. *Phys. Rev. X* **2021**, *11*, 021021. [CrossRef]
18. Tamura, K.; Katsura, H. Quantum many-body scars of spinless fermions with density-assisted hopping in higher dimensions. *Phys. Rev. B* **2022**, *106*, 144306. [CrossRef]
19. Ma, K.K.W.; Volya, A.; Yang, K. Eigenstate thermalization and disappearance of quantum many-body scar states in weakly interacting fermion systems. *Phys. Rev. B* **2022**, *106*, 214313. [CrossRef]
20. Chandran, A.; Iadecola, T.; Khemani, V.; Moessner, R. Quantum many-body scars: A quasiparticle perspective. *arXiv* **2022**, arXiv:2206.11528.
21. Choi, S.; Turner, C.J.; Pichler, H.; Ho, W.W.; Michailidis, A.A.; Papić, Z.; Serbyn, M.; Lukin, M.D.; Abanin, D.A. Emergent SU(2) dynamics and perfect quantum many-body scars. *Phys. Rev. Lett.* **2019**, *122*, 220603. [CrossRef]
22. Izrailev, F.M.; Luna-Acosta, G.A.; Mendez-Bermudez, J.A.; Rendon, M. Amplitude and gradient scattering in billiards with corrugated surfaces. *Phys. Stat. Sol. C* **2003**, 3032. [CrossRef]
23. Izrailev, F.M.; Makarov, N.M.; Rendon, M. Manifestation of the roughness-square-gradient scattering in surface-corrugated billiards. *Phys. Rev. B* **2006**, *73*, 155421. [CrossRef]
24. Rendon, M.; Izrailev, F.M.; Makarov, N.M. Square-gradient mechanism of surface scattering in quasi-one-dimensional rough billiards. *Phys. Rev. B* **2007**, *75*, 205404. [CrossRef]
25. Mendez-Bermudez, J.A.; Izrailev, F.M. Transverse localization in quasi–one–dimensional surface–corrugated billiards. *Mocroelectron. J.* **2008**, *39*, 1376. [CrossRef]
26. Luna-Acosta, G.A.; Na, K.; Reichl, L.E.; Krokhin, A. Band structure and quantum Poincare sections of a classically chaotic quantum rippled channel. *Phys. Rev. E* **1996**, *53*, 3271–3283. [CrossRef]
27. Luna-Acosta, G.A.; Krokhin, A.A.; Rodriguez, M.A.; Hernandez-Tejeda, P.H. Classical chaos and ballistic transport in a mesoscopic channel. *Phys. Rev. B* **1996**, *54*, 11410–11416. [CrossRef]
28. Luna-Acosta, G.A.; Rodriguez, M.A.; Krokhin, A.A.; Na, K.; Méndez, R.A. Quantum and classical ballistic transport in a chaotic 2D electron channel. *Rev. Mex. FíSica* **1998**, *44*, 7–13.

29. Luna-Acosta, G.A.; Mendez-Bermudez, J.A.; Izrailev, F.M. Quantum–classical correspondence for local density of states and eigenstates of a chaotic periodic billiard. *Phys. Lett. A* **2000**, *274*, 192–199. [CrossRef]
30. Luna-Acosta, G.A.; Mendez-Bermudez, J.A.; Izrailev, F.M. Periodic chaotic billiards: Quantum–classical correspondence in energy space. *Phys. Rev. E* **2001**, *64*, 036206. [CrossRef]
31. Luna-Acosta, G.A.; Mendez-Bermudez, J.A.; Izrailev, F.M. Chaotic electron motion in superlattices. Quantum–classical correspondence of the structure of eigenstates and LDOS. *Phys. E* **2002**, *12*, 267–271. [CrossRef]
32. Luna-Acosta, G.A.; Mendez-Bermudez, J.A.; Izrailev, F.M. Classical versus quantum structure of the scattering probability matrix. Chaotic billiards. *Phys. Rev. E* **2002**, *65*, 046605. [CrossRef] [PubMed]
33. Akguc, G.B.; Reichl, L.E. Conductance and statistical properties of chaotic and integrable electron billiards. *J. Stat. Phys.* **2000**, *98*, 813–834. [CrossRef]
34. Huckestein, B.; Ketzmerick, R.; Lewenkopf, C.H. Quantum transport through ballistic cavities: Soft vs hard quantum chaos. *Phys. Rev. Lett.* **2000**, *84*, 5504–5507. [CrossRef] [PubMed]
35. Li, W.; Reichl, L.E.; Wu, B. Quantum chaos in a ripple billiard. *Phys. Rev. E* **2002**, *65*, 056220. [CrossRef] [PubMed]
36. Lichtenberg, A.J.; Lieberman, M.A. *Regular and Chaotic Dynamics*, 2nd ed.; Springer: New York, NY, USA, 1992.
37. Frahm, K.M.; Shepelyansky, D.L. Quantum localization in rough billiards. *Phys. Rev. Lett.* **1997**, *78*, 1440–1443. [CrossRef]
38. Izrailev, F.M.; Mendez-Bermudez, J.A.; Luna-Acosta, G.A. Ballistic localization in quasi–one–dimensional billiards with rough surfaces. *Phys. Rev. E* **2003**, *68*, 066201. [CrossRef]
39. Mendez-Bermudez, J.A.; Luna-Acosta, G.A.; Izrailev, F.M. From chaos to disorder in quasi–1D billiards with corrugated surfaces. *Phys. E* **2004**, *22*, 881. [CrossRef]
40. Adler, R.; Bazin, M.; Schiffer, M. *Introduction to General Relativity*; McGraw-Hill: New York, NY, USA, 1975.
41. Dewitt, B. Dynamical theory in curved spaces I. A review of the classical and quantum action principles. *Rev. Mod. Phys.* **1957**, *29*, 377–397. [CrossRef]
42. Izrailev, F.M. Quantum–classical correspondence for isolated systems of interacting particles: Localization and ergodicity in energy space. *Phys. Scr.* **2001**, *T90*, 95–104. [CrossRef]
43. Flambaum, V.V.; Izrailev, F.M. Statistical theory of finite Fermi systems based on the structure of chaotic eigenstates. *Phys. Rev. E* **1997**, *56*, 5144–5159. [CrossRef]
44. Ingraham, R.L. *A Survey of Nonlinear Dynamics. ("Chaos Theory")*; World Scientific: Singapore, 1992; Section 7.4.
45. Flambaum, V.V.; Izrailev, F.M. Excited eigenstates and strength functions for isolated systems of interacting particles. *Phys. Rev. E* **2000**, *61*, 2539–2542. [CrossRef]
46. Reichl, L.E. *The Transition to Chaos in Conservative Classical Systems: Quantum Manifestations*; Springer: New York, NY, USA, 1992.
47. Berry, M.V.; Howls, C.J. High orders of the Weyl expansion for quantum billiards: Resurgence of periodic orbits, and the Stokes phenomenon. *Proc. R. Soc. Lond. A* **1994**, *447*, 527–555.
48. Feist, J.; Bäcker, A.; Ketzmerick, R.; Rotter, S.; Huckestein, B.; Burgdörfer, J. Nanowires with surface disorder: Giant localization lengths and quantum-to-classical crossover. *Phys. Rev. Lett.* **2006**, *97*, 116804. [CrossRef] [PubMed]
49. Felix, S.; Asch, M.; Filoche, M.; Sapoval, B. Localization and increased damping in irregular acoustic cavities. *J. Sound Vib.* **2007**, *299*, 965–976. [CrossRef]

entropy

MDPI

Article

Generalized Survival Probability

David A. Zarate-Herrada [1], Lea F. Santos [2,†] and E. Jonathan Torres-Herrera [1,*,†]

1 Instituto de Física, Benemérita Universidad Autónoma de Puebla, Apartado Postal J-48, Puebla 72570, Mexico
2 Department of Physics, University of Connecticut, Storrs, CT 06269, USA
* Correspondence: etorresh@ifuap.buap.mx
† These authors contributed equally to this work.

Abstract: Survival probability measures the probability that a system taken out of equilibrium has not yet transitioned from its initial state. Inspired by the generalized entropies used to analyze nonergodic states, we introduce a generalized version of the survival probability and discuss how it can assist in studies of the structure of eigenstates and ergodicity.

Keywords: survival probability; spectral form factor; quench dynamics; many-body quantum chaos; disordered spin model

1. Introduction

The square overlap between a given initial state $|\Psi(0)\rangle$ and its time-evolved counterpart $|\Psi(t)\rangle$,

$$SP(t) = |\langle\Psi(0)|\Psi(t)\rangle|^2, \tag{1}$$

indicates the probability of finding the system still in its initial state at time t. This quantity is known as survival probability, return probability, or simply the fidelity between the initial and the evolved state. This quantity has been extensively investigated since the early decades of quantum mechanics, initially in the context of the uncertainty relation between time and energy [1–3]. As stated by Fock in [3]:

> [the time-energy uncertainty relation] may be viewed as a consequence of the general theorem of Fock and Krylov on the connection between the decay law and the energy distribution function.

The "connection" stated in the quote above refers to the fact that the survival probability (decay law) is the absolute square of the Fourier transform of the energy distribution of the initial state (energy distribution function). That is, for a state evolving according to a Hamiltonian H, whose eigenvalues and eigenstates are given by E_α and $|\alpha\rangle$, one has $|\Psi(t)\rangle = \sum_\alpha C_\alpha^{(0)} e^{-iE_\alpha t}|\alpha\rangle$, and the survival probability can be written as:

$$SP(t) = \left|\sum_\alpha |C_\alpha^{(0)}|^2 e^{-iE_\alpha t}\right|^2 = \left|\int \rho(E) e^{-iEt} dE\right|^2, \tag{2}$$

where $C_\alpha^{(0)} = \langle\alpha|\Psi(0)\rangle$, and

$$\rho(E) = \sum_\alpha |C_\alpha^{(0)}|^2 \delta(E - E_\alpha) \tag{3}$$

is the energy distribution of the initial state. This distribution is also known as the local density of states (LDoS) or strength function, and its mean and variance are [4]:

$$E^{(0)} = \sum_\alpha |C_\alpha^{(0)}|^2 E_\alpha \quad \text{and} \quad \sigma^2 = \sum_\alpha |C_\alpha^{(0)}|^2 (E_\alpha - E^{(0)})^2. \tag{4}$$

Citation: Zarate-Herrada, D.A.; Santos, L.F.; Torres-Herrera, E.J. Generalized Survival Probability. *Entropy* **2023**, *25*, 205. https://doi.org/10.3390/e25020205

Academic Editor: Marko Robnik

Received: 11 December 2022
Revised: 13 January 2023
Accepted: 16 January 2023
Published: 20 January 2023

Both the survival probability and the LDoS are studied in a variety of different fields, from quantum chaos and nuclear physics to localization and quantum information science. These quantities received significant attention from previous researchers of the Budker Institute in Novosibirsk, including those to whom we dedicated the present paper, namely Professor Giulio Casati on the occasion of his 80th birthday in 2022, Professor Felix Izrailev on the occasion of his 80th birthday in 2021, and Professor Vladimir Zelevinsky on the occasion of his 85th birthday in 2022.

Despite the simplicity of Equation (1), the evolution of the survival probability in many-body quantum systems is quite rich, with different behaviors emerging on different time scales, which reveal details about the initial state, the spectrum, and the eigenstates of the considered model. The Taylor expansion of the phase factor in Equation (2) shows that the survival probability, at very short times, $t \ll 1/\sigma$, presents a quadratic and universal behavior, $SP(t) \approx 1 - \sigma^2 t^2$, where σ is the width of the LDoS (see Equation (4)). Beyond this point, but still at short times, $t \lesssim 1/\sigma$, the decay is dictated by the shape of the LDoS. The shape of $\rho(E)$ was investigated in [5,6] in the context of banded random matrices, while in realistic models, the transition from a Lorentzian to a Gaussian form with the increase in the perturbation strength was discussed, as in [7–23]. Depending on the initial state and the model considered, skewed Gaussians and bimodal distributions can also emerge [24]. Beyond the characteristic time for the initial depletion of the initial state, $t \sim 1/\sigma$, the survival probability exhibits a power-law decay $\propto t^{-\gamma}$ with an exponent γ that depends on the level of ergodicity of $|\Psi(0)\rangle$ and $|\alpha\rangle$. When the LDoS is filled ergodically, γ is determined by the bounds of this energy distribution [25–33]. In contrast, when $|\Psi(0)\rangle$ and $|\alpha\rangle$ are non-chaotic states, then γ depends on the level of correlations and multifractality between the states [34–39]. However, this is not yet the end of the story. In chaotic systems, where the energy-level statistics are similar to those of random matrices, the survival probability does not saturate after the algebraic decay. Instead, it reaches a value that is smaller than its infinite-time average,

$$\overline{SP} = \sum_{\alpha} |C_{\alpha}^{(0)}|^4, \tag{5}$$

and then grows in a ramp until it is finally saturated at $\overline{SP}$. The infinite time average is the last term in the equation below, which is obtained from Equation (2),

$$SP(t) = \sum_{\alpha \neq \beta} |C_{\alpha}^{(0)}|^2 |C_{\beta}^{(0)}|^2 e^{-i(E_\alpha - E_\beta)t} + \sum_{\alpha} |C_{\alpha}^{(0)}|^4. \tag{6}$$

The interval in which $SP(t) < \overline{SP}$ is known as the correlation hole [39–56], and there have been different methods proposed with which to measure it experimentally in systems that are out of equilibrium (see [57] and references therein). The correlation hole is a dynamic manifestation of spectral correlations and, as such, can be used to detect many-body quantum chaos in experiments that do not have direct access to the spectrum, such as experiments with cold atoms and ion traps.

In this work, motivated by generalized quantities such as Rényi entropies [58], the inverse participation ratio [59–64], and other similar quantities [65] that play a prominent role in studies of localization and multifractality, we introduce the generalized survival probability, $SP_q(t)$, and its corresponding generalized LDoS, $\rho_q^{(0)}(E)$ (see the definitions below in Equations (10) and (12), respectively). We discuss how they can help to improve our understanding of the structure of the eigenstates.

Using the one-dimensional (1D) disordered spin-1/2 model, which is often employed in the analysis of many-body localization, we compare the results for the generalized survival probability in the chaotic regime and far from it, where the duration of the power-law decay of $SP_q(t)$ becomes dependent on the value of q. We also compare the behavior of $SP_q(t)$ using the chaotic spin model with random matrices from the Gaussian orthogonal ensemble (GOE) and, in the latter case, provide an analytical expression for the entire evolution of the generalized survival probability.

2. Models

Here, we study many-body quantum systems described by the Hamiltonian:

$$H = H_0 + V, \tag{7}$$

where a chosen eigenstate of H_0 corresponds to the initial state, and V is a strong perturbation that takes the system far from equilibrium. We consider initial states that have energies $E^{(0)} = \langle\Psi(0)|H|\Psi(0)\rangle$ close to the middle of the spectrum. Two Hamiltonians H are investigated, of which one is a random matrix from the GOE and the other describes a 1D disordered Heisenberg spin-1/2 model.

2.1. Gaussian Orthogonal Ensemble

The GOE is composed of real and symmetric $\mathcal{D} \times \mathcal{D}$ matrices completely filled with random entries from a Gaussian distribution, with the mean zero and variance given by:

$$\left\langle H_{jk}^2 \right\rangle = \begin{cases} \frac{1}{2}, & \text{for} \quad j \neq k; \\ 1, & \text{for} \quad j = k. \end{cases} \tag{8}$$

We assume that the unperturbed Hamiltonian H_0 is the diagonal part of H and V is the off-diagonal part. The model is non-physical but allows for analytical derivations that can serve as a reference for the study of realistic chaotic many-body quantum systems.

2.2. Disordered Spin-1/2 Model

As a physical model, we consider the 1D Heisenberg spin-1/2 model with onsite disorder, which has been used in studies of many-body localization [66–69]. The Hamiltonian is given by:

$$H = \sum_{k=1}^{L} h_k S_k^z + J \sum_{k=1}^{L} \left(S_k^x S_{k+1}^x + S_k^y S_{k+1}^y + S_k^z S_{k+1}^z \right), \tag{9}$$

where $S^{x,y,z}$ are the spin-1/2 operators, L is the system size, $J = 1$ is the coupling strength, and h_k refers to independent and uniformly distributed random variables in $[-h, h]$, with h being the onsite disorder strength. We assume periodic boundary conditions. The system conserves the total magnetization in the z-direction, $\hat{S}_{\text{tot}}^z = \sum_{k=1}^{L} \hat{S}_k^z$. Throughout this paper, we work in the largest subspace, with $\hat{S}_{\text{tot}}^z = 0$ leading to $\mathcal{D} = L!/(L/2)!^2$. For finite sizes, H shows level statistics comparable to the GOE random matrices when $h \sim 0.5$, while the level repulsion fades away for $h > 1$. We consider the unperturbed Hamiltonian to consist of the terms in the z-direction, $H_0 = \sum_{k=1}^{L} \left(h_k S_k^z + J S_k^z S_{k+1}^z \right)$, and the perturbation to be the flip-flop term, $V = J \sum_{k=1}^{L} \left(S_k^x S_{k+1}^x + S_k^y S_{k+1}^y \right)$.

3. Generalized Survival Probability

We define the generalized survival probability as:

$$SP_q(t) = \frac{1}{\mathcal{N}_q^2} \left| \sum_{\alpha=1}^{\mathcal{D}} |C_\alpha^{(0)}|^q e^{-iE_\alpha t} \right|^2 = \left| \int \rho_q(E) e^{-iE_\alpha t} dE \right|^2, \tag{10}$$

where $\mathcal{N}_q$ is a normalization constant given by:

$$\mathcal{N}_q = \sum_{\alpha=1}^{\mathcal{D}} |C_\alpha^{(0)}|^q, \tag{11}$$

where the parameter $q \geq 0$ is a positive real number, and

$$\rho_q(E) = \frac{1}{\mathcal{N}_q} \sum_{\alpha=1}^{\mathcal{D}} |C_\alpha^{(0)}|^q \delta(E_\alpha - E) \tag{12}$$

is the generalized LDoS (gLDoS), with the mean and variance given, respectively, by:

$$E_q^{(0)} = \frac{1}{\mathcal{N}_q} \sum_{\alpha=1}^{\mathcal{D}} |C_\alpha^{(0)}|^q E_\alpha \quad \text{and} \quad \sigma_q^2 = \frac{1}{\mathcal{N}_q} \sum_{\alpha=1}^{\mathcal{D}} |C_\alpha^{(0)}|^q (E_\alpha - E_q^{(0)})^2. \tag{13}$$

The survival probability, as defined in Equation (2), and the mean and variance given in Equation (4) are recovered when $q = 2$. For $q = 0$, Equation (10) coincides with the spectral form factor [70], which is a quantity used to study level statistics in the time domain. Contrary to the (generalized) survival probability, the spectral form factor is not a dynamical quantity, since it does not depend on the initial state.

If one knows the generalized LDoS, we can obtain the generalized survival probability by performing the Fourier transform in Equation (10). We therefore start our analysis by examining the shape of $\rho_q(E)$.

Generalized LDoS

Figure 1 depicts the generalized LDoS for a single random realization of a GOE matrix and different values of q. We observe that the semicircular shape, typical of random matrices in the limit of large $\mathcal{D}$, and the length of the distribution are conserved independently of the value of q. This is because all eigenstates of GOE matrices are random vectors, and so is the initial state. That is, $C_\alpha^{(0)}$ are random numbers from a Gaussian distribution satisfying the constraint of normalization. Even though for $q > 1$, the larger components $C_\alpha^{(0)}$ become enhanced, leading to the spikes observed in Figure 1c,d, the width of the distribution is not affected by q. This means that after averages over random realizations, one will not notice the differences between the panels. One can then state that the robustness of the generalized LDoS for different values of q is a sign of the ergodicity of the eigenstates of the system.

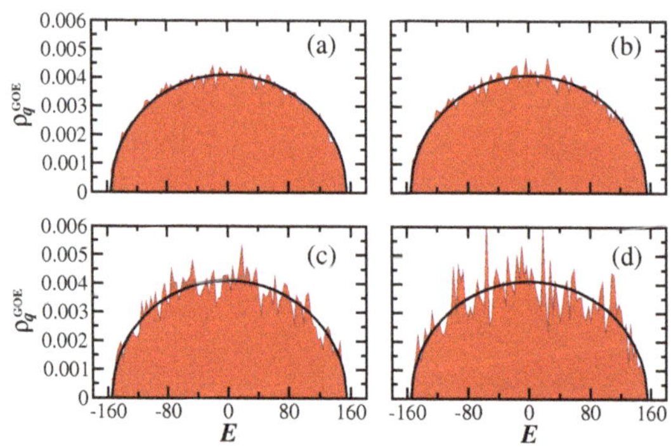

Figure 1. Generalized local density of states for GOE matrices for (**a**) $q = 0.5$, (**b**) $q = 1.0$, (**c**) $q = 2.0$, and (**d**) $q = 3.0$. Shaded areas are numerical results and the solid curves represent the semicircle law in Equation (14). A single disorder realization and a single initial state are considered. The matrix size is $\mathcal{D} = 12{,}000$.

Since the components $|C_\alpha^{(0)}|^q$ are uncorrelated random numbers fluctuating smoothly around the average $\overline{|C_\alpha^{(0)}|^q} = \mathcal{N}_q/\mathcal{D}$, one can see that the gLDoS for the GOE matrices coincides with the normalized density of states $P^{\mathrm{GOE}}(E) = \mathcal{D}^{-1}\sum_\alpha \delta(E_\alpha - E)$. Therefore, for GOE random matrices, we observe that:

$$\rho_q^{\mathrm{GOE}}(E) = P^{\mathrm{GOE}}(E) = \frac{1}{\pi\sigma_q}\sqrt{1-\left(\frac{E}{2\sigma_q}\right)^2}, \tag{14}$$

where the standard deviation $\sigma_q = \sqrt{\mathcal{D}/2}$. In Subfigure (a) in the third figure of this same Section 3 we plot σ_q as a function of q and confirm that σ_q is, indeed, nearly constant for GOE.

For physical many-body quantum systems with two-body interactions, the density of states is Gaussian [71,72]. Thus, the expected shape of the LDoS for a system perturbed far from equilibrium and an initial state in the middle of the spectrum, as considered here, is also Gaussian, as seen in Figure 2c for $q = 2$ and $h = 0.5$.

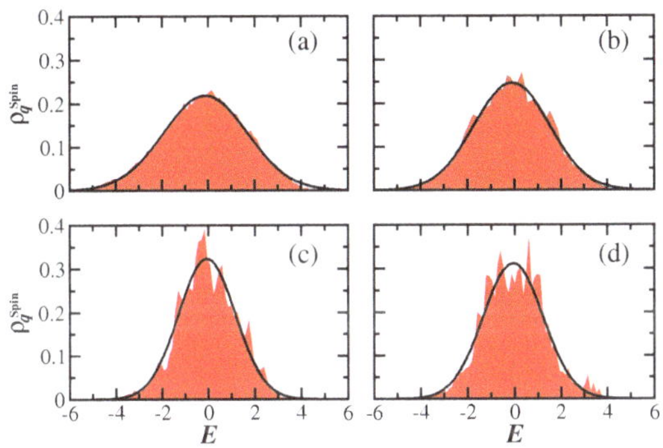

Figure 2. Generalized local density of states for the disordered spin-1/2 model with $h = 0.5$ for (**a**) $q = 0.5$, (**b**) $q = 1.0$, (**c**) $q = 2.0$, and (**d**) $q = 3.0$. Shaded areas are numerical results and the solid curves represent the Gaussian expression in Equation (15). A single disorder realization is considered. The system size is $L = 16$ with $\mathcal{D} = 12{,}870$.

Despite the persistence of the Gaussian shape for different values of q,

$$\rho_q^{\mathrm{Spin}}(E) = \frac{1}{\sqrt{2\pi\sigma_q^2}}\exp\left[-\frac{(E_\alpha - E_q^{(0)})^2}{2\sigma_q^2}\right]. \tag{15}$$

Figure 2 makes it clear that, in contrast to the GOE, the width σ_q depends on q. As q increases and the participation of the larger $|C_\alpha^{(0)}|^q$ becomes amplified, the width of $\rho_q^{\mathrm{Spin}}(E)$ becomes narrower than the density of states. This indicates that the contributions of the components at the tails of the initial-state energy distribution, where chaotic states are nonexistent, are erased.

The dependence of the width of the gLDoS on q reveals the limited degree of ergodicity of physical systems, even those deep in the chaotic regime. The eigenstates of physical systems are not random vectors, as in random matrices, and are not random superpositions of plane waves, as stated by Berry's conjecture [73]. The question of how to define chaotic states in realistic systems is discussed in [10,11,74–78]. Our results add to these studies, providing a way to quantify the level of ergodicity in comparison to random matrices.

In Figure 3, we compare the results for σ_q normalized by the width of the density of states (DoS) as a function of q for the GOE model (Figure 3a) and the spin model (Figure 3b). Each point in Figure 3 is obtained by performing an average over 10 random realizations and a single initial state. The flat curve in Figure 3a indicates the presence of fully ergodic states throughout the spectrum, while in Figure 3b, $\sigma_q^{Spin}/\sigma_{DOS}^{Spin}$ clearly decays as q increases. This occurs in the case of the chaotic model, with $h = 0.5$ (circles), where non-chaotic states are concentrated at the edges of the spectrum, and more abruptly in the case of $h = 2$ (squares), where non-chaotic states are also likely to be found away from the edges of the spectrum.

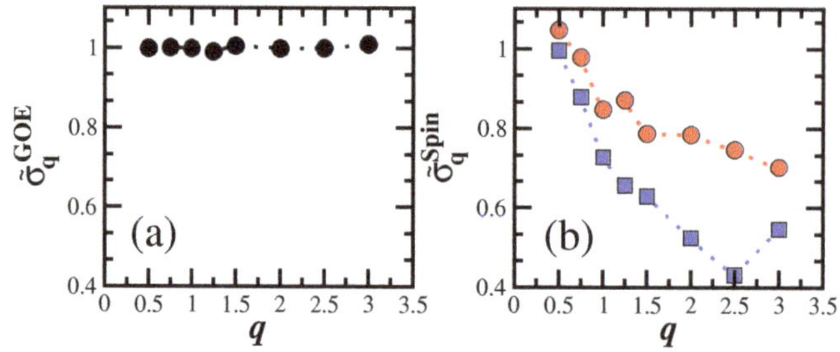

Figure 3. Width of the generalized LDoS normalized by the DoS for (**a**) GOE matrices, $\tilde{\sigma}_q^{\text{GOE}} = \sigma_q^{GOE}/\sigma_{DOS}^{GOE}$, and (**b**) the spin model, $\tilde{\sigma}_q^{\text{Spin}} = \sigma_q^{\text{Spin}}/\sigma_{DOS}^{\text{Spin}}$, with $h = 0.5$ (circles) and $h = 2$ (squares) as a function of q. Each point is an average over 10 disorder realizations and a single initial state. The dotted lines are guides for the eyes. $\mathcal{D} = 12{,}000$ for GOE and $\mathcal{D} = 12{,}870$ ($L = 16$) for the spin model.

The reason for the abrupt decay of σ_q^{Spin} with q for $h = 2$ becomes evident in Figure 4, where we plot $\rho_q(E)$ for different values of q. When $q \leq 1$ (Figure 4a,b), the shape of the LDoS is fragmented, while for $q > 1$ (Figure 4c,d), this structure is nearly erased, and $\rho_q(E)$ indicates a high degree of localization.

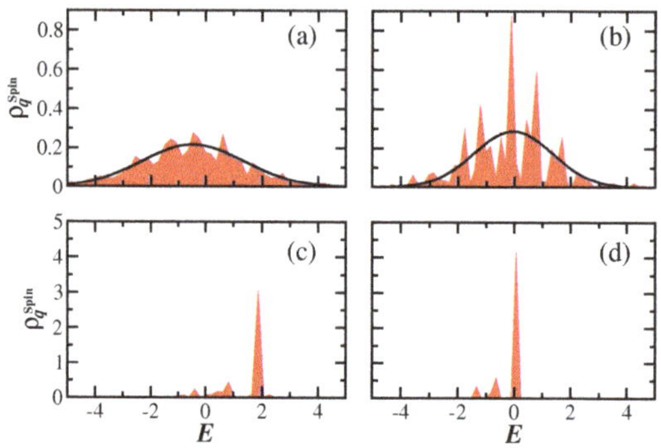

Figure 4. Generalized local density of states for the disordered spin-1/2 model with $h = 2.0$ for (**a**) $q = 0.5$, (**b**) $q = 1.0$, (**c**) $q = 2.0$, and (**d**) $q = 3.0$. Shaded areas are numerical results and solid curves represent the Gaussian expression in Equation (15). A single disorder realization and a single initial state are considered. The system size is $L = 16$ with $\mathcal{D} = 12{,}870$.

For finite-size systems, several numerical studies have supported the notion that the eigenstates of the disordered spin model should become multifractal in its transition to the many-body localized phase [37,64,79–82], although this has not been confirmed in the thermodynamic limit [83]. The patterns observed in Figure 4a,b also suggest fractality.

4. Evolution of the Generalized Survival Probability under the GOE Model: Analytical Expression

According to Equation (10), the survival probability averaged over an ensemble of initial states and random realization is written as:

$$\langle SP_q(t)\rangle = \left\langle \frac{1}{\mathcal{N}_q^2}\sum_{\alpha\neq\beta}|C_\alpha^{(0)}|^q|C_\beta^{(0)}|^q e^{-i(E_\alpha-E_\beta)t}\right\rangle + \left\langle \frac{1}{\mathcal{N}_q^2}\sum_{\alpha}|C_\alpha^{(0)}|^{2q}\right\rangle, \quad (16)$$

where $\langle \ldots \rangle$ denotes the average. The second term on the right-hand side corresponds to the infinite time average, $\overline{SP}_q$, of the generalized survival probability. For GOE random matrices, where $C_\alpha^{(0)}$ are random numbers from a Gaussian distribution,

$$\overline{SP}_q = \frac{1}{\mathcal{N}_q^2} \sum_\alpha |C_\alpha^{(0)}|^{2q} = \frac{\sqrt{\pi}\Gamma\left(q+\frac{1}{2}\right)}{\mathcal{D}\Gamma\left(\frac{q+1}{2}\right)^2}. \tag{17}$$

Since, for random matrices, the eigenvalues and the eigenstates are statistically independent, they can be factorized (see details in [84] and the appendix of [53]). Thus, using

$$\left\langle e^{-i(E_\alpha - E_\beta)t} \right\rangle = \frac{1}{\mathcal{D}-1}\left[\mathcal{D}\frac{\mathcal{J}_1^2(2\sigma t)}{(\sigma t)^2} - b_2\left(\frac{\sigma t}{2\mathcal{D}}\right)\right], \tag{18}$$

we observe that

$$\frac{1}{\mathcal{N}_q^2} \sum_{\alpha \neq \beta} |C_\alpha^{(0)}|^q |C_\beta^{(0)}|^q = 1 - \overline{SP}_q, \tag{19}$$

and due to the requirement that $SP_q(t=0) = 1$, we arrive at the analytical expression:

$$\langle SP_q(t) \rangle = \frac{1 - \langle \overline{SP}_q \rangle}{\mathcal{D}-1}\left[\mathcal{D}\frac{\mathcal{J}_1^2(2\sigma t)}{(\sigma t)^2} - b_2\left(\frac{\sigma t}{2\mathcal{D}}\right)\right] + \langle \overline{SP}_q \rangle. \tag{20}$$

Above, we write $\sigma_q = \sigma$, because σ_q is constant for the GOE. The Fourier transform of the semicircular gLDoS provides the first term on the right-hand side of Equation (20), which involves the Bessel function of the first kind, $\mathcal{J}_1$. This first term describes the initial decay of $\langle SP_q(t) \rangle$, as seen in Figure 5. It presents oscillations with the nth-zeros occurring when the initial state dynamically identifies an orthogonal state at $t_n \sim (\pi n + \sqrt{2}/2)/2\sigma$ with $n = 1, 2, \ldots$. The envelope of the oscillations decays as t^{-3}. The second term on the right-hand side of Equation (20), $b_2(t) = \{t \ln[(2t+1)/(2t-1)] - 1\}\Theta(t-1) + [t \ln(2t+1) - 2t + 1]\Theta(1-t)$, is the so-called two-level form factor that takes $\langle SP_q(t) \rangle$ on a ramp to the saturation value $\langle \overline{SP}_q \rangle$.

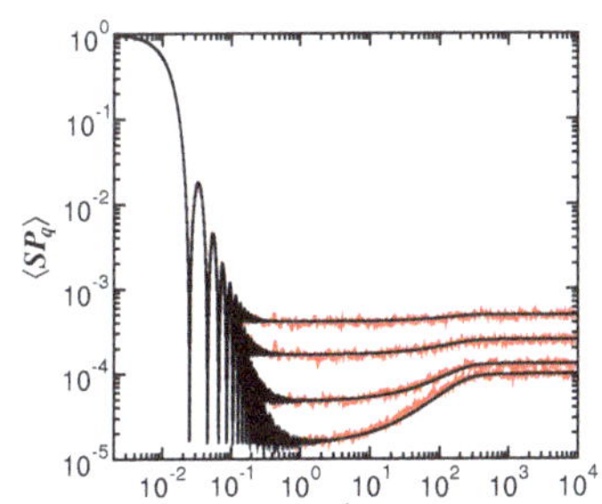

Figure 5. Generalized survival probability evolving under the GOE model for different values of q. Red curves are numerical results and the black lines correspond to the analytical expression in Equation (20). From bottom to top, $q = 0.5$, 1.0, 2.0, and 3.0. Matrix size is $\mathcal{D} = 12{,}000$. Averages over 10^4 samples.

In Figure 5, we compare the numerical results for $\langle SP_q(t) \rangle$ with the analytical expression in Equation (20). The agreement is excellent. The fact that σ_q for the GOE model is independent of q (see Figures 1 and 3) becomes evident, once again, in Figure 5, where the curves for the different values of q coincide at short times, capturing the oscillations of the Bessel function up to the minimum value of $\langle SP_q(t) \rangle$.

To derive the time scale, t_{Th}^{GOE}, where $\langle SP_q(t)\rangle$ reaches the minimum of the correlation hole, we must identify the point where the first and second terms in the square brackets of Equation (20) cross. Following [53], we obtain the long-term expansion of the first term in Equation (20) and expand the two-level form factor for the short times. Combining the two in the derivative of $\langle SP_q(t)\rangle$, we obtain:

$$t_{Th}^{GOE} = \left(\frac{3}{\pi}\right)^{1/4}\frac{\sqrt{\mathcal{D}}}{\sigma} = \left(\frac{3}{\pi}\right)^{1/4}. \tag{21}$$

To obtain the minimum value of $\langle SP_q(t)\rangle$ in the correlation hole, we evaluate Equation (20) at t_{Th}^{GOE}, which results in:

$$\begin{aligned}\langle SP_q(t)\rangle\big|_{t=t_{Th}^{GOE}} &\approx \frac{1-\langle\overline{SP}_q\rangle}{\mathcal{D}-1}\left[\frac{\mathcal{D}}{\pi(\sigma t_{Th}^{GOE})^3} - \left(1-\frac{\sigma}{\mathcal{D}}t_{Th}^{GOE}\right)\right] + \langle\overline{SP}_q\rangle \\ &\approx \frac{1-\langle\overline{SP}_q\rangle}{\mathcal{D}-1}(-1) + \langle\overline{SP}_q\rangle.\end{aligned} \tag{22}$$

Finally, using Equation (17) for $\langle\overline{SP}_q\rangle$, we arrive at:

$$\langle SP_q(t)\rangle\big|_{t=t_{Th}^{GOE}} \approx \frac{\sqrt{\pi}\Gamma\left(q+\frac{1}{2}\right) - \Gamma\left(\frac{q+1}{2}\right)^2}{\mathcal{D}\Gamma\left(\frac{q+1}{2}\right)^2}. \tag{23}$$

For the particular case of $q=2$, Equation (23) leads to the value $2/\mathcal{D}$ previously obtained in [53].

5. Evolution of the Generalized Survival Probability under the Spin Model

In Figure 6, we compare the evolution of $\langle SP_q(t)\rangle$ under the spin model for different values of q. In both panels, Figure 6a for $h=0.5$ and Figure 6b for $h=2$, the initial decay is determined by the envelope of the gLDoS, as seen in Equation (10). Since, according to Figures 2 and 4, the shape of the distribution is Gaussian, one observes in Figure 6 that

$$\langle SP_q(t)\rangle = \exp(-\sigma_q^2 t^2) \tag{24}$$

for $t \lesssim \sigma_q$. The dependence of σ_q on q is noticeable in Figure 6a and evident in Figure 6b.

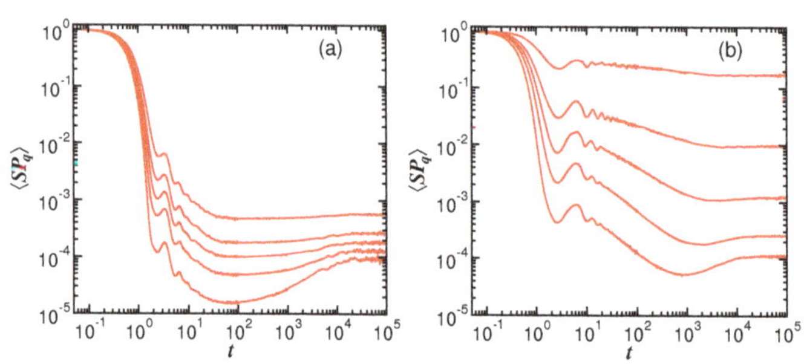

Figure 6. Generalized survival probability evolving under the disordered spin-1/2 model with (**a**) $h=0.5$ and (**b**) $h=2$ for different values of q. From bottom to top, $q=0.5$, 1.0, 1.5, 2.0, and 3.0. The system size is $L=16$. Averages over 3×10^4 samples.

Beyond the Gaussian behavior, a power-law decay emerges:

$$\langle SP_q(t)\rangle \propto t^{-\gamma_q}. \tag{25}$$

In Figure 6a, where the system is chaotic, the power-law exponent should depend on the bounds of the gLDoS. Since the gLDoS for the chaotic model in Figure 2 presents Gaussian tails for any q, we expect the same power-law exponent for all the curves in Figure 6a, which is, indeed, what the nearly parallel lines in the algebraic behavior suggest. In contrast, in Figure 6b, it is clear that γ_q decreases as q increases, and the minimum of the correlation hole takes longer to reach. In this case, the power-law behavior reflects the correlations between the components of the initial state, which become enhanced for larger values of q.

6. Discussion

We introduced the concepts of generalized survival probability, $SP_q(t)$, and the generalized local density of states, $\rho_q(E)$. We showed that the width of the generalized local density of states, σ_q, depends on q, even when the many-body quantum system is deep in the chaotic regime, which stands in contrast with random matrices, where the width is constant and equal to the width of the density of states. Therefore, σ_q may serve as a tool that can be employed to analyze and quantify the level of ergodicity of the states of physical systems with respect to random matrices.

We also showed that the power-law behavior that follows the Gaussian decay of the generalized survival probability is strongly dependent on q when the system is away from the chaotic regime. For a fixed value of the disorder strength, the power-law decay becomes stretched as q increases and the power-law exponent γ_q decreases. This dependence of γ_q on q indicates correlations between the eigenstates. In a future work, we plan to investigate how σ_q and γ_q may be used to study multifractality.

Author Contributions: Conceptualization, L.F.S. and E.J.T.-H.; Investigation, D.A.Z.-H. All authors have read and agreed to the published version of the manuscript.

Funding: D.A.Z.-H. and E.J.T.-H. received financial support from VIEP-BUAP (Mexico), project No. 00270. D.A.Z.-H. thanks CONACyT (Mexico) for the financial support received. L.F.S. is funded by the United States NSF grant No. DMR-1936006. L.F.S. received support from the MPS Simons Foundation Award, ID: 678586.

Institutional Review Board Statement: Not applicable.

Informed Consent Statement: Not applicable.

Data Availability Statement: All data are available upon request.

Acknowledgments: E.J.T.-H. and L.F.S. are grateful to Felix Izrailev for his constant support and invaluable teaching on quantum and classical chaos. D.A.Z.-H. and E.J.T.-H. are grateful to LNS-BUAP for providing access to their supercomputing facility.

Conflicts of Interest: The authors declare no conflict of interest.

Abbreviations

The following abbreviations are used in this manuscript:

DoS	Density of states
LDoS	Local density of states
gLDoS	Generalized local density of states
GOE	Gaussian orthogonal ensemble

References

1. Krylov, N.; Fock, V. On the uncertainity relation between time and energy. *J. Phys. USSR* **1947**, *11*, 112–120.
2. Krylov, N.; Fock, V. On two interpretations of the uncertainty relations between energy and time. *JETP* **1947**, *17*, 93–107.
3. Fock, V.A. Criticism of an attempt to disprove the uncertainty relation between time and energy. *Sov. Phys. JETP* **1962**, *15*, 784.
4. Torres-Herrera, J.; Karp J.; Távora, M.; Santos, L.F. Realistic Many-Body Quantum Systems vs. Full Random Matrices: Static and Dynamical Properties. *Entropy* **2016**, *18*, 359. [CrossRef]

5. Casati, G.; Chirikov, B.V.; Guarneri, I.; Izrailev, F.M. Quantum ergodicity and localization in conservative systems: The Wigner band random matrix model. *Phys. Lett. A* **1996**, *223*, 430. [CrossRef]
6. Fyodorov, Y.V.; Chubykalo, O.A.; Izrailev, F.M.; Casati, G. Wigner Random Banded Matrices with Sparse Structure: Local Spectral Density of States. *Phys. Rev. Lett.* **1996**, *76*, 1603–1606. [CrossRef]
7. Bertulani, C.A.; Zelevinsky, V.G. Excitation of multiphonon giant resonance states in relativistic heavy-ion collisions. *Nuc. Phys. A* **1994**, *568*, 931–952. [CrossRef]
8. Lewenkopf, C.H.; Zelevinsky, V.G. Single and multiple giant resonances: Counterplay of collective and chaotic dynamics. *Nucl. Phys. A* **1994**, *569*, 183–193. [CrossRef]
9. Frazier, N.; Brown, B.A.; Zelevinsky, V. Strength functions and spreading widths of simple shell model configurations. *Phys. Rev. C* **1996**, *54*, 1665–1674. [CrossRef]
10. Zelevinsky, V.; Brown, B.A.; Frazier, N.; Horoi, M. The nuclear shell model as a testing ground for many-body quantum chaos. *Phys. Rep.* **1996**, *276*, 85–176. [CrossRef]
11. Flambaum, V.V.; Izrailev, F.M. Statistical theory of finite Fermi systems based on the structure of chaotic eigenstates. *Phys. Rev. E* **1997**, *56*, 5144. [CrossRef]
12. Flambaum, V.V.; Izrailev, F.M. Excited eigenstates and strength functions for isolated systems of interacting particles. *Phys. Rev. E* **2000**, *61*, 2539. [CrossRef]
13. Flambaum, V.V. Time Dynamics in Chaotic Many-body Systems: Can Chaos Destroy a Quantum Computer? *Aust. J. Phys.* **2000**, *53*, 489–497. [CrossRef]
14. Flambaum, V.V.; Izrailev, F.M. Unconventional decay law for excited states in closed many-body systems. *Phys. Rev. E* **2001**, *64*, 026124. [CrossRef] [PubMed]
15. Flambaum, V.V.; Izrailev, F.M. Entropy production and wave packet dynamics in the Fock space of closed chaotic many-body systems. *Phys. Rev. E* **2001**, *64*, 036220. [CrossRef] [PubMed]
16. Kota, V.K.B.; Sahu, R. Structure of wave functions in (1+2)-body random matrix ensembles. *Phys. Rev. E* **2001**, *64*, 016219. [CrossRef]
17. Chavda, N.; Potbhare, V.; Kota, V. Strength functions for interacting bosons in a mean-field with random two-body interactions. *Phys. Lett. A* **2004**, *326*, 47. [CrossRef]
18. Angom, D.; Ghosh, S.; Kota, V.K.B. Strength functions, entropies, and duality in weakly to strongly interacting fermionic systems. *Phys. Rev. E* **2004**, *70*, 016209. [CrossRef]
19. Kota, V.K.B.; Chavda, N.D.; Sahu, R. Bivariate-*t* distribution for transition matrix elements in Breit-Wigner to Gaussian domains of interacting particle systems. *Phys. Rev. E* **2006**, *73*, 047203. [CrossRef]
20. Kota, V.K.B. *Lecture Notes in Physics*; Springer: Berlin/Heidelberg, Germany, 2014; Volume 884.
21. Santos, L.F.; Borgonovi, F.; Izrailev, F.M. Chaos and Statistical Relaxation in Quantum Systems of Interacting Particles. *Phys. Rev. Lett.* **2012**, *108*, 094102. [CrossRef]
22. Santos, L.F.; Borgonovi, F.; Izrailev, F.M. Onset of chaos and relaxation in isolated systems of interacting spins-1/2: Energy shell approach. *Phys. Rev. E* **2012**, *85*, 036209. [CrossRef] [PubMed]
23. Torres-Herrera, E.J.; Vyas, M.; Santos, L.F. General Features of the Relaxation Dynamics of Interacting Quantum Systems. *New J. Phys.* **2014**, *16*, 063010. [CrossRef]
24. Torres-Herrera, E.J.; Santos, L.F. Nonexponential fidelity decay in isolated interacting quantum systems. *Phys. Rev. A* **2014**, *90*, 033623. [CrossRef]
25. Khalfin, L.A. Contribution to the decay theory of a quasi-stationary state. *Sov. Phys. JETP* **1958**, *6*, 1053.
26. Ersak, I. The number of wave functions of an unstable particle. *Sov. J. Nucl. Phys.* **1969**, *9*, 263.
27. Fleming, G.N. A Unitarity Bound on the Evolution of Nonstationary States. *Il Nuovo Cimento A* **1973**, *16*, 232. [CrossRef]
28. Knight, P. Interaction Hamiltonians, spectral lineshapes and deviations from the exponential decay law at long times. *Phys. Lett. A* **1977**, *61*, 25–26. [CrossRef]
29. Fonda, L.; Ghirardi, G.C.; Rimini, A. Decay theory of unstable quantum systems. *Rep. Prog. Phys.* **1978**, *41*, 587. [CrossRef]
30. Erdélyi, A. Asymptotic Expansions Of Fourier Integrals Involving Logarithmic Singularities. *J. Soc. Indust. Appr. Math.* **1956**, *4*, 38. [CrossRef]
31. Urbanowski, K. General properties of the evolution of unstable states at long times. *Eur. Phys. J. D* **2009**, *54*, 25–29. [CrossRef]
32. Távora, M.; Torres-Herrera, E.J.; Santos, L.F. Inevitable power-law behavior of isolated many-body quantum systems and how it anticipates thermalization. *Phys. Rev. A* **2016**, *94*, 041603. [CrossRef]
33. Távora, M.; Torres-Herrera, E.J.; Santos, L.F. Power-law decay exponents: A dynamical criterion for predicting thermalization. *Phys. Rev. A* **2017**, *95*, 013604. [CrossRef]
34. Ketzmerick, R.; Petschel, G.; Geisel, T. Slow decay of temporal correlations in quantum systems with Cantor spectra. *Phys. Rev. Lett.* **1992**, *69*, 695–698. [CrossRef] [PubMed]
35. Huckestein, B.; Schweitzer, L. Relation between the correlation dimensions of multifractal wave functions and spectral measures in integer quantum Hall systems. *Phys. Rev. Lett.* **1994**, *72*, 713–716. [CrossRef] [PubMed]
36. Huckestein, B.; Klesse, R. Wave-packet dynamics at the mobility edge in two- and three-dimensional systems. *Phys. Rev. B* **1999**, *59*, 9714–9717. [CrossRef]
37. Torres-Herrera, E.J.; Santos, L.F. Dynamics at the many-body localization transition. *Phys. Rev. B* **2015**, *92*, 014208. [CrossRef]

38. Torres-Herrera, E.J.; Távora, M.; Santos, L.F. Survival Probability of the Néel State in Clean and Disordered Systems: An Overview. *Braz. J. Phys.* **2015**, *46*, 239. [CrossRef]
39. Torres-Herrera, E.J.; Santos, L.F. Extended nonergodic states in disordered many-body quantum systems. *Ann. Phys.* **2017**, *529*, 1600284. [CrossRef]
40. Leviandier, L.; Lombardi, M.; Jost, R.; Pique, J.P. Fourier Transform: A Tool to Measure Statistical Level Properties in Very Complex Spectra. *Phys. Rev. Lett.* **1986**, *56*, 2449–2452. [CrossRef]
41. Pique, J.P.; Chen, Y.; Field, R.W.; Kinsey, J.L. Chaos and dynamics on 0.5–300 ps time scales in vibrationally excited acetylene: Fourier transform of stimulated-emission pumping spectrum. *Phys. Rev. Lett.* **1987**, *58*, 475–478. [CrossRef]
42. Guhr, T.; Weidenmüller, H. Correlations in anticrossing spectra and scattering theory. Analytical aspects. *Chem. Phys.* **1990**, *146*, 21–38. [CrossRef]
43. Hartmann, U.; Weidenmüller, H.; Guhr, T. Correlations in anticrossing spectra and scattering theory: Numerical simulations. *Chem. Phys.* **1991**, *150*, 311–320. [CrossRef]
44. Alhassid, Y.; Levine, R.D. Spectral autocorrelation function in the statistical theory of energy levels. *Phys. Rev. A* **1992**, *46*, 4650–4653. [CrossRef]
45. Lombardi, M.; Seligman, T.H. Universal and nonuniversal statistical properties of levels and intensities for chaotic Rydberg molecules. *Phys. Rev. A* **1993**, *47*, 3571–3586. [CrossRef]
46. Michaille, L.; Pique, J.P. Influence of Experimental Resolution on the Spectral Statistics Used to Show Quantum Chaos: The Case of Molecular Vibrational Chaos. *Phys. Rev. Lett.* **1999**, *82*, 2083–2086. [CrossRef]
47. Gorin, T.; Prosen, T.; Seligman, T.H. A random matrix formulation of fidelity decay. *New J. Phys.* **2004**, *6*, 20. [CrossRef]
48. Alhassid, Y.; Fyodorov, Y.V.; Gorin, T.; Ihra, W.; Mehlig, B. Fano interference and cross-section fluctuations in molecular photodissociation. *Phys. Rev. A* **2006**, *73*, 042711. [CrossRef]
49. Leyvraz, F.; García, A.; Kohler, H.; Seligman, T.H. Fidelity under isospectral perturbations: A random matrix study. *J. Phys. A* **2013**, *46*, 275303. [CrossRef]
50. Torres-Herrera, E.J.; Santos, L.F. Dynamical manifestations of quantum chaos: Correlation hole and bulge. *Philos. Trans. R. Soc. A* **2017**, *375*, 20160434. [CrossRef]
51. Torres-Herrera, E.J.; García-García, A.M.; Santos, L.F. Generic dynamical features of quenched interacting quantum systems: Survival probability, density imbalance, and out-of-time-ordered correlator. *Phys. Rev. B* **2018**, *97*, 060303. [CrossRef]
52. Torres-Herrera, E.J.; Santos, L.F. Signatures of chaos and thermalization in the dynamics of many-body quantum systems. *Eur. Phys. J. Spec. Top.* **2019**, *227*, 1897–1910. [CrossRef]
53. Schiulaz, M.; Torres-Herrera, E.J.; Santos, L.F. Thouless and relaxation time scales in many-body quantum systems. *Phys. Rev. B* **2019**, *99*, 174313. [CrossRef]
54. Lerma-Hernández, S.; Villaseñor, D.; Bastarrachea-Magnani, M.A.; Torres-Herrera, E.J.; Santos, L.F.; Hirsch, J.G. Dynamical signatures of quantum chaos and relaxation time scales in a spin-boson system. *Phys. Rev. E* **2019**, *100*, 012218. [CrossRef] [PubMed]
55. Lezama, T.L.; Torres-Herrera, E.J.; Pérez-Bernal, F.; Lev, Y.B.; Santos, L.F. Equilibration time in many-body quantum systems. *Phys. Rev. B* **2021**, *104*, 085117. [CrossRef]
56. Santos, L.F.; Pérez-Bernal, F.; Torres-Herrera, E.J. Speck of chaos. *Phys. Rev. Res.* **2020**, *2*, 043034. [CrossRef]
57. Dag, C.B.; Mistakidis, S.I.; Chan, A.; Sadeghpour, H.R. Many-body quantum chaos in stroboscopically-driven cold atoms. *arXiv* **2022**, arXiv:2210.03840.
58. Rényi, A. On measures of entropy and information. In *Proceedings of the Fourth Berkeley Symposium on Mathematical Statistics and Probability*; University of California Press: Berkeley, CA, USA, 1961; Volume 1, pp. 547–561.
59. Evers, F.; Mirlin, A.D. Anderson transitions. *Rev. Mod. Phys.* **2008**, *80*, 1355–1417. [CrossRef]
60. Wegner, F. Inverse Participation Ratio in 2 + ε Dimensions. *Z. Phys. B* **1980**, *36*, 209. [CrossRef]
61. Soukoulis, C.M.; Economou, E.N. Fractal Character of Eigenstates in Disordered Systems. *Phys. Rev. Lett.* **1984**, *52*, 565–568. [CrossRef]
62. Atas, Y.Y.; Bogomolny, E.; Giraud, O.; Roux, G. Distribution of the Ratio of Consecutive Level Spacings in Random Matrix Ensembles. *Phys. Rev. Lett.* **2013**, *110*, 084101. [CrossRef]
63. Atas, Y.Y.; Bogomolny, E. Calculation of multi-fractal dimensions in spin chains. *Phil. Trans. R. Soc. A* **2013**, *372*. [CrossRef]
64. Solórzano, A.; Santos, L.F.; Torres-Herrera, E.J. Multifractality and self-averaging at the many-body localization transition. *Phys. Rev. Res.* **2021**, *3*, L032030. [CrossRef]
65. Pilatowsky-Cameo, S.; Villaseñor, D.; Bastarrachea-Magnani, M.A.; Lerma-Hernández, S.; F. Santos, L.; Hirsch, J.G. Identification of quantum scars via phase-space localization measures. *Quantum* **2022**, *6*, 644. [CrossRef]
66. Santos, L.F.; Rigolin, G.; Escobar, C.O. Entanglement versus chaos in disordered spin systems. *Phys. Rev. A* **2004**, *69*, 042304. [CrossRef]
67. Santos, L.F.; Dykman, M.I.; Shapiro, M.; Izrailev, F.M. Strong many-particle localization and quantum computing with perpetually coupled qubits. *Phys. Rev. A* **2005**, *71*, 012317. [CrossRef]
68. Dukesz, F.; Zilbergerts, M.; Santos, L.F. Interplay between interaction and (un)correlated disorder in one-dimensional many-particle systems: Delocalization and global entanglement. *New J. Phys.* **2009**, *11*, 043026. [CrossRef]

69. Nandkishore, R.; Huse, D. Many-body localization and thermalization in quantum statistical mechanics. *Annu. Rev. Condens. Matter Phys.* **2015**, *6*, 15. [CrossRef]
70. Mehta, M.L. *Random Matrices*; Academic Press: Boston, MA, USA, 1991.
71. French, J.B.; Wong, S.S.M. Validity of random matrix theories for many-particle systems. *Phys. Lett. B* **1970**, *33*, 449. [CrossRef]
72. Brody, T.A.; Flores, J.; French, J.B.; Mello, P.A.; Pandey, A.; Wong, S.S.M. Random-matrix physics: Spectrum and strength fluctuations. *Rev. Mod. Phys.* **1981**, *53*, 385. [CrossRef]
73. Berry, M.V. Regular and irregular semiclassical wavefunctions. *J. Phys. A* **1977**, *10*, 2083. [CrossRef]
74. Flambaum, V.V.; Gribakina, A.A.; Gribakin, G.F.; Kozlov, M.G. Structure of compound states in the chaotic spectrum of the Ce atom: Localization properties, matrix elements, and enhancement of weak perturbations. *Phys. Rev. A* **1994**, *50*, 267–296. [CrossRef] [PubMed]
75. Flambaum, V.V.; Gribakin, G.F.; Izrailev, F.M. Correlations within eigenvectors and transition amplitudes in the two-body random interaction model. *Phys. Rev. E* **1996**, *53*, 5729–5741. [CrossRef] [PubMed]
76. Flambaum, V.V.; Izrailev, F.M.; Casati, G. Towards a statistical theory of finite Fermi systems and compound states: Random two-body interaction approach. *Phys. Rev. E* **1996**, *54*, 2136–2139. [CrossRef] [PubMed]
77. Borgonovi, F.; Guarnieri, I.; Izrailev, F.M.; Casati, G. Chaos and thermalization in a dynamical model of two interacting particles. *Phys. Lett. A* **1998**, *247*, 140–144. [CrossRef]
78. Borgonovi, F.; Izrailev, F.M.; Santos, L.F.; Zelevinsky, V.G. Quantum chaos and thermalization in isolated systems of interacting particles. *Phys. Rep.* **2016**, *626*, 1. [CrossRef]
79. Luca, A.D.; Scardicchio, A. Ergodicity breaking in a model showing many-body localization. *Europhys. Lett.* **2013**, *101*, 37003. [CrossRef]
80. Luitz, D.J.; Alet, F.; Laflorencie, N. Universal Behavior beyond Multifractality in Quantum Many-Body Systems. *Phys. Rev. Lett.* **2014**, *112*, 057203. [CrossRef]
81. Li, X.; Ganeshan, S.; Pixley, J.; Sarma, S.D. Many-Body Localization and Quantum Nonergodicity in a Model with a Single-Particle Mobility Edge. *Phys. Rev. Lett.* **2015**, *115*, 186601. [CrossRef]
82. Kohlert, T.; Scherg, S.; Li, X.; Lüschen, H.P.; Sarma, S.D.; Bloch, I.; Aidelsburger, M. Observation of Many-Body Localization in a One-Dimensional System with a Single-Particle Mobility Edge. *Phys. Rev. Lett.* **2019**, *122*, 170403. [CrossRef]
83. Macé, N.; Alet, F.; Laflorencie, N. Multifractal Scalings Across the Many-Body Localization Transition. *Phys. Rev. Lett.* **2019**, *123*, 180601. [CrossRef]
84. Santos, L.F.; Torres-Herrera, E.J. Analytical expressions for the evolution of many-body quantum systems quenched far from equilibrium. *AIP Conf. Proc.* **2017**, *1912*, 020015. [CrossRef]

entropy

MDPI

Article

Quantum Bounds on the Generalized Lyapunov Exponents

Silvia Pappalardi * and Jorge Kurchan

Laboratoire de Physique de l'École Normale Supérieure, ENS, Université PSL, CNRS, Sorbonne Université, Université de Paris, F-75005 Paris, France
* Correspondence: silvia.pappalardi@phys.ens.fr

Abstract: We discuss the generalized quantum Lyapunov exponents L_q, defined from the growth rate of the powers of the square commutator. They may be related to an appropriately defined thermodynamic limit of the spectrum of the commutator, which plays the role of a large deviation function, obtained from the exponents L_q via a Legendre transform. We show that such exponents obey a generalized bound to chaos due to the fluctuation–dissipation theorem, as already discussed in the literature. The bounds for larger q are actually stronger, placing a limit on the large deviations of chaotic properties. Our findings at infinite temperature are exemplified by a numerical study of the kicked top, a paradigmatic model of quantum chaos.

Keywords: quantum chaos; generalized Lyapunov exponents; quantum bound to chaos

1. Introduction

Classical chaos is well understood from the sensitivity of the dynamics to respect small changes in the initial conditions, the so-called butterfly effect. This is quantified by the Lyapunov exponent, the rate at which nearby trajectories separate exponentially in time. In the past few years, there has been a lot of attention given to quantum chaos and in particular, on the *quantum Lyapunov exponent* λ_L, defined from the intermediate-time exponential growth of the following square-commutator:

$$\langle |[A(t), B]|^2 \rangle \sim \epsilon^2 e^{\lambda_L t} \,, \tag{1}$$

where ϵ is a small parameter [1]. The interest in this object comes from the fact that λ_L obeys a bound:

$$\lambda_L \leq \frac{2\pi}{\beta\hbar} \,. \tag{2}$$

This result, now known as the *quantum bound to chaos*, was proved within the high energy community [2], driven to the topic because maximal chaos is attained by models of black holes, including the Sachdev–Ye–Kitaev model (SYK) [3–5]. See also Refs. [6,7] for alternative derivations. The interest in these issues has later spread over different communities, from condensed matter to quantum information theory. Recently, the existence of the bound in Equation (2) was physically rationalized as a consequence of the fluctuation–dissipation theorem (FDT) [8,9] since the out of time-order correlators (OTOC) appearing in Equation (1) can be mapped into two-time correlation functions in a duplicated Hilbert space.

Actually, both classically and quantum mechanically, the Lyapunov exponent is a distributed quantity: different starting conditions—or different time intervals after the same starting condition—yield different exponents, and they are peaked on a 'typical' value and have large (and rare) deviations around, for example, when a classical trajectory grazes a regular region. If the distribution is not a delta function, it is referred to as 'multifractal' [10].

A possibility to study the full distribution of Lyapunov exponents is to introduce the generalized Lyapunov exponents (GLE) L_{2q}, defined from the moments of the distribution

Citation: Pappalardi, S.; Kurchan, J. Quantum Bounds on the Generalized Lyapunov Exponents. *Entropy* **2023**, 25, 246. https://doi.org/10.3390/e25020246

Academic Editor: Marko Robnik

Received: 28 December 2022
Accepted: 21 January 2023
Published: 30 January 2023

(This may lead to confusion. The Lyapunov exponents are classified as *first (maximal), second, ...* according to whether they measure linear, area, and in general k-form expansions [11]. Each is also generalized according to the moment q considered. Here we are considering expansions of linear lengths, and all moments q of the 'first, maximum, Lyapunov exponent', thus $L_{2q}^{(k)}$ for all q and $k = 1$). It turns out that the *quantum* generalized exponents also satisfy themselves a bound, stated in Ref. [8], that generalizes Equation (2). These bounds are the subject of this paper. We have two motivations: Firstly, they put limitations on the chaotic properties of *rare* protocols, favoring high and low chaoticity, and also allow us to define a Lyapunov exponent for *typical* protocols, actually different from the usual one considered in quantum mechanics. Secondly, a system that approaches the bounds for all the L_{2q} will turn out to be *mono-fractal*, i.e, the large deviation function becomes peaked on a single value, a property we find intriguing.

The quantum generalized Lyapunov exponents are defined by considering the $2q$-th commutator between two operators at different times, and assuming that they scale exponentially with time as

$$G_{2q}(t) = \langle (i[A(t), B)])^{2q} \rangle_\beta \sim \epsilon^{2q} \, e^{L_{2q}^{(\beta)} t} , \tag{3}$$

where ϵ is a small parameter, two common examples are $\epsilon = \hbar$ (the semiclassical limit) or $\epsilon = N^{-1}$ (as in the SYK model), the latter being more relevant here as we shall be interested in thermodynamic models. The thermal average $\langle \bullet \rangle_\beta$ at inverse temperature β may be defined in various ways, as we shall see below.

The rate $L_{2q}^{(\beta)}$ now defines the *quantum thermal* generalized Lyapunov exponent of order q. The usual (quenched) Lyapunov exponent thus is

$$\lambda_1^{(\beta)} = \lim_{q \to 0} \frac{L_{2q}^{(\beta)}}{2q} \tag{4}$$

On the other hand, the grow rate of the square commutator λ_L actually corresponds to the GLE with $q = 1$: $\lambda_L = L_2^{(\beta)}$.

In the quantum realm, the exponential regime holds only at intermediate times, up to the so-called *Ehrenfest time* $t_{\text{Ehr}} \sim \ln \epsilon^{-1}$. In actual fact, only when $\lim t_{\text{Ehr}} \to \infty$ is the Lyapunov regime unambiguously defined.

Different q-Lyapunovs are dominated by initial conditions having different expansion rates, with larger rates dominating the averages corresponding to larger q. Hence, one must allow for the dependence on the Ehrenfest time itself:

$$t_{\text{Ehr}}^{(2q)} = \frac{2q}{L_{2q}} \ln \epsilon^{-1} \quad \text{such that} \quad G_{2q}(t) \sim e^{L_{2q}^{(\beta)} (t - t_{\text{Ehr}}^{(2q)})} . \tag{5}$$

A crucial assumption we shall make here is that

$$t_{\text{Ehr}}^{(2q)} \leq t_{\text{Ehr}}^{(2q')} \qquad \text{for} \qquad q > q' \tag{6}$$

Under these assumptions, we show that the following holds:

$$\frac{L_{2q}^{(\beta)}}{2q} \leq \frac{\pi}{\beta \hbar} . \tag{7}$$

This bound was already stated in Ref. [8], without the identification of the rate $L_{2q}^{(\beta)}$ as a quantum GLE and the relation to large deviations. We will also clarify some assumptions on which the derivation [8].

Because we are assuming that $\frac{L_{2q}^{(\beta)}}{2q}$ does not decrease, the bounds for larger q (but always of order one with respect to N) are more stringent. The meaning of this, as the classical discussion below will make clear, is that even rarely expanding conditions are bounded.

2. Classical Generalized Lyapunov Exponents

Let us briefly review the classical case [11]. Consider the infinitesimal separation $|\mathbf{\Delta}(t)|$ between two trajectories at time t, starting at a point $\mathbf{x}_0$ and $\mathbf{x}_0 + \mathbf{\Delta}$:

$$R(t) = \frac{|\mathbf{\Delta}(t)|}{|\mathbf{\Delta}(0)|} \sim e^{\lambda t} . \tag{8}$$

The rate λ is a function of the initial condition $\mathbf{x}_0$. This quantity grows in time according to an asymptotic exponential law λt with $(\lambda > 0)$ if the system is chaotic, where λ is a function of time which reaches a finite limit at long times. A very long chaotic trajectory will have explored most of the phase space, and the exponential expansion will be sampled from all regions: the value λ then becomes essentially the same for all initial conditions. This fact is encompassed in the Oseledec theorem [12].

The fact that $\ln R(t)$ is a cumulative process over stretches of time with uncorrelated properties leads to the usual argument for the introduction of a large deviation principle, in this case for the probability of a Lyapunov value given a random initial condition:

$$\ln P(\lambda, t) \sim -t\, S(\lambda) , \tag{9}$$

where S is the large-deviation (Cramér) function. If the system is ergodic, the ensemble of initial conditions may be substituted by the ensemble of initial times along the same trajectory.

The typical Lyapunov exponent λ_1 is given by

$$\lambda_1 \equiv \lim_{t\to\infty} \frac{1}{t} \langle \ln R(t) \rangle_\tau \equiv \lim_{t\to\infty} \int d\lambda\; P(\lambda, t)\lambda \tag{10}$$

(note that $\lambda = \lambda(t)$).

One can study the $2q$-th moments which, for long enough times, shall grow exponentially as

$$\mathcal{R}_{2q}(t) = \langle R(t)^{2q} \rangle \sim e^{L_{2q} t}, \tag{11}$$

where

$$L_{2q} = \lim_{t\to\infty} \frac{1}{t} \ln \langle R(\tau)^{2q} \rangle \tag{12}$$

are called the generalized Lyapunov exponents (GLE) of order $2q$, and characterize the fluctuations of the dynamical system [10]; see also Ref. [13]. They are defined as *annealed averages*:

$$L_{2q} \equiv \frac{1}{t} \ln \lim_{t\to\infty} \int d\lambda\; P(\lambda, t)\, e^{2q\lambda t} \sim \max_{\lambda}\{-S(\lambda) + 2q\lambda\} , \tag{13}$$

where we have evaluated the integral over large t using saddle point. The typical Lyapunov exponent is retrieved as the limit

$$\lambda_1 = \lim_{q\to 0} \frac{L_{2q}}{2q} = \frac{1}{2} \frac{dL_{2q}}{dq}\Big|_{q=0} . \tag{14}$$

The GLE L_{2q} is the Legendre transform of $S(\lambda)$, via Equation (13). As such, also L_{2q} is a convex function of $2q$. As a consequence, one has that L_{2q} obeys two important properties:

1. $L_{2q}/2q$ are an increasing function of the order q:

$$\frac{d}{dq}\left(\frac{L_{2q}}{2q}\right) \geq 0\,; \tag{15}$$

2. The GLEs are always bounded by the linear behavior:

$$L_{2q} \geq 2q\lambda_1\,. \tag{16}$$

These equations are obtained via the property of convex differentiable functions $f(x)$: $f(x) \geq f(y) + f'(y)(x-y)$. For $x = 2q$ and $y = 0$, this yields Equation (16), while for $x = 0$ and $y = 2q$, one has Equation (15).

The equality $L_{2q} = 2q\lambda_1$ in Equation (16) holds only if $P(\lambda) = \delta(\lambda - \lambda_1)$, namely if the Lyapunov exponent is the same and does not fluctuate, we have *mono-fractality*. Interestingly, this characterizes also random matrices of dimension D without any structure and with high connectivity, which satisfy $L_{2q} = 2q\lambda_1 + \mathcal{O}(1/D)$ [10]. Otherwise, the system is characterized by *multifractal behavior*. The higher the moments, the more important the contributions coming from the tales of the distribution. In particular, in the case of a distribution $P(\lambda)$ with a finite support, the limits

$$\lambda_{max/min} = \lim_{q\to\pm\infty} \frac{L_{2q}}{2q} \tag{17}$$

select the maximal and minimal expanding rates.

3. Quantum Generalized Lyapunov Exponents at Infinite Temperature

Our goal is to extend the definition of generalized Lyapunov exponents to the quantum domain to discuss the bound in Equation (7). Systems with a few degrees of freedom do not lend themselves to the implementation of bounds that depend on temperature, as the canonical ensemble is not particularly useful for them. However, we may understand some other features that are also valid in thermodynamic systems by studying infinite-temperature systems of this kind. In this section, we define the quantum generalized Lyapunov exponents at infinite temperature and discuss their "convexity" properties. We will see that it is straightforward to interpret the quantum GLE as a probe of the spectral properties of the square-commutator operator.

3.1. Properties of the Infinite Temperature Quantum GLE

Let us first analyze the infinite temperature $2q$-th commutator Equation (3) at infinite temperature:

$$G_{2q}^{(0)}(t) = \frac{(-1)^q}{Z}\,\mathrm{Tr}\Big([A(t), B]^{2q}\Big), \tag{18}$$

where $Z = \mathrm{Tr}(\mathbb{I}) = \dim\mathcal{H}$ is given by the Hilbert space dimension. This object generalizes the infinite temperature square-commutator in Equation (1), which has been discussed in a variety of models, and it is particularly relevant for dynamical protocols where energy is not conserved (such as with periodic driving or in the open system's scenario).

The infinite temperature quantum GLEs are then defined by the exponential growth at intermediate times:

$$G_{2q}^{(0)}(t) \sim \epsilon^{2q} e^{L_{2q}^{(0)} t}\,. \tag{19}$$

We now show that $L_{2q}^{(0)}$ obey the same properties as the classical ones (e.g., Equation (15)). Let us re-write Equation (18) as

$$G_{2q}^{(0)}(t) = \frac{1}{Z}\sum_i (g_i(t))^{2q}\,, \tag{20}$$

where we defined $g_i(t)$ the eigenvalues of the square-commutator operator, i.e.,

$$-[A(t),B]^2 = \sum_{i_t}(g_i(t))^2|i_t\rangle\langle i_t| \,. \tag{21}$$

and we have made explicit a factor t so that λ_t may have a finite limit. Some properties of this operator have been studied on specific models, see, for example, Refs. [14,15]. If the expectation value of the square commutator grows exponentially (we consider only times before t_{Ehr}), then it is convenient to write each eigenvalue as

$$g_i(t) = e^{\lambda_t^i t} \,. \tag{22}$$

By using $1 = \int d\lambda \delta(\lambda - \lambda_t^i)$, we can re-write Equation (18) as

$$G_{2q}^{(0)}(t) = \int d\lambda \, P(\lambda) e^{2q\lambda t} \tag{23}$$

where we have defined the *distribution of the quantum local Lyapunov exponents*

$$P(\lambda) = \sum_i \delta(\lambda_{i_t} - \lambda) \,. \tag{24}$$

Equations (23) and (24) shows that the $G_{2q}^{(0)}(t)$ are moments since they can be written as an integral of times the powers of a function times a positive function $P(\lambda)$. We can associate to the latter a convex Cramér function $tS(\lambda) \sim \ln P(\lambda)$ as in Equation (9), which gives the Legendre transform of $L_{2q}^{(0)}$. These relations imply the convexity of the quantum GLE at an infinite temperature, which results in the following:

1. $L_{2q}^{(0)}/2q$ is an increasing function of q;
2. The following inequality holds:

$$L_{2q}^{(0)} \geq 2q\lambda_1^{(0)} \,, \tag{25}$$

where $\lambda_1^{(0)} = \lim_{q\to 0} \frac{L_{2q}^{(0)}}{2q}$.

The equality holds in the absence of fluctuations in the spectrum of the square-commutator operator. Such mono-fractal behavior means that—for the appropriate time's range—the square-commutator operator is close to a constant times the identity matrix.

3.2. A Semi-Classical Example: The Quantum Kicked Top

As an illustrative example, we study a driven model: the quantum kicked top. Since the energy is not conserved, this model is equivalent to a system at infinite temperature. We thus show that $L_{2q}^{(0)}$ satisfies the properties of convexity and of large deviation theory. (In this section, we denote $L_{2q}^{(0)} = L_{2q}$, for the sake of clarity.)

The model is described by the time-dependent Hamiltonian:

$$H = \alpha S_x + \frac{J}{N} S_z^2 \sum_{n=-\infty}^{\infty} \delta(t - n\tau) \,, \tag{26}$$

where $S_{x,y,z} = \frac{1}{2}\sum_{i=1}^{N} \sigma_{x,y,z}^i$ are collective spin operators. Due to the collective nature of the interactions, for large N, the classical limit is approached. One can define an effective Planck constant:

$$\hbar = \frac{1}{S} = \frac{2}{N} \tag{27}$$

that vanishes in the thermodynamic limit. The stroboscopic time-evolution operator (namely, the time-evolution operator over one period) reads as

$$\hat{U} = \hat{U}_J \hat{U}_\alpha \text{ with } \quad U_\alpha = e^{-i\alpha S_x} \, , \, \hat{U}_J = e^{-i\frac{J}{N}S_z^2} \, . \tag{28}$$

We fix $\tau = 1$ and $\alpha = \pi/2$. Changing the value of the kicking strength J, this model undergoes a transition between a regular regime and a chaotic one [16,17]. The dynamics of the square commutator (1) have been extensively explored [18–22].

We consider the strongly chaotic limit by choosing $J = 3.5$, and we look at the infinite temperature state with $\rho = \mathbb{I}/\dim H$, with $\dim H = N + 1$. We study the dynamics of the $G_{2q}(t)$ via exact numerical calculations, specifically via exact diagonalization. We compute the stroboscopic time-evolution of the $2q$-th commutator (3) using $\hat{A} = \hat{B} = \hat{S}^z$ at times $t = n\tau = 0, 1, 2, \ldots$.

In Figure 1, we show the dynamics of $G_{2q}(t)$ for different values of $q = 1 \div 19$. The correlators are rescaled by $\hbar^{2q}$ [with $\hbar = 2/N$] to emphasize the scaling of Equation (3). Each commutator grows exponentially before the Ehrenfest times with a different rate that corresponds to the quantum GLE L_{2q}. The value of L_{2q} is then fitted and plotted in Figure 2a, where we display its behavior as a function of q. It is a convex function of q that satisfies $L_{2q} > 2q\lambda_1$ [cf. Equation (25)], being therefore multifractal. The typical Lyapunov exponent λ_1 is computed as in Figure 2b, where $L_{2q}/2q$ is plotted as a function of q. The extrapolation to $q \to 0$ yields $\lambda_1 = 1.1(1)$, which corresponds to the maximum Lyapunov exponent of the classical model in the chaotic phase $\lambda_{class} = 1.12$, as computed via the Benettin et al. algorithm [23,24]; see, for example, the appendix of Ref. [22]. We also extract the maximal expanding rate $\lambda_{max} = 2.4(1)$ [cf. Equation (17)] from the limit $q \to \infty$, signaling that the distribution of Lyapunov has finite support.

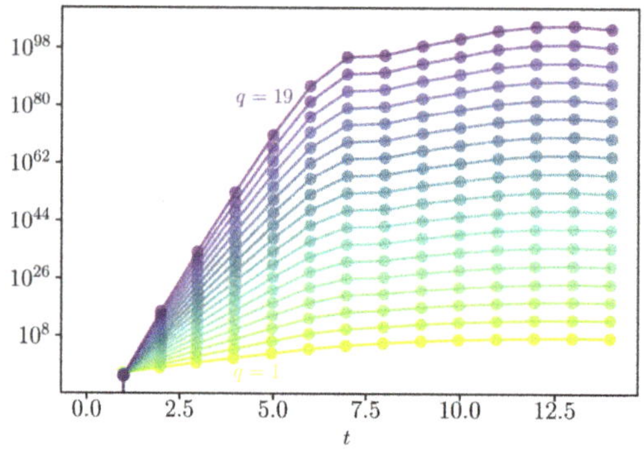

Figure 1. Dynamics of $G_{2q}(t)$ in Equation (3) for different values of $q = 1 \div 19$ as a function of time for $N = 1600$.

Figure 3 (shaded lines) shows the spectrum of the square commutator in Equation (21). Most of the eigenvalues grow exponentially in time before saturation and thus define some local Lyapunov exponents. We compare this behavior with the standard square-commutator expectation $G_2(t)$ [cf. Equation (1)] (blue dots), which grows exponentially at a rate L_2 larger than the maximum Lyapunov λ_1. The figure also shows that λ_{max} as fitted and extracted in Figure 2 (the dashed black line) corresponds to the maximal expanding rate of the local Lyapunov exponents.

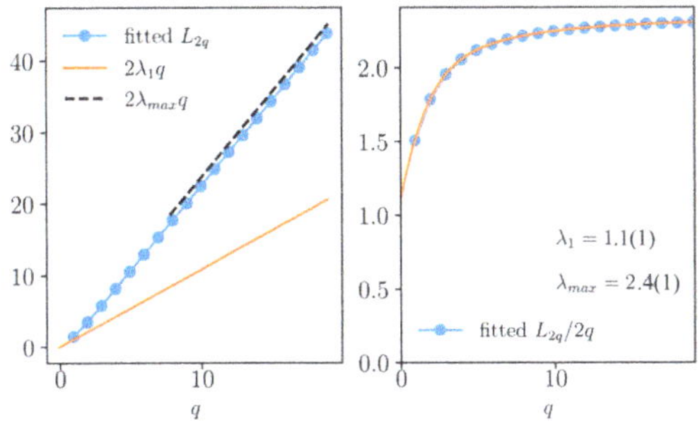

Figure 2. Generalized Lyapunov exponents fitted from Figure 1. (**Left**) L_{2q} as a function of q, contrasted with the actual Lyapunov exponent λ_1 and the maximal expanding rate λ_{max}, obtained by a fit of these data at large q. (**Right**) $L_{2q}/2q$ as a function of the moment q, from which we extract the maximal Lyapunov exponent λ_1.

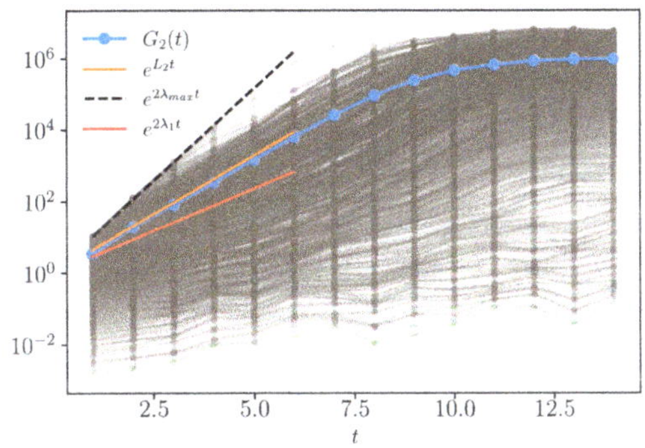

Figure 3. Spectrum of the square commutator $g_i^2(t)$, compared with $G_2(t)$ and the exponential growth with L_2, with the maximal expansion rate λ_{max} and with λ_1 for $N = 1600$.

In Figure 4, we show that the local Lyapunov exponents are a large deviation. We consider the coefficients $\lambda_t^i = \ln(g_i)/2t$ from Figure 3 (we divide everything by a constant factor). On the left, we plot their numerical distribution at different times $t = 3, 4, 5$, which shows that it converges to a distribution at large t. On the right, we plot $-\ln P(\lambda)/t$, which shall correspond to the smooth convex "Cramer" function at large times; see Equation (9).

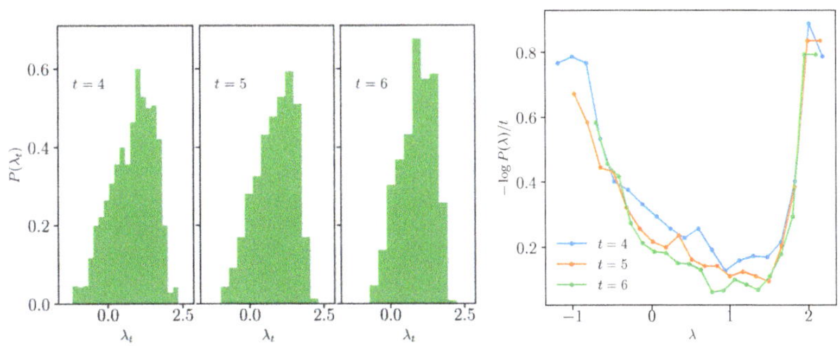

Figure 4. Large deviation properties of the spectrum of the square-commutator for $N = 1600$. (**Left**) Numerical distributions of $\lambda_t^i = \ln(g_i(t))/2t$ with $g_i^2(t)$ the eigenvalues of Equation (20) at different times $t = 3, 4, 5$. (**Right**) $-\ln P(\lambda)/t$ with $P(\lambda)$ the empirical distribution.

4. Thermal Quantum Generalized Exponents

It is useful to consider the different *regularizations* of the $2q$-th commutator as

$$G_{2q}^{(\beta)}(t) = \mathrm{Tr}\left(\left(\rho^{\frac{1}{4q}} i[A(t),B]\rho^{\frac{1}{4q}}\right)^{2q}\right), \tag{29}$$

$$\overline{G}_{2q}^{(\beta)}(t) = \mathrm{Tr}\left(\left(i[\rho^{\frac{1}{8q}} A(t)\rho^{\frac{1}{8q}}, \rho^{\frac{1}{8q}} B\rho^{\frac{1}{8q}}]\right)^{2q}\right). \tag{30}$$

with

$$\rho = e^{-\beta H}/Z_\beta \qquad \text{and} \qquad Z_\beta = \mathrm{Tr} e^{-\beta H} \tag{31}$$

Considering $q = 1$ in Equations (29) and (30), we retrieve the standard regularized square commutators for which the bounds have been proved in Ref. [9]. In Section 4.2 below, we show that the multi-time correlation functions appearing in Equations (29) and (30) can be mapped as two-times functions in a replicated Hilbert space of $2q$ copies. This allows one to rationalize the use of such regularizations—that might seem an artificial construction—as fixing the temperature of the different replicas to be the same.

These regularizations define the thermal average introduced in Equation (3), which defines thermal GLE $L_\beta^{(2q)}$. We consider the situation in which both of the q-th commutators grow exponentially in time as

$$G_{2q}^{(\beta)}(t) \propto \epsilon^{2q} e^{L_{2q}^{(\beta)} t}, \tag{32}$$

$$\bar{G}_{2q}^{(\beta)}(t) \propto \epsilon^{2q} e^{L_{2q}^{(\beta)} t}, \tag{33}$$

which is valid only for an intermediate time regime

$$t_d \ll t \leq t_{\mathrm{Ehr}}^{(2q)} \equiv \frac{L_{2q}^{(\beta)}}{2q} \ln \epsilon^{-1}. \tag{34}$$

4.1. From Commutators to OTOCS

Consider the quantities of Equations (29) and (30). Expanding the commutators, we get a series of OTOC terms containing exactly k times:

$$G_{2q}^{(\beta)}(t) = \sum_{k=1}^{2q} d_k \, [OTOC]_k \tag{35}$$

of the form

$$[OTOC]_k = \langle A_1(t)B_1(0)\dots A_k(t)B_k(0)\rangle_\beta + \text{h.c.}, \tag{36}$$

where $A_1, B_1, \dots$ are powers of A and B respectively and d_k some coefficients. With these notations, $[OTOC]_1 = \langle A^q(t)B^q(0)\rangle_\beta + \text{h.c.}$ is a function of two times. Out-of-time-order correlators between k operators are sometimes referred to as k-OTOC; see Refs. [25–28]. We are here interested in understanding their structure in time for exponential growth. Following [2], we assume that there exists some dissipation time t_d, after which two-point functions factorize as $[OTOC]_1 \sim C_q$.

Each $[OTOC]_{2k}$ may grow at most as fast as the corresponding Lyapunov behavior, during the corresponding Ehrenfest time:

$$C_k - [OTOC]_{2k} \propto \epsilon^{2k} e^{L_{2k} t} = e^{L_{2k}(t - t_{\mathrm{Ehr}}^{(2k)})}. \tag{37}$$

If we evaluate this term at times corresponding to a finite but small fraction of the corresponding Ehrenfest time $t_{\mathrm{Ehr}}^{(2q)}$, we conclude that all the terms with $k < q$ are of lower or equal order, because of the ordering of Ehrenfest times [cf. Equation (6)]. We thus conclude that

$$G_{2q}^{(\beta)}(t) \sim C_q - \text{Re Tr}\left(\rho^{\frac{1}{2q}} A(t) B \rho^{\frac{1}{2q}} A(t) B \dots \rho^{\frac{1}{2q}} A(t) B \right), \quad (38)$$

$$\overline{G}_{2q}^{(\beta)}(t) \sim \overline{C}_q - \text{Tr}\left(\rho^{\frac{1}{4q}} A(t) \rho^{\frac{1}{4q}} B \rho^{\frac{1}{4q}} A(t) \rho^{\frac{1}{4q}} B \dots \rho^{\frac{1}{4q}} A(t) \rho^{\frac{1}{4q}} B \right), \quad (39)$$

where the constants $C_q, \bar{C}_q$ are different given the different regularizations.

4.2. Product Space, Fluctuation-Dissipation Theorem and Bound

In this section, we will show how the multi-time OTOC appearing in the generalized $2q$-th commutators has a simple interpretation as two-time correlation functions in a $2q$-replicated space, see [25]. Focusing on $q = 1$, in Ref. [9] we have stressed that bringing an OTOC into this representation for finite β allows one to write the corresponding fluctuation-dissipation (FDT) relations as a usual KMS one. Here, we will demonstrate it for generic q.

Let us consider the following $4q$ point out of time order correlator:

$$S_{2q}(t) = \frac{1}{Z_\beta} \text{Tr}\left((\rho^{\frac{1}{2q}} A(t) B)^{2q} \right) = \text{Tr}\left(\rho^{\frac{1}{2q}} A(t) B \rho^{\frac{1}{2q}} A(t) B \dots \rho^{\frac{1}{2q}} A(t) B \right). \quad (40)$$

Now, we re-write it in terms of the spectral representation of the Hamiltonian $H|n\rangle = E_n|n\rangle$ as

$$\begin{aligned} S_{2q}(t) &= \frac{1}{Z_\beta} \sum_{n_1 n_2 \dots n_{2q}} e^{-\frac{\beta}{2q}(E_{n_1}+E_{n_2}+\dots E_{n_{2q}})} \times \langle n_1|A(t)B|n_2\rangle\langle n_2|A(t)B|n_3\rangle \dots \langle n_{2q}|A(t)B|n_1\rangle \\ &= \frac{1}{Z_\beta} \sum_{n_1 n_2 \dots n_{2q}} e^{-\frac{\beta}{2q}(E_{n_1}+E_{n_2}+\dots E_{n_{2q}})} \times \langle n_1 n_2 \dots n_{2q-1} n_{2q}|\mathbb{A}(t)\mathbb{B}|n_2 n_3 \dots n_{2q} n_1\rangle, \end{aligned}$$

where we introduced the operators that act in the $2q$-th replicated Hilbert space:

$$\mathbb{A}(t) = A(t) \otimes A(t) \cdots \otimes A(t), \qquad \mathbb{B} = B \otimes B \cdots \otimes B, \quad (41)$$

and the replicated Hamiltonian

$$\mathbb{H} = H \otimes 1 \cdots \otimes 1 + 1 \otimes H \cdots \otimes 1 + \cdots + 1 \otimes 1 \cdots \otimes H. \quad (42)$$

We also define the *cyclic shift operator* $\mathbb{P}$ that permutes cyclically states between the Hilbert spaces as

$$\mathbb{P}|n_1 n_2 \dots n_{2q-1} n_{2q}\rangle = |n_2 n_3 \dots n_q n_1\rangle \quad \text{with} \quad \mathbb{P}^{2q} = 1 \quad , \quad (43)$$

$$\mathbb{P}^\dagger|n_1 n_2 \dots n_{2q-1} n_{2q}\rangle = |n_{2q} n_1 n_2 \dots n_{2q-1}\rangle . \quad (44)$$

Notice that the operator $\mathbb{P}$ is non-Hermitian, but we can define $\tilde{\mathbb{P}} = \frac{\mathbb{P}+\mathbb{P}^\dagger}{2}$ that is. $\tilde{\mathbb{P}}$ also commutes with $\mathbb{A}(t)$, $\mathbb{B}$ and $\mathbb{H}$ so that $\tilde{\mathbb{P}}\mathbb{B}$ is Hermitian.

Let us re-write Equation (41) as

$$\begin{aligned} S_{2q}(t) = \frac{1}{Z_\beta} \sum_{n_1 n_2 \dots n_{2q}} e^{-\frac{\beta}{2q}(E_{n_1}+E_{n_2}+\dots E_{n_{2q}})} \times \frac{1}{2}\Big(&\langle n_1 n_2 \dots n_{2q-1} n_{2q}|\mathbb{A}(t)\mathbb{B}|n_2 n_3 \dots n_q n_1\rangle \\ &+ \langle n_1 n_2 \dots n_{2q-1} n_{2q}|\mathbb{A}(t)\mathbb{B}|n_{2q} n_1 \dots n_{q-2} n_{2q-1}\rangle \Big), \end{aligned}$$

where in the second line, we simply used a different resolution of the identity and a reshuffling of the matrix elements (We use

$$\langle n_1|A(t)B|n_{2q}\rangle\langle n_{2q}|A(t)B|n_{2q-1}\rangle\dots\langle n_3|A(t)B|n_2\rangle\langle n_2|A(t)B|n_1\rangle = \\ \langle n_1|A(t)B|n_{2q}\rangle\langle n_2|A(t)B|n_1\rangle\langle n_3|A(t)B|n_2\rangle\dots\langle n_{2q}|A(t)B|n_{2q-1}\rangle\,.$$

).

Therefore, we can re-write Equation (40) as

$$S_{2q}(t) = \frac{1}{Z_\beta}\mathrm{Tr}\Big(e^{-\beta_{2q}\mathbb{H}}\,\mathbb{A}(t)\mathbb{B}\,\tilde{\mathbb{P}}\Big) \tag{45}$$

which, besides a normalization, is a standard equilibrium expectation value of a two-time function at inverse temperature $\beta_{2q} = \beta/2q$. This result naturally generalizes the one for four times OTOC, for which $\mathbb{P} = \mathbb{P}^\dagger = \tilde{\mathbb{P}}$, as derived in Ref. [9].

4.2.1. Fluctuation–Dissipation in the Replicated Space

We may now write the extended KMS relations. We consider

$$\mathbb{C}_{2q}(t) = \frac{1}{2}\frac{1}{Z}\mathrm{Tr}\Big[e^{-\beta_{2q}\mathbb{H}}\,\{\mathbb{A}(t),\,\mathbb{B}\,\tilde{\mathbb{P}}\}\Big]\,, \tag{46}$$

$$\mathbb{R}_{2q}(t) = \frac{i}{\hbar}\theta(t)\frac{1}{Z}\mathrm{Tr}\Big[e^{-\beta_{2q}\mathbb{H}}\,[\mathbb{A}(t),\,\mathbb{B}\,\tilde{\mathbb{P}}]\Big] \tag{47}$$

$$\mathbb{F}_{2q}(t) = \frac{1}{Z}\mathrm{Tr}\Big[e^{-\frac{\beta_{2q}}{2}\mathbb{H}}\,\mathbb{A}(t)e^{-\frac{\beta_{2q}}{2}\mathbb{H}}\,\mathbb{B}\,\tilde{\mathbb{P}}\Big] \tag{48}$$

where $\mathbb{C}_{2q}$ and $\mathbb{R}_{2q}$ are defined as usual from real and imaginary parts of $S_{2q}(t) = \mathbb{C}_{2q}(t) + \hbar(\mathbb{R}_{2q})''(t)$ and correspond to fluctuations and response functions, respectively. Instead, the (Whiteman) correlation function $\mathbb{F}_{2q}$ in the original space is

$$\mathbb{F}_{2q}(t) = \mathrm{Tr}\Big((\rho^{1/4q}A(t)\rho^{1/4q}B)^{2q}\Big)\,. \tag{49}$$

We remark that the Fourier transforms of the connected parts of $\mathbb{F}_{2q}$, known as free cumulants, directly encode the energy shell correlations appearing in the eigenstate thermalization hypothesis [29,30].

The correlation functions defined in Equations (46)–(48) obey the FDT at a modified temperature β_{2q} [31]. In the frequency domain, the FDT reads

$$\mathbb{C}_{2q}(\omega) = \cosh(\beta_{2q}\hbar\omega/2)\mathbb{F}_{2q}(\omega)\,, \tag{50}$$

$$\hbar(\mathbb{R}_{2q})''(\omega) = \sinh(\beta_{2q}\hbar\omega/2)\mathbb{F}_{2q}(\omega)\,, \tag{51}$$

equivalent to the standard formulation $\hbar(\mathbb{R}_{2q})''(\omega) = \tanh(\beta_{2q}\hbar\omega/2)C_{2q}(\omega)$. We are interested in correlations in the time domain; hence, at the fluctuation–dissipation theorem formulated in the time domain, the *t*-FDT [9]. In particular, we will use the following relations:

$$\mathbb{C}_{2q}(t) = \cos\left(\frac{\beta_{2q}\hbar}{2}\frac{d}{dt}\right)\mathbb{F}_{2q}(t)\,, \tag{52}$$

$$\hbar(\mathbb{R}_{2q})''(t) = \sin\left(\frac{\beta_{2q}\hbar}{2}\frac{d}{dt}\right)\mathbb{F}_{2q}(t)\,. \tag{53}$$

4.2.2. The Bound

At times small but comparable with $t^{(2q)}_{\text{Ehr}}$, the previous arguments showed that the $2q$-OTOC are dominated by the regularized commutators [cf. Equations (38) and (39)]

$$C_q - \mathbb{C}_{2q}(t) \sim G^{(\beta)}_{2q}(t)\,, \qquad \bar{C}_q - \mathbb{F}_{2q}(t) \sim \bar{G}^{(\beta)}_{2q}(t)$$

when the behavior is exponential as $\sim \exp[L^{(\beta)}_{2q}(t - t^{(2q)}_{\text{Ehr}})]$. The t-FDT in Equation (52) implies

$$\frac{C_q - \mathbb{C}_{2q}(t)}{\bar{C}_q - \mathbb{F}_{2q}(t)} = \cos\left(\frac{\beta_{2q}\hbar L^{(\beta)}_{2q}}{2}\right) \tag{54}$$

The positivity of these coefficients—that follows from the fact that the $2q$-th commutators are positive definite—requires that the GLE must be such that $\cos\left(\frac{\beta_{2q}\hbar}{2} L^{(\beta)}_{2q}\right) \geq 0$. We thus conclude

$$\frac{L^{(\beta)}_{2q}}{2q} \leq \frac{\pi}{\beta\hbar}\,. \tag{55}$$

In the models where the Lyapunov depends on temperature [32–34], the cosine above in Equation (54) starts from zero at large temperature and is always in the first quadrant.

The bound on the $2q$-th OTOC rate was previously derived by Tsuji et al. in Ref. [8], by taking Equations (38) and (39) as a working assumption. In Section 4.1 above, we justified it using the ordering of the Ehrenfest times $t^{(2q)}_{\text{Ehr}}$.

4.3. Distribution Functions

The generalized Lyapunov exponents are the moments of a Lyapunov distribution function, as we have seen in the classical case and for the quantum GLE at an infinite temperature. In the case of finite β, the structure is more complex. This is due to the presence of $q-$dependent thermal matrices ρ in the definition of the regularized powers of commutators in Equations (29) and (30).

Nevertheless, one may define the Legendre transform of the thermal GLE as

$$S(\lambda, \beta) = \max_q (2\lambda q - L^{(\beta)}_{2q})\,. \tag{56}$$

In analogy with the previous cases, we may interpret it as the Cramèr function of an associated large deviation function $P(\lambda, \beta) \sim \exp(S(\lambda, \beta)t)$. As such, it shall obey similar properties as discussed above. In particular, the convexity of $S(\lambda, \beta)$ and $L^{(\beta)}_{2q}$ corresponds with the ordering of the Ehrenfest times in Equation (6) assumed at the beginning. The latter is equivalent to the conditions that

$$\frac{L^{(\beta)}_{2q}}{2q} \quad \text{increasing function of } q. \tag{57}$$

It is thus clear that the quantum bound (55) constrains the larger q that are related with the rare large deviations.

5. Discussion and Conclusions

In this work, we studied the quantum generalized Lyapunov exponents that quantify the large deviations of the spectrum of an appropriate operator. First, we discussed their convexity properties at infinite temperatures, which we exemplified on the kicked top. At finite temperatures, the quantum fluctuation–dissipation theorem (KMS) imposes a bound on their value, thus generalizing the celebrated bound to chaos to multipoint correlations. These bounds set a limit on the large deviations of chaotic properties.

A fascinating point is the interpretation of saturating the bound (7) at every q, which implies a form of mono-fractality:

$$L_{2q}^{(\beta)} = \frac{\pi\hbar}{\beta} 2q\,. \tag{58}$$

Classical examples of mono-fractal behavior, *i.e., models for which every trajectory has the same Lyapunov exponent*, are the backer map [35] and the free dynamics on the pseudosphere (the surface with constant negative curvature) [36]. What can we learn about the models that saturate the quantum bound (58)? A natural expectation is that the SYK model would lie in this class. In this case, it would be interesting to explore the meaning of such quantum mono-fractality in connection to the distinct properties of the model.

Author Contributions: Methodology, S.P. and J.K.; Investigation, S.P. and J.K.; Writing—review & editing, S.P. and J.K. All authors have read and agreed to the published version of the manuscript.

Funding: S. P. has received funding from the European Union's Horizon Europe program under the Marie Sklodowska Curie Action VERMOUTH (Grant No. 101059865).

Data Availability Statement: Not applicable.

Conflicts of Interest: The authors declare no conflict of interest.

References

1. Larkin, A.; Ovchinnikov, Y.N. Quasiclassical method in the theory of superconductivity. *Sov. Phys. JETP* **1969**, *28*, 1200–1205.
2. Maldacena, J.; Shenker, S.H.; Stanford, D. A bound on chaos. *J. High Energy Phys.* **2016**, *2016*. [CrossRef]
3. Sachdev, S.; Ye, J. Gapless spin-fluid ground state in a random quantum Heisenberg magnet. *Phys. Rev. Lett.* **1993**, *70*, 3339. [CrossRef]
4. Kitaev, A. Talk Given at the Fundamental Physics Prize Symposium KITP, 07/04/2015. 2015. Available online: https://youtu.be/wFH1huu9Jcs (accessed on 20 January 2023).
5. Chowdhury, D.; Georges, A.; Parcollet, O.; Sachdev, S. Sachdev-Ye-Kitaev models and beyond: Window into non-Fermi liquids. *Rev. Mod. Phys.* **2022**, *94*, 035004. [CrossRef]
6. Murthy, C.; Srednicki, M. Bounds on Chaos from the Eigenstate Thermalization Hypothesis. *Phys. Rev. Lett.* **2019**, *123*. [CrossRef]
7. Kundu, S. Subleading bounds on chaos. *J. High Energy Phys.* **2022**, *2022*, 1–28. [CrossRef]
8. Tsuji, N.; Shitara, T.; Ueda, M. Bound on the exponential growth rate of out-of-time-ordered correlators. *Phys. Rev. E* **2018**, *98*.
9. Pappalardi, S.; Foini, L.; Kurchan, J. Quantum bounds and fluctuation-dissipation relations. *Scipost Phys.* **2022**, *12*, 130. [CrossRef]
10. Crisanti, A.; Paladin, G.; Vulpiani, A. Generalized Lyapunov exponents in high-dimensional chaotic dynamics and products of large random matrices. *J. Stat. Phys.* **1988**, *53*, 583–601. [CrossRef]
11. Vulpiani, A.; Cecconi, F.; Cencini, M. *Chaos: From Simple Models to Complex Systems*; World Scientific: Singapore, 2009; Volume 17, Chapter 5.3. [CrossRef]
12. Oledets, V.I. A multiplicative ergodic theorem. Characteristic Ljapunov, exponents of dynamical systems. *Tr. Mosk. Mat. Obs.* **1968**, *19*, 179–210.
13. Benzi, R.; Paladin, G.; Parisi, G.; Vulpiani, A. On the multifractal nature of fully developed turbulence and chaotic systems. *J. Phys. A* **1984**, *17*, 3521. [CrossRef]
14. Rozenbaum, E.B.; Ganeshan, S.; Galitski, V. Universal level statistics of the out-of-time-ordered operator. *Phys. Rev. B* **2019**, *100*, 035112. [CrossRef]
15. Gharibyan, H.; Hanada, M.; Swingle, B.; Tezuka, M. Quantum Lyapunov spectrum. *J. High Energy Phys.* **2019**, *2019*, 82. [CrossRef]
16. Haake, F. *Quantum Signatures of Chaos*; Springer: Berlin, Germany, 2010. [CrossRef]
17. Haake, F.; Kuś, M.; Scharf, R. Classical and quantum chaos for a kicked top. *Z. für Physik B Condens. Matter.* **1987**, *65*, 381–395. [CrossRef]
18. Pappalardi, S.; Russomanno, A.; Žunkovič, B.; Iemini, F.; Silva, A.; Fazio, R. Scrambling and entanglement spreading in long-range spin chains. *Phys. Rev. B* **2018**, *98*, 134303. [CrossRef]
19. Seshadri, A.; Madhok, V.; Lakshminarayan, A. Tripartite mutual information, entanglement, and scrambling in permutation symmetric systems with an application to quantum chaos. *Phys. Rev. E* **2018**, *98*, 052205. [CrossRef]
20. Pilatowsky-Cameo, S.; Chávez-Carlos, J.; Bastarrachea-Magnani, M.A.; Stránský, P.; Lerma-Hernández, S.; Santos, L.F.; Hirsch, J.G. Positive quantum Lyapunov exponents in experimental systems with a regular classical limit. *Phys. Rev. E* **2020**, *101*, 010202. [CrossRef]
21. Sieberer, L.M.; Olsacher, T.; Elben, A.; Heyl, M.; Hauke, P.; Haake, F.; Zoller, P. Digital quantum simulation, Trotter errors, and quantum chaos of the kicked top. *npj Quantum Inf.* **2019**, *5*, 78. [CrossRef]

22. Lerose, A.; Pappalardi, S. Bridging entanglement dynamics and chaos in semiclassical systems. *Phys. Rev. A* **2020**, *102*, 032404. [CrossRef]
23. Benettin, G.; Galgani, L.; Giorgilli, A.; Strelcyn, J.M. Lyapunov Characteristic Exponents for smooth dynamical systems and for hamiltonian systems: A method for computing all of them. Part 1: Theory. *Meccanica* **1980**, *15*, 9–20. [CrossRef]
24. Benettin, G.; Galgani, L.; Giorgilli, A.; Strelcyn, J.M. Lyapunov Characteristic Exponents for smooth dynamical systems and for hamiltonian systems: A method for computing all of them. Part 2: Numerical application. *Meccanica* **1980**, *15*, 21–30. [CrossRef]
25. Roberts, D.A.; Yoshida, B. Chaos and complexity by design. *J. High Energy Phys.* **2017**, *2017*, 121. [CrossRef]
26. Haehl, F.M.; Rozali, M. Effective field theory for chaotic CFTs. *J. High Energy Phys.* **2018**, *2018*, 118. [CrossRef]
27. Haehl, F.M.; Rozali, M. Fine-Grained Chaos in AdS_2 Gravity. *Phys. Rev. Lett.* **2018**, *120*, 121601. [CrossRef] [PubMed]
28. Bhattacharyya, A.; Chemissany, W.; Haque, S.S.; Yan, B. Towards the web of quantum chaos diagnostics. *Eur. Phys. J. C* **2022**, *82*, 87. [CrossRef]
29. Foini, L.; Kurchan, J. Eigenstate thermalization hypothesis and out of time order correlators. *Phys. Rev. E* **2019**, *99*, 042139. [CrossRef] [PubMed]
30. Pappalardi, S.; Foini, L.; Kurchan, J. Eigenstate Thermalization Hypothesis and Free Probability. *Phys. Rev. Lett.* **2022**, *129*, 170603. [CrossRef]
31. Tsuji, N.; Shitara, T.; Ueda, M. Out-of-time-order fluctuation-dissipation theorem. *Phys. Rev. E* **2018**, *97*, 012101. [CrossRef]
32. Maldacena, J.; Stanford, D. Remarks on the Sachdev-Ye-Kitaev model. *Phys. Rev. D* **2016**, *94*, 106002. [CrossRef]
33. Kurchan, J. Quantum Bound to Chaos and the Semiclassical Limit. *J. Stat. Phys.* **2018**, *171*, 965–979. [CrossRef]
34. Pappalardi, S.; Kurchan, J. Low temperature quantum bounds on simple models. *SciPost Phys.* **2022**, *13*, 006. [CrossRef]
35. Aizawa, Y. Global Aspects of the Dissipative Dynamical Systems. I: Statistical Identification and Fractal Properties of the Lorenz Chaos. *Prog. Theor. Phys.* **1982**, *68*, 64–84. [CrossRef]
36. Balazs, N.; Voros, A. Chaos on the pseudosphere. *Phys. Rep.* **1986**, *143*, 109–240. [CrossRef]

entropy

MDPI

Article

Records and Occupation Time Statistics for Area-Preserving Maps †

Roberto Artuso [1,2,*], Tulio M. de Oliveira [3] and Cesar Manchein [3]

1 Dipartimento di Scienza e Alta Tecnologia and Center for Nonlinear and Complex Systems, Via Valleggio 11, 22100 Como, Italy
2 I.N.F.N, Sezione di Milano, Via Celoria 16, 20133 Milano, Italy
3 Departamento de Física, Universidade do Estado de Santa Catarina, Joinville 89219-710, SC, Brazil
* Correspondence: roberto.artuso@uninsubria.it
† To Giulio, celebrating his birthday and his achievements.

Abstract: A relevant problem in dynamics is to characterize how deterministic systems may exhibit features typically associated with stochastic processes. A widely studied example is the study of (normal or anomalous) transport properties for deterministic systems on non-compact phase space. We consider here two examples of area-preserving maps: the Chirikov–Taylor standard map and the Casati–Prosen triangle map, and we investigate transport properties, records statistics, and occupation time statistics. Our results confirm and expand known results for the standard map: when a chaotic sea is present, transport is diffusive, and records statistics and the fraction of occupation time in the positive half-axis reproduce the laws for simple symmetric random walks. In the case of the triangle map, we retrieve the previously observed anomalous transport, and we show that records statistics exhibit similar anomalies. When we investigate occupation time statistics and persistence probabilities, our numerical experiments are compatible with a generalized arcsine law and transient behavior of the dynamics.

Keywords: area-preserving maps; record statistics; infinite ergodicity

Citation: Artuso, R.; de Oliveira, T.M.; Manchein, C. Records and Occupation Time Statistics for Area-Preserving Maps. *Entropy* **2023**, *25*, 269. https://doi.org/10.3390/e25020269

Academic Editor: Marko Robnik

Received: 18 December 2022
Revised: 26 January 2023
Accepted: 30 January 2023
Published: 1 February 2023

1. Introduction

One of the most remarkable advances in modern dynamics lies in the recognition that deterministic systems may exhibit statistical properties typical of purely stochastic processes: for instance, such systems may display diffusion properties similar to random walks [1–4]. Area-preserving maps (see, for instance, [1]) represent a prominent example of Hamiltonian systems where subtle features of dynamics, such as integrability vs. chaotic properties, may be studied. In this context, one of the most outstanding examples is represented by the Chirikov–Taylor standard map (SM) (see [1,5] and references therein). Even if we are only concerned with classical features, we recall that such a map played an important role in the development of quantum chaos: in particular, early investigations showed how quantum interference suppresses classical diffusion, a phenomenon called quantum dynamical localization [6]. Though the SM has been extensively explored by numerical simulations, very few rigorous results have been proven (see, for instance, the introduction in [7]): however, it is generally believed that for large nonlinearity parameters, this map typically exhibits good stochastic properties and sensitive dependence upon initial conditions. Here a remark is due: such a map can be studied either on a 2-torus or an (unbounded) cylinder: the latter representation is naturally adopted when transport properties are concerned, and analogies with random walks are taken into account [1,3,8,9]. While particular nonlinear parameters in the standard map can be tuned to generate strong anomalous diffusion [10] (see also [11–15]), here we will only deal with the case in which diffusion is normal. Our findings will be compared with those obtained for another area-preserving map, characterized by the lack of exponential instability: the so-called

Casati–Prosen triangle map (TM) [16], introduced by considering, in an appropriate limit, the Birkhoff dynamics of a triangular billiard: apart from its intrinsic interest, such a map is an ideal benchmark to test whether stochasticity properties, exhibited by strongly chaotic systems, are also showcased by systems lacking any exponential instability. It also turns out that many features of the TM are still debated, starting from basic properties, such as ergodicity and mixing (see, for instance, [17,18]). It is also worth mentioning that the TM has been studied in the quantum setting to investigate which features of quantum chaos are displayed in the absence of classical exponential instability [19,20].

In detail, we will compare different indicators for both maps on the cylinder: though, in principle, further complications are added when one considers a non-compact phase space [21,22], this is the appropriate scenario to discuss transport properties and record statistics and to check whether tools from infinite ergodic theory may enrich our understanding of such systems.

Our main findings are that the SM, in its typical chaotic regime, displays all stochastic properties of a purely stochastic system, while, as expected, results are far more complicated for the TM; we believe that some new insight is provided by our analysis, in particular, as regards persistence behavior, occupation time statistics, and the relationship between transport properties and record statistics. When we compare the records statistics for the SM and the TM, the latter exhibits an anomalous growth, which can be related to transport properties. The difference is even more striking when we consider occupation time statistics and survival probability: the SM again behaves like a simple random walk, while the TM numerical experiments suggest transient behavior, coexisting with a generalized arcsine law.

The paper is organized as follows. In Section 2, the Chirikov–Taylor standard map (1) and the triangle map (3), our basic models, are presented, and we also mention the main properties we analyze. Section 3 is dedicated to discussing transport properties, records statistics, and occupation time statistics. We end with a discussion in Section 4.

2. The Basic Setting

We recall the definition of the SM

$$\begin{aligned} p_{n+1} &= p_n + \frac{K}{2\pi}\sin(2\pi x_n), \\ x_{n+1} &= x_n + p_{n+1} \qquad \bmod 1; \end{aligned} \tag{1}$$

where x and p are canonical variables: (1) is the stroboscopic map generated by the time-dependent Hamiltonian $H(p,x,t) = p^2/2 + K/(2\pi)^2\cos(2\pi x)\sum_{m=-\infty}^{+\infty}\delta(t-m)$, describing a periodically kicked rotator, K being the nonlinear parameter: when K is sufficiently big, no KAM invariant circles bound the motion, and one can study moments of the diffusing variable $p \in \mathbb{R}$:

$$\langle |p_n - p_0|^q \rangle \sim n^{q\nu(q)}. \tag{2}$$

The typical behavior observed for the second moment in simulations is normal diffusion $\nu(2) = 1/2$ [23,24], while, for certain parameter values, the existence of stable running orbits (accelerator modes) induces superdiffusion, $\nu(2) > 1/2)$ [11–15,25–27]. We point out that a finer description of anomalous transport is obtained by considering the full spectrum $\nu(q)$: if $\nu(q) = \alpha \cdot q$, for some $\alpha \neq 1/2$ one speaks about weak anomalous diffusion, whereas the case of a nontrivial $\nu(q)$ is dubbed strong anomalous diffusion [10]. As far as the SM is concerned, we will consider the case where transport in the stochastic sea is normal, even if the phase space exhibits a mixture of chaotic and elliptic components (see Figure 1). We will come back to this point in the final discussion.

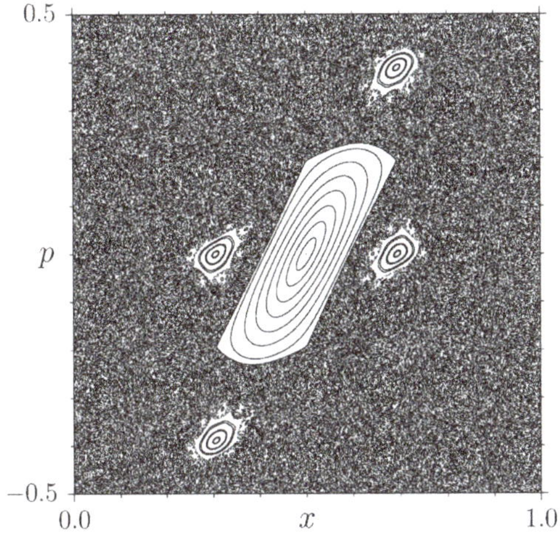

Figure 1. Phase-space portrait for the standard map (1) on the 2-torus, for $K = 2.6$. Here 40 uniformly distributed initial conditions were used for x, while maintaining $p_0 = 0$ fixed: each initial condition is iterated 10^4 times.

On the other side, the TM is defined (on the cylinder) as:

$$\begin{aligned} p_{n+1} &= p_n + 2(x_n - \lfloor x_n \rceil - \mu(-1)^{\lfloor x_n \rceil}), \\ x_{n+1} &= x_n - 2p_{n+1} \qquad \mod 2, \end{aligned} \tag{3}$$

where $\lfloor \cdots \rceil$ denotes the nearest integer. It was introduced in [16] (see also [28]) as an approximate Birkhoff map of irrational triangular billiards: systems lacking exponential instability, whose ergodic properties are subtly related to irrationality properties of the angles [29–32]: in particular, the original derivation refers to an elongated triangular billiard, with a very small angle: x and p are Birkhoff coordinates describing successive collisions with the small side, and the parameter μ is associated with the asymmetry between the two non-small angles. Figure 2 shows the function $\delta p(x) = p_{n+1} - p_n = 2(x_n - \lfloor x_n \rceil - \mu(-1)^{\lfloor x_n \rceil})$. Since the derivative of $\delta p(x)$ (apart from discontinuity points, where it does not exist) is constant $\delta'(x) = 2$, the jacobian of the map is

$$\mathbb{J}(x, p) = \mathbb{J} = \begin{pmatrix} -3 & -2 \\ 2 & 1 \end{pmatrix}. \tag{4}$$

The determinant is 1 (the map is area-preserving, while the trace is -2, corresponding to double degenerate eigenvalues $\lambda = -1$. This implies that the TM lacks any form of exponential instability.

We remark that the polygonal billiards represent both a hard mathematical challenge [33–36] and a natural benchmark when trying to assess which microscopic dynamical features lead to macroscopic transport laws [37–39] (see also [40,41]). In this respect, it is worth mentioning that anomalous transport has been associated with scaling exponents of the spectral measure [42] and that generalized triangle maps have been investigated recently, both as connected to dynamical localization [43] and with respect to slow diffusion [44]. A typical phase portrait (on the torus) of the TM is shown in Figure 3.

Before mentioning the numerical experiments we performed, a crucial observation is in order. When looking at transport properties (and records statistics), considering maps on the cylinder is quite natural, while from the ergodic point of view, this perspective is somehow delicate since no normalizable invariant density exists [21,22], and the appropriate setting is an infinite ergodic theory (we will not employ the full machinery of infinite ergodic theory; however, the lack of an invariant probability measure is essential when properties such as occupation time statistics are examined). When polygonal channels are considered, even establishing recurrent properties of the dynamics is a demanding task [45].

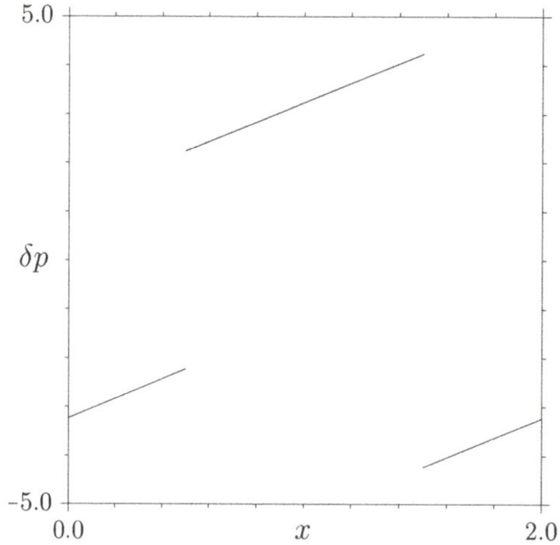

Figure 2. $\delta p = p_{n+1} - p_n$ as a function of x for the triangle map (3), for $\mu = \frac{1+\sqrt{5}}{2}$ (golden mean).

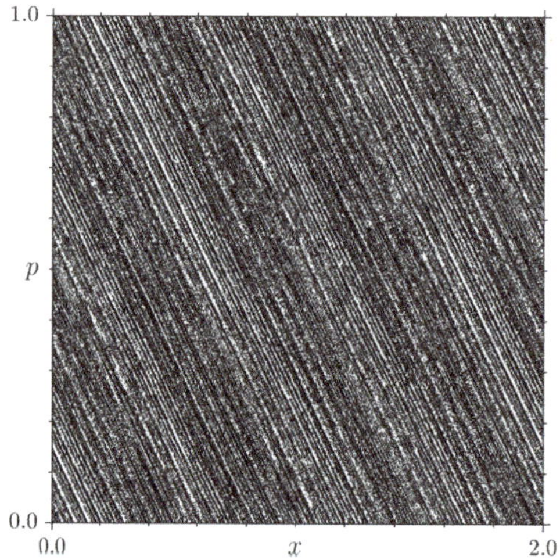

Figure 3. Phase-space dynamics for the triangle map (3), for $\mu = \frac{1+\sqrt{5}}{2}$ (golden mean). Here 100 randomly distributed initial conditions were used for x and p: each initial condition is iterated 5×10^4 times. Notice the typical filament structure in the phase space [30,31].

The first set of properties we investigated is more conventional, and a few results, as we will mention in the next section, have already been considered, especially as far as the SM is concerned. We will look at transport properties, in particular through the first and the second moment of the diffusing variable. We will also study records statistics, which recently have become very popular (see [46,47] and references therein). Then we will scrutinize statistical properties, such as persistence probability and (generalized) arcsine law [48,49]. While motion in the stochastic sea for the SM will exhibit typical properties of a simple stochastic process like a random walk, our findings for the TM suggest both the existence of transient dynamics and anomalous scaling exponents for records statistics and generalized arcsine law.

3. Results

We start by considering properties associated with the spreading of trajectories over the phase space; then we will consider occupation time statistics.

3.1. Diffusion

This is a warm-up exercise since transport properties have been studied both for the SM [1,23,24] and for the TM [32]. We observe *normal* transport for the case of the SM (see panels (a) and (b) in Figure 4, where $\langle (p_n - p_0) \rangle \sim n^{1/2}$ and $\langle (p_n - p_0)^2 \rangle \sim n$, respectively), while the TM results indicate a superdiffusion (see panels (a) and (b) in Figure 5), with

$$\langle (p_n - p_0)^2 \rangle \sim n^{1.86}, \tag{5}$$

in agreement with [32]. We remark that by looking at the power-law exponents of the first two moments, we find that anomalous diffusion is possibly weak [10], namely if we consider the full spectrum of moments' asymptotics:

$$\langle |p_n - p_0|^q \rangle \sim n^{\phi(q)}, \tag{6}$$

we have a single scaling, in the sense that

$$\phi(q) = \alpha \cdot q; \tag{7}$$

where normal diffusion is recovered when $\alpha = 1/2$. This is reasonable since weak anomalous diffusion has been observed in polygonal billiards [50].

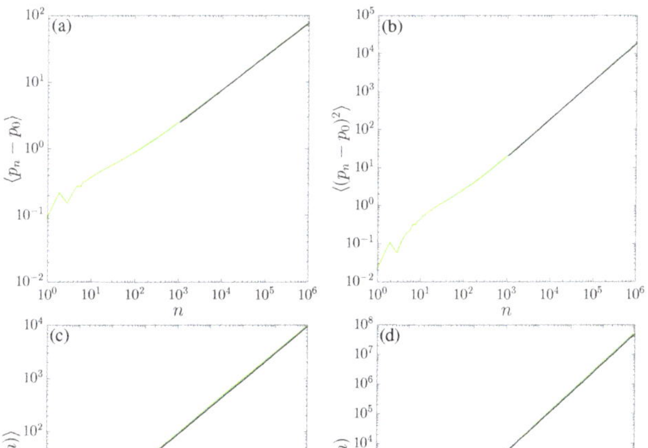

Figure 4. (**a**) Average number of records, (**b**) variance, (**c**) first, and (**b**) second moments of variable p for $K = 2.6$ in the standard map (1), as a function of time. These quantities were computed for 10^6 initial conditions for x_0, arbitrarily chosen in the chaotic sea along the line $p_0 = 0$. Black-continuous lines correspond to power-law asymptotics fit $F(n) = an^{\gamma}$: the fitting parameters are, for (**a**) $a = 0.77(7)$, $\gamma = 0.50(1)$, for (**b**) $a = 0.02(1)$, $\gamma = 0.99(1)$, for (**c**) $a = 0.86(0)$, $\gamma = 0.50(9)$, and for (**d**) $a = 0.50(7)$, $\gamma = 1.00(3)$.

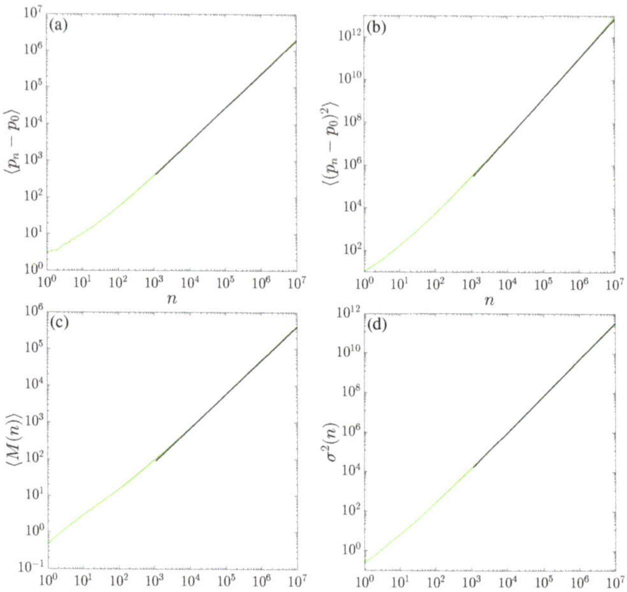

Figure 5. (**a**) Average number of records, (**b**) variance, (**c**) first, and (**b**) second moments of variable p for the golden mean $\mu = \frac{1+\sqrt{5}}{2}$ in the triangular map (3) as a function of time. These quantities were computed for 10^6 initial conditions for x_0, arbitrarily chosen in phase space along the line $p_0 = 0$. Black-continuous lines correspond to power-law asymptotics $F(n) = an^{\gamma}$: the fitting parameters are, for (**a**) $a = 0.65(4)$, $\gamma = 0.92(4)$, for (**b**) $a = 0.67(6)$, $\gamma = 1.86(0)$, for (**c**) $a = 0.13(9)$, $\gamma = 0.92(4)$, and for (**d**) $a = 0.04(0)$, $\gamma = 1.84(9)$.

3.2. Average Number of Records

The statistics of records are very popular in the analysis of correlated and uncorrelated stochastic time sequences [46,47]. Since this subject has not been explored thoroughly in the deterministic setting (with the remarkable exception of [51,52]), we briefly review the basic concepts.

First of all, let us recall the (straightforward) definition of a record: given a sequence of real data $x_0, x_1, \ldots, x_k, \ldots$ the element x_m is a record if

$$x_m > x_j \qquad j = 0, 1, \ldots m-1, \tag{8}$$

(we consider x_0 to be the first record). To the sequence of data points we associate the binary string $\sigma_0, \sigma_1, \ldots, \sigma_k, \ldots$, where

$$\sigma_l = \begin{cases} 1 & \text{if } x_l \text{ is a record} \\ 0 & \text{otherwise} \end{cases} \tag{9}$$

The number of records up to time N is then

$$M_N = \sum_{j=0}^{N} \sigma_j. \tag{10}$$

The properties of the average number of records, $\langle M_N \rangle$, and the corresponding variance

$$Var(M_N) = \langle M_N^2 \rangle - \langle M_N \rangle^2 \tag{11}$$

are important tools to access the nature of the data sequence: as a matter of fact, if the different x_j are independent identically distributed random variables, then, for large N, we have [53,54]:

$$\langle M_N \rangle = \ln N + \gamma_E + \mathcal{O}(N^{-1}), \tag{12}$$

where $\gamma_E = 0.5772\ldots$ is the Euler–Mascheroni constant, and

$$Var(M_N) = \sigma^2(N) = \ln N + \gamma_E - \frac{\pi^2}{6} + \mathcal{O}(N^{-1}). \tag{13}$$

We remark that both quantities are independent of the common distribution of the random variables: this universality is an important feature of record statistics in different contexts.

The results are quite different for a correlated sequence, as when x_j denotes the position of a random walker at time j:

$$x_{j+1} = x_j + \xi_{j+1}, \tag{14}$$

where the jumps are taken from a common distribution $\wp(\xi)$. In this case, the behavior is [46,47]:

$$\langle M_N \rangle \approx \frac{2}{\sqrt{\pi}}\sqrt{N}, \tag{15}$$

and

$$Var(M_N) \approx 2\left(1 - \frac{2}{\pi}\right)N, \tag{16}$$

so that the standard deviation is of the same order of magnitude as the average. Again this is a *universal* result, independent of the particular jump distribution $\wp(\xi)$, as long as the distribution is continuous and symmetric. The crucial ingredient of the proof is that the process renews as soon as a new record is achieved and the appearance of the new record is related to the survival probability for the process, which is universal in view of the Sparre–Andersen theorem [49,55,56] (see also [57]).

Numerical results on records statistics are reported in Figure 4 (for the SM) and Figure 5 (for the TM), panels (c) and (d) for the average number of records $\langle M(n) \rangle$ and variance $\sigma^2(n)$, respectively. For the SM, our results plotted in Figure 4 are consistent with early investigations [51,52] and with the asymptotic behavior of a random walk: our fits are in excellent agreement with $\langle M(n) \rangle \sim n^{1/2}$ and $\sigma^2(n) \sim n$, respectively. For the TM, we observe anomalous scaling with regard to (15) and (16), as plotted in Figure 5: the behavior is related to the transport properties in the sense that data are consistent with the growths.

$$\langle M_N \rangle \sim N^{\phi(1)}, \qquad Var(M_N) \sim N^{\phi(2)}. \tag{17}$$

Similar behavior was observed in [51,52] for the SM in the presence of accelerator modes. We remark that, though in the following we will fix our attention to a particular parameter value for the TM, we checked that reported experiments do not depend on the particular parameter choice, as exemplified in Figure 6, where the growth of the averaged number of records is reported for three different parameters of the TM: $\mu = \sqrt{7}, \sqrt{2}/2$, and $\mu = (\sqrt{5} + e)/12$ plotted with dashed-dotted (green), dashed-dotted-dotted (magenta), and dashed (red) lines. While the actual choice of μ parameter values is somehow arbitrary, the only provision is to choose irrational values. Otherwise, the TM becomes pseudointegrable [32–34]. Figure 6 shows that the asymptotic power-law fitting parameters are almost the same ($a \sim 0.15$ and $\gamma \sim 0.9$) for different irrational values of μ.

We also remark that the number of initial conditions and iteration time plays a role as regards the simulations we will present later on: while our choices are gauged by the stability of the plots upon variations in this parameter, survival probability simulations require extremely large iteration times, as remarked in the caption of Figure 7.

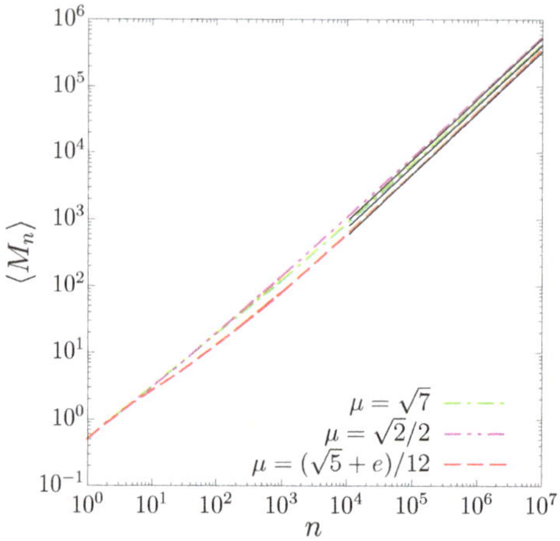

Figure 6. Average number of records for three additional parameters μ in the TM. These quantities were computed for 5×10^5 initial conditions for x_0, arbitrarily chosen in phase space along the line $p_0 = 0$. Black-continuous lines correspond to the power-law asymptotic fitting function $F(n) = an^{\gamma}$, with $(a, \gamma) = [0.20(1), 0.91(8)]$ (magenta), $(a, \gamma) = [0.16(2), 0.91(7)]$ (green), and $(a, \gamma) = [0.11(5), 0.92(4)]$ (red).

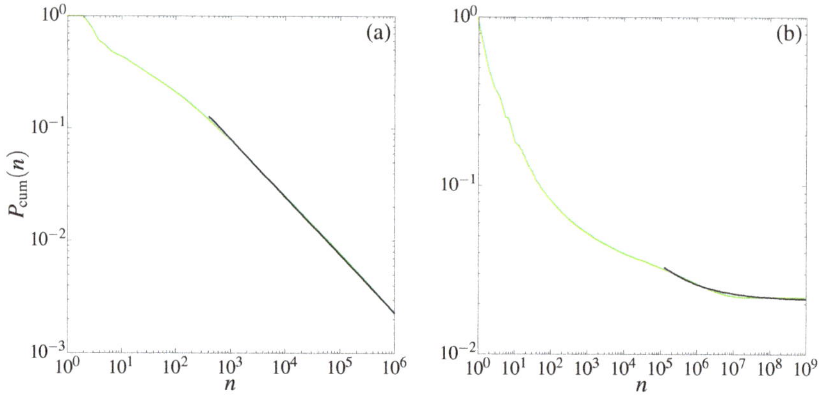

Figure 7. Cumulative distribution function for the survival times obtained for variable p for (**a**) the standard map (1) and (**b**) the triangle map (3) on a logarithmic scale. Data are obtained by simulating 10^6 and 10^5 initial conditions, respectively. Continuous black lines correspond to power-law asymptotic functions $F(n) = a + bn^{-\alpha}$: the fitting parameters are $a = 0$, $b = 2.80(0)$, and $\alpha = 0.51(5)$ in (**a**) and $a = 0.021(0)$, $b = 1.62(6)$, and $\alpha = 0.42(0)$ in (**b**). Notice that to observe the asymptotic plateau for the TM, we have to go to very large iteration times.

While a general, quantitative relationship (if any) between transport exponents and statistical properties of records has not been fully developed, to the best of our knowledge, it is possible in some cases to connect $\phi(1)$ to the expected maximum of the walk [58,59], that, for a random walk with unit jumps, coincides with the number of records. On the other side, we mention that non-homogeneous random walks offer examples where such a relationship does not hold [60–64].

3.3. Occupation Time Statistics

When we consider the evolution of the cylinder, both for the SM and the TM, we are in the presence of infinitely ergodic systems [21,22]. Since, while the Lebesgue measure is preserved, due to area conservation, the (constant) phase space is unbounded, so the invariant density cannot be normalized. This has a series of remarkable consequences, which originally have been considered in the context of stochastic processes, and then explored in the deterministic evolution framework.

One of the most striking properties that has been investigated is the generalized arcsine law (see [48] for the standard formulation for stochastic processes): we briefly recall the main result that lies at the basis of our analysis, namely Lamperti's theorem [65]. The original formulation involves discrete stochastic processes, for which the infinite set of possible states can be separated into two sets, A and B, separated by a single site x_0, such that a transition from one set to the other can only be achieved by passing through x_0, which can be taken as the starting site, and is supposed to be recurrent (namely the probability of returning to it is 1). For instance, we can think of a one-dimensional random walk on an integer lattice, with $x_0 = 0$ and A (B) consisting of strictly positive (negative) lattice sites. We are interested in the limiting distribution of $N(n)/n$, the fraction of time spent in the positive semi-axis up to time n. The theorem states that such a distribution exists in the $n \to \infty$ limit, and it is characterized by two parameters α and η. η is related to the symmetry properties of the process, being the expectation value of the fraction of time spent in $\mathbb{R}_+$:

$$\eta = \lim_{n\to\infty} \mathbb{E}\left(\frac{N(n)}{n}\right) : \tag{18}$$

for a symmetric process $\eta = 1/2$, and from now on, we will only consider such a case.

The other parameter, α, is instead connected to the behavior of the generating function of first return probabilities to the starting site: it can be shown [66] that it can be related to the probability P_n of being at the starting site after n steps in the following way.

$$P_n \sim \frac{H(n)}{n^{1-\alpha}}, \tag{19}$$

where $H(n)$ is a slowly varying function, namely

$$\lim_{n\to\infty} \frac{H(yn)}{n} = 1. \tag{20}$$

Under such conditions, the density of $\varphi = N(n)/n$ in the infinite time limit is given by Lamperti distribution:

$$\mathcal{G}_\alpha(\varphi) = \frac{\sin(\pi\alpha)}{\pi} \frac{\varphi^{1-\alpha}(1-\varphi)^{1-\alpha}}{\varphi^{2\alpha} + 2\varphi^\alpha(1-\varphi)^\alpha\cos(\pi\alpha) + (1-\varphi)^{2\alpha}}, \tag{21}$$

which reproduces the usual arcsine law

$$\mathbb{P}((N_n/n) \le \xi) = \frac{2}{\pi}\arcsin\left(\sqrt{\xi}\right) \tag{22}$$

when $\alpha = 1/2$, in the universality class of the Sparre–Andersen theorem. Deviations from standard arcsine law have been reported in several cases in the framework of deterministic dynamics [67–74], mainly in the context of intermittent maps. Numerical experiments for the SM confirm the validity of the arcsine law, $\alpha = 1/2$, see panel (a) in Figure 8. To the best of our knowledge, this is the first time such an indicator has been considered in the analysis of area-preserving maps.

The results, as expected, are quite different for the TM, and they suggest novel features exhibited by this map. In particular (see panel (b) in Figure 8), numerical results are well fitted by a Lamperti distribution (with $\alpha \approx 0.42$) and thus are different from an ordinary random walk), except for those that present enhanced peaks. Intuitively, such an additional contribution might be due to a fraction of orbits never returning to the origin: this would correspond, in stochastic language, to a transient random walk (we recall that according to Pólya's theorem [75] a simple symmetric random walk is recurrent (so the return to the starting site is sure) in one and two dimensions, and transient in higher dimensions). Such a possibility is indeed not excluded for infinite polygonal channels [45].

Our last set of simulations concerns the survival probability [68]:

$$P_{cum}(n) = \text{prob}(p_n \geq 0 \ldots p_1 \geq 0 | p_0 = 0). \tag{23}$$

When considering recurrent random walks, the asymptotic behavior of the survival probability is again ruled by the Lamperti exponent [65,66] (see also [76]):

$$P_{cum}(n) \sim n^{-\alpha}. \tag{24}$$

Once again, SM simulations (see panel (a) in Figure 7) agree with the expected behavior for simple random walks ($\alpha = 1/2$), while the situation is completely different for the TM, where the survival probability seems to tend to a finite limit for large n, see panel (b) in Figure 7. This is consistent with the transient nature of the TM, which we conjectured in the analysis of generalized arcsine law.

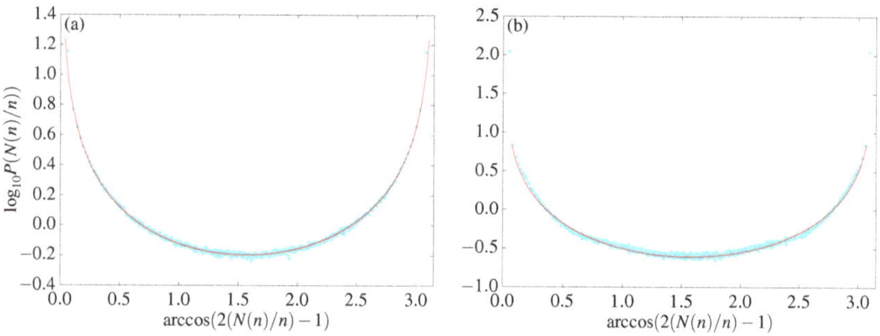

Figure 8. Distribution of the fraction of time spent in the positive axis for the momentum p in the standard (1) (**a**) and triangle (3) (**b**) maps on a semi-logarithmic scale. To enhance the readability of the border values, the transformation $x \to \arccos(2x-1)$ on the horizontal axis. The (light blue) points represent the simulation results, and the (red) line is the Lamperti distribution (21). Data are obtained by computing 10^6 initial conditions iterated 10^6 times for the standard map and 10^6 initial conditions iterated 10^8 times for the triangle map. The fitting parameters are $\alpha = 0.49(9)$ for (**a**) and $\alpha = 0.42(0)$ for (**b**). In the case of the TM, data suggest a superposition of a (rescaled) Lamperti distribution and two Dirac's δ centered on $x = 0$ and $x = 1$ (see text).

4. Discussion

We have performed a set of extensive numerical experiments on two paradigmatic area-preserving maps, the SM and the TM, focusing on the case where such maps are considered on a cylinder, namely a non-compact phase space. First, we reproduced known results about normal diffusion for typical (chaotic) parameters of the SM and superdiffusion for the TM. In particular, for the TM, we recover the anomalous spreading exponent reported in [32]. Then we explored records statistics: numerical simulations again confirm that the SM behaves like a simple random walk in accordance with simulations in [51,52], while anomalous growth is exhibited by the TM. We remark that, in both cases, the exponent that accounts for the growth of the number of records is quite close to the one that determines the asymptotic behavior of the first moment of the transporting variable, see (17). This is also consistent with the results reported in [51,52] for the SM in the presence of accelerator modes. While we only considered the case for which the SM displays normal diffusion (up to the time scale we were able to investigate), it might be possible that accelerator modes (of the possibly very high period) appear for almost every nonlinear parameter K (see the discussion in [12]). This is an issue of conceptual relevance since accelerator modes might eventually determine the true asymptotic behavior, but, on the other side, the time scale on which such behavior is exhibited might be beyond any possible numerical simulation. The most interesting results arise in the analysis of occupation times, such as generalized arcsine law and survival probability: to the best of our knowledge, such properties are investigated here for the first time in the context of area-preserving maps. For the SM, we recover the usual arcsine law and a survival probability in the Sparre–Andersen universality class, while the TM displays different behaviors. In particular, the distribution of occupation times in the positive half-axis (for the transporting variable) is well-fitted by a superposition of a generalized Lamperti distribution (with a different exponent with regard to the SM) and a sum of two δ peaks at the extreme values, which we ascribe to transient orbits. We stress that the analysis of survival probabilities supports such a conjecture since they seem to attain a non-zero limit for very long times. The development of further tools to sustain such a picture is a line of research that we hope will be pursued in further studies, which might also lead to new stochastic modeling of the TM [44]. From a different perspective, it would also be important to consider classes of area-preserving maps for which an analytic approach is feasible. A remarkable example in this respect is provided by the Cerbelli–Giona map [77,78], where jumps in momentum are given by a tent map. Such a map probes the effect of singularities (that are also present in the TM, though without any exponential instability). In particular, in [78], an exact computation of the diffusion coefficient is presented, which bears some similarities to analogous computations for one-dimensional maps [79,80].

Author Contributions: All authors have contributed substantially to the work. All authors have read and agreed to the published version of the manuscript.

Funding: R.A. acknowledges partial support from PRIN Research Project No. 2017S35EHN "Regular and stochastic behavior in dynamical systems" of the Italian Ministry of Education, University and Research (MIUR). C.M. acknowledges the National Council for Scientific and Technological Development—CNPq (Brazilian agency) for partial financial support (Grant Number 310228/2020-4). T.M.d.O. acknowledges the Coordenação de Aperfeiçoamento de Pessoal de Nível Superior—CAPES (Brazilian agency)—Finance Code 001, for partial financial support. Additionally, T.M.d.O. and C.M. also acknowledge the Fundação de Amparo à Pesquisa e Inovação do Estado de Santa Catarina—FAPESC (Brazilian agency) for partial financial support.

Institutional Review Board Statement: Not applicable.

Data Availability Statement: Not applicable.

Acknowledgments: R.A. acknowledges an association to the GNFM group of INDAM. R.A. thanks Gaia Pozzoli for discussions. We thank the referees for suggestions and comments.

Conflicts of Interest: The authors declare no conflict of interest. The funders had no role in the design of the study; in the collection, analyses, or interpretation of data; in the writing of the manuscript, or in the decision to publish the results.

Abbreviations

The following abbreviations are used in this manuscript:

SM	Standard map
TM	Triangle map

References

1. Lichtenberg, A.J.; Lieberman, M.A. *Regular and Chaotic Dynamics*; Springer: Berlin, Germany, 1992.
2. Ott, E. *Chaos in Dynamical Systems*; CUP: Cambridge, UK, 2002.
3. Cvitanović, P.; Artuso, R.; Mainieri, R.; Tanner, G.; Vattay, G. *Chaos: Classical and Quantum*; ChaosBook.org; Niels Bohr Institute: Copenhagen, Denmark, 2020.
4. Artuso, R.; Burioni, R. Anomalous diffusion: Deterministic and stochastic perspectives. In *Large Deviations in Physics*; Vulpiani, A., Cecconi, F., Cencini, M., Puglisi, A., Vergni, D., Eds.; Springer: Berlin/Heidelberg, Germany, 2014; pp. 263–293.
5. Chirikov, B.V.; Shepelyansky, D.L. Chirikov standard map. *Scholarpedia* **2008**, *3*, 3550. [CrossRef]

6. Casati, G.; Chirikov, B.V. (Eds.) *Quantum Chaos;* OUP: Oxford, UK, 1995.
7. Bloor, K.; Luzzatto, S. Some remarks on the geometry of the standard map. *Int. J. Bifurcat. Chaos* **2009**, *19*, 2213–2232. [CrossRef]
8. Chirikov, B.V. A universal instability of many dimensional oscillator systems. *Phys. Rep.* **1979**, *52*, 263–379. [CrossRef]
9. MacKay, R.S.; Meiss, J.D.; Percival, I.C. Stochasticity and transport in hamiltonian systems. *Phys. Rev. Lett.* **1984**, *52*, 697–700. [CrossRef]
10. Castiglione, P.; Mazzino, A.; Muratore-Ginanneschi, P.; Vulpiani, A. On strong anomalous diffusion. *Phys. D* **1999**, *134*, 75–93. [CrossRef]
11. Venegeroles, R. Calculation of superdiffusion for the Chirikov-Taylor map. *Phys. Rev. Lett.* **2008**, *101*, 054102. [CrossRef]
12. Manos, R.; Robnik, M. Survey on the role of accelerator modes for anomalous diffusion: The case of the standard map. *Phys. Rev. E* **2014**, *89*, 022905. [CrossRef]
13. Harsoula, M.; Contopoulos, G. Global and local diffusion in the standard map. *Phys. Rev. E* **2018**, *97*, 022215. [CrossRef]
14. Harsoula, M.; Karamanos, K.; Contopoulos, G. Characteristic times in the standard map. *Phys. Rev. E* **2019**, *99*, 032203. [CrossRef]
15. Moges, H.T.; Manos, T.; Skokos, C. Anomalous diffusion in single and coupled standard maps with extensive chaotic phases *Phys. D* **2022**, *431*, 133120. [CrossRef]
16. Casati, G.; Prosen, T. Triangle map: A model of quantum chaos. *Phys. Rev. Lett.* **2000**, *85*, 4261–4264. [CrossRef]
17. Horvat, M.; Degli Esposti, M.; Isola, S.; Prosen, T.; Bunimovich, L. On ergodic and mixing properties of the triangle map. *Phys. D* **2009**, *238*, 395–415. [CrossRef]
18. Degli Esposti, M.; Galatolo, S. Recurrence near given sets and the complexity of the Casati-Prosen map. *Chaos Solitons Fractals* **2005**, *23*, 1275–1284. [CrossRef]
19. Degli Esposti, M.; O' Keefe, S.; Winn, B. A semi-classical study of the Casati-Prosen map. *Nonlinearity* **2005**, *14*, 1073–1094. [CrossRef]
20. Wang, J.; Benenti, G.; Casati, G.; Wang, W.-G. Statistical and dynamical properties of the quantum triangle map. *J. Phys. A* **2022**, *55*, 234002. [CrossRef]
21. Aaronson, J. *An Introduction to Infinite Ergodic Theory;* AMS: Providence, RI, USA, 1997.
22. Zweimüller, R. Surrey Notes on Infinite Ergodic Theory. Available online: https://mat.univie.ac.at/%7Ezweimueller/PapersAndPreprints.html (accessed on 1 November 2022).
23. Rechester, A.B.; White, R.B. Calculation of turbulent diffusion for the Chirikov-Taylor model. *Phys. Rev. Lett.* **1980**, *44*, 1586–1589. [CrossRef]
24. Dana, I.; Murray, N.W.; Percival, I.C. Resonances and diffusion in periodic Hamiltonian maps. *Phys. Rev. Lett.* **1989**, *62*, 233–236. [CrossRef]
25. Ishizaki, R.; Horita, T.; Kobayashi, T.; Mori, H. Anomalous diffusion due to accelerator modes in the standard map. *Progr. Theor. Phys.* **1991**, *85*, 1013–1022. [CrossRef]
26. Benkadda, S.; Kassibrakis, S.; White, R.B.; Zaslavsky, G.M. Self-similarity and transport in the standard map. *Phys. Rev. E* **1997**, *55*, 4909–4917. [CrossRef]
27. Zaslavsky, G.M.; Edelman, M.; Niyazov, B.A. Self-similarity, renormalization, and phase-space nonuniformity of Hamiltonian chaotic dynamics. *Chaos* **1997**, *7*, 159–181. [CrossRef]
28. Kaplan, L.; Heller, E.J. Weak quantum ergodicity. *Phys. D* **1998**, *121*, 1–18. [CrossRef]
29. Casati, G.; Prosen, T. Mixing properties of triangular billiards. *Phys. Rev. Lett.* **1999**, *83*, 4728–4732. [CrossRef]
30. Artuso, R.; Casati, G.; Guarneri, I. Numerical study on ergodic properties of triangular billiards. *Phys. Rev. E* **1997**, *55*, 6384–6390. [CrossRef]
31. Artuso, R. Correlations and spectra of triangular billiards. *Phys. D* **1997**, *109*, 1–10. [CrossRef]
32. Prosen, T.; Žnidarič, M. Anomalous diffusion and dynamical localization in polygonal billiards. *Phys. Rev. Lett.* **2001**, *87*, 114101. [CrossRef]
33. Gutkin, E. Billiards in polygons. *Phys. D* **1986**, *19*, 311–333. [CrossRef]
34. Gutkin, E. Billiards in polygons: Survey of recent results. *J. Stat. Phys.* **1996**, *83*, 7–26. [CrossRef]
35. Gutkin, E. Billiard dynamics: A survey with the emphasis on open problems. *Regul. Chaotic Dyn.* **2003**, *8*, 1–13. [CrossRef]
36. Gutkin, E. Billiard dynamics: An updated survey with the emphasis on open problems. *Chaos* **2012**, *22*, 026116. [CrossRef]
37. Alonso, D.; Ruiz, A.; De Vega, I. Transport in polygonal billiards. *Phys. D* **2004**, *187*, 184–199. [CrossRef]
38. Jepps, O.G.; Bianca, C.; Rondoni, L. Onset of diffusive behavior in confined transport systems. *Chaos* **2008**, *18*, 013127. [CrossRef]
39. Sanders, D.P.; Larralde, H. Occurrence of normal and anomalous diffusion in polygonal billiard channels. *Phys. Rev. E* **2006**, *73*, 026205. [CrossRef] [PubMed]
40. Cecconi, F.; Del-Castillo-Negrete, D.; Falcioni, M.; Vulpiani, A. The origin of diffusion: The case of non-chaotic systems. *Phys. D* **2003**, *180*, 129–139. [CrossRef]
41. Cecconi, F.; Cencini, M.; Falcioni, M.; Vulpiani, A. Brownian motion and diffusion: From stochastic processes to chaos and beyond. *Chaos* **2005**, *15*, 026102. [CrossRef]
42. Artuso, R.; Guarneri, I.; Rebuzzini, L. Spectral properties and anomalous transport in a polygonal billiard. *Chaos* **2000**, *10*, 189–194. [CrossRef]

43. Guarneri, I.; Casati, G.; Karle, V. Classical dynamical localization. *Phys. Rev. Lett.* **2014**, *113*, 174101. [CrossRef]
44. Yoshida, K.; Casati, G.; Watanabe, S.; Shudo, A. Sublinear diffusion in the generalized triangle map. *Phys. Rev. E* **2022**, *106*, 014206. [CrossRef]
45. Conze, J.-P.; Gutkin, E. On recurrence and ergodicity for geodesic flows on non-compact periodic polygonal surfaces. *Ergod. Theory Dyn. Syst.* **2012**, *32*, 491–515. [CrossRef]
46. Majumdar, M.N. Universal first-passage properties of discrete-time random walks and Lévy flights on a line: Statistics of the global maximum and records. *Phys. A* **2010**, *389*, 4299–4316. [CrossRef]
47. Godrèche, C.; Majumdar, S.N.; Schehr, G. Record statistics of a strongly correlated time series: Random walks and Lévy flights. *J. Phys. A* **2017**, *50*, 333001. [CrossRef]
48. Feller, W. *An Introduction to Probability Theory and Its Applications, Vol. 1*; Wiley: New York, NY, USA, 1968.
49. Feller, W. *An Introduction to Probability Theory and Its Applications, Vol. 2*; Wiley: New York, NY, USA, 1971.
50. Rebuzzini, L.; Artuso, R. Higher order statistics in the annulus square billiard: Transport and polyspectra. *J. Phys. A* **2011**, *44*, 025101. [CrossRef]
51. Srivastava, S.C.L.; Lakshminarayan, A.; Jain, S.R. Record statistics in random vectors and quantum chaos. *Europhys. Lett.* **2013**, *101*, 10003. [CrossRef]
52. Srivastava, S.C.L.; Lakshminarayan, A. Records in the classical and quantum standard map. *Chaos Solitons Fractals* **2015**, *74*, 67–78. [CrossRef]
53. Wergen, G. Records in stochastic processes: Theory and applications. *J. Phys. A* **2013**, *46*, 223001. [CrossRef]
54. Nevzorov, V.B. *Records: Mathematical Theory*; AMS: Providence, RI, USA, 2004.
55. Sparre Andersen, E. On the fluctuations of sums or random variables I. *Math. Scand.* **1953**, *1*, 263–285. [CrossRef]
56. Sparre Andersen, E. On the fluctuations of sums or random variables II. *Math. Scand.* **1954**, *2*, 195–233.
57. Artuso, R.; Cristadoro, G.; Degli Esposti, M.; Knight, G. Sparre-Andersen theorem with spatiotemporal correlations. *Phys. Rev. E* **2014**, *89*, 052111. [CrossRef]
58. Comtet, A.; Majumdar, S.N. Precise asymptotics for a random walker's maximum. *J. Stat. Mech.* **2005**, *2005*, P06013. [CrossRef]
59. Mounaix, P.; Majumdar, S.N.; Schehr, G. Asymptotics for the expected maximum of ramndom walks and Lévy flights with a constant drift. *J. Stat. Mech.* **2018**, *2018*, 083201. [CrossRef]
60. Gillis, J. Centrally biased discrete random walk. *Q. J. Math.* **1956**, *7*, 144–152.
61. Serva, M. Scaling behavior for random walks with memory of the largest distance from the origin. *Phys. Rev. E* **2013**, *88*, 052141. [CrossRef]
62. Radice, M.; Onofri, M.; Artuso, R.; Cristadoro, G. Transport properties and ageing for the averaged Lévy-Lorentz gas. *J. Phys. A* **2020**, *53*, 025701. [CrossRef]
63. Singh, P. Extreme value statistics and arcsine laws for heterogeneous diffusion processes. *Phys. Rev. E* **2022**, *105*, 024113. [CrossRef] [PubMed]
64. Artuso, R.; Onofri, M.; Pozzoli, G.; Radice, M. Extreme value statistics of positive recurrent centrally biased random walks. *J. Stat. Mech.* **2022**, *2022*, 103209. [CrossRef]
65. Lamperti, J. An occupation time theorem for a class of stochastic processes. *Trans. Am. Math. Soc* **1958**, *88*, 380–387. [CrossRef]
66. Radice, M.; Onofri, M.; Artuso, R.; Pozzoli, G. Statistics of occupation times and connection to local properties of nonhomogeneous random walks. *Phys. Rev. E* **2020**, *101*, 042103. [CrossRef]
67. Bel, G.; Barkai, E. Weak ergodicity breaking with deterministic dynamics. *Europhys. Lett.* **2006**, *74*, 16–21. [CrossRef]
68. Bray, A.J.; Majumdar, S.N.; Schehr, G. Persistence and first-passage properties in nonequilibrium systems. *Adv. Phys.* **2013**, *62*, 225–361. [CrossRef]
69. Thaler, M. The Dynkin-Lamperti arc-sine laws for measure preserving transformations. *Trans. Am. Math. Soc.* **1998**, *350*, 4593–4607. [CrossRef]
70. Zweimüller, R. Infinite measure preserving transformations with compact first regeneration. *J. Anal. Math.* **2007**, *103*, 93–131. [CrossRef]
71. Huang, J.; Zhao, H. Ultraslow diffusion and weak ergodicity breaking in right triangular billiards. *Phys. Rev. E* **2017**, *95*, 032209. [CrossRef] [PubMed]
72. Thaler, M. A limit theorem for sojourns near indifferent fixed points of one dimensional maps. *Ergod. Theory Dyn. Syst.* **2002**, *22*, 1289–1312. [CrossRef]
73. Akimoto, T. Generalized arcsine law and stable law in an infinite measure dynamical system. *J. Stat. Phys.* **2008**, *132*, 171–186. [CrossRef]
74. Singh, P.; Kundu, A. Generalized 'arcsine' laws for run-and-tumble particle in one dimension. *J. Stat. Mech.* **2019**, *2019*, 083205. [CrossRef]
75. Hughes, B.D. *Random Walks and Random Environments. Volume I: Random Walks*; Clarendon Press: Oxford, UK, 1995.
76. Barkai, E. Residence time statistics for normal and fractional diffusion in a force field. *J. Stat. Phys.* **2006**, *123*, 883–907. [CrossRef]
77. Corbelli, S.; Giona, M. A continuous archetype of nonuniform chaos in area-preserving dynamical systems. *J. Nonlinear Sci.* **2005**, *15*, 387–421. [CrossRef]
78. MacKay, R.S. Cerbelli and Giona's map is pseudo-Anosov and nine consequences. *J. Nonlinear Sci.* **2006**, *16*, 415–434. [CrossRef]

79. Artuso, R. Diffusive dynamics and periodic orbits of dynamical sustems. *Phys. Lett. A* **1991**, *160*, 528–530. [CrossRef]
80. Cvitanović, P.; Gaspard, P.; Shreiber, T. Investigation of the Lorentz gas in terms of periodic orbits. *Chaos* **1992**, *2*, 85–90. [CrossRef]

entropy

Article

The Metastable State of Fermi–Pasta–Ulam–Tsingou Models

Kevin A. Reiss * and David K. Campbell *

Department of Physics, Boston University, Boston, MA 02215, USA
* Correspondence: kevr@bu.edu (K.A.R.); dkcampbe@bu.edu (D.K.C.)

Abstract: Classical statistical mechanics has long relied on assumptions such as the equipartition theorem to understand the behavior of the complicated systems of many particles. The successes of this approach are well known, but there are also many well-known issues with classical theories. For some of these, the introduction of quantum mechanics is necessary, e.g., the ultraviolet catastrophe. However, more recently, the validity of assumptions such as the equipartition of energy in classical systems was called into question. For instance, a detailed analysis of a simplified model for blackbody radiation was apparently able to deduce the Stefan–Boltzmann law using purely classical statistical mechanics. This novel approach involved a careful analysis of a "metastable" state which greatly delays the approach to equilibrium. In this paper, we perform a broad analysis of such a metastable state in the classical Fermi–Pasta–Ulam–Tsingou (FPUT) models. We treat both the α-FPUT and β-FPUT models, exploring both quantitative and qualitative behavior. After introducing the models, we validate our methodology by reproducing the well-known FPUT recurrences in both models and confirming earlier results on how the strength of the recurrences depends on a single system parameter. We establish that the metastable state in the FPUT models can be defined by using a single degree-of-freedom measure—the spectral entropy (η)—and show that this measure has the power to quantify the distance from equipartition. For the α-FPUT model, a comparison to the integrable Toda lattice allows us to define rather clearly the lifetime of the metastable state for the standard initial conditions. We next devise a method to measure the lifetime of the metastable state t_m in the α-FPUT model that reduces the sensitivity to the exact initial conditions. Our procedure involves averaging over random initial phases in the plane of initial conditions, the P_1-Q_1 plane. Applying this procedure gives us a power-law scaling for t_m, with the important result that the power laws for different system sizes collapse down to the same exponent as $E\alpha^2 \to 0$. We examine the energy spectrum $E(k)$ over time in the α-FPUT model and again compare the results to those of the Toda model. This analysis tentatively supports a method for an irreversible energy dissipation process suggested by Onorato et al.: four-wave and six-wave resonances as described by the "wave turbulence" theory. We next apply a similar approach to the β-FPUT model. Here, we explore in particular the different behavior for the two different signs of β. Finally, we describe a procedure for calculating t_m in the β-FPUT model, a very different task than for the α-FPUT model, because the β-FPUT model is not a truncation of an integrable nonlinear model.

Keywords: metastability; classical statistical mechanics; advanced numerical methods; semiclassical methods and results

Citation: Reiss, K.A.; Campbell, D.K. The Metastable State of Fermi–Pasta–Ulam–Tsingou Models. *Entropy* **2023**, *25*, 300. https://doi.org/10.3390/e25020300

Academic Editor: Marko Robnik

Received: 19 December 2022
Revised: 30 January 2023
Accepted: 30 January 2023
Published: 6 February 2023

1. Introduction

Statistical mechanics, broadly speaking, aims to draw conclusions about the behavior of systems with large numbers of particles without needing to solve the even larger number of equations that the system obeys. This approach was successful in explaining everything from the temperature of a gas to the density of a neutron star, with many stunning discoveries in between [1]. One of the central tenets of this subject is the equipartition theorem [2], which assumes that over time, energy will be shared equally around the system. This assumption has led to many successes, e.g., the ideal gas law, but also some failures, e.g., the ultraviolet catastrophe from the failure of the Rayleigh–Jeans law to describe blackbody radiation [3]. Until recently, it was believed that the resolution of the ultraviolet catastrophe required the quantization of the energy of light into photons. However, more recently, an entirely classical resolution was proposed [4]. By avoiding the assumption of the equipartition theorem, Wang et al. were able to find the Stefan–Boltzmann law through purely classical mechanics, consistent with the results of quantum mechanics. The key was the statistics of a quasi-stationary state in the model, which has the effect of stalling the approach to equilibrium. While the impact of these new results on statistical mechanics and the approach to equilibrium remains to be seen, the suggestion that "metastable" states may play a critical role in the interactions of many-body classical systems is very intriguing and is something that we will study in detail in this paper.

As the background and motivation for our study, we recall that, in the early 1950s, Enrico Fermi, John Pasta, Stanislaw Ulam and Mary Tsingou (FPUT) made the first detailed computational study of the validity of the equipartition theorem. For the parameters used in their studies, instead of equipartition, they observed a similar quasi-stationary state consisting of "recurrences" to the initial state [5]. Their assumption was that adding even a small nonlinear term to the linear couplings between harmonic oscillators would be enough to allow the system to thermalize, i.e., reach a state of equipartition. However, they found that for small enough energies, the system remained localized in the mode space for all the times that were computationally possible to explore with their computer. Further, they found that there were remarkable and entirely unexpected (near) recurrences to the initial state. This discovery opened the door for many important advances in the field of nonlinear dynamical systems: the discovery of solitons [6], q-breathers [7] and many more. Some of the most significant implications of their results were summarized on its 50-year anniversary [8]. The dedicated reader is referred to these major reviews of the FPUT problem: [9–12].

Our interest here is to explore computationally what is referred to as the "metastable state" [13] in the FPUT models. This is a quasi-stationary state which stalls the approach to equipartition. In particular, we are interested in the lifetime of the metastable state, because the system is not able to approach equilibrium until the metastable state has ended. Understanding the lifetime of this state, especially any scaling laws that it exhibits, will likely provide a basis for analyzing other systems with quasi-stationary states. Hence, we will develop and explore some techniques which can standardize the study of metastable states in non-integrable systems. It is our belief that the continued exploration of these states in physical systems has the potential to unlock more equivalencies between quantum mechanics and classical statistical mechanics, as was the case with blackbody radiation [4].

The structure of the remainder of this paper is as follows. First, in Section 2, we introduce the systems we will explore. Then, in Section 3, we lay out the recurrence phenomenon and give an intuitive picture of the metastable state. In Section 4, we explore the metastable state in the α-FPUT model, the primary computational focus of our article. In the next section, we explore the strength of the recurrences (Section 4.1), the lifetime of the metastable state (Section 4.2) and the energy spectrum (Section 4.3). We conclude with a qualitative exploration of the metastable state in the β-FPUT model, in Section 5. We examine a comparison between the two signs of β (Section 5.1), and in Section 5.2, we discuss the possibility of measuring the lifetime of the metastable state in the β model. Section 6 presents a summary of our conclusions.

2. Methods

2.1. Models

The general Hamiltonian for the systems we will consider is that of a chain of oscillators constrained to move in one dimension with nearest neighbor interactions given by a potential $V(r)$, i.e.,

$$H(\boldsymbol{q}, \boldsymbol{p}) = \sum_{n=1}^{N} \frac{p_n^2}{2} + \sum_{n=0}^{N} V(q_{n+1} - q_n). \tag{1}$$

We will consider both the α-FPUT model, with a cubic potential,

$$V_\alpha(r) = \frac{r^2}{2} + \frac{\alpha}{3}r^3, \tag{2}$$

and the β-FPUT model, with a quartic potential,

$$V_\beta(r) = \frac{r^2}{2} + \frac{\beta}{4}r^4, \tag{3}$$

with fixed boundary conditions $q_0 = q_{N+1} = 0$ and $p_0 = p_{N+1} = 0$ such that there are N distinguishable oscillators. The β-FPUT model can be considered as a perturbation of a linear chain of oscillators (with perturbation strength β), while the α-FPUT model behaves as a truncation of the Toda lattice, which has potential energy:

$$V_{\text{Toda}}(r) = V_0\left(e^{\lambda r} - 1 - \lambda r\right), \tag{4}$$

and was shown to be completely integrable [14].

We define the normal modes through the canonical transformation:

$$\begin{bmatrix} q_n \\ p_n \end{bmatrix} = \sqrt{\frac{2}{N+1}} \sum_{k=1}^{N} \begin{bmatrix} Q_k \\ P_k \end{bmatrix} \sin\left(\frac{nk\pi}{N+1}\right). \tag{5}$$

These normal modes have frequencies:

$$\omega_k = 2\sin\left(\frac{k\pi}{2(N+1)}\right). \tag{6}$$

This normal mode transformation diagonalizes the harmonic lattice (i.e., $\alpha = \beta = 0$ only) but leaves off-diagonal terms in the Hamiltonians for the anharmonic models ($\alpha, \beta \neq 0$). These terms lead to the transfer of energy among the modes.

After this normal mode transformation, the Hamiltonian for the α-FPUT model is

$$H_\alpha(\mathbf{Q}, \mathbf{P}) = \sum_{k=1}^{N} \frac{P_k^2 + \omega_k^2 Q_k^2}{2} + \frac{\alpha}{3} \sum_{k,j,l=1}^{N} A_{k,j,l} Q_k Q_j Q_l, \tag{7}$$

while for the β-FPUT:

$$H_\beta(\mathbf{Q}, \mathbf{P}) = \sum_{k=1}^{N} \frac{P_k^2 + \omega_k^2 Q_k^2}{2} + \frac{\beta}{4} \sum_{i,j,l=1}^{N} B_{k,i,j,l} Q_k Q_i Q_j Q_l, \tag{8}$$

where the last (summed) terms in both equations couple the normal modes together, allowing for energy sharing, with coupling constants given by [15,16]:

$$A_{k,j,l} = \frac{\omega_k \omega_j \omega_l}{\sqrt{2(N+1)}} \sum_{\pm} \left[\delta_{k,\pm j \pm l} - \delta_{k \pm j \pm l, 2(N+1)}\right], \tag{9}$$

$$B_{k,i,j,l} = \frac{\omega_k \omega_i \omega_j \omega_l}{2(N+1)} \sum_{\pm} \left[\delta_{k,\pm j \pm l \pm m} - \delta_{k \pm j \pm l \pm m, \pm 2(N+1)}\right], \tag{10}$$

where $\delta_{i,j}$ is the Kronecker delta function and the sums $\sum_\pm$ are overall combination of plus and minus signs in the equation.

The energy E_k of the k-th mode is

$$E_k = \frac{1}{2}\left(P_k^2 + \omega_k^2 Q_k^2\right). \tag{11}$$

This definition is exact only for the harmonic lattice, but serves as a good approximation for weak nonlinearity, because any contributions to the energy coming from coupled modes have a pre-factor of the nonlinear strength (α or β).

Whenever a quantity is time-averaged, we place a line over its symbol (e.g., $\overline{E}$). This represents a time average from time $t = 0$ to $t = T$, i.e.,

$$\overline{E}(T) = \frac{1}{T}\int_0^T E(t)\,dt. \tag{12}$$

2.2. Numerical Methods

For integrations involving the α-FPUT model and β-FPUT model with $\beta < 0$, which were observed to be reasonably stable numerically [17], we use the $SABA_2C$ symplectic integration scheme described in appendix 1 of [18]. This scheme has error $\mathcal{O}([dt]^4)$. For the integration of the Toda lattice, we use the $SABA_2$ scheme, i.e., the same scheme but without the corrector Hamiltonian term, giving error $\mathcal{O}([dt]^2)$, which was determined to provide sufficient accuracy for the range of parameters considered. For $\beta > 0$, the β-FPUT model is known to exhibit exponential numerical instabilities related to instabilities of the soliton solutions to the modified Korteweg–de Vries (mKdV) equation [19], because the mKdV equation arises from the continuum limit of the β-FPUT model. To reduce the need for extremely small time-step sizes, we implement the symplectic integrator $SABA_2Y8\_D$ described in [20] and in Table 2 of [21], which has error $\mathcal{O}([dt]^8)$. In general, we use a time step $dt = 0.1$ unless a failure of time reversal requires us to decrease dt to improve the accuracy.

2.3. Spectral Entropy

We will use spectral entropy to quantify the FPUT system's "distance" from equipartition at a given time. The spectral entropy is similar to Shannon information entropy [22] and is defined as

$$S(t) = -\sum_{k=1}^{N} e_k(t)\ln[e_k(t)], \qquad \text{with: } e_k(t) = \frac{E_k(t)}{\sum_k E_k(t)}, \tag{13}$$

where $e_k(t)$ is the proportion of linear energy in mode k at time t. Spectral entropy ranges from 0, when all the energy is in one mode, to S_{max}, when an equal amount of energy is present in all modes. For the α-FPUT and Toda lattices, equal energy sharing corresponds to $e_k = 1/N\ \forall\ k$; therefore, $S_{max} = \ln(N)$. However, the β-FPUT lattice remains symmetric about its center for initially symmetric excitations, and therefore energy cannot spread from an even-numbered mode number to an odd-numbered mode, or vice versa. Because our initial conditions will include only an odd mode, energy can only be shared among odd modes, so $S_{max} = \ln\lceil \frac{N}{2} \rceil$, where $\lceil\rceil$ is the ceiling function, which rounds a number up to the next highest integer. Because this definition of spectral entropy has a different maximum value for different lattice sizes N, we can rescale it by defining the rescaled spectral entropy (henceforth entropy for short):

$$\eta(t) = \frac{S(t) - S_{max}}{S(0) - S_{max}}. \tag{14}$$

This is a convenient definition because η ranges from 1 at $t = 0$ to 0 when energy is shared equally among all modes (equipartition), regardless of system size N.

3. Phenomena

3.1. FPUT Recurrences

One of the surprising features of the models first explored by FPUT [5] was the presence of what have come to be known as "FPUT recurrences". Indeed, Fermi himself expressed the (understated) opinion that this behavior really constituted a "little discovery in providing limitations that the prevalent beliefs in the universality of "mixing" and "thermalization" in nonlinear systems may not always be justified" [5]. The FPUT recurrences were discovered as follows: when all of the energy was initialized in the first normal mode, this energy was first observed to diffuse to higher order modes, but then the energy began to return to the first normal mode, eventually nearly fully returning at what is called the "recurrence time" (t_r). This phenomenon is shown in Figure 1, which shows the energy in the lowest 5 allowed modes in the α-FPUT and β-FPUT models as a function of time. As noted above, the β-FPUT lattice preserves the symmetry about its center so with the initial energy only in mode 1, only odd modes are allowed. At $t = t_r$, the systems have nearly reproduced their initial conditions. t_r is calculated for the α-FPUT model following [23] and for the β-FPUT model following [24]. The timescale for this recurrent behavior is many orders of magnitude shorter than the Poincaré recurrence time [25], and the recurrences continue quasi-periodically for a long time; indeed, the initial conditions considered by FPUT have yet to be driven to equipartition in any computer simulation. However, for larger initial energy, the FPUT recurrences eventually breakdown, and the system is able to thermalize. Clearly, when most of the energy is quasi-periodically returning near the initial condition, which is extremely localized, the system remains localized while these recurrences continue to occur.

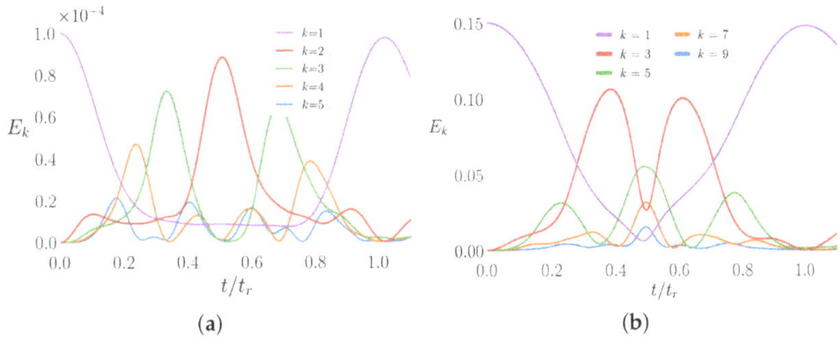

Figure 1. The energy in each normal mode in the α-FPUT and β-FPUT models as a function of time. At $t = t_r$, the first FPUT recurrence is observed, with nearly all energy returning to its initial condition, the first normal mode. The lowest 5 allowed modes in each model are plotted. (**a**) α-FPUT model with initial $E_1 = 10^{-4}$ and $N = 127$ (with $\alpha = 1$). (**b**) β-FPUT model with initial $E_1 = 0.15$ and $N = 127$ (with $\beta = 1$).

FPUT recurrences were used to study ultra-cold Bose gases [26], the nonlinear Schrodinger equation [27] and electron–phonon interactions [28], to mention a few applications. Their study was also extended to higher-order recurrences, such as super-recurrences [18,29]. Their existence was explained in various ways, most notably (1) by using q-breathers [7,30,31] or (2) by the presence of solitons in the KdV (mKdV) equation, which is the continuum limit of the α-FPUT (β-FPUT) model [6,24,32]. The importance of FPUT recurrences is difficult to overstate, but in this paper, we focus primarily on their role in delaying the approach to equipartition.

3.2. Metastable State

The recurrence phenomenon has the effect of stalling the approach to equilibrium by keeping the system's energy localized near its initial condition. This phenomenon was interpreted, as early as 1982 [33], as the system having two distinct "regions" in time: in the first region, the system relaxes into an intermediate quasi-stationary state, which persists for some time, before it again relaxes, this time into its true equilibrium state defined by equipartition, such that $\overline{\eta} = \langle \eta \rangle$. The intermediate or "metastable" state was more recently studied extensively by Giancarlo Benettin [13,34–36]. His work frames the phenomenon as a cross-over between predominantly integrable dynamics to the true non-integrable dynamics of the FPUT models.

In terms of the spectral entropy η, the system is considered to be in equilibrium when $\overline{\eta} = \langle \eta \rangle$, where we calculate $\langle \eta \rangle$ following Danieli [37]:

$$\langle \eta \rangle = \frac{1-\gamma}{S_{max} - S(0)}, \tag{15}$$

where $\gamma \simeq 0.577$ is the Euler–Mascheroni constant. We are interested in the time that the metastable state persists, before its ultimate destruction, and the system's approach to equilibrium. We call this the lifetime t_m of the metastable state. In Figure 2, we illustrate the metastable state in the α-FPUT and β-FPUT models. We can see that their behaviors are qualitatively quite different. While the α-FPUT model appears to be decreasing gradually in $\overline{\eta}$, the β-FPUT model exhibits a clear flat plateau for a long time before some mechanism causes the metastable state to collapse fairly suddenly. The features of the metastable state in the α-FPUT model, at first glance, make it difficult to define where the metastable state ends and the approach to equilibrium begins, but we will show that we can separate these two regions by comparing the α-FPUT model's behavior to that of the Toda lattice and considering the crossover time t_m to be that time at which the behavior of the two systems begins to differ substantially.

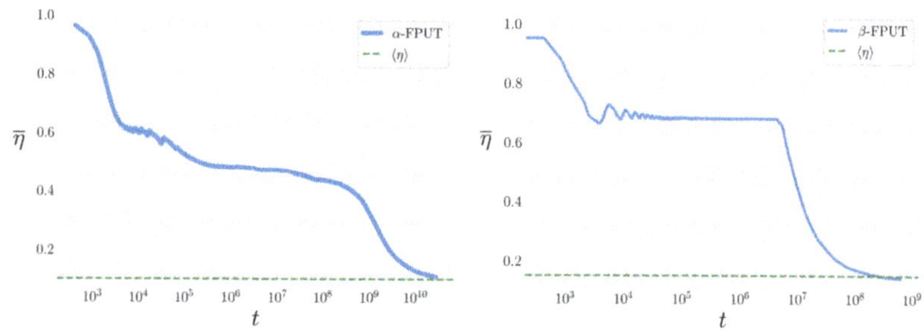

(a) **(b)**

Figure 2. The time-averaged entropy as a function of time in the α-FPUT and β-FPUT models. The ensemble average $\langle \eta \rangle$ (from Equation (15)) is plotted and the agreement $\overline{\eta} = \langle \eta \rangle$ appears to be stalled by a metastable state. (**a**) α-FPUT model with $E\alpha^2 = 0.02$ and $N = 63$. (**b**) β-FPUT model with $E\beta = 0.57$ and $N = 31$.

To make this point more explicitly, we note that up to $\mathcal{O}(r^4)$, the α-FPUT potential (Equation (2)) can be thought of as a truncation of the Toda potential (Equation (4)), through a convenient change in the parameters. By setting $V_0 = \lambda^{-2}$ and $\lambda = 2\alpha$, and Taylor expanding the Toda potential around $r = 0$, we obtain the following series expansion:

$$V_{\text{Toda}}(r) = \frac{r^2}{2} + \frac{\alpha}{3}r^3 + \frac{\alpha^2}{6}r^4 + \frac{\alpha^3}{15}r^5 + \mathcal{O}(r^6) = V_\alpha(r) + \mathcal{O}(r^4). \tag{16}$$

Thus, the α-FPUT model's metastable state can be analyzed by considering its behavior to be similar to the integrable Toda lattice, before it breaks off and exhibits the expected behavior of non-integrable systems [34]. Figure 3 demonstrates the similarity of the evolution of $\overline{\eta}$ between the Toda model and the α-FPUT model up to a certain point in time, after which the α-FPUT model falls to the expected equilibrium value of $\overline{\eta}$: the ensemble average $\langle \eta \rangle$. This comparison to the Toda lattice will allow us to define rather precisely t_m in the α-FPUT model.

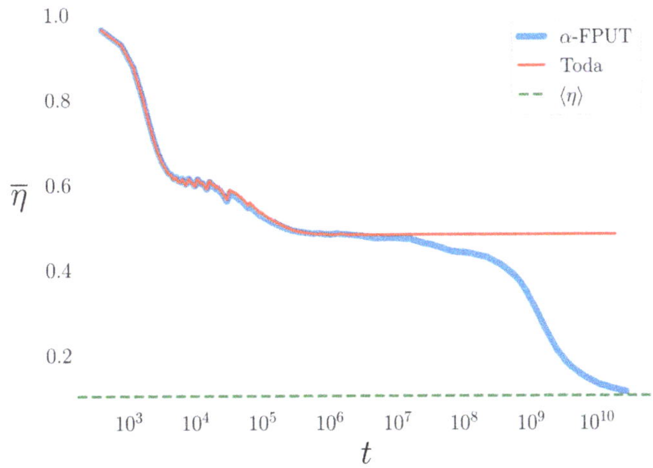

Figure 3. The time-averaged entropy $\overline{\eta}$ as a function of time (note logarithmic time scale) for both the Toda model (red) and α-FPUT model (blue). Both have initial energy $E\alpha^2 = 0.02$ and system size $N = 63$.

4. α-FPUT Model

4.1. Strength of FPUT Recurrences

It was shown [23] that the time to the first FPUT recurrence (t_r) in the α-FPUT model scales as a function of an essential system parameter R, defined as:

$$R = (N+1)^{3/2}\sqrt{E\alpha^2}. \tag{17}$$

Specifically, as was shown in [23], by rescaling the FPUT recurrence time by $(N+1)^3$, then for $R \geq 10$:

$$\frac{t_r}{(N+1)^3} = R^{-1/2}. \tag{18}$$

We use this expected value of the first FPUT recurrence time and look in the region $0.5t_r < t < 1.5t_r$ for the maximum value of the energy in the first normal mode, and name that $E_1(t_r)$. We can then calculate the ratio of this energy to the initial energy and use this as a measure of the relative "strength" of the FPUT recurrence for a given value of R and N. The results are plotted in Figure 4a and demonstrate that the FPUT recurrence strength drops off as R increases, nearly independent of system size N.

Although this discussion is similar in motivation to that describing the behavior of the β-FPUT model (see Section 7 of [24]), the implications are quite different: for the β-FPUT model, FPUT recurrences lose strength as a function of the parameter $E\beta$ independent of N—**NOT** the essential system parameter

$$S = E\beta(N+1), \tag{19}$$

which the FPUT recurrence time scales with. For the α-FPUT model, the strength of the FPUT recurrences scales with R, independent of N, instead of the corresponding energy parameter $E\alpha^2$. It is also worth noting that while in [24] one had to define a parameter called "shareable energy" to compare the quality of the FPUT recurrences between the cases $\beta > 0$ and $\beta < 0$, Figure 4b shows that this is not necessary for the α-FPUT model. This figure plots the quantity:

$$\frac{E_1^{min}}{E} \equiv \min_{0<t<t_r} \frac{E_1(t)}{E}, \tag{20}$$

which quantifies how much energy leaves the first normal mode (the initial condition), before most of it comes back at the recurrence time. Figure 4 demonstrates that for the α-FPUT model, nearly all of the energy consistently leaves the first normal mode before coming back for a recurrence. This appears to be true for all R and N, except in the harmonic limit ($E\alpha^2 \to 0$). However, this sharing of energy is not the case for the β-FPUT model with $\beta < 0$, where roughly 70% of the energy remains in the first normal mode before a recurrence [24].

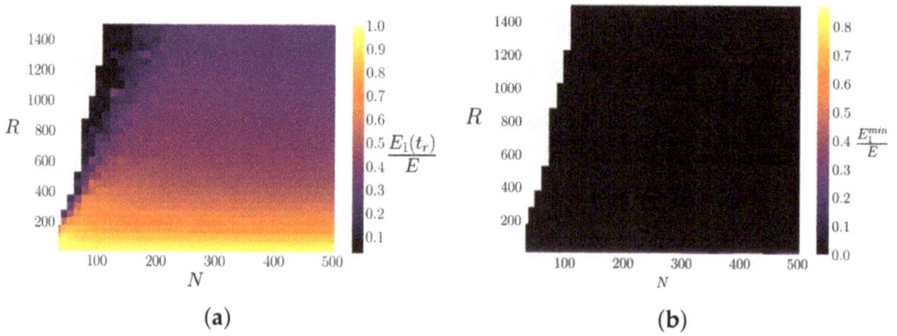

Figure 4. Heatmaps as a function of R (Equation (17)) and system size N in the α-FPUT model. Note that the gray region corresponds to initial conditions which blow up (potential $V(r) \to -\infty$) before $1.5t_r$. (**a**) The "strength" of FPUT recurrences, represented by $E_1(t_r)/E$, the fraction of energy returning to the initial condition at the first recurrence. (**b**) The quantity E_1^{min}/E, which represents the proportion of energy that leaves the 1st normal mode before the first recurrence.

Because Figure 4a seems to show that the strength of recurrences falls off as a function of R, independent of N outside of regimes where blow-up is likely, it helps to look at systems with the same N and plot the recurrence strength $E_1(t_r)/E$ as a function of system parameter R. This is performed in Figure 5, and a nearly exponential decay is found. This result holds for all sufficiently large system sizes, and the results are presented for $N = 502$ to avoid blow-up and other small N behavior [38] (following [39], we only work with systems such that $N + 1$ is a power of 2 or prime). This exponential decay is again in contrast to the β-FPUT model, where recurrence strength appears to be roughly consistent until a cut-off energy $E\beta$ where the recurrence strength falls precipitously [24].

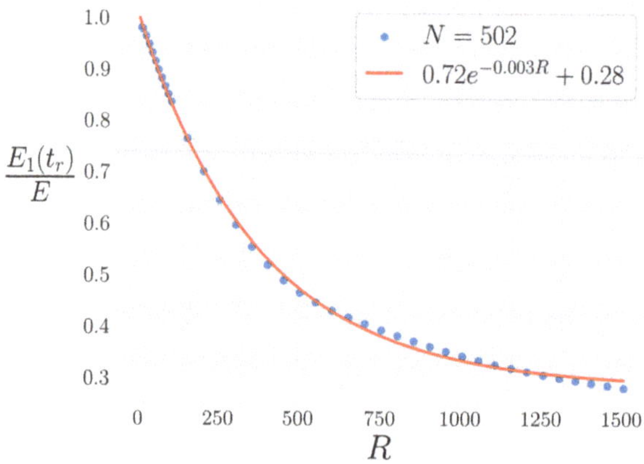

Figure 5. The "strength" of recurrences in the α-FPUT model, $E_1(t_r)/E$ as a function of system parameter R at fixed $N = 502$. An exponential fit is added.

4.2. *Lifetime of Metastable State*

4.2.1. Procedure

We endeavor to find a scaling for the lifetime of the metastable state through a direct comparison to the Toda model, as motivated by Section 3.2. To define t_m by comparing the α-FPUT model's behavior with that of the Toda lattice, the most natural approach is to define some arbitrary tolerance and look for the last time which the α-FPUT model's entropy is within that tolerance of the entropy of the Toda model. Doing so, however, reveals an interesting feature of the metastable state. The results of following this procedure for $N = 63$ are shown in Figure 6. Even though a clear power-law scaling emerges, the data are quite noisy around this scaling. This noisiness appears to be an inherent feature of the chaotic nature of the α-FPUT model around the metastable state.

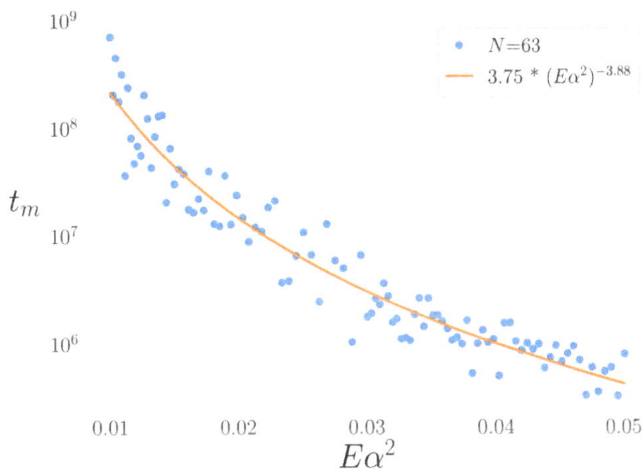

Figure 6. An attempt at defining t_m in the α-FPUT model for $N = 63$ by defining an arbitrary tolerance and waiting for a deviation from the Toda model beyond this tolerance. A power-law best fit is added. Note the logarithmic scaling on the t_m axis.

Note that because the Toda model is integrable, its dynamics can, in theory, be broken down into actions that remain constant in time and angles that evolve periodically in time. The picture of the metastable state of the α-FPUT model presented by Benettin et al. in [36] is that there are two time scales in the system. In the first one, the actions of the Toda model remain nearly constant even in the α-FPUT model, while the corresponding angles evolve on tori, leading to a behavior very similar to that of the Toda model. Eventually, on a longer time scale, the Toda-like actions in the α-FPUT model start to diffuse throughout the phase space, eventually leading to ergodicity and equipartition. The shorter time scale where the Toda-like actions remain nearly constant defines the metastable state. An important aspect of the transition to diffusing actions is that this diffusion behaves chaotically, with positive definite maximal Lyapunov exponents as described in [36]. This leads to an exponential sensitivity to initial conditions when the diffusion of action variables dominates the dynamics, which explains the noise in Figure 6. To quantify the effect of the initial conditions, we next conduct bin averaging over the initial conditions.

Note that from Equation (11), the energy initially given to the first normal mode can be distributed either in a canonical position or momentum. We define the "phase" θ between our canonical coordinates as:

$$\theta = \tan^{-1}\left(\frac{P_1(t)}{\omega_1 Q_1(t)}\right). \tag{21}$$

We can then initialize systems with the same condition $E_1(0) = E$, i.e., the same point in energy space, but slightly separated in the phase space by distributing along the oval of the canonical coordinates defined by rotating θ. In the following, we take 100 random phases for every choice of energy and bin them together to create 10 bins which are each the average of 10 trials with different phases. This bin averaging is meant to calculate an approximation to the ensemble average. An example of the results of this procedure is shown in Figure 7, where each α-FPUT curve represents an average over 10 random phases. Figure 7 demonstrates that each of the α-FPUT trials remains close to the Toda model, up until some time where the entropy starts to decrease below the Toda model entropy (red curve), and then the α-FPUT trajectories start to diverge, not only from the Toda trajectory but also from each other.

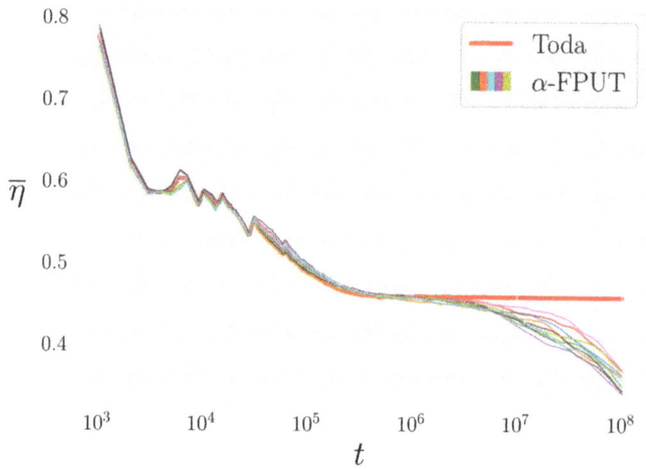

Figure 7. The time-averaged entropy in the Toda model (red curve) compared to 10 bins of α-FPUT trajectories, each made up of the average of 10 random phases. All systems are fixed at energy $E\alpha^2 = 0.028$ and system size $N = 63$.

We gain two advantages from binning in this manner: (1) we now have a natural length to use as a tolerance cut-off to define the separation between the α-FPUT model and Toda model that is not arbitrary: the standard deviation of each bin; and (2) averaging over 10 different bins again gives us an error bar on our measurement of t_m for a given enerzgy. Our procedure is now as follows: Take 10 trials for the α-FPUT model with random phases and average their entropy together. Find the last time that the Toda model's entropy was within one standard deviation of this bin average. Repeat this for 10 total bins, and average those times together to obtain a measurement of t_m with an error bar.

Performing this operation reveals a surprising result, shown in Figure 8. If we look at the bin standard deviation (σ), averaged over bins ($\overline{\sigma}$), there is a feature similar to a phase transition in the plot. The time at which this occurs happens to line up with the time $t = t_m$ as defined in our above procedure. Because our procedure looks at when $\overline{\eta}$ in the α-FPUT model is greater than σ outside of $\overline{\eta}$ in the Toda model, this means $\overline{\eta}$ is falling more quickly than σ is rising in Figure 8, which is significant. This also further validates the point of view that t_m represents a transition from mostly integrable dynamics to chaotic, non-integrable dynamics. The growth in $\overline{\sigma}$ for $t > t_m$ shows that initially nearby systems are deviating in time, whereas for $t < t_m$, $\overline{\sigma}$ is seen to be relatively bounded in time. This also serves to validate our procedure to measure t_m.

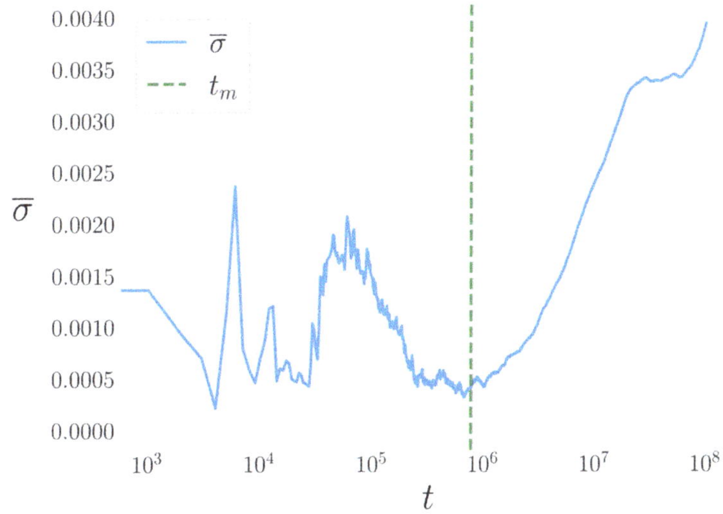

Figure 8. The bin deviation, averaged over bins, again for $E\alpha^2 = 0.028$ and $N = 63$ in the α-FPUT model. The measure value for t_m is marked as a vertical, dashed green line.

4.2.2. Analysis

We apply the procedure described in the previous section and iterate across a range of energies, for $N = 63, 127$ and 255. We determined that system size $N = 31$ was too small and gave erratic results incompatible with the thermodynamic limit. For a discussion of small system size effects in the α-FPUT model, see [38]. We chose system sizes such that $N + 1$ is a power of 2, to avoid resonances discussed in [39]. The results are shown in Figure 9a. The length of each data point is the extent of its bin error. Each system size appears to follow its own trend for high energies. However, for low energies, the data appear to overlap, *regardless of system size*. In this regime, t_m is seen to follow a power law, roughly consistent with an exponent of -4.9, as shown by the red dashed line in Figure 9a. A few simulations indicate that this overlap and scaling is consistent for larger system sizes as well. This result is more significant than that presented in Figure 6, as it both considers the ensemble average and appears to hold in the thermodynamic limit. In particular, this result has significant implications for the $E\alpha^2 \to 0$ limit, which is that originally considered by FPUT.

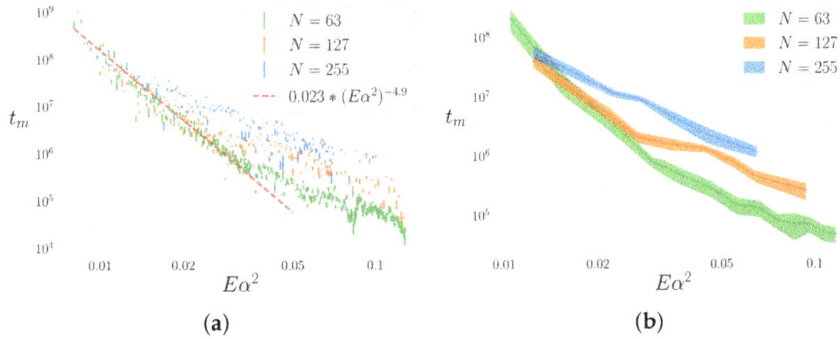

Figure 9. The lifetime of the metastable state as a function of $E\alpha^2$ for $N = 63, 127$, and 255. Note logarithmic scale on all axes. (**a**) The height of each data point represents its bin error. In the $E\alpha^2 \to 0$ limit, the data are seen to follow a power law, in dashed red. (**b**) Nearby energies are binned to obtain an upper bound on the noise as a combination of phase and energy error.

One interesting aspect of Figure 9a is that the error in the noise (shown by the scattering of the data) seems to be larger than the error due to phase averaging and binning (shown by the height of the data points). In order to account for these two possible sources of chaotic noise, we bin data into groups of 20 consecutive energies to estimate the noise in the energy. Then, we assume that the phase noise (σ_θ) and energy noise (σ_E) are independent and add them together as

$$\sigma = \sqrt{\sigma_\theta^2 + \sigma_E^2}, \tag{22}$$

to perform the error propagation and obtain an upper bound on the noise. The results are presented in Figure 9b. This gives a better idea of the noise (inherent because the metastable state signals the onset of chaos) in the lifetime of the metastable state.

4.3. Spectrum

Our use of spectral entropy as the single measure of the destruction of the metastable states gives a qualitative picture, but by plotting the time-averaged energy in each normal mode, at a given time, we have access to many more degrees of freedom than simply looking at the entropy. Therefore, we can obtain a more complete picture. For short times, we expect the spectra of the α-FPUT and Toda models to look essentially identical. This is indeed the case. As time goes on, however, the Toda spectrum flattens out to an exponential tail, which is the shape of the α-FPUT spectrum in the metastable state as well. Nonetheless, some higher modes start to gain energy and spread this energy to the other nearby modes. This process continues until most higher modes are excited and the system approaches equipartition. This behavior is demonstrated in Figure 10. For more discussion on the spectral picture of diffusion in the α-FPUT model compared to the Toda lattice, see [40].

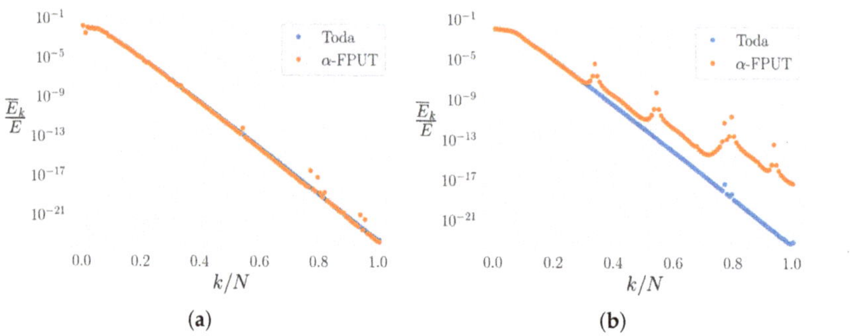

Figure 10. The spectra of the α-FPUT and Toda models for system parameter $R = 75$ and $N = 127$. Spectra are compared at two different times. (**a**) Spectra for $t = 10^5$. (**b**) Spectra for $t = 10^8$.

Figure 10a is plotted at $t = 10^5$ and shows that the spectra of the α-FPUT model and the Toda model largely agree at this time. Figure 10b is plotted at $t = 10^8$, and we can see that resonances have caused local peaks in the α-FPUT spectrum, which diffuse the energy into the modes around them. This has the effect of lifting the spectrum at each resonance, a process which continues until the system reaches equipartition. In [41], Onorato et al. showed that four-wave resonances in the thermodynamic limit of the α-FPUT model lead to irreversible energy mixing. It was also shown that six-wave interactions are always possible and lead to irreversible energy mixing. Despite the appealing possibility that the observed peaks in the spectra might correspond to those predicted by the wave turbulence method of Onorato et al., we have at present been unable to verify this possibility quantitatively. It is possible that the two largest peaks in the spectra of the α-FPUT model (Figure 10b) are actually made of two resonant modes each, so it is unclear if this is an example of a four-wave or six-wave resonance. We are currently investigating this matter further.

Another peculiarity in Figure 10b is the apparent high-k modes in the Toda model which lie well above an exponential tail, even after a long time. The peaks around $k/N \simeq 0.8$ do not appear to be a numerical artifact, so there could possibly be resonances in the integrable limit (which do *not* lead to irreversible energy mixing).

5. β-FPUT Model

5.1. Comparison between $\beta > 0$ and $\beta < 0$

A surprising difference between the FPUT recurrences in the β-FPUT model for the different signs of β was noted in Section 7 of [24]. The difference is qualitatively demonstrated in Figure 11a ($\beta < 0$) and 11b ($\beta > 0$), which show the proportion of the total energy in each of the first 13 modes against time. The results are plotted for the first 50 FPUT recurrences, with the FPUT recurrence time t_r calculated using the results from [24]. For a system with $\beta > 0$, the energy almost entirely leaves the first normal mode before coming back at an FPUT recurrence (as demonstrated by Figure 1, which is essentially a zoom into Figure 11b). When $\beta < 0$, nearly 70% of the energy always remains in the first normal mode during the metastable state. Figure 12b shows that the first normal mode is not isolated when $\beta > 0$ for the relatively small $E\beta = 0.15$, driving the magnitude of $E\beta$ higher for $\beta < 0$, as shown in Figure 12a for $E\beta = -0.35$, which leaves the first normal mode still largely isolated. Figure 6 of [24] shows that this behavior is only a function of the sign of $E\beta$, not its magnitude.

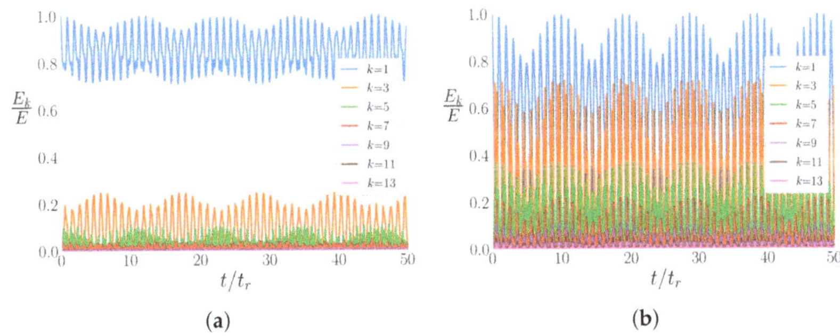

Figure 11. The energy in the lowest 13 modes as a function of time for $N = 127$, and the two choices of the sign of β, in the β-FPUT model when the initial energy is all in the first mode. Recall that in the β model, this means that all even modes have zero energy. Energy in each mode is rescaled by initial energy E and time is rescaled by the FPUT recurrence time t_r. (**a**) $E\beta = -0.35$ (Note: $\beta < 0$). (**b**) $E\beta = 0.15$ (Note: $\beta > 0$).

When the distribution of energy among all the normal modes (energy spectrum) in the metastable state is considered, however, the two systems are relatively similar. Figure 12a ($\beta < 0$) and Figure 12b ($\beta > 0$) plot the spectra of the two β-FPUT systems, i.e., the time-averaged energy in each mode. The time averages are computed after 50 FPUT recurrences have occurred. Both spectra follow an exponential decay, with a few peaks in the spectra raising further questions. In [41], Onorato et al. showed that six-wave resonances lead to irreversible energy mixing; these peaks might correspond to those resonances. This possibility is under further investigation.

Figure 12 shows that the differences noted in Figure 11 are only evident between the 1st and 3rd normal modes, with all the other modes following a qualitatively similar distribution. It is possible that for $\beta < 0$, the $k = 1$ mode engages in the energy-diffusing resonance while $k = 1$ is not a resonant mode for $\beta > 0$. This would explain the lack of energy mixing for $\beta < 0$ and the local peak at $k = 1$ in the spectrum (Figure 12a).

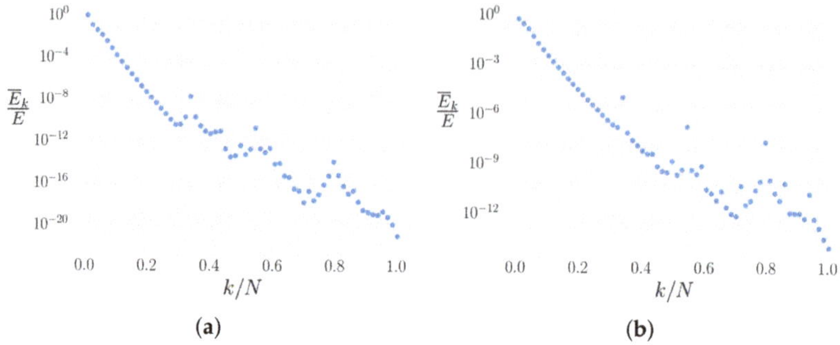

Figure 12. Time-averaged energies in each mode for $N = 127$, as a function of mode number k. The average is taken after 50 FPUT recurrences have taken place. (**a**) $E\beta = -0.35$ (Note: $\beta < 0$). (**b**) $E\beta = 0.15$ (Note: $\beta > 0$).

5.2. Lifetime of Metastable State

As depicted in Figure 2, the metastable state in the β-FPUT model ends much more abruptly than that in the α-FPUT model. However, the α-FPUT model can be considered a truncation of the integrable Toda lattice, so that the point at which the α-FPUT spectral entropy begins to deviate substantially from that of the Toda model is well defined for any set of parameters and can be considered as the end of the metastable state. We showed how to make this even more precise by taking bin averages that effectively approximate an ensemble average. We find that this is very well defined for all the ranges of parameters we have studied in the α-FPUT model. In contrast, the β-FPUT model cannot be viewed as the truncation of any nonlinear integrable model but rather as a perturbation of the linear lattice. This observation coupled with the fact that for $\beta > 0$, the β-FPUT model exhibits well-known exponential numerical instabilities related to the soliton solutions of the modified Korteweg–de Vries (mKdV) equation [19], (see Section 2.2), results in significant convergence issues when we try to adapt our "bin averaging" technique for the β-FPUT model. In particular, it requires very accurate numerics to ensure that we are correctly following the true dynamics of a given trajectory over time in the β-FPUT model, because there are no non-trivial integrable models with which to compare. If we look again at Figure 2 in Section 3.2, these comments seem counter-intuitive, because the α-FPUT model seems to "slide" down from the metastable state to true equilibrium, whereas the β-FPUT model shows a sudden drop-off to the equilibrium value of the spectral entropy. However, in the β-FPUT model, using the binning procedure to establish the true lifetime of the metastable stage requires an enormous number of runs of very high accuracy to ensure that we are not observing a numerical artifact, i.e., an inaccurate calculation of the true trajectory. To determine a fit for t_m against $E\alpha^2$, shown in Figure 9 for the α-FPUT model, roughly 20,000 simulations were needed for each system size. With the aforementioned computational difficulties in the β-FPUT model, this computation is exceedingly expensive.

We will endeavor nonetheless to describe a possible procedure for calculating how t_m in the β-FPUT model scales as a function of $E\beta$, which can be undertaken in future works. As in the α-FPUT model, we run 100 trials with random initial phases and bin them into 10 bins with 10 trials each. We then define $\overline{\eta}_{25}$, the time-averaged entropy after 25 FPUT recurrence times, averaged across the 10 trials in a bin. The recurrence time is calculated following [24]. The number 25 was chosen because this allows the system to relax into its metastable state. After $\overline{\eta}_{25}$, then $\overline{\eta}$ tends to be nearly constant in time up until a critical time where it starts relaxing to equilibrium. We can find this critical time, t_m, by calculating the deviation of $\overline{\eta}$ from $\overline{\eta}_{25}$ by more than the bin standard deviation of the 10 trials. This procedure is visualized in Figure 13, where $\overline{\eta}_{25}$ is plotted in red, and the computed t_m is seen to line up with the end of the plateau in $\overline{\eta}$. Finally, averaging this result over the 10 bins gives a measurement of t_m with associated error bars.

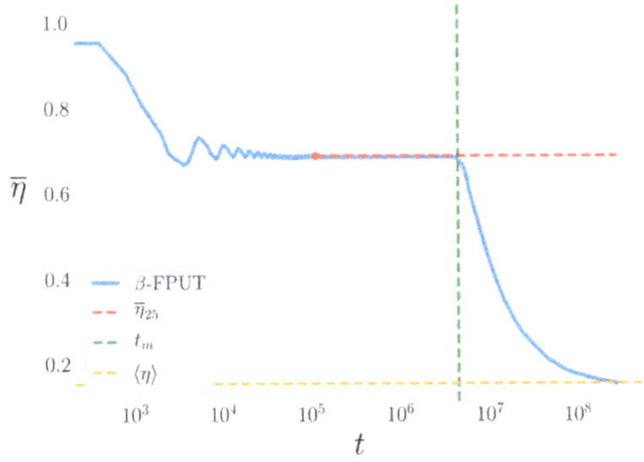

Figure 13. The entropy of the β-FPUT system for $N = 31$, $E\beta = 0.57$. The calculated value of $\overline{\eta}$ in the metastable state, $\overline{\eta}_{25}$, is shown starting at the time of the 25th recurrence as a red dashed line, and the calculated metastable lifetime, t_m, shown as a vertical green line. The ensemble average, $\langle\eta\rangle$, is shown in orange.

6. Conclusions

In this article, we have investigated the metastable state in the α-FPUT and β-FPUT models, both qualitatively and quantitatively. We began with a visualization of the metastable state using spectral entropy (η). This single-degree-of-freedom measure has the power to quantify the distance from the equipartition. This approach allowed us to follow G. Benettin [34] in viewing the α-FPUT model as a truncation of the integrable Toda model.

We next studied the strength of the recurrences in the α-FPUT model, following the results from S. Pace [24] on the β-FPUT model. This yielded the surprising result that the recurrence strength is a function only of the essential system parameter $R = (N+1)^{3/2}\sqrt{E\alpha^2}$ in the α-FPUT model, whereas the strength of the recurrences in the β-FPUT model scale with the energy $E\beta$ and not the essential system parameter $S = E\beta(N+1)$. The strength of the recurrences was shown to decay exponentially with R, independent of the system size N.

We devised a method to measure the lifetime of the metastable state t_m in the α-FPUT model. Our procedure involved averaging over the random initial phases in the P_1-Q_1 plane (at fixed energy $E_1 = \frac{1}{2}(P_1^2 + \omega_k^2 Q_1^2)$). This bin average provided a relevant length distance, the standard deviation, from which we could determine when the α-FPUT model trajectories break off from the entropy of the Toda model. Applying this procedure yielded Figure 9a, which shows t_m as a function of $E\alpha^2$ for different N. Surprisingly, as $E\alpha^2 \to 0$, the data for the different N collapse onto the same power law with the exponent -4.9.

We also explored the spectrum of the α-FPUT model, compared to that of the Toda model. We extended this analysis to relate to a method for an irreversible energy dissipation process suggested by Onorato et al. [41] (four-wave and six-wave resonances in wave turbulence theory). Our preliminary results confirm the presence of resonances in the spectrum, but it is not clear these are those proposed by Oronato et al. [41]. Future work is anticipated on this point.

Turning our attention to the β-FPUT model, we explored the two different signs of β, something which is not interesting in the α-FPUT model because the α-FPUT Hamiltonian is symmetric under $\alpha \to -\alpha$. The spectra for the β-FPUT model suggestively point to resonances leading to equipartition as well. We developed a procedure for calculating t_m in the β-FPUT model, a very different task than for the α-FPUT model because the β-FPUT model is not the truncation of a non-trivial integrable model.

Author Contributions: Conceptualization, K.A.R. and D.K.C.; Writing—original draft, K.A.R. and D.K.C. All authors have read and agreed to the published version of the manuscript.

Funding: This research received no external funding.

Institutional Review Board Statement: Not applicable.

Data Availability Statement: Data available on request due to large file sizes. Contact authors for specific data questions.

Acknowledgments: We are pleased to dedicate this paper to Giulio Casati on the occasion of his eightieth birthday. Casati's profound contributions to nonlinear dynamics, and in particular to chaos, both quantum and classical, have defined these fields for many years and will for many years to come. His more recent work on classical statistical methods [4] was an inspiration for this work and motivated our research. We are grateful to Sal Pace, Kristen Bestavros and Nachiket Karve for the useful discussions. We thank Boston University for the computer support through its Shared Computer Cluster. KAR is grateful to Boston University's Undergraduate Research Program and NSF-sponsored BU-PRO Research Experience for Undergraduates for funding his research.

Conflicts of Interest: The authors declare no conflict of interest.

References

1. Pathria, R.K.; Beale, P.D. *Statistical Mechanics*, 3rd ed.; Elsevier: Amsterdam, The Netherlands, 2011. [CrossRef]
2. Waterson, J.J. On the Physics of media that are composed of free and perfectly elastic molecules in a state of motion. *R. Soc.* **1851**, *5*, 604–604. [CrossRef]
3. Thornton, S.T.; Rex, A.F. *Modern Physics for Scientists and Engineers*, 4th ed.; Cengage Learning: Boston, MA, USA, 2013.
4. Wang, J.; Casati, G.; Benenti, G. Classical Physics and Blackbody Radiation. *Phys. Rev. Lett.* **2022**, *128*, 134101. [CrossRef]
5. Fermi, E.; Pasta, J.R.; Ulam, S.; Tsingou, M. Studies of nonlinear problems, I. In *Los Alamos Report*; Los Alamos National Lab. (LANL): Los Alamos, NM, USA, 1955.
6. Zabusky, N.J.; Kruskal, M.D. Interaction of "Solitons" in a Collisionless Plasma and the Recurrence of Initial States. *Phys. Rev. Lett.* **1965**, *15*, 240–243. [CrossRef]
7. Flach, S.; Ivanchenko, M.V.; Kanakov, O.I. q-Breathers and the Fermi-Pasta-Ulam Problem. *Phys. Rev. Lett.* **2005**, *95*, 064102. [CrossRef]
8. Campbell, D.K.; Rosenau, P.; Zaslavsky, G.M. Introduction: The Fermi–Pasta–Ulam problem—The first fifty years. *Chaos Interdiscip. J. Nonlinear Sci.* **2005**, *15*, 15101. [CrossRef] [PubMed]
9. Gallavotti, G. (Ed.) *The Fermi-Pasta-Ulam Problem: A Status Report*; Number 728 in Lecture Notes in Physics; Springer: New York, NY, USA, 2008.
10. Berman, G.P.; Izrailev, F.M. The Fermi–Pasta–Ulam problem: Fifty years of progress. *Chaos Interdiscip. J. Nonlinear Sci.* **2005**, *15*, 15104. [CrossRef] [PubMed]
11. Porter, M.; Zabusky, N.; Hu, B.; Campbell, D. Fermi, Pasta, Ulam and the Birth of Experimental Mathematics. *Am. Sci.* **2009**, *97*, 214. [CrossRef]
12. Weissert, T.P. *The Genesis of Simulation in Dynamics: Pursuing the Fermi-Pasta-Ulam Problem*; Springer: New York, NY, USA, 1997.
13. Benettin, G.; Carati, A.; Galgani, L.; Giorgilli, A. The fermi—pasta—ulam problem and the metastability perspective. In *The Fermi-Pasta-Ulam Problem*; Gallavotti, G., Ed.; Series Title: Lecture Notes in Physics; Springer: Berlin/Heidelberg, Germany, 2008; Volume 728, pp. 151–189. [CrossRef]
14. Toda, M. Studies of a non-linear lattice. *Phys. Rep.* **1975**, *18*, 1–123. [CrossRef]
15. Sholl, D. Modal coupling in one-dimensional anharmonic lattices. *Phys. Lett. A* **1990**, *149*, 253–257. [CrossRef]
16. Bivins, R.; Metropolis, N.; Pasta, J.R. Nonlinear coupled oscillators: Modal equation approach. *J. Comput. Phys.* **1973**, *12*, 65–87. [CrossRef]
17. Driscoll, C.F.; O'Neil, T.M. Explanation of Instabilities Observed on a Fermi-Pasta-Ulam Lattice. *Phys. Rev. Lett.* **1976**, *37*, 69–72. [CrossRef]
18. Pace, S.D.; Campbell, D.K. Behavior and breakdown of higher-order Fermi-Pasta-Ulam-Tsingou recurrences. *Chaos: Interdiscip. J. Nonlinear Sci.* **2019**, *29*, 23132. [CrossRef]
19. Driscoll, C.; O'Neil, T. Those ubiquitous, but oft unstable, lattice solitons. *Rocky Mt. J. Math.* **1978**, *8*, 211–225. [CrossRef]
20. Danieli, C.; Many Manda, B.; Mithun, T.; Skokos, C. Computational efficiency of numerical integration methods for the tangent dynamics of many-body Hamiltonian systems in one and two spatial dimensions. *Math. Eng.* **2019**, *1*, 447–488. [CrossRef]
21. Yoshida, H. Construction of higher order symplectic integrators. *Phys. Lett. A* **1990**, *150*, 262–268. [CrossRef]
22. Shannon, C.E. A Mathematical Theory of Communication. *Bell Syst. Tech. J.* **1948**, *27*, 45. [CrossRef]
23. Lin, C.; Goedde, C.; Lichter, S. Scaling of the recurrence time in the cubic Fermi-Pasta-Ulam lattice. *Phys. Lett. A* **1997**, *229*, 367–374. [CrossRef]
24. Pace, S.D.; Reiss, K.A.; Campbell, D.K. The β Fermi-Pasta-Ulam-Tsingou recurrence problem. *Chaos Interdiscip. J. Nonlinear Sci.* **2019**, *29*, 113107. [CrossRef]
25. Barreira, L. Poincaré recurrence: Old and new. In Proceedings of the XIVth International Congress on Mathematical Physics, Lisbon, Portugal, 28 July–2 August 2003; World Scientific: Lisbon, Portugal, 2006; pp. 415–422. [CrossRef]
26. Danshita, I.; Hipolito, R.; Oganesyan, V.; Polkovnikov, A. Quantum damping of Fermi-Pasta-Ulam revivals in ultracold Bose gases. *Prog. Theor. Exp. Phys.* **2014**, *2014*, 43103. [CrossRef]

27. Infeld, E. Quantitive Theory of the Fermi-Pasta-Ulam Recurrence in the Nonlinear Schrodinger Equation. *Phys. Rev. Lett.* **1981**, *47*, 717–718. [CrossRef]
28. Kopidakis, G.; Soukoulis, C.M.; Economou, E.N. Electron-phonon interactions and recurrence phenomena in one-dimensional systems. *Phys. Rev. B* **1994**, *49*, 7036–7039. [CrossRef] [PubMed]
29. Drago, G.; Ridella, S. Some more observations on the superperiod of the non-linear FPU system. *Phys. Lett. A* **1987**, *122*, 407–412. [CrossRef]
30. Flach, S.; Ivanchenko, M.V.; Kanakov, O.I. q-breathers in Fermi-Pasta-Ulam chains: Existence, localization, and stability. *Phys. Rev. E* **2006**, *73*, 36618. [CrossRef]
31. Christodoulidi, H.; Efthymiopoulos, C.; Bountis, T. Energy localization on q-tori, long-term stability, and the interpretation of Fermi-Pasta-Ulam recurrences. *Phys. Rev. E* **2010**, *81*, 016210. [CrossRef] [PubMed]
32. Zabusky, J. Nonlinear Lattice Dynamics and Energy Sharing. *J. Phys. Soc. Jpn.* **1969**, *26*, 7.
33. Fucito, F.; Marchesoni, F.; Marinari, E.; Parisi, G.; Peliti, L.; Ruffo, S.; Vulpiani, A. Approach to equilibrium in a chain of nonlinear oscillators. *J. Phys.* **1982**, *43*, 707–713. [CrossRef]
34. Benettin, G.; Ponno, A. Time-Scales to Equipartition in the Fermi–Pasta–Ulam Problem: Finite-Size Effects and Thermodynamic Limit. *J. Stat. Phys.* **2011**, *144*, 793–812. [CrossRef]
35. Benettin, G.; Christodoulidi, H.; Ponno, A. The Fermi-Pasta-Ulam Problem and Its Underlying Integrable Dynamics. *J. Stat. Phys.* **2013**, *152*, 195–212. [CrossRef]
36. Benettin, G.; Pasquali, S.; Ponno, A. The Fermi–Pasta–Ulam Problem and Its Underlying Integrable Dynamics: An Approach Through Lyapunov Exponents. *J. Stat. Phys.* **2018**, *171*, 521–542. [CrossRef]
37. Danieli, C.; Campbell, D.K.; Flach, S. Intermittent many-body dynamics at equilibrium. *Phys. Rev. E* **2017**, *95*, 60202. [CrossRef]
38. Lin, C.Y.; Cho, S.N.; Goedde, C.G.; Lichter, S. When Is a One-Dimensional Lattice Small? *Phys. Rev. Lett.* **1999**, *82*, 4. [CrossRef]
39. Ford, J. Equipartition of Energy for Nonlinear Systems. *J. Math. Phys.* **1961**, *2*, 387–393. [CrossRef]
40. Ponno, A.; Christodoulidi, H.; Skokos, C.; Flach, S. The two-stage dynamics in the Fermi-Pasta-Ulam problem: From regular to diffusive behavior. *Chaos Interdiscip. J. Nonlinear Sci.* **2011**, *21*, 43127. [CrossRef] [PubMed]
41. Onorato, M.; Vozella, L.; Proment, D.; Lvov, Y.V. Route to thermalization in the α-Fermi–Pasta–Ulam system. *Proc. Natl. Acad. Sci. USA* **2015**, *112*, 4208–4213. [CrossRef] [PubMed]

entropy

Article

A Physical Measure for Characterizing Crossover from Integrable to Chaotic Quantum Systems

Chenguang Y. Lyu [1] **and Wen-Ge Wang** [1,2,*]

[1] Department of Modern Physics, University of Science and Technology of China, Hefei 230026, China
[2] CAS Key Laboratory of Microscale Magnetic Resonance, University of Science and Technology of China, Hefei 230026, China
* Correspondence: wgwang@ustc.edu.cn

Abstract: In this paper, a quantity that describes a response of a system's eigenstates to a very small perturbation of physical relevance is studied as a measure for characterizing crossover from integrable to chaotic quantum systems. It is computed from the distribution of very small, rescaled components of perturbed eigenfunctions on the unperturbed basis. Physically, it gives a relative measure to prohibition of level transitions induced by the perturbation. Making use of this measure, numerical simulations in the so-called Lipkin-Meshkov-Glick model show in a clear way that the whole integrability-chaos transition region is divided into three subregions: a nearly integrable regime, a nearly chaotic regime, and a crossover regime.

Keywords: integrability-chaos cross-over; quantum chaos; eigenstate statistics; transition probability; Lipkin-Meshkov-Glick (LMG) model

Citation: Lyu, C.Y.; Wang, W.-g. A Physical Measure for Characterizing Crossover from Integrable to Chaotic Quantum Systems. *Entropy* **2023**, *25*, 366. https://doi.org/10.3390/e25020366

Academic Editor: Marko Robnik

Received: 28 November 2022
Revised: 1 February 2023
Accepted: 3 February 2023
Published: 17 February 2023

1. Introduction

In the past near half century, a large number of topics have been studied in the field of quantum chaos and a huge amount of knowledge about properties of quantum chaotic systems has been accumulated (see, e.g., [1,2]). However, this field is far from being fully explored. For example, although the spectral statistics of quantum chaotic systems have been studied well [3–9], not so much is known about wave functions, particularly about statistical properties of energy eigenfunctions (EFs) as expansions of energy eigenstates in certain bases.

As is well known, in classical mechanics, chaos refers to trajectory sensitivity to initial conditions, characterized by positive (maximum) Lyapunov exponents.The percentage of phase space, which is occupied by chaotic trajectories, supplies a useful quantitative measure in the study of crossover from integrability to chaos, at least for systems in a two-dimensional configuration space. However, the story is much more complicated with quantum systems and the route (from integrability) to quantum chaos is far from being fully understood.

Quantitative characterization of crossover from quantum integrability to quantum chaos is a topic that is of importance in many aspects. Indeed, without a deep understanding of mechanisms that may lead to a break down of integrability, it is unimaginable that a complete understanding of quantum chaos may be achieved; and, it is usually believed that quantum chaos may play a crucial role in establishing a sound foundation for quantum statistical mechanics and for understanding thermalization processes [10–16]. Many quantities have been studied for the characterization; loosely speaking, they may be classified into three classes: spectral statistics (see, e.g., Refs. [2–4,17–21]), statistical properties of EFs (see, e.g., Refs. [14,22–33]), and time evolution properties (see, e.g., Refs. [34–45]). Unfortunately, none of them is as satisfactory as the above-mentioned measure in the classical case.

Regarding spectral statistics, the nearest-level-spacing distribution $P(s)$ is often studied. Its shape is close to the Poison distribution e^{-s} in generic integrable systems [17], while, is close to the Wigner-Dyson distribution $P_W(s) = \frac{\pi}{2}s\exp(-\frac{\pi}{4}s^2)$ in quantum chaotic systems, the latter of which is almost identical to the prediction of random matrix theory (RMT) (for systems with the time-reversal symmetry) [2–4]. Among possible interpolations between Poison and Wigner-Dyson distributions [18–21], the most often studied one is the so-called Brody distribution [18], characterized by a Brody parameter β, with $\beta = 0$ for integrable systems and $\beta = 1$ for chaotic ones. Unfortunately, physical meaning of the Brody parameter is still unclear; in other words, it is regarded merely as a fitting parameter.

Properties of EFs may also be used for characterizing the crossover from quantum integrability to quantum chaos. To compute such properties, usually one needs to take a specific basis, which may be chosen for some physical reason or based on some mathematical consideration [14,25–31]. Recently, two methods were proposed [32,33], which employ intrinsic bases of the systems studied, namely, the eigenbases of their Hamiltonians. The first method makes use of the so-called adiabatic gauge potential (AGP) [32], describing the rate of deformation of eigenstates under infinitesimal perturbation. It was proposed that scaling behavior of the norm of AGP with system size may be used as an indicator for the integrability-chaos crossover. This indicator is basically qualitative and its validity is not completely clear.

The second method also studies the response of eigenstates to small perturbation, but, by making use of the distribution of rescaled components of perturbed states on the unperturbed basis [33]. This distribution has a Gaussian form in quantum chaotic systems, a phenomenon which may be traced back to the so-called Berry's conjecture [22–24,30]; while, it deviates notably from the Gaussian form in quantum integrable systems. This method uses the difference between the distribution of rescaled components and the Gaussian distribution for the purpose of characterizing the integrability-chaos crossover. It suffers from two shortcomings: (i) Although the above-mentioned difference gives a quantitative measure to the distance to chaos, it is irrelevant to integrability; and, (ii) its physical meaning is not directly clear.

In this paper, we go further along the direction of the second method discussed above and make improvements. Specifically, we are to show that the value of the distribution of rescaled components at the origin point is a good candidate, i.e., it is of physical relevance and supplies a measure for characterizing the whole process from integrability to chaos.

The paper is organized as follows. A preliminary discussion is found in Section 2, about the class of systems to be studied, requirements of the type of perturbation to be employed, and some details of the two previous works mentioned above. The proposed measure for integrability-chaos crossover is discussed in Section 3 and its illustration in a model is given in Section 4. Finally, conclusions and discussions are given in Section 5.

2. Preliminary Discussions

In this section, we first discuss basic properties of the systems to be studied (Section 2.1), and then discuss properties of the perturbation to be considered (Section 2.2). Finally, we briefly recall basic contents of the two methods mentioned above, which use intrinsic bases for characterizing crossover from integrable to chaotic quantum systems (Section 2.3).

2.1. Perturbed and Unperturbed Systems

We study the response of Hamiltonian eigenstates to small perturbation. The unperturbed Hamiltonian is denoted by $H^0(\lambda)$, with a running parameter λ, such that it is integrable at $\lambda = 0$ and is chaotic at λ in some region around 1. We use $|k(\lambda)\rangle$ to indicate eigenstates of $H^0(\lambda)$, with eigenenergies $E_k^0(\lambda)$ in the increasing energy order,

$$H^0(\lambda)|k(\lambda)\rangle = E_k^0(\lambda)|k(\lambda)\rangle. \tag{1}$$

Sometimes, when there is no risk of causing confusion, for brevity, the λ-dependence of H^0 and of its eigenstates is not written explicitly.

The perturbed Hamiltonian is written as

$$H = H^0(\lambda) + \epsilon V, \tag{2}$$

where λ-dependence is not written explicitly, where ϵ is a very small parameter and V represents a perturbation operator. Eigenstates of H are denoted by $|\alpha\rangle$,

$$H|\alpha\rangle = E_\alpha |\alpha\rangle, \tag{3}$$

with the eigenenergies E_α also in the increasing energy order. Components of the EF of a perturbed state $|\alpha\rangle$ on the unperturbed basis are written as

$$C_{\alpha k} = \langle k|\alpha\rangle. \tag{4}$$

We assume that the Hilbert space is sufficiently large for meaningful statistical analysis of properties of EFs. For the simplicity in discussion, we consider only systems with the time-reversal symmetry, such that the components $C_{\alpha k}$ are real numbers. (It is straightforward to generalize results to be given below to the generic case without the time-reversal symmetry.)

We are to discuss properties of the distribution of rescaled $C_{\alpha k}$, denoted by $\widetilde{C}_{\alpha k}$,

$$\widetilde{C}_{\alpha k} := \frac{C_{\alpha k}}{\sqrt{\langle |C_{\alpha k}|^2 \rangle}}, \tag{5}$$

where $\langle |C_{\alpha k}|^2 \rangle$ indicates the average shape of EFs. We indicate this distribution by $g(\widetilde{C})$. When computing $\langle |C_{\alpha k}|^2 \rangle$, special attention should be paid to big components of the EFs. For example, for a chaotic system $H^0(\lambda)$, at a sufficiently small parameter ϵ, usually each perturbed state $|\alpha\rangle$ has one big component of $C_{\alpha k}$; its value is close to 1, while other components of $C_{\alpha k}$ are much smaller, proportional to ϵ or smaller as predicted by the perturbation theory. In this case, clearly, if the biggest component is included, it usually makes $\langle |C_{\alpha k}|^2 \rangle$ not smooth. Hence, in the computation of $\langle |C_{\alpha k}|^2 \rangle$, the biggest components of $C_{\alpha k}$ should not be included and, consistently, they are not included in the distribution $g(\widetilde{C})$, either.

Moreover, an average may be taken over those perturbed states that lie within a narrow energy shell, denoted by Γ_α, which is centered at E_α and has a small width δe. To summarize, we write

$$\langle |C_{\alpha k}|^2 \rangle = \frac{1}{\sum'_{E_\beta \in \Gamma_\alpha} 1} \sum_{E_\beta \in \Gamma_\alpha}{}' |C_{\beta k}|^2, \tag{6}$$

where

$$\Gamma_\alpha = [E_\alpha - \delta e/2, E_\alpha + \delta e/2] \tag{7}$$

and the prime over $\sum$ means that big components are excluded. (see Section 3.2 for further discussions on big components to be excluded in computations performed in integrable systems)

2.2. *Properties of the Perturbation V*

Physically, a prominent difference between a quantum integrable system and a quantum chaotic system lies in that the former has at least two independent good quantum numbers, while, the latter has only one which is related to the Hamiltonian. Due to this difference, at least to a certain type of perturbation, the response of an integrable system should be different from that of a chaotic system. Hence, in principle, response to perturbation may be used for the purpose of characterizing the integrability-chaos crossover. In this section, we discuss a type of perturbation V that may be used for this purpose.

It may be useful to provide a little discussion, in comparison with a purely mathematical viewpoint by which an operator V on an arbitrary basis is represented by a matrix with only one restriction — hermiticity. In fact, as is known, if no further restriction is imposed to the form of the matrix of V, there may exist some matrix of V, for which the components $C_{\alpha k}$ in an integrable system may show behaviors qualitatively similar to those in a chaotic system (see, e.g., Ref. [46]). Hence, for the purpose of "detecting" the difference between integrable and chaotic systems, a purely mathematical viewpoint is not enough. In other words, physical considerations should be taken into account, which may give certain restriction to the perturbation V.

To further analyze the above-discussed point, we discuss from the perspective of dynamic groups underlying models studied. Here, a group is called a dynamic group underlying a model, if the model Hamiltonian is a function of generators of the group. (A dynamic group is not necessarily a symmetry group. In other words, it is unnecessary for the model Hamiltonian to possess any symmetry related to the dynamic group.) We note that most physical models of realistic interest have some underlying dynamic Lie groups. Usually, generators of the dynamic group may take the form of raising and lowering operators, which we indicate as $K_\eta^\dagger$ and K_η, respectively, with a label η. For example, for an oscillator in a one-dimensional configuration space, the so-called Weyl-Heisenberg group is the underlying dynamic group, where generators as raising and lowering operators are given by well-known combinations of the position x and momentum p.

On the integrable side of $H^0(\lambda)$ with $\lambda = 0$, two different cases should be treated differently: interacting and noninteracting integrable systems. Let us first discuss a noninteracting integrable system, whose Hamiltonian $H^0(0)$ depends separately on generators of different degrees of freedom. Since there is no interplay between different degrees of freedom, usually the integrable states $|k(0)\rangle$ may be obtained by multiplying raising operators on certain "vacuum state" denoted by $|0\rangle$, i.e.,

$$|k(0)\rangle = \prod_\eta \left(K_\eta^\dagger\right)^{m_\eta} |0\rangle, \tag{8}$$

where m_η are nonnegative integers. Usually, the integers m_η are, or are related to, good quantum numbers in the state $|k(0)\rangle$. On the basis of $\{|k(0)\rangle\}$, elements of V are written as

$$\langle k'(0)|V|k(0)\rangle = \prod_{\eta\eta'} \langle 0| \left(K_{\eta'}\right)^{m'_{\eta'}} V \left(K_\eta^\dagger\right)^{m_\eta} |0\rangle. \tag{9}$$

One useful observation is that, for many types of perturbation V of physical interest, the matrix of V with elements given in Equation (9) has an interesting feature; i.e., it is sparse in the sense of possessing many zero elements of $\langle k'(0)|V|k(0)\rangle$. As one example, one may consider perturbations that are controllable and describable in laboratories. Such a perturbation V is usually represented by some simple function of $\{K_\eta^\dagger, K_\eta\}$ (at least not a complicated function) and, as a result, Equation (9) predicts a sparse matrix of V. As another example, one may consider a perturbation V, which may bring a limited change to the good quantum numbers m_η for most of the states $|k(0)\rangle$. In this case, one also finds a sparse structure of the matrix of $\langle k'(0)|V|k(0)\rangle$.

Then, we discuss an interacting integrable system, for which generators of different degrees of freedom interplay in the Hamiltonian $H^0(0)$. In such system, the states $|k(0)\rangle$ are superpositions of product terms as given on the right-hand side of Equation (8). In this case, the simplicity of the function $V(K_\eta^\dagger, K_\eta)$ does not guarantee a sparse structure of the matrix of $\langle k'(0)|V|k(0)\rangle$. While, for a perturbation V, which may bring a limited change to good quantum numbers of the states $|k(0)\rangle$ (usually not the integers m_η), one still finds a sparse structure of the matrix of $\langle k'(0)|V|k(0)\rangle$. (Mathematically, the simplest example of this type of operator V is written as $V = |k_1(0)\rangle\langle k_2(0)| + |k_2(0)\rangle\langle k_1(0)|$ with some fixed values of k_1 and k_2. Usually, to construct such an operator of physical interest is a model-dependent matter.)

On the chaotic side of $H^0(\lambda)$ with λ around 1, one finds a different story. To be specific in this discussion, let us consider a system that possesses a classical counterpart. To obtain an estimate, one may make use of Berry's conjecture as a semiclassical prediction [22–24,30].

We first discuss $H^0(\lambda)$ of $\lambda \sim 1$, with respect to a noninteracting integrable system $H^0(0)$. Without loss of generality, we assume that the eigenstates $|k(0)\rangle$ in Equation (8) are also eigenstates of action. According to a version of Berry's conjecture given on an action basis [30], the components $\langle k(0)|k(\lambda)\rangle$ have the following expression,

$$\langle k(0)|k(\lambda)\rangle \propto \Omega_{k(0)k(\lambda)} R_{k(0)k(\lambda)}, \tag{10}$$

where $R_{k(0)k(\lambda)}$ are Gaussian random variables (with mean zero and a normal distribution) and $\Omega_{k(0)k(\lambda)}$ indicates the so-called classical analog of the average shape of eigenfunctions, which is given by the overlap of the classical energy surface of $E_k^0(\lambda)$ and a classical torus corresponding to the action given by m_η. On the basis of $|k(\lambda)\rangle$, elements of V are written as

$$\begin{aligned} &\langle k'(\lambda)|V|k(\lambda)\rangle \\ &= \sum_{k(0),k'(0)} \langle k'(\lambda)|k'(0)\rangle\langle k'(0)|V|k(0)\rangle\langle k(0)|k(\lambda)\rangle \end{aligned} \tag{11}$$

Making use of Equations (9) and (10), one finds that, at least for $E_{k'}^0(\lambda)$ not far from $E_k^0(\lambda)$, the elements $\langle k'(\lambda)|V|k(\lambda)\rangle$ are typically nonzero. Hence, the matrix of V on the chaotic basis $\{|k(\lambda)\rangle\}$ has a structure, which is qualitatively different from the previously discussed sparse structure of V on the integrable basis $\{|k(0)\rangle\}$.

Next, we discuss $H^0(\lambda)$ of $\lambda \sim 1$, with respect to an interacting integrable system. In this case, the integrable states $|k(0)\rangle$ are superpositions of the action states. As a consequence, the components $\langle k(0)|k(\lambda)\rangle$ have an expression, which is more complicated than the right-hand side of Equation (10) for the noninteracting case. This implies that the elements $\langle k'(\lambda)|V|k(\lambda)\rangle$ should be also typically nonzero at least for $E_{k'}^0(\lambda)$ not far from $E_k^0(\lambda)$. Moreover, this is also true, even for a perturbation V that brings a limited change to good quantum numbers of the states $|k(0)\rangle$. Hence, as well, the matrix of V on the chaotic basis $\{|k(\lambda)\rangle\}$ has a structure, which is qualitatively different from the previously discussed sparse structure of V on the integrable basis $\{|k(0)\rangle\}$.

To summarize, for a certain perturbation V, its matrix has a sparse structure on an integrable basis, while it does not have a similar structure on a chaotic basis. In Section 3, we are to focus on this type of perturbation V and use it to propose a quantity for characterizing integrability-chaos crossover.

2.3. Two Previously Studied Methods

In this section, we briefly recall basic contents of the two previously studied methods mentioned in the introduction, in which intrinsic bases are employed for characterizing integrability-chaos crossover.

The first method makes use of the AGP, which describes variation of eigenstates under infinitesimal change of the parameter λ. More exactly, denoted by $\mathcal{A}_\lambda$, the AGP generates an adiabatic evolution of the eigenstates,

$$\mathcal{A}_\lambda|k(\lambda)\rangle = i\partial_\lambda|k(\lambda)\rangle. \tag{12}$$

Its offdiagonal elements satisfy

$$\langle k|\mathcal{A}_\lambda|l\rangle = -\frac{i}{E_k^0(\lambda) - E_l^0(\lambda)}\langle k|\partial_\lambda H^0(\lambda)|l\rangle \tag{13}$$

with $k \neq l$; while, its diagonal elements may be set zero due to the freedom in choosing phases of the eigenstates, i.e., $\langle k|\mathcal{A}_\lambda|k\rangle = 0$ for all $|k\rangle$. (In Pandey et al. [32], to avoid a problem that may be caused by degeneracy of the spectrum of $H^0(\lambda)$, on the right-hand side of Equation (13), the term $\frac{1}{E_k^0 - E_l^0}$ is replaced by $\frac{E_k^0 - E_l^0}{(E_k^0 - E_l^0)^2 + \mu^2}$ with some small energy cutoff μ.) The Frobenius norm, also called Hilbert–Schmidt norm, of the AGP operator is written as (This norm is equal to the sum of the so-called fidelity susceptibility of the eigenstates $|k(\lambda)\rangle$ [47,48].)

$$\|\mathcal{A}_\lambda\|^2 = \sum_{k,l} |\langle k|\mathcal{A}_\lambda|l\rangle|^2. \tag{14}$$

It was proposed in Pandey et al. [32] that exponential scaling behavior of the AGP norm with respect to the particle number N of a many-body quantum system may be used as an indicator of quantum chaos, with polynomial behavior expected for integrable systems. For systems satisfying the so-called eigenstate thermalization hypothesis (ETH) (see Equation (24) to be given below), it is not difficult to check that the AGP norm indeed scales exponentially with the system size. However, the reverse statement is not supported by any analytical analysis and this makes the above proposal questionable as a criterion for quantum chaos. In fact, according to numerical simulations given in Pandey et al. [32] in an XXZ chain, the integrability-chaos crossover region predicted by the above-discussed proposal is much lower than that obtained by an ordinary method of employing spectral statistics, by at least one order of magnitude.

In the second method, the difference between the distribution $g(\widetilde{C})$ and the Gaussian distribution is used as a measure for the distance to quantum chaos [33]. Quantitatively, the difference is written as $\Delta_{\rm EF} = \int \left| I(\widetilde{C}) - I_G(\widetilde{C}) \right| d\widetilde{C}$, where $I(\widetilde{C})$ is the cumulative distribution of $g(\widetilde{C})$ and I_G is the cumulative distribution for the Gaussian distribution of

$$g_G(\widetilde{C}) = \frac{1}{\sqrt{2\pi}} \exp(-\widetilde{C}^2/2). \tag{15}$$

This difference is expected to be small in quantum chaotic systems, and large in integrable systems (see the next section for detailed discussions).

Numerical simulations performed in a three-orbital Lipkin-Meshkov-Glick (LMG) model [49] show that the quantity $\Delta_{\rm EF}$ behaves consistently with Δ_W in the regime of integrability-chaos crossover [33]. Here, Δ_W indicates the difference between the nearest-level-spacing distribution $P(s)$ and the Wigner distribution $P_W(s)$, as a measure for the distance to chaos; more precisely,

$$\Delta_W = \int |I(s) - I_W(s)| ds, \tag{16}$$

where "I" also indicate the corresponding cumulative distributions. It was found that integrability is manifested by large values of $\Delta_{\rm EF}$, which are due to high peaks of $g(\widetilde{C})$ at $\widetilde{C} = 0$.

Finally, let us give a brief comparison of the two methods discussed above. The physics lying behind them are in fact related. This point is seen quite clearly, in the special case that $H^0(\lambda) = H^0(0) + \lambda V$. In this case, $\partial_\lambda H^0(\lambda) = V$ and the AGP elements in Equation (13) are directly related to the components $C_{\alpha k}$ divided by c, which are predicted by a first-order perturbation theory [see Equation (19) to be given below].

Meanwhile, the two methods have important differences. First, the second method in fact employs the rescaled components $\widetilde{C}_{\alpha k}$ in Equation (5). Without the rescaling procedure, the distribution of components would not be close to a Gaussian form in quantum chaotic systems. Second, it is not the AGP norm itself that is employed as a measure for integrability-chaos crossover, but it is the AGP scaling behavior with respect to system size. This implies that the first method does not supply a measure for characterizing the whole crossover region, though it might be used for detecting some "crossover point", if in existence.

3. $g(0)$ as a Crossover Measure

In this section, going further along the direction of the second method discussed above, we propose that $g(\widetilde{C} = 0)$ may be used as a measure for characterizing integrability-chaos crossover. Specifically, we discuss its physical meaning in Section 3.1 and discuss its detailed properties in Section 3.2.

3.1. A Physical Meaning of $g(0)$

We first consider a generic case, in which the unperturbed Hamiltonian $H^0(\lambda)$ possesses a nondegenerate spectrum. In this case, under a sufficiently small ϵ, the perturbed states $|\alpha\rangle$ may be approximated by their first-order perturbation expansions. For a given state $|\alpha\rangle$, we use k_α to indicate the label k for which E_k^0 is the closest to E_α. Clearly, $C_{\alpha k_\alpha} \simeq 1$. Then, one writes

$$|\alpha\rangle \simeq |\alpha^{(0)}\rangle + |\alpha^{(1)}\rangle, \tag{17}$$

where $|\alpha^{(0)}\rangle = |k_\alpha\rangle$ and

$$|\alpha^{(1)}\rangle = \sum_{k \neq k_\alpha} \frac{\epsilon V_{kk_\alpha}}{E_{k_\alpha}^0 - E_k^0} |k\rangle. \tag{18}$$

From Equation (18), one sees that

$$C_{\alpha k} \simeq \frac{\epsilon V_{kk_\alpha}}{E_{k_\alpha}^0 - E_k^0} \qquad \text{for } k \neq k_\alpha. \tag{19}$$

To see a physical meaning of $g(0)$, let us discuss the probability of transition from an arbitrary state $|k_0\rangle$ (as an initial state $|\psi(t=0)\rangle$) to an arbitrary state $|k\rangle$ ($k \neq k_0$) under a perturbation ϵV, which is given by $|\langle k|\psi(t)\rangle|^2$. It is easy to find that

$$\langle k|\psi(t)\rangle = \langle k|e^{-iHt}|k_0\rangle = \sum_\alpha C_{\alpha k_0} C_{\alpha k} e^{-iE_\alpha t}. \tag{20}$$

Since $C_{\alpha k} \simeq 1$ for $k = k_\alpha$, one further obtains that

$$\langle k|\psi(t)\rangle \simeq C_{\alpha_1 k_0} e^{-iE_{\alpha_1} t} + C_{\alpha_2 k} e^{-iE_{\alpha_2} t}, \tag{21}$$

where α_1 is determined by the relation of $k_{\alpha_1} = k$ and α_2 by the relation of $k_{\alpha_2} = k_0$. According to Equation (21), the smallness of the components $C_{\alpha_1 k_0}$ and $C_{\alpha_2 k}$, if compared with other unperturbed states, has a clear physical meaning; that is, it implies relative prohibition of the quantum transition from $|k_0\rangle$ to $|k\rangle$. However, a problem is met with quantitative characterization for this, because all the components $C_{\alpha k}$ of $k \neq k_\alpha$ go to zero in the limit of $\epsilon \to 0$.

To circumvent the above-discussed problem, instead, we consider the rescaled components $\widetilde{C}_{\alpha k}$ in Equation (5), which remain finite in the limit of $\epsilon \to 0$. Making use of perturbation theory, it is not difficult to find that the two components $C_{\alpha_1 k_0}$ and $C_{\alpha_2 k}$, which are related to the same unperturbed energy difference of $(E_k^0 - E_{k_0}^0)$, should have similar absolute values on average, i.e., $\langle |C_{\alpha_1 k_0}|^2 \rangle \simeq \langle |C_{\alpha_2 k}|^2 \rangle$. Then, Equation (21) is written as

$$\frac{\langle k|\psi(t)\rangle}{\sqrt{\langle |C_{\alpha_1 k_0}|^2 \rangle}} \simeq \widetilde{C}_{\alpha_1 k_0} e^{-iE_{\alpha_1} t} + \widetilde{C}_{\alpha_2 k} e^{-iE_{\alpha_2} t}. \tag{22}$$

Thus, relative smallness of the rescaled components $\widetilde{C}_{\alpha_1 k_0}$ and $\widetilde{C}_{\alpha_2 k}$ implies relative smallness of the transition probability divided by the average shape of EFs.

From the above discussions, one sees that high population of small rescaled components usually implies strong prohibition of quantum transition. Quantitatively, the population is characterized by the value of the distribution $g(\widetilde{C})$ at $\widetilde{C} = 0$, namely, by $g(0)$.

Next, we discuss the special case, in which $H^0(\lambda)$ possesses a degenerate spectrum. In this case, for a given unperturbed state $|k\rangle$, the components $C_{\alpha k}$ of those perturbed states $|\alpha\rangle$, whose energies E_α are close to E_k^0, may be large. We use S_k to denote the set of the indices α of these perturbed states. For a nondegenerate level E_k^0, the set S_k in fact contains one label α only, such as in the case of nondegenerate spectrum discussed above; while, for a degenerate level, the number of elements of S_k is usually equal to the degeneracy. In most cases, k and k_0 do not belong to a same degenerate level (Study of the case, in which k and k_0 belong to a same degenerate level, is hard and is beyond the scope of perturbative analysis.) and Equation (20) gives that

$$\langle k|\psi(t)\rangle \simeq \sum_{\alpha_1 \in S_k} C_{\alpha_1 k_0} e^{-iE_{\alpha_1} t} + \sum_{\alpha_2 \in S_{k_0}} C_{\alpha_2 k} e^{-iE_{\alpha_2} t}. \quad (23)$$

Again, relative smallness of $C_{\alpha_1 k_0}$ and $C_{\alpha_2 k}$ implies relative prohibition of the transition of $|k_0\rangle \to |k\rangle$. By definition, one may require that the average shape $\langle |C_{\alpha k'}|^2 \rangle$ should change slowly (or remains a constant) within one degenerate subspace. Then, such as in the nondegenerate case discussed above, one still finds that $\langle |C_{\alpha_1 k_0}|^2 \rangle \simeq \langle |C_{\alpha_2 k}|^2 \rangle$ and obtains an equation similar to Equation (23).

When the unperturbed spectrum has a weak degeneracy such that each set S_k contains a small number of elements, the effect induced by the degeneracy is small. Then, according to discussions given above, still, relative smallness of the rescaled components usually implies relative smallness of the transition probability divided by the average shape of EFs. Furthermore, as a result, high population of small rescaled components is related to strong prohibition of quantum transition.

In the rare case that the degeneracy is very high such that the set S_k contains a large number of elements, the situation is more complicated in the quantitative aspect. In fact, the perturbation is extremely strong within each degenerate subspace and this may suppress the number of very small components $\widetilde{C}_{\alpha k}$. Nevertheless, qualitatively, high population of small rescaled components is still related to strong prohibition of quantum transition. (Here, we do not discuss the very special case, in which the spectrum of $H^0(0)$ is completely degenerate. In this case, it is unnecessary for $g(\widetilde{C})$ to be extraordinarily large at $\widetilde{C} = 0$.)

Summarizing the above discussions, we obtain the following result, i.e., the value of $g(0)$ gives a relative measure to the extent of prohibition of quantum transition induced by the perturbation ϵV. Usually, larger value of $g(0)$ may be related to stronger prohibition; while, this expectation may be valid only qualitatively for highly degenerate levels [50]. (As is known, certain elements of the AGP may also be related to transition probability amplitudes. However, the norm of AGP does not show this feature in a direct way.)

3.2. $g(0)$ *in Systems from Integrable and Chaotic*

In this section, we discuss properties of $g(0)$ in chaotic and integrable systems.

First, we discuss $H^0(\lambda)$ as a quantum chaotic system, which satisfies the ETH [13,24,51–56]. (see Ref. [57] for a semiclassical proof of ETH.) Technically, the ETH is written as an ansatz for a special structure of the matrix of an observable O on the energy basis, i.e.,

$$O_{kk'} = f(e_k)\delta_{kk'} + g(e_k, e_{k'}) r_{kk'}, \quad (24)$$

where $f(e)$ and $g(e, e')$ are smooth functions of their variables, $r_{kk'} = r^*_{k'k}$ are independent random variables with normal distribution (zero mean and unit variance), and g^{-2} scales as the density of states ρ_{dos} with the system size. Under a perturbation V that satisfies the ETH ansatz in Equation (24) and under a very small ϵ such that Equation (19) is valid, it is easy to compute the average shape of EFs from Equation (19), obtaining that

$$\langle |C_{\alpha k}|^2 \rangle \simeq \frac{\epsilon\, g(e_k, e_{k_\alpha})}{|E^0_{k_\alpha} - E^0_k|}. \quad (25)$$

Then, noting the randomness of $r_{kk'}$, one sees that the distribution of these rescaled components should have a Gaussian form, as illustrated numerically in Ref. [33]. Hence, $g(0) = 1/\sqrt{2\pi} \approx 0.4$ [cf. Equation (15)] in quantum chaotic systems.

Next, we discuss the integrable case, namely $H^0(0)$. As discussed previously, we consider only those perturbations V, whose matrix $[V_{kk'}]$ in the integrable basis $\{|k(0)\rangle\}$ has a sparse structure, with many zero elements. To figure out properties of $g(0)$ under a very small ϵ, let us consider an arbitrary perturbed state $|\alpha\rangle$. We need to discuss two cases separately, as done below, depending on whether the unperturbed level, which is the closest to E_α, is degenerate or not:

1. In the case that the level is nondegenerate, Equation (19) is valid. Then, clearly, the sparse structure of $[V_{kk'}]$ implies that the distribution $g(\widetilde{C})$ should have a high peak at $\widetilde{C} = 0$, with $g(0) \gg 1$.
2. In the case of a degenerate level, let us use $|k_\alpha^\eta\rangle$ (with $\eta = 1, 2, \ldots$) to indicate those states $|k\rangle$ that correspond to this unperturbed level. It is straightforward to generalize the perturbative treatment given in the above section and obtain an equation similar to Equation (19), but, for $k \neq k_\alpha^\eta$ $(\forall \eta)$. Then, one reaches a similar conclusion that the distribution $g(\widetilde{C})$ should have a high peak at $\widetilde{C} = 0$, usually with $g(0) \gg 1$.

It is useful to give further discussions on the computation of the average shape $\langle |C_{\alpha k}|^2 \rangle$ in the degenerate case. There are two methods that may be adopted. The first method is to exclude all the components $C_{\alpha k}$ for the degenerate subspace, i.e., for $k = k_\alpha^\eta$. This method, though favorable analytically, may face the following subtle problem in numerical simulations, which is induced by the fact that the parameter ϵ always has some lower bound in numerical computations, i.e., there may exist nearly degenerate levels, which are effectively degenerate in numerical simulations with a finite ϵ.

To avoid the above-discussed subtleness in numerical simulations, one may adopt a second method, though it is not so attractive analytically. This method excludes only those components that satisfy $|C_{\alpha k}|^2 > \Lambda_b$, where Λ_b is some (adjusting) parameter that is introduced for obtaining a smooth average shape of EFs. (Meaningful results should be insensitive to the exact value of Λ_b within a reasonable region. We checked this point in our numerical simulations to be discussed later.) When the degeneracy is low, there is usually only a small difference between the values of $g(0)$ obtained by the two methods. While, the difference may be large when the degeneracy is high, with the former value usually being larger than the latter.

In fact, the above-discussed differences corresponding to different situations with degeneracy is not crucial for our purpose here. What is of practical importance is that, in all the cases, the distributions $g(\widetilde{C})$ show high peaks at $\widetilde{C} = 0$, which is the characteristic feature of integrability. Moreover, we recall that, as discussed in the above section, high degeneracy may increase the difficulty in quantitative evaluation of $g(0)$. For these reasons, in characterizing integrability-chaos crossover, the quantity $g^{-1}(0)$ may be practically better than $g(0)$. In fact, most of the indefiniteness, if not all of them, are restrained in $g^{-1}(0)$. The quantity $g^{-1}(0)$ is expected to satisfy $g^{-1}(0) \simeq \sqrt{2\pi}$ for chaotic systems, while, is small for integrable systems. (The exact value of the smallness of $g^{-1}(0)$ is not so important in distinguishing between integrability and chaos, though it may be informative for other purposes).

Based on the above discussions, loosely speaking, one may expect the following picture for behaviors of the quantity $g^{-1}(0)$ in an integrability-chaos crossover. At $\lambda = 0$, one expects that $g^{-1}(0) \approx 0$, except for the case of very high degeneracy. With λ increasing from 0, the EFs of $|k(\lambda)\rangle$ on the integrable basis $\{|k(0)\rangle\}$ become more and more complicated. As a consequence, some previously very small elements $V_{kk'}$ become nonnegligible and, then, as predicted by Equation (19), more and more rescaled components may obtain nonnegligible values. This implies that, with increasing λ, $g^{-1}(0)$ should increase, until it reaches a saturation value of $\sqrt{2\pi}$, which corresponds to quantum chaos.

4. Numerical Simulations

In this section, we present numerical simulations that have been performed in a three-orbital Lipkin-Meshkov-Glick (LMG) model [49,58], illustrating $g(0)$ as a measure for characterizing the integrability-chaos crossover.

4.1. The model

The LMG model consists of Ω fermions, which may occupy three single-particle energy states labeled by $s = 0, 1, 2$. We use η_s to denote the energy of the s-th single-particle state and, for brevity, we set $\eta_0 = 0$. In this paper, we are interested in the collective motion of this model, for which the dimension of the Hilbert space is $\frac{1}{2}(\Omega + 1)(\Omega + 2)$. The classical counterpart of this collective motion has a two-dimensional configuration space.

The Hamiltonian of the model is constructed from the following operators K_{rs},

$$K_{rs} = \sum_{\gamma=1}^{\Omega} a_{r\gamma}^{\dagger} a_{s\gamma}, \quad r, s = 0, 1, 2, \tag{26}$$

where $a_{r\gamma}^{\dagger}$ and $a_{r\gamma}$ are fermionic creation and annihilation operators obeying the usual anti-commutation relations. The operators K_{ss} in fact represent the particle-number operators for single-particle states s, while, K_{sr} with $s \neq r$ are level raising/lowering operators. The Hamiltonian is written as

$$H^0(\lambda) = \eta_1 K_{11} + \eta_2 K_{22} + \lambda V, \tag{27}$$

where

$$V = \sum_{t=1}^{4} \mu_t V^{(t)}. \tag{28}$$

In Equation (28), μ_t are parameters and

$$V^{(1)} = K_{10}K_{10} + K_{01}K_{01}, \tag{29a}$$
$$V^{(2)} = K_{20}K_{20} + K_{02}K_{02}, \tag{29b}$$
$$V^{(3)} = K_{21}K_{20} + K_{02}K_{12}, \tag{29c}$$
$$V^{(4)} = K_{12}K_{10} + K_{01}K_{21}. \tag{29d}$$

For symmetric states in the collective motion, the operators K_{rs} may be written in terms of bosonic creation and annihilation operators denoted by $b_s^{\dagger}$ and b_s [59],

$$K_{rs} = b_r^{\dagger} b_s, \tag{30a}$$
$$K_{r0} = K_{0r}^{\dagger} = b_r^{\dagger} \sqrt{\Omega - b_1^{\dagger} b_1 - b_2^{\dagger} b_2}, \tag{30b}$$

for $r, s = 1, 2$.

The classical counterpart may be obtained by making use of the following transformation,

$$b_s^{\dagger} = \sqrt{\frac{\Omega}{2}}(q_s - ip_s), \quad b_s = \sqrt{\frac{\Omega}{2}}(q_s + ip_s). \tag{31}$$

It is easy to verify that q_r and p_s obey the following commutation relation,

$$[q_r, p_s] = \frac{i}{\Omega}\delta_{rs}. \tag{32}$$

Thus, $1/\Omega$ plays the role of an effective Planck constant, $\hbar_{\text{eff}} = \frac{1}{\Omega}$, and the classical limit of the model is obtained by letting $\Omega \to \infty$. In the classical Hamiltonian, parameters take the form of $\eta_s^{\text{cl}} = \eta_s \Omega$ and $\mu_t^{\text{cl}} = \mu_t \Omega^2$. Quantum systems corresponding to the same values of η_s^{cl} and μ_t^{cl} share a common classical counterpart.

4.2. Numerical Results

As is well known, on the integrable side, a model may behave differently under different winding numbers, particularly, at rational and irrational values. In the LMG model, the winding number is given by the ratio of η_1/η_2. in our numerical simulations, we have studied the crossovers starting from several integrable Hamiltonians, corresponding to different values of η_1 and η_2. We have the freedom of keeping η_1 and η_2 in the order of magnitude of 1. Specifically, we have studied two rational values of η_1/η_2 and two irrational ones: $(\eta_1, \eta_2) = (1, 2), (1, 3), (\sqrt{3}, 2)$ and $(\sqrt{5} - 1, 2)$.

On the chaotic side, parameters in V are chosen such that this observable has essentially the same classical limit as that studied in Ref. [30]; specifically, $\mu_1 = 0.0077625$, $\mu_2 = 0.008775$, $\mu_3 = 0.0095625$, and $\mu_4 = 0.0082125$. Other parameters used in our simulations are $\epsilon = 10^{-4}$ for small perturbation and the total particle number $\Omega = 200$, for which the dimension of the Hilbert space is 20301. Under these parameters, the mean value of $|\epsilon V_{kk'}|$ of nonzero elements of V in the integrable basis is about 2.6×10^{-3}, obviously smaller than the mean level spacing which is about 10^{-2}, satisfying the requirement of weak perturbation.

In the computation of the average shape $\langle |C_{\alpha k}|^2 \rangle$ (Equation (6)), we adopted the second method discussed previously, with $\Lambda_b = 0.1$; i.e., all big components with $|C_{\alpha k}|^2 \geq 0.1$ were excluded. The width δe was adjusted such that each window Γ_α includes 16 levels. In the computation of $g(0)$, for each system $H^0(\lambda)$, 2000 perturbed states $|\alpha\rangle$ lying in the middle energy region were used, and for each $|\alpha\rangle$, 2000 components $\widetilde{C}_{\alpha k}$ with E_k^0 around E_α were used.

Numerical results are shown in Figures 1–4 for the four pairs of (η_1, η_2) mentioned above, respectively. It is seen that the $g^{-1}(0)$-plots in all the four figures show a same pattern of behavior from integrability to chaos. Furthermore, this pattern is much simpler than those of $g(0)$, as explained previously due to the fact that the difference among big values of $g(0)$ is restrained in $g^{-1}(0)$. For the sake of comparison with standard spectral-statistics analysis, we also plot the quantity Δ_W in Equation (16) as a measure for the distance to chaos. Furthermore, moreover, we plot the following quantity as a measure for the distance to integrability, i.e.,

$$\Delta_P = \int |I(s) - e^{-s}| ds, \tag{33}$$

where e^{-s} is the Poison distribution expected for the nearest-level-spacing distribution in generic integrable systems.

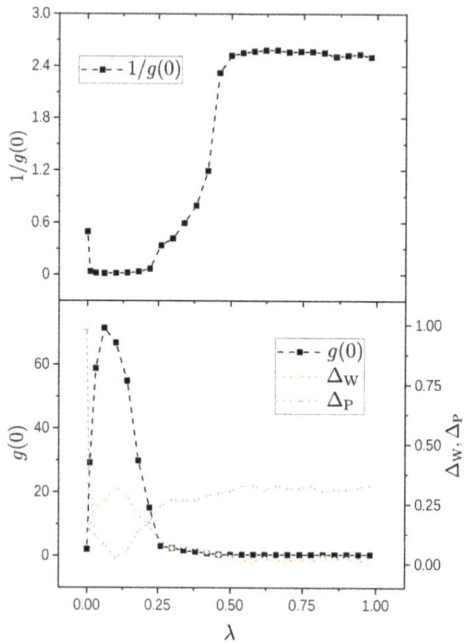

Figure 1. Variation of $g^{-1}(0)$ (upper panel), and $g(0)$ (lower panel) versus λ (solid squares connected by dashed lines) for $\eta_1 = 1, \eta_2 = 2$. Other parameters: $\mu_1 = 0.0077625$, $\mu_2 = 0.008775$, $\mu_3 = 0.0095625$, $\mu_4 = 0.0082125, \Omega = 200$, and $\epsilon = 10^{-4}$. For comparison, variation of Δ_W in Equation (16) (empty circles connected by dotted line (red)) and of Δ_P in Equation (33) (stars connected by dotted line (blue)) are also plotted.

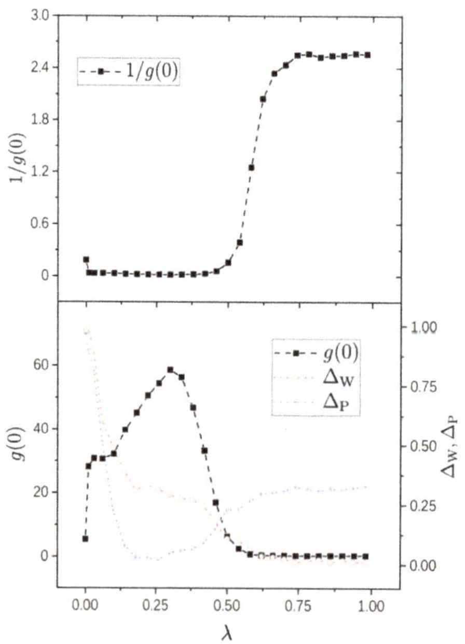

Figure 2. Similar to Figure 1, but for $\eta_1 = 1$ and $\eta_2 = 3$.

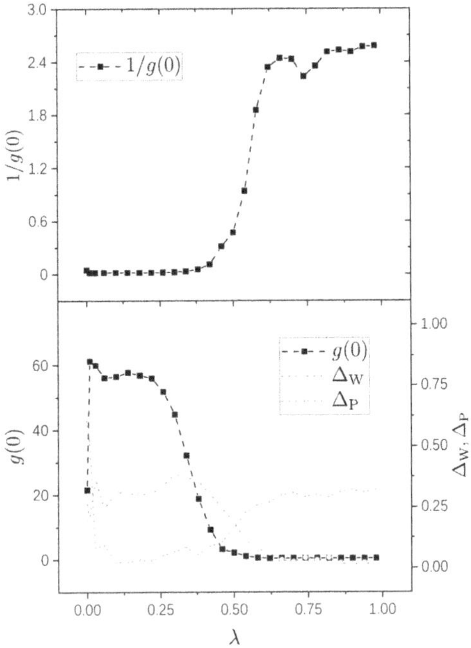

Figure 3. Similar to Figure 1, but for $\eta_1 = \sqrt{3}$ and $\eta_2 = 2$.

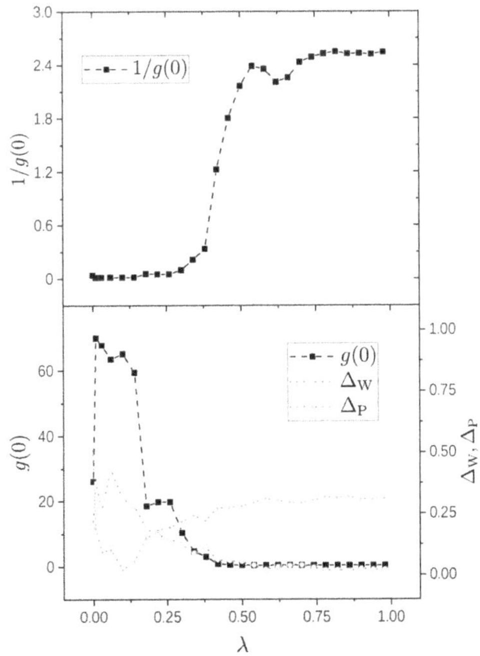

Figure 4. Similar to Figure 1, but for $\eta_1 = \sqrt{5} - 1$ and $\eta_2 = 2$.

As is known, the parameter region $(0, 1)$ for a integrability-chaos crossover may usually be divided into three subregions: a nearly integrable regime, a nearly chaotic regime, and a crossover regime. A merit of the quantity $g^{-1}(0)$ is that it shows this division in a very clear way, the details of which we discuss below.

1. A nearly integrable regime, in which the value of $g^{-1}(0)$ is close to 0. In this regime, the perturbation-induced transition is strongly prohibited between many of the unperturbed states.
2. A nearly chaotic regime, in which $g^{-1}(0) \approx \sqrt{2\pi}$. In this regime, the perturbation-induced transition is not prohibited in a statistical sense.
3. An intermediate (cross-over) regime, in which $g^{-1}(0)$ increases "rapidly", approximately from 0 to 2.5.

Specifically, in the case of $(\eta_1, \eta_2) = (1,2)$, as seen in the upper panel of Figure 1, the above-discussed three regimes occupy approximately the three subregions of $(0, 0.22)$, $(0.5, 1)$ and $(0.22, 0.5)$, respectively. In this figure, the point of $\lambda = 0$ looks quite special, because the value of $g^{-1}(0)$ at $\lambda = 0$ is not small, but, about 0.5. (One should note that, although the value of $g(0)$ at $\lambda = 0$ in this figure is not very large, about 2, it is still obviously larger than the value expected for chaotic systems, the latter of which is about 0.4.) This phenomenon is due to the high degeneracy of the spectrum of the integrable system $H^0(0)$ with $(\eta_1, \eta_2) = (1,2)$. In fact, with $\Lambda_b = 0.1$, most of the components $C_{\alpha k}$ of $|k\rangle$ lying in highly degenerated subspaces were included in the computation of $g(0)$ and, since the perturbation is effectively extremely strong within each degenerate subspace, the number of very small components in them is not large. Please note that this phenomenon disappears with a little increase of λ, which destroys the degeneracy, and as a result the value of $g^{-1}(0)$ becomes quite small, as seen in the figure.

In Figure 2, with $(\eta_1, \eta_2) = (1,3)$, the three regimes are seen clearly, too, approximately occupying $(0, 0.46)$, $(0.74, 1)$, and $(0.46, 0.74)$, respectively. In this figure, the value of $g^{-1}(0)$ at $\lambda = 0$ is not very small, either, about 0.18. This value is smaller than that in Figure 1, because the degeneracy in the system with $(\eta_1, \eta_2) = (1,3)$ is not as high as that in the case of $(\eta_1, \eta_2) = (1,2)$. Consistently, in Figures 3 and 4, the values of $g^{-1}(0)$ are already small at $\lambda = 0$ for integrable systems $H^0(0)$ possessing nondegenerate spectra.

In the lower panels of the four figures, it is seen that the $g(0)$-plots show rich behaviors in the nearly integrable regime. In the case with rational ratios η_1/η_2 (Figures 1 and 2), $g(0)$ is relatively small at $\lambda = 0$ (corresponding to the above-discussed relatively large values of $g^{-1}(0)$) and increases immediately with a little increase of λ. For irrational ratios η_1/η_2 (Figures 3 and 4), although $g(0)$ are already not small at $\lambda = 0$, they also undergo immediate rapid increase with a little increase of λ.

Correspondingly, from the $g(0)$-plots, one also sees three subregions in each figure — nearly integrable, intermediate (crossover), and nearly chaotic. In Figures 1, 2, and 4, the intermediate subregions are close to those obtained from the level-statistics properties (Δ_W). However, notable differences between them are seen in Figures 3, showing the information supplied by $g(0)$ for the crossover is not always in quantitative agreement with that obtained from the level statistics. We are not able to provide further discussion on such a difference, nor on rich behaviors of $g(0)$ on the integrable side, because as discussed previously, further analytical study is needed for a deeper understanding of them.

5. Conclusions

In this paper, a measure has been proposed for characterizing integrability-chaos crossover for a big class of quantum models of practical interest, which possess underlying dynamic Lie groups. To compute this measure, a very small perturbation (ϵV) of physical relevance is applied to the studied system and components of perturbed states on the unperturbed basis are rescaled with respect to the average shape of eigenfunctions. The measure is given by $g(0)$, the value of the distribution of rescaled components at the origin point, or, practically by $g^{-1}(0)$. In a relative sense, $g(0)$ has the physical meaning of describing the extent of prohibition of ϵV-induced transitions between unperturbed states.

As is known, the whole parameter region of integrability-chaos crossover may usually be divided into three subregions: a nearly integrable regime, a nearly chaotic regime, and a crossover regime. Different measures employed usually give divisions with some differences. Numerical simulations performed in the LMG model, which possesses a classical counterpart in a two-dimensional configuration space, show that the behavior of $1/g(0)$ gives such a division in quite a clear way. The division is in qualitative consistency with that obtained from spectral statistics, but not always quantitatively. Future investigations are needed for analytical understanding of the difference.

In principle, the proposed measure may be used in both cases with a noninteracting integrable system and with an interacting integrable system. In this paper, numerical simulations are given only in a simple model of the former type. In future, it would be of interest to see its application to models of the latter type, which might be more complicated than the former type. Moreover, application of the proposed measure to many-body systems, particularly on the nearly integrable side, may be an interesting topic for future study, too.

Author Contributions: Formal analysis, C.Y.L. and W.-G.W.; Writing—original draft, C.Y.L.; Writing—review & editing, W.-G.W.; Supervision, W.-G.W. All authors have read and agreed to the published version of the manuscript.

Funding: This work was partially supported by the Natural Science Foundation of China under Grant Nos. 11535011, 11775210, and 12175222.

Institutional Review Board Statement: Not applicable.

Conflicts of Interest: The authors declare no conflict of interest.

References

1. Casati, G.; Chirikov, B.V.; Guarneri, I.; Izrailev, F.M. Band-random-matrix model for quantum localization in conservative systems. *Phys. Rev. E* **1993**, *48*, R1613–R1616. [CrossRef] [PubMed]
2. Haake, F. *Quantum Signatures of Chaos*; Springer Series in Synergetics; Springer: Berlin/Heidelberg, Germany, 2010; Volume 54; ISBN 978-3-642-05427-3.
3. Casati, G.; Valz-Gris, F.; Guarnieri, I. On the Connection between Quantization of Nonintegrable Systems and Statistical Theory of Spectra. *Lett. Nuovo C.* **1980**, *28*, 279. [CrossRef]
4. Bohigas, O.; Giannoni, M.J.; Schmit, C. Characterization of Chaotic Quantum Spectra and Universality of Level Fluctuation Laws. *Phys. Rev. Lett.* **1984**, *52*, 1–4. [CrossRef]
5. Berry, M.V. Semiclassical theory of spectral rigidity. *Proc. R. Soc. Lond. Ser. A* **1985**, *400*, 229.
6. Sieber, M.; Richter, K. Correlations between periodic orbits and their rôle in spectral statistics. *Phys. Scr.* **2001**, *T90*, 128.; [CrossRef]
7. Sieber, M. Leading off-diagonal approximation for the spectral form factor for uniformly hyperbolic systems. *J. Phys. A Math. Gen.* **2002**, 35, L613. [CrossRef]
8. Heusler, S.; Müller, S.; Braun, P.; Haake, F. Universal spectral form factor for chaotic dynamics. *J. Phys. A: Math. Gen.* **2004**, *37*, L31.; [CrossRef]
9. Müller, S.; Heusler, S.; Braun, P.; Haake, F.; Altland, A. Semiclassical Foundation of Universality in Quantum Chaos. *Phys. Rev. Lett.* **2004**, *93*, 014103. [CrossRef]
10. Polkovnikov, A.; Sengupta, K.; Silva, A.; Vengalattore, M. Colloquium: Nonequilibrium dynamics of closed interacting quantum systems. *Rev. Mod. Phys.* **2011**, *83*, 863–883. [CrossRef]
11. Eisert, J.; Friesdorf, M; Gogolin, C. Quantum many-body systems out of equilibrium. *Nat. Phys.* **2015**, *11*, 124–130. [CrossRef]
12. Gogolin, C.; Eisert, J. Equilibration, thermalisation, and the emergence of statistical mechanics in closed quantum systems. *Rep. Prog. Phys.* **2016**, *79*, 056001. [CrossRef] [PubMed]
13. D'Alessio, L.; Kafri, Y.; Polkovnikov, A.; Rigol, M. From quantum chaos and eigenstate thermalization to statistical mechanics and thermodynamics. *Adv. Phys.* **2016**, *65*, 239–362. [CrossRef]
14. Borgonovi, F.; Izrailev, F.M.; Santos, L.F.; Zelevinsky, V.G. Quantum chaos and thermalization in isolated systems of interacting particles. *Phys. Rep.* **2016**, *626*, 1–58. [CrossRef]
15. Mori, T.; Ikeda, T.N.; Kaminishi, E.; Ueda, M. Thermalization and prethermalization in isolated quantum systems: a theoretical overview. *J. Phys. B: At. Mol. Opt. Phys.* **2018**, *51*, 112001. [CrossRef]
16. Tasaki, H. Typicality of Thermal Equilibrium and Thermalization in Isolated Macroscopic Quantum Systems. *J. Stat. Phys.* **2016**, *163*, 937–997. [CrossRef]
17. Berry, M.V.; Tabor, M. Level Clustering in the Regular Spectrum. *Proc. R. Soc. Lond.* **1977**, *356*, 375–394. [CrossRef]
18. Brody, T.A.; Flores, J.; French, J.B.; Mello, P.A.; Pandey, A.; Wong, S.S.M. Random-matrix physics: spectrum and strength fluctuations. *Rev. Mod. Phys.* **1981**, *53*, 385–479. [CrossRef]
19. Berry, M.V.; Robnik, M. Semiclassical level spacings when regular and chaotic orbits coexist. *J. Phys. A: Math. Gen.* **1984**, *17*, 2413. [CrossRef]

20. Izrailev, F.M. Chaotic stucture of eigenfunctions in systems with maximal quantum chaos. *Phys. Lett. A* **1987**, *125*, 250–252. [CrossRef]
21. Izrailev, F.M. Quantum localization and statistics of quasienergy spectrum in a classically chaotic system. *Phys. Lett. A* **1988**, *134*, 13–18. [CrossRef]
22. Berry, M.V. Regular and irregular semiclassical wavefunctions. *J. Phys. A: Math. Gen.* **1977**, *10*, 2083. [CrossRef]
23. Li, B.; Robnik, M. Sensitivity of the eigenfunctions and the level curvature distribution in quantum billiards. *J. Phys. A: Math. Gen.* **1996**, *29*, 4387. [CrossRef]
24. Srednicki, M. Chaos and quantum thermalization. *Phys. Rev. E* **1994**, *50*, 888–901. [CrossRef] [PubMed]
25. Meredith, D.C.; Koonin, S.E.; Zirnbauer, M.R. Quantum chaos in a schematic shell model. *Phys. Rev. A* **1988**, *37*, 3499–3513. [CrossRef] [PubMed]
26. Blümel, R.; Smilansky, U. Suppression of classical stochasticity by quantum-mechanical effects in the dynamics of periodically perturbed surface-state electrons. *Phys. Rev. A* **1984**, *30*, 1040–1051. [CrossRef]
27. Shapiro, M.; Goelman, G. Onset of Chaos in an Isolated Energy Eigenstate. *Phys. Rev. Lett.* **1984**, *53*, 1714–1717. [CrossRef]
28. Benet, L.; Flores, J.; Hernández-Saldaña, H.; Izrailev, F.M.; Leyvraz, F.; Seligman, T.H. Fluctuations of wavefunctions about their classical average. *J. Phys. A: Math. Gen.* **2003**, *36*, 1289. [CrossRef]
29. Wang, J.; Wang, W. Correlations in eigenfunctions of quantum chaotic systems with sparse Hamiltonian matrices. *Phys. Rev. E* **2017**, *96*, 052221. [CrossRef]
30. Wang, J.; Wang, W. Characterization of random features of chaotic eigenfunctions in unperturbed basis. *Phys. Rev. E* **2018**, *97*, 062219. [CrossRef]
31. Wang, Q.; Robnik, M. Statistical properties of the localization measure of chaotic eigenstates in the Dicke model. *Phys. Rev. E* **2020**, *102*, 032212. [CrossRef]
32. Pandey, M.; Claeys, P.W.; Campbell, D.K.; Polkovnikov, A.; Sels, D. Adiabatic Eigenstate Deformations as a Sensitive Probe for Quantum Chaos. *Phys. Rev. X* **2020**, *10*, 041017. [CrossRef]
33. Xu, Z.; Lyu, Y.-C.; Wang, J.; Wang, W. Sensitivity of energy eigenstates to perturbation in quantum integrable and chaotic systems. *Commun. Theor. Phys.* **2021**, *73*, 15104. [CrossRef]
34. Peres, A. Stability of quantum motion in chaotic and regular systems. *Phys. Rev. A* **1984**, *30*, 1610–1615. [CrossRef]
35. Jalabert, R.A.; Pastawski, H.M. Environment-Independent Decoherence Rate in Classically Chaotic Systems. *Phys. Rev. Lett.* **2001**, *86*, 2490–2493. [CrossRef]
36. Prosen, T.; Znidaric, M. Stability of quantum motion and correlation decay. *J. Phys. A: Math. Gen.* **2002**, *35*, 1455. [CrossRef]
37. Cerruti, N.R.; Tomsovic, S. Sensitivity of Wave Field Evolution and Manifold Stability in Chaotic Systems. *Phys. Rev. Lett.* **2002**, *88*, 054103. [CrossRef]
38. Benenti, G.; Casati, G. Quantum-classical correspondence in perturbed chaotic systems. *Phys. Rev. E* **2002**, *65*, 066205. [CrossRef]
39. Wang, W.; Li, B. Uniform semiclassical approach to fidelity decay: From weak to strong perturbation. *Phys. Rev. E* **2005**, *71*, 066203. [CrossRef]
40. Gorin, T.; Prosen, T.; Seligman, T.H.; Žnidarič, M. Dynamics of Loschmidt echoes and fidelity decay. *Phys. Rep.* **2006**, *435*, 33–156. [CrossRef]
41. Wang, W.; Casati, G.; Li, B. Stability of quantum motion in regular systems: A uniform semiclassical approach. *Phys. Rev. E* **2007**, *75*, 016201. [CrossRef]
42. Leviandier, L.; Lombardi, M.; Jost, R.; Pique, J.P. Fourier Transform: A Tool to Measure Statistical Level Properties in Very Complex Spectra. *Phys. Rev. Lett.* **1986**, *56*, 2449–2452. [CrossRef] [PubMed]
43. Torres-Herrera, E.J.; Santos, L.F. Dynamics at the many-body localization transition. *Phys. Rev. B* **2015**, *92*, 014208. [CrossRef]
44. Torres-Herrera E. J.; Santos Lea F. Dynamical manifestations of quantum chaos: correlation hole and bulgePhil. *Phil. Trans. R. Soc. A* **2016**, *375*, 0434. [CrossRef] [PubMed]
45. Torres-Herrera, E.J.; García-García, A.M.; Santos, L.F. Generic dynamical features of quenched interacting quantum systems: Survival probability, density imbalance, and out-of-time-ordered correlator. *Phys. Rev. B* **2018**, *97*, 060303. [CrossRef]
46. Brenes, M.; Goold, J.; Rigol, M. Low-frequency behavior of off-diagonal matrix elements in the integrable XXZ chain and in a locally perturbed quantum-chaotic XXZ chain. *Phys. Rev. B* **2020**, *102*, 075127. [CrossRef]
47. Gu, S.-J. Fidelity approach to quantum phase transitions. *Int. J. Mod. Phys. B* **2010**, *24*, 4371–4458 [CrossRef]
48. Sierant, P.; Maksymov, A.; Kuś, M.; Zakrzewski, J. Fidelity susceptibility in Gaussian random ensembles. *Phys. Rev. E* **2019**, *99*, 050102. [CrossRef]
49. Lipkin, H.J.; Meshkov, N.; Glick, A.J. Validity of many-body approximation methods for a solvable model: (I). Exact solutions and perturbation theory. *Nucl. Phys.* **1965**, *62*, 188–198. [CrossRef]
50. Kolodrubetz, M.; Sels, D.; Mehta, P.; Polkovnikov, A. Geometry and non-adiabatic response in quantum and classical systems. *Phys. Rep.* **2017**, *697*, 1–87. [CrossRef]
51. Srednicki, M. Thermal fluctuations in quantized chaotic systems. *J. Phys. A: Math. Gen.* **1996**, *29*, L75. [CrossRef]
52. Srednicki, M. The approach to thermal equilibrium in quantized chaotic systems. *J. Phys. A: Math. Gen.* **1999**, *32*, 1163. [CrossRef]
53. Deutsch, J.M. Quantum statistical mechanics in a closed system. *Phys. Rev. A* **1991**, *43*, 2046–2049. [CrossRef] [PubMed]
54. Rigol, M.; Srednicki, M. Alternatives to Eigenstate Thermalization. *Phys. Rev. Lett.* **2012**, *108*, 110601. [CrossRef] [PubMed]

55. De Palma, G.; Serafini, A.; Giovannetti, V.; Cramer, M. Necessity of Eigenstate Thermalization. *Phys. Rev. Lett.* **2015**, *115*, 220401. [CrossRef]
56. Deutsch, J.M. Eigenstate thermalization hypothesis. *Rep. Prog. Phys.* **2018**, *81*, 082001. [CrossRef]
57. Wang, W. Semiclassical proof of the many-body eigenstate thermalization hypothesis. *arXiv* **2022**, arXiv:2210.13183.
58. Wang, W.; Izrailev, F.M.; Casati, G. Structure of eigenstates and local spectral density of states: A three-orbital schematic shell model. *Phys. Rev. E* **1998**, *57*, 323–339. [CrossRef]
59. Gong-ou, X.; Jiang-bin, G.; Wen-ge, W.; Ya-tian, Y.; De-ji, F. Development of quantum nonintegrability displayed in effective Hamiltonians: A three-level Lipkin model. *Phys. Rev. E* **1995**, *51*, 1770–1779. [CrossRef]

entropy

MDPI

Article

Statistical Topology—Distribution and Density Correlations of Winding Numbers in Chiral Systems

Thomas Guhr

Fakultät für Physik, Universtät Duisburg–Essen, 47048 Duisburg, Germany; thomas.guhr@uni-due.de

Abstract: Statistical Topology emerged as topological aspects continue to gain importance in many areas of physics. It is most desirable to study topological invariants and their statistics in schematic models that facilitate the identification of universalities. Here, the statistics of winding numbers and of winding number densities are addressed. An introduction is given for readers with little background knowledge. Results that my collaborators and I obtained in two recent works on proper random matrix models for the chiral unitary and symplectic cases are reviewed, avoiding a technically detailed discussion. There is a special focus on the mapping of topological problems to spectral ones as well as on the first glimpse of universality.

Keywords: statistical topology; random matrices; chirality; winding numbers

Citation: Guhr, T. Statistical Topology—Distribution and Density Correlations of Winding Numbers in Chiral Systems. *Entropy* **2023**, *25*, 383. https://doi.org/10.3390/e25020383

Academic Editor: Marko Robnik

Received: 18 January 2023
Revised: 15 February 2023
Accepted: 16 February 2023
Published: 20 February 2023

1. Introductory Remarks

Statistical Topology aims at combining, in a generalizing form, topological questions appearing in physics with the powerful concepts of Random Matrix Theory (RMT) which is capable of describing spectral statistics in a huge number of systems, stemming from different areas of physics and beyond. The focus in this work is exclusively on winding numbers and associated statistical quantities studied in the framework of a random matrix model; other topological invariants, such as the Chern numbers, which are also of considerable interest, are not discussed. The long-term aim is to study the emergence of universalities whose identification and usage is always, in all branches of statistical physics, the most rewarding enterprise. I have two goals. First, I want to present an introduction to Statistical Topology, restricted to statistical problems which are related to winding numbers, for readers without a pertinent background. Neither physics expert jargon, nor heavy mathematics and mathematical physics terminology are used. Second, I want to review and summarize results that my collaborators and I obtained in two recent studies [1,2]. We calculated for a chiral unitary random matrix model correlators of winding number densities and the winding number distribution. We also computed generators for these correlators in a chiral unitary and a chiral symplectic random matrix model. Furthermore, we made first steps towards finding universalities.

The paper is organized as follows: In Section 2, the salient features of winding numbers and chiral symmetry are presented. In Section 3, a schematic model with the necessary mathematical setup is formulated. Results are reviewed in Section 4, discussion and conclusions are given in Section 5.

2. Winding Numbers and Chirality

After briefly revisiting the occurrence of winding numbers in complex analysis in Section 2.1, the Kitaev chain is discussed in Section 2.2 and the statistical ansatz is motivated in Section 2.3. The research is placed in the framework of Quantum Chromodynamics (QCD) and Condensed Matter Physics in Section 2.4, summarizing the corresponding remarks in Refs. [1,2].

2.1. A Simple Topological Invariant in Complex Analysis

The winding number is a topological concept encountered in complex analysis. Before discussing applications in physics, we briefly sketch the mathematical background. The winding number $W = W(z_i)$ counts how many times a point z_i in the complex plane $\mathbb{C}$ is encircled by a closed contour γ, where counterclockwise or clockwise give a positive or a negative contribution, respectively. An example is shown in Figure 1, we have $W(z_1) = 0$, $W(z_2) = 1$ and $W(z_3) = 2$. Obviously, the winding number $W(z_i)$ is a topological constant or, in physics terminology, a quantum number. It is invariant under all deformations of γ that do not cross the point z_i in question. In particular, the winding number is always a positive or negative integer, $W \in \mathbb{Z}$. It may be written as the contour integral

$$W(z_i) = \frac{1}{2\pi i} \oint_{\gamma} \frac{d\zeta}{\zeta - z_i} \,. \tag{1}$$

One easily establishes the link to Cauchy's argument principle: Consider a meromorphic function $f(z)$ and a closed contour Γ, encircling some zeros and poles of $f(z)$ in the complex plane $\mathbb{C}$ as shown in the example in Figure 1. The integral along this contour Γ over the logarithmic derivative of $f(z)$ yields the difference of the number N_Z of zeros and the number N_P of poles, hence

$$\frac{1}{2\pi i} \oint_{\Gamma} \frac{f'(z)}{f(z)} dz = N_Z - N_P \,. \tag{2}$$

The close relation to the winding number is found by making the change of variable $\zeta = f(z)$ and in accordance with the contour, $\Gamma \to f(\Gamma)$,

$$N_Z - N_P = \frac{1}{2\pi i} \oint_{\Gamma} \frac{f'(z)}{f(z)} dz = \frac{1}{2\pi i} \oint_{f(\Gamma)} \frac{d\zeta}{\zeta} = W(0) \,. \tag{3}$$

We conclude that $N_Z - N_P$ is the winding number $W(0)$ of the closed contour $f(\Gamma)$ around the origin $z = 0$. As, from now on, all winding numbers will refer to the origin, we drop the argument and simply write W.

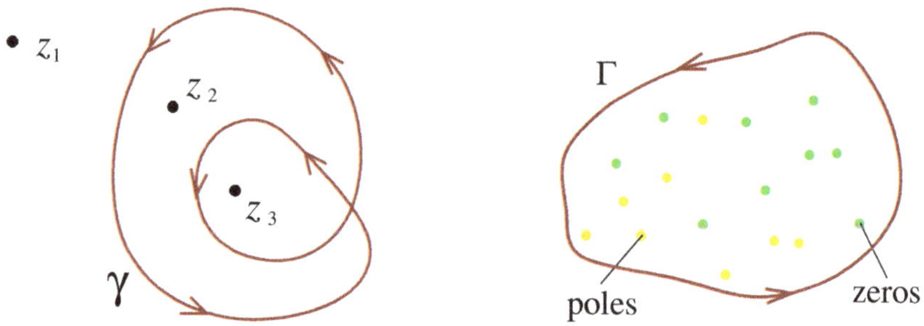

Figure 1. Left: Three points z_i, $i = 1, 2, 3$ in the complex plane $\mathbb{C}$ and a closed contour γ. **Right**: A closed contour Γ encircling zeros and poles of a meromorphic function $f(z)$.

2.2. Kitaev Chain and Winding Numbers

To illustrate the occurrence of topological invariants in physics, we look at the Kitaev chain [3,4] as a prominent example. It consists of spinless electrons with next-neighbor hopping and superconductive pairing. The Hamiltonian reads, in a slightly simplified form sufficient for the present discussion,

$$\hat{H} = \sum_n \left(t\left(\hat{c}_n^\dagger \hat{c}_{n+1} + \hat{c}_{n+1}^\dagger \hat{c}_n\right) + \mu \hat{c}_n^\dagger \hat{c}_n + \frac{\Delta}{2}\left(\hat{c}_{n+1}^\dagger \hat{c}_n^\dagger + \hat{c}_n \hat{c}_{n+1}\right)\right), \tag{4}$$

where $\hat{c}_n$ and $\hat{c}_n^\dagger$ are annihilation and creation operators, respectively, at position n on the chain. Moreover, μ and Δ are chemical and pairing potentials and t is the hopping strength. The Hamiltonian may be reformulated in terms of Majorana fermions whose number is twice that of the electrons. Remarkably, depending on the parameters, there are two possibilities, as schematically depicted in Figure 2. Either all Majorana fermions are paired or, at the ends of the chain, two of them are unpaired [5]. In the former case, the chain is in a normal or trivial superconducting phase, in the latter, in a topological one. This aspect deserves further discussion.

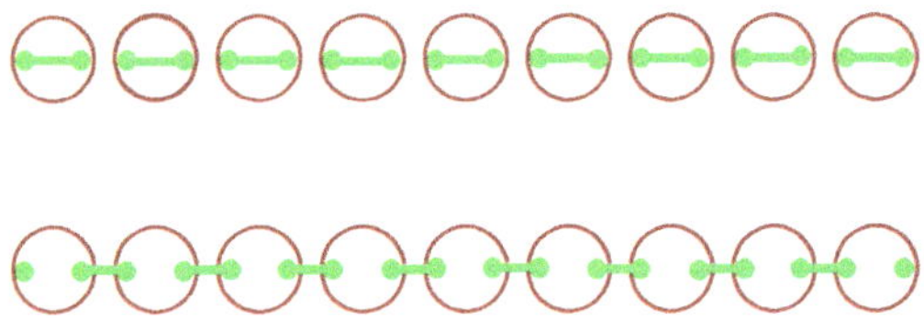

Figure 2. Kitaev chain, electrons as larger open circles (red), Majorana fermions as small dots (green) with the pairing indicated by connecting lines (green). Top: All Majorana fermions are paired, normal or trivial superconducting phase. Bottom: Unpaired Majorana fermions at the ends of the chain, topological superconducting phase.

In Fourier space, the Kitaev chain corresponds to the Bloch–Bogolyubov–de Gennes Hamiltonian matrix $H(k)$. It is a crucial that this 2×2 matrix satisfies chiral symmetry,

$$\{H(k), \mathcal{C}\} = 0 \qquad \text{with} \qquad \mathcal{C} = \begin{bmatrix} 1 & 0 \\ 0 & -1 \end{bmatrix}. \tag{5}$$

The matrix $\mathcal{C}$ is the chiral operator in its proper basis and $\{\ ,\ \}$ is the anticommutator. It is then possible to write the Hamiltonian matrix in the form

$$H(k) = \vec{d}(k) \cdot \vec{\sigma} \qquad \text{with} \qquad \vec{d}(k) = (0, \Delta \sin k, \mu + 2t \cos k). \tag{6}$$

Hence, using the three-component vector $\vec{\sigma}$ of the 2×2 Pauli matrices, $H(k)$ is found to be a scalar product with all physics encoded in the vector $\vec{d}(k)$ that depends on the wave number k and the three parameters μ, Δ and t. Importantly, the first component is zero, $d_x = 0$. This restriction to effectively only two dimensions can be shown to be a consequence of chiral symmetry (5).

To see how topology enters, we notice that the vector $\vec{d}(k)$ describes an ellipse with parameter k on the curve, μ determines the position of its center, Δ and t determine its shape. In Figure 3, we depict $\vec{d}(k)$ for fixed values of $\mu = 1, \Delta = 1$ and three different values $t = 0.25, 0.5, 1$ with the corresponding energy dispersion relation $E(k)$. If the origin of the (y, z) plane is included in the closed contour that the ellipse describes, its winding number is one, $W = 1$. If not, the winding number is zero, $W = 0$. These are two topologically separated scenarios, reflecting the distinctly different role of the Majorana fermions in the top and bottom parts of Figure 2. For $W = 0$, the superconducting phase is the normal or trivial one, while it is topological for $W = 1$. A special situation occurs if the ellipse just touches the x axis, the band gap disappears, marking the phase transition point.

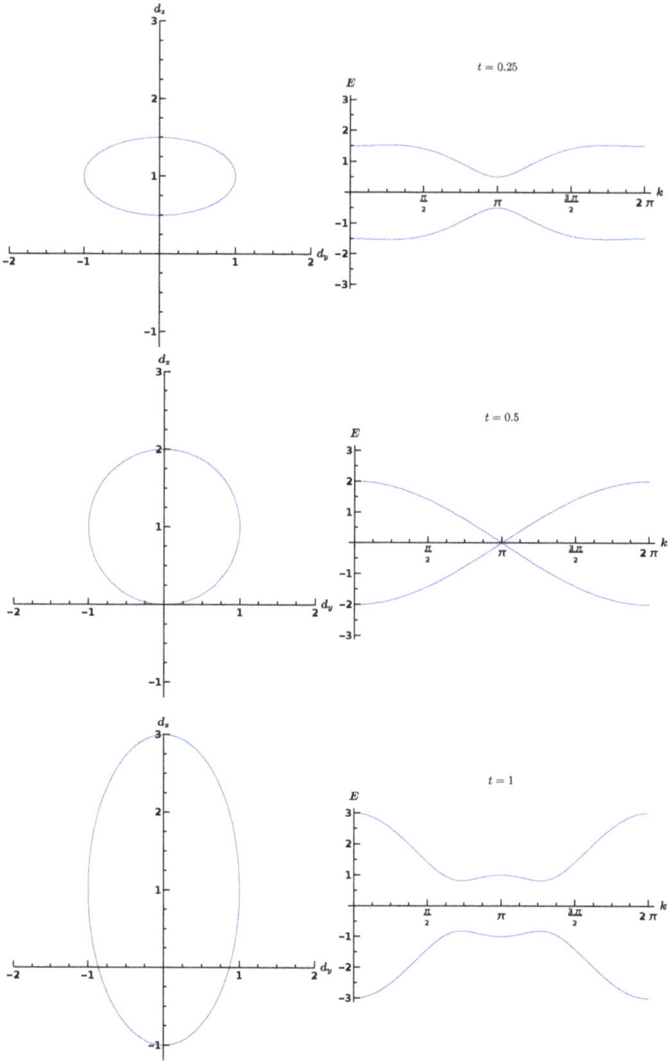

Figure 3. Ellipses described by $\vec{d}(k)$ (**left**) and corresponding dispersion relations $E(k)$ (**right**). **Top**: $t = 0.25$, normal superconducting phase, $W = 0$. **Center**: $t = 0.5$, phase transition point. **Bottom**: $t = 1$, topological superconducting phase, $W = 1$. Courtesy of Nico Hahn.

2.3. *Chirality, Random Winding Numbers and Modelling Aspects*

When studying such topological invariants in statistical physics, the closed contour might be a random quantity, for example, generated by a proper ensemble of Hamiltonians. In the case of the Kitaev chain, this ensemble may be realized by choosing the parameters μ, Δ and t from probability distributions. Hence, the contour can be different for a particular choice, i.e., it becomes random, and the winding number W will be random as well. In general, the dynamics of a system under consideration, described by the Hamiltonian, and the distributions of its parameters will determine the distribution $P(W)$. Are there universalities when comparing different systems? If yes, in which quantities do these universalities manifest? In the distributions $P(W)$ on their original scales or on some scales which make these systems comparable? These are the guiding questions for our research.

Universalities are best identified in random schematic systems that only contain the most basic ingredients needed for the relevant physics, in the present case for the occurence of winding numbers. Random Matrix Theory (RMT) [6,7] is known to be a powerful concept in this spirit when studying universalities in spectral correlations as well as in the correlations of parametric level motion [8,9]. The chiral symmetry (5) and, thus, the restriction to two dimensions are essential for the interpretation of the two superconducting phases in the Kitaev chain in terms of the winding number. Hence, when setting up a schematic random matrix model, we need to employ chirality.

2.4. Connections to Quantum Chromodynamics and Condensed Matter Physics

In Quantum Chromodynamics, the chiral symmetry of the Dirac operator is broken spontaneously as well as explicitly by the quark masses. The chiral condensate is the order parameter of the phase transition that occurs at a high temperature and that restores chiral symmetry, which is related to the confinement–deconfinement transition. To investigate statistical properties of lattice gauge calculations, chiral RMT [10–17] is remarkably successful. As in the original RMT, presence or absence of time-reversal invariance combined with spin-rotation symmetries results in three classes of chiral random matrices: orthogonal, unitary and symplectic. It was then shown that altogether ten RMT symmetry classes [18–22] exist, referred to as the tenfold way. The three original and the three chiral ones comprise six of these ten classes, the remaining four emerge when particle-hole symmetry is also considered, see Refs. [23,24]. In condensed matter physics, chiral symmetry is realized by sublattice symmetry (see early work in Ref. [25]) or as a combination of time reversal and particle-hole symmetry [24].

In the terminology of condensed matter physics, the winding number comes in as characterization of translationally invariant one-dimensional chiral systems that are gapped at the centre of the spectrum. The winding number is the integer topological index with respect to the bundle of negative-energy bands. A non-zero winding number W indicates the topologically non-trivial situation with $|W|$ modes at each boundary [26–29]. The winding number differs for different realization of the disorder, i.e., it becomes random. Our research on the winding number was inspired by studies of systems with energy bands in two dimensions, allowing for a topological classification by the (first) Chern number. A random matrix model [30,31] revealed a Gaussian distribution of Chern numbers with a universal covariance.

Another intriguing direction might be the application of Statistical Topology to classical wave phenomena such as microwaves or acoustics and, furthermore, to photonics where topological issues are already in focus [32].

3. Formulation of the Problem and Mathematical Setup

After introducing chiral random matrix ensembles with a parameter dependence in Section 3.1, the statistical quantities of interest are defined in Section 3.2. In Section 3.3, a crucial step for all of our mathematical investigations is presented, namely, the mapping of the topological problem addressed to a spectral one which greatly facilitates the computations.

3.1. Chiral Random Matrix Ensembles with Parametric Dependence

We derived results [1,2] for the chiral unitary and the chiral symplectic symmetry classes labeled AIII and CII, respectively, see Ref. [18]. The latter case is mathematically much more demanding than the former, but not as involved as the orthogonal case, labeled BDI. Only very recently have we been able to solve it, this will be published elsewhere. The cases BDI and CII describe time-reversal invariant systems, while this invariance does not exist in the case AIII. We refer to the matrices as Hamiltonians H, as most of the present application of winding numbers seem to stem from Condensed Matter Physics. The matrices H are complex Hermitean or quaternion real, i.e., self-adjoint, with even dimension $\beta N \times \beta N$ where we employ the Dyson indices $\beta = 2$ and $\beta = 4$ for AIII and CII. Chiral symmetry manifests in the relation

$$\{\mathcal{C}, H\} = 0 \tag{7}$$

where in the chiral basis

$$\mathcal{C} = \begin{bmatrix} \mathbf{1}_{\beta N/2} & 0 \\ 0 & -\mathbf{1}_{\beta N/2} \end{bmatrix} . \tag{8}$$

The Hamiltonians thus take the block off-diagonal form

$$H = \begin{bmatrix} 0 & K \\ K^\dagger & 0 \end{bmatrix} , \tag{9}$$

where the $\beta N/2 \times \beta N/2$ matrices K have no further symmetries. We draw the matrices H from the chiral Gaussian Unitary, respectively, Symplectic Ensembles (chGUE, chGSE). To study questions of topology, we give these random matrices a parametric dependence $K = K(p)$ and thus, $H = H(p)$, where the real variable p lies on the unit circle. The winding number corresponding to these Hamiltonians is then [33,34]

$$W = \frac{1}{2\pi i} \int_0^{2\pi} w(p)\, dp \, , \tag{10}$$

with the winding number density

$$w(p) = \frac{d}{dp} \ln \det K(p) = \frac{1}{\det K(p)} \frac{d}{dp} \det K(p) \, . \tag{11}$$

Cauchy's argument principle applies to the integral (10), provided $\det K$ is a non-zero analytic function of p, see Section 2.1 and particularly, Equation (3).

To produce explicit results, we choose a particular realization of the parameter dependence. With two smooth and 2π periodic scalar functions $a(p)$ and $b(p)$, we set

$$K(p) = a(p)K_1 + b(p)K_2 \, , \tag{12}$$

where the matrices K_1 and K_2 have dimensions $\beta N/2 \times \beta N/2$. The associated Hamiltonians

$$H(p) = a(p)H_1 + b(p)H_2 \qquad \text{with} \qquad H_m = \begin{bmatrix} 0 & K_m \\ K_m^\dagger & 0 \end{bmatrix} , \; m = 1, 2 \, , \tag{13}$$

define parametric combinations of either two chGUE's or two chGSE's. Averages over these combined ensembles have to be performed. It is convenient to introduce the vector

$$v(p) = (a(p), b(p)) \in \mathbb{C}^2 \, . \tag{14}$$

Time-reversal invariance imposes the condition $v^*(p) = v(-p)$ in the chiral symplectic case CII.

3.2. Statistical Quantities Considered

Considering k different points p_i, $i = 1, \ldots, k$, on the unit circle, we are interested in the k-point correlators of winding number densities

$$C_k^{(\beta,N)}(p_1, \ldots, p_k) = \langle w(p_1) \cdots w(p_k) \rangle \tag{15}$$

The precise meaning of the angular brackets indicating the ensemble average will be given later on. In the chiral unitary case AIII, we computed these correlators directly [1], see Section 4.1. As, first, this approach becomes forbiddingly complicated in the chiral symplectic case CII, and, second, results in cumbersome expressions for larger k, we calculated the generators

$$Z_{k|l}^{(\beta,N)}(q,p) = \left\langle \frac{\prod_{j=1}^{l} \det K(p_j)}{\prod_{j=1}^{k} \det K(q_j)} \right\rangle \tag{16}$$

for two sets of variables $p_1, \ldots, p_l$ and $q_1, \ldots, q_k$ in Ref. [2], see Section 4.4. Only the case $k = l$ is needed, but the more general definition (16) for k and l being different has technical advantages. We notice that k and l are the numbers of determinants in denominator and numerator, respectively. The k-fold derivative

$$C_k^{(\beta,N)}(p_1, \ldots, p_k) = \frac{\partial^k}{\prod_{j=1}^{k} \partial p_j} Z_{k|k}^{(\beta,N)}(q,p)\Bigg|_{q=p} \tag{17}$$

of the generator (16) for $k = l$ at $q = p$ yields the correlator (15). Anticipating the later discussion, we emphasize that the generators for both Dyson indices $\beta = 2, 4$ will exhibit a remarkably clear structure [2] which is an important reason to address them here. It is worth mentioning that the correlators (15) and the generators (16) are very different from those for the parametric level motion considered in Refs. [8,9].

Furthermore, we also computed the distribution of winding numbers $P(W)$ in the chiral unitary case AIII [1], see Section 4.2.

3.3. Mapping a Topological to a Spectral Problem

At first sight, the computation of the correlators (15) and the generators (16) appears as a formidable task, requiring the development of completely new techniques in RMT. Luckily, one can establish a link between the topological problem set up above and spectral problems in RMT for which a wealth of literature exists. This amounts to a tremendous simplification, even though the calculations to be performed are still involved and quite demanding, particularly in the chiral symplectic case. The key observation is that a combination of the two matrices K_1 and K_2 in Equation (12) encodes all the statistical information needed. Pulling out K_1, say, one has

$$K(p) = a(p)K_1 + b(p)K_2 = b(p)K_1\left(\kappa(p)\mathbf{1}_{\beta N/2} + K_1^{-1}K_2\right) \tag{18}$$

with the ratio

$$\kappa(p) = \frac{a(p)}{b(p)} . \tag{19}$$

Since the winding number density (11) is the derivative of the logarithm

$$\ln\det K(p) = \ln\det K_1 + \beta N \ln b(p) + \ln\det\left(\kappa(p) + K_1^{-1}K_2\right) , \tag{20}$$

the first term $\ln\det K_1$ does not contribute and, remarkably, only the combination $Y = K_1^{-1}K_2$ is relevant. Using Equation (18), the generators acquire the form

$$Z_{k|k}^{(\beta,N)}(q,p) = \left(\prod_{j=1}^{k} \frac{b(p_j)}{b(q_j)}\right)^{\beta N} \left\langle \prod_{j=1}^{k} \frac{\det(\kappa(p_j)\mathbf{1}_{\beta N/2} + Y)}{\det(\kappa(q_j)\mathbf{1}_{\beta N/2} + Y)} \right\rangle , \tag{21}$$

which as well only contains the matrix Y.

The task to be solved is the derivation of the probability density for the random matrices $Y = K_1^{-1}K_2$ from the independent Gaussian distributions for the random matrices K_1 and K_2. Once again, luckily, the results are known as spherical [35,36] ensembles and their probability densities read explicitly

$$\widetilde{G}^{(\beta)}(Y) = \frac{1}{\pi^{\beta N^2/2}} \prod_{j=1}^{N} \frac{(\beta(N+j)/2-1)!}{(\beta j/2-1)!} \frac{1}{\det^{2N}\left(\mathbf{1}_{\beta N/2} + YY^{\dagger}\right)} . \tag{22}$$

These ensembles are referred to as complex spherical and quaternion spherical for $\beta = 2, 4$. In the complex case, the probability density (22) can be reduced to a joint probability density of the N complex eigenvalues $z = \mathrm{diag}\,(z_1, \ldots, z_N)$ of Y and reads

$$G^{(2)}(z) = \frac{1}{c^{(2)}} |\Delta_N(z)|^2 \prod_{j=1}^{N} \frac{1}{(1+|z_j|^2)^{N+1}} \tag{23}$$

with the the Vandermonde determinant

$$\Delta_N(z) = \prod_{j<l} (z_j - z_l)\,. \tag{24}$$

In the quaternion case, however, each eigenvalue z_j of Y has a complex conjugate z_j^*, which is also an eigenvalue. The corresponding joint probability density of the eigenvalues $z = \mathrm{diag}\,(z_1, z_1^*, z_2, z_2^*, \ldots, z_N, z_N^*)$ is given by

$$G^{(4)}(z) = \frac{1}{c^{(4)}} \Delta_{2N}(z) \prod_{j=1}^{N} \frac{z_j - z_j^*}{(1+|z_j|^2)^{2N+2}}\,. \tag{25}$$

The normalization constants are

$$c^{(\beta)} = \left(\frac{\beta\pi}{2}\right)^N N! \prod_{j=1}^{N} \mathrm{B}\left(\frac{\beta j}{2}, \frac{\beta(N+1-j)}{2}\right), \tag{26}$$

where $\mathrm{B}(x, y)$ is Euler's Beta function. The question whether the integrals to be calculated are well-defined for $\beta = 4$ arises, but the answer is affirmative [2]. Hence, the ensemble average over a function $f(z)$ to be performed amounts to carrying out the integral

$$\langle f(z) \rangle = \int_{\mathbb{C}} d[z_1] \cdots \int_{\mathbb{C}} d[z_N]\, G^{(\beta)}(z)\, f(z) \tag{27}$$

over all complex eigenvalues. Hence, by reducing the two chiral ensembles to a single spherical one for either β, all information of the topological problem is contained in the determinants $\det(\kappa(p)\mathbf{1}_{\beta N/2} + Y)$ or their derivatives. Most advantageously, this is equivalent to a spectral problem where Y and $\kappa(p)$ formally play the roles of a (complex or quaternion) "Hamiltonian" and of the corresponding "energy", respectively.

4. Results

The correlators for the unitary case are addressed in Section 4.1, the distribution is given in in Section 4.2. Aspects of universality are discussed in Section 4.3. The generators in the chiral unitary and symplectic cases are presented in Section 4.4.

4.1. Winding Number Correlators in the Chiral Unitary Case

In Ref. [1], we calculated the winding number correlators $C_k^{(2,N)}(p_1, \ldots, p_k)$ as defined in Equation (15) in the unitary case directly. We chose

$$a(p) = \cos p \qquad \text{and} \qquad b(p) = \sin p\,. \tag{28}$$

Using Equations (11) and (20) as well as the complex eigenvalues of Y, one has

$$w(p) = N \cot p + y(p) \qquad \text{with} \qquad y(p) = -\frac{1}{\sin^2 p} \sum_{n=1}^{N} \frac{1}{\cot p + z_n}\,. \tag{29}$$

Only the k-fold products of $y(p)$ have to be ensemble averaged with the joint probability density (23), the presence of the inconvenient term $N \cot p$ implies that the correlator

$C_k^{(2,N)}(p_1,\ldots,p_k)$ of the k winding number densities $w(p_j)$ becomes a combinatorial sum of the $y(p_j)$ correlators. Furthermore, the latter themselves turn out to be rather involved combinatorial expressions. Eventually, $C_k^{(2,N)}(p_1,\ldots,p_k)$ is found to be a combinatorial sum of determinants with the entries

$$L_{nml}(q_l) = \frac{(-1)^{m-n}\pi}{q_l^{m-n+1}} \mathrm{B}(m, N-m+1) \begin{cases} u_m(N, q_l^2) & m \geq n \\ -v_m(N, q_l^2) & m < n \end{cases} \tag{30}$$

with the properly normalized incomplete Beta functions

$$\begin{aligned} u_m(N, q_l^2) &= \frac{2}{\mathrm{B}(m, N-m+1)} \int_0^{q_l} d\rho \frac{\rho^{2m-1}}{(1+\rho^2)^{N+1}} \\ v_m(N, q_l^2) &= \frac{2}{\mathrm{B}(m, N-m+1)} \int_{q_l}^{\infty} d\rho \frac{\rho^{2m-1}}{(1+\rho^2)^{N+1}} \end{aligned} \tag{31}$$

that satisfy $u_m(N, q_l^2) + v_m(N, q_l^2) = 1$. Even though $\mathrm{B}(m, N-m+1)$ drops out in the $L_{nml}(q_l)$, this normalization has advantages, see Ref. [1]. The first two correlators read

$$\begin{aligned} C_1^{(2,N)}(p_1) &= 0 \\ C_2^{(2,N)}(p_1, p_2) &= -\frac{1-\cos^{2N}(p_1-p_2)}{1-\cos^2(p_1-p_2)} \,. \end{aligned} \tag{32}$$

The at-first-sight surprising vanishing of the averaged winding number density is actually quite natural, as the winding number W must have a symmetric distribution with vanishing first moment. The integral of $C_1^{(2,N)}(p_1)$ over p_1 is this first moment.

4.2. Winding Number Distribution

In Ref. [1], we also computed the winding number distribution $P(W)$ in the unitary case for the choice (28). Using Cauchy's argument principle, we derive the discrete probability distribution

$$P(W) = r\left(\frac{W+N}{2}\right)\binom{N}{(W+N)/2} \tag{33}$$

on the integers W between $-N$ and $+N$ for arbitrary, finite matrix dimension N. Here, $r(m)$ is the probability that m eigenvalues are inside the unit circle and the remaining ones outside which may be written as

$$r(m) = \int_{|z_1|<1} d[z_1] \cdots \int_{|z_m|<1} d[z_m] \int_{|z_{m+1}|>1} d[z_{m+1}] \cdots \int_{|z_N|>1} d[z_N]\, G^{(2)}(z)\,. \tag{34}$$

Calculating the integrals yields

$$r(m) = \frac{1}{N!} \sum_{\omega \in \mathbb{S}_N} \left(\prod_{i=1}^{m} u_{\omega(i)}(N,1)\right)\left(\prod_{i=m+1}^{N} v_{\omega(i)}(N,1)\right), \tag{35}$$

in terms of the functions (31). The combinatorial factor in Formula (33) takes into account the permutation invariance of the eigenvalues inside, respectively, outside, the unit circle. The sum runs over all permutations, $\mathbb{S}_N$ is the permutation group.

4.3. Aspects of Universality

The quest for universality is twofold, first, there is the question of whether the same statistical effects, distributions or scalings, etc, can be identified in empirical or experimental data of different physical systems. Second, there is the theoretical and mathematical side concerned with often schematic models and their ability to describe or even predict the results from data analysis. In the case of spectral correlations, universal statistics is found on the local scale of the mean level spacing, i.e., universalities are revealed after a rescaling of the energies, referred to as unfolding. The unfolded correlators of, on the one hand, RMT for infinite level number and of, on the other hand, numerous physical systems of very different nature with large number of levels coincide, see the discussion in Refs. [6,7]. The theoretical and mathematical challenge is non-trivial as it amounts to showing that a most general class of probability densities for the random matrices yields after unfolding the same statistical quantities. Put differently, it suffices to consider Gaussians because the resulting statistics is, always after unfolding, universal.

In the case of statistical topology, universality is of equally high importance, but it appears to be considerably more complicated. Already, on the theoretical and mathematical side, there are several natural questions to be posed: First, is there an unfolding scale comparable to the local mean level spacing and how is it related to the scale of the level velocity as in the parametric correlations [8,9,37]? Second, which probability densities for the random matrices yield in the model set up in Section 3.1 the same statistics? Third, what are the conditions on the functions $a(p)$ and $b(p)$ or, more precisely, the combined conditions on these functions and the probability densities that yield in the model universal statistics? Fourth, is it possible to find universal statistics for models more general than the one in Section 3.1?

In Ref. [1], we started addressing these issues in the unitary case for the choice (28). Of course, this limits our discussion, a future fully fledged investigation ought to also consider the impact of different choices for these functions. Guided by unfolding in spectral statistics, we rescaled the arguments p_i in the correlation functions $C_k^{(2,N)}(p_1,\ldots,p_k)$ according to

$$\psi_i = N^\alpha p_i \,. \tag{36}$$

The power α should be positive because we want to zoom into the parametric dependence in the limit $N \to \infty$. Consider the two-point function (32) and the limit

$$\lim_{N\to\infty} C_2^{(2,N)}\left(\frac{\psi_1}{N^\alpha}, \frac{\psi_2}{N^\alpha}\right)\frac{d\psi_1}{N^\alpha}\frac{d\psi_2}{N^\alpha} = f_2^{(\alpha)}(\psi_1,\psi_2)d\psi_1 d\psi_2 \tag{37}$$

defining the function $f_2^{(\alpha)}$, if existing. A straightforward calculation yields

$$f_2^{(\alpha)}(\psi_1,\psi_2) = \begin{cases} -\dfrac{1}{(\psi_1-\psi_2)^2} & \alpha < \frac{1}{2} \\ -\dfrac{1-\exp(-(\psi_1-\psi_2)^2)}{(\psi_1-\psi_2)^2} & \alpha = \frac{1}{2} \\ 0 & \alpha > \frac{1}{2} \end{cases} . \tag{38}$$

We notice $C_2^{(2,N)}(p_1,p_1) = -1$, see Equation (32), implying that $\psi_1 \neq \psi_2$ when taking the limit for arbitrary α. The result (38) reveals different regimes, the one for $\alpha = 1/2$ involves the same scale as in Refs. [8,9]. Figure 4 shows results for two values of α and various values of N, the unfolded two-point function approaches the limit (38) when N increases. We conjectured that the function $f_2^{(\alpha)}(\psi_1,\psi_2)$ is universal [1].

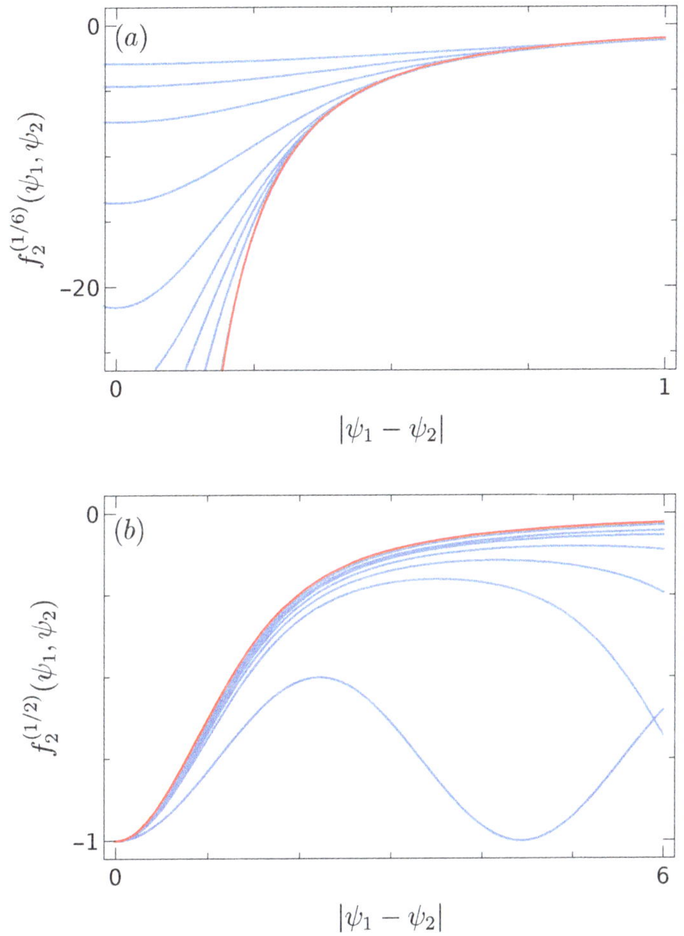

Figure 4. Unfolded two-point function after the rescaling (36) for different values of N (blue). In (**a**), we used $N = 5, 10, 20, 50, 100, 150, 200, 300, 1000$ and $\alpha = 1/6$, in (**b**) $N = 2, 5, 7, 10, 15, 20, 50, 100$ and $\alpha = 1/2$. For comparison, the limit (37) is presented (red). Taken from Ref. [1].

In Ref. [1], we also showed that the winding number distribution (33) becomes Gaussian for large N. More precisely, its second moment behaves like $\langle W^2 \rangle \sim \sqrt{N}$, suggesting an unfolding of the form $W/N^{1/4}$, i.e., different from the rescaling above. It then follows that $P(W)$ approaches a Gaussian with variance $2\sqrt{N/\pi}$ for large N.

4.4. Generators in the Chiral Unitary and Symplectic Cases

We computed the generators (16), respectively, (21) exactly for $\beta = 2$ and $\beta = 4$ in Ref. [2]. To this end, we used the method proposed some years ago in Refs. [38,39]. It identifies and employs, in ordinary space, supersymmetric structures deeply rooted in the ensemble averages. As there is no mapping performed of the ensemble averages to superspace, the method is often referred to, jokingly, but not deceptively, as "supersymmetry without supersymmetry". In the chiral unitary case $\beta = 2$, we found a ratio of two determinants,

$$Z_{k|k}^{(2,N)}(q,p) = \frac{\det\left[\frac{1}{v^T(q_m)\sigma_2 v(p_n)}\left(\frac{v^\dagger(q_m)v(p_n)}{v^\dagger(q_m)v(q_m)}\right)^N\right]_{1\leq m,n\leq k}}{\det\left[\frac{1}{v^T(q_m)\sigma_2 v(p_n)}\right]_{1\leq m,n\leq k}}, \tag{39}$$

where σ_2 is the second 2×2 Pauli matrix and $v(p_n)$ is the vector defined in Equation (14). In the chiral symplectic case $\beta = 4$, we arrived at a ratio of a Pfaffian and a determinant,

$$Z_{k|k}^{(4,N)}(q,p) = \frac{\mathrm{Pf}\begin{bmatrix}\widehat{K}_1(p_m,p_n) & \widehat{K}_2(p_m,q_n)\\ -\widehat{K}_2(p_n,q_m) & \widehat{K}_3(q_m,q_n)\end{bmatrix}_{1\leq m,n\leq k}}{\det\left[\frac{1}{iv^T(q_m)\sigma_2 v(p_n)}\right]_{1\leq m,n\leq k}}. \tag{40}$$

The three kernel functions $\widehat{K}_l(p_m,p_n)$, $l = 1,2,3$ are quite complicated and can be found explicitly in Ref. [2]. Considering the complexity of the problem and of its mathematical structure, these are remarkably compact results, even in the chiral symplectic case. This compactness is the reason why we present these results here. Their form is intimately connected with the mapping of the topological to a spectral problem discussed in Section 3.3 because such determinant and Pfaffian expressions are ubiquitous for the generators in spectral statistics. Importantly, this carries over, at least for the model considered, to the generators for the correlators of winding number densities.

5. Discussion and Conclusions

Statistical Topology is an emerging branch in statistical physics, with connections to various branches of mathematics. It is triggered by the identification of topological questions in many areas of physics, ranging from quantum mechanics and quantum field theory over semiclassics to QCD and Condendsed Matter Physics. First, I tried to give an introduction to winding number statistics for newcomers who do not have any background, avoiding usage of expert jargon and of burying the key ideas under the adavanced terminology developed in mathematics and mathematical physics. Second, I reviewed results that my collaborators and I obtained in two recent works. We studied winding numbers and associated statistical quantities in a random matrix model. There are, of course, also other topological invariants of considerable interest in physics, most notably, the Chern numbers.

I presented our first, probably awkward, steps towards looking at universal behavior. In my opinion, the most fascinating challenge for the future is the further study of universality in statistical topology, more precisely, of both of its aspects, the experimental–empirical as well as the theoretical–mathematical one.

Funding: This work was funded by the German–Israeli Foundation within the project Statistical Topology of Complex Quantum Systems, grant number GIF I-1499-303.7/2019.

Data Availability Statement: No new data were created in this study.

Acknowledgments: I thank Omri Gat, Nico Hahn, Mario Kieburg and Daniel Waltner, my collaborators of Refs. [1,2], I deeply regret that I cannot thank anymore Petr Braun who passed away in late 2020. I am greatful to Nico Hahn for Figure 3.

Conflicts of Interest: I declare no conflict of interest.

Abbreviations

The following abbreviations are used in this manuscript:

RMT	Random Matrix Theory
QCD	Quantum Chromodynamics
chGUE	chiral Gaussian Unitary Ensemble
chGSE	chiral Gaussian Symplectic Ensemble
$\mathbb{S}_N$	permutation group of N objects

References

1. Braun, P.; Hahn, N.; Waltner, D.; Gat, O.; Guhr, T. Winding number statistics of a parametric chiral unitary random matrix ensemble. *J. Phys. A Math. Theor.* **2022**, *55*, 224011.
2. Hahn, N.; Kieburg, M.; Gat, O.; Guhr, T. Winding Number Statistics for Chiral Random Matrices: Averaging Ratios of Determinants with Parametric Dependence. *arXiv* **2022**, arXiv:2207.08612.
3. Kitaev, A.Y. Unpaired Majorana fermions in quantum wires. *Physics-Uspekhi* **2001**, *44*, 131–136. [CrossRef]
4. Kitaev, A.Y. Anyons in an exactly solved model and beyond. *Ann. Phys.* **2006**, *321*, 2–111. [CrossRef]
5. Pattanayak, A.; Pujari, S.; Dixit, G. Role of Majorana fermions in high-harmonic generation from Kitaev chain. *Sci. Rep.* **2022**, *12*, 6722. [CrossRef]
6. Mehta, M.L. *Random Matrices*; Academic Press: Cambridge, MA, USA, 2004.
7. Guhr, T.; Müller-Groeling, A.; Weidenmüller, H.A. Random-matrix theories in quantum physics: Common concepts. *Phys. Rep.* **1998**, *299*, 189–425. [CrossRef]
8. Simons, B.D.; Altshuler, B.L. Universal velocity correlations in disordered and chaotic systems. *Phys. Rev. Lett.* **1993**, *70*, 4063–4066. [CrossRef]
9. Simons, B.D.; Altshuler, B.L. Universalities in the spectra of disordered and chaotic systems. *Phys. Rev. B* **1993**, *48*, 5422–5438. [CrossRef]
10. Verbaarschot, J. Spectrum of the QCD Dirac operator and chiral random matrix theory. *Phys. Rev. Lett.* **1994**, *72*, 2531–2533. [CrossRef]
11. Verbaarschot, J.; Wettig, T. Random Matrix Theory and Chiral Symmetry in QCD. *Annu. Rev. Nucl. Part. Sci.* **2000**, *50*, 343–410. [CrossRef]
12. Shuryak, E.; Verbaarschot, J. Random matrix theory and spectral sum rules for the Dirac operator in QCD. *Nucl. Phys. A* **1993**, *560*, 306–320. [CrossRef]
13. Wettig, T.; Schäfer, A.; Weidenmüller, H. The chiral phase transition and random matrix models. *Nucl. Phys. A* **1996**, *610*, 492–499. [CrossRef]
14. Wettig, T.; Schäfer, A.; Weidenmüller, H. The chiral phase transition in a random matrix model with molecular correlations. *Phys. Lett. B* **1996**, *367*, 28–34. [CrossRef]
15. Jackson, A.D.; Verbaarschot, J.J.M. Random matrix model for chiral symmetry breaking. *Phys. Rev. D* **1996**, *53*, 7223–7230. [CrossRef] [PubMed]
16. Verbaarschot, J.J.M.; Zahed, I. Spectral density of the QCD Dirac operator near zero virtuality. *Phys. Rev. Lett.* **1993**, *70*, 3852–3855. [CrossRef]
17. Guhr, T.; Wilke, T.; Weidenmüller, H.A. Stochastic Field Theory for a Dirac Particle Propagating in Gauge Field Disorder. *Phys. Rev. Lett.* **2000**, *85*, 2252–2255. [CrossRef]
18. Altland, A.; Zirnbauer, M.R. Nonstandard symmetry classes in mesoscopic normal-superconducting hybrid structures. *Phys. Rev. B* **1997**, *55*, 1142–1161. [CrossRef]
19. Heinzner, P.; Huckleberry, A.; Zirnbauer, M.R. Symmetry Classes of Disordered Fermions. *Commun. Math. Phys.* **2005**, *257*, 725–771. [CrossRef]
20. Kitaev, A. Periodic table for topological insulators and superconductors. *AIP Conf. Proc.* **2009**, *1134*, 22–30. [CrossRef]
21. Schnyder, A.P.; Ryu, S.; Furusaki, A.; Ludwig, A.W.W. Classification of topological insulators and superconductors in three spatial dimensions. *Phys. Rev. B* **2008**, *78*, 195125. [CrossRef]
22. Chiu, C.K.; Teo, J.C.Y.; Schnyder, A.P.; Ryu, S. Classification of topological quantum matter with symmetries. *Rev. Mod. Phys.* **2016**, *88*, 035005. [CrossRef]
23. Oppermann, R. Anderson localization problems in gapless superconducting phases. *Phys. A Stat. Mech. Its Appl.* **1990**, *167*, 301–312. [CrossRef]
24. Zirnbauer, M.R. Particle–hole symmetries in condensed matter. *J. Math. Phys.* **2021**, *62*, 021101. [CrossRef]
25. Gade, R. Anderson localization for sublattice models. *Nucl. Phys. B* **1993**, *398*, 499–515. [CrossRef]
26. Prodan, E.; Schulz-Baldes, H. *Bulk and Boundary Invariants for Complex Topological Insulators: From K-Theory to Physics*; Mathematical Physics Studies; Springer International Publishing: Cham, Switzerland, 2016. [CrossRef]
27. Chen, B.H.; Chiou, D.W. An elementary rigorous proof of bulk-boundary correspondence in the generalized Su-Schrieffer-Heeger model. *Phys. Lett. A* **2020**, *384*, 126168. [CrossRef]
28. Shapiro, J. The bulk-edge correspondence in three simple cases. *Rev. Math. Phys.* **2020**, *32*, 2030003. [CrossRef]

29. Alicea, J. New directions in the pursuit of Majorana fermions in solid state systems. *Rep. Prog. Phys.* **2012**, *75*, 076501. [CrossRef] [PubMed]
30. Walker, P.N.; Wilkinson, M. Universal Fluctuations of Chern Integers. *Phys. Rev. Lett.* **1995**, *74*, 4055–4058. [CrossRef] [PubMed]
31. Gat, O.; Wilkinson, M. Correlations of quantum curvature and variance of Chern numbers. *SciPost Phys.* **2021**, *10*, 149. [CrossRef]
32. Lu, L.; Joannopoulos, J.; Soljačić, M. Topological Photonics. *Nat. Photonics* **2014**, *8*, 821–829. [CrossRef]
33. Maffei, M.; Dauphin, A.; Cardano, F.; Lewenstein, M.; Massignan, P. Topological characterization of chiral models through their long time dynamics. *New J. Phys.* **2018**, *20*, 013023. [CrossRef]
34. Asbóth, J.K.; Oroszlány, L.; Pályi, A. *A Short Course on Topological Insulators*; Springer International Publishing: Berlin/Heidelberg, Germany, 2016. [CrossRef]
35. Forrester, P.J.; Krishnapur, M. Derivation of an eigenvalue probability density function relating to the Poincaré disk. *J. Phys. Math. Theor.* **2009**, *42*, 385204. [CrossRef]
36. Mays, A. A Real Quaternion Spherical Ensemble of Random Matrices. *J. Stat. Phys.* **2013**, *153*, 48–69. [CrossRef]
37. Beenakker, C.; Rejaei, B. Random-matrix theory of parametric correlations in the spectra of disordered metals and chaotic billiards. *Phys. A Stat. Mech. Its Appl.* **1994**, *203*, 61–90. [CrossRef]
38. Kieburg, M.; Guhr, T. Derivation of determinantal structures for random matrix ensembles in a new way. *J. Phys. A* **2010**, *43*, 075201. [CrossRef]
39. Kieburg, M.; Guhr, T. A new approach to derive Pfaffian structures for random matrix ensembles. *J. Phys. A* **2010**, *43*, 135204. [CrossRef]

entropy

Article

Towards the Resolution of a Quantized Chaotic Phase-Space: The Interplay of Dynamics with Noise

Domenico Lippolis [1] and Akira Shudo [2,*]

[1] Institute for Applied Systems Analysis, Jiangsu University, Zhenjiang 212013, China
[2] Department of Physics, Tokyo Metropolitan University, Minami-Osawa, Hachioji 192-0397, Tokyo, Japan
* Correspondence: shudo@tmu.ac.jp

Abstract: We outline formal and physical similarities between the quantum dynamics of open systems and the mesoscopic description of classical systems affected by weak noise. The main tool of our interest is the dissipative Wigner equation, which, for suitable timescales, becomes analogous to the Fokker–Planck equation describing classical advection and diffusion. This correspondence allows, in principle, to surmise a finite resolution, other than the Planck scale, for the quantized state space of the open system, particularly meaningful when the latter underlies chaotic classical dynamics. We provide representative examples of the quantum-stochastic parallel with noisy Hopf cycles and Van der Pol-type oscillators.

Keywords: quantum dissipation; stochasticity; nonlinearity; chaos; Wigner equation; Fokker–Planck equation

Citation: Lippolis, D.; Shudo, A. Towards the Resolution of a Quantized Chaotic Phase-Space: The Interplay of Dynamics with Noise. *Entropy* **2023**, *25*, 411. https://doi.org/10.3390/e25030411

Academic Editor: Marko Robnik

Received: 30 December 2022
Revised: 6 February 2023
Accepted: 12 February 2023
Published: 24 February 2023

1. Introduction

Efforts to reconcile classical and quantum mechanics are just about as old as quantum mechanics itself. While the formulation in Hilbert space makes it difficult to establish a direct correspondence between the two, a projection of the wave function to phase space may reveal some formal affinities between the quantum evolution of probability density and the traditional Liouville formalism of classical mechanics. The closest one can get to relate the two is by projecting the Liouville–von Neumann equation onto a suitable state space. For example, choosing the traditional phase space, we may obtain the so-called Wigner representation, which shares similarities with the aforementioned classical density evolution.

Yet, there are also notable differences, as it stands to reason. The Wigner function, which is the projection of the density operator onto the phase space, may also take on negative values; its evolution is governed by an equation plagued with an infinity of derivatives, and, as an indirect consequence of that, it may attain scales smaller than Planck's constant [1]. This is especially true in systems whose underlying classical dynamics exhibit chaotic behavior.

In reality, however, no system is perfectly and eternally isolated, and exchanges of matter or energy with the surrounding environment are inevitable, whether due to measurements, thermal interactions, or shot noise [2–4]. That brings dissipation into the picture and, with that, decoherence.

The effect of the environment on the evolution of a density matrix in a phase-space representation was first studied by Feynman and Vernon [5], who extended the path-integral formalism to dissipative quantum dynamics. Later, Caldeira and Leggett [6] derived an equivalent partial differential equation for the density matrix, which bears diffusive and dissipative terms similar to the Fokker–Planck equation. The latter analogy was then thought to hold in the semiclassical limit until a new wave of contributions [7–10] reexamined the problem in a quantum chaotic setting. A most remarkable outcome of those works is the identification of a decoherence time, beyond which the Wigner equation is, in all, a Fokker–Planck equation since the higher-order derivatives may be safely neglected, and the quantized phase space may be resolved only up to a finite scale. Such resolution does not depend on the Planck constant, but rather emerges from the balance of the phase-space contraction rate (Lyapunov exponent) with the coupling of the system with an Ohmic environment.

More recent contributions have focussed on the efficiency of Wigner evolution for general types of dissipation [11], and on obtaining a Lindblad-based dissipative Wigner equation to tackle quantum friction [12,13].

Once it is established that, under suitable conditions and after a sufficiently long time of evolution, the Wigner equation has the form of a Fokker–Planck equation, the quantum dissipative problem is cast into a classical stochastic process. Moreover, if the underlying classical dynamics of the quantum system in exam are chaotic, the limiting resolution of the phase space postulated in refs. [8,10] is not expected to be uniform, but will depend on the local interplay of the stretching/contraction with the dissipation. In the equivalent classical noisy problem, it is the 'Brownian' diffusion that plays the role of dissipation.

In the past decade, significant steps [14–18] were taken to determine the resolution of a chaotic state space in the presence of weak noise and reduce the dynamics to a Markov process of finite degrees of freedom in the form of a connectivity matrix. Low-dimensional discrete-time dynamical systems such as logistic- or Hénon-type maps have been treated in a non-Hamiltonian setting, whose quantum analogs are, in principle, difficult to identify. The optimal resolution hypothesis should be extended to continuous time flows as well, and the starting point of that roadmap is a thorough comprehension of the steady-state solutions of the Fokker–Planck equation around the building blocks of chaos: periodic orbits.

Here, we intend to lay the foundation of that understanding, by solving the Fokker–Planck equation of nonlinear paradigmatic dynamical systems, classical and with weak noise. We examine two-dimensional flows featuring nonlinearities but not yet chaos, where the competition between contraction and noise around a limit cycle results in a stationary density, which characterizes the steady state, and, as shown at the very end of the manuscript, shares common traits with the steady-state Wigner function of a case study in quantum dissipative dynamics.

The article is structured as follows: in Section 2 we review the basic tools of the phase-space representation of quantum dynamics, both in closed and open systems. We follow up in Section 3 by discussing the main issues related to the evolution of the Wigner function in a quantum chaotic setting, the effects of dissipation, and the correspondence of the Wigner with the Fokker–Planck equation. In Section 4, a novel methodology is introduced to evaluate the steady-state solution of the Fokker–Planck equation around a periodic orbit, which casts the partial differential equation into an ordinary differential equation for the covariance matrix, known as the Lyapunov equation. We first present a proof of concept on the simplest limit cycle, of circular shape as from a Hopf bifurcation, to be followed in Section 5 by the treatment of the nonlinear oscillators, which are the main object of the current study. At the end of the section, the results on the Fokker–Planck steady-state densities are paralleled to those obtained for the Wigner function in a recent study of a quantum-dissipative model of the same oscillators. Summary and discussion close the paper.

2. Density Matrix, Wigner Function, and Dissipation

Given a collection of physical systems, the ensemble average of an observable A is given by

$$\langle A \rangle = \sum_i \rho_i \langle \psi_i | A | \psi_i \rangle \,, \tag{1}$$

or, using

$$\rho = \sum_i \rho_i | \psi_i \rangle \langle \psi_i | \,, \tag{2}$$

one can simply write

$$\langle A \rangle = \mathrm{Tr}[\rho\, A]\,, \tag{3}$$

so that, if the observable A is time-independent, knowing ρ at all times means solving the problem of dynamics. That is the motivation for studying the density matrix ρ in the first place.

2.1. *Quantum Dynamics in the Phase Space*

Now, the density ρ evolves according to the Liouville–von Neumann equation

$$i\hbar\rho_t = [\rho, H]\,, \tag{4}$$

the quantum analog of the well-known Liouville equation

$$\rho_t = \{\rho, H\} \tag{5}$$

of classical dynamics, which we spell out in phase space:

$$\partial_t\rho = -\frac{p}{m}\partial_x\rho + (\partial_x V(x)\partial_p)\rho\,, \tag{6}$$

assuming a Hamiltonian of the form $H = \frac{p^2}{2m} + V(x)$.

In order to integrate the Liouville–von Neumann Equation (4), we need to project it onto some basis, and several representations are already available to us, for instance, the P- or the Q-representation (a.k.a. Husimi's)

$$Q(\alpha, \alpha^*) = \frac{1}{\pi}\langle\alpha|\rho|\alpha\rangle\,, \tag{7}$$

with $|\alpha\rangle$ a coherent state. Studying quantum-to-classical correspondence, especially of a system that exhibits chaotic behavior, is generally best achieved by using the Wigner representation [19]

$$W(x,p) = \frac{1}{2\pi\hbar}\int e^{-ipy/\hbar}\psi\left(x+\frac{y}{2}\right)\psi^*\left(x-\frac{y}{2}\right)dy\,, \tag{8}$$

which can also be expressed in terms of the density matrix, as

$$W(x,p) = \frac{1}{2\pi\hbar}\int e^{-ipy/\hbar}\langle x+y/2|\rho|x-y/2\rangle dy\,, \tag{9}$$

the operation being called Weyl transform. An operator A may also be projected onto phase space, by applying a Weyl transform:

$$\tilde{A}(x,p) = \frac{1}{2\pi\hbar}\int e^{-ipy/\hbar}\langle x+y/2|A|x-y/2\rangle dy\,, \tag{10}$$

which can prove handy in the evaluation of expectation values, that is

$$\langle A \rangle = \mathrm{Tr}[\rho\, A] = \int W(x,p)\tilde{A}(x,p)dxdp\,, \tag{11}$$

since, in general,

$$\mathrm{Tr}[AB] = \int \tilde{A}(x,p)\tilde{B}(x,p)dxdp\,. \tag{12}$$

Thus, expectation values of observables are determined by means of phase-space averages, and the problem of quantum mechanics boils down to that of the time evolution of the Wigner function. It has been shown [19] that $W(x,p)$ obeys the Wigner equation

$$\partial_t W(x,p) = -\frac{p}{m}\partial_x W(x,p) + \sum_{s=0}^{\infty} c_s(-\hbar^2)^s \partial_x^{2s+1} V(x) \partial_p^{2s+1} W(x,p), \tag{13}$$

(with $c_s = \frac{2^{-2s}}{(2s+1)!}$) that, in general, bears an infinite number of terms. In reality, integrating Equation (13) can already be impractical if there are just a few nontrivial terms in the summation [20]. If the potential $V(x)$ is at most quadratic, the Wigner equation reduces to Liouville's, as in (6). Otherwise, Equation (13) is still not easy to deal with, and, importantly, it may not be truncated in the semiclassical limit since the terms $\partial_p^{2s+1} W(x,p)$ bring down powers of $\hbar^{-1-2s}$, so that

$$\hbar^{2s} \cdot \frac{1}{\hbar^{2s+1}} \sim \hbar^{-1},$$

and $O\left(\hbar^{-1}\right)$ does grow in the limit $\hbar \to 0$, making no terms in the Wigner equation negligible, in principle.

2.2. Open Systems

On the other hand, let us suppose the system is connected to an environment, whose interaction produces two additional terms on the right-hand side of the Wigner Equation (13), that is [8]

$$2\gamma\partial_p[pW(x,p)] + D\partial_{pp}^2 W(x,p).$$

The first term produces relaxation, due to the exchange of energy with the environment, and γ is the relaxation rate. The second term means diffusion, responsible for the so-called decoherence process, where one sets $D = 2\gamma M k_B T$, with M mass of the system, and T temperature of the environment. The dissipation and diffusion terms are obtained from a path-integral formulation of the system-environment interaction, which traces back to the works of Feynman and Vernon [5], and, later, of Caldeira and Leggett [6].

If the potential $V(x)$ is, at most, quadratic, one recovers the Fokker–Planck equation, which describes the classical evolution of the density of trajectories produced by a particle subject to Brownian motion:

$$\partial_t W(x,p) = -\frac{p}{m}\partial_x W(x,p) + \partial_x V(x)\partial_p W(x,p) + 2\gamma\partial_p[pW(x,p)] + D\partial_{pp}^2 W(x,p). \tag{14}$$

This equation is fully quantum mechanical, and $W(x,p)$ may take on negative values, unlike the classical phase-space density of a Brownian particle.

Yet, for a general potential $V(x)$, the evolution of the dissipative system is ruled by the full-fledged Wigner Equation (13) plus the terms (14) due to the environment:

$$\begin{aligned}\partial_t W(x,p) = -\frac{p}{m}\partial_x W(x,p) + \sum_{s=0}^{\infty} c_s(-\hbar^2)^s \partial_x^{2s+1} V(x) \partial_p^{2s+1} W(x,p) + \\ + 2\gamma\partial_p[pW(x,p)] + D\partial_{pp}^2 W(x,p).\end{aligned} \tag{15}$$

The resulting equation is still plagued with an infinite number of derivatives, and is thus of impractical integration. In the next section, we discuss whether and how it is safe to neglect the higher-order terms in Equation (15), in the context of quantum chaos.

3. Stretching, Contracting, and Zaslavsky's Time

Let us examine some aspects of the evolution of the Wigner function, when the underlying classical dynamics of the system is chaotic. By established knowledge [21], the two main features of chaos are:

1. Nearby trajectories diverge exponentially fast, meaning that, letting $\mathbf{x} = (x, p)$,

$$\lambda = \lim_{t \to \infty} \ln \left| \frac{\delta \mathbf{x}(t)}{\delta \mathbf{x}(0)} \right| > 0\,, \tag{16}$$

in other words, the difference $\delta \mathbf{x}(t)$ between any two nearby trajectories grows exponentially fast for any initial condition. This feature is also described as extreme sensitivity of the system to initial conditions.
2. The number M of qualitatively distinct orbits ('configurations', tagged by symbolic sequences) scales exponentially with their length, so that the topological entropy is positive:

$$S = \lim_{t \to \infty} \frac{1}{t} \ln M(t) > 0\,. \tag{17}$$

3.1. Chaos and the Wigner Function

Chaos is the result of a stretching and folding process mainly due to nonlinearities. For a Hamiltonian system, volumes in the phase space are conserved (by Liouville's theorem), so that the amount of stretching (diverging trajectories) in some directions must be compensated by an equal amount of contraction in others.

As a result, the inconvenient higher-order terms in the Wigner Equation (13) can be estimated to evolve as

$$\partial_p^{2s+1} W(x, p) \propto \frac{W(x, p)}{\delta p^{2s+1}(t)} \sim \frac{W(x, p)}{\delta p(0)\, e^{-(2s+1)\lambda t}}\,, \tag{18}$$

for smooth enough $W(x, p)$. Thus, the inherent problem is, in principle, not with the smoothness of the density, but rather with the fact that the phase space contracts at an exponential rate, and therefore the contribution of higher-order derivatives in Equation (13) is more and more important, as time proceeds. To better illustrate that, let us compare the terms in the Poisson brackets of the Liouville Equation (6) (also present in the full-fledged Wigner equation), with the higher-order terms in Equation (13):

$$\frac{\partial_x V(x) \partial_p W(x, p)}{c_s \partial_x^{2s+1} V(x) \partial_p^{2s+1} W(x, p)} \sim \frac{1}{c_s} \frac{\partial_x V(x)}{\partial_x^{2s+1} V(x)} \delta p^{2s} \sim \frac{1}{c_s} \frac{\partial_x V(x)}{\partial_x^{2s+1} V(x)} \delta p(0)\, e^{-2s\lambda t} \gg 1 \tag{19}$$

is a condition for the higher-order terms to be negligible with respect to the lower-order 'Liouville' terms. The above inequality can be inverted, and turned into a condition for the time t:

$$t \ll \frac{1}{\lambda} \ln \left[\frac{\partial_x V(x)}{\partial_x^{2s+1} V(x)} \frac{\delta p(0)}{c_s} \right]^{-1/2s}\,. \tag{20}$$

Identifying the quantity $\mathcal{X}_V\, \delta p(0) = \frac{\partial_x V(x) \delta p(0)}{\partial_x^{2s+1} V(x)}$ with the typical action of the system, we can now understand

$$t^* \sim \frac{1}{\lambda} \ln \frac{\mathcal{X}_V\, \delta p(0)}{\hbar} \tag{21}$$

as the time scale within which the inconvenient higher-order terms ($s \geq 1$) of the Wigner Equation (13) may be neglected. Some literature refers to t^* as Zaslavsky's time [22]. Its meaning is somehow related to the more commonly mentioned Ehrenfest time, as in fact, t^* is longer the larger the ratio of the typical action to $\hbar$, and longest in the semiclassical limit. Still, the basic idea of this correspondence time does not relate directly with interference or need 'semiclassical' dynamics, but rather implies a finite resolution for the quantized phase space within a certain time scale, irrespective of the scale of the action. In general, the smoother the potential $V(x)$, the longer t^*, whereas the larger the Lyapunov exponent λ, the shorter t^*.

3.2. A Resolution for the Quantized Phase Space

In a simplified but physically meaningful description, that will then prove more accurate as a local model, we may recognize and estimate the competing effects of dynamical contraction on the one hand, and of dissipation-induced diffusion on the other hand, by quantizing the Hamiltonian $H = \lambda xp$. A wave packet of the form

$$W(x,p) \sim e^{-x^2/\sigma^2-\sigma^2 p^2} \tag{22}$$

evolves separately along the stretching x-direction, and the contracting p-direction. In momentum space, we have the Schrödinger equation

$$\partial_t u(p,t) = \lambda p \partial_p u(p,t) \Rightarrow u(p,t) = u_0\left(pe^{\lambda t}\right), \tag{23}$$

that maps the wave packet in the p-direction as

$$e^{-\sigma^2 p^2/2} \to e^{e^{2\lambda t}\sigma^2 p^2/2}, \tag{24}$$

and thus the width σ^{-2} shrinks by a factor of $e^{-2\lambda t}$ after time t. Identifying σ^{-1} with the uncertainty $\delta p(t)$ of the momentum, we may say that

$$\delta p(t) \sim \delta p(0) e^{-\lambda t} \tag{25}$$

along the contracting direction. On the other hand, connecting the system to an environment brings about diffusion, and

$$\partial_t u(p,t) = D\partial_{pp} u(p,t) \Rightarrow u(p,t) \sim e^{-p^2/2(\delta p(0)+Dt)}, \tag{26}$$

whose variance evolves as $\sqrt{Dt}$:

$$\delta p(t) \sim [\delta p(0) + Dt]^{1/2}. \tag{27}$$

Then, intuitively, there must be some minimal scale in the contracting direction, set by

$$\delta p_{\min} \sim \left[\frac{D}{2\lambda}\right]^{1/2}. \tag{28}$$

The full picture is called the Ornstein–Uhlenbeck problem [23]

$$\partial_t u(p,t) = D\partial_{pp} u(p,t) - \lambda \partial_p u(p,t). \tag{29}$$

In particular, the larger $\delta p_{\min}$, the closer the evolution of the Wigner function to a stochastic process. More precisely, the regime where we may neglect the higher-order derivatives in the Wigner equation is deduced from Equation (19) as

$$\frac{\mathcal{X}_V\, \delta p_{\min}}{\hbar} \gg 1, \tag{30}$$

and that requires a relatively smooth potential, and the coefficient of the decoherence term in Equation (14), D, to be comparable to the Lyapunov exponent λ. Now, if $\delta p_{\min}$ is an 'equilibrium' value as argued above, the chaotic contraction is no longer shrinking the scale of phase-space probability exponentially and indefinitely as in the non-dissipative setting (recall $\delta p(t) \sim \delta p(0)e^{-\lambda t}$). Hence, in principle, there would be no Zaslavsky's time t^*, but rather, the quantum dissipative evolution may be well described by the Fokker–Planck type of Equation (14) at all times, provided that the initial condition is smooth enough. Importantly, the semiclassical limit is not required for this approximation to work.

4. Contraction vs. Diffusion in Stochastic Dynamics

Equation (14) and the discussion from the previous section suggest that, under suitable conditions, the problem of the dynamics of a quantum system connected to an environment may be cast into the classical evolution of a density according to a Fokker–Planck equation. As a consequence, studying the interplay of stretching/contracting dynamics with weak noise may also help shed light on quantum dissipation. Particularly interesting scenarios arise when the deterministic dynamics exhibits chaotic behavior. It is, in fact, well known that the phase space of a chaotic system has a self-similar (fractal) structure of infinite resolution. However, in reality, every system experiences noise, coming from experimental uncertainties, neglected degrees of freedom, or roundoff errors, for example. No matter how weak, noise smooths out fractals, giving the system a finite resolution. The consequences are dramatic for the computation of long-time dynamical averages, such as diffusion coefficients or escape rates, since infinite-dimensional operators describing the evolution of the system (such as Fokker–Planck) effectively become finite matrices. With the aim of efficiently estimating long-time averages of observables for a chaotic dynamical system affected by background noise, a recent endeavor carried on over the past decade has achieved a technique to partition the chaotic phase space up to its optimal resolution, using unstable periodic orbits. The benchmark models already treated range from one-dimensional discrete-time repellers [14], and general unimodal maps [15], to two-dimensional chaotic attractors [17,18]. Most importantly, a finite resolution for the state space of these models has effectively changed the dimensionality of the Fokker–Planck operator from infinite to inherently finite. Consequently, computations of the desired long-time averages become simpler and more efficient. On a more intuitive note, the present results also bear physical significance because, even when the external noise is uncorrelated, additive, isotropic, and homogenous, the interplay of noise and nonlinear dynamics always results in a local stochastic neighborhood, whose covariance depends on both the past and the future noise integrated and nonlinearly convolved with deterministic evolution along the trajectory. In that sense, noise is effectively never 'white' in nonlinearity, and thus, the optimal resolution varies from neighborhood to neighborhood and has to be computed locally.

As stated in the introduction, here we attack continuous-time dynamical systems, and begin by studying the evolution of noisy neighborhoods of periodic orbits. The simplest yet meaningful models are two-dimensional limit cycles, that can serve as a testbed for parsing the interaction of deterministic dynamics with noise.

4.1. *The Lyapunov Equation around a Cycle*

Consider the Fokker–Planck equation

$$\partial_t \rho(\mathbf{x},t) = -\partial_{\mathbf{x}}(v(\mathbf{x})\rho(\mathbf{x},t)) + \Delta\partial_{\mathbf{xx}}\rho(\mathbf{x},t)\,, \tag{31}$$

where Δ is the diffusion tensor, whose entries are the noise amplitudes along each direction (Δ is diagonal with identical entries for isotropic noise). If we look at the dynamics in the neighborhood of a particular deterministic trajectory, we may linearize the velocity field $v(\mathbf{x})$ locally, and replace it with $A_a(\mathbf{x} - \mathbf{x}_a)$, where $A_a = \frac{\partial v(\mathbf{x})}{\partial \mathbf{x}}\Big|_{\mathbf{x}=\mathbf{x}_a}$ is the so-called matrix of variations. Moreover, we may switch to a co-moving reference frame in the desired neighborhood, say $z_a = \mathbf{x} - \mathbf{x}_a$.

Suppose we start off with an initial density of trajectories of Gaussian shape, that is $\rho_a(z_a) = \frac{1}{C_a}\exp\left(-z_a^\top \frac{1}{Q_a} z_a\right)$. The short-time solution to (31) can then be written in the path-integral form

$$\begin{aligned} \rho_{a+1}(z_{a+1}) &= \frac{1}{C_a}\int [dz_a]\, e^{-\frac{1}{2}(z_{a+1}-(\mathbf{1}+A_a\delta t)z_a)^\top \frac{1}{\Delta\delta t}(z_{a+1}-(\mathbf{1}+A_a\delta t)z_a)-\frac{1}{2}z_a^\top \frac{1}{Q_a} z_a} \\ &= \frac{1}{C_{a+1}}\, e^{-\frac{1}{2}z_{a+1}^\top \frac{1}{(\mathbf{1}+A_a\delta t)Q_a(\mathbf{1}+A_a\delta t)^T+\Delta\delta t} z_{a+1}}\,. \end{aligned} \tag{32}$$

One can then infer the relation between input and output quadratic forms in the exponential

$$Q_{a+1} = \Delta\delta t + (1 + A_a\delta t)Q_a(1 + A_a\delta t)^\top, \tag{33}$$

and, neglecting terms of order δt^2, recover the time-dependent Lyapunov equation

$$\dot{Q} = A(t)Q + QA^\top(t) + \Delta, \qquad Q(t_0) = Q_0. \tag{34}$$

Following the theory of time-dependent ordinary differential equations [24], we may write the solution of (34) as

$$Q(t) = J(t,t_0)\left[Q(t_0) + \int_{t_0}^{t} J^{-1}(s,t_0)\Delta\left(J^{-1}(s,t_0)\right)^\top ds\right]J^\top(t,t_0). \tag{35}$$

Here, $J(t,t_0)$ is the Jacobian along a flow $\mathbf{x} = \mathbf{x}(t)$:

$$\frac{d}{dt}J(t,t_0) = A(\mathbf{x})J(t,t_0), \qquad J(t_0,t_0) = \mathbf{1}. \tag{36}$$

One can verify this by just plugging the solution above into the equation. Alternatively, one can write Equation (35) in the simpler form

$$Q(t) = J(t,t_0)Q(t_0)J^\top(t,t_0) + \int_{t_0}^{t} J(t,s)\Delta J^\top(t,s)ds, \tag{37}$$

where the notation $J(t,s)$ means that the Jacobian is computed following a trajectory that starts at time s and ends at time t, consistently with Equation (36).

4.2. Noisy Circle

Next, consider one of the simplest two-dimensional dynamical systems, a pair of ODEs with a circular limit cycle of radius r_c, together with additive isotropic white noise of strength $2D$:

$$\begin{aligned} \dot{x} &= \lambda(r_c - \sqrt{x^2+y^2})x - \omega y + \sqrt{2D}\xi_x \\ \dot{y} &= \lambda(r_c - \sqrt{x^2+y^2})y + \omega x + \sqrt{2D}\xi_y \end{aligned} \tag{38}$$

where

$$< \xi_x(t)\xi_x(\tau) >= \delta(t-\tau), \qquad < \xi_x(t)\xi_y(\tau) >= 0. \tag{39}$$

In polar coordinates, this is written

$$\begin{aligned} \dot{r} &= \lambda(r_c - r)r + \sqrt{2D}\xi_x\cos\theta + \sqrt{2D}\xi_y\sin\theta \\ \dot{\theta} &= \omega - \sqrt{2D}\xi_x\frac{\sin\theta}{r} + \sqrt{2D}\xi_y\frac{\cos\theta}{r} \end{aligned} \tag{40}$$

This Langevin-type equation produces the drift and diffusion coefficients [23]

$$\begin{aligned} D_r &= \lambda(r_c - r)r + \frac{2D}{r} \\ D_\theta &= \omega \\ D_{rr} &= 2D \\ D_{\theta\theta} &= \frac{2D}{r^2} \end{aligned} \tag{41}$$

which then determine the Fokker–Planck equation for the system:

$$\partial_t P + \frac{1}{r}\partial_r[\lambda(r_c - r)rP] + \partial_\theta\omega P - \frac{D}{r}\partial_r(r\partial_r P) - \frac{D}{r^2}\partial_{\theta\theta}P = 0 \tag{42}$$

The limit cycle $r = r_c$ can be either stable (Figure 1) or unstable depending on the sign of λ. Let us consider the stable case.

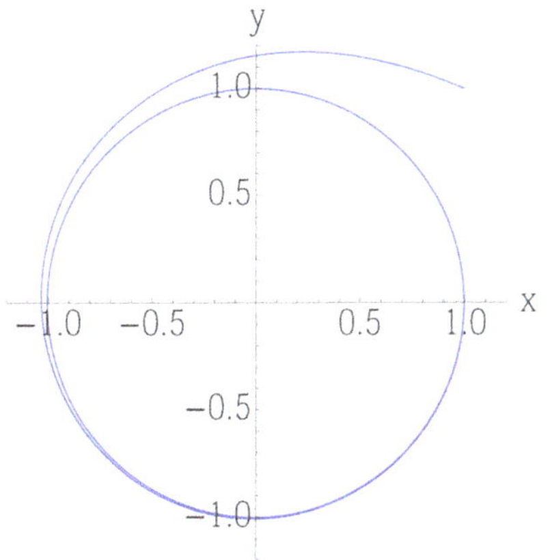

Figure 1. Solution of the numerically integrated Equation (38) without noise. Any initial condition converges to the circular limit cycle.

The first thing to look for is a stationary solution to the asymptotic form of (42):

$$\partial_r[\lambda(r_c - r)rP_\infty] - D\partial_r(r\partial_r P_\infty) = 0 \tag{43}$$

A solution is

$$P_\infty(r) = Ce^{-\frac{\lambda}{2D}(r-r_c)^2}, \tag{44}$$

which implies that P_∞ is a Gaussian of width $2\sqrt{D/\lambda}$ in the neighborhood of the limit cycle. The general solution to (42) is [23]

$$P(r,\theta,t) = e^{-\frac{\lambda}{2D}(r-r_c)^2} \sum_{n=0}^{\infty} \sum_{\nu=-\infty}^{\infty} A_n^\nu e^{-s_n^\nu t}(r-r_c)^{|\nu|} L_n^{|\nu|}(r-r_c)e^{i\nu\theta} \tag{45}$$

where $L_n^{|\nu|}(r-r_c)$ are generalized Laguerre polynomials and both the eigenvalues s_n^ν and the coefficients A_n^ν can be found numerically.

Neighborhood and Coordinates

This problem has an obvious symmetry, which allows us to guess the right (nonlinear!) change of coordinates, as well as the stationary solution, independent of the angular coordinate. The result is that the noisy neighborhood of the limit cycle is determined by the variance of the stationary solution (44). In general, however, we might not be so lucky, and guessing a suitable, possibly nonlinear, change of coordinates is probably beyond our reach. One way to identify a neighborhood for a periodic or any other orbit is to integrate the time-dependent Lyapunov Equation (34) in the original (Cartesian) coordinates, but in a co-moving frame defined by the local coordinates $z_a = \mathbf{x} - \mathbf{x}_a$ introduced in Section 4.1.

Figure 2 illustrates this second approach: the forward Lyapunov Equation (34) is numerically solved along an orbit that converges to the circular limit cycle (here we take the diffusion tensor $\Delta = \begin{pmatrix} 2D & 0 \\ 0 & 2D \end{pmatrix}$), and its solution (37) is sampled along the trajectory and *inverted* to obtain $Q^{-1}(t)$, the covariance matrix of the Gaussian density, that produces a tube (in the figure in light yellow) along the orbit. The eigenvalues of $Q^{-1}(t)$ are found to converge to $\Lambda_1 = \frac{\lambda}{2D}$, consistently with the result for the width of the stationary-state solution (44) of the full-blown radial Fokker–Planck Equation (43): that determines the width of the tube, $\sigma = \sqrt{1/2\Lambda_1}$. The second eigenvalue of Q^{-1} is $\Lambda_2 = 0$, as it appears from Figure 2b. The latter eigenvalue is to be read as follows: while the forward Lyapunov equation converges to a finite limit in the stable (radial) direction, where noise balances contracting dynamics, it diverges along the marginal (tangent) direction, and therefore its inverse converges to zero, asymptotically.

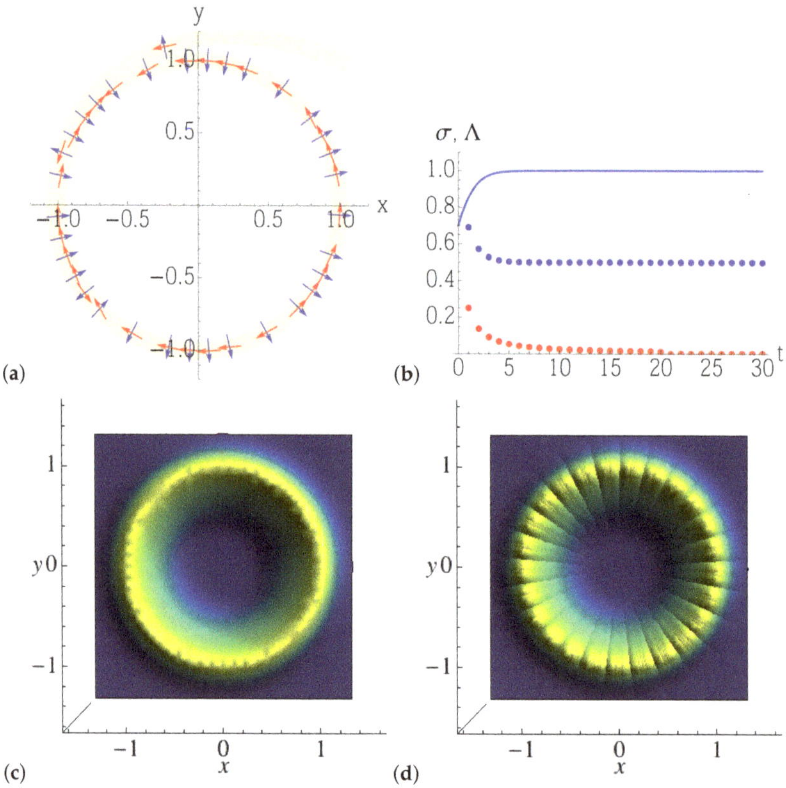

Figure 2. (**a**) Solution of the numerically integrated Equation (38), together with the eigenvectors (arrows) of the covariant matrix Q^{-1}, as given by the solution (37) of the forward Lyapunov equation, for noise amplitude $2D = 0.1$. The light yellow stripe is a pictorial representation of the Gaussian tube along the limit cycle (plan view). (**b**) The eigenvalues Λ_1 (blue dots), Λ_2 (red dots) of Q^{-1}, and the width σ (solid line) of the evolved density versus time t. (**c**) The exact steady-state solution (44) to the Fokker–Planck equation for a noisy circular limit cycle. (**d**) The approximation to the same steady state, obtained by piecing together solutions (37) to the Lyapunov equation around the limit cycle.

As shown in Figure 2c,d, the exact solution (44) of the Fokker–Planck equation is well reproduced by piecing together Gaussian tubes of covariance $Q^{-1}(t)$, each computed around a definite point of the noiseless limit cycle (the spurious lines orthogonal to the circle in Figure 2d are due to the finite sampling of the Gaussian tubes, that should ideally be a continuum).

As we may mostly be interested in the solution of the Lyapunov equation near unstable periodic orbits (like in a chaotic system), we then need to solve the same problem backwards in time, otherwise said by studying adjoint evolution, or the adjoint Lyapunov equation.

4.3. Adjoint Lyapunov Equation

The backward evolution is described by the adjoint Fokker–Planck equation

$$\partial_t \rho(\mathbf{x}, t) = v(\mathbf{x}) \partial_{\mathbf{x}} \rho(\mathbf{x}, t) + \Delta \partial_{\mathbf{xx}} \rho(\mathbf{x}, t) \,. \tag{46}$$

Following the line of thought of Section 4.1, we can write the path-integral evolution of a Gaussian density in the neighborhood of an orbit

$$\begin{aligned} \rho_a(z_a) &= \frac{1}{C_{a+1}} \int [dz_{a+1}]\, e^{-\frac{1}{2}(z_{a+1}-(\mathbf{1}+A_a\delta t)z_a)^\top \frac{1}{\Delta\delta t}(z_{a+1}-(\mathbf{1}+A_a\delta t)z_a)-\frac{1}{2}z_{a+1}^\top \frac{1}{Q_{a+1}} z_{a+1}} \\ &= \frac{1}{C_a} e^{-\frac{1}{2}z_a^\top \frac{1}{Q_a} z_a}, \end{aligned} \tag{47}$$

where

$$Q_a = (\mathbf{1}+A_a\delta t)^{-1}(Q_{a+1}+\Delta\delta t)\left[(\mathbf{1}+A_a\delta t)^\top\right]^{-1}. \tag{48}$$

Analogously to the forward evolution, we can take the limit of infinitesimal time intervals and get the differential equation

$$\dot{Q} = \Delta - A(t)Q - QA^\top(t), \tag{49}$$

the adjoint (or backward-) Lyapunov equation. Compared to the forward Lyapunov Equation (34), the adjoint evolution (49) features the 'time reversal' operation $A(t) \to -A(t)$, and therefore we can still use Equation (37) as a solution, as long as the Jacobian along the orbit is computed as

$$\frac{d}{dt}J(t,t_0) = -A(\mathbf{x})J(t,t_0), \qquad J(t_0,t_0) = \mathbf{1}, \tag{50}$$

and its computation follows the time reversed flow, that is the solution to the dynamical system $\dot{\mathbf{x}} = -v(\mathbf{x})$.

5. Non-Circular Limit Cycles: Classical Noise vs. Quantum Dissipation

We now turn our attention to non-circular limit cycles with background noise, and determine the steady-state density distribution yielded by the Fokker–Planck equation. We do so by integrating the Lyapunov equation in the neighborhood of a trajectory that eventually converges to the limit cycle.

The paradigmatic models of our choice both come from the nonlinear oscillator

$$\ddot{x} + \omega_0^2 x + \mu\, g(x,\dot{x}) = 0, \tag{51}$$

with g, a nonlinear function of position and velocity. Equation (51) may be reduced to a dynamical system as the Van der Pol oscillator, by taking $g = (x^2 - b)\dot{x}$, or as the Rayleigh model, by taking $g = \frac{\dot{x}^3}{3} - b\dot{x}$ in Equation (51). In what follows, we shall set $b = 3$, $\omega_0 = 1$, and we will tweak μ. The Van der Pol oscillator takes the form

$$\begin{aligned} \dot{x} &= y \\ \dot{y} &= -\mu\left(x^2-3\right)y - x, \end{aligned} \tag{52}$$

while the Rayleigh model reads (it is conventional to swap x and y)

$$\begin{aligned} \dot{x} &= y - \mu\left(\frac{1}{3}x^3 - 3x\right) \\ \dot{y} &= -x. \end{aligned} \tag{53}$$

In both classical systems, the dynamics converges to a limit-cycle of non-circular shape (Figure 3), which depends on the parameters, and it is characterized by fast and slow motions. One, therefore, expects the interplay of noise with the inherent nonlinear contraction to be non-uniform, unlike in the circular limit cycle examined in the previous section, and to give rise to a stationary density distribution of varying covariance along the cycle. We consider both models (52) and (53) for different values of the parameter μ, so as to gradually increase the eccentricity of the limit cycle, from a deformed circle (Figure 4a) to a nearly rectangular orbit (Figure 5b), where the deterministic stretching/contraction are most inhomogeneous along the cycle. The most notable feature is the oscillation of the covariance of the Gaussian solution of the linearized Fokker–Planck equation along the direction orthogonal to that of noiseless motion, denoted by σ in Figures 4c,d and 5c,d: the more eccentric the limit cycle, the more widely and rapidly σ oscillates. That translates to a Gaussian steady-state density featuring a width that increasingly depends on the position along the orbit with the parameter μ, as we can see in the three-dimensional/density plots of Figures 4e,f and 5e,f. As anticipated, these monodromic Gaussian distributions computed by means of the Lyapunov equation and portrayed in the figures would be, in a chaotic setting, the building blocks of a partition of the noisy phase space, whose non-uniform resolution is determined by their overlaps.

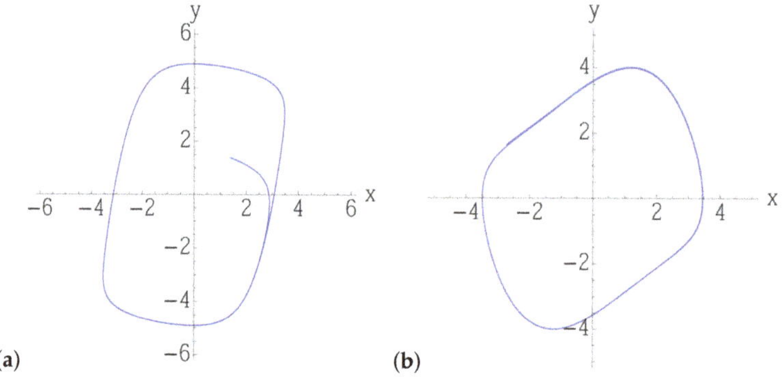

(a) (b)

Figure 3. Solution of the numerically integrated (a) Equation (53), and (b) Equation (52), without noise. Any initial condition converges to a limit cycle.

The Gaussian solutions of the Lyapunov equation computed and illustrated here share common traits with the steady-state Wigner function of the two oscillators (52) and (53), as obtained from a fully quantum mechanical computation that has recently appeared in the literature [25]. In that work, the quantization is performed by means of creation/annihilation operators

$$\hat{a} = \frac{1}{2}(\hat{x} + i\hat{y}), \tag{54}$$

and its adjoint $\hat{a}^\dagger$, while dissipative terms are added to the Liouville–von Neumann Equation (4), in the spirit of Lindblad's formalism (in units of $\hbar$):

$$\rho_t = -i\left[\hat{H}, \rho\right] + \sum_j \alpha_j \mathcal{D}\left[f_j(\hat{a}^\dagger, \hat{a})\right]\rho, \tag{55}$$

where

$$\mathcal{D}[\hat{c}]\rho = \hat{c}\rho\hat{c}^\dagger - \frac{1}{2}\hat{c}^\dagger\hat{c}\rho - \frac{1}{2}\rho\hat{c}^\dagger\hat{c}, \tag{56}$$

while the coefficients α_j are functions of the parameters ω_0 and μ. The Hermitian Hamiltonian $\hat{H}$ has two distinct expressions for the Rayleigh and the Van der Pol oscillators, and, like f_j, it is a function of linear (e.g., $\hat{a}^\dagger$), bilinear ($\hat{a}^\dagger\hat{a}$), and nonlinear ($\hat{a}^2$) terms involving the creation/annihilation operators (see ref. [25] for details). The above Equation (55) was numerically integrated, and the Wigner function was then found to eventually concentrate around the classical limit cycles, that feature similar eccentricities to the ones considered in the present work and plotted in Figures 4 and 5. In particular (Figure 6), the steady-state Wigner distribution is enhanced along 'tubes' of varying width, as it can be noticed in the more eccentric density plots of the Rayleigh model (Figure 6c,d). This feature is especially apparent in Figure 6d, where the high-density region is narrower along the vertical segments of the limit cycle (faster classical motion), and wider along its horizontal segments (slower classical motion). It compares directly with the Gaussian solution of the Lyapunov equation portrayed in Figure 5f.

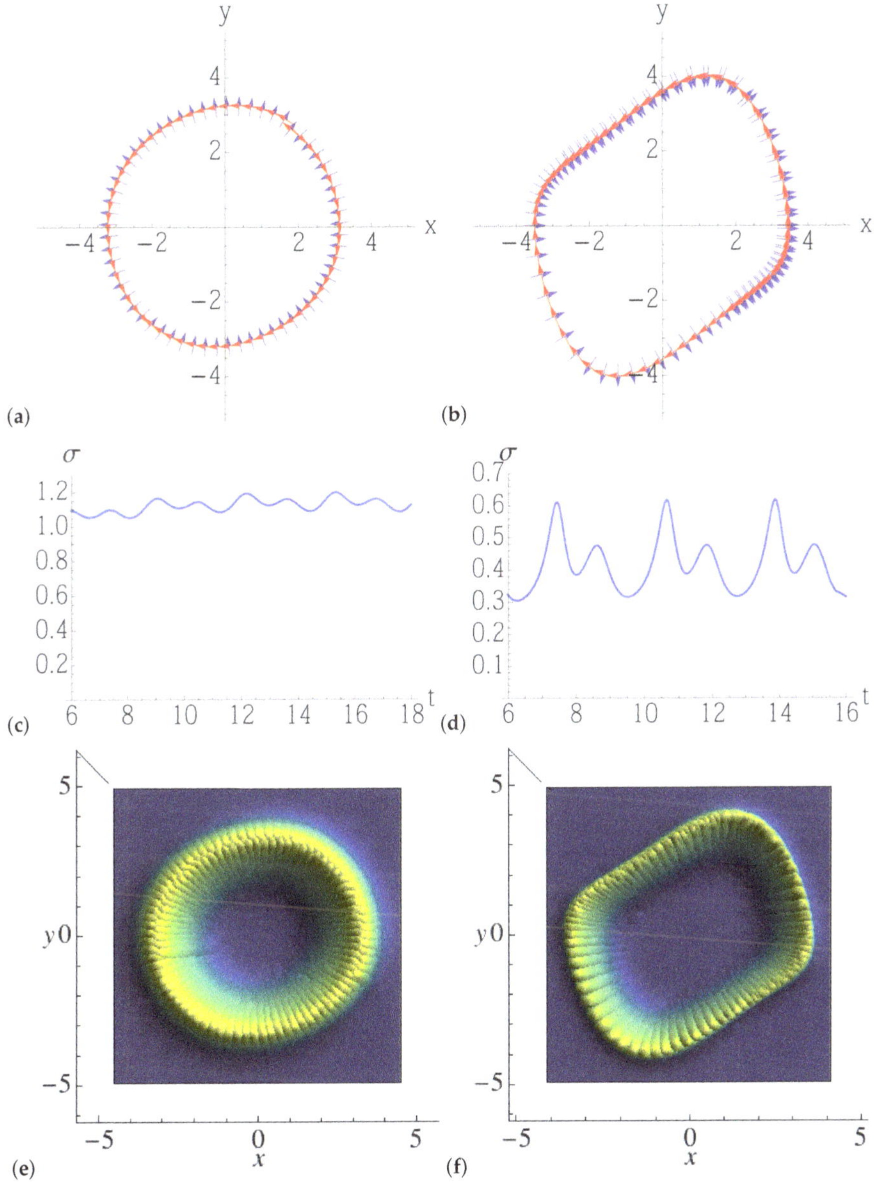

Figure 4. Top: solution of the numerically integrated Equation (52), together with the eigenvectors (arrows) of the covariant matrix Q^{-1}, as given by the solution (37) of the forward Lyapunov equation. In Equation (52), we take (**a**,**c**,**e**) $\mu = 0.03$, and (**b**,**d**,**f**) $\mu = 0.2$, while the amplitude of the noise is set to $2D = 0.1$. The light yellow stripe represents the width σ of the Gaussian density around the limit cycle. Middle: width σ of the Lyapunov tube, determined by the non-zero eigenvalue of $Q^{-1}(t)$ vs. time t. The cycle period is $t_p \approx 7$ time units. Bottom: Gaussian solutions of the linearized Fokker–Planck equation along the limit cycle.

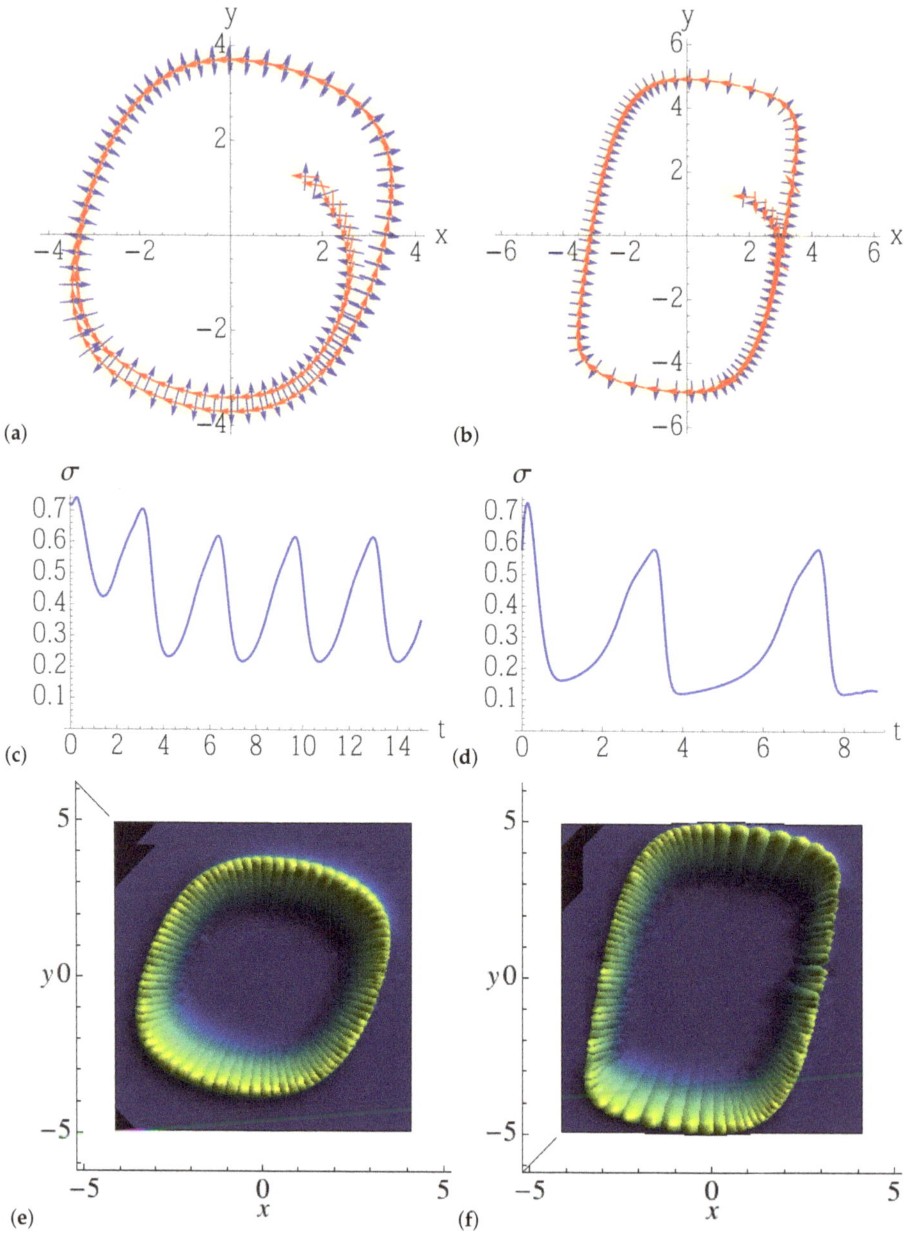

Figure 5. Top: solution of the numerically integrated Equation (53), together with the eigenvectors (arrows) of the covariant matrix Q^{-1}, as given by the solution (37) of the forward Lyapunov equation. In Equation (53), we take (**a**,**c**,**e**) $\mu = 0.3$, and (**b**,**d**,**f**) $\mu = 0.8$, while the amplitude of the noise is set to $2D = 0.1$. The light yellow stripe represents the width σ of the Gaussian density around the limit cycle. Middle: width σ of the Lyapunov tube, determined by the non-zero eigenvalue of $Q^{-1}(t)$ vs. time t. The cycle period is $t_p \approx 7$ time units. Bottom: Gaussian solutions of the linearized Fokker–Planck equation along the limit cycle.

It is noted that in the cited work, the authors did not integrate the Wigner equation, but the Lindblad equation with a full-fledged quantum-mechanical algorithm. In particular, the localization of the Wigner function around the classical limit cycles is not to be taken for granted, and it legitimates the parallel between their steady-state solutions and the local Gaussian tubes obtained in the present work from the noisy classical system.

On the other hand, the numerical steady-state solutions to Equation (56) obtained by the authors of [25] are clearly not Gaussians centered at the limit cycles (except in Figure 6a, the case of least eccentricity), as demonstrated by their varying intensities along the periodic orbits. In that sense, the Gaussian Ansatz that turns the Fokker–Planck into the Lyapunov equation carries limited information on the phase-space density distributions at equilibrium. Therefore, the analogy proposed here should be taken with a grain of salt, and only considered as a hint for the noisy-classical to quantum-dissipative correspondence in a particular system with nontrivial interplay of contraction and diffusion.

Finally, we would like to briefly comment on the difference between classical and quantum dissipation in the present models featuring limit cycles. In the classical system the dissipation produces damping, which is then balanced by the noise-induced diffusion. Instead, the quantum dissipation, generated by the characteristic Lindblad terms in Equation (55), is responsible for both the 'friction' that drives densities to localize along the classical limit cycles, and the diffusion that spreads out the steady-state Wigner density distribution in the same region of the attracting orbits. This is consistent with the more general picture of Section 2.2, where the quantum dissipation brings about both a damping and a diffusive term in the Wigner equation.

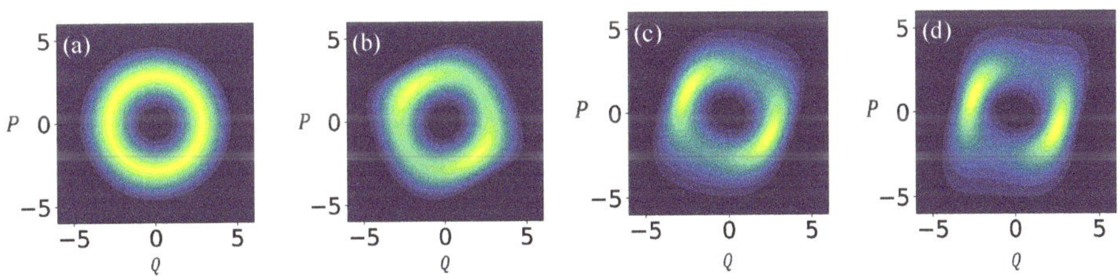

Figure 6. Quantum (**a**,**b**) Van der Pol and (**c**,**d**) Rayleigh oscillators from Reference [25]. The steady-state Wigner function (density plot) localizes around limit cycles of increasing eccentricity (controlled by the parameter μ in the classical models (52) and (53)). Courtesy of A. Chia.

6. Summary and Discussion

Having reviewed the parallels between the problem of dynamical evolution of a quantum system subject to dissipation and that of a stochastic process ruled by the Fokker–Planck equation, we have narrowed our attention down to chaos, and, in particular, to the problem of an inherent scale resolution of the phase space. The issue is measuring the conjugated variables down to a certain precision, which may be set by the balance of the contraction rate of the classical chaos with the coupling to the environment, the source of dissipation.

Using the analogy with the problem of classical chaotic dynamics with background noise, we consider the Fokker–Planck equation, and study its local solutions in the neighborhood of a periodic orbit, that effectively give the latter a finite width, in the phase space. Solving the problem for two-dimensional limit cycles, as done here, is the starting point: in a chaotic setting, a number of periodic tubes of finite width that proliferate exponentially with their length must end up overlapping, and thus determine the finest resolution for the noisy/quantized state space, that is expected to be non-uniform, as chaos interacts differently with diffusion/dissipation everywhere, in general.

The analysis performed in this venue shows that the problem is tractable, and it provides the basic technology to attack it. Complications and obstacles are ahead for higher-dimensional systems, where stable, unstable, and marginal directions coexist along the same orbit, and where the solution to the adjoint Fokker–Planck operator introduced here will almost certainly be instrumental to the method. Still, the progress already achieved by periodic orbit theory in such complex models as the Kuramoto–Sivashinsky or the Navier–Stokes equation gives us confidence in the feasibility of the optimal partition hypothesis in higher-dimensional chaos.

Author Contributions: Conceptualization, D.L.; Methodology, D.L.; Validation, D.L.; Writing—original draft, D.L.; Writing—review & editing, A.S.; Supervision, A.S. All authors have read and agreed to the published version of the manuscript.

Funding: DL acknowledges NSF China—Grant No. 11750110416-1601190090 for partial support.

Conflicts of Interest: The authors declare no conflict of interest.

References

1. Zurek, W.H. Sub-Planck structure in phase space and its relevance for quantum decoherence. *Nature* **2001**, *412*, 6848. [CrossRef] [PubMed]
2. Breuer, H.-P.; Petruccione, F. *The Theory of Open Quantum Systems*; Oxford University Press: London, UK, 2007.
3. Zubairy, M.S.; Scully, M.O. *Quantum Optics*; Cambridge University Press: Cambridge, UK, 1997.
4. Gardiner, C.W.; Zoller, P. *Quantum Noise*; Springer: Berlin, Germany, 2004.
5. Feynman, R.P.; Vernon, F.L. The theory of a general quantum system interacting with a linear dissipative system. *Ann. Phys.* **1963**, *24*, 118. [CrossRef]
6. Caldeira, A.O.; Leggett, A.J. Path integral approach to quantum brownian motion. *Physica A* **1983**, *121*, 587–616. [CrossRef]
7. Dittrich, T.; Graham, R. Effects of weak dissipation on the long-time behaviour of the quantized standard map. *Europhys. Lett.* **1988**, *7*, 287–291. [CrossRef]
8. Zurek, W.H.; Paz, J.P. Decoherence, Chaos, and the Second Law. *Phys. Rev. Lett.* **1993**, *72*, 2508–2511. [CrossRef] [PubMed]
9. Kolovsky, A.R. A remark on the problem of quantum-classical correspondence in the case of chaotic dynamics. *Europhys. Lett.* **1994**, *27*, 79–84. [CrossRef]
10. Kolovsky, A.R. Quantum coherence, evolution of the Wigner function, and transition from quantum to classical dynamics for a chaotic system. *Chaos* **1996**, *6*, 534–542. [CrossRef] [PubMed]
11. Cabrera, R.; Bondar, D.I.; Jacobs, K.; Rabitz, H.A. Efficient method to generate time evolution of the Wigner function for open quantum systems. *Phys. Rev. A* **2015**, *92*, 042122. [CrossRef]
12. Bondar, D.I.; Cabrera, R.; Campos, A.; Mukamel, S.; Rabitz, H.A. Correction to Wigner-Lindblad equations for quantum friction. *J. Phys. Chem. Lett.* **2016**, *7*, 1632. [CrossRef] [PubMed]
13. Carlo, G.G.; Ermann, L.; Rivas, A.M.F. Effects of chaotic dynamics on quantum friction. *Phys. Rev. E* **2019**, *99*, 042214. [CrossRef] [PubMed]
14. Lippolis, D.; Cvitanović, P. How well can one resolve the state space of a chaotic map? *Phys. Rev. Lett.* **2010**, *104*, 014101. [CrossRef] [PubMed]
15. Cvitanović, P.; Lippolis, D. Knowing when to stop: How noise frees us from determinism. *AIP Conf. Proc.* **2012**, *1468*, 82–126.
16. Heninger, J.M.; Lippolis, D.; Cvitanović, P. Perturbation theory for the Fokker-Planck operator in chaos. *Commun. Nonlinear Sci. Numer. Simulat.* **2018**, *55*, 16–28. [CrossRef]
17. Lippolis, D. Mapping densities in a noisy state space. In Proceedings of the 2013 International Symposium on Nonlinear Theory and Its Applications, Santa Fe, NM, USA, 8–11 September 2013; Volume 2, p. 318.
18. Heninger, J.M.; Lippolis, D.; Cvitanović, P. Neighborhoods of periodic orbits and the stationary distribution of a noisy chaotic system. *Phys. Rev. E* **2015**, *92*, 062922. [CrossRef] [PubMed]
19. Case, W.B. Wigner functions and Weyl transforms for pedestrians. *Am. J. Phys.* **2008**, *76*, 937–946. [CrossRef]
20. Altland, A.; Haake, F. Quantum Chaos and Effective Thermalization. *Phys. Rev. Lett.* **2012**, *108*, 073601. [CrossRef] [PubMed]
21. Gaspard, P. *Chaos, Scattering and Statistical Mechanics*; Cambridge University Press: Cambridge, UK, 1998.
22. Berman, G.P.; Zaslavsky, G.M. Condition of stochasticity in quantum nonlinear systems. *Physica A* **1978**, *91*, 450–460. [CrossRef]
23. Risken, H. *The Fokker-Planck Equation*; Springer: Berlin, Germany, 1996.
24. Amann, H. *Ordinary Differential Equations: An Introduction to Nonlinear Analysis*; De Gruyter: Berlin, Germany, 1990.
25. Chia, A.; Kwek, L.C.; Noh, C. Relaxation oscillations and frequency entrainment in quantum mechanics. *Phys. Rev. E* **2020**, *102*, 042213.

entropy

Article

The Effect of On-Site Potentials on Supratransmission in One-Dimensional Hamiltonian Lattices

Tassos Bountis [1,2,*] and Jorge E. Macías-Díaz [3,4]

1 Department of Mathematics, University of Patras, 26500 Patras, Greece
2 Center for Integrable Systems, P. G. Demidov Yaroslavl State University, 150003 Yaroslavl, Russia
3 Department of Mathematics and Didactics of Mathematics, School of Digital Technologies, Tallinn University, Narva Rd. 25, 10120 Tallinn, Estonia
4 Departamento de Matemáticas y Física, Universidad Autónoma de Aguascalientes, Avenida Universidad 940, Ciudad Universitaria, Aguascalientes 20100, Mexico
* Correspondence: tassosbountis@gmail.com

Abstract: We investigated a class of one-dimensional (1D) Hamiltonian N-particle lattices whose binary interactions are quadratic and/or quartic in the potential. We also included on-site potential terms, frequently considered in connection with localization phenomena, in this class. Applying a sinusoidal perturbation at one end of the lattice and an absorbing boundary on the other, we studied the phenomenon of supratransmission and its dependence on two ranges of interactions, $0 < \alpha < \infty$ and $0 < \beta < \infty$, as the effect of the on-site potential terms of the Hamiltonian varied. In previous works, we studied the critical amplitude $A_s(\alpha, \Omega)$ at which supratransmission occurs, for one range parameter α, and showed that there was a sharp threshold above which energy was transmitted in the form of large-amplitude nonlinear modes, as long as the driving frequency Ω lay in the forbidden band-gap of the system. In the absence of on-site potentials, it is known that $A_s(\alpha, \Omega)$ increases monotonically the longer the range of interactions is (i.e., as $\alpha \longrightarrow 0$). However, when on-site potential terms are taken into account, $A_s(\alpha, \Omega)$ reaches a maximum at a low value of α that depends on Ω, below which supratransmission thresholds *decrease* sharply to lower values. In this work, we studied this phenomenon further, as the contribution of the on-site potential terms varied, and we explored in detail their effect on the supratransmission thresholds.

Keywords: system with long-range interactions; on-site potential; supratransmission; Hamiltonian lattices

MSC: 65M06; 65M12

Citation: Bountis, T.; Macías-Díaz, J.E. The Effect of On-Site Potentials on Supratransmission in One-Dimensional Hamiltonian Lattices. *Entropy* **2023**, *25*, 423. https://doi.org/10.3390/e25030423

Academic Editor: Marko Robnik

Received: 28 December 2022
Revised: 21 February 2023
Accepted: 23 February 2023
Published: 26 February 2023

1. Introduction

Supratransmission was first investigated in chains of coupled nonlinear oscillators [1], when the first oscillator is forced sinusoidally at a frequency in the so-called forbidden band gap, so that the excitation of linear normal modes is avoided. The theoretical prediction of the critical driving amplitude, above which the boundary begins to transmit energy, was first established experimentally, and later approximated analytically, using known solutions of the sine-Gordon equation [2]. Further studies showed that this phenomenon is also present in well-known nonlinear models, such as the classical Fermi–Pasta–Ulam–Tsingou (FPUT) chain [3] and the Klein–Gordon and double sine-Gordon models [4]. Various investigations have led to the development of reliable computational techniques to approximate the occurrence of this phenomenon [5,6]. In this direction, various works have been devoted to the design and analysis of novel numerical methods using variational properties, while new computational methodologies have been proposed based on these approaches [7].

Initially, most studies focused on mathematical models with only nearest neighbor interactions. However, more recently, new results have emerged focusing on global interactions with a long range of applicability [8–10]. Indeed, globally interacting systems have led to a better understanding of nonlinear problems in physics [11]. More generally, it is known that systems with long-range interactions (LRIs) can be expressed as Riesz-type fractional models in the continuous limit [12–14]. Such approaches provide a convenient bridge between spatially discrete and continuous systems, and they lead to a better understanding of globally interacting systems [15,16].

Regarding the study of supratransmission in globally interacting systems, its presence has been established in Riesz space fractional sine-Gordon equations [17], fractional models of Josephson transmission lines [18] and in FPUT chains with different ranges of interactions [19]. More recently, a general 1D Hamiltonian lattice consisting of quadratic global interactions, which included an on-site potential (of the Klein–Gordon type), was investigated in [20], where it was found that supratransmission was also present in that system. However, the critical amplitude $A_s(\alpha)$ in that model exhibited a surprising non-monotonic behavior: as expected, it occurred at *higher* amplitudes the *longer* the range of interactions, but reached a maximum at a value $\alpha = \alpha_{max}$ that depended on Ω!. Below this value, supratransmission thresholds *decreased* sharply to values lower than the nearest neighbor $\alpha = \infty$ limit.

In [21], the authors considered a different Hamiltonian form of the model studied in [20], in which on-site potentials were combined with LRIs of purely quartic order. One of the reasons was to examine whether the presence of quadratic interactions and their associated linear spectrum are needed for supratransmission, as suggested by the seminal work in [1]. The conclusion in [21] was that this feature does *not* constitute a necessary condition for supratransmission. However, as in the case of [20], the critical amplitudes $A_s(\alpha)$ again exhibited the same behavior of reaching a maximum at $\alpha \approx 2$ that depended on the value of Ω, and then as $\alpha \longrightarrow 0$ fell to values even below their $\alpha = \infty$ limit .

In this work, we investigated the dependence of critical supratransmission amplitudes on two LRI scale parameters, α and β (of the quadratic and quartic interactions, respectively), as well as the frequency of oscillations Ω. To this end, we employed a 1D Hamiltonian lattice with both quadratic and quartic interactions, in which the on-site potential terms were gradually switched off, as a multiplicative parameter $c > 0$ tended to zero. The results were quite surprising and revealed the role of the driving frequency Ω on the critical supratransmission amplitudes of these lattices as the on-site potential terms were gradually switched off.

The paper is organized as follows. In Section 2, we introduce our Hamiltonian model and its associated equations of motion in the form of an initial–boundary value problem. Our Hamiltonian system, in its general form, describes globally interacting particles with both quadratic and quartic interactions, as well as on-site terms of the sine-Gordon type. Section 4 describes the fully discretized model we used to approximate the solutions of the equations of motion and the Hamiltonian functional. In Section 5, we describe the computational experiments we performed to study the occurrence of supratransmission in three cases: Case I with both quartic and quadratic LRIs, Case II with only quartic LRIs and Case III with only quadratic LRIs, in the limit where the on-site potentials vanish. Next, in Section 6 we study the dependence of supratransmission amplitudes on different values of α and β, determining ranges of interactions. Finally, in Section 7 we end our study with a discussion of our results and concluding remarks.

2. Mathematical Model

We consider N-particle one-dimensional Hamiltonian lattices, with $N \in \mathbb{N}$ sufficiently large, whose $N+2$ particles are governed by the Hamiltonian

$$\begin{aligned} \mathcal{H} = &\sum_{n=1}^{N} \left(\frac{1}{2} p_n^2 + cV(x_n) \right) + \frac{a^2}{2M_1} \sum_{n=0}^{N} \sum_{m=n+1}^{N+1} \frac{(x_m - x_n)^2}{(m-n)^\alpha} \\ &+ \frac{b^2}{4M_2} \sum_{n=0}^{N} \sum_{m=n+1}^{N+1} \frac{(x_m - x_n)^4}{(m-n)^\beta} \end{aligned} \tag{1}$$

where α and β are non-negative real numbers (including ∞) determining the range of pair particle interactions, $V : \mathbb{R} \to \mathbb{R}$ is a continuously differentiable function representing the on-site potential of the system and $a, b, c \in \mathbb{R}^+ \cup \{0\}$ are constants. We also use the scaling factors

$$M_1 = \frac{1}{N+1} \sum_{n=0}^{N} \sum_{m=n+1}^{N+1} \frac{1}{(m-n)^\alpha}, \quad M_2 = \frac{1}{N+1} \sum_{n=0}^{N} \sum_{m=n+1}^{N+1} \frac{1}{(m-n)^\beta}, \tag{2}$$

to ensure that all terms in our Hamiltonians are *extensive*, i.e., proportional to the number of particles N, see [20] and references therein. The constants $a \geq 0$, $b \geq 0$ are fixed as follows: in Case I, both a, b are nonzero; in Case II $a = 0$ and $b \neq 0$; and in Case III $a \neq 0$ $b = 0$. In all cases, the parameter $c \geq 0$ is taken to gradually tend to zero.

Each $x_n(t) : \mathbb{R} \to \mathbb{R}$ is a function such that $x_n(t) \in \mathcal{C}^2(\mathbb{R})$, for each $n \in \{1, \dots, N\}$, and $p_n(t)$ is the time derivative of $x_n(t)$. In this context, we define, respectively, the kinetic and potential energy of the nth particle by

$$\mathcal{K}_n = \frac{1}{2} p_n^2 \tag{3}$$

and

$$\mathcal{P}_n = cV(x_n) + \frac{a^2}{2M_1} \sum_{m=n+1}^{N+1} \frac{(x_m - x_n)^2}{(m-n)^\alpha} + \frac{b^2}{4M_2} \sum_{m=n+1}^{N+1} \frac{(x_m - x_n)^4}{(m-n)^\beta}. \tag{4}$$

The on-site potential function $V(x_n)$ is taken to be of the sine-Gordon type, $V(u) = 1 - \cos(u)$. Alternatively, we can also use the Klein–Gordon potential

$$V(u) = \frac{1}{2!} u^2 - \frac{1}{4!} u^4 + \frac{1}{6!} u^6. \tag{5}$$

Clearly, the values of $\alpha = \beta = \infty$ represent the case of nearest neighbor interactions. If $\alpha, \beta \in [0, 1]$, the mutual particle interactions have comparable contributions and the system is said to have *long-range interactions* (LRIs). The case where $\alpha, \beta \in (1, \infty)$, is said to correspond to *short-range interactions* (SRIs).

Thus, the equations of motion associated with our Hamiltonian (1) are as follows:

$$\begin{aligned} &\ddot{x}_n = \frac{a^2}{M_1} \sum_{\substack{m=1 \\ m \neq n}}^{N} \frac{x_m - x_n}{|m-n|^\alpha} + \frac{b^2}{M_2} \sum_{\substack{m=1 \\ m \neq n}}^{N} \frac{(x_m - x_n)^3}{|m-n|^\beta} - cV'(x_n) - \gamma_n \dot{x}_n, \\ &\text{with b.c.} \begin{cases} x_0(t) = A\sin(\Omega t), & \forall t \geq 0, \\ x_{N+1}(t) = x_N(t), & \forall t \geq 0, \\ x_n(0) = \dot{x}_n(0) = 0, & \forall n \in \{1, \dots, N+1\}, \end{cases} \end{aligned} \tag{6}$$

where a "damping" term is added at each node (with coefficient $\gamma_n > 0$), simulating an absorbing boundary at the right end of the system. This avoids the generation of returning shock waves at the right end of the chain [22,23]. In fact, it is convenient to use

$$\gamma_n = 0.5\left[1 + \tanh\left(\frac{2n - N_0 + N}{6}\right)\right], \tag{7}$$

for each $n \in \{1, \ldots, N+1\}$, assuming that $N_0 < N$.

A simpler form of this model was investigated in [20], which corresponds to (6) with $b = 0$. An interesting question, therefore, arises as to whether supratransmission is still present if the interaction terms in the potentials are quartic instead of quadratic (i.e., $a = 0$). As was shown in [21], supratransmission neither requires the presence of harmonic interactions, nor the absence of on-site terms in the potential, but it is rather a generic phenomenon present in this class of systems.

Regarding the role of the damping term in (6), we note that for relatively large N, we let $1 \ll N_0 \ll N$. Thus, the form of (7) guarantees that the nodes before $N - N_0$ will have a damping coefficient that is approximately equal to zero. On the other hand, the γ_n monotonically increase near the right end of the chain, thus ensuring that the waves reflected back in the system are rapidly attenuated, see [24]. We have found empirically that if $N_0 \approx N/4$ wave reflection at the right boundary is effectively negligible; thus, in all our simulations we used $N = 100$ and $N_0 = 25$.

3. Calculation of the Phonon Spectrum

It is important to determine the phonon (or forbidden) band of frequencies for our system, so as to be able to interpret the results of our computations in later sections. To this end, we considered the linear terms in the position variables on the right-hand side (RHS) of (6) and obtained an $N \times N$ symmetric matrix, whose eigenvalues were difficult to obtain analytically for general values of α and N.

Note, however, that in the limit $\alpha \to 0$ and large N, we have $M_1 = N/2$, while our symmetric matrix has a multiplicative factor $2a^2/N$, with diagonal terms equal to $-N(1 + c/2a^2) + 1$ and all off-diagonal terms equal to 1. Fortunately, this matrix can be diagonalized and has the eigenvalues $-c$ and $-(c + 2a^2)$ (multiplicity $N - 1$). Now, for $\alpha > 0$ (and small), all the terms that were unity in the matrix slightly decrease, which suggests that the phonon frequencies become non-degenerate and lie within an interval approximated by $[\sqrt{c}, \sqrt{c + 2a^2}]$.

4. Numerical Methodology

The solutions of system (6) will be approximated using an explicit finite difference scheme, which possesses similar Hamiltonian properties as our mathematical model and efficiently approximates the solutions of (6) over the time interval $[0, T]$, for large T.

We first fix a uniform partition

$$0 = t_0 < t_1 < \ldots < t_j < \ldots < t_K = T \tag{8}$$

of the interval $[0, T]$, with τ sufficiently small. Next, we define the time step $t_{j+\frac{1}{2}} = t_j + \frac{1}{2}\tau$, for each $j \in \{0, 1, \ldots, K-1\}$, set $x_n^j = x_n(t_j)$, for $n \in \{0, 1, \ldots, N+1\}$ and $j \in \{0, 1, \ldots, K\}$, and introduce the discrete linear operators

$$\mu_t x_n^j = \frac{1}{2}(x_n^{j+1} + x_n^j), \tag{9}$$

$$\delta_t^{(1)} x_n^j = \frac{x_n^{j+1} - x_n^j}{\tau}, \tag{10}$$

$$\delta_t^{(2)} x_n^j = \frac{x_n^{j+1} - 2x_n^j + x_n^{j-1}}{\tau^2}, \tag{11}$$

and

$$\delta_{t,x} V(x_n^j) = \begin{cases} \dfrac{V(x_n^{j+1}) - V(x_n^j)}{x_n^{j+1} - x_n^j}, & \text{if } x_n^{j+1} \neq x_n^j, \\ V'(x_n^{j+\frac{1}{2}}), & \text{if } x_n^{j+1} = x_n^j. \end{cases} \tag{12}$$

Thus, the fully discretized form of (6) is given by the following system of equations:

$$\mu_t \delta_t^{(2)} x_n^j = \frac{a^2}{M_1} \sum_{\substack{m=1 \\ m \neq n}}^{N} \frac{\mu_t (x_m^j - x_n^j)}{|m-n|^\alpha} + \frac{b^2}{M_2} \sum_{\substack{m=1 \\ m \neq n}}^{N} \frac{\mu_t (x_m^j - x_n^j)^3}{|m-n|^\beta} - c\delta_{t,x} V(x_n^j) - \gamma_n \delta_t^{(1)} x_n^j,$$

$$\text{such that} \begin{cases} x_0^j = A\sin(\Omega t_j), & \text{if } 0 \leq j \leq K, \\ x_{N+1}^j = x_N^j, & \text{if } 0 \leq j \leq K, \\ x_n^0 = x_n^1 = x_n^2 = 0, & \text{if } 1 \leq n \leq N. \end{cases} \tag{13}$$

It follows that the discrete form of the Hamiltonian functional is given by the following:

$$\begin{aligned} H^j = \sum_{n=1}^{N} \left[\frac{1}{2} (\delta_t^{(1)} x_n^j)(\delta_t^{(1)} x_n^{j-1}) + cV(x_n^j) \right] + \frac{a^2}{2M_1} \sum_{n=0}^{N} \sum_{n=n+1}^{N+1} \frac{(x_m^j - x_n^j)^2}{(x-n)^\alpha} \\ + \frac{b^2}{4M_2} \sum_{n=0}^{N} \sum_{n=n+1}^{N+1} \frac{(x_m^j - x_n^j)^4}{(x-n)^\beta}, \end{aligned} \tag{14}$$

for each $j \in \{1, \ldots, K-1\}$. Thus, the discrete kinetic and potential energies of the nth particle at time t_j are given, respectively, by

$$K_n^j = \frac{1}{2} (\delta_t^{(1)} x_n^j)(\delta_t^{(1)} x_n^{j-1}), \tag{15}$$

and

$$P_n = cV(x_n^j) + \frac{a^2}{2M_1} \sum_{n=n+1}^{N+1} \frac{(x_m^j - x_n^j)^2}{(x-n)^\alpha} + \frac{b^2}{4M_2} \sum_{n=n+1}^{N+1} \frac{(x_m^j - x_n^j)^4}{(x-n)^\beta}. \tag{16}$$

This method is an adaptation of a four-step scheme that approximates the solutions of Riesz space fractional hyperbolic partial differential equations with Hamiltonian structures [25]. It uses spatial step sizes equal to 1 and is both stable and quadratically convergent [23].

In the following sections, we describe how we conducted a series of computational experiments to study the behavior of the critical amplitude A_s for supratransmission in 1D lattices governed by (6), for various values of a, b and range parameters α and β. For comparison purposes, we employed the approach followed in [20] for 1D lattices consisting of $N = 100$ particles. Regarding the absorbing boundary, we let $N_0 = 25$ and set $\tau = 0.05$ in all our simulations. Moreover, we used the sine-Gordon on-site potential, i.e.,

$$V(x) = 1 - \cos x, \qquad \forall x \in \mathbb{R}. \tag{17}$$

5. Computational Results

The purpose of this section is to carry out simulations in order to study the behavior of critical amplitudes for supratransmission as a function of the model parameters, particularly $c \geq 0$ measuring the contribution of the on-site potential terms. Throughout, we consider $N = 100$ nodes and final time $T = 100$. The computation time step everywhere is $\tau = 0.05$.

5.1. Case I for Different Frequencies

As our first example, we considered the Case I parameters $a = b = 1$, with maximal LRIs $\alpha = \beta = 0$, and selected various values for the driving frequency Ω. Under these circumstances, Figure 1 shows the graph of the total energy (6) (measured at the steady state of the system) as a function of the driving amplitude A and the coefficient c, for the following values of the frequencies: (a) $\Omega = 3.5$, (b) $\Omega = 2.5$, (c) $\Omega = 1.5$ and (d) $\Omega = 0.9$.

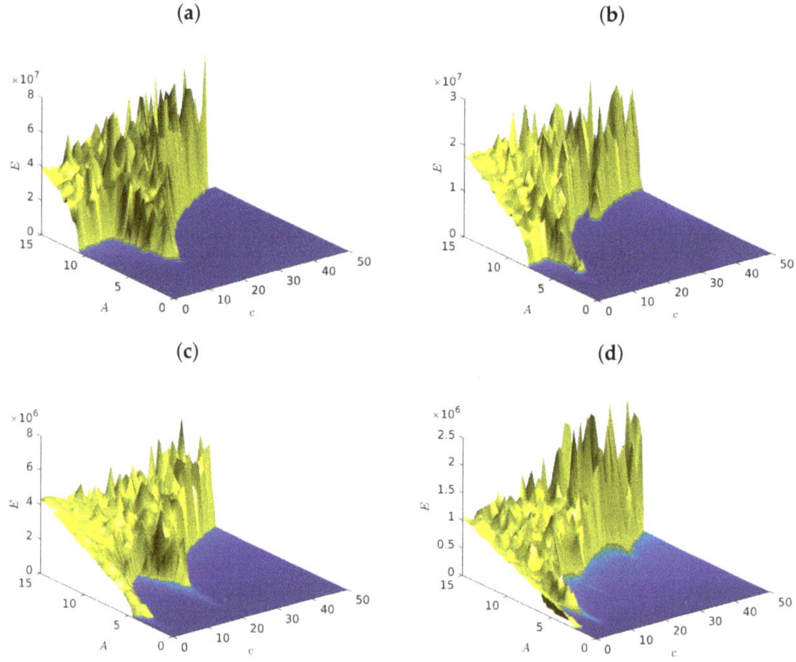

Figure 1. Graphs of the total energy of system (6) versus the driving amplitude A and c, for **(a)** $\Omega = 3.5$, **(b)** $\Omega = 2.5$, **(c)** $\Omega = 1.5$ and **(d)** $\Omega = 0.9$. We used $\alpha = \beta = 0$, fixed $a = b = 1$ and the time step size was equal to 0.05.

Let us first observe in this figure the presence of supratransmission in all cases, as exemplified by the sudden, upward surge of the energy from the blue $E = 0$ level of inactivity to the yellow/green "mountain ranges" at higher values of A. Note now in Figure 1a that, as c decreased, the critical amplitude $A_s(c)$ for supratransmission experienced a minimum around $c = 10$ and then increased to a higher value as c tended to zero. A similar picture also occurs in Figure 1b with the minimum of A_s located near $c = 5$.

According to the analysis in Section 3, the phonon band in this case was $[\sqrt{c}, \sqrt{c+2}]$, which yielded $[3.16, 3.46]$ for (a) and $[2.23, 2.65]$ for (b), which are close to the driving frequencies $\Omega = 3.5$ and $\Omega = 2.5$ respectively. A similar observation regarding the first minimum also held for Figure 1c,d, where the phonon bands were $[1.4, 2$ and $[1, 1.73]$, respectively. Regarding the secondary minima in the last two figures, we suggest that they correspond to higher order resonances. For example, the second minimum in Figure 1c occurred at $c \approx 19$, which would excite a $\sqrt{19} \approx 4.5 = 3\Omega$ resonance.

5.2. Case II for Different Frequencies

We next examined a Hamiltonian (1) where the quadratic interactions were absent, setting $a = 0$ and $b = 1$ and allowing the range parameters to vary from $0 < \alpha = \beta$ up to $\alpha = \beta = 9$ and higher. Remarkably, the results obtained, for the same frequency values as considered in Case I, led to figures that *differed very little* from Figure 1 (and hence are not plotted here) provided we kept the same frequency values used in the plots of Figure 1.

Next, we proceeded to study, in Figure 2, the same Case II example with $a = 0$ and $b = 1$, with all other parameters the same as in Figure 1. The important difference here was that we chose driving frequencies *smaller than 1* in the different plots: (a) $\Omega = 0.9$, (b) $\Omega = 0.8$, (c) $\Omega = 0.7$ and (d) $\Omega = 0.6$.

The difference between the results of Figures 1 and 2 was striking: the energy landscapes for these frequencies were dramatically smoother, and the A_s values exhibited a single minimum, $0 < c_{min} < 1$, whose value diminished monotonically as Ω decreased. Interestingly, since the only quadratic term in the potential was cx_n^2, the minimum of the A_s was smaller than 1 and decreased as Ω decreased. Below this minimum, A_s *significantly increased* as $c \to 0$, which meant that the lattice became more resilient against supratransmission in the absence of on-site potential terms.

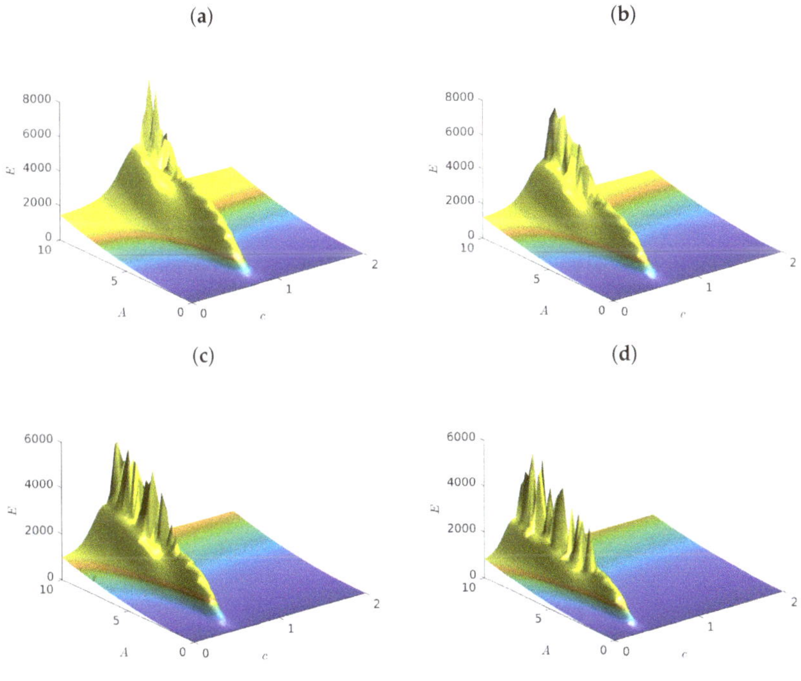

Figure 2. Graphs of the total energy of system (6) versus A_s and c, for (**a**) $\Omega = 0.9$, (**b**) $\Omega = 0.8$, (**c**) $\Omega = 0.7$ and (**d**) $\Omega = 0.6$. We used $\alpha = \beta = 0$, and fixed $a = 0$ and $b = 1$.

5.3. Case III: Absence of Quartic Terms

Next, we turn to an example of Case III, with $a = 2$ and $b = 0$, and examine in Figure 3 the behavior of critical amplitudes A_s as functions of the range parameter α, for diminishing $c = 1, 0.8, 0.4, 0.2$. In all plots of this figure we used frequencies $\Omega < 1$, and observed no supratransmission whatsoever. One might think, therefore, that this was due to resonances with the phonon band and would not occur if we used $\Omega > 1$. However, when we did apply higher driving frequencies, all we observed was harmonic waves traveling down the lattice, showing no sudden increase in their amplitudes. This suggested that quartic interactions in the potential were crucial for the occurrence of supratransmission.

In the next section, where we discuss a case in which quartic interaction terms were present, the plots of Figure 3 are similar to those of Figure 4a,c,e, for $\Omega < 1$, and supratransmission was absent. However, when $\Omega > 1$ in that system, as Figure 4b,d,f show, resonances with the phonon band were avoided and supratransmission did occur!

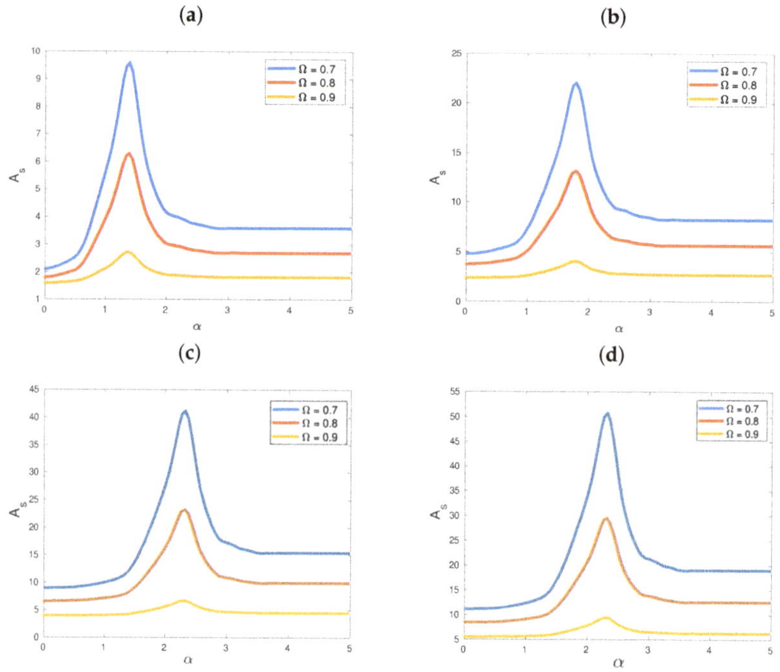

Figure 3. Graphs of the supratransmission threshold A_s as a function of α, for various values of Ω (see legends). We let $a = 2$, $b = 0$, $N = 100$, $T = 200$ and $V(x) = 1 - \cos x$. Various values of C were employed, namely (**a**) $c = 1$, (**b**) $c = 0.8$, (**c**) $c = 0.4$ and (**d**) $c = 0.2$.

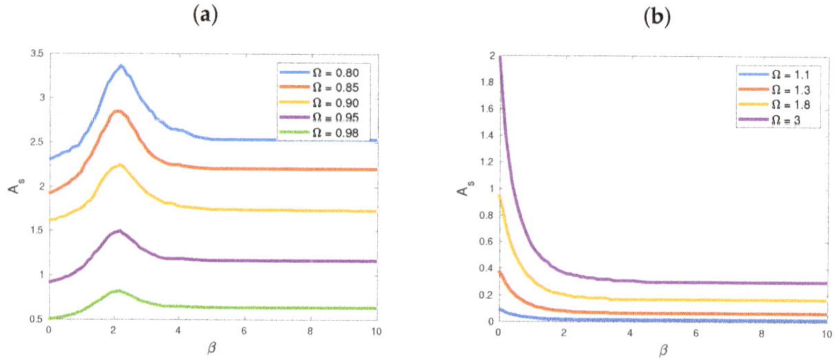

Figure 4. *Cont.*

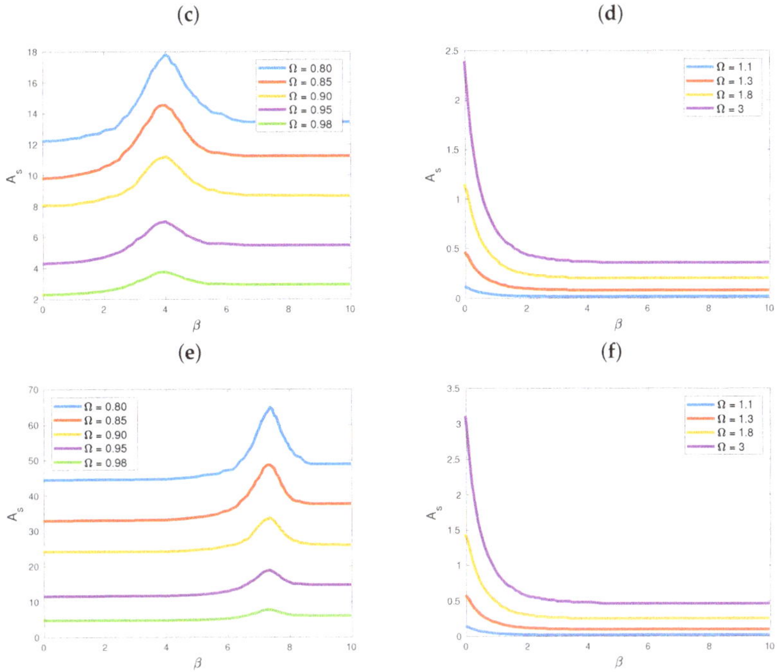

Figure 4. Graphs of the supratransmission threshold A_s as a function of the parameter β, using various values of Ω. We let $a = 0$, $b = 4$, $N = 100$, $T = 200$ and $V(x) = 1 - \cos x$. Decreasing values of c were employed, namely $c = 0.9$ (**top row**), $c = 0.4$ (**middle row**) and $c = 0.2$ (**bottom row**). The subfigures in (**a–f**) give the value of Ω that corresponds to the curve of the figure with the same color. In the right column figures, all critical amplitudes A_s attained finite values at $\beta = 0$, which increased linearly with Ω.

6. A Case II Study with Different β Parameters

Finally, we examined the dependence of critical amplitudes A_s on the range parameter β, for different values of the driving frequency Ω. We did this for Case II of our study, corresponding to $a = 0$, $b = 4$ and examined the behavior of the results as the value of c is decreased, plotting the results of our computations in Figure 4.

Observe first the general behavior of A_s vs. β for $\Omega < 1$ in Figure 4a,c,e: note the presence of a local maximum in the graphs, which shifted to higher values of β as c diminished from 0.9 (top row) to 0.4 (middle row) and 0.2 (bottom row). Clearly, for these frequencies, the location of the maximum increased for higher β. Note first that, in all these figures, all critical amplitudes *increased* as c decreased, in agreement with the results of previous sections.

By contrast, in Figure 4b,d,f, the behavior of A_s vs. β monotonically increased as β tended to zero, converging to constant values, which increased as c diminished. This was very similar to what we already encountered in Figure 3b,d and was in agreement with the results of Section 3, as there were no resonances with the phonon band in this case; hence, supratransmission occurred in this case also, with the A_s values rising linearly with Ω as $\beta \to 0$.

7. Discussion and Conclusions

The process of supratransmission has become very popular in recent years since its discovery nearly two decades ago. The main reason for this is due to its appeal as a mechanism that allows one to transmit a large amount of energy through a nonlinear medium by simply driving one end of the medium by $A\sin(\Omega t)$ at the appropriate frequencies and amplitudes. Thus, it has inspired many authors due to its obvious applications in many physical situations.

Almost all studies so far have concentrated on one-dimensional (1D) chains of nonlinear oscillators, as a first step towards understanding supratransmission in higher dimensions. An additional advantage of 1D systems is the fact that, at least under nearest neighbor coupling, they have been extensively analyzed for more than 60 years, numerically as well as analytically. Thus, the first papers on supratrasmission focused on a number of classical conservative 1D models, for which a continuum limit analysis yielded partial differential equations that connected supratransmission to nonlinear wave solutions of the systems and, thus, permitted the derivation of analytical formulas connecting A_s and Ω to the emergence of such solutions.

More recently, a number of papers appeared verifying that supratransmission is also ubiquitous in models that involve long-range interactions (LRIs) among the particles by multiplying the interaction between the nth and mth particles by a factor of the form $1/(n-m)^{\alpha}$, with $0 \leq \alpha \leq \infty$. In this context, $0 \leq \alpha \leq 1$ represents LRIs, while short range interactions (SRIs) correspond to the $\alpha > 1$ values, with $\alpha = \infty$ being the nearest neighbor case. The authors of the present paper and their collaborators analyzed in detail the occurrence of supratransmission in 1D systems using the FPUT 1D lattice as an example [9,19], and they later included in their studies the effect of on-site terms in the potential [20].

In the absence of on-site potentials, one finds that $A_s(\alpha, \Omega)$ grows monotonically the longer the range of interactions (i.e., as $\alpha \longrightarrow 0$). However, when on-site potential terms are introduced, $A_s(\alpha, \Omega)$ is seen to attain a maximum at low α values depending on Ω, below which supratransmission thresholds sharply decrease. In this paper, we made a first attempt to further explore these phenomena by introducing a parameter $c > 0$ before the on-site potential, and letting $c \longrightarrow 0$.

We first examined the case where both quadratic and quartic global LRIs were present and found that, the higher $\Omega > 1$ is, the clearer it became that as $c \longrightarrow 0$, supratransmition critical amplitudes increased significantly. This picture became even clearer when only quartic terms were included in the potential. Next, we examined the dependence of critical amplitudes on the range parameters, α and β, in a Case II and a Case III study for one set of frequencies with $\Omega < 1$ and one with $\Omega > 1$, for decreasing values of c. When $\Omega < 1$, in both studies, A_s had a single maximum at a β value that increased as c decreased, hence no supratransmission. However, for $\Omega > 1$, in the case where quartic interaction terms were present, supratransmission did occur, as there was no local maximum, and the A_s monotonically increased at $\beta = 0$ values, which also grew steadily as $c \longrightarrow 0$.

These findings were explained by carrying out an analysis of the phonon band of our system. This allowed us to explain the reason why supratransmission failed to occur, as the range parameters α and β decreased, due to resonances of the driving frequency with the phonons, while it did occur when Ω lay above the phonon band.

It would, of course, be desirable to connect analytically the occurrence of supratransmission with the excitation of specific nonlinear solutions of the problem, as has been conducted in a number of references, where it was possible to derive a PDE in the continuum limit, see, e.g., [3,4]. This was mainly accomplished in the continuum limit of 1D lattices with nearest neighbor interactions, which appeared difficult to extend to the LRI systems studied in this paper. Perhaps it could be conducted for cases where *a fractional PDE* can be derived for our system, but this is a matter that we would prefer to address in a future publication.

We also looked at snapshots of the energy distribution along the chain to guess the type of nonlinear excitation activated at the supratransmission state. Interestingly, the graphs we obtained strongly resembled a *travelling breather*, similar to what one finds in the continuum limit of a lattice with only nearest neighbor interactions! It would thus be very interesting, in a future publication, to attempt to find an analogous analytical solution and study its stability for a lattice that involves long range interactions, such as with the one studied in the present paper.

In closing, we believe that the results we have described in this paper provide a motivation to search for physical arguments to justify and perhaps explain our findings. One main result is that the presence of on-site potentials often hinders the occurrence of supratransmission. The relationships we have uncovered between the fundamental parameters of our system and the phenomenon of supertransmission are interesting, and, in our opinion, quite remarkable. They, therefore, merit further study before we can say that we understand this phenomenon and we are ready to investigate it in higher dimensional settings.

Author Contributions: Conceptualization, T.B. and J.E.M.-D.; data curation, J.E.M.-D.; formal analysis, J.E.M.-D.; funding acquisition, J.E.M.-D. and T.B.; investigation, T.B. and J.E.M.-D.; methodology, T.B. and J.E.M.-D.; project administration, T.B. and J.E.M.-D.; resources, J.E.M.-D.; software, J.E.M.-D.; supervision, T.B. and J.E.M.-D.; validation, J.E.M.-D. and T.B.; visualization, J.E.M.-D.; writing—original draft, T.B. and J.E.M.-D.; writing—review and editing, T.B. and J.E.M.-D. All authors have read and agreed to the published version of the manuscript.

Funding: T.B.: The work in Sections 1,3,6,7 was funded under the scientific project No. 21-71-30011 of the Russian Science Foundation. J.E.M.-D.: The work in Sections 2,4,5 reports on a set of final results of the research project "Conservative methods for fractional hyperbolic systems: analysis and applications", funded by the National Council for Science and Technology of Mexico (CONACYT) through grant A1-S-45928.

Institutional Review Board Statement: Not applicable.

Informed Consent Statement: Not applicable.

Data Availability Statement: Not applicable.

Acknowledgments: The authors express their gratitude to the two reviewers for their stimulating comments and remarks that helped significantly improve the presentation and results of the present paper.

Conflicts of Interest: The authors declare no conflict of interest.

References

1. Caputo, J.G.; Leon, J.; Spire, A. Nonlinear energy transmission in the gap. *Phys. Lett. A* **2001**, *283*, 129–135. [CrossRef]
2. Khomeriki, R.; Leon, J. Bistability in the sine-Gordon equation: The ideal switch. *Phys. Rev. E* **2005**, *71*, 056620. [CrossRef] [PubMed]
3. Khomeriki, R.; Lepri, S.; Ruffo, S. Nonlinear supratransmission and bistability in the Fermi-Pasta-Ulam model. *Phys. Rev. E* **2004**, *70*, 066626. [CrossRef] [PubMed]
4. Geniet, F.; Leon, J. Nonlinear supratransmission. *J. Phys. Condens. Matter* **2003**, *15*, 2933. [CrossRef]
5. Ervin, V.J.; Macías-Díaz, J.E.; Ruiz-Ramírez, J. A positive and bounded finite element approximation of the generalized Burgers–Huxley equation. *J. Math. Anal. Appl.* **2015**, *424*, 1143–1160. [CrossRef]
6. Macías-Díaz, J.E.; Medina-Ramírez, I.E. An implicit four-step computational method in the study on the effects of damping in a modified α-Fermi–Pasta–Ulam medium. *Commun. Nonlinear Sci. Numer. Simul.* **2009**, *14*, 3200–3212. [CrossRef]
7. Macías-Díaz, J.; Puri, A. A numerical method for computing radially symmetric solutions of a dissipative nonlinear modified Klein–Gordon equation. *Numer. Methods Partial. Differ. Equ. Int. J.* **2005**, *21*, 998–1015. [CrossRef]
8. Christodoulidi, H.; Bountis, A.; Drossos, L. The effect of long-range interactions on the dynamics and statistics of 1D Hamiltonian lattices with on-site potential. *Eur. Phys. J. Spec. Top.* **2018**, *227*, 563–573. [CrossRef]
9. Christodoulidi, H.; Bountis, T.; Tsallis, C.; Drossos, L. Dynamics and statistics of the Fermi–Pasta–Ulam β-model with different ranges of particle interactions. *J. Stat. Mech. Theory Exp.* **2016**, *2016*, 123206. [CrossRef]
10. Dauxois, T.; Ruffo, S.; Arimondo, E.; Wilkens, M. Dynamics and thermodynamics of systems with long-range interactions: An introduction. In *Dynamics and Thermodynamics of Systems with Long-Range Interactions*; Springer: Berlin/Heidelberg, Germany, 2002; pp. 1–19.
11. Viana, R.; Batista, A.; Batista, C.; Iarosz, K. Lyapunov spectrum of chaotic maps with a long-range coupling mediated by a diffusing substance. *Nonlinear Dyn.* **2017**, *87*, 1589–1601. [CrossRef]
12. Tarasov, V.E. Continuous limit of discrete systems with long-range interaction. *J. Phys. A Math. Gen.* **2006**, *39*, 14895. [CrossRef]
13. Tarasov, V.E.; Zaslavsky, G.M. Fractional dynamics of systems with long-range interaction. *Commun. Nonlinear Sci. Numer. Simul.* **2006**, *11*, 885–898. [CrossRef]

14. Tarasov, V.E.; Zaslavsky, G.M. Conservation laws and Hamilton's equations for systems with long-range interaction and memory. *Commun. Nonlinear Sci. Numer. Simul.* **2008**, *13*, 1860–1878. [CrossRef]
15. Dong, B.Q.; Wu, J.; Ye, Z. Global regularity for a 2D tropical climate model with fractional dissipation. *J. Nonlinear Sci.* **2019**, *29*, 511–550. [CrossRef]
16. Wu, G.C.; Deng, Z.G.; Baleanu, D.; Zeng, D.Q. New variable-order fractional chaotic systems for fast image encryption. *Chaos Interdiscip. J. Nonlinear Sci.* **2019**, *29*, 083103. [CrossRef]
17. Macías-Díaz, J.E. Numerical study of the process of nonlinear supratransmission in Riesz space-fractional sine-Gordon equations. *Commun. Nonlinear Sci. Numer. Simul.* **2017**, *46*, 89–102. [CrossRef]
18. Macías-Díaz, J.E. Persistence of nonlinear hysteresis in fractional models of Josephson transmission lines. *Commun. Nonlinear Sci. Numer. Simul.* **2017**, *53*, 31–43. [CrossRef]
19. Macías-Díaz, J.E.; Bountis, A. Supratransmission in β-Fermi–Pasta–Ulam chains with different ranges of interactions. *Commun. Nonlinear Sci. Numer. Simul.* **2018**, *63*, 307–321. [CrossRef]
20. Macías-Díaz, J.E.; Bountis, A.; Christodoulidi, H. Energy transmission in Hamiltonian systems of globally interacting particles with Klein-Gordon on-site potentials. *Math. Eng.* **2019**, *1*, 343–358. [CrossRef]
21. Macías-Díaz, J.E.; Bountis, A. Nonlinear Supratransmission in quartic Hamiltonian systems with globally interacting particles and on-site potentials. *J. Comput. Nonlinear Dyn.* **2021**, *16*, 021001. [CrossRef]
22. Macías-Díaz, J.E. Numerical study of the transmission of energy in discrete arrays of sine-Gordon equations in two space dimensions. *Phys. Rev. E* **2008**, *77*, 016602. [CrossRef] [PubMed]
23. Macías-Díaz, J.E. Numerical simulation of the nonlinear dynamics of harmonically driven Riesz-fractional extensions of the Fermi–Pasta–Ulam chains. *Commun. Nonlinear Sci. Numer. Simul.* **2018**, *55*, 248–264. [CrossRef]
24. Macías-Díaz, J.E.; Puri, A. On the propagation of binary signals in damped mechanical systems of oscillators. *Phys. D Nonlinear Phenom.* **2007**, *228*, 112–121. [CrossRef]
25. Macías-Díaz, J.E. An explicit dissipation-preserving method for Riesz space-fractional nonlinear wave equations in multiple dimensions. *Commun. Nonlinear Sci. Numer. Simul.* **2018**, *59*, 67–87. [CrossRef]

entropy

Article

Spectral Form Factor and Dynamical Localization

Črt Lozej

Max Planck Institute for the Physics of Complex Systems, Nöthnitzer Str. 38, 01187 Dresden, Germany; crt@pks.mpg.de

Abstract: Quantum dynamical localization occurs when quantum interference stops the diffusion of wave packets in momentum space. The expectation is that dynamical localization will occur when the typical transport time of the momentum diffusion is greater than the Heisenberg time. The transport time is typically computed from the corresponding classical dynamics. In this paper, we present an alternative approach based purely on the study of spectral fluctuations of the quantum system. The information about the transport times is encoded in the spectral form factor, which is the Fourier transform of the two-point spectral autocorrelation function. We compute large samples of the energy spectra (of the order of 10^6 levels) and spectral form factors of 22 stadium billiards with parameter values across the transition between the localized and extended eigenstate regimes. The transport time is obtained from the point when the spectral form factor transitions from the non-universal to the universal regime predicted by random matrix theory. We study the dependence of the transport time on the parameter value and show the level repulsion exponents, which are known to be a good measure of dynamical localization, depend linearly on the transport times obtained in this way.

Keywords: quantum chaos; spectral form factor; dynamical localization; billiards; stadium

Citation: Lozej, Č. Spectral form Factor and Dynamical Localization. *Entropy* **2023**, *25*, 451. https://doi.org/10.3390/e25030451

Academic Editor: Lawrence Horwitz

Received: 30 January 2023
Revised: 1 March 2023
Accepted: 3 March 2023
Published: 4 March 2023

1. Introduction

One of the central areas of study in quantum chaos is that of the spectral statistics of quantum chaotic systems and how they relate to classical chaos and random matrix theory (RMT) [1,2]. The spectral form factor (SFF) is one of the most widely used spectral statistics due to the stark contrast in behaviour between the chaotic and integrable regimes. However, the SFF is not a self-averaging quantity [3], meaning that the typical value may be far from the average value. Because of this, its numerical computation remains challenging, and its practical evaluation requires some sort of smoothing procedure, either by computing disorder averages (only possible when considering systems with disorder) or local time averages. Nevertheless, the SFF has been used as the fundamental indicator of quantum chaos in many of the central rigorous results. A heuristic proof of the quantum chaos (Bohigas–Giannoni–Schmit) conjecture [4,5], which was initiated by Berry [6], developed by Sieber and Richter [7], and later completed by Haake's group [8–10], clearly relates random matrix spectral correlations to correlations among classical unstable (hyperbolic) orbits by computing the orbit contributions to the SFF. Recently, much attention has been given to the SFF in many-body settings [11–16]. Rigorous proofs of quantum chaos by computing the SFF have been preformed in kicked spin chains [17,18] and more generally in dual-unitary circuits [19,20]. In the high-energy physics context, for example, studies of the SFF have been performed in Sachdev–Ye–Kitaev-type models [21–24] and using hydrodynamic theories [25]. Pioneering experimental studies of the SFF were carried out on excitation spectra of molecules [26] and microwave billiards [27]. The SFF has recently been used to probe the many-body localization (MBL) transition [28,29]. In this paper, we will adapt a similar methodology to that in Ref. [28] to study the dynamical localization transition in single-body systems on the example of stadium billiards. Even though quantum billiards are ubiquitous in the field of quantum chaos, not many theoretical studies of the SFF in chaotic billiards are to be found in the literature. Previous studies focus mainly on the (pseudo)integrable and closely related regime, such as rectangular billiards [30], including perturbations [31], barrier billiards [32–34] and Veech triangular [35] billiards. Recently, the SFF has also been computed in the case of generic triangular billiards [36], where it was demonstrated that the spectral statistics follow RMT, thereby extending the quantum chaos conjecture to strongly mixing systems without classical Lyapunov chaos.

The origin of the study of dynamical localization in the stadium billiards can be traced to the pioneering work of Borgonovi, Casati and Li [37], later continued by Casati and Prosen [38,39]. Quantum dynamical localization (DL) occurs when quantum interference stops the diffusion of wave packets. The phenomenon is analogous to the famous Anderson localization, but occurring in momentum space instead of the configuration space. The two can be explicitly related in the example of the quantum kicked rotor system [40]. The following heuristic argument explains when dynamical localization may be expected. The transition is governed by the ratio of two typical time scales, namely the transport time t_T, controlling the typical rate of diffusion, and the Heisenberg time t_H, which is the inverse of the mean level spacing. The discreetness of the quantum spectrum may only be resolved on time scales greater than the Heisenberg time. If $t_T > t_H$, we expect the interference will localize the wave packets in only part of the momentum space. On the other hand, if $t_T < t_H$, we expect the wave packet to encompass the full extent of the momentum space before any interference effects might stop the diffusion. The transition from the dynamically localized regime to the fully delocalized ergodic regime has been extensively studied in the quantum kicked rotor system (see the review articles [41,42]), billiard systems [37–39,43–48], Dicke model [49], etc. In particular, our previous studies of DL in the stadium billiard [45] show the functional dependence of the localization measures and level repulsion exponents on the ratio $\alpha = t_H/t_T$. However, to ascertain the transport times, a separate classical computation of the transport times was necessary. This also introduces some ambiguity in defining the transport time because of the complex inhomogeneous diffusion that occurs (see Ref. [50] for details). Furthermore, in generic billiards with divided regular/chaotic phase space the diffusion process is even more complex because of the hierarchical structures of islands of stability in the phase space and the stickiness phenomenon (see Refs. [48,51–53] and references therein). In the present paper, we present an alternative definition of the transport time based on the timescale of the onset of RMT spectral statistics in the SFF. The definition is inspired by the methodology used to extract the Thouless time of spin chains in Ref. [28]. We show the transport time extracted from the spectral form factor can be used to describe the transition from the DL regime to the ergodic regime.

2. Definitions and Methods

2.1. Quantum Billiards

Quantum billiards are archetypical models of both classical and quantum chaos. In the quantum billiard problem, we consider a quantum particle trapped inside a region $\mathcal{B} \subset \mathbb{R}^2$ referred to as the billiard table. The eigenfunctions ψ_n are given by the solutions of the Helmholtz equation

$$\left(\nabla^2 + k_n^2\right)\psi_n = 0, \quad (1)$$

and Dirichlet b.c. $\psi_n|_{\partial\mathcal{B}} = 0$, with eigenenergies $E_n = k_n^2$, where k_n is the wave number of the n-th eigenstate. We use a system of units where $\hbar = 1$, and the mass of the particle is $m = 1/2$. The very efficient scaling method, devised by Vergini and Saraceno [54] and extensively studied by Barnett [55], allows us to compute very large spectra of the order of 10^6 states (the implementation is available as part of [56]). The spectral staircase function counts the number of eigenstates (or modes) up to some energy $N(E) := \#\{n|E_n < E\}$. The asymptotic mean of the spectral staircase for billiards is given by the well known generalized Weyl's law [57]

$$N_{\mathrm{Weyl}}(E) = (\mathcal{A}E - \mathcal{L}\sqrt{E})/4\pi + c \quad (2)$$

where $\mathcal{A}$ is the area of the billiard, $\mathcal{L}$ is the circumference, and c is a constant corner and curvature correction. The asymptotic density of states is then

$$\rho(E) = \frac{\mathcal{A}}{4\pi} - \frac{\mathcal{L}}{8\pi\sqrt{E}}. \tag{3}$$

The Heisenberg time is defined as the inverse of the mean level spacing or

$$t_H = 2\pi\rho(E). \tag{4}$$

To compare the universal statistical fluctuations it is convenient to unfold the spectra. This is performed by inserting the numerically computed billiard spectrum into Weyl's formula $e_n := N_{\mathrm{Weyl}}(E_n)$. The resulting unfolded spectrum e_n has a uniform mean level density equal to one. In the unfolded spectrum $t_H = 2\pi$.

One of the paradigmatic examples is the stadium billiard of Bunimovich [58,59]. The stadium is constructed from two semicircles separated by a rectangular region. We fix the radius of the semicircles to one. The family of stadium billiards is characterized by the width of the separation ε. The stadium is classically chaotic for any value of ε. Because of the two reflection symmetries, it is sufficient to consider the quarter stadium in the quantum case, corresponding to the odd–odd symmetry sector of the full stadium. Two examples of stadium eigenstates are shown in Figure 1. In panel (a), we show a typical dynamically localized eigenstate in the $\varepsilon = 0.02$ stadium. The localization is evident in the distinctly regular nodal patterns that are similar in appearance to very strong scarring. Although the probability density function extends over all the configuration space, it is visibly depleted in the inner part of the billiard near the origin (note the colour scale is logarithmic). In (b), we show a typical eigenstate in the $\varepsilon = 0.5$ stadium. The state is practically uniformly extended, with the typical chaotic nodal patterns of random superpositions of plane waves, with some scarring visible around an unstable (bow-tie-shaped) periodic orbit.

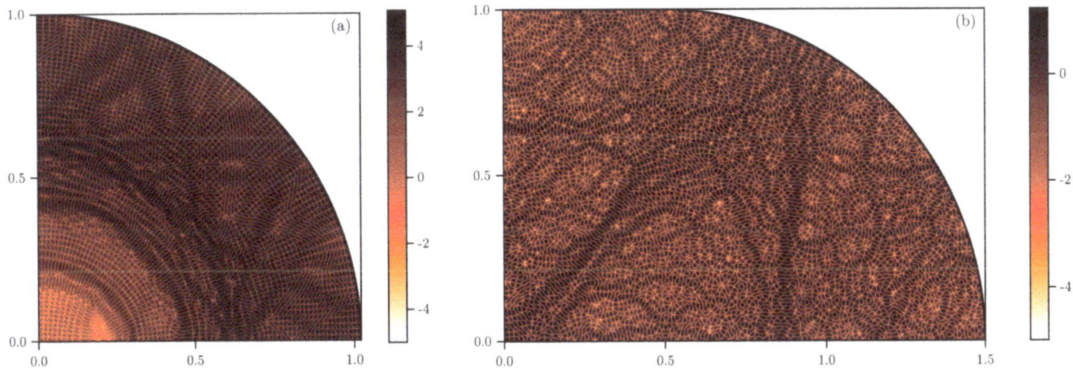

Figure 1. Representativeeigenstates in the (quarter) stadium billiards. (**a**) Localized eigenstate at $k = 302.60195$ and $\varepsilon = 0.02$. (**b**) Extended state at $k = 302.6037$ and $\varepsilon = 0.5$. We plot the probability density in the logarithmic scale, $\log_{10}(|\psi|^2)$.

2.2. *Spectral Form Factor*

The SFF is loosely defined as the Fourier transform of the spectral two-point correlation function and may be written as

$$K(\tau) = \left\langle \left| \sum_n \exp(2\pi i e_n \tau) \right|^2 \right\rangle, \tag{5}$$

where the sum goes over the unfolded energy levels. The time τ is measured in units of Heisenberg time $\tau_H = 1$. The SFF is not a self-averaging quantity [3], it exhibits erratic fluctuations with time. This means a separate averaging must be performed, represented by $\langle \cdots \rangle$. This is commonly an average over different realizations when considering random matrices or disordered systems. For clean single-body systems, we instead perform a moving time average to smooth out the fluctuations [26,27]. This is achieved by convolving the SFF with a Gaussian function in time,

$$K(\tau) = \int_0^\infty \left| \sum_n \exp(2\pi i e_n \tau) \right|^2 \frac{1}{\sqrt{2\pi\sigma^2}} \exp(-\frac{1}{2}\frac{(\tau - t)^2}{\sigma^2}) dt. \tag{6}$$

This introduces an additional numerical parameter σ. It is further useful to decompose the SFF into the connected and disconnected part $K = K_{\text{conn}} + K_{\text{disc}}$. The disconnected part is given by the diagonal terms from definition (5) and depends solely on the density of states (see Ref. [25] for more details). It is also evident from definition (5) that the SFF behaves as a delta distribution at $t = 0$. This narrow peak is produced by the disconnected part of the SFF. The spectral fluctuations are encoded in the connected part of the SFF, which we obtain by subtracting the disconnected part $K_{\text{conn}} = K - K_{\text{disc}}$. Since we are only interested in spectral fluctuations, we will only consider the connected part of the SFF in all further instances.

The stadium billiards are classically chaotic systems with time-inversion symmetry. Their universal spectral statistics are therefore expected to follow the Gaussian orthogonal ensemble (GOE) of RMT [1,2]. In the infinite dimensional GOE case, the SFF has the following analytical form,

$$K_{\text{GOE}}(\tau) = \begin{cases} 2\tau - \tau\ln(2\tau + 1) & \tau < 1 \\ 2 - \tau\ln(\frac{2\tau+1}{2\tau-1}) & \tau > 1 \end{cases}. \tag{7}$$

This has the basic anatomy of a so-called "ramp" followed by a saturation regime after reaching the Heisenberg time. This contrasts well with the integrable case, where an immediate saturation is expected. Since all stadium billiards are ergodic chaotic systems, we expect the SFF will follow the universal GOE prediction. However, when ε is small, the transport times become very large and should even diverge as we approach the limit $\varepsilon \to 0$ (the limiting case is the integrable circle billiard, where the momentum becomes a strictly conserved quantity). Classically, the fact that the system is ergodic becomes apparent only after the transport time is reached, and the dynamics are able to explore all the phase space. We expect the SFF of the stadia will follow the GOE prediction only after the transport elapses. We will therefore define the quantum transport time τ_T as the time at which the SFF of the numerically computed billiard spectrum begins to follow the RMT prediction. The procedure that is used to extract τ_T is described in more detail in Appendix A. The transport time may either be greater or smaller than $\tau_H = 1$ (note that by definition (5) we measure time in the SFF in units of Heisenberg time). Following the argument from the introduction, this means we expect localization when $\tau_T > 1$, and no localization (extendedness) when $\tau_T < 1$.

2.3. Dynamical Localization and Level Repulsion

We will measure the localization of the eigenstates indirectly by computing the level of the repulsion exponent of the spectra. The level repulsion exponent is defined by using the nearest-neighbour level spacing. An intuitive understanding of the connection between level repulsion and localization may be gained from the following heuristic picture. The eigenstates of chaotic systems in the non-localized regime are extended in the phase space, and there is a great deal of overlap between them. In the extreme case of full extendedness, the differences in the overlaps will stem purely from local fluctuations of the wave (or Husimi) functions. This means strong couplings between the consecutive states are possible and indeed very probable, leading to a gap in the eigenenergies. For an extremely simplified case, one may consider a two-level system, where the gap of the eigenenergies (avoided crossing) may be computed directly. On the other hand, dynamically localized states only occupy a smaller area of the phase space. If the consecutive states occupy different areas of the phase space, there will be essentially no overlap between them, with the couplings exponentially suppressed. This is more likely to happen if the eigenstates are more severely localized, leading to a much weaker level repulsion. The connection between localization and level repulsion has a strong foundation in our previous works and also related studies in different systems. In particular, in Ref. [45] we showed that the level repulsion exponents in the stadium billiards are proportional to the mean values of localization measures based on the Husimi representation of the eigenstates (for a recent study of the localization measures in more general divided phase space systems, see also [60]).

The level spacing is defined as the difference in energy between two consecutive levels in the unfolded spectrum $s_i = e_{i+1} - e_i$. The unfolding procedure guarantees that the mean level spacing is in unity. We also studied the probability density distribution $P(s)$. The level repulsion is given by the behaviour of $P(s)$ at small s, namely $P(s) \propto s^\beta$, where β is called the level repulsion exponent. Following the quantum chaos conjecture, the level spacing distribution of chaotic quantum systems is well described by the Wigner surmise obtained from RMT. In the GOE case, $\beta = 1$ indicates linear level repulsion. On the other hand, integrable systems are expected to show Poissonian level statistics (Berry–Tabor conjecture) and no level repulsion $\beta = 0$. In the localized regime, the distribution is not known analytically. Empirically, the level repulsion exponent changes from 0 to 1 as we transition from the severely localized to the delocalized chaotic regime. One of the most popular ways of describing the level spacing distribution in the transition region is to use the Brody distribution [61], which interpolates the two regimes

$$P_B(S) = cS^\beta \exp\left(-dS^{\beta+1}\right), \tag{8}$$

where the normalization constants are given by $c = (\beta + 1)d$, and $d = \left(\Gamma\left(\frac{\beta+2}{\beta+1}\right)\right)^{\beta+1}$. Alternatively, another popular choice is the Izrailev distribution [41]; however, we opted for the Brody distribution due to the simpler expression and empirically good description of the numerical results in previous papers [43–46,62]. The level repulsion exponent β is the indicator of dynamical localization, which we compare to τ_T across the transition.

3. Results

3.1. Transport Times

To compute the quantum transport times, we computed the spectra of the stadium billiards at 22 values of $\varepsilon \in (0.01, 0.07)$. Each spectrum contains around 10^6 levels with $k_n \in (640, 4000)$. The lowest levels start at around the 10^4-th eigenstate. Because the scaling method computes the eigenvalues in only small intervals, some levels are lost in the computation due to numerical errors. By comparing with Weyl's law, we estimate that less than 0.1% are lost.Since the SFF is a linear spectral statistic, we expect this to have a negligible effect on the result. Even with the great efficiency of the numerical method, collecting the spectra and computing the SFF takes considerable computational effort due to the large spectra required to obtain good results.

The connected SFF of the selected stadia are shown in Figure 2a. The numerical results are compared with the GOE curve (7). We see the $\varepsilon = 0.5$ result, where the transport time is expected to be very short, nicely follows the GOE curve from start to finish. When ε is decreased, the numerical SFF detaches from the GOE curve at some point. This point is by our definition the transport time. We see the transport time increases as ε is decreased, eventually becoming longer than the Heisenberg time. We note the SFF still exhibits some fluctuations, even though each of the spectra contains many levels—approximately 10^6. The smoothing parameter in the presented case is $\sigma = 0.01$, which we find is the optimal compromise between fine resolution and the intensity of fluctuations. We extract the transport times, including some error estimates (shown with the error bars), as described in Appendix A. The result is presented in Figure 2b. In the inset, we show the same graph in the decadic log–log scale. The transport times appear to roughly follow a power law decay $\tau_T \propto \varepsilon^{-\gamma}$, with a transition from $\gamma = 1$ to $\gamma = 1/2$ above $\varepsilon_c \approx 0.04$. The caveat is that the power laws should not be seen as a definitive result, since the range of the parameter values is within one decade. In Ref. [50], we computed the classical transport times of the stadia in the space of conjugated momenta and discrete time (the conjugated momenta of the billiard mapping, describing the classical dynamics, are $p = \sin\theta$, where θ is the angle of reflection when the particle hits the boundary). There, we found $N_T \propto \varepsilon^{-\gamma}$ with a transition from $\gamma = 5/2$ to $\gamma = 2$ above $\varepsilon_c \approx 0.05$; however, the transitional value is not sharply defined. We note that considering the transport in the flow of the stadium billiard (real time) instead of the billiard map (discrete time) might give different results, because the slow decay of correlations in the classical stadium billiard is caused by special types of bouncing ball and boundary glancing orbits [63]. The difference in the decay rates indicates the quantum transport time extracted from the SFF is not directly proportional to the discrete transport time in momentum space. However, both are monotonic functions (within some fluctuations) of the parameter ε, and both seem to exhibit a transition in the power law behaviour at roughly the same parameter range.

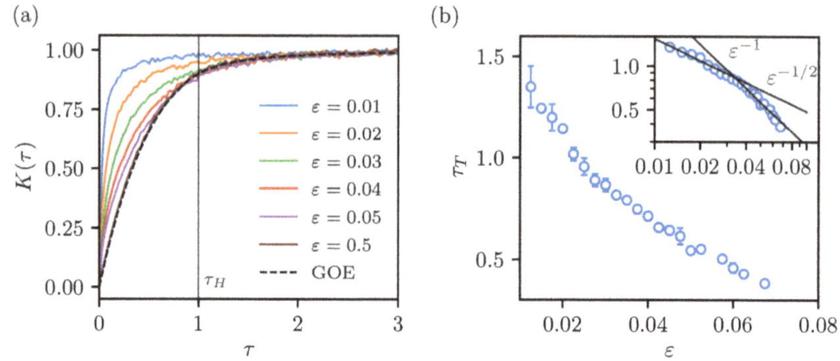

Figure 2. (**a**) Connected spectral form factors of stadium billiards in units of Heisenberg time. The GOE curve, expected in chaotic systems, is shown with the black dashed line. (**b**) Dependence of the quantum transport times (in units of Heisenberg time), extracted from the SFF, on the billiard parameter ε. The error bars show the estimated errors due to the fluctuations in crossing the threshold value. The inset shows the same plot in the decadic log–log scale.

3.2. Level Repulsion

To determine the level repulsion exponents β, we fit the level spacing distributions of the computed spectra with the Brody distribution (8). In Figure 3a, we show some examples of the fits. We see the level spacings are indeed described well by the Brody distribution. In panel (b), we show β as a function of τ_T. We observe the transition from the extended to the localized regime as the transport time increases, empirically confirming the heuristic argument that the transition should happen when the transport time is close to the Heisenberg time. Quantitatively, the mid-point of the transition $\beta = 0.5$ occurs already at $\tau_T \approx 0.8$. The relation between the two quantities appears to be close to linear. In Ref. [45], we found a nonlinear functional relation between β and the parameter $\alpha = t_H/t_T$ (the denominator is the classical transport time) that would be analogous to $1/\tau_T$. This indicates that the quantum transport times are not exactly analogous to the classical transport times. Nevertheless, we clearly establish a functional relation between the level repulsion exponent and the quantum transport times. Because the level repulsion exponents are a linear function of localization measures (see Refs. [45,46]), this demonstrates the link to dynamical localization and potentially also a more general relation (a similar non-universal but characteristic behaviour) between level spacing distributions and spectral form factors in other contexts.

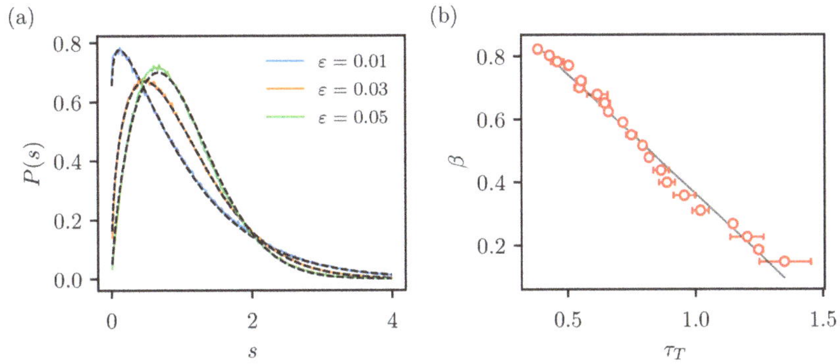

Figure 3. (**a**) Representative examples of nearest-neighbour level spacing distributions (coloured lines) fitted by the Brody distribution (black dashed lines). (**b**) Dependence of the level repulsion exponent (Brody parameter) on the quantum transport times (in units of Heisenberg time).

4. Discussion

We have presented a numerical study of the spectral form factors of the stadium billiards in relation to dynamical localization. The main result is the computation of the connected spectral form factors and extraction of the quantum transport times τ_T (in units of Heisenberg time) from the SFF. By relating τ_T to the level repulsion exponent β, we show that the transition from the localized to the delocalized regime is governed by the ratio between the transport time and the Heisenberg time. The novelty of the presented approach compared with previous studies of the dynamical localization transition is that all computations are based on the quantum spectral statistics alone. No classical computations of the transport times are needed. This might be especially beneficial in cases where the classical transport processes are very complex and the definition of the relevant transport time might be ambiguous, such as, for instance, in systems with divided phase space and, as already demonstrated in Ref. [28], in many-body systems without a classical limit. The relationship between β and τ_T is close to linear. This is different from the nonlinear relation with the analogous quantity $\alpha = t_H/t_T$ found in Ref. [45], where the transport times t_T were computed from the classical momentum diffusion in discrete time. Nevertheless, both definitions of the transport time exhibit a power law regime change at roughly the same value of ε. Since quantum billiards may be considered a generic example of Hamiltonian systems, the results are widely applicable. Further research directions might include a similar study of the SFF in systems with divided phase space, such as, for instance, the limaçon billiards (see Refs. [46,64,65] and references therein).

Funding: The author acknowledges the Max Planck Institute for the Physics of Complex Systems (MPIPKS) for financial support. The APC was funded by The Max Planck Society for the Advancement of Science.

Institutional Review Board Statement: Not applicable.

Data Availability Statement: The data presented in this study are available upon reasonable request from the corresponding author.

Acknowledgments: This paper is dedicated to the celebration of the 80th birthday of Giulio Casati. His work was a great inspiration at the start of my research career and indeed for the contents of the present paper. I am very privileged to be counted as one of his collaborators. I thank him for the kind words of encouragement and for inviting me to contribute to solving some long-standing questions of quantum chaos. I thank the Max Planck Society for its hospitality and M.T. Eiles and M. Robnik for carefully proofreading the manuscript.

Conflicts of Interest: The authors declare no conflict of interest.

Abbreviations

The following abbreviations are used in this manuscript:

RMT	Random matrix theory
SFF	Spectral form factor
GOE	Gaussian orthogonal ensemble
DL	Dynamical localization

Appendix A. Extracting the Quantum Transport Times

The appendix describes the details of how we extract the quantum transport time from the SFF data. We follow the procedure outlined in Ref. [28]. The objective is to find the point in time when the connected SFF starts to follow the GOE curve (7). Let us define the following quantity

$$g(\tau) = \left| \log_{10}\left(\frac{K(\tau)}{K_{\mathrm{GOE}}(\tau)} \right) \right|. \tag{A1}$$

This measures the ratio between the numerical data and the GOE curve, and the logarithm gives the order of magnitude. When $g \to 0$, the two quantities are exactly equal. To determine the quantum transport time, we select a threshold value g_0 and define τ_T as the time at which $g(\tau) < g_0$. Because the SFF fluctuates even after the smoothing procedure, pinpointing the exact value of τ_T remains challenging. A local fluctuation exactly at the threshold may obscure the result. To estimate the errors incurred by the fluctuations, we vary g_0 in a small interval and compute the mean and standard deviation of the obtained τ_T. In Figure A1, we show $g(\tau)$ for a few different stadium billiards. The fluctuations of $g(\tau)$ in the final stationary regime are of the order of 0.01, i.e., about 2% relative difference. We therefore opted for $g_0 = 0.02$, or about 5% relative difference, and varied it down to $g_0 = 0.015$ to obtain the error estimates. The final results are shown in Figure 2b in Section 3.1.

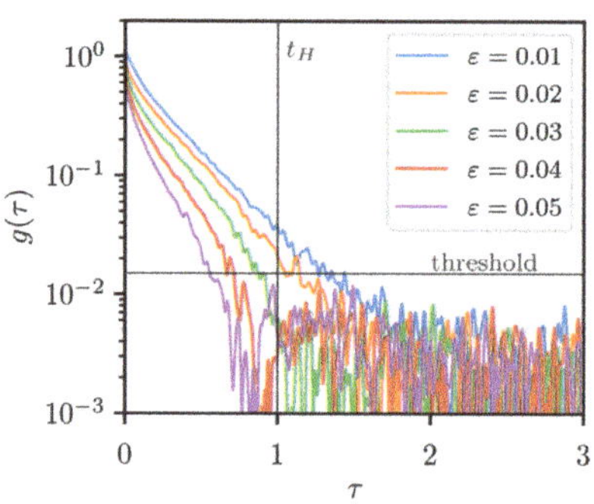

Figure A1. Dependence of the quantity $g(\tau)$ for representative values of ε. The transport time is determined by the point where the curve crosses the threshold for the first time.

References

1. Stöckmann, H.J. *Quantum Chaos—An Introduction*; Cambridge University Press: Cambridge, UK, 1999.
2. Haake, F. *Quantum Signatures of Chaos*; Springer: Berlin, Germany, 2001.
3. Prange, R. The spectral form factor is not self-averaging. *Phys. Rev. Lett.* **1997**, *78*, 2280. [CrossRef]
4. Bohigas, O.; Giannoni, M.J.; Schmit, C. Characterization of chaotic quantum spectra and universality of level fluctuation laws. *Phys. Rev. Lett.* **1984**, *52*, 1–4 . [CrossRef]

5. Casati, G.; Valz-Gris, F.; Guarnieri, I. On the connection between quantization of nonintegrable systems and statistical theory of spectra. *Lett. Nuovo C.* **1980**, *28*, 279–282. [CrossRef]
6. Berry, M.V. Semiclassical theory of spectral rigidity. *Proc. R. Soc. Lond. A Math. Phys. Sci.* **1985**, *400*, 229–251.
7. Sieber, M.; Richter, K. Correlations between periodic orbits and their role in spectral statistics. *Phys. Scr.* **2001**, *T90*, 128. [CrossRef]
8. Müller, S.; Heusler, S.; Braun, P.; Haake, F.; Altland, A. Semiclassical Foundation of Universality in Quantum Chaos. *Phys. Rev. Lett.* **2004**, *93*, 014103. [CrossRef]
9. Heusler, S.; Müller, S.; Braun, P.; Haake, F. Universal spectral form factor for chaotic dynamics. *J. Phys. Math. Gen.* **2004**, *37*, L31. [CrossRef]
10. Müller, S.; Heusler, S.; Braun, P.; Haake, F.; Altland, A. Periodic-orbit theory of universality in quantum chaos. *Phys. Rev. E* **2005**, *72*, 046207. [CrossRef] [PubMed]
11. Roy, D.; Prosen, T. Random matrix spectral form factor in kicked interacting fermionic chains. *Phys. Rev. E* **2020**, *102*, 060202. [CrossRef]
12. Roy, D.; Mishra, D.; Prosen, T. Spectral form factor in a minimal bosonic model of many-body quantum chaos. *Phys. Rev. E* **2022**, *106*, 024208. [CrossRef]
13. Chan, A.; De Luca, A.; Chalker, J.T. Solution of a minimal model for many-body quantum chaos. *Phys. Rev. X* **2018**, *8*, 041019. [CrossRef]
14. Chan, A.; De Luca, A.; Chalker, J. Spectral statistics in spatially extended chaotic quantum many-body systems. *Phys. Rev. Lett.* **2018**, *121*, 060601. [CrossRef]
15. Moudgalya, S.; Prem, A.; Huse, D.A.; Chan, A. Spectral statistics in constrained many-body quantum chaotic systems. *Phys. Rev. Res.* **2021**, *3*, 023176. [CrossRef]
16. Chan, A.; Shivam, S.; Huse, D.A.; De Luca, A. Many-body quantum chaos and space-time translational invariance. *Nat. Commun.* **2022**, *13*, 7484. [CrossRef]
17. Kos, P.; Ljubotina, M.; Prosen, T. Many-body quantum chaos: Analytic connection to random matrix theory. *Phys. Rev. X* **2018**, *8*, 021062. [CrossRef]
18. Bertini, B.; Kos, P.; Prosen, T. Exact spectral form factor in a minimal model of many-body quantum chaos. *Phys. Rev. Lett.* **2018**, *121*, 264101. [CrossRef]
19. Kos, P.; Bertini, B.; Prosen, T. Correlations in perturbed dual-unitary circuits: Efficient path-integral formula. *Phys. Rev. X* **2021**, *11*, 011022. [CrossRef]
20. Bertini, B.; Kos, P.; Prosen, T. Random matrix spectral form factor of dual-unitary quantum circuits. *Commun. Math. Phys.* **2021**, *387*, 597–6204. [CrossRef]
21. Cotler, J.S.; Gur-Ari, G.; Hanada, M.; Polchinski, J.; Saad, P.; Shenker, S.H.; Stanford, D.; Streicher, A.; Tezuka, M. Black holes and random matrices. *J. High Energy Phys.* **2017**, *2017*, 1–54. [CrossRef]
22. Gharibyan, H.; Hanada, M.; Shenker, S.H.; Tezuka, M. Onset of random matrix behavior in scrambling systems. *J. High Energy Phys.* **2018**, *2018*, 1–62. [CrossRef]
23. Khramtsov, M.; Lanina, E. Spectral form factor in the double-scaled SYK model. *J. High Energy Phys.* **2021**, *2021*, 1–38. [CrossRef]
24. Cáceres, E.; Misobuchi, A.; Raz, A. Spectral form factor in sparse SYK models. *J. High Energy Phys.* **2022**, *2022*, 1–22. [CrossRef]
25. Winer, M.; Swingle, B. Hydrodynamic theory of the connected spectral form factor. *Phys. Rev. X* **2022**, *12*, 021009. [CrossRef]
26. Delon, A.; Jost, R.; Lombardi, M. NO_2 jet cooled visible excitation spectrum: Vibronic chaos induced by the $X^2A_1 - A^2B_2$ interaction. *J. Chem. Phys.* **1991**, *95*, 5701–5718. [CrossRef]
27. Alt, H.; Gräf, H.D.; Guhr, T.; Harney, H.; Hofferbert, R.; Rehfeld, H.; Richter, A.; Schardt, P. Correlation-hole method for the spectra of superconducting microwave billiards. *Phys. Rev. E* **1997**, *55*, 6674. [CrossRef]
28. Šuntajs, J.; Bonča, J.; Prosen, T.; Vidmar, L. Quantum chaos challenges many-body localization. *Phys. Rev. E* **2020**, *102*, 062144. [CrossRef]
29. Prakash, A.; Pixley, J.; Kulkarni, M. Universal spectral form factor for many-body localization. *Phys. Rev. Res.* **2021**, *3*, L012019. [CrossRef]
30. Marklof, J. Spectral form factors of rectangle billiards. *Commun. Math. Phys.* **1998**, *199*, 169–202. [CrossRef]
31. Rahav, S.; Fishman, S. Spectral statistics of rectangular billiards with localized perturbations. *Nonlinearity* **2002**, *15*, 1541. [CrossRef]
32. Wiersig, J. Spectral properties of quantized barrier billiards. *Phys. Rev. E* **2002**, *65*, 046217. [CrossRef]
33. Giraud, O. Periodic orbits and semiclassical form factor in barrier billiards. *Commun. Math. Phys.* **2005**, *260*, 183–201. [CrossRef]
34. Bogomolny, E. Level compressibility of certain random unitary matrices. *Entropy* **2022**, *24*, 795. [CrossRef]
35. Bogomolny, E.; Giraud, O.; Schmit, C. Periodic orbits contribution to the 2-point correlation form factor for pseudo-integrable systems. *Commun. Math. Phys.* **2001**, *222*, 327–369. [CrossRef]
36. Lozej, Č.; Casati, G.; Prosen, T. Quantum chaos in triangular billiards. *Phys. Rev. Res.* **2022**, *4*, 013138. [CrossRef]
37. Borgonovi, F.; Casati, G.; Li, B. Diffusion and localization in chaotic billiards. *Phys. Rev. Lett.* **1996**, *77*, 4744. [CrossRef]
38. Casati, G.; Prosen, T. Quantum localization and cantori in the stadium billiard. *Phys. Rev. E* **1999**, *59*, R2516. [CrossRef]
39. Casati, G.; Prosen, T. The quantum mechanics of chaotic billiards. *Phys. D Nonlinear Phenom.* **1999**, *131*, 293–310. [CrossRef]
40. Grempel, D.; Fishman, S.; Prange, R. Localization in an incommensurate potential: An exactly solvable model. *Phys. Rev. Lett.* **1982**, *49*, 833. [CrossRef]

41. Izrailev, F.M. Simple models of quantum chaos: Spectrum and eigenfunctions. *Phys. Rep.* **1990**, *196*, 299–392. [CrossRef]
42. Santhanam, M.; Paul, S.; Kannan, J.B. Quantum kicked rotor and its variants: Chaos, localization and beyond. *Phys. Rep.* **2022**, *956*, 1–87. [CrossRef]
43. Batistić, B.; Robnik, M. Dynamical localization of chaotic eigenstates in the mixed-type systems: Spectral statistics in a billiard system after separation of regular and chaotic eigenstates. *J. Phys. A Math. Theor.* **2013**, *46*, 315102. [CrossRef]
44. Batistić, B.; Robnik, M. Quantum localization of chaotic eigenstates and the level spacing distribution. *Phys. Rev. E* **2013**, *88*, 052913. [CrossRef]
45. Batistić, B.; Lozej, Č.; Robnik, M. The Level Repulsion Exponent of Localized Chaotic Eigenstates as a Function of the Classical Transport Time Scales in the Stadium Billiard. *Nonlinear Phenom. Complex Syst.* **2018**, *21*, 225–236.
46. Batistić, B.; Lozej, Č.; Robnik, M. Statistical properties of the localization measure of chaotic eigenstates and the spectral statistics in a mixed-type billiard. *Phys. Rev. E* **2019**, *100*, 062208. [CrossRef]
47. Batistić, B.; Lozej, Č.; Robnik, M. The distribution of localization measures of chaotic eigenstates in the stadium billiard. *Nonlinear Phenom. Complex Syst.* **2020**, *23*, 17–32. [CrossRef]
48. Lozej, Č. Transport and Localization in Classical and Quantum Billiards. Ph.D. Thesis, University of Maribor, Maribor, Slovenia, 2020.
49. Wang, Q.; Robnik, M. Statistical properties of the localization measure of chaotic eigenstates in Dicke model. *Phys. Rev. E* **2020**, *102*, 032212. [CrossRef] [PubMed]
50. Lozej, Č.; Robnik, M. Aspects of diffusion in the stadium billiard. *Phys. Rev. E* **2018**, *97*, 012206. [CrossRef] [PubMed]
51. Lozej, Č.; Robnik, M. Structure, size, and statistical properties of chaotic components in a mixed-type Hamiltonian system. *Phys. Rev. E* **2018**, *98*, 022220. [CrossRef]
52. Lozej, Č. Stickiness in generic low-dimensional Hamiltonian systems: A recurrence-time statistics approach. *Phys. Rev. E* **2020**, *101*, 052204. [CrossRef] [PubMed]
53. Lozej, Č.; Lukman, D.; Robnik, M. Effects of stickiness in the classical and quantum ergodic lemon billiard. *Phys. Rev. E* **2021**, *103*, 012204. [CrossRef]
54. Vergini, E.; Saraceno, M. Calculation by scaling of highly excited states of billiards. *Phys. Rev. E* **1995**, *52*, 2204–2207. [CrossRef]
55. Barnett, A. Dissipation in Deforming Chaotic Billiards. Ph.D. Thesis, Harvard University, Cambridge, MA, USA, 2001.
56. Lozej, Č.; Batistić, B.; Lukman, D. Quantum Billiards. Available online: https://github.com/clozej/quantum-billiards/tree/crt_public (accessed on 30 January 2023).
57. Baltes, H.P.; Hilf, E.R. *Spectra of Finite Systems*; BI-Wissenschafts: Mannheim, Germany, 1976.
58. Bunimovich, L.A. On billiards close to dispersing. *Mat. Sb.* **1974**, *136*, 49–73.
59. Bunimovich, L.A. On the ergodic properties of nowhere dispersing billiards. *Commun. Math. Phys.* **1979**, *65*, 295–312. [CrossRef]
60. Lozej, Č.; Lukman, D.; Robnik, M. Phenomenology of quantum eigenstates in mixed-type systems: Lemon billiards with complex phase space structure. *Phys. Rev. E* **2022**, *106*, 054203. [CrossRef]
61. Brody, T. A statistical measure for the repulsion of energy levels. *Lett. Nuovo C. (1971–1985)* **1973**, *7*, 482–484. [CrossRef]
62. Manos, T.; Robnik, M. Dynamical localization in chaotic systems: Spectral statistics and localization measure in the kicked rotator as a paradigm for time-dependent and time-independent systems. *Phys. Rev. E* **2013**, *87*, 062905. [CrossRef] [PubMed]
63. Vivaldi, F.; Casati, G.; Guarneri, I. Origin of long-time tails in strongly chaotic systems. *Phys. Rev. Lett.* **1983**, *51*, 727. [CrossRef]
64. Robnik, M. Classical dynamics of a family of billiards with analytic boundaries. *J. Phys. A Math. Gen.* **1983**, *16*, 3971. [CrossRef]
65. Robnik, M. Quantising a generic family of billiards with analytic boundaries. *J. Phys. A Math. Gen.* **1984**, *17*, 1049. [CrossRef]

entropy

MDPI

Review

Quantum Chaos and Level Dynamics

Jakub Zakrzewski [1,2]

[1] Institute of Theoretical Physics, Faculty of Physics, Astronomy and Applied Computer Science, Jagiellonian University, Łojasiewicza 11, 30-348 Kraków, Poland; jakub.zakrzewski@uj.edu.pl
[2] Mark Kac Complex Systems Research Center, Jagiellonian University, 30-348 Kraków, Poland

Abstract: We review the application of level dynamics to spectra of quantally chaotic systems. We show that the statistical mechanics approach gives us predictions about level statistics intermediate between integrable and chaotic dynamics. Then we discuss in detail different statistical measures involving level dynamics, such as level avoided-crossing distributions, level slope distributions, or level curvature distributions. We show both the aspects of universality in these distributions and their limitations. We concentrate in some detail on measures imported from the quantum information approach such as the fidelity susceptibility, and more generally, geometric tensor matrix elements. The possible open problems are suggested.

Keywords: 80th birthday of Giulio Casati; quantum chaos; level dynamics

Citation: Zakrzewski, J. Quantum Chaos and Level Dynamics. *Entropy* **2023**, *25*, 491. https://doi.org/10.3390/e25030491

Academic Editor: Marko Robnik

Received: 12 February 2023
Revised: 8 March 2023
Accepted: 9 March 2023
Published: 13 March 2023

1. Introduction

It is a great pleasure to be able to contribute to the volume of Giulio Casati's 80th birthday. Since the very beginning of my encounter with quantum chaos, Giulio Casati was one of those whose works inspired younger people. As an example, let me mention a contribution of late Prot Pakoński with whom I had a pleasure to consider the Kepler map (one of the toy models of Giulio), extending it to arbitrary polarization of the microwaves [1]. In this review I will discuss, however, a different topic—the statistical measures related to level dynamics in quantally chaotic systems. This is where we met scientifically, contributing to a single paper I had an honour to coauthor with Giulio [2]. The subject of level dynamics developed mainly in the 1980s and 1990s of the former millennium, yet it has recently found interesting extensions and applications in the modern many-body physics.

Level dynamics, described in a pedagogical way in the books of late Fritz Haake [3] and Hans-Jürgen Stöckmann [4], considers the motion of levels as a function of some arbitrary scalar parameter λ which characterizes the Hamiltonian $H(\lambda)$ of the system. It may be viewed as the motion of interacting fictitious particles (represented by levels) with λ being the effective time, as described originally by Pechukas [5] and followed by Yukawa [6] who built a corresponding statistical mechanics picture. This was applied in different ways to either justify the random matrix theory application to quantum chaotic spectra (this aim has been, in fact, accomplished by a beautiful periodic orbit theory application by the late Fritz Haake and collaborators—for a review see the Ref. [3]) or to define new statistical measures of energy spectra and find their distributions. Without any claim for completeness, some of these applications will be reviewed below.

We begin with defining the notation for level dynamics and the corresponding statistical mechanics picture in Section 2 showing how standard results from this approach provide a prediction for level statistics in a chaotic integrable transition. The resulting distribution seems to work surprisingly well for the data in many-body localization crossovers [7]. To somehow complete the picture, we review in the next Section several other models for the statistics in the transition, notably the banded matrix model, developed by Casati and coworkers [8–12]. In Section 4 we review the universality conjecture in level dynamics [13] showing how it is reflected in the so-called curvature distributions [14] in Section 5. We mention the velocity correlations in Section 6 and the avoided crossings statistics [15,16] in Section 7. Applications of these measures are discussed, stressing their limitations in real systems. Section 8 describes, on the other hand, very recent findings on the distribution of fidelity susceptibility [17], while the extensions to many parameter dynamics with geometric tensor matrix elements distributions are reviewed in Section 9. We mention briefly the parametric measures in the transition to localized regimes in Section 10. We finish with conclusions discussing future perspectives.

While the presentation that follows is only theoretical, we should mention that the predictions concerning level dynamics were tested, to some extent, in experiments. As typical for quantum chaos, those experiments were not carried out on eigenvalues of the Schrödinger equation, but rather on related models of quasi-2D microwave cavities or propagation of acoustic waves. We provide an incomplete list of references to beautiful experiments [18–24] stressing the work of the Stöckmann group [18] where a rather complete comparison of different measures with experimental microwave resonance data was carried out.

2. Level Dynamics

Let us recall some basic details on level dynamics, to fix the notation. Let the Hamiltonian, $H(\lambda) = H_0 + \lambda V$, depend on some parameter λ for arbitrary H_0 and V. The eigenvalue equation

$$H(\lambda)|a(\lambda)\rangle = E(\lambda)_a|a(\lambda)\rangle, \tag{1}$$

(where $E(\lambda)_a$ is the eigenvalue corresponding to eigenvector $|a(\lambda)\rangle$) upon differentiation with respect to λ gives

$$\frac{d}{d\lambda}E_a \equiv \dot{E}_a = \langle a|V|a\rangle \equiv V_{aa}. \tag{2}$$

Let us call $p_a \equiv \dot{E}_a$ a velocity of level E_a where λ becomes a fictitious time. Such a "motion" for a set of levels $\{E_a(\lambda)\}$ is visualized in Figure 1. A λ derivative (denoted by a dot over the variable) of p_a yields

$$\dot{p}_a = 2\sum_{b\neq a}\frac{V_{ab}V_{ba}}{E_a - E_b} = 2\sum_{b\neq a}\frac{|f_{ab}|^2}{(E_a - E_b)^3}, \tag{3}$$

where we introduced $f_{ab} = V_{ab}(E_a - E_b)$. Following the procedure one step further, we find equations for $\dot{f_{ab}}$:

$$\dot{f}_{ab} = \sum_{r\neq a,b} f_{ar}f_{rb}\left[\frac{1}{(E_a - E_r)^2} - \frac{1}{(E_b - E_r)^2}\right]. \tag{4}$$

One notices that no new quantities appear, and the set of Equations (2)–(4) is closed. It is sometimes called the Pechukas–Yukawa equations following original works [5,6]. This set of nonlinear equations is integrable—this is not surprising as a problem is equivalent to a diagonalization of the Hamiltonian matrix as noted in the Ref. [3].

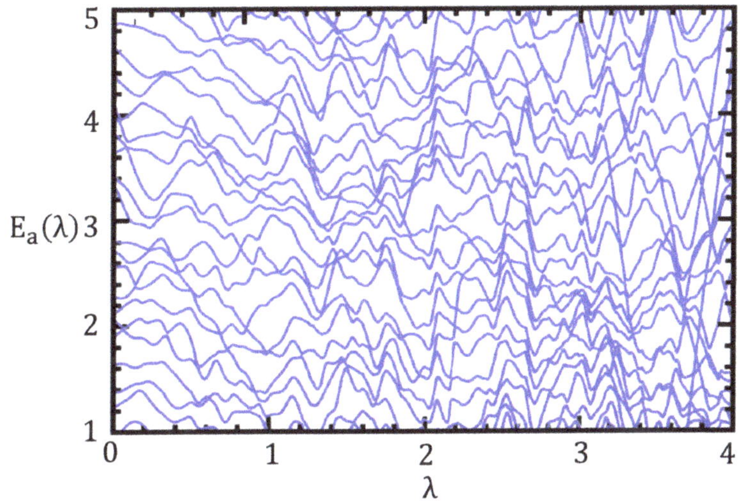

Figure 1. Levels of a model system as a function of the parameter λ. Each level may be visualized as a particle with the position given by the energy, $E_a(\lambda)$, the velocity given by the level slope $dE_a/d\lambda$, and acceleration (curvature of the level, $d^2E_a/d\lambda^2$.

The eigenvalues of $H = H_0 + \lambda V$ are V-dominated for large λ and the dynamics become trivial. Haake [3] introduces a different λ dependence (equivalent for small λ): $H(\lambda) = \sqrt{f}(H_0 + \lambda V)$ with $f = (1+\lambda^2)^{-1}$ while we shall follow the "trigonometric choice" [4,14,16]

$$H = H_0 \cos(\lambda) + V \sin(\lambda). \tag{5}$$

This results in an additional harmonic binding of eigenvalues which prevents them from escaping to infinity. In effect, the equations of motion become:

$$\dot{E}_a = \langle a|\dot{H}|a\rangle = p_a \tag{6}$$

and

$$\dot{p}_a = -E_a + 2\sum_{b\neq a} \frac{|f_{ab}|^2}{(E_a - E_b)^3} \tag{7}$$

with $f_{ab} = \langle a|\dot{H}|b\rangle(E_a - E_b)$.

Since the dynamics are integrable, the appropriate statistical description should involve all possible constants of the motion. Such an approach would be a formidable task. The Yukawa simplified way is just to consider the simplest integrals of motion, the total energy H and the trace of the square of f_{ab} matrix, $Q = \frac{1}{2}\mathrm{Tr}(F^2)$ [4] with

$$H = \frac{1}{2}\sum_n (p_n^2 + E_n^2) + \frac{1}{2}\sum_{n,m} \frac{|f_{nm}|^2}{(E_n - E_m)^2}. \tag{8}$$

The phase space density, according to Gibbs, is:

$$\rho = \frac{1}{Z}\exp(-\alpha H - \gamma Q). \tag{9}$$

The reader may be surprised that we use α as an effective inverse temperature. The simple reason is that, in order to follow the sacred quantum chaos notation, we reserve β for a level repulsion parameter with $\beta = 1, 2, 4$ characterizing different universality classes of Dyson and corresponding, for Gaussian ensembles to Gaussian Orthogonal, Unitary, and

Symplectic Ensembles (denoted as GOE, GUE, and GSE, respectively). The density ρ may be explicitly written out as

$$\rho = \frac{1}{Z}\exp\Big(-\alpha\Big[\frac{1}{2}\sum_n(p_n^2 + E_n^2) + \frac{1}{2}\sum_{n,m}\frac{|f_{nm}|^2}{(E_n - E_m)^2}\Big] - \gamma\frac{1}{2}\sum_{n,m}\Big|f_{nm}\Big|^2\Big). \tag{10}$$

By integrating out the variables p_n and f_{nm}, we can compute the joint probability distribution (JPD) of eigenvalues [4]

$$P(E_1, E_2, \ldots\ldots, E_n) \sim \prod_{n<m}\left|\frac{(E_n - E_m)^2}{1 + \frac{\gamma}{\alpha}(E_m - E_n)^2}\right|^{\beta/2} \exp\Big(-\frac{\alpha}{2}\sum_n E_n^2\Big), \tag{11}$$

with $\beta = 1, 2, 4$ corresponding to three universality classes. The β appears in (11) as the structure of the integrated variables f_{nm} depends on the universality class with F being orthogonal, unitary or symplectic. The similar ensemble was considered by Gaud in the Ref. [25], as well as Forrester [26] and Hasegawa and Ma [27]. They considered mainly two point correlation functions for the unitary ensemble. We rather concentrate on the time-reversal invariant case, as most commonly met in current many-body localization studies.

Equation (11) is simplified in different limiting cases. The Poissonian distribution was reached in the $\gamma/\alpha >> 1$ limit. The distribution becomes

$$P(E_1, E_2, \ldots\ldots, E_n) \sim \exp\big(-\frac{\alpha}{2}\sum_n E_n^2\big). \tag{12}$$

On the other hand, for $\gamma/\alpha \ll 1$ the distribution yields the Gaussian ensemble result. We have

$$P(E_1, E_2, \ldots\ldots, E_n) \sim \prod_{n>m}\Big|E_n - E_m\Big|^{\beta} \exp\big(-\frac{\alpha}{2}\sum_n E_n^2\big). \tag{13}$$

Finally, to reach (GOE) in this limit, we fix $\beta = 1$, and we also fix, for simplicity, $\alpha = 1$ (the latter choice affects the global energy scale only). We also denote $\gamma/\alpha = 10^p$. The distribution (11) takes the final form

$$P(E_1, E_2, \ldots\ldots, E_n) \sim \prod_{n<m}\left|\frac{(E_n - E_m)^2}{1 + 10^p(E_m - E_n)^2}\right|^{1/2} \exp\Big(-\frac{1}{2}\sum_n E_n^2\Big), \tag{14}$$

where $p = \log_{10}\frac{\gamma}{\beta}$ is the single parameter interpolating between GOE ($p \to -\infty$) and Poisson ($p \to \infty$) limits. The first term in (14) represents the pairwise interaction between the particles and the exponential term provides the harmonic binding of the eigenvalues. The resulting distribution, obtained using Monte-Carlo sampling for different p, was shown to faithfully reproduce statistics of eigenvalues on the transition between ergodic and many body localized situations [7].

3. Other Interpolating Ensembles

It is interesting to review several interpolating statistics models proposed in the past. An early work of Rosenzweig and Porter [28] is certainly worth mentioning. In this model the variance of off-diagonal elements in random matrices is controlled by a matrix dimension-dependent parameter. Its value interpolates between the Gaussian orthogonal ensemble (GOE) value to vanishing values for the Poissonian case. The other approach was proposed on the basis of a Wigner-inspired 2×2 matrix approach by Lenz and Haake [29]. Yet another was an ad hoc expression known as Brody distribution [30] which fit to low-resolution experimental data surprisingly well. Berry and Robnik [31] proposed the distribution based on sound physical assumption of the separation between "chaotic" wave functions faithful to GOE and those localized in the regular part of the phase space. The corresponding distribution was shown to work well in the so-called deep semiclassical limit [32]. Another proposition due to Seligman and coworkers [33] assumed that the variance of off-diagonal elements a_{ij}should scale as $\exp\left[-(i-j)^2/\sigma^2\right]$. For $\sigma \to 0$ one recovers the Poisson case while $\sigma \to \infty$ becomes GOE. Yet another well-known approach is that of Guhr [34] who used supersymmetric techniques to express the two-level correlation function in the Poisson-GOE ensemble in terms of a double integral. It is worth stressing that another popular proposition was advocated by Giulio Casati and coworkers [8,9]. They considered banded Gaussian random matrices as a useful tool in describing the transition, where the corresponding parameter was $y = b^2/N$ with b being the matrix bandwidth and N its rank.

While the (unfolded) level spacing statistics was the main object of quantum chaos studies, in a many-body localization (MBL) context Huse and Oganesyan [35] introduced a new measure, the gap ratio, defined as $r_n = min[\delta_n, \delta_{n-1}]/max[\delta_n, \delta_{n-1}]$, where $\delta_n = E_{n+1} - E_n$ is the energy gap between the consecutive energy levels. The dimensionless gap ratio does not require unfolding. The MBL transition description was addressed by Serbyn and Moore [36] who proposed two stages of the GOE–Poisson transition: (1) A Short Range Plasma Model (SRPM) and (2) a semi-Poissonian level statistics [37,38]. Recent efforts worth mentioning are a β-Gaussian ($\beta - G$) model [39]. A comparison of the performance of different models was given in the Ref. [40], while the Ref. [41] proposes a more complicated, two-parameter $\beta - h$ model, where the pairwise interaction between the levels is limited to a number h.

Comparison of some of these models with numerics for interacting disordered spin systems modelling ergodic to the MBL transition is given in the Ref. [7]. The interested reader should consult the Ref. [7] for details, but it suffices to say here that the single-parameter Yukawa-like model described above compared favorably with other single-parameter models and quite faithfully reproduced the disordered spin data for the MBL–ergodic crossover.

4. Universality of Parametric Dynamics

A simple inspection of Equation (10) shows that the velocities, p_n, have, in this approach, a Gaussian distribution with the variance determined by the "inverse temperature" α. This is the essence of level dynamics universality, as was determined and thoroughly studied by Simons and Altschuler [13,42]. The level spacings have a single scale—the mean level spacing, Δ. The unfolding then corresponds to rescaling the energy levels $\epsilon_i = E_i/\Delta$. Level dynamics introduces a novel scale determining how fast the eigenvalues change with the parameter λ. The original definition [13] involves the velocity–velocity correlation function for unfolded levels

$$C(\lambda) = \langle p_n(0)p_n(\lambda)\rangle/\Delta^2 \tag{15}$$

averaged over eigenstates n. Then $C(0)$ yields the second, apart from the mean level spacing, important scale. When the levels are unfolded using mean spacing, Δ, and the parametric dependence is "unfolded" using $C(0)$ as [13]

$$x = \sqrt{C(0)}\lambda \tag{16}$$

the spectral properties of different systems should be similar. Clearly, $C(0)$ in our notation is directly related, modulo-unfolding to the "inverse temperature" α in the Gibbs ensemble.

There is a hidden assumption in the level dynamics universality as formulated by Simons and Altschuler [13,42] and apparent in the Pechukas–Yukava statistical approach—the parameter change is global in a sense that both H_0 and V are of similar "strength" i.e., belong to the same ensemble. This does not have to be so, as nicely exemplified in the Ref. [43] where it is shown that small local perturbation modifies strongly the velocity distribution in the billiard example studied. Assuming random wavefunctions it is derived that in such a case $P(v) \propto K_0(A|v|)$ where A is a constant and $K_0(.)$ denotes a modified

Bessel function. The distinction between local and global perturbations in the context of unviersality were further studied in the Ref. [44].

It seems natural to review now the properties of velocity correlation function. For the reasons that become obvious later, it is more convenient to consider first the second derivatives of energy levels with respect to the parameter, the so-called curvatures.

5. Curvature Distributions

The curvatures of levels $K_n = \dot{p}_n$ as derivatives of velocities should be called in the dynamics language "level accelerations". We stick to curvatures as this is a commonly used name. Large curvatures appear in the vicinities of avoided crossings in the system. Then, essentially only two levels are involved. Following this strategy [45] showed that the large curvature tail behaves as $|K|^{-(\beta+2)}$ for all three universality classes.

The full analysis of curvature distributions, not limited to large curvature tail, was carried out in the Ref. [14]. Large numerical data collected for all three ensembles suggested the following simple and analytic form:

$$P(K) = \mathcal{N}_\beta \frac{1}{\left(1 + (K/\gamma_\beta)^2\right)^{\frac{\beta+2}{2}}} \tag{17}$$

(with $\beta = 1, 2, 4$ for GOE, GUE and GSE, respectively) and

$$\gamma_\beta = \pi\beta C(0)\Delta \tag{18}$$

where, recall, Δ is the mean level spacing (i.e., an inverse of the mean density of states). Defining the dimensionless curvature, k, as

$$k = \frac{K}{\gamma_\beta} \tag{19}$$

we have explicitly

$$P_O(K) = \frac{1}{2}\frac{1}{(1+k^2)^{3/2}} \tag{20}$$

$$P_U(K) = \frac{2}{\pi}\frac{1}{(1+k^2)^2} \tag{21}$$

$$P_S(K) = \frac{8}{3\pi}\frac{1}{(1+k^2)^3}. \tag{22}$$

These expressions, which could be claimed as being determined via Monte-Carlo integration and inspired guess, were soon proven analytically for all three ensembles of Gaussian random matrices [46,47] and by an alternative technique in the Refs. [48,49].

Let us remark that the above definition of k differs from the form suggested by the universality rule, (16), $d^2\epsilon/dx^2$ by a multiplicative factor $\pi\beta$ which simplifies (20)–(22).

The distributions (20)–(22) appear to work well for circular ensembles as well as some quantally chaotic systems such as kicked tops [14] or periodic band random matrices in the metalic regime, as shown by Casati and coworkers [11]. The question remains, however, to what extent these RMT predictions are universal and to what extend the particular quantally chaotic systems are faithful to them. The first aspect was clarified by Li and Robnik [50] who pointed out that a nonlinear transformation from λ to some other parameter $\mu(\lambda)$ leads to a different curvature distribution as the transformation between curvatures is nonlinear. It reads [50]:

$$k_\mu = k_\lambda - \frac{p_\lambda}{\pi\beta\langle p_\lambda^2\rangle}\frac{\mu''}{\mu'}. \tag{23}$$

In the expressions above k_μ and k_λ are normalized curvatures calculated with respect to the corresponding parameters, p_λ-the slope and prime denotes derivative with respect to λ. As Li and Robnik [50] point out since velocities are Gaussian distributed (fast decaying) the universality of curvatures may be restored for large curvatures but the nonlinearity of the transformation (23) prevents universality at all scales, see also [51]. The same argument shows, however, that for any "local" linear transformation the universality may hold. As long as changes of $H(\lambda)$ are linear in λ, as assumed in the derivation above, one might expect the universality to hold.

There is, however, another origin for the lack of universality which gives us insight into the physics involved. Already Takami and Hasegawa [52] suggested that the presence of scarring, i.e., strong localization of eigenstates in the space where unstable periodic orbits exist in the classical limit [53,54] may affect curvatures. Numerical studies of several examples such as the hydrogen atom in a magnetic field or quantum billiard proved that this is indeed the case [14]. While referring the reader to an original paper for numerical details it suffices to say that strong scarring leads to a peculiar level dynamics with some levels (scarred eigenstates) have quite different slope than the rest and interact with other levels only locally in narrow avoided crossings. Those levels behave like solitons and may be described as such [55,56]. Their behavior leads to an excess of small curvatures (outside of these avoided crossings) as well as very large curvatures (at the centers of avoided crossings).

It is worth stressing (which we just do with a single sentence) that the curvatures are strongly linked with transport and conductance [11,57]. Particularly interesting for this case are situations where the parameter breaks time reversal invariance as it happens for twisted boundary conditions.

6. Velocity Correlations

Let us come back to the level slopes, i.e., velocity correlations. Simons and Altschuler [13,42] in their analysis considered the autocorrelator of velocities at a some energy difference, ω, $\tilde{c}(x,\omega)$ (note—this is a different object than $C(x)$ which involves correlations for the same level n) that involves all level velocities in a given interval studied. We refer the reader to original papers for details [13,42]. The $C(\lambda)$ or rather $C(x)$ was studied numerically [58] for all three unitary classes. A simple analytic approximation for $C(x)$ was proposed in terms of the plasma error function, see [58].

The large x limits was elegantly solved [13,42] yielding $C(x) = -2/\beta\pi^2x^2$ for GUE. Interesting information may be obtained from a small x limit when $\tilde{c}(x,\omega)$ (for $\omega = 0$) and $C(x)$ behave similarly.

Application of Taylor series expansion of $C(x)$ allows one to link the velocity correlator to the variance of the rescales curvatures. Explicitly, one obtains [58]

$$C(x) = C(0)(1 - \frac{1}{2}\beta^2\pi^2x^2\langle k^2\rangle). \tag{24}$$

Defining normalized correlation $c(x) = C(x)/C(0)$ one reproduces the results [42] for GUE: $c_{\mathrm{GUE}}(x) = 1 - 2\pi^2x^2$ and gets $c_{\mathrm{GSE}}(x) = 1 - \frac{8}{3}\pi^2x^2 + \ldots$ for the symplectic ensemble [58]. Interestingly, for the most common orthogonal universality class one encounters the problem as the variance of curvatures, following (20), does not exist. This indicates that the small x behavior may be singular and the Taylor expansion is questionable.

This issue has been studied further in the Ref. [2] where it was shown that in fact $c(x)$ reveals singularities around $x = 0$. Taking the parametric dependence (5) one may show that Fourier components of the Fourier transform of $c(x)$ have algebraic tails which directly indicates singularities at $x = 0$ of the velocity correlator. Again we just quote the the final result which shows that

$$c_{\text{GOE}}(x) \sim 1 + b_1 x^2 |\ln(x)|,$$
$$c_{\text{GUE}}(x) \sim 1 - 2\pi^2 x^2 + b_2 |x^3|, \quad (25)$$
$$c_{\text{GSE}}(x) \sim 1 - \frac{8}{3}\pi^2 x^2 + b_3 x^4 + b_5 |x^5|,$$

with b_i being coefficients of the order of unity. One may observe that the singularity at $x = 0$ becomes weaker with growing level repulsion β, being most severe for GOE. An even more in depth analysis of singularities appears in the Ref. [59] where explicit values for the parameters, b_i are found.

As mentioned in Section 4 local (instead of global) perturbation affect strongly velocity distributions [43,44]. This has also a pronounced effect on velocity correlations [43,44] as further analysed in detail for the unitary ensemble [60].

7. Avoided Crossings Distributions

Another statistical property with interesting links to level spacings is the distribution of avoided crossings, i.e., minima of distances between neighboring levels. The problem of finding the corresponding distribution was formulated by Wilkinson [61] who has shown that integrated distributions for small minimal distances D for GOE (GUE) show similar repulsion as present in spacing distributions. Avoided crossings for billiard models were numerically studied by Goldberg and Schweizer [62]. While in a general case the exact distributions are not known in some analytic form, a well working approximations based on two-levels approximation may be easily derived following the Wigner approach for level spacings themselves [15]. For GOE case it is written down immediately as the two-level Hamiltonian $H = H_0 + \lambda V$ may be written in the eigenbasis of V as

$$H = \begin{bmatrix} a & d \\ d & b \end{bmatrix} + \lambda \begin{bmatrix} v_1 & 0 \\ 0 & v_2 \end{bmatrix}. \quad (26)$$

The minimal distance between levels is simply $2|d|$. Since H_0 is assumed to correspond to GOE, d is Gaussian distributed, so we get the distribution (for $D = 2|d|$) normalized to unit mean avoided crossing:

$$P(D) = \frac{2}{\pi} \exp\left[-\frac{D^2}{\pi}\right], \ D > 0. \quad (27)$$

Situation is only slightly more complicated for other ensembles. For GUE d in (26) should be complex, $d = d_1 + id_2$, with independently Gaussian distributed (with the same variance) d_i. A simple integral leads to a normalized distribution

$$P(D) = \frac{\pi D}{2} \exp\left[-\frac{\pi}{4} D^2\right], \quad (28)$$

which is identical to the so-called Wigner surmise for spacings for GOE. We observe a simple rule that the avoided crossings in two-level approximation share the same distribution as the nearest neighbor spacings but the the repulsion parameter β reduced by unity. So for GUE with $\beta = 2$ we get the Wigner formula corresponding to spacings for $\beta = 1$. This is in full agreement with small D perturbative prediction of the Ref. [61]. Similarly, an explicit calculus shows that for $\beta = 4$ GSE the avoided crossing distribution behaves as D^3 for small D.

Numerical tests (which have to be carefully done to correctly estimate and avoided crossing values [16]) show excellent agreement between two-level approximate formulae and numerical data for random matrices of larger sizes. The agreement is in fact better than the spacings and the Wigner surmise. The reason is simple—the two-level approximation works better for minimal distances between levels.

8. Fidelity Susceptibility

Rapidly developing in the past 20 years, the quantum information field brought yet another measure which may be related to level dynamics, the fidelity, $\mathcal{F}$ [63]. While generally defined for mixed states, a pure-state definition [64] suffices for our purposes

$$\mathcal{F} = |\langle \psi(0)|\psi(\lambda)\rangle|. \tag{29}$$

Here, $\psi(\lambda)$ is an eigenstate at the value of the parameter equal to λ. Taylor expansion for small λ leads to the definition of fidelity susceptibility, χ

$$\mathcal{F} = 1 - \frac{1}{2}\chi\lambda^2 + O(\lambda^3). \tag{30}$$

Fidelity susceptibility became an indicator of quantum phase transitions. At the transition point, the ground state properties change, leading to the enhancement of χ [64–66]. Apart from ground states, thermal states were also considered [67–69]). The fidelity statistics, taking into account a set of eigenstates, was discussed for the first time in the Ref. [17]; note that an attempt to identify many-body localization transition was due to the Ref. [70]. We shall briefly review the results of the Ref. [17] that provide one of the rare situations when exact analytic results are available for the arbitrary size of random matrices.

Consider $H = H_0 + \lambda V$ with both H_0, V belonging to GOE or GUE. Fidelity susceptibility of state $|n\rangle$ with energy E_n of H_0 is easily derived as

$$\chi_n = \sum_{m \neq n} \frac{|V_{nm}|^2}{(E_n - E_m)^2}, \tag{31}$$

showing some similarity to curvatures $\dot{p}_n$ (3)—the difference is just a power in the denominator.

The probability distribution of the fidelity susceptibility at energy E reads:

$$P(\chi, E) = \frac{1}{N\rho(E)}\left\langle \sum_{n=1}^{N} \delta(\chi - \chi_n)\delta(E - E_n), \right\rangle \tag{32}$$

which we consider at the center of the spectrum $E = 0$. Following the technique developed in the Refs. [48,71,72], one arrives [17] at, for the GOE case,

$$P_N^O(\chi) = \frac{C_N^O}{\sqrt{\chi}}\left(\frac{\chi}{1+\chi}\right)^{\frac{N-2}{2}}\left(\frac{1}{1+2\chi}\right)^{\frac{1}{2}}\left[\frac{1}{1+2\chi} + \frac{1}{2}\left(\frac{1}{1+\chi}\right)^2 \mathcal{I}_{N-2}^O\right],$$

where C_N^O is a normalization constant and

$$\mathcal{I}_N^O = \begin{cases} N\frac{N+2}{N+3/2}, & N \quad \text{even}, \\ N+1/2, & N \quad \text{odd}. \end{cases} \tag{33}$$

Equation (33) is an exact result for an arbitrary rank N of the random matrix from GOE. This is one of the rare situations when analytic formulae for arbitrary N and not only for the $N = 2$ or $N \to \infty$ limit are known.

The $N \to \infty$ limit is interesting. As $\mathcal{I}_N^O$ is asymptotically proportional to N, one can define a scaled fidelity susceptibility $y = \chi/N$. Its distribution, in the $N \to \infty$ limit takes a rather simple form

$$P^O(y) = \frac{1}{6}\frac{1}{y^2}\left(1 + \frac{1}{y}\right)\exp\left(-\frac{1}{2y}\right). \tag{34}$$

As tested numerically, this expression works well for $N \sim 200$ already.

Similarly, see the Ref. [17] for the derivation, where one obtains an analytic, valid for arbitrary N results, for GUE. We quote here just the $N \to \infty$ limit for the scaled fidelity susceptibility

$$P^U(y) = \frac{1}{3\sqrt{\pi}} \frac{1}{y^{5/2}} \left(\frac{3}{4} + \frac{1}{y} + \frac{1}{y^2} \right) \exp\left(-\frac{1}{y} \right), \tag{35}$$

while for a full expression for χ valid for arbitrary N as well as for a comparison with numerical data, we refer to the Ref. [17].

The distributions (33) and (34) for GOE and GUE, respectively, are presented in Figure 2. For large scaled fidelities y a power law decay is observed $1/y^2$ for GOE and $1/y^{5/2}$ for GUE.

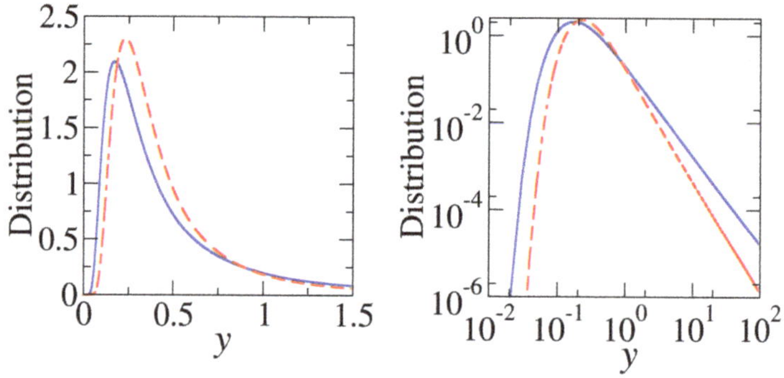

Figure 2. Scaled fidelity susceptibility distributions in the linear (**left**) and logarithmic (**right**) scales for GOE, Equation (33) (blue lines) and GUE, Equation (34). Observe a power law tail at large scaled fidelities, y.

9. Generalization to More Parameters

A natural extension of parametric level dynamics occurs in the presence of more than one parameter. One may define

$$H = H_0 + \sum_i \lambda_i V_i \tag{36}$$

where H_0, V_i are statistically independent and drawn from the same (as in this review) or different universality classes. Obviously, novel problems appear in that case which we describe briefly only. Probably it was Michael Wilkinson and his student, Elisabeth Austin, who addressed first such a situation in their study of density of degeneracies, for example, diabolical points [73]. A three-parameter family was considered for Chern integer fluctuations [59,74]. Steuwer and Simons [75] found the distribution of adiabatic curvature (related to Berry phase) for GUE. The multiparameter dynamics was recently revisited by Berry and Shukla who discussed Berry curvature deriving two-level and three-level distributions [76–78]. They found the large curvature scaling, $P(c) \sim c^{-2}$ for the orthogonal and $P(c) \sim c^{-5/2}$ for the unitary class, the result already present in the Ref. [75].

Importantly, however, Berry and Shukla [78] linked the problem with the quantum information concept of geometric tensor and the distance between quantum states [79–81]. The Fubini–Study distance in the Hilbert space between states differing by a small change of parameters form $\vec{\lambda} \equiv (\lambda_1, \lambda_2, \ldots, \lambda_n)$ to $\vec{\lambda} + d\vec{\lambda}$ is

$$ds^2 = 1 - |\langle n(\vec{\lambda}) | n(\vec{\lambda} + d\vec{\lambda}) \rangle|^2 = \sum_{ij} \mathrm{Re} g_{ij}^{(n)}(\vec{\lambda}) d\lambda_i d\lambda_j, \tag{37}$$

where $g_{ij}^{(n)}(\vec{\lambda})$ is the so-called geometric tensor [79–81] which governs the quenches (in λ) of the system. For (36)

$$g_{ij}^{(n)} = \sum_{m(\neq n)} \frac{\langle n|V_i|m\rangle\langle m|V_j|n\rangle}{(E_n - E_m)^2}. \tag{38}$$

Note that the distance between states is determined by the real part of the geometric tensor only—(37). The imaginary part, $\mathrm{Im}g_{ij}^{(n)}$, gives the Berry curvature [78,82] related to changes of two parameters λ_i, λ_j. For a single-parameter problem the geometric tensor reduces to a scalar proportional to the fidelity susceptibility discussed in the previous Section. Additionally, if V_i belongs to the same ensembles, then the distributions of g_{ii} are equal. One may consider, therefore, the trace $G = \mathrm{Tr}g_{ij}^{(n)}$ as an equivalent of the fidelity susceptibility while the distribution of the imaginary part of the geometric tensor reduces to the Berry curvature distribution.

The distribution of trace, G, is, therefore, given by (33) discussed above, valid for arbitrary N [17] after taking into account the fact that, for equal diagonal elements the trace is simply the matrix rank multiplying the diagonal element. The alternative derivation using supersymmetric techniques is provided in the Ref. [82] for GUE in the $N \to \infty$ limit. The same authors obtained for the Berry curvature the result derived earlier by the Ref. [75].

10. Towards the Localization Limit

While in the introductory part we have considered the transition between an integrable (localized) regime in the context of level spacings and the dynamics of Pechukas–Yukawa gas, later we mainly described the results for the ergodic regime well-described by Gaussian random ensembles. Here we shall briefly mention some of the results for level dynamics measures that involve the transition.

Here the seminal contributions were provided by the Como group centered around Giulio Casati [8–11]. The banded random matrix ensemble provided a natural tool to study the transition from ergodic (metallic) to localized transition by varying the width of the band. The team addressed also curvature distributions [2,11]. The analytic approach to the problem was pursued by Yan Fyodorov, who, starting around 1994, considered comprehensively level dynamics features close to the localization transition studying velocity correlations [83,84] or curvature distributions [85]. In particular, the velocity distribution for one-dimensional disordered wire is derived using the supersymmetric approach to be

$$P(v_s) = \frac{\pi}{\sinh^2(\pi v_s)}\{\pi v_s \coth(\pi v_s) - 1\}, \tag{39}$$

for the scale velocity v_s. The curvature distributions were more recently addressed in the context of MBL studies [86,87]. The level dynamics across the many-body localization transition for the paradigmatic XXZ spin model was considered in the Ref. [88]. Velocity, curvature and fidelity susceptibility distributions were considered. Interestingly, while velocities depended on the choice of the parameter (being the interaction strength or the kinetic tunneling), curvature exhibited universal behavior in the delocalized regime. In the localized regime, curvature distributions reveal system specific characteristics that exemplify the presence of local integrals of motion in the localized phase. Large curvature or large fidelity susceptibility tails change their slope when entering the localized regime. Such behavior is well-understood qualitatively and linked to weaker level repulsion. One may also mention here a proposition to use adiabatic eigenstate deformation as a probe of entering the localized regime [89].

Recent studies address avoided crossing distributions and level dynamics on the localized side of the MBL transition [90–92]. Such an analysis, however, has to face the problem of assumed integrability which leads necessarily to exact crossing of levels associated with different quantum numbers.

11. Conclusions—Where Do We Stand

With this traipse through the last 40 years, starting with the Pechukas model [5], we hope to have shown that many fascinating results have been obtained in the studies of level dynamics of complex systems, but that there are still many open questions and unsolved problems. There are at least two areas that await a more decisive attack and, hopefully, solutions. One is the transition between different ensembles [93]. We have not, on purpose, reviewed only a few studies in this domain for the reader to formulate their own problems. A second related area with several white spots lies in multiparameter level dynamics. The latter has been mostly limited to studies within unitary ensembles, notoriously easiest to treat. We do not know the Berry curvature distribution for the orthogonal ensemble. We do not know about the geometric tensor properties when different parameters induce different transitions. Even the simplest questions remain unanswered. For example, simple analysis shows that fidelity susceptibility of Berry curvature decays with reverse quadratic power for GOE while the corresponding power is $-5/2$ for GUE. Can we say something when we couple GOE and GUE ensembles by some parameter, such as weakly breaking the time reversal invariance? Can we generalize the findings to the symplectic ensemble? What about the ten-fold method [94]? It is my belief that we may expect in the future some very interesting results coming from new people entering the subject. I already anticipate the excitement.

Funding: This research has been funded by National Science Centre (Poland) under project 2019/35/B/ST2/00034. The support by a grant from the Priority Research Area DigiWorld under the Strategic Programme Excellence Initiative at Jagiellonian University is also acknowledged.

Institutional Review Board Statement: Not applicable.

Data Availability Statement: No original numerical or experimental data were produced for the present contribution.

Acknowledgments: I would like to thank several of the colleagues with whom some of the results presented were obtained. Special thanks are due to Dominique Delande, the work on curvature [14] and avoided crossing distributions [16] done during my stay in Paris formed just the beginning of many years of collaboration with about 50 papers. I profited a lot from discussions with Marek Kuś with whom we first treated avoided crossing distributions using the two-level approach [15] and then considered fidelity susceptibility recently [17]. Parametric correlations of velocities were studied in collaboration with Italo Guarneri and Luca Molinari under the guidance of Giulio Casati [2] with contributions of Karol Życzkowski who, at that period, collaborated with the Como gang. I want also to mention the recent collaboration with Artur Maksymov and Bitan De. Finally, I would like to thank Piotr Sierant, my former student for his contribution to fidelity susceptibility distribution derivation [17] and many discussions. Without him this work could not be completed. I am also grateful to Eugene Bogomolny, Yan Fyodorov, Marko Robnik, and Ben Simons for suggestions on the literature of the subject. The support of PL-Grid Infrastructure is acknowledged.

Conflicts of Interest: The authors declare no conflict of interest.

References

1. Pakoński, P.; Zakrzewski, J. Kepler Map for H Atom Driven by Microwaves with Arbitrary Polarization. *Acta Phys. Polon B* **2001**, *32*, 2801.
2. Guarneri, I.; Życzkowski, K.; Zakrzewski, J.; Molinari, L.; Casati, G. Parametric spectral correlations of disordered systems in the Fourier domain. *Phys. Rev. E* **1995**, *52*, 2220–2235. [CrossRef] [PubMed]
3. Haake, F. *Quantum Signatures of Chaos*; Springer: Berlin, Germany, 2010.
4. Stöckmann, H. *Quantum Chaos: An Introduction*; Cambridge University Press: Cambridge, UK, 1999.
5. Pechukas, P. Distribution of Energy Eigenvalues in the Irregular Spectrum. *Phys. Rev. Lett.* **1983**, *51*, 943–946. [CrossRef]
6. Yukawa, T. New Approach to the Statistical Properties of Energy Levels. *Phys. Rev. Lett.* **1985**, *54*, 1883–1886. [CrossRef] [PubMed]
7. De, B.; Sierant, P.; Zakrzewski, J. On intermediate statistics across many-body localization transition. *J. Phys. A Math. Theor.* **2021**, *55*, 014001. [CrossRef]
8. Casati, G.; Molinari, L.; Izrailev, F. Scaling properties of band random matrices. *Phys. Rev. Lett.* **1990**, *64*, 1851–1854. [CrossRef]

9. Casati, G.; Izrailev, F.; Molinari, L. Scaling properties of the eigenvalue spacing distribution for band random matrices. *J. Phys. A Math. Gen.* **1991**, *24*, 4755–4762. [CrossRef]
10. Casati, G.; Chirikov, B.V.; Guarneri, I.; Izrailev, F.M. Band-random-matrix model for quantum localization in conservative systems. *Phys. Rev. E* **1993**, *48*, R1613–R1616. [CrossRef]
11. Casati, G.; Guarneri, I.; Izrailev, F.M.; Molinari, L.; Życzkowski, K. Periodic band random matrices, curvature, and conductance in disordered media. *Phys. Rev. Lett.* **1994**, *72*, 2697–2700. [CrossRef]
12. Casati, G.; Chirikov, B.; Guarneri, I.; Izrailev, F. Quantum ergodicity and localization in conservative systems: The Wigner band random matrix model. *Phys. Lett. A* **1996**, *223*, 430–435. . [CrossRef]
13. Simons, B.D.; Altshuler, B.L. Universal velocity correlations in disordered and chaotic systems. *Phys. Rev. Lett.* **1993**, *70*, 4063–4066. [CrossRef] [PubMed]
14. Zakrzewski, J.; Delande, D. Parametric motion of energy levels in quantum chaotic systems. I. Curvature distributions. *Phys. Rev. E* **1993**, *47*, 1650–1664. [CrossRef]
15. Zakrzewski, J.; Kuś, M. Distributions of avoided crossings for quantum chaotic systems. *Phys. Rev. Lett.* **1991**, *67*, 2749–2752. [CrossRef] [PubMed]
16. Zakrzewski, J.; Delande, D.; Kuś, M. Parametric motion of energy levels in quantum chaotic systems. II. Avoided-crossing distributions. *Phys. Rev. E* **1993**, *47*, 1665–1676. [CrossRef] [PubMed]
17. Sierant, P.; Maksymov, A.; Kuś, M.; Zakrzewski, J. Fidelity susceptibility in Gaussian random ensembles. *Phys. Rev. E* **2019**, *99*, 050102. [CrossRef]
18. Stöckmann, H.J.; Stoffregen, U.; Kollmann, M. A relation between billiard geometry and the temperature of its eigenvalue gas. *J. Phys. A Math. Gen.* **1997**, *30*, 129. [CrossRef]
19. Bertelsen, P.; Ellegaard, C.; Guhr, T.; Oxborrow, M.; Schaadt, K. Measurement of Parametric Correlations in Spectra of Resonating Quartz Blocks. *Phys. Rev. Lett.* **1999**, *83*, 2171–2174. [CrossRef]
20. Dietz, B.; Heine, A.; Richter, A.; Bohigas, O.; Leboeuf, P. Spectral statistics in an open parametric billiard system. *Phys. Rev. E* **2006**, *73*, 035201. [CrossRef]
21. Poli, C.; Dietz, B.; Legrand, O.; Mortessagne, F.; Richter, A. Avoided-level-crossing statistics in open chaotic billiards. *Phys. Rev. E* **2009**, *80*, 035204. [CrossRef]
22. Hul, O.; Šeba, P.; Sirko, L. Investigation of parameter-dependent properties of quantum graphs with and without time-reversal symmetry. *Phys. Scr.* **2009**, *2009*, 014048. [CrossRef]
23. Hul, O.; Šeba, P.; Sirko, L. Departure of some parameter-dependent spectral statistics of irregular quantum graphs from random matrix theory predictions. *Phys. Rev. E* **2009**, *79*, 066204. [CrossRef]
24. Ławniczak, M.; Nicolau-Kuklińska, A.; Hul, O.; Masiak, P.; Bauch, S.; Sirko, L. Experimental and numerical determination of the correlation function of level velocities for microwave networks simulating quantum graphs. *Phys. Scr.* **2013**, *2013*, 014041. [CrossRef]
25. Gaudin, M. Une famille à un paramètre d'ensembles unitaires. *Nucl. Phys.* **1966**, *85*, 545. [CrossRef]
26. Forrester, P. Statistical properties of the eigenvalue motion of Hermitian matrices. *Phys. Lett. A* **1993**, *173*, 355–359. [CrossRef]
27. Hasegawa, H.; Ma, J.Z. Intermediate level statistics with one-parameter random matrix ensembles. *J. Math. Phys.* **1998**, *39*, 2564–2583.
28. Rosenzweig, N.; Porter, C.E. "Repulsion of Energy Levels" in Complex Atomic Spectra. *Phys. Rev.* **1960**, *120*, 1698–1714. [CrossRef]
29. Lenz, G.; Haake, F. Reliability of small matrices for large spectra with nonuniversal fluctuations. *Phys. Rev. Lett.* **1991**, *67*, 1–4. [CrossRef] [PubMed]
30. Brody, T.A. A statistical measure for the repulsion of energy levels. *Lett. Nuovo C. (1971–1985)* **1973**, *7*, 482–484. [CrossRef]
31. Berry, M.V.; Robnik, M. Semiclassical level spacings when regular and chaotic orbits coexist. *J. Phys. A Math. Gen.* **1984**, *17*, 2413–2421. [CrossRef]
32. Prosen, T. Berry-Robnik level statistics in a smooth billiard system. *J. Phys. A Math. Gen.* **1998**, *31*, 7023–7029. [CrossRef]
33. Seligman, T.H.; Verbaarschot, J.J.M.; Zirnbauer, M.R. Quantum Spectra and Transition from Regular to Chaotic Classical Motion. *Phys. Rev. Lett.* **1984**, *53*, 215–217. [CrossRef]
34. Guhr, T. Transitions toward Quantum Chaos: With Supersymmetry from Poisson to Gauss. *Ann. Phys.* **1996**, *250*, 145–192. [CrossRef]
35. Oganesyan, V.; Huse, D.A. Localization of interacting fermions at high temperature. *Phys. Rev. B* **2007**, *75*, 155111. [CrossRef]
36. Serbyn, M.; Moore, J.E. Spectral statistics across the many-body localization transition. *Phys. Rev. B* **2016**, *93*, 041424. [CrossRef]
37. Bogomolny, E.B.; Gerland, U.; Schmit, C. Models of intermediate spectral statistics. *Phys. Rev. E* **1999**, *59*, R1315–R1318. [CrossRef]
38. Bogomolny, E.; Gerland, U.; Schmit, C. Short-range plasma model for intermediate spectral statistics. *Eur. Phys. J. B* **2001**, *19*, 121–132. [CrossRef]
39. Buijsman, W.; Cheianov, V.; Gritsev, V. Random Matrix Ensemble for the Level Statistics of Many-Body Localization. *Phys. Rev. Lett.* **2019**, *122*, 180601. [CrossRef]
40. Sierant, P.; Zakrzewski, J. Level statistics across the many-body localization transition. *Phys. Rev. B* **2019**, *99*, 104205. [CrossRef]
41. Sierant, P.; Zakrzewski, J. Model of level statistics for disordered interacting quantum many-body systems. *Phys. Rev. B* **2020**, *101*, 104201. [CrossRef]

42. Simons, B.D.; Altshuler, B.L. Universalities in the spectra of disordered and chaotic systems. *Phys. Rev. B* **1993**, *48*, 5422–5438. [CrossRef]
43. Barth, M.; Kuhl, U.; Stöckmann, H.J. Global versus Local Billiard Level Dynamics: The Limits of Universality. *Phys. Rev. Lett.* **1999**, *82*, 2026–2029. [CrossRef]
44. Marchetti, F.M.; Smolyarenko, I.E.; Simons, B.D. Universality of parametric spectral correlations: Local versus extended perturbing potentials. *Phys. Rev. E* **2003**, *68*, 036217. [CrossRef] [PubMed]
45. Gaspard, P.; Rice, S.A.; Mikeska, H.J.; Nakamura, K. Parametric motion of energy levels: Curvature distribution. *Phys. Rev. A* **1990**, *42*, 4015–4027. [CrossRef] [PubMed]
46. von Oppen, F. Exact distribution of eigenvalue curvatures of chaotic quantum systems. *Phys. Rev. Lett.* **1994**, *73*, 798–801. [CrossRef] [PubMed]
47. von Oppen, F. Exact distributions of eigenvalue curvatures for time-reversal-invariant chaotic systems. *Phys. Rev. E* **1995**, *51*, 2647–2650. [CrossRef]
48. Fyodorov, Y.V.; Sommers, H.J. Universality of "level curvature" distribution for large random matrices: Systematic analytical approaches. *Z. Phys. B Condens. Matter* **1995**, *99*, 123–135. [CrossRef]
49. Fyodorov, Y.V. Level Curvature Distribution: From Bulk to the Soft Edge of Random Hermitian Matrices. *Acta Phys. Pol. A* **2011**, *120*, A-100. [CrossRef]
50. Li, B.; Robnik, M. Sensitivity of the eigenfunctions and the level curvature distribution in quantum billiards. *J. Phys. A Math. Gen.* **1996**, *29*, 4387. [CrossRef]
51. Lebœuf, P.; Sieber, M. Universality in quantum parametric correlations. *Phys. Rev. E* **1999**, *60*, 3969–3972. [CrossRef]
52. Takami, T.; Hasegawa, H. Curvature distribution of chaotic quantum systems: Universality and nonuniversality. *Phys. Rev. Lett.* **1992**, *68*, 419–422. [CrossRef]
53. Heller, E.J. Bound-State Eigenfunctions of Classically Chaotic Hamiltonian Systems: Scars of Periodic Orbits. *Phys. Rev. Lett.* **1984**, *53*, 1515–1518. [CrossRef]
54. Bogomolny, E. Smoothed wavefunctions of chaotic quantum systems. *Physica D* **1988**, *31*, 169. [CrossRef]
55. Gaspard, P.; Rice, S.A.; Nakamura, K. Solitonlike structure in the parametric distortions of bounded-system energy spectra. *Phys. Rev. Lett.* **1989**, *63*, 930–933. [CrossRef]
56. Nakamura, K. *Quantum Chaos, a New Paradigm of Nonlinear Dynamics*; Cambridge University Press: Cambridge, UK, 1993.
57. Braun, D.; Montambaux, G. Universal spectral correlations in diffusive quantum systems. *Phys. Rev. B* **1994**, *50*, 7776–7785. [CrossRef]
58. Zakrzewski, J. On "Universal" correlations in disordered and chaotic systems. *Z. Phys. B Condens. Matter* **1995**, *98*, 273–277. [CrossRef]
59. Walker, P.N.; Sánchez, M.J.; Wilkinson, M. Singularities in the spectra of random matrices. *J. Math. Phys.* **1996**, *37*, 5019–5032.
60. Smolyarenko, I.E.; Simons, B.D. Parametric spectral statistics in unitary random matrix ensembles: From distribution functions to intra-level correlations. *J. Phys. A Math. Gen.* **2003**, *36*, 3551. [CrossRef]
61. Wilkinson, M. Statistics of multiple avoided crossings. *J. Phys. A Math. Gen.* **1989**, *22*, 2795. [CrossRef]
62. Goldberg, J.; Schweizer, W. Distribution of multiple avoided crossings: Numerical evaluation. *J. Phys. A Math. Gen.* **1991**, *24*, 2785. [CrossRef]
63. Uhlmann, A. The "transition probability" in the state space of a $*$-algebra. *Rep. Math. Phys.* **1976**, *9*, 273–279. [CrossRef]
64. Zanardi, P.; Paunković, N. Ground state overlap and quantum phase transitions. *Phys. Rev. E* **2006**, *74*, 031123. [CrossRef]
65. You, W.L.; Li, Y.W.; Gu, S.J. Fidelity, dynamic structure factor, and susceptibility in critical phenomena. *Phys. Rev. E* **2007**, *76*, 022101. [CrossRef] [PubMed]
66. Invernizzi, C.; Korbman, M.; Campos Venuti, L.; Paris, M.G.A. Optimal quantum estimation in spin systems at criticality. *Phys. Rev. A* **2008**, *78*, 042106. [CrossRef]
67. Zanardi, P.; Quan, H.T.; Wang, X.; Sun, C.P. Mixed-state fidelity and quantum criticality at finite temperature. *Phys. Rev. A* **2007**, *75*, 032109. [CrossRef]
68. Sirker, J. Finite-Temperature Fidelity Susceptibility for One-Dimensional Quantum Systems. *Phys. Rev. Lett.* **2010**, *105*, 117203. [CrossRef] [PubMed]
69. Rams, M.M.; Sierant, P.; Dutta, O.; Horodecki, P.; Zakrzewski, J. At the Limits of Criticality-Based Quantum Metrology: Apparent Super-Heisenberg Scaling Revisited. *Phys. Rev. X* **2018**, *8*, 021022. [CrossRef]
70. Hu, T.; Xue, K.; Li, X.; Zhang, Y.; Ren, H. Fidelity of the diagonal ensemble signals the many-body localization transition. *Phys. Rev. E* **2016**, *94*, 052119. [CrossRef] [PubMed]
71. Fyodorov, Y.V.; Savin, D.V. Statistics of Resonance Width Shifts as a Signature of Eigenfunction Nonorthogonality. *Phys. Rev. Lett.* **2012**, *108*, 184101. [CrossRef]
72. Fyodorov, Y.V.; Nock, A. On Random Matrix Averages Involving Half-Integer Powers of GOE Characteristic Polynomials. *J. Stat. Phys.* **2015**, *159*, 731–751. [CrossRef]
73. Wilkinson, M.; Austin, E.J. Densities of degeneracies and near-degeneracies. *Phys. Rev. A* **1993**, *47*, 2601–2609. [CrossRef]
74. Walker, P.N.; Wilkinson, M. Universal Fluctuations of Chern Integers. *Phys. Rev. Lett.* **1995**, *74*, 4055–4058. [CrossRef]
75. Steuwer, A.; Simons, B.D. Adiabatic curvature distribution in disordered and chaotic structures. *Phys. Rev. B* **1998**, *57*, 9186–9192. [CrossRef]

76. Berry, M.V.; Shukla, P. Geometric phase curvature for random states. *J. Phys. A Math. Theor.* **2018**, *51*, 475101. [CrossRef]
77. Berry, M.V.; Shukla, P. Geometric Phase Curvature Statistics. *J. Stat. Phys.* **2020**, *180*, 297–303. [CrossRef]
78. Berry, M.V.; Shukla, P. Quantum metric statistics for random-matrix families. *J. Phys. A Math. Theor.* **2020**, *53*, 275202. [CrossRef]
79. Provost, J.P.; Vallee, G. Riemannian structure on manifolds of quantum states. *Commun. Math. Phys.* **1980**, *76*, 289–301. [CrossRef]
80. Campos Venuti, L.; Zanardi, P. Quantum Critical Scaling of the Geometric Tensors. *Phys. Rev. Lett.* **2007**, *99*, 095701. [CrossRef]
81. Kolodrubetz, M.; Sels, D.; Mehta, P.; Polkovnikov, A. Geometry and non-adiabatic response in quantum and classical systems. *Phys. Rep.* **2017**, *697*, 1–87. . [CrossRef]
82. Penner, A.G.; von Oppen, F.; Zaránd, G.; Zirnbauer, M.R. Hilbert Space Geometry of Random Matrix Eigenstates. *Phys. Rev. Lett.* **2021**, *126*, 200604. [CrossRef]
83. Fyodorov, Y.V. Distribution of "Level Velocities" in Quasi-1D Disordered or Chaotic Systems with Localization. *Phys. Rev. Lett.* **1994**, *73*, 2688–2691. [CrossRef]
84. Fyodorov, Y.V.; Mirlin, A.D. Mesoscopic fluctuations of eigenfunctions and level-velocity distribution in disordered metals. *Phys. Rev. B* **1995**, *51*, 13403–13409. [CrossRef]
85. Titov, M.; Braun, D.; Fyodorov, Y.V. Log-normal distribution of level curvatures in the localized regime: Analytical verification. *J. Phys. A Math. Gen.* **1997**, *30*, L339–L345. [CrossRef]
86. Filippone, M.; Brouwer, P.W.; Eisert, J.; von Oppen, F. Drude weight fluctuations in many-body localized systems. *Phys. Rev. B* **2016**, *94*, 201112. [CrossRef]
87. Monthus, C. Many-body-localization transition: Sensitivity to twisted boundary conditions. *J. Phys. A Math. Theor.* **2017**, *50*, 095002. [CrossRef]
88. Maksymov, A.; Sierant, P.; Zakrzewski, J. Energy level dynamics across the many-body localization transition. *Phys. Rev. B* **2019**, *99*, 224202. [CrossRef]
89. Pandey, M.; Claeys, P.W.; Campbell, D.K.; Polkovnikov, A.; Sels, D. Adiabatic Eigenstate Deformations as a Sensitive Probe for Quantum Chaos. *Phys. Rev. X* **2020**, *10*, 041017. [CrossRef]
90. Garratt, S.J.; Roy, S.; Chalker, J.T. Local resonances and parametric level dynamics in the many-body localized phase. *Phys. Rev. B* **2021**, *104*, 184203. [CrossRef]
91. Garratt, S.J.; Roy, S. Resonant energy scales and local observables in the many-body localized phase. *Phys. Rev. B* **2022**, *106*, 054309. [CrossRef]
92. Grabarits, A. Level dynamics and avoided level crossings in driven disordered quantum dots. *Phys. Rev. B* **2023**, *107*, 014206. [CrossRef]
93. Kunstman, P.; Życzkowski, K.; Zakrzewski, J. Nonuniversality in level dynamics. *Phys. Rev. E* **1997**, *55*, 2446–2454. [CrossRef]
94. Altland, A.; Zirnbauer, M.R. Nonstandard symmetry classes in mesoscopic normal-superconducting hybrid structures. *Phys. Rev. B* **1997**, *55*, 1142–1161. [CrossRef]

entropy

MDPI

Article

Quantization of Integrable and Chaotic Three-Particle Fermi–Pasta–Ulam–Tsingou Models

Alio Issoufou Arzika [1,2], Andrea Solfanelli [2,3], Harald Schmid [4] and Stefano Ruffo [2,3,5,*]

1 LCEMR, Faculty of Science and Technology, Abdou Moumouni University, Niamey BP 10662, Niger
2 SISSA, Via Bonomea 265, 34136 Trieste, Italy
3 INFN, Sezione di Trieste, Via Valerio 2, 34127 Trieste, Italy
4 Dahlem Center for Complex Quantum Systems, Freie Universität Berlin, 14195 Berlin, Germany
5 Istituto dei Sistemi Complessi, Via Madonna del Piano 10, 50019 Sesto Fiorentino, Italy
* Correspondence: ruffo@sissa.it

Abstract: We study the transition from integrability to chaos for the three-particle Fermi–Pasta–Ulam–Tsingou (FPUT) model. We can show that both the quartic β-FPUT model ($\alpha = 0$) and the cubic one ($\beta = 0$) are integrable by introducing an appropriate Fourier representation to express the nonlinear terms of the Hamiltonian. For generic values of α and β, the model is non-integrable and displays a mixed phase space with both chaotic and regular trajectories. In the classical case, chaos is diagnosed by the investigation of Poincaré sections. In the quantum case, the level spacing statistics in the energy basis belongs to the *Gaussian orthogonal ensemble* in the chaotic regime, and crosses over to Poissonian behavior in the quasi-integrable low-energy limit. In the chaotic part of the spectrum, two generic observables obey the *eigenstate thermalization hypothesis*.

Keywords: integrable systems; chaotic Hamiltonian systems; quantum chaos; eigenstate thermalization hypothesis

1. Introduction

The Fermi–Pasta–Ulam–Tsingou (FPUT) model [1] has widely been investigated with the aim of understanding the approach to thermodynamic equilibrium in an isolated system [2,3]. Since the beginning of these studies, the relation with integrability in Hamiltonian mechanics and the Kolmogorov–Arnold–Moser theorem [4–6] has been stressed [7]. More recently, the "closeness" of the model to the integrable Toda lattice has carefully been examined [8].

Here, we take a different approach to integrability in the FPUT model, with the aim of discussing its quantization. We study the N-particle model for periodic boundary conditions and we introduce a specific Fourier transform, which allows us to conveniently write the nonlinear interaction terms among the Fourier mode variables [9]. We then consider the three-particle case $N = 3$ and prove that it is integrable when either the cubic nonlinearity vanishes or the quartic nonlinearity does. Interestingly, the model with only quartic nonlinearity, dubbed β-FPUT model, has an additional integral of the motion besides energy and momentum. This integral, previously discovered in Ref. [10] using an analogy with celestial mechanics, can be associated to a cylindrical symmetry, which is only manifested in Fourier modes. Already for $N = 4$, this integral is no longer present, and what remains for higher values of N, even in the thermodynamic limit, is the presence of invariant subspaces in the Fourier mode space [9], whose stability has carefully been examined by Chechin and coworkers (see Ref. [11] and Refs. therein). It should be here mentioned that a study of the Einstein–Brillouin–Keller (EBK) quantization of the integrable $N = 3$ Toda lattice was undertaken by Isola, Kantz and Livi [12]. The complete $(\alpha + \beta)$-FPUT model, with both non-zero cubic and quartic nonlinearity is non-integrable and displays strong chaotic motion in an intermediate energy range, as we show numerically by displaying Poincaré sections.

Citation: Arzika, A.I.; Solfanelli, A.; Schmid, H.; Ruffo, S. Quantization of Integrable and Chaotic Three-Particle Fermi–Pasta–Ulam–Tsingou Models. *Entropy* **2023**, *25*, 538. https://doi.org/10.3390/e25030538

Academic Editor: Marko Robnik

Received: 25 January 2023
Revised: 13 March 2023
Accepted: 14 March 2023
Published: 21 March 2023

The quantization of classical chaotic systems is still an extremely active field of research, which has its roots in the discovery of the phenomenon of *dynamical localization* by Casati, Chirikov, Ford and Izrailev [13] and in the pioneering investigation of level statistics for Sinai billiards by Berry [14]. This latter topic was further investigated by Berry and Tabor [15], who characterized specifically the Poisson statistics in the integrable limit. These studies led to the famous *universality hypothesis* by Bohigas, Giannoni and Schmit [16], who first suggested that the spectra of quantum chaotic systems are in agreement with the predictions of random matrix theory (an early account of these developments can be found in Tabor's book [17]). The analysis of spectral statistics for smooth Hamiltonians continued with the work of Seligman, Verbaarschot and Zirnbauer [18], who studied the transition from regular to chaotic phase-space and fitted the levels with Poisson or Gaussian orthogonal ensemble (GOE) statistics, respectively, for an even parity potential.

In this paper, we quantize the three-particle FPUT model in the formalism of first quantization, and discuss its transition from integrability to chaos using methods from random matrix theory. Moreover, we check the *eigenstate thermalization hypothesis* (ETH) [19,20] for two relevant observables, kinetic energy and the additional integral of the motion. This complements works about the bosonic FPUT model [21–23], and in particular a recent study [24] investigating the localization transition in the thermodynamic limit.

The paper is organized as follows. In Section 2, we introduce the appropriate Fourier mode variables and prove the integrability of both the α- and β- FPUT $N = 3$ model. In Section 3, we quantize the three-particle FPUT model and study its level statistics. Section 4 is devoted to conclusions.

2. The FPUT Model and the Integrable Three-Particle Case

The Hamiltonian of the $(\alpha + \beta)$-Fermi–Pasta–Ulam–Tsingou coupled oscillator model is

$$H = \sum_{i=1}^{N} \left[\frac{p_i^2}{2} + \frac{1}{2}(q_{i+1} - q_i)^2 + \frac{\alpha}{3}(q_{i+1} - q_i)^3 + \frac{\beta}{4}(q_{i+1} - q_i)^4 \right] = H_0 + H_\alpha + H_\beta, \quad (1)$$

where q_i is the relative displacements of the i-th oscillator from its equilibrium position and p_i the conjugate momentum. We choose periodic boundary conditions $q_{N+1} = q_1$, $p_{N+1} = p_1$. The unperturbed Hamiltonian can be written in quadratic form

$$H_0 = H(\alpha = 0, \beta = 0) = \frac{1}{2}\left[{}^t\mathbf{pp} +{}^t \mathbf{qAq}\right], \quad (2)$$

where ^{t} denotes the transposed of a column vector and $\mathbf{A}$ is the NxN symmetric matrix whose elements are

$$A_{i,j} = -\delta_{i-1,j} + 2\delta_{i,j} - \delta_{i+1,j} - \delta_{i,1}\delta_{j,N} - \delta_{i,N}\delta_{j,1}. \quad (3)$$

Due to the periodic boundary conditions, matrix $\mathbf{A}$ is degenerate. Its eigenvalues are $\mu_k = \omega_k^2$, where

$$\omega_k = 2\sin\left(\frac{\pi k}{N}\right), \; k = 0, \ldots, N-1 \quad (4)$$

are the harmonic frequencies of the linear oscillator chain. The normalized eigenvectors are $u_n^0 = 1/\sqrt{N}$, for $k = 0$, $u_n^k = \sqrt{2/N}\sin(2\pi kn/N + \gamma)$, for $k = 1, \ldots, [(N-1)/2]$, $u_n^{N/2} = (-1)^n/\sqrt{N}$, for $k = N/2$ (N even), $u_n^{N-k} = \sqrt{2/N}\sin(2\pi kn/N + \gamma + \pi/2)$, $u_n^{N-k} = \sqrt{2/N}\sin(2\pi kn/N + \gamma + \pi/2)$, for $k = \lfloor N/2 \rfloor + 1, \ldots, N-1$. Here, $\lfloor \bullet \rfloor$ denotes the integer part, the index n runs from 1 to N and the free parameter γ can be fixed arbitrarily, due to the degeneracy of $\mathbf{A}$. The NxN matrix $\mathbf{S}$, which maps Fourier modes $Q_0, \ldots, Q_{N-1}$ into $q_1, \ldots, q_N$, is orthogonal, and its columns are given by the eigenvectors of $\mathbf{A}$. The momenta in the two dual spaces, $P_0, \ldots, P_{N-1}$ and $p_1, \ldots, p_N$, are related by the same linear transformation $\mathbf{S}$. Although the choice of γ does not affect the quadratic part of the Hamiltonian in Fourier modes, it does have consequences on the expression of the nonlinear interaction terms among the Fourier modes. A particularly convenient choice is $\gamma = \pi/4$, for which all columns of the transformation matrix **S** take the same expression $1/\sqrt{N}[\sin(2\pi kn/N) + \cos(2\pi kn/N)]$, with $n = 1, \ldots, N$ as the row index and $k = 0, \ldots, N-1$ as the column index, respectively. Using this transformation, the nonlinear cubic term of the Hamiltonian is transformed into

$$H_\alpha(\mathbf{Q}) = \frac{\alpha}{3}\left(\frac{2}{N}\right)^{3/2} \sum_{n=1}^{N} \sum_{k_1,k_2,k_3=0}^{N-1} U^{\pi/4}_{n,k_1} U^{\pi/4}_{n,k_2} U^{\pi/4}_{n,k_3} \omega_{k_1}\omega_{k_2}\omega_{k_3} Q_{k_1} Q_{k_2} Q_{k_3}\,, \tag{5}$$

where

$$U^\gamma_{n,k} = \cos\left(\frac{\pi k(2n+1)}{N} + \gamma\right). \tag{6}$$

The sum over n in Equation (5) can be explicitly performed using the properties of the $U's$ (see Appendix A). The result is

$$H_\alpha(\mathbf{Q}) = \frac{\alpha}{6}\left(\frac{\sqrt{2}}{N\sqrt{N}}\right) \sum_{k_1,k_2,k_3=0}^{N-1} B_{k_1,k_2,k_3} \omega_{k_1}\omega_{k_2}\omega_{k_3} Q_{k_1} Q_{k_2} Q_{k_3}\,, \tag{7}$$

where

$$B_{k_1,k_2,k_3} = -\Delta_{k_1+k_2+k_3} + \Delta_{k_1+k_2-k_3} + \Delta_{k_1-k_2+k_3} + \Delta_{k_1-k_2-k_3}\,, \tag{8}$$

and $\Delta_r = (-1)^m$ if $r = mN$ and 0 otherwise, with m an integer. The B_{k_1,k_2,k_3} coupling coefficients incorporate momentum conservation. The same procedure can be used to derive the quartic term in the Hamiltonian, obtaining

$$H_\beta(\mathbf{Q}) = \frac{\beta}{8N} \sum_{k_1,k_2,k_3,k_4=0}^{N-1} C_{k_1,k_2,k_3,k_4} \omega_{k_1}\omega_{k_2}\omega_{k_3}\omega_{k_4} Q_{k_1} Q_{k_2} Q_{k_3} Q_{k_4}\,, \tag{9}$$

where

$$C_{k_1,k_2,k_3,k_4} = -\Delta_{k_1+k_2+k_3+k_4} + \Delta_{k_1+k_2-k_3-k_4} + \Delta_{k_1-k_2+k_3-k_4} + \Delta_{k_1-k_2-k_3+k_4} \tag{10}$$

again reflecting momentum conservation. This method is fully general and can be applied to derive the Hamiltonian in Fourier modes **Q** for all values of N.

The first nontrivial value of N is $N = 3$, since the $N = 2$ case, being a two-body problem, is totally integrable. For $N = 3$, we obtain the following potential:

$$V(Q_1, Q_2) = \frac{3}{2}(Q_1^2 + Q_2^2) + \frac{\alpha}{2}(Q_1 + Q_2)^3 + \frac{9\beta}{8}(Q_1^2 + Q_2^2)^2 \tag{11}$$

where, for convenience, the Hamiltonian is rewritten in slightly different variables where Q_1 is swapped with Q_2 and, then, the sign of Q_1 is changed, leaving the Hamiltonian invariant. Therefore, we can represent the dynamics in Fourier space as the motion of a particle in the two-dimensional potential (11).

Let us first discuss the potential (11) with $\alpha = 0$ and $\beta \neq 0$. The kinetic term of the Hamiltonian is

$$K = \frac{1}{2}(P_0^2 + P_1^2 + P_2^2)\,, \tag{12}$$

and hence, the center of mass motion is free, ($P_0 = const.$). The potential is invariant under rotations in the (Q_1, Q_2) plane; therefore, the pseudo-angular momentum

$$L_0 = Q_1 P_2 - Q_2 P_1 \tag{13}$$

is conserved. Going back to the original coordinates with an inverse Fourier transform, it can be shown that

$$L_0 = \frac{1}{\sqrt{3}}[p_1(q_3 - q_2) + p_2(q_1 - q_3) + p_3(q_2 - q_1)] \tag{14}$$

which was known to be a constant of motion, see [10]. In Fourier coordinates, this unexpected constant of motion acquires an explicit physical meaning, by associating it to a rotational symmetry. Thinking of the three coordinates (q_1, q_2, q_3) as positions of a particle in three dimensions, this symmetry corresponds to a cylindrical rotational symmetry of the potential around the axis $(u_1^0, u_2^0, u_3^0) = (1, 1, 1)/\sqrt{3}$, the first eigenvector of $\mathbf{A}$. This symmetry is related to the component of the angular momentum, L_0, along this axis. The two other constants of motion for this three-dimensional phase space are energy and the momentum along the axis $P_0 = (p_1 + p_2 + p_3)/\sqrt{3}$.

The case $\alpha \neq 0$, $\beta = 0$ is also integrable. This can be shown by introducing the variables $s = Q_1 + Q_2$ and $d = Q_1 - Q_2$. In terms of these variables, the potential is written as $V(s, d) = (3/4)(s^2 + d^2) + (\alpha/2)s^3$ and is therefore separable along the directions s and d. Along d, the motion is harmonic, while along s, the motion is the one of a cubic oscillator. The motion in the full space is therefore a composition of free, harmonic and cubic anharmonic motions along the three directions.

The intermediate case where both $\alpha \neq 0$ and $\beta \neq 0$ is non-integrable. A typical potential surface is displayed in Figure 1: it shows two valleys of different depths. If the motion is within each valley, at low energy, it is quasi-regular and very weakly chaotic, close to integrable. On the contrary, when the motion runs across the two valleys, at higher energies, it is strongly chaotic, displaying intersections on the Poincaré plane that cover an area, see Figures 2 and 3 with two different section planes (Q_1, P_1) and (Q_1, Q_2) (the numerical integration algorithm is described in Appendix B). More precisely, the potential minima have coordinates $Q_1^- = Q_2^- = -1/(\alpha - \sqrt{\alpha^2 - 3\beta})$ and $Q_1 = Q_2 = 0$, while the saddle point is located at $Q_1^+ = Q_2^+ = -1/(\alpha + \sqrt{\alpha^2 - 3\beta})$ and the corresponding values of the potential energy are

$$V(Q_1^{\pm}, Q_2^{\pm}) = \frac{4\alpha\left(\alpha \pm \sqrt{\alpha^2 - 3\beta}\right) - 9\beta}{2\left(\alpha \pm \sqrt{\alpha^2 - 3\beta}\right)^4}, \quad V(0,0) = 0\,. \tag{15}$$

Two valleys exist for weak to moderate quartic interactions ($\beta < 3\alpha^2$). The boundary value of the potential separating motions within the valleys and above them is given by $V_b = V_+$. Notice that in the case of strong quartic interactions ($\beta > 3\alpha^2$), only one minimum occurs, and the motion is again quasi-integrable.

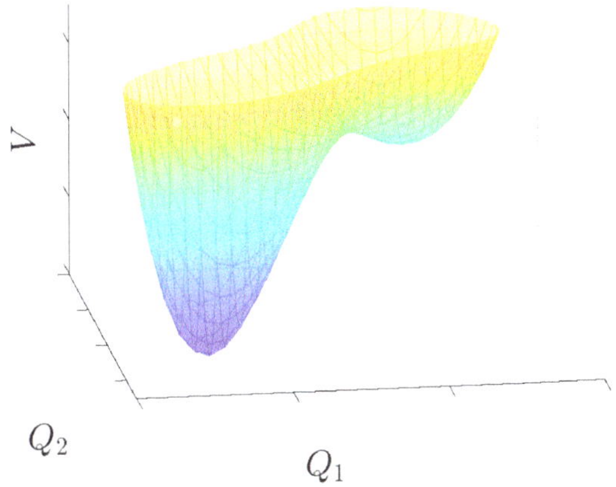

Figure 1. Effective potential surface $V(Q_1, Q_2) = (3/2)(Q_1^2 + Q_2^2) + (\alpha/2)(Q_1 + Q_2)^3 + (9\beta/8)(Q_1^2 + Q_2^2)^2$ for $\alpha = \sqrt{2}$ and $\beta = 0.5$.

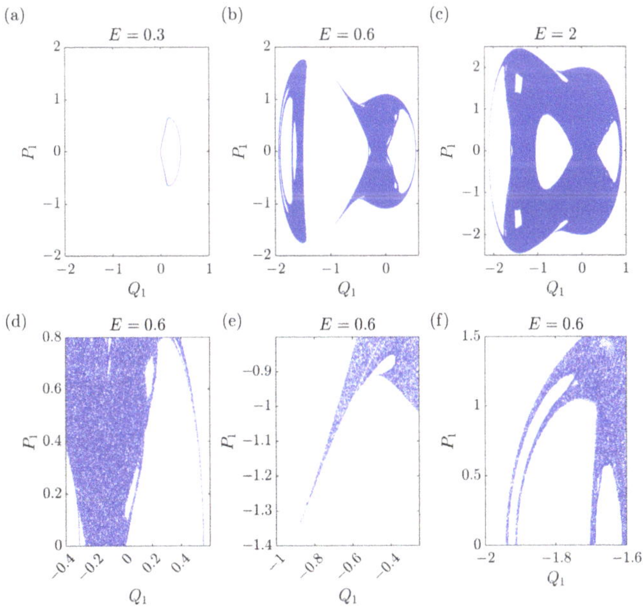

Figure 2. Poincaré maps $(Q_1(t^*), P_1(t^*))$ for the particle in the potential (11) at fixed times t^* defined by $P_2(t^*) = 0$. The particle is initially at rest, and in the crater of the upper potential valley, i.e., $Q_1(0) > 0, Q_2(0) = 0$. (**a**) At low energies, the particle does not escape from the valley. The motion describes a closed curve and it is hence quasi-integrable. (**b**,**c**) At higher energies, the particle visits both potential valleys and a single orbit fills an area in the Poincaré section: the motion shows strong chaotic behavior. (**d**–**f**) Zooms of (**b**). Parameters: $\alpha = \sqrt{2}$ and $\beta = 0.5$ and 250,000 points in the Poincare sections.

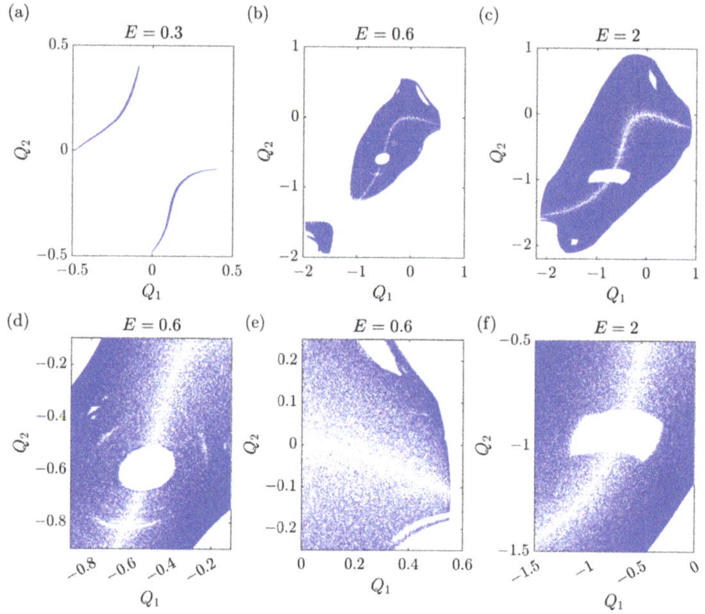

Figure 3. Poincaré maps $(Q_1(t^*), Q_2(t^*))$ for fixed times t^* defined by $P_2(t^*) = 0$. Parameters are the same as in Figure 2. (**a**) Integrable regime. (**b**,**c**) Chaotic regime. (**d**,**e**) Zooms of (**b**). (**f**) Zoom of (**c**).

Summarizing, this analysis shows that the FPUT three-particle model is integrable in the two limits $\alpha \to 0, \beta \neq 0$ and $\alpha \neq 0, \beta \to 0$. Moreover, the model displays a strongly chaotic behavior for intermediate values and α and β and for energies that are close to a saddle in the two-dimensional phase-space of a fictitious particle, which represents the motion in Fourier space. These results call for an analysis of the quantum behavior of such a model, which is pursued in the following section.

3. The Quantum Three-Particle FPUT Model

In the remainder of the paper, we extend the study to a quantum mechanical version of the three-particle FPUT model. To this end, we promote classical coordinates p_j, q_j to quantum operators $\hat{p}_j$, $\hat{q}_j$, which satisfy canonical commutation relations $[\hat{q}_i, \hat{p}_j] = i\delta_{i,j}$ ($\hbar = 1$). The quantum Hamiltonian of the three-particle FPUT model is then given by

$$H = \sum_{i=1}^{3} \left[\frac{\hat{p}_i^2}{2} + \frac{1}{2}(\hat{q}_{i+1} - \hat{q}_i)^2 + \frac{\alpha}{3}(\hat{q}_{i+1} - \hat{q}_i)^3 + \frac{\beta}{4}(\hat{q}_{i+1} - \hat{q}_i)^4 \right]. \quad (16)$$

Analogously to the classical part, we can reduce the degrees of freedom by defining normal modes $\hat{Q}_0$, $\hat{Q}_1$ and $\hat{Q}_2$. The mode $\hat{Q}_0$, corresponding to center of mass motion, decouples from the other two, and we are thus left with the problem of obtaining the eigenvalues and eigenstates for the relevant normal modes $\hat{Q}_1$ and $\hat{Q}_2$ of the potential

$$\hat{V} = \frac{3}{2}(\hat{Q}_1^2 + \hat{Q}_2^2) + \frac{\alpha}{2}(\hat{Q}_1 + \hat{Q}_2)^3 + \frac{9\beta}{8}(\hat{Q}_1^2 + \hat{Q}_2^2)^2. \quad (17)$$

3.1. The Integrable Cases

As for the classical FPUT model, if α or β are equal to zero, the model is integrable. In particular, if $\alpha = 0$, as already discussed in Section 2, the model has a manifest cylindrical symmetry with respect to the Q_0 axis. Therefore, it is convenient to pass to cylindrical coordinates (r, ϕ, z) defined as

$$\begin{cases} Q_1 = r\cos\phi \\ Q_2 = r\sin\phi \\ Q_0 = z \end{cases} . \tag{18}$$

In terms of these new variables, the time-independent Schrödinger equation reads

$$\left[-\frac{1}{2r}\frac{\partial}{\partial r}\left(r\frac{\partial}{r\partial r}\right) - \frac{1}{2r^2}\frac{\partial^2}{\partial\phi^2} - \frac{1}{2}\frac{\partial^2}{\partial z^2} + \frac{3}{2}r^2 + \frac{9}{8}\beta r^4\right]\psi = E\psi\,, \tag{19}$$

where we introduced the wave function $\psi = \psi(r, \phi, z)$. Moreover, the Hamiltonian commutes with the operators corresponding to the two integrals of motion $\hat{P}_0$ and $\hat{L}_0 = \hat{Q}_1\hat{P}_2 - \hat{Q}_2\hat{P}_1$. As a consequence, the energy levels can be labeled by three quantum numbers $E = E_{n,l_0,p_0}$ corresponding to the three commuting observables $\hat{H}, \hat{P}_0, \hat{L}_0$, and we can search for the solutions of the Schrödinger equation having the separable form

$$\psi_{n,p_0,l_0}(r, \phi, z) = \frac{e^{ip_0 z}}{\sqrt{L}}\frac{e^{il_0\phi}}{\sqrt{2\pi}}f_{n,l_0}(r)\,. \tag{20}$$

The normalized eigenfunctions of the $\hat{P}_0$ and $\hat{L}_0$ operators factorize, and $f_{n,l_0}(r)$ is the radial wave function which remains to be found. L is the size of the system along the axial direction z. We notice that square modulus of the total wave function must be normalized to unity by requiring that

$$\int \mathrm{d}r\mathrm{d}\phi\mathrm{d}z r^2\left|\psi_{n,p_0,l_0}(r, \phi, z)\right|^2 = \int \mathrm{d}r r^2\left|f_{n,l_0}(r)\right|^2 = 1\,. \tag{21}$$

It is then convenient to introduce the new radial function $u_{n,l_0}(r) = \sqrt{r}f_{n,l_0}(r)$. Finally, inserting this ansatz in Equation (19), we obtain the radial equation

$$-\frac{1}{2}\frac{\mathrm{d}^2u_{n,l_0}(r)}{\mathrm{d}r^2} + V_{\mathrm{eff}}(r)u_{n,l_0}(r) = E_{n,l_0}u_{n,l_0}(r)\,, \tag{22}$$

where $E_{n,l_0} = E_{n,l_0,p_0} - p_0^2/2$, and we introduced the effective potential

$$V_{\mathrm{eff}}(r) = \frac{4l_0^2 - 1}{8r^2} + \frac{3}{2}r^2 + \frac{9}{8}\beta r^4\,, \tag{23}$$

which, in addition to the interaction potential $V(r) = (3/2)r^2 + (9/8)\beta r^4$, also contains a centrifugal barrier term. Summarizing, we mapped the problem into a single particle moving in a one-dimensional potential. We are not aware of the exact analytical solutions of Equation (22) with given boundary conditions, but the equation could be of course solved numerically to determine the radial spectrum quantitatively. This concludes the proof of the integrability of the quantum version of the three-particle β-FPUT model.

The quantum integrability of the α-FPUT model is even simpler because the corresponding Schrödinger equation is separable in two independent one-dimensional Schrödinger equations, describing a harmonic and an anharmonic oscillator, respectively.

3.2. The Non-Integrable (α+β)-FPUT Three-Particle Model

The rotational symmetry of potential (11) for $\alpha = 0$, leading to integrability, is lost when the general case $\alpha \neq 0$ is considered. In the classical case, we hence found chaotic behavior, as shown by the Poincarè sections. Quantum mechanically, we should expect a situation similar to the one studied in Ref. [18], where the Poisson level spacing statistics was found in the integrable limit, while these statistics were seen to belong to the GOE universality class in the case of the mixed regular–chaotic phase space. In the quantum FPUT model, we should therefore find the GOE statistics at intermediate energies and Poisson statistics at both low and high energy.

We adopt a numerical diagonalization procedure of Hamiltonian Equation (16) [25]. We discretize the Hamiltonian on a two-dimensional square lattice with grid size $[-L/2, L/2]^2$ and N_e grid points assuming open boundary conditions. The grid is carefully chosen such that the minima of the potential are well resolved, and the boundaries of the grid are completely dominated by the quartic term. Rescaled numerical spectra are presented in Figure 4, with ascending order of energy levels. Eigenvalues stabilize as N_e is increased, and we checked that the uncertainty in the numerical value of each eigenvalue is small compared to the mean level spacing Δ.

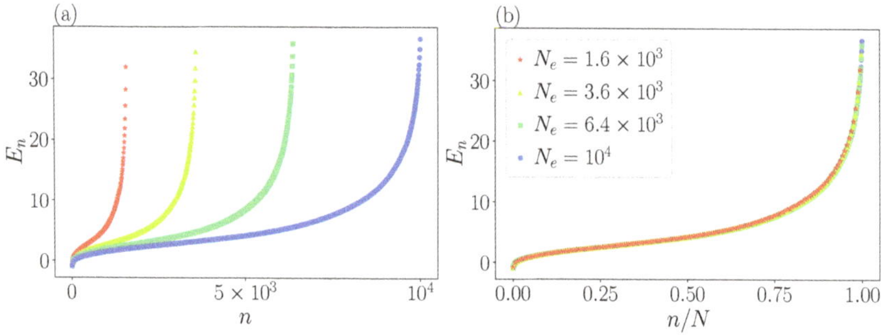

Figure 4. Energy levels of the three-particle FPUT model for $\alpha = \sqrt{2}$ and $\beta = 1/2$. In panel (**a**), the levels are ordered in increasing value and for different grid points N_e of the lattice. In panel (**b**), we scale the level order by the number of grid points, obtaining a perfect data collapse.

The central object in our analysis is the distribution function

$$P(s) = \delta(s - \delta_n)\,, \tag{24}$$

which is obtained from consecutive level spacings

$$\delta_n = \frac{E_{n+1} - E_n}{\Delta}\,. \tag{25}$$

Here, Δ denotes the mean level spacing. The spectral statistics we present are in line with the *Gaussian orthogonal ensemble* (GOE) of random matrices with distribution function [17,26,27]

$$P_{\mathrm{GOE}}(s) = \frac{\pi}{2} s e^{-\frac{\pi}{4}s^2}\,. \tag{26}$$

This poses a clear signature of a chaotic quantum system with real, normally distributed entries of the Hamiltonian. In contrast, integrability will be associated with an absence of level repulsion and a Poissonian level distribution function

$$P_{\mathrm{P}}(s) = e^{-s}\,. \tag{27}$$

Another measure for chaos is given by the level-statistics ratio

$$\bar{r} = \frac{\min(\delta_n, \delta_{n+1})}{\max(\delta_n, \delta_{n+1})}\,. \tag{28}$$

Figure 5a shows the level distribution function $P(s)$ for an energy window $E \in [1.9, 2.2]$ with $n = 300$ levels. With apparent level repulsion in place, level spacings are distributed according to the GOE prediction, indicating chaotic behavior. This is contrasted with a distribution obtained in an energy window $E \in [7, 9]$ ($n = 300$ levels) in Figure 5b, with the

Poissonian level statistic. The two discussed cases confirm the physical picture presented above, though with noticeable fluctuations due to the number of considered levels.

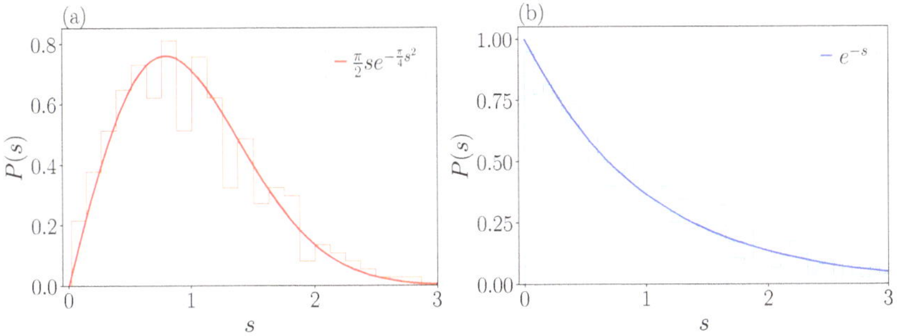

Figure 5. Level spacing distribution of the energy spectrum of the FPUT model $P(s)$ with $\alpha = \sqrt{2}$, $\beta = 1/2$. In panel (**a**) the distribution is well fitted with GOE statistics, while in panel (**b**) the statistics is clearly Poissonian.

The level spacing statistics in different regions of the spectrum is strongly affected by the shape of the potential. We will consider representative parameter values of α and β for which the potential landscape has two non-equivalent minima separated by a barrier of height V_b, see Figure 6. At very low energies, the wave function localizes in one of the potential minima, which gives rise to harmonic levels $E_n \simeq \omega(n + \frac{1}{2})$. At very high energies, the quartic term dominates over the cubic, effectively restoring rotational symmetry, and therefore integrability. From a semi-classical point of view, we expect most of the chaos to appear near the separatrix for levels at energies $E \gtrsim V_b$.

The transition from Poissonian to GOE statistics as a function of energy is traced via the level ratio $\overline{r}$, see Figure 7. For this, we sweep the energy window with n eigenvalues centered around an energy $\overline{E}$. At high energies ($\overline{E} \gg V_b$), we find a level ratio consistent with the Poisson distribution, $\overline{r} \approx 0.38$. At energies near to the barrier height but sufficiently high so that it is not strongly affected by the presence of the potential minima ($V_b \leq \overline{E} \leq 20V_b$), the level ratio peaks at the GOE value $\overline{r} \approx 0.52$, signaling a transition to a fully quantum chaotic behavior of the system. The transition between these two extremal cases occurs on the same energy scale on which rotational symmetry is broken, and hence appears visible in Figure 7. For low energies, the $\overline{r}$-value decreases again but not fully to the Poissonian value since the minima support harmonic oscillator eigenfunctions. This behavior should be compared with the one of the integrable model ($\alpha = 0$) shown in Figure 7b. In this case, the level ratio is never compatible with chaotic statistics independently from the energy interval considered.

Real-space wave functions support the picture of chaos in distinct regions of the spectra, see Figure 6a–c. At low energies ($E < V_b$), back to the picture of quasi-integrability, wave functions in the potential minimum attain the shape of two-dimensional orbitals which can be labeled by radial and angular quantum numbers, Figure 6a. In contrast, the chaotic behavior for states with intermediate energies ($V_b < E < 20V_b$) is reflected in irregularly shaped wavefunctions with random direction of the nodal curves, Figure 6b. At high energy ($E \gg V_b$), close to the other integrable limit, wavefunctions have a structure similar to the one observed at low energy, Figure 6c.

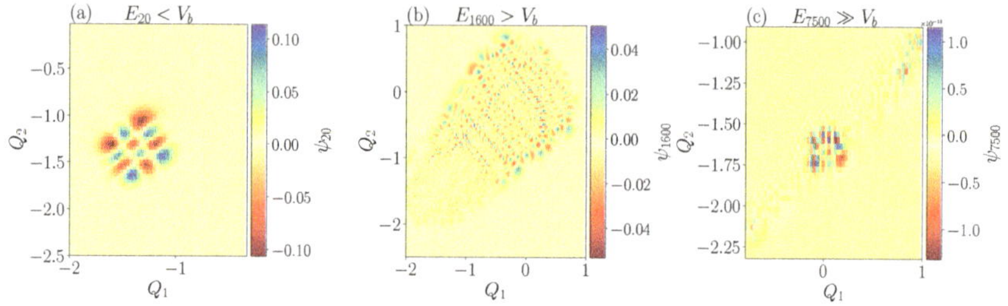

Figure 6. Representative wavefunctions at low $E_{20} \simeq -0.43 < V_b$ (panel (**a**)), $V_b < E_{1600} \simeq 2.0 < 20V_b$ intermediate (panel (**b**)) and high $E_{7500} \simeq 7.7 \gg V_b$ (panel (**c**)) energy.

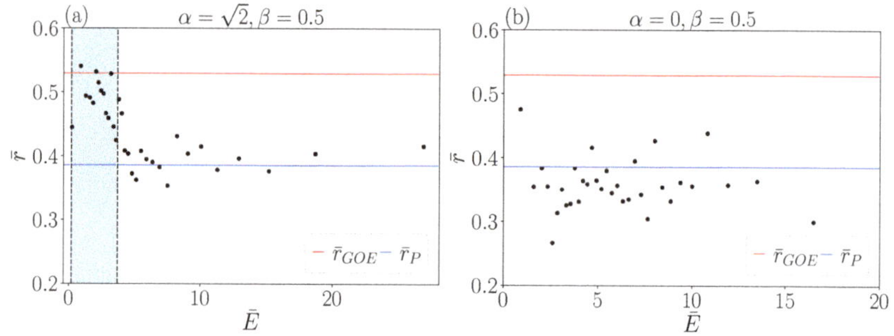

Figure 7. Window averaged level–statistic ratio (28) as a function of the window averaged energy for $\alpha = \sqrt{2}$, $\beta = 0.5$ (panel **a**) and $\alpha = 0$, $\beta = 0.5$ (panel **b**). The GOE and the Poisson level statistics values are shown by horizontal lines. The vertical light blue stripe indicates the region of average energy $V_b \leq \bar{E} \leq 20V_b$.

In fact, the nodal curves strongly repel each other, and do not cross, up to our numerical resolution, see Figure 8b. This was predicted by Berry [14] and first verified for a free particle in a stadium [28].

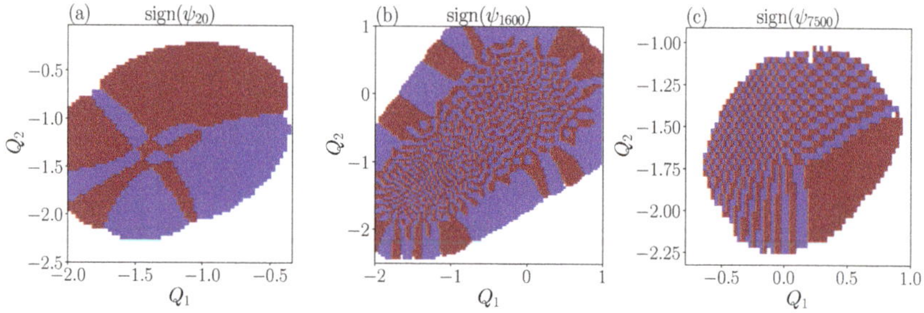

Figure 8. Sign of the wavefunctions, showing nodal lines in different energy regions: low energy (panel (**a**)), intermediate energy (panel (**b**)) and high energy (panel (**c**)).

3.3. *Eigenstate Thermalization Hypothesis*

According to the *eigenstate thermalization hypothesis* (ETH) of Deutsch [19] and Srednicki [20] (see Ref. [29] for a review), if a quantum system is chaotic, the average of a generic observable taken on the eigenstates of the energy is a *smooth* function of the energy eigenvalue itself. In connection with ETH, and focusing on the relation between quantum and classical systems (e.g., for the FPUT model), the adiabatic gauge potential is introduced to characterize suppressed heating by a slow drive [30–32]. We checked the ETH property, restricting to the stationary solutions of the Schrödinger equation, without any attempt to analyze the time evolution. Indeed, it is also guessed that the relaxation time to the average is controlled by the size of the system, i.e., the number of particles. Therefore, we cannot expect a fast relaxation in such a low-dimensional phase space as the one of the three-particle FPUT model. The averages on the energy eigenstates of kinetic energy and angular momentum are shown in Figures 9 and 10 as a function of energy in three different regions of the spectrum. Both at low and at high energy, the kinetic energy and angular momentum are irregularly scattered in a wide region of values, and change abruptly as the energy is varied. It is only in the central chaotic region of the spectrum that they appear to fluctuate stochastically around an average. We believe that this result is a particularly transparent numerical verification of the ETH hypothesis.

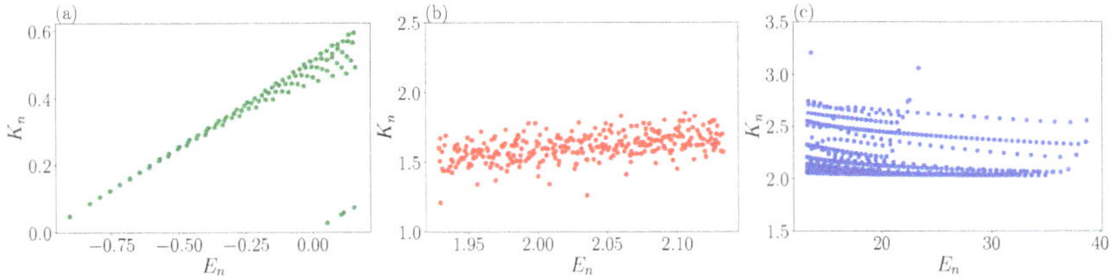

Figure 9. Average kinetic energy vs. energy for (**a**) the lower part of the spectrum ($E_n < V_b$), (**b**) the central part of the spectrum where eigenstates are chaotic ($E_n \approx 10V_b$), (**c**) upper part of the spectrum ($E_n > 50V_b$).

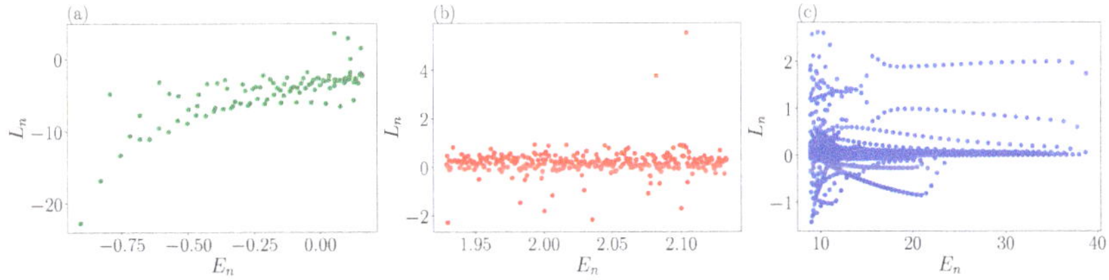

Figure 10. Average angular momentum vs. energy for (**a**) the lower part of the spectrum ($E_n < V_b$), (**b**) the central part of the spectrum where eigenstates are chaotic ($E_n \approx 10V_b$), (**c**) upper part of the spectrum ($E_n > 50V_b$).

4. Conclusions

In this paper, we investigated the classical and quantum Fermi–Pasta–Ulam–Tsingou model with three particles. Classically, the model is integrable in the two limits where the cubic or the quartic term of the Hamiltonian vanishes. In the intermediate region, the model shows a mixed phase space with both regular and chaotic trajectories, in agreement with the picture suggested by the Kolmogorov–Arnold–Moser theorem. We computed the energy spectrum and showed that, as expected from the Bohigas–Giannoni–Schmit universality hypothesis, the level spacing satisfies the *Gaussian orthogonal ensemble* statistics in the chaotic part of the spectrum. In the two integrable limits, the statistics become Poissonian. We also showed how two generic observables obey the *eigenstate thermalization hypothesis* of Deutsch and Srednicki in the chaotic part of the spectrum.

Author Contributions: Conceptualization, A.I.A., H.S., A.S. and S.R.; methodology, A.I.A., H.S., A.S. and S.R.; software, A.I.A., H.S. and A.S.; validation, A.I.A., H.S., A.S. and S.R.; formal analysis, A.I.A., H.S., A.S. and S.R.; investigation, A.I.A., H.S., A.S. and S.R.; resources, S.R.; data curation, H.S. and A.S.; writing—original draft preparation, A.I.A., H.S., A.S. and S.R.; writing—review and editing, A.I.A., H.S., A.S. and S.R.; supervision, S.R.; project administration, S.R.; funding acquisition, S.R. All authors have read and agreed to the published version of the manuscript.

Funding: This work is funded by MUR Italy under the PRIN 2017 project Coarse-grained description for non-equilibrium systems and transport phenomena (CO-NEST) No. 201798CZL.

Institutional Review Board Statement: Not applicable.

Data Availability Statement: Data available upon reasonable request.

Acknowledgments: SR thanks Luca Celardo for useful suggestions. AIA thanks the TWAS-SISSA-Lincei program for supporting his visit to SISSA.

Conflicts of Interest: The authors declare no conflict of interest.

Appendix A. Properties of the U Symbols

We here summarize some useful properties of the U symbols of Formula (6). It is straightforward to prove that

$$U^{\phi}_{n,r}U^{\delta}_{n,s} = \frac{1}{2}\left(U^{\phi+\delta}_{n,r+s} + U^{\phi-\delta}_{n,r-s}\right), \tag{A1}$$

and

$$U^{\delta+\pi}_{n,r} = -U^{\delta}_{n,r}\,. \tag{A2}$$

Using these properties, one derives the following identities when restricting to $\phi = \delta = \pi/4$:

$$U^{\pi/4}_{n,k_1}U^{\pi/4}_{n,k_2}U^{\pi/4}_{n,k_3} = \frac{1}{4}\left(-U^{-\pi/4}_{n,k_1+k_2+k_3} + U^{\pi/4}_{n,k_1+k_2-k_3} + U^{\pi/4}_{n,k_1-k_2+k_3} + U^{\pi/4}_{n,k_1-k_2-k_3}\right), \tag{A3}$$

which is used for the derivation of the expression of the cubic term of the Hamiltonian of the FPUT model in terms of Fourier modes. Moreover,

$$U^{\pi/4}_{n,k_1}U^{\pi/4}_{n,k_2}U^{\pi/4}_{n,k_3}U^{\pi/4}_{n,k_4} = \tag{A4}$$

$$\frac{1}{8}(-U^{0}_{n,k_1+k_2+k_3+k_4} + U^{0}_{n,k_1+k_2-k_3-k_4} + U^{0}_{n,k_1-k_2+k_3-k_4} + U^{0}_{n,k_1-k_2-k_3+k_4} \tag{A5}$$

$$+U^{\pi/2}_{n,k_1+k_2+k_3-k_4} + U^{\pi/2}_{n,k_1+k_2-k_3+k_4} + U^{\pi/2}_{n,k_1-k_2+k_3+k_4} - U^{\pi/2}_{n,k_1-k_2-k_3-k_4}), \tag{A6}$$

which is used in the derivation of the quartic term.

Appendix B. Details on Numerical Integration of the Classical Equations of Motion

We integrated the classical equations of motion for the effective two-dimensional potential using a fourth-order Runge–Kutta scheme [33]. A stepsize $\Delta t = 0.01$ proved to be stable and sufficiently accurate for the purposes of this paper. We monitored the relative error in energy conservation of 10^{-8} for the longest simulation times reported. Our main goal is of a qualitative nature, i.e., showing integrability and chaos in different regions of phase space. We cross-checked some results with a velocity Verlet integrator.

References

1. Fermi, E.; Pasta, J.; Ulam, S.; Tsingou, M. *Studies of the Nonlinear Problems;* Los Alamos Internal Report, Document LA-1940; Los Alamos National Lab. (LANL): Los Alamos, NM, USA, 1955. [CrossRef]
2. Ford, J. The Fermi-Pasta-Ulam problem: Paradox turns discovery. *Phys. Rep.* **1992**, *213*, 271–310. [CrossRef]
3. Gallavotti, G. *The Fermi-Pasta-Ulam Problem: A Status Report*; Lecture Notes in Physics; Springer: Berlin/Heidelberg, Germany, 2008.
4. Kolmogorov, A.N. On conservation of conditionally periodic motions for a small change in Hamilton's function. *Dokl. Akad. Nauk.* **1954**, *98*, 527.
5. Arnold, A.N. Invariant Tori and Cylinders for a Class of Perturbed Hamiltonian Systems. *Usp. Mat. Nauk.* **1963**, *18*, 13.
6. Moser, J. On invariant curves of area-preserving mappings of annulus. *Matematika* **1962**, *6*, 51–68.
7. Chirikov, B. A universal instability of many-dimensional oscillator systems. *Phys. Rep.* **1979**, *52*, 263. [CrossRef]
8. Benettin, G.; Christodoulidi, H.; Ponno, A. The Fermi-Pasta-Ulam Problem and Its Underlying Integrable Dynamics. *J. Stat. Phys.* **2013**, *152*, 195–212. [CrossRef]
9. Poggi, P.; Ruffo, S. Exact solutions in the FPU oscillator chain. *Physica D* **1997**, *103*, 251–272. [CrossRef]

10. Choodnovsky, G.V.; Choodnovsky, D.V. Novel first integrals for the Fermi-Pasta-Ulam lattice with cubic nonlinearity and for other many-body systems in one and three dimensions. *Lett. Nuovo C* **1977**, *19*, 291. [CrossRef]
11. Chechin, G.M.; Ryabov, D.S. Stability of nonlinear normal modes in the Fermi-Pasta-Ulam β chain in the thermodynamic limit. *Phys. Rev. E* **2012**, *85*, 056601. [CrossRef]
12. Isola, S.; Kantz, H.; Livi, R. On the quantization of the three-particle Toda lattice. *J. Phys. A* **1991**, *24*, 3061–3076. [CrossRef]
13. Casati, G.; Ford, J. *Stochastic Behavior in Classical and Quantum Hamiltonian Systems*; Lecture Notes in Physics; Springer: Berlin/Heidelberg, Germany, 1979; Volume 93, p. 334.
14. Berry, M.V.; Tabor, M. Level clustering in the regular spectrum *Proc. R. Soc. A Lond. Math. Phys. Sci.* **1977**, *356*, 375–394. [CrossRef]
15. Berry, M.V. Quantizing a classically ergodic system: Sinai's billiard and the KKR method. *Ann. Phys.* **1981**, *131*, 163–216. [CrossRef]
16. Bohigas, O.; Giannoni, M.J. Chaotic motion and random matrix theories. In *Mathematical and Computational Methods in Nuclear Physics*; Springer: Berlin/Heidelberg, Germany, 1984; pp. 1–99.
17. Tabor, M. *Chaos and Integrability in Nonlinear Dynamics: An Introduction*; Wiley: Hoboken, NJ, USA, 1989.
18. Seligman, T.H.; Verbaarschot, J.J.M.; Zirnbauer, M.R. Quantum Spectra and Transition from Regular to Chaotic Classical Motion. *Phys. Rev. Lett.* **1984**, *53*, 215–217. [CrossRef]
19. Deutsch, J.M. Quantum statistical mechanics in a closed system. *Phys. Rev. A* **1991**, *43*, 2046. [CrossRef] [PubMed]
20. Srednicki, M. Chaos and quantum thermalization. *Phys. Rev. E* **1994**, *50*, 888. [CrossRef]
21. Ivić, Z.; Tsironis, G.P. Biphonons in the β-Fermi-Pasta-Ulam model. *Phys. D Nonlinear Phenom.* **2006**, *216*, 200–206. [CrossRef]
22. Berman, G.; Tarkhanov, N. Quantum Dynamics in the Fermi–Pasta–Ulam Problem. *Int. J. Theor. Phys.* **2006**, *45*, 1846–1868. [CrossRef]
23. Riseborough, P.S. Phase transition arising from the underscreened Anderson lattice model: A candidate concept for explaining hidden order in URu_2Si_2. *Phys. Rev. E* **2012**, *85*, 11129. [CrossRef]
24. Burin, A.L.; Maksymov, A.O.; Schmidt, M.; Polishchuk, I.Y. Chaotic Dynamics in a Quantum Fermi–Pasta–Ulam Problem. *Entropy* **2019**, *21*, 51. [CrossRef]
25. Press, W.H.; Teukolsky, S.A.; Vetterling, W.T.; Flannery, B.P. *Numerical Recipes: The Art of Scientific Computing*; Cambridge University Press: Cambridge, UK, 2007.
26. Beenakker, C.W. Random-matrix theory of quantum transport. *Rev. Mod. Phys.* **1997**, *69*, 731. [CrossRef]
27. Dieplinger, J.; Bera, S.; Evers, F. Emergent Relativistic Effects in Condensed Matter. *Ann. Phys.* **2021**, *435*, 168503. [CrossRef]
28. McDonald, S.W.; Kaufman, A.N. Spectrum and Eigenfunctions for a Hamiltonian with Stochastic Trajectories. *Phys. Rev. Lett.* **1979**, *42*, 1189–1191. [CrossRef]
29. Reimann, P. Eigenstate thermalization: Deutsch's approach and beyond. *New J. Phys.* **2015**, *17*, 055025. [CrossRef]
30. Gjonabalaj, M.O.; Campbell, D.K.; Polkovnikov, A. Counterdiabatic driving in the classical β-Fermi-Pasta-Ulam-Tsingou chain. *Phys. Rev. E* **2022**, *106*, 014131. [CrossRef]
31. Claeys, P.W.; Pandey, M.; Sels, D.; Polkovnikov, A. Floquet-Engineering Counterdiabatic Protocols in Quantum Many-Body Systems. *Phys. Rev. Lett.* **2019**, *123*, 090602. [CrossRef]
32. Kolodrubetz, M.; Sels, D.; Mehta, P.; Polkovnikov, A. Geometry and non-adiabatic response in quantum and classical systems. *Phys. Rep.* **2017**, *697*, 1. [CrossRef]
33. Schmid, H.; Succi, S.; Ruffo, S. Nonlinearity accelerates the thermalization of the quartic FPUT model with stochastic baths. *J. Stat. Mech.* **2021**, *2021*, 053205. [CrossRef]

entropy

MDPI

Article

On Two Non-Ergodic Reversible Cellular Automata, One Classical, the Other Quantum †

Tomaž Prosen

Faculty of Mathematics and Physics, University of Ljubljana, Jadranska 19, SI-1000 Ljubljana, Slovenia; tomaz.prosen@fmf.uni-lj.si

† To Prof. Giulio Casati, on the occasion of his 80th birthday.

Abstract: We propose and discuss two variants of kinetic particle models—cellular automata in 1 + 1 dimensions—that have some appeal due to their simplicity and intriguing properties, which could warrant further research and applications. The first model is a deterministic and reversible automaton describing two species of quasiparticles: stable massless *matter* particles moving with velocity ± 1 and unstable standing (zero velocity) *field* particles. We discuss two distinct continuity equations for three conserved charges of the model. While the first two charges and the corresponding currents have support of three lattice sites and represent a lattice analogue of the conserved energy–momentum tensor, we find an additional conserved charge and current with support of nine sites, implying non-ergodic behaviour and potentially signalling integrability of the model with a highly nested R-matrix structure. The second model represents a quantum (or stochastic) deformation of a recently introduced and studied charged hardpoint lattice gas, where particles of different binary charge (± 1) and binary velocity (± 1) can nontrivially mix upon elastic collisional scattering. We show that while the unitary evolution rule of this model does not satisfy the full Yang–Baxter equation, it still satisfies an intriguing related identity which gives birth to an infinite set of local conserved operators, the so-called glider operators.

Keywords: cellular automata; interacting dynamics; ergodic theory; ergodicity breaking; integrability; conservation laws

Citation: Prosen, T. On Two Non-Ergodic Reversible Cellular Automata, One Classical, the Other Quantum. *Entropy* **2023**, *25*, 739. https://doi.org/10.3390/e25050739

Academic Editor: Marko Robnik

Received: 31 March 2023
Revised: 28 April 2023
Accepted: 29 April 2023
Published: 30 April 2023

1. Introduction

This article is dedicated to my friend and mentor, Giulio Casati. For that reason I will also take the liberty to write the paper from a very personal perspective. One of the most important things that Giulio taught me, was to passionately appreciate extremely simple models of dynamics. Constructing and solving simple models of (lately, most often many-body) dynamics with carefully chosen physical properties has thus become my personal obsession throughout my career. The other thing which I owe Giulio is a pragmatic but deep appreciation of ergodic theory. Together we have been responsible for a few endeavours in experimental mathematics, which sometimes aroused the attention of experts. For instance, we pointed out intriguing ergodic properties of classical and quantum polygonal billiards and related non-hyperbolic and non-integrable dynamical systems [1–4].

Crudely speaking, ergodic theory [5] divides dynamical systems into ergodic and non-ergodic ones. The former are characterized by the property that time averages of (relevant) observables can be computed in terms of an ensemble average over the entire space of states (phase space), while the latter exhibit memory of the initial condition in typical (say, physically relevant) observables. Additionally, ergodic dynamics can have various degrees of dynamical complexity or chaos, while non-ergodic ones can conform to some algebraic tools of exact solvability, or integrability. However, it is not clear to what extent the breaking of ergodicity is connected to any form of integrability, as it can precisely be defined in terms of Lax pairs (in classical, deterministic systems) or the Yang–Baxter equation (in a quantum or stochastic setting). In the physics literature, these questions have been extensively discussed recently in the context of many-body dynamics, where the mathematics of ergodic theory is much less developed, say on the lattice and having local interactions, both in the classical and quantum realms (see, e.g., some recent reviews [6–9]). Besides integrability, many different forms of ergodicity breaking have been suggested, such as phase-space or Hilbert-space fragmentation, due to various forms of kinetic constraints, the so-called many-body localization due to static disorder, many-body scarring due to hidden weakly broken non-abelian symmetries, etc.

In this paper, we propose two very simple many-body locally interacting dynamical systems defined on a discrete 1 + 1 space-time lattice. The first model, which we call a matter–field automaton, for reasons that will hopefully appear clear to the reader, is an example of a deterministic particle kinematics with two distinct types of degrees of freedom (matter and field), which is non-ergodic for a nontrivial reason: existence of a highly nontrivial but local conserved charge. Inspired by brute-force empirical computer-algebra calculations, we suggest that the model may be integrable, but if it is, the corresponding R-matrix should be highly nontrivial. The second model, which is a quantum deformation of the previously studied charged hardpoint lattice gas [10–12], obeys a remarkable reduced Yang–Baxter-like identity for a non-abelian spectral parameter. This again ensures manifest ergodicity breaking in the model through an existence of an infinite set of local translationally invariant conserved operators, the so-called gliders.

2. Matter–Field Automaton

2.1. *Definition of the Automaton*

Let us define a deterministic reversible cellular automaton on $\{0,1\}^{\mathbb{Z}} \ni \underline{s}$ as follows. Even lattice sites correspond to the *matter* variable $s_{2x} \in \{0,1\}$, while odd lattice sites correspond to the *field* variable $s_{2x+1} \in \{0,1\}$, $x \in \mathbb{Z}$. The instantaneous system configuration is thus specified by an infinite binary sequence $\underline{s}$, on which we specify deterministic dynamics.

The dynamics is defined in a staggered fashion, in even and odd time layers. Specifically, for so-called *even* time steps we update the matter–field–matter triples $(s_{4x}, s_{4x+1}, s_{4x+2})$ for all $x \in \mathbb{Z}$ with a deterministic, reversible rule $(ss's'') \longrightarrow (rr'r'')$, while for *odd* time-steps, the triples $(s_{4x-2}, s_{4x-1}, s_{4x})$, are updated by the same rule. The rule is specified as

$$\begin{aligned} (000) &\longleftrightarrow (000), \\ (001) &\longleftrightarrow (100), \\ (011) &\longleftrightarrow (110), \\ (101) &\longleftrightarrow (010), \\ (111) &\longleftrightarrow (111), \end{aligned} \tag{1}$$

or, graphically and perhaps more intuitively, by the following diagrams

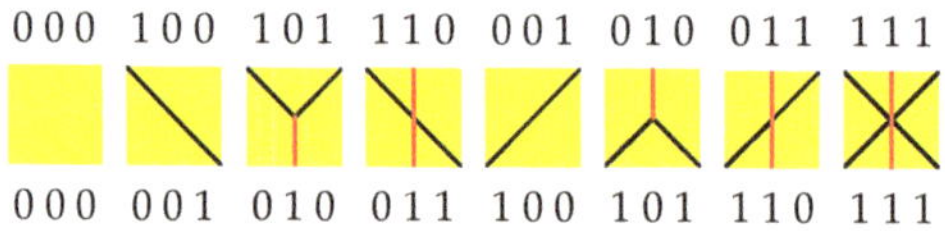

which indicate mappings of all eight lower matter–field–matter triples to corresponding upper matter–field–matter triples, or vice versa due to time-reversal symmetry. The black and red lines thus represent the worldlines of matter and field particles, respectively. In Figure 1, we show an example of a piece of a many-body trajectory $\{s_x^t; x, t \in \mathbb{Z}\}$, where time t runs in the vertical direction (upwards). The construction of the trajectory can be understood as a checkerboard tiling of the rule's yellow squares, which is uniquely prescribed by the lowest two lines of tiles (the initial condition).

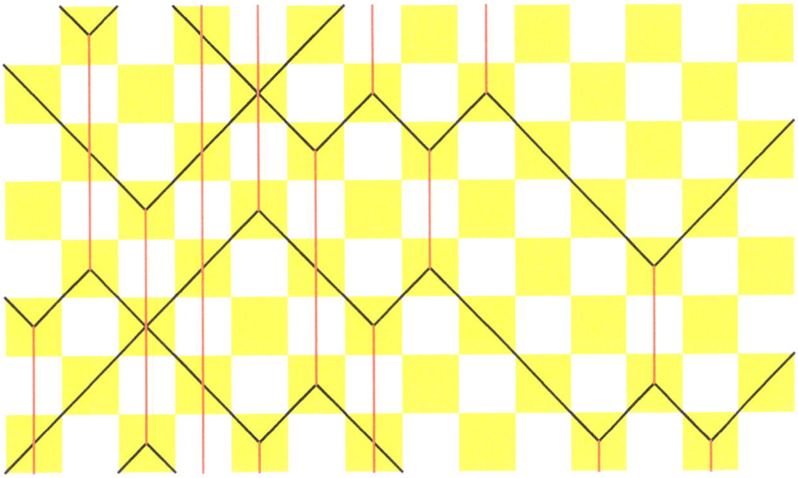

Figure 1. Example of an automaton's trajectory (time runs in the vertical direction, say upwards). Yellow squares denote application of the local rule, or equivalently, of the three-bit permutation gate Y. Black/red lines denote the worldlines of matter/field quasiparticles.

Note that this dynamical system has a rich set of three independent $\mathbb{Z}^2$ symmetries, specifically C, P, and T:

1. "Charge" conjugation C, $s \to 1 - s$, i.e., if s^t_x is a valid trajectory, then $1 - s^t_x$ is a valid trajectory,
2. Parity P, $x \to -x$, i.e., if s^t_x is a valid trajectory, then s^t_{-x} is a valid trajectory,
3. Time reversal T, $t \to -t$, i.e., if s^t_x is a valid trajectory, then s^{-t}_x is a valid trajectory.

In Figures 2–6, we depict some interesting characteristic trajectories of the matter–field automaton. In Figure 2, we prepare a maximum entropy initial state in a finite box of size L, where both field and matter are sampled randomly and uniformly in the set $\{0, 1\}$, and plot an emerging deterministic trajectory. Since the initial condition vanishes outside the box, $s^{t=0}_x = 0$, for $x < 0$ or $x > L$, we see a slow 'evaporation' of matter particles from the edges of the box. In Figure 3, we plot a similar trajectory, where initially we have only matter (at maximum entropy) and no field, while in Figure 4 we plot the reverse situation, where initially there is only the field at maximum entropy and no matter. The fact that spatio-temporal patterns of these trajectories seem qualitatively distinct depending on the initial conditions, and in particular the relative ratio of field versus matter drastically differs, suggests that the dynamics may be non-ergodic. We shall establish this more precisely below by constructing nontrivial algebraic conservation laws. While the previous plots represent cartoons of equilibrium dynamics, we plot a typical trajectory representing far-from-equilibrium dynamics in Figure 5: a head-on collision of two streams of matter (with no field in the initial data), which experience a nontrivial scattering process via formation of a complex field pattern, finally resulting in two oppositely moving scattered streams of matter.

Finally, we demonstrate that the matter–field system can be understood as a very general automaton which can also mimic the dynamics of the popular Rule 54 reversible cellular automaton [13] for a dynamically closed (invariant) subset of initial configurations. More generally, we can show it can simulate dynamics for classes of negative-length hard-rod systems with commensurate initial data. For instance, if we take initial condition $\underline{s}$ where only every sixth site can be occupied, i.e., $s_x \equiv 0$ for $x \neq 0 \pmod 6$, then one can show that no field worldline can be crossed by another matter particle, and hence each field particle decays at the immediate next time step (the field is maximally short-lived). This means that scattering of matter particles always experiences a fixed phase-shift, which makes the dynamics qualitatively similar (yet not precisely equivalent) to Rule 54 on a certain reduced lattice (see Figure 6 for a demonstration). In Rule 54 dynamics, for instance, the minimal distance between movers/kinks (=2) is twice the time-tag in kink-scattering (=1), while in this case, the minimal distance between (left/right) movers (=6) is three times the scattering time-lag (=2), i.e., free field lifetime. A similar commensurate negative-length hard-rod dynamics is obtained for the subset of initial data $\{\underline{s}\}$ of the form $s_{x'} \equiv 0$, $x' \neq 0 \pmod{2(2k+1)}$, for any $k = 1, 2 \ldots$, where the ratio between the minimal distance between parallel kinks to scattering-lag is $2k+1$.

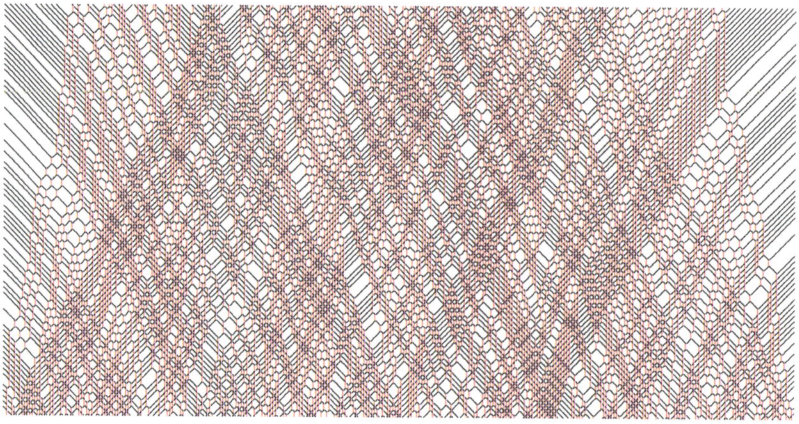

Figure 2. Example of a typical trajectory in a maximum entropy initial state, where $s_x^{t=0}$ are sampled randomly and uniformly in $\{0,1\}$, for a section of $0 \leq x \leq 600$ and duration $t_{\max} = 150$, $0 \leq t \leq t_{\max}$. Note the evaporation of matter particles from the sides, and reduction in the field intensity, which is due to initial vacuum in the complement region, $s_{x'}^{t=0} = 0$ for $x' < 0$ or $x' > 600$.

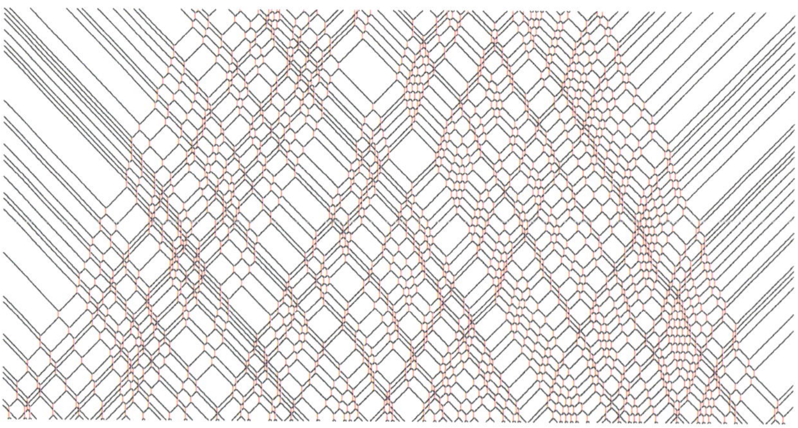

Figure 3. Example of a typical trajectory from the initial state where the matter is in a maximal entropy state, i.e., s_{2x} are sampled randomly and uniformly for $0 \leq 2x \leq 600$, and there is no field, i.e., $s_{2x+1} \equiv 0$, of duration $t_{\max} = 150$, $0 \leq t \leq t_{\max}$.

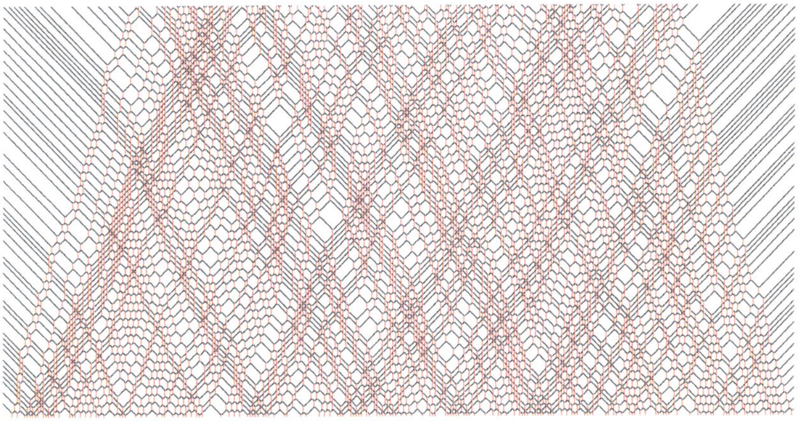

Figure 4. Example of a typical trajectory from the initial state where the field is in a maximal entropy state, i.e., s_{2x+1} are sampled randomly and uniformly for $0 \leq 2x+1 \leq 600$, and there is no matter, i.e., $s_{2x} \equiv 0$, of duration $t_{\max} = 150$, $0 \leq t \leq t_{\max}$.

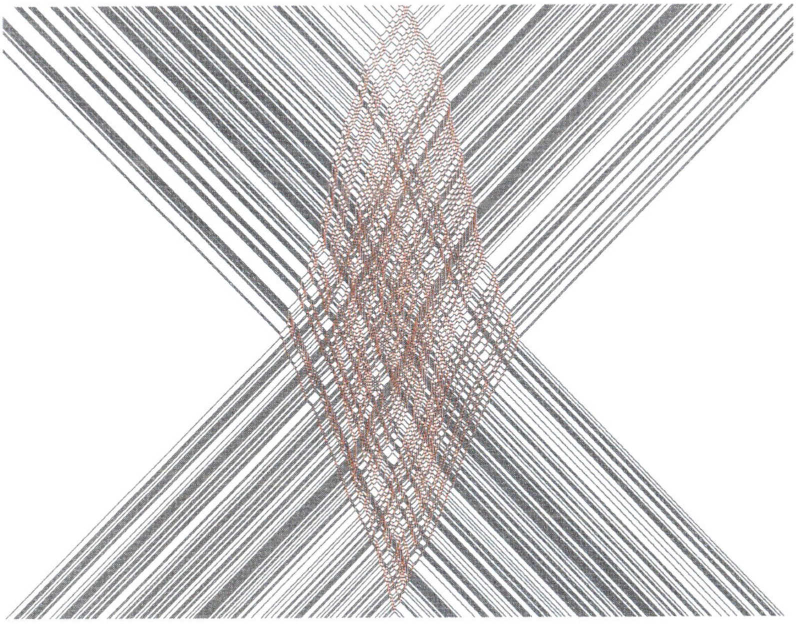

Figure 5. Head-on collision of a random (maximum entropy) sample of left movers with a random sample of right movers. The total width of each initial stream of left/right movers is 800 and the duration is $t_{\max} = 600$.

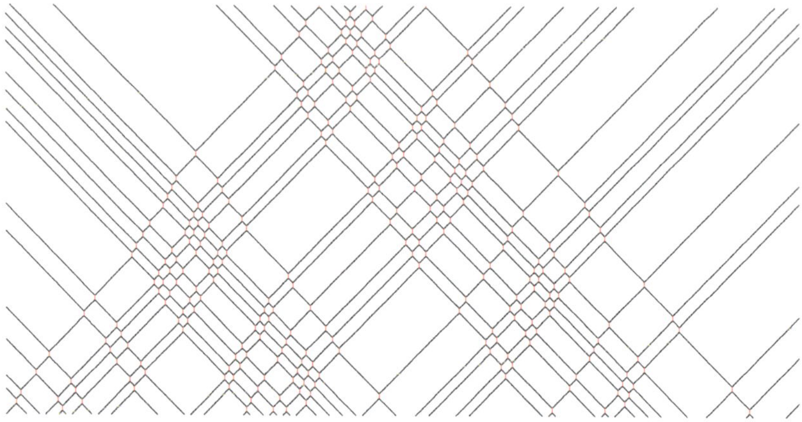

Figure 6. Trajectory starting from a random initial condition where only s_{6x} ($0 \leq x \leq 180$) are chosen at random, uniform in $\{0,1\}$, while all other $s_{x'} = 0$. Such a trajectory features only maximally short-lived field particles, which provide 'unit' phase shifts for left/right movers upon scattering, and corresponds to dynamics of negative-length hard-rods, such as, e.g., in the Rule 54 system [13].

2.2. *Time Evolution over Abelian Algebra of Local Observables*

Let us now reinterpret the dynamics of the matter–field automaton in terms of an abelian C^*-algebraic dynamical system, and then study its conservation laws. We begin by defining a quasilocal algebra of observables, i.e., real functions over space of configurations $\{0,1\}^{\mathbb{Z}}$. For more details on a fully analogous but more detailed construction in a related model of Rule 54 reversible cellular automaton, see ref. [14]. Let $[0]_{2x}, [1]_{2x}$ define a local basis of observables, checking if $s_{2x} = 0$ (or 1) for the matter degrees of freedom, and correspondingly $[0]_{2x+1}, [1]_{2x+1}$ for the field:

$$[r]_x(\underline{s}) = \delta_{r,s_x}. \tag{2}$$

We have $[0]_x + [1]_x = \mathbb{1}$ (unit element in the algebra of observables), and denote $[\bullet]_x = [1]_x - [0]_x$, for x either even or odd. We note that $[0]_x, [1]_x$ (or $\mathbb{1}, [\bullet]_x$) generate the entire (quasi)local algebra under multiplication and addition (and closure). We denote local observables with larger support as $[s_0, s_1, \ldots, s_r]_x = [s_0]_x[s_1]_{x+1}\cdots[s_r]_{x+r}$.

The fundamental three site operator—local propagator—in the algebra of observables implementing the rules

$$\begin{aligned}
[000]_{2x} &\longleftrightarrow [000]_{2x}, \\
[001]_{2x} &\longleftrightarrow [100]_{2x}, \\
[011]_{2x} &\longleftrightarrow [110]_{2x}, \\
[101]_{2x} &\longleftrightarrow [010]_{2x}, \\
[111]_{2x} &\longleftrightarrow [111]_{2x},
\end{aligned} \tag{3}$$

is encoded in terms of the 8×8 permutation matrix

$$Y = \begin{pmatrix}
1 & 0 & 0 & 0 & 0 & 0 & 0 & 0 \\
0 & 0 & 0 & 0 & 1 & 0 & 0 & 0 \\
0 & 0 & 0 & 0 & 0 & 1 & 0 & 0 \\
0 & 0 & 0 & 0 & 0 & 0 & 1 & 0 \\
0 & 1 & 0 & 0 & 0 & 0 & 0 & 0 \\
0 & 0 & 1 & 0 & 0 & 0 & 0 & 0 \\
0 & 0 & 0 & 1 & 0 & 0 & 0 & 0 \\
0 & 0 & 0 & 0 & 0 & 0 & 0 & 1
\end{pmatrix}. \tag{4}$$

The full time step of dynamical automorphism on the algebra of observables $a^t(\underline{s}) \equiv a(\underline{s}^{-t})$,

$$a^{2(t+1)} = Ua^{2t}, \qquad a^{2t+1} = U_e a^{2t}, \quad a^{2t+2} = U_o a^{2t+1}, \tag{5}$$

is constructed as follows

$$U = U_o U_e, \tag{6}$$

where even and odd time steps are generated as

$$U_e = \prod_x Y_{4x,4x+1,4x+2}, \quad U_o = \prod_x Y_{4x-2,4x-1,4x}, \tag{7}$$

where three indices denote the positions in the string, or in the tensor product of local algebras ($\bigotimes_x \text{span}\{[0]_x, [1]_x\}$), where the three site operator (4) acts nontrivially.

2.3. Conservation Laws

Let $a_{[x,y]}$ be any local observable supported in the interval of sites $[x,y] \subset \mathbb{Z}$. We define a shift automorphism Π

$$\Pi a_{[x,y]} = a_{[x+2,y+2]}, \tag{8}$$

which shifts an observable by 'one lattice unit', so that the identity of matter/field sites is preserved. Note that

$$\Pi U^e = U^o \Pi, \quad \Pi U^o = U^e \Pi, \quad [\Pi^2, U^e] = 0, \quad [\Pi^2, U^o] = 0. \tag{9}$$

A local observable q is called a conserved charge if, together with an appropriate conserved current p, it satisfies a conservation law (space–time discrete continuity equation):

$$U_e q - U_o q + \Pi p - \Pi^{-1} p = 0, \tag{10}$$

where we note that

$$U_o^2 = U_e^2 = \mathbb{1}, \tag{11}$$

since $Y^2 = \mathbb{1}$. Then, extensive local observable is globally conserved

$$Q = \sum_{x \in \mathbb{Z}} q_x, \quad q_x := \Pi^{2x} q, \qquad UQ = Q. \tag{12}$$

Importantly, it holds in general that if (q,p) is a local conservation law, then $(\Pi q, -\Pi p)$ is a conservation law as well. This can be proven by multiplying Equation (10) by Π and using (9).

Limiting to local observables on clusters of five sites (beginning and ending with a matter site), covering one 'unit cell' (periodicity block of 4 sites) of the automaton, we find (using exact computer algebra) exactly two inequivalent conservation laws

$$q_x^{(1)} = [\bullet]_{4x} - [\bullet]_{4x+2}, \tag{13}$$

$$p_x^{(1)} = \frac{1}{2}[\bullet]_{4x} + \frac{1}{2}[\bullet]_{4x+1} + \frac{1}{2}[\bullet]_{4x+2} + \frac{1}{2}[\bullet]_{4x}[\bullet]_{4x+1}[\bullet]_{4x+2}, \tag{14}$$

$$q_x^{(2)} = [\bullet]_{4x+1} + [\bullet]_{4x+2} + [\bullet]_{4x+3}, \tag{15}$$

$$p_x^{(2)} = -[\bullet]_{4x+2}, \tag{16}$$

with local densities and current supported on up to three consecutive sites. $q_x^{(1)}$ can be interpreted as the *momentum* density, while $q_x^{(2)}$ is the analog of the *energy* density. Indeed, writing $P = Q^{(1)} = \sum_x q_x^{(1)}$, $E = Q^{(2)} + \Pi Q^{(2)} = \sum_x (q_x^{(2)} + \Pi q_x^{(2)})$ we have intuitive expressions for the total momentum and total energy of a configuration:

$$P(\underline{s}) = \sum_x (-1)^x s_{2x}, \quad E(\underline{s}) = \sum_x (s_{2x} + 2s_{2x+1}), \tag{17}$$

meaning that the field particle carries no momentum and has an energy content (rest energy) equal to the energy of two matter particles. Note that a pair (P, E) is equivalent to a pair $(Q^{(1)}, Q^{(2)})$, since $\Pi Q^{(1)} = \sum_x \Pi q_x^{(1)} = -P$ and $Q^{(2)} - \Pi Q^{(2)} = \sum_x (q_x^{(2)} - \Pi q_x^{(2)}) = -\sum_x q_x^{(1)} = -P$.

If we then increase the support size to clusters of nine $(2 \times 4 + 1)$ consecutive sites, we obtain precisely one new additional conservation law:
(using compact notation: $[\circ] \equiv \mathbb{1}$, $[r_0 r_1 r_2 \ldots]_{4x} \equiv [r_0 r_1 r_2 \ldots]$)

$$\begin{aligned}
q^{(3)} = {} & 2[\circ\circ\circ\circ\circ\circ\bullet\circ\bullet] - 2[\circ\circ\circ\circ\circ\circ\bullet\bullet\circ] - 4[\circ\circ\circ\circ\circ\bullet\circ\circ\bullet] - 2[\circ\circ\circ\circ\circ\bullet\bullet\circ\circ] \\
& + 2[\circ\circ\circ\circ\circ\bullet\bullet\bullet\bullet] - 8[\circ\circ\circ\circ\bullet\circ\circ\circ\bullet] - 4[\circ\circ\circ\circ\bullet\circ\circ\bullet\circ] + 2[\circ\circ\circ\circ\bullet\circ\bullet\circ\circ] \\
& - 2[\circ\circ\circ\circ\bullet\circ\bullet\bullet\bullet] + 4[\circ\circ\circ\circ\bullet\bullet\circ\bullet\bullet] - 2[\circ\circ\circ\circ\bullet\bullet\bullet\circ\bullet] + 2[\circ\circ\circ\circ\bullet\bullet\bullet\bullet\circ] \\
& - 2[\circ\circ\circ\bullet\circ\circ\circ\circ\bullet] - 2[\circ\circ\circ\bullet\circ\circ\circ\bullet\circ] + 2[\circ\circ\circ\bullet\circ\circ\bullet\circ\circ] - 2[\circ\circ\circ\bullet\circ\circ\bullet\bullet\bullet] \\
& - 4[\circ\circ\circ\bullet\circ\bullet\circ\circ\circ] + 2[\circ\circ\circ\bullet\bullet\bullet\circ\circ\bullet] + 2[\circ\circ\circ\bullet\bullet\bullet\circ\bullet\circ] - 2[\circ\circ\circ\bullet\bullet\bullet\bullet\circ\circ] \\
& + 2[\circ\circ\circ\bullet\bullet\bullet\bullet\bullet\bullet] + [\circ\circ\bullet\circ\circ\circ\circ\circ\bullet] + [\circ\circ\bullet\circ\circ\circ\circ\bullet\circ] - [\circ\circ\bullet\circ\circ\circ\bullet\circ\circ] \\
& + [\circ\circ\bullet\circ\circ\circ\bullet\bullet\bullet] + 2[\circ\circ\bullet\circ\circ\bullet\circ\circ\circ] - [\circ\circ\bullet\circ\bullet\bullet\circ\circ\bullet] - [\circ\circ\bullet\circ\bullet\bullet\circ\bullet\circ] \\
& + [\circ\circ\bullet\circ\bullet\bullet\bullet\circ\circ] - [\circ\circ\bullet\circ\bullet\bullet\bullet\bullet\bullet] + [\circ\circ\bullet\bullet\circ\bullet\circ\circ\bullet] + [\circ\circ\bullet\bullet\circ\bullet\circ\bullet\circ] \\
& - [\circ\circ\bullet\bullet\circ\bullet\bullet\circ\circ] + [\circ\circ\bullet\bullet\circ\bullet\bullet\bullet\bullet] - [\circ\circ\bullet\bullet\bullet\circ\circ\circ\bullet] - [\circ\circ\bullet\bullet\bullet\circ\circ\bullet\circ] \\
& + [\circ\circ\bullet\bullet\bullet\circ\bullet\circ\circ] - [\circ\circ\bullet\bullet\bullet\circ\bullet\bullet\bullet] - 2[\circ\circ\bullet\bullet\bullet\bullet\circ\circ\circ] - [\circ\bullet\circ\circ\circ\circ\circ\circ\bullet] \\
& - [\circ\bullet\circ\circ\circ\circ\circ\bullet\circ] + [\circ\bullet\circ\circ\circ\circ\bullet\circ\circ] - [\circ\bullet\circ\circ\circ\circ\bullet\bullet\bullet] - 2[\circ\bullet\circ\circ\circ\bullet\circ\circ\circ] \\
& + [\circ\bullet\circ\circ\bullet\bullet\circ\circ\bullet] + [\circ\bullet\circ\circ\bullet\bullet\circ\bullet\circ] - [\circ\bullet\circ\circ\bullet\bullet\bullet\circ\circ] + [\circ\bullet\circ\circ\bullet\bullet\bullet\bullet\bullet] \\
& - [\circ\bullet\circ\bullet\circ\bullet\circ\circ\bullet] - [\circ\bullet\circ\bullet\circ\bullet\circ\bullet\circ] + [\circ\bullet\circ\bullet\circ\bullet\bullet\circ\circ] - [\circ\bullet\circ\bullet\circ\bullet\bullet\bullet\bullet] \\
& + [\circ\bullet\circ\bullet\bullet\circ\circ\circ\bullet] + [\circ\bullet\circ\bullet\bullet\circ\circ\bullet\circ] - [\circ\bullet\circ\bullet\bullet\circ\bullet\circ\circ] + [\circ\bullet\circ\bullet\bullet\circ\bullet\bullet\bullet] \\
& + 2[\circ\bullet\circ\bullet\bullet\bullet\circ\circ\circ] - [\bullet\circ\circ\circ\circ\circ\circ\circ\bullet] - [\bullet\circ\circ\circ\circ\circ\circ\bullet\circ] + [\bullet\circ\circ\circ\circ\circ\bullet\circ\circ] \\
& - [\bullet\circ\circ\circ\circ\circ\bullet\bullet\bullet] - 2[\bullet\circ\circ\circ\circ\bullet\circ\circ\circ] + [\bullet\circ\circ\circ\bullet\bullet\circ\circ\bullet] + [\bullet\circ\circ\circ\bullet\bullet\circ\bullet\circ] \\
& - [\bullet\circ\circ\circ\bullet\bullet\bullet\circ\circ] + [\bullet\circ\circ\circ\bullet\bullet\bullet\bullet\bullet] - [\bullet\circ\circ\bullet\circ\bullet\circ\circ\bullet] - [\bullet\circ\circ\bullet\circ\bullet\circ\bullet\circ] \\
& + [\bullet\circ\circ\bullet\circ\bullet\bullet\circ\circ] - [\bullet\circ\circ\bullet\circ\bullet\bullet\bullet\bullet] + [\bullet\circ\circ\bullet\bullet\circ\circ\circ\bullet] + [\bullet\circ\circ\bullet\bullet\circ\circ\bullet\circ] \\
& - [\bullet\circ\circ\bullet\bullet\circ\bullet\circ\circ] + [\bullet\circ\circ\bullet\bullet\circ\bullet\bullet\bullet] + 2[\bullet\circ\circ\bullet\bullet\bullet\circ\circ\circ] - [\bullet\bullet\bullet\circ\circ\circ\circ\circ\bullet] \\
& - [\bullet\bullet\bullet\circ\circ\circ\circ\bullet\circ] + [\bullet\bullet\bullet\circ\circ\circ\bullet\circ\circ] - [\bullet\bullet\bullet\circ\circ\circ\bullet\bullet\bullet] - 2[\bullet\bullet\bullet\circ\circ\bullet\circ\circ\circ] \\
& + [\bullet\bullet\bullet\circ\bullet\bullet\circ\circ\bullet] + [\bullet\bullet\bullet\circ\bullet\bullet\circ\bullet\circ] - [\bullet\bullet\bullet\circ\bullet\bullet\bullet\circ\circ] + [\bullet\bullet\bullet\circ\bullet\bullet\bullet\bullet\bullet] \\
& - [\bullet\bullet\bullet\bullet\circ\bullet\circ\circ\bullet] - [\bullet\bullet\bullet\bullet\circ\bullet\circ\bullet\circ] + [\bullet\bullet\bullet\bullet\circ\bullet\bullet\circ\circ] - [\bullet\bullet\bullet\bullet\circ\bullet\bullet\bullet\bullet] \\
& + [\bullet\bullet\bullet\bullet\bullet\circ\circ\circ\bullet] + [\bullet\bullet\bullet\bullet\bullet\circ\circ\bullet\circ] - [\bullet\bullet\bullet\bullet\bullet\circ\bullet\circ\circ] + [\bullet\bullet\bullet\bullet\bullet\circ\bullet\bullet\bullet] \\
& + 2[\bullet\bullet\bullet\bullet\bullet\bullet\circ\circ\circ]
\end{aligned} \tag{18}$$

with the current

$$
\begin{aligned}
p^{(3)} = {} & 2[\circ\circ\circ\circ\circ\bullet\bullet\circ\circ] - 4[\circ\circ\circ\circ\bullet\circ\circ\circ\bullet] - 4[\circ\circ\circ\circ\bullet\circ\circ\bullet\circ] + 2[\circ\circ\circ\circ\bullet\circ\bullet\circ\circ] \\
& -4[\circ\circ\circ\circ\bullet\circ\bullet\bullet\bullet] - 2[\circ\circ\circ\bullet\circ\circ\circ\circ\bullet] - 2[\circ\circ\circ\bullet\circ\circ\circ\bullet\circ] - 2[\circ\circ\circ\bullet\circ\circ\bullet\bullet\bullet] \\
& +2[\circ\circ\circ\bullet\bullet\bullet\circ\circ\bullet] + 2[\circ\circ\circ\bullet\bullet\bullet\circ\bullet\circ] + 2[\circ\circ\circ\bullet\bullet\bullet\bullet\bullet\bullet] + [\circ\circ\bullet\circ\circ\circ\circ\circ\bullet] \\
& +[\circ\circ\bullet\circ\circ\circ\circ\bullet\circ] + [\circ\circ\bullet\circ\circ\circ\bullet\bullet\bullet] - [\circ\circ\bullet\circ\bullet\bullet\circ\circ\bullet] - [\circ\circ\bullet\circ\bullet\bullet\circ\bullet\circ] \\
& -[\circ\circ\bullet\circ\bullet\bullet\bullet\bullet\bullet] + [\circ\circ\bullet\bullet\circ\bullet\circ\circ\bullet] + [\circ\circ\bullet\bullet\circ\bullet\circ\bullet\circ] + [\circ\circ\bullet\bullet\circ\bullet\bullet\bullet\bullet] \\
& -[\circ\circ\bullet\bullet\bullet\circ\circ\circ\bullet] - [\circ\circ\bullet\bullet\bullet\circ\circ\bullet\circ] - [\circ\circ\bullet\bullet\bullet\circ\bullet\bullet\bullet] - [\circ\bullet\circ\circ\circ\circ\circ\circ\bullet] \\
& -[\circ\bullet\circ\circ\circ\circ\circ\bullet\circ] - [\circ\bullet\circ\circ\circ\circ\bullet\bullet\bullet] + [\circ\bullet\circ\circ\bullet\bullet\circ\circ\bullet] + [\circ\bullet\circ\circ\bullet\bullet\circ\bullet\circ] \\
& +[\circ\bullet\circ\circ\bullet\bullet\bullet\bullet\bullet] - [\circ\bullet\circ\bullet\circ\bullet\circ\circ\bullet] - [\circ\bullet\circ\bullet\circ\bullet\circ\bullet\circ] - [\circ\bullet\circ\bullet\circ\bullet\bullet\bullet\bullet] \\
& +[\circ\bullet\circ\bullet\bullet\circ\circ\circ\bullet] + [\circ\bullet\circ\bullet\bullet\circ\circ\bullet\circ] + [\circ\bullet\circ\bullet\bullet\circ\bullet\bullet\bullet] - [\bullet\circ\circ\circ\circ\circ\circ\circ\bullet] \\
& -[\bullet\circ\circ\circ\circ\circ\circ\bullet\circ] - [\bullet\circ\circ\circ\circ\circ\bullet\bullet\bullet] + [\bullet\circ\circ\circ\bullet\bullet\circ\circ\bullet] + [\bullet\circ\circ\circ\bullet\bullet\circ\bullet\circ] \\
& +[\bullet\circ\circ\circ\bullet\bullet\bullet\bullet\bullet] - [\bullet\circ\circ\bullet\circ\bullet\circ\circ\bullet] - [\bullet\circ\circ\bullet\circ\bullet\circ\bullet\circ] - [\bullet\circ\circ\bullet\circ\bullet\bullet\bullet\bullet] \\
& +[\bullet\circ\circ\bullet\bullet\circ\circ\circ\bullet] + [\bullet\circ\circ\bullet\bullet\circ\circ\bullet\circ] + [\bullet\circ\circ\bullet\bullet\circ\bullet\bullet\bullet] - [\bullet\bullet\bullet\circ\circ\circ\circ\circ\bullet] \\
& -[\bullet\bullet\bullet\circ\circ\circ\circ\bullet\circ] - [\bullet\bullet\bullet\circ\circ\circ\bullet\bullet\bullet] + [\bullet\bullet\bullet\circ\bullet\bullet\circ\circ\bullet] + [\bullet\bullet\bullet\circ\bullet\bullet\circ\bullet\circ] \\
& +[\bullet\bullet\bullet\circ\bullet\bullet\bullet\bullet\bullet] - [\bullet\bullet\bullet\bullet\circ\bullet\circ\circ\bullet] - [\bullet\bullet\bullet\bullet\circ\bullet\circ\bullet\circ] - [\bullet\bullet\bullet\bullet\circ\bullet\bullet\bullet\bullet] \\
& +[\bullet\bullet\bullet\bullet\bullet\circ\circ\circ\bullet] + [\bullet\bullet\bullet\bullet\bullet\circ\circ\bullet\circ] + [\bullet\bullet\bullet\bullet\bullet\circ\bullet\bullet\bullet].
\end{aligned} \tag{19}
$$

If we furthermore increase this to a 3-cell cluster of $13 = 3 \times 4 + 1$ consecutive sites, we obtain only one additional conservation law which, however, is trivially related to the previous one:

$$q_x^{(4)} = \Pi q_x^{(3)}, \qquad p_x^{(4)} = -\Pi p_x^{(3)}. \tag{20}$$

Scanning for the existence of higher conservation laws with densities supported over four or more unit cells seems impossible with brute-force computer algebra.

Furthermore, note that the conservation law property (10) is invariant under local space–time discrete gauge transformation, i.e., taking any local observable a

$$
\begin{aligned}
q &\longrightarrow q + \Pi a - \Pi^{-1} a, && (21) \\
p &\longrightarrow p - U_{\mathrm{e}} a + U_{\mathrm{o}} a. && (22)
\end{aligned}
$$

In the explicit expressions listed above, the gauge has been fixed by right-alignment of the densities.

One might expect that matter–field automaton hosts an infinite tower of local conserved quantities of increasing support size, however, these may be very difficult to find empirically. This hypothesis would suggest also that the matter–field automaton may be a completely integrable system, in a similar spirit as Rule 54 [15], as it also displays similar negative-length hard-rod dynamics for a class of initial data, but preliminary attempts to find a Yang–Baxter or Lax structure failed. The other option is that the matter–field system has only a few, perhaps finitely many (or an incomplete set) conserved local charges, and would then represent a paradigmatic theory between integrable and ergodic dynamics.

Deterministic cellular automata on discrete phase space (space of configurations) often admit continuous, single (or few) parameter modifications, which render the models quantum or stochastic (depending on the type of deformation which turns a deterministic evolution to a unitary or Markov map). For instance, in the definition of the matter–field automaton, one may in each step allow for the triple (010) to remain (010) with probability α and to map to (101) with probability $1 - \alpha$. Similarly, (101) remains in (101) with probability β and maps to (010) with probability $1 - \beta$. All the definitions of the previous section remain valid, except that the permutation matrix Y is now replaced by a stochastic (Markov) matrix:

$$Y_{\text{stoch}} = \begin{pmatrix} 1 & 0 & 0 & 0 & 0 & 0 & 0 & 0 \\ 0 & 0 & 0 & 0 & 1 & 0 & 0 & 0 \\ 0 & 0 & \alpha & 0 & 0 & 1-\beta & 0 & 0 \\ 0 & 0 & 0 & 0 & 0 & 0 & 1 & 0 \\ 0 & 1 & 0 & 0 & 0 & 0 & 0 & 0 \\ 0 & 0 & 1-\alpha & 0 & 0 & \beta & 0 & 0 \\ 0 & 0 & 0 & 1 & 0 & 0 & 0 & 0 \\ 0 & 0 & 0 & 0 & 0 & 0 & 0 & 1 \end{pmatrix}. \tag{23}$$

Interestingly, the first two conservation laws, the momentum and energy (17), see also (13), (14), survive the stochastic deformation, while, as indicated by computer algebra, the third conservation law $(q^{(3)}, p^{(3)})$ is *broken*. Hence, the deformation of the matter–field automaton likely no longer posses nontrivial ergodicity breaking.

3. Deformed (Quantized) Hardpoint Lattice Gas

Nevertheless, we show briefly in this section how continuous deformations of some other reversible deterministic cellular automata can lead to interesting non-ergodic dynamics. For this purpose, we use the so-called charged hardpoint lattice gas [10] (also referred to as XXC model in [16] since it is a deterministic limit of models introduced and studied by Maassarani [17]).

In comparison to the matter–field automaton, the model has a richer local configuration space, i.e., three states per site, and a simpler local interaction map, applying only to a pair of neighbouring sites instead of three. In physical terms, the model describes the dynamics of free point particles which move with speeds ± 1, with initial positions placed at commensurate (multiples of integer) coordinates, but which may carry an internal (say binary) degree of freedom (charge). Denoting the three local states as $s \in \{0, +, -\}$, with 0 referred to as a vacancy, the local reversible and deterministic rule, which we apply sequentially to pairs $(2x, 2x+1)$ and $(2x+1, 2x)$, reads

$$(00) \longleftrightarrow (00), \quad (0\alpha) \longleftrightarrow (\alpha 0), \quad (\alpha\beta) \longleftrightarrow (\alpha\beta), \quad \alpha, \beta \in \{+, -\}. \tag{24}$$

The model allows for a remarkable set of exact results on equilibrium and non-equilibrium dynamics, ranging from dynamical correlations in equilibrium and the transport coefficients [10], to an analytic *matrix-product-state* description of time-dependent quenches and non-equilibrium steady states in a boundary driven setup [11], to a recent exact solution of full counting statistics [12]. One may wonder if the model still allows for some degree of solvability, if hard-core collision is relaxed to a general stochastic, or unitary scattering

$$(+-) \longrightarrow u_{11}(+-) + u_{12}(-+), \qquad (-+) \longrightarrow u_{21}(+-) + u_{22}(-+), \tag{25}$$

where u may be a 2×2 stochastic, or unitary matrix. In the latter case, we think of the quantum lattice gas model and quantum superposition states as elaborated precisely below. We note that this system is closely related to a semiclassically quantized sine-Gordon model, recently studied in [18].

Consider a local Hilbert space $\mathcal{H}_1 = \mathbb{C}^3$, with states $|s\rangle_{s=0,+,-}$. We refer to states $|\pm\rangle$ as *charged particles* and we denote them by the Greek index $|\alpha\rangle_{\alpha=\pm}$, and to the state $|0\rangle$ as a *vacancy*. Let us define a local propagator over $\mathcal{H}_1 \otimes \mathcal{H}_1$ as

$$\begin{aligned} U[u] &= |00\rangle\langle 00| + \sum_{\alpha\in\{\pm\}} (|0\alpha\rangle\langle\alpha 0| + |\alpha 0\rangle\langle 0\alpha| + |\alpha\alpha\rangle\langle\alpha\alpha|) + \\ &+ u_{11}|+-\rangle\langle+-| + u_{12}|+-\rangle\langle-+| + u_{21}|-+\rangle\langle+-| + u_{22}|-+\rangle\langle-+|, \end{aligned} \tag{26}$$

where u can be in principle any invertible matrix, $u \in GL(2)$, for the properties that will be discussed below, while for quantum physics applications we will think of u as a unitary mixing matrix. Such $U[u]$ satisfies two remarkable identities, valid for any u:

$$U_{12}[u]U_{23}[\mathbb{1}]U_{12}[u^{-1}] = U_{23}[u^{-1}]U_{12}[\mathbb{1}]U_{23}[u], \tag{27}$$
$$U[u]U[u^{-1}] = \mathbb{1}, \tag{28}$$

which can be checked by straightforward computation. Note that identity (27) bears some similarity to the braid group form of the Yang–Baxter equation if u is considered as a non-abelian spectral parameter. Specifically, $U[\mathbb{1}]$ obeys the braid relation—it constitutes the so-called Yang–Baxter map—and provides a deterministic update rule for the charged hardpoint lattice gas. However, $U[u]$ does not have the full Yang–Baxter property, but ust enough to render the existence of an infinite set of conserved local operators, as shown below.

For simplicity we can now assume a finite system of an even number of sites L and define its many-body Hilbert space as a tensor product $\mathcal{H} = \mathcal{H}_1^{\otimes L}$. Then, we define a locality-preserving discrete time dynamical system (a quantum cellular automaton) over it, in terms of a brickwork circuit, i.e.,

$$\mathcal{U} = \mathcal{U}_{\rm o}\mathcal{U}_{\rm e}, \qquad \mathcal{U}_{\rm e} = \prod_{x=1}^{L/2} U_{2x-1,2x}[u], \qquad \mathcal{U}_{\rm o} = \prod_{x=1}^{L/2} U_{2x,2x+1}[u], \quad L+1 \equiv 1, \tag{29}$$

and write the evolution of observables as

$$a^{2t} = \mathcal{U}_{\rm e} a^{2t-1}\mathcal{U}_{\rm e}^{-1}, \quad a^{2t+1} = \mathcal{U}_{\rm o} a^{2t}\mathcal{U}_{\rm o}^{-1}. \tag{30}$$

Let us define the following pair of local operators (note periodic boundary conditions if needed):

$$\begin{aligned} g_x^{(+)} &= U_{x,x+1}[u]U_{x-1,x}U[\mathbb{1}]U_{x,x+1}[u^{-1}], \\ g_x^{(-)} &= U_{x-1,x}[u]U_{x,x+1}U[\mathbb{1}]U_{x-1,x}[u^{-1}]. \end{aligned} \tag{31}$$

Observing identities (27) and (28), we can straightforwardly derive a few remarkable properties of these operators:

$$\mathcal{U}_{\rm e} g_{2x}^{(+)}\mathcal{U}_{\rm e}^{-1} = g_{2x+1}^{(+)}, \qquad \mathcal{U}_{\rm o} g_{2x-1}^{(+)}\mathcal{U}_{\rm o}^{-1} = g_{2x}^{(+)}, \tag{32}$$
$$\mathcal{U}_{\rm e} g_{2x}^{(-)}\mathcal{U}_{\rm e}^{-1} = g_{2x-1}^{(-)}, \qquad \mathcal{U}_{\rm o} g_{2x+1}^{(-)}\mathcal{U}_{\rm o}^{-1} = g_{2x}^{(-)}, \tag{33}$$

which justify the term glider operators [19] (see Figure 7 for a simple graphical proof of the properties (32) and (33) in terms of quantum circuit representation). We note that our gliders would be trivial $g_x^{(\pm)} = \mathbb{1}$ for *integrable trotterization* [20] where $U(\mathbb{1}) = \mathbb{1}$.

These are particular kinds of quantum conservation laws, where the formal charge and current densities are identical. Consequently, we can build products of such operators, separated by at least two pairs of sites, which again behave in the same way

$$g_{x_1,x_2\ldots x_m}^{(\pm)} = \prod_{n=1}^{m} g_{2x_n}^{(\pm)}, \quad \mathcal{U} g_{x_1,x_2\ldots x_m}^{(\pm)}\mathcal{U}^{-1} = g_{x_1\pm1,x_2\pm1\ldots x_m\pm1}^{(\pm)} \tag{34}$$

under constraints $x_{j+1} - x_j \geq 2$. We thus have a large set of extensive charges

$$Q_{r_1,r_2\ldots r_m}^{(\pm)} = \sum_{x=1}^{L/2} g_{x,x+r_1\ldots x+r_m}^{(\pm)}, \quad r_1 \geq 2,\ r_{m+1} - r_m \geq 2, \qquad Q_{\underline{r}}^{(\pm)} \equiv \mathcal{U} Q_{\underline{r}}^{(\pm)}\mathcal{U}^{-1}. \tag{35}$$

These conserved operators concisely encapsulate the non-ergodic properties of the deformed hardpoint lattice gas. The non-ergodicity should not be considered surprising, as the pattern of left and right movers is evolving regularly according to ballistic (free) dynamics. However, the internal (charge) degrees of freedom undergo interacting dynamics, which is nontrivially constrained by conserving $Q_{\underline{r}}^{\pm}$.

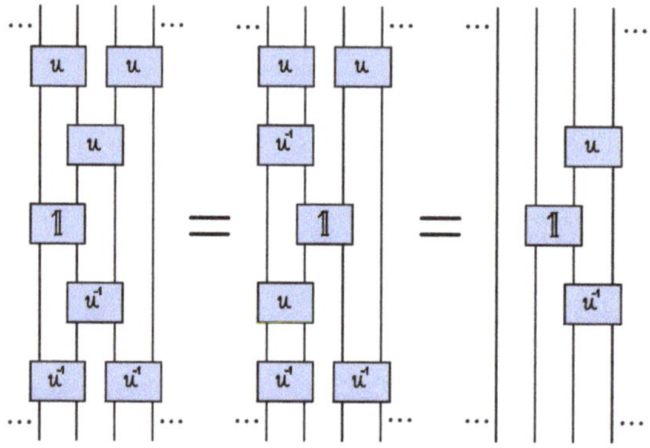

Figure 7. Diagrammatic proof of the glider property (32) in terms of equivalence of quantum circuits, where gates $U[u]$, $U[\mathbb{1}]$, and $U[u^{-1}]$ are indicated by blue boxes. The proof of (33) is analogous (via left–right reflection of the above diagrams).

The fact that the number of distinct glider charges $Q_{\underline{r}}^{(\pm)}$ increases exponentially with the size of support, say ℓ such that all $r_j \leq \ell$, is reminiscent of superintegrable quantum cellular automata [13,15,16], yet the absence of an exact Yang–Baxter property suggests that the dynamical system is not completely integrable (i.e., there should be no Bethe ansatz!?). We are thus suggesting a new class of non-ergodic many-body quantum dynamics.

4. Conclusions

The dynamical systems introduced and discussed in this short contribution have been constructed for the sole amusement of the author. Yet, we hope these constructions may at some point find some nontrivial applications in the context of non-equilibrium statistical mechanics. Specifically, one of the long-standing key goals in the field is to derive irreversible macroscopic transport laws, such as Fick's law of diffusion or Fourier's law of heat transport, from microscopic reversible equations of motion. This question has also been among the main research focuses of Giulio Casati [21], as well as stimulating the beginning of my research career [22], and some of our joint work investigated Fourier's law in systems very similar to the models studied here: hardpoint colliding masses in one dimension [23].

The coexistence of diffusive and ballistic transport has been rigorously demonstrated in the Rule 54 automaton (reviewed in ref. [13], see also ref. [24]), and one may hope that such results can be extended to the matter–field automaton, which seems richer and perhaps more generic (for example, Rule 54 dynamics is superintegrable with an exponentially growing number of conserved local charges, while the matter–field automaton is perhaps at most integrable, as it has much rarer nontrivial local charges).

Funding: This work has been supported by the Slovenian Research Agency (ARRS) under the Program P1-0402 and grants N1-0233 and N1-0219.

Data Availability Statement: All data shown or used in this paper have been generated by *Mathematica* encoding the procedures explicitly described above.

Acknowledgments: I acknowledge enjoying numerous discussions with Balazs Pozsgay on related problems.

Conflicts of Interest: The authors declare no conflicts of interest.

References

1. Casati, G.; Prosen, T. Mixing property of triangular billiards. *Phys. Rev. Lett.* **1999**, *83*, 4729–4732. [CrossRef]
2. Casati, G.; Prosen, T. Triangle map: A model of quantum chaos. *Phys. Rev. Lett.* **2000**, *85*, 4261–4264. [CrossRef] [PubMed]
3. Casati, G.; Prosen, T. Nonergodicity and localization of invariant measure for two colliding masses. *Phys. Rev. E* **2014**, *89*, 042918.
4. Lozej, Č.; Casati, G.; Prosen, T. Quantum chaos in triangular billiards. *Phys. Rev. Res.* **2022**, *4*, 013138. [CrossRef]
5. Cornfeld, I.P.; Fomin, S.V.; Sinai, Y.G. *Ergodic Theory*; Springer: New York, NY, USA, 1982.
6. D'Alessio, L.; Kafri, Y.; Polkovnikov, A.; Rigol, M. From quantum chaos and eigenstate thermalization to statistical mechanics and thermodynamics. *Adv. Phys.* **2016**, *65*, 239–362. [CrossRef]
7. Abanin, D.A.; Altman, E.; Bloch, I.; Serbyn, M. Many-body localization, thermalization, and entanglement. *Rev. Mod. Phys.* **2019**, *91*, 021001. [CrossRef]
8. Moudgalya, S.; Bernevig, B.A.; Regnault, N. Quantum many-body scars and Hilbert space fragmentation: A review of exact results. *Rep. Prog. Phys.* **2022**, *85*, 086501. [CrossRef]
9. Spohn, H. Hydrodynamic scales of integrable many-particle systems. *arXiv* **2023**, arXiv:2301.08504.
10. Medenjak, M.; Klobas, K.; Prosen, T. Diffusion in Deterministic Interacting Lattice Systems. *Phys. Rev. Lett.* **2017**, *119*, 110603. [CrossRef]
11. Medenjak, M.; Popkov, V.; Prosen, T.; Ragoucy, R.; Vanicat, M. Two-species hardcore reversible cellular automaton: Matrix ansatz for dynamics and nonequilibrium stationary state. *SciPost Phys.* **2019**, *6*, 074. [CrossRef]
12. Krajnik, Ž.; Schmidt, J.; Pasquier, V.; Ilievski, I.; Prosen, T. Exact Anomalous Current Fluctuations in a Deterministic Interacting Model. *Phys. Rev. Lett.* **2022**, *128*, 160601. [CrossRef] [PubMed]
13. Buča, B.; Klobas, K.; Prosen, T. Rule 54: Exactly solvable model of nonequilibrium statistical mechanics. *J. Stat. Mech.* **2021**, *7*, 074001. [CrossRef]
14. Klobas, K.; Medenjak, M.; Prosen, T.; Vanicat, M. Time-Dependent Matrix Product Ansatz for Interacting Reversible Dynamics. *Commun. Math. Phys.* **2019**, *371*, 651–688. [CrossRef]
15. Gombor, T.; Pozsgay, B. Integrable deformations of superintegrable quantum circuits. *arXiv* **2022**, arXiv:2205.02038.
16. Gombor, T.; Pozsgay, B. Superintegrable cellular automata and dual unitary gates from Yang-Baxter maps. *SciPost Phys.* **2022**, *12*, 102. [CrossRef]
17. Maassarani, Z. The XXC models. *Phys. Lett. A* **1998**, *244*, 160–164. [CrossRef]
18. Kormos, M.; Vörös, D.; Zaránd, G. Finite temperature dynamics in gapped 1D models in the sine-Gordon family. *Phys. Rev. B* **2022**, *106*, 205151. [CrossRef]
19. Gütschow, J.; Uphoff, S.; Werner, R.F.; Zimborás, Z. Time Asymptotics and Entanglement Generation of Clifford Quantum Cellular Automata. *J. Math. Phys.* **2010**, *51*, 015203. [CrossRef]
20. Vanicat, M.; Zadnik, L.; Prosen, T. Integrable Trotterization: Local Conservation Laws and Boundary Driving. *Phys. Rev. Lett.* **2018**, *121*, 030606. [CrossRef]
21. Casati, G.; Ford, J.; Vivaldi, F.; Visscher, W.M. One-Dimensional Classical Many-Body System Having a Normal Thermal Conductivity. *Phys. Rev. Lett.* **1984**, *52*, 1861. [CrossRef]
22. Prosen, T.; Robnik, M. Energy transport and detailed verification of Fourier heat law in a chain of colliding harmonic oscillators. *J. Phys. A Math. Gen.* **1992**, *25*, 3449. [CrossRef]
23. Casati, G.; Prosen, T. Anomalous heat conduction in a one-dimensional ideal gas. *Phys. Rev. E* **2003**, *67*, 015203(R). [CrossRef] [PubMed]
24. Gopalakrishnan, S.; Huse, D.A.; Khemani, V.; Vasseur, R. Hydrodynamics of operator spreading and quasiparticle diffusion in interacting integrable systems. *Phys. Rev. B* **2018**, *98*, 220303(R). [CrossRef]

entropy

MDPI

Article

Density of Avoided Crossings and Diabatic Representation

Anatoly E. Obzhirov [1,*] and Eric J. Heller [2,*]

1 Max Planck Institute for the Structure and Dynamics of Matter and Center for Free-Electron Laser Science, 22761 Hamburg, Germany
2 Department of Physics and Chemistry and Chemical Biology, Harvard University, Cambridge, MA 02138, USA
* Correspondence: anatoly.obzhirov@mpsd.mpg.de (A.E.O.); eheller@fas.harvard.edu (E.J.H.)

Abstract: Electronic structure theory describes the properties of solids using Bloch states that correspond to highly symmetrical nuclear configurations. However, nuclear thermal motion destroys translation symmetry. Here, we describe two approaches relevant to the time evolution of electronic states in the presence of thermal fluctuations. On the one hand, the direct solution of the time-dependent Schrodinger equation for a tight-binding model reveals the diabatic nature of time evolution. On the other hand, because of random nuclear configurations, the electronic Hamiltonian falls into the class of random matrices, which have universal features in their energy spectra. In the end, we discuss combining two approaches to obtain new insights into the influence of thermal fluctuations on electronic states.

Keywords: time evolution of quantum systems; structure of eigenstates and energy spectra; random matrix theory; semiclassical methods and results; atomic, molecular and solid-state systems

Citation: Obzhirov, A.E.; Heller, E.J. Density of Avoided Crossings and Diabatic Representation. *Entropy* **2023**, *25*, 751. https://doi.org/10.3390/e25050751

Academic Editor: Marko Robnik

Received: 31 March 2023
Revised: 2 May 2023
Accepted: 2 May 2023
Published: 4 May 2023

1. Introduction

Bloch's theorem and translational symmetry are two of the main bricks in the foundation of solid state physics. Their use in crystal lattices makes an implicit assumption, not emphasized in the texts we are aware of. Bloch's theorem is mathematically correct of course for a fixed, periodic potential, but real nuclei move under the influence of thermal fluctuations, so that they have random configuration without translational symmetry at any moment in time. This is clearest in the coherent state representation of the lattice, which assigns a mean position and momentum to each atom [1].

The success of solid state theory suggests there should be a physical justification for using electronic states corresponding to fixed highly symmetrical positions. Recently, it has been found that for graphene π-bands, thermal fluctuations turn out to be too fast for the adiabatic Born–Oppenheimer approximation (ABO) to be valid, so that the other limit—the diabatic limit (DBO)—holds [2].

If we define the Born–Oppenheimer approximation as supposing that the electron wavefunction returns to its starting form if the nuclei return to their starting positions after an arbitrary journey, this leaves the possibility of adiabatic and diabatic motion of the electron along the journey. We allow the caveat of possible Berry phases within the Born–Oppenheimer realm.

During thermal fluctuations, atomic positions have a random component for any arbitrary moment in time. The electronic Hamiltonian, which is a function of atomic positions, acquires this random character. Properties of simple prototype random matrix models are well studied in random matrix theory; some of them are universal and can be extended to more general cases. Here, we are interested in ensembles of parameter-dependent random matrices. In addition to level statistics, statistics of singularities in their spectrum including degeneracies and avoided crossings is known. We believe these results can be used to characterize electronic time evolution in the presence of phonons.

Indeed, the ABO breaks down when degeneracies occur; for the case of avoided crossings, Landau and Zener found that the probability of adiabatic dynamics depends on level spacing and sensitivity to perturbations. Random matrix theory combined with Landau–Zener theory makes it possible to characterize the effects of thermal fluctuations on electronic states.

The paper is organized as follows. In Section 2.1, we describe parameter-dependent random matrices; in Sections 2.2 and 2.3, we discuss statistical properties of singularities in their spectrum, namely, conical intersections and avoided crossings. In Section 3, we describe results on graphene thermal fluctuations that suggest diabatic representation. In Section 4, we sketch future directions for describing electronic dynamics in the presence of phonons using random matrix theory.

2. Singularities in the Spectra of Random Matrices

2.1. Parameter-Dependent Random Matrices

Random matrices provide a pathway to universal behavior for complex quantum systems [3]. One example of such a system is the chaotic dynamics of a kicked top [4,5]. In this paper, we consider only Gaussian ensembles, which are denoted by the Dyson index β [3], where $\beta = 1$ for GOE, $\beta = 2$ for GUE, and $\beta = 4$ for GSE. Corresponding matrices H_β can be represented as a combination of symmetrical S and antisymmetrical A matrices, whose elements are independent Gaussian distributed random variables with mean zero and variance $(1 \pm \delta_{ij})/\beta$:

$$H_1 = S, \quad H_2 = S + iA, \quad H_4 = Se_0 + A^1 e_1 + A^2 e_2 + A^3 e_3 \tag{1}$$

where e_i are 2×2 matrix representations of quaternion algebra bases:

$$e_0 = \begin{pmatrix} 1 & 0 \\ 0 & 1 \end{pmatrix}, e_1 = \begin{pmatrix} 0 & i \\ i & 0 \end{pmatrix}, e_2 = \begin{pmatrix} 0 & -1 \\ 1 & 0 \end{pmatrix}, e_3 = \begin{pmatrix} i & 0 \\ 0 & -i \end{pmatrix} \tag{2}$$

and the direct product is assumed for $A^n e_n$.

We study the cases when a Hamiltonian $\hat{H}$ depends on a set of parameters, which, in turn, depend on time. The idea of parameter-dependent random matrices first appears in [6] to quantify the dissipation rate of a driven complex quantum system. We use the following parametrization $H(x_1, x_2, \ldots, x_n)$ suggested by Wilkinson and Austin [7]

$$H(x_1, x_2, \ldots) = \cos x_1 H_1 + \sin x_1 H_2 + \cos x_2 H_3 + \sin x_2 H_4 + \ldots. \tag{3}$$

where matrices H_1, H_2, ... belong to the same ensemble and $\mathbf{x} = (x_1, x_2, \ldots, x_n)$. These matrices may also belong to different ensembles to study symmetry-breaking perturbations as Fritz Haake suggested [8–10].

This parametrization has a number of advantages. Distribution of matrix elements is the same for all $\mathbf{x}$, and matrices H and $\partial H/\partial x_j$ are independent, therefore leading to rather simple analytical averaging within the degenerate perturbation theory; the extension to the cases when $\partial H/\partial X_1$ and $\partial H/\partial X_2$ have non-zero correlation was made by Wilkinson [11,12]. Equation (3) provides an ergodic property [13]—averaging over energy levels leads to the same level spacing distribution as averaging over parameter space for a given pair of levels—enabling, for example, geometrical considerations to determine scaling laws for the density of avoided crossings [11].

An alternative to Equation (3) is the linear parametrization

$$H(x_1, x_2, \ldots) = H + H_1 x_1 + H_2 x_2 + \ldots. \tag{4}$$

However, now singularities in the spectrum happen only in the vicinity of $\mathbf{x} = 0$, and the distribution of eigenvalue curvatures $\partial^2 \epsilon_i / \partial x_j^2$ tends to zero for $x_j \to \infty$.

2.2. Geometrical Properties of Conical Intersections

For a Hamiltonian $\hat{H}$ to have a doubly degenerate energy level, it is necessary to satisfy two independent conditions. We reproduce here the argument of Teller [14]. Let us assume that all but two of the $\hat{H}$ eigenstates have been found. Expressing $\hat{H}$ in the basis formed by these eigenstates and two arbitrary states i and j, forming altogether a complete set of functions, the condition for the degeneracy is

$$H_{ij} = 0, H_{ii} = H_{jj}. \tag{5}$$

For ensemble β, it leads to a system of $\beta + 1$ linear equations. In two-dimensional parameter space, the levels are degenerate in isolated points called conical intersections.

The geometrical properties of conical intersections have been established in a number of papers. Longuett-Higgins found that if a wavefunction changes sign when transported adiabatically round a given loop in parameter space, then the state must become degenerate with another one at some point within the loop [15]. This topological property is one consequence of the Berry phase [16].

The distribution of conical intersections in the parameter space is characterized by density $\mathcal{D}_n^{(\mathrm{ci})}$, which is the number of points, in which energy levels n and $n+1$ are degenerate, per unit area. Wilkinson and Austin proposed the idea for its analytical derivation [11]. For a small element dA of the parameter space, the probability dp to find a conical intersection is $\mathcal{D}_n^{(\mathrm{ci})} dA$. At the same time, dp is determined by the distribution $P[R]$ of the distances R from a random point to the nearest conical intersection

$$dp = P[R]dR = \mathcal{D}_n^{(\mathrm{ci})} dA. \tag{6}$$

Hence, the problem is reduced to analytical calculation of $P[R]$. It is possible to express R by using the degenerate perturbation theory. Averaging over the corresponding ensemble leads to well-known integrals [17–19]. $\mathcal{D}_n^{(\mathrm{ci})}$ depends on the density of states $\rho_n = \rho(E_n)$ and the variance of the off-diagonal matrix elements σ^2 and has the general form $C_\beta(\rho\sigma)^{\beta+1}$:

$$\mathcal{D}_n^{(\mathrm{ci})} = \frac{\pi}{3}\rho_n^2\sigma^2,\ \mathcal{D}_n^{(\mathrm{ci})} = \frac{2\sqrt{\pi}}{3}\rho_n^3\sigma^3,\ \mathcal{D}_n^{(\mathrm{ci})} = \frac{16\sqrt{2}\pi^{3/2}}{45}\rho_n^5\sigma^5 \tag{7}$$

for GOE [11], GUE, and GSE [20], respectively. We note that prefactors C_β were found only after corresponding scaling had been revealed in numerical calculations [21,22]. Such an order of discoveries is typical in this field, for example, the exact distribution for energy level curvatures for GUE was found only after numerical experiments [23–26].

By putting the Wigner semicircle distribution [3] in Equation (7) and integrating the resulting expression over energy E, one gets the density of degeneracies summed over all pairs of levels. Multiplied by the volume of the parameter space, this gives the total number of degeneracies $\mathcal{N}$. For $M \times M$ GOE matrices parametrized as Equation (3), Wilkinson and Austin found [11]

$$\mathcal{N} = \frac{\pi}{2}M^2. \tag{8}$$

For GOE-parametrized matrices, the potential energy surfaces $E_n(x_1, x_2)$ and $E_{n+1}(x_1, x_2)$ form a cone near conical intersections. The corresponding energy level lines are ellipses. Wilkinson and Austin [11] calculated the distribution of eccentricities and cone slopes. We illustrate them in Figure 1, which is taken from their paper. We note that ellipses are randomly oriented and mostly elongated in accordance with the distribution

$$P[e] = \left(\frac{2e}{2-e^2}\right)^3 \tag{9}$$

coming from the joint probability distribution of slopes of the cone s and eccentricity e:

$$P[t,d] = \frac{d}{256\sigma^6}e^{-t/8\sigma^2}, \tag{10}$$

where d and t unambiguously determine e and s [11].

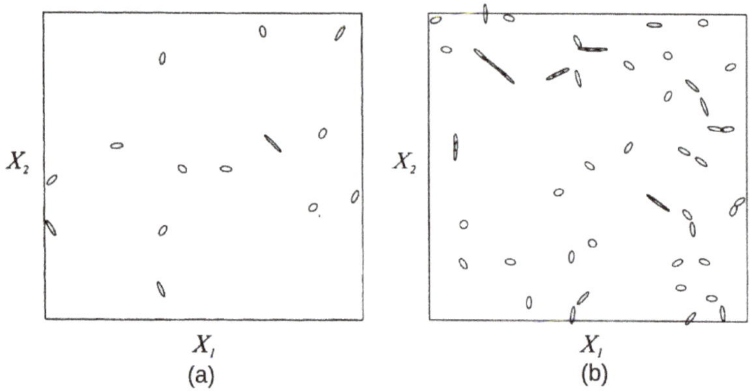

Figure 1. Energy level lines near the corresponding conical intersections between (**a**) the 1st and 2nd and (**b**) the 9th and the 10th energy levels for a 20×20 GOE-parametrized matrix. We note that ellipses of nearby conical intersections are correlated but still degeneracies do not exhibit repulsion. According to Equation (8), the total anticipated number of degeneracies is $\frac{\pi}{2}M^2 = 628$; in real calculations [11], it is 564. (Adapted with permission from ref. [11]. Copyright 1993 American Institute of Physics.)

One might expect that, similar to the spacings between adjacent energy levels, conical intersections between the same pair of levels would tend to avoid each other. It turns out that this is not the case for the GOE-parametrized model, Equation (3); there, degeneracies can be considered as randomly distributed points [11]. However, as one can see on Figure 1, their eccentricities and orientations are correlated.

2.3. Geometrical Properties of Avoided Crossings

Consider the case when the Hamiltonian depends on only one parameter x. This is one of a wide range of systems with an arbitrary large number of time-dependent parameters. The noncrossing rule follows from Equation (5) [27]; when x is varied, two energy levels never cross. Still, they can approach each other with an arbitrary small spacing. A local minimum in the distance between adjacent energy levels is called an avoided crossing. We refer to the difference between energy levels at the minimum as a gap. The distribution of crossing gaps is known for the GOE [28–30], GUE, and GSE [29,30]. Surprisingly, the ratio of mean crossing to mean spacing is not universal and varies from 0.42 to 0.52 for different realizations of the same ensemble β [30]. Furthermore, known as level repulsion, the noncrossing rule has been generalized to non-hermitian Hamiltonians that describe dissipative systems [31].

Energy profiles near an avoided crossing have the hyperbolic form

$$E_{n,n+1} = B(x - x_0) \pm \frac{1}{2}\sqrt{\epsilon^2 + A^2(x - x_0)^2} \tag{11}$$

which is a solution for the Hamiltonian

$$H(x) = \begin{pmatrix} (B + A/2)(x - x_0) & \epsilon/2 \\ \epsilon/2 & (B - A/2)(x - x_0) \end{pmatrix}. \tag{12}$$

When the Hamiltonian depends on a set of time-dependent parameters, such avoided crossings appear in the vicinity of degeneracies (the conical intersection geometry). Using this property together with ergodicity, Wilkinson and Austin have shown in a simple manner that the number of avoided crossings with gaps less than Δ is proportional to Δ in the limit $\Delta \to 0$ [11].

The density of avoided crossings $\mathcal{D}_n^{(\mathrm{ac})}$ is the number of avoided crossings between n and $n+1$ energy levels per unit length as we traverse a line by varying x. The density $\mathcal{D}_\beta^{(\mathrm{ac})}(A, B, \epsilon)$ of avoided crossings with gap ϵ and parameters A and B of the hyperbola Equation (11) was found for β-parameterized models [6,20]

$$\mathcal{D}_\beta^{(\mathrm{ac})}(A, B, \epsilon) = C_\beta^{(\mathrm{ac})} \frac{\rho^{\beta+1}}{\sigma^{\beta+1}} A^{\beta+1} \epsilon^{\beta-1} \exp\left(-\frac{\beta A^2}{8\sigma^2}\right) \exp\left(-\frac{\beta B^2}{2\sigma^2}\right) \frac{\sqrt{\beta}}{\sqrt{2\pi}\sigma} dA dB d\epsilon \tag{13}$$

with the prefactors [20]

$$C_1^{(\mathrm{ac})} = \frac{\pi}{24}, C_2^{(\mathrm{ac})} = \frac{\pi^{3/2}}{12}, C_4^{(\mathrm{ac})} = \frac{8\pi^{7/2}}{135\sqrt{2}}. \tag{14}$$

3. Thermal Fluctuations in Solids

3.1. Born–Oppenheimer Approximation

All information about the properties of a system containing N_n nuclei and N_e electrons is contained in the eigenstates of the many-body Hamiltonian

$$\hat{H} = \sum_{I=1}^{N_n} \frac{\mathbf{P}_I^2}{2M_I} + \sum_{i=1}^{N_e} \frac{\mathbf{p}_i^2}{2m} + \sum_{1 \le I < J}^{N_n} \frac{Z_I Z_J e^2}{|\mathbf{R}_I - \mathbf{R}_J|} + \sum_{1 \le i < j}^{N_e} \frac{e^2}{|\mathbf{r}_i - \mathbf{r}_j|} - \sum_{I=1}^{N_n} \sum_{i=1}^{N_e} \frac{Z_I e^2}{|\mathbf{r}_i - \mathbf{R}_I|}, \tag{15}$$

where m is the mass of an electron, e is the charge of an electron, $\mathbf{r}_i$ and $\mathbf{p}_i$ are the position and the momentum of the i-th electron, respectively, M_I is the mass of I-th nucleus, Z_I is the charge of I-th nucleus, and $\mathbf{R}_I$ and $\mathbf{P}_I$ are the position and the momentum of I-th nucleus, respectively. The Hamiltonian describes kinetic energy of electrons and nuclei, Coulomb interaction between nuclei, Coulomb interaction between electrons, and finally Coulomb interaction between electrons and nuclei.

Eigenstates of Equation (15) can be found analytically only for the hydrogen atom. Numerical solutions are possible for systems containing several atoms and electrons. If there are M points in grids representing the positions of electrons and nuclei, $M^{3(N_n+N_e)}$ real numbers would be required to store one state of the system. The exponential complexity quickly hits the memory limits of modern machines, so approximations are necessary.

The Born–Oppenheimer approximation [32,33] is the first simplification of the problem. It separates nuclear and electronic degrees of freedom. The separation is justified by the fact that nuclei are much heavier than electrons, and as a result electrons adjust to nuclear motion.

Within the Born–Oppenheimer approximation, atomic positions $\mathbf{R}_I$ are fixed to give the electronic Hamiltonian

$$\hat{H}_e(\mathbf{R}_1, \ldots, \mathbf{R}_{N_n}) = \sum_{i=1}^{N_e} \frac{\mathbf{p}_i^2}{2m} + \sum_{1 \le I < J}^{N_n} \frac{Z_I Z_J e^2}{|\mathbf{R}_I - \mathbf{R}_J|} + \sum_{1 \le i < j}^{N_e} \frac{e^2}{|\mathbf{r}_i - \mathbf{r}_j|} - \sum_{I=1}^{N_n} \sum_{i=1}^{N_e} \frac{Z_I e^2}{|\mathbf{r}_i - \mathbf{R}_I|}. \tag{16}$$

The electronic Hamiltonian is then diagonalized to obtain eigenvalues

$$\hat{H}_{el}(\mathbf{R}_1, \ldots, \mathbf{R}_{N_n}) |\psi_n\rangle = E_n(\mathbf{R}_1, \ldots, \mathbf{R}_{N_n}) |\psi_n\rangle, \tag{17}$$

which are functions of nuclear positions. These functions are then regarded as potentials in which nuclei move

$$H_n = \sum_{I=1}^{N_n} \frac{\mathbf{P}_I^2}{2M_I} + E_n(\mathbf{R}_1, \ldots, \mathbf{R}_{N_n}). \tag{18}$$

This Hamiltonian includes only nuclear degrees of freedom. It can be used to obtain both quantum and classical solutions. In the classical treatment, the motion of a nuclei is approximated as

$$M_I \ddot{\mathbf{R}}_I = -\frac{\partial E_n(\mathbf{R}_1, \ldots, \mathbf{R}_{N_n})}{\mathbf{R}_I}. \tag{19}$$

In many cases, the collection of nuclei can be represented as classical harmonic oscillators, which is justified by the Schrodinger correspondence principle [34,35].

A more restrictive simplification is the clamped nuclei approximation [36], which assumes that nuclei have fixed positions corresponding to the equilibrium structure. The original problem is then reduced to finding the ground state of the electronic Hamiltonian for a fixed set of parameters, Equation (16). This problem still has exponential complexity, since M^{3N_e} real numbers are required to store an electronic state. This complexity is overcome by a single particle picture, in which an electron moves independently of all other electrons. In this picture, the interaction of a single electron with other electrons is represented as the interaction of an electron with some average external potential due to the remaining electrons. Density functional theory provides the most accurate description of electron–electron interactions in this way [37]. Another method commonly used to solve the electronic Hamiltonian is the tight-binding method [38,39].

3.2. Supercell Technique

Within the Born–Oppenheimer approximation, clamped nuclei approximation, and single-particle picture, the microscopic properties of a solid can be found by solving the Schrodinger equation $\hat{H}\,|\psi_n\rangle = E_n\,|\psi_n\rangle$ for a single electron in the external potential set by equilibrium atomic positions $\{\mathbf{R}_I\}$. The problem is still too complex, since solids consist of an infinite number of atoms and electrons, making the electronic Hamiltonian a function of an infinite number of parameters. The use of translational symmetry together with periodic boundary conditions reduces the problem to a limited number of electrons inside a primitive cell, and the Hamiltonian is hence a function of a finite set of parameters—vectors that define the primitive unit cell and atomic coordinates inside the cell. Furthermore, Bloch's theorem gives eigenstates of the resulting Hamiltonian. Since atomic positions are fixed in this description, this solution provides only the static properties of a solid. A central property is the total energy of the system as a function of its structure. Minimizing it, one can calculate equilibrium lattice constants, the bulk modulus, and the equation of state $E(V)$ or $P(V)$, which is now a standard methodology in computational materials science [37].

In real crystals, of course, atoms are constantly fluctuating around equilibrium positions. At an arbitrary moment of time, atomic positions are random and there is no translational symmetry. Bloch's theorem is no longer applicable. The problem becomes intractable since the Hamiltonian is again a function of an infinite number of parameters. In some cases, the complexity can still be harnessed by using the supercell technique [40,41]. In this technique, a supercell is constructed of N repeating primitive cells and the atoms inside it are allowed to move freely. Periodic boundary conditions are then imposed on this structure. It therefore becomes an infinite solid, where the supercell plays the role of a primitive cell. Translational symmetry is restored and Bloch's theorem can be applied. The atomic motion inside the supercell allows access to dynamical properties such as phonons or relaxation times. Strictly speaking, the thus calculated dynamical properties would be different from those of the real solid; they would rather represent the properties of a large "molecule" consisting of N primitive cells whose motion is repeated throughout the space. However, upon an increase in N, the calculated properties would more closely resemble those of the real solid; in other words, by increasing N, it is possible to achieve numerical convergence.

Surprisingly, in many cases, the size of the supercell N necessary for numerical convergence is not exceedingly high in terms of numerical calculations. Physically, this means that in such cases interactions inside a solid quickly decay over distance. For example, for α-quartz, phononic dispersion can be accurately calculated by displacing atoms in a supercell consisting of only 27 unit cells. The reason is that force constants between atoms in the unit cells separated by more than three lattice spacings are weak [42]. In general, of course, the value of N necessary for convergence depends both on the properties and the solid, and it is possible that numerical convergence may not be achieved due to the limited computational resources of modern computers.

The dynamical properties of solids calculated using the supercell technique also rely on the assumption that electronic dynamics has a Born–Oppenheimer adiabatic character. The electronic Hamiltonian for the supercell is set by positions of the atoms inside it. Supposing the atoms are moving classically, the Hamiltonian becomes time dependent. Therefore, to describe electronic dynamics $|\psi(t)\rangle$, it is necessary to solve the time-dependent Schrodinger equation starting from the n-th eigenstate

$$i\hbar\frac{\partial\,|\psi(t)\rangle}{\partial t} = \hat{H}(t)\,|\phi(t)\rangle\,, \tag{20}$$

$$|\psi(0)\rangle = |\psi_n(0)\rangle\,. \tag{21}$$

The time-dependent wavefunction $|\psi(t)\rangle$ often used in supercell calculations corresponds to the n-th eigenstate of the Hamiltonian $\hat{H}(t)$ at time t

$$\hat{H}(t)\,|\psi_n(t)\rangle = E_n(t)\,|\psi_n(t)\rangle\,, \tag{22}$$

$$|\psi(t)\rangle = |\psi_n(t)\rangle\,. \tag{23}$$

It is different from the wavefunction that one obtains by solving the time-dependent Schrödinger equation, but according to the adiabatic theorem, they coincide if the Hamiltonian changes slowly enough (the adiabatic Born–Oppenheimer approximation (ABO)).

The adiabatic assumption can be checked by comparing the time-dependent solution with the adiabatic one. Alternatively, the adiabatic assumption can be indirectly validated by the fact that calculated dynamical observables coincide with experimental values. In most applications, there is no need for direct validation, since, for solids, theoretical predictions turn out to correctly describe the experimental behavior.

The case of graphene is unique in this regard. Graphene is a two-dimensional honeycomb structure of carbon atoms. Each carbon atom has four valence orbitals, three of which hybridize to form strong in-plane σ-bonds between adjacent atoms. The fourth electron is in a p_z orbital perpendicular graphene's plane. Out-of-plane p_z electrons form π-bonds that are significantly weaker than σ-bonds, and therefore their impact on atomic motion is weak. Although atomic motion is primarily determined by σ-bonds, electrons in σ-bonds do not contribute to electronic transport. Therefore, the fact that numerical calculations reproduce experimental values for atomic vibrations confirms the validity of the adiabatic assumption for σ-bonds only. One can check the validity of the adiabatic theorem for π-electrons by directly comparing time-dependent electronic wavefunctions to the adiabatic ones. This comparison was performed by Mohanty and Heller [2], which we describe below.

3.3. Simulating Graphene Thermal Fluctuations

To model the thermal fluctuations of graphene, the authors constructed a 4×4 supercell with 32 carbon atoms. To check that qualitative conclusions hold for larger sizes, they repeated calculations for a 5×10 supercell. Within a supercell, atoms were allowed to move freely under the harmonic σ-bonded force field, with atomic motion taken as a superposition of graphene phonons that satisfied the periodic boundary conditions of a given supercell. In turn, graphene phononic frequencies were computed by considering only force constants between adjacent atoms, and the corresponding value was taken from infrared spectroscopy of the C=C aromatic bond. The amplitude of each phonon corresponded to a temperature $T = 300$ K. Eventually, atomic positions were predetermined for each moment of time according to

$$\mathbf{r}(t) = \sum_n \sqrt{\frac{2k_BT}{M\omega_n^2}}\mathbf{q}_n \cos\omega_n t, \tag{24}$$

where $\mathbf{q}_n$ is the normal mode for the n-th phonon with frequency ω_n. π-electrons do not directly influence this motion. This is a simplification, but it is expected that the qualitative conclusions hold since the stiff σ-bonds contribute most to the force.

At each moment of time, atomic positions determine the nearest-neighbor tight-binding Hamiltonian for π-electrons. The hopping integral between adjacent atomic sites depends on the distance $r_{ij}(t)$ between the atoms

$$H(t) = -\sum_{\langle ij\rangle} \beta \exp[-\alpha r_{ij}(t)](\hat{c}_i^+\hat{c}_j + H.c.), \tag{25}$$

where $\langle ij\rangle$ stand for the indices of adjacent atoms.

The time-dependent Hamiltonian $\hat{H}(t)$ is used to obtain two types of wavefunction. The first corresponds to the adiabatic solution $|\psi_n(t)\rangle$, Equation (22). Adiabatic diagonalization based on the instantaneous atomic positions provides the eigenenergies $E_n(t)$. The second type of wavefunction is the actual electronic wavefunction that is given by Equation (20).

To study the validity of the ABO, the true wavefunction $|\psi(t)\rangle$ is compared with the adiabatic solution $|\psi_n(t)\rangle$ by calculating the overlap between them

$$a_n(t) = |\langle\psi_n(t)|\psi(t)\rangle|^2, \tag{26}$$

which is called the adiabatic correlation.

The adiabatic correlation reveals whether the adiabatic theorem, and consequently the ABO, is valid for this process. However, it does not say how strong the changes in the wavefunctions $|\psi_n(t)\rangle$ and $|\psi(t)\rangle$ are. For this, the overlap of the actual wavefunction with the initial wavefunction was calculated. The corresponding quantity is called the diabatic autocorrelation

$$d(t) = |\langle\psi(0)|\psi(t)\rangle|^2. \tag{27}$$

3.4. *Breakdown of the ABO for Thermally Fluctuating Graphene*

Eigenenergies of the time-dependent Hamiltonian $E_n(t)$ for a 4×4 supercell, Equation (22), are shown in Figure 2a. Time is given in the units of the shortest vibrational period, which is 191 $\hbar/\mathrm{Ry}$ for the case of the 4×4 supercell. There are 32 eigenenergies because the supercell accommodates 32 atoms; for any moment of time, eigenvalues are not degenerate because atomic positions do not correspond to a highly symmetrical configuration including the initial moment of time $t = 0$, Equation (24). Eigenenergies form clusters that are separated from each other by gaps of several eV.

The time evolution of the eigenenergies of one of the clusters is seen to fluctuate around the average values in Figure 2b. Adjacent levels constantly approach each other; however, spacings between them never go to zero. If the dynamics were adiabatic, $E_n(t)$ would describe how the expected values of the time-dependent states evolve. To reveal the character of the dynamics, adiabatic correlations and diabatic autocorrelations are calculated for the states starting from the ground state and from the $n = 18$ state at $t = 0$.

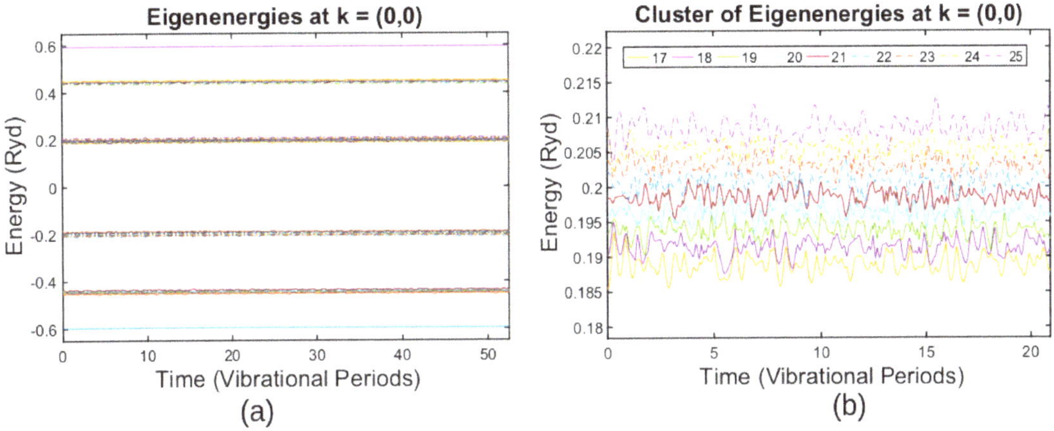

Figure 2. (**a**) Time-dependent energy spectrum at the Γ point. (**b**) Time evolution of the $n = 18$th to the $n = 25$th bands. (Reprinted from ref. [2].)

For the ground state, the adiabatic correlation remains one $a_1(t) = 1$ during the simulation time equal to tens of the shortest vibrational period. The diabatic autocorrelation remains one $d(t) = 1$ as well. Preservation of adiabatic correlation means that the solution of Equation (20) $|\psi(t)\rangle$ and the solution of Equation (22) $|\psi_1(t)\rangle$ coincide up to an arbitrary phase, $|\psi(t)\rangle = |\psi_1(t)\rangle$; therefore, the adiabatic theorem and consequently the ABO hold for time evolution of this particular state. Since the diabatic autocorrelation remains one as well, we conclude that the thermal motion of atoms has no effect on the ground state, and neither the adiabatic nor the true wavefunctions change.

The behavior of higher states that form the clusters is different. Figure 3a shows adiabatic correlations for the state $|\psi(t)\rangle$, starting from the $n = 18$ eigenstate at $t = 0$. In contrast to the ground state, the overlap probability with the corresponding adiabatic state sharply decays to zero during the first vibrational period. For the first four vibrational periods, it remains zero most of the time, occasionally sharply rising to $a_{18} \approx 0.5$ and then sharply decaying to zero again three times during the aforementioned time. In addition to this, we can also observe the overlap probability with other adiabatic states in the Figure. These states form a cluster of eigenenergies, Figure 2b. The overlap with each of these states is constantly changing, rising from zero to a maximum value and then decaying back to zero. The maximum value varies among states of the cluster and lies within the range from $a_{25} \approx 0.25$ for the $n = 25$ adiabatic eigenstate to $a_{21} = 0.8$ for the $n = 21$ adiabatic eigenstate. For certain states, for example, $n = 21$, peaks in the adiabatic correlation merge, while for the others, peaks in the adiabatic correlation stand separately.

The diabatic autocorrelation for the $n = 18$ state also has a different behavior to the one of the ground state, Figure 3b. In contrast to the ground state case, it does not remain one throughout the simulation. It gradually decays to zero within 60 periods of the shortest vibration. The decay time of the diabatic correlation is several orders of magnitude larger than the decay time of the adiabatic correlation. After dropping to zero, the diabatic correlation function grows.

From the quick decay of the adiabatic correlations, it is clear that in the presence of thermal fluctuations, the adiabatic theorem and the ABO do not hold for states that form clusters. The adiabatic theorem and the ABO are still valid for the ground state that is separated from all the others by gaps of the order of several eV. Such a sharp decay of the adiabatic correlation can be attributed to the near degeneracy of the states, suffering frequent avoided crossings. As we discuss in the next section, these avoided crossings indeed have parameters that lead to the breakdown of the adiabatic theorem.

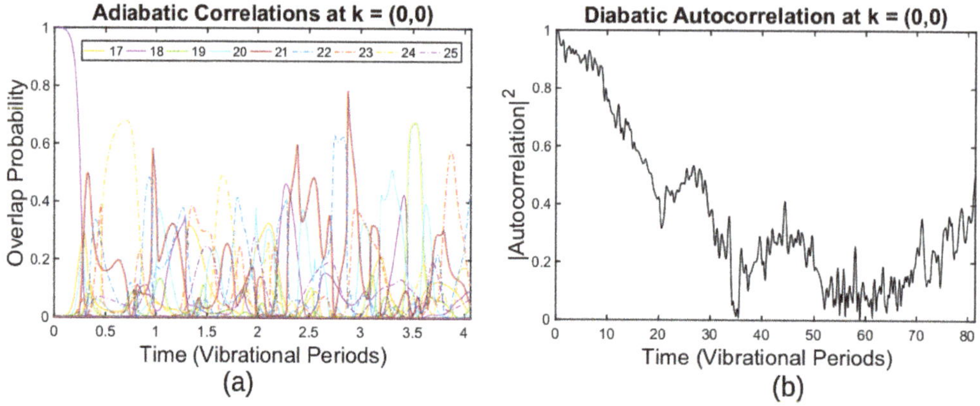

Figure 3. (**a**) Adiabatic autocorrelation function for the band $n = 18$ and overlap probability between the time-dependent solution starting from the $n = 18$th state at $t = 0$ and other adiabatic eigenstates. (**b**) Diabatic autocorrelation function for the band $n = 18$. (Reprinted from ref. [2].)

Secondly, the fact that the diabatic autocorrelation decays on a timescale several orders larger than adiabatic correlations means that during the very first oscillations, electronic wavefunctions stay intact in the sense that the atomic orbital coefficients remain nearly unchanged as they are carried around by their respective nuclei. Since adiabatic correlations drop sharply to zero in the meantime, one can conclude that even the smallest changes in atomic positions lead to enormous changes in the adiabatic wavefunctions $|\psi_n(t)\rangle$ that form a cluster and that are solutions of Equation (22). This is in sharp contrast to the ground state, where both adiabatic and diabatic wavefunctions did not change during the whole simulation time. From the perspective of the zero-th order degenerate perturbation theory, almost degenerate states form a complete basis set to describe changes in adiabatic electronic wavefunctions induced by thermal motion of atoms. Quick decay of adiabatic correlations and preservation of the initial wavefunction means that even slight atomic displacements lead to strong mixing between initial adiabatic states. This is a characteristic feature of states that form clusters.

Finally, the decay of the diabatic autocorrelation at much larger times than adiabatic correlations means that electrons cannot adjust to nuclear motion. The ABO ansatz is based on this assumption. Therefore, it is not optimal for description of such dynamical processes. Qualitatively, one may rather say that electronic states are preserved during dynamics. Since time evolution has diabatic character, then another ansatz, which the authors called the diabatic Born–Oppenheimer approximation, would be more suitable than the ABO for a description of such processes.

Qualitatively, the diabatic character is in accordance with experiments. Indeed, electronic structure theory and Bloch functions are built assuming atoms form perfect symmetrical structures. The success of electronic structure theory indicates that during thermal fluctuations, electronic wavefunctions are preserved by obeying a diabatic solution near avoided crossings. Therefore, there is an implicit diabatic assumption in electronic structure theory, otherwise it would be impossible to exploit translational symmetry. The diabatic representation is that the ABO solution for the highest symmetry is used no matter what the actual positions of the nuclei are.

4. Discussion and Further Work

In this section, we suggest future directions in the study of time-dependent dynamics of electrons in the presence of phonons. Mohanty and Heller [2] have established its diabatic character by running time-dependent simulations. It is interesting, however, that the lowest frequency phonons have not been studied, owing to the the finite size of the supercell. They are slow; thus, they may be again in the adiabatic realm. It would be very constructive to determine this.

We now suggest a new perspective based on avoided crossings. We outline how one can potentially come to the same conclusions by using Landau–Zener theory in tandem with random matrix theory without running time-dependent calculations.

The idea of using Landau–Zener theory in tandem with random matrix theory belongs to Wilkinson [6]. They considered a driven finite-sized system and calculated the rate of increase in energy of the system using a distribution of gap sizes and the slopes of avoided crossings, similar to Equation (13). It was noted in reference [6] that this idea could be applied in the context of the Born–Oppenheimer approximation to characterize vibronic coupling in complex molecules.

Diabatic preservation of electronic character suggests a breakdown of the adiabatic approximation. Adiabaticity breaks down when the Hamiltonian changes too rapidly. One can estimate a timescale of adiabatic motion by using Heisenberg's uncertainty principle; for a process with the minimal spacing between energy levels ΔE, adiabatic evolution can take place only for time τ

$$\tau \geq \hbar/\Delta E. \tag{28}$$

From Equation (28), it follows that avoided crossings or local minima in the distance between adjacent levels determine the timescale for adiabatic motion. We have considered their statistical properties in Section 2.3.

Figure 2b shows how energy levels vary over time for graphene thermal fluctuations. Each state experiences several avoided crossings during one cycle of the shortest vibrational mode. We conclude that the presence of multiple avoided crossings is an inherent feature of this motion, and therefore should be explicitly incorporated into theoretical treatment. Specifically, avoided crossings determine a characteristic energy gap, which in turn determines a timescale for adiabatic evolution. This timescale is to be compared with the vibrational period, which is primarily set by σ-bonds and only weakly depends on p_z-orbitals.

For the case [2], the shortest vibrational period is $\tau \approx 191\hbar/\text{Ry}$, so the energy gap should be at least $\Delta E = 1/191 \approx 0.005$ Ry. As one can see in Figure 2b, avoided crossings have a much smaller gap, leaving no hope for adiabatic motion. Indeed, in Figure 3a, the adiabatic autocorrelation of the 18th state drops to zero at $t = 0.25$, which is exactly the first avoided crossing one can observe in Figure 2b.

Landau and Zener gave a more accurate estimation for adiabatic timescales [43]. They considered a two-level system with the Hamiltonian in the form of Equation (12) for the case $B = 0$, $x_0 = 0$, and $x \equiv t$. The probability of adiabatic time evolution by the end of the motion is

$$p = 1 - \exp\left(-\frac{\pi\epsilon^2}{2A\hbar}\right). \tag{29}$$

The estimation of adiabatic A_{adia} based on the exact solution $A_{\text{adia}} = \frac{\pi}{2}\epsilon^2/\hbar$ is close to the one based on Heisenberg's uncertainty principle, $A = \epsilon^2/\hbar$. In the case of graphene thermal fluctuations, the electronic structure of the p_z orbitals determines ϵ, while the temperature and σ-bonds determine A.

Thus, instead of explicitly calculating the time-dependent wavefunction, one can more simply and more insightfully study the evolution of electronic character by finding avoided crossings in the time-dependent electronic spectrum, extracting A and ϵ from the fit to Equation (11) and putting these parameters into Equation (29).

The procedure we have just outlined still requires performing numerical calculations to obtain a time-dependent electronic spectrum. Further analytical insights at much less effort might have been obtained using the theoretical results from Section 2.3. As a result of the randomness of thermal fluctuations, one can model the electronic Hamiltonian by a random matrix. This theory allows determining the distribution of ϵ, A, and B for avoided crossings, Equation (13). In tandem with Landau–Zener theory, it is possible to understand the diabatic character of time evolution even without running simulations. In order to make random matrix theory applicable, it is necessary to adjust the density of states and the sensitivity to perturbations by rescaling energy and time; this procedure has been described by Wilkinson [22].

There is one caveat: Landau–Zener theory considers only two coupled states, whereas all states that form a cluster turn out to be coupled. There is a close and suggestive analogy with the Wigner surmise for the nearest neighbor level spacing distribution; a very accurate result was obtained by Wigner using only an ensemble of 2×2 random matrices instead of the full $N \times N$ case.

One can see a manifestation of the full N-state cluster coupling in Figure 3a for the first avoided crossing of the 18th state; the state makes a transition to the 21st but not to the adjacent 19th or 17th state. A general solution to the multi-level Landau–Zener problem is necessary here, which has not been obtained yet [44–46].

We have suggested a new way to analyze electronic dynamics in the presence of phonons, which enables to use random matrix theory and may shed new light on this problem.

5. Summary

The main purpose of this paper is to show that two seemingly distinct subjects, random matrix theory and electronic dynamics in the presence of thermal fluctuations, can be brought together through considering statistically known properties of the electronic structure spectrum. This perspective can potentially give more information than just the preservation of diabatic electronic character recently found in the time-dependent simulations of graphene thermal fluctuations.

We first introduced parameter-dependent random matrices and the reasons for selecting a particular parametrization. Then, we considered singularities in their spectra, i.e., conical intersections and avoided crossings. We have described the geometrical properties of conical intersections, along with a typical workflow for calculating their statistical distributions analytically. Qualitatively, analytical predictions were in agreement with the illustrated numerical results. Then, we quoted the statistical distribution of fitting parameters for energy levels near an avoided crossing, which is important for the perspective we suggest.

We next examined the time evolution of graphene π-band electronic states in the presence of phonons. A direct solution of time-dependent Schrodinger equation shows that a projection of the time-dependent state on an instantaneous adiabatic state drops to zero during the very first vibrational cycle, whereas overlap with the initial electronic state decays at a much longer timescale. Moreover, it tends to recover at an even larger timescale. That indicates preservation of the electronic character and the diabatic nature of time evolution.

Finally, we sketched a possible way to treat the dynamics of electrons in the presence of phonons based on random matrix theory and avoided crossings. We started with the observation that numerous avoided crossings arise in the supercell electronic bandstructure because of nuclear motion. Overlap of the time-dependent wavefunction with the instantaneous adiabatic states suggests avoided crossings are responsible for the rapid decay of the adiabatic states. A simple estimation based on the uncertainty principle confirms that avoided crossings have gaps that are too small for the adiabatic theorem to be valid. We then outlined a more sophisticated analytical treatment based on Landau–Zener theory. All in all, the suggested perspective may enable the use of random matrix theory to characterize the time evolution of electrons in the presence of phonons. This could be a powerful new perspective in this important problem.

Author Contributions: Conceptualization, A.E.O. and E.J.H.; writing—original draft, A.E.O. and E.J.H. All authors have read and agreed to the published version of the manuscript.

Funding: This work was supported by the NSF Center for Integrated Quantum Materials (CIQM), grant no. DMR-1231319. A.O. acknowledges a fellowship from the Zimin Foundation SMTB Alumni summer research program. Open Access funding was provided by the Max Planck Society.

Institutional Review Board Statement: Not applicable.

Informed Consent Statement: Not applicable.

Data Availability Statement: The data presented in this study are available on request from the corresponding author.

Conflicts of Interest: The authors declare no conflict of interest.

Abbreviations

The following abbreviations are used in this manuscript:

ABO	Adiabatic Born–Oppenheimer approximation
DBO	Diabatic Born–Oppenheimer approximation
GOE	Gaussian orthogonal ensemble
GUE	Gaussian unitary ensemble
GSE	Gaussian symplectic ensemble

References

1. Donghwan, K.; Aydin, A.; Daza, A.; Avanaki, K.N.; Keski-Rahkonen, J.; Heller, E.J. Coherent charge carrier dynamics in the presence of thermal lattice vibrations. *Phys. Rev. B* **2022**, *106*, 054311.
2. Mohanty, V.; Heller, E.J. Lazy electrons in graphene. *Proc. Natl. Acad. Sci. USA* **2019**, *116*, 18316–18321. [CrossRef] [PubMed]
3. Haake, F. *Quantum Signatures of Chaos*; Springer: Berlin/Heidelberg, Germany, 2010.
4. Scharf, R.; Dietz, B.; Kuś, M.; Haake, F.; Berry, M.V. Kramers' Degeneracy and Quartic Level Repulsion. *Europhys. Lett.* **1988**, *5*, 383–389. [CrossRef]
5. Grobe, R.; Haake, F.; Sommers, H.-J. Quantum Distinction of Regular and Chaotic Dissipative Motion. *Phys. Rev. Lett.* **1988**, *61*, 1899–1902. [CrossRef] [PubMed]
6. Wilkinson, M. Statistical aspects of dissipation by Landau-Zener transitions. *J. Phys. A Math. Gen.* **1988**, *21*, 4021–4037. [CrossRef]
7. Wilkinson, M.; Austin, E.J. Suppression of dissipation by localization. *J. Phys. A Math. Gen.* **1990**, *23*, L957–L963. [CrossRef]
8. Lenz, G.; Haake, F. Transitions between universality classes of random matrices. *Phys. Rev. Lett.* **1990**, *65*, 2325–2328. [CrossRef]
9. Lenz, G.; Haake, F. Reliability of small matrices for large spectra with nonuniversal fluctuations. *Phys. Rev. Lett.* **1991**, *67*, 1. [CrossRef]
10. Lenz, G.; Haake, F. Classical Hamiltonian Dynamics of Rescaled Quantum Levels. *Europhys. Lett.* **1990**, *13*, 577–582.
11. Wilkinson, M.; Austin, E.J. Densities of degeneracies and near-degeneracies. *Phys. Rev. A* **1993**, *47*, 2601–2609. [CrossRef]
12. Wilkinson, M. A semiclassical sum rule for matrix elements of classically chaotic systems. *J. Phys. A* **1987**, *20*, 2415–2423. [CrossRef]
13. Pandey, A. Statistical properties of many-particle spectra: III. Ergodic behavior in random-matrix ensembles. *Ann. Phys.* **1979**, *119*, 170–191. [CrossRef]
14. Teller, E. The Crossing of Potential Surfaces. *J. Phys. Chem.* **1937**, *41*, 109–116. [CrossRef]
15. Longuett-Higgins, H.C. The intersection of potential energy surfaces in polyatomic molecules. *Proc. R. Soc. Lond. A* **1975**, *344*, 147–156.
16. Berry, M.V. Quantal phase factors accompanying adiabatic changes. *Proc. R. Soc. Lond. A* **1984**, *392*, 45–57.
17. Selberg, A. Remarks on a multiple integral. *Nor. Mat. Tidsskr.* **1944**, *26*, 71–78.
18. Aomoto, K. Jacobi Polynomials Associated with Selberg Integrals. *SIAM J. Math. Anal.* **1987**, *18*, 545–549. [CrossRef]
19. Mehta, M.L. *Random Matrices*; Academic Press: New York, NY, USA, 1990.
20. Walker, P.N.; Sanchez, M.J.; Wilkinson, M. ingularities in the Spectra of Random Matrices. *J. Math. Phys.* **1996**, *37*, 5019–5032. [CrossRef]
21. Berry, M.V.; Wilkinson, M. Diabolical points in the spectra of triangles. *Proc. R. Soc. Lond. A* **1984**, *392*, 15–43.
22. Walker, P.N.; Wilkinson, M. Universal Fluctuations of Chern Integers. *Phys. Rev. Lett.* **1995**, *74*, 4055–4058. [CrossRef]
23. Kuś, M.; Scharf, R.; Haake, F. Symmetry versus degree of level repulsion for kicked quantum systems. *Z. Phys. B* **1987**, *66*, 129–134. [CrossRef]
24. Haake, F.; Kuś, M.; Scharf, R. Classical and Quantum Chaos for a Kicked Top. *Z. Phys. B* **1987**, *65*, 381–395. [CrossRef]
25. Zakrzewski, J.; Delande, D. Parametric motion of energy levels in quantum chaotic systems. I. Curvature distributions. *Phys. Rev. E* **1993**, *47*, 1650–1664. [CrossRef] [PubMed]
26. von Oppen, F. Exact distribution of eigenvalue curvatures of chaotic quantum systems. *Phys. Rev. Lett.* **1994**, *73*, 798–801. [CrossRef] [PubMed]

27. von Neumann, J.; Wigner, E. Uber merkwürdige diskrete Eigenwerte. Uber das Verhalten von Eigenwerten bei adiabatischen Prozessen. *Phys. Z.* **1929**, *30*, 467–470.
28. Austin, E.J.; Wilkinson, M. Statistical properties of parameter-dependent classically chaotic quantum systems. *Nonlinearity* **1992**, *5*, 1137–1150. [CrossRef]
29. Zakrzewski, J.; Kuś, M. Distributions of avoided crossings for quantum chaotic systems. *Phys. Rev. Lett.* **1991**, *67*, 2749–2752. [CrossRef]
30. Zakrzewski, J.; Delande, D.; Kuś, M. Parametric motion of energy levels in quantum chaotic systems. II. Avoided-crossing distributions. *Phys. Rev. E* **1993**, *47*, 1665–1676. [CrossRef]
31. Grobe, R.; Haake, F. Universality of cubic-level repulsion for dissipative quantum chaos. *Phys. Rev. Lett.* **1989**, *62*, 2893–2896. [CrossRef]
32. Born, M.; Oppenheimer, R. Zur quantentheorie der Molekeln. *Ann. Phys.* **1927**, *389*, 457–484. [CrossRef]
33. Heller, E.J. *The Semiclassical Way to Dynamics and Spectroscopy*; Princeton University Press: Princeton, NJ, USA, 2018.
34. Born, M. Über die Serienspektra der Elemente. *Z. Phys.* **1920**, *2*, 423–469.
35. Heller, E.J.; Kim, D. Schrödinger correspondence applied to crystals. *J. Phys. Chem. A* **2019**, *123*, 4379–4388. [CrossRef] [PubMed]
36. Ruggenthaler, M.; Tancogne-Dejean, N.; Flick, J.; Appel, H.; Rubio, A. From a quantum-electrodynamical light–matter description to novel spectroscopies. *Nat. Rev. Chem.* **2018**, *2*, 0118. [CrossRef]
37. Payne, M.C.; Teter, M.P.; Allan, D.C.; Arias, T.A.; Joannopoulos, J.D. Iterative minimization techniques for ab initio total-energy calculations: Molecular dynamics and conjugate gradients. *Rev. Mod. Phys.* **1992**, *64*, 1045–1097. [CrossRef]
38. Hoffmann, R. *Solids and Surfaces: A Chemist's View of Bonding in Extended Structures*; VCH: New York, NY, USA, 1988.
39. Turchi, P.E.A.; Gonis, A.; Colombo, L. *Tight-Binding Approach to Computational Materials Science*; Materials Research Society: Warrendale, PA, USA, 1998.
40. Makov, G.; Payne, M.C. Periodic boundary conditions in Ab Initio Calculations. *Phys. Rev. B* **1995**, *51*, 4014–4022. [CrossRef]
41. Makov, G.; Shah, R.; Payne, M.C. Periodic boundary conditions in *ab initio* calculations. II. Brillouin-zone sampling for aperiodic systems. *Phys. Rev. B* **1996**, *53*, 15513–15517. [CrossRef]
42. Mizokami, K.; Togo, A.; Tanaka, I. Lattice thermal conductivities of two SiO_2 polymorphs by first-principles calculations and the phonon Boltzmann transport equation. *Phys. Rev. B* **2018**, *97*, 224306. [CrossRef]
43. Landau, L.D. On the Theory of Transfer of Energy at Collisions II. *Phys. Z. Sowjetunion* **1932**, *2*, 46–51.
44. Brundobler, S.; Elser, V. S-matrix for generalized Landau-Zener problem. *J. Phys. A Math. Gen.* **1993**, *26*, 1211–1227. [CrossRef]
45. Sinitsyn, N.A. Counterintuitive transitions in the multistate Landau–Zener problem with linear level crossings. *J. Phys. A Math. Gen.* **2004**, *37*, 10691–10697. [CrossRef]
46. Sinitsyn, N.A.; Chernyak, V.Y. The quest for solvable multistate Landau-Zener models. *J. Phys. A Math. Theor.* **2017**, *50*, 255203. [CrossRef]

entropy

Article

Semi-Poisson Statistics in Relativistic Quantum Billiards with Shapes of Rectangles

Barbara Dietz

Center for Theoretical Physics of Complex Systems, Institute for Basic Science (IBS), Daejeon 34126, Republic of Korea; barbara@ibs.re.kr

Abstract: Rectangular billiards have two mirror symmetries with respect to perpendicular axes and a twofold (fourfold) rotational symmetry for differing (equal) side lengths. The eigenstates of rectangular neutrino billiards (NBs), which consist of a spin-1/2 particle confined through boundary conditions to a planar domain, can be classified according to their transformation properties under rotation by π ($\pi/2$) but not under reflection at mirror-symmetry axes. We analyze the properties of these symmetry-projected eigenstates and of the corresponding symmetry-reduced NBs which are obtained by cutting them along their diagonal, yielding right-triangle NBs. Independently of the ratio of their side lengths, the spectral properties of the symmetry-projected eigenstates of the rectangular NBs follow semi-Poisson statistics, whereas those of the complete eigenvalue sequence exhibit Poissonian statistics. Thus, in distinction to their nonrelativistic counterpart, they behave like typical quantum systems with an integrable classical limit whose eigenstates are non-degenerate and have alternating symmetry properties with increasing state number. In addition, we found out that for right triangles which exhibit semi-Poisson statistics in the nonrelativistic limit, the spectral properties of the corresponding ultrarelativistic NB follow quarter-Poisson statistics. Furthermore, we analyzed wave-function properties and discovered for the right-triangle NBs the same scarred wave functions as for the nonrelativistic ones.

Keywords: quantum chaos; relativistic quantum chaos; quantum billiards; relativistic quantum billiard

Citation: Dietz, B. Semi-Poisson Statistics in Relativistic Quantum Billiards with Shapes of Rectangles. *Entropy* **2023**, *25*, 762. https://doi.org/10.3390/e25050762

Academic Editor: Marko Robnik

Received: 28 February 2023
Revised: 29 April 2023
Accepted: 4 May 2023
Published: 6 May 2023

1. Introduction

This paper is a contribution to the special issue *Quantum Chaos*, which is dedicated to the 80th birthday of Giulio Casati who is a leading expert in the fields of classical and quantum chaos. He, actually, already expressed in [1] the conjecture that the spectral properties of quantum systems with a chaotic classical dynamics coincide with those of random matrices from the Gaussian ensembles (GEs) of corresponding universality class [2], that is, before Bohigas, Gianoni and Schmit formulated the famous BGS conjecture [3]. According to the BGS conjecture, they are well described by random marix theory (RMT) [2,4–7], where the appropriate GE for quantum systems with preserved and violated time-reversal invariance are the Gaussian orthogonal ensemble (GOE) and the Gaussian unitary ensemble (GUE), respectively. Criteria for its validity were identified in Ref. [8] based on the semiclassical periodic orbit (PO) theory, which was pioneered by Gutzwiller [9,10]. Casati considered a billiard with the shape of a stadium. Billiards provide a particularly suited model for studies in the context of quantum chaos. The dynamics of classical billiards (CBs), consisting of a point particle which moves freely inside a bounded two-dimensional domain and is reflected specularly at the boundary, can be engineered through the choice of their shape [11–13]. The eigenstates of the corresponding nonrelativistic quantum billiard (QB) are determined by solving the Schrödinger equation for a free particle and imposing the Dirichlet boundary condition (BC) on the resulting wave functions. Berry and Tabor showed in [14] based on action-angle variables that the spectral properties of typical integrable systems [15] agree well with those of Poissonian random numbers. However, there are numerous examples for 'untypical' integrable systems, a paradigm one being the harmonic oscillator [16]. Another example is polygonal billiards [17–19], such as triangular billiards, which have been investigated by Casati and his collaborators for nearly three decades [20–23] and the rectangular billiard [24]. Billiards with the shapes of rational polygonals of which the boundary comprises diffractive corners with angles $\alpha_i = \frac{n_i}{m_i}\pi$ where m_i, n_i are integers and $n_i \neq 1$ [17,25–32] are neither integrable nor chaotic. Therefore, they are referred to as pseudointegrable systems. Their phase space trajectories propagate on invariant surfaces that are topologically equivalent to multihandled, two-dimensional tori with genus $2 \leq g < \infty$, implying that they are not ergodic in phase space. Yet, the motion in such billiards has a chaotic component arising from the diffractive corners, which are of measure zero in classical phase space. The spectral properties of the corresponding QB differ considerably from those of typical integrable systems [26,28–31]. To be more explicit, they exhibit features that are intermediate between those of Poissonian random numbers and those of random matrices from the GOE in the sense that the levels repel each other linearly like in chaotic systems, and their nearest-neighbor spacing distribution decreases exponentially for large spacings, which is typical for integrable systems [33]. Depending on the choice of angles of a right-triangle QB, the properties are well described by those of semi-Poissonian numbers, which are obtained by deleting every second one in a sequence of Poissonian ones [23,34]. Such cases are considered in the present work.

We present results obtained for the spectral properties, momentum distributions [35] and Husimi functions [36,37] of the rectangular and square-shaped relativistic neutrino billiards (NBs) and their symmetry-projected eigenstates. These are obtained by separating their eigenstates according to their transformation properties under rotation by π and $\pi/2$, respectively. Neutrino billiards were introduced by Berry and Mondragon [38]. They are governed by the Weyl equation [39] for a non-interacting, massless spin-1/2 particle—commonly referred to as the Dirac equation in this context [38]—with the BC that the outward current vanishes. In distinction to QBs, NBs and the relativistic quantum systems considered in Ref. [40] do not have a well-defined classical counterpart. Insight into their behavior in the semiclassical limit can be obtained based on a semiclassical approximation for the fluctuating part of their spectral density in terms of a trace formula [41,42], which is applicable from the ultrarelativistic limit for massless neutrinos to the nonrelativistic limit of large mass m_0 where the energy is close to the rest energy $E_0 = m_0c^2$ [43]. The trace formula is a sum over periodic orbits of the CB of corresponding shape, where in the ultrarelativistic limit, those with an odd number of reflections at the boundary are missing.

An alternative type of billiards exhibiting relativistic features in part of their eigenvalue spectrum are graphene billiards, which are constructed by cutting a honeycomb lattice out of their shape. They are used to model properties of artificial graphene flakes based on a tight-binding model [44,45]. Finite-size sheets of graphene [46–49], referred to as graphene quantum dots, have the advantage that they can be studied experimentally. The first experiments were presented in [50–52]. In the vicinity of the corners of the hexagonal Brillouin zone, where the conduction and valence band touch each other conically, the energy excitations are governed by the Dirac equation for massless spin-1/2 particles [53]. The occurrence of the conical structure solely originates from the honeycomb structure of graphene, which is formed by two interpenetrating triangular lattices. This led to the realization of numerous experimental 'artificial-graphene' realizations [54]. Boundary conditions on the spinor components in a graphene billiard were formulated in [55–57]. We modeled rectangular, Africa-shaped and threefold-symmetric graphene billiards experimentally with flat superconducting microwave photonic crystals [58–62] and found agreement with the spectral properties of massive neutrino billiards only beyond a certain mass. In addition, theoretical studies of rectangular graphene quantum dots yielded deviations from those of massless neutrino billiards [63]. Their origin is explained in Ref. [64] and may be attributed to the presence of the boundary and differing BCs. The extraordinary features presented in this work are only observed in the ultrarelativistic limit for massless neutrino billiards. Similar studies will be performed for graphene billiards for varying BCs in a separate work.

The work was motivated by results obtained in [65] for the equilateral triangle and in [66] for sectors of the circle and ellipse NB. The boundary of the equilateral triangle has a C_{3v} symmetry [67], that is, threefold rotational symmetry and mirror symmetries with respect to its main axes: that of the circle belongs to the U(1) symmetry class, which comprises all M-fold rotational symmetries with $M \geq 2$, and that of the ellipse has mirror symmetries with respect to its minor and major axes and a twofold rotational symmetry. Generally, the eigenfunctions of a QB with a mirror symmetry are separated into eigenfunctions which are either symmetric or antisymmetric with respect to the symmetry axes, and they fulfill either Neumann or Dirichlet BCs along them. This is not possible for NBs. However, as will be outlined in Section 2, the eigenstates of NBs whose boundary has a M-fold rotational symmetry can be separated according to their transformation properties under rotation by $\frac{2\pi}{M}$ into symmetry-projected eigenstates. Sectors of NBs are constructed by cutting these along the borders of a fundamental domain associated with a discrete rotational symmetry and imposing the same boundary conditions along the cutlines as along the outer boundary, that is, as for the full NB. These NBs are referred to as symmetry-reduced NBs in the following. The circle and ellipse sector CBs have an integrable classical dynamic [68,69], and the corresponding QBs exhibit Poissonian statistics. The symmetry-projected eigenstates of the circle and ellipse NBs have been determined analytically in Refs. [38,41,65,70]. Their spectral properties also exhibit Poissonian statistics, that is, they agree with those of the corresponding QB. The spinor components of the eigenfunctions of an NB are linked through the BCs at the boundary [38]. Furthermore, for NBs with a discrete rotational symmetry, they transform differently under the associated rotation [65,71], implying that the symmetry classes are intermingled when cutting an NB into symmetry-reduced NBs. We demonstrated in [66] that consequently, the symmetry-reduced and full NBs cannot have any common eigenstates and that the spectral properties of symmetry-reduced circle NBs with an arbitrary inner angle smaller than 2π and the quarter-ellipse NBs with sufficiently small eccentricity agree with GOE after extracting the contributions from librational modes. This was also attributed to the discontinuity of the BC at the corners, where straight and curved parts are connected.

In [65], we computed the eigenstates of massive equilateral triangle NBs and their symmetry-projected eigenstates analytically and found that they coincide with those of the corresponding QB. Their short-range correlations exhibit the nontypical behavior expected for rectangular billiards whose side lengths are commensurable; however, otherwise, the spectral properties agree with semi-Poisson statistics. In contrast, those of the massless right-triangle NB, which is obtained by cutting the equilateral triangle NB along a mirror-symmetry axis, agree with Poisson [66]. These results are in contrast to those for the ellipse and circle NBs and corresponding sector NBs. They are attributed to the fact that the equilateral triangle NB has no curved boundary parts. In the present work, we investigate properties of the symmetry-projected and symmetry-reduced eigenstates of rectangular billiards with commensurable and incommensurable side length. The central question was whether the symmetry-projected eigenstates of rectangular NBs show a similar behavior as those of the equilateral triangle.

2. Review of Characteristic Features of Neutrino Billiards

In the two-dimensional plane $\boldsymbol{r} = (x, y)$, the Dirac equation for a free spin-1/2 particle with mass m_0 and momentum $\hat{\boldsymbol{p}} = -i\hbar\boldsymbol{\nabla}$ reads

$$\hat{\boldsymbol{H}}_D\boldsymbol{\psi} = \left(c\hat{\boldsymbol{\sigma}} \cdot \hat{\boldsymbol{p}} + m_0c^2\hat{\sigma}_z\right)\boldsymbol{\psi} = E\boldsymbol{\psi},\ \boldsymbol{\psi} = \begin{pmatrix}\psi_1\\ \psi_2\end{pmatrix}. \tag{1}$$

Here, $\hat{\boldsymbol{H}}_D$ denotes the Dirac Hamiltionian, $\hat{\boldsymbol{\sigma}} = (\hat{\sigma}_x, \hat{\sigma}_y)$, $\hat{\sigma}_{x,y,z}$ are the Pauli matrices, and $E = \hbar ck_E = \hbar ck\sqrt{1+\beta^2}$ is the energy of the particle, where k is the free-space wave vector and $\beta = \frac{m_0 c}{\hbar k}$ is the ratio of the rest-energy momentum and free-space momentum. The NB was introduced in Ref. [38] for the ultrarelativistic, i.e., massless case $m_0 = 0$. It is characterized by the way the particle is confined to the billiard domain Ω without destroying the self-adjointness of the Hamiltonian. This is ensured by imposing along its boundary $\partial\Omega$ on the solutions of Equation (1) the BC that the normal component of the local current, which is given by the expectation value of the current operator $\hat{\boldsymbol{u}} = \boldsymbol{\nabla}_p \hat{\boldsymbol{H}}_D = c\hat{\boldsymbol{\sigma}}$, $\boldsymbol{u}(\boldsymbol{r}) = c\boldsymbol{\psi}^\dagger \hat{\boldsymbol{\sigma}} \boldsymbol{\psi}$, vanishes, yielding independently of the mass [38,42],

$$\psi_2(t) = i\mu e^{i\alpha(t)} \psi_1(t). \tag{2}$$

Here, the boundary $\boldsymbol{r}(t) = [x(t), y(t)]$ or in the complex plane $w(t) = x(t) + iy(t)$ is parameterized in terms of t, and $\mu = \pm 1$ determines the rotational direction of the flow at the boundary, where it is unidirectional [38]. We chose $\mu = 1$. The parameter is defined for rectangular billiards with side lengths $2a$ and $2b$ in a coordinate system, whose origin is at its center and with the x-axis parallel to the sides with lengths $2a$ where $w(t) = t \pm ib$, $t \in [-a, a]$, while the y-axis is parallel to those with lengths $2b$ where $w(t) = \pm a + it$, $t \in [-b, b]$. Furthermore, $\alpha(t)$ is the angle of the outward-pointing normal vector $\boldsymbol{n}(t)$ at $\boldsymbol{r}(t)$ with respect to the x-axis. Another choice of parameter which is commonly used is the arc-length $s \in [0, \mathcal{L}]$ with $\mathcal{L}$ denoting the perimeter, which essentially corresponds to t plus the sum of the lengths of the sides that have been passed when moving along the boundary where we set $s = 0$ at $\boldsymbol{r} = (0, -b)$. In a local coordinate system (n, s) which moves counterclockwise along the boundary $(n = 0, s)$ and whose coordinate axes are in the directions of the tangential vector $\boldsymbol{t}(s)$ to $\partial\Omega$ at $\boldsymbol{r}(s)$ and the normal vector $\boldsymbol{n}(s)$, respectively, the combination of the Dirac Equation (1) at the boundary and Equation (2) yields with $(\partial_x \pm i\partial_y) = e^{\pm i\alpha}(\partial_n \pm i\partial_s)$ BCs in terms of separate equations for the wave-function components [42],

$$\begin{aligned} (\partial_n + i\partial_s)\psi_1(n,s)\big|_{n\to 0^-} &= -k\mathcal{K}^{-1}\psi_1(s), \\ (\partial_n - i\partial_s)\psi_2(n,s)\big|_{n\to 0^-} &= k\mathcal{K}\psi_2(s). \end{aligned} \tag{3}$$

Here, $\mathcal{K} = \sqrt{\frac{1-\sin\theta_\beta}{1+\sin\theta_\beta}}$ with $\sin\theta_\beta = \frac{\beta}{\sqrt{1+\beta^2}}$. Note that when introducing

$$\boldsymbol{\psi} = \begin{pmatrix} \sqrt{\frac{1+\sin\theta_\beta}{2}}\tilde{\psi}_1 \\ \sqrt{\frac{1-\sin\theta_\beta}{2}}\tilde{\psi}_2 \end{pmatrix}, \tag{4}$$

the Dirac equation Equation (1) takes the form of that for massless neutrinos with modified BCs,

$$k\tilde{\boldsymbol{\psi}}(\boldsymbol{r}) + i\hat{\boldsymbol{\sigma}} \cdot \nabla \tilde{\boldsymbol{\psi}}(\boldsymbol{r}) = 0 \tag{5}$$

$$\mathcal{K}\tilde{\psi}_2(s) = ie^{i\alpha(s)}\tilde{\psi}_1(s). \tag{6}$$

The nonrelativistic limit is reached when the energy is close to the rest energy, $E \simeq m_0 c^2$ [43], that is, for sufficiently large $\beta \to \infty$, corresponding to $\mathcal{K} \simeq \frac{1}{2\beta} \to 0$ and $\theta_\beta \to \pi/2$. Conversely, in the ultrarelativistic case $m_0 = 0$, they equal $\mathcal{K} = 1$ and $\theta_\beta = 0$. The BC imposes a phase relation on the wave-function components $\psi_{1,2}(s)$ at $\partial\Omega$ and provides a quantization condition whose solutions are the eigenstates of the Hamiltonian $\hat{\boldsymbol{H}}_{NB}$ associated with the NB. Alternative BCs for the confinement of relativistic particles to a bounded domain are proposed in [70,72] and based on the 'MIT' bag model [73].

Exact analytical solutions were derived for the equilateral triangle NB [65,70], the circle NB [38] and the ellipse NB [41] based on plane wave expansions. In [74–78], the eigenenergies of Dirac particles confined to a one- and a three-dimensional box were computed based on a plane-wave ansatz employing the MIT bag model. However, even though the BCs depend either on x or on y, a complete quantization of the rectangular NB with such an ansatz is not possible, because the x and y parts of the kinetic energy term do not commute [38]. A plane wave expansion

$$\psi_1(\mathbf{r}) = \sum_j a_j(k)e^{i\mathbf{k}_j\cdot\mathbf{r}} \tag{7}$$

$$\psi_2(\mathbf{r}) = \sum_j e^{i\theta_j}a_j(k)e^{i\mathbf{k}_j\cdot\mathbf{r}} \tag{8}$$

with $\mathbf{k}_j = k(\cos\theta_j, \sin\theta_j)$ and BC Equation (2) or, equivalently, Equation (3) yields only one-dimensional solutions corresponding to eigenmodes propagating parallel to the x or y axis, for which either $\theta_j = 0, \pi$ or $\theta_j = \frac{\pi}{2}, \frac{3\pi}{2}$. Therefore, we employed an extension of the boundary-integral equations (BIEs) derived in Ref. [38] for massless NBs to massive ones [42,79].

The derivation is based on Green's theorem, which provides exact integral equations for the eigenvalues and the spinor components in the interior of the billiard in terms of those on the boundary. An advantage of the boundary integral approach is that the eigenvalue problem is reduced from a two-dimensional differential equation to a one-dimensional boundary integral. The BIE is given by

$$(1-\sin\theta_\beta)\tilde{\psi}_1^*(\phi') = \frac{ik}{4}\int_0^{2\pi} |w'(\phi)|d\phi Q_1(\phi',\phi)\tilde{\psi}_1^*(\phi) \tag{9}$$

where the integration variable ϕ is related to s by $ds = |w'(\phi)|d\phi$ and

$$\begin{aligned} Q_1(\phi',\phi) =& \cos\theta_\beta\left[e^{i(\alpha(\phi')-\alpha(\phi))}-1\right]H_0^{(1)}(k\rho) \\ &+\left\{\left[1-\sin\theta_\beta\right]e^{i(\xi(\phi,\phi')-\alpha(\phi))}+\left[1+\sin\theta_\beta\right]e^{-i(\xi(\phi,\phi')-\alpha(\phi'))}\right\}H_1^{(1)}(k\rho), \end{aligned} \tag{10}$$

with

$$e^{i\xi(\phi,\phi')} = \frac{w(\phi)-w(\phi')}{|w(\phi)-w(\phi')|},\ \rho(\phi,\phi') = |w(\phi)-w(\phi')|. \tag{11}$$

Here, $H_m^{(1)}(k\rho) = J_m(k\rho) + iY_m(k\rho)$ is the Hankel function of the first kind of order m. At $\phi = \phi'$, i.e., $\rho = 0$, $H_0^{(1)}(k\rho)$ and $H_1^{(1)}(k\rho)$ have a logarithmic and a $1/\rho$ singularity. The integral over these singularities leads to the $\sin\theta_\beta$ term on the left-hand side [79]. Accordingly, an interval $[\phi'-\delta\phi, \phi'+\delta\phi]$, where $\delta\phi$ is arbitrarily small, is excluded from the integration range on the right-hand side. The corresponding BIE for $\tilde{\psi}_2^*(\phi')$ is obtained by employing in Equation (10) the BC in Equation (6) and those for the spinor components $\psi_{1,2}(\phi)$ are obtained with Equation (4). In the nonrelativistic limit $\sin\theta_\beta \to 1$, the left-hand side approaches zero, implying that the evaluation of the BIE becomes a numerical challenge. We would like to note that for $\sin\theta_\beta \simeq 1$, the BIE in Equation (10) does not provide a suitable quantization procedure since the spinor components decouple, $\psi_2(\mathbf{r})$ becomes vanishingly small and the BC for $\psi_1(\mathbf{r})$ turns into Robin BCs, which become Dirichlet BCs for $\sin\theta_\beta = 1$. For that reason, the BIEs are replaced in the nonrelativistic limit by one for its normal derivative [80].

We are interested in the spectral properties and properties of the wave functions of the associated symmetry-projected eigenstates of rectangular NBs which have mirror symmtries with respect to the x and y axes and a twofold rotational symmetry—for the square NB, ($a = b$) even a fourfold symmetry. Yet, in this respect, there is a crucial difference between QBs and NBs, which has its origin in the fact that the BCs for NBs connect the spinor components $\psi_1(s)$ and $\psi_2(s)$ at $\partial\Omega$ and may lead to distinct spectral properties, as outlined in [66,71,81]. The characteristics of NBs under a reflection or a rotation operator can be summarized as follows.

Applying on the coordinate vector an orthogonal transformation, $\mathbf{r}' = \hat{\mathbf{R}}\mathbf{r}$ corresponds to applying a unitary transformation $\hat{\mathbf{U}}$ to the Dirac Hamiltonian $\hat{H}_D(\mathbf{r})$. If $\tilde{\boldsymbol{\psi}}(\mathbf{r})$ is a solution of the Dirac equation in Equation (1), then the eigenfunction of the transformed Dirac Hamiltonian $\hat{\boldsymbol{H}}_D(\boldsymbol{r}') = \hat{\boldsymbol{U}}^\dagger \hat{\boldsymbol{H}}_D \hat{\boldsymbol{U}}$ is given by $\bar{\boldsymbol{\psi}}(\boldsymbol{r}') = \hat{\boldsymbol{U}}^\dagger \tilde{\boldsymbol{\psi}}(\boldsymbol{r})$. The unitary operators for a mirror reflection $\hat{\boldsymbol{R}}_x = \hat{\sigma}_z$ at the x-axis or $\hat{\boldsymbol{R}}_y = -\hat{\sigma}_z$ at the y-axis are $\hat{\boldsymbol{U}}_x = \hat{\sigma}_x$ or $\hat{\boldsymbol{U}}_y = i\hat{\sigma}_y$, respectively. However, application of the reflection operator $\hat{R}_X$ with $X = x$ or $X = y$ on a spinor eigenfunction $\tilde{\boldsymbol{\psi}}(\boldsymbol{r})$ of $\hat{\boldsymbol{H}}_{NB}(\boldsymbol{r})$ does not yield an eigenfunction of $\hat{\boldsymbol{H}}_{NB}(\boldsymbol{r}')$ [82], because $\bar{\boldsymbol{\psi}}(\boldsymbol{r}' = \hat{\boldsymbol{R}}_X \boldsymbol{r})$ does not fulfill the BC in Equation (2).

The unitary operator $\hat{\boldsymbol{U}}_M$ corresponding to a counterclockwise rotation by $\frac{2\pi}{M}$, $\boldsymbol{r}' = \hat{\boldsymbol{R}}_M \boldsymbol{r}$ with

$$\hat{\boldsymbol{R}}_M = \begin{pmatrix} \cos\left(\frac{2\pi}{M}\right) & -\sin\left(\frac{2\pi}{M}\right) \\ \sin\left(\frac{2\pi}{M}\right) & \cos\left(\frac{2\pi}{M}\right) \end{pmatrix}, \tag{12}$$

reads

$$\hat{\boldsymbol{U}}_M = \begin{pmatrix} e^{i\frac{\pi}{M}} & 0 \\ 0 & e^{-i\frac{\pi}{M}} \end{pmatrix}. \tag{13}$$

The eigenfunctions of the transformed Dirac Hamiltonian are $\bar{\boldsymbol{\psi}}^T_M(\boldsymbol{r}') = [e^{-i\frac{\pi}{M}}\tilde{\psi}_1(\boldsymbol{r}), e^{i\frac{\pi}{M}}\tilde{\psi}_2(\boldsymbol{r})]$. For NBs whose boundary has an M-fold rotational symmetry $w(s') = e^{i\frac{2\pi}{M}} w(s)$ and $e^{i\alpha(s')} = e^{i\frac{2\pi}{M}} e^{i\alpha(s)}$, the BC in Equation (2) is fulfilled for $\bar{\boldsymbol{\psi}}(\boldsymbol{r}')$ if $\tilde{\boldsymbol{\psi}}(\boldsymbol{r})$ is an eigenfunction of $\hat{\boldsymbol{H}}_{NB}$. Thus, the Hamiltonian $\hat{\boldsymbol{H}}_{NB}(\boldsymbol{r})$ can be brought to a block diagonal form according to the M one-dimensional irreducible representations labeled by $l = 0, \dots, M-1$; that is, its eigenstates can be grouped into M subspaces defined by their transformation properties under a rotation by $\frac{2\pi}{M}$ [83–86], yielding the symmetry-projected eigenstates

$$\psi^{(l)}_{1,2}(\hat{\boldsymbol{R}}^{-\lambda}_M \boldsymbol{r}) = e^{i\lambda\frac{2l\pi}{M}} \tilde{\psi}^{(l)}_{1,2}(\boldsymbol{r}), \lambda = 0, \dots M-1. \tag{14}$$

The wave-function components corresponding to $l = 0$ are invariant under rotation by $\frac{2\pi}{M}$, whereas for $l \neq 0$, a rotation by 2π is needed to recover the original ones. The Dirac equation in Equation (1) and the BC in Equation (6) relate components of the spinor eigenfunction with different transformation properties under rotation by $\frac{2\pi}{M}$ [65,71]. Namely, if the first component belongs to the subspace labeled by l,

$$\tilde{\psi}^{(l)}_1(\boldsymbol{r}') = e^{-il\frac{2\pi}{M}} \tilde{\psi}^{(l)}_1(\boldsymbol{r}) \tag{15}$$

where $\boldsymbol{r}' = \hat{\boldsymbol{R}}_M \boldsymbol{r}$, then inserting this property into the Dirac equation yields for the second one

$$\tilde{\psi}_2(\boldsymbol{r}') = e^{-i(l-1)\frac{2\pi}{M}} \tilde{\psi}_2(\boldsymbol{r}) \equiv \tilde{\psi}^{(l-1)}_2(\boldsymbol{r}'), \tag{16}$$

where $l = -1$ corresponds to $l = M-1$. Similarly, using the M-fold symmetry of the boundary of the NB, i.e., $e^{i\alpha(s')} = e^{i\left(\alpha(s) + \frac{2\pi}{M}\right)}$ and inserting Equation (15) into the BC in Equation (6) implies that [65,71]

$$\tilde{\psi}_2(s') = i e^{i\alpha(s')} \mathcal{K}^{-1} \tilde{\psi}^{(l)}_1(s') = e^{-i(l-1)\frac{2\pi}{M}} \tilde{\psi}_2(s) \equiv \tilde{\psi}^{(l-1)}_2(s').$$

Thus, the spinor components cannot be simultaneously rotationally invariant, implying that the spinor eigenfunctions cannot be rotationally invariant. This intermingling of symmetry properties has its origin in the additional spin degree of freedom [65,71].

For a billiard whose boundary has an M-fold rotational symmetry, the corresponding symmetry-reduced one is constructed by cutting it along the sides of a fundamental domain into sectors with inner angle $\phi_0 = \frac{2\pi}{M}$. The corresponding symmetry-reduced QB and NB are obtained by imposing the same BCs along the cutlines as along the outer bondary, i.e., as for the full QB, respectively, NB. These are Dirichlet BCs for the QB. Accordingly, their eigenstates coincide with rotationally invariant ones of the full QB. For the corresponding symmetry-reduced NB, the BCs in Equation (2) are imposed along the cutlines on the outgoing current, which is in the opposite direction to the current in the full NB at one of the cutlines [66]. Thus, the intermingling of symmetries in Equations (15) and (16) of the wave-function components implies that the eigenstates of an NB with a rotationally symmetric boundary cannot be eigenstates of the corresponding symmetry-reduced NB. Accordingly, their spectral properties do not necessarily coincide as demonstrated for circle and ellipse sectors in [66]. In the following, we will refer to the symmetry class of the spinor component $\tilde{\psi}_1^{(l)}(\boldsymbol{r})$ when specifying the value for l.

We would like to remark that the symmetry-projected eigenfunctions of a QB with M-fold rotational symmetry are complex for $l \neq 0, M/2$, implying that they are not invariant under application of the time-reversal operator $\hat{T}$ [83]. Indeed, the eigenfunctions with symmetry class l are transformed by $\hat{T}$ into eigenstates with $\tilde{l} = M - l$. Time-reversal invariance of the QB implies that the corresponding eigenvalues are doubly degenerate. It was demonstrated in [83] that the spectral properties of the states with $l \neq 0, M/2$ typically show GUE statistics, whereas those of the states with $l = 0, M/2$ are well described by the GOE if the shape of the QB generates chaotic classical dynamics. On the other hand, the Dirac Hamiltonian in Equation (1) itself does not commute with the time-reversal operator $\hat{T}$. Consequently, if the shape has no mirror symmetry, for all l, the eigenvalues of the corresponding NB are not degenerate, and the spectral properties of massless NBs typically coincide with those of the GUE.

In the present work, we investigate the spectral properties of symmetry-projected and symmetry-reduced rectangular NBs. To compute the eigenstates, we applied the BIE. It can be separated into individual BIEs for each symmetry class. To take into account the M-fold symmetry of the boundary $w(\phi)$ of an NB, we use the periodicity properties

$$w\left(\phi + \lambda\frac{2\pi}{M}\right) = e^{i\lambda\frac{2\pi}{M}} w(\phi), \tag{17}$$

$$w'\left(\phi + \lambda\frac{2\pi}{M}\right) = e^{i\lambda\frac{2\pi}{M}} w'(\phi), \tag{18}$$

$$e^{i\alpha\left(\phi + \lambda\frac{2\pi}{M}\right)} = e^{i\lambda\frac{2\pi}{M}} e^{i\alpha(\phi)}, \tag{19}$$

with $\lambda = 0, 1, 2, \dots, M-1$. Restricting the range of ϕ' to one fundamental domain, $\phi \in [0, \frac{2\pi}{M})$, the BIEs are given as

$$\tilde{\psi}_1^{(l)*}(\phi') = \int_0^{\frac{2\pi}{M}} d\phi \tilde{M}^{(l)}(k; \phi, \phi') \tilde{\psi}_1^{(l)*}(\phi) \tag{20}$$

with $l = 0, 1, , 2, \dots, M-1$ and

$$\tilde{M}_1^{(l)}(k; \phi, \phi') = \sum_{\lambda=0}^{M-1} e^{i\frac{2l\pi}{M}\lambda} M_{1,\lambda}(k; \phi, \phi'), \tag{21}$$

where

$$M_{1,\lambda}(k; \phi, \phi') = Q_1\left(k; \phi + \lambda\frac{2\pi}{M}, \phi'\right). \tag{22}$$

3. Tools Employed for the Analysis of Properties of the Eigenstates

We analyzed the spectral properties in terms of the nearest-neighbor spacing distribution $P(s)$, the integrated nearest-neighbor spacing distribution $I(s)$, the number variance $\Sigma^2(L)$ and the Dyson–Mehta statistic $\Delta_3(L)$, which provides a measure for the rigidity of a spectrum [2]. For this, the ordered eigenvalues $k_m = \sqrt{E_m}$ with $k_1 \leq k_2 \leq \dots$ were unfolded to mean spacing unity by replacing them with the smooth part of the integrated spectral density, $\epsilon_m = N^{smooth}(k_m)$, which is given for QBs by Weyl's formula [87] $N^{Weyl}(k_m) = \frac{\mathcal{A}}{4\pi}k_m^2 - \frac{\mathcal{L}}{4\pi}k_m + C_0$, with $\mathcal{A}$ denoting the area of the billiard, whereas for massless NBs, the perimeter contribution cancels out [38]. Furthermore, we analyzed distributions of the ratios [88,89] of consecutive spacings between nearest neighbors, $r_j = \frac{\epsilon_{j+1}-\epsilon_j}{\epsilon_j - \epsilon_{j-1}}$, which are dimensionless so that unfolding is not needed [88–90].

We also computed Husimi functions which are defined in classical phase space [5,36]. The classical dynamics of a CB is determined by the shape of its boundary $\partial\Omega$. Similarly, the eigenstates of the corresponding QB or NB are obtained by employing the BIEs originating from Green's theorem and thus are fully determined by their values along $\partial\Omega$. Therefore, an appropriate choice of the Poincaré surface of section (PSOS) is obtained by restricting the phase space to $\partial\Omega$, which commonly is defined in terms of Poincaré–Birkhoff coordinates (s, p), where $p = \sin\chi(s)$ with $\chi(s)$ denoting the angle of the particle trajectory with respect to the normal vector $\boldsymbol{n}(s)$. In Ref. [37], Poincaré–Husimi functions are defined for nonrelativistic QBs as the projection of the normal derivative of the associated wave function $\psi(s)$ at the boundary onto a coherent state [37], which is localized at $\partial\Omega$ and periodic with period $\mathcal{L}$,

$$\mathrm{H}_j(s,p) = \frac{1}{2\pi k_j}\frac{1}{\int_0^{\mathcal{L}} ds'|\partial_{n'}\psi(s')|^2}\left|\int_0^{\mathcal{L}} ds'\partial_{n'}\psi(s')C^{\delta}_{(s',p)}(s';k_j)\right|^2.$$

Here, $n' = n(s')$ and

$$C^{\delta}_{(s,p)}(s';k_j) = \left(\frac{k_j}{\pi\delta^2}\right)^{1/4}\sum_{m=-\infty}^{\infty}\exp\left(ipk_j(s'-s+m\mathcal{L}) - \frac{k_j}{2\delta^2}(s'-s+m\mathcal{L})^2\right). \quad (23)$$

The parameter δ controls the resolution of the Husimi plots. The Poincaré–Husimi function provides a probability density of the eigenstates on the Poincaré surface of section [37] and accordingly is commonly referred to as a quantum Poincaré surface of section. We computed Husimi functions for the NB by replacing $\psi(s)$ by the symmetry-projected spinor eigenfunctions $\psi^{(l)}_{1,2}(s)$, where we employed Equation (3) to determine their normal derivatives.

We, furthermore, computed momentum distributions [35], i.e., the Fourier transform of the spinor eigenfunctions $\boldsymbol{\psi}_n(\boldsymbol{r})$ from coordinate space (x, y) to momentum space (q_x, q_y),

$$\boldsymbol{\psi}_n(\boldsymbol{q}) = \iint_{\Omega} d\boldsymbol{r}e^{i\boldsymbol{q}\boldsymbol{r}}\boldsymbol{\psi}_n(\boldsymbol{r}). \quad (24)$$

They are localized on the energy shell, that is, at values $q = \sqrt{q_x^2 + q_y^2} = k_n$, and provide information on the directions of the plane waves that form the eigenmode, which is especially useful when it is scarred along periodic orbits or localized.

4. Numerical Results for the Symmetry-Projected Eigenstates of Rectangular NBs

In this section, we present numerical results for the spectral properties and properties of the wave functions, Husimi functions and momentum distributions of the symmetry-projected eigenstates of rectangular NBs with different side-length ratios,

$$\begin{aligned}
R_1(T_1) &: \frac{a}{b} = 1 \\
R_2(T_2) &: \frac{a}{b} = 2 \\
R_3(T_3) &: \frac{a}{b} = \frac{\sqrt{5}+1}{2} \\
R_4(T_4) &: \frac{a}{b} = \frac{\sqrt{2}+\sqrt{\pi}}{2} \\
R_5(T_5) &: \frac{a}{b} = \sqrt{3}
\end{aligned} \quad (25)$$

and their symmetry-reduced counterparts, namely triangles T_i, that are obtained by cutting the corresponding rectangle R_i along its diagonal. The triangles T_2, T_3, T_4 have in common that their angles have the structure $\left\{\alpha_1\pi, \alpha_2\pi, \frac{\pi}{2}\right\}$ with α_1, α_2 irrational numbers.

Furthermore, in [66], we considered a triangle, named T_5, which is obtained by cutting the equilatarал triangle with side lengths 2 along a mirror axis into two halves. Its inner angles are $\{\frac{\pi}{2}, \frac{\pi}{3}, \frac{\pi}{6}\}$, implying that it is integrable like the R_1 and T_1 billiards. We computed 5000 eigenvalues for each symmetry class and NB. These studies were motivated by recent results which we obtained in Ref. [66] for the semi-circle NB, the quarter-ellipse NB for various values of its eccentricity, and the T_5 triangle NB [65]. For an ellipse with semimajor and semiminor axes lengths $a = \cosh\mu_0$ and $b = \sinh\mu_0$, respectively, the eccentricity is $\epsilon = \frac{1}{\cosh\mu_0}$. The classical dynamics of a particle in an ellipse CB and the semi- and quarter-ellipse CBs, which are obtained by cutting the ellipse along the semiminor and semimajor axes, respectively, is integrable because the product of the angular momenta with respect to the two focal points is a constant of motion. The eigenfunctions of the corresponding QBs are given by products of the radial and the periodic Mathieu functions [69,91–94], and the eigenvalues are the zeroes of the radial Mathieu function at the boundary. Their spectral properties coincide with those of Poissonian random numbers. The orbits of the ellipse CB and the eigenmodes of the ellipse QB can be separated into librational modes and rotational modes. With decreasing ϵ, the ellipse turns into a circle, and the librational modes turn into the diameter orbit, whereas with increasing eccentricity, the modes resemble those in the rectangular billiard. The eigenfunctions of the ellipse NB are given in terms of superpositions of products of the radial and periodic Mathieu functions [41]. It has a twofold symmetry so that the spinor components ψ_1 are either symmetric with respect to a rotation by π and the associated second component ψ_2 is antisymmetric, or vice versa.

We showed in [66] that the spectral properties of circle-sector NBs and quarter ellipse-sector NBs with small eccentricity, i.e., nearly circular shape, are well described by the GOE after the extraction of contributions from the diameter orbit in the former one and librational modes in the latter one that bounce back and forth parallel to the semiminor axis and have a vanishing support at the corners connecting straight and curved parts. This behavior was attributed to the intermingling of the symmetry properties associated with the rotational symmetry of the corresponding full NBs and the effects of these corners. Indeed, for the T_5 NB, the spectral properties are close to Poisson statistics [66], whereas we found good agreement with semi-Poissonian statistics for the symmetry-projected eigenstates of the equilateral triangle NB and QB [65].

In Figure 1, we present some results for the semi-ellipse NB for different eccentricities. Shown are the spectral properties of the ellipse NB and its symmetry-projected eigenstates for $\epsilon = 0.265$ and for the semi-ellipse NBs with $\epsilon = 0.1, 0.5, 0.65$. Those of the symmetry-projected eigenstates are close to those of the semi-ellipse NB with $\epsilon = 0.65$, which is close to Poisson, whereas with decreasing ϵ, they approach semi-Poisson statistics, as is clearly visible in the ratio distributions which are shown in the right part of Figure 1.

In Figure 2, we show results for the square NB (R_1 in Equation (25)), its symmetry-projected eigenstates labeled by $l = 0, 1, 2, 3$ and the 45° triangle (T_1 in Equation (25)).

The spectral properties of the symmetry-projected eigenstates agree well with semi-Poisson for the short-range correlations up to a certain number of mean spacings L for the long-range correlations. The size of the deviations is similar to that of a QB with integrable classical dynamics [20,24] for a similar number of eigenvalues, such as, e.g., the ellipse and circle QBs and NBs [41,66]. The overshooting of those for the square NB with respect to Poisson statistics originates from non-systematic, occasionally occurring degeneracies of eigenvalues associated with different symmetry classes, which have also been observed for the equilateral triangle NB in Ref. [65]. The ratio distribution of the R_1 QB is shown together with that of the R_1 NB in the right part of Figure 2 in (c). It exhibits the nongeneric behavior commonly observed for rectangular QBs with rational ratios of side lengths, whereas that of the NB agrees well with Poisson statistics.

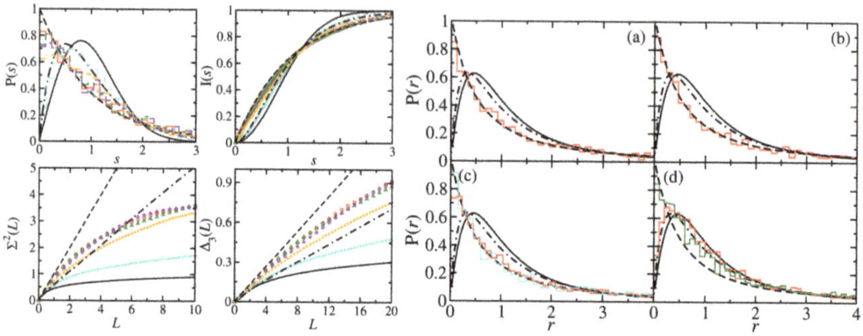

Figure 1. (**Left**) Spectral properties of the symmetry-projected eigenstates of the ellipse NB with $\epsilon = 0.65$ (red histogram and circles: $l = 0$; green histogram and triangles: $l = 1$), and the semi-ellipse NB with $\epsilon = 0.65$ (violet histogram and diamonds), $\epsilon = 0.5$ (orange histogram and plus) and $\epsilon = 0.1$ (turquoise histogram and stars). (**Right**) Ratio distributions for (**a**) the symmetry-projected eigenstates of the ellipse NB with $l = 0$, (**b**) $l = 1$, (**c**) all eigenvalues of the ellipse NB (red) and the corresponding ellipse QB (turquoise) and (**d**) the semi-ellipse NB for $\epsilon = 0.5$ (green) and $\epsilon = 0.1$ (red). They are compared to the GOE (black solid line), Poisson (black dashed line) and semi-Poisson (black dashed–dotted lines) statistics.

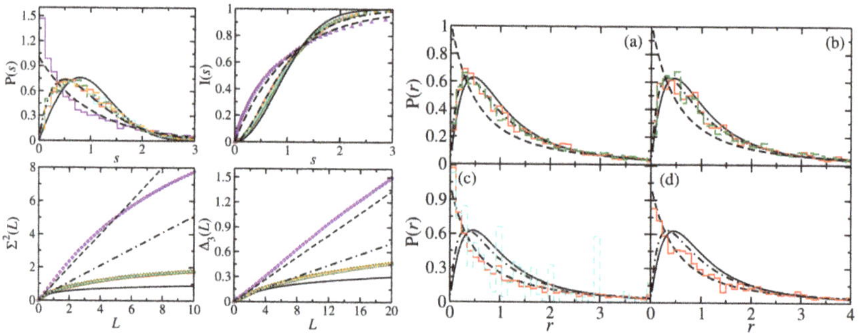

Figure 2. (**Left**) Spectral properties of the R_1, i.e., square NB (violet histogram and diamonds) and its symmetry-projected eigenstates (red histogram and circles: $l = 0$; green histogram and triangles: $l = 1$; orange histogram and plus: $l = 2$; turquoise histogram and stars: $l = 3$). (**Right**) Ratio distributions for (**a**) the symmetry-projected eigenstates of the R_1 NB with $l = 0$ (red) and $l = 2$ (green), (**b**) $l = 1$ (red) and $l = 3$ (green), (**c**) all eigenvalues of the square NB (red) and the corresponding square QB (turquoise) and (**d**) the 45°-triangle, i.e., T_1 NB (red). They are compared to the GOE (black solid line), Poisson (black dashed line) and semi-Poisson (black dashed–dotted lines) statistics.

In the left part of Figure 3, we compare length spectra, that is, the modulus of the Fourier transform $|\tilde{\rho}^{fluc}(l)|$ of the fluctuating part of the spectral density, $\rho^{fluc}(k)$, from wave number k to length l for the full square NB, the symmetry-projected ones and the 45° triangle NB. The length spectra of the QB and NB exhibit peaks at the lengths of its periodic orbits. Generally, in the length spectra of NBs, peaks at lengths, which correspond to periodic orbits with an odd number of reflections, are missing [42]. Such orbits are absent in rectangular CBs. Those of the symmetry-projected eigenstates show additional peaks at lengths which correspond to pseudo-orbits, that is, orbits that are periodic in the fundamental domains but not in the full system [85,86].

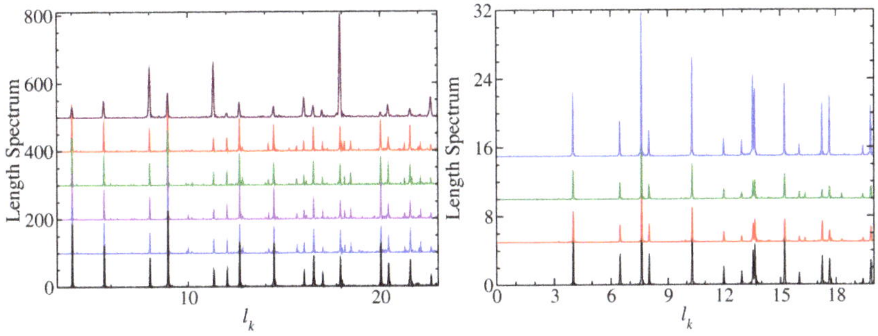

Figure 3. (**Left**) Comparison of the length spectra of, from bottom to top, the full square NB (black line), for $l = 0$ (blue line), $l = 1$ (violet line), $l = 2$ (green), $l = 3$ (red), and the square QB (maroon). (**Right**) Comparison of the length spectra of, from bottom to top, the rectangular R_3 NB (black line), for $l = 0$ (red line), $l = 1$ (green line), and the corresponding QB (blue).

In Figure 4, we show examples for the momentum distributions, real parts of the spinor components $\psi_1(\boldsymbol{r})$ and $\psi_2(\boldsymbol{r})$, the local current $|\boldsymbol{u}(\boldsymbol{r})| \propto \left|\left(\Re[\psi_1^*(\boldsymbol{r})\psi_2(\boldsymbol{r})], \Im[\psi_1^*(\boldsymbol{r})\psi_2(\boldsymbol{r})]\right)\right|$ and Husimi functions for the symmetry-projected eigenstates of the R_1 NB with $l = 0$ (a) and $l = 1$ (b). The wave-function patterns are invariant under rotation by $\frac{\pi}{2}$ for $\psi_1(\boldsymbol{r})$ and $l = 0$, and for $\psi_2(\boldsymbol{r})$ and $l = 1$. For the other components, they need to be rotated by 2π to recover the original patterns. The momentum distributions are peaked around values $(q_x, q_y) = (\pm k_x, \pm k_y)$ and $(q_x, q_y) = (\pm k_y, \pm k_x)$ along a circle, whose radius is defined by the eigenwavenumber $q = \sqrt{q_x^2 + q_y^2} = k_n$ corresponding to the eigenstate number n. For $n = 68$ and $l = 0$ or for $n = 75$ and $l = 1$, it is peaked at $(\pm k_n, 0)$ and $(0, \pm k_n)$ and exhibit chessboard structures which resemble those of the square QB, except that their intensities decrease with the distance from the center of the square. Otherwise, the wave-function patterns are distinct and more complex. In addition, from the pattern structure of the local currents, we may conclude that the eigenfunctions are not given by simple superpositions of plane waves. Some of the Husimi functions are localized either in the upper or the lower part of the PSOS, indicating that the corresponding eigenmodes propagate in a preferred direction. This may be attributed to the fact that the BCs in Equation (2) lead to a unidirectionality of the local current along the boundary. The wave-function patterns are more complex than for the corresponding QB; however, they exhibit a regular structure, so that the GOE-like behavior of the spectral properties is not expected.

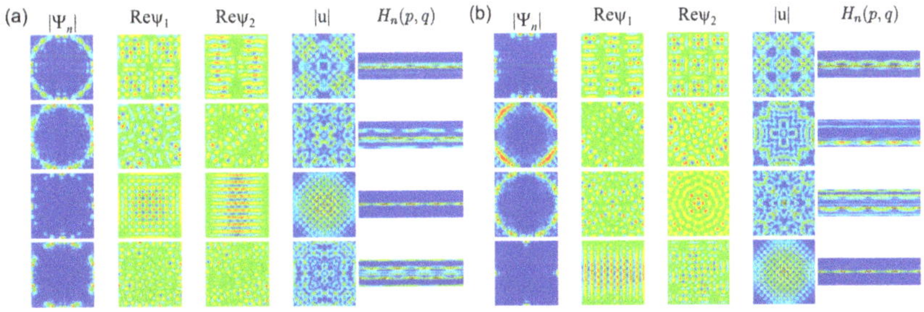

Figure 4. (**a**) From left to right, the momentum distribution in the (q_x, q_y) plane, the real parts of the spinor components $\psi_1(\boldsymbol{r})$ and $\psi_2(\boldsymbol{r})$ function in the (x, y) plane, the local current in the (x, y) plane and the Husimi functions in the Birkhoff coordinate plane (s, p), where $s = 0$ at the center of the lower horizontal side and increases in counterclockwise direction, for the symmetry-projected eigenstates states of the square NB with $l = 0$ and, from top to bottom, numbers $n = 49, 51, 68, 75$. (**b**) Same as left for $l = 1$.

Similar results are obtained for the spectral properties of the rectangular R_2 and R_3 NBs and their symmetry-projected eigenstates, shown in the left and right part of Figure 5, respectively, and in the right part of Figure 6. The gaps observed in the nearest-neighbor spacing distribution of the R_2 QB are typical for rectangular QBs whose ratios of side lengths are no irrational numbers [95]. The long-range correlations, on the other hand, approach Poisson statistics with an increasing number of eigenvalues [24,96]. The spectral properties of the R_2 NB exhibit Poisson statistics, whereas those of its symmetry-projected eigenstates are well described by semi-Poisson statistics. In the right part of Figure 3, we show length spectra for the R_3 QB and NB and the symmetry-projected eigenstates of the R_3 NB. The length spectra of the QB and NB exhibit peaks at the same lengths, as there are only periodic orbits with an even number of reflections. The length spectra of the symmetry-projected eigenstates also exhibit peaks at the lengths of the classical periodic orbits and a few additional ones at the lengths of peudo-orbits.

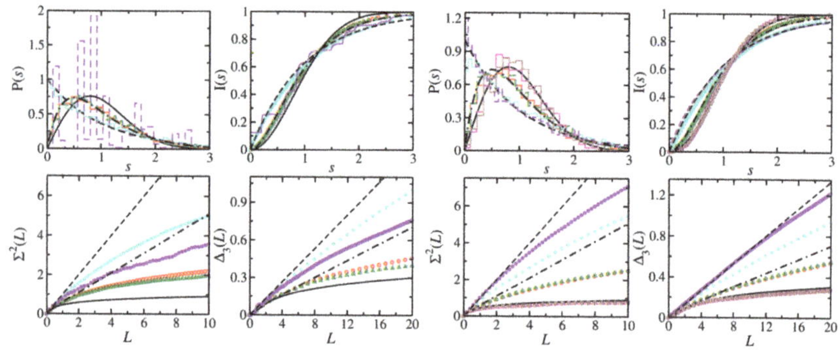

Figure 5. **(Left)** Spectral properties of the symmetry-projected eigenstates of the rectangular R_2 NB (red histogram and circles: $l = 0$; green histogram and triangles: $l = 1$), the R_2 QB (violet histogram, lines and dots), and the R_2 NB (turquoise histogram and diamonds). **(Right)** Spectral properties of the symmetry-projected eigenstates of the R_3 NB (red histogram and circles: $l = 0$; green histogram and triangles: $l = 1$), the R_3 QB (violet histogram, lines and dots), the R_3 NB (turquoise histogram and diamonds) and the triangular T_3 (magenta histogram and squares) and T_4 NB (brown histogram and dots). They are compared to the GOE (black solid line), Poisson (black dashed line) and semi-Poisson (black dashed–dotted lines) statistics.

In the right part of Figure 5 and in Figure 6, we include the results for the triangular T_3 and T_4 NBs which are constructed by cutting the rectangular R_3 and R_4 NBs along the diagonals. Their spectral properties are close to GOE. We found out that they, actually, agree well with those of a quarter-Poissonian sequence which is obtained by taking from an ordered sequence of Possonian random numbers each fourth number or, equivalently, every second one from a sequence of semi-Poisson numbers, such as those of the corresponding symmetry-projected eigenvalue sequences. The spectral properties of the T_3 and T_4 QBs agree well with semi-Poisson statistics [34]. This is illustrated in the left part of Figure 6. Those of the T_5 NB coincide with Poissonian statistics [66]. Right triangles, that have only one angle which is rational with respect to π, have been studied in detail in Refs. [21,23,34,97,98]. It was shown that triangles, whose angles are all irrational with respect to π, exhibit GOE-like spectral properties, whereas those of right triangles are non-Poissonian but differ from GOE. Deviations from GOE were shown to originate from the presence of wave functions that are scarred along bouncing-ball orbits that exhibit a regular pattern, such as the periodic-orbit family consisting of orbits that are reflected perpendicular to the tilted side of the triangle and form a periodic-orbit channel (POC) whose maximum width extends from one diffractive corner to the other one (see below) [19,99,100]. Their contributions, in fact, can be extracted for irrational triangles by proceeding as, e.g., in [101], whereas for the T_3 and T_4 QBs and NBs, their number is too large. This explains the agreement with semi-Poisson statistics for the T_3 and T_4 QBs [34], which corresponds to a linear level repulsion and is a special case of intermediate statistics for which analytical expressions exist. These are shown as dashed–dotted lines, e.g., in Figure 6. For the corresponding NBs, the level repulsion is cubic [33]. After the extraction of contributions from bouncing-ball orbits, the spectral properties of the T_2 QB are close to GOE behavior, whereas those of the T_2 NB agree well with GUE statistics, as illustrated in Figure 7, implying that the effect of scarred wave functions is stronger for the QB than it is for the NB. Indeed, scarred wave functions occur more rarely for the T_2 NB than for the T_2 QB. The ratio distributions, shown in the right part of Figure 7, agree well with GOE and GUE for the T_2 QB and NB, respectively, even though contributions from periodic orbits that lead to scarred wave functions were not extracted, implying that it is insensitive to scarred states. Note that no unfolding is needed for the analysis of ratio distributions. The left part of Figure 8 shows the fluctuating part of the spectral density (black dots). The contributions from bouncing-ball orbits lead to slow oscillations. To determine them, we used the procedure introduced in [101], that is, we computed the Fourier transform of the fluctuating part of the spectral density, $\tilde{\rho}^{fluc}(l)$, and then the inverse Fourier transform over those parts that correspond to lengths of these orbits. The associated peaks are plotted as orange and turquoise dashed lines for the NB and QB, respectively, in the length spectrum shown in the right part of Figure 8. The resulting oscillating part of $N(k)$, $N^{osc}(k)$, is shown as red dots connected by a red dashed line in the left part of Figure 8. To extract the contributions of this part to the spectral properties, the eigenvalues k_m were unfolded by replacing them with the sum of the smooth and oscillating part of $N(k)$, $\epsilon_m = N^{smooth}(k_m) + N^{osc}(k_m)$ [102].

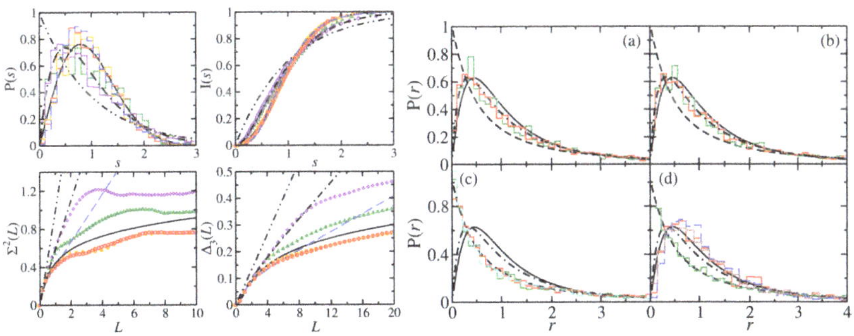

Figure 6. (**Left**) Spectral properties of the triangular T_3 NB (red histogram and circles) and T_3 QB (green histogram and triangles) and the triangular T_4 NB (orange histogram and dots) and T_4 QB (violet histogram and diamonds). They are compared to the GOE (black solid line), Poisson (black dashed–dotted–dotted line), semi-Poisson (black dashed–dotted lines) and quarter-Poisson (blue dashed lines) statistics. (**Right**) Ratio distributions for (**a**) the $l = 0$ states of the R_2 (green histogram) and R_3 (red histogram) NBs, (**b**) same as (**a**) for the states with $l = 1$, (**c**) the rectangular R_2 (green histogram) and R_3 (red histogram) NBs and for the rectangular R_3 QB (turquoise histogram), (**d**) the T_5 (green histogram), T_3 (red histogram) and T_4 (brown histogram) NBs. They are compared to the GOE (black solid line), Poisson (black dashed line), semi-Poisson (black dashed–dotted lines) and quarter-Poisson (blue dashed lines) statistics.

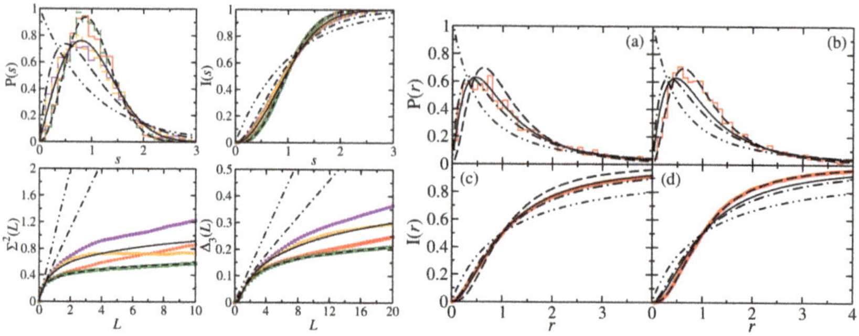

Figure 7. (**Left**) Spectral properties of the triangular T_2 QB and NB before ((violet histogram and stars) and (red histogram and dots)) and after extraction of contributions from orbits that lead to scarred wave functions ((orange dashed-line histogram and crosses) and (green histogram and squares)). They are compared to the GOE (black solid line), Poisson (black dashed–dotted–dotted line), semi-Poisson (black dashed–dotted lines) and GUE (black dashed lines) statistics. (**Right**) Ratio distributions $P(r)$ (upper panels) and integrated ratio distributions $I(r)$ (lower panels) for the T_2 QB (**a**,**c**) and NB (**b**,**d**). They are compared to the GOE (black solid line), Poisson (black dashed–dotted–dotted line), semi-Poisson (black dashed–dotted lines) and GUE (black dashed lines) statistics.

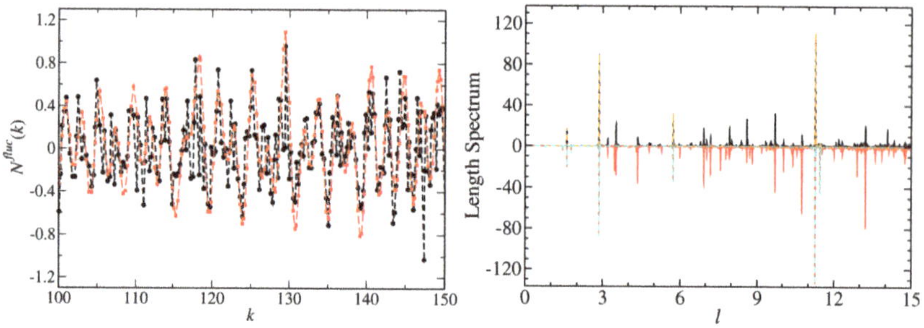

Figure 8. (**Left**) Fluctuating part of the integrated spectral density (black) and the contributions of bouncing-ball orbits that are extracted to obtain the results shown in Figure 7 (red) for the T_2 NB. (**Right**) Comparison of the length spectra of the T_2 NB (upper part) and the T_2 QB (lower part).

In Figure 9, we show momentum distributions, real parts of the spinor components and Husimi functions for a few symmetry-projected eigenstates of the R_3 NB ((a): $l = 0$; (b): $l = 1$) and in Figure 10 for the corresponding right triangle T_3 NB (a) and T_3 QB (b). The momentum distributions are peaked at values $(q_x, q_y) = (\pm k_x, \pm k_y)$ along the circle defined by the eigenwavenumbers $k_n = \sqrt{q_x^2 + q_y^2}$. In the example shown for $l = 1$ and $n = 57$, $\boldsymbol{q}$ takes the values $(q_x, q_y) = (\pm k_n, 0)$. This case, in fact, can be derived based on a plane-wave ansatz, since the propagation of the eigenmodes is one-dimensional [77]. The wave-function patterns shown for $l = 0$ in the second row are similar to corresponding ones in the R_3 QB. In these examples, the Husimi functions are localized along the $p = 0$ axis. For the other cases, they again exhibit a preferred direction of propagation, but they are distributed over the (s, p) plane. Scarred wave functions of the type presented in [23] are observed in the corresponding symmetry-reduced billiard, i.e., the T_3 QB and NB. The examples shown in Figure 10 correspond to relatively low-lying states. Here, we chose for the QB eigenstates for which the wave functions and Husimi functions exhibit similar patterns to those of the NB. We show them because for higher excitations, the wave function and local current patterns become complex and are no longer discernible. Still, some of them exhibit scarred eigenstates, such as those shown in the third row. In Figure 11, we show examples of scarred wave functions for higher-lying eigenstates of the T_3 NB (upper row) and similar ones for the T_3 QB (lower row). They are scarred along bouncing ball orbits from the family of periodic orbits that hit the tilted side perpendicularly and are reflected with constant angles from the sides to the left and right of the $\frac{\pi}{2}$ corner.

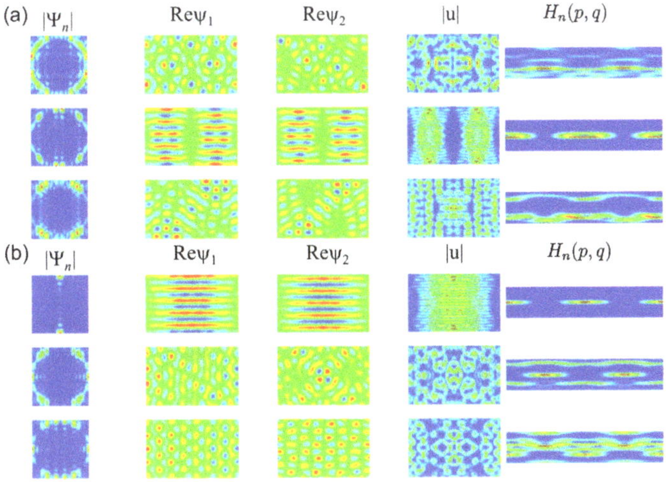

Figure 9. (**a**) Same as in Figure 4 for the symmetry-projected eigenstates of the rectangular R_3 NB with $l = 0$ and, from top to bottom, numbers $n = 57$–59. (**b**) Same as (**a**) for $l = 1$.

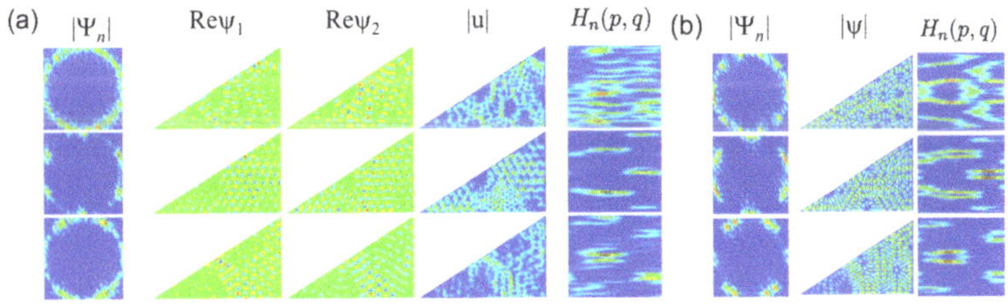

Figure 10. (**a**) Same as Figure 4 for the triangular T_3 NB with, from top to bottom, numbers $n = 255, 288, 264$. Here, the Borkhoff coordinates are chosen such that $s = 0$ at the left corner of the triangle, and it increases in the counterclockwise direction. (**b**) From left to right, momentum distributions, intensity distributions and Husimi functions for the corresponding QB with, from top to bottom, numbers $n = 266, 259, 267$.

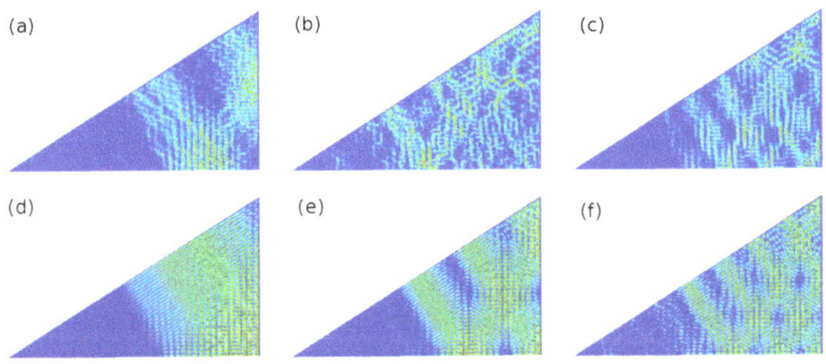

Figure 11. First row: Local current for the eigenstates of the T_3 NB for, from left to right, numbers n =1460 (**a**), 1453 (**b**), 1471 (**c**). Second row: intensity distribution for the eigenstates of the T_3 QB for, from left to right, numbers n =1459 (**d**), 1463 (**e**), 1469 (**f**).

5. Conclusions

We analyzed the properties of symmetry-projected eigenstates of rectangular NBs whose side lengths are either commensurable or incommensurable. Independently of the choice of the ratio of the side lengths, we find very good agreement of the spectral properties of the symmetry-projected eigenstates with semi-Poisson statistics. In distinction to the eigenvalues of rectangular QBs, those of the corresponding NBs are non-degenerate and exhibit Poisson statistics. The fact that the complete spectra of rectangular NBs exhibit Poisson statistics—whereas after their separation into two (or four) sequences of eigenvalues corresponding to the symmetry-projected eigenstates, the spectral properties follow semi-Poisson statistics—indicates that rectangular NBs behave like typical quantum systems with an integrable counterpart, whose eigenstates are non-degenerate and have alternating symmetry properties with increasing state number. Note that this is not strictly true, because when two eigenvalues come close to each other, then the ordering of the symmetry classes may change. Generally, Poisson statistics is expected for the symmetry-projected states if they are uncorrelated, as is the case for the corresponding nonrelativistic QB; however, in the ultrarelativistic case, they are linked through the different symmetry properties of the spinor components. Namely, if the first component belongs to class l, then the second one belongs to class $(l-1)$. Furthermore, we found out that the eigenvalues of triangular NBs, which are constructed by cutting the rectangular NB with incommensurable side lengths along the diagonal and follow semi-Poisson statistics in the nonrelativistic limit, exhibit a cubic-level repulsion and spectral properties that agree well with those of random numbers. These random numbers are composed of every fourth number from a Poisson sequence or every second one from a semi-Poisson one. That is, we observe a hierarchy, induced by the variation of the BCs along the diagonal of the R_3 and R_4 NBs, in the spectral properties in the sense that they follow Poisson statistics for the full rectangle NB, semi-Poisson statistics for the symmetry-projected eigenstates and corresponding triangle QB and quarter-Poisson statistics for the symmetry-reduced NB, which would correspond to taking every second level from successive eigenvalue sequences. For the T_2 QB and NB, we find good agreement with GOE and GUE statistics, respectively, after the extraction of contributions from bouncing-ball orbits, that is, from eigenstates whose wave functions are scarred along these orbits. An important difference between a QB and the corresponding NB is that the associated Hamiltonian preserves and violates time-reversal invariance, respectively. Thus, the results for the T_2 NB are not surprising, and accordingly, we may deduce from our results for the T_3 and T_4 QBs and NBs that when inducing time-reversal invariance violation in a QB which exhibits semi-Poisson statistics, it will yield quarter-Poisson statistics. We chose these triangles because their spectral properties are particular and only considered the ultrarelativistic and the nonrelativistic limits. The transition region was investigated in detail in Refs. [65,71] for the symmetry-projected eigenstates of NBs whose boundaries have a threefold rotational symmetry. The results confirm that, as was shown in [42], with increasing mass, the spectral properties approach those of the nonrelativistic QB.

Funding: BD acknowledges financial support from the Institute for Basic Science in Korea through the project IBS-R024-D1.

Institutional Review Board Statement: Not applicable.

Informed Consent Statement: Not applicable.

Data Availability Statement: All data are contained within this article.

Conflicts of Interest: The author declares no conflict of interest.

References

1. Casati, G.; Valz-Gris, F.; Guarnieri, I. On the connection between quantization of nonintegrable systems and statistical theory of spectra. *Lett. Nuovo Cim.* **1980**, *28*, 279–282. [CrossRef]
2. Mehta, M.L. *Random Matrices*; Academic Press: London, UK, 1990.

3. Bohigas, O.; Giannoni, M.J.; Schmit, C. Characterization of Chaotic Quantum Spectra and Universality of Level Fluctuation Laws. *Phys. Rev. Lett.* **1984**, *52*, 1. [CrossRef]
4. Berry, M. *Structural Stability in Physics*; Pergamon Press: Berlin, Germany, 1979.
5. Giannoni, M.; Voros, A.; Zinn-Justin, J. (Eds.) *Chaos and Quantum Physics*; Elsevier: Amsterdam, The Netherlands, 1989.
6. Guhr, T.; Müller-Groeling, A.; Weidenmüller, H.A. Random-matrix theories in quantum physics: Common concepts. *Phys. Rep.* **1998**, *299*, 189. [CrossRef]
7. Haake, F.; Gnutzmann, S.; Kuś, M. *Quantum Signatures of Chaos*; Springer: Berlin/Heidelberg, Germany, 2018.
8. Heusler, S.; Müller, S.; Altland, A.; Braun, P.; Haake, F. Periodic-orbit theory of level correlations. *Phys. Rev. Lett.* **2007**, *98*, 044103. [CrossRef]
9. Gutzwiller, M.C. Periodic Orbits and Classical Quantization Conditions. *J. Math. Phys.* **1971**, *12*, 343–358. . 1.1665596. [CrossRef]
10. Gutzwiller, M.C. *Chaos in Classical and Quantum Mechanics*; Springer: Berlin/Heidelberg, Germany, 1990.
11. Sinai, Y.G. Dynamical systems with elastic reflections. *Russ. Math. Surv.* **1970**, *25*, 137. [CrossRef]
12. Bunimovich, L.A. On the Ergodic Properties of Nowhere Dispersing Billiards. *Commun. Math. Phys.* **1979**, *65*, 295. [CrossRef]
13. Berry, M.V. Regularity and chaos in classical mechanics, illustrated by three deformations of a circular 'billiard'. *Eur. J. Phys.* **1981**, *2*, 91. [CrossRef]
14. Berry, M.V.; Tabor, M. Calculating the bound spectrum by path summation in actionangle variables. *J. Phys. A* **1977**, *10*, 371. [CrossRef]
15. Robnik, M.; Veble, G. On spectral statistics of classically integrable systems. *J. Phys. A* **1998**, *31*, 4669. [CrossRef]
16. Drożdż, S.; Speth, J. Near-ground-state spectral fluctuations in multidimensional separable systems. *Phys. Rev. Lett.* **1991**, *67*, 529–532. [CrossRef]
17. Gutkin, E. Billiards in polygons. *Physica D* **1986**, *19*, 311–333. [CrossRef]
18. Gutkin, E. Billiards in polygons: Survey of recent results. *J. Stat. Phys.* **1996**, *83*, 7–26. [CrossRef]
19. Bogomolny, E. Formation of superscar waves in plane polygonal billiards. *J. Phys. Commun.* **2021**, *5*, 055010. [CrossRef]
20. Casati, G.; Chirikov, B.V.; Guarneri, I. Energy-Level Statistics of Integrable Quantum Systems. *Phys. Rev. Lett.* **1985**, *54*, 1350–1353. [CrossRef]
21. Artuso, R.; Casati, G.; Guarneri, I. Numerical study on ergodic properties of triangular billiards. *Phys. Rev. E* **1997**, *55*, 6384–6390. [CrossRef]
22. Casati, G.; Prosen, T. Mixing Property of Triangular Billiards. *Phys. Rev. Lett.* **1999**, *83*, 4729–4732. [CrossRef]
23. Lozej, Č.; Casati, G.; Prosen, T. Quantum chaos in triangular billiards. *Phys. Rev. Res.* **2022**, *4*, 013138. [CrossRef]
24. Marklof, J. Spectral Form Factors of Rectangle Billiards. *Comm. Math. Phys.* **1998**, *199*, 169. [CrossRef]
25. Zemlyakov, A.N.; Katok, A.B. Topological transitivity of billiards in polygons. *Mat. Notes* **1975**, *18*, 291. [CrossRef]
26. Richens, P.J.; Berry, M.V. Pseudointegrable systems in classical and quantum mechanics. *Physica D* **1981**, *2*, 495. [CrossRef]
27. Mirbach, B.; Korsch, H.J. Longlived states and irregular dynamics in inelastic collisions: Analysis of a polygon billiard model. *Nonlinearity* **1989**, *2*, 327. [CrossRef]
28. Życzkowski, K. Classical and quantum billiards-integrable, nonintegrable and pseudo-integrable. *Act. Phys. Pol. B* **1992**, *49*, 245–270.
29. Biswas, D.; Jain, S.R. Quantum description of a pseudointegrable system: The $\pi/3$-rhombus billiard. *Phys. Rev. A* **1990**, *42*, 3170–3185. [CrossRef] [PubMed]
30. Shudo, A.; Shimizu, Y. Extensive numerical study of spectral statistics for rational and irrational polygonal billiards. *Phys. Rev. E* **1993**, *47*, 54–62. [CrossRef]
31. Shudo, A.; Shimizu, Y.; Šeba, P.; Stein, J.; Stöckmann, H.J.; Życzkowski, K. Statistical properties of spectra of pseudointegrable systems. *Phys. Rev. E* **1994**, *49*, 3748–3756. [CrossRef]
32. Hasselblatt, B.; Katok, A. (Eds.) *Handbook of Dynamical Systems*; Elsevier: Amsterdam, The Netherlands, 2002; Volume 1A.
33. Bogomolny, E.B.; Gerland, U.; Schmit, C. Models of intermediate spectral statistics. *Phys. Rev. E* **1999**, *59*, R1315–R1318. [CrossRef]
34. Gorin, T. Generic spectral properties of right triangle billiards. *J. Phys. A Math. Gen.* **2001**, *34*, 8281. [CrossRef]
35. Bäcker, A.; Schubert, R. Chaotic eigenfunctions in momentum space. *J. Phys. A Math. Gen.* **1999**, *32*, 4795. [CrossRef]
36. Husimi, K. Some formal properties of the density matrix. *Proc. Phys. Math. Soc. Jpn.* **1940**, *22*, 264.
37. Bäcker, A.; Fürstberger, S.; Schubert, R. Poincaré Husimi representation of eigenstates in quantum billiards. *Phys. Rev. E* **2004**, *70*, 036204. [CrossRef] [PubMed]
38. Berry, M.V.; Mondragon, R.J. Neutrino Billiards: Time-Reversal Symmetry-Breaking Without Magnetic Fields. *Proc. R. Soc. London A* **1987**, *412*, 53.
39. Weyl, H. Elektron und Gravitation. I. *Z. Physik* **1929**, *56*, 330. [CrossRef]
40. Bolte, J.; Keppeler, S. A Semiclassical Approach to the Dirac Equation. *Ann. Phys.* **1999**, *274*, 125–162. [CrossRef]
41. Dietz, B. Circular and Elliptical Neutrino Billiards: A Semiclassical Approach. *Act. Phys. Pol. A* **2019**, *136*, 770. [CrossRef]
42. Dietz, B.; Li, Z.Y. Semiclassical quantization of neutrino billiards. *Phys. Rev. E* **2020**, *102*, 042214. [CrossRef]
43. Baym, G. *Lectures on Quantum Mechanics*; CRC Press: Boca Raton, FL, USA, 2018.
44. Dresselhaus, M.; Dresselhaus, G.; Eklund, P. *Science of Fullerenes and Carbon Nanotubes*; Academic Press: San Diego, CA, USA, 1996. [CrossRef]
45. Reich, S.; Maultzsch, J.; Thomsen, C.; Ordejón, P. Tight-binding description of graphene. *Phys. Rev. B* **2002**, *66*, 035412. [CrossRef]

46. Novoselov, K.S.; Geim, A.K.; Morozov, S.V.; Jiang, D.; Zhang, Y.; Dubonos, S.V.; Grigorieva, I.V.; Firsov, A.A. Electric Field Effect in Atomically Thin Carbon Films. *Science* **2004**, *306*, 666–669. [CrossRef]
47. Geim, A.; Novoselov, K. The rise of graphene. *Nat. Mater.* **2007**, *6*, 183. [CrossRef]
48. Beenakker, C.W.J. Colloquium: Andreev reflection and Klein tunneling in graphene. *Rev. Mod. Phys.* **2008**, *80*, 1337. [CrossRef]
49. Castro Neto, A.H.; Guinea, F.; Peres, N.M.R.; Novoselov, K.S.; Geim, A.K. The electronic properties of graphene. *Rev. Mod. Phys.* **2009**, *81*, 109. [CrossRef]
50. Ponomarenko, L.A.; Schedin, F.; Katsnelson, M.I.; Yang, R.; Hill, E.W.; Novoselov, K.S.; Geim, A.K. Chaotic Dirac Billiard in Graphene Quantum Dots. *Science* **2008**, *320*, 5874. [CrossRef] [PubMed]
51. Güttinger, J.; Stampfer, C.; Hellmüller, S.; Molitor, F.; Ihn, T.; Ensslin, K. Charge detection in graphene quantum dots. *Appl. Phys. Lett.* **2008**, *93*, 212102. [CrossRef]
52. Güttinger, J.; Frey, T.; Stampfer, C.; Ihn, T.; Ensslin, K. Spin States in Graphene Quantum Dots. *Phys. Rev. Lett.* **2010**, *105*, 116801. [CrossRef]
53. Wallace, P.R. The Band Theory of Graphite. *Phys. Rev.* **1947**, *71*, 622–634. [CrossRef]
54. Polini, M.; Guinea, F.; Lewenstein, M.; Manoharan, H.C.; Pellegrini, V. Artificial graphene as a tunable Dirac material. *Nat. Nanotechnol.* **2013**, *8*, 625. [CrossRef] [PubMed]
55. Akhmerov, A.R.; Beenakker, C.W.J. Detection of Valley Polarization in Graphene by a Superconducting Contact. *Phys. Rev. Lett.* **2007**, *98*, 157003. [CrossRef]
56. Akhmerov, A.R.; Beenakker, C.W.J. Boundary conditions for Dirac fermions on a terminated honeycomb lattice. *Phys. Rev. B* **2008**, *77*, 085423. [CrossRef]
57. Wurm, J.; Richter, K.; Adagideli, İ. Edge effects in graphene nanostructures: From multiple reflection expansion to density of states. *Phys. Rev. B* **2011**, *84*, 075468. [CrossRef]
58. Bittner, S.; Dietz, B.; Miski-Oglu, M.; Richter, A. Extremal transmission through a microwave photonic crystal and the observation of edge states in a rectangular Dirac billiard. *Phys. Rev. B* **2012**, *85*, 064301. [CrossRef]
59. Dietz, B.; Iachello, F.; Miski-Oglu, M.; Pietralla, N.; Richter, A.; von Smekal, L.; Wambach, J. Lifshitz and excited-state quantum phase transitions in microwave Dirac billiards. *Phys. Rev. B* **2013**, *88*, 104101. [CrossRef]
60. Dietz, B.; Klaus, T.; Miski-Oglu, M.; Richter, A. Spectral properties of superconducting microwave photonic crystals modeling Dirac billiards. *Phys. Rev. B* **2015**, *91*, 035411. [CrossRef]
61. Dietz, B.; Klaus, T.; Miski-Oglu, M.; Richter, A.; Wunderle, M.; Bouazza, C. Spectral Properties of Dirac Billiards at the van Hove Singularities. *Phys. Rev. Lett.* **2016**, *116*, 023901. [CrossRef] [PubMed]
62. Zhang, W.; Zhang, X.; Che, J.; Miski-Oglu, M.; Dietz, B. Properties of the eigenmodes and quantum-chaotic scattering in a superconducting microwave Dirac billiard with threefold rotational symmetry. *arXiv* **2023**, arXiv:2302.10094v1.
63. Libisch, F.; Stampfer, C.; Burgdörfer, J. Graphene quantum dots: Beyond a Dirac billiard. *Phys. Rev. B* **2009**, *79*, 115423. [CrossRef]
64. Wurm, J.; Rycerz, A.; Adagideli, İ.; Wimmer, M.; Richter, K.; Baranger, H.U. Symmetry Classes in Graphene Quantum Dots: Universal Spectral Statistics, Weak Localization, and Conductance Fluctuations. *Phys. Rev. Lett.* **2009**, *102*, 056806. [CrossRef]
65. Dietz, B. Relativistic quantum billiards with threefold rotational symmetry: Exact, symmetry-projected solutions for the equilateral neutrino billiard. *Act. Phys. Pol. A* **2021**, *140*, 473. [CrossRef]
66. Yu, P.; Zhang, W.; Dietz, B.; Huang, L. Quantum signatures of chaos in relativistic quantum billiards with shapes of circle- and ellipse-sectors. *J. Phys. Math. Theor.* **2022**, *55*, 224015. [CrossRef]
67. McIsaac, P. Symmetry-Induced Modal Characteristics of Uniform Waveguides - I: Summary of Results. *IEEE Trans. Microw. Theory Tech.* **1975**, *23*, 421–429. [CrossRef]
68. Sieber, M. Semiclassical transition from an elliptical to an oval billiard. *J. Phys. A* **1997**, *30*, 4563–4596. [CrossRef]
69. Waalkens, H.; Wiersig, J.; Dullin, H.R. Elliptic Quantum Billiard. *Ann. Phys.* **1997**, *260*, 50–90. [CrossRef]
70. Gaddah, W.A. Exact solutions to the Dirac equation for equilateral triangular billiard systems. *J. Phys. A Math. Theor.* **2018**, *51*, 385304. [CrossRef]
71. Zhang, W.; Dietz, B. Microwave photonic crystals, graphene, and honeycomb-kagome billiards with threefold symmetry: Comparison with nonrelativistic and relativistic quantum billiards. *Phys. Rev. B* **2021**, *104*, 064310. [CrossRef]
72. Gaddah, W.A. Discrete symmetry approach to exact bound-state solutions for a regular hexagon Dirac billiard. *Phys. Script.* **2021**, *96*, 065207. [CrossRef]
73. Greiner, W.; Schäfer, A. (Eds.) *Quantum Chromodynamics*; Springer: New York, NY, USA, 1994.
74. Alberto, P.; Fiolhais, C.; Gil, V.M.S. Relativistic particle in a box. *Eur. J. Phys.* **1996**, *17*, 19. [CrossRef]
75. Alonso, V.; Vincenzo, S.D. General boundary conditions for a Dirac particle in a box and their non-relativistic limits. *J. Phys. A Math. Gen.* **1997**, *30*, 8573. [CrossRef]
76. Alonso, V.; Vincenzo, S.D.; Mondino, L. On the boundary conditions for the Dirac equation. *Eur. J. Phys.* **1997**, *18*, 315. [CrossRef]
77. Alberto, P.; Das, S.; Vagenas, E.C. Relativistic particle in a three-dimensional box. *Phys. Lett. A* **2011**, *375*, 1436–1440. [CrossRef]
78. Yusupov, J.; Otajanov, D.; Eshniyazov, V.; Matrasulov, D. Classical and quantum dynamics of a kicked relativistic particle in a box. *Phys. Lett. A* **2018**, *382*, 633–638. [CrossRef]
79. Dietz, B. Unidirectionality and Husimi functions in constant-width neutrino billiards. *J. Phys. A Math. Theor.* **2022**, *55*, 474003. [CrossRef]

80. Bäcker, A. Numerical Aspects of Eigenvalue and Eigenfunction Computations for Chaotic Quantum Systems. In *The Mathematical Aspects of Quantum Maps*; Esposti, M.D., Graffi, S., Eds.; Springer: Berlin/Heidelberg, Germany, 2003; pp. 91–144. [CrossRef]
81. Yu, P.; Dietz, B.; Xu, H.Y.; Ying, L.; Huang, L.; Lai, Y.C. Kac's isospectrality question revisited in neutrino billiards. *Phys. Rev. E* **2020**, *101*, 032215. [CrossRef] [PubMed]
82. Yu, P.; Dietz, B.; Huang, L. Quantizing neutrino billiards: An expanded boundary integral method. *New J. Phys.* **2019**, *21*, 073039. [CrossRef]
83. Leyvraz, F.; Schmit, C.; Seligman, T.H. Anomalous spectral statistics in a symmetrical billiard. *J. Phys. A* **1996**, *29*, L575. [CrossRef]
84. Keating, J.P.; Robbins, J.M. Discrete symmetries and spectral statistics. *J. Phys. A* **1997**, *30*, L177. [CrossRef]
85. Robbins, J.M. Discrete symmetries in periodic-orbit theory. *Phys. Rev. A* **1989**, *40*, 2128–2136. [CrossRef] [PubMed]
86. Joyner, C.H.; Müller, S.; Sieber, M. Semiclassical approach to discrete symmetries in quantum chaos. *J. Phys. A* **2012**, *45*, 205102. [CrossRef]
87. Weyl, H. Über die Abhängigkeit der Eigenschwingungen einer Membran und deren Begrenzung. *J. Reine Angew. Math.* **1912**, *141*, 1. [CrossRef]
88. Oganesyan, V.; Huse, D.A. Localization of interacting fermions at high temperature. *Phys. Rev. B* **2007**, *75*, 155111. [CrossRef]
89. Atas, Y.Y.; Bogomolny, E.; Giraud, O.; Roux, G. Distribution of the Ratio of Consecutive Level Spacings in Random Matrix Ensembles. *Phys. Rev. Lett.* **2013**, *110*, 084101. [CrossRef]
90. Atas, Y.; Bogomolny, E.; Giraud, O.; Vivo, P.; Vivo, E. Joint probability densities of level spacing ratios in random matrices. *J. Phys. A* **2013**, *46*, 355204. [CrossRef]
91. McLachlan, N. (Ed.) *Theory and Application of Mathieu Functions*; Oxford University Press: London, UK, 1947.
92. Morse, P.; Feshbach, H. (Eds.) *Methods of Theoretical Physics*; MacGraw-Hill: New York, NY, USA, 1953.
93. Abramowitz, M.; Stegun, I.A. (Eds.) *Handbook of Mathematical Functions with Formulas, Graphs and Mathematical Tables*; Dover: New York, NY, USA, 2013.
94. Dietz, B.; Smilansky, U. A scattering approach to the quantization of billiards—The inside—Outside duality. *Chaos* **1993**, *3*, 581. [CrossRef] [PubMed]
95. Elkies, N.D.; McMullen, C.T. Gaps in $\sqrt{n}$ mod 1 and ergodic theory. *Duke Math. J.* **2004**, *123*, 95–139. [CrossRef]
96. El-Baz, D.; Marklof, J.; Vinogradov, I. The two-point correlation function of the fractional parts of sqrt(n) is Poisson. *Proc. Am. Math. Soc.* **2015**, *143*, 2815–2828.
97. Wang, J.; Casati, G.; Prosen, T. Nonergodicity and localization of invariant measure for two colliding masses. *Phys. Rev. E* **2014**, *89*, 042918. [CrossRef] [PubMed]
98. Huang, J.; Zhao, H. Ultraslow diffusion and weak ergodicity breaking in right triangular billiards. *Phys. Rev. E* **2017**, *95*, 032209. [CrossRef] [PubMed]
99. Bäcker, A.; Schubert, R.; Stifter, P. On the number of bouncing ball modes in billiards. *J. Phys. Math. Gen.* **1997**, *30*, 6783. [CrossRef]
100. Bogomolny, E.; Schmit, C. Structure of Wave Functions of Pseudointegrable Billiards. *Phys. Rev. Lett.* **2004**, *92*, 244102. [CrossRef]
101. Dietz, B.; Heine, A.; Heuveline, V.; Richter, A. Test of a numerical approach to the quantization of billiards. *Phys. Rev. E* **2005**, *71*, 026703. [CrossRef]
102. Sieber, M.; Smilansky, U.; Creagh, S.C.; Littlejohn, R.G. Non-generic spectral statistics in the quantized stadium billiard. *J. Phys. A* **1993**, *26*, 6217. [CrossRef]

entropy

MDPI

Article

Quantum Entropies and Decoherence for the Multiparticle Quantum Arnol'd Cat †

Giorgio Mantica [1,2,3]

1 Center for Non-Linear and Complex Systems Università dell'Insubria, Via Valleggio 11, 22100 Como, Italy; giorgio.mantica@uninsubria.it
2 Istituto Nazionale di Alta Matematica "F. Severi", GNFM Gruppo Nazionale per la Fisica Matematica, P. le Aldo Moro 5, 00185 Rome, Italy
3 I.N.F.N. Gruppo Collegato di Como, Sezione di Milano, Via Celoria 16, 20133 Milan, Italy
† Dedicated to Giulio Casati on his eightieth.

Abstract: I study the scaling behavior in the physical parameters of dynamical entropies, classical and quantum, in a specifically devised model of collision-induced decoherence in a chaotic system. The treatment is fully canonical and no approximations are involved or infinite limits taken. I present this model in a detailed way, in order to clarify my views in the debate about the nature, definition, and relevance of quantum chaos.

Keywords: classical and quantum entropies; chaos; quantum to classical transition

Citation: Mantica, G. Quantum Entropies and Decoherence for the Multiparticle Quantum Arnol'd Cat. *Entropy* **2023**, *25*, 1004. https://doi.org/10.3390/e25071004

Academic Editor: Marko Robnik

Received: 30 March 2023
Revised: 14 June 2023
Accepted: 21 June 2023
Published: 29 June 2023

1. Preface: Forty Years With Giulio Casati

Forty years ago, in December 1982, I defended my master's thesis in Pavia under the guidance of Italo Guarneri. He had introduced me to his coworker from the university of Milano, Giulio Casati, who at the same time was turning forty. Ever since then, my scientific career has been marked by this encounter, in repeated periods of proximity and elongation. After a brief permanence in Milano, Giulio managed to place me in contact with Joe Ford, with whom I was a postdoc in Atlanta for many years, before taking a three-year turn to France with Daniel Bessis, a fourth important figure in my education.

However, now back to Giulio: while I was in Paris, using his unrivaled political ability, he was setting up nothing less than a new university, in Como, where I later joined him (Figure 1) and Italo Guarneri in what at the beginning was a heroic endeavor: we shared a single office with a modem by which I connected to computers in Paris and Milano. We worked hard to establish the course in physics—later also in mathematics—and we brought an intense activity to the newly instituted Center for Non-linear and Complex Systems. We organized conferences and workshops; we invited many visitors, hosting postdocs and graduate students. The focus of this activity was chaos in classical and quantum systems, and also orthogonal polynomials of singular measures and their relevance for the phenomenon of quantum intermittency [1], on which I will not dwell here.

Figure 1. Como, early 1990s. Picture by Felix Izrailev.

In this paper I will instead present a reflection on the problem of quantum chaos and the quantum–classical correspondence. Notwithstanding the dedicatory goal, I will not hide the differences between my views and those of Giulio [2–6], the *extremist of the middle*, as Joe Ford jokingly called him, not without a hint of affection. The nickname implied that such views, Giulio's, were widely shared. Indeed, to the question *Does chaos pose a problem to quantum mechanics?* few would answer in the positive, and many would quote the notions of quantum chaology [7,8] (Berry) or quantum pseudo-chaos [9,10] (Chirikov). Joe Ford was convinced of the contrary and with some differences I shared his views [11] (but notice the differences between [12] and [13]). This celebration is now the occasion for me to place on the record what I have learned in these years and what I believe is still to be investigated.

2. Introduction and Summary of the Paper

As I mentioned, Giulio introduced me to Joe Ford, who had learned the basic tenets of algorithmic complexity theory [14] from his colleague and friend Boris Chirikov (Figure 2) from the Soviet Union, where this theory had flourished thanks to the school of Kolmogorov. It leaves a sour taste to remember the many liaisons that ensued, in a moment when we fear for many dear friends and colleagues. This theory quantifies the amount of information contained in a sequence of symbols [15–18]. Now, symbolic dynamics is a tool to translate motions into sequence of symbols, in such a way that one can equate chaotic trajectories of a system to random symbolic sequences—roughly speaking, sequences that cannot be significantly compressed without losing information. In this perspective, chaos is the uninterrupted production of information by the motion of a dynamical system, just in the same way as a stochastic process could act; hence, in Ford's wording [19], *chaos is deterministic randomness*. In a historical/theoretical introduction to classical chaos coauthored with Giulio for the Journal of the Italian Physical Society [20], we adopted this perspective: that instability of trajectories, transverse homoclinic intersections, and horseshoes are the dynamical *mechanisms* by which chaos *arises*, but not its *essence*, which is the algorithmic complexity of symbolic trajectories.

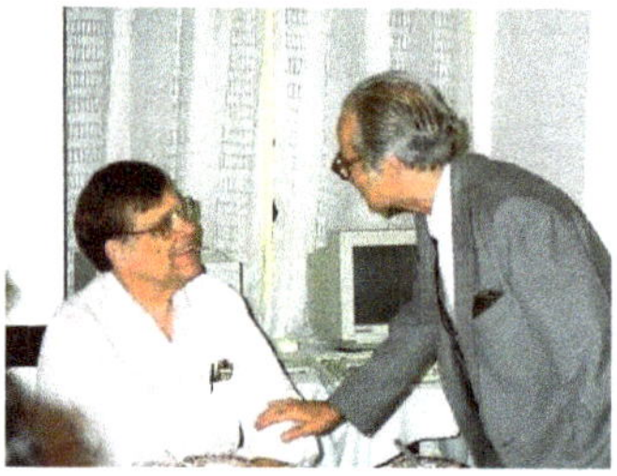

Figure 2. Joe Ford and Boris Chirikov. Como, early 1990s. Picture by Felix Izrailev.

Similarly, in quantum dynamics, a variety of indicators have been introduced to unveil the mechanisms behind instability—or its lack. It is clearly not my aim here to list all contributions, early or recent, to this research, but just to name a few, for the sake of argument. Phase space representations, such as the Wigner function, have been employed since long ago for comparison with classical dynamics, as in [21–23]. In [11], this analysis was instrumental to find the eigenfunctions of the evolution operator and to expose the exponential propagation of mode number in the classical Liouville evolution, which impairs stability and practical reversibility, a technique later revisited by Benenti and Casati [2]. Quantum Lyapunov exponents have also been defined, see, e.g., [24,25], and more recently, much attention has been devoted to the out-of-time autocorrelation functions; see the paper by Kurchan in this volume [26] and references therein.

At a more general level than these mechanisms there is the study of the algorithmic complexity of the quantum motion. Many years ago, Berry and Hannay quantized the

Arnol'd cat map [27] via semiclassical means [28]. Ford, myself, and Ristow [11] developed a fully Hamiltonian quantization and set out to estimate the algorithmic complexity of the evolution, finding that it grows similar to the logarithm of the dimension of the Hilbert space of the system plus the logarithm of the time–length of the evolution. Prior to saturation of such upper bounds (see the description in Section 4), the quantum behavior reproduces the classical, as it has been repeatedly reported using the other indicators mentioned above. Yet, it was recognized that the *scaling* of the range of such *correspondence* with the physical parameters is fully unsatisfactory to justify a sensible approach to the classical limit (a sort of *logarithmic catastrophe* [12]).

As is well known, a possible solution to this problem was proposed at the end of the last century [29–36] by considering that systems are coupled to an external environment. I believe that this approach still need to be completed, or at least polished, in two respects. Firstly, by casting it in a fully Hamiltonian approach, as I shall perform by introducing a model system of collision-induced decoherence that avoids recurse to mean field approximations or infinite baths of oscillators, by which nondynamical randomness might creep in uncontrolled. Secondly, by framing it into an information theoretical perspective, employing the so-called Alicky–Fannes (AF) entropy [37–39], which offers computational advantages and avoids some of the difficulties of algorithmic complexity. It must be noted that quantum dynamical entropies generalizing the classical KS notion [40] were already developed in the early days of "quantum chaos" [41–43].

The scheme of this paper is as follows: in the next section I briefly recall the notion of dynamical entropy. Section 4 discusses the importance of dynamical entropies in the notion of chaos and the fundamental problem that one encounters in taking the classical limit of chaotic quantum systems. Section 5 gives a review of the quantum version of the classical Arnol'd cat map [27], as the motion of a *single* kicked particle, so as to open the way to a *multi*particle generalization. The Alicki–Fannes entropy of such a single particle system has been computed in a series of papers. The formalism is presented in Section 6, again in view of its generalization to the multiparticle Arnol'd cat [44], described in Section 7. This generalization aims at introducing in a fully controllable, dynamical way the concept of decoherence induced by scattering by light particles, following the paradigm of Joos and Zeh [45]. In a small aside, Section 8, I briefly describe recent results on the *Von Neumann* entropy of the multiparticle Arnol'd cat [46], which confirm its suitability as a model of quantum decoherence. In Section 9 I return to the AF entropy, of which in my view is theoretically more relevant than Von Neumann's, to the same degree as it is harder to treat theoretically and compute numerically. I explain how a combined limit in the number of particles and their mass *could* provide a physically sound means to understand the classical limit. Final comments are presented in the conclusions.

3. Dynamical Entropy, Classical and Quantum

Consider a classical dynamical system [27,40] consisting of a phase-space X on which a transformation ϕ acts in such a way to preserve a measure μ defined on a suitable sigma-algebra of subsets of X: $\mu(\phi^{-1}(A)) = \mu(A)$ for any A in the sigma-algebra. The map ϕ generates the dynamics, that we assume to evolve in discrete time instants, denoted by the symbol n throughout the paper. Consider a finite partition $\mathcal{P}$ of X by measurable subsets $P_l, l = 0, \ldots, L-1, \bigcup_l P_l = X$. The Shannon entropy of this partition, $S(\mathcal{P})$, is defined as

$$S(\mathcal{P}) = -\sum_{l=0}^{L-1} \mu(P_l) \log \mu(P_l). \tag{1}$$

This value is the average amount of information obtained by determining the element of the partition that contains a given point in phase space: this action can be physically interpreted as the result of a measurement experiment.

Next, consider the partition obtained by intersecting all elements of $\mathcal{P}$ and all their preimages by the map ϕ: indicate this new partition by

$$\mathcal{P} \vee \phi^{-1}(\mathcal{P}) = \{P \cap \phi^{-1}(Q), \text{ for } P, Q \in \mathcal{P}\}, \quad (2)$$

where obviously some of the intersections at right hand side may be empty. If we label sets by the *alphabet* composed of the *symbols* 0 to $L-1$, as in Equation (1), the elements of the partition in Equation (2) are labeled as $P_{lk} = P_l \cap \phi^{-1}(P_k)$. These sets comprise all points that at time zero are located in P_l and at the successive time instant in P_k. Therefore, the Shannon entropy of $\mathcal{P} \vee \phi^{-1}(\mathcal{P})$ measures the average amount of information provided by knowledge of the *symbolic dynamics* of a point over two time instants.

In the lore, a *word* is a sequence of symbols $\sigma = \sigma_0, \sigma_1, \ldots, \sigma_{n-1}$ in the alphabet. We use the notation $|\sigma|$ for the length of a word. Let Σ_n be the set of all words of length n in the alphabet. Clearly, the cardinality of this set is L^n. Every word defines a *cylinder* P_σ in X: this is the set of all points x for which $\phi^i(x) \in P_{\sigma_i}$, for $i = 0, \ldots, n-1$. The set of cylinders associated with words of length n composes the partition $\vee_{i=0}^{n-1} \phi^{-i}(\mathcal{P})$, whose entropy is therefore

$$S(\mathcal{P}, n) = S(\vee_{i=0}^{n-1} \phi^{-i}(\mathcal{P})) = -\sum_{\sigma \in \Sigma_n} \mu(P_\sigma) \log(\mu(P_\sigma)). \quad (3)$$

Consider the difference $S(\mathcal{P}, n+1) - S(\mathcal{P}, n)$. This quantifies the average amount of new information obtained by observing the state of the motion at time n, on top of what is already obtained by the observations at the previous n instants, from time 0 to time $n-1$. It could also be described as a *surprise* that the dynamical evolution of the system is capable of producing at such time. It is then understandable that one can define as dynamical entropy the asymptotic rate of information production as follows:

$$h(\mathcal{P}) = \lim_{n \to \infty} S(\mathcal{P}, n+1) - S(\mathcal{P}, n) = \lim_{n \to \infty} \frac{1}{n} S(\mathcal{P}, n). \quad (4)$$

This leads to the definition of the Kolmogorov Sinai entropy of the dynamical system as a supremum over partitions:

$$h_{KS}(X, \phi, \mu, \mathcal{A}) = \sup\{h(\mathcal{P}), \mathcal{P} \text{ finite measurable partition of } X\}. \quad (5)$$

Chaotic classical dynamical systems are then properly defined as those having positive KS entropy, and Brudno's theorem [47] links this concept to positive algorithmic complexity.

Let us now turn to quantum dynamics for comparison. Robert Alicki and Mark Fannes constructed a definition of entropy following the same steps as in the classical construction [37,38,48]. A quantum dynamical system is composed of a Hilbert space $\mathcal{H}$ and of a unitary evolution U that leaves an invariantdensity matrix ρ. The analogue of the partition $\mathcal{P}$ is a finite family of orthogonal projection operators P_l, from $l = 0$ to $L-1$, such that $\sum_l P_l = I$, with I being the identity. We use the same symbol P_l for both the classical and the quantum case, since this highlights the analogy.

For illustration, suppose that $\rho = |\psi><\psi|$, even if generically this is not invariant. The wave function $|\psi>$ represents the initial state of the system. Projecting $|\psi>$ by P_l is the analogue of the classical *measurement* of the partition element which the dynamical state belongs to. In quantum dynamics, the square modulus $\|P_l|\psi>\|^2$ is the probability that the result of this measurement yields the value l. Clearly, because of normalization, $\sum_l \|P_l|\psi>\|^2 = 1$. Following suit, the analogues of the elements of $\mathcal{P} \vee \phi^{-1}(\mathcal{P})$ are $P_{\sigma_1} U P_{\sigma_0}|\psi>$, which shows that the classical action of ϕ^{-1} is replaced by the quantum unitary evolution operator U. This leads to defining the quantum dynamical partitions:

$$\vee_{i=0}^{n-1} U^i(\mathcal{P}) = \{P_{\sigma_{n-1}} U \circ P_{\sigma_{n-2}} U \circ \cdots \circ P_{\sigma_1} U \circ P_{\sigma_0}, \ \sigma \in \Sigma_n\} = \{P_\sigma, \ \sigma \in \Sigma_n\}. \quad (6)$$

We used the notation P_σ for the operators in the above set. Then, the quantum analogue of $\mu(P_\sigma)$ takes the form of the trace of a product of operators:

$$\mu(P_\sigma) = \mathrm{Tr}(P_\sigma^\dagger \rho P_\sigma). \tag{7}$$

Alas, it is not possible to use this quantity to define a dynamical entropy using Equation (3). In fact, suppose again for simplicity that $\rho = |\psi><\psi|$. Then,

$$\mu(P_\sigma) = \| P_\sigma |\psi> \|^2. \tag{8}$$

While it is true that $\sum_\sigma \mu(P_\sigma) = 1$, observe that Equation (8) can be rewritten as follows:

$$\| P_\sigma |\psi> \|^2 = < \psi P_\sigma | P_\sigma \psi > . \tag{9}$$

The word σ appears twice in the above equation. Replace the first appearance with a different word $\theta \in \Sigma_n$, to obtain

$$< \psi P_\theta | P_\sigma \psi > = < \psi | P_\theta^\dagger P_\sigma \psi > = < \psi | P_\theta^\dagger P_{\sigma_{n-1}} U \circ \cdots \circ P_{\sigma_1} U \circ P_{\sigma_0} \psi > =$$
$$= < \psi | P_{\theta_0} \circ U^{-1} P_{\theta_1} \circ \cdots \circ U^{-1} P_{\theta_{n-1}} \circ P_{\sigma_{n-1}} U \circ \cdots \circ P_{\sigma_1} U \circ P_{\sigma_0} \psi > .$$

It helps intuition to read the last line from right to left: in sequence, the initial function $|\psi>$ is projected by P_{σ_0}, acted upon by U, projected by P_{σ_1}, acted upon by U, and so on until it is projected by $P_{\sigma_{n-1}}$. Then, the adjoint of P_θ operates by projecting by $P_{\theta_{n-1}}$, acting by the *inverse* of U (that is, by reversing time), projecting by $P_{\theta_{n-2}}$, acting by U^{-1}, and so on until the projection by P_{θ_0}, and finally taking the scalar product with the initial function $|\psi>$.

Clearly, the result of this operation is null unless $\sigma_{n-1} = \theta_{n-1}$. In the classical case where ϕ^{-1} takes the place of U, the result is also null unless the *full* word θ coincides with σ. In fact, it is enlightening to consider the above process as a sequence of many-slits experiments: a dynamical trajectory must pass through the slits σ_0 to σ_{n-1}. Then, time is reversed and it is required to pass from the slits θ_{n-1} to θ_0. Since trajectories in classical dynamical systems are time-reversible, this is only possible if $\theta = \sigma$. On the contrary, in quantum dynamics, two different *paths* from $\sigma_0 = \theta_0$ to $\sigma_{n-1} = \theta_{n-1}$, the end points being fixed, may *not* be orthogonal because of *quantum interference*: $< \psi P_\theta | P_\sigma \psi > = < \psi_\theta | \psi_\sigma > \neq 0$, where we introduced the notation $|\psi_\sigma > = P_\sigma \psi >$.

If we now take for ρ an invariant density matrix (which in this paper will be the normalized identity matrix I), we define the *decoherence matrix* D with entries $D_{\theta,\sigma}(n)$

$$D_{\theta,\sigma}(n) = \mathrm{Tr}(P_\theta^\dagger \rho P_\sigma) = \mathrm{Tr}(P_\theta^\dagger P_\sigma) = \frac{1}{\mathcal{N}} \sum_{j=1}^{\mathcal{N}} < \psi_\theta^j | \psi_\sigma^j >, \tag{10}$$

where $\mathcal{N}$ is the dimension of the Hilbert space and $\psi^j >$ is an orthonormal basis of $\mathcal{H}$. In this paper I shall consider a notable example of such finite-dimensional Hilbert space. The decoherence matrix is a fundamental object of the *coherent histories* formalism [49–54]. Observe that $D(n)$ is an $L^n \times L^n$ square matrix, Hermitean, of unit-trace and is non-negative. In the classical case, this matrix is diagonal. In the quantum case, one defines the partition entropy as

$$S(\mathcal{H}, U, \mathcal{P}, n) = -Tr(D(n) \log D(n)) \tag{11}$$

and can therefore proceed to define the AF entropy operating as in Equations (4) and (5) [48]. To suit our purposes we do not take the infinite n limit or the supremum over partitions, but, rather, consider the scaling behavior of the quantity $S(\mathcal{H}, U, \mathcal{P}, n)$, concisely denoted $S(n)$, that we will call AF entropy for short (Shannon AF entropy at finite resolution and time being a more precise term) for physically accessible partitions, ranges of time, and physical parameters.

4. Dynamical Entropy And Quantum Chaos

On the same theoretical basis as in classical dynamics, we shall consider as chaotic a quantum system capable of producing new information at an inferiorly bounded rate indefinitely in time when measured by a sufficiently refined partition $\mathcal{P}$. For this, we adopt the AF quantum entropy defined in the previous section. It helps to remark that Benatti derived a quantum version of the classical Brudno theorem, linking AF entropy and algorithmic complexity [55], so that the parallelism with classical dynamics is completed.

At a superficial inspection it may seem that there should be no limitation for the entropy $S(\mathcal{H}, U, \mathcal{P}, n)$ in Equation (11) to grow linearly with n: as mentioned, the matrix D has size $L^n \times L^n$. Yet, thanks to a result in [38] the spectrum of nonzero eigenvalues of D coincides with that of a second matrix, $\Omega(n)$, of size $\mathcal{N}^2 \times \mathcal{N}^2$ independent of n, of unit trace and Hermitean. This matrix is the basis of a technique to numerically *compute* $Tr[D(n) \log D(n)]$, combining the theory of orthogonal polynomials with a sampling method from lattice gauge theory [56]. It appears to be the only method proposed so far capable of attacking nontrivial Hilbert space and word lengths dimension. While this technique is rather complicated, one consequence of its is immediate: $S(\mathcal{H}, U, \mathcal{P}, n)$ is the Shannon entropy of a probability distribution over $\mathcal{N}^2$ discrete states, whose number is *independent* of n. Therefore, it is bounded by $2 \log \mathcal{N}$ and the limit in Equation (4) is forcefully null.

Why this is a serious problem can be easily understood: consider a classical dynamical system that can be *quantized* via the standard procedure of replacing classical variables by quantum operators. Further, assume that the resulting Hilbert space has finite dimension $\mathcal{N}$. Consider a classical partition $\mathcal{P}$ that corresponds to a quantum family of projectors. One can compute the classical probabilities $\mu(\mathcal{P}_\sigma)$ as well as the eigenvalues of the quantum decoherence matrix $D(n)$ and then compare the associated Shannon entropies at variable duration n of the experiment. The *correspondence principle* leads us to expect the two results to be comparable, in the *classical* limit. Yet, recall that the quantum system is endowed with a finite *information reservoir*. In the initial instants of the evolution, the classical system is outputting information and the quantum analogue follows suit, keeping pace with the classical. Yet, this can last as long as the quantum system has not fully exhausted its information reservoir.

Suppose now that the classical system is *not* chaotic, in the sense that $S(\mathcal{P}, n)$ grows *less* than linearly, say logarithmically, with n. In this case, the value of n at which the quantum dynamics can provide the same information output as the classical is roughly proportional to $\mathcal{N}$, and therefore is sufficiently large to justify a physical relevance of correspondence. On the contrary, when the classical dynamics is chaotic, i.e., $S(\mathcal{P}, n) \sim h(\mathcal{P})\ n$, the finite quantum information reservoir is eaten away at a rate sustainable only up to $n \sim \log \mathcal{N}$, after which the two evolutions qualitatively and quantitatively differ. This is, in essence, the logarithmic catastrophe in the correspondence principle mentioned before.

From a physical point of view, the last estimate corresponds to what has been defined the Zaslavsky time scale [57] (also termed Ehrenfest time scale [58]) $\tau = \log(\mathcal{I}/\hbar)$, where $\mathcal{I}$ is a typical action of the system, a fact that yields the analogy $\mathcal{N} \sim (\mathcal{I}/\hbar)$ and prompts matter for thought. In this paper, $\hbar$ will be constant, while the *mass* of a quantum particle will determine $\mathcal{N}$.

From a theoretical point of view, advocated by Berry, among others [7,8], it is clear that dynamical entropies are properly defined only in the limit when n tends to infinity, while the classical limit is $\mathcal{I}/\hbar \to \infty$ and the two limits do not commute. At the same time, it is apparent that this attitude ignores the unphysically low value of the Zaslavsky time for, e.g., the chaotic three-body problem. Moreover, as I first suggested in [12], rather than considering the limit for n tending to infinity, one should study the *scaling relations* with respect to the physical parameters of the information–theoretical quantities (4) and (11) in realistic ranges of the physical variables. In other words: Will a sequence of physically realizable experiments at finite precision be able to reveal a significant difference among the two dynamics in the semiclassical region? A model of this situation is found in the dynamics of classical billiards [59]: a difference emerges clearly in the behavior of the finite time algorithmic complexity of symbolic trajectories, when polygonal tables approach chaotic or integrable curved billiards.

In the next section I briefly recap the details of a much-investigated system, the quantum Arnol'd cat, where this situation emerges with clarity. This system is also particularly useful to study the arguably most significant way to explain the emergence of classical properties (among which chaos) in quantum dynamics: decoherence. In this approach, a quantum system is not isolated, but coupled to an *environment* that is responsible for the decay of the off-diagonal matrix elements of D. Yet, in the usual application of this approach, the environment is mathematically represented by a random source, or an infinite bath of oscillators: in any case, a system possessing an *infinite* amount of information, and not, to the best of my knowledge, by a quantum system with finite dimensional Hilbert space. In previous works [44], I proposed a scheme by which the environment is composed of the tensor product of simple low-dimensional systems, so that the dimension of the full, purely quantum, isolated system–environment pairgrows exponentially in the number of the elements of the latter and can therefore provide a physically reasonable solution to the logarithmic catastrophe described above. The multiparticle Arnol'd cat introduced in this paper is a simple model to implement this scheme. Due to technical difficulties, both theoretical and numerical, much is still to be understood on the information flow in this system. But first, let us review the original quantum Arnol'd cat.

5. Short Review of the Quantum Arnol'd Cat

Despite the fact that this quantum system has now become *classical* (pun), it is interesting to present a derivation that opens the way to its decoherent generalization. We study a point particle of mass M free to rotate on a ring, without gravity. This ring is a one-dimensional torus of length L and a position on it is $X \in [0, L)$. Most of the time, the particle evolves freely at constant velocity. At discrete times, it is subject to a periodic impulsive force, of period T, the relative Hamiltonian being:

$$H_{cat}(X, Y, t) = \frac{Y^2}{2M} - \alpha \frac{X^2}{2} \sum_{j=-\infty}^{\infty} \delta(t/T - j). \tag{12}$$

Y is the conjugate momentum to X and α is a coupling constant. According to Hamilton's equations, the effect of the force is to instantaneously change the particle's momentum.

Let us show that H_{cat} is the Hamiltonian that yields the Arnol'd cat mapping. Consider the classical period evolution of the dynamical variables observed at instants immediately following the action of the impulsive force:

$$\begin{cases} X \to X + \frac{T}{M}Y \\ Y \to Y(1 + \frac{\alpha T}{M}) + \alpha T X. \end{cases} \tag{13}$$

Introduce the rescaled momentum $\tilde{Y} := \frac{T}{M}Y$, which has the dimensions of a length. The period evolution operator in the new variables is

$$\begin{cases} X \to X + \tilde{Y} \\ \tilde{Y} \to \frac{\alpha T^2}{M}X + (1 + \frac{\alpha T^2}{M})\tilde{Y}. \end{cases} \tag{14}$$

We require that the rescaled momentum variable $\tilde{Y}$ also be periodic, of the same period L as of X. Choosing $L = 1$ and $\frac{\alpha T^2}{M}$ to be an integer fulfills this requirement. Finally, the Arnol'd cat mapping on the unit two-dimensional torus is obtained by further imposing that

$$\frac{\alpha T^2}{M} = 1. \tag{15}$$

A rigorous mathematical framework for the quantum dynamics of this map can be found in [60]. I now follow the treatment in [11], with results in agreement with the original work of Schwinger [61] on the quantum dynamics on a torus. Because of the X periodicity, the wavefunction takes the form

$$\psi(X) = \sum_k c_k \phi_k(X) = \sum_k c_k e^{i2\pi kX/L}. \tag{16}$$

In the above, $\phi_k(X) := e^{i2\pi kX/L}$ are momentum eigenfunctions and c_k, with k integer, are the expansion coefficients. Next, we consider periodicity in Y with period ML/T. The momentum operator is $\mathbf{Y} = -i\hbar\partial_X$, so that $\mathbf{Y}\phi_k(X) = \frac{kh}{L}\phi_k(X)$; h is Planck's constant, therefore, periodicity in momentum implies that

$$c_k = c_{k+\mathcal{N}} \tag{17}$$

for any k, where $\mathcal{N}$ is an integer number:

$$\frac{ML^2}{T} = \mathcal{N}h. \tag{18}$$

One parameter can be freely chosen and we set $L = 1$, with no loss of generality. One can explain the rationale behind Equation (18): to embed quantum dynamics in a two-dimensional torus of unit periodicity in the X direction, the periodicity in Y must be an integer multiple of the Planck constant. Finally, using $c_k = c_{k+\mathcal{N}}$ in Equation (16) implies that $\psi(X)$ takes the form of a periodic train of delta functions at the points X_j:

$$X_j = \frac{j}{\mathcal{N}} + s,\ j = 0, \ldots, \mathcal{N} - 1 \tag{19}$$

where s is an arbitrary constant shift. We shall use the notation $|j>$ for the delta at such positions.

The Hilbert space of the system is so isomorphic to $\mathbf{C}^{\mathcal{N}}$. In the position representation, wavefunctions can be written as

$$\psi(X_j) = \frac{1}{\sqrt{\mathcal{N}}} \sum_{k=0}^{\mathcal{N}-1} c_k \phi_k(X_j) = \frac{1}{\sqrt{\mathcal{N}}} \sum_{k=0}^{\mathcal{N}-1} c_k e^{-i2\pi kj/\mathcal{N}} \tag{20}$$

and the discrete Fourier transform maps the above to the momentum representation:

$$c_k = \hat{\psi}(Y_k) = \frac{1}{\sqrt{\mathcal{N}}} \sum_{j=0}^{\mathcal{N}-1} \psi(X_j) e^{i2\pi kj/\mathcal{N}}, \tag{21}$$

where $Y_k = kh$.

An important remark at this point is the following: via Equation (18), the dimension of the Hilbert space is directly proportional to the mass M of the particle. Therefore, we can take the classical limit in its correct physical form, by keeping $\hbar$ to its real physical value and by considering a particle of larger and larger mass M.

Quantum motion induced by the Hamiltonian Equation (12) becomes a unitary operator in $\mathcal{H} = \mathbf{C}^{\mathcal{N}}$ that was computed in [11]. In the position representation, where ψ is the vector $\psi(X_j)$, $j = 0, \ldots, \mathcal{N}$, the unitary evolution U^0 induced by the free rotation $\frac{Y^2}{2M}$ has matrix elements

$$U^0_{kl} = \frac{1}{\sqrt{\mathcal{N}}} e^{-(\pi i l^2/\mathcal{N})} e^{2\pi ikl/\mathcal{N}}. \tag{22}$$

The instantaneous change in momentum induced by the kick is rendered by the operator K with matrix elements:

$$K_{kl} = \frac{1}{\sqrt{\mathcal{N}}} e^{i\pi l^2/\mathcal{N}} \delta_{k,l}, \tag{23}$$

where $\delta_{k,l}$ is the Kronecker delta. To complete the review of the formulae in [11], the quantum Arnol'd cat evolution operator is the product $U = KU^0$.

6. The Alicki–Fannes Entropy of the Quantum Arnol'd Cat

The theory of the Alicki–Fannes entropy can be applied to the dynamics of the quantum Arnol'd cat, using projections operators P_l measuring the position coordinate only: letting $|j>$ for $j = 0, \cdots, \mathcal{N} - 1$ be the canonical basis vectors of $\mathbf{C}^{\mathcal{N}}$ and letting, for convenience, $\mathcal{N} = 2^q$, $Q = 2^{q-p}$ with p and q positive integers, $p < q$, we define

$$P_l = \sum_{j=lQ}^{(1+l)Q-1} |j>< j|, \quad l = 0, \ldots, 2^p - 1. \tag{24}$$

For illustration, P_0 is the projection on the first $Q = 2^{q-p}$ basis states, which corresponds to the classical projection on the phase-space set $0 \leq X < 1/2^p$, that is, the leftmost $1/2^p$ section of the unit two-dimensional torus (notice that all values of Y are allowed).

Even though the adaptation of the formalism of Section 3 to the case of the quantum Arnol'd cat is theoretically straightforward and physically transparent, its practical implementation, in terms of numerical analysis, is particularly challenging. Details of the sophisticated technique required for its implementation can be found in [56]. Following the approach presented in [38], this technique permits to compute the entropy for arbitrarily long times, avoiding the exponential increase of the dimension of the decoherence matrix. In Figure 3, the quantity $S(n)$ (shorthand notation for $S(\mathcal{H}, \mathcal{U}, \mathcal{P}, n)$) is plotted versus n for different values of q (recall, $\mathcal{N} = 2^q$) and $p = 2$. We see that the bound $S(n) \leq 2\log(\mathcal{N})$ is approached at a rapid pace in n for any finite value of q. Before this saturation, which happens when n is proportional to q, that is $\log(\mathcal{N})$, there is a small interval when $S(n)$ increases linearly with a slope dictated by the classical Lyapunov exponent. After this brief lapse of time, output of new information fades.

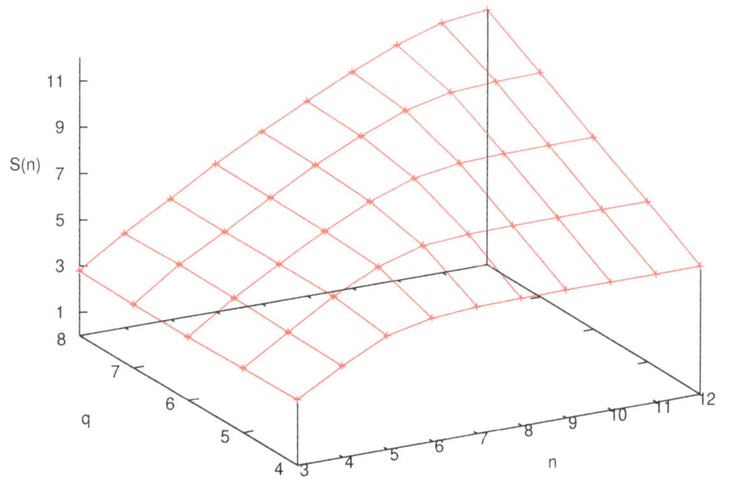

Figure 3. AF entropy $S(n)$ as a function of n and q.

This is the logarithmic catastrophe described above. Indeed, to *linearly* increase the length of this span of chaotic freedom, one has to *exponentially* increase the dimension of phase space, i.e., the *mass* of the particle. It is clear that, while mathematically the classical complexity is fully regained in the limit, this is quite unsatisfactory from a physical viewpoint.

A detailed analysis of the decoherence matrix yielding these results further reveals their physical content. In the numerical experiments that follow, we fix $p = 2$, meaning that phase space and Hilbert space are partitioned in four sets (see Equation (24) and discussion) and, consequently, the decoherence matrix is the direct sum of sixteen diagonal blocks, of which we consider the block $\sigma_0 = \theta_0 = 0$, $\sigma_{n-1} = \theta_{n-1} = 0$, with the other blocks providing similar results. In Figure 4, left panel, the matrix elements $|D_{\sigma,\theta}|$ are plotted as blue bars. The word length is three, so that interference between the paths $(0, \sigma_1, 0)$ and $(0, \theta_1, 0)$, as in a four-slit experiment, is clearly evident. Because of this, the diagonal values compare poorly with the classical probabilities $\mu(\sigma)$ (a shorthand notation for $\mu(P_\sigma)$), plotted as red bars on the diagonal. The forest of off-diagonal blue bars is even more evident when considering words of length four $(0, \sigma_1, \sigma_2, 0)$ (there are 16 of them), in the right panel of Figure 4.

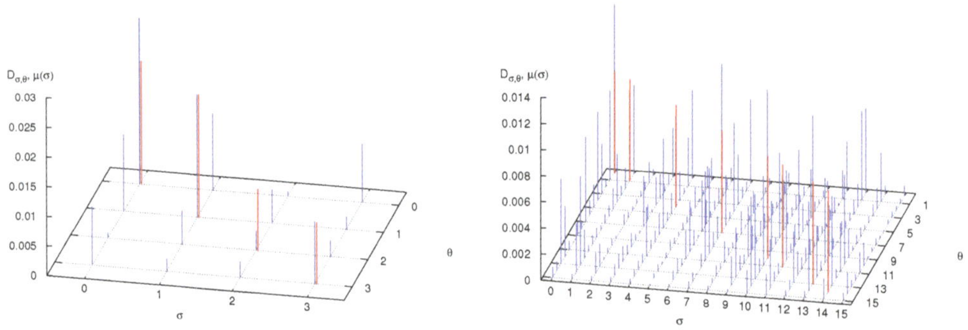

Figure 4. Absolute value of the decoherence matrix elements $D_{\sigma,\theta}$ (blue bars) and classical word probabilities $\mu(\sigma)$ (red bars, plotted on the diagonal for comparison), when $q = 2$ so that $\mathcal{N} = 2^q = 4$, and $p = 2$. In the left panel $|\sigma| = |\theta| = 3$, in the right panel $|\sigma| = |\theta| = 4$. See text for further detail.

How, then, does the classical limit arise in this context? In Figure 5 the *diagonal* matrix elements $D_{\sigma,\sigma}$ are plotted versus q, and compared with the classical probabilities $\mu(\sigma)$. Since the mass M is proportional to $\mathcal{N} = 2^q$, increasing q by one unit means doubling the mass of the particle. Convergence to the classical values is observed as well as the progressive quenching of off-diagonal matrix elements. In fact, in Figure 6, left panel, it is seen that when q increases, the decay of the sum of the absolute values of the off-diagonal matrix elements $D_{\sigma,\theta}$, $\sigma \neq \theta$, trends in pair with convergence to the classical entropy S_{cl} of the *diagonal* entropy $S_{diag} = -\sum_\sigma D_{\sigma,\sigma} \log D_{\sigma,\sigma}$ and of the AF entropy S_{AF} in Equation (11). The fitting line in the figure reveals that these phenomena scale as a power law with the mass M (proportional to 2^q), with exponent between minus one and minus one half.

The right panel of Figure 6 reveals, in even clearer detail, the logarithmic catastrophe alluded to before. The symbolic distance, $d_{\text{symb}} = \sum_\sigma |D_{\sigma,\sigma} - \mu(\sigma)|$, at fixed word length $|\sigma|$, is plotted when both $|\sigma|$ and q are varied. One observes that classical word probabilities of fixed length $|\sigma|$ are recovered when increasing the mass of the particle (i.e., q), while quantum interference destroys this convergence when, at fixed mass, longer words are considered. This is the well-known noncommutativity of the classical limit and the long-time limit [7,8]. Yet, the graph shows that quantum effects spoil the *correspondence* at a word length that is *logarithmic* in the mass of the particle. One so recovers from the entropic point of view the results obtained via algorithmic complexity [11].

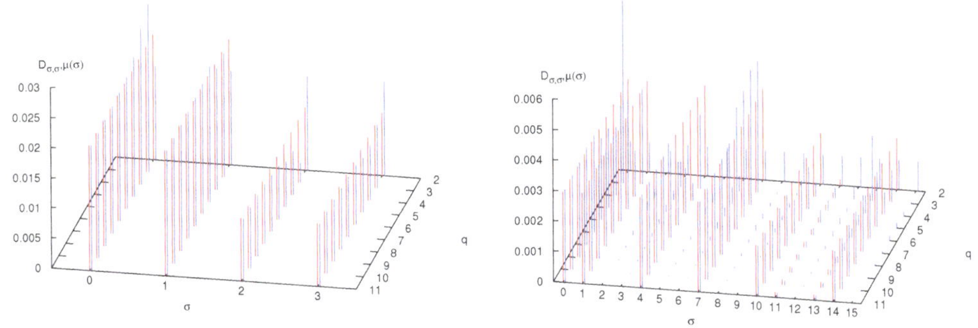

Figure 5. Absolute value of the diagonal decoherence matrix elements $D_{\sigma,\sigma}$ (blue bars) as function of q, compared with the classical word probabilities $\mu(\sigma)$ (red bars, obviously independent of q and plotted sideways for comparison). In the left panel $|\sigma| = 3$, in the right panel $|\sigma| = 4$. See text for further detail.

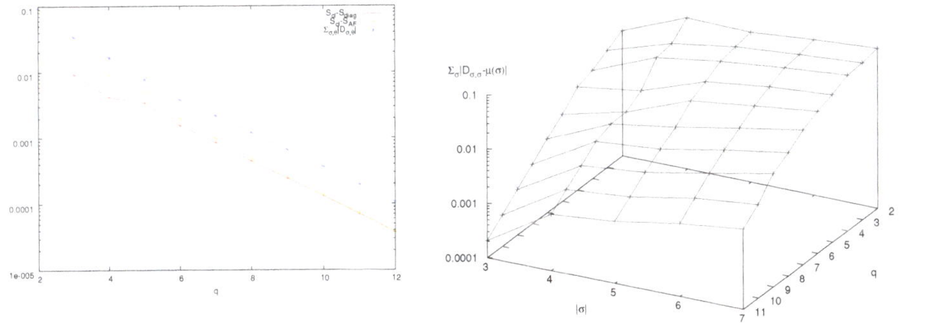

Figure 6. Left panel: for $|\sigma| = 3$, sum of the absolute values of the off-diagonal matrix elements $D_{\sigma,\theta}$ (blue curve), absolute difference between classical entropy S_{cl} and diagonal entropy S_{diag} (red curves), absolute difference between classical entropy S_{cl} and AF entropy S_{AF} (green curves). Right panel: symbolic distance $d_{\text{symb}} = \sum_\sigma |D_{\sigma,\sigma} - \mu(\sigma)|$ versus word length $|\sigma|$ and q. See text for further detail.

As mentioned, decoherence has been proposed as a way to overcome this *impasse*. It is the purpose of the remaining part of this paper to study decoherence in a controlled setting, with the aid of a model system.

7. Decoherence and the Multiparticle Arnol'd Cat

Decoherence, that is, the loss of quantum coherence of a system, is today principally studied because it is a major obstacle in the performance of a quantum computer [62]. In the 1990s of the XX century it was invoked for the explanation of the puzzling behavior of quantum Schrödinger cats [63] and the emergence of classical properties in quantum mechanics [29–33,45]. In actuality, as early as 1984, Guarneri [64] proved that a random perturbation of the motion of a kicked rotor prevents quantum localization from setting in. This induces a diffusive quantum motion, whose diffusion coefficient D was computed by Ott et al. [65]. In the same perspective, Dittrich and Graham [66], Kolovsky [21,23], and Sundaram et al. [67] considered systems (both classical and quantum) randomly perturbed or coupled to an environment: the resulting motion is also diffusive. The quantum dynamics of these systems were shown to be similar to the classical when $\sqrt{D/\lambda}$ is larger than $\hbar$ times a dimensional constant (λ is the Lyapunov exponent). The heuristic explanation of this inequality stems from the observation that since diffusion *smears* classical phase space to the finest discernable resolution of $\sqrt{D/\lambda}$, when this latter is larger than $\hbar$, the quantum smeariness, the two dynamics appear similar. To quote a famous musical, *"The way she sees and the way he looks is a perfect match"*: the reader is asked to interpret the analogy. We shall return to this point in the conclusions.

Let us now describe a system that I have introduced to model *exactly* the interaction of a system with the environment, avoiding recourse to master equations, Lindblad operators, or similar, with the goal of having a fully Hamiltonian system [44]. The model draws upon the original view of decoherence by Joos and Zeh [45,68] as due to collisions of particles.

Consider adding I smaller particles of mass m in the ring that confines the particle of mass M, freely rotating except for collisions with the large particle. Let x_i and y_i, $i = 1, \ldots, I$ the coordinates and momenta of these particles, respectively. The Hamiltonian of the classical system is, therefore,

$$H = H_{cat}(X, Y, t) + \sum_{i=1}^{I} \frac{y_i^2}{2m} + \kappa \sum_{i=1}^{I} V(x_i - X), \tag{25}$$

where H_{cat} was introduced in Equation (12), κ is a coupling constant, and the potential function $V(\cdot)$ will be described momentarily.

Let us first define the quantum kinematics. In addition to the $L = 1$ periodicity in the variables x_i, we also impose periodicity $\frac{mL}{T}$ in the momenta y_i. Therefore, performing a similar analysis to the one presented above, we obtain

$$\frac{mL^2}{T} = \nu h, \tag{26}$$

where ν is the dimension of the Hilbert space of a single small particle, an integer that we also choose to be a power of two: $\nu = 2^r$. Therefore, the many-particle wavefunction reads

$$\Psi(X, x_1, \ldots, x_I) = \mathcal{N}^{-1/2} \nu^{-I/2} \sum_{k_0=0}^{\mathcal{N}-1} \sum_{k_1=0}^{\nu-1} \cdots \sum_{k_I=0}^{\nu-1} c_{k_0,k_1,\ldots,k_I} e^{2\pi i (k_0 X + \sum_i k_i x_i)}. \tag{27}$$

In the same way as the position X is restricted to a lattice, so are the x_is. We further require that small particles' positions x_i, $i = 1, \ldots, I$, be a subset of those of the large one:

$$x_i^{j_i} = j_i \frac{1}{\nu} + s_i \frac{1}{\mathcal{N}}, \quad j_i = 0, \ldots, \nu - 1. \tag{28}$$

where s_i an integer ranging from zero to $\mathcal{N}/\nu - 1$. Introducing the multi-index $j = j_0, j_1, \ldots, j_I$, in which the label 0 refers to the large particle and $i > 0$ to the i-th small one, we let $|j >$ denote the delta function at the particles' position given in Equations (19) and (28). A different representation for the wave function is, therefore,

$$\Psi = \sum_{j_0=0}^{\mathcal{N}-1} \sum_{j_1=0}^{\nu-1} \cdots \sum_{j_I=0}^{\nu-1} d_{j_0,j_1,\ldots,j_I} |j_0, j_1, \ldots, j_I > . \tag{29}$$

The Hilbert space of the system $\mathcal{H}$ is the tensor product of the Hilbert spaces of the single particles; therefore, it is isometric to a power of $\mathbf{C}$:

$$\mathcal{H} = \mathcal{H}_{big} \otimes \mathcal{H}_{small}^{I} = \mathbf{C}^{\mathcal{N} I^{\nu}}. \tag{30}$$

The potential V must represent a short-range interaction occurring only when the large particle and a small one occupy the same position. In the representation Equation (29), its matrix elements are, therefore,

$$< j'|Vj >= V_{j_0',j_1',\ldots,j_I';j_0,j_1,\ldots,j_I} = \prod_{i=0}^{I} \delta_{j_i',j_i} \sum_{i=1}^{I} \delta_{j_0, \frac{\mathcal{N}}{\nu} j_i + s_i}. \tag{31}$$

8. The Von Neumann Entropy

In a pair of papers I have investigated the dynamics of the quantum system introduced in the previous section [44,46]. Treating the large, zeroth particle as the system and the small particles as the environment, the reduced density matrix $\hat{\rho}$ is defined by tracing over the small particles degrees of freedom:

$$\hat{\rho}_{j_0',j_0} = \sum_{j_1,\ldots,j_I} \rho_{j_0',j_1,\ldots,j_I;j_0,j_1,\ldots,j_I} \qquad (32)$$

Firstly, by choosing an initial density matrix as a pure state in a Schrödinger cat configuration [63], I showed that as time progresses, interaction with small particles wipes out the outdiagonal components of the reduced density matrix $\hat{\rho}$, obtained by tracing over the small particles degrees of freedom, exactly as in the experimental results. This dynamics is shown in the position representation in Figure 7. Here, $\alpha = 0$, that is, there is no cat kick. These results show that the multiparticle cat is a *bona fide* model of decoherence.

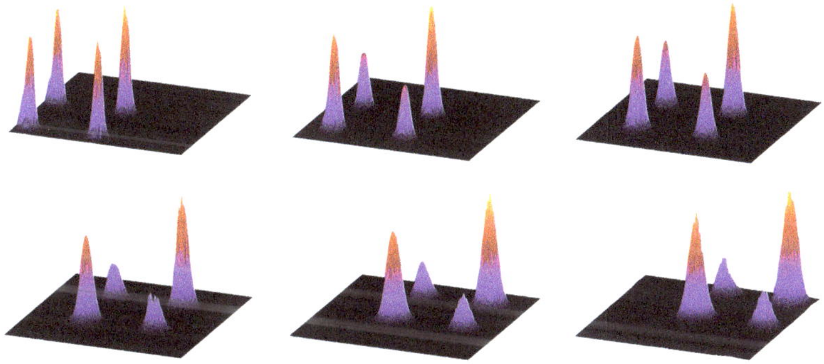

Figure 7. Absolute value of the reduced density matrix, in the position representation, of a system with $q = 7$, $I = 2$, $\nu = 1$, $\kappa = 50$. Time snapshot ordered from left to right, top to bottom. The cat particle has a positive velocity, so that peaks drift on the torus.

I then considered the Von Neumann entropy, $-\mathrm{Tr}\, \hat{\rho} \log \hat{\rho}$. This entropy quantifies the amount of information in the eigenvalues of the reduced density matrix. The analysis was first performed for the free colliding particles case, given by the Hamiltonian (13), (25) with $\alpha = 0$. The initial increase of the Von Neumann entropy was observed to be logarithmic. On the contrary, when $\alpha T^2/M = 1$, as in the Arnol'd cat map, the initial increase of entropy is linear, with a slope that equals the Kolmogorov Sinai entropy *provided* the coupling constant κ Equation (25) takes a *specific* value, which scales according to the physical parameters of the system. Clearly, when κ is null there is no decoherence and the Von Neumann entropy is null. For small values of κ, entropy increases linearly at a small rate before saturation, while above the aforementioned value the rate is larger than the classical dynamical entropy. This parameter dependence is in my view a *caveat* for the physical interpretation of similar results in different systems. For details, the interested reader is referred to [46].

9. The AF Entropy of the Multiparticle Arnol'd Cat

Even though interestingly, results about the Von Neumann entropy lack the generality and theoretical significance provided by the Alicki–Fannes entropy. In particular, the former is bounded above by the logarithm of the dimension of the Hilbert space of the *reduced* system, in this case $\log \mathcal{N}$, *independently* of the dimension of the *full* Hilbert space. The latter, on the contrary is bound by $2\log\mathcal{N} + 2I\log(\nu)$, which scales *linearly* with the number of particles I and therefore *might* provide a physically reasonable classical limit.

Moreover, the decoherence matrix introduced in Section 3 permits us to compare analytically, i.e., one by one, the classical probabilities $\mu(P_\sigma)$ with the diagonal elements

$D_{\sigma,\sigma}$ and to gauge the importance of nondiagonal matrix elements, which encode quantum coherence. Observe that the projectors P_l in Equation (24) can be generalized to act on the coordinate of the large particle only, labeled by the index zero:

$$P_l = \sum_{j_0=lQ}^{(1+l)Q-1} \sum_{j_1=0}^{\nu-1} \cdots \sum_{j_l=0}^{\nu-1} |j_0, j_1, \ldots, j_l >< j_l, \ldots, j_1, j_0|, \quad l = 0, \ldots, 2^p - 1. \tag{33}$$

In so doing, we are observing the symbolic dynamics of the large particle only, in the presence of the external environment provided by the other, smaller, particles.

9.1. The Decohering Effect of the Perturbation

Firstly, let us consider a system composed of a large particle, $q = 4$, coupled to two small particles, $r = 1$ (mass ratio 1:8) and let us vary the coupling constant κ. In Figure 8 we plot, as before, the decoherence matrix entries for words of length 5, beginning and ending with the symbol 0. In the left panel $\kappa = 0$ (that is, small particles do not interact with the large one), while in the right panel $\kappa = 10$. Nondiagonal matrix elements are clearly smaller in the second case, so that diagonal values become closer to the eigenvalues of the decoherence matrix. Nonetheless, we do not expect them to approach *in the limit of very large* κ the classical word probabilities, displayed as red bars in both panels. This is evident in Figure 9 where the diagonal matrix elements $D_{\sigma,\sigma}$ are plotted versus κ and compared to $\mu(\sigma)$. The symbolic distance $d_{\text{symb}} = \sum_\sigma |D_{\sigma,\sigma} - \mu(\sigma)|$ quantifies the agreement: in the right panel of Figure 9 it is seen that d_{symb} does not significantly diminishes when increasing κ, differently from the sum of the absolute values of the off-diagonal matrix elements, $\sum_{\sigma \neq \theta} |D_{\sigma,\theta}|$.

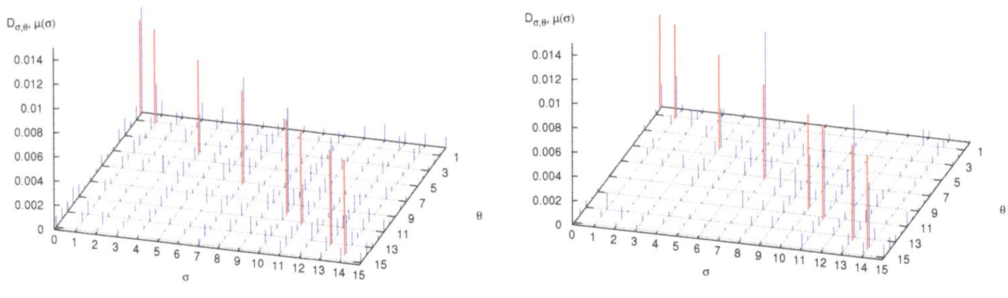

Figure 8. Absolute value of the decoherence matrix elements $D_{\sigma,\theta}$ (blue bars) and classical word probabilities $\mu(\sigma)$ (red bars, plotted on the diagonal for comparison) when $q = 4$, $r = 1$, $I = 2$, and $p = 2$. In the left and right panels, $\kappa = 0$ and $\kappa = 10$, respectively. See text for further detail.

I believe that these results are general, albeit obtained in a restricted range of parameters. Firstly, increasing κ reduces the off-diagonal matrix elements and therefore makes the system less coherent, i.e., more classical. Secondly, observe that the sums $\sum_\sigma D_{\sigma,\sigma}$ and $\sum_\sigma \mu(\sigma)$ over words of fixed length $|\sigma|$ are constant and equal to the probabilities, quantum and classical, of the sector σ_0, σ_{n-1}, (0, 0 in the figures) so that changing the value of κ results in a transfer of probability among all words of length $n = |\sigma|$ while keeping their sum approximately constant (because transfer of probability among different blocks σ_0, σ_{n-1} is small). Therefore, the fact that increasing κ at fixed $|\sigma|$ does not reduce the symbolic distance means that small particles are rendering the *observation* of the large one more classical, but at the same time are perturbing it too much. Finally, the increase of the symbolic distance with $|\sigma|$ (an evident proxy for time) at fixed κ shows, again, the rapidly emerging difference between classical and quantum dynamics. Figure 10 confirms this conclusion: we see that in the presence of decoherence (that is, low off-diagonal matrix elements), AF and diagonal entropies approach each other, but differ from the classical value.

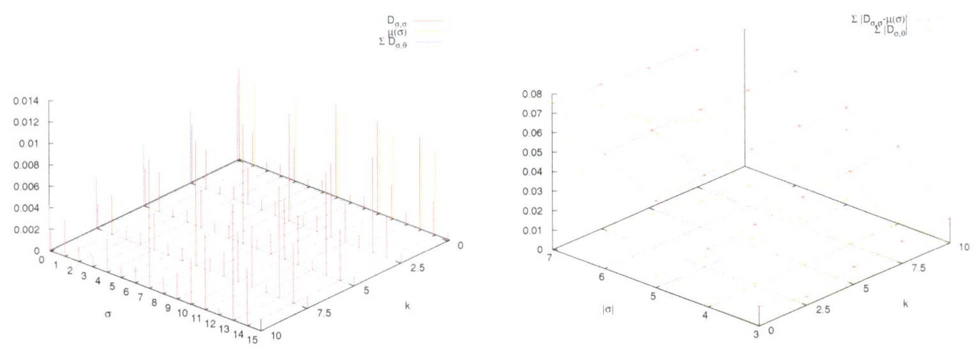

Figure 9. Left panel: Parameters as in Figure 8. Diagonal matrix elements $D_{\sigma,\sigma}$ (red bars) versus κ, classical word probabilities $\mu(\sigma)$ (green bars), and sum of the absolute value of outdiagonal matrix elements $D_{\sigma,\theta}$ (blue bars, rescaled by a factor to improve readability of the graph). Right panel: sum of the absolute differences between diagonal matrix elements and classical probabilities (red), sum of the absolute values of off-diagonal matrix elements (green, rescaled by a factor).

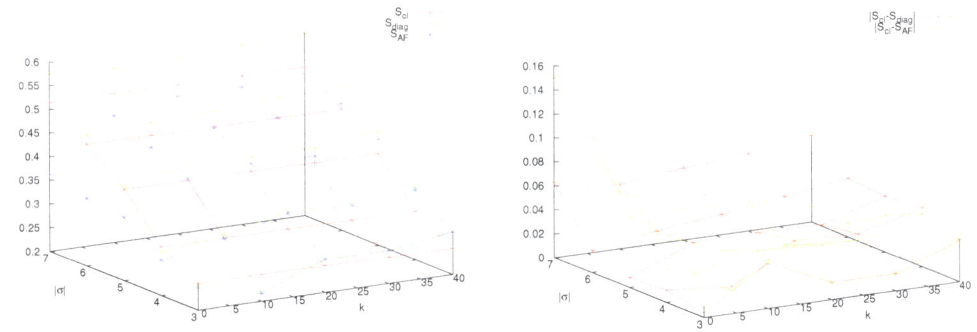

Figure 10. Parameters as in Figure 8, wider range of the coupling constant $\kappa \in [0, 40]$. Left panel: classical entropy S_{cl} (red curves), diagonal entropy S_{diag} (green curves), AF entropy S_{AF} (blue curves). Right panel: absolute difference $|S_{cl} - S_{diag}|$ (red curves) and $|S_{cl} - S_{AF}|$ (green curves).

9.2. The Limit of Large Mass with Decoherence

The previous results show that there exist two ways to induce decoherence, defined as quenching of the off-diagonal entries of the matrix D: increasing the mass of the particle (that is, the dimension of its Hilbert space), or coupling it to a larger, yet still purely quantum, system. The former leads in the limit to the classical results, but is plagued by its very slow, logarithmic convergence, which renders its application unphysical. The second cannot provide by itself the hardly-sought classical limit, but it may feature an exponential increase in Hilbert space dimension, simply augmenting the number of small particles. I made this observation in [44], where I surmised that a combination of the two approaches might yield scaling relations (in the number of small particles, coupling, and mass of the large particle) by which complexity of the motion, as a linear increase of the Shannon Alicki–Fannes entropy $S(n)$, could be observed over physically significant time spans.

This investigation is subtle, both from the theoretical and the numerical viewpoint. In this paper we can add a few experiments in this direction. In Figure 11, we consider two families of systems, the former composed of a single particle, of increasing mass (different values of q), already studied in Section 6; the second, composed again of such large particle, but coupled to two small particles, with coupling constant κ that increases proportionally to the mass of the large one. We compute the AF entropies S_{AF} of these systems and compare them with the classical entropy. We observe that, while both converge to the last quantity, the multiparticle case provides a better approximation for small values of q. Considering that the mass M is proportional to the exponential of q, this is encouraging in the quest outlined above.

To investigate this phenomenon, recall that, as described above, for any finite value of $n = |\sigma|$ and for any finite sector σ_0, σ_{n-1}, (0, 0 in the figures) the AF, diagonal, and classical entropies are the sum of L^{n-1} contributions of the form $s_j = -\zeta_j \log(\zeta_j)$, when ζ_j is an eigenvalue of the matrix D, or a diagonal entry of this matrix, or a classical word probability, respectively. We can therefore compare different entropies by first sorting the values s_j in descending order, from the largest to the smallest, and then computing their integrated distribution $S_i = \sum_{j=0}^{i} s_j$. These distributions are plotted in Figure 12, in the same case of Figure 11. Different panels are related to increasing values of q, from 2 to 7. The horizontal segments in the plots of the AF partial entropies are due to the Hilbert space dimensionality bound, which imposes a maximum number of non-null eigenvalues of D. We can observe that the effect of the small particles is to free the AF entropy from this bound, assuring a faster convergence to the classical values.

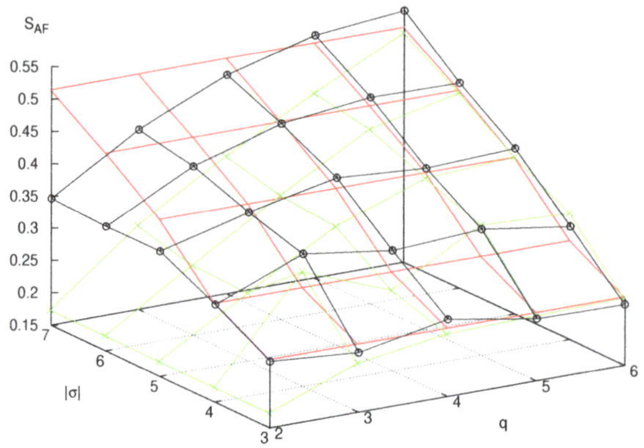

Figure 11. Classical entropy S_{cl} (red curves) compared with AF entropy S_{AF} of systems with variable q and $I = 0$ (green curves) and $I = 2$, $r = 1$, and $\kappa = 0.125 \times 2^q$ (black curves).

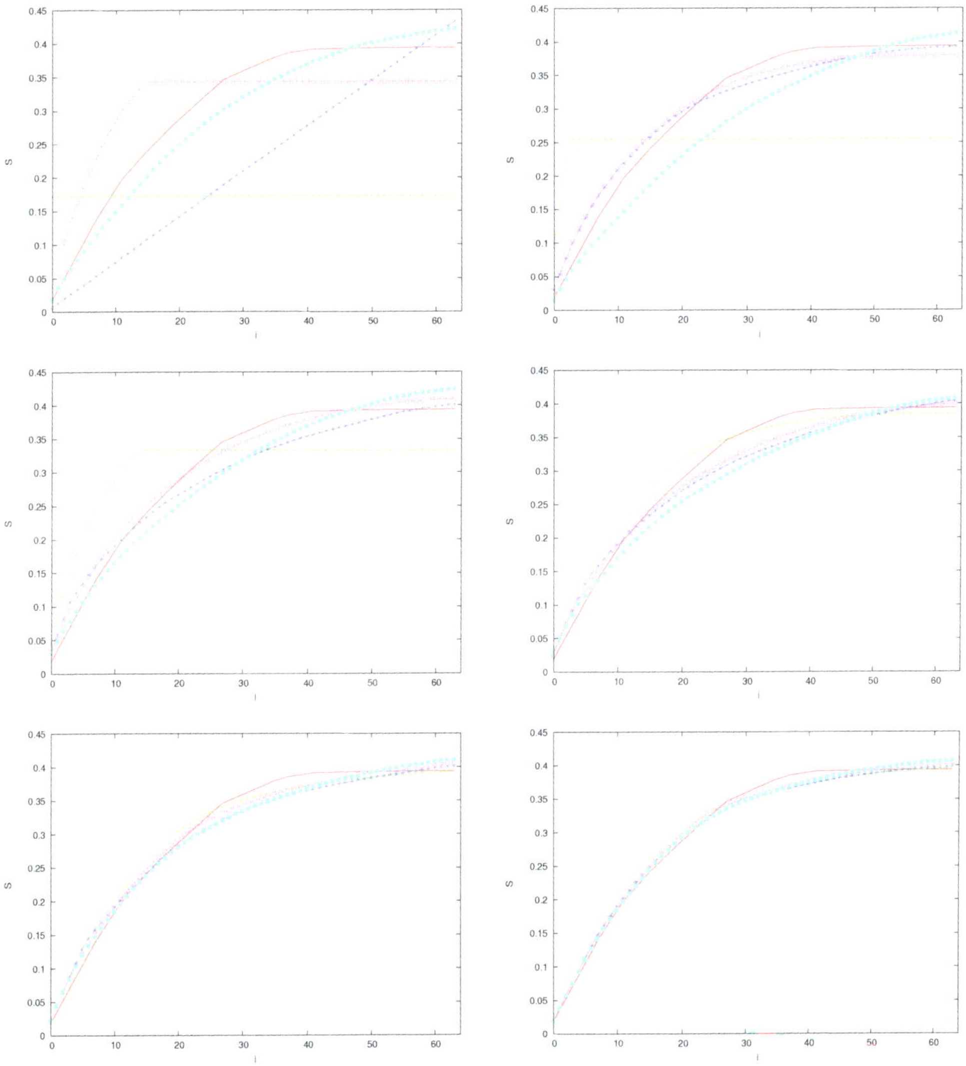

Figure 12. Increasing partial entropies S_i, $|\sigma| = 5$, of systems with variable q, from 2 to 7, left to right and top to bottom. The classical entropy S_i is denoted by a red curve in all panels. In the case $I = 0$, the green curve marks the AF partial entropy, and the blue curve the diagonal partial entropy. In the case $I = 2$, $r = 1$, and $\kappa = 0.125 \times 2^q$, the AF partial entropy is marked in magenta and the diagonal partial entropy in light blue.

10. Conclusions

The multiparticle Arnol'd cat is a model system particularly suited to investigate the nature, definition and relevance of quantum chaos. In my works I have followed the original suggestion of Chirikov and Ford that this problem should be considered within the framework of information theory. Dynamical entropies provide a major tool for this investigation and their formal aspects have been fully developed in the last years [48]. In this respect, the model described herein permits to free dynamics, in a controlled, fully canonical way, from the bound induced by the finite dimensionality of the Hilbert space.

This bound is due, in this particular model, by the finiteness of the mass of a particle, and therefore carries a deep physical significance.

I considered herein the classical limit as the limit of increasing mass of such particle, while keeping the value of the Planck constant fixed. Chaotic behavior has been defined, for a finite time span, as the production of information at an inferiorly bounded, positive rate. Consideration of the decoherence matrix from the consistent histories formalism has also permitted the comparison, word by word, of classical and quantum symbolic dynamics. It came as no surprise that, as I presented in earlier works, in the single-particle case, the length of agreement between the two, and consequently the length of the span of the chaotic freedom of the quantum system, grows only logarithmically with the mass of the particle.

As reviewed in this paper, coupling to an external system has been shown to provide quantum motions (via the Wigner function, or similar phase space tools) that *correspond* to a classical *noisy* evolution, within the finite phase-space resolution of this latter, for longer times. In a discrete lattice model, mimicking the quantum cat Hamiltonian with noise [69], we have observed information production over long time scales. Returning now to many particles quantum systems, in [44], I proposed the hypothesis that a particular scaling of parameters—number of particles, their mass, and the observational time interval—could extend, in a physically reasonable range, the span of chaotic behavior. The results of this paper are admittedly not conclusive, but suggest that this endeavor could be feasible. In this view, the (fully quantum) environment is to be considered as a source of *disorganized* information that, when input to a regular or chaotic quantum system, is *organized* by this latter in such a way to agree, within observational limits, to the classical description. This is, in essence, the program outlined in [69]: *"For a long time, research in quantum chaos has looked for quantum characteristics related to classical chaotic motion. The fact that none of these could be properly called chaos led to the concept of pseudochaos and cast doubts on the very existence of chaos in nature. [...] one might try to reverse this approach and consider classical dynamics as an effective theory that, via truly chaotic deterministic dynamical systems, models a randomly perturbed quantum motion under observational coarse graining."*

Funding: This research was partially funded from grant PRIN 2017S35EHN *Regular and stochastic behavior in dynamical systems*.

Institutional Review Board Statement: Not applicable.

Data Availability Statement: No experimental data involved.

Acknowledgments: Computations for this paper were performed on the Zefiro cluster at the INFN computer center in Pisa. Enrico Mazzoni is warmly thanked for assistance.

Conflicts of Interest: The author declares no conflict of interest.

References

1. Mantica, G. Fourier-Bessel functions of singular continuous measures and their many asymptotics, ETNA. *Electron. Trans. Numer. Anal.* **2006**, *25*, 409–430.
2. Benenti, G.; Casati, G. How complex is quantum motion? *Phys. Rev. E* **2009**, *79*, 025201(R). [CrossRef] [PubMed]
3. Sokolov, V.V.; Zhirov, O.V.; Benenti, G.; Casati, G. Complexity of quantum states and reversibility of quantum motion. *Phys. Rev. E* **2008**, *78*, 046212. [CrossRef] [PubMed]
4. Casati, G.; Chirikov, B.V. Comment on "Decoherence, chaos and the second law". *Phys. Rev. Lett.* **1995**, *75*, 350. [CrossRef]
5. Casati, G.; Chirikov, B.V. The Legacy of Chaos in Quantum Mechanics. In *Quantum Chaos*; Cambridge University Press: Cambridge, UK, 1995.
6. Casati, G.; Chirikov, B.V. Quantum Chaos: Unexpected complexity. *Physica D* **1995**, *89*, 220. [CrossRef]
7. Berry, M.V. The Bakerian Lecture, 1987. Quantum chaology. *Proc. Roy. Soc. Lond. A* **1987**, *413*, 183.
8. Berry, M.V. Quantum chaology, not quantum chaos. *Phys. Scr.* **1989**, *40*, 335–336. [CrossRef]
9. Chirikov, B.V. *The Problem of Quantum Chaos*; Lecture Notes Phys; Springer: Berlin/Heidelberg, Germany, 1992.
10. Chirikov, B.V.; Vivaldi, F. An algorithmic view of pseudo–chaos. *Physica D* **1999**, *129*, 223. [CrossRef]
11. Ford, J.; Mantica, G.; Ristow, G. Does Quantum Mechanics Obey the Correspondence Principle? Is It Complete? *Physica D* **1991**, *50*, 493–520. [CrossRef]

12. Mantica, G.; Ford, J. On the Completeness of the Classical Limit of Quantum Mechanics. In *Quantum Chaos–Quantum Measurement*; Cvitanović, P., Percival, I., Wirzba, S., Eds.; Kluwer: Boston, MA, USA, 1992.
13. Ford, J.; Mantica, G. Does Quantum Mechanics Obey the Correspondence Principle? Is It Complete? *Am. J. Phys.* **1992**, *60*, 1086–1098. [CrossRef]
14. Ford, J. How random is a coin toss? *Phys. Today* **1983**, *36*, 40. [CrossRef]
15. Kolmogorov, A.N. Three Approaches to the Definition of the "Amount of information" Concept". *Probl. Peredachi Informatsii* **1965**, *1*, 1–7.
16. Chaitin, G.J. On the Length of Programs for Computing Finite Binary Sequences. *J. ACM* **1966**, *13*, 547–569. [CrossRef]
17. Solomonoff, R.J. A formal theory of inductive inference. Part I and II. *Inf. Control* **1964**, *7*, 224–254. [CrossRef]
18. Chaitin, G.J. Algorithmic Information Theory. In *Encyclopedia of Statistical Sciences*; Wiley: Hoboken, NJ, USA, 2004.
19. Ford, J. What is Chaos, That We Should be Mindful of It? In *The New Physics*; Davies, P., Ed.; Cambridge University Press: Cambridge, UK, 1989; pp. 348–371.
20. Casati, G.; Mantica, G. Caos-Chaos. *G. Fis. Soc. Ital. Fis.* **2022**, *62*, 461–486.
21. Kolovsky, A.R. A Remark on the Problem of Quantum-Classical Correspondence in the Case of Chaotic Dynamics *Europhys. Lett.* **1994**, *27*, 79.
22. Kolovsky, A.R. Quantum coherence, evolution of the Wigner function, and transition from quantum to classical dynamics. *Chaos* **1996**, *6*, 534–542. [CrossRef]
23. Kolovsky, A.R. Condition of Correspondence between Quantum and Classical Dynamics for a Chaotic System. *Phys. Rev. Lett.* **1996**, *76*, 340.
24. Majewski, W.A.; Kuna, M. On quantum characteristic exponents. *J. Math. Phys.* **1993**, *34*, 5007–5015. [CrossRef]
25. Man'ko, V.I.; Mendes, R.V. Lyapunov exponent in quantum mechanics. A phase-space approach. *Physica D* **2000**, *145*, 330–348 [CrossRef]
26. Pappalardi, S.; Kurchan, J. Quantum Bounds on the Generalized Lyapunov Exponents. *Entropy* **2023**, *25*, 246. [CrossRef]
27. Arnold, V.I.; Avez, A. *Ergodic Problems of Classical Mechanics*; Benjamin: New York, NY, USA, 1968.
28. Hannay, J.H.; Berry, M.V. Quantization of linear maps on the torus - Fresnel diffraction by a periodic grating. *Physica D* **1980**, *1*, 267–291. [CrossRef]
29. Brun, T.A.; Percival, I.; Schack, R. Quantum chaos in open systems: A quantum state diffusion analysis. *J. Phys. A Math. Gen.* **1996**, *28*, 2077–2090. [CrossRef]
30. Halliwell, J.; Zoupas, A. Quantum state diffusion, density matrix diagonalization, and decoherent histories: A model. *Phys. Rev. D* **1995**, *52*, 7294–7307. [CrossRef] [PubMed]
31. Percival, I. *Quantum State Diffusion*; Cambridge University Press: Cambridge, UK, 1998.
32. Zurek, W.H.; Paz, J.P. Decoherence, chaos and the 2nd law. *Phys. Rev. Lett.* **1994**, *72*, 2508. [CrossRef]
33. Zurek, W.H.; Paz, J.P. Quantum chaos. A decoherent definition. *Physica D* **1995**, *83*, 300. [CrossRef]
34. Zurek, W.H. Decoherence, einselection, and the quantum origins of the classical. *Rev. Mod. Phys.* **2003**, *75*, 715–775. [CrossRef]
35. Monteoliva, D.; Paz, J.P. Decoherence and the rate of entropy production in chaotic quantum systems. *Phys. Rev. Lett.* **2000**, *85*, 3373. [CrossRef]
36. Bianucci, P.; Paz, J.P.; Saraceno, M. Decoherence for classically chaotic quantum maps. *Phys. Rev. E* **2002**, *65*, 46226 [CrossRef]
37. Alicki, R.; Fannes, M. Defining quantum dynamical entropy. *Lett. Math. Phys.* **1994**, *32*, 75–82. [CrossRef]
38. Alicki, R.; Makowiec, D.; Miklaszewski, W. Quantum chaos in terms of entropy for a periodically kicked top. *Phys. Rev. Lett.* **1996**, *77*, 838–841. [CrossRef]
39. Alicki, R.; Lozinski, A.; Pakoński, P.; Życzkowski, K. Quantum dynamical entropy and decoherence rate. *J. Phys. A. Math. Gen.* **2004**, *37*, 5157–5172.
40. Katok, A.; Hasselblatt, B. *Introduction to the Modern Theory of Dynamical Systems*; Cambridge University Press: Cambridge, UK, 1995.
41. Kosloff, R.; Rice, S.A. The Influence of Quantization on the Onset of Chaos in Hamiltonian Systems: The Kolmogorov Entropy Interpretation. *J. Chem. Phys.* **1980**, *74*, 1340–1349. [CrossRef]
42. Pechukas, P. Kolmogorov entropy and quantum chaos. *J. Phys. Chem.* **1982**, *86*, 2239. [CrossRef]
43. Helton, J.W.; Tabor, M. On classical and quantal Kolmogorov entropies. *J. Phys. A Math. Gen.* **1985**, *18*, 2743. [CrossRef]
44. Mantica, G. The Multiparticle Quantum Arnol'd Cat: A test case for the decoherence approach to quantum chaos. *J. Sib. Fed. Univ.* **2010**, *3*, 369–380.
45. Joos, E.; Zeh, H.D. The emergence of classical properties through interaction with the environment. *Z. Phys. B* **1985**, *59*, 223–243. [CrossRef]
46. Mantica, G. Many-Body Systems and Quantum Chaos: The Multiparticle Quantum Arnol'd Cat. *Condens. Matter* **2019**, 4 72. [CrossRef]
47. Alekseev, V.M.; Yakobson, M.V. Symbolic dynamics and hyperbolic dynamic systems. *Phys. Rep.* **1981**, *75*, 290–325. [CrossRef]
48. Benatti, F. *Dynamics, Information and Complexity in Quantum Systems*; Springer: Berlin/Heidelberg, Germany, 2009.
49. Griffiths, R.B. Consistent Histories and the Interpretation of Quantum Mechanics. *J. Stat. Phys.* **1984**, *36*, 219. [CrossRef]
50. Griffiths, R.B. Consistent Interpretations of Quantum Mechanics Using Quantum Trajectories. *Phys. Rev. Lett.* **1993**, *70*, 2201. [CrossRef]
51. Gell-Mann, M.; Hartle, J.B. Classical Equations for Quantum Systems. *Phys. Rev. D* **1993**, *47*, 3345. [CrossRef]

52. Omnès, R. From Hilbert Space to Common Sense: A Synthesis of Recent Progress in the Interpretation of Quantum Mechanics. *Ann. Phys.* **1990**, *201*, 354. [CrossRef]
53. Omnès, R. Consistent Intepretations of Quantum Mechanics. *Rev. Mod. Phys.* **1992**, *64*, 339. [CrossRef]
54. Halliwell, J.J. A Review of the Decoherent Histories Approach to Quantum Mechanics. *Ann. N. Y. Acad. Sci.* **1995**, *755*, 726–740. [CrossRef]
55. Benatti, F.; Krueger, T.; Mueller, M.; Siegmund-Schultze, R.; Szkola, A. Entropy and Quantum Kolmogorov Complexity: A Quantum Brudno's Theorem. *Commun. Math. Phys.* **2006**, *265*, 437–461. [CrossRef]
56. Mantica, G. Quantum Dynamical Entropy and an Algorithm by Gene Golub. *Electron. Trans. Numer. Anal.* **2007**, *28*, 190–205.
57. Berman, G.P.; Zaslavsky, G.M. Condition of stochasticity in quantum nonlinear systems. *Phys. Stat. Mech. Appl.* **1978**, *91*, 450–460. [CrossRef]
58. Chirikov, B.V.; Izrailev, F.M.; Shepelyansky, D.L. Quantum chaos: Localization vs. ergodicity. *Physica D* **1988**, *33*, 77. [CrossRef]
59. Mantica, G. Quantum Algorithmic Integrability: The Metaphor of Rational Billiards. *Phys. Rev. E* **2000**, *61*, 6434–6443. [CrossRef]
60. Esposti, M.D.; Graffi, S.; Isola, S. Classical limit of the quantized hyperbolic toral automorphisms. *Comm. Math. Phys.* **1995**, *167*, 471–507. [CrossRef]
61. Schwinger, J. Unitary operator bases. *Proc. Natl. Acad. Sci. USA* **1960**, **46**, 570. [CrossRef]
62. Benenti, G.; Casati, G.; Strini, G. *Principles of Quantum Computation and Information*; World Scientific: London, UK, 2004.
63. Brune, M.; Hagley, E.; Dreyer, J.; Maitre, X.; Maali, A.; Wunderlich, C.; Raimond, J.M.; Haroche, S. Observing the progressive decoherence of the meter in a quantum measurement. *Phys. Rev. Lett.* **1996**, **77**, 4887. [CrossRef]
64. Guarneri, I. Energy growth in a randomly kicked quantum rotator. *Lett. Nuovo C.* **1984**, *40*, 171–175. [CrossRef]
65. Ott, E.; Antonsen, T.M.; Hanson, J.D. Effect of Noise on Time-Dependent Quantum Chaos. *Phys. Rev. Lett.* **1984**, *53*, 2187. [CrossRef]
66. Dittrich, T.; Graham, R. Continuous quantum measurements and chaos. *Phys. Rev. A* **1990**, *42*, 4647. [CrossRef]
67. Pattanayak, A.K.; Sundaram, B.; Greenbaum, B.J. Parameter scaling in the decoherent quantum-classical transition for chaotic systems. *Phys. Rev. Lett.* **2003**, *90*, 14103. [CrossRef]
68. Carlone, R.; Figari, R.; Teta, A. The Joos-Zeh formula and the environment induced decoherence. *Int. J. Mod. Phys. B* **2004**, *18*, 667–674.
69. Falcioni, M.; Mantica, G.; Pigolotti, S.; Vulpiani, A. Coarse Grained Probabilistic Automata Mimicking Chaotic Systems. *Phys. Rev. Lett.* **2003**, *91*, 044101. [CrossRef]

entropy

MDPI

Article

Chaos Detection by Fast Dynamic Indicators in Reflecting Billiards

Gabriele Gradoni [1,*], Giorgio Turchetti [2,3] and Federico Panichi [2]

[1] Department of Electrical and Electronic Engineering, School of Mathematical Sciences, University of Nottingham, University Park, Nottingham NG7 2RD, UK
[2] Department of Physics and Astronomy, Alma Mater Studiorum, University of Bologna, Viale Berti Pichat 6/2, 40127 Bologna, Italy
[3] INdAM Gruppo Nazionale per la Fisica Matematica Piazzale Aldo Moro, 00185 Roma, Italy
* Correspondence: Gabriele.Gradoni@nottingham.ac.uk

Abstract: The propagation of electromagnetic waves in a closed domain with a reflecting boundary amounts, in the eikonal approximation, to the propagation of rays in a billiard. If the inner medium is uniform, then the symplectic reflection map provides the polygonal rays' paths. The linear response theory is used to analyze the stability of any trajectory. The Lyapunov and reversibility error invariant indicators provide an estimate of the sensitivity to a small initial random deviation and to a small random deviation at any reflection, respectively. A family of chaotic billiards is considered to test the chaos detection effectiveness of the above indicators.

Keywords: Lyapunov error; reversibility error; Gibbs entropy; wave chaos

Citation: Gradoni, G.; Turchetti, G.; Panichi, F. Chaos Detection by Dynamic Indicators in Absorbing Billiards. *Entropy* **2023**, *25*, 1251. https://doi.org/10.3390/e25091251

Academic Editor: Marko Robnik

Received: 31 March 2023
Revised: 26 July 2023
Accepted: 2 August 2023
Published: 23 August 2023

1. Introduction

Electromagnetic cavities exhibit wave chaos that can be predicted by a semi-classical analysis and random matrix theory; see [1] for a two-dimensional open electromagnetic cavity, ref. [2] for closed three-dimensional cavities and [3] for a review. These powerful prediction tools hold true and can be extended to include coupling through antennas and wave-guides, as well as interconnected cavities [4]. Furthermore, the semi-classical treatment of electromagnetic cavities has been tackled extensively in [5,6]. The experimental verification of wave chaos in microwave billiards has been addressed by several international research groups over the last few decades. This extensive work led to the observation of a rich phenomenology originating from complex wave dynamics [7], including: Wave-function scars [8], chaotic dynamics in superconducting billiards [9], time-reversal symmetry breaking [10], nodal domains in rough billiards [11], time-invariance violation at and around exceptional points [12] as well as electromagnetic reverberation [13]. Concerning optical resonators, it has been shown that chaotic dynamics are a key mechanism in asymmetric geometries [14–16]. This effect is important in the generation of lasers [17] with directional emissions [18] of eccentric cavities. Chaotic cavities have been employed in a plethora of electromagnetic engineering applications, including time reversal energy focusing [19] and electromagnetic compatibility (EMC) testing [20]. The equations for the propagation of acoustic and electromagnetic waves in resonant cavities are similar and coincide when the medium within the cavity is uniform. Usually, the cavities have a cylindrical or spherical symmetry and the resonant modes can be computed in a closed mathematical form. When the eikonal approximation is applicable, for sufficiently short wavelengths, the symplectic reflection map is integrable [21]. Deformations of the boundary cause the loss of integrability and the emergence of chaotic behavior in the particles or rays trajectories [22]. Even if the dynamics of integrable billiards are still an exciting research topic that continuously unveils novel phenomena, a growing interest has been developing within the scientific community for the properties of chaotic billiards [23–27].

For a sphere, the trajectory of any ray develops in a plane, which is determined by the initial ray position and velocity. The intersection of the sphere with the invariant plane is a circle, where the 2D area preserving the reflection map is integrable.

The symplectic 4D reflection map on a sphere is no longer integrable if the sphere is deformed. Thus, as opposed to a 4D flow, a first integral is no longer available and the 2D Poincaré map cannot be computed, preventing the analysis of the dynamical structures in phase plane portraits. For a cylinder, the trajectory of any ray develops in a plane, which intersects it on an ellipse, where the 2D reflection map is integrable. When the cylinder is deformed, the orbits no longer belong to a plane, the 4D symplectic map is not integrable and the phase portraits of the reflection map cannot be drawn [28]. This is important also in wave propagation within deformed cavities as the evolution of quantum states show footprints of classical trajectories [29].

The Lyapunov error ($E_n(\mathbf{x})$) and the reversibility error ($E_n^R(\mathbf{x})$) are dynamic indicators, based on the linear response theory, and allow the testing of the sensitivity of the orbits to small random deviations (see, for instance, [30]). For any fixed number of iterations, this sensitivity can be compared on a set of initial conditions chosen in a phase plane (see, for instance, [31]).

In Section 2, we present a brief introduction of the fast indicator used, their properties and relations. In Appendix B, we provide the definition of Lyapunov and reversibility errors. For a single orbit, the dependence on n can be investigated and the limit $n \to \infty$, limit of $n^{-1} \log E_n(\mathbf{x})$, provides the maximum Lyapunov exponent. $E_n^R(\mathbf{x})$ has been shown to be very sensitive to multi-dimensional problems, such as chaos detection in planetary systems (see, for instance, [32]) and in 2D and 3D waveguides (see [33]). In Section 3, we present a numerical analysis of the 2D reflection map in a convex domain, given by a deformed circle. We compare the phase portraits with the color plots of the Lyapunov and reversibility errors computed in a regular grid of phase space.

The mathematical description of the convex billiard and its new parametrization are presented in Appendix A.

To conclude, we consider the transport of particles and rays within the billiard. Given a source of particles or rays within the billiard, the time evolution of the probability density of particles or the energy density of rays is analyzed. Such a density can hardly be determined analytically, even for integrable billiards, and a numerical strategy is presented.

The dynamic indicators could be evaluated at time t for the inner points of the billiard after performing an average with respect to the initial ray or particle direction. A detailed analysis of 2D and 3D billiards can be worked out by using the algorithms described here.

2. Lyapunov and Reversibility Error Indicators

The dynamic stability of the reflection map can be analyzed with the Lyapunov error indicator (i.e., $E_n(\mathbf{x})$) or with the Reversibility error indicator (i.e., $E_n^R(\mathbf{x})$). The former measures the sensitivity of the initial conditions to small random deviations, and the latter measures the sensitivity to small deviations at each reflection. To implement the Reversibility error indicator, we iterate both the randomly perturbed and exact maps' n times, and compute the distance of the final point of the perturbed orbit with respect to the one obtained from the exact orbit. Letting ϵ be the amplitude of the random perturbations, we consider the linear response given by the $\epsilon \to 0$ limit. Given a symplectic or measure preserving map $M(\mathbf{x})$, defined as a compact manifold of $\mathbb{R}^{2d}$, we denote with $DM(\mathbf{x})$ the tangent map where $(DM)_{ij}(\mathbf{x}) = \partial M_i(\mathbf{x})/\partial x_j$. The orbit $\mathbf{x}_n$ is obtained by iterating n times the map M. We introduce the matrix L_n obtained by taking the products of the tangent map along the orbit:

$$\begin{aligned} &\mathbf{x}_n = M(\mathbf{x}_{n-1}) && \mathbf{x}_0 = \mathbf{x} && \mathbf{x}_n = M^{\circ n}(\mathbf{x}) \\ &\mathsf{L}_n(\mathbf{x}) = DM(\mathbf{x}_{n-1})\,\mathsf{L}_{n-1}(\mathbf{x}) && \mathsf{L}_0 = \mathsf{I} && \mathsf{L}_n(\mathbf{x}) = DM^{\circ n}(\mathbf{x}) \end{aligned} \tag{1}$$

For the initial condition $\mathbf{y} = \mathbf{x} + \epsilon\boldsymbol{\xi}$, where $\boldsymbol{\xi}$ is a random vector with a zero mean and unit covariance matrix $\langle \boldsymbol{\xi}\,\boldsymbol{\xi}^T \rangle = \mathsf{I}$, we compute the orbit $\mathbf{y}_n = M(\mathbf{y}_{n-1})$ where $\mathbf{y}_0 = \mathbf{y}$. We compare the perturbed and the reference orbit by defining the linear response Ξ_n according to

$$\Xi_n(\mathbf{x}) = \lim_{\epsilon\to 0} \frac{\mathbf{y}_n - \mathbf{x}_n}{\epsilon} = \lim_{\epsilon\to 0} \frac{M^{\circ n}(\mathbf{x}+\epsilon\boldsymbol{\xi}) - M^{\circ n}(\mathbf{x})}{\epsilon} = DM^{\circ n}(\mathbf{x})\,\boldsymbol{\xi} \equiv \mathsf{L}_n(\mathbf{x})\,\boldsymbol{\xi} \tag{2}$$

The square of $E_n(\mathbf{x})$ is defined as the trace of the covariance matrix $\Sigma_n^2(\mathbf{x})$ of the random vector $\Xi_n(\mathbf{x})$:

$$\Sigma_n^2(\mathbf{x}) = \langle\, \Xi_n(\mathbf{x})\, \Xi_n^T(\mathbf{x})\, \rangle = \mathsf{L}_n(\mathbf{x})\, \mathsf{L}_n^T(\mathbf{x}) \qquad E_n^2(\mathbf{x}) = \mathrm{Tr}\,(\Sigma_n^2(\mathbf{x})) = \mathrm{Tr}\left(\mathsf{L}_n^T(\mathbf{x})\mathsf{L}_n(\mathbf{x})\right) \tag{3}$$

The matrix $\mathsf{L}_n\,\mathsf{L}_n^T$ has the same eigenvalues as the Lyapunov matrix $\mathsf{L}_n^T\mathsf{L}_n$ but their eigenvectors are different. The invariants $I_n^{(k)}(\mathbf{x})$ of the Lyapunov matrix are equal to the sum of the $\binom{d}{k}$ products of the eigenvalues and can be computed with the Faddeev–Leverrier algorithm. From a geometrical viewpoint, the invariants are equal to the sum of the squared volumes of the $\binom{d}{k}$ parallelotopes whose edges are $\mathsf{L}_n(\mathbf{x})\mathbf{e}_j$, and $\mathbf{e}_j$ are any orthonormal base vectors.According to the Oseledec theorem, by writing the Lyapunov matrix as $\mathsf{L}_n^T(\mathbf{x})\mathsf{L}_n(\mathbf{x}) = \mathsf{W}_n(\mathbf{x})\, e^{2n\,\Lambda_n(\mathbf{x})}\, \mathsf{W}_n(\mathbf{x})$, the diagonal matrix $\Lambda_n(\mathbf{x})$ and the eigenvectors matrix $\mathsf{W}_n(\mathbf{x})$ have a limit, for $n \to \infty$ is equal to $\Lambda(\mathbf{x})$ and $\mathsf{W}(\mathbf{x})$, respectively. As a consequence, letting $\Lambda(\mathbf{x}) = \mathrm{diag}\,(\lambda_1(\mathbf{x}),\ldots,\lambda_{2d}(\mathbf{x}))$, we have

$$\lim_{n\to\infty} \frac{1}{2n} \log I_n^{(k)}(\mathbf{x}) = \lambda_1(\mathbf{x}) + \lambda_2(\mathbf{x}) + \ldots + \lambda_k(\mathbf{x}) \tag{4}$$

If the map M is symplectic, the exponents are pairwise opposite $\lambda_{d+j} = -\lambda_{d-j+1}$. If the map is integrable, then the first d eigenvalues of the Lyapunov matrix grow as n^2, while the last d decrease as n^{-2}, so that all the Lyapunov exponents vanish. For a generic map, the power law growth or exponential growth with n of the first d invariants guarantee the classification of the phase space regions of regular and chaotic evolution. Below, we briefly introduce the reversibility error. For a complete description of the indicators and their properties, see Appendix B.

Reversibility Error

We consider the Backward–Forward process (BF) as one implementation of the Reversibility error indicator. In this case, we iterate n times the map with a random perturbation of amplitude ϵ first, and then the inverse map. The recurrence is given by

$$\mathbf{y}_{n'} = M(\mathbf{y}_{n'-1}) + \epsilon\boldsymbol{\xi}_{n'} \qquad 1 \le n' \le n \qquad \mathbf{y}_{n'} = M^{-1}(\mathbf{y}_{n'-1}) \qquad n+1 \le n' \le 2n \tag{5}$$

and the linear response is defined by

$$\Xi_{2n}^{BF}(\mathbf{x}) = \lim_{\epsilon\to 0} \frac{\mathbf{y}_{2n} - \mathbf{x}}{\epsilon} = \sum_{k=1}^{n} \mathsf{L}_k^{-1}(\mathbf{x})\,\boldsymbol{\xi}_k \tag{6}$$

where L_k^{-1} is the inverse of the matrix L_k defined by (1). In the absence of perturbation, we are back to the initial condition $\mathbf{y}_{2n} \equiv \mathbf{x}_{2n} = \mathbf{x}$. The covariance matrix of the random vector Ξ_{2n}^{BF} is given by

$$\Sigma_{BF\,n}^2(\mathbf{x}) = \left\langle \Xi_{2n}^{BF}(\mathbf{x}) \left(\Xi_{2n}^{BF}\right)^T \right\rangle = \sum_{n'=1}^{n} \left(\mathsf{L}_{n'}^T(\mathbf{x})\mathsf{L}_{n'}(\mathbf{x})\right)^{-1} \tag{7}$$

The square of the BF reversibility error $E_{BF\,n}$ is defined as the trace of $\Sigma_{BF\,n}^2$. The asymptotic limit of the invariants $I_{BF\,n}^{(k)}(\mathbf{x})$ is the same as (4) for $k \le d$. The log of the last invariant

$I^{(2d)}_{BF\ n}(\mathbf{x})$ is the Gibbs entropy of the process with the covariance matrix $\Sigma^2_{BF\,n}(\mathbf{x})$ and its asymptotic behavior is the same as the Kolmogorov–Sinai entropy.

If the map is symplectic, then the matrices $\mathsf{L}_{n'}\mathsf{L}^T_{n'}$ and $\mathsf{L}^T_{n'}\mathsf{L}_{n'}$ are symplectic. Therefore, the trace of $\mathsf{L}^T_{n'}\mathsf{L}_{n'}$ and its inverse are equal. In this case, the *BF* reversibility error is simply related to the Lyapunov error E_n as

$$E^2_{BF\ n}(\mathbf{x}) = \sum_{n'=1}^{n} E^2_{n'}(\mathbf{x}) \tag{8}$$

Previously, the *BF* process has been considered with the noise applied to both the B and F iterations. The covariance matrix of the linear response in in this case is given by

$$\Sigma^2_{BF\ n}(\mathbf{x}) = \frac{1}{2}\Big\langle \Xi^{BF}_{2n}(\mathbf{x}) \left(\Xi^{BF}_{2n}\right)\Big\rangle = \frac{1}{2}\,\mathsf{I} + \sum_{n'=1}^{n-1} \left(\mathsf{L}^T_{n'}(\mathbf{x})\mathsf{L}_{n'}(\mathbf{x})\right)^{-1} + \frac{1}{2}\left(\mathsf{L}^T_{n}(\mathbf{x})\mathsf{L}_{n}(\mathbf{x})\right)^{-1} \tag{9}$$

and proof of (9) can be found in [3] Section 2.3. The asymptotic behavior of the invariants of the reversibility error covariance matrix Σ^2_{nBF} is determined by the positive Lyapunov exponents. Indeed, the limit of $(2t)^{-1}\log(I^{(k)}_{nBF})$ for $t \to \infty$ is the sum of the positive exponents among the first k. As a consequence, $\frac{1}{2}\log(I^{(2d)}_{nBF})$, which corresponds to the Gibbs entropy of the BF random process where $I^{(2d)}_{nBF} = det(\Sigma^2_{nBF})$, is the sum of all the positive Lyapunov exponents (the first d for a symplectic map), just as the Kolmogorov–Sinai entropy. Notice that the difference with the previous definition and (7) is negligible for a large n. Such a definition was initially proposed to compare the reversibility error due to a small random displacement with respect to the reversibility error due to the round off. The Reversibility Error indicator due to round off, denoted by *REM*, is a numerical implementation of $E^R_n(\mathbf{x})$ and numerical equivalence has been proven for simple maps [34].

Denoting with M_ϵ the map evaluated with round off and M^{-1}_ϵ the inverse map evaluated with the round off, the *REM* indicator is then

$$REM_{BF\ n}(\mathbf{x}) = \frac{\|M_\epsilon^{\circ\ -n} \circ M_\epsilon^{\circ\ n}(\mathbf{x}) - \mathbf{x}\|}{\sqrt{2}\,\epsilon} \tag{10}$$

The round off is a pseudo-random process in which just one realization is available. When different values for n or $\mathbf{x}$ are used, it is evident that $REM_{BF\ n}$ exhibits significant fluctuations with respect to $E_{BF\ n}$, which is the result of an averaging process over the random displacements. No higher invariants can be defined for the reversibility error due to round off.

We can also consider the specular implementation, namely the Forward–Backward REM_{FB}, which consists of iterating n times first the perturbed inverse map M^{-1} and then the map M. For an autonomous symplectic map however, all the invariants are the same as for the BF process. Moreover, for a Hamiltonian flow, equivalence of the *FB* and *BF* invariants is a consequence of the time reversal invariance.

The linear approximation given by $\mathbf{y}_n = \mathbf{x}_n + \epsilon\Xi_n$ can be considered; however, since the remainder of order ϵ^2 can generically be neglected only for a number of iterations, small with respect to $\log(1/\epsilon)$, it is of no practical use. The linear response being based on the $\epsilon \to 0$ limit is valid for any number n of iterations.

3. Numerical Results for the 2D Billiard

In this section, we introduce numerical results for a 2D billiard; the reflection map for a convex billiard defined by (A12) with $f(x) = x^2/3$ is presented. The billiard analyzed in this section has a closed boundary for $\epsilon \le 1/\sqrt{3} \sim 0.577$ as one can easily show. Indeed, letting $V(x) = x^2/2 + \epsilon x^3/3$, the maximum occurs for $x = -1/\epsilon$ and we require $V(-1/\epsilon) = 1/(6\epsilon^2) \ge 1/2$ to have a closed boundary so that $\epsilon \le 1/\sqrt{3}$. We have restricted our analysis to $\epsilon \le /1/2$, observing that for $\epsilon \le 0.1$, the map is almost integrable; for

$\epsilon = 0.2$, the area of chaotic regions is significant and already for $\epsilon = 0.4$, a large fraction of the phase space is chaotic. We have used the coordinates (ϕ, p) in the phase space even though the map preserves the measure but is not area preserving. We have compared the phase portraits with the Lyapunov error $E_n(\mathbf{x})$, the reversibility error E_n^{BF} and the round-off induced reversibility error REM_n^{BF}, computed according to (A28), (A38) and (A41), respectively. In Figure 1, we show the billiard for:

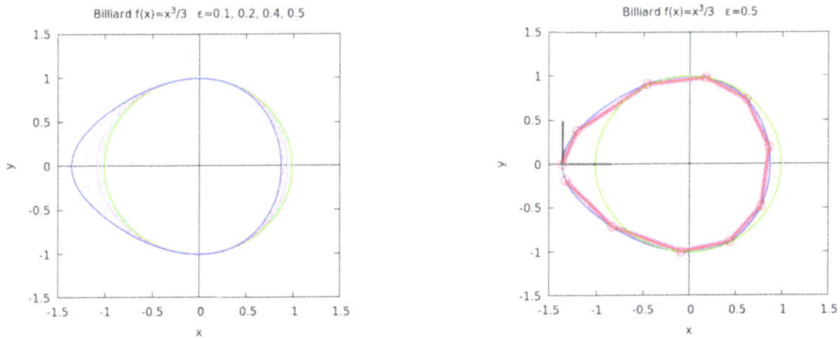

Figure 1. Left panel: billiard boundary defined by (A12) with $f(x) = x^2/3$ and ϵ=0, 0.1, 0.2, 0.4, 0.5 (green, cyan, purple, grey, blue). **Right panel**: rays trajectory for $\epsilon = 0.5$ and $n = 10$ (purple line) and the reversed ray trajectory (red line). At the initial point, the tangent and inner normal vectors are shown (black lines).

We have analyzed the orbits for the billiard, comparing them with the color map of the Lyapunov error $E_n(\mathbf{x})$, reversibility error $E_n^{BF}(\mathbf{x})$ and the round-off induced reversibility error $REM_n^{BF}(\mathbf{x})$, where $\mathbf{x} = (\phi, p)^T$ is chosen on a regular grid of the $N_g \times N_g$ points. A logarithmic color map is used to show the results. By defining the tangent vector $\boldsymbol{\tau}(s) = \partial \mathbf{r}(s)/\partial s$ and the ray velocity as $\|\mathbf{v}(s)\|$, then the range of the phase ϕ of the position vector is $[0, 2\pi]$, and the range of the momentum $p = \boldsymbol{\tau} \cdot \mathbf{v}$ is $[-1, 1]$. Since the orbits are symmetric, by changing p into $-p$, the chosen range for p is then in the range $[0, 1]$. For a deeper discussion on the properties of the billiard, see Appendix A.

It is important to notice that the indicators give additional information with regard to the phase portrait, since they provide a quantitative measurement of the chaos for an orbit. Additionally, for 4D billiards Poincare' sections are not available and the indicators are the only methods to investigate the stability of the phase space.

In Figures 2, we compare the phase portrait against the $E_n(\mathbf{x})$ for the billiard with $\epsilon = 0.1$. The correspondence between the phase portrait and $E_n(\mathbf{x})$ color plot is good and in both cases, a thin layer of chaotic orbits is seen at the boundary of the main chains of islands. In the interior of the chains of islands, the error tends to be zero because the de-tuning (derivative of the frequency with respect to the action) is low. By approaching the separatrix, the de-tuning grows, diverging on the separatrix itself, where the error growth with n, changing from a power law to an exponential one.

In Figure 3, $E_n^R(\mathbf{x})$ and $REM_n^{BF}(\mathbf{x})$ show a similar behavior. REM is similar to the error induced by the small random displacements, but exhibits higher fluctuations because the averaging over the random process is missing. It is worth noticing that $\left(E_n^R\right)^2$ constitutes the sum along the orbit of $(E_n)^2$, provided (s, p) are used as canonical coordinates. The Lyaponov error oscillates with n in the regions of quasi-integrable motion and for fixed n, oscillations are observed in phase space. These oscillations can be eliminated by using the MEGNO average (see, for instance, Ref. [35]) and they disappear for $E_n^R(\mathbf{x})$.

In Figures 4 and 5, the Poincare section, $E_n^R(\mathbf{x})$, $E_n(\mathbf{x})$ and $REM_n^{BF}(\mathbf{x})$ are shown for a larger perturbation $\epsilon = 0.2$. The larger deformation with respect to the integrable billiard increases the area of the chaotic orbits. For $\epsilon = 0.3$, almost one half of the unit phase space area—in the $(\phi_0/2pi, p_0)$ initial coordinates used in the figures—is filled with chaotic orbits and this fraction increases approaching 1, when ϵ tends to the limit value $\epsilon = 1/\sqrt{3}$. For a

2D map, the phase space portraits provide the required information on the orbits stability; however, one advantage of the proposed indicators is that they provide a quantitative value to discriminate between regular and chaotic orbits. To explore small details in the phase space, one can analyze a smaller region and increase the number of iterations. Moreover, the indicators become the unique stability analysis tool when the dimensionality of the problem is increased. This is the case for the 3D billiard which leads to a 4D reflection map. In this case, the 2D phase portraits are no longer available.

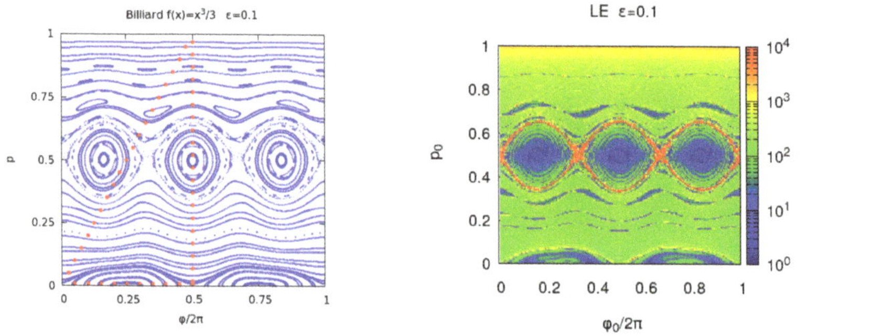

Figure 2. Left panel: phase portrait of the billiard with $\epsilon = 0.1$. Each orbit is computed for $n = 1000$ and the initial points of each orbit are the red dots. **Right panel**: Lyapunov error $E_n(\mathbf{x})$ where $\phi_0/(2\pi)$ and p are chosen in a regular grid of the unit square with $N_g \times N_g$ points and $N_g = 200$. The iterations number is $n = 200$ and the results are shown in a logarithmic color scale.

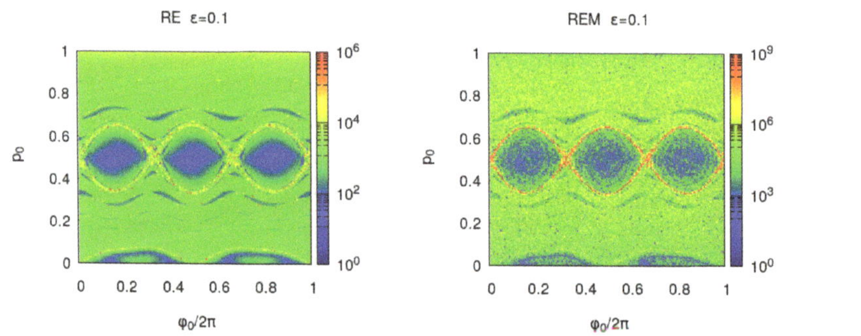

Figure 3. Left panel: billiard with $\epsilon = 0.1$ reversibility BF error $E_n^{BF}(\mathbf{x})$ where $\phi_0/(2\pi)$ and p are chosen in a regular grid of the unit square with $N_g \times N_g$ points and $N_g = 200$ and the iterations number is $n = 200$. **Right panel**: round-off induced BF reversibility error $REM_n^{BF}(\mathbf{x})$.

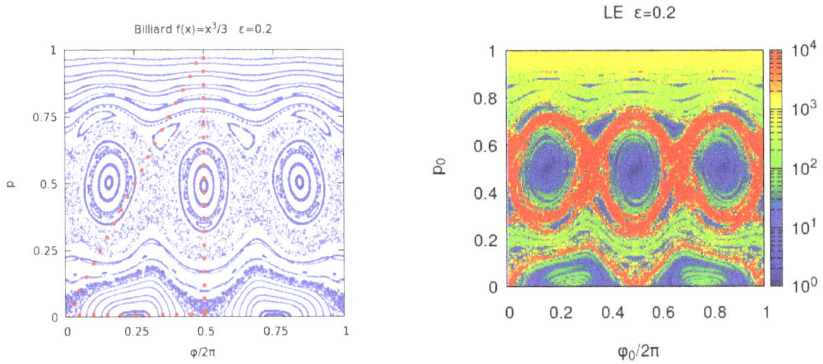

Figure 4. Left panel: phase portrait of the billiard with $\epsilon = 0.2$. Each orbit is computed for $n = 1000$ and the initial points of each orbit are the red dots. **Right panel**: Lyapunov error $E_n(\mathbf{x})$ where $\phi_0/(2\pi)$ and p are chosen in a regular grid of the unit square with $N_g \times N_g$ points and $N_g = 200$. The iterations number is $n = 200$ and the results are shown in a logarithmic color scale.

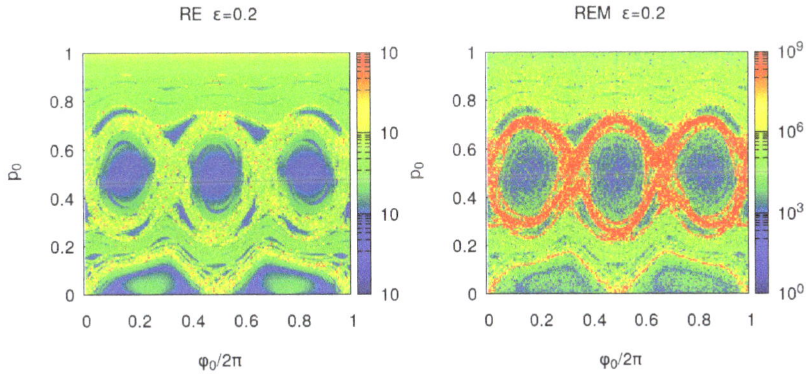

Figure 5. Left panel: billiard with $\epsilon = 0.2$ reversibility BF error $E_n^{BF}(\mathbf{x})$ where $\phi_0/(2\pi)$ and p are chosen in a regular grid of the unit square with $N_g \times N_g$ points and $N_g = 200$ and the iterations number is $n = 200$. **Right panel**: round-off induced reversibility error $REM_n^{BF}(\mathbf{x})$.

4. Conclusions

We have considered the motion of particles on a 2D billiard with a convex reflecting boundary using a parametrization different from the one proposed by [36]. Our implementation can be easily extended to a 3D billiard. The computation of the arc length s on the boundary requires a numerical integration, which can be avoided choosing the phase ϕ of the position vector $\mathbf{r}$ rather then the curvilinear abscissa s, even though, in this case, the map is only measure preserving. The stability of the orbits has been analyzed using the Lyapunov and Reversibility error fast indicators. Both indicators are invariant with respect to the choice of the initial deviation and of the orthogonal reference frame. We have shown that the logarithm of the second reversibility error invariant is the Gibbs entropy of the random vector defining the deviation from reversibility. The numerical results of the Lyapunov and reversibility error indicators, presented for a selected family of billiards, confirm the reliability of the proposed methods to explore the sensitivity of ray propagation to small random perturbations.

Author Contributions: Methodology, G.G. and G.T.; Software, F.P.; Validation, F.P.; Formal analysis, G.G., G.T. and F.P.; Writing—original draft, G.G., G.T. and F.P. All authors have read and agreed to the published version of the manuscript.

Funding: This research received no external funding.

Institutional Review Board Statement: Not applicable.

Data Availability Statement: Not applicable.

Conflicts of Interest: The authors declare no conflict of interest.

Abbreviations

The following abbreviations are used in this manuscript:

2D	Two Dimensional
3D	Three Dimensional
4D	Four Dimensional
EMC	ElectroMagnetic Compatibility
BF	Backward-Forward
REM	Reversibility Error Method
RE	Reversibility Error
LE	Lyapunov Error
MEGNO	Mean Exponential Growth Factor of Nearby Orbits

Appendix A. Convex Billiards

In this appendix, we consider a plane billiard with a convex boundary and the reflections map, which is area preserving. Letting s be the arc length of the boundary, whose length is ℓ, the boundary is parameterized by $\mathbf{r} = \mathbf{r}(s)$, where $\mathbf{r} = (x, y)^T$ and $x(s)$, $y(s)$ are periodic functions of s with period ℓ [37]. Choose two points $\mathbf{r}_0 = \mathbf{r}(s_0)$ and $\mathbf{r}_2 = \mathbf{r}(s_2)$ and consider a ray starting at $\mathbf{r}_0 = \mathbf{r}(s_0)$, hitting the boundary at $\mathbf{r}(s)$ and again at $\mathbf{r}_2 = \mathbf{r}(s_2)$. Since $\mathbf{r}_0$ and $\mathbf{r}_2$ are given, the intermediate point $\mathbf{r}$ is determined by the Maupertuis–Fermat principle which requires the length $H(s)$ of the trajectory to be a minimum

$$H(s) = h(s_0, s) + h(s, s_2) \qquad h(s, s_0) = \|\mathbf{r}(s) - \mathbf{r}(s_0)\| \qquad h(s_2, s) = \|\mathbf{r}(s_2) - \mathbf{r}(s)\| \tag{A1}$$

At any point of the curve s, we define the tangent vector $\boldsymbol{\tau}(s) = \partial\mathbf{r}(s)/\partial s$ and the inner normal $\mathbf{n}(s) = \boldsymbol{\tau}(s) \times \mathbf{e}_z$. Denoting $\dot{x} = dx/ds$ and $\dot{y} = dy/ds$, we have $\boldsymbol{\tau}(s) = (\dot{x}, \dot{y})$ and $\mathbf{n}(s) = (\dot{y}, -\dot{x})$. The ray velocity $\|\mathbf{v}(s)\|$ is constant and is chosen to be equal to 1. Denoting with $\mathbf{v}(s) \equiv \mathbf{v}(s + 0)$ the velocity of the outgoing ray at the point $\mathbf{r}(s)$, we have

$$\mathbf{v}_0 \equiv \mathbf{v}(s_0) = \frac{\mathbf{r}(s) - \mathbf{r}_0}{h(s, s_0)} \qquad \mathbf{v} \equiv \mathbf{v}(s) = \frac{\mathbf{r}_2 - \mathbf{r}(s)}{h(s_2, s)} \tag{A2}$$

Letting ψ_0 and ψ represent the angles that the ray forms at s_0 and s with the tangents $\boldsymbol{\tau}(s_0)$ and $\boldsymbol{\tau}(s)$, the condition that $H(s)$ is a minimum for s_0 and s_2 fixed then gives

$$\begin{aligned} \frac{dH(s)}{ds} &= \frac{\partial}{\partial s}h(s_0, s) + \frac{\partial}{\partial s}h(s, s_2) \\ &= \frac{\mathbf{r}(s) - \mathbf{r}(s_0)}{h} \cdot \frac{\partial\mathbf{r}(s)}{\partial s} - \frac{\mathbf{r}(s_2) - \mathbf{r}(s)}{h} \cdot \frac{\partial\mathbf{r}(s)}{\partial s} \\ &= \mathbf{v}_0 \cdot \boldsymbol{\tau} - \mathbf{v} \cdot \boldsymbol{\tau} = \cos\psi_0 - \cos\psi = 0 \end{aligned} \tag{A3}$$

The minimum condition is satisfied for $s = s_1$, where $\psi = \psi_1$ and setting $\mathbf{v}_1 = \mathbf{v}(s_1)$ as the reflection condition $\psi_1 = \psi_0$ is fulfilled. Indeed, the angles that the rays form with the normal are $\theta_0 = \pi/2 - \psi_0$ and $\theta_1 = \pi/2 - \psi_1$ and the standard condition $\theta_1 = \theta_0$ is recovered. In order to determine the reflection map, we notice that $p_0 = \cos\psi = \mathbf{v}_0 \cdot \boldsymbol{\tau}_0$ and $p_1 = \cos\psi_1 = \mathbf{v}_1 \cdot \boldsymbol{\tau}_1$ are the momenta conjugated to s_0 and s_1. To this end, we consider $h(s_0, s_1)$ as the generating function of the map. The moments p_0 and p_1 conjugated to s_0 and s_1 are given by

$$p_1 = \frac{\partial}{\partial s_1} h(s_0, s_1) = \mathbf{v}_1 \cdot \boldsymbol{\tau}_1 = \cos\psi_1 \qquad p_0 = \frac{\partial}{\partial s_0} h(s_0, s_1) = -\mathbf{v}_0 \cdot \boldsymbol{\tau}_0 = -\cos\psi_0 \tag{A4}$$

As a consequence, the map M from $\mathbf{x}_0 = (s_0, p_0)^T$ to $\mathbf{x}_1 = (s_1, p_1)^T$ generated by $h(s_0, s_1)$ is area preserving and s, p are canonical coordinates.

Appendix A.1. Berry's Parametrization

A parametrization $\mathbf{r} = \mathbf{r}(\phi)$ which allows the analytical computation of the curvilinear abscissa $s = s(\phi)$ was proposed in [36]. The basic idea is to use $\rho = \rho(\phi)$ to represent the curve where ρ is the curvature radius and ϕ is the angle which the tangent forms with one of the coordinates' axes. We choose the origin on the curve at the point where $x = x_{\text{min}}$ and we move clock-wise. In the case of a circle of radius R, we have $s = R\phi$ and the tangent and the inner normal are given by

$$\mathbf{r} = (-\cos\phi, \sin\phi)^T \qquad \frac{d\mathbf{r}}{ds} = \boldsymbol{\tau} = (\sin\phi, \cos\phi)^T \qquad \frac{d\boldsymbol{\tau}}{ds} = \frac{1}{R}(\cos\phi, -\sin\phi)^T = \frac{\mathbf{r}}{R} \tag{A5}$$

Letting a generic closed curve parameterize according to $\mathbf{r} = \mathbf{r}(\phi)$, where ϕ is the angle that the tangent forms with the y axis, and $\rho(\phi)$ the curvature radius, it follows, from the standard formula $d\boldsymbol{\tau}/ds = \mathbf{n}/\rho$, where $\mathbf{n}$ is the unit inner normal, that

$$\boldsymbol{\tau} = \frac{d\mathbf{r}}{ds} = (\sin\phi, \cos\phi)^T \qquad \frac{d\boldsymbol{\tau}}{d\phi} = \frac{d\boldsymbol{\tau}}{ds}\frac{ds}{d\phi} = \mathbf{n}\frac{1}{\rho}\frac{ds}{d\phi} \tag{A6}$$

Taking the norm of the last equation, where $||d\boldsymbol{\tau}/d\phi|| = 1$, we have

$$\frac{ds}{d\phi} = \rho(\phi) \qquad \mathbf{r}(\phi) - \mathbf{r}(0) = \int_0^{\phi} \boldsymbol{\tau}(\phi')\frac{ds}{d\phi'}\,d\phi' = \int_0^{\phi} (\sin\phi', \cos\phi')^T \rho(\phi')\,d\phi' \tag{A7}$$

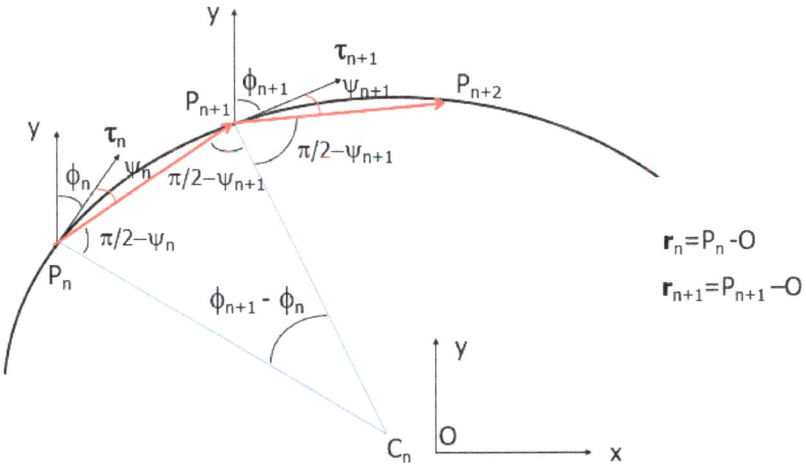

Figure A1. Angles and vectors defining the ray propagation.

The functions $\mathbf{r}(\phi)$ and $\rho(\phi)$ are periodic with period 2π and can be expanded in a Fourier series. Since $\mathbf{r}(0) = \mathbf{r}(2\pi)$, Equation (A7) implies that the integral of $\rho\,\cos\phi$ and $\rho\,\sin\phi$ in $[0, 2\pi]$ vanishes. The simplest deformation of the circle is given by

$$\rho(\phi) = 1 + \epsilon\cos(2\phi) \tag{A8}$$

but any trigonometric polynomial in $\cos(k\phi)$ and $\sin(k\phi)$ with $k \geq 2$ might be considered. In this case, the curve parametrization becomes

$$\begin{aligned} s(\phi) &= \phi + \frac{\epsilon}{2}\,\sin(2\phi) \\ x(\phi) &= -\cos\phi + \frac{\epsilon}{6}\,\sin(3\phi) - \frac{\epsilon}{2}\sin\phi \\ y(\phi) &= \sin\phi - \frac{\epsilon}{6}\,\cos(3\phi) - \frac{\epsilon}{2}\cos\phi \end{aligned} \tag{A9}$$

In order to find the map, we first determine the phase ϕ_{n+1} corresponding to the new reflection at $\mathbf{r}_{n+1} = \mathbf{r}(\phi_{n+1})$. To this end, we notice that the direction of the ray emerging at $\mathbf{r}_n$ is given by the vector $\mathbf{v}_n$, forming an angle ψ_n with τ_n which in turn forms an angle ϕ_n with the y axis. As a consequence, $\mathbf{v}_n$ forms an angle $\phi_n + \psi_n$ with the y axis. Recalling that $\mathbf{v}_n = (\mathbf{r}_{n+1} - \mathbf{r}_n)/\|\mathbf{r}_{n+1} - \mathbf{r}_n)\|$, the ratio of the x and y components is given by

$$\frac{v_{x\,n}}{v_{y\,n}} = \frac{x_{n+1} - x_n}{y_{n+1} - y_n} = \frac{\int_{\phi_n}^{\phi_{n+1}} \rho(\phi')\,\sin\phi\,d\phi'}{\int_{\phi_n}^{\phi_{n+1}} \rho(\phi')\,\cos\phi\,d\phi'} = \tan(\phi_n + \psi_n) \tag{A10}$$

This equation implicitly determines ϕ_{n+1} as a function of ϕ_n and ψ_n. Next, we compute $s_{n+1} = s(\phi_{n+1})$ and $\mathbf{r}_{n+1} = \mathbf{r}(\phi_{n+1})$ via (A9). The angles entering the reflections at $\mathbf{r}_n$ and $\mathbf{r}_{n+1}$ are shown in Figure A1. The last step is to determine ψ_{n+1}; to this end, we consider the triangle whose vertices are P_n, P_{n+1} and C_n where the half lines directed along the normals $\mathbf{n}_n$ and $\mathbf{n}_{n+1}$ intersect. Figure A1 shows the displacement vectors with respect to the origin being $\mathbf{r}_n$, $\mathbf{r}_{n+1}$, $\mathbf{r}_c$. Taking into account the reflection condition, the angles of this triangle are $\pi/2 - \psi_n$, $\pi/2 - \psi_{n+1}$ and $\phi_{n+1} - \phi_n$ from which we obtain

$$\psi_{n+1} = \phi_{n+1} - \phi_n - \psi_n \qquad\qquad p_{n+1} = \cos\psi_{n+1} \tag{A11}$$

Appendix A.2. Alternative Parametrization and Reflection Map

We propose another equation for a deformed circle given by

$$F(\mathbf{r}) = \frac{y^2}{2} + \frac{x^2}{2} + \epsilon f(x) - \frac{1}{2} = 0 \qquad f(x) = O(x^3) \tag{A12}$$

with $\epsilon > 0$ and sufficiently small so that the billiard is convex. A parametrization $\mathbf{r} = \mathbf{r}(\phi)$ can be easily obtained but requires the solution of an implicit equation. This parametrization was already obtained in [38] and re-derived here. The advantage is that the present form can be easily extended to a deformed sphere. Indeed, at a given point, the curvatures of a surface depend on the two lines traced on it and therefore, on the corresponding tangents which are usually not orthogonal. It is not trivial to represent the curves by assigning the principal curvatures because this implies the knowledge of two families of curves that are mutually orthogonal. The parametrization we propose for the billiard boundary, defined by (A12), is the following:

$$\mathbf{r} = \mathbf{r}(\phi) = r(\phi)(-\cos\phi,\,\sin\phi)^T \qquad\qquad r(\phi) = \|\mathbf{r}(\phi)\| \tag{A13}$$

In this case, $r(\phi)$ is the distance from the origin and we denote by $\rho(\phi)$, the radius of curvature. Setting $\dot{r} = dr/d\phi$, the curvilinear abscissa $s(\phi)$ is given by

$$\frac{dx}{d\phi} = r\sin\phi - \dot{r}\cos\phi \qquad \frac{dy}{d\phi} = r\cos\phi + \dot{r}\sin\phi$$
$$ds = (r^2 + \dot{r}^2)^{1/2} d\phi \qquad s(\phi) = \int_0^{\phi} (r^2(\phi') + \dot{r}^2(\phi'))^{1/2}\, d\phi' \tag{A14}$$

The tangent vector $\boldsymbol{\tau}$, the inner normal vector $\mathbf{n}$, and the radius of curvature ρ are given by

$$\boldsymbol{\tau}(\phi) = \frac{d\mathbf{r}}{ds} = \frac{d\mathbf{r}}{d\phi}\left(\frac{ds}{d\phi}\right)^{-1} = \frac{(r\sin\phi - \dot{r}\cos\phi,\, r\cos\phi + \dot{r}\sin\phi)^T}{(r^2(\phi) + \dot{r}^2(\phi))^{1/2}} \tag{A15}$$

$$\rho(\phi) = \left(\frac{ds}{d\phi}\right)^2 \left\| \frac{d^2\mathbf{r}}{d\phi^2} - \frac{d\mathbf{r}}{d\phi}\frac{d^2 s}{d\phi^2}\left(\frac{ds}{d\phi}\right)^{-1} \right\|^{-1} \qquad \mathbf{n}(\phi) = (\tau_y(\phi),\ -\tau_x(\phi)) \tag{A16}$$

The expressions of $\boldsymbol{\tau}$ and $\mathbf{n}$ are still relatively simple, unlike the radius of curvature, which never enters in the computation of the reflection map.

Notice that ϕ is no longer the phase of the vector $\boldsymbol{\tau}$ with respect to the y axis. Denoting with ϕ_τ the phase of $\boldsymbol{\tau}$, which we choose to be 0 when $\boldsymbol{\tau} = (1, 0)$ and increasing clock-wise as ϕ, we have

$$\boldsymbol{\tau}(\phi) = (\sin\phi_\tau,\ \cos\phi_\tau)^T \tag{A17}$$

As a consequence, given $\boldsymbol{\tau}$, its phase is given by

$$\phi_\tau = \arccos(\tau_y) \quad \text{if} \quad \tau_x \geq 0 \qquad \phi_\tau = 2\pi - \arccos(\tau_y) \quad \text{if} \quad \tau_x < 0 \tag{A18}$$

When the equation of the billiard boundary is given by (A13), we determine $r = r(\phi)$ and $\dot{r} = \dot{r}(\phi)$ by solving the equations

$$r = (1 - 2\epsilon f(-r\cos\phi))^{1/2} \qquad \dot{r} = \frac{\epsilon}{r} f'(-r\cos\phi)(\dot{r}\cos\phi - r\sin\phi) \tag{A19}$$

iteratively staring with $r = 1$ and $\dot{r} = 0$. At first order, $r(\phi) = 1 - \epsilon f(-\cos\phi) + O(\epsilon^2)$. The solution $r = r(\phi)$ is obtained iteratively and the convergence is very fast for $\epsilon << 1$. The iterations compute the fixed point stops when machine accuracy is reached. Then, $s = s(\phi)$ is obtained by numerical integration according to (A14).

To compute the reflection map, we start from a point $\mathbf{r}_0$ with ray velocity $\mathbf{v}_0$. At the point $\mathbf{r}_0$, the tangent and inner normal vectors are $\boldsymbol{\tau}_0$ and $\mathbf{n}_0$. We define the phase of a vector $\mathbf{b}$ with respect to a vector $\mathbf{a}$ as the angle between $\mathbf{a}$ and $\mathbf{b}$ counted clock-wise and its range is $[0, 2\pi]$. To the initial position, we associate, according to (A13), the angle ϕ_0 defined as the phase of $-\mathbf{r}_0$ with respect to the x axis. To the initial tangent vector, we associate, according to (A18), the angle $\phi_{\tau\,0}$ defined as the phase of $\boldsymbol{\tau}_0$ with respect to the y axis. If $y_0 = 0$ and $x_0 < 0$, we have $\phi_0 = \phi_{\tau\,0} = 0$. The phase of $\mathbf{v}_0$ with respect to $\boldsymbol{\tau}_0$ is ψ_0 and its range is $[0,\ \pi]$ whereas the range of ϕ_0 and $\phi_{\tau\,0}$ is $[0, 2\pi]$. As a consequence, the phase of $\mathbf{v}_0$ with respect to the y axis is $\phi_{\tau\,0} + \psi_0 \bmod 2\pi$. The tangential component of the velocity is $p_0 = \mathbf{v}_0 \cdot \boldsymbol{\tau}_0 \equiv \cos\psi_0$ and if $p_0 > 0$, the reflection is clock-wise; if $p_0 < 0$, it is counter clock-wise. Alternatively, we can start with $(\phi_0,\ p_0)$ and determine the vectors $\mathbf{r}_0,\ \mathbf{v}_0$.

After n iterations, the point $\mathbf{r}_n$ on the boundary is reached and the velocity of the outgoing ray is $\mathbf{v}_n$. The angles $\phi_n,\ \phi_{\tau\,n},\ \psi_n$ and the momentum $p_n = \cos\psi_n$ are determined and eventually the curvilinear abscissa s_n is obtained with a numerical integration.

To compute the reflection map, we determine first the unique intersection $\mathbf{r}_{n+1}$ of the ray emerging from $\mathbf{r}_n$ with the boundary. Letting $\mathbf{r}(t) = \mathbf{r}_n + t\mathbf{v}_n$, we obtain the unique positive solution of $F(\mathbf{r}(t)) = 0$ by iteratively solving the equation

$$t = -2\mathbf{v}_n \cdot \mathbf{r}_n - 2\epsilon \frac{f(x_n + t\, v_{x\,n}) - f(x_n)}{t} \tag{A20}$$

As a consequence, we have $\mathbf{r}_{n+1} = \mathbf{r}_n + t_* \mathbf{v}_n$. The phase ϕ_{n+1} of $-\mathbf{r}_{n+1}$ with respect to the x axis, according to (A13), is given by

$$\begin{aligned} \phi_{n+1} &= \arccos\left(-\frac{x_{n+1}}{r_{n+1}}\right) \qquad \text{if} \quad y_n \geq 0 \\ \phi_{n+1} &= 2\pi - \arccos\left(-\frac{x_{n+1}}{r_{n+1}}\right) \qquad \text{if} \quad y_n < 0 \end{aligned} \tag{A21}$$

We compute the tangent vector and inner normal vector $\tau_{n+1} = \tau(\phi_{n+1})$ and $\mathbf{n}_{n+1} = \mathbf{n}(\phi_{n+1})$ according to (A16). The direction of the emerging ray is computed using the reflection condition:

$$\mathbf{v}_{n+1} = \mathbf{v}_n - 2(\mathbf{v}_n \cdot \mathbf{n}_{n+1})\, \mathbf{n}_{n+1} \tag{A22}$$

$$p_{n+1} = \mathbf{v}_{n+1} \cdot \tau_{n+1} = \mathbf{v}_n \cdot \tau_{n+1} \tag{A23}$$

From the momentum $p_{n+1} = \mathbf{v}_n \cdot \tau_{n+1}$, we determine the phase $\psi_{n+1} = \arccos(p_{n+1})$ of $\mathbf{v}_{n+1}$ with respect to τ_{n+1}. Recalling that the phase of $\mathbf{v}_{n+1}$ with respect to the y axis is $\phi_{\tau\, n+1} + \psi_{n+1}$, we can check that

$$\mathbf{v}_{n+1} = \big(\sin(\phi_{\tau\, n+1} + \psi_{n+1}),\ \cos(\phi_{\tau\, n+1} + \psi_{n+1})\big)^T \tag{A24}$$

We recall that the map $(\phi_{n+1}, p_{n+1}) = M(\phi_n, p_n)$ is measure preserving.

The map $(s_{n+1}, p_{n+1}) = M(s_n, p_n)$ is symplectic (i.e. area preserving) and can be obtained from the previous map by computing the function $s = s(\phi)$ according to (A14). This can be done through achieving machine accuracy by storing the results of an integration on a sufficiently fine uniform grid by using Simpson or Gauss integration schemes in each interval, and a linear interpolation.

Finally, the strategy is the following: given an initial value ϕ_0 and p_0, we determine $\mathbf{r}_0$ and $\mathbf{v}_0$. Notice the component of $\mathbf{v}_0$ along the inner normal is positive $(1 - p_0^2)^{1/2}$ so that $\mathbf{v}_0$ always points towards an inner point of the billiard. As a consequence, the ray is reflected at a point $\mathbf{r}_1$ on the boundary, the motion being clock-wise if $p_0 > 0$ and counter clock-wise if $p_0 < 0$. After N reflections, the point $\mathbf{r}_N$ is reached. The reversed trajectory $\mathbf{r}*_n$ is obtained by choosing $\mathbf{r}_0^* = \mathbf{r}_N$ and $\mathbf{v}_0^* = -\mathbf{v}_{N-1}$. It is not difficult to show that $\mathbf{r}_n^* = \mathbf{r}_{N-n}$ and $\mathbf{v}_n^* = -\mathbf{v}_{N-n-1}$. As a consequence, we have $\phi_n^* = \phi_{N-n}$ and for the momentum,

$$p_n^* = \mathbf{v}_n^* \cdot \tau_n^* = -\mathbf{v}_{N-n-1} \cdot \tau_{N-n} = -\mathbf{v}_{N-n} \cdot \tau_{N-n-1} = -p_{N-n} \tag{A25}$$

These relations imply that $\phi_N^* = \phi_0$ and $p_N^* = -p_0$ and prove that the initial condition is recovered after N iterations of the inverse map. If the map is slightly perturbed, then, after N iterations of the map, the initial conditions are no longer recovered and the deviation from reversibility is measured by $\big((\phi_N^* - \phi_0)^2 + ((p_N^* + p_0)^2\big)^{1/2}$.

Appendix B. Lyapunov and Reversibility Error Indicators

In this appendix, we derive the equivalence between the Lyapunov and reversibility error indicators. Those indicators have been previously defined and used to analyze the stability of other Hamiltonian systems (see, for instance, Ref. [34]). Those indicators measure the growth of a small random perturbation induced on an initial condition or at each reflection. In the latter case, it is convenient to perturb the orbit during the first n iterations and integrate the unperturbed inverse map.

These indicators are computed according to the linear response theory, namely when the limit of the perturbation ϵ tends to 0. For this reason, the linear approximation is valid only for a limited number of iterations n. The number n of iterations up to which the linear approximation is valid depends on $\mathbf{x}$ and ϵ and uniform estimates are typically $n \sim \log(1/\epsilon)$. The linear response is valid for any iteration number N since it involves the limit to zero of the noise amplitude ϵ. Given a symplectic or measure preserving map $M(\mathbf{x})$, defined in $\mathbb{R}^{2d}$, or a compact manifold of dimension $2d$, we denote with $\mathbf{x}_n$ the point of the orbit obtained by iterating the map n times, starting from an initial condition $\mathbf{x}$, and by $\mathsf{L}_n(\mathbf{x})$, its tangent map. Denoting the composition of the map with $M^{\circ 2}(\mathbf{x}) = M \circ M(\mathbf{x}) \equiv M(M(\mathbf{x}))$ we have

$$\begin{aligned} &\mathbf{x}_{n+1} = M(\mathbf{x}_{n-1}) \qquad && \mathbf{x}_0 = \mathbf{x} \qquad && \mathbf{x}_n = M^{\circ n}(\mathbf{x}) \\ &\mathsf{L}_{n+1}(\mathbf{x}) = DM(\mathbf{x}_n)\,\mathsf{L}_n(\mathbf{x}) && \mathsf{L}_0 = \mathsf{I} && \mathsf{L}_n(\mathbf{x}) = DM^{\circ n}(\mathbf{x}) \\ &(DM)_{ij}(\mathbf{x}) = \frac{\partial}{\partial x_j} M_i(\mathbf{x}) \end{aligned} \tag{A26}$$

Consider a nearby initial condition $\mathbf{y} = \mathbf{x} + \epsilon\boldsymbol{\xi}$ where $\boldsymbol{\xi}$ is a random vector with a zero mean and unit covariance matrix $\langle\, \boldsymbol{\xi}\,\boldsymbol{\xi}^T \,\rangle = \mathsf{I}$. The orbit $\mathbf{y}_n = M(\mathbf{y}_{n-1})$ is compared with the unperturbed orbit $\mathbf{x}_n = M(\mathbf{x}_{n-1})$ and the linear response is given by a random vector Ξ_n defined by

$$\Xi_n(\mathbf{x}) = \lim_{\epsilon\to 0} \frac{\mathbf{y}_n - \mathbf{x}_n}{\epsilon} = \lim_{\epsilon\to 0} \frac{M^{\circ n}(\mathbf{x}+\epsilon\boldsymbol{\xi}) - M^{\circ n}(\mathbf{x})}{\epsilon} = DM^{\circ n}(\mathbf{x})\,\boldsymbol{\xi} \equiv \mathsf{L}_n(\mathbf{x})\,\boldsymbol{\xi} \tag{A27}$$

The square of $E_n(\mathbf{x})$ is defined as the trace of the covariance matrix:

$$\begin{aligned} \Sigma_n^2(\mathbf{x}) &= \langle\, \Xi_n(\mathbf{x})\,\Xi_n^T(\mathbf{x})\,\rangle = \mathsf{L}_n(\mathbf{x})\,\mathsf{L}_n^T(\mathbf{x}) \\ E_n(\mathbf{x})^2 &= \operatorname{Tr}(\Sigma_n^2(\mathbf{x})) = \operatorname{Tr}\left(\mathsf{L}_n^T(\mathbf{x})\mathsf{L}_n(\mathbf{x})\right) \end{aligned} \tag{A28}$$

The matrix $\mathsf{L}_n\,\mathsf{L}_n^T$ has the same eigenvalues of the Lyapunov matrix $\mathsf{L}_n^T\mathsf{L}_n$ but the eigenvectors are different. The invariants $I_n^{(k)}(\mathbf{x})$ of $\mathsf{L}_n\,\mathsf{L}_n^T$ or $\mathsf{L}_n^T\,\mathsf{L}_n$ are the sum of the $\binom{d}{k}$ products of the eigenvalues. From a geometric viewpoint, the invariant $I_n^{(k)}$ is the sum of the squared volumes of the $\binom{d}{k}$ parallelotopes, whose edges are all the k distinct choices among the vectors $\mathsf{L}_n\mathbf{e}_1, \dots, \mathsf{L}_n\mathbf{e}_{2d}$, having denoted with $\mathbf{e}_j$ the base vectors $(\mathbf{e}_j)_i = \delta_{ij}$. From the polar decomposition $\mathsf{L}_n = \mathsf{R}_n\, e^{n\,\Lambda(n)}\,\mathsf{W}_n^T$, where R_n and W_n are orthogonal matrices, and the fact that $\Lambda(n)$ is a real diagonal matrix whose elements are $\lambda_1(n) \ge \lambda_2(n) \ge \ldots \ge \lambda_{2d}(n)$, is it possible to show that

$$\mathsf{L}_n^T\mathsf{L}_n = \mathsf{W}_n\, e^{2n\,\Lambda(n)}\,\mathsf{W}_n^T \qquad I_n^{(k)} = \sum_{j_1<j_2<\ldots<j_k} \exp\Big(2n\left(\lambda_{j_1}(n) + \ldots + \lambda_{j_k}(n)\right)\Big) \tag{A29}$$

If the matrix is symplectic, letting $\mathbf{r}^T = (\mathbf{q}, \mathbf{p})$ with $\mathbf{q} = (q_1, \dots, q_d)$ and $\mathbf{p} = (p_1, \dots, p_d)$, we have $\lambda_{d+j}(n) = -\lambda_{d-j+1}(n)$. The first invariant is $I_n^{(1)} = \operatorname{Tr}(\mathsf{L}_n^T\mathsf{L}_n) = E_n(\mathbf{x})^2$ and the next invariants can be recursively computed with the Faddeev–Leverrier formula, involving the traces of the powers of $\mathsf{L}_n^T\mathsf{L}_n$. The last invariant is given by the determinant.

Since the eigenvalues $\lambda_j(n)$ have a limit λ_j for $n \to \infty$, the asymptotic limit of the invariants is given by

$$\lim_{n\to\infty} \frac{1}{2n} \log I_n^{(k)}(\mathbf{x}) = \lambda_1(\mathbf{x}) + \lambda_2(\mathbf{x}) + \ldots + \lambda_k(\mathbf{x}) \tag{A30}$$

If a symplectic map M is integrable, the eigenvalues grow according to a power law

$$e^{2n\,\lambda_j(n)} \sim (1+\alpha_j^2\, n^2) \qquad e^{2n\,\lambda_{d+j}(n)} \sim \frac{1}{(1+\alpha_{d-j+1}^2\, n^2)} \qquad 1 \le j \le d \tag{A31}$$

As a consequence, the invariants behave as $I^{(k)}(n) \sim n^{2k}$ for $1 \le k \le d$. In the fully chaotic regions, all the first d Lyapunov exponents are positive; in the regions of regular motion, they vanish.

Appendix B.1. Reversibility Error

We consider the Backward–Forward process (BF), in which at each iteration of the reflection map, we add a random perturbation of amplitude ϵ up to the iteration n and then iterate n times with the inverse map. The linear response Ξ_n^{BF} for the BF process is obtained by first computing the perturbed orbit:

$$\begin{aligned} \mathbf{y}_{n'} &= M(\mathbf{y}_{n'-1}) + \epsilon\boldsymbol{\xi}_{n'} \qquad 1 \le n' \le n \\ \mathbf{y}_{n'} &= M^{-1}(\mathbf{y}_{n'-1}) \qquad n+1 \le n' \le 2n \end{aligned} \tag{A32}$$

and observing that the unperturbed orbit, after $2n$ iterations, comes back to the initial condition $\mathbf{x}_{2n} = M^{\circ -n} \circ M^{\circ n}(\mathbf{x}) = \mathbf{x}$:

$$\Xi_{2n}^{BF}(\mathbf{x}) = \lim_{\epsilon\to 0} \frac{\mathbf{y}_{2n} - \mathbf{x}}{\epsilon} = \sum_{k=1}^{n} \mathsf{L}_k^{-1}(\mathbf{x})\, \boldsymbol{\xi}_k \tag{A33}$$

The proof of this relation can be easily obtained by observing that $\Xi_{n'}$ satisfies the following linear recurrence equation

$$\begin{aligned} &\Xi_{n'}^{BF}(\mathbf{x}) = \lim_{\epsilon\to 0} \frac{M(\mathbf{x}_{n'-1}) + \epsilon\boldsymbol{\xi}_{n'-1} - M(\mathbf{x}_{n'-1})}{\epsilon} = DM(\mathbf{x}_{n'-1})\, \Xi_{n'-1}^{BF}(\mathbf{x}) + \boldsymbol{\xi}_{n'} \\ &\Xi_0 = 0 \qquad 1 \le n' \le n \end{aligned} \tag{A34}$$

for $n' \le n$ so that for $n' = n$ we obtain

$$\Xi_n^{BF}(\mathbf{x}) = \boldsymbol{\xi}_n + \sum_{k=1}^{n-1} DM(\mathbf{x}_{n-1})DM(\mathbf{x}_{n-2}) \cdots DM(\mathbf{x}_k)\, \boldsymbol{\xi}_k = \boldsymbol{\xi}_n + \sum_{k=1}^{n-1} \mathsf{L}_n(\mathbf{x})\, \mathsf{L}_k^{-1}(\mathbf{x})\, \boldsymbol{\xi}_k \tag{A35}$$

Since there is no perturbation when iterating with the inverse map M^{-1} for $n+1 \le n' \le 2n$, we have

$$\Xi_{2n}^{BF}(\mathbf{x}) = \lim_{\epsilon\to 0} \frac{M^{-n}(\mathbf{y}_n) - M^{-n}(\mathbf{x}_n)}{\epsilon} = DM^{-n}(\mathbf{x}_n)\, \Xi_n^{BF}(\mathbf{x}) = \mathsf{L}_n^{-1}(\mathbf{x})\Xi_n^{BF}(\mathbf{x}) \tag{A36}$$

Indeed, differentiating $M^{-n} \circ M^n(\mathbf{x}) = \mathbf{x}$, we obtain $DM^{-n}(\mathbf{x}_n)\, DM^n(\mathbf{x}) = \mathsf{I}$ so that $DM^{-n}(\mathbf{x}_n) = \mathsf{L}_n^{-1}(\mathbf{x})$. Replacing (A36) into (A34), the equation (A33) follows. The covariance matrices for the BF processes are given by

$$\Sigma_n^{BF\,2}(\mathbf{x}) = \left\langle \Xi_{2n}^{BF}(\mathbf{x})\Xi_{2n}^{BF}(\mathbf{x}) \right\rangle = \sum_{n'=1}^{n} \mathsf{L}_{n'}^{-1}(\mathbf{x}) \left(\mathsf{L}_{n'}^{-1}(\mathbf{x})\right)^{-T} = \sum_{n'=1}^{n} \left(\mathsf{L}_{n'}^{T}(\mathbf{x})\mathsf{L}_{n'}(\mathbf{x})\right)^{-1} \tag{A37}$$

The square of the $E_n^R(\mathbf{x})$ is defined as the first invariant, namely, the trace of the covariance matrix Σ_n^{BF}:

$$E_n^{BF\;2}(\mathbf{x}) = \operatorname{Tr}\left(\Sigma_n^{BF\;2}(\mathbf{x})\right) = \sum_{n'=1}^{n} \operatorname{Tr}\left(\left(\mathsf{L}_{n'}^T(\mathbf{x})\mathsf{L}_{n'}(\mathbf{x})\right)^{-1}\right) \tag{A38}$$

The next invariants of $\Sigma_n^{BF\;2}$ can be computed. If the map is symplectic, then $\mathsf{L}_{n'}\mathsf{L}_{n'}^T$ and $\mathsf{L}_{n'}^T\mathsf{L}_{n'}$ are symplectic matrices. Therefore, the traces of $\mathsf{L}_{n'}^T\mathsf{L}_{n'}$ and its inverse are equal and the RF reversibility error is given by a quadratic average of the $E_n(\mathbf{x})$.

$$E_n^{BF\;2}(\mathbf{x}) = \sum_{n'=1}^{n} E_n^2(\mathbf{x}) \tag{A39}$$

Previously, the BF process has been considered with the noise applied to both the backward and forward iterations. In this case, the covariance matrix is given by

$$\Sigma_n^{BF\;2}(\mathbf{x}) = \frac{1}{2}\left\langle \Xi_{2n}^{BF}(\mathbf{x})\left(\Xi_{2n}^{BF}\right)\right\rangle = \frac{1}{2}\,\mathsf{I} + \sum_{n'=1}^{n-1}\left(\mathsf{L}_{n'}^T(\mathbf{x})\mathsf{L}_{n'}(\mathbf{x})\right)^{-1} + \frac{1}{2}\left(\mathsf{L}_n^T(\mathbf{x})\mathsf{L}_n(\mathbf{x})\right)^{-1} \tag{A40}$$

A proof of (A40) can be found in [3] Section 2.3. This definition, whose difference with the previous one (A38) is negligible for large n, was initially proposed to compare $E_n^R(\mathbf{x})$, due to a small random displacement, against REM. Denoting with M_ϵ the map and with M_ϵ^{-1} its inverse, both computed by including the round-off, then the REM is defined by

$$REM_n^{BF}(\mathbf{x}) = \frac{\| M_\epsilon^{\circ\;-n} \circ M_\epsilon^{\circ\;n}(\mathbf{x}) - \mathbf{x}\|}{\sqrt{2}\;\epsilon} \tag{A41}$$

The round-off is a pseudo-random process of which just one realization is available. It is evident that, as opposed to $E_n^{BF}(\mathbf{x})$, REM_n^{BF}, it exhibits significant fluctuations when n or $\mathbf{x}$ vary. No higher invariants can be defined for the reversibility error due to round-off.

The specular process FB, in which we first iterate n times the perturbed inverse map M^{-1} and then the map M, can be considered. The previous formulae hold where $\mathsf{L}_{n'}(\mathbf{x}) = DM^{n'}(\mathbf{x})$ is replaced with $\mathsf{L}_{-n'}(\mathbf{x}) = DM^{-n'}(\mathbf{x}) = DM^{n'}(M^{-n'})) = \left(\mathsf{L}_{n'}(\mathbf{x}_{-n'})\right)^{-1}$ and therefore, the covariance matrix for the FB process is

$$\Sigma_n^{FB\;2}(\mathbf{x}) = \sum_{n'=1}^{n}\left(\mathsf{L}_{-n'}^T(\mathbf{x})\mathsf{L}_{-n'}(\mathbf{x})\right)^{-1} = \sum_{n'=1}^{n} \mathsf{L}_{n'}(\mathbf{x}_{-n'})\mathsf{L}_{n'}^T(\mathbf{x}_{-n'}) \qquad \mathbf{x}_{-n'} = M^{-n'}(\mathbf{x}) \tag{A42}$$

This formula can be proved just as the formula (A40) for the BF covariance matrix. One should take into account exchanging F and B amounts to change the map M with M^{-1} so that $\mathsf{L}_{n'}^T\mathsf{L}_{n'}$ is replaced by $\mathsf{L}_{-n'}^T\mathsf{L}_{-n'}$ since $\mathsf{L}_{n'} = DM^{\circ n'}$ and $\mathsf{L}_{-n'} = DM^{\circ\,-n'} \equiv D(M^{-1})^{\circ n'}$. For a symplectic map, the BF and FB covariance matrix are equivalent, provided that one exchanges the sign on the momentum, whereas for a dissipative map, they are intrinsically different. In this case, their different behavior reflects the break-up of the time reversal invariance for the unperturbed system. As a consequence, for a symplectic map, the FB reversibility error is related to the Lyapunov error by

$$E_n^{FB\;2}(\mathbf{x}) = \sum_{n'=1}^{n} E_n^2(\mathbf{x}_{-n'}) \tag{A43}$$

The corresponding round-off induced reversibility error REM is defined by iterating n times the inverse map M_ϵ^{-1} first, and then the map M_ϵ. If $\mathbf{x}$ belongs to an ergodic component, then the asymptotic behavior of $E_n(\mathbf{x})$ is the same for any $\mathbf{x}$ except for a subset of zero measure. As a consequence, the asymptotic of $E_n^{BF}(\mathbf{x})$ and $E_n^{FB}(\mathbf{x})$ is the same almost everywhere.

This is the reason why, for a symplectic map, we can ignore the FB error and consider just the BF reversibility error, which we usually denoted with $E_n^R(\mathbf{x})$.

References

1. Doron, E.; Smilansky, U.; Frenkel, A. Experimental demonstration of chaotic scattering of microwaves. *Phys. Rev. Lett.* **1990**, *65*, 3072–3075. [CrossRef] [PubMed]
2. Dembowski, C.; Dietz, B.; Gräf, H.D.; Heine, A.; Papenbrock, T.; Richter, A.; Richter, C. Experimental Test of a Trace Formula for a Chaotic Three-Dimensional Microwave Cavity. *Phys. Rev. Lett.* **2002**, *89*, 064101. [CrossRef] [PubMed]
3. Turchetti, G.; Panichi, F. Fast Indicators for Orbital Stability: A Survey on Lyapunov and Reversibility Errors. In *Progress in Relativity*; Buzea, C.G., Agop, M., Butler, L., Eds.; IntechOpen: Rijeka, Croatia, 2019; Chapter 10. [CrossRef]
4. Gradoni, G.; Yeh, J.H.; Xiao, B.; Antonsen, T.M.; Anlage, S.M.; Ott, E. Predicting the statistics of wave transport through chaotic cavities by the random coupling model: A review and recent progress. *Wave Motion* **2014**, *51*, 606–621. [CrossRef]
5. Balian, R.; Duplantier, B. Electromagnetic waves near perfect conductors. I. Multiple scattering expansions. Distribution of modes. *Ann. Phys.* **1977**, *104*, 300–335. [CrossRef]
6. Balian, R.; Duplantier, B. Electromagnetic waves near perfect conductors. II. Casimir effect. *Ann. Phys.* **1978**, *112*, 165–208 [CrossRef]
7. Stöckmann, H.J.; Kuhl, U. Microwave studies of the spectral statistics in chaotic systems. *J. Phys. Math. Theor.* **2022**, *55*, 383001. [CrossRef]
8. Sridhar, S. Experimental observation of scarred eigenfunctions of chaotic microwave cavities. *Phys. Rev. Lett.* **1991**, *67*, 785–788. [CrossRef]
9. Gräf, H.D.; Harney, H.L.; Lengeler, H.; Lewenkopf, C.H.; Rangacharyulu, C.; Richter, A.; Schardt, P.; Weidenmüller, H.A. Distribution of eigenmodes in a superconducting stadium billiard with chaotic dynamics. *Phys. Rev. Lett.* **1992**, *69*, 1296–1299. [CrossRef]
10. So, P.; Anlage, S.M.; Ott, E.; Oerter, R.N. Wave Chaos Experiments with and without Time Reversal Symmetry: GUE and GOE Statistics. *Phys. Rev. Lett.* **1995**, *74*, 2662–2665. [CrossRef]
11. Savytskyy, N.; Hul, O.; Sirko, L. Experimental investigation of nodal domains in the chaotic microwave rough billiard. *Phys. Rev. E* **2004**, *70*, 056209. [CrossRef]
12. Dietz, B.; Harney, H.L.; Kirillov, O.N.; Miski-Oglu, M.; Richter, A.; Schäfer, F. Exceptional Points in a Microwave Billiard with Time-Reversal Invariance Violation. *Phys. Rev. Lett.* **2011**, *106*, 150403. [CrossRef] [PubMed]
13. Gros, J.B.; Legrand, O.; Mortessagne, F.; Richalot, E.; Selemani, K. Universal behavior of a wave chaos based electromagnetic reverberation chamber. *Wave Motion* **2014**, *51*, 664–672. [CrossRef]
14. Nöckel, J.U.; Stone, A.D. Ray and wave chaos in asymmetric resonant optical cavities. *Nature* **1997**, *385*, 45–47. [CrossRef]
15. Stone, A.D. Wave-chaotic optical resonators and lasers. In *Quantum Chaos Y2K*; World Scientific: Singapore 2001; pp. 248–262.
16. Chernov, N.; Markarian, R. *Chaotic billiards*; Number 127; American Mathematical Soc.: Providence, RI, USA, 2006.
17. Liu, C.; Di Falco, A.; Molinari, D.; Khan, Y.; Ooi, B.S.; Krauss, T.F.; Fratalocchi, A. Enhanced energy storage in chaotic optical resonators. *Nat. Photonics* **2013**, *7*, 473–478. [CrossRef]
18. Creagh, S.C.; White, M.M. Differences between emission patterns and internal modes of optical resonators. *Phys. Rev.* **2012**, *85*, 015201. [CrossRef]
19. Draeger, C.; Aime, J.C.; Fink, M. One-channel time-reversal in chaotic cavities: Experimental results. *J. Acoust. Soc. Am.* **1999**, *105*, 618–625. [CrossRef]
20. Serra, R.; Gradoni, G.; Andrieu, G.; Primiani, V.M.; Magdowski, M.; Legrand, O.; Ahmed, M. Reverberation Chambers at the Edge of Chaos: Discussion Forum at EMC Europe 2020. *IEEE Electromagn. Compat. Mag.* **2022**, *11*, 73–88. [CrossRef]
21. Kaloshin, V.; Sorrentino, A. On the integrability of Birkhoff billiards. *Philos. Trans. R. Soc. Math. Phys. Eng. Sci.* **2018**, *376*, 20170419. [CrossRef]
22. Mikoss, I.; Garcia, P. An exact map for a chaotic billiard. *Int. J. Mod. Phys.* **2011**, *25*, 673–681. [CrossRef]
23. Artuso, R.; Casati, G.; Guarneri, I. Numerical experiments on billiards. *J. Stat. Phys.* **1996**, *83*, 145–166. [CrossRef]
24. Dragović, V.; Radnović, M. Periodic ellipsoidal billiard trajectories and extremal polynomials. *Commun. Math. Phys.* **2019**, *372*, 183–211. [CrossRef]
25. Reznik, D.; Garcia, R.; Keller, J. Can elleptic billiards still surprise us? *Math. Intell.* **2020**, *42*, 6–17. [CrossRef]
26. Dettmann, C.P.; Rahman, M.R. Spherical billiards with almost complete escape. *Chaos Interdiscip. J. Nonlinear Sci.* **2021**, *31*, 123119. [CrossRef]
27. Vasconcelos, J.; Miranda-Filho, L.; de Souza, A.; de C Romaguera, A.; Barbosa, A. Interaction between a robot and Bunimovich stadium billiards. *Sci. Rep.* **2022**, *12*, 4983. [CrossRef] [PubMed]
28. M. Firmbach, S. Lange, R. Ketzmerick and A. Backer. 3D billiards: visualization of regular structures and trapping of chaotic trajectories. *Phys. Rev. E* **2018**, *98*, 022214. [CrossRef]
29. Richter, M.; Lange, S.; Bäcker, A.; Ketzmerick, R. Visualization and comparison of classical structures and quantum states of four-dimensional maps. *Phys. Rev. E* **2014**, *89*, 022902. [CrossRef]
30. G. Turchetti, F. Panichi, S. Sinigardi, and S. Vaienti. Errors, Correlations and Fidelity for noisy Hamilton flows. Theory and numerical examples. *J. Phys. Math. Theor.* **2015**, *50*, 064001. [CrossRef]

31. F. Panichi, K. Goździewski and G. Turchetti. The reversibility error method (REM): a new, dynamical fast indicator for planetary dynamics. *MNRAS* **2017**, *468*, 469–491. [CrossRef]
32. F. Panichi, L. Ciotti and G. Turchetti. Fidelity and reversibility in the restricted three body problem. *Commun. Nonlinear Sci. Numer. Simul.* **2016**, *35*, 53–68. [CrossRef]
33. Gradoni, G.; Federico, P.; Giorgio, T. Propagation of rays in 2D and 3D waveguides: a stability analysis with Lyapunov and Reversibility fast indicators. *Chaos* **2021**, *31*, 043138. [CrossRef]
34. Turchetti, G.; Panichi, F. Birkhoff normal forms and stability indicators for betatronic motion. *Nonlinear Dyn. Collect. Eff. Part. Beam Phys.* **2019**, 47–69. [CrossRef]
35. Cincotta, P.M.; Giordano, C.M.; Simó, C. Phase space structure of multi-dimensional systems by means of the mean exponential growth factor of nearby orbits. *Phys. Nonlinear Phenom.* **2003**, *182*, 151–178. [CrossRef]
36. Berry, M.V. Regularity and chaos in classical mechanics, illustrated by three deformations of a circular 'billiard'. *Eur. J. Phys.* **1981**, *2*, 91. [CrossRef]
37. Markus Himmelstrand. A Survey of Dynamical Billiards. Available online: https://www.diva-portal.org/smash/get/diva2:650284/FULLTEXT01.pdf (accessed on 1 August 2023)
38. Dietz, B.; Smilansky, U. A scattering approach to the quantization of billiards- The inside-outside duality. *Chaos* **1993**, 3, 581–589. [CrossRef] [PubMed]

MDPI
St. Alban-Anlage 66
4052 Basel
Switzerland
www.mdpi.com

Entropy Editorial Office
E-mail: entropy@mdpi.com
www.mdpi.com/journal/entropy

www.ingramcontent.com/pod-product-compliance
Lightning Source LLC
Chambersburg PA
CBHW061448220926
51850CB00004B/43
9783036588582